**Electron Flow Version/Sixth Edition**

D0216816

# Introductory Electronic Devices and Circuits

## Robert T. Paynter

Prentice
Hall

Upper Saddle River, New Jersey ■ Columbus, Ohio

*To Susan, for encouraging me to write when I didn't want to
and making me laugh when I couldn't*

**Library of Congress Cataloging-in-Publication Data**

Paynter, Robert T.
    Introductory electronic devices and circuits: electron flow version/Robert T. Paynter.—6th ed.
        p. cm.
    Includes index.
    ISBN 0-13-061750-4 (alk. paper)
    1.  Solid state electronics.    2.  Operational amplifiers.
    I. Title.
    TK7871.85 .P395 2003
        621.381—dc21                                                                 2002035493

**Editor in Chief:** Stephen Helba
**Acquisitions Editor:** Dennis Williams
**Development Editor:** Kate Linsner
**Production Editor:** Rex Davidson
**Cover Designer:** Jason Moore
**Cover Image:** Corbis
**Design Coordinator:** Karrie Converse-Jones
**Interior Design:** Rebecca Bobb
**Interior Design Coordinator:** Diane Ernsberger
**Illustrations:** Steve Botts
**Production Manager:** Pat Tonneman
**Marketing Manager:** Ben Leonard

The book was set in Times Roman by The Clarinda Company and was printed and bound by R. R.
Donnelley & Sons Company. The cover was printed by Phoenix Color Corp.

Pearson Education Ltd.
Pearson Education Australia Pty. Limited
Pearson Education Singapore Pte. Ltd.
Pearson Education North Asia Ltd.
Pearson Education Canada, Ltd.
Pearson Educación de Mexico, S. A., de C. V.
Pearson Education—Japan
Pearson Education Malaysia Pte. Ltd.
Pearson Education, *Upper Saddle River, New Jersey*

**Copyright © 2003, 2000, 1997, 1994, 1991, 1989 by Pearson Education, Inc., Upper Saddle
River, New Jersey 07458**. All rights reserved. Printed in the United States of America. This
publication is protected by Copyright and permission should be obtained from the publisher prior to
any prohibited reproduction, storage in a retrieval system, or transmission in any form or by any
means, electronic, mechanical, photocopying, recording, or likewise. For information regarding
permission(s), write to: Rights and Permissions Department.

10 9 8 7 6 5 4 3 2 1
ISBN: 0-13-061750-4

# Preface: To the Instructor

If you compare the sixth edition of *Introductory Electronic Devices and Circuits* with its predecessor, you'll immediately see the changes in style and appearance that are part of this revision. The enhanced illustrations, combined with subtle changes in wording, are intended to make it easier for your students to study and comprehend the material being presented.

Users of previous editions will notice some changes in presentation and content. For example:

- The presentation on emitter bias in Chapter 7 (*DC Biasing Circuits*) has been reduced and moved into a new section (Section 7.4, *Other Transistor Biasing Circuits*) with the feedback bias circuits.
- The approach to transformer-coupled amplifier analysis in Chapter 11 (*Power Amplifiers*) has been modified to bring it in line with *RC*-coupled circuit analysis.
- The component specification sheets have been updated to reflect changes in component ratings and availability.

Several of the learning aids from previous editions have undergone revision (or relocation) as well. For example:

- The summary illustrations, in most cases, have been modified to include the primary component and/or circuit equations.
- Critical Thinking questions, which previously appeared in the margins, have been added to the Section Review questions.
- The glossary (Appendix E) has been updated and revised extensively.

## Learning Aids

From the start, my goal has been to produce a text that students can really *use* in their studies. As a result, many of the learning aids developed in the previous editions of *Introductory Electronic Devices and Circuits* have been retained:

① **Performance-based objectives** enable students to measure their progress by telling them what they are expected to be able to do as a result of their studies.
② **Chapter outlines** provide a handy overview of the chapter organization.
③ **Objective identifiers** in the margins cross-reference the objectives with the chapter material. This helps students to locate the material that will enable them to fulfill any objective.
④ **Margin notes** (which are color coded in this edition) include:
   - A running glossary of new terms
   - Notes that highlight the differences between theory and practice
   - Reminders of principles covered in earlier sections or chapters

⑤ **In-chapter practice problems** are included in the examples to provide students with an immediate opportunity to apply the principles being demonstrated. The **answers** to these problems appear at the end of each chapter.

⑥ **Summary illustrations** provide a convenient review of circuit operating principles, analysis equations, and applications. Many also provide comparisons between two or more related components or circuits.

⑦ **Highlighted lab references** help to tie the material in the text to the exercises in the accompanying lab manual.

Examples of these learning aids are shown on the following pages.

The following learning aids have also been retained from previous editions:

1. **Section review** questions at the end of each section. Most of these reviews now include Critical Thinking questions.

2. Each chapter ends with an **equation summary**, a **key terms list**, and an **extended chapter summary** (written in list form).

3. An extensive set of practice problems at the end of each chapter.[1] In addition to standard practice problems, the problem sets include:
   - **Troubleshooting Practice Problems**
   - **Pushing the Envelope** (challenging questions)
   - **Suggested Computer Applications Problems**

## MULTISIM APPLICATIONS PROBLEMS

In response to reviewer input, applications problems incorporating **EWB**® software were integrated throughout the previous edition of this text. These files have been upgraded to **MultiSim**® and incorporated here so that instructors can decide (on an individual basis) whether or not to include them in their courses. The CD-ROM packaged with the text contains MultiSim applications problems developed by **George Shaiffer** (Pikes Peak Community College, Colorado Springs, CO). Various figures throughout the text are marked with an EWB icon. The file associated with each figure can be accessed from the CD-ROM using the figure number.

Many instructors see MultiSim as a valuable learning tool. Others believe that its classroom use should be limited to solving circuit problems encountered by more advanced students. I believe that the method used to integrate MultiSim into this text will make it valuable to those who wish to use it while keeping it unobtrusive to those who do not.

## COMPANION WEBSITE

*Introductory Electronic Devices and Circuits* has a companion website designed to provide additional review materials, questions, and practice problems. The website provides the following for each chapter in the text:

- A list of chapter objectives
- A chapter summary (written in a different form than the summary provided in the text)
- Multiple-choice review questions and problems
- Fill-in-the-blank review questions and problems

These items combine to provide a valuable tool for reviewing every chapter in the book.

## Lab Manual to Accompany *Introductory Electronic Devices and Circuits*

The lab manual that accompanies this text has also gone through extensive revision. The circuit schematics have been revised to better illustrate the test equipment connections called for in the exercises. Optional MultiSim procedures have also been added to each exercise.

---

[1] This does not apply to Chapter 1, which has no practice problems.

chapter **7**

# DC Biasing Circuits

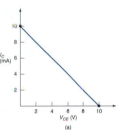

**FIGURE 7.5**

These combinations (which were chosen at random) are verified by equation (6.12) as follows:

For $I_C = 1$ mA,

$$V_{CE} = V_{CC} - I_C R_C = 10\text{ V} - (1\text{ mA})(1\text{ k}\Omega) = 10\text{ V} - 1\text{ V} = \mathbf{9\text{ V}}$$

For $I_C = 2$ mA,

$$V_{CE} = V_{CC} - I_C R_C = 10\text{ V} - (2\text{ mA})(1\text{ k}\Omega) = 10\text{ V} - 2\text{ V} = \mathbf{8\text{ V}}$$

For $I_C = 5$ mA,

$$V_{CE} = V_{CC} - I_C R_C = 10\text{ V} - (5\text{ mA})(1\text{ k}\Omega) = 10\text{ V} - 5\text{ V} = \mathbf{5\text{ V}}$$

These calculations verify the values obtained from the dc load line.

⑦ **Lab Reference:** A dc load line is plotted and used to predict circuit values in Exercise 7.

**PRACTICE PROBLEM 7.2**

A circuit like the one shown in Figure 7.4 has values of $V_{CC} = +16$ V and $R_C = 2$ k$\Omega$. Plot the dc load line for the circuit, and determine the values of $V_{CE}$ for $I_C = 2$ mA, 4 mA, and 6 mA. Then, verify your values using equation (6.12).

### 7.1.1 The Q-Point

When a transistor does not have an input signal, its output rests at specific dc values of $I_C$ and $V_{CE}$. As you have seen, these values correspond to a specific point on the dc load line. This point is called the **Q-point**. The letter $Q$ comes from the word **quiescent**, meaning *at rest*. A quiescent amplifier is one that has no input signal applied and, therefore, has constant dc values of $I_C$ and $V_{CE}$.

When the dc load line of an amplifier is superimposed on the collector curves for the transistor, the $Q$-point value can easily be determined. This point is illustrated in Figure 7.6.

Assume that the collector curves shown in Figure 7.6 are the curves for the transistor in Figure 7.4. The load line found in Example 7.2 has been superimposed over the collector curves. The $Q$-point is the point where the load line intersects the appropriate collector curve. For example, if the amplifier is operated at $I_B = 20$ μA, the $Q$-point is located at the point where the dc load line intersects the $I_B = 20$ μA curve, as shown in the illustration. From the load line, we can then determine that the circuit has $Q$-point values of $I_C = 4$ mA and $V_{CE} = 6$ V (the coordinates that correspond to the $Q$-point location).

◄ **OBJECTIVE 3**

**Q-point**
A point on the dc load line that indicates the values of $V_{CE}$ and $I_C$ for an amplifier at rest.

**Quiescent**
At rest.

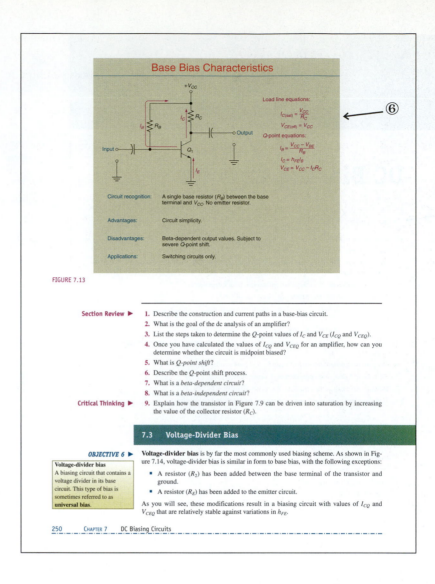

**Base Bias Characteristics**

Load line equations:

$$I_{C(sat)} = \frac{V_{CC}}{R_C}$$

$$V_{CE(off)} = V_{CC}$$

Q-point equations:

$$I_B = \frac{V_{CC} - V_{BE}}{R_B}$$

$$I_C = h_{FE}I_B$$

$$V_{CE} = V_{CC} - I_C R_C$$

| Circuit recognition: | A single base resistor ($R_B$) between the base terminal and $V_{CC}$. No emitter resistor. |
| Advantages: | Circuit simplicity. |
| Disadvantages: | Beta-dependent output values. Subject to severe Q-point shift. |
| Applications: | Switching circuits only. |

FIGURE 7.13

**Section Review ▶**

1. Describe the construction and current paths in a base-bias circuit.
2. What is the goal of the dc analysis of an amplifier?
3. List the steps taken to determine the Q-point values of $I_C$ and $V_{CE}$ ($I_{CQ}$ and $V_{CEQ}$).
4. Once you have calculated the values of $I_{CQ}$ and $V_{CEQ}$ for an amplifier, how can you determine whether the circuit is midpoint biased?
5. What is *Q-point shift*?
6. Describe the Q-point shift process.
7. What is a *beta-dependent circuit*?
8. What is a *beta-independent circuit*?

**Critical Thinking ▶**

9. Explain how the transistor in Figure 7.9 can be driven into saturation by increasing the value of the collector resistor ($R_C$).

### 7.3 Voltage-Divider Bias

**OBJECTIVE 6 ▶**

> **Voltage-divider bias**
> A biasing circuit that contains a voltage divider in its base circuit. This type of bias is sometimes referred to as **universal bias**.

**Voltage-divider bias** is by far the most commonly used biasing scheme. As shown in Figure 7.14, voltage-divider bias is similar in form to base bias, with the following exceptions:

- A resistor ($R_2$) has been added between the base terminal of the transistor and ground.
- A resistor ($R_E$) has been added to the emitter circuit.

As you will see, these modifications result in a biasing circuit with values of $I_{CQ}$ and $V_{CEQ}$ that are relatively stable against variations in $h_{FE}$.

250    CHAPTER 7    DC Biasing Circuits

## Acknowledgments

A project of this size cannot be completed without help from a variety of capable and concerned individuals. First and foremost, I am indebted (as always) to **Toby Boydell**, Seva Electronics (formerly with Conestoga College, Ontario). Once again, Toby provided input and editing throughout the revision process. He also directed the revision of the accompanying lab manual, serving as its primary contributing author. I am also indebted to the following individuals for the reviews they provided prior to the start of this revision:

**Jhones Bastardo**, ITT Technical Institute

**Mohammad A. Dabbas**, ITT Technical Institute

**D. J. D'Stefan**, County College of Morris

**Albert Gerth**, Corning Community College

**Tony Hearn**, Community College of Philadelphia

**Philip J. Regalbuto**, Trident Technical College

Their input helped determine the direction of the revision.

I would also like to thank the following for their contributions to the project:

**Wesley Morrison**, Copy Editor, for his thorough and extremely helpful copy editing.

**Steve Botts**, Technical Artist, for his fantastic job on upgrading the illustrations.

**Dennis Williams**, Acquisitions Editor, for promoting my work at every turn and providing the support needed to get it done.

**Kate Linsner**, Development Editor, for overseeing the review process and the initial development of this revision.

**Rex Davidson**, Production Editor, for overseeing the production process and helping to keep things on track.

**Steve Helba**, Editor-in-Chief, for believing in the project and providing the support needed to make it a reality.

**Lara Dimmick**, Editorial Assistant, for always knowing where to get the answers.

Finally, a special thanks goes out to my family and friends for their love and support—and to Susan . . . for helping to put it all together again.

*Bob Paynter*

# Preface: To the Student

## "Why Am I Learning This?"

Have you ever found yourself asking this question? If you have, then take a moment to read further.

I believe that any subject is easier to learn if you know *why* you are learning it. For this reason, we're going to take a moment to discuss:

- Why the study of electronic devices is important
- How this area of study relates to the other areas of electronics
- How you can get the most out of your study of electronic devices

Each electronics course serves, in part, as a foundation for the next. For example, you were taught about *resistors* in your fundamentals course. If you take a moment to flip through this book, you'll see that very few circuits do not contain at least one resistor. So, it should make sense that a thorough understanding of resistors is necessary to learn the principles and circuits discussed in this book.

*You are studying electronic devices at this point because it is serves as a foundation for the courses that will follow.* Just as the knowledge of basic components and circuit principles is essential for understanding electronic devices and circuits, you must successfully learn the material in this book to be prepared for later courses.

What *are* electronic devices? They are components with *dynamic* resistance characteristics. That is, they are components with resistance characteristics that are *current-controlled* or *voltage-controlled* (depending on the component). These fairly complex components are used in virtually every type of electronic system. They are used extensively in *communications systems* (such as televisions, stereos, and cellular phones), *digital systems* (such as PCs and calculators), *industrial systems* (such as process control systems), and *avionics* (aviation electronics).

As you can see, the study of electronic devices is essential if your knowledge is to advance beyond where it is now.

## "What Can I Do to Get the Most Out of This Course?"

There are several steps that you can take to help you successfully complete your study of electronics. The first is to realize that *learning electronics requires that you take an active role in your education*. It's like learning to ride a bicycle—you have to hop on and take a few spills. You can't learn how to ride a bike just by "reading the book," and the same can be said about learning electronics. You must be *actively involved in the learning process*.

How do you get actively involved in the learning process? Here are some guidelines worth following:

1. *Attend class on a regular basis.* The book provides information. Insight (which is just as important) is gained through classroom and lab experience.
2. *Take part in classroom problem-solving sessions.* Get out your calculator, and solve the problems along with your classmates.

3. *Do all the assigned homework.* Circuit analysis is a skill. As with any skill, you gain competency only through practice.

4. *Take part in classroom discussions.* Classroom discussions can clarify points that otherwise may be confusing, and they can help you to better understand how the various principles tie together to form a complete picture.

5. *Actively study the material in your textbook.*

Actively studying the material in the textbook means that you must do more than simply read it. When you are reading material for the first time, there are several things you should do:

1. *Learn the terminology.* You are taught new terms so that you will know what they mean and how to use them. When you come across a new term in the text, commit the new term to memory. How do you know when a new term is being introduced? Throughout this text, new terms are identified in the margins. When you see a new term and its definition in the margin, stop and learn the term before going on to the next section.

2. *Use your calculator to work through the examples.* When you come across an example, get your calculator and work the calculations in the example for yourself. When you do this, you develop the skills needed to solve the circuit problems on your own.

3. *Solve the example practice problems.* Most of the examples in this book are followed by a practice problem that is identical in nature to the example. When you see these problems, solve them using the model provided by the example. Then, check your solutions by looking up the answers at the end of the chapter.

4. *Use the chapter objectives to measure your learning.* Each chapter begins with an extensive list of performance-based objectives. These objectives tell you what you should be able to do as a result of learning the material.

This book also contains *objective identifiers* that are located in the margins of the text. For example, if you look at page 6 of the text, you'll see *OBJECTIVE 3* printed in the margin. This identifier tells you that this is the point where the material associated with objective 3 (in the list on page 1) is located. These objective identifiers can be used to help you with your studies. If you are unsure about how to perform a specific objective, flip through the chapter until you see the appropriate identifier. At that point, you will find the information that you need.

## One Final Note

It has been said that *success is not an entitlement . . . you have to work for it.* There is a lot of work involved in learning. However, the extra effort will pay off in the end. Your understanding of electronic devices will be more thorough as a result of your efforts, and learning the areas of electronics that follow will be easier. I wish you the best of success.

*Bob Paynter*

# Contents

## chapter 1

# Fundamental Solid-State Principles

## Objectives

*After studying the material in this chapter, you should be able to:*

1. Describe the makeup of the atom, and state the relationship between the number of valence electrons and its conductivity.
2. List the principles that govern the association between electrons and orbital shells.
3. Describe the relationship between conduction and temperature.
4. Contrast trivalent and pentavalent elements.
5. List the similarities and differences between *n*-type and *p*-type semiconductors.
6. Describe *diffusion current*.
7. Explain how a *depletion layer* is formed around a *pn* junction.
8. Explain the source of *barrier potential*, and list the barrier potential values for silicon and germanium.
9. Describe the relationship between depletion layer width, junction resistance, and junction current.
10. Define *bias*.
11. Describe the different methods of forward and reverse biasing a *pn* junction.
12. Explain why silicon is used more commonly than germanium in the production of solid-state devices.

## Outline

# Little Did They Know . . . .

In the March 1948 edition of the *Bell Laboratories Record,* an article appeared that was buried between stories of company promotions and seminars. The article carried the headline *Queen of Gems May Have Role in Telephony* and described a discovery that may well have been the predecessor to the transistor. Here is an excerpt from the article:

A radically new method of controlling the flow and amplification of electronic current—one that may have far-reaching influence on the future of electronics—was described on January 31 before the American Physical Society by K. G. McKay of Physical Electronics. Part of the Laboratories' broad investigations in communications for Bell System, this research has grown out of earlier work by D. E. Wooldridge, A. J. Ahearn, and J. A. Burton.

The method is based on the discovery that when beams of electrons are shot at certain insulators, in this case a diamond chip, electric currents are produced in the insulator which may be as much as 500 times as large as the current in the original electron beam.

The technique holds promise, after engineering development, of opening up an entirely new approach to the design and use of certain types of electron tubes. *It is not expected to replace existing electronic technologies but rather to supplement them.*

In fact, there are indications that if certain engineering problems can be overcome, the technique may lead to the development of important new electron tubes which do not exist today. For example, the technique *might* profitably be applied to the development of an entirely new means of obtaining extremely high amplification. . . . Methods of amplifying currents in gas or vacuum tubes have been known since the development of the first practical high-vacuum tube, also at Bell Telephone Laboratories, approximately 35 years ago, but this has never been done previously in solids.

---

Electronic systems, such as radios, televisions, and computers, were originally constructed using *vacuum tubes.* Vacuum tubes were generally used to increase the strength of ac signals (*amplify*) and to convert ac energy to dc energy (*rectify*). Although they were able to perform these critical operations very well, vacuum tubes had several characteristic problems. They were large and fragile, and they wasted tremendous amounts of power through heat loss.

In the 1940s, a team of scientists working for Bell Labs developed the *transistor,* the first solid-state device capable of amplifying an ac signal. The term *solid-state* was coined because the transistor was solid rather than hollow like the vacuum tube it was designed to replace. The transistor was also smaller, more rugged, and wasted much less power than did its vacuum-tube counterpart. Since the development of the transistor, solid-state components have replaced vacuum tubes in nearly every application.

Solid-state components are made from elements that are classified as *semiconductors.* A **semiconductor** element is one that is neither a conductor nor an insulator but, rather, lies halfway between the two. Under certain circumstances, the resistive properties of a semiconductor can be varied between those of a conductor and those of an insulator. As you will see later, it is *this* characteristic of semiconductor elements that makes them useful as amplifiers and rectifiers.

> **Semiconductor**
> An element that is neither an insulator nor a conductor.

## 1.1    Atomic Theory

Before we get started, let's discuss the reason for reviewing atomic theory. *Some* coverage of atomic theory is essential to understanding the characteristics of semiconductors. However, our goal at this time is to establish *what* happens on the atomic level, not *why* it happens. As you read through this section, realize that the material is being presented to give you a *basic* understanding of what happens on the atomic level. Any physics text should provide more information if you wish to learn more about these principles.

### 1.1.1    The Atom

OBJECTIVE 1 ▶ The atom has been shown to contain three basic particles: the *protons* and *neutrons* that make up the nucleus (core) of the atom, and *electrons* that orbit about the nucleus. The simplest model of the atom, called the *Bohr model*, is illustrated in Figure 1.1. The orbital paths, or *shells*, are identified using the letters $K$ through $Q$. The innermost shell is the K shell, followed by the L shell, and so on. The outermost shell for a given atom is called the **valence shell**. The valence shell of an atom is critical because it determines the conductivity of the atom.

> **Valence shell**
> The outermost shell that determines the conductivity of an atom.

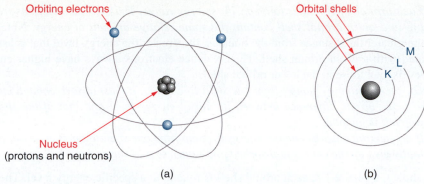

FIGURE 1.1   Bohr model of the atom.

The valence shell of an atom can contain up to eight electrons. The conductivity of the atom depends on the number of electrons that are in the valence shell. When an atom has one valence electron, it is a nearly perfect conductor. When an atom has eight valence electrons, the valence shell is said to be *complete*, and the atom is an insulator. Therefore, *conductivity decreases with an increase in the number of valence electrons.*

## 1.1.2   Semiconductors

Semiconductors are atoms that contain four valence electrons. Because the number of valence electrons in a semiconductor is halfway between one (for a conductor) and eight (for an insulator), a semiconductor atom is neither a good conductor nor a good insulator.

Three of the most commonly used semiconductor materials are *silicon* (Si), *germanium* (Ge), and *carbon* (C). These atoms are all represented in Figure 1.2. Note that each of these elements contains four valence electrons. Of the semiconductors shown, silicon and germanium are used in the production of solid-state components. Carbon is used mainly in the production of resistors and potentiometers.

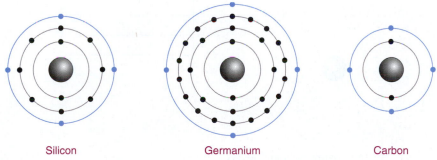

Silicon                     Germanium                     Carbon

FIGURE 1.2   Semiconductor atoms.

## 1.1.3   Charge and Conduction

When no outside force causes conduction, the number of electrons in a given atom equals ◄ *OBJECTIVE 2* the number of protons. Since the charges of electrons (−) and protons (+) are equal and opposite, the *net charge* on the atom is zero. If an atom *loses* one valence electron, the atom then contains fewer electrons than protons, and the net charge on the atom is *positive.* If an atom with an incomplete valence shell *gains* one valence electron, the atom contains more electrons than protons, and the net charge on the atom is *negative.*

Some fundamental laws regarding the relationship between electrons and orbital shells have been shown to be true:

1. *Electrons travel in orbital shells. They cannot orbit the nucleus in the space that exists between any two orbital shells.*

2. *Each orbital shell relates to a specific energy range. Thus, all the electrons traveling in a given orbital shell contain the same relative amount of energy.* Note that the greater the distance from the nucleus, the greater the energy level that is associated with a given orbital shell. Thus, valence electrons always have higher energy levels than those in the lower orbital shells.

3. *For an electron to "jump" from one shell to another, it must absorb enough energy to make up the difference between its initial energy level and that of the shell to which it is jumping.*

4. *If an electron absorbs enough energy to jump from one shell to another, it will eventually give up the energy it absorbed and return to a lower-energy shell.*

As shown in Figure 1.3, each orbital shell is related to a specific energy level. The difference between the energy levels of any two orbital shells is referred to as an **energy gap**. For an electron to jump from one orbital shell to another, it must absorb enough energy to overcome the energy gap between the shells. For example, in Figure 1.3, the valence shell, or **band**, is shown to have an energy level of approximately 0.7 **electron-volt (eV)**. The **conduction band** is shown to have an energy level of 1.8 eV. Thus, for an electron to jump from the valence band to the conduction band, it would have to absorb an amount of energy equal to

$$1.8 \text{ eV} - 0.7 \text{ eV} = 1.1 \text{ eV}$$

For conductors, semiconductors, and insulators, the valence to conduction-band energy gaps are approximately 0.4, 1.1, and 1.8 eV, respectively. The higher this energy gap, the harder it is to cause conduction, because more energy must be absorbed for an electron to jump to the conduction band.

**Energy gap**
The difference between the energy levels of any two orbital shells.

**Band**
Another name for an orbital shell.

**eV (electron-volt)**
The energy absorbed by an electron when it is subjected to a 1 V difference of potential.

**Conduction band**
The band outside the valence shell.

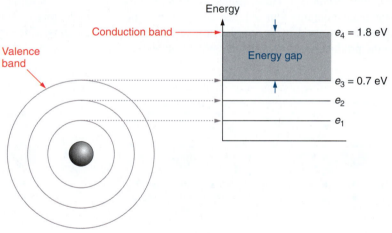

**FIGURE 1.3**  Silicon energy gaps and levels.

When an electron absorbs enough energy to jump from the valence band to the conduction band, the electron is said to be in an *excited* state. An excited electron will eventually give up the energy it absorbed and return to its original energy level. The energy given up by the electron is in the form of *light* or *heat*.

### 1.1.4   Covalent Bonding

**Covalent bonding** is the method by which atoms complete their valence shells by "sharing" valence electrons with other atoms. The covalent bonding of a group of silicon atoms is represented in Figure 1.4. To help illustrate the bonding process, each atom has been represented by an octagon (eight-sided figure) containing a square. The octagon represents the valence shell of the atom, and the square is used to identify the electrons that belong to the particular atom. As you can see, the center atom has eight valence electrons,

**Covalent bonding**
A means of holding atoms together by sharing valence electrons.

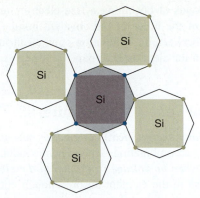

FIGURE 1.4    Silicon covalent bonding.

four that belong to the center atom plus one that belongs to each of the four surrounding atoms. These atoms, in turn, share electrons with four surrounding atoms, and so on. The results of this bonding are:

1. The atoms are held together, forming a solid substance.
2. The atoms are all electrically stable, because their valence shells are complete.
3. The completed valence shells cause the silicon to act as an *insulator*. Thus, **intrinsic** (pure) silicon is a very poor conductor. The same principle holds true for intrinsic germanium.

**Intrinsic**
Another word for *pure*.

When semiconductor atoms bond together in a set pattern like the one shown in Figure 1.4, the resulting material is called a *crystal*. A crystal is a smooth glassy solid. You have probably seen quartz crystals at some time. Silicon and germanium commonly crystallize in the same manner. At room temperature, a silicon crystal has fewer free electrons than a germanium crystal. Therefore, silicon is a better insulator than germanium at room temperature (which is a desirable characteristic). This characteristic is one of the reasons why silicon is used more often than germanium to make semiconductor components.

Carbon can crystallize in the same fashion as silicon and germanium, but carbon crystals are too expensive to use in solid-state component production. (Carbon crystals are more commonly known as *diamonds*.)

## 1.1.5    Conduction

When a valence electron absorbs enough energy, it jumps from the valence band to the conduction band. One result of this action is that a *gap* is left in the covalent bond. This gap is referred to as a **hole**. It would follow that *for every conduction-band electron, there must exist a valence-band hole*. The term used to describe this combination is **electron-hole pair**. The basic concept of the electron-hole pair is illustrated in Figure 1.5.

**Hole**
A gap in a covalent bond.

**Electron-hole pair**
A free electron and its matching valence band hole.

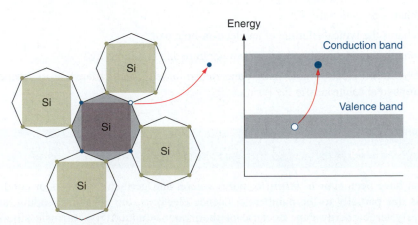

FIGURE 1.5    Generation of an electron-hole pair.

**Recombination**
When a free electron returns to the valence shell.

**Lifetime**
The time from electron-hole pair generation to recombination.

Within a few microseconds of becoming a free electron, an electron will give up its energy and fall into one of the holes in the covalent bond. This process is known as **recombination**. The time between the generation of an electron-hole pair and recombination is called the **lifetime** of the electron-hole pair.

### 1.1.6 Conduction Versus Temperature

*OBJECTIVE 3* ▶ At room temperature, *thermal energy* (heat) causes the constant creation of electron-hole pairs, with their subsequent recombination. Thus, a semiconductor always has some number of free electrons *even when no voltage is applied to the element*. As you increase the temperature, more electrons in the element absorb enough energy to break free of their covalent bonds, and the number of free electrons increases.

On the other hand, if you *decrease* the temperature, there is less thermal energy to release electrons from their covalent bonds, and the number of free electrons decreases. This relationship continues until the temperature reaches *absolute zero*. Absolute zero is, by definition, the temperature at which there is no thermal energy and occurs at $-273.16°C$ ($-459.69°F$). Since there is no thermal energy at this temperature, there are no free electrons. (There is no energy for the electrons to absorb, so they cannot jump to the conduction band.)

The important relationship here is the fact that *conductivity in a semiconductor varies directly with temperature*. This is why the current through a semiconductor increases as the material warms up and decreases when the material cools down.

Section Review ▶

1. What are the three particles that make up the atom?
2. What is the relationship between the number of valence electrons and the conductivity of a given element?
3. How many valence electrons are there in a conductor? An insulator? A semiconductor?
4. What three semiconductor elements are most commonly used in electronics?
5. What are the relationships between electrons and orbital shells that were listed in this section?
6. What is an *energy gap*? What are the energy gap values for insulators, semiconductors, and conductors?
7. What forms of energy are given off by an electron that is falling into the valence band from the conduction band?
8. What is *covalent bonding*?
9. What are the effects of covalent bonding on intrinsic semiconductor materials?
10. What is an *electron-hole pair*?
11. What is *recombination*?
12. What is the typical lifetime of an electron-hole pair?
13. What is the relationship between temperature and conductivity?
14. Why aren't electron-hole pairs generated in a semiconductor when its temperature drops to absolute zero?

## 1.2 Doping

*OBJECTIVE 4* ▶ As you have been shown, *intrinsic* (pure) silicon and germanium are poor conductors. This is due partially to the number of valence electrons, the covalent bonding, and the relatively large energy gap. Because of their poor conductivity, intrinsic silicon and germanium are of little use.

**Doping** is the process of *adding impurity atoms to intrinsic silicon or germanium to improve the conductivity of the semiconductor*. The term *impurity* is used to describe the doping elements, because the silicon or germanium is no longer pure once the doping has occurred. Since a doped semiconductor is no longer pure, it is called an **extrinsic** semiconductor.

Two element types are used for doping: **trivalent** and **pentavalent**. A trivalent element is one that has three valence electrons. A pentavalent element is one that has five valence electrons. When trivalent atoms are added to intrinsic semiconductors, the resulting material is called a *p*-type material. When pentavalent impurity atoms are used, the resulting material is called an *n*-type material. The most commonly used doping elements are listed in Table 1.1.

**Doping**
Adding impurity elements to intrinsic semiconductors.

**Extrinsic**
Another word for *impure*.

**Trivalent**
Elements with three valence shell electrons.

**Pentavalent**
Elements with five valence shell electrons.

***p*-type material**
A semiconductor that has added trivalent impurities.

***n*-type material**
A semiconductor that has added pentavalent impurities.

TABLE 1.1   Commonly Used Doping Elements

| Trivalent Impurities | Pentavalent Impurities |
|---|---|
| Aluminum (Al) | Phosphorus (P) |
| Gallium (Ga) | Arsenic (As) |
| Boron (B) | Antimony (Sb) |
| Indium (In) | Bismuth (Bi) |

## 1.2.1   *N*-Type Materials

When pentavalent impurities are added to silicon or germanium, the result is an excess of electrons *in the covalent bonds*. This principle is illustrated in Figure 1.6. As you can see, the pentavalent arsenic atom is surrounded by four silicon atoms. The silicon atoms bond with the arsenic atom, each sharing one arsenic electron. However, the fifth arsenic electron is not bound to any of the surrounding silicon atoms. As a result, the fifth electron requires little energy to break free and enter the conduction band. If literally millions of arsenic atoms are added to pure silicon or germanium, then millions of electrons are not part of the covalent bonds. All these electrons can be made to flow through the material with little difficulty.

One point that needs to be understood is this: Even though millions of electrons are not part of the covalent bonding, *the material is still electrically neutral*. This is because each arsenic atom has the same number of protons as electrons, just like the silicon or germanium atoms. Since the overall numbers of protons and electrons in the material are equal, the net charge on the material is zero.

Because there are more conduction-band electrons than valence-band holes in an *n-type material*, the electrons are called *majority carriers*, and the valence-band holes are called *minority carriers*. The relationship between majority and minority carriers in an *n*-type material can be better understood by taking a look at Figure 1.7. The valence band is shown to contain *some* holes. These holes are caused by thermal energy excitation of electrons, as was discussed earlier. With the excess of conduction-band electrons, however, the lifetime of an electron-hole pair is shortened significantly, because the hole is filled almost immediately by one of the excess electrons in the conduction band.

Why are *n*-type materials electrically neutral?

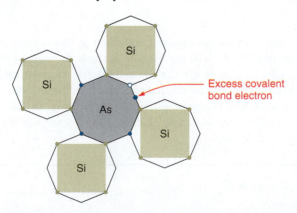

FIGURE 1.6   *n*-Type material.

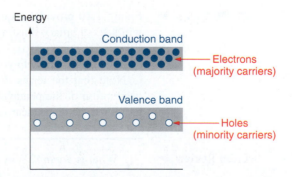

FIGURE 1.7   Energy diagram (*n*-type material).

Since the only holes that exist in the covalent bonding are those caused by thermal energy, the number of holes is far less than the number of conduction-band electrons. This is where the terms *majority* and *minority* come from.

Note that the term *n-type* implies an excess of electrons. As you will see, a *p-type* material is one with an excess of holes and relatively few free electrons.

### 1.2.2    *P-*Type Materials

When intrinsic silicon or germanium is doped with a trivalent element, the resulting material is called a *p-type material.* The use of a trivalent element causes the existence of a hole in the covalent bonding structure. This point is illustrated in Figure 1.8. As you can see, the aluminum atom is shown to be surrounded by four silicon atoms. This time, however, there is a gap in the covalent bond caused by the lack of a fourth valence electron in the aluminum atom. Now, instead of an excess of electrons, we have an excess of holes in the covalent bonds. This condition is represented in Figure 1.9. The *p*-type material is shown to have an excess of holes in the valence band. At the same time, there are some electrons in the conduction band. Again, the free electrons in the conduction band are there because of thermal energy. Since there are many more valence-band holes than conduction-band electrons, the holes are the majority carriers and the electrons are the minority carriers.

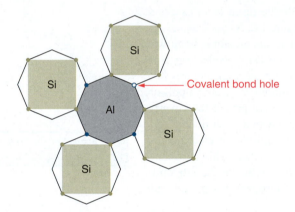

FIGURE 1.8   *p-*Type material.

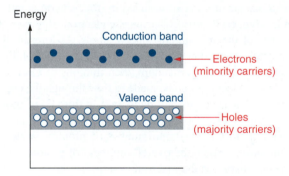

FIGURE 1.9   Energy diagram (*p*-type material).

Why are *p*-type materials electrically neutral?

Again, remember that the number of electrons in the *p*-type material is equal to the number of protons. Even though there are gaps in the covalent bonds, the proton-electron balance still exists. Therefore, the net charge on the *p*-type material is zero.

### 1.2.3    Summary

OBJECTIVE 5 ▶

Figure 1.10 provides a quick summary of the material presented in this section. As you will see, summary illustrations like Figure 1.10 are used extensively throughout the text. When you come across one, take time to study the illustration and the relationships it contains. This will help you to remember the points that were made in the section.

Note that the terms *donor atom* and *acceptor atom* are used in Figure 1.10 in the description of the pentavalent and trivalent doping elements. The meanings of these two terms will be made clear in the next section.

Section Review ▶

1. What is *doping*? Why is it necessary?

2. What is an *impurity element*?

3. What are *trivalent* and *pentavalent* elements?

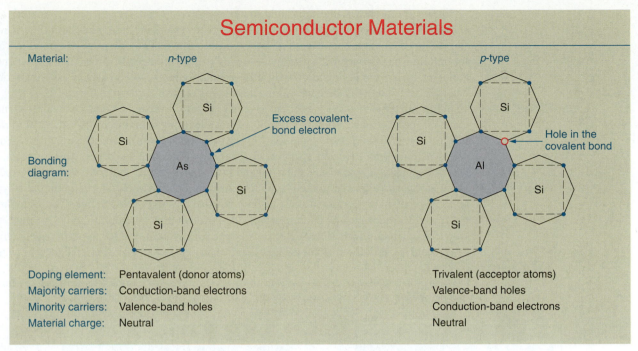

**Semiconductor Materials**

FIGURE 1.10

4. Despite their respective characteristics, *n*-type and *p*-type materials are still electrically neutral. Why?

5. In what ways are *n*-type and *p*-type materials similar? In what ways are they different?

## 1.3    The *PN* Junction

An *n*-type material and a *p*-type material become extremely useful when *joined* together to form a *pn junction*. Figure 1.11 illustrates the relative conditions of the *n* and *p* materials before and at the moment they are joined together.

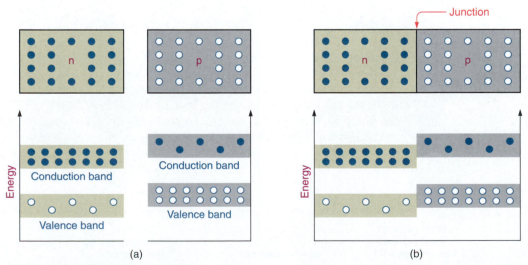

FIGURE 1.11    *pn*-Junction initial energy levels.

Figure 1.11a shows the individual materials. As you can see, the *n*-type material is shown (on top) as containing an excess of electrons (solid circles), and the *p*-type material is shown as having an excess of holes (open circles). The energy diagrams illustrate the relationship between the energy levels of the two materials. Note that the valence bands of the two materials are at slightly different energy levels, as are the conduction bands. This is due to differences in the atomic makeup of the two materials.

**OBJECTIVE 6 ▶**    When the two materials are joined together (Figure 1.11b), the conduction and valence bands of the materials overlap. This allows free electrons from the *n*-type material to *diffuse* (wander) to the *p*-type material. This action and its results are illustrated in Figure 1.12. When a free electron wanders from the *n*-type material across the junction, it becomes trapped in one of the valence-band holes in the *p*-type material. As a result, there is one net *positive* charge in the *n*-type material and one net *negative* charge in the *p*-type material. This may seem confusing at first, but it really isn't that strange. Take a look at Figure 1.13. Under normal circumstances, the covalent bond shown in the *n*-type material would have a total of 21 valence electrons (4 per silicon atom + 5 for the pentavalent atom = 21) and 21 protons that offset the negative valence charges.* This balance between

*To simplify things, we consider only *valence* electrons and their "matching" protons. The actual number of total electrons and protons is not considered, because the electrical activity is determined by the valence shell.

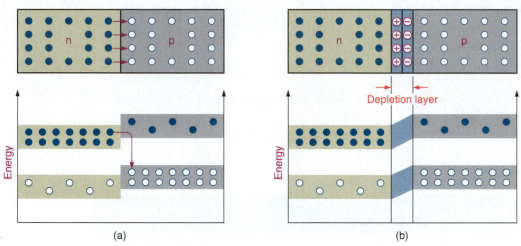

(a)                                      (b)

**FIGURE 1.12**   The forming of the depletion layer.

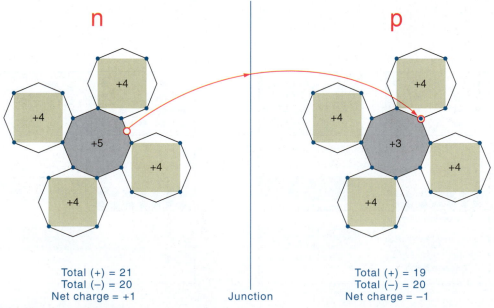

**FIGURE 1.13**   Depletion layer charges.

valence electrons and protons results in a net charge of zero. When an electron wanders into the *p*-type material as shown:

1. The bond shown in the *n*-type material has lost an electron. Now the bond has only 20 electrons but *still* has 21 offsetting protons. As a result, the overall charge is positive (+1).
2. The bond in the *p*-type material has gained an electron. Now the bond has 20 electrons but *still* has 19 offsetting protons. As a result, the overall charge is negative (−1).

Now consider what happens on a larger scale. When looking at the large-scale picture, there are several things to remember:

◀ *OBJECTIVE 7*

1. Each electron that diffuses across the junction leaves one positively charged bond in the *n*-type material and produces one negatively charged bond in the *p*-type material.
2. Both conduction-band electrons and valence shell holes are needed for conduction through the materials. When an electron diffuses across the junction, the *n*-type material has lost a conduction-band electron. When the electron falls into a hole in the *p*-type material, that material has lost a valence-band hole. At this point, both bonds have been *depleted* of charge carriers.

Since the action just described happens on a large scale, the junction ends up with a layer (on both sides) that is depleted of charge carriers. This layer is called the **depletion layer** and is represented in Figure 1.12b. Note that the overall charge of the layer is shown to be positive on the *n* side of the junction and negative on the *p* side of the junction.

With the buildup of (−) charges on the *p* side of the junction and of (+) charges on the *n* side of the junction, there is a natural *difference of potential* between the two sides of the junction. This potential is referred to as the **barrier potential**. The barrier potential for a *pn* junction is typically in the millivolt range.

**Depletion layer**
The area around a *pn* junction that is depleted of charge carriers. Note that the depletion layer is also referred to as the *depletion region*.

◀ *OBJECTIVE 8*

**Barrier potential**
The natural potential across a *pn* junction.

## 1.3.1    Summary

When an *n*-type material is joined with a *p*-type material:

1. A small amount of *diffusion* occurs across the junction. The amount of diffusion is limited by the difference between the conduction-band energy levels of the two materials.
2. When electrons diffuse into the *p* region, they give up their energy and "fall" into the holes in the valence-band covalent bonds.
3. Since the pentavalent atoms (near the junction) in the *n* region have lost an electron, they have an overall *positive charge*.
4. Since the trivalent atoms (near the junction) in the *p* region have gained an electron, they have an overall *negative charge*.
5. The difference in charges on the two sides of the junction is called the *barrier potential*. This barrier potential is typically in the millivolt range.

Note that since the pentavalent atoms in the *n* material are giving up electrons, they are often referred to as **donor atoms**. The trivalent atoms, on the other hand, are accepting electrons from the pentavalent atoms and thus are referred to as **acceptor atoms**.

**Donor atoms**
Another name for *pentavalent atoms*.

**Acceptor atoms**
Another name for *trivalent atoms*.

---

1. What is the overall charge on an *n-type* covalent bond that has just given up a conduction-band electron?
2. What is the overall charge on a *p-type* covalent bond that has just accepted an extra valence-band electron?
3. Describe the forming of the *depletion layer*.
4. What is *barrier potential*? What causes it?

◀ **Section Review**

**OBJECTIVE 9** ▶    A *pn* junction is useful because we can control the width of its depletion layer. By controlling the width of its depletion layer, we are able to control the resistance of the *pn* junction and thus the amount of current that can pass through the device. The relationship between the width of the depletion layer and the junction current is summarized as:

| Depletion Layer Width | Junction Resistance | Junction Current |
|---|---|---|
| Minimum | Minimum | Maximum |
| Maximum | Maximum | Minimum |

**OBJECTIVE 10** ▶    **Bias** is a potential applied to a *pn* junction to obtain a desired mode of operation. This potential is used to control the width of the depletion layer. The two types of bias are *forward bias* and *reverse bias*. **Forward bias** is a potential used to reduce the resistance of a *pn* junction. A forward-biased *pn* junction has minimum depletion layer width and junction resistance. **Reverse bias** is a potential used to increase the resistance of a *pn* junction. A reverse-biased *pn* junction has maximum depletion layer width and junction resistance. In this section, we look at the two types of bias.

> **Bias**
> A potential applied to a *pn* junction to obtain a desired mode of operation.
>
> **Forward bias**
> A potential used to reduce the resistance of a *pn* junction.
>
> **Reverse bias**
> A potential that causes a *pn* junction to have a high resistance.

### 1.4.1    Forward Bias

A *pn* junction is *forward biased* when the applied potential causes the *n*-type material to be more *negative* than the *p*-type material. When forward biased, a *pn* junction allows current to pass with little opposition. The effects of forward bias are illustrated in Figure 1.14. Figure 1.14a shows a *pn* junction connected to a voltage source (*V*) and an open switch (*SW₁*). The energy diagram is included to show the initial energy states of the junction materials.

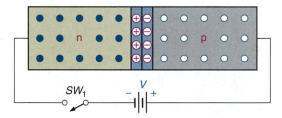

(a) An unbiased *pn* junction

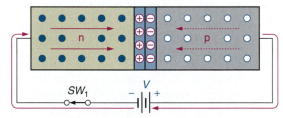

(b) Charge motion at the moment *SW₁* is closed

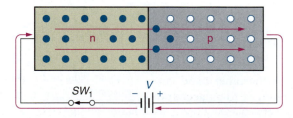

(c) Conduction increases as the depletion layer breaks down.

(d) Bulk resistance

**FIGURE 1.14    The effects of forward bias.**

When *SW₁* is closed, the voltage source is connected to the *pn* junction as shown in Figure 1.14b. As you can see, a negative potential is applied to the *n*-type material and a positive potential is applied to the *p*-type material. As a result:

1. The conduction band electrons in the *n*-type material are pushed toward the junction.
2. The valence band holes in the *p*-type material are pushed toward the junction.

*Assuming that V is greater than the barrier potential of the junction,* the electrons in the *n*-type material break through the depletion layer. When this happens, the electrons are free to recombine with the holes in the *p*-type material, and conduction occurs. Conduction through a *pn* junction is illustrated in Figure 1.14c. Once a *pn* junction begins to conduct, it provides a slight opposition to current. This opposition to current is referred to as **bulk resistance**. Bulk resistance (illustrated in Figure 1.14d) is the combined resistance of the *n*-type and *p*-type materials as follows:

$$R_B = R_p + R_n$$

**Bulk resistance**
The combined resistance of the *n*-type and *p*-type materials in a forward-biased *pn* junction.

The value of $R_B$ is typically in the range of 25 Ω or less. The exact value of $R_B$ for a given junction depends on the dimensions of the *n*-type and *p*-type materials, the amount of doping used to produce the materials, and the operating temperature.

Since the value of $R_B$ is extremely low, very little voltage is dropped across this resistance. For this reason, the voltage drop across $R_B$ is usually ignored in circuit calculations.

When a forward-biased *pn* junction begins to conduct, the **forward voltage ($V_F$)** across the junction is slightly greater than the barrier potential for the device. The values of $V_F$ are approximated as

**Forward voltage ($V_F$)**
The voltage across a forward-biased *pn* junction.

$$V_F \cong 0.7 \quad \text{(for silicon)}$$

and

$$V_F \cong 0.3 \quad \text{(for germanium)}$$

Any difference between the measured and the approximated values of $V_F$ is a result of the current through the bulk resistance of the device. This point is discussed further in Chapter 2.

It should be noted that there are two ways in which a *pn* junction can be forward biased:

◄ *OBJECTIVE 11*

1. By applying a potential to the *n*-type material that drives it more *negative* than the *p*-type material.
2. By applying a potential to the *p*-type material that drives it more *positive* than the *n*-type material.

These two biasing methods are illustrated in Figure 1.15. In each case, one of the two materials is connected to ground. In Figure 1.15a, a positive voltage is applied to the *p*-type material. In Figure 1.15b, a *negative voltage* is applied to the *n*-type material. Assuming the magnitude of the applied voltage in each case is sufficient, the depletion layer breaks down and allows conduction. Thus, we can cause a *pn* junction to conduct by driving the *n*-type material more negative than the *p*-type material or by driving the *p*-type material more positive than the *n*-type material. Both methods of forward biasing a *pn* junction are used in practice.

### 1.4.2   Reverse Bias

A *pn* junction is *reverse biased* when the applied potential causes the *n*-type material to be more *positive* than the *p*-type material. When a *pn* junction is reverse biased, the depletion layer becomes wider, and junction current is reduced to almost zero. Reverse bias and its effects are illustrated in Figure 1.16. Figure 1.16a shows a junction that is forward biased. As you have already been shown, the junction is allowing current to pass with little opposition. If the biasing potential returns to zero, the depletion layer re-forms as electrons diffuse across the junction. This is illustrated in Figure 1.16b.

When we apply a voltage with the polarity shown in Figure 1.16c, the electrons in the *n*-type material head toward the positive terminal of the source. At the same time, the holes in the *p*-type material move toward the negative source terminal. The electron motion away from the *n* side of the junction further depletes the material of free

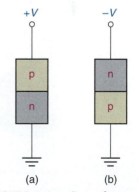

FIGURE 1.15   Some forward-biased *pn* junctions.

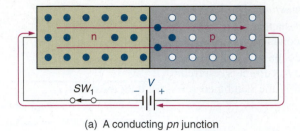
(a) A conducting *pn* junction

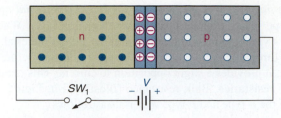
(b) A depletion layer forms when the current path is broken.

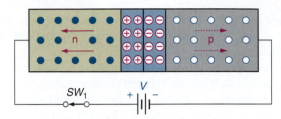
(c) When the polarity of *V* is reversed, the depletion layer widens as charge carriers move away from the junction.

**FIGURE 1.16**   The effects of reverse bias.

carriers. The depletion layer has effectively been widened. The same principle holds true for the *p* side of the material. Since there are fewer holes near the junction, the depletion layer has grown. *The overall effect of the widening of the depletion layer is that the resistance of the junction is drastically increased, and conduction drops to near zero.*

Figure 1.16 helps to demonstrate another point about reverse bias. During the time that the depletion layer is forming and growing, there is *still* majority carrier current in both materials. This *diffusion current* lasts only as long as it takes the depletion layer to reach its maximum width. Diffusion current is undesirable in high-frequency circuits, so special diodes have been developed for these type of circuits. (Some of these diodes are discussed in Chapter 5.)

Just as there are two ways to forward bias a junction, there are two ways to reverse bias a junction:

1. By applying a potential to the *n*-type material that drives it more *positive* than the *p*-type material.
2. By applying a potential to the *p*-type material that drives it more *negative* than the *n*-type material.

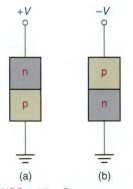

The junctions shown in Figure 1.17 are reverse biased. Again, try to relate the diagrams to the statements made above.

### 1.4.3   The Bottom Line

Bias is a potential applied to a *pn* junction that determines the operating characteristics of the device. Bias polarities and effects are summarized as follows:

| Bias Type | Junction Polarities | Junction Resistance |
|---|---|---|
| Forward | *n*-type material is more (−) than *p*-type material | Extremely *low* |
| Reverse | *p*-type material is more (−) than *n*-type material | Extremely *high* |

**FIGURE 1.17**   Some reverse-biased *pn* junctions.

The voltage across a forward-biased *pn* junction will be approximately equal to 0.7 V for silicon and 0.3 V for germanium. When a *pn* junction is reverse biased, it acts essentially as an *open* circuit. These points are summarized in Figure 1.18.

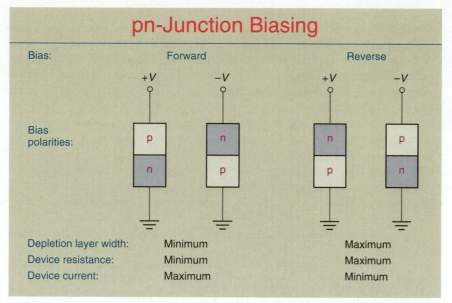

FIGURE 1.18

## 1.4.4    A Final Note

Up to this point, we have included both silicon and germanium in our discussions of semiconductor materials and junctions. Of the two, silicon is almost always preferred for several reasons:

◀ *OBJECTIVE 12*

1. Silicon is more tolerant of heat.
2. Germanium oxide is water soluble, making germanium more difficult to process than silicon.
3. At a given value of reverse voltage, germanium allows more leakage current than does silicon.

Since it is the most commonly used semiconductor, we will limit future discussions to silicon. Just remember that germanium-based circuits work the same way as silicon-based circuits. The primary difference is in the value of $V_F$ for germanium versus silicon.

---

1. What are the resistance and current characteristics of a *pn* junction when its depletion layer is at its *maximum* width? Its *minimum* width?

◀ Section Review

2. What purpose is served by the use of *bias*?
3. What effect does forward bias have on the depletion layer width of a *pn* junction?
4. What is the junction resistance of a forward-biased *pn* junction?
5. What are the approximate values of $V_F$ for forward-biased silicon and germanium *pn* junctions?
6. List the commonly used methods for forward biasing a *pn* junction.
7. What effect does reverse bias have on the depletion layer width of a *pn* junction?
8. What is the junction resistance of a reverse-biased *pn* junction?
9. List the commonly used methods for reverse biasing a *pn* junction.
10. Why is silicon more commonly used than germanium in the production of solid-state components?

---

**CHAPTER SUMMARY**

Here is a summary of the major points made in this chapter:

1. Early electronic systems used *vacuum tubes* to amplify and rectify ac signals. These tubes were large, fragile, and wasted a tremendous amount of power through heat loss.

2. Vacuum tubes (for the most part) have been replaced by *solid-state* components. These components, which are made using semiconductor materials, are smaller, more rugged, and more efficient than their vacuum tube counterparts.

3. A *semiconductor* element is one that is neither a conductor nor an insulator, but rather, lies halfway between the two. Under certain circumstances, the resistive properties of a semiconductor material can be varied between those of a conductor and those of an insulator.

4. The outermost orbital shell of an atom is referred to as the *valence shell*. This shell is important because it determines the conductivity of the atom.
   a. Elements with one valence shell electron are conductors.
   b. Elements with eight valence shell electrons are insulators.
   c. Semiconductor elements have four valence shell electrons.

5. The most commonly used semiconductor elements are *carbon*, *silicon*, and *germanium.*
   a. Silicon and germanium are commonly used in the production of solid-state components.
   b. Carbon is used primarily in the production of resistors and potentiometers.

6. Many semiconductor operating characteristics are based on the following principles:
   a. Electrons travel in orbital shells. They cannot orbit the nucleus in the space between shells.
   b. Each orbital shell has a specific energy range; that is, the electrons orbiting in a given shell have the same relative amount of energy.
   c. For an electron to jump from one shell to another, it must absorb enough energy to make up the difference between its initial energy level and that of the shell to which it is jumping.
   d. When an electron jumps to a higher-energy shell, it eventually releases the energy it absorbed and returns to a lower-energy orbit. (Electrons always seek the lowest possible energy shell.)

7. When an electron drops from one shell to a lower-energy shell, it releases energy in the form of *light* or *heat*.

8. *Covalent bonding* is the method by which atoms complete their valence shells when they "share" electrons with other atoms (see Figure 1.4).

9. The results of covalent bonding are:
   a. Atoms are held together, forming a solid substance.
   b. The atoms are electrically stable, because their valence shells are complete.
   c. The complete valence shells cause intrinsic (pure) silicon or germanium to act as an insulator.

10. Silicon is a better insulator than germanium at room temperature. This desirable characteristic is one of the reasons that most solid-state components are made using silicon.

11. When a valence electron jumps to a higher-level energy shell, it leaves a *hole* in the covalent bond. The combination of the electron and the hole is called an *electron-hole pair*.

12. Within a few microseconds of becoming a free electron, the particle will give up its energy and fall into a nearby valence-shell hole. This process is called *recombination*.

13. The time span from the generation of an electron-hole pair until recombination occurs is called the *lifetime* of the electron-hole pair.

14. Thermal energy (heat) causes the constant creation of electron-hole pairs. This means that a semiconductor always has some free electrons even when no voltage is applied to the element.

15. Conductivity in a semiconductor is directly proportional to temperature.

16. Because of their poor conductivity, intrinsic silicon and germanium are of little use.

17. *Doping* is the process of adding impurities to semiconductor elements to improve their conductivity ratings. An *impurity* (in this case) is an element other than silicon or germanium.

18. *Trivalent* and *pentavalent* elements are commonly used as doping elements. (A list of the commonly used impurity elements is provided in Table 1.1.)

    **a.** A *trivalent* element has *three* valence electrons.

    **b.** A *pentavalent* element has *five* valence electrons.

**19.** When an intrinsic semiconductor element is doped using a *pentavalent* element, an *n-type material* is produced (see Figure 1.6).

    **a.** *n*-Type materials contain electrons that are not part of the covalent bonds. Relatively little energy is required to force these excess electrons into the conduction band.

    **b.** Although *n*-type materials contain an excess of conduction band electrons, each is offset by an extra proton in the parent atom. Therefore, *n*-type materials are electrically neutral.

    **c.** Conduction band electrons in an *n*-type material are referred to as *majority carriers.*

    **d.** Valence band holes in an *n*-type material are referred to as *minority carriers.*

**20.** When an intrinsic semiconductor element is doped using a *trivalent* element, a *p-type material* is produced (see Figure 1.8).

    **a.** Trivalent doping elements do not provide enough electrons to complete the covalent bonds when added to silicon (or germanium). As a result, *p*-type material contains an excess of valence band holes.

    **b.** Although *p*-type materials contain an excess of valence band holes, the total numbers of protons and electrons are equal. Therefore, *p*-type materials are electrically neutral.

    **c.** Valence-band holes in a *p*-type material are referred to as *majority carriers.*

    **d.** Conduction-band electrons in a *p*-type material are referred to as *minority carriers.*

**21.** *p*-Type and *n*-type materials are compared in Figure 1.10.

**22.** *p*-Type and *n*-type materials become useful when joined together to form a *pn junction* (see Figure 1.12).

**23.** When a *pn* junction is formed:

    **a.** A small amount of *diffusion current* passes between the materials.

    **b.** Electrons that diffuse into the *p*-type material fall into the excess valence band holes.

    **c.** The atoms near the junction in the *n*-type material have given up electrons, so they have a net *positive* charge.

    **d.** The atoms near the junction in the *p*-type material have gained electrons, so they have a net *negative* charge.

    **e.** There is a difference of potential between the charges on the two sides of the junction. This difference of potential, called the *barrier potential*, is typically in the millivolt range.

    **f.** The barrier potential for silicon is assumed to be approximately 0.7 V. The barrier potential for germanium is assumed to be approximately 0.3 V.

**24.** The pentavalent atoms in the *n*-type material that lose electrons (when the junction is formed) are referred to as *donor atoms.*

**25.** The trivalent atoms in the *p*-type material that gain electrons (when the junction is formed) are referred to as *acceptor atoms.*

**26.** *Bias* is a potential applied to a *pn* junction to obtain a desired mode of operation.

**27.** *Forward bias* is used to reduce the width of the depletion layer. This reduces the resistance of the junction and allows more current (per volt) to pass through the component.

    **a.** Forward bias is achieved when the applied potential causes the *n*-type material to be more *negative* than the *p*-type material (see Figures 1.14 and 1.15).

    **b.** When conducting, the forward voltage across a *pn* junction is approximately 0.7 V (for silicon) or 0.3 V (for germanium).

**28.** *Reverse bias* is used to increase the width of the depletion layer. This increases the resistance of the junction and decreases the current (per volt) through the component.

    **a.** Reverse bias is achieved when the applied potential causes the *n*-type material to be more *positive* than the *p*-type material (see Figures 1.16 and 1.17).

    **b.** A reverse-biased *pn* junction acts (essentially) as an open circuit, preventing the flow of charge (current) through the circuit.

**29.** Forward bias and reverse bias are compared in Figure 1.18.

## KEY TERMS

acceptor atoms  11
band  4
barrier potential  11
bias  12
bulk resistance  13
conduction band  4
covalent bonding  4
depletion layer  11
diffusion current  14
donor atoms  11
doping  7
electron-hole pair  5

electron-volt (eV)  4
energy gap  4
extrinsic  7
forward bias  12
forward voltage ($V_F$)  13
germanium  3
hole  5
intrinsic  5
lifetime  6
majority carrier  7
minority carrier  7

negative voltage  13
*n*-type material  7
pentavalent  7
*p*-type material  7
recombination  6
reverse bias  12
reverse voltage ($V_R$)  15
semiconductor  2
silicon  3
trivalent  7
valence shell  2

## chapter 2

# Diodes

## Objectives

*After studying the material in this chapter, you should be able to:*

1. Identify the terminals of a *pn*-junction diode, given the schematic symbol of the component.
2. Analyze the schematic diagram of a simple diode circuit and determine:
   a. Whether the diode is conducting.
   b. The direction of current through a conducting diode.
3. List and describe the three diode models and the applications for each.
4. List the main parameters of the *pn*-junction diode, and explain how each limits the use of the component.
5. Determine the suitability of a given diode for a given application using diode spec sheets and/or selector guides.
6. Identify the schematic symbol of the zener diode, and determine the direction of current through the device.
7. Discuss the basic operating principles of the zener diode.
8. List the main zener diode parameters, and explain how each limits the use of the component.
9. Discuss the basic operating principles of the light-emitting diode (LED).
10. Calculate the value of the *current-limiting* resistor needed for an LED in a given circuit.
11. Determine whether a given *pn*-junction diode, zener diode, or LED is good or faulty.

## Outline

## Which Came First?

In later chapters, you will learn about a type of electronic device called a *transistor*. While the transistor is similar in construction to a *pn*-junction diode, it is actually a more complex component whose operation is based on more complex concepts.

It would seem to the casual observer that the *pn*-junction diode was developed before the transistor. After all, most complex devices are developed as outgrowths of similar—but simpler—devices. This, however, is not the case.

The *pn*-junction diode was actually developed almost six years *after* the first transistor. In fact, the transistor had already been in commercial use for two years when Bell Laboratories announced the development of the *pn*-junction diode! In this instance, the chicken definitely came before the egg.

The diode is the most basic solid-state component. There are many diode types, each with its own operating characteristics and applications. The various diode types are easily identified by name, circuit application, and schematic symbol. It should be noted that the term *diode*, used by itself, refers to the basic *pn*-junction diode. All other diodes have other identifying names, such as *zener diode*, *light-emitting diode*, and so on.

A **diode** is a *two-electrode* (two-terminal) device that acts as a *one-way conductor*. The most basic type of diode is the *pn-junction diode*, which is nothing more than a *pn* junction with a lead connected to each of the semiconductor materials. When forward biased, this type of diode conducts. When reverse biased, diode conduction drops to nearly zero.

In this chapter, we look at the three most commonly used types of diodes; the *pn-junction diode*; the *zener diode*; and the *light-emitting diode*, or *LED*. Several additional types of diodes are covered in Chapter 5.

> **Diode**
> A two-terminal device that acts as a one-way conductor.

---

### 2.1   Introduction to the *PN*-Junction Diode

**OBJECTIVE 1** ▶

The schematic symbol for the *pn*-junction diode is shown in Figure 2.1. The *n*-type material is called the **cathode**, and the *p*-type material is called the **anode**.

> **Cathode**
> The *n*-type terminal of a diode.
>
> **Anode**
> The *p*-type terminal of a diode.

Anode (A)
*p* type

Cathode (K)
*n* type

**FIGURE 2.1**   *pn*-Junction diode schematic symbol.

Recall that a *pn* junction conducts when the *n*-type material (cathode) is more negative than the *p*-type material (anode). Relating this characteristic to the schematic symbol of the diode, we can say that a diode conducts when the following conditions are met:

1. The arrow points to the more negative of the diode potentials; that is, the *cathode* is more negative than the *anode*.
2. The difference of potential (voltage) across the anode and cathode leads exceeds the barrier potential of the device.
3. The diode conducts fully when the *forward voltage* ($V_F$) across the component is approximately:
   **a.** 0.7 V for a silicon diode.
   **b.** 0.3 V for a germanium diode.

**OBJECTIVE 2** ▶

These conditions are illustrated in Figure 2.2, which shows several forward-biased (conducting) diodes. Note that each diode symbol points to the more negative potential. Since the diode symbol points to the more negative potential when the component is conducting, the direction of diode forward current is *against the arrow*, as shown in Figure 2.2.

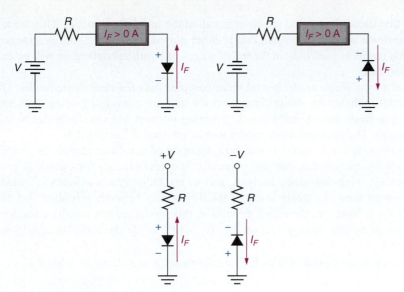

FIGURE 2.2    Forward-biased diodes.

A *pn*-junction diode is reverse biased when the *n*-type material (cathode) is more positive than the *p*-type material (anode). This causes the depletion layer to widen and block
current. Relating this characteristic to the schematic symbol for the diode, we can make
the following statement: *A diode does not conduct when the symbol points to the more
positive of the diode potentials.* This point is illustrated in Figure 2.3, which shows several reverse-biased (nonconducting) diodes. Note that each diode symbol points to the
more positive potential.

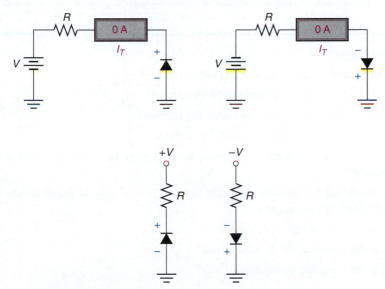

FIGURE 2.3    Reverse-biased diodes.

## 2.1.1    Diode Models

In this chapter, you will be introduced to three diode *models*. A **model** is a representation
of a component or circuit that demonstrates one or more of the characteristics of that
component or circuit. For example, the dc model of a capacitor may represent the component as an open circuit, since a capacitor blocks dc. At the same time, the ac model of a
capacitor may represent the component as a variable reactance, since the reactance of a
capacitor varies inversely with frequency.

Component models are usually used to represent the component under specific circumstances or in specific applications. As you will see, the diode model that you use
depends on what you are trying to do.

> **Model**
> A representation of a
> component or circuit that
> demonstrates one (or more) of
> its characteristics.

**Troubleshooting**
The process of locating faults
in electronic equipment.

The first diode model that we cover is called the *ideal diode model*. This model represents the diode as a simple switch that is either *closed* (conducting) or *open* (nonconducting). This model is used only in the initial stages of **troubleshooting**, as will be explained in Section 2.2.

The *practical diode model* is a bit more complex than the ideal diode model. The practical model includes the diode characteristics that are considered when mathematically analyzing a diode circuit and when determining whether a given diode can be used in a given circuit. The practical diode model will be covered in Section 2.3.

The *complete diode model* is the most accurate of the diode models. It includes several diode characteristics that are generally considered only for circuit development (engineering), high-frequency analysis, and so on. Other characteristics included in this model can be used to explain many of the differences between predicted and measured circuit values. Note that the characteristics in this model are not usually considered on a regular basis by the average technician. We will look at the complete diode model in Section 2.5.

The three diode models and their applications are summarized in Table 2.1.

**TABLE 2.1** Diode Model Summary

| Diode Model | Applications |
|---|---|
| **Ideal** | Circuit troubleshooting |
| **Practical** | Mathematical circuit analysis |
| | Determining suitability of a diode for a given application |
| **Complete** | Circuit development (engineering) |
| | Special-case circuit analysis |
| | Explaining differences between predicted and measured circuit values |

**Section Review ▶**

1. What is a *diode*?

2. What type of bias causes a diode to *conduct*?

3. What type of bias causes a diode to *block* current?

4. Draw the symbol for a diode, and label the terminals.

5. What polarity is required to *forward bias* a diode?

6. When analyzing a schematic diagram, how do you determine whether a diode is conducting?

7. When analyzing a schematic diagram, how do you determine the *direction* of diode current? Explain your answer.

8. When is the *ideal diode model* used?

9. When is the *practical diode model* used?

10. When is the *complete diode model* used?

## 2.2   The Ideal Diode

The ideal diode is viewed as a switch.

The *ideal* diode is viewed as an open switch when it is reverse biased and as a closed switch when forward biased. You may recall that a switch has the following characteristics:

| Switch Condition | Switch Characteristics |
|---|---|
| Open | Infinite resistance, allowing no current |
| | Full applied voltage dropped across the component terminals |
| Closed | No resistance, allowing maximum current |
| | No voltage dropped across the component terminals |

Based on the characteristics of a switch, we can make the following statements about the ideal diode:

What are the characteristics of the *ideal* diode?

1. When *reverse biased (open switch)*:
   a. The diode has infinite resistance.
   b. The diode does not pass current.
   c. The diode drops the applied voltage across its terminals.
2. When *forward biased (closed switch)*:
   a. The diode has no resistance.
   b. The diode does not limit the current through it.
   c. The diode has no voltage drop across its terminals.

The graph in Figure 2.4 illustrates the characteristics of the ideal diode model. Diode forward voltage ($V_F$) and reverse voltage ($V_R$) are measured along the positive and negative x-axes, respectively. Diode forward current ($I_F$) and reverse current ($I_R$) are measured along the positive and negative y-axes, respectively. Quadrant I of the graph is labeled as the *forward operating region*, because every combination of $V_F$ and $I_F$ falls within this region of the graph. By the same token, Quadrant III of the graph is labeled as the *reverse operating region*.

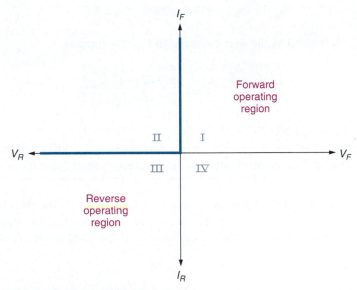

FIGURE 2.4    Characteristics of the ideal diode.

The reverse operating region in Figure 2.4 illustrates the reverse-bias characteristics of the ideal diode. Note that as **reverse voltage** ($V_R$) increases, **reverse current** ($I_R$) remains at zero. This implies that the reverse-biased diode is an *open circuit* (just like an open switch), because there is no current through the device regardless of the value of the applied voltage. *Since the reverse-biased diode is acting as an open, the applied voltage is dropped across the terminals of the device.* This point is illustrated in Example 2.1.

**Reverse voltage ($V_R$)**
The voltage across a reverse-biased diode.

**Reverse current ($I_R$)**
The current through a reverse-biased diode.

### EXAMPLE 2.1

Determine the values of $V_{D1}$, $I_T$, and $V_{R1}$ for the circuit shown in Figure 2.5.

*Solution:*    The arrow in the schematic symbol is pointing toward the positive terminal of the source, so we know that the diode is reverse biased. Therefore:
   a. The full applied voltage is dropped across $D_1$.

$$V_{D1} = V_S = 5\text{ V}$$

   b. $D_1$ does not allow conduction. Therefore, $I_T = 0$ A.
   c. Since there is no current through $R_1$, there is no voltage across the component ($V_{R1} = 0$ V).

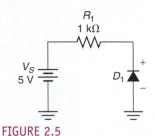

FIGURE 2.5

## PRACTICE PROBLEM 2.1

A series circuit consists of a 12 V source, a 470 Ω resistor ($R_1$), a 330 Ω resistor ($R_2$), and a diode. If the diode is reverse biased, what is the value of $V_{R1}$? What is the value of $V_{R2}$?

---

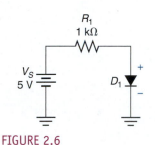

**FIGURE 2.6**

| What determines the value of diode forward current? |
| --- |

The forward operating region in Figure 2.4 illustrates the forward-bias characteristics of the ideal diode. Note that the forward voltage ($V_F$) across the diode is assumed to be 0 V for this diode model, while forward current ($I_F$) is shown to be at some measurable value. When the ideal diode is forward biased, the value of $I_F$ is determined by the voltage and resistance values that are external to the component. This point is illustrated in Example 2.2.

## EXAMPLE 2.2

Determine the values of $V_{D1}$, $V_{R1}$, and $I_T$ for the circuit shown in Figure 2.6.

*Solution:* The arrow in the schematic symbol is pointing toward the negative terminal of the source, so we know that the diode is forward biased. Therefore:

    **a.** $V_{D1} = 0$ V, leaving the total applied voltage across $R_1$.
    **b.** $V_{R1} = V_S = 5$ V.
    **c.** $I_T$ is determined by the source voltage and $R_1$. By formula,

$$I_T = \frac{V_R}{R_1} = \frac{5 \text{ V}}{1 \text{ k}\Omega} = \textbf{5 mA} \quad (ideal)$$

## PRACTICE PROBLEM 2.2

A series circuit consists of a 12 V source, a 470 Ω resistor, a 330 Ω resistor, and a diode. If the diode is forward biased, what is the ideal value of $I_T$ for the circuit?

---

If you compare the values found in Examples 2.1 and 2.2 with the switch characteristics described earlier, you will see that the diode was treated as an ideal switch in both cases. The forward and reverse characteristics of the ideal diode are summarized in Figure 2.7.

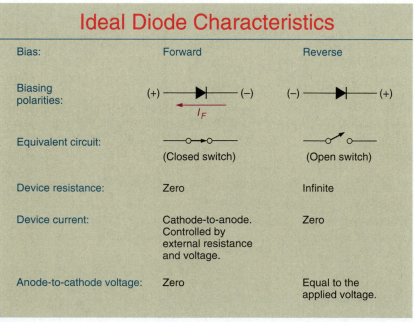

**FIGURE 2.7**

### 2.2.1 When Do We Use the Ideal Diode Model?

Normally, the ideal model of the diode is used in the initial stages of circuit troubleshooting. When troubleshooting most diode circuits, your initial concern is only whether a given diode is acting as a one-way conductor. If it is, the component is assumed to be good. If it is not, the component is faulty and must be replaced.

◄ Section Review

1. What are the *forward characteristics* of the ideal diode model?
2. What are the *reverse characteristics* of the ideal diode model?
3. When do we use the *ideal* diode model?

## 2.3   The Practical Diode Model

In our discussion of the ideal diode, we did not consider many of the diode characteristics that must be dealt with by working technicians on a regular basis. One of these characteristics, *forward voltage*, is normally considered in the mathematical analysis of a diode circuit. Many other practical diode characteristics are used when determining whether one diode may be used in place of another or in a specific circuit. These characteristics include *peak reverse voltage*, *average forward current*, and *forward power dissipation*.

In this section, we will take a look at *forward voltage* ($V_F$) and the effect that it has on the mathematical analysis of basic diode circuits. The other characteristics listed are covered in detail in Section 2.4.

### 2.3.1   Forward Voltage ($V_F$)

In Chapter 1, we established the fact that a slight voltage is developed across a forward-biased *pn* junction. The effect of this *forward voltage* ($V_F$) on the diode characteristic curve is illustrated in Figure 2.8.

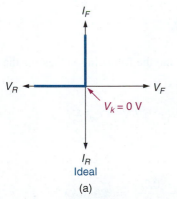

FIGURE 2.8   Diode characteristic curves.

Figure 2.8a is the same ideal diode characteristic curve that was shown in Figure 2.4. Note that the point where $I_F$ suddenly increases is labeled $V_k$ in the figure. This label is commonly used to identify what is called the *knee voltage* on a voltage-versus-current graph. The term **knee voltage ($V_k$)** is often used to describe the point on a voltage-versus-current graph where current suddenly increases or decreases. As you can see, the ideal diode model assumes a knee voltage of 0 V.

In Figure 2.8b, we see the practical diode characteristic curve. The only difference between this curve and the one shown for the ideal diode model is the value of $V_F$. In the practical diode curve, the value of $V_k$ is shown to be equal to the approximated value of $V_F$ for a silicon *pn* junction, 0.7 V. In an actual circuit, the $V_F$ may fall between 0.7 V and 1.1 V, depending on the current through the device.

**Knee voltage ($V_k$)**
The voltage at which device current suddenly increases or decreases.

Using the curve shown in Figure 2.8b, we can make the following statements about the forward operating characteristics of the practical diode:

1. Diode current remains at zero until the knee voltage is reached.
2. Once the applied voltage reaches the value of $V_k$, the diode turns *on*, and forward conduction occurs.
3. As long as the diode is conducting, the value of $V_F$ is approximately equal to $V_k$. In other words, $V_F$ is assumed to be approximately 0.7 V, regardless of the value of $I_F$.

## 2.3.2    The Effect of $V_F$ on Circuit Analysis

So, how does including the value of $V_F$ change the analysis of a diode circuit? To answer this question, let's take a look at the circuit shown in Figure 2.9. According to Kirchhoff's voltage law, the sum of the component voltages in the circuit must equal the applied voltage. By formula,

$$V_S = V_F + V_{R1}$$

If we substitute the value of $V_k$ (0.7 V) for $V_F$ and rearrange the equation to solve for $V_{R1}$, we get

$$V_{R1} = V_S - 0.7 \text{ V} \tag{2.1}$$

According to Ohm's law,

$$I_T = \frac{V_{R1}}{R_1}$$

Substituting equation (2.1) in place of $V_{R1}$ in the above equation, we get

$$I_T = \frac{V_S - 0.7 \text{ V}}{R_1} \tag{2.2}$$

Using equations (2.1) and (2.2), we can determine the following values for the circuit in Figure 2.9:

$$V_{R1} = V_S - 0.7 \text{ V} = 5 \text{ V} - 0.7 \text{ V} = \mathbf{4.3 \text{ V}}$$

and

$$I_T = \frac{V_S - 0.7 \text{ V}}{R_1} = \frac{5 \text{ V} - 0.7 \text{ V}}{1 \text{ k}\Omega} = \mathbf{4.3 \text{ mA}}$$

If you compare these values with those obtained for the same circuit in Example 2.2, you'll see the effect of including $V_F$ on the results of the circuit analysis. The current and voltage values calculated for the circuit in Figure 2.9 are summarized as follows:

| Value | Ideal | Practical |
|-------|-------|-----------|
| $V_F$ | 0 V | 0.7 V |
| $V_{R1}$ | 5 V | 4.3 V |
| $I_T$ | 5 mA | 4.3 mA |

Examples 2.3 and 2.4 further demonstrate the effect of $V_F$ on circuit calculations.

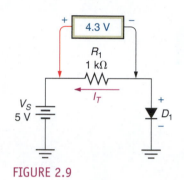

FIGURE 2.9

## EXAMPLE 2.3

Predict the voltmeter reading in Figure 2.10.

*Solution:* The voltage across the diode is assumed to be 0.7 V. Thus, the voltage across the resistor will be equal to the difference between the source voltage and the value of $V_F$. By formula,

$$V_{R1} = V_S - 0.7 \text{ V} = 6 \text{ V} - 0.7 \text{ V} = \textbf{5.3 V}$$

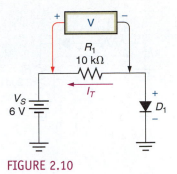

**PRACTICE PROBLEM 2.3**

A circuit like the one in Figure 2.10 has a 15 V source. Determine the voltage across the resistor.

**FIGURE 2.10**

## EXAMPLE 2.4

Determine the total current in the circuit shown in Figure 2.10.

*Solution:* The total circuit current is found as

$$I_T = \frac{V_S - 0.7 \text{ V}}{R_1} = \frac{6 \text{ V} - 0.7 \text{ V}}{10 \text{ k}\Omega} = \textbf{530 } \boldsymbol{\mu}\textbf{A}$$

**PRACTICE PROBLEM 2.4**

A circuit like the one shown in Figure 2.10 has a 5 V source and a 510 $\Omega$ resistor. Determine the value of $I_T$ for the circuit.

### 2.3.3 Percentage of Error

In most practical analysis problems, a calculated value is considered to be accurate enough if it is within ±10% of the actual measured value. The percentage of error in a given calculation is determined using

$$\% \text{ of error} = \frac{|X - X'|}{X} \times 100 \qquad \textbf{(2.3)}$$

**Lab Reference:** Percentage of error calculations appear throughout the lab manual, beginning in Exercise 2.

where $X$ = the measured value
$\quad\;\; X'$ = the calculated value

Using the ideal diode model in circuit calculations can introduce a percentage of error in the results that is not acceptable (that is, not within ±10%). For example, we used the ideal and practical diode models to determine the value of $V_{R1}$ for the circuit shown in Figure 2.9. The percentage of error introduced by using the ideal diode model in this case would be found as

$$\% \text{ of error} = \frac{|4.3 \text{ V} - 5 \text{ V}|}{4.3 \text{ V}} \times 100 = \textbf{16.28}\%$$

As you can see, the percentage of error is greater than 10% and therefore is not acceptable. This is why we use the practical diode model in most circuit analysis problems.

When more than one resistor is in series with a diode, the *total* resistance ($R_T$) must be used in determining the value of $I_T$, as was the case in all the circuits you studied in basic electronics. This point is illustrated in Example 2.5.

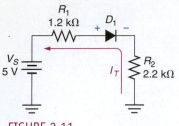

FIGURE 2.11

## EXAMPLE 2.5

Determine the value of $I_T$ for the circuit shown in Figure 2.11.

**Solution:** For the circuit shown, we can calculate the total circuit current using

$$I_T = \frac{V_S - 0.7\text{ V}}{R_T} = \frac{5\text{ V} - 0.7\text{ V}}{3.4\text{ k}\Omega} = \textbf{1.26 mA}$$

### PRACTICE PROBLEM 2.5

A circuit like the one in Figure 2.11 has the following values: $R_1 = 330\ \Omega$, $R_2 = 220\ \Omega$, and $V_S = 2$ V. Calculate the value of the circuit current.

Note that we used $R_T$ in place of $R_1$ in equation (2.2) to solve the circuit in Example 2.5.

Just as you must consider the total resistance in the analysis of a diode circuit, you must consider the *sum of the diode voltage drops* if there are several *series-connected* diodes in a circuit. This point is illustrated in Example 2.6.

## EXAMPLE 2.6

Determine the value of $I_T$ for the circuit shown in Figure 2.12.

**Solution:** With two diodes in the circuit, the total value of $V_F$ is assumed to be 1.4 V. Using this value in the place of 0.7 V in equation (2.2) allows us to determine the value of $I_T$ as follows:

$$I_T = \frac{V_S - 1.4\text{ V}}{R_1} = \frac{4\text{ V} - 1.4\text{ V}}{5.1\text{ k}\Omega} = \textbf{509.8 } \boldsymbol{\mu}\textbf{A}$$

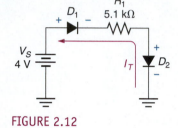

FIGURE 2.12

### PRACTICE PROBLEM 2.6

A series circuit consists of two forward-biased diodes, a 470 $\Omega$ resistor, a 330 $\Omega$ resistor, and a 6 V source. What is the value of $I_T$ for the circuit?

Now, here's one more example to tie everything together.

## EXAMPLE 2.7

Determine the value of $I_T$ for the circuit shown in Figure 2.13 *using the ideal diode model*. Then recalculate the value *using the practical diode model*. What is the percentage of error introduced by using the ideal diode model?

**Solution:** The ideal diode model assumes that $V_F = 0$ V. Therefore, the total applied voltage is dropped across the two resistors, and $I_T$ is found as

$$I_T = \frac{V_S}{R_T} = \frac{10\text{ V}}{3.3\text{ k}\Omega} = \textbf{3.03 mA}$$

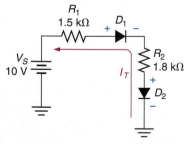

FIGURE 2.13

The practical diode model assumes that $V_F = 0.7$ V for each diode. Therefore, the value of $I_T$ is found as

$$I_T = \frac{V_S - 1.4\text{ V}}{R_T} = \frac{10\text{ V} - 1.4\text{ V}}{3.3\text{ k}\Omega} = \textbf{2.61 mA}$$

The percentage of error between the two calculations is found as

$$\% \text{ of error} = \frac{|2.61 \text{ mA} - 3.03 \text{ mA}|}{2.61 \text{ mA}} \times 100 = \textbf{16.1\%}$$

Note that the value of $I_T$ found using the *practical diode model* was used in the denominator of the fraction in the percentage of error calculation. This value was used because the practical value is always assumed to be closer than the ideal to the actual value of $I_T$.

**PRACTICE PROBLEM 2.7**
Refer to Practice Problem 2.6. Recalculate the value of total circuit current using the *ideal diode model*. Then determine the percentage of error introduced by using this diode model.

### 2.3.4   One Final Note

It can be argued that it isn't always necessary to include the diode forward voltage in circuit calculations. For example, if a forward-biased diode and a resistor are in series with a 100 V source, ignoring the 0.7 V across the diode causes an error of less than 1% in the circuit calculations (which is well within acceptable limits). While this argument is valid, we are always interested in getting the most accurate results possible (within reason), so we always use the practical diode model in our circuit calculations.

In the next section, we will look at the factors that must be considered when determining whether a specific diode can be used in a given situation. While these factors are not normally considered in voltage and current calculations, they are very important when it comes to actually working with diode circuits.

1. What diode characteristic must be considered in the mathematical analysis of a diode circuit?  ◀ **Section Review**

2. Which diode characteristics are normally considered when replacing one diode with another?

3. What is the assumed value of a knee voltage for a silicon diode?

4. List the characteristics of the practical forward-biased diode.

5. How does including the value of $V_F$ affect the accuracy of circuit calculations?

6. How accurate must practical calculations be before they are considered to be acceptable?

7. A circuit voltage is calculated to be 10 V. The actual value of this voltage is 12.2 V. What is the percentage of error in the calculation?

8. Why do you think the *measured* value is used (as opposed to the *calculated* value) in  ◀ **Critical Thinking**
the denominator of equation (2.3)?

## 2.4   Other Practical Considerations

While troubleshooting a circuit, you find that a diode in the circuit is faulty and must be  ◀ *OBJECTIVE 4*
replaced. Now you discover that you have a wide variety of diodes in stock, but none of them has the same part number as the one that needs replacing. How can you determine whether a specific diode can be used in place of the faulty one? Several diode characteristics must be considered when determining whether a specific diode can be used in a given circuit. These characteristics are *peak reverse voltage*, *average forward current*, and *forward power dissipation*.

### 2.4.1 Peak Reverse Voltage ($V_{RRM}$)

Any insulator will conduct if an applied voltage is high enough to cause the insulator to break down. For a reverse-biased diode, the *maximum* reverse voltage that *won't* force the diode to conduct is called the **peak reverse voltage ($V_{RRM}$)**. When $V_{RRM}$ is exceeded, the depletion layer may break down and allow the diode to conduct in the reverse direction. Typical values of $V_{RRM}$ range from a few volts (for zener diodes) to thousands of volts.

The effect that $V_{RRM}$ has on the diode characteristic curve is illustrated in Figure 2.14. Note that the value of reverse current ($I_R$) is shown to be zero until the value of $V_{RRM}$ ($-70$ V, in this case) is exceeded. When $V_R > V_{RRM}$, the value of $I_R$ increases rapidly as the depletion layer breaks down. Normally, when a *pn*-junction diode is forced to conduct in the reverse direction, the device is destroyed. A *zener diode*, on the other hand, is designed to work in the reverse direction without harming the diode. This point is discussed in detail later in this chapter.

**Peak reverse voltage ($V_{RRM}$)**
The *maximum* reverse voltage that won't force a *pn*-junction to conduct.

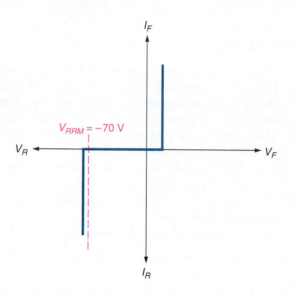

**FIGURE 2.14**

The current generated through a *pn*-junction diode when $V_R > V_{RRM}$ is called **avalanche current**. This name describes a process where one free electron bumps other electrons in the diode, causing them to break free from their covalent bonds, which then rapidly causes even more electrons to be broken free, and so on. The result is that the diode is destroyed by excessive current and the heat it produces.

Peak reverse voltage is a very important **parameter** (limit). When you are considering whether to use a specific diode in a given application, you must ensure that the component's $V_{RRM}$ rating is greater than the maximum reverse voltage in the circuit. For example, consider the circuit in Figure 2.15. During the negative alternation of the input, the peak value of the source (50 V) is dropped across the reverse-biased diode. Therefore, the $V_{RRM}$ rating for $D_1$ must be greater than 50 V to ensure that the component isn't damaged or destroyed by the voltage source.

**Avalanche current**
The current that occurs when $V_{RRM}$ is exceeded. Avalanche current can generate sufficient heat to destroy a *pn*-junction diode.

**Parameter**
A limit.

*A Practical Consideration:*
Most peak reverse voltage ratings for *pn*-junction diodes are multiples of 50 or 100 V. Common values are 50, 100, 150, 200, and so on.

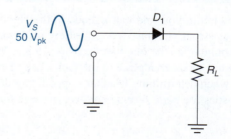

**FIGURE 2.15**

In practice, the diode used in a given circuit should have a $V_{RRM}$ that is at least 20% greater than the maximum voltage it is expected to block. For example, with a peak reverse voltage $(V_{R(pk)})$ of 50 V, the diode used in Figure 2.15 would require a *minimum* $V_{RRM}$ rating of

$$V_{RRM} = 1.2V_{R(pk)} = (1.2)(50 \text{ V}) = \textbf{60 V} \quad (minimum)$$

As long as the diode used in the circuit has a $V_{RRM}$ rating that is equal to (or greater than) 60 V, it will be able to handle minor variations in the source amplitude without being driven beyond its reverse voltage limit.

One point should be made at this time: As long as the $V_{RRM}$ rating of a diode is at least 20% greater than the maximum reverse voltage produced in a circuit, we are not really concerned about its exact value. For example, if the diode in Figure 2.15 had to be replaced, we could use one with a $V_{RRM}$ rating of 100, 200, or even 1000 V. This point is demonstrated in Example 2.8.

### EXAMPLE 2.8

The diode in the circuit shown in Figure 2.16 is faulty and must be replaced. When checking the parts bin, you see that you have the following diodes in stock:

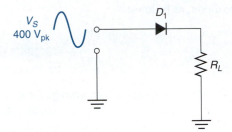

FIGURE 2.16

| Diode Part Number | Peak Reverse Voltage |
|---|---|
| 1N4001 | 50 |
| 1N4002 | 100 |
| 1N4003 | 200 |
| 1N4004 | 400 |
| 1N4005 | 600 |
| 1N4006 | 800 |
| 1N4007 | 1000 |

Which of these components could be used to replace the faulty diode?

*Solution:* Since the peak value of the source is 400 $V_{pk}$, there is a full 400 V across the diode terminals whenever the device is reverse biased. Therefore, the $V_{RRM}$ rating of the replacement part must be *at least 20% greater than* 400 V. Therefore, the only diodes we can use safely are the 1N4005, 1N4006, or 1N4007.

#### PRACTICE PROBLEM 2.8
A circuit like the one in Figure 2.16 has a 175 $V_{pk}$ signal source. Which of the diodes listed in Example 2.8 could be used safely in the circuit?

---

The value of $V_{RRM}$ for a given *pn*-junction diode can be obtained from the *specification sheet* (or *data sheet*) for the component. A **specification sheet** (spec sheet) *lists all the important parameters and operating characteristics of a device or circuit*. The specification sheet for a given component can usually be obtained from the manufacturer's website or hard-copy data books. You will be shown in Section 2.6 how to find the value of $V_{RRM}$ (and many other important parameters) on a spec sheet.

**Specification sheet**
A listing of all the important parameters and operating characteristics of a device or circuit.

Peak reverse voltage is only one of several parameters that must be considered before trying to use a specific diode in a given application. Another parameter that must be considered is the *average forward current* rating of the diode.

## 2.4.2 Average Forward Current ($I_0$)

**Average forward current**
The maximum allowable value of dc forward current for a diode.

The **average forward current** rating of a diode is the *maximum allowable value of dc forward current*. For example, the 1N4001 diode has an average forward current rating of 1 A. This means that the dc forward current through the diode must never be allowed to exceed 1 A. If the dc forward current through the diode is allowed to exceed 1 A, the diode may be destroyed from excessive heat. The average forward current rating is a parameter found on the spec sheet for the component.

When you are considering whether to use a specific diode in a specific circuit, you must determine the average value of forward current ($I_F$) for the circuit. Then you must make sure that the diode you want to use has an average forward current rating that is at least 20% greater than the value of $I_F$ for the circuit. This point is illustrated in Example 2.9.

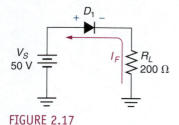

FIGURE 2.17

### EXAMPLE 2.9

Determine the *minimum* average forward current rating that would be required for the diode in Figure 2.17.

*Solution:* The forward current in the circuit is calculated (as always) using the practical model of the diode, as follows:

$$I_F = \frac{V_S - 0.7\text{ V}}{R_L} = \frac{50\text{ V} - 0.7\text{ V}}{200\ \Omega} = \textbf{246.5 mA}$$

Thus, any diode used in the circuit must have an average forward current rating that is greater than 246.5 mA. In practice, you would use a diode with an average forward current rating that is *at least 20% greater than* the calculated value of $I_F$, as follows:

$$I_0 = (1.2)(246.5\text{ mA}) = \textbf{295.8 mA} \quad (\textit{minimum})$$

### PRACTICE PROBLEM 2.9

A circuit like the one shown in Figure 2.17 has a 100 V source. If the resistor has a value of 51 $\Omega$, what is the minimum allowable average forward current rating for the diode in the circuit?

*A Practical Consideration:*
Let's say that you calculate the dc forward current in a circuit to be 1 A. If you use a diode that has an average forward current rating of 1 A in the circuit, you will be pushing the diode to its limit, which will shorten the lifetime of the component. If, on the other hand, you use a diode with an average forward current rating of 10 A, the diode will not be pushed to its limit and will last much longer.

In Chapter 3, you will be shown how to determine the *dc equivalent current* for an ac source. When dealing with ac circuits, you will need to determine the *average* (or *dc equivalent*) current for the circuit. Then this dc equivalent current value must be compared to the average forward current rating of the diode to see if it can be used in the circuit.

## 2.4.3 Forward Power Dissipation ($P_{D(\text{max})}$)

**Forward power dissipation ($P_{D(\text{max})}$)**
A diode rating that indicates the maximum possible power dissipation of the foward-biased diode.

Many diodes have a **forward power dissipation** rating. This rating indicates the *maximum possible power dissipation of the device when it is forward biased.*

Recall from your study of basic electronics that power is found as

$$P = IV$$

where $P$ = the power dissipated by a component
   $I$ = the device current
   $V$ = the voltage across the device

Using the basic power equation and the values of $V_F$ and $I_F$ for a diode, you can determine the required forward power dissipation rating for a replacement component, as demonstrated in Example 2.10.

## EXAMPLE 2.10

Calculate the *minimum* forward power dissipation rating for any diode that would be used in the circuit shown in Figure 2.18.

*Solution:* First, we have to calculate the total circuit current. This current is found as

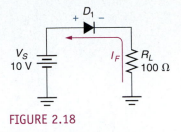

FIGURE 2.18

$$I_F = \frac{10 \text{ V} - 0.7 \text{ V}}{100 \text{ }\Omega} = \textbf{93 mA}$$

Using $I_F = 93$ mA and $V_F = 0.7$ V, the diode power dissipation is found as

$$P_F = I_F V_F = (93 \text{ mA})(0.7 \text{ V}) = \textbf{65.1 mW}$$

To provide a safety margin, the diode used should have a forward power dissipation rating that is *at least 20% greater than* the value of $P_F$. Thus, the minimum value of $P_{D(max)}$ is found as

$$P_{D(max)} = 1.2P_F = (1.2)(65.1 \text{ mW}) = \textbf{78.12 mW} \quad (minimum)$$

### PRACTICE PROBLEM 2.10

A circuit like the one shown in Figure 2.18 has a 20 V source and a 68 $\Omega$ series resistor. What is the minimum required forward power dissipation rating for any diode used in the circuit?

Some diode specification sheets provide a forward power dissipation rating rather than an average forward current rating. When this is the case, the average forward current rating for the diode can be found using

$$I_0 = \frac{P_{D(max)}}{V_F} \qquad (2.4)$$

where $I_0$ = the limit on the average forward current
$P_{D(max)}$ = the forward power dissipation rating of the diode
$V_F$ = the forward voltage across the diode, assumed to be 0.7 V for a silicon *pn*-junction diode

This equation is simply a variation on the standard power dissipation equation. Its use is demonstrated in Example 2.11.

## EXAMPLE 2.11

A diode has a forward power dissipation rating of 500 mW. What is the maximum allowable value of forward current for the device?

*Solution:* The value of $I_0$ is found as

$$I_0 = \frac{P_{D(max)}}{V_F} = \frac{500 \text{ mW}}{0.7 \text{ V}} = \textbf{714.29 mA}$$

To provide a safety margin, the forward current ($I_F$) is normally restricted to 80% of the calculated value of $I_0$. Therefore, the maximum allowable forward current is found as

$$I_{F(max)} = 0.8I_0 = (0.8)(714.29 \text{ mA}) = \textbf{571.43 mA}$$

This value indicates that the diode described in this example can be used safely in any circuit with a maximum forward current that is less than (or equal to) 571.43 mA.

**PRACTICE PROBLEM 2.11**

Show that the power dissipation rating of 500 mW in Example 2.11 will be exceeded if the diode forward current equals 750 mA.

### 2.4.4    Summary

When can one diode be
replaced with another?

When you are trying to replace one diode with another, three main parameters must be considered. Before substituting one diode for another, ask yourself:

1. Is the $V_{RRM}$ rating of the replacement diode at least 20% greater than the maximum reverse voltage in the circuit?
2. Is the *average forward current* rating of the replacement diode at least 20% greater than the average (dc) value of $I_F$ in the circuit?
3. Is the *forward power dissipation* rating of the replacement diode at least 20% greater than the value of $P_F$ in the circuit?

If the answer to all these questions is *yes*, then you can use the diode in the circuit.

**Section Review ▶**

1. What is *peak reverse voltage*?
2. Why do you need to consider the $V_{RRM}$ rating for a diode before attempting to use the diode in a specific circuit?
3. A diode circuit has a 29 $V_{pk}$ ac source. Which of the diodes listed in Example 2.8 could be used in the circuit without having reverse breakdown problems? Explain your answer.
4. What is *average forward current*?
5. How do you determine whether the *average forward current* rating of a diode will be exceeded in a given circuit?
6. What is the *forward power dissipation* rating of a diode?
7. How do you determine whether a given diode has a *forward power dissipation* rating that is high enough to allow the device to be used in a specific circuit?
8. When you know the *forward power dissipation* rating for a given diode, how can you determine the limit on its average forward current?

## 2.5    The Complete Diode Model

The complete diode model most accurately represents the true operating characteristics of the diode. Two factors that make this model so accurate are *bulk resistance* and *reverse current*. When these factors are taken into account, we get the diode characteristic curve shown in Figure 2.19. This illustration will be referred to throughout our discussion on the complete diode model.

### 2.5.1    Bulk Resistance ($R_B$)

As you learned in Chapter 1, *bulk resistance* is the natural resistance of the diode *p*-type and *n*-type materials. The effect of bulk resistance on diode operation can be seen in the *forward operation region* of the curve. As Figure 2.19 illustrates, $V_F$ is *not* constant, but rather, varies with the value of $I_F$. The change in $V_F$ ($\Delta V_F$) is caused by the diode current passing through the bulk resistance of the diode. This concept is illustrated by the diode

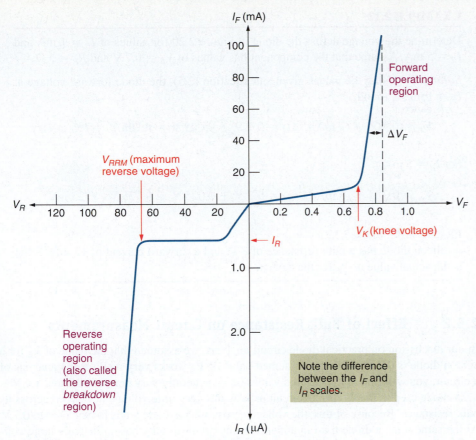

**Lab Reference:** The forward region of the diode curve shown here is plotted (using measured values) in Exercise 1.

FIGURE 2.19    Complete-model diode curve.

equivalent circuit shown in Figure 2.20. The voltage source in the diode ($V_B$) represents the voltage required to overcome the barrier potential of the component. The resistor ($R_B$) represents the bulk resistance of the component. The diode current ($I_F$) passing through the bulk resistance develops a voltage equal to $I_F R_B$. As shown in the meter, the diode forward voltage equals the sum of $V_B$ and $I_F R_B$. By formula,

$$V_F = V_B + I_F R_B \qquad (2.5)$$

Note that as $I_F$ increases, so does $I_F R_B$. Therefore, the total voltage across a diode varies with the value of $I_F$. This is shown in Example 2.12.

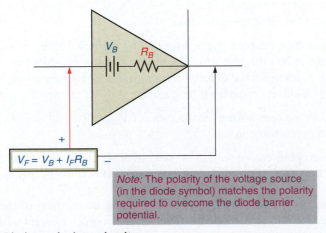

FIGURE 2.20    Diode equivalent circuit.

## EXAMPLE 2.12

Determine the voltage across the diode in Figure 2.20 for values of $I_F = 1$ mA and $I_F = 5$ mA. Assume that the component has values of $V_B = 0.7$ V and $R_B = 5\ \Omega$.

*Solution:* Using the values given and equation (2.5), the diode forward voltage at $I_F = 1$ mA is found as

$$V_F = 0.7\ \text{V} + (1\ \text{mA})(5\ \Omega) = 0.7\ \text{V} + 0.005\ \text{V} = \mathbf{0.705\ V} \quad (705\ \text{mV})$$

For $I_F = 5$ mA,

$$V_F = 0.7\ \text{V} + (5\ \text{mA})(5\ \Omega) = 0.7\ \text{V} + 0.025\ \text{V} = \mathbf{0.725\ V} \quad (725\ \text{mV})$$

**PRACTICE PROBLEM 2.12**

A silicon diode has a bulk resistance of 8 $\Omega$ and a forward current of 12 mA. What is the actual value of $V_F$ for the device?

### 2.5.2 Effect of Bulk Resistance on Circuit Measurements

In our discussion on practical diode circuit analysis, we assumed that the value of $V_F$ for a silicon diode is 0.7 V. While this assumed value of $V_F$ works very well for the analysis of a circuit, you will find that measured values of $V_F$ generally vary between 0.7 and 1.1 V.

A diode used in a low-current circuit usually has very little voltage developed across its bulk resistance. Because of this, the voltage across such a diode tends to be closer to 0.7 V. At the same time, a diode used in a high-current circuit usually has a relatively large voltage developed across its bulk resistance, resulting in a value of $V_F$ that is closer to 1.1 V.

How does bulk resistance affect $V_F$ readings?

### 2.5.3 Reverse Current ($I_R$)

When a diode is reverse biased, the depletion layer reaches its maximum width, and conduction through the diode (ideally) stops. In reality, however, there is a very small amount of current through a reverse-biased diode. This *reverse current* is illustrated in the *reverse operating region* of the diode curve (Figure 2.19). Reverse current is made up of two independent currents: *reverse saturation current, $I_S$*, and *surface-leakage current, $I_{SL}$*. By formula,

$$I_R = I_S + I_{SL} \tag{2.6}$$

where $I_R$ = the diode reverse current
$\quad I_S$ = the reverse saturation current
$\quad I_{SL}$ = the surface-leakage current

**Reverse saturation current ($I_S$)**
A current caused by thermal activity in a reverse-biased diode. $I_S$ is temperature dependent.

**Reverse saturation current ($I_S$)** is produced by thermal activity in the diode materials. This current *varies directly with temperature* and is *independent of the value of reverse voltage*. $I_S$ accounts for a major portion of diode reverse current, so reverse current ($I_R$) remains relatively constant as long as:

1. Reverse voltage does not exceed the value of $V_{RRM}$.
2. Temperature remains constant.

The first of these points can be seen in the reverse operating region of the curve in Figure 2.19. The effect of temperature on reverse current is described in greater detail later in this section.

**Surface-leakage current ($I_{SL}$)**
A current along the surface of a reverse-biased diode. $I_{SL}$ is $V_R$ dependent.

**Surface-leakage current ($I_{SL}$)** is generated along the surface of the diode by the reverse voltage across the component. This current *varies directly with the value of reverse voltage* and is *independent of temperature*. However, since $I_{SL} \ll I_S$, there is no significant change in $I_R$ when $I_{SL}$ changes.

The total value of $I_R$ for most diodes is typically in the microampere range or less. However, as temperature increases, $I_S$ and, consequently, $I_R$ increase. This point is discussed in detail later in this section.

What is the typical range of $I_R$ values?

### 2.5.4 Effect of Reverse Current on Circuit Measurements

Until now, we have assumed that the value of $I_R$ is zero. As you now know, a small amount of reverse current occurs in a diode circuit. This reverse current causes a slight voltage to be developed across any resistance in the circuit. An example of this is shown in Figure 2.21. We have an applied voltage of 25 V and a diode reverse current of 20 µA. We can calculate the component voltage values as follows: The voltage across the load resistor is found as

$$V_{RL} = I_R R_L = 20 \text{ µA} \times 10 \text{ k}\Omega = \mathbf{200 \text{ mV}}$$

Now, the voltage across the diode is found as

$$V_{D1} = V_S - V_{RL} = 25 \text{ V} - 200 \text{ mV} = \mathbf{24.8 \text{ V}}$$

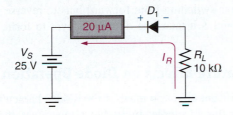

**FIGURE 2.21**

The voltage developed across any series resistance by $I_R$ is usually insignificant. For example, only 0.8% of the supply voltage in Figure 2.21 is developed across $R_1$. However, knowing the potential effects of reverse current is important for several reasons. First, any voltage across the series resistance may increase dramatically with increases in operating temperature (and the resulting increases in diode reverse current). Second, it explains why we often do not measure the full applied voltage across a reverse-biased diode. (In most cases, the voltage across a reverse-biased diode is slightly lower than the applied voltage.)

### 2.5.5 Diode Capacitance

An insulator placed between two closely spaced conductors forms a capacitor. When a diode is *reverse biased* as shown in Figure 2.22, it forms a depletion layer (insulator) between two semiconductor materials (which are conductors by comparison). Therefore,

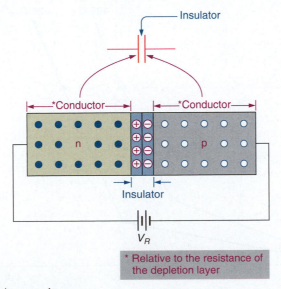

**FIGURE 2.22** Diode capacitance.

**Junction capacitance**

The capacitance of a reverse-biased *pn* junction, formed by the depletion layer (insulator) and the semiconductor materials.

a reverse-biased diode has some measurable amount of **junction capacitance**. Under normal circumstances, junction capacitance can be ignored because of its extremely low value, typically in the low pF range. However, when the *high-frequency* operation of a diode is analyzed, junction capacitance can become extremely important. The effects of junction capacitance on high-frequency operation are discussed in detail in Chapter 14.

Incidentally, there is a type of diode designed to make use of its junction capacitance. This diode, called a *varactor*, is introduced in Chapter 5.

### 2.5.6    Diffusion Current

Refer to the forward operating region of the diode curve in Figure 2.19. The 0.7 V point on the diode curve is labeled *knee voltage* ($V_k$). Below the knee voltage, you will notice that $I_F$ does *not* instantly drop to the zero point. The reason for this is simple: When $V_F$ goes below the barrier potential of the diode, the device starts to form a depletion layer. However, until the diode is actually reverse biased, the depletion layer will not reach its maximum width and thus will not reach its maximum resistance. In other words, as long as there is *some* forward voltage, the depletion layer will not be at its maximum width, and some amount of $I_F$ will occur. This small amount of $I_F$ is called **diffusion current**. Note that when a diode is switching from forward bias to reverse bias, the diffusion current will last only as long as it takes the depletion layer to form, generally a few milliseconds or less.

**Diffusion current**

The $I_F$ below the knee voltage.

### 2.5.7    Temperature Effects on Diode Operation

Temperature has a significant effect on most of the diode characteristics discussed in this section. The reason for this is simple: Increased temperature means increased thermal activity and decreased diode resistance, as shown in Chapter 1. This holds true for both forward and reverse diode operation.

How does a change in temperature affect $I_F$ and $V_F$?

The effects of increased temperature on forward diode operation are illustrated in Figure 2.23. As you can see, there are two forward operation curves. One represents the diode operation at 25°C, and the other represents the diode operation at 100°C. From the graph, we can draw two conclusions regarding forward diode operation and temperature:

1. *As temperature increases, $I_F$ increases at a specified value of $V_F$.* This is illustrated in Figure 2.23 by the two points labeled $I_1$ and $I_2$. As you can see, both of these points fall on the $V_F = 0.7$ V line. However, $I_2$ is greater than $I_1$ due to the increased thermal activity in the diode. As temperature has increased from 25°C to 100°C, the value of $I_F$ has increased from 5 mA ($I_1$) to 25 mA ($I_2$).

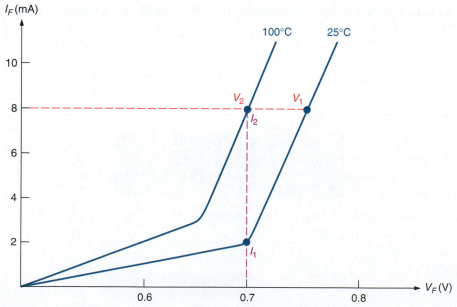

**FIGURE 2.23**    Temperature effects on forward operation.

**2.** *As temperature increases, $V_F$ decreases at a specified value of $I_F$. This is illustrated by the $V_1$ and $V_2$ points on the curve. Note that both points correspond to a value of $I_F = 8$ mA. As temperature increases, $V_F$ decreases from approximately 0.75 V ($V_1$) to 0.7 V ($V_2$).*

*In practice, a rise in temperature usually results in both a slight increase in $I_F$ and a slight decrease in $V_F$.*

The effect of temperature on diode reverse current ($I_R$) is basically the same as the effect on $I_F$. The effect of temperature on $I_R$ is illustrated in Figure 2.24. You may recall that $I_R$ is equal to the sum of reverse saturation current ($I_S$) and surface-leakage current ($I_{SL}$). Since $I_S$ is normally much greater than $I_{SL}$, it is safe to assume that $I_R$ is approximately equal to $I_S$.

How does a change in temperature affect $I_R$?

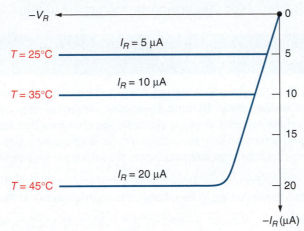

**FIGURE 2.24** Temperature effects on reverse operation.

## 2.5.8    The Bottom Line

Generally, you will not have to deal with most of the characteristics discussed in this section unless you are designing a circuit with very low variation tolerances or very high operating frequencies. Even in most practical circuit designs, such as rectifiers and amplifiers, knowing the diode's parameters is more important than knowing the *exact* values of diode voltage, current, resistance, and capacitance.

When you are involved in practical circuit analysis and troubleshooting, you need to consider only the more practical aspects of diode operation. These aspects are summarized as follows:

A summary of *practical* diode characteristics.

1. When the forward voltage across a diode reaches the barrier potential, the diode will start to conduct. From that point on, the value of $V_F$ will be approximately equal to the barrier potential.
2. As $V_F$ decreases below the barrier potential, the diode will start to turn off and begin to act like an open switch. This "open-switch" characteristic will continue for all reverse voltage values up to peak reverse voltage, $V_{RRM}$.
   **a.** As long as $V_R \leq V_{RRM}$, the total applied voltage will be dropped across the diode.
   **b.** When $V_R > V_{RRM}$, the diode may break down and conduct in the reverse direction. When this happens, you will usually end up replacing the diode.
3. The forward current through a diode is limited by the applied voltage and the resistance values that are external to the diode. For all practical purposes, the reverse current through a diode is zero.

◀ **Section Review**

1. What is *bulk resistance*?
2. What effect does bulk resistance have on circuit current measurements?
3. What is the relationship between $I_F$ and $V_F$?
4. What is reverse current?
5. Which component of reverse current is affected by temperature?

6. Which component of reverse current is affected by the amount of reverse bias?

7. Which component of reverse current makes up a majority of that current?

8. What effect does $I_R$ have on the measured value of voltage across any series resistance?

9. Describe the reverse-biased diode as a capacitor.

10. What is diffusion current?

11. What effect does an increase in temperature have on $I_F$ and $V_F$?

12. What effect does an increase in temperature have on $I_R$?

**Critical Thinking** ▶ 13. Determine the reverse resistance of the diode in Figure 2.21.

14. Which do you think would be of greater concern: the effect of an increase in temperature on diode *forward* operation or diode *reverse* operation? Why?

## 2.6 Diode Specification Sheets

The *specification sheet*, or *spec sheet*, for any component lists the *parameters* and *operating characteristics* of the device. Because parameters are limits, they are almost always designated as *maximum* values. Operating characteristics, on the other hand, may be designated as *minimum*, *maximum*, or *typical* values, depending on the rating.

Why are parameters important?

Diode operating characteristics and parameters are important for several reasons:

1. They indicate whether a given diode can be used for a specific application.
2. They establish the operating limits of any circuit designed to use the diode.

You have already been introduced to most of the commonly used diode characteristics and parameters. In this section, we look at how these characteristics and parameters may be found on the spec sheet of a given diode.

### 2.6.1 Spec Sheet Organization

Diode spec sheets are commonly divided into two sections: *maximum ratings* and *electrical characteristics*. This can be seen in Figure 2.25, which contains the spec sheet for the 1N4001–1N4007 (or 1N400X) series diodes.

The **maximum ratings** table contains the diode parameters that must not be exceeded under any circumstances. If any of these parameters are exceeded, you will more than likely have to replace the diode.

The **electrical characteristics** table contains the guaranteed operating characteristics of the device. As long as the *maximum ratings* limits are observed, the diode is guaranteed to work within the limits shown in the *electrical characteristics* table.

Confused? Let's take a look at the *maximum reverse current* rating in the *electrical characteristics* table. The 1N400X series of diodes is guaranteed to have a *maximum reverse current* ($I_R$) of 10 μA at 25°C, with a *typical* value of 0.05 μA. These values assume that the magnitude of $V_R$ is not greater than the *reverse breakdown rating* listed on the spec sheet. If it is, the diode reverse current may exceed the maximum value listed. If any diode parameter is exceeded, the *electrical characteristics* values of the device cannot be guaranteed.

### 2.6.2 Diode Maximum Ratings

Several of the parameters listed in the *maximum ratings* table were introduced earlier in this chapter. Table 2.2 summarizes the ratings listed and their meanings.

A few notes:

Maximum ratings<br>
Device parameters that must not be exceeded under any circumstances.<br><br>
Electrical characteristics<br>
The guaranteed operating characteristics of the device.

1. **Peak repetitive reverse voltage** ($V_{RRM}$) indicates the maximum allowable reverse voltage that can be applied to the device. The 1N4001 has a $V_{RRM}$ rating of 50 V. If you apply a reverse voltage to the device that is *greater than* 50 V, the device *may* go into reverse breakdown. Note that this rating is also called *working peak reverse voltage* or *dc blocking voltage*.

Peak repetitive reverse voltage ($V_{RRM}$)<br>
The maximum allowable value of diode reverse voltage.

# 1N4001, 1N4002, 1N4003, 1N4004, 1N4005, 1N4006, 1N4007

**1N4004 and 1N4007 are Preferred Devices**

## Axial Lead Standard Recovery Rectifiers

This data sheet provides information on subminiature size, axial lead mounted rectifiers for general–purpose low–power applications.

**Mechanical Characteristics**

- Case: Epoxy, Molded
- Weight: 0.4 gram (approximately)
- Finish: All External Surfaces Corrosion Resistant and Terminal Leads are Readily Solderable
- Lead and Mounting Surface Temperature for Soldering Purposes: 220°C Max. for 10 Seconds, 1/16″ from case
- Shipped in plastic bags, 1000 per bag.
- Available Tape and Reeled, 5000 per reel, by adding a "RL" suffix to the part number
- Available in Fan–Fold Packaging, 3000 per box, by adding a "FF" suffix to the part number
- Polarity: Cathode Indicated by Polarity Band
- Marking: 1N4001, 1N4002, 1N4003, 1N4004, 1N4005, 1N4006, 1N4007

**ON Semiconductor™**

http://onsemi.com

**LEAD MOUNTED RECTIFIERS 50–1000 VOLTS DIFFUSED JUNCTION**

CASE 59–03
AXIAL LEAD
PLASTIC

**MARKING DIAGRAM**

AL
1N
400x
YYWW

| AL | = Assembly Location |
|---|---|
| 1N400x | = Device Number |
| x | = 1, 2, 3, 4, 5, 6 or 7 |
| YY | = Year |
| WW | = Work Week |

## MAXIMUM RATINGS

| Rating | Symbol | 1N4001 | 1N4002 | 1N4003 | 1N4004 | 1N4005 | 1N4006 | 1N4007 | Unit |
|---|---|---|---|---|---|---|---|---|---|
| *Peak Repetitive Reverse Voltage Working Peak Reverse Voltage DC Blocking Voltage | $V_{RRM}$ $V_{RWM}$ $V_R$ | 50 | 100 | 200 | 400 | 600 | 800 | 1000 | Volts |
| *Non–Repetitive Peak Reverse Voltage (halfwave, single phase, 60 Hz) | $V_{RSM}$ | 60 | 120 | 240 | 480 | 720 | 1000 | 1200 | Volts |
| *RMS Reverse Voltage | $V_{R(RMS)}$ | 35 | 70 | 140 | 280 | 420 | 560 | 700 | Volts |
| *Average Rectified Forward Current (single phase, resistive load, 60 Hz, $T_A$ = 75°C) | $I_O$ | | | | 1.0 | | | | Amp |
| *Non–Repetitive Peak Surge Current (surge applied at rated load conditions) | $I_{FSM}$ | | | | 30 (for 1 cycle) | | | | Amp |
| Operating and Storage Junction Temperature Range | $T_J$ $T_{stg}$ | | | | –65 to +175 | | | | °C |

*Indicates JEDEC Registered Data

## ELECTRICAL CHARACTERISTICS*

| Rating | Symbol | Typ | Max | Unit |
|---|---|---|---|---|
| Maximum Instantaneous Forward Voltage Drop ($i_F$ = 1.0 Amp, $T_J$ = 25°C) | $v_F$ | 0.93 | 1.1 | Volts |
| Maximum Full–Cycle Average Forward Voltage Drop ($I_O$ = 1.0 Amp, $T_L$ = 75°C, 1 inch leads) | $V_{F(AV)}$ | – | 0.8 | Volts |
| Maximum Reverse Current (rated dc voltage) ($T_J$ = 25°C) ($T_J$ = 100°C) | $I_R$ | 0.05 1.0 | 10 50 | μA |
| Maximum Full–Cycle Average Reverse Current ($I_O$ = 1.0 Amp, $T_L$ = 75°C, 1 inch leads) | $I_{R(AV)}$ | – | 30 | μA |

*Indicates JEDEC Registered Data

**FIGURE 2.25** The 1N400X series specifications. (Copyright of Semiconductor Component Industries, LLC. Used by permission.)

**TABLE 2.2** Diode Maximum Ratings

| Rating | Discussion |
|---|---|
| Peak repetitive reverse voltage, $V_{RRM}$ | This is the maximum allowable value of diode reverse voltage. This rating applies to both dc and peak signal voltages. |
| Nonrepetitive peak reverse voltage, $V_{RSM}$ | This is the maximum allowable value of a single-event reverse voltage. For any diode, this value is always greater than $V_{RRM}$. |
| RMS reverse voltage, $V_{R(rms)}$ | This is simply the rms equivalent of $V_{RRM}$. Its value can be found as $V_{R(rms)} = 0.707 V_{RRM}$. |
| Average rectified forward current, $I_0$ | This is the value of average forward current described earlier in this chapter. For the 1N400X series, this value is 1 A when the ambient temperature is 75°C. Note that the limit on forward current decreases as temperature increases. |
| Nonrepetitive peak surge current, $I_{FSM}$ | This is the maximum allowable value of forward current *surge* (a high-value current or voltage that typically does not occur at regular intervals). The 1N400X series diodes are designed to withstand a single 30 A surge. |
| Operating and storage junction temperature, $T_J$ or $T_{stg}$ | The range of temperatures that the diode can withstand. |

2. The reverse voltage ratings are the only ratings that distinguish the diodes in the 1N400X series from each other. In other words, the diodes listed have the same maximum ratings and electrical characteristics *outside their reverse voltage characteristics*. This is typical for a given series of diodes.
3. The *average forward current* ($I_0$) rating shows that the maximum allowable average forward current is measured at a specific temperature. The reason is simple: The current through a diode generates *heat*, so the maximum allowable current through a diode depends on *how much heat the component can dissipate*. When the air that surrounds a diode is hot, the diode cannot dissipate as much heat. Therefore, the limit on device current decreases.
4. The *peak surge current* ($I_{FSM}$) indicates the diode's ability to handle a *surge* (short duration, extremely high value) of current. We will cover surge current and one of its common sources in Chapter 3.

### 2.6.3 Diode Terminal Identification

A number of diode packages are shown in Figure 2.26. Whenever possible, use the spec sheet diagram to identify the anode and cathode leads of a given diode. In most cases, the terminals are identified as shown in Figure 2.26.

### 2.6.4 Diode Parameters and Device Substitution

*OBJECTIVE 5* ▶ It was stated earlier in the chapter that *average forward current* and *peak reverse voltage* ratings are two of the primary considerations when substituting one diode for another. To make component substitutions easier, diode data books and manufacturer websites sometimes contain *selector guides*. These guides group diodes by *average forward current* (or *forward power dissipation*) and *peak reverse voltage*. An example of a selector guide is shown in Figure 2.27.

The selector guide allows you to select a diode based on circuit requirements. For example, let's say that you need a diode with an average forward current rating of 1.5 A

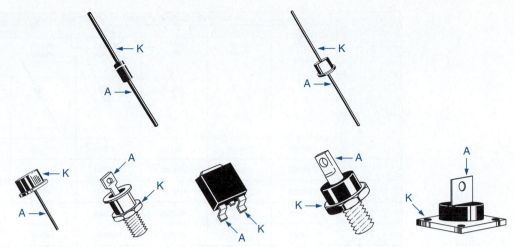

FIGURE 2.26  Common types of diodes.

and a peak reverse voltage of 100 V. You would locate the column that corresponds to 1.5 A and cross-match it with the row that has a $V_{RRM}$ of 100 V. The block that corresponds to both values contains the number 1N5392. Thus, the 1N5392 would be suitable for the desired application.

### 2.6.5  Electrical Characteristics

As stated earlier, the values provided under the *electrical characteristics* heading indicate the guaranteed operating characteristics of the diode. Table 2.3 gives a brief explanation of the electrical characteristics listed in Figure 2.25. Note that the spec sheet lists both *typical* and *maximum* values for many of the characteristics. When analyzing the operation of the component in a circuit, the value listed as *typical* is used. When determining circuit tolerances for circuit development purposes, the *maximum* value is generally used. When only one value is given (whether typical or maximum), it is used for all circuit analyses.

> *A Practical Consideration:* Any diode in the selector guide to the right and/or below the 1N5392 can be used in the application. As long as the forward current and reverse voltage ratings are *at least* 1.5 A and 100 V, the diode can be used. (This assumes the diode has the proper body type.)

TABLE 2.3  Diode Electrical Characteristics

| | |
|---|---|
| Maximum forward voltage drop, $V_F$ | This is the maximum value that $V_F$ will ever reach. All $V_F$ values are guaranteed to be 1.1 V or less. Note that this value was determined at a temperature of 25°C. On the average, $V_F$ will decrease by about 1.8 mV for every 1°C rise in temperature above 25°C. |
| Maximum full-cycle average forward voltage drop, $V_{F(AV)}$ | This is the maximum *average* forward voltage ($V_F$). For 1N400X, this value is 0.8 V. Note that this parameter is also temperature dependent and was measured at 75°C. |
| Maximum reverse current, $I_R$ | These ratings, 10 and 50 μA, were given for 25° and 100°C, respectively. The spec sheet lists this parameter "*at rated dc voltages.*" This means that the rating is valid for all dc values of $V_R$ at or below the 50 V $V_{RRM}$ rating. |
| Maximum full-cycle average reverse current, $I_{R(AV)}$ | This is the maximum average value of $I_R$. Note that this rating is at a temperature of 75°C. At all temperatures below 75°C, the average value of $I_R$ will be less than 30 μA. |

| | $I_O$, AVERAGE RECTIFIED FORWARD CURRENT (Amperes) | | | | | |
|---|---|---|---|---|---|---|
| | 1.0 | 1.5 | | 3.0 | | 6.0 |
| VRRM (Volts) | 59-03 (DO-41) Plastic  | 59-04 Plastic  | 60-01 Metal  | 267-03 Plastic  | 267-02 Plastic  | 194-04 Plastic  |
| 50 | †1N4001 | **1N5391 | 1N4719 | **MR500 | 1N5400 | MR750 |
| 100 | †1N4002 | **1N5392 | 1N4720 | **MR501 | 1N5401 | MR751 |
| 200 | †1N4003 | 1N5393 *MR5059 | 1N4721 | **MR502 | 1N5402 | MR752 |
| 400 | †1N4004 | 1N5395 *MR5060 | 1N4722 | **MR504 | 1N5404 | MR754 |
| 600 | †1N4005 | 1N5397 *MR5061 | 1N4723 | **MR506 | 1N5406 | MR756 |
| 800 | †1N4006 | 1N5398 | 1N4724 | MR508 | | MR758 |
| 1000 | †1N4007 | 1N5399 | 1N4725 | MR510 | | MR760 |
| IFSM (Amps) | 30 | 50 | 300 | 100 | 200 | 400 |
| TA @ Rated IO (°C) | 75 | $T_L = 70$ | 75 | 95 | $T_L = 105$ | 60 |
| TC @ Rated IO (°C) | | | | | | |
| TJ (Max) (°C) | 175 | 175 | 175 | 175 | 175 | 175 |

† Package Size: 0.120" Max Diameter by 0.260" Max Length.
* 1N5059 series equivalent Avalanche Rectifiers.
** Avalanche versions available, consult factory.

| | $I_O$, AVERAGE RECTIFIED FORWARD CURRENT (Amperes) | | | | | | | |
|---|---|---|---|---|---|---|---|---|
| | 12 | 20 | 24 | 25 | 30 | | 40 | 50 |
| VRRM (Volts) | 245A-02 (DO-203AA) Metal  | | 339-02 Plastic Note 1  | 193-04 Plastic Note 2  | 43-02 (DO-21) Metal  | | 42A-01 (DO-203AB) Metal 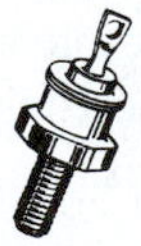 | 43-04 Metal  |
| 50 | MR1120 1N1199,A,B | MR2000 | MR2400 | MR2500 | 1N3491 | 1N3659 | 1N1183A | MR5005 |
| 100 | MR1121 1N1200,A,B | MR2001 | MR2401 | MR2501 | 1N3492 | 1N3660 | 1N1184A | MR5010 |
| 200 | MR1122 1N1202,A,B | MR2002 | MR2402 | MR2502 | 1N3493 | 1N3661 | 1N1186A | MR5020 |
| 400 | MR1124 1N1204,A,B | MR2004 | MR2404 | MR2504 | 1N3495 | 1N3663 | 1N1188A | MR5040 |
| 600 | MR1126 1N1206,A,B | MR2006 | MR2406 | MR2506 | | Note 3 | 1N1190A | Note 3 |
| 800 | MR1128 | MR2008 | | MR2508 | | Note 3 | Note 3 | Note 3 |
| 1000 | MR1130 | MR2010 | | MR2510 | | Note 3 | Note 3 | Note 3 |
| IFSM (Amps) | 300 | 400 | 400 | 400 | 300 | 400 | 800 | 600 |
| TA @ Rated IO (°C) | | | | | | | | |
| TC @ Rated IO (°C) | 150 | 150 | 125 | 150 | 130 | 100 | 150 | 150 |
| TJ (Max) (°C) | 190 | 175 | 175 | 175 | 175 | 175 | 190 | 195 |

**Note 1.** Meets mounting configuration of TO-220 outline.
**Note 2.** Request Data Sheet for Mounting Information.
**Note 3.** Available on special order.

FIGURE 2.27 (Copyright of Semiconductor Component Industries, LLC. Used by permission.)

### 2.6.6    Finding Data on Spec Sheets

When you are looking for data on a spec sheet, follow a three-step procedure:

1. Determine whether the data you are looking for is a *maximum rating* (parameter) or an *electrical characteristic* (guaranteed minimum performance).
2. Look in the appropriate table for the desired data. If it isn't there, try the alternative-search technique covered in step 3. If you still can't find the data, contact the manufacturer of the component.
3. Sometimes it is difficult to locate a particular parameter or characteristic because of the wording used in the spec sheet. When you can't locate a particular bit of information, try this:
   a. Determine the unit of measure for the desired parameter or characteristic. For example, *average forward current* would typically be measured in *mA* or *A*, while *reverse current* would typically be measured in *nA* or *μA*.
   b. Search the *unit* column in the spec sheet tables to find the appropriate unit of measure.
   c. If you find the unit of measure, look at the data name to see if it is the value you are trying to locate.

The use of step 3 can be illustrated using Figure 2.25. Let's say that we are interested in the temperature range of the 1N400X series of diodes. We know that temperature is measured in °C. Looking under the *unit* column, we see that only one parameter is measured in °C. Looking to the left side of that measurement, we see the parameter is identified as *operating and storage junction temperature*. This is the parameter that we are trying to locate.

While the spec sheet for the 1N400X series of diodes is relatively short and simple, the spec sheets for many other devices are relatively complex. When dealing with these more complex spec sheets, you will find the steps listed under step 3 very useful for quickly finding the information that you need.

### 2.6.7    One Final Note

Not all operating parameters and characteristics used to describe diodes are listed on the 1N400X spec sheet. However, the most critical of the operating parameters and characteristics have been covered. Some parameters that were *not* included in this sheet are *junction capacitance* (rated in *pF*), *forward power dissipation* (rated in *mW* or *W*), and *maximum switching frequency* (rated in *kHz* or *MHz*). Also, as mentioned earlier, not all spec sheets use the same terminology to describe various parameters and characteristics. However, the wording is usually close enough that you will be able to find the information you need.

---

1. Why are parameters important?
2. What is a *maximum rating*? Give an example.
3. What is an *electrical characteristic*? Give an example.
4. What is the difference between a *maximum rating* and an *electrical characteristic*?
5. Why does the limit on forward current decrease when temperature increases?
6. What is the procedure for locating information on a diode specification sheet?

◄ **Section Review**

## 2.7    Zener Diodes

The **zener diode** is a special type of diode that is designed to work in the reverse breakdown region of its characteristic curve. A *pn*-junction diode operated in this region is usually destroyed by the excessive reverse current and the heat it produces. This is not the case for the zener diode.

> **Zener diode**
> A diode that is designed to work in the reverse breakdown region.

The purpose served by a component designed to operate in the reverse breakdown region can be seen in Figure 2.28. This figure shows the reverse breakdown region of the diode characteristic curve. As the curve illustrates, two things happen when the **reverse breakdown voltage** $V_{BR}$ is reached:

1. The diode current increases drastically.
2. The reverse voltage across the diode ($V_R$) remains relatively constant.

In other words, *the voltage across a diode operated in this region is relatively constant over a range of component current values.* This characteristic makes the zener diode useful as one type of *voltage regulator.* A **voltage regulator** is a circuit designed to maintain a relatively constant voltage despite anticipated variations in load current or input voltage. A simple zener regulator is introduced in Chapter 3. More practical (and more complex) voltage regulators are covered in Chapter 21.

**Reverse breakdown voltage ($V_{BR}$)**
The minimum $V_R$ that causes a device to break down and conduct in the reverse direction.

**Voltage regulator**
A circuit designed to maintain a constant voltage despite anticipated variations in load current or input voltage.

**Lab Reference:** The reverse operating curve of a zener diode is plotted (using measured values) in Exercise 2.

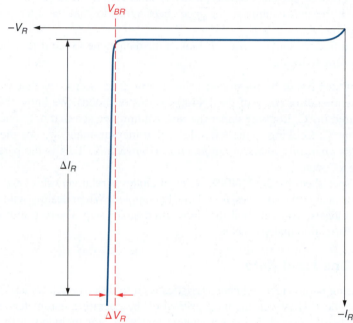

FIGURE 2.28   Reverse breakdown characteristics.

◀ **OBJECTIVE 6** ▶

**Zener voltage ($V_Z$)**
The approximate voltage across a zener when operated in reverse breakdown.

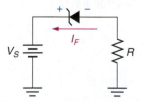

FIGURE 2.29   Zener current.

Since the zener diode is normally reverse biased, it should not surprise you that zener current is normally *in the direction of the arrow*, as shown in Figure 2.29. As the polarity signs in the figure show, the cathode is more positive than the anode. Thus, the component is reverse biased, and device current is in the direction shown.

Does a zener ever conduct in the same direction as a conventional diode? Is it ever operated in the forward operating region? Not very often. Even though a zener diode has forward operating characteristics similar to those of a *pn*-junction diode, the value of the zener lies in its application as a voltage regulator. This application is lost in a forward-biased zener, so it is rarely used in the forward operating region.

When a zener is operating in its reverse operating region, the voltage across the device is *nearly* constant and equal to its **zener voltage ($V_Z$)** rating. Zener diodes have a range of $V_Z$ ratings from about 1.8 V to several hundreds of volts, with power dissipation ratings from 500 mW to 50 W.

### 2.7.1   Diode Breakdown

There are two types of reverse breakdown. We have already briefly discussed the first type, *avalanche breakdown.* The other type of breakdown is called *zener breakdown.*

**Zener breakdown** is a type of reverse breakdown that occurs at relatively low reverse voltages. The *n*-type and *p*-type materials of a zener diode are heavily doped, resulting in a relatively narrow depletion layer. This depletion layer can break down at a lower reverse

◀ **OBJECTIVE 7** ▶

**Zener breakdown**
A type of reverse breakdown that occurs at relatively low reverse voltages.

voltage ($V_R$) than the depletion layer in a typical *pn*-junction diode. Note that zener diodes with low $V_Z$ ratings experience zener breakdown, while those with high $V_Z$ ratings usually experience avalanche breakdown.

## 2.7.2 Zener Operating Characteristics

A zener diode maintains a near-constant reverse voltage for a *range* of reverse current values. These values are identified in Figure 2.30. The minimum current required to maintain voltage regulation (constant voltage) is the **zener knee current ($I_{ZK}$)**. When the zener is used as a voltage regulator, the current through the diode must never be allowed to drop below this value.

**Zener knee current ($I_{ZK}$)**
The minimum value of zener reverse current required to maintain voltage regulation.

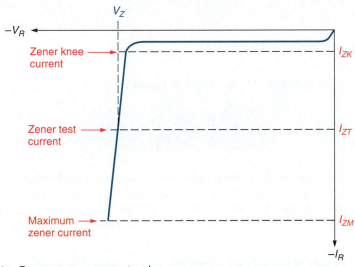

**FIGURE 2.30**   Zener reverse current values.

The **maximum zener current ($I_{ZM}$)** is the maximum amount of current the diode can handle without being damaged or destroyed. The **zener test current ($I_{ZT}$)** is the current level at which the $V_Z$ rating of the diode is measured. For example, if a given zener diode has values of $V_Z = 9.1$ V and $I_{ZT} = 20$ mA, the diode has a reverse voltage of 9.1 V when the test current is 20 mA. At other current values, the value of $V_Z$ will vary *slightly* above or below the rated value.

$I_{ZT}$ is a *test* current value and, thus, is not a critical value for circuit analysis. However, it is an important value in circuit design. *If you need to have a zener voltage that is as close to the nominal (rated) value of $V_Z$ as possible, you need to design the circuit to have a current value equal to $I_{ZT}$.* The further your circuit current is from $I_{ZT}$, the further $V_Z$ will be from its rated value. $I_{ZK}$ and $I_{ZM}$ are important for both circuit analysis and component substitution. This point will be demonstrated in our discussion on zener voltage regulators in Chapter 3.

**Zener impedance ($Z_Z$)** is the zener diode's opposition to *any change in current*. The spec sheet for the 1N746–1N759 series zener diodes lists the following test conditions for measuring $Z_Z$:

**Maximum zener current ($I_{ZM}$)**
The maximum allowable value of zener reverse current.

**Zener test current ($I_{ZT}$)**
The value of zener current at which the nominal values of the component are measured.

$$I_{ZT} = 20 \text{ mA} \qquad I_{zt} = 2 \text{ mA}$$

This means that $Z_Z$ is measured while *varying* zener current by 2 mA around the value of $I_{ZT}$ (20 mA). This variation in zener current is illustrated in Figure 2.31. Note that the 2 mA variation in $I_Z$ causes a 56 mV variation in $V_Z$. From this information, $Z_Z$ is determined using

**Zener impedance ($Z_Z$)**
The zener diode's opposition to a change in current.

$$Z_Z = \frac{\Delta V_Z}{\Delta I_Z} \qquad (2.7)$$

**Lab Reference:** Zener impedance is measured in Exercise 2.

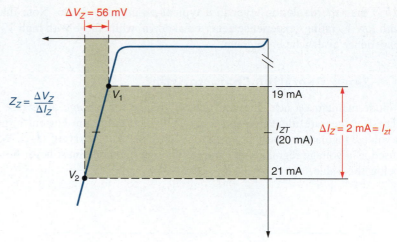

**FIGURE 2.31**  Determining zener impedance.

Thus, for the diode represented by the curve in Figure 2.31:

$$Z_Z = \frac{\Delta V_Z}{\Delta I_Z} = \frac{56 \text{ mV}}{2 \text{ mA}} = 28 \text{ }\Omega$$

> **Static reverse current ($I_R$)**
> The reverse current through a
> diode when $V_R$ is less than the
> component's reverse breakdown
> voltage.

**Static reverse current ($I_R$)** is the reverse current through a diode when $V_R$ is less than the component's reverse breakdown voltage. In other words, this is the reverse leakage current through the diode when it is off. For the 1N746, this value is 10 μA at 25°C and 30 μA at 150°C.

### 2.7.3    Zener Equivalent Circuits

> The *ideal* and *practical* zener
> models.

There are basically two equivalent circuits for the zener diode. Both are shown in Figure 2.32. The *ideal* model simply considers the zener to be a voltage source equal to $V_Z$. When placed in a circuit, this voltage source *opposes* the applied circuit voltage.

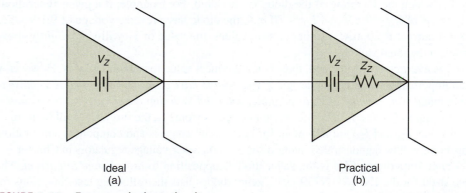

Ideal
(a)

Practical
(b)

**FIGURE 2.32**  Zener equivalent circuits.

The *practical* model of the zener includes a series resistor, labeled $Z_Z$. This model of the zener is used mainly for predicting the response of the diode to a change in circuit current. This point is demonstrated in our discussion on zener voltage regulators (in Chapter 3).

---

**Section Review ▶**

1. What is the primary difference between zener diodes and *pn*-junction diodes?

2. What characteristic of a zener diode makes it useful as a voltage regulator?

3. How do you determine the direction of zener current in a schematic diagram?

4. What is the significance of the zener voltage rating?

5. Name and define the following symbols: $V_Z$, $Z_Z$, $I_{ZK}$, $I_{ZM}$, $I_{ZT}$, $I_{zt}$, and $I_R$.

6. Which rated zener currents limit the total current in a zener diode circuit?

---

## 2.8    Zener Diode Specification Sheets

As you will see in this section, zener diode spec sheets differ somewhat from those for *pn*-junction diodes. Figure 2.33 shows a portion of the spec sheet for the 1N4370A series zener diodes. We will refer to this figure throughout this discussion.

### 2.8.1    Maximum Ratings

The primary parameters for the zener diode are the *maximum steady-state power dissipation* and the *power derating factor*. While the operating and storage temperature range is also listed in the maximum ratings section, the average technician would not be concerned with this parameter under normal circumstances.

The **maximum steady-state power dissipation rating** is the maximum allowable average power dissipation ($P_D$) for a zener diode that is operating in reverse breakdown. This rating is used (along with the nominal zener voltage rating) to determine the value of maximum zener current ($I_{ZM}$) when $I_{ZM}$ is not listed on the component spec sheet. When the value of $I_{ZM}$ is not listed on the spec sheet for a given zener diode, its value can be approximated using

◀ *OBJECTIVE 8*

> **Maximum steady state power dissipation rating**
> The maximum allowable average power dissipation for a zener diode that is operating in reverse breakdown.

$$I_{ZM} = \frac{P_{D(max)}}{V_Z} \qquad (2.8)$$

Example 2.13 demonstrates the use of this equation.

#### EXAMPLE 2.13

A 1N754A zener diode has a dc power dissipation rating of 500 mW and a nominal zener voltage of 6.8 V. What is the value of $I_{ZM}$ for the device?

*Solution:*    Using equation (2.8), the value of $I_{ZM}$ is found as

$$I_{ZM} = \frac{P_{D(max)}}{V_Z} = \frac{500 \text{ mW}}{6.8 \text{ V}} = \textbf{73.5 mA}$$

*Remember, the value of $I_{ZM}$ is important because it determines the maximum current the diode can tolerate!* If the average current through this diode exceeds 73.5 mA, you'll end up replacing the diode.

#### PRACTICE PROBLEM 2.13

A zener diode has a dc power dissipation rating of 1 W and a nominal zener voltage of 27 V. What is the value of $I_{ZM}$ for the device?

---

The **power derating factor** (listed in the maximum ratings table) tells you how much the maximum power dissipation rating *decreases* when the operating temperature increases above a specified value. For example, the spec sheet in Figure 2.33 shows a derating factor of *4 mW/°C at temperatures above 75°C*. This means that the maximum power dissipation rating for the device decreases by 4 mW for every 1°C rise in operating temperature above 75°C. Example 2.14 demonstrates the use of this derating factor.

> **Power derating factor**
> The rate at which the maximum power rating decreases per 1°C rise above a specified temperature.

## 1N4370A Series

## 500 mW DO-35 Hermetically Sealed Glass Zener Voltage Regulators

This is a complete series of 500 mW Zener diodes with limits and excellent operating characteristics that reflect the superior capabilities of silicon–oxide passivated junctions. All this in an axial–lead hermetically sealed glass package that offers protection in all common environmental conditions.

**Specification Features:**
- Zener Voltage Range – 2.4 V to 12 V
- ESD Rating of Class 3 (>16 KV) per Human Body Model
- DO–204AH (DO–35) Package – Smaller than Conventional DO–204AA Package
- Double Slug Type Construction
- Metallurgical Bonded Construction

**Mechanical Characteristics:**
**CASE:** Double slug type, hermetically sealed glass
**FINISH:** All external surfaces are corrosion resistant and leads are readily solderable
**MAXIMUM LEAD TEMPERATURE FOR SOLDERING PURPOSES:**
230°C, 1/16″ from the case for 10 seconds
**POLARITY:** Cathode indicated by polarity band
**MOUNTING POSITION:** Any

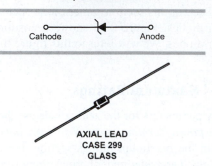

**ON Semiconductor™**

http://onsemi.com

Cathode    Anode

**AXIAL LEAD
CASE 299
GLASS**

**MARKING DIAGRAM**

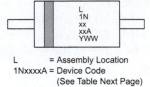

L
1N
xx
xxA
YWW

| L | = Assembly Location |
| 1NxxxxA | = Device Code |
| | (See Table Next Page) |
| Y | = Year |
| WW | = Work Week |

**MAXIMUM RATINGS (Note 1.)**

| Rating | Symbol | Value | Unit |
|--------|--------|-------|------|
| Max. Steady State Power Dissipation @ $T_L \leq 75°C$, Lead Length = 3/8″ | $P_D$ | 500 | mW |
| Derate above 75°C | | 4.0 | mW/°C |
| Operating and Storage Temperature Range | $T_J$, $T_{stg}$ | −65 to +200 | °C |

1. Some part number series have lower JEDEC registered ratings.

FIGURE 2.33    (Copyright of Semiconductor Component Industries, LLC. Used by permission.)

### EXAMPLE 2.14

The *power derating curve* in Figure 2.33 (below the maximum ratings table) shows that the 1N4370A has a maximum power dissipation rating of 400 mW at $T = 100°C$. Verify this value using the component's derating factor and $P_D$ rating.

*Solution:*    The first step is to determine the total derating value at $T = 100°C$. This value is found as

$$\textbf{Derating value} = (4 \text{ mW/°C})(100°C - 75°C) = \textbf{100 mW}$$

Subtracting this value from the maximum power dissipation rating of 500 mW, we get

$$\boldsymbol{P_D} = 500 \text{ mW} - 100 \text{ mW} = \textbf{400 mW}$$

which agrees with the value shown on the curve.

### PRACTICE PROBLEM 2.14
Using Figure 2.33, determine the power dissipation rating of the 1N746A at 125°C.

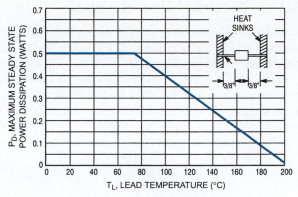

Figure 1. Steady State Power Derating

## 1N4370A Series

**ELECTRICAL CHARACTERISTICS** ($T_A$ = 25°C unless otherwise noted, $V_F$ = 1.5 V Max @ $I_F$ = 200 mA for all types)

| Symbol | Parameter |
|--------|-----------|
| $V_Z$ | Reverse Zener Voltage @ $I_{ZT}$ |
| $I_{ZT}$ | Reverse Current |
| $Z_{ZT}$ | Maximum Zener Impedance @ $I_{ZT}$ |
| $I_{ZM}$ | Maximum DC Zener Current |
| $I_R$ | Reverse Leakage Current @ $V_R$ |
| $V_R$ | Reverse Voltage |
| $I_F$ | Forward Current |
| $V_F$ | Forward Voltage @ $I_F$ |

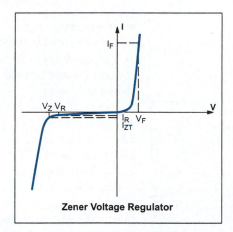

**Zener Voltage Regulator**

**ELECTRICAL CHARACTERISTICS** ($T_A$ = 25°C unless otherwise noted, $V_F$ = 1.5 V Max @ $I_F$ = 200 mA for all types)

| Device (Note 2.) | Device Marking | Zener Voltage (Note 3.) | | | | $Z_{ZT}$ (Note 4.) @ $I_{ZT}$ | $I_{ZM}$ (Note 5.) | $I_R$ @ $V_R$ = 1 V | |
|---|---|---|---|---|---|---|---|---|---|
| | | $V_Z$ (Volts) | | | @ $I_{ZT}$ | | | $T_A$ = 25°C | $T_A$ = 150°C |
| | | Min | Nom | Max | (mA) | (Ω) | (mA) | (µA) | (µA) |
| 1N4370A | 1N4370A | 2.28 | 2.4 | 2.52 | 20 | 30 | 150 | 100 | 200 |
| 1N4371A | 1N4371A | 2.57 | 2.7 | 2.84 | 20 | 30 | 135 | 75 | 150 |
| 1N4372A | 1N4372A | 2.85 | 3.0 | 3.15 | 20 | 29 | 120 | 50 | 100 |
| *1N746A* | *1N746A* | *3.14* | *3.3* | *3.47* | *20* | *28* | *110* | *10* | *30* |
| 1N747A | 1N747A | 3.42 | 3.6 | 3.78 | 20 | 24 | 100 | 10 | 30 |
| 1N748A | 1N748A | 3.71 | 3.9 | 4.10 | 20 | 23 | 95 | 10 | 30 |
| 1N749A | 1N749A | 4.09 | 4.3 | 4.52 | 20 | 22 | 85 | 2 | 30 |
| 1N750A | 1N750A | 4.47 | 4.7 | 4.94 | 20 | 19 | 75 | 2 | 30 |
| *1N751A* | *1N751A* | *4.85* | *5.1* | *5.36* | *20* | *17* | *70* | *1* | *20* |
| *1N752A* | *1N752A* | *5.32* | *5.6* | *5.88* | *20* | *11* | *65* | *1* | *20* |
| *1N753A* | *1N753A* | *5.89* | *6.2* | *6.51* | *20* | *7* | *60* | *0.1* | *20* |
| 1N754A | 1N754A | 6.46 | 6.8 | 7.14 | 20 | 5 | 55 | 0.1 | 20 |
| 1N755A | 1N755A | 7.13 | 7.5 | 7.88 | 20 | 6 | 50 | 0.1 | 20 |
| 1N756A | 1N756A | 7.79 | 8.2 | 8.61 | 20 | 8 | 45 | 0.1 | 20 |
| 1N757A | 1N757A | 8.65 | 9.1 | 9.56 | 20 | 10 | 40 | 0.1 | 20 |
| 1N758A | 1N758A | 9.50 | 10 | 10.5 | 20 | 17 | 35 | 0.1 | 20 |
| 1N759A | 1N759A | 11.40 | 12 | 12.6 | 20 | 30 | 30 | 0.1 | 20 |

2. **TOLERANCE AND TYPE NUMBER DESIGNATION ($V_Z$)**
   The type numbers listed have a standard tolerance on the nominal zener voltage of ±5%.

3. **ZENER VOLTAGE ($V_Z$) MEASUREMENT**
   Nominal zener voltage is measured with the device junction in the thermal equilibrium at the lead temperature ($T_L$) at 30°C ± 1°C and 3/8″ lead length.

4. **ZENER IMPEDANCE ($Z_Z$) DERIVATION**
   $Z_{ZT}$ and $Z_{ZK}$ are measured by dividing the ac voltage drop across the device by the ac current applied. The specified limits are for $I_{Z(ac)}$ = 0.1 $I_{Z(dc)}$ with the ac frequency = 60 Hz.

5. **MAXIMUM ZENER CURRENT RATINGS ($I_{ZM}$)**
   Values shown are based on the JEDEC rating of 400 mW where the actual zener voltage ($V_Z$) is known at the operating point, the maximum zener current may be increased and is limited by the derating curve.

FIGURE 2.33  *(continued)*

### 2.8.2 Electrical Characteristics

The *electrical characteristics* table in Figure 2.33 contains the zener voltage, current, and impedance ratings for each diode in the series. If you look closely at the values in the **nominal zener voltage** column (shaded), you will notice something interesting: *The values listed follow the same basic progression as standard resistor values.* Note that each nominal zener value is measured at the listed value of zener test current ($I_{ZT}$).

**Nominal zener voltage**

The rated value of $V_Z$ for a given zener diode.

Each component listed in the electrical characteristics table has a tolerance in its $V_Z$ rating of $\pm 5\%$. For example, the 1N758A has a nominal zener rating of 10 V. The actual value of the component could fall anywhere between

$$V_Z = 10 \text{ V} - 5\% = 10 \text{ V} - 0.5 \text{ V} = \mathbf{9.5 \text{ V}}$$

and

$$V_Z = 10 \text{ V} + 5\% = 10 \text{ V} + 0.5 \text{ V} = \mathbf{10.5 \text{ V}}$$

when the device current equals 20 mA. As you can see, these minimum and maximum values of $V_Z$ are listed in the table.

The $Z_{ZT}$ rating is the value of *zener impedance* at the listed value of $I_{ZT}$. The value of $Z_{ZT}$ listed would be used in any circuit calculations involving zener impedance, as will be shown in Chapter 3.

Each value in the $I_{ZM}$ (*maximum zener current*) column is based on a JEDEC rating of 400 mW and the rated zener voltage of the component (as indicated by Note 5 below the table). In each case, the product of $I_{ZM}$ and $V_Z$ is approximately 360 mW, which is 10% lower than the JEDEC rated value of $P_D = 400$ mW. Thus, there is an added safety margin built into the rating.

The values in the $I_R$ column represent the *maximum reverse leakage current* (or *static reverse current*) values for the components, measured at a value of $V_R = 1$ V. As the spec sheet indicates, this value is temperature dependent, just like the reverse current rating of the *pn*-junction diode.

The JEDEC Solid State Technology Association is a branch of the Electronics Industries Alliance (EIA) that deals with the standardization of semiconductor devices and integrated circuits. *JEDEC* is an acronym for **J**oint **E**lectron **D**evice **E**ngineering **C**ouncil.

### 2.8.3 Zener Diode Selector Guides

The selector guides for zener diodes are very similar to those used for the *pn*-junction diodes. A zener diode selector guide is shown in Figure 2.34.

What parameters are critical for component substitution?

As you can see, the critical parameters for device substitution are *nominal zener voltage* (vertical listing) and *dc power dissipation* (horizontal listing). Examples 2.15 and 2.16 show how the selector guide is used.

#### EXAMPLE 2.15

We need a substitute component for a zener diode that is faulty. The substitute component must have a nominal zener voltage of 12 V and be capable of dissipating 1.2 W. Which of the zeners listed in Figure 2.34 can be used?

***Solution:*** First, the zener diode must have a rating of 12 V. Therefore, our diode is listed in the row that corresponds to $V_Z = 12$ V. Two diodes are listed in the 12 V row that have power dissipation ratings greater than 1.2 W. These are the 1N5927A (1.5 W) and the 1N5349A (5 W). Either of these diodes could be used as a substitute component.

#### PRACTICE PROBLEM 2.15

We need a zener diode with a nominal zener voltage of 75 V and a power dissipation capability of 4 W. Which zener diode(s) shown in Figure 2.34 can be used in this case?

When we know the zener voltage and current requirements of a component and need a substitute, we have to do a little calculating to find the right substitute part. This point is illustrated in the following example.

*A Practical Consideration:* When we were dealing with substituting one *pn*-junction diode for another, we were concerned only with whether the current and peak reverse voltage ratings were high enough to survive in the circuit. When dealing with zener diodes, the power rating of the substitute diode may be higher than needed, but the $V_Z$ rating of the substitute must *equal* that of the component it is replacing. We *cannot* use a diode with a higher (or lower) $V_Z$ rating than the original component.

**FIGURE 2.34** Zener selector guide.

| Nominal Zener Voltage (*Note 1) | 500 mW Cathode = Polarity Mark (*Notes 4,11) | 500 mW Cathode = Polarity Mark (*Notes 9,11) | 1 Watt Cathode = Polarity Mark (*Note 6) | 1 Watt Cathode = Polarity Mark (*Notes 6,12) | 1 Watt Cathode to Case (*Note 7) | 1.5 Watt Cathode = Polarity Mark (*Note 8) | 5 Watt Cathode = Polarity Mark (*Note 8) |
|---|---|---|---|---|---|---|---|
|  | Glass Case 362-01 | | Glass Case 59-04 (DO-41) | Glass Case 362B-01 | Metal Case 52-03 (DO-13) | Sumetic 30 Case 59-03 (DO-41) | Sumetic 40 Case 17-02 |
| 1.8 | | | | | | | |
| 2.0 | | | | | | | |
| 2.2 | | | | | | | |
| 2.4 | MLL4370 | MLL5221A | | | | | |
| 2.5 | | MLL5222A | | | | | |
| 2.7 | MLL4371 | MLL5223A | | | | | |
| 2.8 | | MLL5224A | | | | | |
| 3.0 | MLL4372 | MLL4225A | | | | | |
| 3.3 | MLL746 | MLL5226A | 1N4728 | MLL4728 | 1N3821 | 1N5913A | 1N5333A |
| 3.6 | MLL747 | MLL5227A | 1N4729 | MLL4729 | 1N3822 | 1N5914A | 1N5334A |
| 3.9 | MLL748 | MLL5228A | 1N4730 | MLL4730 | 1N3823 | 1N5915A | 1N5335A |
| 4.3 | MLL749 | MLL5229A | 1N4731 | MLL4731 | 1N3824 | 1N5916A | 1N5336A |
| 4.7 | MLL750 | MLL5230A | 1N4732 | MLL4732 | 1N3825 | 1N5917A | 1N5337A |
| 5.1 | MLL751 | MLL5231A | 1N4733 | MLL4/33 | 1N3826 | 1N5918A | 1N5338A |
| 5.6 | MLL752 | MLL5232A | 1N4734 | MLL4734 | 1N3827 | 1N5919A | 1N5339A |
| 6.0 | | MLL5233A | | | | | |
| 6.2 | MLL753 | MLL5234A | 1N4735 | MLL4735 | 1N3828 | 1N5920A | 1N5341A |
| 6.8 | MLL754 MLL957A | MLL5235A | 1N4736 | MLL4736 | 1N3829 1N3016A | 1N5921A | 1N5342A |
| 7.5 | MLL755 MLL958A | MLL5236A | 1N4737 | MLL4737 | 1N3830 1N3017A | 1N5922A | 1N5343A |
| 8.2 | MLL756 MLL959A | MLL5237A | 1N4738 | MLL4738 | 1N3018A | 1N5923A | 1N5344A |
| 8.7 | | MLL5238A | | | | | 1N5345A |
| 9.1 | MLL757 MLL960A | MLL5239A | 1N4739 | MLL4739 | 1N3019A | 1N5924A | 1N5346A |
| 10 | MLL758 MLL961A | MLL5240A | 1N4740 | MLL4740 | 1N3020A | 1N5925A | 1N5347A |
| 11 | MLL962A | MLL5241A | 1N4741 | MLL4741 | 1N3021A | 1N5926A | 1N5348A |
| 12 | MLL759 MLL963A | MLL5242A | 1N4742 | MLL4742 | 1N3022A | 1N5927A | 1N5349A |
| 13 | MLL964A | MLL5243A | 1N4743 | MLL4743 | 1N3023A | 1N5928A | 1N5350A |
| 14 | | MLL5244A | | | | | 1N5351A |
| 15 | MLL965A | MLL5245A | 1N4744 | MLL4744 | 1N3024A | 1N5929A | 1N5352A |
| 16 | MLL966A | MLL5246A | 1N4745 | MLL4745 | 1N3025A | 1N5930A | 1N5353A |
| 17 | | MLL5247A | | | | | 1N5354A |
| 18 | MLL967A | MLL5248A | 1N4746 | MLL4746 | 1N3026A | 1N5931A | 1N5355A |
| 19 | | MLL5249A | | | | | 1N5356A |
| 20 | MLL968A | MLL5250A | 1N4747 | MLL4747 | 1N3027A | 1N5932A | 1N5357A |
| 22 | MLL969A | MLL5251A | 1N4748 | MLL4748 | 1N3028A | 1N5933A | 1N5358A |
| 24 | MLL970A | MLL5252A | 1N4749 | MLL4749 | 1N3029A | 1N5934A | 1N5359A |
| 25 | | MLL5253A | | | | | 1N5360A |
| 27 | MLL971A | MLL5254A | 1N4750 | MLL4750 | 1N3030A | 1N5935A | 1N5361A |
| 28 | | MLL5255A | | | | | 1N5362A |
| 30 | MLL972A | MLL5256A | 1N4751 | MLL4751 | 1N3031A | 1N5936A | 1N5363A |
| 33 | MLL973A | MLL5257A | 1N4752 | MLL4752 | 1N3032A | 1N5937A | 1N5364A |
| 36 | MLL974A | MLL5258A | 1N4753 | MLL4753 | 1N3033A | 1N5938A | 1N5365A |
| 39 | MLL975A | MLL5259A | 1N4754 | MLL4754 | 1N3034A | 1N5939A | 1N5366A |
| 43 | MLL976A | MLL5260A | 1N4755 | MLL4755 | 1N3035A | 1N5940A | 1N5367A |
| 47 | MLL977A | MLL5261A | 1N4756 | MLL4756 | 1N3036A | 1N5941A | 1N5368A |
| 51 | MLL978A | MLL5262A | 1N4757 | MLL4751 | 1N3037A | 1N5942A | 1N5369A |
| 56 | MLL979A | MLL5263A | 1N4758 | MLL4758 | 1N3038A | 1N5943A | 1N5370A |
| 60 | | MLL5264A | | | | | 1N5371A |
| 62 | MLL980A | MLL5265A | 1N4759 | MLL4759 | 1N3039A | 1N5944A | 1N5372A |
| 68 | MLL981A | MLL5266A | 1N4760 | MLL4760 | 1N3040A | 1N5945A | 1N5373A |
| 75 | MLL982A | MLL5267A | 1N4761 | MLL4761 | 1N3041A | 1N5946A | 1N5374A |
| 82 | MLL983A | MLL5268A | 1N4762 | MLL4762 | 1N3042A | 1N5947A | 1N5375A |
| 87 | | MLL5269A | | | | | 1N5376A |
| 91 | MLL984A | MLL5270A | 1N4763 | MLL4763 | 1N3043A | 1N5958A | 1N5377A |
| 100 | MLL985A | | 1N4764 | MLL4764 | 1N3044A | 1N5949A | 1N5378A |
| 110 | MLL986A | | | | 1N3045A | 1N5950A | 1N5379A |
| 120 | | | | | 1N3046A | 1N5951A | 1N5380A |
| 130 | | | | | 1N3047A | 1N5952A | 1N5831A |
| 150 | | | | | 1N3048A | 1N5953A | 1N5383A |
| 160 | | | | | 1N3049A | 1N5954A | 1N5384A |
| 170 | | | | | | | 1N5385A |
| 175 | | | | | | | |
| 180 | | | | | 1N3050A | 1N5955A | 1N5386A |
| 200 | | | | | 1N3051A | 1N5956A | 1N5388A |

**FIGURE 2.34** Zener selector guide. (Copyright of Semiconductor Component Industries, LLC. Used by permission.)

## EXAMPLE 2.16

We need to replace a faulty zener diode. The substitute component must have a nominal zener voltage of 20 V and be able to handle the power generated by a maximum current of 150 mA. Which diode(s) listed in Figure 2.34 can be used in this application?

*Solution:* First, we need to determine the power dissipation requirements of the substitute diode. The dc power requirement is found as

$$P_{D(max)} = I_{ZM}V_Z = (150 \text{ mA})(20 \text{ V}) = 3 \text{ W}$$

Therefore, the substitute component must be able to dissipate *at least* 3 W. Checking the 20 V/5 W location on the selector guide, we see that the only component we can use is the 1N5357A.

### PRACTICE PROBLEM 2.16

We need a zener diode with a nominal zener voltage of 6.8 V that can handle a maximum zener current of 175 mA. Which diode(s) listed in Figure 2.34 can be used for this application?

**Section Review ▶**

1. When a spec sheet does not list the value of $I_{ZM}$, how do you determine its value?

2. What is a *power derating factor*? How is it used?

3. Explain how you would determine whether one zener diode can be used in place of another.

## 2.9 Light-Emitting Diodes (LEDs)

**OBJECTIVE 9 ▶**

Anode

Cathode

FIGURE 2.35 The LED schematic symbol.

LEDs are diodes that emit light when biased properly. LEDs are available in a variety of colors, such as red, yellow, and green. Some, such as *infrared* LEDs, emit light that is not visible to the naked eye. The schematic symbol for the LED is shown in Figure 2.35. Note that nothing in the symbol indicates the color of the component or the light that it emits.

The construction of a typical LED is illustrated in Figure 2.36. Since LEDs have clear (or semiclear) cases, there is normally no label on the case to identify the leads. Rather, the leads are normally identified in one of two ways:

1. The leads may have different lengths, as shown in Figure 2.37a. When this scheme is used, the shorter of the two leads is usually the *cathode*.

2. One side of the case may be flattened, as shown in Figure 2.37b. When this scheme is used, the lead closest to the flattened side is usually the *cathode*.

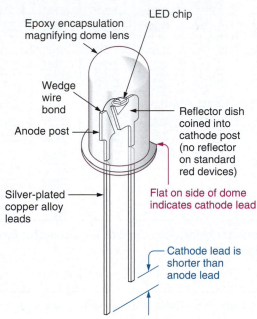

FIGURE 2.36 Construction features of T-1³⁄₄ plastic LED lamp. (Courtesy of Agilent Technologies, Inc.)

### 2.9.1 LED Characteristics

LEDs have characteristic curves that are very similar to those for *pn*-junction diodes. However, LEDs have higher forward voltage ($V_F$) values and lower reverse breakdown voltage ($V_{BR}$) ratings. Typically:

- Forward voltage ($V_F$) is between 1.4 and 3.6 V (at $I_F = 20$ mA).
- Reverse breakdown voltage ($V_{BR}$) is between −3 and −10 V.

The color emitted by a given LED depends on the combination of elements used to produce the component. LEDs are generally produced using gallium (Ga) and one or more of the following elements: arsenic (As), aluminum (Al), indium (In), phosphorus (P), and nitrogen (N). Some common element combinations are identified in Table 2.4.

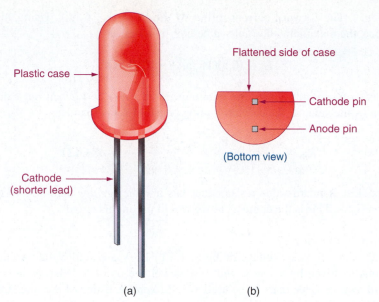

Plastic case →

Flattened side of case

Cathode pin

Anode pin

(Bottom view)

Cathode
(shorter lead) →

(a)

(b)

FIGURE 2.37

TABLE 2.4    Common LEDs

| Elements | Forward Voltage ($V_F$) | Color Emitted |
|----------|------------------------|---------------|
| GaAs | 1.5 V @ $I_F$ = 20 mA | Infrared (invisible) |
| AlGaAs | 1.8 V @ $I_F$ = 20 mA | Red |
| GaP | 2.4 V @ $I_F$ = 20 mA | Green |
| AlGaInP | 2.0 V @ $I_F$ = 20 mA | Amber (yellow) |
| AlGaInN | 3.6 V @ $I_F$ = 20 mA | Blue |

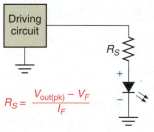

$$R_S = \frac{V_{out(pk)} - V_F}{I_F}$$

FIGURE 2.38    An LED needs a current-limiting resistor.

Most LEDs have a forward current ($I_F$) limit of 100 mA or less. For that reason, LEDs typically require the use of a series current-limiting resistor.

## 2.9.2    Current-Limiting Resistors

When used in any practical application, an LED requires the use of a series **current-limiting resistor**, as shown in Figure 2.38. The resistor ensures that the maximum current rating of the LED cannot be exceeded by the circuit current. As the illustration shows, the value of the limiting resistor ($R_S$) is determined using the following equation:

**Current-limiting resistor**
A resistor in series with a component to limit the current through the component.

$$R_S = \frac{V_{out(pk)} - V_F}{I_F} \qquad (2.9)$$

where $V_{out(pk)}$ = the peak output voltage of the driving circuit
$V_F$ = the *minimum* value of $V_F$ for the LED
$I_F$ = the desired value of $I_F$ for the LED

To provide a safety margin, the value of $I_F$ used in equation (2.9) should be no greater than 80% of the rated maximum LED forward current. Example 2.7 demonstrates the process of determining the needed value for a current-limiting resistor.

◄ *OBJECTIVE 10*

### EXAMPLE 2.17

The driving circuit shown in Figure 2.38 has a peak output of 8 V. The LED has ratings of $V_F$ = 1.8 to 2.0 V, and the diode maximum forward current rating is 16 mA. What value of current-limiting resistor is needed in the circuit?

**Solution:** The forward current rating of the diode is 16 mA. Using the 80% guideline, the maximum allowable value of $I_F$ is found as

$$I_F = (0.8)(16 \text{ mA}) = \textbf{12.8 mA}$$

Now, using the peak voltage, the minimum rated value of $V_F$, and our calculated limit on $I_F$, the value of the current limiting resistor ($R_S$) is found as

$$R_S = \frac{V_{\text{out(pk)}} - V_F}{I_F} = \frac{8 \text{ V} - 1.8 \text{ V}}{12.8 \text{ mA}} = \textbf{484.4 } \boldsymbol{\Omega}$$

The smallest standard-value resistor that has a value *greater than* 484.4 $\Omega$ is the 510 $\Omega$ resistor. This is the component we would use in this circuit.

**PRACTICE PROBLEM 2.17**
An LED with a forward voltage rating of 1.4 to 1.8 V and a forward current rating of 12 mA is driven by a source with a peak voltage of 14 V. What is the smallest standard resistor value that can be used as $R_S$? (*Note:* A listing of the standard resistor values appears in Appendix A.)

### 2.9.3    Multicolor LEDs

> **Multicolor LED**
> An LED that emits different colors when the polarity of the supply voltage changes.

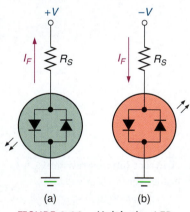

**FIGURE 2.39** Multicolor LED.

Some LEDs emit one color of light when the biasing voltage has one polarity and a second color when the polarity of the biasing voltage reverses. By rapidly switching the polarity of the biasing voltage, a **multicolor LED** can be made to appear to generate a third color. One schematic symbol for a multicolor LED is shown in Figure 2.39.

Multicolor LEDs contain two *pn* junctions, one for each color of light that the component emits. Because the junction in an LED emits light only when it is forward biased, the two diode junctions are connected in reverse parallel, meaning that the anode of each is connected to the cathode of the other.

If a positive potential is applied to the LED (as shown in Figure 2.39a), the *pn* junction on the *left* emits light. Note that the device current passes through the left *pn* junction. If the polarity of the voltage source is reversed (as shown in Figure 2.39b), the *pn* junction on the right emits light. Note that the direction of diode current has reversed and is now passing through the right *pn* junction.

Multicolor LEDs are typically *red* when biased in one direction and *green* when biased in the other. If a multicolor LED is switched fast enough between the two polarities, the LED appears to produce a *third* color. For example, a red/green LED appears to produce a *yellow* light when rapidly switched back and forth between biasing polarities.

**Section Review ▶**

1. How are the leads on an LED usually identified?
2. How do the electrical characteristics of LEDs differ from those of *pn*-junction diodes?
3. Why do LEDs need series current-limiting resistors?

## 2.10  Diode Testing

> What causes most diode failures?

Most diode failures are caused by component aging, excessive forward current, or exceeding the component's $V_{RRM}$ rating. When a diode has been damaged by excessive current, the problem is usually easy to diagnose. In most cases, the diode cracks or falls apart completely. Burned connection points and copper traces on a printed circuit board are also symptoms of excessive diode current.

When a diode does not show obvious signs of damage (like those listed above), some simple tests will tell you whether or not the component is faulty.

## 2.10.1    Testing *PN*-Junction Diodes

A *pn*-junction diode can be tested using a digital ohmmeter, as shown in Figure 2.40. When connected as shown in Figure 2.40a, the diode is forward biased by the ohmmeter. As a result, the component resistance reading should be very low, typically lower than 1 kΩ. When connected as shown in Figure 2.40b, the ohmmeter reverse biases the diode. As a result, the component resistance should be extremely high, typically in the high MΩ range. In most cases, the ohmmeter gives an *out-of-range* indication as a result of the high diode reverse resistance. *When you test a diode, a high forward resistance or low reverse resistance indicates that the diode is faulty and must be replaced.*

Several notes of caution need to be made regarding this procedure:

1. When set on low resistance scales, some ohmmeters can generate enough current to destroy a low-current diode. To prevent component damage, use only the resistance scales rated in the low kΩ or higher. For example, the minimum resistance setting for the meter represented in Figure 2.40 would be 2 kΩ. (When in doubt, check the current limit of the diode under test, and compare it to the current limits of the meter.)

◀ **OBJECTIVE 11**

*A Practical Consideration:*
Some meters provide an output voltage that is less than $V_F$ for resistance measurements. These meters will give an open indication on a good diode. You need to check your meter documentation before using it for diode testing.

**Lab Reference:** Diode testing with an ohmmeter is demonstrated in Exercise 1.

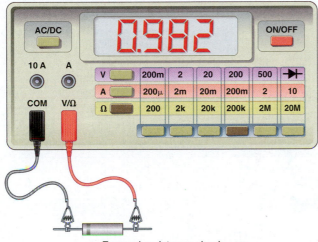

Forward resistance check

(a)

*Note:* Typical forward resistance readings for a good diode are around 1kΩ or lower.

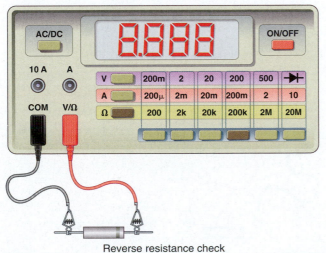

Reverse resistance check

(b)

*Note:* Typical reverse resistance for a good diode results in an "out of range" indication on the meter.

**FIGURE 2.40**   Diode testing.

**2.** Some meters supply current from the common lead, but others supply it from the "ohms" lead. Check the documentation on your meter to be sure that you know the lead polarities before testing any diodes. Otherwise, you may think you have a bad diode when you don't.

Many DMMs have a diode test function, as illustrated in Figure 2.41. Note that the meter function and range switches are set to select the diode symbol. When the meter is connected as shown in Figure 2.41a, the diode is forward biased by the meter. The reading indicates the approximate value of $V_F$ across the diode, or 0.7 V in this case. When the meter is connected as shown in Figure 2.41b, the diode is reverse biased. In this case, the meter reading indicates the value of $V_R$ across the diode, which is approximately equal to the voltage provided by the internal meter power supply (3.0 V in this case). If the meter reads 3.0 V for a forward voltage, or less than 1 V for a reverse voltage, the component is faulty and must be replaced.

An ohmmeter can also be used to identify the anode and the cathode of an unmarked diode. Diodes are usually marked to indicate which end of the component is the cathode. However, with age and heat, these markings may fade. When the ohmmeter is connected to the diode so that it is forward biased, the positive meter lead is connected to the anode and the negative lead is connected to the cathode.

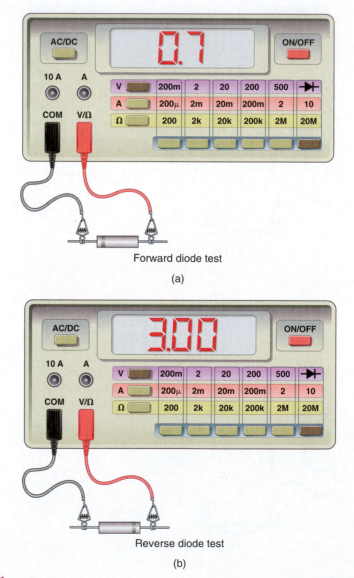

Forward diode test

(a)

Reverse diode test

(b)

**FIGURE 2.41**

## 2.10.2  Testing Zener Diodes

In most cases, a zener diode cannot be tested as shown in Figures 2.40 and 2.41, simply because the component is designed to break down and conduct in the reverse direction when the magnitude of the component voltage equals $V_Z$.

The simplest test of a zener diode is to measure the voltage across its terminals while the component is in the circuit under test. If the voltage across the zener is within tolerance, the component is good. If not, there is a strong possibility that the component is faulty. In this case, simply replace the component. If the circuit operation returns to normal, then the zener diode was the source of the problem; if not, continued testing of the other circuit components is required.

## 2.10.3  Testing LEDs

Normally you will not need to test an LED to see if it is defective. The usual cause of a faulty LED is excessive forward current, which destroys the *pn* junction. When this occurs, the discoloration of the LED due to the burnt junction is easy to recognize. In Figure 2.42, the LED on the right is normal, and the LED on the left has been damaged by excessive current.

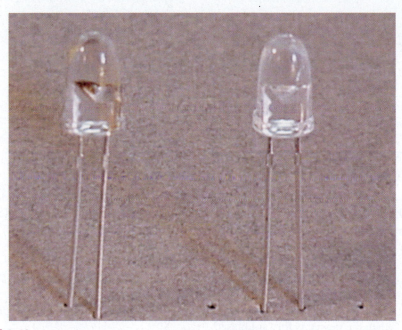

FIGURE 2.42

---

1. What are the common causes of diode failure?

2. What steps are involved in testing a *pn*-junction diode with an ohmmeter?

3. What precautions should be taken before using an ohmmeter to test a *pn*-junction diode?

4. How do you test a zener diode?

◀ **Section Review**

## 2.10.4  Diodes: A Comparison

You have been shown quite a lot in this chapter. The summary illustration in Figure 2.43 will help you to remember the primary points about each diode covered. If you have difficulty remembering any of the points listed, review the appropriate section of the chapter or the chapter summary.

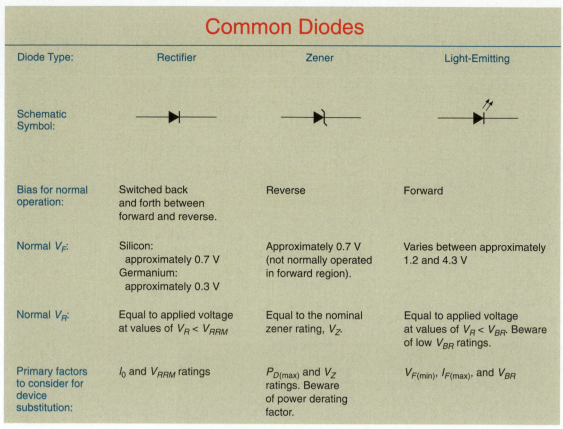

| Diode Type: | Rectifier | Zener | Light-Emitting |
|---|---|---|---|
| **Schematic Symbol:** | | | |
| **Bias for normal operation:** | Switched back and forth between forward and reverse. | Reverse | Forward |
| **Normal $V_F$:** | Silicon: approximately 0.7 V Germanium: approximately 0.3 V | Approximately 0.7 V (not normally operated in forward region). | Varies between approximately 1.2 and 4.3 V |
| **Normal $V_R$:** | Equal to applied voltage at values of $V_R < V_{RRM}$ | Equal to the nominal zener rating, $V_Z$. | Equal to applied voltage at values of $V_R < V_{BR}$. Beware of low $V_{BR}$ ratings. |
| **Primary factors to consider for device substitution:** | $I_0$ and $V_{RRM}$ ratings | $P_{D(max)}$ and $V_Z$ ratings. Beware of power derating factor. | $V_{F(min)}$, $I_{F(max)}$, and $V_{BR}$ |

FIGURE 2.43

## CHAPTER SUMMARY

Here is a summary of the major points made in this chapter:

1. A *diode* is a two-electrode (two-terminal) device that acts as a one-way conductor.
2. The *pn-junction* diode is nothing more than a *pn* junction with a lead connected to each of the semiconductor materials.
    a. The *n-type* material is referred to as the *cathode*.
    b. The *p-type* material is referred to as the *anode*.
3. A *pn*-junction diode is *forward biased* when the cathode is *more negative than the anode*.
    a. The arrow in the schematic symbol points toward the more negative potential when the device is forward biased.
    b. Electron flow through a forward-biased diode is *against* the arrow.
4. A *pn*-junction diode is *reversed biased* when the cathode is *more positive* than the anode.
    a. The device is reverse biased when the arrow in the schematic symbol points toward the more positive of the diode potentials.
    b. Device current is (ideally) zero when the component is reverse biased.
5. There are three diode models.
    a. The *ideal model* is the simplest of the three and is used primarily in the initial stages of troubleshooting.
    b. The *practical model* includes several diode characteristics that must be considered for circuit analysis problems and component substitution.
    c. The *complete model* includes characteristics that are considered in circuit development (design) and high-frequency analysis. It also contains several characteristics that account for the differences between predicted and measured circuit values.

6. The *ideal* diode is viewed as an open switch when reverse biased and as a closed switch when forward biased.
7. When reverse biased, the ideal diode model assumes that the component:
   a. Has infinite resistance.
   b. Does not pass current.
   c. Drops the entire applied voltage across its terminals.
8. When forward biased, the ideal diode model assumes that the component:
   a. Has no resistance.
   b. Has no control over diode forward current.
   c. Has no voltage drop across its terminals.
9. When the ideal diode is forward biased, diode current is determined by the voltage and resistance values that are external to the component.
10. The ideal diode model is used in the initial stages of troubleshooting.
11. The practical diode model includes diode forward voltage (among other characteristics).
12. The term *knee voltage* is commonly used to describe the voltage on a voltage-versus-current graph where current suddenly increases (or decreases).
13. The knee voltage for a practical forward-biased diode is assumed to be approximately:
    a. 0.7 V for silicon (Si).
    b. 0.3 V for germanium (Ge).
14. In the forward operating region, the practical diode model assumes that:
    a. Diode current remains at zero until the knee voltage is reached.
    b. The diode turns on and forward conduction occurs when $V_F = V_k$.
    c. Forward voltage remains fixed regardless of the value of $I_F$.
15. Including $V_F$ in any circuit analysis reduces the percentage of error in the circuit calculations.
16. The maximum reverse voltage that *won't* force a diode to conduct is called the *peak reverse voltage* ($V_{RRM}$).
17. If the reverse voltage applied to a diode exceeds its $V_{RRM}$ rating, the component may break down and allow conduction in the reverse direction.
18. If forced to conduct in the reverse direction, *avalanche current* may generate sufficient heat to destroy the diode.
19. $V_{RRM}$ is a diode *parameter* (limit). If any diode parameter is exceeded, the component may be damaged or destroyed.
    a. As a precaution, the value of $V_{RRM}$ for a diode should be at least 20% greater than the maximum anticipated reverse voltage in a circuit.
    b. The value of $V_{RRM}$ for a given diode is always provided on the component *specification sheet*.
20. The *average forward current* ($I_0$) rating for a diode is the maximum allowable value of dc forward current.
    a. $I_0$ is a diode parameter.
    b. As a precaution, the $I_0$ rating for a diode should be at least 20% greater than the maximum anticipated value of $I_F$ in a circuit.
21. The *power dissipation* ($P_{D(max)}$) rating of a diode indicates its maximum possible power dissipation when forward biased.
    a. $P_{D(max)}$ is a diode parameter.
    b. As a precaution, the value of $P_{D(max)}$ for a diode should be at least 20% greater than the maximum anticipated value of $P_F$ in a circuit.
22. When determining whether to use a diode in a given circuit, you must consider:
    a. Is the $V_{RRM}$ rating of the component at least 20% greater than the maximum value of reverse voltage in the circuit?
    b. Is the $I_0$ rating of the component at least 20% greater than the maximum value of forward current in the circuit?
    c. Is the $P_{D(max)}$ rating of the component at least 20% greater than the maximum anticipated value of $P_F$ in the circuit?
    If the answer to all of these questions is *yes,* you can use the diode in the circuit.

23. *Bulk resistance* is the natural resistance of the diode p- and n-type materials.
24. The bulk resistance of a diode causes $V_F$ to increase slightly as $I_F$ increases. The effect of bulk resistance on diode forward operation is illustrated in Figure 2.19 and Example 2.12.
25. When reverse biased, a diode passes a very low-value *reverse current*.
    a. Reverse current is made up of *reverse saturation current* ($I_S$) and *surface-leakage current* ($I_{SL}$).
    b. Reverse current is normally in the low microamp ($\mu$A) range or lower.
26. Diode reverse current remains relatively constant as long as the $V_{RRM}$ rating of the diode is not exceeded and the temperature remains constant.
    a. Reverse saturation current is temperature dependent.
    b. Surface leakage current varies directly with the amount of reverse bias applied to a diode.
27. The presence of reverse current causes a slight voltage to be developed across any resistance in series with a diode when the component is reverse biased.
28. When reverse biased, a diode has measurable *junction capacitance*.
29. The forward current through a diode at values of $V_F < V_k$ is referred to as *diffusion current*.
30. An increase in diode operating temperature causes:
    a. A decrease in $V_F$ at a given value of $I_F$.
    b. An increase in $I_F$ at a given value of $V_F$.
31. Typically, an increase in diode operating temperature results in a slight increase in $I_F$ and a slight decrease in $V_F$.
32. The specification (spec) sheet for a component lists its parameters and operating characteristics (see Figure 2.25).
    a. *Parameters* (or *maximum ratings*) are limits that are not to be exceeded under any circumstances.
    b. *Operating characteristics* (or *electrical characteristics*) are guaranteed device characteristics. As long as the parameters of the device are observed, the operating characteristics are guaranteed to fall within the ranges listed.
33. When substituting one diode for another, *average forward current* ($I_0$) and *peak reverse voltage* ($V_{RRM}$) are the primary considerations. Each of these ratings must be at least 20% greater than the expected values of forward current and reverse voltage in the circuit.
34. Diode *selector guides* are references that are used to locate suitable diodes for specific circuit requirements (see Figure 2.27).
35. The *zener diode* is designed to work in the reverse breakdown region of the diode characteristic curve.
36. When operated in the reverse breakdown region of its operating curve, the voltage across a zener diode remains relatively constant over a specified range of current values.
    a. The specified current range is defined by *zener knee current* ($I_{ZK}$) and *maximum zener current* ($I_{ZM}$).
    b. The voltage across a zener that is operated between $I_{ZK}$ and $I_{ZM}$ is nearly constant and equal to its *zener voltage* ($V_Z$) rating.
37. The $V_Z$ rating of a zener is measured at a specific test current value ($I_{ZT}$).
38. Zener impedance ($Z_Z$) is the diode's opposition to a *change* in current.
39. The *dc power dissipation* rating of a zener can be used (with $V_Z$) to determine the value of $I_{ZM}$.
40. The *power derating factor* tells you how much the power dissipation rating of a component decreases with increases in temperature above a specified value. (See Example 2.14.)
41. You can substitute one zener diode for another when:
    a. The $V_Z$ ratings of the components are equal.
    b. The $P_{D(\text{max})}$ rating of the substitute component is equal to (or greater than) that of the original component.

**42.** Zener diode selector guides list the components by $V_Z$ and power dissipation ratings (see Figure 2.34).

**43.** Light-emitting diodes (LEDs) emit light when biased properly.

**44.** LEDs typically have the following ranges of values:
   **a.** $V_F = 1.4$ to $3.6$ V (rated at $I_F = 20$ mA)
   **b.** $V_{BR} = -3$ to $-10$ V

**45.** LEDs have relatively low maximum forward current ratings, typically 100 mA or less. As a result, they require the use of a series current-limiting resistor. (The required value of a current-limiting resistor is determined as shown in Example 2.17.)

**46.** Multicolor LEDs emit one color when forward biased and another when reverse biased. When rapidly switched between polarities, these two colors combine to produce a third color.

**47.** Most diode failures are caused by excessive current, component age, or surpassing the $V_{RRM}$ rating of the component.

**48.** Many faulty diodes show visible signs of damage, such as damaged casings and/or damage to surrounding copper traces on a PC board.

**49.** A *pn*-junction diode can be tested with a DMM as shown in Figures 2.40 and 2.41.
   **a.** When connected as shown in Figure 2.40, the DMM is used to measure the component resistances. High forward resistance or low reverse resistance indicates a faulty diode.
   **b.** When connected as shown in Figure 2.41, the DMM performs a diode function test. A low voltage reading (less than 1 V) should be displayed when the component is forward biased. A higher voltage reading (greater than 2 V) should be displayed when the component is reverse biased.
   **c.** Only high resistance scales on an ohmmeter should be used, because the meter output current (when on a low resistance scale) may be sufficient to damage a good diode.

**50.** The simplest test of a zener diode is to check the voltage across its terminals while it is in the circuit. If the voltage across the component is not at (or near) its rated value, there is a strong possibility that the component is faulty.

**51.** Faulty LEDs show obvious signs of damage and need no testing (see Figure 2.42).

## EQUATION SUMMARY

| Equation Number | Equation | Section Number |
|:---:|:---:|:---:|
| **(2.1)** | $V_{R1} = V_S - 0.7\text{V}$ | 2.3 |
| **(2.2)** | $I_T = \dfrac{V_S - 0.7\text{V}}{R_1}$ | 2.3 |
| **(2.3)** | $\% \text{ of error} = \dfrac{\lvert X - X' \rvert}{X} \times 100$ | 2.3 |
| **(2.4)** | $I_0 = \dfrac{P_{D(max)}}{V_F}$ | 2.4 |
| **(2.5)** | $V_F = V_B + I_F R_B$ | 2.5 |
| **(2.6)** | $I_R = I_S + I_{SL}$ | 2.5 |
| **(2.7)** | $Z_Z = \dfrac{\Delta V_Z}{\Delta I_Z}$ | 2.7 |
| **(2.8)** | $I_{ZM} = \dfrac{P_{D(max)}}{V_Z}$ | 2.8 |
| **(2.9)** | $R_S = \dfrac{V_{out(pk)} - V_F}{I_F}$ | 2.9 |

## PRACTICE PROBLEMS

### Section 2.1

1. Draw a circuit containing a dc voltage source, a resistor, and a forward-biased diode.

2. Add an arrow to the circuit you drew in Practice Problem 1 to indicate the direction of diode current.

3. Draw a circuit containing a dc voltage source, a resistor, and a reverse-biased diode.

4. For each of the circuits shown in Figure 2.44, determine the direction (if any) of diode forward current.

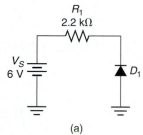
(a)

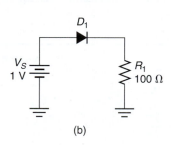
(b)

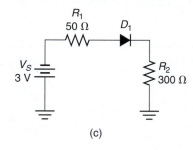
(c)

**FIGURE 2.44**

5. For each of the circuits shown in Figure 2.45, determine the direction (if any) of diode forward current.

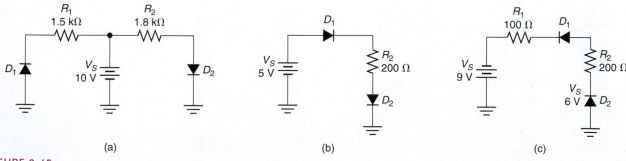
(a)       (b)       (c)

**FIGURE 2.45**

## Section 2.2

6. Using the *ideal diode model*, determine the voltage across each of the diodes shown in Figure 2.44.

7. Using the *ideal diode model*, determine the voltage across each of the components shown in Figure 2.45a.

## Section 2.3

8. Using the *practical diode model*, determine the values of $V_{D1}$, $V_{R1}$, and $I_T$ for the circuit shown in Figure 2.44a.

9. Using the *practical diode model*, determine the values of $V_{D1}$, $V_{R1}$, and $I_T$ for the circuit shown in Figure 2.44b.

10. Using the *practical diode model*, determine the values of $V_{D1}$, $V_{R1}$, $V_{R2}$, and $I_T$ for the circuit shown in Figure 2.44c.

11. Determine the values of $V_{D1}$, $V_{R1}$, $I_1$, $V_{D2}$, $V_{R2}$, and $I_2$ for the circuit shown in Figure 2.45a.*

12. Determine the values of $V_{D1}$, $V_{D2}$, $V_{R1}$, and $I_T$ for the circuit shown in Figure 2.45b.

13. Determine the values of $V_{D1}$, $V_{D2}$, $V_{R1}$, $V_{R2}$, and $I_T$ for the circuit shown in Figure 2.45c.

14. A voltage is calculated to be 12.8 V. The measured voltage is 13.2 V. What is the percentage of error in the calculation?

15. A current is calculated to be 750 μA. The measured current is 880 μA. Determine whether the percentage of error in the calculation is acceptable.

16. A voltage is calculated to be 144 mV. The measured voltage is 160 mV. Determine whether the percentage of error in the calculation is acceptable.

17. Ignoring the presence of the meter, calculate the value of $V_{R2}$ for the circuit shown in Figure 2.46. Then, determine the percentage of error between your calculated value and the meter reading shown.

18. Ignoring the presence of the meter, calculate the value of $V_{R2}$ for the circuit shown in Figure 2.47. Then, determine the percentage of error between your calculated value and the meter reading shown.

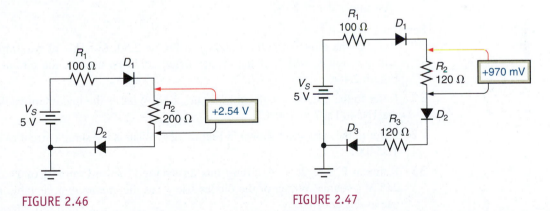

FIGURE 2.46                    FIGURE 2.47

## Section 2.4

19. What is the minimum required peak reverse voltage rating for the diode in Figure 2.48a?

20. What is the minimum required peak reverse voltage rating for the diode in Figure 2.48b? (*Hint*: Don't forget how to work with voltage dividers!)

21. It was stated in Section 2.4 that practical $V_{RRM}$ ratings are usually multiples of 50 or 100 V. With this in mind, what would be the minimum acceptable $V_{RRM}$ rating for the diode in Figure 2.48c?

*From now on, the practical diode model will be assumed unless indicated otherwise.

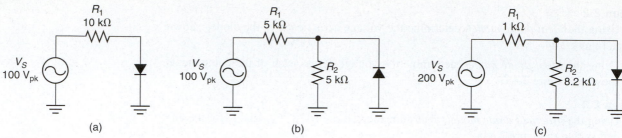

FIGURE 2.48

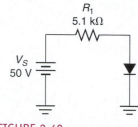

FIGURE 2.49

**22.** What is the minimum acceptable average forward current rating for the diode shown in Figure 2.49?

**23.** What is the minimum acceptable average forward power dissipation rating for the diode shown in Figure 2.49?

**24.** A diode has a $P_{D(max)}$ rating of 1.2 W. What is the maximum allowable value of forward current for the device?

**25.** A diode has a $P_{D(max)}$ rating of 750 mW. What is the limit on the value of average forward current for the device?

### Section 2.5

**26.** A small-signal diode (silicon) has a forward current of 10 mA and a bulk resistance $(R_B)$ of 5 $\Omega$. What is the actual value of $V_F$ for the device?

**27.** A small-signal diode (silicon) has a forward current of 8.2 mA and a bulk resistance $(R_B)$ of 12 $\Omega$. What is the actual value of $V_F$ for the device?

**28.** A diode (silicon) has a bulk resistance of 20 $\Omega$. At what value of $I_F$ will the value of $V_F$ actually *equal* 0.8 V?

**29.** Refer to Figure 2.48a. The diode in the circuit has a maximum rated value of $I_R = 10 \ \mu A$ at 25°C. Assuming that $I_R$ reaches its maximum value at each negative peak of the input cycle, what value of voltage will be measured across $R_1$ when $I_R$ peaks? (Assume that $T = 25$°C.)

### Section 2.6

**30.** Refer to the diode spec sheet shown in Figure 2.50. In terms of maximum reverse voltage ratings, which of the diodes listed could be used in the circuit shown in Figure 2.48a?

**31.** Refer to the spec sheet shown in Figure 2.50. What is the maximum value of $I_R$ for the 1N5400 at $T = 150$°C?

**32.** Refer to the spec sheet shown in Figure 2.50. What is the surge current rating for the 1N5400?

**33.** Refer to Figure 2.27. A circuit has an average forward current of 24.5 A and a 225 $V_{pk}$ source. Which of the diodes listed has the *minimum* acceptable ratings for use in this circuit?

**34.** Refer to Figure 2.27. A circuit has an average forward current of 3.6 A and a 170 $V_{pk}$ source. Which diode has the *minimum* acceptable ratings for use in this circuit?

**35.** Refer to Figure 2.27. A circuit has an average forward power dissipation (for the diode) of 2.8 W and a 470 $V_{pk}$ source. Which diode has the *minimum* acceptable ratings for use in this circuit?

### Section 2.7

**36.** A zener diode spec sheet lists values of $I_{ZT} = 20$ mA and $I_{zt} = 1$ mA. If the measured change in $V_Z$ (at $I_{zt}$) is 25 mV, what is the value of zener impedance for the device?

**37.** Refer to Figure 2.51. In each circuit shown, determine whether the biasing voltage has the correct polarity for *normal* zener operation.

**38.** For each of the properly biased zener diodes shown in Figure 2.51, draw an arrow indicating the direction of zener current.

### Section 2.8

**39.** A 6.8 V zener diode has a $P_{D(max)}$ rating of 1 W. What is the value of $I_{ZM}$ for the device?

# 1N5400 thru 1N5408

1N5404 and 1N5406 are Preferred Devices

## Axial-Lead Standard Recovery Rectifiers

Lead mounted standard recovery rectifiers are designed for use in power supplies and other applications having need of a device with the following features:

- High Current to Small Size
- High Surge Current Capability
- Low Forward Voltage Drop
- Void–Free Economical Plastic Package
- Available in Volume Quantities
- Plastic Meets UL 94V–0 for Flammability

**Mechanical Characteristics**

- Case: Epoxy, Molded
- Weight: 1.1 gram (approximately)
- Finish: All External Surfaces Corrosion Resistant and Terminal Leads are Readily Solderable
- Lead and Mounting Surface Temperature for Soldering Purposes: 220°C Max. for 10 Seconds, 1/16″ from case
- Polarity: Cathode Indicated by Polarity Band
- Marking: 1N5400, 1N5401, 1N5402, 1N5404, 1N5406, 1N5407, 1N5408

## ON Semiconductor™

**http://onsemi.com**

**STANDARD RECOVERY RECTIFIERS 50–1000 VOLTS 3.0 AMPERES**

**AXIAL LEAD CASE 267–05 STYLE 1**

**MARKING DIAGRAM**

```
AL
1N
540x
YYWW
```

| AL | = Assembly Location |
| 1N540x | = Device Number |
| x | = 0, 1, 2, 4, 6, 7 or 8 |
| YY | = Year |
| WW | = Work Week |

### MAXIMUM RATINGS

| Rating | Symbol | 1N5400 | 1N5401 | 1N5402 | 1N5404 | 1N5406 | 1N5407 | 1N5408 | Unit |
|---|---|---|---|---|---|---|---|---|---|
| Peak Repetitive Reverse Voltage Working Peak Reverse Voltage DC Blocking Voltage | $V_{RRM}$ $V_{RWM}$ $V_R$ | 50 | 100 | 200 | 400 | 600 | 800 | 1000 | Volts |
| Non–repetitive Peak Reverse Voltage | $V_{RSM}$ | 100 | 200 | 300 | 525 | 800 | 1000 | 1200 | Volts |
| Average Rectified Forward Current (Single Phase Resistive Load, 1/2″ Leads, $T_L$ = 105°C) | $I_O$ | 3.0 | | | | | | | Amp |
| Non–repetitive Peak Surge Current (Surge Applied at Rated Load Conditions) | $I_{FSM}$ | 200 (one cycle) | | | | | | | Amp |
| Operating and Storage Junction Temperature Range | $T_J$ $T_{stg}$ | − 65 to +170 − 65 to +175 | | | | | | | °C |

### THERMAL CHARACTERISTICS

| Characteristic | Symbol | Typ | Unit |
|---|---|---|---|
| Thermal Resistance, Junction to Ambient (PC Board Mount, 1/2″ Leads) | $R_{\theta JA}$ | 53 | °C/W |

### ELECTRICAL CHARACTERISTICS

| Characteristic | Symbol | Min | Typ | Max | Unit |
|---|---|---|---|---|---|
| Forward Voltage ($I_F$ = 3.0 Amp, $T_A$ = 25°C) | $v_F$ | – | – | 1.0 | Volts |
| Reverse Current (Rated dc Voltage) $T_A$ = 25°C $T_A$ = 150°C | $I_R$ | – – | – – | 10 100 | μA |

Ratings at 25°C ambient temperature unless otherwise specified.
60 Hz resistive or inductive loads.
For capacitive load, derate current by 20%.

**FIGURE 2.50** (Copyright of Semiconductor Component Industries, LLC. Used by permission.)

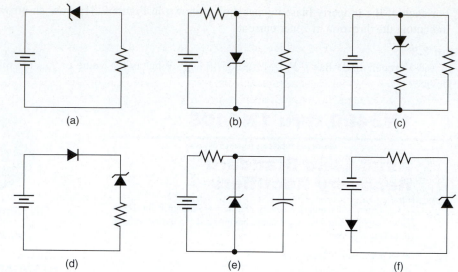

FIGURE 2.51

**40.** A 24 V zener diode has a $P_{D(max)}$ rating of 10 W. What is the value of $I_{ZM}$ for the device?

**41.** A zener diode with a $P_{D(max)}$ rating of 5 W has a derating factor of 8 mW/°C above 50°C. What is the maximum allowable value of $P_D$ for the device if it is operating at 120°C?

**42.** The MLL4678 zener diode has a $P_{D(max)}$ rating of 250 mW and a derating factor of 1.67 mW/°C above 50°C. What is the maximum allowable value of $P_D$ for the device if it is operating at 150°C?

**43.** Refer to Figure 2.34. Which of the diodes could be used in place of a 28 V zener diode that has a maximum power dissipation of 1.8 W?

**44.** Refer to Figure 2.34. Which of the diodes listed could be used in place of a 6.8 V zener diode that has a maximum power dissipation of 1.2 W?

**45.** Refer to Figure 2.34. Which of the diodes listed could be used in place of a 12 V zener diode that has a maximum operating current of 150 mA?

**Section 2.9**

**46.** An LED has a range of $V_F = 1.5$ to 1.8 V and $I_F = 18$ mA (desired). If the LED is driven by a 20 $V_{pk}$ source, what standard value of current-limiting resistor is needed to protect the LED?

**47.** An LED has a range of $V_F = 1.6$ to 2.0 V and $I_F = 20$ mA (desired). Determine the minimum standard resistor value that could be used as a current-limiting resistor if the LED is driven by a 32 $V_{pk}$ source.

**TROUBLESHOOTING PRACTICE PROBLEMS**

**48.** The following table lists the results of testing several diodes. In each case, determine whether the diode is good, open, or shorted.

|  | *Forward Resistance* | *Reverse Resistance* |
|---|---|---|
| **a.** | 1200 MΩ | 1200 MΩ |
| **b.** | 15 Ω | 3500 MΩ |
| **c.** | 75 Ω | 175 MΩ |
| **d.** | 30 Ω | 50 Ω |

**49.** Refer to Figure 2.48a. When the output from the source is positive, the peak voltage across $R_1$ is approximately 100 V. When the output from the source is negative, the peak voltage across $R_1$ is approximately 0 V. Is the diode good, open, or shorted? Explain your answer.

**50.** Refer to Figure 2.48a. The voltage across $R_1$ is always equal to the peak source voltage. Is the diode good, open, or shorted? Explain your answer.

**51.** Refer to Figure 2.48a. The voltage across $R_1$ is 0 V, regardless of the value of the source voltage. Is the diode good, open, or shorted? Explain your answer.

---

**PUSHING THE ENVELOPE**

**52.** Calculate the total current for the circuit shown in Figure 2.52.

**53.** Calculate the power being dissipated by the zener diode shown in Figure 2.53.

**54.** Explain how you could measure the value of $R$ in Figure 2.54 without disconnecting the diode.

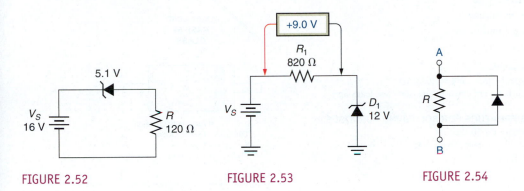

FIGURE 2.52          FIGURE 2.53          FIGURE 2.54

**55.** The spec sheet for the MLL755 zener diode is shown in Figure 2.55. What is the value of $I_{ZM}$ for the device when it is operated at 150°C?

**56.** The MLL756 zener diode *cannot* be used in the circuit shown in Figure 2.56. Why not? (*Note:* The temperature range shown is the normal operating temperature for the circuit.)

---

**SUGGESTED COMPUTER APPLICATIONS PROBLEMS**

**57.** Write a program to determine the total voltage drop across a forward-biased *pn*-junction diode. The program must take into account the type of semiconductor material used and the values of $V_k$, $I_F$, and $R_B$.

**58.** Write a program to determine the total voltage drop across a zener diode when $V_Z$, $Z_Z$, and $I_Z$ are known.

**59.** Write a program that will determine whether a given *pn*-junction diode is good, open, or shorted when provided with forward and reverse resistance readings.

**60.** Write a program that will determine the value of $I_{ZM}$ at specified values of $V_Z$, $P_{D(max)}$, power derating factor, and temperature.

---

**ANSWERS TO THE EXAMPLE PRACTICE PROBLEMS**

**2.1** The diode acts as an open, and an open drops the full applied voltage. Therefore, $V_{R1} = V_{R2} = 0$ V.

**2.2** Using the ideal diode model, $I_T = 15$ mA.

**2.3** 14.3 V

**2.4** $I_T = 8.43$ mA.

**2.5** $I_T = 2.36$ mA.

**2.6** $I_T = 5.75$ mA.

**2.7** $I_T$ (ideal) $= 7.5$ mA; % of error $= 30.4$%.

# 1N4728A - 1N4764A Series

## 1 Watt DO-41 Hermetically Sealed Glass Zener Voltage Regulator Diodes

This is a complete series of 1 Watt Zener diode with limits and excellent operating characteristics that reflect the superior capabilities of silicon–oxide passivated junctions. All this in an axial–lead hermetically sealed glass package that offers protection in all common environmental conditions.

**Specification Features:**
- Zener Voltage Range — 3.3 V to 91 V
- ESD Rating of Class 3 (>16 KV) per Human Body Model
- DO–41 (DO–204AL) Package
- Double Slug Type Construction
- Metallurgical Bonded Construction
- Oxide Passivated Die

**Mechanical Characteristics:**

**CASE:** Double slug type, hermetically sealed glass

**FINISH:** All external surfaces are corrosion resistant and leads are readily solderable

**MAXIMUM LEAD TEMPERATURE FOR SOLDERING PURPOSES:** 230°C, 1/16″ from the case for 10 seconds

**POLARITY:** Cathode indicated by polarity band

**MOUNTING POSITION:** Any

**ON Semiconductor™**

http://onsemi.com

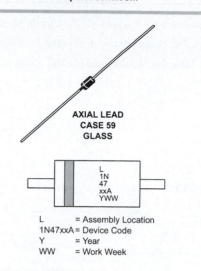

**AXIAL LEAD
CASE 59
GLASS**

L = Assembly Location
1N47xxA = Device Code
Y = Year
WW = Work Week

Cathode     Anode

### MAXIMUM RATINGS

| Rating | Symbol | Value | Unit |
|---|---|---|---|
| Max. Steady State Power Dissipation @ $T_L \leq 50°C$, Lead Length = 3/8″ | $P_D$ | 1.0 | Watt |
| Derated above 50°C | | 6.67 | mW/°C |
| Operating and Storage Temperature Range | $T_J, T_{stg}$ | − 65 to +200 | °C |

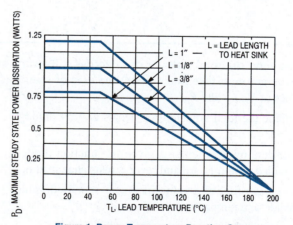

**Figure 1. Power Temperature Derating Curve**

(a)

**FIGURE 2.55** (Copyright of Semiconductor Component Industries, LLC. Used by permission.)

**ELECTRICAL CHARACTERISTICS** ($T_A$ = 25°C unless otherwise noted, $V_F$ = 1.2 V Max, $I_F$ = 200 mA for all types)

| JEDEC Device [2] | Zener Voltage [3][4] | | | @ $I_{ZT}$ | Zener Impedance [5] | | | Leakage Current | | $I_r$ [6] |
| | $V_Z$ (Volts) | | | | $Z_{ZT}$ @ $I_{ZT}$ | $Z_{ZK}$ @ $I_{ZK}$ | | $I_R$ @ $V_R$ | | |
| | Min | Nom | Max | (mA) | (Ω) | (Ω) | (mA) | (µA Max) | (Volts) | (mA) |
|---|---|---|---|---|---|---|---|---|---|---|
| 1N4728A | 3.14 | 3.3 | 3.47 | 76 | 10 | 400 | 1 | 100 | 1 | 1380 |
| 1N4729A | 3.42 | 3.6 | 3.78 | 69 | 10 | 400 | 1 | 100 | 1 | 1260 |
| 1N4730A | 3.71 | 3.9 | 4.10 | 64 | 9 | 400 | 1 | 50 | 1 | 1190 |
| 1N4731A | 4.09 | 4.3 | 4.52 | 58 | 9 | 400 | 1 | 10 | 1 | 1070 |
| 1N4732A | 4.47 | 4.7 | 4.94 | 53 | 8 | 500 | 1 | 10 | 1 | 970 |
| 1N4733A | 4.85 | 5.1 | 5.36 | 49 | 7 | 550 | 1 | 10 | 1 | 890 |
| 1N4734A | 5.32 | 5.6 | 5.88 | 45 | 5 | 600 | 1 | 10 | 2 | 810 |
| 1N4735A | 5.89 | 6.2 | 6.51 | 41 | 2 | 700 | 1 | 10 | 3 | 730 |
| 1N4736A | 6.46 | 6.8 | 7.14 | 37 | 3.5 | 700 | 1 | 10 | 4 | 660 |
| 1N4737A | 7.13 | 7.5 | 7.88 | 34 | 4 | 700 | 0.5 | 10 | 5 | 605 |
| 1N4738A | 7.79 | 8.2 | 8.61 | 31 | 4.5 | 700 | 0.5 | 10 | 6 | 550 |
| 1N4739A | 8.65 | 9.1 | 9.56 | 28 | 5 | 700 | 0.5 | 10 | 7 | 500 |
| 1N4740A | 9.50 | 10 | 10.50 | 25 | 7 | 700 | 0.25 | 10 | 7.6 | 454 |
| 1N4741A | 10.45 | 11 | 11.55 | 23 | 8 | 700 | 0.25 | 5 | 8.4 | 414 |
| 1N4742A | 11.40 | 12 | 12.60 | 21 | 9 | 700 | 0.25 | 5 | 9.1 | 380 |

**TOLERANCE AND TYPE NUMBER DESIGNATION**

2. The JEDEC type numbers listed have a standard tolerance on the nominal zener voltage of ±5%.

**SPECIALS AVAILABLE INCLUDE:**

3. Nominal zener voltages between the voltages shown and tighter voltage tolerances. For detailed information on price, availability, and delivery, contact your nearest ON Semiconductor representative.

**ZENER VOLTAGE ($V_Z$) MEASUREMENT**

4. ON Semiconductor guarantees the zener voltage when measured at 90 seconds while maintaining the lead temperature ($T_L$) at 30°C ± 1°C, 3/8″ from the diode body.

**ZENER IMPEDANCE ($Z_Z$) DERIVATION**

5. The zener impedance is derived from the 60 cycle ac voltage, which results when an ac current having an rms value equal to 10% of the dc zener current ($I_{ZT}$ or $I_{ZK}$) is superimposed on $I_{ZT}$ or $I_{ZK}$.

**SURGE CURRENT ($I_R$) NON-REPETITIVE**

6. The rating listed in the electrical characteristics table is maximum peak, non-repetitive, reverse surge current of 1/2 square wave or equivalent sine wave pulse of 1/120 second duration superimposed on the test current, $I_{ZT}$, per JEDEC registration; however, actual device capability is as described in Figure 5 of the General Data — DO-41 Glass.

(b)

**ELECTRICAL CHARACTERISTICS** ($T_A$ = 25°C unless otherwise noted, $V_F$ = 1.2 V Max, $I_F$ = 200 mA for all types) (continued)

| JEDEC Device [2] | Zener Voltage [3][4] | | | @ $I_{ZT}$ | Zener Impedance [5] | | | Leakage Current | | $I_r$ [6] |
| | $V_Z$ (Volts) | | | | $Z_{ZT}$ @ $I_{ZT}$ | $Z_{ZK}$ @ $I_{ZK}$ | | $I_R$ @ $V_R$ | | |
| | Min | Nom | Max | (mA) | (Ω) | (Ω) | (mA) | (µA Max) | (Volts) | (mA) |
|---|---|---|---|---|---|---|---|---|---|---|
| 1N4743A | 12.4 | 13 | 13.7 | 19 | 10 | 700 | 0.25 | 5 | 9.9 | 344 |
| 1N4744A | 14.3 | 15 | 15.8 | 17 | 14 | 700 | 0.25 | 5 | 11.4 | 304 |
| 1N4745A | 15.2 | 16 | 16.8 | 15.5 | 16 | 700 | 0.25 | 5 | 12.2 | 285 |
| 1N4746A | 17.1 | 18 | 18.9 | 14 | 20 | 750 | 0.25 | 5 | 13.7 | 250 |
| 1N4747A | 19.0 | 20 | 21.0 | 12.5 | 22 | 750 | 0.25 | 5 | 15.2 | 225 |
| 1N4748A | 20.9 | 22 | 23.1 | 11.5 | 23 | 750 | 0.25 | 5 | 16.7 | 205 |
| 1N4749A | 22.8 | 24 | 25.2 | 10.5 | 25 | 750 | 0.25 | 5 | 18.2 | 190 |
| 1N4750A | 25.7 | 27 | 28.4 | 9.5 | 35 | 750 | 0.25 | 5 | 20.6 | 170 |
| 1N4751A | 28.5 | 30 | 31.5 | 8.5 | 40 | 1000 | 0.25 | 5 | 22.8 | 150 |
| 1N4752A | 31.4 | 33 | 34.7 | 7.5 | 45 | 1000 | 0.25 | 5 | 25.1 | 135 |
| 1N4753A | 34.2 | 36 | 37.8 | 7 | 50 | 1000 | 0.25 | 5 | 27.4 | 125 |
| 1N4754A | 37.1 | 39 | 41.0 | 6.5 | 60 | 1000 | 0.25 | 5 | 29.7 | 115 |
| 1N4755A | 40.9 | 43 | 45.2 | 6 | 70 | 1500 | 0.25 | 5 | 32.7 | 110 |
| 1N4756A | 44.7 | 47 | 49.4 | 5.5 | 80 | 1500 | 0.25 | 5 | 35.8 | 95 |
| 1N4757A | 48.5 | 51 | 53.6 | 5 | 95 | 1500 | 0.25 | 5 | 38.8 | 90 |
| 1N4758A | 53.2 | 56 | 58.8 | 4.5 | 110 | 2000 | 0.25 | 5 | 42.6 | 80 |
| 1N4759A | 58.9 | 62 | 65.1 | 4 | 125 | 2000 | 0.25 | 5 | 47.1 | 70 |
| 1N4760A | 64.6 | 68 | 71.4 | 3.7 | 150 | 2000 | 0.25 | 5 | 51.7 | 65 |
| 1N4761A | 71.3 | 75 | 78.8 | 3.3 | 175 | 2000 | 0.25 | 5 | 56 | 60 |
| 1N4762A | 77.9 | 82 | 86.1 | 3 | 200 | 3000 | 0.25 | 5 | 62.2 | 55 |
| 1N4763A | 86.5 | 91 | 95.6 | 2.8 | 250 | 3000 | 0.25 | 5 | 69.2 | 50 |

(c)

FIGURE 2.55    (*continued*)

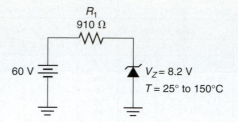

$R_1$
910 Ω

60 V

$V_Z$ = 8.2 V
$T$ = 25° to 150°C

FIGURE 2.56

**2.8** Any diode with $V_{RRM} > 210$ V. Therefore, you could use any diode from 1N4004–1N4007.

**2.9** The value of $I_T$ for the circuit is 1.95 A. The rating for the diode would have to be 20% greater than this value, or 2.34 A.

**2.10** $P_{D(min)} = 198.7$ mW $\times$ 1.20 = 238.4 mW

**2.11** $P_D = (750$ mA$)(0.7$ V$) = 525$ mW. This exceeds the 500 mW limit.

**2.12** 0.796 V (or 796 mV)

**2.13** $I_{ZM} = 37$ mA

**2.14** $P_D$ (at 125°C) = 300 mW

**2.15** The 1N5374A

**2.16** $P_D$ for the device is 1.19 W. Therefore, any diode in the 6.8 V row with a $P_D$ of 1.5 W or higher could be used. (The 1N5921A or 1N5342A could be used.)

**2.17** $R_S = 1.5$ kΩ minimum standard value; the calculated value is 1313 Ω.

## chapter 3

# Common Diode Applications

## Basic Power Supply Circuits

## Objectives

*After studying the material in this chapter, you should be able to:*

1. Briefly describe the purpose served by a power supply and the function of each circuit it contains.

2. Calculate the peak and dc (average) load voltage and current values for a *positive* half-wave rectifier.

3. Describe the operation of the full-wave rectifier.

4. Calculate the values of peak and dc (average) load voltage and current for any full-wave rectifier.

5. Describe the operation of the bridge rectifier.

6. Calculate the peak and dc (average) load voltage and current values for a bridge rectifier.

7. Discuss the effects that *filtering* has on the output of a rectifier.

8. Describe the operation of the basic capacitive filter.

9. Discuss the reason why full-wave rectifiers are preferred over half-wave rectifiers.

10. Calculate the values of $I_Z$, $I_L$, and $I_T$ for a zener voltage regulator, given the required circuit values and diode ratings.

11. Calculate the values of $V_{dc}$, $I_L$, and $V_{r(out)}$ for a basic power supply.

12. List the faults that commonly occur in a basic power supply and the symptoms of each.

## Outline

## The Role of Power Supplies in Semiconductor Development

Power supplies are the most commonly used circuits in electronics. Virtually every electronic system requires the use of a power supply to convert an ac line voltage to the dc voltages needed for the system's internal operation.

In addition to being the most commonly used type of circuit, the power supply also played a major role in the development of today's electronic devices. Early power supplies used vacuum tubes to *rectify* ac; that is, to convert ac to pulsating dc. These vacuum tubes wasted a tremendous amount of power.

Early semiconductor research centered on the use of *germanium*, a semiconductor material that cannot withstand any significant amount of current and heat. With the development of the commercial *pn*-junction diode in 1954, researchers turned to the problem of developing rectifier diodes, that is, diodes that could withstand large current values and the heat produced by those currents.

In 1955, *silicon pn*-junction rectifier diodes had been developed that could handle current values up to 2 A. From that point on, research centered on the use of silicon rather than germanium. Even now, silicon is used for most semiconductor applications. Germanium is rarely used in the production of semiconductor devices.

**OBJECTIVE 1** ▶

**Power supply**
A group of circuits used to convert ac to dc.

It would take several volumes to discuss every diode application in modern electronics. In this chapter, we will focus on the various roles that diodes play in the operation of basic power supply circuits. In Chapter 4, we will look at several additional diode applications.

The **power supply** of an electronic system is a group of circuits that convert the ac energy provided by the wall outlet to dc energy. The power cord of an electronic system provides the input to the power supply, which then provides all the internal dc voltages needed for proper circuit operation.

There are two basic types of power supplies. The *linear* power supply is the original and simpler of the two, and it is the focus of this chapter. The *switching* power supply is a newer and more complex circuit that is used in a wide variety of electronic systems. Switching supplies are covered in Chapter 21.

**Rectifier**
A circuit that converts ac to pulsating dc.

**Filter**
A circuit that reduces the variations in the output of a rectifier.

**Voltage regulator**
A circuit designed to maintain a constant power supply output voltage.

A basic linear power supply can be broken down into three circuit groups, as shown in Figure 3.1a. The ac input is applied to a rectifier. A **rectifier** is a diode circuit that converts the ac to what is called *pulsating dc*. This pulsating dc is then applied to a *filter*. The **filter** is a circuit that reduces the variations in the output from the rectifier. The final stage of the power supply is the *voltage regulator*. The **voltage regulator** is designed to maintain a constant power supply output voltage.

At one time, voltage regulators were designed using zener diodes as the regulating element. However, the development of IC (integrated-circuit) voltage regulators led to the replacement of zener diodes as regulating elements. IC voltage regulators are far more efficient than zener diodes, so power supply design emphasizes their use. At the same time, zener regulators are easier to understand and serve as valuable educational circuits. We will therefore concentrate on a zener regulator in this chapter. IC voltage regulators are covered in detail in Chapter 21.

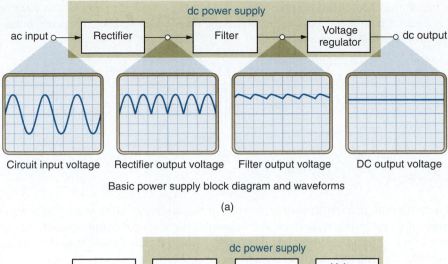

Basic power supply block diagram and waveforms

(a)

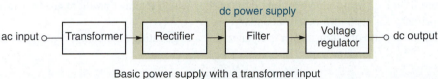

Basic power supply with a transformer input

(b)

**FIGURE 3.1** Basic power supply block diagram and waveforms.

In many cases, a *transformer* is used to connect the power supply to the ac line input, as shown in Figure 3.1b. In this circuit, the ac line input is applied to the transformer, which may be a *step-up*, *step-down*, or *isolation* transformer, depending on the needs of the power supply. Because they are commonly used in conjunction with power supply circuits, we will begin our discussion by reviewing some basic transformer principles.

## 3.1 Transformers

Transformers are not considered to be solid-state devices, but they do play an integral role in the operation of most power supplies. The basic schematic symbol for the transformer is shown in Figure 3.2. The component consists of two windings, called the *primary* and the *secondary*. The input to the transformer is applied to the primary, and the output is taken from the secondary.

Transformers are made up of inductors that are in close proximity to each other, yet are not electrically connected. An alternating voltage applied to the primary induces an alternating voltage in the secondary. At the same time, the primary and secondary are electrically isolated, so no actual current is transferred between the two circuits. Therefore, a transformer provides ac coupling from primary to secondary while providing physical isolation between the two circuits.

There are three types of transformers: *step-up*, *step-down*, and *isolation*. These components are described as follows:

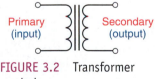

**FIGURE 3.2** Transformer symbol.

1. The *step-up transformer* provides a secondary voltage that is *greater than* the primary voltage. For example, a step-up transformer may provide a 240 $V_{ac}$ output with a 120 $V_{ac}$ input.
2. The *step-down transformer* provides a secondary voltage that is *less than* the primary voltage. For example, a step-down transformer may provide a 30 $V_{ac}$ output with a 120 $V_{ac}$ input.
3. An *isolation transformer* provides an output voltage that is *equal to* the input voltage. This type of transformer is used to isolate the power supply electrically from the ac power line, which helps to protect the power supply (and the technician who is working on it) from the line voltage.

The transformers described here are represented in Figure 3.3. As you can see, each symbol contains a ratio that is printed just above the transformer symbol. This ratio is known as the *turns ratio* of the component.

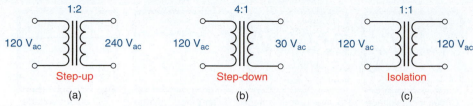

**FIGURE 3.3**

The turns ratio of a transformer is *the ratio of the number of turns in the primary to the number of turns in the secondary*. For example, the step-down transformer shown in Figure 3.3b is shown to have a turns ratio of 4:1, which means that there are four turns in the primary for each turn in the secondary.

The turns ratio of a transformer is equal to the voltage ratio of the component. By formula,

$$\frac{N_S}{N_P} = \frac{V_S}{V_P} \tag{3.1}$$

where $N_S$ = the number of turns in the secondary
$\quad\quad N_P$ = the number of turns in the primary
$\quad\quad V_S$ = the secondary voltage
$\quad\quad V_P$ = the primary voltage

Thus, the primary voltage of the step-down transformer shown in Figure 3.3b is four times as great as the secondary voltage.

### 3.1.1 Calculating Secondary Voltage

When the turns ratio and primary voltage of a transformer are known, the secondary voltage can be found as

$$V_S = \frac{N_S}{N_P} V_P \tag{3.2}$$

For example, let's say that the step-down transformer shown in Figure 3.3b has a 120 $V_{ac}$ input. The secondary voltage for the component would be found as

$$V_S = \frac{N_S}{N_P} V_P = \frac{1}{4}(120 \ V_{ac}) = \mathbf{30 \ V_{ac}}$$

As you can see, the primary voltage (120 $V_{ac}$) is four times the secondary voltage (30 $V_{ac}$).

### 3.1.2 Calculating Secondary Current

Ideally, transformers are 100% efficient. This means that the ideal transformer transfers 100% of its input power to the secondary. By formula,

$$P_S = P_P$$

Since power equals the product of voltage and current,

$$V_S I_S = V_P I_P$$

and

$$\frac{I_P}{I_S} = \frac{V_S}{V_P} \qquad (3.3)$$

As you can see, the current ratio is the inverse of the voltage ratio. This means that

**1.** For a step-down transformer, $I_S > I_P$.
**2.** For a step-up transformer, $I_S < I_P$.

In other words, current varies (from primary to secondary) in the opposite way that voltage varies. If voltage increases, current decreases, and vice versa.

Since the voltage ratio of a transformer is equal to its turns ratio, equation (3.3) can be rewritten as

$$\frac{I_P}{I_S} = \frac{N_S}{N_P}$$

or

$$I_S = \frac{N_P}{N_S}I_P \qquad (3.4)$$

The following example demonstrates a practical application of this relationship.

### EXAMPLE 3.1

The fuse shown in Figure 3.4 is used to limit the current in the primary of the transformer. Assuming that the fuse limits the value of $I_P$ to 1 A, what is the limit on the value of the secondary current?

*Solution:* The maximum secondary current is found using the limit on $I_P$ and the turns ratio of the transformer as follows:

$$I_S = \frac{N_P}{N_S}I_P = \frac{1}{4}(1 \text{ A}) = 250 \text{ mA}$$

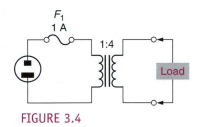

**FIGURE 3.4**

If the secondary current tries to exceed the 250 mA limit, the primary current will exceed its limit and blow the fuse.

#### PRACTICE PROBLEM 3.1
A circuit like the one shown in Figure 3.4 has a turns ratio of 1:12 and a fuse that limits the primary current to 250 mA. Calculate the maximum allowable value of $I_S$.

The current relationships developed here have been based on the idea that a transformer is 100% efficient. In reality, there are a number of losses within a transformer that cause the secondary power to be somewhat lower than the primary power. However, the difference between primary power and secondary power is small enough to have little effect on the relationships covered in this section.

## 3.1.3 Transformer Ratings

Some manufacturers' catalogs rate transformers by turns ratios, while others list them by *secondary voltage ratings*. For example, a transformer may be listed as a 40 $V_{ac}$ transformer. When this rating is used, *it indicates the ac secondary voltage produced by a 120 $V_{ac}$ input to the primary*. In other words, it gives you the rms output from the transformer when it is supplied by a standard 120 $V_{ac}$ line input. Throughout this chapter, we will use both of these methods of rating transformers.

**Section Review** ▶

1. What are the names of the transformer input and output circuits?

2. List and describe the three types of transformers.

3. What is the *turns ratio* of a transformer?

4. Describe the relationship between the turns ratio of a transformer and its input and output voltages.

5. Describe the relationship between the transformer primary and secondary power.

6. Describe the relationship between the voltage ratio of a transformer and its current ratio.

7. Describe the two means by which transformers are normally listed in parts catalogs.

## 3.2  Half-Wave Rectifiers

There are three basic types of rectifier circuits: the *half-wave*, *full-wave*, and *bridge* rectifiers. Of the three, the bridge rectifier is the most commonly used, followed by the full-wave rectifier. Our discussion on rectifiers begins with the half-wave rectifier simply because it is the easiest to understand.

> What does a half-wave rectifier do?

The *half-wave rectifier* is simply a *diode* that is connected in series between a transformer (or ac line input) and its load. A half-wave rectifier with a resistive load ($R_L$) is shown in Figure 3.5. The diode is used to eliminate either the negative alternation of the input or the positive alternation of the input. As you will see, the diode direction determines which half-cycle is eliminated.

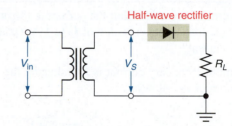

**FIGURE 3.5**   Half-wave rectifier.

Because the half-wave rectifier configuration requires the least number of parts, it is the cheapest to produce. However, it is the least efficient of the three types of rectifiers. As a result, it is normally used for noncritical, low-current applications.

### 3.2.1  Basic Circuit Operation

The negative half-cycle of the input to the rectifier shown in Figure 3.5 is eliminated by the one-way conduction of the diode. Figure 3.6 details the operation of the circuit for one complete cycle of the input signal. During the positive half-cycle of the input, $D_1$ is forward biased and provides a path for current. This allows a voltage ($V_L$) to be developed across $R_L$ that is approximately equal to the voltage across the secondary of the transformer ($V_S$).

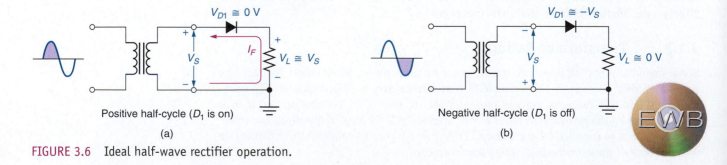

**FIGURE 3.6**   Ideal half-wave rectifier operation.

When the polarity of the input signal reverses, $D_1$ is reverse biased, preventing conduction in the circuit. With no current through $R_L$, no voltage is developed across the load. In this case, the output voltage remains at approximately 0 V, and the voltage across the diode ($V_D$) is approximately equal to $V_S$.

If all this seems confusing, remember the *ideal diode model.* This diode model was shown in Chapter 2 to represent the component as either an *open* (reverse-biased) or *closed* (forward-biased) switch. When forward biased, this ideal switch drops no voltage. When reverse biased, this ideal switch drops all the applied voltage.

Ideally, the diode in Figure 3.6a can be viewed as a closed switch. Therefore,

$$V_L \cong V_S \quad \text{(forward operation)} \tag{3.5}$$

This is because no voltage is dropped across a closed switch. By the same token, if we view the diode in Figure 3.6b as an *open* switch, we see that

$$V_{D1} = V_S \quad \text{(reverse operation)} \tag{3.6}$$

This is because an open switch drops all the applied voltage. This would leave no voltage to be dropped across $R_L$, and $V_L$ would equal 0 V, as shown in the figure. The *ideal* circuit operating characteristics are summarized as follows:

| Diode Condition | $V_{D1}$ (ideal) | $V_L$ (ideal) |
|---|---|---|
| Forward biased | 0 V | Equal to $V_S$ |
| Reverse biased | Equal to $V_S$ | 0 V |

Using these relationships, it is easy to understand the input/output waveforms shown in Figure 3.7. During $t_1$, the diode is forward biased, and the output ($V_L$) is approximately equal to the input ($V_S$). During $t_2$, the diode is reverse biased, and the output drops to 0 V. This is how half-wave rectifiers work.

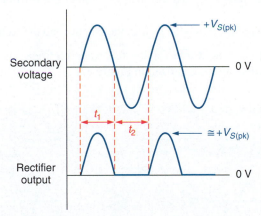

**FIGURE 3.7** Ideal input and output waveforms.

**Lab Reference:** These waveforms are observed in Exercise 3.

## 3.2.2 Negative Half-Wave Rectifiers

Figure 3.8 shows a *negative* half-wave rectifier. If you compare this circuit with the one shown in Figure 3.6, you'll see that the only physical difference between the circuits is the orientation of the diode.

During the positive half-cycle of the input, $D_1$ is reverse biased (as shown in Figure 3.8a), and equation (3.6) applies. In this case, it is the positive half-cycle of the input that is blocked by the diode. During the negative half-cycle of the input, $D_1$ is forward biased (as shown in Figure 3.8b), and equation (3.5) applies. Note the direction of the secondary current and the polarity of the load voltage.

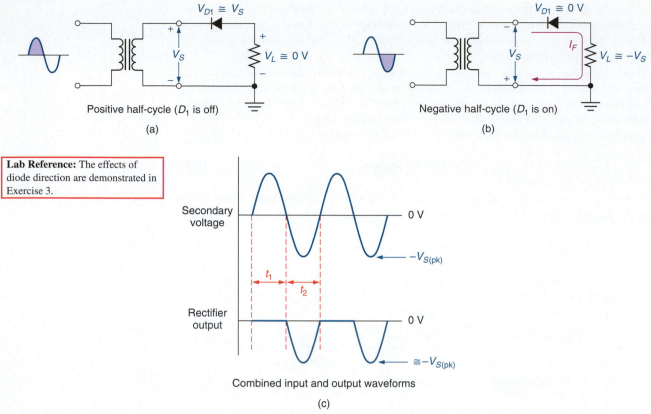

$V_{D1} \cong V_S$

$V_S$

$V_L \cong 0\ V$

Positive half-cycle ($D_1$ is off)

(a)

$V_{D1} \cong 0\ V$

$V_S$

$I_F$

$V_L \cong -V_S$

Negative half-cycle ($D_1$ is on)

(b)

**Lab Reference:** The effects of diode direction are demonstrated in Exercise 3.

Secondary voltage

0 V

$-V_{S(pk)}$

$t_1$

$t_2$

Rectifier output

0 V

$\cong -V_{S(pk)}$

Combined input and output waveforms

(c)

FIGURE 3.8   Negative half-wave rectifier.

How do you determine the output polarity of a half-wave rectifier?

As you can see, the *direction* of the diode determines whether the output from the rectifier is positive or negative. For circuit recognition, the following statements generally hold true:

1. When the diode points toward the load ($R_L$), the output from the rectifier will be *positive*.
2. When the diode points toward the source, the output from the rectifier will be *negative*.

These two statements also hold true for the full-wave rectifier. The points made so far about half-wave rectifiers are summarized in Figure 3.9.

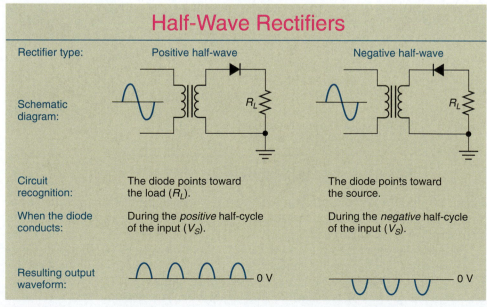

## Half-Wave Rectifiers

| Rectifier type: | Positive half-wave | Negative half-wave |
|---|---|---|
| Schematic diagram: | $R_L$ | $R_L$ |
| Circuit recognition: | The diode points toward the load ($R_L$). | The diode points toward the source. |
| When the diode conducts: | During the *positive* half-cycle of the input ($V_S$). | During the *negative* half-cycle of the input ($V_S$). |
| Resulting output waveform: | 0 V | 0 V |

FIGURE 3.9

## 3.2.3    Calculating Load Voltage and Load Current

In our discussion of the *ideal* half-wave rectifier, we ignored the value of $V_F$ for the diode. When we take this value into account, the *peak load voltage*, $V_{L(pk)}$, is found as

◄ OBJECTIVE 2

$$V_{L(pk)} = V_{S(pk)} - V_F \tag{3.7}$$

You shouldn't have any difficulty with this equation. It is simply a variation on equation (2.1). $V_{S(pk)}$ is the *peak secondary voltage* of the transformer and is found as

> Equation (2.1):
>
> $V_{R1} = V_S - 0.7\text{ V}$

$$V_{S(pk)} = \frac{N_S}{N_P} V_{P(pk)} \tag{3.8}$$

where $\dfrac{N_S}{N_P}$ = the ratio of transformer secondary turns to primary turns

$V_{P(pk)}$ = the peak transformer primary voltage

*A word of caution*: Equation (3.8) assumes that the input to the transformer is given as a *peak* value. More often than not, source voltages are given as rms values. When this is the case, the source voltage can be converted to a peak value as follows:

$$V_{pk} = \frac{V_{rms}}{0.707} \tag{3.9}$$

Example 3.2 illustrates the procedure for calculating the peak output voltage from a positive half-wave rectifier.

### EXAMPLE 3.2

Determine the peak load voltage for the circuit shown in Figure 3.10.

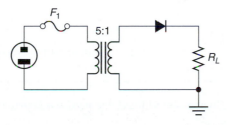

> Figure 3.10 contains the symbol for an ac plug (in the transformer primary). Whenever this symbol is used, we will assume that the circuit input is a 120 $V_{ac}$ sine wave.

FIGURE 3.10

**Solution:**    First, the ac input to the transformer is converted to a peak value as follows:

$$V_{P(pk)} = \frac{V_{P(rms)}}{0.707} = \frac{120\text{ V}}{0.707} = \mathbf{169.7\text{ V}}$$

Now, the voltage values in the secondary circuit are found as

$$V_{S(pk)} = \frac{N_S}{N_P} V_{P(pk)} = \frac{1}{5}(169.7\text{ V}) = \mathbf{33.94\text{ V}}$$

and

$$V_{L(pk)} = V_{S(pk)} - V_F = 33.94\text{ V} - 0.7\text{ V} = \mathbf{33.24\text{ V}}$$

**PRACTICE PROBLEM 3.2**
A half-wave rectifier has values of $N_P = 10$, $N_S = 1$, and $V_{P(pk)} = 180$ V. What is the peak load voltage for the circuit?

Here is another practical situation: Most transformers are rated for a specific rms output voltage. For example, a 25 $V_{ac}$ transformer would have an rms output of 25 V when supplied from a 120 V wall outlet. When a transformer has an output voltage rating, simply divide the rated output voltage by 0.707 to obtain the value of $V_{S(pk)}$, as shown in Example 3.3.

### EXAMPLE 3.3

Determine the peak load voltage for the circuit shown in Figure 3.11.

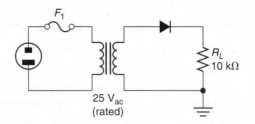

25 $V_{ac}$
(rated)

FIGURE 3.11

*Question and Answer:*
At this point, you may be wondering why we are so interested in *peak* values. Wouldn't it be easier to stick with *effective* (rms) values? We are interested in peak values for several reasons. First, knowing the peak values of the circuit voltages allows us to effectively analyze the circuit operation with an oscilloscope. Second (and most important) is the fact that *average* (dc) voltage and current values are commonly defined in terms of (and are calculated using) peak values.

**Solution:** The transformer is shown to have a 25 $V_{ac}$ rating. This value of $V_S$ is converted to peak form as follows:

$$V_{S(pk)} = \frac{V_{S(rms)}}{0.707} = \frac{25 \text{ V}}{0.707} = \mathbf{35.36 \text{ V}}$$

Now, the value of $V_{L(pk)}$ is found as

$$V_{L(pk)} = V_{S(pk)} - V_F = 35.36 \text{ V} - 0.7 \text{ V} = \mathbf{34.66 \text{ V}}$$

**PRACTICE PROBLEM 3.3**
A 12 $V_{ac}$ transformer is being used in a positive half-wave rectifier. What is the peak load voltage for the circuit?

Once the peak load voltage is determined, the *peak load current* is found as

$$I_{L(pk)} = \frac{V_{L(pk)}}{R_L} \qquad\qquad (3.10)$$

Example 3.4 demonstrates the calculation of peak load current.

### EXAMPLE 3.4

What is the peak load current for the circuit shown in Figure 3.12?

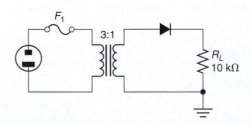

FIGURE 3.12

**Solution:** The circuit has a 120 $V_{ac}$ input. This value is converted to a *peak* value as follows:

$$V_{P(pk)} = \frac{V_{P(rms)}}{0.707} = \frac{120\ V}{0.707} = \textbf{169.7 V}$$

Now, the value of $V_{S(pk)}$ is found as

$$V_{S(pk)} = \frac{N_S}{N_P} V_{P(pk)} = \frac{1}{3}(169.7\ V) = \textbf{56.6 V}$$

Finally, the load voltage and current values are found as

$$V_{L(pk)} = V_{S(pk)} - V_F = 56.6\ V - 0.7\ V = \textbf{55.9 V}$$

and

$$I_{L(pk)} = \frac{V_{L(pk)}}{R_L} = \frac{55.9\ V}{10\ k\Omega} = \textbf{5.59 mA}$$

**PRACTICE PROBLEM 3.4**

A circuit like the one shown in Figure 3.12 has values of $N_P = 12$, $N_S = 1$, and $R_L = 8.2\ k\Omega$. What is the peak load current for the circuit?

### 3.2.4    Average Voltage and Current

The value of **average voltage ($V_{ave}$)** for an ac (or other) waveform is *the value that would be measured with a dc voltmeter*. In other words, $V_{ave}$ is the *dc equivalent* value of the waveform. In most cases, $V_{ave}$ and $V_{dc}$ are used to describe the same value. Since rectifiers are used to convert ac to dc, $V_{ave}$ is a very important value. For a half-wave rectifier, $V_{ave}$ is found as

> **Average voltage ($V_{ave}$)**
> The dc equivalent of an ac (or other) waveform. $V_{ave}$ is measured with a dc voltmeter.

$$V_{ave} = \frac{V_{(pk)}}{\pi} \quad \text{(half-wave rectified)} \tag{3.11}$$

Another form of this equation is

> Equation (3.11) is typically introduced in an ac circuits course. The derivation of this equation is provided (as a review) in Appendix D.

$$V_{ave} = 0.318 V_{pk} \quad \text{(half-wave rectified)} \tag{3.12}$$

where $0.318 \cong 1/\pi$. Either of these equations can be used to determine the *dc equivalent* load voltage for a *half-wave* rectifier. Example 3.5 demonstrates the process for determining the value of $V_{ave}$ for a half-wave rectifier.

**EXAMPLE 3.5**

Determine the value of $V_{ave}$ for the circuit shown in Figure 3.13.

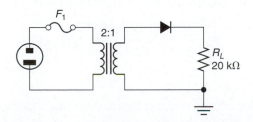

> **Lab Reference:** These values are calculated and measured as part of Exercise 3.

FIGURE 3.13

**Solution:**

$$V_{P(pk)} = \frac{V_{P(rms)}}{0.707} = \frac{120 \text{ V}}{0.707} = \textbf{169.7 V}$$

$$V_{S(pk)} = \frac{N_S}{N_P}V_{P(pk)} = \frac{1}{2}(169.7 \text{ V}) = \textbf{84.85 V}$$

$$V_{L(pk)} = 84.85 \text{ V} - 0.7 \text{ V} = \textbf{84.15 V}$$

$$V_{ave} = \frac{V_{L(pk)}}{\pi} = \frac{84.15 \text{ V}}{\pi} = \textbf{26.8 V}$$

**PRACTICE PROBLEM 3.5**

A half-wave rectifier like the one in Figure 3.13 has values of $N_P = 14$ and $N_S = 1$. What is the dc load voltage for the circuit?

---

**Average current ($I_{ave}$)**
The dc equivalent of an alternating current. $I_{ave}$ is measured with a dc ammeter.

Just as we can convert a *peak* voltage to *average* voltage, we can also convert a *peak* current to an *average* current. The value of the **average current ($I_{ave}$)** for an ac waveform is the value that would be measured with a *dc ammeter*. Thus, the value of $I_{ave}$ for a waveform gives us an *equivalent dc current*. This is another very important value for any circuit used to convert ac to dc. The value of $I_{ave}$ can be calculated in one of two ways:

**1.** We can determine the value of $V_{ave}$ and then use Ohm's law as follows:

$$I_{ave} = \frac{V_{ave}}{R_L}$$

**2.** We can calculate the value of $I_{pk}$ as demonstrated in Example 3.4. Then, we can convert this peak value to average form using equations similar to those we used to convert $V_{pk}$ to $V_{ave}$ (equations 3.11 and 3.12). The current forms of these equations are

Equation (3.13) can be derived in the same fashion as equation (3.11). The derivation of that equation is provided in Appendix D.

$$I_{ave} = \frac{I_{pk}}{\pi} \quad \text{(half-wave rectified)} \tag{3.13}$$

and

$$I_{ave} = 0.318 I_{pk} \quad \text{(half-wave rectified)} \tag{3.14}$$

Example 3.6 demonstrates the first method for determining the value of $I_{ave}$.

**EXAMPLE 3.6**

Determine the value of $I_{ave}$ for the circuit shown in Figure 3.13.

**Solution:** In Example 3.5, we determined the value of $V_{ave}$ to be 26.8 V. Using this value and the value of $R_L$, the value of $I_{ave}$ is found to be

$$I_{ave} = \frac{V_{ave}}{R_L} = \frac{26.8 \text{ V}}{20 \text{ k}\Omega} = \textbf{1.34 mA}$$

**PRACTICE PROBLEM 3.6**

A half-wave rectifier has an average output voltage that is equal to 24 V. The load resistance is 2.2 kΩ. What is the value of the dc load current for the circuit?

---

Example 3.7 demonstrates the second method of determining the value of $I_{ave}$.

**EXAMPLE 3.7**

Determine the dc load current for the rectifier shown in Figure 3.14.

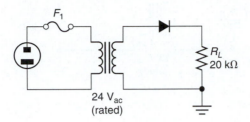

FIGURE 3.14

*Solution:* The transformer has a 24 $V_{ac}$ rating. Thus, the peak secondary voltage is found as

$$V_{S(pk)} = \frac{24 \ V_{ac}}{0.707} = \mathbf{33.9 \ V}$$

The peak load voltage is now found as

$$V_{L(pk)} = V_{S(pk)} - 0.7 \ V = \mathbf{33.2 \ V}$$

The peak load current is found as

$$I_{L(pk)} = \frac{V_{L(pk)}}{R_L} = \frac{33.2 \ V}{20 \ k\Omega} = \mathbf{1.66 \ mA}$$

Finally,

$$I_{ave} = \frac{I_{(pk)}}{\pi} = \frac{1.66 \ mA}{\pi} = \mathbf{528.39 \ \mu A}$$

**PRACTICE PROBLEM 3.7**

A half-wave rectifier is fed by a 48 $V_{ac}$ transformer. If the load resistance for the circuit is 12 $k\Omega$, what is the dc load current for the circuit?

## 3.2.5  Negative Half-Wave Rectifiers

The mathematical analysis of a *negative* half-wave rectifier is nearly identical to that for a positive half-wave rectifier. The only difference is that all the voltage polarities are reversed.

Here is a simple method you can use to perform the mathematical analysis of a negative half-wave rectifier:

1. Analyze the circuit as if it were a positive half-wave rectifier.
2. After completing your calculations, change all your voltage polarity signs from positive to negative.

This approach to analyzing a negative half-wave rectifier is demonstrated in Example 3.8.

**EXAMPLE 3.8**

Determine the dc output voltage for the circuit shown in Figure 3.15.

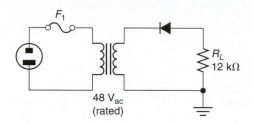

**FIGURE 3.15**

*Solution:* We'll start by solving the circuit as if it were a positive half-wave rectifier. First,

$$V_{S(pk)} = \frac{48 \text{ V}_{ac}}{0.707} = \textbf{67.9 V}$$

and

$$V_{L(pk)} = V_{S(pk)} - 0.7 \text{ V} = \textbf{67.2 V}$$

Finally,

$$V_{ave} = \frac{V_{L(pk)}}{\pi} = \frac{67.2 \text{ V}}{\pi} = \textbf{21.39 V}$$

Now, we simply convert all the positive voltage values to negative voltage values. Thus, for the circuit shown in Figure 3.15,

$$V_{S(pk)} = -67.9 \text{ V}$$
$$V_{L(pk)} = -67.2 \text{ V}$$
$$V_{ave} = -21.39 \text{ V}$$

**PRACTICE PROBLEM 3.8**

A negative half-wave rectifier is fed by a 36 $V_{ac}$ transformer. Using the method illustrated in this example, calculate the values of $V_{S(pk)}$, $V_{L(pk)}$, and the dc output voltage.

As you can see, there isn't really a whole lot of difference between the mathematical analysis of a negative half-wave rectifier and that of a positive half-wave rectifier.

### 3.2.6    Component Substitution

What role does the value of $I_{ave}$ play in component substitution?

The value of $I_{ave}$ is important for another reason. You may recall from Chapter 2 that the *maximum dc forward current* that can be drawn through a diode is equal to the *average forward current* ($I_0$) rating of the device. When working with rectifiers, you may need at some point to substitute one diode for another. When this is the case, you must make sure that the value of $I_{ave}$ for the diode in the circuit is less than the $I_0$ rating of the substitute component. This point is demonstrated in Section 3.5.

### 3.2.7 Peak Inverse Voltage (PIV)

The maximum amount of reverse bias that will be applied to a diode in a given circuit is called the **peak inverse voltage**, or **PIV**, of the circuit. For the half-wave rectifier, the PIV is found as

$$\text{PIV} = V_{S(pk)} \quad \text{(half-wave rectified)} \tag{3.15}$$

**Peak inverse voltage (PIV)**
The maximum reverse bias that will be applied to a diode in a given circuit.

The basis for this equation can be seen by referring to Figure 3.6. When the diode is reverse biased (Figure 3.6b), no voltage is dropped across the load. Therefore, all of $V_S$ is dropped across the diode.

The PIV of a given circuit is important because it determines the *minimum allowable value* of $V_{RRM}$ for any diode used in the circuit. This point is demonstrated in Section 3.5.

Why is PIV important?

◄ **Section Review**

1. Briefly explain the forward and reverse operation of a half-wave rectifier.
2. Describe the difference between the output waveforms of a positive and a negative half-wave rectifier.
3. How can you determine the output polarity of a half-wave rectifier?
4. List, in order, the steps you would take to calculate the dc output voltage from a rectifier if you were given the turns ratio of the transformer and the rms primary voltage.
5. List, in order, the steps you would take to calculate the dc output voltage from a rectifier if you knew the rated rms secondary voltage.
6. What test equipment is used to measure the average load voltage ($V_{ave}$) of a half-wave rectifier?
7. What test equipment is used to measure the average load current ($I_{ave}$) of a half-wave rectifier?
8. What is the relationship between circuit PIV and the $V_{RRM}$ rating of a diode?
9. How do you determine the PIV for a half-wave rectifier?
10. Based on the waveforms shown in Figure 3.7, why do you think the half-wave rectifier is viewed as being inefficient?

◄ **Critical Thinking**

11. Using only the information provided in this section, derive an equation that converts an *effective* (rms) voltage to an *average* (dc) value.

## 3.3 Full-Wave Rectifiers

The *full-wave rectifier* consists of *two diodes* that are connected as shown in Figure 3.16a. The result of this change in circuit construction is illustrated in Figure 3.16b.

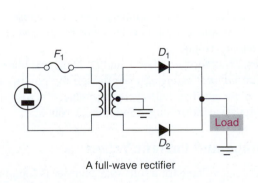

A full-wave rectifier

(a)

Input signal

Half-wave rectifier output

Full-wave rectifier output

$t_0$

Typical rectifier waveforms

(b)

**FIGURE 3.16** Full-wave rectifier.

In Figure 3.16b, the output from the full-wave rectifier is compared with that from a half-wave rectifier. Note that the full-wave rectifier produces two half-cycles out for every one produced by the half-wave rectifier.

**Center-tapped transformer**

A transformer with an output lead connected to the center of the secondary winding.

The transformer shown in Figure 3.16 is a **center-tapped transformer**. This type of transformer has a lead connected to the center of the secondary winding. The voltage from the center tap to each of the outer winding terminals is equal to half the secondary voltage. For example, let's say that we have a 24 V center-tapped transformer. The voltage from the center tap to each of the outer winding terminals is 12 V.

As you will see, the operation of the center-tapped transformer plays a major role in the operation of the full-wave rectifier. For this reason, the full-wave rectifier cannot be *line operated*; that is, it cannot be connected directly to the ac line input.

### 3.3.1    Basic Circuit Operation

**OBJECTIVE 3** ▶ Figure 3.17 shows the operation of the full-wave rectifier during one complete cycle of the input signal. During the positive half-cycle of the input, $D_1$ is forward biased, and $D_2$ is reverse biased. Note the direction of current through the load ($R_L$). *Using the ideal diode model*, the peak load voltage can be found as

$$V_{L(pk)} \cong \frac{V_{S(pk)}}{2} \tag{3.16}$$

**Lab Reference:** Full-wave operation is demonstrated in Exercise 3.

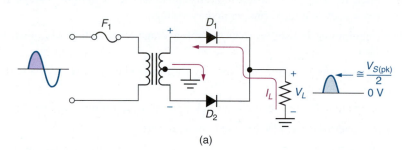

(a)

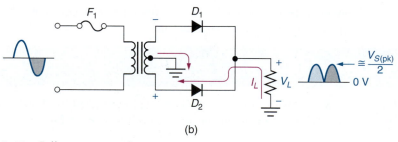

(b)

FIGURE 3.17    Full-wave operation.

Why is load voltage only half the value of the transformer secondary voltage?

The load voltage is approximately equal to half the secondary voltage because the transformer is center tapped. The voltage from one end of a center-tapped transformer to the center tap is *always* half the total secondary voltage.

When the polarity of the input reverses, $D_2$ is forward biased, and $D_1$ is reverse biased. Note that the direction of the load current *has not changed*, even though the polarity of the transformer secondary has. Thus, another positive half-cycle is produced across the load, producing the full-wave rectifier output waveform shown in Figure 3.16b.

### 3.3.2    Calculating Load Voltage and Current Values

**OBJECTIVE 4** ▶ Using the *practical diode model*, the peak load voltage for a full-wave rectifier is found as

$$V_{L(pk)} = \frac{V_{S(pk)}}{2} - 0.7 \text{ V} \tag{3.17}$$

The full-wave rectifier produces twice as many output pulses (per input cycle) as the half-wave rectifier. In other words, for every output pulse produced by a half-wave rectifier, two are produced by a full-wave rectifier. For this reason, the *average load voltage* for the full-wave rectifier is found as

The full-wave rectifier has twice the output frequency of the half-wave rectifier.

$$V_{ave} = \frac{2V_{L(pk)}}{\pi} \tag{3.18}$$

or

$$V_{ave} = 0.637 V_{L(pk)} \tag{3.19}$$

where $0.637 \cong 2/\pi$. Note that the value 0.637 used in equation (3.19) is approximately twice the value of 0.318, used in equation (3.12) to find the value of $V_{ave}$ for a half-wave rectifier. The procedure for determining the dc load voltage for a full-wave rectifier is determined in Example 3.9.

## EXAMPLE 3.9

Determine the dc load voltage for the circuit shown in Figure 3.18.

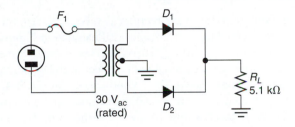

**Lab Reference:** These values are calculated and measured as part of Exercise 3.

**FIGURE 3.18**

*Solution:* The transformer is rated at 30 $V_{ac}$. Therefore, the value of $V_{S(pk)}$ is found as

$$V_{S(pk)} = \frac{30\ V_{ac}}{0.707} = \textbf{42.4 V}$$

The peak load voltage is now found as

$$V_{L(pk)} = \frac{V_{S(pk)}}{2} - 0.7\ V = 21.2\ V - 0.7\ V = \textbf{20.5 V}$$

Finally, the dc load voltage is found as

$$V_{ave} = \frac{2V_{L(pk)}}{\pi} = \frac{41\ V}{\pi} = \textbf{13.05 V}$$

## PRACTICE PROBLEM 3.9

A full-wave rectifier is fed by a 24 $V_{ac}$ center-tapped transformer. What is the dc load voltage for the circuit?

Once the peak and average load voltage values are known, it is easy to determine the values of $I_{L(pk)}$ and $I_{ave}$. Just use the known values and Ohm's law, as demonstrated in Example 3.10.

**EXAMPLE 3.10**

Determine the values $I_{L(pk)}$ and $I_{ave}$ for the circuit shown in Figure 3.18.

***Solution:*** In Example 3.9, we calculated the peak and average output voltages for the circuit. Using these calculated values and the value of $R_L$ shown in the circuit, we determine the circuit current values as follows:

$$I_{L(pk)} = \frac{V_{L(pk)}}{R_L} = \frac{20.5 \text{ V}}{5.1 \text{ k}\Omega} = \textbf{4.02 mA}$$

and

$$I_{ave} = \frac{V_{ave}}{R_L} = \frac{13.05 \text{ V}}{5.1 \text{ k}\Omega} = \textbf{2.56 mA}$$

**PRACTICE PROBLEM 3.10**

The circuit described in Practice Problem 3.9 has a 2.2 k$\Omega$ load. What are the peak and dc load current values for the circuit?

### 3.3.3    Peak Inverse Voltage

When one of the diodes in a full-wave rectifier is reverse biased, the voltage across that diode is approximately equal to $V_S$. This point is illustrated in Figure 3.19. With a 24 $V_{ac}$ rating, there is approximately 34 $V_{pk}$ across the secondary of the transformer. As a result, the secondary voltages (measured with respect to the center tap) are $+17$ $V_{pk}$ and $-17$ $V_{pk}$. With the polarities shown, $D_1$ is conducting, and $D_2$ is reverse biased. If we assume $D_1$ to be ideal, then $V_{D1} = 0$ V, and the cathode of $D_1$ is also at $+17$ $V_{pk}$. Since the cathode of $D_1$ is connected directly to the cathode of $D_2$, its cathode is also at $+17$ $V_{pk}$. With $-17$ $V_{pk}$ applied to the anode of $D_2$, the difference of potential across the component is 34 $V_{pk}$.

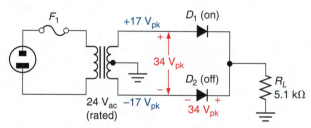

**FIGURE 3.19**   Full-wave rectifier PIV.

The peak load voltage supplied by the full-wave rectifier is approximately half the secondary voltage, $V_S$. Therefore, the reverse voltage across either diode is approximately twice the peak load voltage. By formula,

$$\text{PIV} \cong 2V_{L(pk)} \tag{3.20}$$

As shown in Figure 3.19, the PIV across the reverse-biased diode is approximately equal to the peak secondary voltage (when the other diode is assumed to be *ideal*). Therefore, the PIV applied to either diode can be approximated as

$$\text{PIV} \cong V_{S(pk)} \tag{3.21}$$

Though it provides a fairly accurate value of diode PIV, equation (3.21) ignores the practical characteristics of the other diode in the full-wave rectifier. In Figure 3.19, $D_1$ is on and has a voltage drop of 0.7 V. Because $D_1$ is in series with $D_2$, the PIV across $D_2$ is reduced by the voltage drop across $D_1$. Therefore, a more accurate value of PIV can be obtained using

$$\text{PIV} \cong V_{S(pk)} - 0.7 \text{ V} \qquad (3.22)$$

*A Practical Consideration:*
It may seem confusing to have three different equations for PIV. However, you should keep two points in mind:

1. Equations (3.20) and (3.21) are effectively saying the same thing.

2. The 0.7 V drop included in equation (3.22) may become important when you are dealing with low-voltage rectifiers (like those typically found in personal computers).

### 3.3.4   Negative Full-Wave Rectifiers

If we reverse the directions of the diodes in the positive full-wave rectifier, we have a *negative* full-wave rectifier. The negative full-wave rectifier and its output waveform are shown in Figure 3.20. As you can see, the main differences between the positive and negative full-wave rectifiers are the direction in which the diodes are pointing and the polarity of the output voltage.

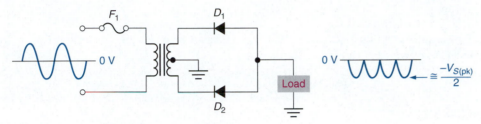

FIGURE 3.20   A negative full-wave rectifier.

The mathematical analysis of a negative full-wave rectifier can be performed using the same approach we used for the negative half-wave rectifier; that is, the circuit voltage values are calculated using the relationships given in this section. Then, the polarities of the results are changed from positive to negative. We will not go through this analysis procedure again, but you should have no problem making the transition from the negative half-wave to the negative full-wave rectifier. If you don't remember how to analyze a negative rectifier, refer back to Section 3.2.

### 3.3.5   Full-Wave Versus Half-Wave Rectifiers

There are quite a few similarities between full-wave and half-wave rectifiers. Figure 3.21 summarizes the relationships that you have been shown for these two circuits.

It would seem (at first) that the only similarity between half-wave and full-wave rectifiers is the method of finding the PIV values for the two circuits. However, there is another similarity that may not be as obvious. Let's assume for a moment that both of the rectifiers shown in Figure 3.21 are fed by 24 $V_{ac}$ transformers. If you were to calculate the dc output voltages for the two circuits, you would get the following values:

$$V_{ave} = 10.58 \text{ V}_{dc} \quad \text{(for the half-wave rectifier)}$$
$$V_{ave} = 10.36 \text{ V}_{dc} \quad \text{(for the full-wave rectifier)}$$

As you can see, the two circuits produce nearly identical dc output voltages *for identical values of transformer secondary voltage*. In fact, if the values of $R_L$ for the two circuits are equal, the dc output current values for the circuits are also nearly identical.

So why do we bother with the full-wave rectifier when we can get the same dc output values with the half-wave rectifier? There are a couple of reasons. First, *if the peak load voltages for the two circuits are equal, the full-wave rectifier will have twice the dc load voltage and power efficiency of the half-wave rectifier.* The second reason deals with the operation of filters. As you will be shown in Section 3.6, the full-wave rectifier has twice the output frequency of the half-wave rectifier, which has an impact on the filtering of the rectifier output. When we add filters to the half-wave and full-wave rectifiers, the advantages of the full-wave rectifier will become clear.

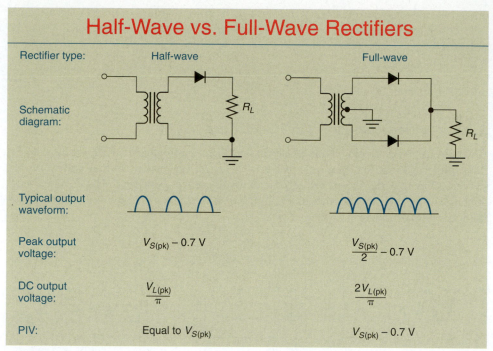

| Rectifier type: | Half-wave | Full-wave |
|---|---|---|
| Schematic diagram: | | |
| Typical output waveform: | | |
| Peak output voltage: | $V_{S(pk)} - 0.7\ V$ | $\dfrac{V_{S(pk)}}{2} - 0.7\ V$ |
| DC output voltage: | $\dfrac{V_{L(pk)}}{\pi}$ | $\dfrac{2V_{L(pk)}}{\pi}$ |
| PIV: | Equal to $V_{S(pk)}$ | $V_{S(pk)} - 0.7\ V$ |

**FIGURE 3.21**

**Section Review ▶**

1. Briefly explain the operation of a full-wave rectifier.
2. How do you determine the PIV across each diode in a full-wave rectifier?
3. Briefly discuss the similarities and differences between the half-wave rectifier and the full-wave rectifier.

**Critical Thinking ▶**

4. Why does a full-wave rectifier have twice the efficiency of a half-wave rectifier with the same peak load voltage?

## 3.4    Full-Wave Bridge Rectifiers

> Why are bridge rectifiers preferred over other full-wave rectifiers?

The *bridge rectifier* is the most commonly used full-wave rectifier for several reasons:

1. It does not require the use of a center-tapped transformer and therefore can be coupled directly to the ac power line (if desired).
2. When connected to a transformer with the same secondary voltage, it produces nearly double the peak output voltage of the full-wave center-tapped rectifier. This results in a higher dc output voltage from the supply.

The bridge rectifier consists of *four diodes* connected as shown in Figure 3.22.

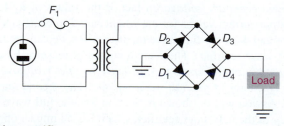

**FIGURE 3.22**    Bridge rectifier.

### 3.4.1 Basic Circuit Operation

◄ OBJECTIVE 5

In Figure 3.17, you were shown that the full-wave rectifier produces its output by alternating circuit conduction between the two diodes. When one diode is *on* (conducting), the other diode is *off* (not conducting). The bridge rectifier works in the same basic fashion. However, the bridge rectifier alternates conduction between two diode *pairs*. This principle is illustrated in Figure 3.23. During the positive half-cycle of the input, $V_S$ has the polarity shown in Figure 3.23a, causing $D_1$ and $D_3$ to conduct. Note the direction of current through the load resistor and the polarity of the resulting load voltage. During the negative half-cycle of the input, $V_S$ has the polarity shown in Figure 3.23b. $D_2$ and $D_4$ now conduct (rather than $D_1$ and $D_3$). However, the current direction through the load has not changed, nor has the resulting polarity of the load voltage.

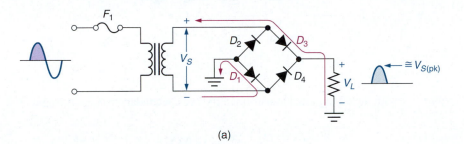

(a)

Lab Reference: Bridge rectifier operation is demonstrated in Exercise 3.

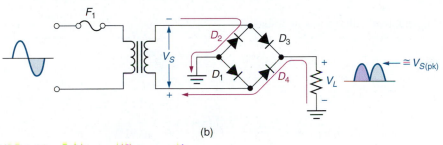

(b)

FIGURE 3.23  Bridge rectifier operation.

### 3.4.2 Calculating Load Voltage and Current Values

◄ OBJECTIVE 6

The full-wave rectifier has an output voltage that is approximately half the secondary voltage (assuming that the conducting diode is ideal). This relationship was expressed in equation (3.16). As you know, the output voltage in a full-wave rectifier is reduced to half the secondary voltage by the center tap on the transformer secondary. The center-tapped transformer is essential for the full-wave rectifier to work, but it cuts the output voltage to approximately half the value of $V_S$.

Equation (3.16):
$$V_{L(pk)} \cong \frac{V_{S(pk)}}{2}$$

The bridge rectifier does not require the use of a center-tapped transformer. Assuming the diodes in the bridge to be ideal, the rectifier has a peak output voltage of

$$V_{L(pk)} \cong V_{S(pk)} \quad \text{(ideal)} \qquad (3.23)$$

Refer to Figure 3.23. If you consider the diodes to be ideal, the cathode and anode voltages are equal for each diode. If you view the conducting diodes as being shorted connections to the transformer secondary, you can see that the voltage across the load resistor is equal to the voltage across the secondary.

When calculating circuit output values, you will get more accurate results if you take the voltage drops across the two conducting diodes into account. Since there are two conducting diodes in series with the load, the peak load voltage is found as:

$$V_{L(pk)} = V_{S(pk)} - 1.4 \text{ V} \qquad (3.24)$$

The 1.4 V value represents the sum of the diode voltage drops. The rest of the load voltage and current values are found using the same equations as those used for the full-wave rectifier. This is demonstrated in Example 3.11.

### EXAMPLE 3.11

Determine the dc load voltage and current values for the circuit shown in Figure 3.24.

**Lab Reference:** These voltages are calculated and measured as part of Exercise 3.

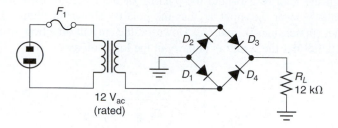

FIGURE 3.24

*Solution:* With the 12 $V_{ac}$ rated transformer, the peak secondary voltage is found as

$$V_{S(pk)} = \frac{12\ V_{ac}}{0.707} = \mathbf{16.97\ V}$$

The peak load voltage is now found as

$$V_{L(pk)} = V_{S(pk)} - 1.4\ V = \mathbf{15.57\ V}$$

The dc load voltage is found as

$$V_{ave} = \frac{2V_{L(pk)}}{\pi} = \frac{(2)(15.57\ V)}{\pi} = \mathbf{9.91\ V}$$

Finally, the dc load current is found as

$$I_{ave} = \frac{V_{ave}}{R_L} = \frac{9.91\ V}{12\ k\Omega} = \mathbf{825.8\ \mu A}$$

### PRACTICE PROBLEM 3.11

A bridge rectifier is fed by an 18 $V_{ac}$ transformer. Determine the dc load voltage and current for the circuit when it has a 1.2 k$\Omega$ load.

## 3.4.3   Bridge Versus Full-Wave Rectifiers

Let's analyze one more full-wave rectifier. This will give us some values for comparing the outputs from the full-wave and bridge rectifiers.

### EXAMPLE 3.12

A full-wave rectifier has a 12 $V_{ac}$ transformer and a 12 k$\Omega$ load. Determine the dc load voltage and current values for the circuit.

*Solution:* The peak secondary voltage for the circuit is found as

$$V_{S(pk)} = \frac{12\ V_{ac}}{0.707} = \mathbf{16.97\ V}$$

The peak load voltage is now found as

$$V_{L(pk)} = \frac{V_{S(pk)}}{2} - 0.7V = \frac{16.97\ V}{2} - 0.7\ V = \mathbf{7.79\ V}$$

The dc load voltage is found as

$$V_{ave} = \frac{2V_{L(pk)}}{\pi} = \frac{(2)(7.79\ V)}{\pi} = \mathbf{4.96\ V}$$

Finally, the dc load current is found as

$$I_{ave} = \frac{V_{ave}}{R_L} = \frac{4.96\ V}{12\ k\Omega} = \mathbf{413.3\ \mu A}$$

Now, let's compare the results from Examples 3.11 and 3.12. The only difference between these two circuits is that one is a bridge rectifier and the other is a full-wave rectifier. For convenience, the circuit values are summarized as follows:

| Value | Bridge Rectifier | Full-Wave Rectifier |
|---|---|---|
| Peak load voltage | 15.57 V | 7.79 V |
| dc load voltage | 9.91 V | 4.96 V |
| dc load current | 825.8 μA | 413.3 μA |

As you can see, the bridge rectifier has output values that are twice as high as those of a comparable full-wave rectifier. This higher output is the primary advantage of using a bridge rectifier. The bridge rectifier also has higher power efficiency than the full-wave rectifier and does not require use of a center-tapped transformer.

> What are the primary advantages of using a bridge rectifier in place of a full-wave rectifier?

### 3.4.4 Peak Inverse Voltage

Using the ideal diode model, the PIV of each diode in the bridge rectifier is equal to $V_S$. This is the same voltage that was applied to the diodes in the full-wave center-tapped rectifier. Figure 3.25 helps to illustrate this point.

> Since each "off" diode is in series with a conducting diode, the actual PIV is equal to $V_{S(pk)} - 0.7\ V$.

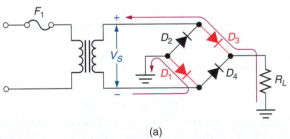
(a)

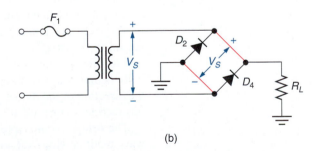
(b)

FIGURE 3.25   Bridge rectifier PIV.

Figure 3.25a shows the conduction through the secondary when $V_S$ has the polarity shown. In Figure 3.25b, the conducting diodes ($D_1$ and $D_3$) have been replaced by wires representing the *ideal* characteristics of the conducting diodes. As shown, $D_2$ and $D_4$ are connected across the transformer secondary. Therefore, the voltage across each of these nonconducting diodes equals the secondary voltage ($V_S$). The same holds true for $D_1$ and $D_3$ when they are reverse biased (during the negative half-cycle of the input).

### 3.4.5    Putting It All Together

The three commonly used rectifiers are the *half-wave rectifier, full-wave (center-tapped) rectifier*, and the *bridge rectifier*. The output characteristics of these circuits are summarized in Figure 3.26.

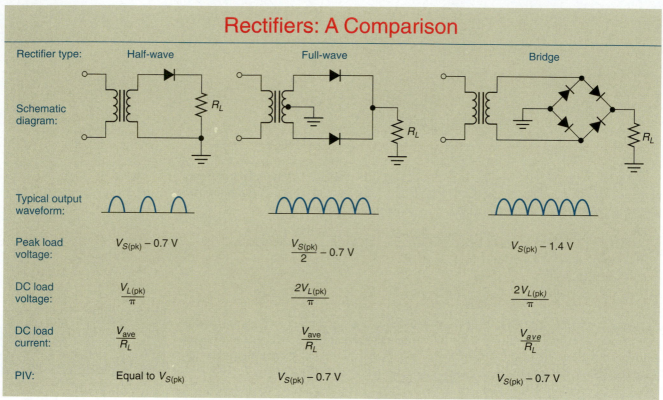

| Rectifier type: | Half-wave | Full-wave | Bridge |
|---|---|---|---|
| Schematic diagram: | | | |
| Typical output waveform: | | | |
| Peak load voltage: | $V_{S(pk)} - 0.7$ V | $\dfrac{V_{S(pk)}}{2} - 0.7$ V | $V_{S(pk)} - 1.4$ V |
| DC load voltage: | $\dfrac{V_{L(pk)}}{\pi}$ | $\dfrac{2V_{L(pk)}}{\pi}$ | $\dfrac{2V_{L(pk)}}{\pi}$ |
| DC load current: | $\dfrac{V_{ave}}{R_L}$ | $\dfrac{V_{ave}}{R_L}$ | $\dfrac{V_{ave}}{R_L}$ |
| PIV: | Equal to $V_{S(pk)}$ | $V_{S(pk)} - 0.7$ V | $V_{S(pk)} - 0.7$ V |

FIGURE 3.26

As you know, the *half-wave rectifier* is the simplest of the three circuits. For each input cycle, the half-wave rectifier produces a single half-cycle output. The polarity of the half-cycle output depends on the direction of the diode in the circuit. The half-wave rectifier is normally used in conjunction with a transformer. However, the circuit can be directly coupled to an ac line input.

The *full-wave (center-tapped) rectifier* uses two diodes in conjunction with a center-tapped transformer to convert an ac input to a pulsating dc output. For each input cycle, this rectifier produces two output half-cycles (as shown in Figure 3.26). As is the case with the half-wave rectifier, the output polarity of the full-wave rectifier is determined by the direction of the diodes in the circuit. Since a center-tapped transformer is required for this rectifier to operate, it *cannot* be directly coupled to an ac line input.

The *bridge rectifier* produces an output that is similar to that of a center-tapped full-wave rectifier. However, the diode configuration in this rectifier eliminates the need for a center-tapped transformer. As a result, the bridge rectifier has two distinct advantages over its full-wave counterpart:

1. The peak output (for a given peak input) is twice the value of the output produced by a center-tapped full-wave rectifier.
2. The bridge rectifier can be directly coupled to an ac line input, just like the half-wave rectifier.

Figure 3.26 summarizes the output relationships for the rectifiers we have discussed. Whenever you need to briefly review the relationships listed, you can refer to this illustration. For more detailed information, refer to the appropriate section of this chapter.

1. Describe the operation of the bridge rectifier.
2. List the advantages that the bridge rectifier has over the full-wave rectifier.
3. Describe the method for determining the PIV for a diode in a bridge rectifier.

## 3.5 Working with Rectifiers

Several practical aspects of working with rectifiers have been ignored up to this point. Most of these factors explain the differences between *theory* and *practice.*

### 3.5.1 Effects of Bulk Resistance and Reverse Current on Circuit Measurements

In Chapter 2, we discussed the effects of bulk resistance and reverse current on circuit voltage measurements. In that chapter, the following points were made:

1. The current through the bulk resistance of a diode can affect the actual value of $V_F$ for the device.
2. When a diode is in series with a resistor, reverse current can cause a voltage to be developed across the resistor when the diode is biased *off.*

While the effects of bulk resistance and reverse current are not normally very drastic, they do serve to explain a few things:

1. Rectifiers tend to be high-current circuits. Thus, the effects of $I_F R_B$ on the value of $V_F$ can be fairly significant. Many rectifier diodes will have a value of $V_F$ that is closer to 1 V than to 0.7 V.
2. It is fairly common for *power rectifiers* (those with high-current capabilities) to have high reverse current ratings. For example, the spec sheet shown in Figure 3.27 shows that the MBR1045 has a *maximum instantaneous reverse current* rating of 15 mA at $T = 125°C$. This value of reverse current could severely impact the operation of a power supply. To keep the reverse current to a minimum, power supplies are kept as cool as possible.

> *A Practical Consideration:* Diode forward voltage values in high-current rectifiers can be as high as 1.4 V.

Many rectifier diodes are mounted to a heat sink to keep them cool when used in high-current circuits. The heat sink draws the heat away from the diode, keeping the *pn*-junction cooler. There are many configurations for heat sinks, from flat, aluminum stock to those with elaborate cooling fins. Many times, heat sinks have forced air flowing through and around them to increase their power dissipation capabilities. Many power diodes, such as the one shown in Figure 3.27, have a drilled metal plate that allows them to be bolted directly to a heat sink.

> Reverse current is one of the reasons that many power supplies are fan cooled.

Integrated-circuit (IC) bridge rectifiers, like the one illustrated in Figure 3.28, are also designed to be mounted onto a heat sink. There is a hole in the middle of the component to allow a mounting bolt to pass through. The bottom of the bridge may be aluminum, which is electrically isolated from all diode junctions and serves as an excellent thermal conductor. (IC bridge rectifiers are discussed further at the end of this section.)

### 3.5.2 Transformer Rating Tolerance

The tolerance of a transformer's output rating can have a major effect on circuit voltage measurements. Depending on the quality of the transformer, the output voltage may (at times) be above or below the rated value by as much as 20%! This can have a major impact on any voltage measured in the circuit.

Later in this chapter, we will discuss *zener voltage regulators*. These circuits provide varying degrees of **line regulation**; that is, they have *the ability to maintain a constant load voltage despite anticipated variations in rectifier output voltage.* In other words, as long as the variations in the output from a rectifier remain within an anticipated range, the

> **Line regulation**
> The ability of a voltage regulator to maintain a constant load voltage despite anticipated variations in rectifier output voltage.

# MBR1035, MBR1045

**MBR1045 is a Preferred Device**

## SWITCHMODE™ Power Rectifiers

. . . using the Schottky Barrier principle with a platinum barrier metal. These state–of–the–art devices have the following features:

- Guardring for Stress Protection
- Low Forward Voltage
- 150°C Operating Junction Temperature
- Epoxy Meets UL94, VO at 1/8″

**Mechanical Characteristics:**
- Case: Epoxy, Molded
- Weight: 1.9 grams (approximately)
- Finish: All External Surfaces Corrosion Resistant and Terminal Leads are Readily Solderable
- Lead Temperature for Soldering Purposes: 260°C Max. for 10 Seconds
- Shipped 50 units per plastic tube
- Marking: B1035, B1045

**ON Semiconductor™**

http://onsemi.com

---

### SCHOTTKY BARRIER RECTIFIERS 10 AMPERES 35 to 45 VOLTS

---

3 o———▷|———o 1, 4

---

**TO–220AC CASE 221B PLASTIC**

**MARKING DIAGRAM**

B10x5

B10x5 = Device Code
x = 3 or 4

## MAXIMUM RATINGS

| Rating | Symbol | Value | Unit |
|---|---|---|---|
| Peak Repetitive Reverse Voltage Working Peak Reverse Voltage DC Blocking Voltage     MBR1035     MBR1045 | $V_{RRM}$ $V_{RWM}$ $V_R$ | 35 45 | V |
| Average Rectified Forward Current (Rated $V_R$, $T_C$ = 135°C) | $I_{F(AV)}$ | 10 | A |
| Peak Repetitive Forward Current, (Rated $V_R$, Square Wave, 20 kHz, $T_C$ = 135°C) | $I_{FRM}$ | 20 | A |
| Non–Repetitive Peak Surge Current (Surge Applied at Rated Load Conditions Halfwave, Single Phase, 60 Hz) | $I_{FSM}$ | 150 | A |
| Peak Repetitive Reverse Surge Current (2.0 μs, 1.0 kHz) See Figure 12. | $I_{RRM}$ | 1.0 | A |
| Storage Temperature Range | $T_{stg}$ | –65 to +175 | °C |
| Operating Junction Temperature | $T_J$ | –65 to +150 | °C |
| Voltage Rate of Change (Rated $V_R$) | dv/dt | 10,000 | V/μs |

## THERMAL CHARACTERISTICS

| Characteristic | Symbol | Value | Unit |
|---|---|---|---|
| Maximum Thermal Resistance, Junction to Case | $R_{\theta JC}$ | 2.0 | °C/W |

## ELECTRICAL CHARACTERISTICS

| | Symbol | Value | Unit |
|---|---|---|---|
| Maximum Instantaneous Forward Voltage (Note 1.)   ($i_F$ = 10 Amps, TC = 125°C)   ($i_F$ = 20 Amps, $T_C$ = 125°C)   ($i_F$ = 20 Amps, $T_C$ = 25°C) | $v_F$ | 0.57 0.72 0.84 | Volts |
| Maximum Instantaneous Reverse Current (Note 1.)   (Rated dc Voltage, $T_C$ = 125°C)   (Rated dc Voltage, $T_C$ = 25°C) | $i_R$ | 15 0.1 | mA |

1. Pulse Test: Pulse Width = 300 μs, Duty Cycle ≤ 2.0%.

FIGURE 3.27  (Copyright of Semiconductor Component Industries, LLC. Used by permission.)

voltage regulator will prevent those variations from affecting the operation of the load. Thus, transformer rating tolerances may cause a measured rectifier voltage to be off by a considerable margin, but they will have little effect on dc load voltage when a voltage regulator is used. The effect of voltage regulation on the operation of a power supply is covered in detail later in this chapter.

### 3.5.3 Power Rectifiers

Many power supplies require the use of rectifier diodes that have extremely high forward current and/or power dissipation ratings. The characteristics of these **power rectifiers** are illustrated in Figure 3.27, which contains the spec sheet for the MBR1045 power rectifier.

> **Power rectifier**
> A diode with extremely high forward current and/or power dissipation ratings.

The spec sheet shown illustrates the primary differences between power rectifiers and small-signal diodes. These differences are:

1. Power rectifiers have high forward current ratings. The MBR1045 is capable of handling an average forward current of 10 A. (Average forward current ratings of 50 A or greater are not uncommon.)
2. Power rectifiers can handle extremely high forward surge currents. The spec sheet for the MBR1045 shows that the component can handle a forward surge of up to 150 A!
3. Power diodes tend to have relatively high reverse current ratings. The instantaneous reverse current rating for the MBR1045 ranges from 100 $\mu$A (at $T = 25°C$) to 15 mA (at $T = 125°C$).

**FIGURE 3.28** Integrated rectifier package.

If you compare these ratings to those for the 1N4001 (Figure 2.25), you can see that these current ratings are significantly higher.

### 3.5.4 Integrated Rectifiers

Advances in semiconductor device manufacturing have made it possible to construct entire circuits on a single piece of semiconductor material. A circuit that is constructed entirely on a single piece of semiconductor material is called an **integrated circuit (IC)**.

> **Integrated circuit (IC)**
> An entire circuit constructed on a single piece of semiconductor material.

There are a number of advantages to using integrated rectifiers. Among them are:

1. Reduced cost (it takes fewer components to construct a power supply).
2. Troubleshooting is made easier.
3. All the diodes in the bridge operate at the same temperature. As a result, they have equal values of $V_F$ and leakage current.
4. The bridge can easily be mounted to a heat sink.

The component case illustrated in Figure 3.28 contains an entire bridge rectifier. Typically, IC bridge rectifiers can handle average currents of 25 A (and higher) and surge currents in the hundreds of amperes. In these respects, IC bridge rectifiers are like power rectifiers. At the same time, they have high $V_{RRM}$ and low $I_R$ ratings, like many small-signal diodes.

The IC bridge rectifier can be represented using a block diagram like the one shown in Figure 3.29. This figure shows how a single component would be wired to replace the bridge circuit shown in Figure 3.24.

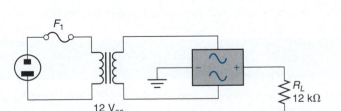

FIGURE 3.29

When using an IC bridge rectifier, the same calculations are made as when working with a bridge made up of individual diodes. Be aware, however, that the $V_F$ rating given on the component's spec sheet is the *forward voltage per diode*, so you still need to consider the effect of *two* diode voltages in the circuit analysis. This point is demonstrated in Example 3.13.

**EXAMPLE 3.13**

Calculate the value of $V_{ave}$ for the load in Figure 3.30.

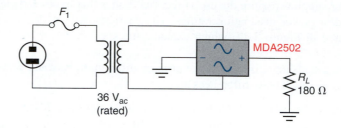

FIGURE 3.30

*Solution:*   First, the peak secondary voltage is found as

$$V_{S(pk)} = \frac{36 \text{ V}_{ac}}{0.707} = \textbf{50.9 V}$$

According to the spec sheet for the MDA2502 bridge rectifier, the forward voltage ($V_F$) rating for the component is 0.95 V per diode. With two diodes in the conduction path, the peak load voltage is found as

$$V_{L(pk)} = V_{S(pk)} - 2V_F = 50.9 \text{ V} - (2)(0.95 \text{ V}) = 50.9 \text{ V} - 1.9 \text{ V} = \textbf{49 V}$$

Finally, the average load voltage is found as

$$V_{ave} = \frac{2V_{L(pk)}}{\pi} = \frac{(2)(49 \text{ V})}{\pi} = \textbf{31.2 V}$$

**PRACTICE PROBLEM 3.13**

The MDA2502 is used in a circuit with a 12 $V_{ac}$ transformer and a 150 $\Omega$ load. Determine the average load voltage for the circuit.

**Section Review ▶**

1. What effect does bulk resistance have on the measured value of $V_F$ for a rectifier diode?
2. What effect can reverse current have on measured load voltages?
3. How does cooling a power supply reduce the effects of reverse current?
4. What is the typical tolerance range for transformers?
5. What is *line regulation*?
6. What are the primary differences between power rectifiers and small-signal diodes?
7. What advantages of using IC rectifiers were listed in this section?

## 3.6   Filters

The circuit that follows the rectifier in a power supply is the *filter* (see Figure 3.1). Power supply filters are used to *reduce the variations in the rectifier output signal*. Since our goal is to produce a *constant* dc output voltage, it is necessary to remove as much of the rectifier output variation as possible.

Filters reduce the variations in the rectifier output.

The overall result of using a filter is illustrated in Figure 3.31. Here, we see the output from the half-wave rectifier, both before and after filtering. Note that there are still voltage variations after filtering. However, the *amount* of variation has been greatly reduced.

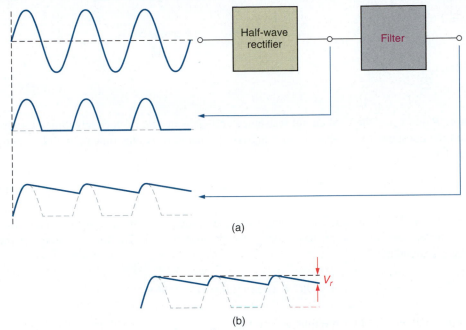

**FIGURE 3.31**  The effects of filtering on the output of a half-wave rectifier.

The variation in the output voltage of a filter is called **ripple voltage** ($V_r$). As you will see, the amount of ripple voltage in the output from a given filter depends on the rectifier used, the filter component values, and the load resistance.

Power supplies are designed to produce as little ripple voltage as possible. Too much ripple in the output can have different adverse effects, depending on the application of the power supply. In an audio amplifier, excessive power supply ripple can produce an annoying hum at 60 or 120 Hz, depending on the type of rectifier used. In video circuits, excessive ripple can produce video "hum" bars in the picture. In digital circuits, it can result in erroneous outputs from logic gates. Therefore, it is important for the filter output to contain as little ripple as possible.

> **Ripple voltage**
> The variation in the output voltage from a filter.

### 3.6.1    Basic Capacitive Filter

The *capacitive filter* is the most basic filter type and the most commonly used. The simplest capacitive filter is simply a capacitor connected in parallel with the load resistance, as shown in Figure 3.32. The filtering action is based on the charge/discharge action of the capacitor. During the positive half-cycle of the input, $D_1$ conducts and the capacitor charges rapidly (Figure 3.32a). As the input starts to go negative, $D_1$ turns off, and the capacitor slowly discharges through the load resistance (Figure 3.32b). As the output from the rectifier drops below the charged voltage of the capacitor, the capacitor acts as the voltage source for the load. It is the difference between the charge and discharge times of the capacitor that reduces the variations in the rectifier output voltage.

The difference between the charge and discharge times of the capacitor is caused by two distinct *RC time constants* in the circuit. You may recall from your study of basic electronics that a capacitor will charge (or discharge) in *five* time constants. One time constant (represented by the Greek letter *tau*) is found as

$$\tau = RC \qquad\qquad \textbf{(3.25)}$$

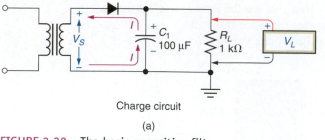

Charge circuit

(a)

Discharge circuit

(b)

**FIGURE 3.32**   The basic capacitive filter.

where $R$ and $C$ are the total circuit resistance and capacitance, respectively. Since it takes five time constants for a capacitor to charge or discharge fully, this time period ($T$) can be found as

$$T = 5RC \qquad\qquad (3.26)$$

Now refer to Figure 3.32a. The capacitor charges through the diode. For the sake of discussion, let's assume that $D_1$ has a forward resistance of 5 Ω. The time constant for the circuit is found as

$$\tau = RC = (5\ \Omega)(100\ \mu F) = 500\ \mu s$$

and the total capacitor charge time is found as

$$T = 5RC = (5)(500\ \mu s) = 2.5\ ms$$

Thus, the capacitor charges to the peak input voltage in 2.5 ms. The discharge path for the capacitor is through the resistor (Figure 3.32b). For this circuit, the time constant is found as

$$\tau = RC = (1\ k\Omega)(100\ \mu F) = 100\ ms$$

and the total capacitor discharge time is found as

$$T = 5RC = (5)(100\ ms) = 500\ ms$$

*A Practical Consideration:* Ideally, there would be no ripple at the output of a filter. However, this can occur only under *open-load* conditions in any practical filter circuit. If the load is open, the capacitor does not have a discharge path, and the time constant of the *RC* circuit is (theoretically) infinite.

Therefore, the capacitor in Figure 3.32 has a charge time of 2.5 ms and a discharge time of 500 ms. This is why it charges almost instantly yet barely starts to discharge before another charging voltage is provided by the rectifier.

You have seen that the values of filter capacitance and load resistance determine the discharge time of the capacitor. As such, the values of filter capacitance and load resistance also affect the amplitude of the ripple voltage at the filter output.

*The amplitude of the ripple voltage at the output of a filter varies inversely with the values of filter capacitance and load resistance,* as illustrated in Figure 3.33. The values to the left of the waveforms indicate what happens to ripple voltage when $C_F$ is constant and $R_L$ increases. The values to the right indicate what happens when $R_L$ is constant and $C_F$ increases. In either case, the amplitude of the filter output ripple decreases.

Ripple in a power supply is minimized by having a high-value filter capacitance combined with a high resistance load. However, these two values are limited by other considerations. The load on a given power supply may consist of any number of circuits, each with its own input resistance characteristics. (The load resistor used in our schematics represents the combined resistance of these circuits.) Thus, it is impractical to vary the

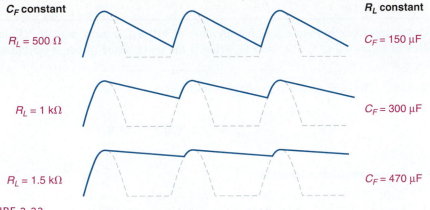

| $C_F$ constant | | $R_L$ constant |
| --- | --- | --- |
| $R_L = 500\ \Omega$ | | $C_F = 150\ \mu F$ |
| $R_L = 1\ k\Omega$ | | $C_F = 300\ \mu F$ |
| $R_L = 1.5\ k\Omega$ | | $C_F = 470\ \mu F$ |

**Lab Reference:** The effects of changing capacitance on filtering are demonstrated in Exercise 4.

FIGURE 3.33

resistance of the power supply load. In other words, a practical power supply is designed to accommodate a specific load resistance, not the other way around. By the same token, the value of the filter capacitor is limited by three factors:

**1.** The maximum allowable charge time for the component.
**2.** The amount of *surge current* ($I_{surge}$) that the rectifier diodes can withstand.
**3.** The cost of "larger-than-needed" filter capacitors.

> What limits the value of a filter capacitor?

The capacitor is not only involved in the discharge action, it is also involved in the charging action. If you make the value of $C$ too high, your discharge time is greatly increased, but so is the charge time. This ties in with the second factor listed, *surge current*. We'll take a look at surge current and its causes and effects now.

## 3.6.2 Surge Current

When you first turn on a power supply, the filter capacitor has no accumulated charge to oppose $V_S$. For the first instant, the discharged capacitor acts as a short circuit, as shown in Figure 3.34. As a result, the diode current is initially limited only by the resistance of the transformer secondary and the bulk resistance of the diode. Since these resistances are usually very low, the initial current tends to be extremely high. This high initial current is referred to as **surge current** and is calculated as follows:

> **Surge current**
> The high initial current in a power supply.

$$I_{surge} = \frac{V_{S(pk)}}{R_W + R_B}$$ (3.27)

where $V_{S(pk)}$ = the peak secondary voltage
$R_W$ = the resistance of the secondary windings
$R_B$ = the diode bulk resistance

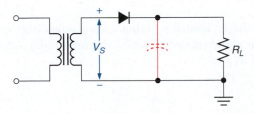

FIGURE 3.34

Example 3.14 demonstrates the calculation of surge current for a filtered rectifier.

## EXAMPLE 3.14

Assume the circuit in Figure 3.32 has an input of 170 V$_{pk}$, a turns ratio of 2:1, and values of $R_W = 0.8\ \Omega$ and $R_B = 5\ \Omega$. What is the initial value of surge current for the circuit?

*Solution:*   The peak secondary voltage is found as

$$V_{S(pk)} = \frac{N_S}{N_P}V_{P(pk)} = \frac{1}{2} \times 170\ \text{V} = \mathbf{85\ V}$$

Now, the surge current is found as

$$I_{\text{surge}} = \frac{V_{S(pk)}}{R_W + R_B} = \frac{85\ \text{V}}{0.8\ \Omega + 5\ \Omega} = \mathbf{14.66\ A}$$

### PRACTICE PROBLEM 3.14

Assume the circuit described above has values of $R_W = 0.5\ \Omega$ and $R_B = 8\ \Omega$. What is the initial surge current for the circuit?

---

The surge current value found in Example 3.14 may seem to be extremely high, but it shouldn't cause any problems. As you may recall, most rectifier diodes have relatively high surge current ratings. An example of this is the 1N400X series of diodes. As listed on the spec sheet in Figure 2.25, the nonrepetitive surge current ($I_{FSM}$) is rated at 30 A. The 14.655 A surge in Example 3.14 is well below the $I_{FSM}$ rating of the 1N400X series diodes.

When the amount of surge current produced by a circuit is more than the rectifier diodes can handle, the problem can be resolved by using a series *current-limiting* resistor, as shown in Figure 3.35. The current-limiting resistor is usually a low-resistance, high-wattage component. The resistor limits the surge current, but it also reduces the output voltage from the circuit. This is because the output voltage from the rectifier is divided between the current-limiting resistor and the load resistance.

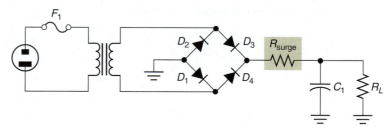

FIGURE 3.35

Surge current can also be limited by using a lower-value filter capacitor. Lower-value capacitors charge in a shorter period of time. The relationship among capacitance, current, and time is given as

$$C = \frac{I(t)}{\Delta V_C} \tag{3.28}$$

where $C$ = the capacitance, in farads
$\quad\quad I$ = the dc (average) charge/discharge current
$\quad\quad t$ = the charge/discharge time
$\quad\quad \Delta V_C$ = the change in capacitor voltage during charge/discharge

Equation (3.28) can be rearranged to produce the following:

$$t = \frac{C(\Delta V_C)}{I}$$

(3.29)

As equation (3.29) shows, the time required for a capacitor to charge to a specified value is directly proportional to the value of the capacitor. Thus, a lower-value capacitor charges faster, reducing the duration of the surge. At the same time, lower-value capacitors produce more ripple voltage (as was shown in Figure 3.33).

### 3.6.3 Filter Output Voltages

Ideally, a filter capacitor would charge to a peak value and remain at that value; that is, there would be no variations in the filter output voltage. However, we know this isn't the case. As shown in Figure 3.36, the output from a filter has *peak, average* (dc), and *ripple* voltage values. The dc output voltage ($V_{dc}$) is shown to equal the peak voltage ($V_{pk}$) minus half the peak-to-peak value of the ripple voltage. By formula,

$$V_{dc} = V_{pk} - \frac{V_r}{2}$$

(3.30)

where $V_{pk}$ = the peak rectifier output voltage
$V_r$ = the peak-to-peak value of ripple voltage

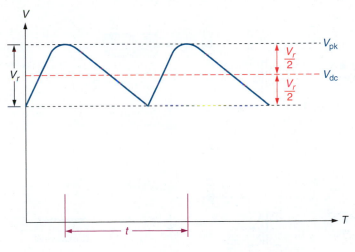

**Lab Reference:** The effects of filtering on dc load voltage are demonstrated in Exercise 4.

$t \cong 16.67$ ms (for half-wave rectifiers)
$t \cong 8.33$ ms (for full-wave rectifiers)

FIGURE 3.36

The ripple voltage from the filter can be found using a variation of equation (3.28), as follows:

$$V_r = \frac{I_L t}{C}$$

(3.31)

where $I_L$ = the dc load current
$t$ = the time between charging peaks
$C$ = the capacitance, in farads

The values of $t$ listed in Figure 3.36 were obtained using the frequency of the rectifier output alternations. Assuming that the line frequency into a *half-wave rectifier* is 60 Hz, the frequency of the output alternations is also 60 Hz. Therefore, the time between the charging peaks for a half-wave rectifier can be found as

$$t = \frac{1}{f} = \frac{1}{60 \text{ Hz}} = \textbf{16.67 ms} \quad \text{(for a } half\text{-}wave \text{ rectifier)}$$

The full-wave rectifier produces two output peaks for every one produced by a half-wave rectifier. Therefore, the full-wave rectifier has twice the output frequency (120 Hz). Using $f = 120$ Hz, the time between the charging peaks for a full-wave rectifier is found as

$$t = \frac{1}{f} = \frac{1}{120 \text{ Hz}} = \textbf{8.33 ms} \quad \text{(for a } full\text{-}wave \text{ rectifier)}$$

With half the time between charging peaks, the output from a filtered full-wave rectifier contains half the ripple of a comparable filtered half-wave rectifier. This point is demonstrated in Example 3.15.

### EXAMPLE 3.15

Determine the ripple output from each of the circuits shown in Figure 3.37. As the meters show, the load current is 20 mA for each circuit.

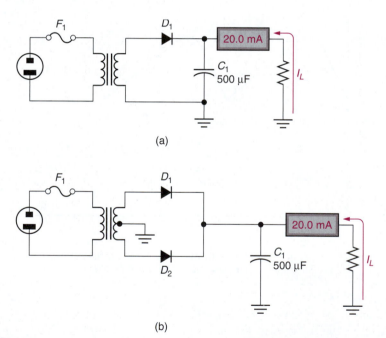

(a)

(b)

**FIGURE 3.37**

***Solution:*** The half-wave rectifier (Figure 3.37a) has a time of 16.67 ms between charging pulses. Therefore,

$$V_r = \frac{I_L t}{C} = \frac{(20 \text{ mA})(16.67 \text{ ms})}{500 \text{ }\mu\text{F}} = \textbf{666.7 mV}_{\text{pp}}$$

For the full-wave rectifier (Figure 3.37b), $t = 8.33$ ms. Therefore,

$$V_r = \frac{I_L t}{C} = \frac{(20 \text{ mA})(8.33 \text{ ms})}{500 \text{ } \mu\text{F}} = \mathbf{333.3 \text{ mV}_{pp}}$$

As Example 3.15 demonstrates, the full-wave rectifier has exactly half the ripple output produced by the half-wave rectifier. This is due to the shortened time period between capacitor charging pulses. Figure 3.38 compares the capacitor discharge times of the filtered half-wave and full-wave rectifiers. Although the discharge slopes are the same for the circuits, the time between the charging peaks accounts for the difference in the ripple voltage between the two.

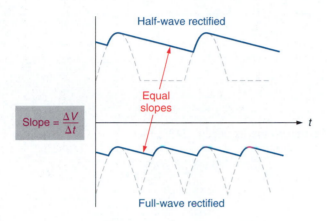

FIGURE 3.38

Lab Reference: An oscilloscope is used to measure $V_r$ in Exercise 4.

An oscilloscope can be used to measure the ripple voltage that is riding on a dc level. Because the ripple voltage is normally very small compared to the dc voltage, the oscilloscope must be set for *ac coupling* to block the dc component of the ripple. This allows you to decrease the V/div setting of the scope so that you can measure low amplitude signal voltages.

You were told earlier in the chapter that the primary advantage of using full-wave rectifiers rather than half-wave rectifiers would be seen when we discussed filtering. As Example 3.15 has shown, a full-wave rectifier will have half the output ripple of a comparable half-wave rectifier. Since our goal is to have a steady dc voltage that has as little ripple voltage as possible, the full-wave rectifier gets us much closer to our goal than does the half-wave rectifier.

◄ *OBJECTIVE 9*

### 3.6.4   Filter Effects on Rectifier Analysis

Refer back to Figure 3.36. As the figure shows, you can find the value of $V_{dc}$ for a filtered rectifier by subtracting $V_r/2$ from the peak rectifier output voltage. There is only one problem: To determine the value of $V_r$, we have to know the value of the dc load current ($I_L$), and to determine the value of $I_L$, we have to know the value of $V_{dc}$.

What we have here is referred to as a *loop*. We have to know the value of $V_{dc}$ to find the value of $V_r$, but we have to know the value of $V_r$ to find the value of $V_{dc}$.

So, what is the solution? If you look closely at Figure 3.36, you'll see that the final value of $V_{dc}$ is *very close* to the value of $V_{pk}$. Thus, we can start our mathematical analysis of the circuit by making the following assumption:

$$V_{dc} \cong V_{L(pk)}$$

Then, using this *assumed* value of $V_{dc}$, we can calculate the approximate value of $I_L$. From there, we can find the approximate value of $V_r$ and an even closer value of $V_{dc}$. As Example 3.16 demonstrates, the final value of $V_{dc}$ is usually very close to the value that was originally assumed.

### EXAMPLE 3.16

Determine the value of $V_{dc}$ for the circuit shown in Figure 3.39.

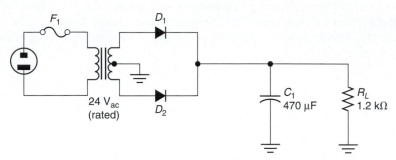

FIGURE 3.39

*Solution:* The transformer is rated at $24\ V_{ac}$, so

$$V_{S(pk)} = \frac{24\ V_{ac}}{0.707} = 33.95\ V$$

$$V_{L(pk)} = \frac{V_{S(pk)}}{2} - 0.7\ V = 16.28\ V$$

Now, we will *assume* that

$$V_{dc} \cong V_{L(pk)} = 16.28\ V$$

Using this assumed value of $V_{dc}$, the dc load current is found as

$$I_L = \frac{V_{dc}}{R_L} = \frac{16.28\ V}{1.2\ k\Omega} = 13.6\ mA$$

Now, the peak-to-peak value of the ripple voltage is found as

$$V_r = \frac{I_L t}{C} = \frac{(13.6\ mA)(8.33\ ms)}{470\ \mu F} = 241\ mV$$

Finally, the calculated value of $V_{dc}$ is found as

$$V_{dc} = V_{L(pk)} - \frac{V_r}{2} = 16.28\ V - 120.5\ mV = 16.16\ V$$

### PRACTICE PROBLEM 3.16

A full-wave rectifier has a $24\ V_{ac}$ transformer, a $330\ \mu F$ filter capacitor, and a $1.5\ k\Omega$ load. Calculate the values of $V_r$ and $V_{dc}$ for the circuit.

The percentage of error between our assumed and calculated values of $V_{dc}$ is 0.74%. This shows that the initial assumption of $V_{dc} \cong V_{L(pk)}$ for the circuit is valid.

### 3.6.5 Filter Effects on Diode PIV

For the full-wave and bridge rectifiers, the filter does not have any significant effect on the peak inverse voltage across each diode. However, when filtered, the half-wave rectifier diode produces a peak inverse voltage that is twice the secondary voltage. By formula,

$$\text{PIV} = 2V_{S(pk)} \quad \text{(half-wave, filtered)} \qquad (3.32)$$

The basis for this relationship is shown in Figure 3.40. When the diode conducts, $C_1$ charges to the value of $V_{S(pk)}$, as shown in Figure 3.40a (assuming, of course, that the diode is an ideal component). When the polarity of the secondary voltage reverses (Figure 3.40b), $D_1$ turns off. For an instant, the reverse voltage across the diode is equal to the sum of the secondary voltage and the capacitor voltage. Since these two voltages are equal, the peak inverse voltage across the diode equals $2V_{S(pk)}$. Note that the diode reverse voltage decreases as the capacitor discharges through the load resistance. However, the diode must have a maximum reverse voltage rating that is greater than $2V_{S(pk)}$.

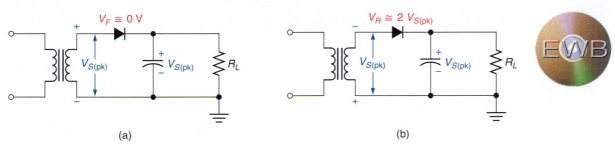

FIGURE 3.40   The effects of filtering on diode PIV.

### 3.6.6 Other Filter Types

There are several other types of filters. The *LC* filters shown in Figure 3.41 make use of the reactance properties of inductors and capacitors. Each has high inductive (series) reactance and low capacitive (shunt) reactance at the power supply ripple frequency. These reactances combine to form a voltage divider, which greatly reduces the amplitude of the ripple voltage. Since the inductors also oppose any rapid change in current, the filters in Figure 3.41 provide surge current protection. Even so, capacitive filters (like those discussed in this section) are more commonly used because of cost and size factors.

> Inductive filters provide protection against surge current problems.

   The operation of each filter in Figure 3.41 is relatively simple. In each case, the inductor aids in keeping the series current constant, while one or more capacitors short any voltage changes to ground. The result is a very stable output in terms of both current and voltage. As a result, *LC* filters are often used in circuits that require extremely low variations in power supply output.

### 3.6.7 One Final Note

The use of a filter will greatly reduce the variations in the output from a rectifier. At the same time, the *ideal* power supply would provide a stable dc output voltage with no ripple voltage at all. While there will always be *some* ripple voltage at the output of a power supply, the use of a voltage regulator will reduce the filter output ripple even further. This point will be demonstrated in Section 3.7.

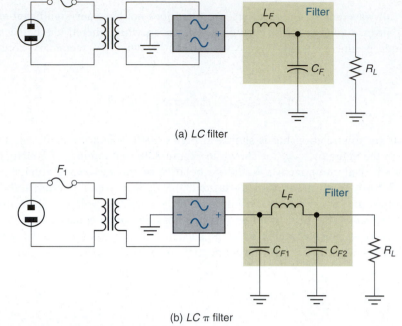

(a) *LC* filter

(b) *LC* π filter

FIGURE 3.41 Some other filter circuits.

**Section Review ▶**

1. What are filters used for?

2. What is *ripple voltage*?

3. Describe the operation of the basic capacitive filter in terms of *charge/discharge time constants*.

4. What limits the value of a filter capacitor?

5. What causes *surge current* in a power supply?

6. What are the *two* devices shown in this section that limit surge current? How does each limit $I_{\text{surge}}$?

7. Describe the relationship between $V_r$ and $V_{\text{dc}}$ for a filtered rectifier.

8. Why are full-wave rectifiers preferred over half-wave rectifiers?

9. Describe the process used to calculate the values of $V_r$ and $V_{\text{dc}}$ for a filtered rectifier.

10. Explain the effect that filtering has on the PIV across the diode in a half-wave rectifier.

## 3.7    Zener Voltage Regulators

The final circuit in a basic power supply is the *voltage regulator* (as shown in Figure 3.1). There are many types of voltage regulators. The more complex and practical voltage regulators contain a number of transistors and/or ICs (as will be discussed in Chapter 21). In this chapter, we are going to concentrate on the simple zener regulator shown in Figure 3.42a.

As you may recall, a zener diode operated in reverse breakdown maintains a relatively constant voltage across its terminals as long the zener current remains between its knee current ($I_{ZK}$) and maximum current ($I_{ZM}$) ratings. This operating characteristic is illustrated in 3.42b. Since the load resistance is in parallel with $D_1$, the load voltage ($V_L$) always equals the zener voltage ($V_Z$), as shown in Figure 3.42c. As long as the zener current remains within its allowable range of values, zener voltage and load voltage remain relatively constant. Therefore, *the key to keeping the load voltage constant is to keep the zener current within its specified range* (between $I_{ZK}$ and $I_{ZM}$).

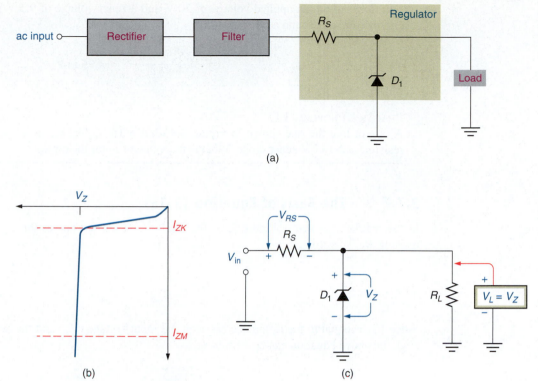

(a)

(b)                    (c)

**FIGURE 3.42**  The basic zener voltage regulator.

## 3.7.1    Total Circuit Current

For a zener regulator like the one shown in Figure 3.42c, the source current is found as         ◀ *OBJECTIVE 10*

$$I_T = \frac{V_{in} - V_Z}{R_S}$$         **(3.33)**

where $I_T$ = the total current drawn through $R_S$
    $V_{in}$ = the input voltage
    $V_Z$ = the the nominal (rated) zener voltage
    $R_S$ = the series resistor

Example 3.17 demonstrates the use of this equation.

### EXAMPLE 3.17

Determine the total circuit current for Figure 3.43.

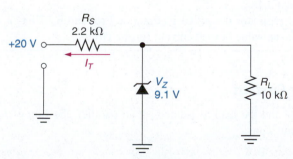

**FIGURE 3.43**

*Solution:* With an applied voltage of 20 V and a zener voltage of 9.1 V, the total circuit current is found as

$$I_T = \frac{V_{in} - V_Z}{R_S} = \frac{20\ V - 9.1\ V}{2.2\ k\Omega} = \textbf{4.95 mA}$$

**PRACTICE PROBLEM 3.17**
A circuit like the one shown in Figure 3.43 has a 15 $V_{dc}$ source, a 1.5 k$\Omega$ series resistor, and a 12 V zener diode. What is the value of $I_T$ for the circuit?

### 3.7.2 The Basis of Equation (3.33)

In the regulator shown in Figure 3.43, the total circuit current passes through the series resistor, $R_S$. Therefore,

$$I_T = \frac{V_{RS}}{R_S}$$

Since $V_{RS}$ is equal to the difference between the input voltage ($V_{in}$) and the zener voltage ($V_Z$), the above equation can be written as

$$I_T = \frac{V_{in} - V_Z}{R_S}$$

### 3.7.3 Load Current

Since the load resistance ($R_L$) is in parallel with the zener diode, the voltage across the load always equals $V_Z$. Thus, the load current can be found as

$$I_L = \frac{V_Z}{R_L} \qquad\qquad (3.34)$$

Example 3.18 demonstrates the use of this equation.

**EXAMPLE 3.18**

Determine the load current for the circuit shown in Figure 3.43.

*Solution:* With a 9.1 V zener and a 10 k$\Omega$ resistor, the total load current is found as

$$I_L = \frac{V_Z}{R_L} = \frac{9.1\ V}{10\ k\Omega} = \textbf{910 } \boldsymbol{\mu}\textbf{A}$$

**PRACTICE PROBLEM 3.18**
The zener voltage regulator described in Practice Problem 3.17 has a 12 k$\Omega$ load resistance. What is the value of load current for the circuit?

### 3.7.4 Zener Current

Since the zener diode and the load resistance are in parallel, the sum of $I_Z$ and $I_L$ equals the total circuit current. Thus,

$$I_Z = I_T - I_L \qquad\qquad (3.35)$$

where $I_Z$ = the total zener current
$I_T$ = the total circuit current
$I_L$ = the load current

Example 3.19 demonstrates the procedure for determining the value of $I_Z$.

### EXAMPLE 3.19

Determine the load current and zener current for the circuit shown in Figure 3.43.

*Solution:* In Example 3.17, we determined $I_T$ to be 4.95 mA. In Example 3.18, we determined the value of $I_L$ to be 910 μA. Therefore,

$$I_Z = I_T - I_L = 4.95 \text{ mA} - 910 \text{ μA} = \textbf{4.04 mA}$$

### PRACTICE PROBLEM 3.19

Determine the value of the zener current for the circuit described in Practice Problems 3.17 and 3.18.

## 3.7.5 Load Variations

It was stated earlier a zener regulator maintains a relatively constant output voltage as long as zener current stays between $I_{ZK}$ and $I_{ZM}$. The possible effects of load resistance variations on zener current are illustrated in Figure 3.44. If load resistance is reduced to $0 \text{ Ω}$ (Figure 3.44a), all circuit current passes through the load. In this case, zener current drops below $I_{ZK}$, and the diode stops regulating the output voltage. If the load resistance is increased to $\infty \text{ Ω}$ (Figure 3.44b), total circuit current passes through the zener diode. In this case, the diode may be destroyed unless the value of $R_S$ is high enough to keep the value of $I_Z$ from exceeding $I_{ZM}$.

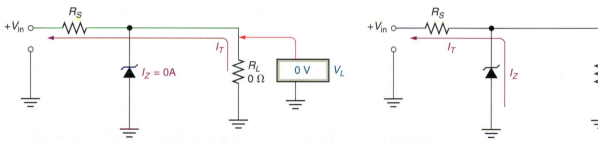

Regulation is lost when the load is shorted.

(a)

Total circuit current is drawn through the zener when the load is open.

(b)

FIGURE 3.44 The effects of load variation on the operation of a zener voltage regulator.

What are the practical limits on the value of $R_L$? The minimum value of $R_L$ is determined by the zener voltage and the value of $I_{ZK}$. To maintain zener regulation, the minimum zener current must be equal to $I_{ZK}$. Therefore,

$$I_{L(\text{max})} = I_T - I_{ZK}$$

Since $I_{L(\text{max})}$ occurs when the load resistance is at a minimum,

$$R_{L(\text{min})} = \frac{V_Z}{I_{L(\text{max})}} \qquad \textbf{(3.36)}$$

This relationship is illustrated in Example 3.20.

## EXAMPLE 3.20

The zener diode shown in Figure 3.45 has values of $I_{ZK} = 3$ mA and $I_{ZM} = 100$ mA. What is the minimum allowable value of $R_L$?

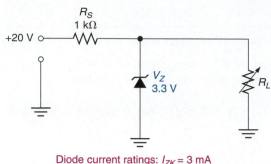

Diode current ratings: $I_{ZK} = 3$ mA
$I_{ZM} = 100$ mA

FIGURE 3.45

*Solution:* First, the total circuit current is found as

$$I_T = \frac{V_{\text{in}} - V_Z}{R_S} = \frac{20\text{ V} - 3.3\text{ V}}{1\text{ k}\Omega} = \mathbf{16.7\text{ mA}}$$

Now, $I_{L(\text{max})}$ is found as

$$I_{L(\text{max})} = I_T - I_{ZK} = 16.7\text{ mA} - 3\text{ mA} = \mathbf{13.7\text{ mA}}$$

and

$$R_{L(\text{min})} = \frac{V_Z}{I_{L(\text{max})}} = \frac{3.3\text{ V}}{13.7\text{ mA}} = \mathbf{241\ \Omega}$$

**PRACTICE PROBLEM 3.20**
If the diode shown in Figure 3.45 has values of $V_Z = 5.1$ V and $I_{ZK} = 5$ mA, what is the minimum allowable value of $R_L$ for the circuit?

For the circuit shown in Figure 3.45, the zener maintains a relatively constant output voltage as long as $R_L$ does not go below 241 $\Omega$. If $R_L$ goes below this value, $I_L$ will increase above its maximum allowable value, and $I_Z$ will drop below $I_{ZK}$. The diode will then stop regulating the output voltage. Note that a load that draws *maximum* current is referred to as a **full load**.

| **Full load** |
|---|
| A minimum load resistance that draws maximum current. |

### 3.7.6 Load Regulation

You have been shown that the zener regulator can maintain a relatively constant load voltage for a range of load current values. The ability of a regulator to maintain a constant load voltage despite anticipated changes in load current demand is called **load regulation**. Load regulation is discussed in detail in Chapter 21.

| **Load regulation** |
|---|
| The ability of a regulator to maintain a constant load voltage despite anticipated variations in load current demand. |

### 3.7.7 Zener Reduction of Ripple Voltage

The zener regulator provides an added bonus: It reduces the amount of ripple voltage present at the filter output. The effect that the regulator has on ripple voltage is easy to understand when you consider the equivalent circuit of the diode. The basic zener regulator and its equivalent circuit are shown in Figure 3.46. You may recall from Chapter 2 that zener impedance ($Z_Z$) is a dynamic value; that is, it is the opposition that a zener diode presents to *a change in voltage or current*. Since ripple voltage is a changing quantity, it is affected by $Z_Z$.

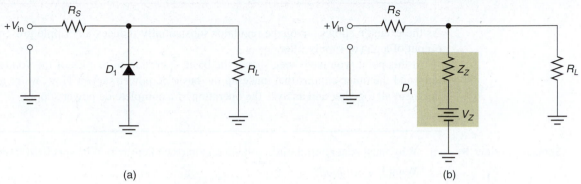

**FIGURE 3.46**

To the ripple waveform, there is a voltage divider present in the regulator. This voltage divider is made up of the series resistance ($R_S$) and the parallel combination of $Z_Z$ and the load. The ripple output from the regulator can be found as

$$V_{r(\text{out})} = \frac{(Z_Z \| R_L)}{(Z_Z \| R_L) + R_S} V_r \tag{3.37}$$

where $V_{r(\text{out})}$ = the ripple present at the regulator output
$(Z_Z \| R_L)$ = the parallel combination of $Z_Z$ and the load resistance
$R_S$ = the regulator series resistance
$V_r$ = the peak-to-peak ripple voltage present at the regulator input

Example 3.21 demonstrates the use of this equation.

**EXAMPLE 3.21**

The filtered output from a full-wave rectifier has a peak-to-peak ripple voltage of 1.5 V. If this signal is applied to the circuit shown in Figure 3.47, what will the ripple at the load equal?

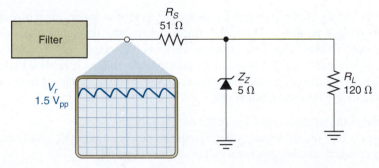

**FIGURE 3.47**

**Solution:** The zener regulator is shown to have values of $R_L = 120\ \Omega$ and $Z_Z = 5\ \Omega$. With a 51 $\Omega$ series resistor and a ripple input of 1.5 $V_{pp}$, the ripple at the output of the regulator is found as

$$V_{r(\text{out})} = \frac{(Z_Z \| R_L)}{(Z_Z \| R_L) + R_S} V_r = \frac{4.8\ \Omega}{4.8\ \Omega + 51\ \Omega} (1.5\ V_{pp}) = \mathbf{129\ mV_{pp}}$$

**PRACTICE PROBLEM 3.21**

A zener regulator has a 91 Ω series resistance, a 200 Ω load resistance, and a zener impedance that equals 25 Ω. If the input ripple to the circuit is 1.2 $V_{pp}$, what is the amount of load ripple?

As the example shows, a voltage regulator substantially reduces any ripple present at the output of a power supply filter.

At this point, you have been shown the basic operating principles of the voltage regulator and the other circuits that make up the basic dc power supply. Now, we're going to put them all together and analyze the operation of a complete dc power supply.

**Section Review ▶**

1. Why must zener current in a voltage regulator be kept within its specified limits?
2. What is a *full load*?
3. How does a zener voltage regulator reduce the ripple voltage from a filter?

## 3.8   Putting It All Together

We have discussed the operation of transformers, rectifiers, filters, and zener regulators in detail. Now, it is time to put them all together into a basic working power supply. In this section, we will analyze the basic power supply shown in Figure 3.48.

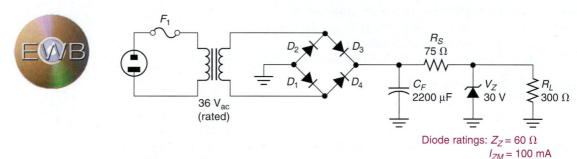

**FIGURE 3.48**

The power supply shown contains a transformer, bridge rectifier, capacitive filter, and zener diode voltage regulator. The transformer converts the incoming line voltage to a lower secondary voltage. The bridge rectifier converts the transformer secondary ac voltage into a *positive* pulsating dc voltage. This pulsating dc voltage is applied to the capacitive filter, which reduces the variations in the rectifier dc output voltage. Finally, the zener voltage regulator performs two functions:

1. It reduces the ripple (variations) in the output voltage.
2. It ensures that the dc output voltage from the power supply ($V_{dc}$) will remain relatively constant despite variations in load current demand.

Thus, the combination of the four circuits has converted an ac line voltage to a steady dc supply voltage that remains relatively constant when load current demands change.

**OBJECTIVE 11 ▶**

Our goal in analyzing a basic power supply is to determine the values of dc output voltage ($V_{dc}$), ripple voltage ($V_r$), and load current ($I_L$). The procedure for determining these values is as follows:

1. Determine the rms value of the transformer secondary voltage.
2. Determine the value of $V_{S(pk)}$.
3. Determine the peak rectifier output voltage.

**4.** Determine the total current through the series resistor. This current value (designated as $I_R$) will be used when calculating the value of ripple voltage.

**5.** Determine the value of ripple voltage from the filter.

**6.** Find $V_{dc}$ at the output. This value equals the $V_Z$ rating of the zener diode under normal circumstances.

**7.** Using the rated value of $Z_Z$, approximate the final ripple output voltage.

**8.** Using $V_Z$ and $R_L$, determine the value of load current.

Example 3.22 illustrates the process for analyzing the schematic of a basic power supply.

### EXAMPLE 3.22

Determine the values of $V_{dc}$, $V_{r(out)}$, and $I_L$ for the power supply shown in Figure 3.48.

*Solution:* First, we must convert the rated value of the transformer secondary voltage to a peak value as follows:

$$V_{S(pk)} = \frac{36 \text{ V}_{ac}}{0.707} = \textbf{51 V}$$

Now, we determine the value of peak voltage at the filter input.

$$V_{pk} = V_{S(pk)} - 1.4 \text{ V} = \textbf{49.6 V}$$

Next, we assume that we have a dc source voltage of 49.6 V. Using 49.6 V as our value for $V_{in}$, we determine the value of the current through the series resistor as follows:

$$I_R = \frac{V_{in} - V_Z}{R_S} = \frac{49.6 \text{ V} - 30 \text{ V}}{75 \text{ }\Omega} = \textbf{261.3 mA}$$

We now use the value of $I_R = 261.3$ mA to determine the value of $V_r$ as follows:

$$V_r = \frac{I_R t}{C} = \frac{(261.3 \text{ mA})(8.33 \text{ ms})}{2200 \text{ }\mu\text{F}} = \textbf{989 mV}_{pp}$$

As stated earlier, the dc output voltage from the power supply equals the value of $V_Z$. By formula,

$$V_{dc} = V_Z = \textbf{30 V}$$

and

$$I_L = \frac{V_Z}{R_L} = \frac{30 \text{ V}}{300 \text{ }\Omega} = \textbf{100 mA}$$

Finally, the value of $V_{r(out)}$ is found as

$$V_{r(out)} = \frac{(Z_Z \| R_L)}{(Z_Z \| R_L) + R_S} V_r = \frac{50 \text{ }\Omega}{125 \text{ }\Omega} (989 \text{ mV}_{pp}) = \textbf{396 mV}_{pp}$$

### PRACTICE PROBLEM 3.22

A power supply like the one shown in Figure 3.48 has the following values: $V_S = 24 \text{ V}_{ac}$ (rated), $C = 470 \text{ }\mu\text{F}$, $R_S = 500 \text{ }\Omega$, $V_Z = 10 \text{ V}$, $Z_Z = 20 \text{ }\Omega$, and $R_L = 5.1 \text{ k}\Omega$. Determine the values of $V_{dc}$, $I_L$, and $V_{r(out)}$ for the circuit.

Earlier in the chapter, you were told that zener voltage regulators are not commonly used. One reason is that zener diodes waste a significant amount of power. Consider the circuit we analyzed in Example 3.22. The current through $R_S$ was found to be 261.3 mA. With 100 mA being drawn through the load, the zener current is found as

$$I_Z = 261.3 \text{ mA} - 100 \text{ mA} = 161.3 \text{ mA}$$

and the power dissipated by the zener is found as

$$P_Z = I_Z V_Z = (161.3 \text{ mA})(30 \text{ V}) = 4.84 \text{ W}$$

which is greater than the load power (3 W).

---

**Section Review ▶**

1. What is the goal of power supply analysis? In other words, what values are you trying to determine?

2. Why are transistor and/or IC voltage regulators preferred over zener voltage regulators?

## 3.9    Power Supply Troubleshooting

**OBJECTIVE 12 ▶**

Power supply faults may occur in the transformer, rectifier diodes, filter, or voltage regulator. When a fault develops in a power supply, the type of symptom and a few simple tests will tell you where the fault is located. In this section, we will take a look at some common power supply fault symptoms and the tests used to isolate a faulty component.

### 3.9.1    Primary Fuse

Every power supply contains a fuse that is located in the primary circuit of the supply. This fuse is normally placed in series with the primary of the transformer, as shown in Figure 3.49.

**Lab Reference:** Several rectifier faults are simulated in Exercise 3.

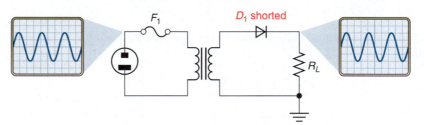

(a)  A half-wave rectifier with a shorted diode

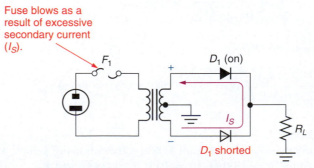

(b)  A full-wave rectifier with a shorted diode

**FIGURE 3.49**

If a fault in a power supply causes an extremely high amount of current to be drawn from the transformer, the primary fuse will blow (open). When this happens, the ac line input from the wall outlet is prevented from reaching the power supply. This eliminates the excessive current, protecting the power supply and the technician who is working on it.

Many power supplies use *slow-blow* fuses. Slow-blow fuses are designed to handle the high surge current produced when a power supply is first turned on. At the same time, if the current demand on a slow-blow fuse is too high for a long enough period of time (usually around 2 to 4 seconds), the slow-blow fuse opens to protect the circuits.

When a power supply fuse needs to be replaced, *you must use the same value and type of fuse.* Never, under any circumstances, use a fuse that has a higher current rating than the one you are replacing. Using a higher-value fuse will defeat the purpose of the fuse (protecting the circuit from excessive current) and create a safety hazard.

As you will see throughout this section, many power supply faults will cause the primary fuse to blow, while many others will not. When a fault causes the primary fuse to blow, simply replace the fuse after the fault is diagnosed and corrected.

*A Word of Caution:*
When you begin to work in the field, you may meet some technicians who will use a wire to "defeat" (bypass) the primary fuse in a faulty power supply. This allows the power supply to operate under circumstances that would normally cause a system shutdown. *Never defeat the fuse in a power supply.* You not only risk starting a fire, but you could also be injured or killed if you should accidentally come into contact with any high-current components in the circuit.

### 3.9.2    Transformer Faults

The transformer in a power supply may develop one of several possible faults:

1. A shorted primary or secondary winding.
2. An open primary or secondary winding.
3. A short between the primary or secondary winding and the transformer frame.

In most cases, a shorted primary or secondary winding will cause the fuse to blow. If the fuse does not blow, the dc output from the power supply will be extremely low, and the transformer itself will get extremely hot.

If the primary or the secondary winding of the transformer opens, the output from the power supply will drop to zero. In this case, the primary fuse will not blow. If you believe that either transformer winding is open, a simple resistance check of the winding will verify your suspicions. If either winding reads a very high resistance, the winding is open.

If either winding shorts to the transformer casing, the result will be a blown fuse. This fault is isolated by checking the resistance from the winding leads to the transformer casing. A low resistance measurement indicates that a winding-to-case short circuit exists.

With any of the problems above, the repair procedure is simple. You must replace the transformer.

### 3.9.3    Rectifier Faults

The *half-wave rectifier* is the easiest rectifier to troubleshoot. If the diode in the rectifier *shorts*, the output from the rectifier will be a sine wave that is identical to $V_S$. This is illustrated in Figure 3.49a. Since the diode is shorted, it acts as a straight piece of wire. Therefore, neither half-cycle of the input ($V_S$) is eliminated, and the output is a replica of $V_S$.

If the diode in the half-wave rectifier opens, the output from the circuit will drop to zero. If the rectifier diode is either shorted or open, simply replace the diode.

In a *full-wave rectifier*, a shorted diode will cause the power supply fuse to blow. Figure 3.49b shows the effects of a shorted $D_2$ on circuit operation. Note that the diode has been replaced by a straight wire. When $D_1$ is forward biased by $V_S$, the transformer secondary is shorted through $D_1$. This produces excessive current in the secondary, causing the primary fuse to blow.

If you suspect that either diode in a full-wave rectifier has shorted, simply measure the reverse resistance of both diodes. In each case, the diode should have a very high reverse resistance. If the reverse resistance of either diode is extremely low, replace the diode.

If a diode in a full-wave rectifier opens, the output from the rectifier will resemble the output from a half-wave rectifier. The output ripple voltage will approximately double, and the ripple frequency will go from 120 to 60 Hz. In this case, measure the diode resistances, and look for a large value of forward resistance. When you have such a reading, replace the diode.

*A Practical Consideration:*
If one diode in a rectifier goes bad, it will usually damage or destroy one or more of the other rectifier diodes in the process. If you determine the fault to be in the rectifier of a power supply, you should replace *all* the diodes.

*A Practical Consideration:*
Be sure to disconnect either end of any diode before measuring its reverse resistance. Otherwise, you may obtain a faulty reading (caused by the components that are connected to the diode).

The symptoms for shorted and open diodes in the bridge rectifier are the same as those for full-wave rectifiers. In the case of the bridge rectifier, you simply have more diodes that must be tested.

### 3.9.4    Filter Faults

When working with filter capacitors, technician safety is very important. The hazard of electrical shock or heat burns is always present. Capacitors store an electrical charge, and they can retain that charge even after the power switch has been turned off or the ac plug has been disconnected. As a safety precaution, capacitors should be discharged by shorting the terminals with a 50 to 100 Ω resistor before desoldering them or taking any measurements. Besides being a danger to the technician, test equipment can be damaged if any capacitors remained charged. Many power supply circuits have bleeder resistors connected directly across the capacitor terminals so that, if the load is disconnected from the power supply, the capacitors still discharge. Otherwise, the capacitor could stay charged until the voltage bleeds off internally through leakage.

When a filter capacitor *shorts*, the primary fuse will blow. The reason for this is illustrated in Figure 3.50. When the filter capacitor shorts, it shorts out the load resistance. This has the same effect as wiring the two sides of the bridge together (Figure 3.50a). If you trace from the high side of the bridge to the low side, you will see that the only resistance across the secondary of the transformer is the forward resistance of the two conducting diodes. This effectively shorts out the transformer secondary, causing excessive secondary current and a blown fuse in the primary.

When checking for a shorted filter capacitor, simply measure the resistance of the component, as shown in Figure 3.50b. If the capacitor is shorted, you will measure a very low resistance. Be sure to disconnect the capacitor from the line before measuring its resistance. This will prevent you from forward biasing a rectifier diode with the meter, which will give you a faulty reading.

> *Caution:* High-value capacitors (like those typically used in commercial power supplies) can store a lethal charge. Always consult the documentation on such circuits before attempting to discharge or handle any capacitors they contain.

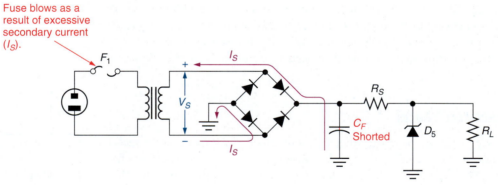

(a) The faulty capacitor shorts out the transformer secondary

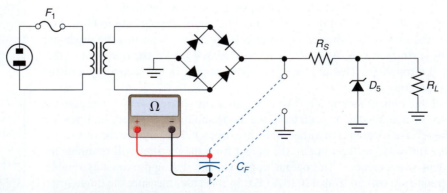

(b) Testing the filter capacitor with an ohmmeter

FIGURE 3.50    Shorted *C* effects and testing.

When electrolytic capacitors get old, the electrolyte tends to dry out. This process is accelerated when the capacitor is exposed to high temperatures, such as those in a power supply. As the electrolyte dries out, the capacitor tends to lose capacity. For example, a 1000 µF capacitor may drop to 700 µF. This reduced capacitance causes the ripple voltage to increase. As the capacitor continues to dry out, it eventually becomes an *open*. There are several ways to check the quality of a capacitor. The best way is to check its value with a capacitance checker. Another way of checking for reduced capacitance is to place an identical good capacitor in parallel with the suspected bad one. If the ripple decreases by more than 50%, the suspect capacitor is probably bad and should be replaced. An *open* filter capacitor will display similar symptoms and must also be replaced.

> *Caution: Be very careful when replacing electrolytic capacitors.* If you connect an electrolytic capacitor backward in a high-current circuit (such as a power supply), it will either explode or become hot enough to cause serious burns.

### 3.9.5  Zener Regulator Faults

If a zener diode *shorts*, the symptoms are the same as those for a shorted filter capacitor. When the output symptoms indicate a short, simply disconnect the zener diode from the line and connect a voltmeter across the output. If the zener diode is the source of the short, the power supply output will resemble that of a filtered rectifier. If the short circuit symptoms remain, the zener diode is not the cause of the problem.

If the zener diode *opens*, the peak output voltage and ripple will both increase. At the same time, voltage regulation will be lost. Thus, you will see a significant amount of variation in the output when the load current demands change.

### 3.9.6  Secondary Fuses

Many power supply circuits have fuses installed in the secondary dc outputs. One purpose for multiple fuses in the secondary is that, if one section of a system fails, causing a blown fuse, the rest of the system will remain functional. Multiple fuses also allow different areas of a system to be fused at different current levels. For example, a videocassette recorder (VCR) normally has a separate fuse for each power supply output. This allows the remaining parts of the system to operate when one part fails.

### 3.9.7  Troubleshooting Applications

It is easy to tell you about common power supply faults and their causes. However, it is not always easy initially to apply these fault/cause relationships to actual circuit problems. In this section, we go through several example troubleshooting cases to show how a little thought and some simple testing can help you to diagnose basic power supply faults.

---

**SYMPTOMS**

When the power supply shown in Figure 3.51 is first turned on, the output is normal. After a few minutes, the dc output voltage drops to almost zero, and then the primary fuse blows.

Application 3.1

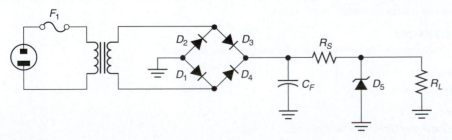

FIGURE 3.51

### OBSERVATION

Some components operate as they should when cool and develop the symptoms of a short when they get hot. Electrolytic capacitors are extremely susceptible to this type of problem. For this reason, the capacitor is suspected of being the cause of the problem.

### TESTING

To diagnose the possible capacitor problem, the capacitor must be heated in one way or another. This is because the problem is heat related. A soldering iron connected to one of the capacitor leads will heat the component enough for testing purposes.

An ohmmeter is connected in the circuit as shown in Figure 3.50b. When first connected, the ohmmeter reads a relatively high resistance. When the hot soldering iron is connected to the capacitor lead, the component starts to get hot. After a moment, the resistance reading on the meter drops to a very low value.

### CONCLUSION

Since the capacitor resistance dropped when temperature increased, the component is *leaky* (partially shorted) and must be replaced. ■

There is an alternative method to test a capacitor suspected of being leaky. Replace the primary fuse, and turn on the power supply. When the output voltage first drops, spray the capacitor with a specially made aerosol coolant. Such coolants are available at most electronic parts stores. If the output voltage goes back to normal when the capacitor is cooled, the capacitor is leaky and must be replaced.

## Application 3.2

### SYMPTOMS

When the power supply shown in Figure 3.51 is turned on, the transformer becomes extremely hot, and the dc output voltage is nearly zero. After a moment, the primary fuse blows.

### OBSERVATION

The symptoms listed are classic for a shorted transformer primary or secondary. For this reason, the transformer is tested.

### TESTING

The typical power supply transformer has a primary resistance around 10 to 50 $\Omega$ and a secondary resistance of 10 $\Omega$ or less. When tested, the transformer gives resistance readings of $R_P = 5\ \Omega$ and $R_S = 8\ \Omega$.

### CONCLUSION

Since the primary resistance is extremely low, the transformer primary must be shorted. Replacing the transformer corrects the problem. ■

Some transformers have a secondary resistance rating printed on the side of the component. To determine the approximate value of primary resistance, multiply the secondary resistance by the turns ratio. For example, a transformer with a secondary resistance of 1.5 $\Omega$ and a turns ratio of 20:1 has a primary resistance of approximately 1.5 $\Omega \times 20 = 30\ \Omega$ (assuming the component isn't faulty).

## Application 3.3

### SYMPTOMS

The power supply shown in Figure 3.51 has a dc output voltage that is approximately half its rated value and a large amount of ripple voltage.

### OBSERVATION

These symptoms can be caused only by an open filter capacitor.

### TESTING

None is required.

### CONCLUSION

The capacitor filter must be replaced. ■

**SYMPTOMS**

The cassette drive motor of a VCR does not operate when the play function is selected. Other system circuits, such as the electronic clock and channel tuner, are functioning normally.

**OBSERVATION**

When a problem develops in any electronics system, one of the first tests is to check the power supply input(s) to the circuits that appear to be faulty.

**TESTING**

The normal dc supply input to the motor drive circuit is 12 $V_{dc}$. When this input is checked, it measures 0 V. The circuit input is traced back to a power supply secondary fuse, which is visually determined to be open. The secondary fuse must be replaced. ∎

An important point should be made at this time: Although it may seem obvious from earlier discussions, fuses don't open arbitrarily. They are *caused* to open by faults in their loads. In other words, the secondary fuse in Application 3.4 most likely opened because there is a fault in the motor drive circuit. This fault must be diagnosed and corrected, or the new fuse will blow in a relatively short time.

### 3.9.8 What the Fuse Tells You

If you look back through this section, you'll notice that there is a common thread to troubleshooting the circuits in the power supply: the fuse in the transformer primary (or secondary). The condition of the fuse indicates whether the problem in the power supply is a *short* or an *open*. If the fuse is blown, a short exists somewhere in the power supply. If it is not blown, an open exists somewhere in the supply. This, of course, assumes that there is a problem in the supply to begin with. (The problem *may* be in the load.)

### 3.9.9 One Final Note

Commercially produced power supplies are far more complex than the ones we have seen in this chapter. The main difference lies in the voltage regulator circuitry, which usually contains one or more transistors and/or IC voltage regulators. However, the basic principles covered in this chapter apply to all power supplies. While the exact circuitry used varies from one power supply to another, the basic principles of operation do not.

You will need to learn a great deal about transistors and IC voltage regulators before you will be ready to take on some of the more complex power supply circuits. However, by the time you get to Chapter 21, you will be ready to deal with the complex circuits covered.

◀ **Section Review**

1. List the common transformer faults and their symptoms.
2. What should you do when you diagnose a rectifier diode fault?
3. List the common filter faults and their symptoms.
4. List the common voltage regulator faults and their symptoms.
5. What does the condition of the primary (or secondary) fuse in a dc power supply tell you?

**CHAPTER SUMMARY**

Here is a summary of the major points made in this chapter:

1. The power supply of an electronic system is used to convert the ac energy provided by the wall outlet to dc energy.
2. There are two basic types of power supplies: *linear* and *switching*. Basic linear power supply circuits are the focus of this chapter.

3. A basic linear power supply consists of a *rectifier*, a *filter*, and a *voltage regulator* (see Figure 3.1).
   a. In many cases, the ac line input is coupled to the power supply via a transformer (as shown in Figure 3.1b).
   b. The rectifier is a diode circuit that converts ac to *pulsating dc.*
   c. The filter reduces the variations in the output voltage from the rectifier.
   d. The voltage regulator maintains a relatively constant output voltage from the power supply.
4. The three basic types of transformers are listed below, along with their voltage and current characteristics:

| Transformer Type | Voltage | Current |
|---|---|---|
| Step-down | $V_S < V_P$ | $I_S > I_P$ |
| Step-up | $V_S > V_P$ | $I_S < I_P$ |
| Isolation | $V_S \cong V_P$ | $I_S \cong I_P$ |

5. The *turns ratio* of a transformer is the ratio of primary turns to secondary turns.
   a. The input-to-output voltage ratio of a transformer is equal to its turns ratio.
   b. The input-to-output current ratio equals the reciprocal of the turns ratio.
6. Many transformers have secondary voltage ratings. These ratings indicate the ac output voltage produced by the component with a 120 $V_{ac}$ input.
7. The *half-wave* rectifier is simply a diode that is placed in series between a transformer and its load.
8. The diode in the half-wave rectifier eliminates either the negative or positive alternations of the transformer output.
   a. A *positive* half-wave rectifier eliminates the negative alternations of the transformer output. As a result, the rectifier output contains only positive alternations.
   b. A *negative* half-wave rectifier eliminates the positive alternations of the transformer output. As a result, the rectifier output contains only negative alternations.
9. Half-wave rectifiers are inexpensive to produce but very inefficient. As a result, they are typically used in noncritical, low-current applications.
10. The operation of a typical half-wave rectifier is illustrated in Figures 3.6 and 3.7.
11. The direction of the rectifier diode determines the output polarity for half-wave rectifiers, as follows:

| Rectifier Type | Connections |
|---|---|
| Positive | Anode → transformer <br> Cathode → load |
| Negative | Cathode → transformer <br> Anode → load |

   These connections are compared in Figure 3.9.
12. The *average load voltage* from an ac circuit is the reading you get when the voltage is measured with a dc voltmeter.
13. The *average load current* from an ac circuit is the reading you get when the current is measured with a dc ammeter.
14. One rectifier diode can be substituted for another when:
   a. The $I_0$ rating of the diode is greater than the circuit value of $I_{ave}$.
   b. The $V_{RRM}$ rating of the diode is greater than the circuit PIV.
15. A *full-wave rectifier* produces a single-polarity output by converting negative alternations to positive alternations (or vice versa).
   a. A full-wave rectifier requires the use of a *center-tapped* transformer.
   b. The operation of a full-wave rectifier is illustrated in Figure 3.17.

16. The output polarity of a full-wave rectifier is determined by the direction of the diodes as follows:

| Rectifier Type | Connections |
|---|---|
| Positive | Anodes → transformer |
| | Cathodes → load |
| Negative | Cathodes → transformer |
| | Anodes → load |

17. The PIV produced by a full-wave rectifier is approximately twice the peak load voltage.
18. Half-wave and full-wave rectifiers are compared in Figure 3.21.
19. Half-wave and full-wave rectifiers produce nearly identical dc load voltages for equal values of transformer secondary voltage.
20. The bridge rectifier is a four-diode full-wave rectifier. It is the most commonly used because:
    a. It does not require the use of a center-tapped transformer.
    b. Using transformers with equal secondary voltages, it produces nearly twice the peak load voltage, average load voltage, and average load current of a full-wave center-tapped rectifier.
21. The bridge rectifier works on the principle of alternating the conduction of two diode pairs.
22. The operation of a bridge rectifier is illustrated in Figure 3.23.
23. Figure 3.26 provides a comparison of half-wave, full-wave, and bridge rectifiers.
24. Rectifiers are generally high-current circuits. As a result, the voltage developed across the bulk resistance of a rectifier diode can produce a forward voltage of 1 V (or higher).
25. Many rectifier diodes are mounted to a heat sink to keep them cool when used in high-power circuits. This is required (in part) because of the high reverse current values that are typical for rectifer diodes operated at high temperatures.
26. Transformer tolerances can be in the range of 20% (or greater).
27. Power rectifiers are components with:
    a. Extremely high forward current and power dissipation ratings.
    b. Relatively high foward voltage values, typically 1 V or greater.
    c. High reverse current ratings.
    d. Extremely high nonrepetitive surge current ratings.
28. Integrated circuit (IC) rectifiers are complete rectifier circuits housed in a single casing. The advantages of this technology include:
    a. Reduced cost.
    b. Ease of troubleshooting.
    c. Identical operating temperature of all components in the rectifier, resulting in equal values of $V_F$ and reverse current.
    d. Ease of mounting the circuit on a heat sink.
29. The output from a rectifier is normally connected to a *filter* (see Figure 3.1).
    a. A filter is a circuit that significantly reduces the variations in the pulsating dc output from the rectifer.
    b. The most commonly used filter is the *capacitive* filter.
30. The change in voltage that remains in the output from a filter is referred to as *ripple voltage*.
31. A capacitive filter works on the principle of *switching time constants* (see Figure 3.32).
    a. The capacitor charges through the diode(s). The charge circuit has a short time constant.
    b. The capacitor discharges through the load. The discharge circuit has a long time constant.

32. A brief *surge current* is generated through a filtered rectifier when power is first applied to the circuit.

33. The greater the value of a filter capacitor, the greater the magnitude and duration of the surge current. Therefore, surge current places a practical limit on the value of any filter capacitor.

34. When a filter is connected to the output of a *half-wave* rectifier, the PIV applied to the rectifier diode doubles to $2V_{S(pk)}$.

35. Adding a filter to a full-wave or bridge rectifier does not affect the PIV applied to the rectifier diodes.

36. Surge current problems can be eliminated by using a filter that contains an inductor placed in series between the rectifier output and the load (see Figure 3.41).

37. *LC* filters provide the best overall protection from surge currents and excessive ripple but are the most expensive to use.

38. A zener diode acts as a *voltage regulator* when placed in parallel with a power supply load.
    a. A zener maintains a relatively constant terminal voltage as long as the device current stays within specified limits.
    b. Zener regulators are rarely used because they are extremely inefficient. However, they are commonly used to teach the concept of voltage regulation.

39. A zener regulator is in parallel with the load, so $V_L = V_Z$.

40. As load resistance varies (within limits), the zener maintains a constant load voltage.

41. *Load regulation* is the ability of a regulator to maintain a constant output voltage despite changes in load current demand.

42. As shown in Figure 3.46, the combined impedance of a zener regulator ($Z_Z$) and the load forms a voltage divider with any series resistance ($R_S$). This voltage divider reduces the ripple at the output from the filter (see Example 3.21).

43. Figure 3.48 shows a complete basic dc power supply.
    a. The transformer reduces the line voltage to a lower secondary voltage (36 $V_{ac}$).
    b. The bridge rectifier converts the 36 $V_{ac}$ to pulsating dc.
    c. The filter reduces the variations in the output from the rectifier.
    d. The zener regulator further reduces the ripple produced by the rectifier and helps to ensure that a steady load voltage is maintained over a range of load current demands.

44. The most common transformer faults are:
    a. A shorted primary or secondary winding.
    b. An open primary or secondary winding.
    c. A short between the primary or secondary and the transformer case.

45. Any transformer short generally causes the power supply fuse to blow.

46. Any transformer open causes the output from the transformer to drop to zero.

47. Any suspected fault in a transformer can be verified using a resistance check of the circuit in question.

48. For a half-wave rectifier:
    a. A shorted rectifier diode passes the ac secondary voltage to the load.
    b. An open rectifier diode drops the load voltage to 0 V.

49. For a full-wave (or bridge) rectifier:
    a. A shorted diode will blow the power supply fuse.
    b. An open diode causes the rectifier output signal to resemble that of a half-wave rectifier.

50. A shorted filter capacitor will blow the power supply fuse.

51. An open filter capacitor causes a significant increase in the ripple present at the load.

52. A shorted zener regulator has the same symptoms as a shorted filter capacitor.

53. An open zener regulator causes the dc load voltage and output ripple to increase.

| Equation Number | Equation | Section Number |
|:---|:---:|:---:|
| (3.1) | $\dfrac{N_S}{N_P} = \dfrac{V_S}{V_P}$ | 3.1 |
| (3.2) | $V_S = \dfrac{N_S}{N_P} V_P$ | 3.1 |
| (3.3) | $\dfrac{I_P}{I_S} = \dfrac{V_S}{V_P}$ | 3.1 |
| (3.4) | $I_S = \dfrac{N_P}{N_S} I_P$ | 3.1 |
| (3.5) | $V_L \cong V_S$   (forward operation) | 3.2 |
| (3.6) | $V_{D1} = V_S$   (reverse operation) | 3.2 |
| (3.7) | $V_{L(pk)} = V_{S(pk)} - V_F$ | 3.2 |
| (3.8) | $V_{S(pk)} = \dfrac{N_S}{N_P} V_{P(pk)}$ | 3.2 |
| (3.9) | $V_{pk} = \dfrac{V_{rms}}{0.707}$ | 3.2 |
| (3.10) | $I_{L(pk)} = \dfrac{V_{L(pk)}}{R_L}$ | 3.2 |
| (3.11) | $V_{ave} = \dfrac{V_{pk}}{\pi}$   (half-wave rectified) | 3.2 |
| (3.12) | $V_{ave} = 0.318 V_{pk}$   (half-wave rectified) | 3.2 |
| (3.13) | $I_{ave} = \dfrac{I_{pk}}{\pi}$   (half-wave rectified) | 3.2 |
| (3.14) | $I_{ave} = 0.318 I_{pk}$   (half-wave rectified) | 3.2 |
| (3.15) | $PIV = V_{S(pk)}$   (half-wave rectified) | 3.2 |
| (3.16) | $V_{L(pk)} \cong \dfrac{V_{S(pk)}}{2}$ | 3.3 |
| (3.17) | $V_{L(pk)} = \dfrac{V_{S(pk)}}{2} - 0.7\ V$ | 3.3 |
| (3.18) | $V_{ave} = \dfrac{2 V_{L(pk)}}{\pi}$ | 3.3 |
| (3.19) | $V_{ave} = 0.637 V_{L(pk)}$ | 3.3 |
| (3.20) | $PIV \cong 2 V_{L(pk)}$ | 3.3 |
| (3.21) | $PIV \cong V_{S(pk)}$ | 3.3 |
| (3.22) | $PIV = V_{S(pk)} - 0.7\ V$ | 3.3 |
| (3.23) | $V_{L(pk)} \cong V_{S(pk)}$   (ideal) | 3.4 |
| (3.24) | $V_{L(pk)} = V_{S(pk)} - 1.4\ V$ | 3.4 |
| (3.25) | $\tau = RC$ | 3.6 |
| (3.26) | $T = 5RC$ | 3.6 |
| (3.27) | $I_{surge} = \dfrac{V_{S(pk)}}{R_W + R_B}$ | 3.6 |

| Equation Number | Equation | Section Number |
|---|---|---|
| **(3.28)** | $$C = \frac{I(t)}{\Delta V_C}$$ | 3.6 |
| **(3.29)** | $$t = \frac{C(\Delta V_C)}{I}$$ | 3.6 |
| **(3.30)** | $$V_{dc} = V_{pk} - \frac{V_r}{2}$$ | 3.6 |
| **(3.31)** | $$V_r = \frac{I_L t}{C}$$ | 3.6 |
| **(3.32)** | $$PIV = 2V_{S(pk)} \quad \text{(half-wave, filtered)}$$ | 3.6 |
| **(3.33)** | $$I_T = \frac{V_{in} - V_Z}{R_S}$$ | 3.7 |
| **(3.34)** | $$I_L = \frac{V_Z}{R_L}$$ | 3.7 |
| **(3.35)** | $$I_Z = I_T - I_L$$ | 3.7 |
| **(3.36)** | $$R_{L(min)} = \frac{V_Z}{I_{L(max)}}$$ | 3.7 |
| **(3.37)** | $$V_{r(out)} = \frac{(Z_Z \| R_L)}{(Z_Z \| R_L) + R_S} V_r$$ | 3.7 |

**KEY TERMS**

average current ($I_{ave}$) 84
average voltage ($V_{ave}$) 83
bridge rectifier 92
capacitive filter 101
center-tapped
   transformer 88
filter 74
full load 114
full-wave rectifier 87

half-wave rectifier 78
inductive filter 109
integrated circuit (IC) 99
integrated rectifier 99
linear power supply 74
line regulation 97
load regulation 114
peak inverse voltage 87
power rectifier 99

power supply 74
rectifier 74
ripple voltage 101
surge current 103
switching power
   supply 74
voltage regulator 74

**PRACTICE PROBLEMS**

### Section 3.1

**1.** Determine the ac secondary voltage for the circuit shown in Figure 3.52.

**2.** Determine the ac secondary voltage for the circuit shown in Figure 3.53.

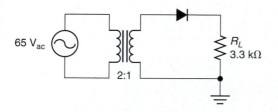

**FIGURE 3.52**

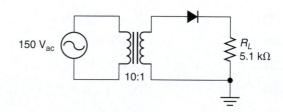

**FIGURE 3.53**

**3.** Determine the peak secondary voltage for the circuit shown in Figure 3.52.

**4.** Determine the peak secondary voltage for the circuit shown in Figure 3.53.

**5.** A transformer has values of $V_P = 40$ V and $V_S = 320$ V. What is the component's turns ratio?

**6.** A transformer with a 12:1 turns ratio has a 250 mA primary current. What is the value of the secondary current?

## Section 3.2
**7.** Determine the peak load voltage for the circuit shown in Figure 3.52.

**8.** Determine the peak load voltage for the circuit shown in Figure 3.53.

**9.** Determine the peak load voltage for the circuit shown in Figure 3.54.

**10.** Determine the peak load current for the circuit shown in Figure 3.54.

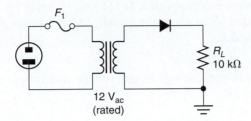

**FIGURE 3.54**

**11.** Determine the average (dc) load voltage for the circuit shown in Figure 3.52.

**12.** Determine the average (dc) load voltage for the circuit shown in Figure 3.53.

**13.** Determine the average (dc) load voltage for the circuit shown in Figure 3.54.

**14.** Assume that the diode in Figure 3.52 is reversed. Determine the new values of $V_{L(pk)}$, $V_{ave}$, and $I_{ave}$ for the circuit.

**15.** Assume that the diode in Figure 3.53 is reversed. Determine the new values of $V_{L(pk)}$, $V_{ave}$, and $I_{ave}$ for the circuit.

**16.** Assume that the diode in Figure 3.54 is reversed. Determine the new values of $V_{L(pk)}$, $V_{ave}$, and $I_{ave}$ for the circuit.

**17.** Determine the PIV for the diode in Figure 3.53.

**18.** A negative half-wave rectifier with a 12 k$\Omega$ load is driven by a 20 V$_{ac}$ transformer. Draw the schematic for the circuit, and determine the following values: PIV, $V_{L(pk)}$, $V_{ave}$, and $I_{ave}$.

## Section 3.3
**19.** Determine the values of $V_{L(pk)}$, $V_{ave}$, and $I_{ave}$ for the circuit shown in Figure 3.55.

**20.** Determine the values of $V_{L(pk)}$, $V_{ave}$, and $I_{ave}$ for the circuit shown in Figure 3.56.

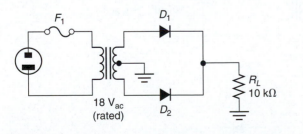

**FIGURE 3.55**

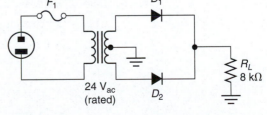

**FIGURE 3.56**

**21.** Determine the values of $V_{L(pk)}$, $V_{ave}$, and $I_{ave}$ for the circuit shown in Figure 3.57.

**22.** Determine the PIV of the circuit shown in Figure 3.55.

**23.** Determine the PIV of the circuit shown in Figure 3.56.

**24.** Determine the PIV of the circuit shown in Figure 3.57.

**25.** Assume that the diodes in Figure 3.55 are both reversed. Determine the new values of $V_{L(pk)}$, $V_{ave}$, and $I_{ave}$ for the circuit.

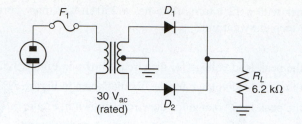

FIGURE 3.57

**26.** A negative full-wave rectifier with a 910 Ω load is driven by a 16 $V_{ac}$ transformer. Draw the schematic diagram of the circuit, and determine the following values: $V_{L(pk)}$, PIV, $V_{ave}$, and $I_{ave}$.

**Section 3.4**

**27.** Determine the peak and average output voltage and current values for the circuit shown in Figure 3.58.

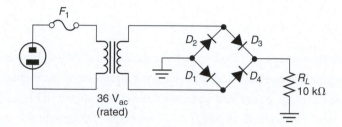

FIGURE 3.58

**28.** Repeat Problem 27. Assume that the transformer is a 16 $V_{ac}$ transformer.

**29.** What is the minimum allowable value of $V_{RRM}$ for each of the diodes in Problem 28?

**30.** A bridge rectifier with a 1.2 kΩ load is driven by a 48 $V_{ac}$ transformer. Draw the schematic diagram for the circuit, and calculate the dc load voltage and current values. Also, determine the PIV for each diode in the circuit.

**Section 3.6**

**31.** The circuit shown in Figure 3.59a has values of $R_W = 1\ \Omega$ and $R_B = 6\ \Omega$. What is the value of surge current for the circuit?

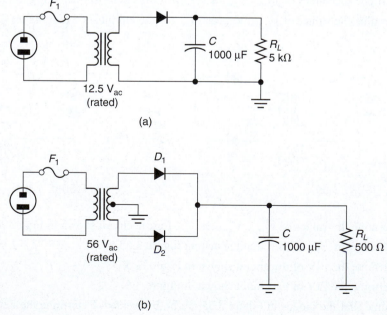

FIGURE 3.59

**32.** A half-wave rectifier has values of $R_W = 2\ \Omega$, $R_B = 12\ \Omega$, and $V_S = 36\ V_{ac}$. What is the value of surge current for the circuit?

**33.** What are the values of $V_{dc}$ and $V_r$ for the circuit shown in Figure 3.59a?

**34.** What are the values of $V_{dc}$ and $V_r$ for the circuit shown in Figure 3.59b?

**35.** The circuit shown in Figure 3.60 has the following values: $V_S = 18\ V_{ac}$ (rated), $C = 470\ \mu F$, and $R_L = 820\ \Omega$. Determine the values of $V_r$, $V_{dc}$, and $I_L$ for the circuit.

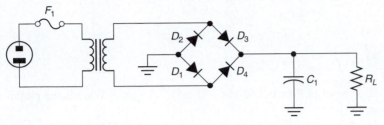

FIGURE 3.60

**36.** The circuit shown in Figure 3.60 has the following values: $V_S = 24\ V_{ac}$ (rated), $C = 1200\ \mu F$, and $R_L = 200\ \Omega$. Determine the values of $V_r$, $V_{dc}$, and $I_L$ for the circuit.

**37.** What is the PIV for the circuit described in Problem 36?

**38.** What is the PIV for the circuit shown in Figure 3.59a?

**Section 3.7**

**39.** For the circuit shown in Figure 3.61, determine the total circuit current.

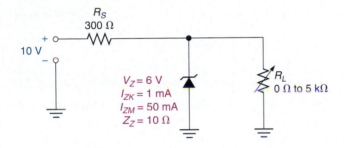

FIGURE 3.61

**40.** For the circuit shown in Figure 3.61, determine the value of $I_L$ for $R_L = 2\ k\Omega$.

**41.** For the circuit shown in Figure 3.61, determine the value of $I_Z$ for $R_L = 2\ k\Omega$.

**42.** For the circuit shown in Figure 3.61, determine the value of $I_Z$ for $R_L = 3\ k\Omega$.

**43.** For the circuit shown in Figure 3.61, determine the minimum allowable value of $R_L$.

**44.** For the circuit shown in Figure 3.62, determine the total circuit current.

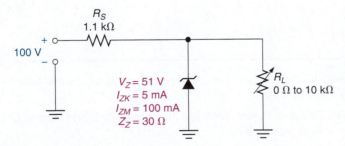

FIGURE 3.62

**45.** For the circuit shown in Figure 3.62, determine the value of $I_Z$ for $R_L = 5\ k\Omega$.

**46.** For the circuit shown in Figure 3.62, determine the minimum allowable value of $R_L$.

**47.** Determine the values of $I_T$, $I_L$, and $I_Z$ for the circuit shown in Figure 3.63.

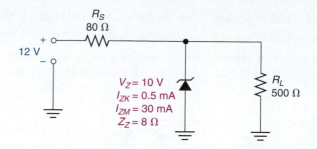

**FIGURE 3.63**

**48.** The 12 V input to Figure 3.63 has 770 mV$_{pp}$ of ripple. What is the output ripple for the circuit?

**Section 3.8**

**49.** Calculate the dc output voltage and current values for the circuit shown in Figure 3.64.

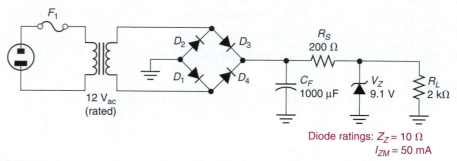

**FIGURE 3.64**

**50.** Calculate the output ripple voltage for the circuit shown in Figure 3.64.

**51.** Calculate the dc output values and output ripple voltage for the circuit shown in Figure 3.65.

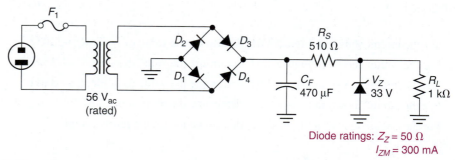

**FIGURE 3.65**

**52.** The circuit shown in Figure 3.66a has the output signal shown. Discuss the possible cause(s) of the problem.

**53.** The circuit shown in Figure 3.66b has the output waveform shown. Discuss the possible cause(s) of the problem.

**54.** The circuit in Figure 3.64 has an output that equals 0 V$_{dc}$. The primary fuse is not blown. Discuss the possible cause(s) of the problem.

**55.** The load resistance in Figure 3.67 opens. Will this cause the zener diode to be destroyed? Use circuit calculations to explain your answer.

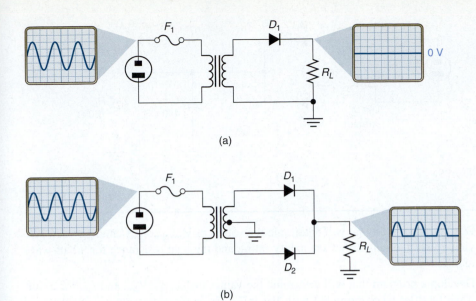

(a)

(b)

FIGURE 3.66

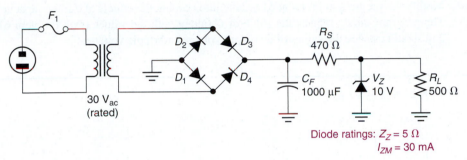

30 V$_{ac}$
(rated)

$R_S$
470 Ω

$C_F$
1000 μF

$V_Z$
10 V

$R_L$
500 Ω

Diode ratings: $Z_Z = 5$ Ω
$I_{ZM} = 30$ mA

FIGURE 3.67

PUSHING THE ENVELOPE

**56.** Assume the zener diode in Figure 3.67 has opened. Which of the diodes in Figure 2.34 could be used as a replacement component?

**57.** The circuit shown in Figure 3.68 has measured output values of $V_{dc} = 12$ V and $V_r = 1.22$ V$_{pp}$. Determine whether or not there is a problem in the circuit. (*Remember:* Transformer tolerances can be as high as ±20%.)

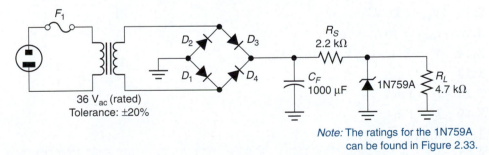

36 V$_{ac}$ (rated)
Tolerance: ±20%

$R_S$
2.2 kΩ

$C_F$
1000 μF

1N759A

$R_L$
4.7 kΩ

*Note:* The ratings for the 1N759A
can be found in Figure 2.33.

FIGURE 3.68

**58.** The rectifier diodes in Figure 3.69 must be replaced. Using Figure 2.25, determine which, if any, of the 1N4001–7 series diodes can be used as a replacement component in the circuit.

**59.** The primary fuse in Figure 3.68 must be replaced. Determine the minimum acceptable value for the fuse, assuming that its value must be 20% greater than the primary current. Assume you have the following rated fuses: ⅛ A, ¼ A, ½ A, 1 A, 2 A, and 5 A.

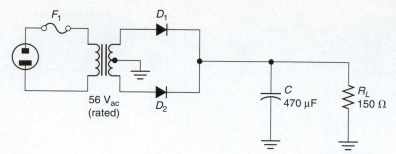

FIGURE 3.69

**SUGGESTED COMPUTER APPLICATIONS PROBLEMS**

**60.** Develop a program that will determine the values of $V_{out(pk)}$, $V_{ave}$, and $I_{ave}$ when provided with the transformer secondary rating and the load resistance for a half-wave rectifier.

**61.** Develop a program that will determine the values of $V_{out(pk)}$, $V_{ave}$, and $I_{ave}$ for a full-wave rectifier when provided with the transformer secondary rating and the load resistance.

**62.** Modify the program in Problem 61 to take into account the effects of a filter capacitor. The program should request the value of $C$ (along with the values from Problem 61) and should provide the values of dc output voltage and ripple voltage.

**ANSWERS TO THE EXAMPLE PRACTICE PROBLEMS**

**3.1** $I_S = 20.8$ mA (maximum)

**3.2** $V_{L(pk)} = 17.3$ V$_{pk}$

**3.3** $V_{L(pk)} = 16.27$ V$_{pk}$

**3.4** $I_{L(pk)} = 1.63$ mA

**3.5** $V_{ave} = 3.64$ V

**3.6** $I_{ave} = 10.9$ mA

**3.7** $I_{ave} = 1.78$ mA

**3.8** $V_{S(pk)} = -50.92$ V, $V_{L(pk)} = -50.22$ V$_{pk}$, $V_{ave} = -15.99$ V$_{dc}$

**3.9** $V_{ave} = 10.36$ V

**3.10** $I_{L(pk)} = 7.4$ mA, $I_{ave} = 4.71$ mA

**3.11** $V_{ave} = 15.3$ V, $I_{ave} = 12.77$ mA

**3.13** $V_{ave} = 9.6$ V

**3.14** $I_{surge} = 10$ A

**3.16** $V_r = 274$ mV$_{pp}$, $V_{dc} = 16.14$ V

**3.17** $I_T = 2$ mA

**3.18** $I_L = 1$ mA

**3.19** $I_Z = 1$ mA

**3.20** $R_{L(min)} = 515$ Ω

**3.21** $V_{r(out)} = 235$ mV$_{pp}$

**3.22** $V_{dc} = 10$ V, $I_L = 1.96$ mA, $V_{r(out)} = 30.7$ mV$_{pp}$

## chapter 4

# Common Diode Applications

## Clippers, Clampers, Voltage Multipliers, and Displays

### Objectives

*After studying the material in this chapter, you should be able to:*

1. State the purposes served by clippers, clampers, and voltage multipliers.
2. Describe the operation of *series clippers*.
3. Describe and analyze the operation of *shunt clippers*.
4. Describe and analyze the operation of *biased shunt clippers*.
5. Describe the effects of *negative clampers* and *positive clampers* on an input waveform.
6. Describe the circuit operation of a clamper.
7. Describe and analyze the operation of the *half-wave voltage doubler*.
8. Describe and analyze the operation of the *full-wave voltage doubler*.
9. Describe the use of the LED as a power-level indicator.
10. Describe the use of the LED in multisegment displays.
11. Describe the common fault symptoms that occur in clippers and clampers.
12. Describe the procedure for troubleshooting voltage multipliers.

### Outline

**4.4** Voltage Multipliers

**4.5** LED Applications

**4.6** Diode Circuit Troubleshooting

Chapter Summary

Diodes are among the primary components in power supplies, as we saw in Chapter 3, but they have many other common applications as well. In this chapter, we will look at several additional diode circuits and applications.

**OBJECTIVE 1** ▶

**Clipper** (or **limiter**)
A diode circuit used to eliminate some portion(s) of a waveform.

**Clamper** (or **dc restorer**)
A diode circuit that is used to set (or restore) the dc reference of a waveform.

**Voltage multipliers** produce a dc output voltage that is some multiple of a peak input voltage.

**Multisegment displays** are used to display **alphanumeric symbols** (numbers, letters, and punctuation marks).

The first type of circuit we will cover is the **clipper** (or **limiter**). A clipper is a diode circuit that is used to eliminate some portion(s) of a waveform. As you will see, there are many types of clippers.

The second type of circuit we will cover is the **clamper** (or **dc restorer**). A clamper is a diode circuit that is used to set (or restore) the dc reference of a waveform. As you will see, it is not uncommon for waveforms to be centered around some *dc reference voltage*. For example, you might see a 12 $V_{pp}$ sinusoidal waveform that varies equally above and below $+10$ $V_{dc}$. A clamper could be used to set (or change) that $+10$ $V_{dc}$ reference.

The third type of circuit we will discuss is the **voltage multiplier**. A voltage multiplier produces a dc output voltage that is some multiple of an ac peak input voltage.

Finally, we will look at the most common LED application: the **multisegment display**. Multisegment displays are used to display **alphanumeric symbols**; that is, they are used to display numbers, letters, and punctuation marks.

## 4.1 Clippers

You have already been introduced to the basic operating principles of the clipper. The half-wave rectifier is basically a clipper that eliminates one of the alternations of an ac signal.

There are *four* basic clipper configurations, as shown in Figure 4.1. Each *series clipper* contains a diode that is in series with the load. Each *shunt clipper* contains a diode that is in parallel with the load.

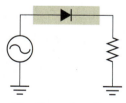
Negative series clipper

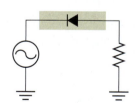
Positive series clipper

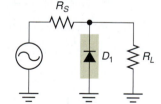
Negative shunt clipper

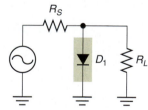
Positive shunt clipper

FIGURE 4.1

### 4.1.1 Series Clippers

The *series clipper*, as shown in Figure 4.2, has the same circuit operating characteristics as the half-wave rectifier. In fact, if you compare the circuits shown in Figure 3.9 to those shown in Figure 4.2, you will see that the half-wave rectifier is nothing more than a series clipper.

**OBJECTIVE 2** ▶

Since the operation of the series clipper is identical to that of the half-wave rectifier, we will not go into any great detail on series clipper operation at this point. As a review, the following points are made about series clippers:

1. When the diode in a *negative series clipper* is forward biased by the input signal, it conducts, and the load voltage is found as

$$V_L = V_{in} - 0.7 \text{ V}$$ 

(4.1)

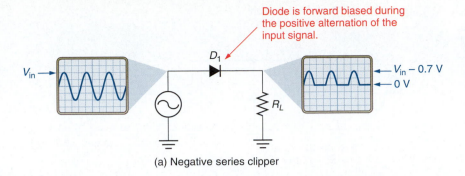

Diode is forward biased during the positive alternation of the input signal.

$V_{in} - 0.7$ V
0 V

(a) Negative series clipper

As shown in Figure 4.2, a *negative clipper* eliminates the *negative* alternation of its input. The *positive clipper* eliminates the *positive* alternation of its input.

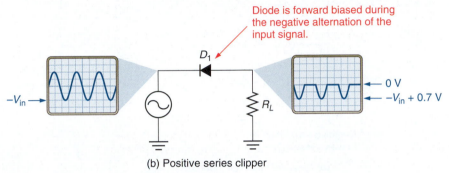

Diode is forward biased during the negative alternation of the input signal.

0 V
$-V_{in} + 0.7$ V

(b) Positive series clipper

**FIGURE 4.2**   Series clipper operation.

2. When the diode in the negative series clipper is reverse biased by the input signal, it does not conduct. Therefore,

$$V_{D1} = V_{in} \qquad \textbf{(4.2)}$$

and

$$V_L = 0 \text{ V} \qquad \textbf{(4.3)}$$

A negative series clipper and its associated waveforms are shown in Figure 4.2a.

3. The *positive series clipper* operates in the same fashion. The only differences are:
   **a.** The output voltage polarities are reversed.
   **b.** The current directions through the circuit are reversed.

A positive series clipper and its associated waveforms are shown in Figure 4.2b.

For a complete review of series clippers, refer to the coverage of half-wave rectifiers in Chapter 3.

## 4.1.2   Shunt Clippers

The operation of the *shunt clipper* is exactly opposite that of the series clipper. The *series clipper* provides an output when the diode is forward biased and no output when the diode is reverse biased. The *shunt clipper* provides an output when the diode is reverse biased and shorts the output signal to ground when the diode is forward biased. This operation is illustrated in Figure 4.3.

◄ *OBJECTIVE 3*

The circuit shown in Figure 4.3 is a *negative shunt clipper.* When the diode in the negative shunt clipper is reverse biased, it is effectively removed from the circuit. This is shown in Figure 4.3a. With the diode reversed biased, the resistors form a voltage divider, and the load voltage can be found by

$$V_L = \frac{R_L}{R_L + R_S} V_{in} \qquad \textbf{(4.4)}$$

While the output signal resembles the positive alternation of the input, the peak output voltage is somewhat less than the peak input voltage. This point is illustrated in Example 4.1.

**Lab Reference:** The operation illustrated in Figure 4.3 is demonstrated in Exercise 5.

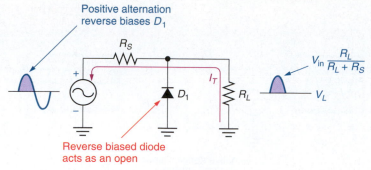

Positive alternation reverse biases $D_1$

Reverse biased diode acts as an open

(a) Positive half-cycle operation

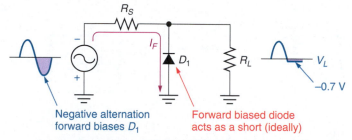

Negative alternation forward biases $D_1$

Forward biased diode acts as a short (ideally)

(b) Negative half-cycle operation

**FIGURE 4.3** Shunt clipper operation.

## EXAMPLE 4.1

The negative shunt clipper shown in Figure 4.4 has a peak input voltage of $+12$ V. What is the peak load voltage for the circuit?

**FIGURE 4.4**

***Solution:*** When the input is positive, the diode is reverse biased and does not conduct. Therefore, the peak load voltage is found as

$$V_L = \frac{R_L}{R_L + R_S} V_{in} = \frac{5.1 \text{ k}\Omega}{6.1 \text{ k}\Omega} (+12 \text{ V}_{pk}) = \mathbf{10 \text{ V}_{pk}}$$

### PRACTICE PROBLEM 4.1

A negative shunt clipper has values of $R_L = 510 \ \Omega$ and $R_S = 100 \ \Omega$. If the input voltage is $+15 \text{ V}_{pk}$, what is the peak load voltage?

*Just a Reminder:*
A diode doesn't conduct fully until its cathode is 0.7 V *more negative* than its anode. Thus, the diode in the negative shunt clipper conducts fully when the cathode is at $-0.7$ V. That is why $V_F$ is given as a negative value in equation (4.5).

During the negative alternation of the input signal, the diode in the negative shunt clipper is forward biased and conducts as shown in Figure 4.2b. With the diode conducting, the voltage across the component equals the diode forward voltage ($V_F$). Since the diode and load are in parallel, $V_L$ also equals the diode forward voltage. By formula,

$$V_L = -V_F \tag{4.5}$$

Since the output side of $R_S$ is held to approximately $-0.7$ V when the diode is forward biased, the voltage across $R_S$ (which is designated $V_{RS}$) is equal to the *difference between* $V_{in}$ and the value $V_F$. By formula,

$$V_{RS} = -V_{in} + 0.7 \text{ V} \qquad\qquad (4.6)$$

when the diode is forward biased. Example 4.2 illustrates the circuit conditions that exist when the diode in a negative shunt clipper is forward biased.

### EXAMPLE 4.2

The circuit described in Example 4.1 has a $-12$ $V_{pk}$ input. Determine the values of $V_L$ and $V_{RS}$ for the circuit.

*Solution:* Since the diode is forward biased, the load voltage equals the value of $V_F$ for the diode. Thus,

$$V_L = -0.7 \text{ V}$$

The voltage across the series resistor ($R_S$) is now found as

$$V_{RS} = -V_{in} + 0.7 \text{ V} = -12 \text{ V}_{pk} + 0.7 \text{ V} = \mathbf{-11.3\ V_{pk}}$$

### PRACTICE PROBLEM 4.2

The circuit described in Practice Problem 4.1 has a $-15$ $V_{pk}$ input. Determine the values of $V_L$ and $V_{RS}$ for the circuit.

So far, we have worked with the *negative shunt clipper.* As Example 4.3 illustrates, the positive shunt clipper works according to the same principles and is analyzed in the same fashion.

### EXAMPLE 4.3

The *positive shunt clipper* shown in Figure 4.5 has the input waveform shown. Determine the value for $V_L$ for each of the input alternations.

The diode in a *positive* shunt clipper conducts when the input is *positive.* Thus, for the positive shunt clipper, equations (4.5) and (4.6) are modified as follows:

$$V_L = V_F$$
$$V_{RS} = V_{in} - 0.7 \text{ V}$$

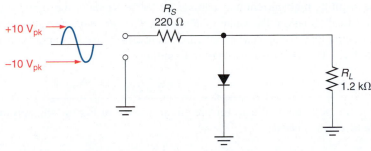

FIGURE 4.5

*Solution:* When the input is *positive,* the diode is forward biased. Thus,

$$V_L = 0.7 \text{ V}$$

and

$$V_{RS} = V_{in} - 0.7 \text{ V} = 10 \text{ V}_{pk} - 0.7 \text{ V} = \mathbf{9.3\ V_{pk}}$$

When the input to the circuit is *negative*, the diode is reverse biased and effectively removed from the circuit. Thus,

$$V_L = \frac{R_L}{R_L + R_S} V_{in} = \frac{1.2 \text{ k}\Omega}{1.42 \text{ k}\Omega}(-10 \text{ V}_{pk}) = -8.45 \text{ V}_{pk}$$

### PRACTICE PROBLEM 4.3

A *positive shunt clipper* with values of $R_S = 100 \ \Omega$ and $R_L = 1.1 \text{ k}\Omega$ has a $\pm 12 \text{ V}_{pk}$ input signal. Determine the value of $V_L$ for each alternation of the input. Also, determine the value of $V_{RS}$ when the diode is forward biased.

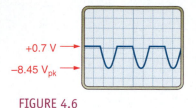

+0.7 V →

−8.45 V$_{pk}$ →

FIGURE 4.6

The load waveform for the circuit shown in Figure 4.5 is illustrated in Figure 4.6. As you can see, the load voltage has peak values of +0.7 and −8.45 V.

### 4.1.3 The Purpose Served by $R_S$

$R_S$ is included in the shunt clipper as a *current-limiting* resistor. Its purpose becomes clear when you consider what would happen if the input signal forward biased the diode and $R_S$ was not in the circuit. This operating condition is illustrated in Figure 4.7.

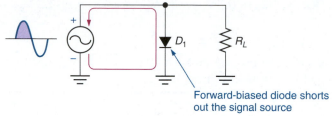

Forward-biased diode shorts out the signal source

FIGURE 4.7   Without $R_S$, $D_1$ shorts the source.

Without $R_S$ in the circuit, the diode shorts the signal source to ground during the positive alternation of the input signal. This probably results in one of the following:

1. The diode being destroyed by excessive forward current.
2. One or more components in the signal source being destroyed by the excessive current demand of the conducting diode.

For example, assume that the circuit input signal shown in Figure 4.7 has a value of $+V_{pk} = 12 \text{ V}$. When the input signal is at its positive peak, the diode shorts the +12 V to ground. The resulting high current may damage either the diode or the signal source.

In any practical situation, the value of $R_S$ is *much lower than* the value of $R_L$. When this is the case, the load voltage (when the diode is reverse biased) is approximately equal to the value of $V_{in}$. This point is illustrated in Example 4.4.

### EXAMPLE 4.4

Determine the peak load voltage for the circuit shown in Figure 4.8. Assume that the diode is reverse biased.

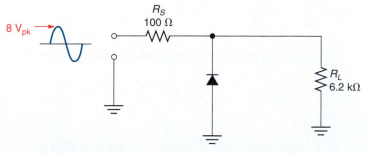

8 V$_{pk}$ →

$R_S$
100 Ω

$R_L$
6.2 kΩ

FIGURE 4.8

**Solution:** The peak load voltage is found using equation (4.4) as follows:

$$V_L = \frac{R_L}{R_L + R_S} V_{in} = \frac{6.2 \text{ k}\Omega}{6.3 \text{ k}\Omega} (8 \text{ V}_{pk}) = \mathbf{7.87 \text{ V}_{pk}}$$

Since $R_L \gg R_S$ in this circuit, we could have assumed that $V_{in}$ and $V_L$ were approximately equal in any practical circuit analysis.

### 4.1.4  Biased Clippers

A **biased clipper** uses a dc biasing source to set the limit(s) on the circuit output voltage. As shown in Figure 4.9, this allows the circuit to clip input waveforms at values other than the diode $V_F$ of 0.7 V. In each circuit shown, the bias voltage ($V_B$) is in series with the shunt diode. As a result, the diode conducts (and clips the input waveform) when the signal voltage equals the sum of $V_F$ and $V_B$.

◀ **OBJECTIVE 4**

**Biased clipper**
A shunt clipper that uses a dc biasing source to set the limit(s) on the circuit output voltage.

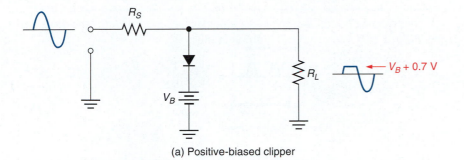

(a) Positive-biased clipper

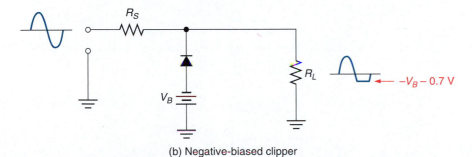

(b) Negative-biased clipper

**FIGURE 4.9**  Biased shunt clippers.

The *positive-biased clipper* (Figure 4.9a) clips its input signal at $V_B$ + 0.7 V. The actual value at which the circuit clips the input signal depends on the value of the biasing voltage ($V_B$). If $V_B$ is 2 V, the input signal is clipped at 2.7 V. If the value of $V_B$ is 5 V, the input signal is clipped at 5.7 V, and so on.

The *negative-biased clipper* (Figure 4.9b) works in the same fashion, but it clips its input signal at $-V_B$ − 0.7 V. If $V_B$ is −2 V, the input signal is clipped at −2.7 V. If the value of $V_B$ is −5 V, the input signal is clipped at −5.7 V, and so on.

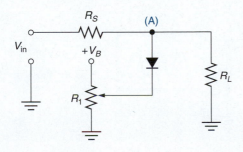

**FIGURE 4.10**

In practice, a potentiometer is used to provide an adjustable value of $V_B$, as shown in Figure 4.10. In this circuit, the biasing voltage ($+V_B$) is connected to the diode via the potentiometer ($R_1$). $R_1$ is adjusted in this circuit to provide the desired clipping limit at point $A$ in the circuit. By reversing the direction of the diode and the polarity of $V_B$, the circuit in Figure 4.10 can be modified to work as a negative-biased clipper.

### 4.1.5    One Final Note

As you can see, there are many different clipper configurations. These configurations, which are used for a variety of applications, are summarized in Figure 4.11. What are they used for? We will answer this question in the next section.

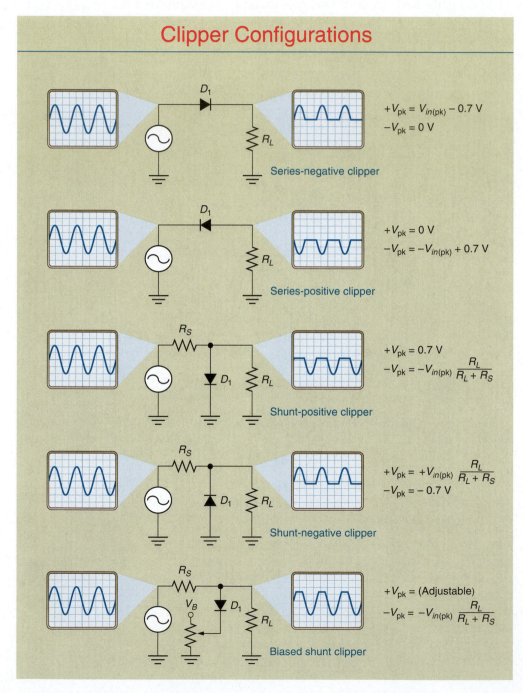

**Clipper Configurations**

$+V_{pk} = V_{in(pk)} - 0.7$ V
$-V_{pk} = 0$ V

Series-negative clipper

$+V_{pk} = 0$ V
$-V_{pk} = -V_{in(pk)} + 0.7$ V

Series-positive clipper

$+V_{pk} = 0.7$ V
$-V_{pk} = -V_{in(pk)} \dfrac{R_L}{R_L + R_S}$

Shunt-positive clipper

$+V_{pk} = +V_{in(pk)} \dfrac{R_L}{R_L + R_S}$
$-V_{pk} = -0.7$ V

Shunt-negative clipper

$+V_{pk} = $ (Adjustable)
$-V_{pk} = -V_{in(pk)} \dfrac{R_L}{R_L + R_S}$

Biased shunt clipper

FIGURE 4.11

1. What purpose is served by a *clipper*?
2. What is another name for a clipper?
3. What purpose is served by a *clamper*?
4. What is another name for a clamper?
5. Discuss the differences between series and shunt clippers.
6. What purpose is served by $R_S$ in the shunt clipper?
7. Describe the operation of a biased clipper.

## 4.2 Clipper Applications

Clippers are used in a wide variety of electronic systems. They are generally used to perform one of two functions:

1. Alter the shape of a waveform.
2. Provide circuit transient protection.

You have already seen one example of the first function in the half-wave rectifier. This circuit alters the shape of an ac signal, changing it to pulsating dc. You will see another application of this type in this section, along with several transient protection circuits.

### 4.2.1 Transient Protection

A **transient** is an abrupt current or voltage spike that has an extremely short duration. A *current surge* would be one type of transient.

Transients can do serious damage to circuits whose inputs must stay within certain voltage or current limits. For example, many digital circuits have inputs that can handle only voltages that fall within a specified range. For these circuits, a voltage transient that goes outside the specified voltage range cannot be allowed to reach the input. A clipper can be used in this case to prevent such a transient from reaching a digital circuit, as illustrated in Figure 4.12a.

> **Transient**
> An abrupt current or voltage spike.

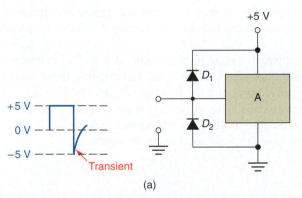

(a)

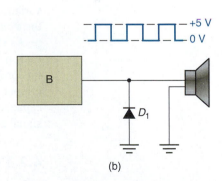

(b)

**FIGURE 4.12** Transient-protection circuits.

The block labeled A represents a digital circuit whose input voltage must not be allowed to go outside the range of 0 to +5 V. The clipper will protect the circuit from any voltage transients outside this range. For example, let's assume that the transient shown in the figure occurs on the input line. The transient will forward bias $D_2$, causing the diode to conduct. With $D_2$ conducting, the transient will be shorted to ground, protecting the input to circuit A. If the input square wave should go above 5 V, $D_1$ (which is located between the +5 V supply and the input) will be forward biased. This will short any input greater than 5 V back to the +5 V supply.

> The description of the circuit in Figure 4.12a assumes that the diodes are ideal. When you consider the diode voltage drops, the input signal is clipped at +5.7 and −0.7 V.

*A Practical Consideration:*
Let's say that circuit B in Figure 4.12b goes bad. When it is replaced, the new circuit works fine until it tries to drive the speaker. After driving the speaker for a few seconds, the new circuit B also goes bad. The problem in this case is likely an open diode. If the diode is open, the counter emf produced by the speaker isn't shorted to ground and destroys each new circuit that is used to replace the previous one.

The circuit in Figure 4.12b includes a clipper designed to protect the *output* of circuit B from transients produced when a square wave is used to drive a speaker (a common practice with computer sound effects). A speaker is essentially a big coil (inductor). When a square wave is used to drive the coil in a speaker, that coil produces a *counter emf*. This counter emf is produced each time that the output voltage makes the transition from +5 to 0 V. The counter emf is greater in magnitude than the original voltage, and opposite in polarity. Thus, each time circuit B tries to drive the output line to 0 V, the speaker tries to force the line to continue to some negative value. This counter emf could destroy the driving circuit. However, the clipping diode ($D_1$) shorts the counter emf produced by the speaker coil to ground before it can harm the driving circuit.

### 4.2.2    The AM Detector

Another common application for the diode clipper can be found in a typical AM receiver. In this case, the clipper is part of a circuit called an **AM detector**. A simple detector is shown in Figure 4.13.

**AM detector**
A diode clipper that converts a varying amplitude ac input to a varying dc level.

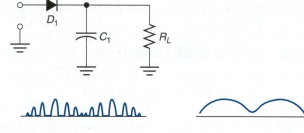

FIGURE 4.13    $C_1$ charges to $V_{ave}$ of the clipped signal.

The purpose of the detector is to produce dc output voltage that varies with the peak variations in the input signal. This output is produced by the capacitor, which charges to the average of the input signal.

As shown, the input signal to the circuit has an average value of 0 V. This is because each positive peak has an equal and opposite negative peak. The clipping diode solves this problem by eliminating the negative portion of the input waveform. (In this case, the clipper alters the shape of the waveform.) With the negative portion eliminated, the capacitor charges and discharges at the rate of the peak input variations. This provides a signal at the load that is a reproduction of the peak variations at the input.

### 4.2.3    One Final Note

The applications introduced in this section are intended only to give you an idea of clipper applications. To cover every possible circuit would require several volumes. Even though we have covered only a few clipper circuits, you should now have a good idea of the purposes they serve.

---

**Section Review ▶**

1. What is a *transient*?

2. Why must digital circuit inputs be protected from voltage transients?

3. Describe the operation of the AM detector.

## 4.3    Clampers (DC Restorers)

A *clamper* is a circuit designed to shift a waveform either above or below a given reference voltage without distorting the waveform. There are two types of clampers: the *positive clamper* and the *negative clamper*. The input/output characteristics of these two circuits are illustrated in Figure 4.14.

◀ *OBJECTIVE 5*

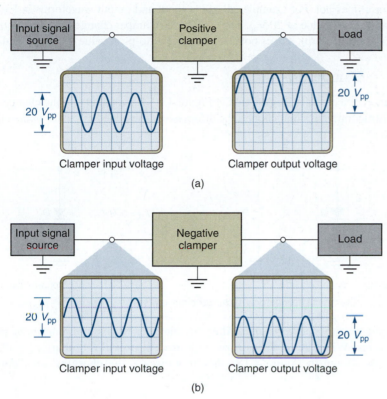

**FIGURE 4.14**

A **positive clamper** shifts its input waveform so that the *negative* peak of the waveform is approximately equal to the clamper dc reference voltage. For example, Figure 4.14a shows what happens when a 20 $V_{pp}$ sine wave is applied to a positive clamper with a dc reference of 0 V. As you can see, the input and output waveforms have a value of 20 $V_{pp}$. However, the clamper output waveform has peak values of +20 and 0 V (the negative peak). The positive clamper has shifted the entire waveform so that its negative peak value is approximately equal to the circuit's dc reference voltage.

A **negative clamper** shifts its input waveform so that the *positive* peak of the waveform is approximately equal to the clamper dc reference voltage. For example, Figure 4.14b shows what happens when a 20 $V_{pp}$ sine wave is applied to a negative clamper with a dc reference of 0 V. In this case, the clamper output waveform has peak values of 0 V (the positive peak) and −20 V. The negative clamper has shifted the entire waveform so that its positive peak value is approximately equal to the circuit's dc reference voltage.

> **Positive clamper**
> A circuit that shifts an entire input signal *above* a dc reference voltage.

> **Negative clamper**
> A circuit that shifts an entire input signal *below* a dc reference voltage.

### 4.3.1    $V_{ave}$ Shift

You were shown in Chapter 3 how to determine the dc average ($V_{ave}$) value of a *rectified* sine wave. When we are dealing with a sine wave that is *not* rectified, the value of $V_{ave}$ falls halfway between the positive and negative peak voltage values. For example, if we were to measure the input waveform in Figure 4.14a with a dc voltmeter, we would get a reading of 0 V. Note that the dc average of the waveform (0 V) falls halfway between +10 and −10 $V_{pk}$.

How does a clamper affect the value of $V_{ave}$ for a waveform?

When a waveform is shifted by a clamper, the value of $V_{ave}$ for the waveform changes. For example, consider the output waveform shown in Figure 4.14a. This waveform has peak values of +20 and 0 V. The dc average, which falls halfway between the peak values, is +10 V. Thus, if we were to measure this output waveform with a dc voltmeter, we would get a reading of 10 $V_{dc}$. By the same token, if we were to measure the output waveform shown in Figure 4.14b with a dc voltmeter, we would get a reading of −10 $V_{dc}$.

Do clampers affect the peak-to-peak value of a waveform?

An interesting point can be made at this time. While the value of $V_{ave}$ for a waveform normally changes when the waveform goes through the clamper, the peak-to-peak value of the waveform does not. For example, all of the input and output waveforms in Figure 4.14 have a peak-to-peak value of 20 $V_{pp}$. Thus, while the clamper changes the peak and average (dc) values of a waveform, it does not change the peak-to-peak value of the original signal.

### 4.3.2    Clamper Operation

The operation of a clamper is illustrated in Figure 4.15. As you can see, the clamper is similar (in construction) to a shunt clipper; the difference is the added capacitor in the clamper.

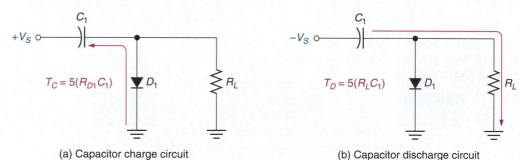

(a) Capacitor charge circuit                    (b) Capacitor discharge circuit

FIGURE 4.15    Clamper *charge* and *discharge* time constants.

Like the capacitive power supply filter, the clamper works on the basis of *switching time constants*. When the diode is forward biased, it provides a charging path for the capacitor, as shown in Figure 4.15a. For this circuit, the charging time constant is found as

$$\tau = R_{D1}C_1$$

and the approximate total charge time is found as

$$T_C = 5(R_{D1}C_1) \tag{4.7}$$

where $R_{D1}$ is the bulk resistance of the diode. When the diode is reverse biased, the capacitor starts to discharge *through the resistor*, as shown in Figure 4.15b. Thus, the discharge time constant for the circuit is found as

$$\tau = R_L C_1$$

and the approximate total discharge time is found as

$$T_D = 5(R_L C_1) \tag{4.8}$$

Note the different resistance values that appear in equations (4.7) and (4.8). As Example 4.5 demonstrates, the difference between the capacitor charge and discharge times is significant.

### EXAMPLE 4.5

Determine the capacitor charge and discharge times for the circuit represented in Figure 4.15. Assume that the forward resistance of the diode is 10 $\Omega$, $R_L = 10$ k$\Omega$, and $C_1 = 1$ $\mu$F.

*Solution:* The capacitor charges through the diode. Therefore, the charge time is found as

$$T_C = 5(R_{D1}C_1) = 5(10\ \Omega \times 1\ \mu F) = \mathbf{50\ \mu s}$$

The capacitor discharges through the resistor. Thus, the discharge time is found as

$$T_D = 5(R_LC_1) = 5(10\ k\Omega \times 1\ \mu F) = \mathbf{50\ ms}$$

As you can see, it takes a lot longer for the capacitor to discharge than it does for it to charge. This is the basis of clamper operation.

### PRACTICE PROBLEM 4.5
A clamper has values of $R_{D1} = 8\ \Omega$, $C_1 = 4.7\ \mu F$, and $R_L = 1.2\ k\Omega$. Determine the charge and discharge times for the capacitor.

Using the values obtained in Example 4.5, we will discuss the overall operation of the ◄ **OBJECTIVE 6** clamper. For ease of discussion, we will look at the circuit's response to a *square-wave* input. (When designed properly, clampers work for a variety of input waveforms.) We will also assume that the diodes are *ideal* components. The response of a clamper to a square-wave input is illustrated in Figure 4.16.

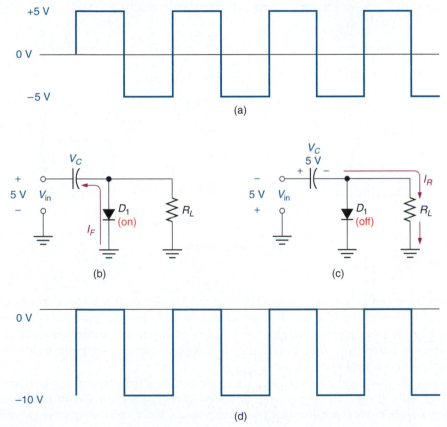

**FIGURE 4.16**

Figure 4.16a shows the input waveform. When the input goes to its positive peak (+5 V), $D_1$ is forward biased. This provides a low-resistance current path for charging $C_1$, as shown in Figure 4.16b. Assuming that the charge time of the capacitor ($5R_{D1}C_1$) is very *short*, $C_1$ rapidly charges to the full value of the peak input voltage, 5 V. With the applied voltage dropped across $C_1$, the load voltage has a positive peak value of 0 V.

When the input waveform begins to go negative, the output side of the capacitor (which is initially at 0 V) also begins to go negative. This turns $D_1$ off, and the capacitor is forced to discharge through the resistor, as shown in Figure 4.16c. Assuming that the discharge time of the capacitor $(5R_LC_1)$ is very *long*, $C_1$ will lose very little of its charge. With the input voltage and $V_C$ having the values and polarities shown in Figure 4.16c, the load voltage has a negative peak value of $-10$ V. Thus, with the input shown, the clamper provides an output square wave that varies between 0 and $-10$ V, as shown in Figure 4.16d.

Remember that the clamper illustrated in Figure 4.16 is described using the ideal diode model. In reality, the $V_F$ of the diode (0.7 V) affects the output waveform. The capacitor can charge only to the difference between $V_{\text{in(pk)}}$ and the $V_F$ of the diode. This causes the output waveform shown in Figure 4.16d to be shifted by approximately $+0.7$ V.

### 4.3.3    Negative Clampers Versus Positive Clampers

As was the case with clippers, *the difference between the negative clamper and the positive clamper is simply the direction of the diode.* This point is illustrated in Figure 4.17. Figure 4.17a shows a negative clamper and its effect on a sine-wave input. By reversing the diode, we get the circuit operation shown in Figure 4.17b. Here, the diode direction is reversed, making it a positive clamper. Since the diode is returned to ground (0 V), the circuit shifts the input waveform until the negative peak voltage of the waveform is approximately equal to 0 V.

**Lab Reference:** Both positive and negative clampers are analyzed in Exercise 5.

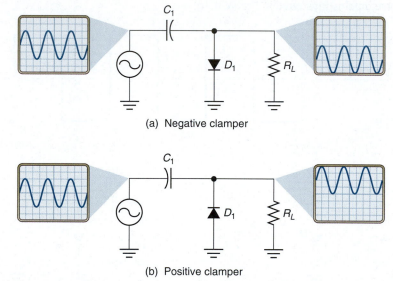

(a)  Negative clamper

(b)  Positive clamper

**FIGURE 4.17**

It is important to remember that, when reversing the diode, the capacitor (if polarized) must also be reversed. The positive side of a polarized capacitor must always be connected to the more positive voltage. In a negative clamper, the positive terminal of the capacitor is connected to the signal source. In a positive clamper, the positive side of the capacitor is connected to the output.

There is a quick and easy memory trick for determining what type of clamper you are dealing with: *If the diode is pointing up (away from ground), the circuit is a positive clamper. If the diode is pointing down (toward ground), the circuit is a negative clamper.*

### 4.3.4    Biased Clampers

**Biased clamper**

A clamper that allows a waveform to be shifted above or below a dc reference other than 0 V.

**Biased clampers** allow a waveform to be shifted so that it falls above or below a dc reference other than 0 V. Several biased clampers are shown in Figure 4.18.

The circuit shown in Figure 4.18a is easy to analyze using the procedure we just established: The diode is pointing down, so we know the circuit is a *negative clamper.* The dc reference voltage for the circuit depends on the value of the dc biasing source $(V_B)$ and the setting of the potentiometer $(R_1)$. If $V_B$ is negative, the potential applied to $D_1$ falls

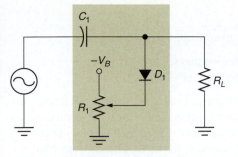

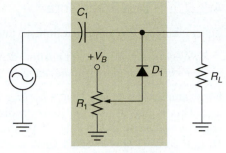

(a) Negative-biased clamper

(b) Positive-biased clamper

FIGURE 4.18  Biased clampers.

somewhere between $-V_B$ and 0 V. For example, let's say that $V_B$ is $-20$ V and $R_1$ is set so that the potential applied to $D_1$ is $-10$ V. In this case, the input waveform would be shifted so that its positive peak voltage is approximately $-10$ V.

The circuit shown in Figure 4.18b is a positive-biased clamper. It works according to the same principles as the negative-biased clamper. However, the circuit shifts its input waveform so that the negative peak of the output waveform is approximately equal to the potential applied to $D_1$. $R_1$ is adjustable, so the dc reference of the circuit can be adjusted to any voltage between $V_B$ and 0 V.

The circuits shown in Figure 4.19 require a bit more explaining. These circuits are referred to as **zener clampers**. The zener diodes are used to set the dc reference voltages of the circuits. For example, if a 10 V zener is used in Figure 4.19a, the dc reference voltage is approximately $+10$ V. If a 10 V zener is used in Figure 4.19b, the dc reference voltage for that circuit is approximately $-10$ V.

> **Zener clamper**
> A clamper that uses zener diodes to establish the dc reference voltages.

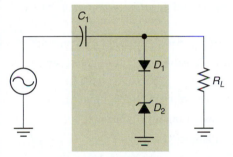

(a) Negative zener clamper

(b) Positive zener clamper

FIGURE 4.19  Zener clampers.

Why are the *pn*-junction diodes included? The operation of any clamper depends on the *one-way* conduction of its diode (as illustrated in Figure 4.16). As you learned in Chapter 2, a zener conducts in both the reverse (zener) and forward operating regions. By including $D_1$ in Figure 4.19a, the circuit allows diode conduction in one direction only, as follows:

- During the positive alternation of the input, the diode branch conducts, clamping the positive peak to $(V_Z + 0.7 \text{ V})$.

- During the negative alternation of the input, $D_1$ is reversed biased, blocking conduction through the diode branch.

By preventing zener forward conduction during the negative alternation of the input cycle, $D_1$ established the one-way conduction needed for proper clamper operation.

It is easy to tell the difference between a positive or negative clamper circuit that uses a zener diode as a voltage reference. The direction of the *pn*-junction diode determines the polarity of the clamper because the diode determines the polarity of the charge on the capacitor.

### 4.3.5    One Final Note

Zener clampers are somewhat limited; that is, they come in only two varieties:

1. *Negative* clampers with *positive* dc reference voltages.
2. *Positive* clampers with *negative* dc reference voltages.

The term *common-cathode* is used to describe components that are connected cathode-to-cathode. By the same token, the term *common-anode* is used to describe components that are connected anode-to-anode.

For example, you cannot have a *positive* zener clamper with a *positive* dc reference voltage. This is because the *pn*-junction diode and the zener diode must be connected as either *common-cathode* (Figure 4.19a) or *common-anode* (Figure 4.19b) components. If $D_1$ in Figure 4.19a is reversed, there is nothing to prevent the diode branch from conducting during the negative alternation of the input signal, eliminating the circuit clamping action. This defeats the purpose of including the diode in the first place. By the same token, reversing $D_1$ in Figure 4.19b allows the diode branch in that circuit to conduct during the positive alternation of the input, eliminating the circuit clamping action.

### 4.3.6    Summary

The clamper shifts a waveform to a predetermined dc reference voltage. The dc reference voltage for a clamper is determined by the potential to which the diode in the circuit is referenced.

When the diode in a clamper is pointing up (away from ground), the circuit is a positive clamper. The positive clamper shifts the input waveform so that its negative peak voltage is approximately equal to the dc reference voltage of the circuit. When the diode in a clamper is pointing down (toward ground), the circuit is a negative clamper. The negative clamper shifts the input waveform so that its positive peak voltage is approximately equal to the dc reference voltage of the circuit.

---

**Section Review ▶**

1. Describe the difference between *positive clampers* and *negative clampers* in terms of input/output relationships.
2. What effect does a clamper have on the value of $V_{ave}$ for a given input waveform?
3. What test equipment would you use to measure the value of $V_{ave}$ for a clamper output signal?
4. What effect does a clamper have on the rms voltage of a sine-wave input?
5. Explain the operation of a clamper in terms of switching time constants.
6. What determines the dc reference voltage of a clamper?
7. How can you tell whether a given clamper is a *positive* or *negative* clamper?
8. Why can't you have a positive zener clamper with a positive dc reference voltage?

## 4.4    Voltage Multipliers

Voltage multipliers are circuits that provide a *dc output* that is a multiple of a *peak input voltage*. For example, a *voltage doubler* provides a dc output voltage that is *twice* its peak input voltage, and so on.

*A Practical Consideration:* Since the diodes in a voltage multiplier will dissipate some power, the output power for the circuit is actually *less than* the input power.

While voltage multipliers provide an output voltage that is much greater than the peak input voltage, they are not power generators. When a voltage multiplier increases the peak input voltage by a given factor, the peak input current is *decreased* by approximately the same factor. Thus, the actual output power from a voltage multiplier is *never* greater than the input power.

Because the output current capability of a voltage multiplier is reduced every time the voltage is increased, the device normally ends up with very low current capability. As a result, voltage multipliers are usually used in high-voltage, low-current applications. One typical application is to supply the high-voltage, low-current dc voltage required to operate the cathode-ray tube (a.k.a. the picture tube) in a television.

In this section, we will take a look at several types of voltage multipliers. For ease of discussion, the multiplier circuits are shown as being driven by a simple voltage source, $V_S$. It should be noted that, in practice, multiplier circuits are generally used in power supply applications. Thus, they are usually connected to the secondary of a power supply transformer.

## 4.4.1    Half-Wave Voltage Doublers

The **half-wave voltage doubler** consists of two diodes and two capacitors. *Electrolytic* capacitors are normally used because of their high capacity ratings. A half-wave voltage doubler is shown (along with its source and load) in Figure 4.20.

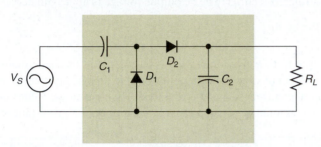

Half-wave voltage doubler

**FIGURE 4.20**

> **Half-wave voltage doubler**
> Provides a dc output voltage that is approximately twice its peak input voltage. The name is derived from the fact that the output capacitor is charged during half the input cycle.

The operation of the half-wave doubler is illustrated in Figure 4.21. During the negative alternation of the input cycle, $D_1$ is forward biased, and $D_2$ is reverse biased, as shown in Figure 4.21a. During this half cycle:

◄ **OBJECTIVE 7**

- $C_1$ charges until its plate-to-plate voltage (ideally) equals the peak value of the input signal.
- $C_2$ discharges through the load.

During the positive alternation of the input cycle, $D_1$ is reverse biased, and $D_2$ is forward biased (as shown in Figure 4.21b). During this half cycle:

- The voltage source ($V_S$) and $C_1$ are effectively connected in series with $C_2$.
- $V_S$ and $C_1$ act as *series-aiding* voltage sources, charging $C_2$ until its plate-to-plate voltage is approximately equal to $2V_{S(pk)}$.

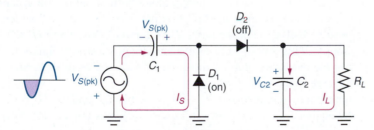

(a)  $C_1$ charges and $C_2$ discharges during the negative alternation of the input.

> **Lab Reference:** Half-wave voltage doubler operation is demonstrated in Exercise 5.

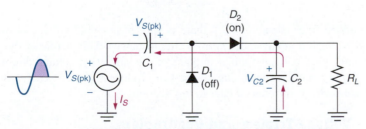

(b)  The source and $C_1$ charge $C_2$ during the positive alternation of the input.

**FIGURE 4.21**

> What prevents the capacitor from discharging rapidly?

When $V_S$ returns to its original polarity, $D_2$ is again turned off. With $D_2$ off, the only discharge path for $C_2$ is through the load resistance. Normally, the time constant of this circuit is such that $C_2$ has little time to lose any of its charge before the input reverses polarity again. In other words, during the negative alternation of the input, $C_2$ discharges *slightly*. Then, during the positive alternation, $D_2$ is turned on, and $C_2$ is recharged until its plate-to-plate voltage again equals $2V_{S(pk)}$.

Since $C_2$ barely discharges between input cycles, the output waveform of the half-wave voltage doubler closely resembles that of a filtered half-wave rectifier. Typical input and output waveforms for a half-wave voltage doubler are shown in Figure 4.22. As the figure shows, the circuit has a dc output voltage and ripple voltage that closely resemble the output from a filtered rectifier. The dc output voltage is approximated as

$$V_{dc} \cong 2V_{S(pk)} \qquad (4.9)$$

The output ripple from a given multiplier is calculated in the same manner as for a filtered half-wave rectifier. As such, the amount of ripple depends primarily on the values of the capacitors and the current demand of the load. High capacitor values and low current demands reduce the amount of ripple present at the output of a given multiplier.

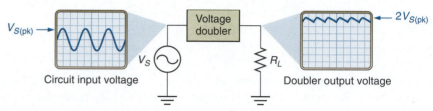

FIGURE 4.22

> Reversing the directions of the diodes and capacitors reverses the polarity of the output from the half-wave doubler.

If the directions of the diodes in Figure 4.20 are reversed, along with the polarity of the two capacitors, we then have a negative voltage doubler. The circuit operates in the same way as the positive doubler except that $C_1$ charges to a negative voltage, which adds to the incoming negative peak voltage. The circuit output is a negative voltage at twice the value of $-V_{S(pk)}$.

### 4.4.2    Full-Wave Voltage Doublers

◀ OBJECTIVE 8 ▶

**Full-wave voltage doubler**
Provides a dc output voltage that is approximately twice its peak input voltage. The name is derived from the fact that the output capacitors are charged during alternate half-cycles of the input.

The **full-wave voltage doubler** closely resembles the half-wave doubler. It contains two diodes and two capacitors ($C_1$ and $C_2$), as shown in Figure 4.23a. In the figure, the voltage source and load resistance are both shown, along with an added filter capacitor, $C_3$. This filter capacitor is used to reduce the ripple output from the voltage doubler.

During the positive half-cycle of the input, $D_1$ is forward biased and $D_2$ reverse biased by the voltage source. This gives us the equivalent circuit shown in Figure 4.23b. Again, we have idealized the diodes to simplify the circuit. Using the equivalent circuit, you can see that $C_1$ will charge to the value of $V_{S(pk)}$.

When the input polarity is reversed, $D_1$ is reverse biased, and $D_2$ is forward biased. This gives us the equivalent circuit shown in Figure 4.23c. As this circuit shows, $C_2$ now charges to the value of $V_{S(pk)}$. Since $C_1$ and $C_2$ are in series, the total voltage across the two components is found as

$$V_{dc} \cong 2V_{S(pk)}$$

With the added filter capacitor ($C_3$), there is very little ripple voltage at the output of the full-wave voltage doubler under normal circumstances. This is one of the advantages of using the full-wave voltage doubler in place of the half-wave voltage doubler. Another advantage of this circuit is that it can be used to produce a *dual-polarity power supply*. This application will be introduced later in this section.

### 4.4.3    Voltage Triplers and Quadruplers

*Voltage triplers* and *voltage quadruplers* are both variations on the basic half-wave voltage doubler. The schematic diagram for the voltage tripler is shown in Figure 4.24a.

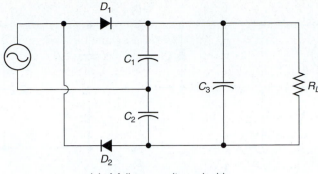

(a) A full-wave voltage doubler

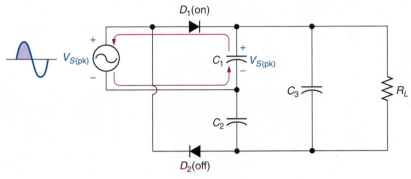

(b) $C_1$ is charged (via $D_1$) during the positive alternation of the input cycle.

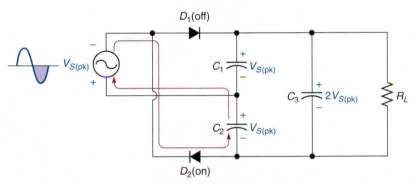

(c) $C_2$ is charged (via $D_2$) during the negative alternation of the input cycle.

**FIGURE 4.23**   Full-wave voltage doubler operation.

As you can see, the **voltage tripler** is very similar to the half-wave voltage doubler shown in Figure 4.21. In fact, the $D_1$, $D_2$, $C_1$, and $C_2$ circuitry forms a half-wave doubler. The added components ($D_3$, $C_3$, and $C_4$) form the rest of the voltage tripler.

There are two keys to the operation of this circuit. The first is that the half-wave voltage doubler works exactly as described earlier in this section. The second is that $D_3$ conducts whenever $D_1$ conducts.

When $V_S$ is negative, as shown in Figure 4.24b, $D_1$ and $D_3$ both conduct, allowing $C_1$ and $C_3$ to charge up to $V_{S(pk)}$. Figure 4.24c shows what happens when $V_S$ goes positive: $D_2$ is turned on, allowing $C_2$ to charge up to $2V_{S(pk)}$. The voltages across $C_2$ and $C_3$ now add up to $3V_{S(pk)}$.

$C_4$ is a filter capacitor that is added to reduce the ripple in the dc output voltage. Since this capacitor is in parallel with the $C_2/C_3$ series circuit, it charges to a plate-to-plate voltage equal to $3V_{S(pk)}$. Note that the value of $C_4$ is chosen so that the time constant formed with the load resistance is extremely long. Thus, $C_4$ barely discharges between charging cycles.

The **voltage quadrupler** provides a dc output that is four times its peak input voltage. As Figure 4.25 shows, the voltage quadrupler is simply made up of two half-wave voltage

**Voltage tripler**
Produces a dc output voltage that is three times its peak input voltage.

**Voltage quadrupler**
Produces a dc output voltage that is four times its peak input voltage.

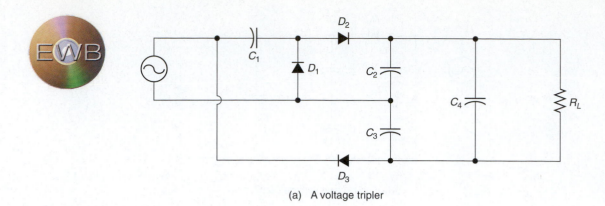

(a) A voltage tripler

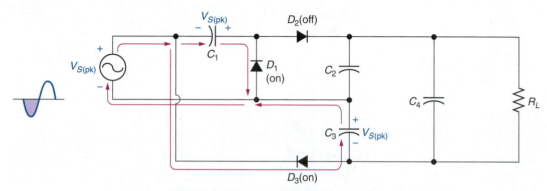

(b) $C_1$ and $C_3$ are charged during the negative alternation of the input cycle.

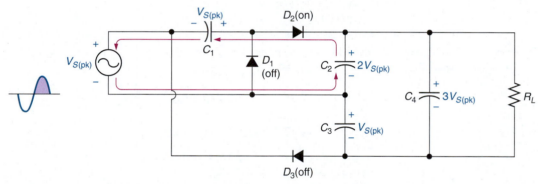

(c) $C_2$ is charged during the positive alternation of the input cycle.

FIGURE 4.24 Voltage tripler operation.

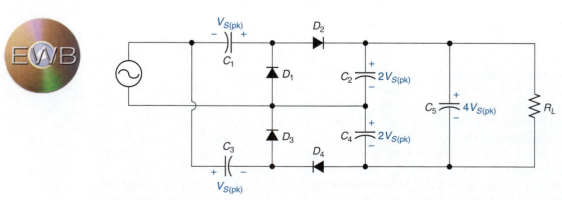

FIGURE 4.25

doublers. The half-wave doubler that is made of $D_1$, $D_2$, $C_1$, and $C_2$ charges $C_2$ until its plate-to-plate voltage is equal to $2V_{S(pk)}$. The half-wave doubler formed by the other components charges $C_4$ until its plate-to-plate voltage is also equal to $2V_{S(pk)}$. The series combination of $C_2$ and $C_4$ charges the filter capacitor ($C_5$) until its plate-to-plate voltage is equal to $4V_{S(pk)}$. Again, the value of $C_5$ is chosen so that it retains most of its charge between charging cycles.

### 4.4.4 A Dual-Polarity Power Supply

One application for the voltage multiplier can be seen in a basic dual-polarity dc power supply. A **dual-polarity power supply** is one that provides both positive and negative dc output voltages. One such supply is shown in Figure 4.26.

**Dual-polarity power supply**
A dc supply that provides both positive and negative dc output voltages.

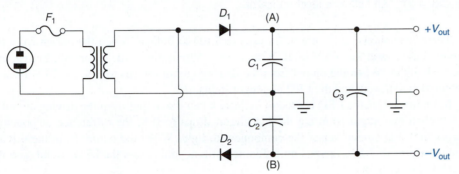

FIGURE 4.26

The transformer in the figure is connected to a full-wave voltage doubler. The common point for the capacitors ($C_1$ and $C_2$) is used as the power supply ground, meaning that all circuit voltages are measured with respect to this point. Thus, point A will be *positive* with respect to ground, and point B will be *negative* with respect to ground. Note that the two dc output voltages will be approximately equal in magnitude to $V_{S(pk)}$. For example, if $V_{S(pk)}$ is 12 V, the power supply will have outputs approximately equal to $+12$ $V_{dc}$ and $-12$ $V_{dc}$. The filter capacitor ($C_3$) is added to reduce the ripple in the dc output voltages.

A dual-voltage supply such as this is used only in low-current, noncritical applications where exact voltages are not a concern. A practical, dual-polarity power supply is more complicated and includes more filtering and some type of voltage regulation for each polarity. You will see one such power supply in Chapter 21.

1. Describe the operation of the half-wave voltage doubler.

2. Describe the operation of the full-wave voltage doubler.

3. Why are full-wave voltage doublers preferred over half-wave voltage doublers?

4. What is the similarity between the voltage tripler and the half-wave voltage doubler?

5. What is the similarity between the voltage quadrupler and the half-wave voltage doubler?

◀ **Section Review**

## 4.5 LED Applications

The most common application for the LED is as a *power indicator* that can replace a neon or incandescent lamp. The power indicator that lights on the front of your stereo whenever the system is on is probably an LED.

Another common application for the LED is as a *level indicator* in switching circuits. Given a circuit whose output will always be at either a *high dc voltage level* or a *low dc voltage level*, the LED can be used to indicate the output *state* at any given time. This application is illustrated in Figure 4.27.

◀ *OBJECTIVE 9*

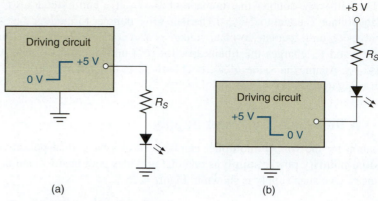

(a)                                                  (b)

FIGURE 4.27    An LED as a level indicator.

The circuit shown in Figure 4.27a uses an LED to indicate when the output from the driving circuit is at the +5 V level. With a +5 V output, the diode is forward biased, and the LED lights. When the output from the driving circuit returns to 0 V, the LED biasing potential is removed, and the LED no longer lights.

In Figure 4.27b, the LED is used to indicate when the output from the driving circuit is 0 V. When the output from the driving circuit drops to 0 V, the difference of potential across the LED circuit causes the component to light. When the output from the driving circuit returns to +5 V, there is no difference of potential across the LED circuit, and the LED no longer lights.

## 4.5.1    Multisegment Displays

OBJECTIVE 10 ▶

LEDs are used as the primary components in *multisegment displays*. Multisegment displays are used to display alphanumeric characters. One type of multisegment display is the **seven-segment display** shown in Figure 4.28. The seven-segment display uses seven individual LEDs that are arranged in a figure-8 configuration. By lighting a combination of individual LEDs, any number from 0 to 9, along with several letters, can be displayed. For example, if LEDs b, c, f, and g are lit, the display will show the number 4.

**Seven-segment display**
A device made up of seven LEDs shaped in a figure-8 that is used to display numbers.

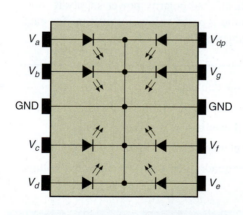

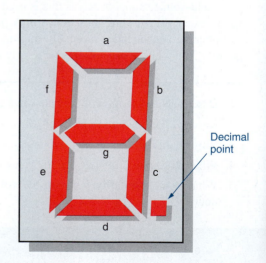

FIGURE 4.28

**Common-cathode display**
A display in which all LED cathodes are tied to a single pin.

The seven-segment display shown in Figure 4.28 is called a **common-cathode display**. This term means that the cathodes of the LEDs are tied together. Because of this, the display can function with only one ground connection. The individual LEDs are lit by applying a *high* voltage at the appropriate pins. Normally, this high voltage is within the range of +5 to +10 V. Note that each LED in the seven-segment display must have its own series current-limiting resistor.

Another type of display is the **common-anode display**. This type of display has a single +$V$ input that is common to all the LEDs. Then individual LEDs are lit by applying a *ground* to the appropriate cathode.

Variations on the basic seven-segment display include those with added LEDs and built-in decoders. Figure 4.29a shows a display that allows any letter or number to be produced. Figure 4.29b represents a 5 × 7 dot matrix readout that can display the digits 0 to 9, the alphabet, and other symbols by turning on a combination of the individual LEDs.

**Common-anode display**
A display in which all LED anodes are tied to a single pin.

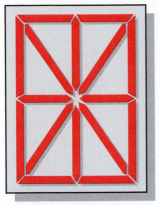

(a) 16-Segment alphanumeric display

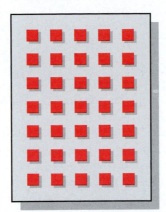

(b) 5 × 7 Dot matrix display

FIGURE 4.29

## 4.5.2 Liquid-Crystal Displays (LCDs)

**Liquid-crystal displays (LCDs)** are low-power displays that are used in all kinds of electronic systems, from digital watches and calculators to laptop computers and flat screen TVs. LCD displays use far less power than comparable LED displays because they block or pass the light from other sources rather than emit their own.

The simplest LCD displays consist of clear segments on a reflective background material. When a voltage is applied across a given segment, that segment does not pass (or reflect) light, giving it a dark appearance. When the voltage is removed, the segment clears, allowing the background material to reflect any ambient (or artificial) light. Characters and numbers are produced by activating the appropriate segments.

More advanced LCD displays (like those in a laptop computer) are *backlit*. These displays contain built-in fluorescent tubes that light the display surface. The average lifetime of these tubes is approximately 15,000+ hours. (At 6 hours a day, that comes to just under 7 years.)

**Liquid-crystal display (LCD)**
A display consisting of segments that reflect (or do not reflect) ambient light.

◀ Section Review

1. Describe the use of an LED as a power-level indicator.

2. Describe the use of LEDs in multisegment displays.

3. Describe the LCD and its operation.

## 4.6 Diode Circuit Troubleshooting

Now that we have covered the operating principles of clippers, clampers, voltage multipliers, and multisegment displays, we will take a look at some of the fault symptoms that commonly occur in these circuits.

## 4.6.1 Clipper Faults

The faults that normally occur in series clippers are the same as those of the half-wave rectifier that were discussed in Chapter 3. The fault–symptom relationships shown in Table 4.1 are those for the basic shunt clipper (shown in Figure 4.30) and the biased clipper.

◀ *OBJECTIVE 11*

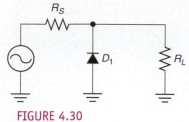

FIGURE 4.30

**TABLE 4.1** Shunt Clipper Faults

| Fault | Symptom(s) |
|---|---|
| $R_S$ open | Since the source is isolated from the load, the load voltage and current both drop to zero. |
| $D_1$ open | Clipping action is lost, so the output waveform is nearly identical in shape to the input waveform. |
| $D_1$ shorted | The symptoms are the same as those for $R_S$ open because the full applied voltage is dropped across $R_S$. The shorted diode shorts out the load. |

### 4.6.2    Clamper Faults

The fault–symptom relationships shown in Table 4.2 are those for the basic clamper shown in Figure 4.31a.

The *biased clamper* (Figure 4.31b) may develop any of the fault–symptom combinations listed in Table 4.2. In addition, it can develop any of the fault–symptom combinations given in Table 4.3.

**TABLE 4.2** Clamper Faults

| Fault | Symptom(s) |
|---|---|
| $C_1$ open | The voltage source is isolated from the rest of the circuit, so load voltage and current drop to zero. |
| $C_1$ shorted | The circuit resembles a shunt clipper without a current-limiting resistor. Either the diode will be destroyed from excessive current or the signal source will be shorted to ground and damaged. |
| $C_1$ leaky | The capacitor attempts to charge but is not able to hold the charge for any period of time. This causes the dc reference of the output signal to change constantly. The waveform itself is also extremely distorted. |
| $D_1$ open | All clamping action is lost and the output waveform is centered around 0 V. |
| $D_1$ shorted | The voltage source is shorted to ground by the short $RC$ time constant of the capacitor and the shorted diode. The output from the voltage source is loaded down, and the voltage source itself may be damaged. |

> *A Practical Consideration:* Carbon resistors do not internally short. If a short exists, the short is normally caused by a solder bridge or by accidentally using the wrong value of resistor. A visual inspection will find either of these two problems.

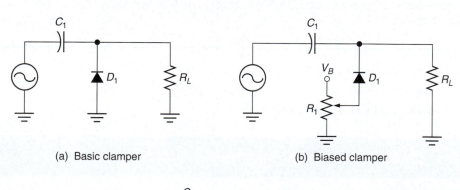

(a) Basic clamper          (b) Biased clamper

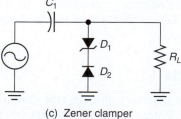

(c) Zener clamper

FIGURE 4.31

**TABLE 4.3   Additional Biased-Clamper Faults**

| Fault | Symptom(s) |
|---|---|
| $V_B$ missing | If the dc biasing voltage is missing (drops to 0 V), there is still some dc offset in the circuit output. This dc offset, which will not equal the normal dc offset for the circuit, is produced by the diode current through $R_1$. |
| $R_1$ open | If $R_1$ opens, the results are the same as for $D_1$ opening. All clamping action is lost, and the output waveform is centered around 0 V. |

Additional faults that may develop in *biased clampers*.

The *zener clamper* (Figure 4.31c) may develop any of the capacitor fault–symptom combinations listed in Table 4.2. Also, if either diode *opens*, the results will be the same as those shown in Table 4.2 for an open diode. If either diode in the zener clamper *shorts*, the results are slightly different from those listed in Table 4.2. The results of either diode shorting in the zener clamper are shown in Table 4.4.

**TABLE 4.4   Additional Zener Clamper Faults**

| Fault | Symptom(s) |
|---|---|
| $D_1$ shorted | If $D_1$ shorts, the clamper acts as an unbiased clamper. As you can see, shorting $D_1$ in Figure 4.31c gives us the exact same circuit as the one shown in Figure 4.31a. |
| $D_2$ shorted | The zener diode turns on during both alternations of the input waveform. The result is an extremely distorted output waveform. |

Additional faults that may develop in the *zener clamper*.

As always, there is a strong possibility that both diodes in the zener clamper will be destroyed if either diode fails. You may recall from our discussion on rectifier circuits that it is not uncommon for a failing diode to destroy any series diodes in the process. The best thing to do whenever you find a faulty diode is to replace any diodes in series with the component.

### 4.6.3   Voltage-Multiplier Faults

Of all the circuits covered in this chapter, voltage multipliers are by far the most difficult to troubleshoot. The maze of diodes and capacitors that make up the voltage multiplier can develop many faults, each having symptoms that may closely resemble those of some other fault.

The best approach to troubleshooting the voltage multiplier is to start by reading the voltage across the output filter capacitor. The voltage across the filter capacitor may help to isolate the area where the fault is located. For example, let's say that we are testing the voltage tripler shown in Figure 4.32 and that the voltage across the filter capacitor ($C_4$) is close to $2V_{S(pk)}$. This would indicate that $C_3$ is not charging since the voltage across this component is usually equal to $V_S$. We would then test $D_3$ and $C_3$, the two components in the $C_3$ charging path.

◀ *OBJECTIVE 12*

*A Practical Consideration:* When working with voltage multipliers, you should be extremely careful because of the high voltages that may be present within the circuit. All capacitors should be discharged before any component is removed from the circuit. When replacing capacitors in voltage multipliers, the replacement capacitors must have the same or higher voltage ratings as the originals. Using a lower-rated capacitor (or reversing the capacitor polarity) can result in circuit damage and/or injury to you.

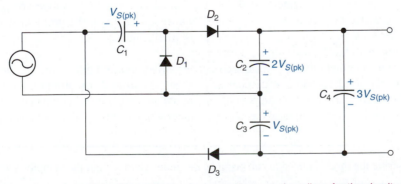

*Note:* The voltages shown are the normal readings for the circuit.

**FIGURE 4.32**

If the voltage across $C_4$ were approximately equal to $V_{S(pk)}$, we would then check the voltage across $C_2$ since the voltage across this component is normally equal to $2V_{S(pk)}$. If the voltage across the component was not correct, we would check the voltage across $C_1$. If the voltage across $C_1$ was correct, we would test $D_2$ and $C_2$. If the voltage across $C_1$ was not correct, we would test $D_1$ and $C_1$.

When testing voltage multipliers, it helps if you have a quick reference showing you the normal capacitor voltage readings for the various multiplier circuits. Such a reference is supplied in Table 4.5.

*A Practical Consideration:*
Don't expect the capacitor voltages to be exactly equal to the peak values we have assumed in our discussions. The diodes each drop some voltage, and the capacitors are actually charging and discharging continually. However, the average capacitor voltages should be close to the appropriate multiples of $V_{S(pk)}$ that you were shown earlier.

TABLE 4.5   Normal Capacitor Voltage Readings for Various Multiplier Circuits

| Capacitor | Half-Wave Doubler | Full-Wave Doubler | Tripler | Quadrupler |
|---|---|---|---|---|
| $C_1$ | $V_{S(pk)}$ | $V_{S(pk)}$ | $V_{S(pk)}$ | $V_{S(pk)}$ |
| $C_2$ | $2V_{S(pk)}$ | $V_{S(pk)}$ | $2V_{S(pk)}$ | $2V_{S(pk)}$ |
| $C_3$ | | $2V_{S(pk)}$ | $V_{S(pk)}$ | $V_{S(pk)}$ |
| $C_4$ | | | $3V_{S(pk)}$ | $2V_{S(pk)}$ |
| $C_5$ | | | | $4V_{S(pk)}$ |

When testing a voltage multiplier, use the listing shown in Table 4.5 and the following procedure:

1. *Measure the voltage across the output capacitor. The reading may direct you to test one of the parallel charging capacitors. For example, if the voltage across $C_4$ (Figure 4.32) is low by an amount equal to $V_S$, you would check $C_3$. If it is low by an amount equal to $2V_{S(pk)}$, you would check $C_2$.*
2. Measure the voltage across any capacitor that lies in the charge path of an uncharged capacitor. For example, if $C_2$ (Figure 4.32) is not charged, you would measure the voltage across $C_1$.
3. Continue the process until you find the uncharged capacitor that is closest to the source. Then test all the components in the charge path for that capacitor. For example, if $C_2$ (Figure 4.32) is not charged and the voltage across $C_1$ is normal, check $D_2$ and $C_2$. If the voltages across $C_1$ and $C_2$ are both wrong, check $D_1$ and $C_1$.

If none of the capacitor charges is correct, and there is no problem with either $C_1$ or $D_1$, the voltage source is most likely the problem.

## 4.6.4   Display Faults

**Decoder–driver**
An IC used to drive a multisegment display.

Multisegment displays are driven by ICs called **decoder–drivers**. The most common symptom of a fault in the driver-display circuit is the failure of one or more segments to light.

The testing of a driver-display circuit is simple. When a segment should be *on*, check the driving pin for that segment. If the voltage there is correct (a positive voltage for a *common-cathode* display and 0 V for a *common-anode* display), the problem is the display. If the voltage is not there, the problem is likely the decoder–driver. This assumes that the inputs to the decoder–driver are correct.

Some multisegment displays have the decoder–driver circuit built within the unit. This eliminates the need for external decoder and driver circuits and for current-limiting resistors. When testing this type of component, the inputs are checked and compared to the output of the display. If the output of the display is incorrect, the entire unit should be replaced.

**Section Review ▶**

1. Describe the diode faults that can occur in the zener clamper and the symptoms of each.
2. Why should you replace both diodes in a zener clamper if one of them shorts?
3. Describe the process used to troubleshoot a voltage multiplier.

Here is a summary of the major points made in this chapter:

1. A *clipper* (or *limiter*) is a circuit used to eliminate a portion of an ac signal.
2. A *clamper* (or *dc restorer*) is a circuit used to change the dc reference of an ac signal.
3. A *voltage multiplier* is a circuit used to produce a dc output voltage that is some multiple of an ac peak input voltage.
4. A *multisegment display* is an LED circuit used to display *alphanumeric symbols* (numbers, letters, punctuation marks, etc.).
5. A *series clipper* contains a diode that is positioned in series between a source and its load.
   a. The circuit input passes through to the output when the diode is forward biased (conducting).
   b. When the diode is reverse biased by the input signal, it does not conduct. In this case, there is no voltage developed across the load.
6. A *negative* series clipper eliminates the negative portion of its input signal. A *positive* series clipper eliminates the positive portion of its input signal (see Figure 4.2).
7. A *shunt* clipper contains a diode that is in parallel with the load.
   a. When the diode is reverse biased (not conducting), the output waveform resembles the input waveform.
   b. When the diode is forward biased (conducting), the load is shorted. Therefore, the voltage across the load equals the drop across the diode.
8. Shunt clipper operation is illustrated in Figure 4.3. The series *current-limiting* resistor ($R_S$) is included to prevent the diode from shorting the signal source to ground when forward biased.
9. A *biased clipper* is a shunt clipper that uses a dc voltage source to bias the diode. This allows the circuit to clip the input at values other than the diode $V_F$.
10. A comparison of the various clipper circuits is provided in Figure 4.11.
11. Clippers are commonly used to:
    a. Alter the shape of a waveform.
    b. Protect a circuit from *transients*.
12. A *transient* is an abrupt current or voltage spike that has an extremely short duration. Transients can seriously damage any circuit not designed to handle them.
13. Several transient protection circuits are shown in Figure 4.12.
14. An *AM detector* is a diode clipper that converts a varying amplitude ac input to a varying dc level. AM detector operation is illustrated in Figure 4.13.
15. A *positive clamper* shifts its entire input signal *above* a dc reference voltage.
16. A *negative clamper* shifts its entire input signal *below* a dc reference voltage.
17. Clamper input/output relationships are illustrated in Figure 4.14.
18. A clamper changes the peak and dc average values of its input waveform. However, it does not affect the waveform's rms and peak-to-peak values.
19. Clampers work on the principle of *switching time constants* (see Figure 4.15).
    a. The capacitor charge time is extremely short when compared to the input cycle time.
    b. The capacitor discharge time is extremely long when compared to the input cycle time.
20. Clamper operation is illustrated in Figures 4.15 and 4.16.
21. *Biased clampers* allow us to shift a waveform above (or below) a dc reference other than the approximate value of 0 V. Several biased clampers are shown in Figure 4.18.
22. Zener clampers use a zener diode in series with the *pn*-junction diode to establish a dc reference voltage (that is approximately equal to $V_Z$). Several zener clampers are shown in Figure 4.19.
23. There are only two types of zener clampers:
    a. Negative clampers with positive dc reference voltages.
    b. Positive clampers with negative dc reference voltages.
24. The term *common-cathode* is often used to describe two diodes that are connected cathode-to-cathode. The term *common-anode* is used to describe two diodes that are connected anode-to-anode.

25. A *voltage multiplier* provides a dc output voltage that is some multiple of its peak input voltage.
   a. As voltage is increased by a given factor, current is decreased by approximately the same factor.
   b. Voltage multipliers are generally used in high-voltage, low-current applications.
   c. Since all the diodes in a voltage multiplier use some amount of power, the output power from a multiplier is always lower than its input power.
26. A *half-wave voltage doubler* provides a dc output that is approximately twice its peak input voltage. The name is derived from the fact that the output capacitor is charged during one-half of the circuit input cycle.
27. The operation of a half-wave voltage doubler is illustrated in Figure 4.21.
28. A *full-wave voltage doubler* uses two capacitors to produce an output that is approximately twice its peak input voltage. The name is derived from the fact that the two series capacitors are charged during alternate half-cycles of the circuit input cycle.
29. The configuration of a full-wave voltage doubler allows a *filter capacitor* to be placed across the two series capacitors. This filter capacitor reduces the ripple in the circuit output.
30. Full-wave voltage doubler operation is illustrated in Figure 4.23.
31. A *voltage tripler* produces a dc output voltage that is approximately three times its peak input voltage. Voltage tripler operation is illustrated in Figure 4.24.
32. A *voltage quadrupler* produces a dc output voltage that is approximately four times its peak input voltage. A voltage quadrupler can be made using two half-wave voltage doublers, as shown in Figure 4.25.
33. LEDs are commonly used as *power indicators* and *level indicators*.
   a. A power indicator lights whenever a system (like a stereo) is turned on.
   b. A level indicator is used to indicate whether a dc voltage is at one level or another. Several level indicators are shown in Figure 4.27.
34. LEDs are used as the primary components in *multisegment displays*.
   a. A *seven-segment display* (Figure 4.28) is made up of seven LEDs shaped in a figure-8.
   b. A *common-cathode* display is one in which all LED cathodes are tied to a single pin.
   c. A *common-anode* display is one in which all LED anodes are tied to a single pin.
35. A *liquid-crystal display* (LCD) is a low-power display made of segments that can be made to pass or block light.
   a. When no voltage is applied to a segment, that segment is clear, allowing light to pass through the segment.
   b. When a voltage is applied to a segment, that segment darkens, blocking light.
36. LCDs are backlit (like those in a laptop computer) or simply reflective (like those in a digital watch).
37. The fault symptoms for *series clippers* are the same as those given in Chapter 3 for half-wave rectifiers (see Section 3.9).
38. The fault symptoms for *shunt clippers* are listed in Table 4.1.
39. The fault symptoms for *clampers* are listed in Tables 4.2 through 4.4.
40. Care must be taken when troubleshooting voltage multipliers since many faults can prevent the circuit capacitors from discharging.
41. The normal capacitor voltages for various multiplier circuits are listed in Table 4.5.
42. Most multisegment displays are driven by external ICs called *decoder–drivers*.

## EQUATION SUMMARY

| Equation Number | Equation | Section Number |
|---|---|---|
| **(4.1)** | $V_L = V_{in} - 0.7 \text{ V}$ | 4.1 |
| **(4.2)** | $V_{D1} = V_{in}$ | 4.1 |

| Equation Number | Equation | Section Number |
|:---:|:---:|:---:|
| (4.3) | $V_L = 0 \text{ V}$ | 4.1 |
| (4.4) | $V_L = \dfrac{R_L}{R_L + R_S} V_{in}$ | 4.1 |
| (4.5) | $V_L = -V_F$ | 4.1 |
| (4.6) | $V_{RS} = -V_{in} + 0.7 \text{ V}$ | 4.1 |
| (4.7) | $T_C = 5(R_{D1}C_1)$ | 4.3 |
| (4.8) | $T_D = 5(R_L C_1)$ | 4.3 |
| (4.9) | $V_{dc} \cong 2V_{S(pk)}$ | 4.4 |

## KEY TERMS

alphanumeric
  symbols 136
AM detector 144
biased clamper 148
biased clipper 141
clamper 136
clipper 136
common-anode
  display 157
common-cathode
  display 156
dc reference
  voltage 136
dc restorer 136

decoder–driver 160
dual-polarity power
  supply 155
full-wave voltage
  doubler 152
half-wave voltage
  doubler 151
limiter 136
liquid-crystal
  display 157
multisegment
  display 136
negative-biased
  clipper 141

negative clamper 145
positive-biased
  clipper 141
positive clamper 145
series clipper 136
seven-segment
  display 156
shunt clipper 137
transient 143
voltage multiplier 136
voltage quadrupler 153
voltage tripler 153
zener clamper 149

## PRACTICE PROBLEMS

### Section 4.1

1. Determine the positive peak load voltage for the circuit shown in Figure 4.33a.

2. Draw the output waveform for the circuit shown in Figure 4.33a. Label the peak voltage values on the waveform.

3. Determine the negative peak load voltage for the circuit shown in Figure 4.33b.

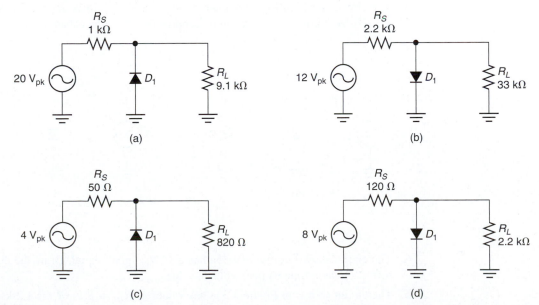

FIGURE 4.33

4. Draw the output waveform for the circuit shown in Figure 4.33b. Label the peak voltage values on the waveform.

5. Determine the peak voltage values for the output from Figure 4.33c. Then, draw the waveform, and include the voltage values in the drawing.

6. Repeat Problem 5 for the circuit shown in Figure 4.33d.

7. The potentiometer in Figure 4.34a is set so that the anode of $D_1$ is at $-2$ V. If the value of $V_S$ is 14 $V_{pp}$, what are the peak load voltages for the circuit?

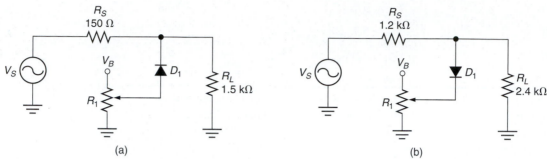

FIGURE 4.34

8. The potentiometer in Figure 4.34a is set so that the anode voltage of $D_1$ is $-8$ V. Assuming that $V_S = 24$ $V_{pp}$, determine the peak load voltages for the circuit, and draw the output waveform.

9. The potentiometer in Figure 4.34b is set so that the cathode voltage of $D_1$ is $+4$ V. Assuming that $V_S = 22$ $V_{pp}$, determine the peak load voltages for the circuit, and draw the output waveform.

10. The potentiometer in Figure 4.34b is set so that the cathode voltage of $D_1$ is $+2$ V. Assuming that $V_S = 4$ $V_{pp}$, determine the peak load voltages for the circuit, and draw the output waveform. *(Be careful on this one!)*

### Section 4.3

11. A clamper with a 24 $V_{pp}$ input shifts the waveform so that its peak voltages are 0 and $-24$ V. Determine the dc average of the output waveform.

12. A clamper with a 14 $V_{pp}$ input shifts the waveform so that its peak voltages are 0 and $+14$ V. Determine the dc average of the output waveform.

13. The circuit shown in Figure 4.35a has values of $R_{D1} = 24$ Ω, $R_L = 2.2$ kΩ, and $C_1 = 4.7$ μF. Determine the charge and discharge times for the capacitor.

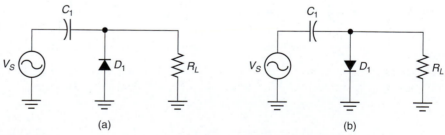

FIGURE 4.35

14. The circuit shown in Figure 4.35a has a 14 $V_{pp}$ input signal. Draw the output waveform, and determine its peak voltage values.

15. The circuit shown in Figure 4.35a has values of $R_{D1} = 8$ Ω, $R_L = 1.2$ kΩ, and $C = 33$ μF. Determine the charge and discharge times for the capacitor.

**16.** The circuit shown in Figure 4.35b has values of $R_{D1} = 14 \, \Omega$, $R_L = 1.5 \, \text{k}\Omega$, and $C = 1 \, \mu\text{F}$. Determine the charge and discharge times for the capacitor.

**17.** The input to the circuit shown in Figure 4.35b is a 12 $V_{pp}$ signal. Determine the peak load voltages for the circuit, and draw the output waveform.

**18.** The potentiometer in Figure 4.36a is set so that the voltage at the anode of $D_1$ is $+3$ V. The value of $V_S$ is 9 $V_{pp}$. Draw the output waveform, and determine its peak voltage values.

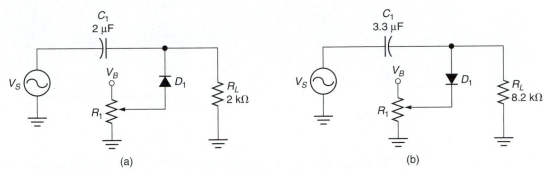

(a)    (b)

**FIGURE 4.36**

**19.** The potentiometer in Figure 4.36b is set so that the cathode voltage of $D_1$ is $+6$ V. The value of $V_S$ is 30 $V_{pp}$. Draw the output waveform for the circuit, and determine its peak voltage values.

**20.** Draw the output waveform for the circuit shown in Figure 4.37a, and determine its peak voltage values.

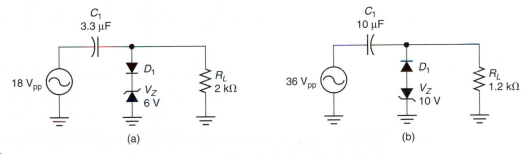

(a)    (b)

**FIGURE 4.37**

**21.** Draw the output waveform for the circuit shown in Figure 4.37b, and determine its peak voltage values.

**Section 4.4**

**22.** The circuit shown in Figure 4.38 has a 15 $V_{pk}$ input signal. Determine the values of $V_{C1}$ and $V_{C2}$ for the circuit. Assume that the diodes are ideal components.

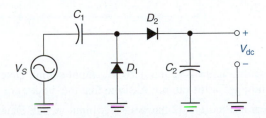

**FIGURE 4.38**

**23.** The circuit shown in Figure 4.38 has a 48 $V_{pk}$ input signal. Determine the values of $V_{C1}$ and $V_{C2}$ for the circuit. Assume that the diodes are ideal components.

**24.** The circuit shown in Figure 4.39 has a 24 $V_{pk}$ input signal. Determine the values of $V_{C1}$, $V_{C2}$, and $V_{C3}$ for the circuit. Assume that the diodes are ideal components.

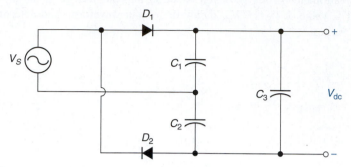

FIGURE 4.39

**25.** The circuit shown in Figure 4.39 has a 25 $V_{ac}$ input signal. Determine the values of $V_{C1}$, $V_{C2}$, and $V_{C3}$ for the circuit. Assume that the diodes are ideal components. *(Be careful on this one!)*

**26.** Determine the dc output voltage for the circuit shown in Figure 4.40. Assume that the diodes are ideal components.

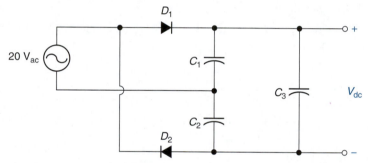

FIGURE 4.40

**27.** Determine the dc output voltage for the circuit shown in Figure 4.41. Assume that the diodes are ideal components.

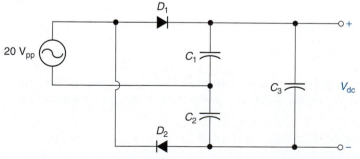

FIGURE 4.41

**28.** The circuit shown in Figure 4.42 has a 20 $V_{pk}$ input signal. Determine the values of $V_{C1}$, $V_{C2}$, $V_{C3}$, and $V_{C4}$ for the circuit. Assume that the diodes are ideal components.

**29.** The circuit shown in Figure 4.42 has a 15 $V_{ac}$ input signal. Determine the values of $V_{C1}$, $V_{C2}$, $V_{C3}$, and $V_{C4}$ for the circuit. Assume that the diodes are ideal components.

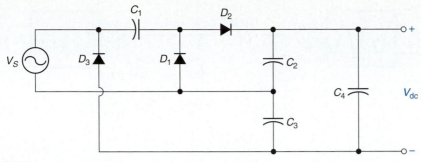

FIGURE 4.42

30. The circuit shown in Figure 4.43 has an 18 $V_{pk}$ input signal. Determine the values of $V_{C1}$, $V_{C2}$, $V_{C3}$, $V_{C4}$, and $V_{C5}$ for the circuit. Assume that the diodes are ideal components.

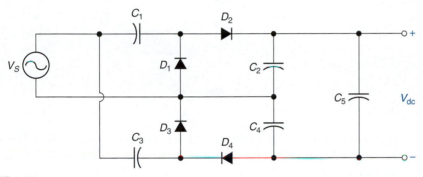

FIGURE 4.43

31. The circuit shown in Figure 4.43 has a 36 $V_{ac}$ input signal. Determine the values of $V_{C1}$, $V_{C2}$, $V_{C3}$, $V_{C4}$, and $V_{C5}$ for the circuit. Assume that the diodes are ideal components.

32. Determine the dc output voltage values for the dual-polarity power supply in Figure 4.44. Assume that the diodes are ideal components.

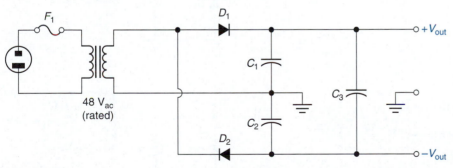

FIGURE 4.44

33. Assume that the transformer in Figure 4.44 is a 36 $V_{ac}$ transformer and that the diodes in the circuit are ideal components. Determine the dc output voltage values for the circuit.

---

34. The circuit shown in Figure 4.45a has the input/output waveforms illustrated. Determine whether there is a problem in the circuit. If there is, discuss the possible cause(s) of the problem.

**TROUBLESHOOTING
PRACTICE PROBLEMS**

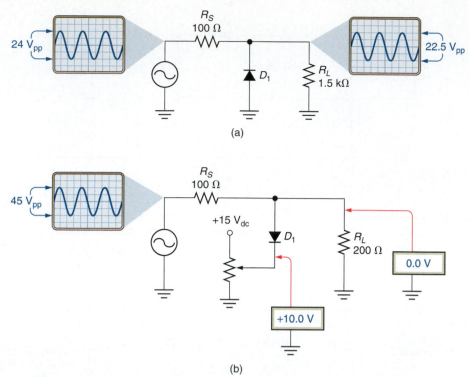

(a)

(b)

**FIGURE 4.45**

**35.** The circuit shown in Figure 4.45b has the input/output values shown. Determine whether there is a problem in the circuit. If there is, discuss the possible cause(s) of the problem.

**36.** The circuit shown in Figure 4.46a has the input/output waveforms illustrated. Determine whether there is a problem in the circuit. If there is, discuss the possible cause(s) of the problem.

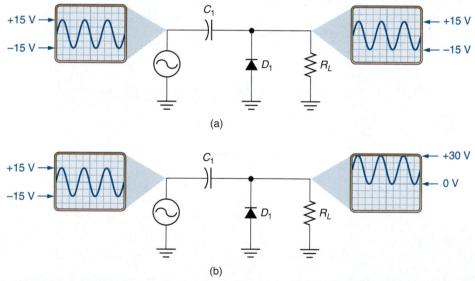

(a)

(b)

**FIGURE 4.46**

**37.** The circuit shown in Figure 4.46b has the input/output waveforms illustrated. Determine whether there is a problem in the circuit. If there is, discuss the possible cause(s) of the problem.

**38.** The circuit shown in Figure 4.47a has the input/output waveforms illustrated. Determine whether there is a problem in the circuit. If there is, discuss the possible cause(s) of the problem.

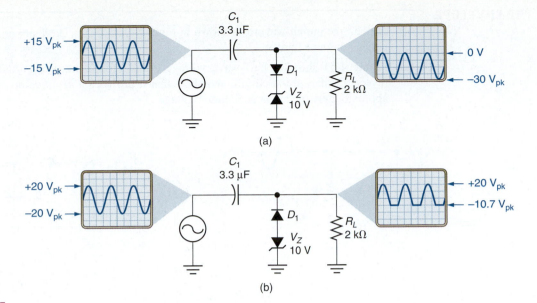

(a)

(b)

**FIGURE 4.47**

39. The circuit shown in Figure 4.47b has the input/output waveforms illustrated. Determine whether there is a problem in the circuit. If there is, discuss the possible cause(s) of the problem.

40. The circuit shown in Figure 4.48 has the capacitor voltages illustrated. Determine whether there is a problem in the circuit. If there is, list the steps you would take to find the faulty component.

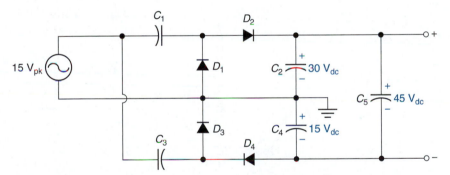

**FIGURE 4.48**

41. The circuit shown in Figure 4.49 has the capacitor voltages illustrated. Determine whether there is a problem in the circuit. If there is, list the steps you would take to find the faulty component.

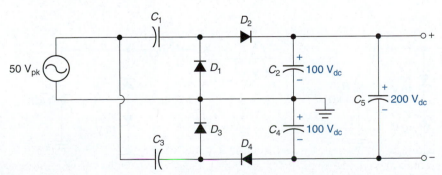

**FIGURE 4.49**

**42.** Explain the output waveform shown in Figure 4.50. (*Note:* There are no faulty components in the circuit.)

**43.** Figure 4.51 shows a half-wave voltage doubler with the diodes and capacitors reversed. Analyze the circuit, and show that it provides a *negative* dc output that is approximately twice the peak input voltage.

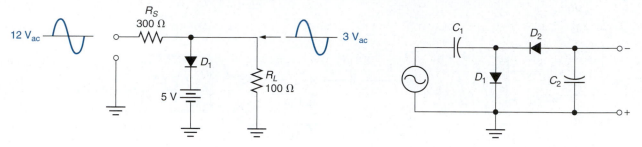

FIGURE 4.50                    FIGURE 4.51

**44.** Schematics can be drawn in many different ways. For example, the circuit shown in Figure 4.52 is actually a voltage quadrupler. However, one of the components has accidentally been reversed. Which one is it? Explain your answer.

**45.** The circuit shown in Figure 4.53 is a *zener clipper.* Using your knowledge of diode operation, determine the shape of the circuit output waveform and its peak values.

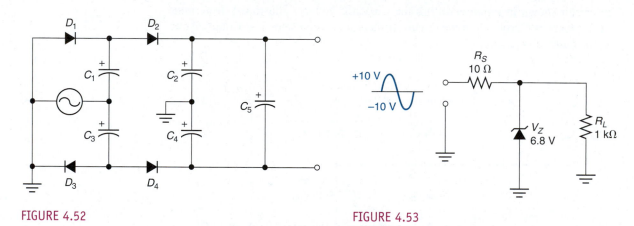

FIGURE 4.52                    FIGURE 4.53

**46.** Repeat Problem 45 for the *dual-zener clipper* shown in Figure 4.54.

**47.** If we reverse the zener diodes in Figure 4.54, we get the circuit shown in Figure 4.55. Explain what difference (if any) results from the change.

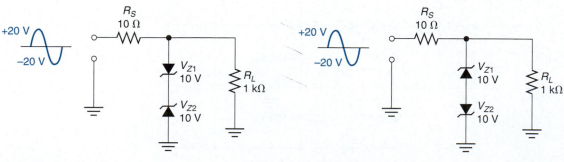

FIGURE 4.54                    FIGURE 4.55

48. Write a program that will determine the peak output voltages for a biased shunt clipper. The program should allow the user to input the value of $V_{S(pk)}$ and the value of the diode biasing voltage. It should also be able to solve both negative and positive clippers.

49. Modify the program in Problem 48 to accept the value of $V_S$ in peak, peak-to-peak, or rms form.

50. Write a program that will determine the peak output voltages and dc output voltage for either a positive or negative unbiased clamper. The program should provide for $V_S$ inputs in any standard voltage form (peak, rms, and so on).

4.1  12.54 $V_{pk}$

4.2  $V_L = -0.7$ V, $V_{RS} = -14.3$ $V_{pk}$

4.3  Positive alternation: $V_L = 0.7$ V, $V_{RS} = 11.3$ $V_{pk}$; negative alternation: $V_L = -11$ $V_{pk}$

4.5  $T_C = 188$ μs, $T_D = 28.2$ ms

## chapter 5

# Special Applications Diodes

## Objectives

*After studying the material in this chapter, you should be able to:*

1. State the purpose served by a *varactor*.
2. Describe the relationship between varactor bias and junction capacitance.
3. Describe the method by which a varactor can be used as the tuning component in a parallel (or series) *LC* circuit.
4. Describe a *surge* and the danger it presents to an electronic system.
5. List the required characteristics of every surge-protection circuit.
6. Describe the differences between constant-current diodes and *pn*-junction diodes.
7. Describe the operation of the *tunnel* diode.
8. Describe the construction and operation of the Schottky diode.
9. Describe the construction and operation of the PIN diode.
10. Compare and contrast the forward operation of the PIN and *pn*-junction diodes.

## Outline

In the first four chapters, we discussed the operation, common circuit applications, and troubleshooting of *pn*-junction diodes, zener diodes, and LEDs. In this chapter, we take a relatively brief look at several other types of diodes. The diodes covered in this chapter are called *special applications diodes* because they are used to perform functions other than those performed by the *pn*-junction diode, zener diode, and LED. The diodes covered in this chapter are rarely used for any purposes other than those presented here.

The diodes we will be discussing have very little in common with each other. For this reason, each section in this chapter reads as a *stand-alone* section. This means that you do not have to cover the sections in this chapter in any particular order. You can choose those sections you wish to cover in any order.

## 5.1    Varactor Diodes

**OBJECTIVE 1 ▶**

> **Varactor**
> A diode that has relatively high junction capacitance when reverse biased.

The **varactor** is a type of *pn*-junction diode that has relatively high junction capacitance when reverse biased. The capacitance of the junction is controlled by the amount of reverse voltage applied to the device. This makes the component very useful as a *voltage-controlled capacitor*. Note that the varactor is also referred to as a *varicap, tuning diode*, or *epicap*. Two commonly used schematic symbols for the varactor are shown in Figure 5.1.

**OBJECTIVE 2 ▶**

The ability of a varactor to act as a voltage-controlled capacitor is easy to understand when you consider the reverse bias characteristics of the device. When a *pn*-junction is reverse biased, the depletion layer acts as an insulator between the *p*-type and *n*-type materials, as shown in Figure 5.2a. As you know, a capacitor is made up of an insulator (called the *dielectric*) that separates two conductors (called the *plates*). If we view the *p*-type and *n*-type materials in the varactor as the plates and the depletion layer as the dielectric, it is easy to view the reverse-biased component as a capacitor.

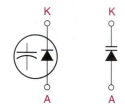

FIGURE 5.1   Varactor schematic symbols.

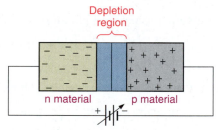

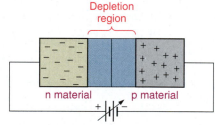

(a)  *pn* Junction with small reverse bias voltage applied

(b)  *pn* Junction with increased reverse bias voltage applied

FIGURE 5.2

The capacitance of a reverse-biased varactor junction is found as

$$C_t = \epsilon \frac{A}{W_d}$$

(5.1)

where $C_t$ = the total junction capacitance
$\epsilon$ = the permittivity of the semiconductor material
$A$ = the cross-sectional area of the junction
$W_d$ = the *width* of the depletion layer

Equation (5.1) is not used in any practical applications because a lot of intricate mathematical analyses are required to determine the values needed to solve the equation. However, it is introduced here because it does illustrate one important concept: The value of $C_t$ is *inversely proportional* to the width of the depletion layer. Since the width of the depletion layer increases when the amount of reverse bias increases, as shown in Figure 5.2b, we can say that *the junction capacitance of the varactor increases as diode reverse bias decreases, and vice versa*. This relationship between $V_R$ and $C_t$ is further illustrated in Figure 5.3. As the graph shows, you increase or decrease the junction capacitance by

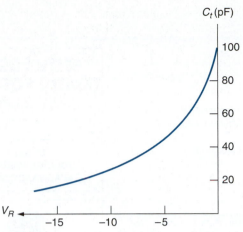

**FIGURE 5.3**   Varactor bias versus capacitance curve.

simply changing the amount of reverse bias applied to the device. This makes the varactor ideal for use in circuits that require voltage-controlled tuning. One such application is introduced later in this section and again in Chapter 17.

## 5.1.1   Varactor Specifications

The spec sheet for a given varactor provides the information you need for any circuit analysis. Figure 5.4 shows the spec sheet and operating curves for the Motorola MV209 series varactors. We will use this spec sheet for our discussion on the commonly used varactor parameters and characteristics.

The *maximum ratings* for the varactor are the same as those used for the *pn*-junction diode and the zener diode. The same holds true for the *reverse breakdown voltage* and *reverse leakage current* ratings that appear in the *electrical characteristics* section of the spec sheet. Since you are already familiar with these ratings, we will not discuss them here. If you need to review any of them, refer back to the appropriate discussions in Chapter 2.

The **diode capacitance temperature coefficient (*TC_C*)** rating tells you how much the component's capacitance changes for each 1°C rise in temperature above 25°C. The $TC_C$ rating for the MV209 series varactors is 300 ppm/°C. This means that the varactor capacitance *increases* by 300 *parts per million* for each 1°C rise in temperature above 25°C.

What is meant by *parts per million*? It means that the capacitance increases by *300 millionths (0.0003) of its rated value*. If a capacitor's rated value is 1 pF, the change is found as

$$\Delta C = (0.0003)(1 \text{ pF}) = \textbf{0.0003 pF}$$

Since the *nominal* capacitance listed on the MV209 spec sheet is 29 pF, its increase in capacitance per 1°C rise in temperature is found as

$$\Delta C = (0.0003)(29 \text{ pF}) = \textbf{0.0087 pF}$$

Thus, the MV209 varactor must experience a temperature increase of nearly 115°C to produce an increase in capacitance of 1 pF. A graph that shows the relationship between the ambient temperature and the actual capacitance of the varactor is also shown in Figure 5.4. This graph plots the *normalized* capacitance versus temperature. As you may know, a *normalized value* is one that has been set to 1 (one). When a value is normalized, the result indicates *the factor by which the value changes*. For example, the capacitance versus temperature graph in Figure 5.4 shows approximate normalized values of

$$C = 0.978 \text{ at } T = -50°C \quad \text{to} \quad C = 1.016 \text{ at } T = 100°C$$

> **Diode capacitance temperature coefficient (*TC_C*)**
> The amount by which varactor capacitance changes when temperature changes.

> How do you determine the $\Delta C$ for a temperature change of 1°C?

> Temperature has little effect on the capacitance rating of most varactors.

## Silicon Epicap Diodes

Designed for general frequency control and tuning applications; providing solid–state reliability in replacement of mechanical tuning methods.

- High Q with Guaranteed Minimum Values at VHF Frequencies
- Controlled and Uniform Tuning Ratio
- Available in Surface Mount Package

**MBV109T1
MMBV109LT1\*
MV209\***

\* ON Semiconductor Preferred Devices

**26–32 pF
VOLTAGE VARIABLE
CAPACITANCE DIODES**

CASE 419–04, STYLE 3
SC–70/SOT–323

CASE 318–08, STYLE 6
SOT–23 (TO–236AB)

CASE 182–06, STYLE 1
TO–92 (TO–226AC)

### MAXIMUM RATINGS

| Rating | Symbol | MBV109T1 | MMBV109LT1 | MV209 | Unit |
|--------|--------|----------|------------|-------|------|
| Reverse Voltage | $V_R$ | | 30 | | Vdc |
| Forward Current | $I_F$ | | 200 | | mAdc |
| Forward Power Dissipation<br>@ $T_A$ = 25°C<br>Derate above 25°C | $P_D$ | 280<br>2.8 | 200<br>2.0 | 200<br>1.6 | mW<br>mW/°C |
| Junction Temperature | $T_J$ | | +125 | | °C |
| Storage Temperature Range | $T_{stg}$ | | −55 to +150 | | °C |

### DEVICE MARKING

MBV109T1 = J4A, MMBV109LT1 = M4A, MV209 = MV209

### ELECTRICAL CHARACTERISTICS ($T_A$ = 25°C unless otherwise noted.)

| Characteristic | Symbol | Min | Typ | Max | Unit |
|----------------|--------|-----|-----|-----|------|
| Reverse Breakdown Voltage ($I_R$ = 10 µAdc) | $V_{(BR)R}$ | 30 | — | — | Vdc |
| Reverse Voltage Leakage Current ($V_R$ = 25 Vdc) | $I_R$ | — | — | 0.1 | µAdc |
| Diode Capacitance Temperature Coefficient ($V_R$ = 3.0 Vdc, f = 1.0 MHz) | $TC_C$ | — | 300 | — | ppm/°C |

| | $C_t$, Diode Capacitance $V_R$ = 3.0 Vdc, f = 1.0 MHz pF | | | Q, Figure of Merit $V_R$ = 3.0 Vdc f = 50 MHz | $C_R$, Capacitance Ratio $C_3/C_{25}$ f = 1.0 MHz (Note 1) | |
|--------|-----|-----|-----|-----|-----|-----|
| Device | Min | Nom | Max | Min | Min | Max |
| MBV109T1, MMBV109LT1, MV209 | 26 | 29 | 32 | 200 | 5.0 | 6.5 |

1. $C_R$ is the ratio of $C_t$ measured at 3 Vdc divided by $C_t$ measured at 25 Vdc.

**MMBV109LT1** is also available in bulk packaging. Use **MMBV109L** as the device title to order this device in bulk.

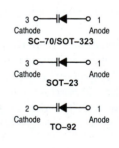

3 ○—▷|—○ 1
Cathode     Anode
**SC–70/SOT–323**

3 ○—▷|—○ 1
Cathode     Anode
**SOT–23**

2 ○—▷|—○ 1
Cathode     Anode
**TO–92**

(a)

**FIGURE 5.4** Spec sheet and operating curves for the MV209 series varactors. (Copyright of Semiconductor Component Industries, LLC. Used by permission.)

If we assume that the varactor has an actual capacitance of 29 pF, the range of values for $C_t$ is found as

$$C_t = (0.978)(29 \text{ pF}) = \mathbf{28.36\ pF} \quad (\text{at } T = -50°\text{C})$$

to

$$C_t = (1.016)(29 \text{ pF}) = \mathbf{29.46\ pF} \quad (\text{at } T = 100°\text{C})$$

Thus, a 150°C increase in temperature causes a 1.1 pF (3.79%) increase in capacitance.

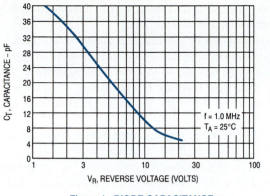

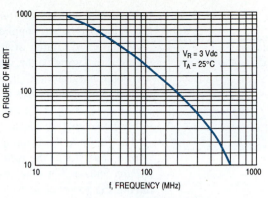

Figure 1. DIODE CAPACITANCE

Figure 2. FIGURE OF MERIT

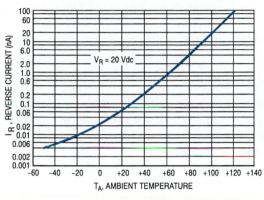

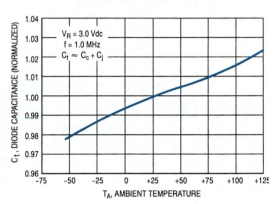

Figure 3. LEAKAGE CURRENT

Figure 4. DIODE CAPACITANCE

(b)

FIGURE 5.4   *(continued)*

The **diode capacitance ($C_t$)** rating is self-explanatory. For the MV209 series varactors, the $C_t$ rating is 26 to 32 pF when $V_R$ is 3 $V_{dc}$. Note that the *nominal* value (29 pF) is the *rated* value of the component. In other words, the MV209 would be rated as a 29 pF varactor diode even though the actual value could fall anywhere within the specified range at $V_R = 3\,V_{dc}$.

The **capacitance ratio ($C_R$)** rating tells you how much the junction capacitance varies over the given range of voltages. For the MV209 series, the $C_R$ rating ranges from 5.0 to 6.5 for $V_R = 3$ to 25 $V_{dc}$. This means that the capacitance at $V_R = 3\,V_{dc}$ will be 5.0 to 6.5 times as high as it is at $V_R = 25\,V_{dc}$.

The *higher* the value of $C_R$, the wider the range of capacitance values for a given varactor. For example, let's say that we have two varactors, each having a $C_t$ rating of 51 pF when $V_R = 5\,V_{dc}$. Let's also assume that $D_1$ has a rating of $C_R = 1.5$ for $V_R = 5$ to 10 $V_{dc}$ and $D_2$ has a rating of $C_R = 4$ for $V_R = 5$ to 10 $V_{dc}$. To find the capacitance of each diode at $V_R = 10\,V_{dc}$, we would divide the value of $C_t$ by the value of $C_R$ as follows:

$$C_{D1} = \frac{51\ pF}{1.5} = \textbf{34 pF} \quad (\text{at } V_R = 10\,V_{dc})$$

and

$$C_{D2} = \frac{51\ pF}{4} = \textbf{12.75 pF} \quad (\text{at } V_R = 10\,V_{dc})$$

Thus, $D_1$ would have a range of capacitance from 34 to 51 pF, and $D_2$ would have a range from 12.75 to 51 pF.

The $C_R$ rating of a varactor diode is important when designing circuits. A varactor with a *high* $C_R$ rating could be used in a *coarse-tuning* circuit, while a varactor with a *low* $C_R$

> **Diode capacitance ($C_t$)**
> The rated value (or range) of $C$ for a varactor at a specific value of $V_R$.

> **Capacitance ratio ($C_R$)**
> The factor by which $C$ changes from one specified value of $V_R$ to another.

> Why is the varactor $C_R$ rating important?

rating could be used in a *fine-tuning* circuit. Tuning will be discussed further when we take a look at varactor applications later in this section.

*An Important Point:*
You can use $C_R$ to find the range of $C_t$ values. However, you must use the *capacitance versus reverse voltage* graph to find the value of $C$ at various values of $V_R$.

While the $C_R$ rating of a varactor can be used to determine its capacitance *range*, it is not used to determine the specific diode capacitance at a specific value of $V_R$. For example, we would not use the $C_R$ rating of the MV209 to determine its capacitance at $V_R = 6$ V$_{dc}$. Instead, you need to use the *capacitance versus reverse voltage* curve that is shown in Figure 5.4. Using this curve, we could determine the value of capacitance at $V_R = 6$ V$_{dc}$ to be approximately 23 pF.

The *figure of merit* rating is the $Q$ of the junction capacitance. You may recall that the $Q$ of a capacitor is the *ratio of energy stored in the capacitor to the energy lost through leakage current*. In other words, it is a measure of how close the component comes to having the power characteristics of an *ideal* capacitor. An ideal capacitor would be 100% energy efficient; that is, all the energy stored by the component would be returned to the circuit when the capacitor discharges. Therefore, the ideal capacitor would have *infinite Q*.

The higher the $Q$ of a capacitor, the better the quality of the component. The MV209 is rated at $Q = 200$ (minimum). This means that the energy returned to the circuit by the capacitor (when it discharges) is at least 200 times the energy lost through leakage current.

### 5.1.2    Varactor Applications

**OBJECTIVE 3** ▶    Varactors are used almost exclusively in *tuned circuits*, which are discussed in detail in Chapter 17. A tuned *LC* circuit that contains a varactor is shown in Figure 5.5. Note that the capacitance of the varactor is in *parallel* with the inductor. Thus, the varactor and the inductor form a *parallel LC circuit*, or *LC tank circuit*.

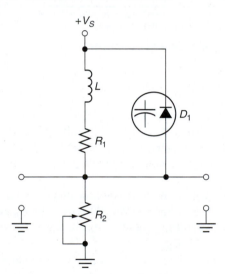

**FIGURE 5.5**  *LC* tank circuit.

Before we analyze the operation of the varactor in the circuit, a few points should be made. First, note the direction of the varactor. Since the varactor is pointing toward the positive source voltage, we know that it is reverse biased. Therefore, it is acting as a voltage-variable capacitance. *For normal operation, a varactor diode is operated in its reverse operating region.* (A forward-biased varactor diode would serve no special purpose because it has the same forward characteristics as a standard *pn*-junction diode.)

How is a varactor biased for normal operation?

Second, $R_1$ and $R_2$ form a voltage divider that determines the amount of reverse bias across $D_1$ and, therefore, its capacitance. By adjusting the setting of $R_2$, we can vary the diode capacitance. This, in turn, varies the *resonant frequency* of the *LC* circuit.

The resonant frequency of the *LC* tank circuit is found using

$$f_r = \frac{1}{2\pi\sqrt{LC}}$$    (5.2)

If the amount of varactor reverse bias is *decreased*, the value of *C* for the component *increases*. The increase in *C* will cause the resonant frequency of the circuit to *decrease*. Thus, *a decrease in reverse bias causes a decrease in resonant frequency*. By the same token, an increase in varactor reverse bias causes an increase in the value of $f_r$. This point is illustrated in Example 5.1.

### EXAMPLE 5.1

The *LC* tank circuit shown in Figure 5.5 has a 1 mH inductor. The varactor has the following specifications: $C_t = 100$ pF when $V_R = 5$ V, and $C_R = 2.5$ for $V_R = 5$ to 10 V. Determine the resonant frequency for the circuit at $V_R = 5$ V and $V_R = 10$ V.

*Solution:*  When the varactor reverse bias is 5 V, the value of *C* is 100 pF. For this condition, the value of $f_r$ is found as

$$f_r = \frac{1}{2\pi\sqrt{LC}} = \frac{1}{2\pi\sqrt{(1\text{ mH})(100\text{ pF})}} = 503.29\text{ kHz}$$

The value of *C* at $V_R = 10$ V is found by dividing the $C_t$ rating by the value of $C_R$ as follows:

$$C = \frac{100\text{ pF}}{2.5} = 40\text{ pF}$$

Now, the value of $f_r$ at $V_R = 10$ V is found as

$$f_r = \frac{1}{2\pi\sqrt{LC}} = \frac{1}{2\pi\sqrt{(1\text{ mH})(40\text{ pF})}} = 795.77\text{ kHz}$$

Thus, when the varactor reverse bias increases from 5 to 10 V, the value of $f_r$ increases from 503.29 to 795.77 kHz.

#### PRACTICE PROBLEM 5.1

A circuit like the one shown in Figure 5.5 has a 3.3 mH inductor and a varactor with the following specifications: $C_t = 51$ pF at $V_R = 4$ V and $C_R = 1.8$ for $V_R = 4$ to 10 V. Determine the value of $f_r$ for $V_R = 4$ V and $V_R = 10$ V.

You were told in our discussion on varactor parameters that high-$C_R$ varactors are used in *coarse-tuning* circuits, while low-$C_R$ varactors are used in *fine-tuning* circuits. This point can be illustrated with the aid of Example 5.2.

### EXAMPLE 5.2

The diode in Figure 5.5 is replaced with one that has ratings of $C_t = 100$ pF when $V_R = 5$ V and $C_R = 1.02$ for $V_R = 5$ to 10 V. Determine the frequency range of the circuit for $V_R = 5$ to 10 V.

*Solution:*  In Example 5.1 we determined the value of $f_r$ at $V_R = 5$ V to be 503.29 kHz. This value has not changed. For the new circuit, the value of *C* at $V_R = 10$ V is found as

$$C = \frac{100\text{ pF}}{1.02} = 98\text{ pF}$$

The value of $f_r$ at $V_R = 10$ V is now found as

$$f_r = \frac{1}{2\pi\sqrt{LC}} = \frac{1}{2\pi\sqrt{(1 \text{ mH})(98 \text{ pF})}} = 508.40 \text{ kHz}$$

**PRACTICE PROBLEM 5.2**

The varactor diode in Practice Problem 5.1 has been replaced with a diode that has ratings of $C_t = 51$ pF when $V_R = 4$ V and $C_R = 1.07$ for $V_R = 4$ to 10 V. Determine the frequency range of the circuit for $V_R = 4$ to 10 V.

Table 5.1 summarizes the values found in Examples 5.1 and 5.2. As you can see, the range of frequencies for the circuit in Example 5.2 is much smaller than that for the circuit in Example 5.1.

TABLE 5.1   Results from Examples 5.1 and 5.2

| *Example* | $f_r$ *at* $V_R = 5$ V | $f_r$ *at* $V_R = 10$ V | $\Delta f_r$ |
|-----------|------------------------|-------------------------|--------------|
| 5.1 | 503.29 kHz | 795.77 kHz | 292.48 kHz |
| 5.2 | 503.29 kHz | 508.4  kHz | 5.11 kHz |

> What is the difference between coarse tuning and fine tuning?

The circuit in Example 5.1 would be a *coarse-tuning circuit;* that is, it would be used to vary the value of $f_r$ over a wide range of values. The circuit in Example 5.2, on the other hand, would be a *fine-tuning* circuit. It would be used to select a frequency within a smaller range.

**Section Review ▶**

1. A varactor acts as what type of capacitance?
2. What is the relationship between the amount of reverse bias applied to a varactor and its capacitance?
3. What is the *diode capacitance temperature coefficient* rating? What is its unit of measure?
4. What is the relationship between varactor capacitance and temperature?
5. What is the *capacitance ratio* of a varactor?
6. Why is the $C_R$ rating of a varactor important?
7. What is the $Q$ of a capacitor?
8. Explain the operation of the circuit shown in Figure 5.5.
9. What is the relationship between varactor reverse bias and the resonant frequency of its tuned circuit?
10. What is the difference between *coarse tuning* and *fine tuning*?
11. What is the relationship between the $C_R$ rating of a varactor and the type of tuning provided?

**Critical Thinking ▶**

12. What type of voltage (fixed or varying) do you think you'd most often find applied to a varactor, and why?

## 5.2   Transient Suppressors and Constant-Current Diodes

> **Transient suppressor**
> A zener diode with extremely high surge-handling capabilities.

In this section, we will discuss two diodes that are very similar in operation to the standard zener diode. The first is the **transient suppressor**. Transient suppressors are zener diodes that have extremely high *surge-handling* capabilities. These diodes are used to protect voltage-sensitive circuits from surges that can occur under a variety of circumstances. The second is the **constant-current diode**. The constant-current diode is an extremely high-impedance diode that maintains a relatively constant device current over a wide range of forward operating voltages. Thus, the constant-current diode can be viewed as a "current version" of the zener diode. Note that the constant-current diode is not actually a zener diode, despite the similarities between the two. Rather, it is a variation on the *pn*-junction diode.

**Constant-current diode**
A diode that maintains a relatively constant device current over a wide range of forward operating voltages.

## 5.2.1    Transient Suppressors

In Chapter 4, you were introduced to the idea of using a shunt clipper to protect a circuit from a *surge* (or *transient*). The *surge-protection* circuits shown in Figure 5.6 use transient suppressors, configured as shunt clippers, to protect the input of a power supply from any ac line surges. Each circuit has a pair of transient suppressors wired in a *common-cathode* configuration between each pair of lines.

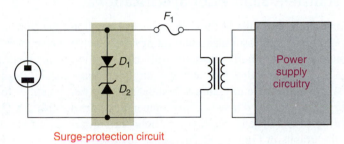

Surge-protection circuit

(a) Surge protection between the hot and neutral lines

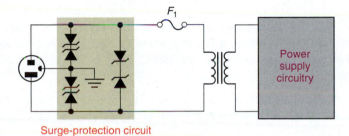

Surge-protection circuit

(b) Full surge protection between all input lines

FIGURE 5.6

The surge protectors in Figure 5.6 use *common-cathode* transient suppressors to eliminate the problem of surge polarity. For example, a positive surge on the hot line (or negative surge on the neutral line) forward biases $D_1$ and causes $D_2$ to operate in reverse breakdown. The opposite surge polarity causes $D_1$ to operate in reverse breakdown and forward biases $D_2$. In either case, the surge is shorted back to the wall outlet through the suppressor circuit.

You may recall that a *transient* (or *surge*) is an abrupt, high-voltage (or current) condition that lasts for a very brief time, usually in the microsecond or millisecond range. Left unchecked, a surge in the ac power lines can cause serious damage to the power supply of any electronic system (such as a television or personal computer). Transients and surges are caused by various conditions. Most often, they are generated by electric motors, air-conditioning and heating units, arcing switches, and lightning. The circuits shown in Figure 5.6 protect the power supply from such surges by shorting out any voltages greater than the $V_Z$ ratings of the diodes.

Transient suppressors have many other uses besides conditioning the ac voltage that goes into a power supply. They are used in many telecommunication, automotive, and consumer devices. Because of the wide range of applications, transient suppressors are available in a wide range of voltages.

For a surge-protection circuit to operate properly, it must have several characteristics:

◄ *OBJECTIVE 4*

◄ *OBJECTIVE 5*

1. The diodes used must have extremely high power dissipation ratings. This is because most ac power line surges contain a relatively high amount of power, generally in the hundreds of watts or higher.
2. The diodes must be able to turn on very rapidly. If the diodes in the surge-protection circuit are too slow, the power supply could be damaged before they have a chance to turn on.

These requirements for the surge-protection circuit are easily fulfilled by using transient suppressors.

Transient suppressors have the same general operating characteristics as standard zener diodes. In fact, the schematic symbol for the transient suppressor is identical to that of a standard zener diode. The main difference between the transient suppressor and the standard zener diode is the suppressor's *surge-handling capability*. The transient suppressor is designed to dissipate extremely high amounts of power for a very limited time. For example, the 1N5908 transient suppressor can dissipate up to 1.5 kW for a period of slightly less than 10 ms. This amount of power, even for a short period of time, would destroy any standard zener diode.

You might think that the time limit on the power rating of a transient suppressor could be a problem, but it isn't. Remember, surges generally last only a few milliseconds. Thus, transient suppressors can generally handle any surges that occur.

## 5.2.2 Transient Suppressor Specifications

Figure 5.7 shows a portion of the 1N5908 spec sheet. We will use the values shown in our discussion on transient suppressor maximum ratings and electrical characteristics.

The maximum ratings are fairly standard, with the exception of the **peak power dissipation rating ($P_{PK}$)**. This rating indicates the surge-handling capability of the component. As note 1 below the table in Figure 5.7a indicates, this rating applies to a *nonrepetitive* current surge and must be derated at lead temperatures greater than 25°C.

The $P_{PK}$ rating of a transient suppressor is both *temperature* and *time* dependent, as illustrated by the graphs in Figure 5.8. The power derating curve (labeled *Figure 2*) is a standard power derating curve, indicating the decrease in $P_{PK}$ that occurs (as a percentage) as temperature increases. The use of this curve is demonstrated in Example 5.3.

### EXAMPLE 5.3

A 1N5908 surge suppressor has a $P_{PK}$ of 1500 W at 25°C. It is being used in a circuit that has an ambient temperature of 100°C. What is the value of $P_{PK}$ at this temperature?

*Solution:* The point where the curve intersects the $T = 100°C$ line corresponds to a derating percentage of 50%. Thus, $P_{PK}$ will be 50% of its maximum value at this temperature, or 750 W.

### PRACTICE PROBLEM 5.3
What is the $P_{PK}$ rating of the 1N5908 at $T = 150°C$?

You may be wondering why we bothered with the temperature curve when the spec sheet for the 1N5908 series suppressors lists a derating value. *The derating value listed in the maximum ratings table applies to the $P_D$ rating of the components, not their $P_{PK}$ rating.* To derate the $P_{PK}$ rating, you must use the power versus temperature curve.

The time dependency of the $P_{PK}$ rating is illustrated in Figure 5.8. As you can see, $P_{PK}$ *and surge duration (pulse width) vary in opposite directions.* As the duration of the surge increases, the power-dissipating capability of the suppressors decreases, and vice versa.

Another maximum rating listed in Figure 5.7 is *forward surge current.* As shown, the 1N5908 can handle a forward surge current of 200 A. As note 2 below the table in part (a) indicates, this rating assumes a maximum duration of 8.3 ms and a limit of four pulses per minute.

Many of the *electrical characteristics* listed in Figure 5.7 are identified on the diode curve in part (b). As shown on the curve, there are three reverse voltage ratings of interest:

- The **working peak reverse voltage ($V_{RWM}$)** is the *maximum peak or dc reverse voltage that will not drive the component into its reverse breakdown (zener) region of operation.* In other words, this is the highest reverse voltage that will not trigger the device into conduction. The importance of this rating can be explained using the circuit shown in Figure 5.6a. With a line input of 120 $V_{rms}$, the peak primary voltage is approximately 170 V. Thus, any transient suppressors in the circuit must have

**How do transient suppressors differ from standard zener diodes?**

**Peak power dissipation rating ($P_{PK}$)**
Indicates the amount of surge power that the suppressor can dissipate.

**Power derating**

**Working peak reverse voltage ($V_{RWM}$)**
The maximum peak or dc reverse voltage that will not drive a transient suppressor into its reverse breakdown (zener) region of operation.

# 1N5908

## 1500 Watt Mosorb™ Zener Transient Voltage Suppressors

### Unidirectional*

Mosorb devices are designed to protect voltage sensitive components from high voltage, high–energy transients. They have excellent clamping capability, high surge capability, low zener impedance and fast response time. These devices are ON Semiconductor's exclusive, cost-effective, highly reliable Surmetic™ axial leaded package and are ideally-suited for use in communication systems, numerical controls, process controls, medical equipment, business machines, power supplies and many other industrial/consumer applications, to protect CMOS, MOS and Bipolar integrated circuits.

**Specification Features:**
- Working Peak Reverse Voltage Range – 5 V
- Peak Power – 1500 Watts @ 1 ms
- Maximum Clamp Voltage @ Peak Pulse Current
- Low Leakage < 5 μA Above 10 V
- Response Time is Typically < 1 ns

**Mechanical Characteristics:**
**CASE:** Void-free, transfer-molded, thermosetting plastic
**FINISH:** All external surfaces are corrosion resistant and leads are readily solderable
**MAXIMUM LEAD TEMPERATURE FOR SOLDERING PURPOSES:** 230°C, 1/16″ from the case for 10 seconds
**POLARITY:** Cathode indicated by polarity band
**MOUNTING POSITION:** Any

### ON Semiconductor™

**http://onsemi.com**

Cathode      Anode

**AXIAL LEAD
CASE 41A
PLASTIC**

L
1N
5908
YYWW

L = Assembly Location
1N5908 = JEDEC Device Code
YY = Year
WW = Work Week

### MAXIMUM RATINGS

| Rating | Symbol | Value | Unit |
|---|---|---|---|
| Peak Power Dissipation (Note 1.)<br>@ $T_L \leq 25°C$ | $P_{PK}$ | 1500 | Watts |
| Steady State Power Dissipation<br>@ $T_L \leq 75°C$, Lead Length = 3/8″<br>Derated above $T_L = 75°C$ | $P_D$ | 5.0<br><br>50 | Watts<br><br>mW/°C |
| Thermal Resistance, Junction–to–Lead | $R_{\theta JL}$ | 20 | °C/W |
| Forward Surge Current (Note 2.)<br>@ $T_A = 25°C$ | $I_{FSM}$ | 200 | Amps |
| Operating and Storage<br>Temperature Range | $T_J, T_{stg}$ | – 65 to<br>+175 | °C |

1. Nonrepetitive current pulse per Figure 4 and derated above $T_A = 25°C$ per Figure 2.
2. 1/2 sine wave (or equivalent square wave), PW = 8.3 ms, duty cycle = 4 pulses per minute maximum.

\* Bidirectional device will not be available in this device

(a)

**FIGURE 5.7** 1N5908 transient suppressor ratings and characteristics. (Copyright of Semiconductor Component Industries, LLC. Used by permission.)

$V_{RWM}$ ratings *greater than* 170 V. Otherwise, the diodes will turn on during the normal operation of the circuit.

- The **breakdown voltage ($V_{BR}$)** rating is *the peak or dc reverse voltage that will drive the transient suppressor into its reverse breakdown (zener) operating region.* In other words, this is the reverse voltage that will trigger the device into conduction. As shown in the diode curve, $V_{BR}$ is greater in magnitude than $V_{RWM}$ and is measured at a higher value of device current.

- The **clamping voltage ($V_C$)** is *the rated voltage across the component when it is conducting, measured at the indicated current.* Note that $V_C$ decreases as device current increases.

> **Breakdown voltage ($V_{BR}$)**
> The peak or dc reverse voltage that will drive a transient suppressor into its reverse breakdown (zener) operating region.
>
> **Clamping voltage ($V_C$)**
> The rated reverse voltage across a transient suppressor when it is conducting.

**ELECTRICAL CHARACTERISTICS** ($T_A$ = 25°C unless otherwise noted, $V_F$ = 3.5 V Max. @ $I_F$ (Note 3.) = 100 A)

| Symbol | Parameter |
|--------|-----------|
| $I_{PP}$ | Maximum Reverse Peak Pulse Current |
| $V_C$ | Clamping Voltage @ $I_{PP}$ |
| $V_{RWM}$ | Working Peak Reverse Voltage |
| $I_R$ | Maximum Reverse Leakage Current @ $V_{RWM}$ |
| $V_{BR}$ | Breakdown Voltage @ $I_T$ |
| $I_T$ | Test Current |
| $I_F$ | Forward Current |
| $V_F$ | Forward Voltage @ $I_F$ |

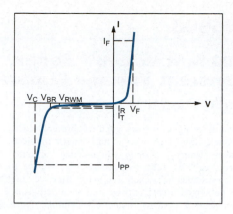

**Uni–Directional TVS**

**ELECTRICAL CHARACTERISTICS** ($T_A$ = 25°C unless otherwise noted, $V_F$ = 3.5 V Max. @ $I_F$ (Note 3.) = 53 A)

| Device (Note 4.) | $V_{RWM}$ (Note 5.) (Volts) | $I_R$ @ $V_{RWM}$ (µA) | Breakdown Voltage | | | | $V_C$ (Volts) (Note 7.) | | |
|---|---|---|---|---|---|---|---|---|---|
| | | | $V_{BR}$ (Note 6.) (Volts) | | | @ $I_T$ | | | |
| | | | Min | Nom | Max | (mA) | @ $I_{PP}$ = 120 A | @ $I_{PP}$ = 60 A | @ $I_{PP}$ = 30 A |
| 1N5908 | 5.0 | 300 | 6.0 | – | – | 1.0 | 8.5 | 8.0 | 7.6 |

NOTES:
3. Square waveform, PW = 8.3 ms, Non–repetitive duty cycle.
4. 1N5908 is JEDEC registered as a unidirectional device only (no bidirectional option)
5. A transient suppressor is normally selected according to the maximum working peak reverse voltage ($V_{RWM}$), which should be equal to or greater than the dc or continuous peak operating voltage level.
6. $V_{BR}$ measured at pulse test current $I_T$ at an ambient temperature of 25°C and minimum voltages in $V_{BR}$ are to be controlled.
7. Surge current waveform per Figure 4 and derate per Figure 2 of the General Data – 1500 W at the beginning of this group

(b)

FIGURE 5.7    *(continued)*

**1N5908**

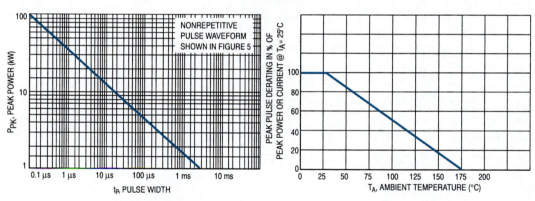

Figure 1.  Pulse Rating Curve          Figure 2.  Pulse Derating Curve

FIGURE 5.8    1N5908 transient suppressor derating curves. (Copyright of Semiconductor Component Industries, LLC. Used by permission.)

### 5.2.3    Selector Guides

Selector guides for transient suppressors contain all the ratings that we have discussed in this section. The selector guides for a given series of transient suppressors can be obtained from the series manufacturer.

### 5.2.4    Back-to-Back Suppressors

The surge protection circuit in Figure 5.9a contains two transient suppressors in a *common-cathode* configuration, meaning that the cathodes of the two components are connected to each other. Assuming that each transient suppressor has a reverse breakdown rating of

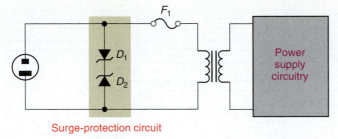

Surge-protection circuit

(a) Surge protection between the hot and neutral lines

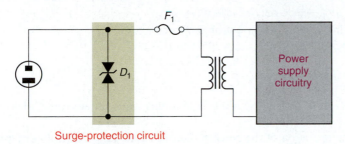

Surge-protection circuit

(b) Surge protection using a single back-to-back suppressor

FIGURE 5.9    Surge-protection circuits.

200 V, one of two things will happen if a transient in the ac line exceeds this value. Depending on the polarity of the transient:

- $D_1$ will be forward biased, and $D_2$ will be driven into its reverse operating region. In this case, the voltage across $D_2$ equals its *clamping voltage* ($V_{C2}$), and the primary voltage is held to the sum of $V_{F1}$ and $V_{C2}$.

- $D_2$ will be forward biased, and $D_1$ will be driven into its reverse operating region. In this case, the primary voltage is held to the sum of $V_{F2}$ and $V_{C1}$.

Assuming that $D_1$ and $D_2$ have the same ratings, the maximum voltage across the transformer primary cannot exceed the sum of $V_C$ and $V_F$.

Surge protection can be accomplished using a single **back-to-back suppressor**, as shown in Figure 5.9b. This type of suppressor, whose terminals are identified as anode 1 and anode 2, actually contains two transient suppressors that are connected *internally* like the components in Figure 5.9a. The obvious advantages to using this type of component are reduced circuit manufacturing costs and simpler circuitry. Note that back-to-back suppressors have no $V_F$ ratings, since they are designed to break down at the rated value of $V_{BR}$ in both directions.

> **Back-to-back suppressor**
> A single package containing two transient suppressors in either a common-cathode or common-anode configuration.

### 5.2.5    Constant-Current Diodes

Constant-current diodes are drastically different from any of the diodes you have seen so far. As stated earlier, a constant-current diode *maintains a relatively constant device current over a wide range of forward operating voltages.* As such, a constant-current diode is often referred to as a **current regulator diode**. The forward operating curves for five current regulator diodes are shown in Figure 5.10.

> **Current regulator diode**
> Another name for the constant-current diode.

When a conventional *pn*-junction diode is forward biased, $V_F$ is approximately 0.7 V, and $I_F$ is determined by the components in the diode circuit. This does not hold true for the constant-current diode. As the forward curve in Figure 5.10 shows, the value of $V_F$ for a constant-current diode can have a wide range of values, in this case anywhere from 0.1 to 100 V. At the same time, the value of $I_F$ for the constant-current diode is limited by the diode itself, not by the components in the diode circuit. For example, the 1N5290 curve shows that the value of $I_F$ for the device increases as $V_F$ increases from 0.15 V to approximately 1.5 V. At that point, the diode regulates the value of $I_F$ to around 500 μA (0.5 mA).

◀ *OBJECTIVE 6*

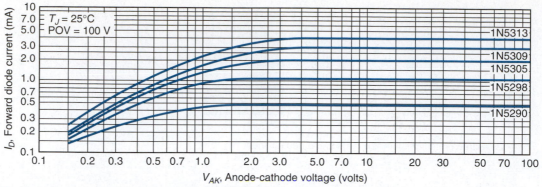

FIGURE 5.10   Current regulator forward operating curves. (Copyright of Semiconductor Component Industries, LLC. Used by permission.)

FIGURE 5.11   Constant-current diode schematic symbol.

The value of $I_F$ through the 1N5290 is held at this value for any value of $V_F$ between 1.5 and 100 V.

Because the operation of the constant-current diode is so radically different from that of any other diode, it has its own schematic symbol. This symbol is shown in Figure 5.11.

### 5.2.6   Constant-Current Diode Specifications

The spec sheet in Figure 5.12 shows the maximum ratings and electrical characteristics for the 1N5283–1N5314 series current regulator diodes. The maximum ratings should all seem familiar, with the exception of the **peak operating voltage (POV)** rating. The peak operating voltage of a current regulator is *its maximum allowable value of* $V_F$. As shown, current regulator diodes can have $V_F$ ratings that are significantly greater than 0.7 V (up to 100 V in this case).

For reasons that you'll see in a moment, we start our coverage of the *electrical characteristics* with the last rating given, the **maximum limiting voltage ($V_L$)**. The $V_L$ rating indicates *the voltage at which the diode begins to regulate current.* For the 1N5290, the value of $V_L$ is 1.05 V. This is lower than the value of $V_L$ that we obtained from the current versus voltage graph in Figure 5.10 because it is measured at $0.8I_P$.

The **regulator current ($I_P$)** rating is the *regulated value of forward current for values of* $V_F$ *that are between* $V_L$ *and POV.* As long as the forward voltage is kept between the rated values of $V_L$ and POV, the current through the diode is maintained at the value of $I_P$. Note that the *nominal* value of $I_P$ for the 1N5290 is shown to be 470 µA (0.47 mA). This is very close to the value provided by the current versus voltage graph in Figure 5.10.

The **minimum dynamic impedance ($Z_T$)** rating of the constant-current diode demonstrates another big difference between this diode and the *pn*-junction diode. You may recall that the bulk resistance of a *pn*-junction diode is typically very small, around 10 Ω or less. The impedance of the constant-current diode is extremely high, typically in the high-kΩ to low-MΩ range. The **minimum knee impedance ($Z_K$)** is in the low-kΩ to low-MΩ range.

**Peak operating voltage (POV)**
The maximum allowable value of $V_F$.

**Maximum limiting voltage ($V_L$)**
The voltage at which the diode starts to limit current.

**Regulator current ($I_P$)**
The regulated value of forward current for forward voltages that are between $V_L$ and POV.

**Minimum dynamic impedance ($Z_T$)**
The minimum forward impedance of a constant-current diode when operated in the current-limiting region of operation.

**Minimum knee impedance ($Z_K$)**
The minimum forward impedance of a constant-current diode when operated at the knee voltage of the operating curve.

**Series current regulator**
A circuit that maintains a constant circuit input current over a wide range of input voltages.

### 5.2.7   Series Current Regulators: A Constant-Current Diode Application

A **series current regulator** is a circuit used to maintain a constant circuit or load current. A basic series current regulator is shown in Figure 5.13. In this circuit, the current regulator diode is placed in series between a source and its load. Despite the variations in the source output (shown in the display), the load current is maintained at the value of $I_P$ for the diode.

# Current Regulator Diodes

Field-effect current regulator diodes are circuit elements that provide a current essentially independent of voltage. These diodes are especially designed for maximum impedance over the operating range. These devices may be used in parallel to obtain higher currents.

## MAXIMUM RATINGS

| Rating | Symbol | Value | Unit |
|--------|--------|-------|------|
| Peak Operating Voltage ($T_J = -55°C$ to $+200°C$) | POV | 100 | Volts |
| Steady State Power Dissipation @ $T_L = 75°C$ Derate above $T_L = 75°C$ Lead Length = 3/8 (Forward or Reverse Bias) | $P_D$ | 600 4.8 | mW mW/°C |
| Operating and Storage Junction Temperature Range | $T_J$, $T_{stg}$ | −55 to +200 | °C |

## ELECTRICAL CHARACTERISTICS  ($T_A = 25°C$ unless otherwise noted)

| Type No. | Regulator Current $I_P$ (mA) @ $V_T = 25$ V | | | Minimum Dynamic Impedance @ $V_T = 25$ V $Z_T$ (MΩ) | Minimum Knee Impedance @ $V_K = 6.0$ V $Z_K$ (MΩ) | Maximum Limiting Voltage @ $I_L = 0.8$ Ip (min) $V_L$ (Volts) |
|----------|------|-----|-----|---|---|---|
| | Nom | Min | Max | | | |
| *1N5283* | 0.22 | 0.198 | 0.242 | 25.0 | 2.75 | 1.00 |
| 1N5284 | 0.24 | 0.216 | 0.264 | 19.0 | 2.35 | 1.00 |
| 1N5285 | 0.27 | 0.243 | 0.297 | 14.0 | 1.95 | 1.00 |
| 1N5286 | 0.30 | 0.270 | 0.330 | 9.00 | 1.60 | 1.00 |
| *1N5287* | 0.33 | 0.297 | 0.363 | 6.60 | 1.35 | 1.00 |
| 1N5288 | 0.39 | 0.351 | 0.429 | 4.10 | 1.00 | 1.05 |
| 1N5289 | 0.43 | 0.387 | 0.473 | 3.30 | 0.870 | 1.05 |
| *1N5290* | 0.47 | 0.423 | 0.517 | 2.70 | 0.750 | 1.05 |
| *1N5291* | 0.56 | 0.504 | 0.616 | 1.90 | 0.560 | 1.10 |
| 1N5292 | 0.62 | 0.558 | 0.682 | 1.55 | 0.470 | 1.13 |
| 1N5293 | 0.68 | 0.612 | 0.748 | 1.35 | 0.400 | 1.15 |
| 1N5294 | 0.75 | 0.675 | 0.825 | 1.15 | 0.335 | 1.20 |
| 1N5295 | 0.82 | 0.738 | 0.902 | 1.00 | 0.290 | 1.25 |
| 1N5296 | 0.91 | 0.819 | 1.001 | 0.880 | 0.240 | 1.29 |
| *1N5297* | 1.00 | 0.900 | 1.100 | 0.800 | 0.205 | 1.35 |
| *1N5298* | 1.10 | 0.990 | 1.21 | 0.700 | 0.180 | 1.40 |
| *1N5299* | 1.20 | 1.08 | 1.32 | 0.640 | 0.155 | 1.45 |
| 1N5300 | 1.30 | 1.17 | 1.43 | 0.580 | 0.135 | 1.50 |
| 1N5301 | 1.40 | 1.26 | 1.54 | 0.540 | 0.115 | 1.55 |
| *1N5302* | 1.50 | 1.35 | 1.65 | 0.510 | 0.105 | 1.60 |
| 1N5303 | 1.60 | 1.44 | 1.76 | 0.475 | 0.092 | 1.65 |
| 1N5304 | 1.80 | 1.62 | 1.98 | 0.420 | 0.074 | 1.75 |
| *1N5305* | 2.00 | 1.80 | 2.20 | 0.395 | 0.061 | 1.85 |
| *1N5306* | 2.20 | 1.98 | 2.42 | 0.370 | 0.052 | 1.95 |
| *1N5307* | 2.40 | 2.16 | 2.64 | 0.345 | 0.044 | 2.00 |
| 1N5308 | 2.70 | 2.43 | 2.97 | 0.320 | 0.035 | 2.15 |
| *1N5309* | 3.00 | 2.70 | 3.30 | 0.300 | 0.029 | 2.25 |
| 1N5310 | 3.30 | 2.97 | 3.63 | 0.280 | 0.024 | 2.35 |
| 1N5311 | 3.60 | 3.24 | 3.96 | 0.265 | 0.020 | 2.50 |
| *1N5312* | 3.90 | 3.51 | 4.29 | 0.255 | 0.017 | 2.60 |
| *1N5313* | 4.30 | 3.87 | 4.73 | 0.245 | 0.014 | 2.75 |
| *1N5314* | 4.70 | 4.23 | 5.17 | 0.235 | 0.012 | 2.90 |

FIGURE 5.12  1N5283–1N5314 Current regulator maximum ratings and electrical characteristics. (Copyright of Semiconductor Component Industries, LLC. Used by permission.)

Two important points need to be made about the circuit shown in Figure 5.13:

**1.** Because the current regulator diode maintains a constant value of $I_P$, the sinusoidal waveform at the output of the source does not reach the load. In other words, the voltage on the load side of the diode remains constant as long as the load resistance does not change.

**2.** For the circuit to operate properly, the voltage across the diode, which equals the difference between the source and load voltages, must remain between the diode's $V_L$ and POV ratings.

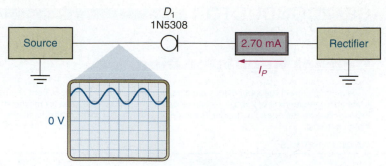

**FIGURE 5.13** Series current regulator operation.

---

**Section Review ▶**

1. What is a *transient suppressor*?

2. What is a *constant-current diode*?

3. What is a *surge*? How are ac power line surges commonly generated?

4. Why is a surge in the ac power line hazardous?

5. What are the required characteristics for a surge-protection circuit?

6. How does a *transient suppressor* differ from a standard zener diode?

7. Define each of the following transient suppressor ratings:
   a. Peak power dissipation ($P_{PK}$)
   b. Working peak reverse voltage ($V_{RWM}$)
   c. Clamping voltage ($V_C$)

8. Why don't back-to-back suppressors have $V_F$ ratings?

9. What are the differences between constant-current diodes and *pn*-junction diodes?

10. Draw the schematic symbol for the constant-current diode, and label its terminals.

11. Define each of the following constant-current diode specifications:
    a. Peak operating voltage (POV)
    b. Maximum limiting voltage ($V_L$)
    c. Regulator current ($I_P$)

12. What purpose is served by the series current regulator?

13. Describe the operation of the series current regulator.

**Critical Thinking ▶**

14. Based on the curves in Figure 5.10, what is the relationship between the rated forward current ($I_D$) of a current regulator diode and the range of $V_{AK}$ over which $I_D$ is constant?

## 5.3 Tunnel Diodes

**Tunnel diode**
A heavily doped diode used in high-frequency communications circuits.

**Ultrahigh frequency (UHF)**
The band of frequencies between 300 MHz and 3 GHz.

**OBJECTIVE 7 ▶**

**Tunnel diodes** are components used in the **ultrahigh frequency (UHF)** and microwave-frequency range. They have many applications in high-frequency communication electronics. Applications using tunnel diodes include amplifiers, oscillators, modulators, and demodulators. Because of the way that they are manufactured, they exhibit a unique characteristic curve, unlike any of the diodes that we have already studied. The schematic symbol and operating curve for the tunnel diode are shown in Figure 5.14a. The operating curve is a result of the extremely heavy doping used in the manufacturing of the tunnel diode. In fact, the tunnel diode is doped approximately 1000 times as heavily as standard *pn*-junction diodes. Figure 5.14b provides a comparison of the tunnel diode and *pn*-junction diode curves.

In the forward operating region of the tunnel diode, we are interested in the area between the **peak voltage ($V_P$)** and the **valley voltage ($V_V$)**. At $V_F = V_P$, forward current is called **peak current ($I_P$)**. As $V_F$ is increased to the value of $V_V$, $I_F$ *decreases* to its min-

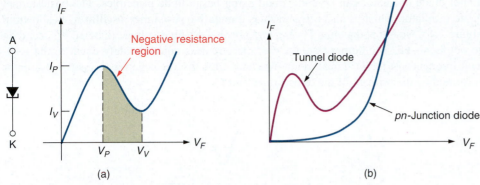

FIGURE 5.14  Tunnel diode symbol and characteristic curve.

imum value, called **valley current** ($I_V$). As you can see, *forward voltage and current are inversely proportional when the diode is operated between the values of $V_P$ and $V_V$.*

The region of operation between the peak and valley voltage is referred to as the *negative resistance region*. The term **negative resistance** describes the *dynamic* resistance of the component over the range of $V_P$ to $V_V$, which is found as

$$R_d = \frac{\Delta V}{\Delta I} = \frac{V_V - V_P}{I_V - I_P} \qquad (5.3)$$

Since the dynamic resistance of the component is defined using *inversely changing values* of voltage and current, the calculated value of $R_d$ is *negative*. This point is illustrated in Example 5.4.

**Peak and valley ratings**
The peak and valley voltage and current values for a tunnel diode can be obtained from the component spec sheet. When operated between $V_P$ and $V_V$, tunnel diode forward voltage and current are inversely proportional.

**Negative resistance**
A term used to describe any device with current and voltage values that are inversely related.

#### EXAMPLE 5.4

A tunnel diode has the following values: $I_P$ = 2 mA at $V_P$ = 150 mV and $I_V$ = 100 µA at $V_V$ = 500 mV. Calculate the dynamic resistance of the device.

*Solution:*  Using the values given, the dynamic resistance of the component is found as

$$R_d = \frac{\Delta V}{\Delta I} = \frac{V_V - V_P}{I_V - I_P} = \frac{500\ \text{mV} - 150\ \text{mV}}{100\ \text{µA} - 2\ \text{mA}} = \frac{350\ \text{mV}}{-1.9\ \text{mA}} = \boldsymbol{-184.2\ \Omega}$$

#### PRACTICE PROBLEM 5.4
A tunnel diode has the following values: $I_P$ = 5 mA at $V_P$ = 100 mV and $I_V$ = 250 µA at $V_V$ = 200 mV. Calculate the dynamic resistance of the component.

As you can see, the dynamic resistance of the tunnel diode has a *negative* value. This is why the operating region between the peak and valley voltages is referred to as the *negative resistance region.*

It should be noted that *negative resistance* is a mathematical concept that describes the change that takes place over a range of values. At any given point on the curve, the ratio of voltage to current still yields a positive value.

Tunnel diodes are operated almost exclusively in the negative resistance region. One common use of the tunnel diode is as the active component in an oscillator.

### 5.3.1    Tunnel Diode Oscillator

An **oscillator** is a circuit that is used to convert dc to an ac signal. The ac signal created by the tunnel diode oscillator could be used for many applications that require a high-frequency sine wave. Oscillators are discussed in detail in Chapter 18; however, the

**Oscillator**
A circuit that converts dc to ac. An ac signal generator.

**Negative resistance oscillator**

An oscillator whose operation is based on a negative resistance device.

tunnel diode oscillator can be understood using basic diode principles. The basic tunnel diode oscillator, which is also referred to as a **negative resistance oscillator**, is shown in Figure 5.15. Note that the circuit is shown to have a dc input voltage (labeled $+V_S$) and an ac output signal. The output frequency of the circuit is approximately equal to the resonant frequency of the $LC$ tank circuit. The tank circuit in the negative resistance oscillator is formed by the primary of the transformer and $C_2$.

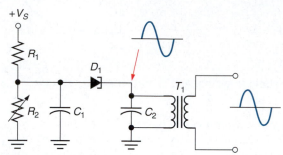

**FIGURE 5.15**  Tunnel diode oscillator.

The negative resistance oscillator in Figure 5.15 uses a tunnel diode to generate and sustain a sinusoidal output with only a dc input (as shown in the figure). To understand this principle, we need to take a brief look at what happens when a current pulse is provided to a parallel $LC$ circuit.

In Figure 5.16a, an $LC$ tank circuit is shown in series with a switch. If the switch were closed momentarily, we'd get the waveform shown in Figure 5.16b. This waveform is generated by the back-and-forth current between the inductor and the capacitor. Note that the waveform loses amplitude from one cycle to the next because of a continual loss of energy in the circuit.

Oscillator operation is discussed in detail in Chapter 18. The description provided here is intended only to give you a basic idea of how the negative resistance oscillator works.

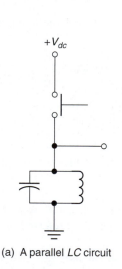

(a) A parallel $LC$ circuit

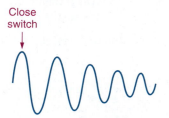

(b) Current waveform produced by the $LC$ circuit after a single switch closing

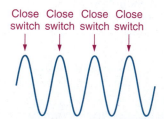

(c) Current waveform produced by the $LC$ circuit when the switch is closed at each current peak

**FIGURE 5.16**

The key to sustaining a sinusoidal output from an $LC$ tank circuit is to provide it with additional current at the peak of each cycle, as shown in Figure 5.16c. When current is provided to the tank circuit on each cycle, the energy lost is restored to the circuit and the cycles continue at a constant amplitude. With this in mind, let's go back to the circuit shown in Figure 5.15.

The tunnel diode in the circuit provides additional current to the tank circuit during the positive alternations of the sine wave. $R_1$ and $R_2$ are used to bias the tunnel diode. The biasing of the diode is set (by design) so that the following conditions are met:

1. When the tank circuit waveform is at its positive peak, the difference between this voltage and the biasing voltage equals the $V_P$ rating of the diode. Thus, diode current is at its maximum value when the tank circuit waveform is at its positive peak.
2. As the tank circuit waveform decreases toward zero, the difference between this voltage and the biasing voltage approaches the $V_V$ rating of the diode. Thus, diode current is decreasing as the sine wave voltage decreases.

As these statements indicate, *maximum current is supplied to the tank circuit when the circuit waveform is at its positive peak*. This enables the circuit to sustain oscillations, as shown in Figure 5.16c.

You may be wondering how the oscillations start in the first place. When power is first applied to the circuit, the dc voltage source starts charging $C_1$ (via $R_1$). As the voltage across the capacitor increases, the change in voltage (and, therefore, current) is coupled to the tank circuit through the diode. The change in tank circuit current is sufficient to start the oscillations.

The negative resistance oscillator has one major drawback. While the circuit works very well at high frequencies (in the upper megahertz range and higher), it cannot be used efficiently at lower frequencies. Lower-frequency oscillators are generally made with transistors and other integrated circuits.

> Negative resistance oscillators cannot be used efficiently at low output frequencies.

◀ **Section Review**

1. What is a tunnel diode?
2. What is the relationship between tunnel diode forward current and voltage when the component is operated in the negative resistance region?
3. Refer to the curves in Figure 5.14b. What tunnel diode characteristic accounts for the difference in the operating curves?
4. What is meant by the term *negative resistance*?
5. What is an *oscillator*?
6. What determines the output frequency for a negative resistance oscillator?
7. Briefly explain the operation of the circuit in Figure 5.15.

## 5.4    Other Diodes

The diodes covered in this section are used less often than the other diodes we have discussed. Except for this fact, they are not necessarily related to each other in characteristics or applications.

### 5.4.1    Schottky Diodes

The **Schottky diode** has very little junction capacitance. Because of this, it can be operated at much higher frequencies than the typical *pn*-junction diode. The reduced junction capacitance also results in a much faster *switching time*. For this reason, Schottky devices are used more and more in digital switching applications.

> **Schottky diode**
> A high-speed diode with very little junction capacitance.

The Schottky diode is often referred to by any of the following names: *Schottky barrier diode*, *hot-carrier diode*, and *surface-barrier diode*. The schematic symbol and characteristic curve for this component are shown in Figure 5.17. As the characteristic curve indicates, the Schottky diode has lower $V_F$ and $V_{RRM}$ values than those of the *pn*-junction diode. Typically, the Schottky diode has a $V_F$ of approximately 0.3 V and a $V_{RRM}$ of less than 50 V. These are much lower than the typical *pn*-junction ratings of $V_F = 0.7$ V and $V_{RRM} = 150$ V.

◀ *OBJECTIVE 8*

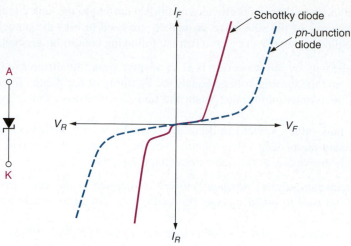

FIGURE 5.17  Schottky diode symbol and characteristic curve.

FIGURE 5.18  Schottky diode construction.

**Propagation delay**
The time required for a signal to get from the input to the output of a component or circuit.

The low junction capacitance and high-switching-speed capability of the Schottky diode are the result of the construction of the component. Schottky diodes have a junction that uses metal in place of the *p*-type material, as shown in Figure 5.18. As a result, the component has no depletion layer to break down or rebuild.

By forming a junction with a semiconductor and metal, you still have a junction, but now there is very little junction capacitance. With very little junction capacitance, the Schottky diode can be switched back and forth very rapidly between forward and reverse operation. While many components are capable of switching at this frequency, most of them are *low-current* devices. The Schottky diode is a relatively high-current device, capable of switching rapidly while providing forward currents in the neighborhood of 50 A. In sinusoidal and *low-current* switching circuits, the Schottky diode is capable of operating at frequencies of 20 GHz and more.

Schottky diodes are also used in the manufacture of integrated-circuit chips to decrease the **propagation delay** time of the internal circuits. Shorter propagation delay time increases the maximum operating speed of integrated circuits. This relationship is discussed further in Chapter 19.

### 5.4.2  PIN Diodes

**OBJECTIVE 9** ▶

The *PIN diode* is made up of *three* semiconductor materials. The construction of the PIN diode is illustrated in Figure 5.19. The center material is made up of *intrinsic* (pure) silicon. The *p*- and *n*-type materials are very heavily doped and, therefore, have very low resistances.

FIGURE 5.19  PIN diode.

When reverse biased, the PIN diode acts as a capacitor.

When reverse biased, the PIN diode acts as a capacitor. The reason for this is illustrated in Figure 5.20. You may recall that an intrinsic semiconductor acts as an insulator. Thus, the intrinsic material in the PIN diode can be viewed as the dielectric of a capacitor.

By comparison to the intrinsic material, the heavily doped *p*- and *n*-type materials can be viewed as *conductors*. Therefore, we have a dielectric (the intrinsic material) sandwiched between two conductors (the *p*- and *n*-type materials). This forms the PIN diode capacitor.

The capacitance of a reverse-biased PIN diode remains relatively constant over a wide range of reverse voltages.

The capacitance of a reverse-biased PIN diode is relatively constant over a wide range of reverse voltages. For example, the diode capacitance curve for the MPN300 PIN diode is shown in Figure 5.21. Note that the diode capacitance remains at approximately

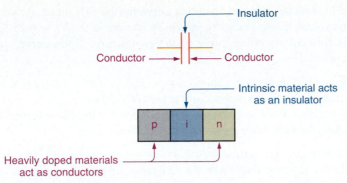

FIGURE 5.20  PIN diode capacitance.

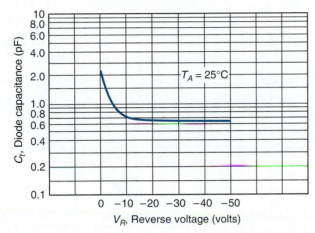

FIGURE 5.21  MPN3700 diode capacitance curve. (Copyright of Semiconductor Component Industries, LLC. Used by permission.)

0.65 pF over the $V_R$ range of $-30$ to $-50$ V. Over the $V_R$ range of 0 to $-30$ V, the component has a capacitance curve similar to that of a varactor. The capacitance curve in Figure 5.21 is typical for PIN diodes.

When forward biased, the intrinsic material is forced into conduction. As the number of free carriers in the intrinsic material increases, the resistance of the material decreases. Thus, when forward biased, the PIN diode acts as a *current-controlled resistance*. This point is illustrated in Figure 5.22. Note that the *series resistance* (diode resistance) *decreases* as forward current increases. This is due to the increase in the number of free carriers that are in the intrinsic material.

> When forward biased, the PIN diode acts as a *current-controlled* resistor.

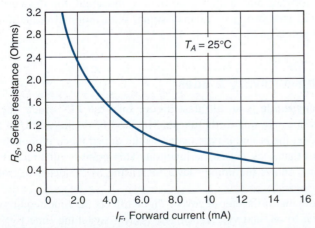

FIGURE 5.22  MPN3700 diode series resistance curve. (Copyright of Semiconductor Component Industries, LLC. Used by permission.)

Figure 5.23 shows the forward operating curve for the PIN diode. As you can see, the current versus voltage curve gradually increases, starting at the $V_F = 0.75$ V point. If you compare this forward operating curve to the *pn*-junction diode curve, you'll see two major differences:

1. The *pn*-junction diode curve shows conduction to start at nearly 0 V, while the PIN diode *starts* conducting (in this case) at 0.75 V (750 mV).
2. The *pn*-junction diode has a distinct turning point in the curve (called the *knee voltage*), while the PIN diode shows no distinct knee voltage.

> PIN diode operation versus *pn*-junction diode operation.

These two differences are caused by the construction of the PIN diode. For the PIN diode to conduct, $V_F$ must overcome the resistance of the insulating intrinsic material. For the MMBV3700, $V_F$ must be at least 750 mV before the intrinsic material will allow conduction.

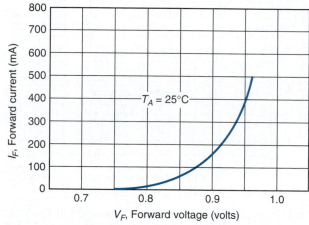

**FIGURE 5.23**   MPN3700 diode forward voltage curve. (Copyright of Semiconductor Component Industries, LLC. Used by permission.)

The lack of a knee voltage, or turning point in the curve, is because the PIN diode does not have a *pn* junction. Without a *pn* junction, the device does not have any sudden *turn-on* point. Rather, conduction increases at a less abrupt rate.

The PIN diode is used primarily in UHF and microwave applications. They are also used as RF switches in many amateur radio systems. The low reverse capacitance and current-controlled resistance make it ideal for high-frequency communication circuits.

> **Modulator**
> A circuit that combines two signals of different frequencies into a single signal.

PIN diodes are commonly used as switches or **modulators**. If the $V_F$ value is below 0.75 V, the diode has virtually no current leakage because the intrinsic material acts as an insulator. When forward biased above the 0.75 V threshold, the device has a low-value resistance and will pass a high-frequency signal with minimal reduction in amplitude, thus closely resembling an ideal switch.

### 5.4.3   Step-Recovery Diodes

> **Step-recovery diode**
> A heavily doped diode with an ultrafast switching time.

The **step-recovery diode** is an ultrafast diode. Like the PIN diode, the step-recovery diode's characteristics are due to the unusual method of doping used. In the case of the step-recovery diode, the *p*- and *n*-type materials are doped much more heavily at the ends of the component than they are at the junction. This point is illustrated by the graph in Figure 5.24, which shows the doping level increasing as the distance from the junction increases.

The unusual doping of the step-recovery diode affects the time required for the device to switch from *off* to *on*, and vice versa. The typical switching time for a step-recovery diode is in the low-picosecond range. This makes them ideal for switching applications in the VHF frequency range and above.

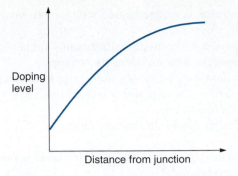

Doping level

Distance from junction

FIGURE 5.24

### 5.4.4 One Final Note

There are too many types of diodes to cover adequately in one chapter. However, the diodes covered in this chapter have been selected to give you a basis for understanding some of the other diode types.

◄ **Section Review**

1. What are the commonly used names for the Schottky diode?
2. Why can the Schottky diode be operated at much higher frequencies than the typical *pn*-junction diode?
3. How does the construction of the Schottky diode differ from that of the *pn*-junction diode?
4. Describe the construction of the PIN diode.
5. Explain the reverse characteristics of the PIN diode.
6. Explain the forward characteristics of the PIN diode.
7. Why is the forward operating curve of the PIN diode so different from that of a *pn*-junction diode?
8. Describe the doping of the step-recovery diode.
9. What is the result of the step-recovery diode doping scheme?

**CHAPTER SUMMARY**

Here is a summary of the major points made in this chapter:

1. The *varactor* is a type of *pn*-junction diode that has relatively high junction capacitance when reverse biased.
   a. Varactors are used primarily as *voltage-controlled capacitors*.
   b. Varactors are also referred to as *varicaps*, *tuning diodes*, and *epicaps*.
2. Junction capacitance ($C_t$) is inversely proportional to depletion layer width.
   a. An increase in reverse bias causes the width of the depletion layer to increase.
   b. $C_t$ decreases as reverse bias increases.
   c. The bias versus $C_t$ curve for the varactor is shown in Figure 5.3.
3. The *diode capacitance temperature coefficient* ($TC_C$) rating tells you how much a varactor's capacitance increases per 1°C rise in temperature (above 25°C).
4. The *capacitance ratio* ($C_R$) rating tells you how much the junction capacitance of a varactor varies over the given range of voltages.
5. Varactors are used almost exclusively in tuning circuits.
   a. A high-$C_R$ varactor is best suited for *coarse-tuning* applications.
   b. A low-$C_R$ varactor is best suited for *fine-tuning* applications.
6. The $Q$ of any capacitor is a measure of how close the component comes to having the power characteristics of an ideal capacitor.
   a. The ideal capacitor would have infinite $Q$.
   b. The higher the $Q$ of a capacitor, the closer it comes to the ideal component.
   c. Since they are voltage-controlled capacitors, varactors have $Q$ ratings.

7. A *transient suppressor* is a zener diode with an extremely high surge-handling capability.

8. A *constant-current diode* is a diode that maintains a relatively constant device current over a wide range of forward operating voltages.

9. Transient suppressors are commonly used to protect power supply inputs from surges that may occur on the ac power line (see Figure 5.6).

10. Any surge protector must have:
    a. An extremely high power dissipation rating (so that it can handle the surge power).
    b. A rapid turn-on time (so that it can short out a surge before it has time to damage the circuit being protected).

11. The *peak power dissipation rating* ($P_{PK}$) indicates the surge-handling capability of a transient suppressor.
    a. The $P_{PK}$ rating of a transient suppressor must be derated at ambient temperatures above 25°C (see Figure 5.8a).
    b. The power-handling capability of a transient suppressor decreases as the pulse width of a surge increases (see Figure 5.8b).

12. Transient suppressors have extremely high forward surge current ratings.

13. The *working peak reverse voltage* ($V_{RWM}$) rating indicates the maximum reverse voltage that will *not* turn on a transient suppressor.

14. The *breakdown voltage* ($V_{BR}$) rating is the peak or dc reverse voltage that will drive a transient suppressor into its reverse breakdown (zener) operating region.

15. The *clamping voltage* ($V_C$) is the rated reverse voltage across a transient suppressor when it is conducting.

16. A *back-to-back* suppressor contains two components in a single case that are connected as common-cathode (or common-anode) components.

17. A *constant-current* diode maintains a relatively constant device current as long as the component voltage stays within established limits.
    a. The *maximum limiting voltage* ($V_L$) rating indicates the value of forward voltage at which the component starts to limit current.
    b. The *peak operating voltage* (POV) rating is the maximum allowable value of diode forward voltage.
    c. As long as forward voltage is kept between the limits set by the diode $V_L$ and POV ratings, diode current remains relatively constant (see Figure 5.10).
    d. The *regulator current* ($I_P$) rating is the value of diode forward current over the range of rated component voltages.

18. Constant-current diodes are typically used as *series current regulators* (see Figure 5.13).

19. A *tunnel diode* is a heavily doped diode that is used in high-frequency communications circuits.
    a. In the forward operating region, the diode exhibits a unique characteristic between two rated voltages, called the *peak voltage* ($V_P$) and *valley voltage* ($V_V$).
    b. Between $V_P$ and $V_V$, an increase in component voltage causes a *decrease* in component current.

20. The region of tunnel diode operation between $V_P$ and $V_V$ is referred to as the *negative resistance region*.
    a. The term *negative resistance* describes the *dynamic* resistance ($R_d$) of the component.
    b. The dynamic resistance of the component equals the ratio of $\Delta V$ to $\Delta I$ over the range between $V_P$ and $V_V$. Since current decreases over this range, the value of $R_d$ is *negative* (see Example 5.4).

21. An *oscillator* is a circuit that converts dc energy to ac energy.

22. The tunnel diode is the active component in a *negative resistance oscillator* (see Figure 5.15).

23. A *Schottky diode* is a high-speed diode with very little junction capacitance.
    a. The Schottky diode consists of an *n*-type material and a metal plate. As such, it has no *pn* junction (see Figure 5.18).
    b. Since a Schottky diode does not have a *pn*-junction, it has no junction capacitance. As a result, it is capable of switching at extremely high frequencies.
24. The Schottky diode has lower forward and maximum reverse voltages than a *pn*-junction diode (see Figure 5.17).
25. The PIN diode is made up of three semiconductor materials.
    a. An *intrinsic* (pure) semiconductor is sandwiched between heavily doped *p*- and *n*-type materials.
    b. When reverse biased, the PIN diode acts as a capacitor.
    c. When forward biased, the PIN diode acts as a current-controlled resistor.
26. The *step-recovery diode* is a heavily doped diode with an ultrafast switching time.

## EQUATION SUMMARY

| Equation Number | Equation | Section Number |
|---|---|---|
| (5.1) | $$C_T = \epsilon \frac{A}{W_d}$$ | 5.1 |
| (5.2) | $$f_r = \frac{1}{2\pi\sqrt{LC}}$$ | 5.1 |
| (5.3) | $$R_d = \frac{\Delta V}{\Delta I} = \frac{V_V - V_P}{I_V - I_P}$$ | 5.3 |

## KEY TERMS

back-to-back suppressor 185
breakdown voltage ($V_{BR}$) 183
capacitance ratio ($C_R$) 177
clamping voltage ($V_C$) 183
constant-current diode 181
current-controlled resistance 193
current regulator diode 185
diode capacitance ($C_t$) 177
diode capacitance temperature coefficient ($TC_C$) 175
epicap 174
hot-carrier diode 191
maximum limiting voltage ($V_L$) 186

minimum dynamic impedance ($Z_T$) 186
minimum knee impedance ($Z_K$) 186
modulator 194
negative resistance 189
negative resistance oscillator 190
oscillator 189
peak current ($I_P$) 189
peak operating voltage (POV) 186
peak power dissipation rating ($P_{PK}$) 182
peak voltage ($P_V$) 189
PIN diode 192
propagation delay 192
regulator current ($I_P$) 186

Schottky barrier diode 191
Schottky diode 191
series current regulator 186
step-recovery diode 194
surface-barrier diode 191
transient suppressor 180
tunnel diode 188
tunnel diode oscillator 190
ultrahigh frequency (UHF) 188
valley current ($I_V$) 189
valley voltage ($V_V$) 189
varactor 174
varicap 174
working peak reverse voltage ($V_{RWM}$) 182

## PRACTICE PROBLEMS

### Section 5.1

1. A varactor has ratings of $C_t$ = 50 pF at 25°C and $TC_C$ = 500 ppm/°C. Determine the change in capacitance for each 1°C rise in temperature.

2. A varactor has ratings of $C_t$ = 48 pF at 25°C and $TC_C$ = 800 ppm/°C. Determine the change in capacitance for each 1°C rise in temperature.

3. The varactor in Figure 5.25 has the following values: $C_t$ = 48 pF at $V_R$ = 3 $V_{dc}$, and $C_R$ = 4.8 for $V_R$ = 3 to 12 $V_{dc}$. Determine the resonant frequency for the circuit at $V_R$ = 3 $V_{dc}$ and $V_R$ = 12 $V_{dc}$.

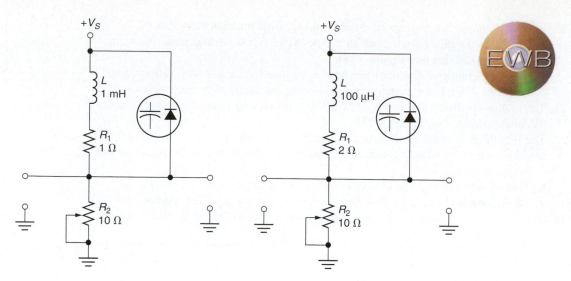

FIGURE 5.25                          FIGURE 5.26

**4.** The varactor in Figure 5.26 has the following values: $C_t = 68$ pF at $V_R = 4$ $V_{dc}$, and $C_R = 1.12$ for $V_R = 4$ to $10$ $V_{dc}$. Determine the resonant frequency for the circuit at $V_R = 4$ $V_{dc}$ and $V_R = 10$ $V_{dc}$.

### Section 5.2

**5.** Figure 5.27 shows the surge power curves for the Motorola MPZ-16 series transient suppressors. How much power can these diodes handle if the surge duration is 8 ms and $T_C = 35°C$?

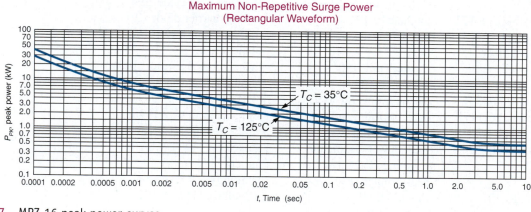

Maximum Non-Repetitive Surge Power
(Rectangular Waveform)

FIGURE 5.27    MPZ-16 peak power curves.

**6.** Refer to Figure 5.27. Determine the maximum power dissipation for the MPZ-16 series diodes when the surge duration is 15 ms and $T_C = 35°C$.

**PUSHING THE ENVELOPE**

**7.** The varactor in Figure 5.28 has the characteristic curves shown. What value of $V_R$ is required to set the resonant frequency of the tank circuit to 100 kHz?

**8.** The 1N6303A transient suppressor has the characteristics shown in the selector guide in Figure 5.29. Determine the dynamic zener impedance ($Z_Z$) of this suppressor. (*Hint*: Review the method used to determine $Z_Z$ that was shown in Chapter 2.)

**9.** The circuit shown in Figure 5.30 cannot tolerate an input voltage surge that is greater than 30% above the rated peak value of $V_{in}$. Which of the suppressors listed in Figure 5.29 would be best suited for use in this circuit?

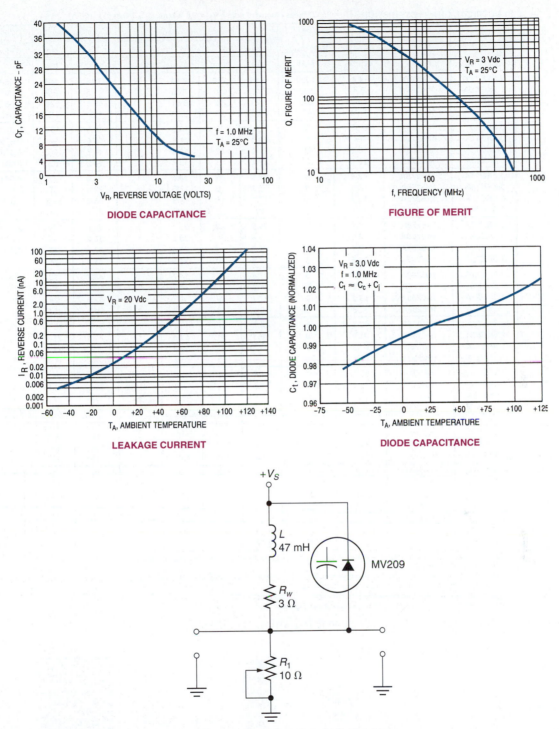

FIGURE 5.28 (Copyright of Semiconductor Component Industries, LLC. Used by permission.)

**ELECTRICAL CHARACTERISTICS**  (T$_A$ = 25°C unless otherwise noted, V$_F$ = 3.5 V Max. @ I$_F$ (Note 1) = 100 A)

| Device | JEDEC Device (Note 2) | V$_{RWM}$ (Note 3) (Volts) | I$_R$ @ V$_{RWM}$ (μA) | Breakdown Voltage V$_{BR}$ (Note 4) (Volts) Min | Nom | Max | @ I$_T$ (mA) | V$_C$ @ I$_{PP}$ (Note 5) V$_C$ (Volts) | I$_{PP}$ (A) | ΘV$_{BR}$ (%/°C) |
|---|---|---|---|---|---|---|---|---|---|---|
| *1.5KE6.8A* | *1N6267A* | *5.8* | *1000* | *6.45* | *6.8* | *7.14* | *10* | *10.5* | *143* | *0.057* |
| 1.5KE7.5A | 1N6268A | 6.4 | 500 | 7.13 | 7.5 | 7.88 | 10 | 11.3 | 132 | 0.061 |
| 1.5KE8.2A | 1N6269A | 7.02 | 200 | 7.79 | 8.2 | 8.61 | 10 | 12.1 | 124 | 0.065 |
| 1.5KE9.1A | 1N6270A | 7.78 | 50 | 8.65 | 9.1 | 9.55 | 1 | 13.4 | 112 | 0.068 |
| 1.5KE10A | 1N6271A | 8.55 | 10 | 9.5 | 10 | 10.5 | 1 | 14.5 | 103 | 0.073 |
| 1.5KE11A | 1N6272A | 9.4 | 5 | 10.5 | 11 | 11.6 | 1 | 15.6 | 96 | 0.075 |
| 1.5KE12A | 1N6273A | 10.2 | 5 | 11.4 | 12 | 12.6 | 1 | 16.7 | 90 | 0.078 |
| 1.5KE13A | 1N6274A | 11.1 | 5 | 12.4 | 13 | 13.7 | 1 | 18.2 | 82 | 0.081 |
| *1.5KE15A* | *1N6275A* | *12.8* | *5* | *14.3* | *15* | *15.8* | *1* | *21.2* | *71* | *0.084* |
| 1.5KE16A | 1N6276A | 13.6 | 5 | 15.2 | 16 | 16.8 | 1 | 22.5 | 67 | 0.086 |
| 1.5KE18A | 1N6277A | 15.3 | 5 | 17.1 | 18 | 18.9 | 1 | 25.2 | 59.5 | 0.088 |
| 1.5KE20A | 1N6278A | 17.1 | 5 | 19 | 20 | 21 | 1 | 27.7 | 54 | 0.09 |
| *1.5KE22A* | *1N6279A* | *18.8* | *5* | *20.9* | *22* | *23.1* | *1* | *30.6* | *49* | *0.092* |
| *1.5KE24A* | *1N6280A* | *20.5* | *5* | *22.8* | *24* | *25.2* | *1* | *33.2* | *45* | *0.094* |
| *1.5KE27A* | *1N6281A* | *23.1* | *5* | *25.7* | *27* | *28.4* | *1* | *37.5* | *40* | *0.096* |
| *1.5KE30A* | *1N6282A* | *25.6* | *5* | *28.5* | *30* | *31.5* | *1* | *41.4* | *36* | *0.097* |
| *1.5KE33A* | *1N6283A* | *28.2* | *5* | *31.4* | *33* | *34.7* | *1* | *45.7* | *33* | *0.098* |
| 1.5KE36A | 1N6284A | 30.8 | 5 | 34.2 | 36 | 37.8 | 1 | 49.9 | 30 | 0.099 |
| *1.5KE39A* | *1N6285A* | *33.3* | *5* | *37.1* | *39* | *41* | *1* | *53.9* | *28* | *0.1* |
| 1.5KE43A | 1N6286A | 36.8 | 5 | 40.9 | 43 | 45.2 | 1 | 59.3 | 25.3 | 0.101 |
| 1.5KE47A | 1N6287A | 40.2 | 5 | 44.7 | 47 | 49.4 | 1 | 64.8 | 23.2 | 0.101 |
| *1.5KE51A* | *1N6288A* | *43.6* | *5* | *48.5* | *51* | *53.6* | *1* | *70.1* | *21.4* | *0.102* |
| *1.5KE56A* | *1N6289* | *47.8* | *5* | *53.2* | *56* | *58.8* | *1* | *77* | *19.5* | *0.103* |
| 1.5KE62A | 1N6290A | 53 | 5 | 58.9 | 62 | 65.1 | 1 | 85 | 17.7 | 0.104 |
| 1.5KE68A | 1N6291A | 58.1 | 5 | 64.6 | 68 | 71.4 | 1 | 92 | 16.3 | 0.104 |
| 1.5KE75A | 1N6292A | 64.1 | 5 | 71.3 | 75 | 78.8 | 1 | 103 | 14.6 | 0.105 |
| 1.5KE82A | 1N6293A | 70.1 | 5 | 77.9 | 82 | 86.1 | 1 | 113 | 13.3 | 0.105 |
| 1.5KE91A | 1N6294A | 77.8 | 5 | 86.5 | 91 | 95.5 | 1 | 125 | 12 | 0.106 |
| 1.5KE100A | 1N6295A | 85.5 | 5 | 95 | 100 | 105 | 1 | 137 | 11 | 0.106 |
| 1.5KE110A | 1N6296A | 94 | 5 | 105 | 110 | 116 | 1 | 152 | 9.9 | 0.107 |
| 1.5KE120A | 1N6297A | 102 | 5 | 114 | 120 | 126 | 1 | 165 | 9.1 | 0.107 |
| 1.5KE130A | 1N6298A | 111 | 5 | 124 | 130 | 137 | 1 | 179 | 8.4 | 0.107 |
| 1.5KE150A | 1N6299A | 128 | 5 | 143 | 150 | 158 | 1 | 207 | 7.2 | 0.108 |
| 1.5KE160A | 1N6300A | 136 | 5 | 152 | 160 | 168 | 1 | 219 | 6.8 | 0.108 |
| 1.5KE170A | 1N6301A | 145 | 5 | 162 | 170 | 179 | 1 | 234 | 6.4 | 0.108 |
| 1.5KE180A | 1N6302A | 154 | 5 | 171 | 180 | 189 | 1 | 246 | 6.1 | 0.108 |
| 1.5KE200A | 1N6303A | 171 | 5 | 190 | 200 | 210 | 1 | 274 | 5.5 | 0.108 |
| 1.5KE220A | | 185 | 5 | 209 | 220 | 231 | 1 | 328 | 4.6 | 0.109 |
| 1.5KE250A | | 214 | 5 | 237 | 250 | 263 | 1 | 344 | 5 | 0.109 |

1.  1/2 sine wave (or equivalent square wave), PW = 8.3 ms, duty cycle = 4 pulses per minute maximum.
2.  Indicates JEDEC registered data
3.  A transient suppressor is normally selected according to the maximum working peak reverse voltage (V$_{RWM}$), which should be equal to or greater than the dc or continuous peak operating voltage level.
4.  V$_{BR}$ measured at pulse test current I$_T$ at an ambient temperature of 25°C
5.  Surge current waveform per Figure 5 and derate per Figures 1 and 2.

FIGURE 5.29   (Copyright of Semiconductor Component Industries, LLC. Used by permission.)

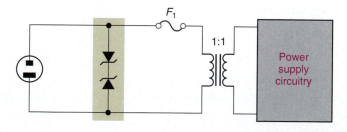

FIGURE 5.30

**ANSWERS TO THE EXAMPLE PRACTICE PROBLEMS**

5.1   387.95 kHz at $V_R = 4$ V, 520.49 kHz at $V_R = 10$ V

5.2   387.95 kHz at $V_R = 4$ V, 401.30 kHz at $V_R = 10$ V

5.3   $P_{PK} = 270$ W at $T = 150°C$ (assuming a derating value of 18%)

5.4   $R_d = -21.05 \ \Omega$

## chapter 6

# Bipolar Junction Transistors

## Objectives

*After studying the material in this chapter, you should be able to:*

1. Name and identify (by schematic symbol) each terminal of the bipolar junction transistor (BJT), and explain its relationship to the other terminals of the transistor.

2. Describe the construction of a BJT.

3. Describe the characteristics of a BJT in the *cutoff*, *saturation*, and the *active* regions of operation.

4. Discuss the transistor as a current-controlled device, and state the relationship among the three terminal currents.

5. Define *beta* and use its value in transistor current calculations.

6. Define *alpha* and use its value in transistor current calculations.

7. Calculate the maximum allowable base current for a transistor given the maximum allowable value of collector current and the maximum beta rating of the device.

8. List and describe the various transistor voltage ratings.

9. Describe the characteristic curves of a BJT.

10. Describe the relationship among *beta*, *temperature*, and *dc collector current*.

11. Describe *thermal resistance* and its effect on the operating temperature of a transistor junction.

12. Describe the five ohmmeter checks used to test for a faulty BJT.

13. Explain the difference between *discrete* and *integrated* transistors.

14. Describe the characteristics of *high-current*, *high-voltage*, and *high-power* transistors.

15. Describe the basic construction of *surface-mount components*, and list the advantages they have over other ICs.

# Outline

## A Common Belief

Popular belief holds that the *bipolar junction transistor*, or BJT, was developed by Schockley, Brattain, and Bardeen in 1948. However, this is not entirely accurate.

The transistor developed by the Bell Laboratories team in 1948 was a *point-contact transistor*. This device consisted of a thin germanium wafer connected to two extremely thin wires. The wires were spaced only a few thousandths of an inch apart. A current introduced to one of the wires was amplified by the germanium wafer, and the larger output current was taken from the other wire.

The BJT wasn't actually developed until late 1951. The component was developed by Dr. Schockley and another Bell Laboratories team. The first time the component was used in any type of commercial venture was in October 1952, when the Bell System employed transistor circuits in the telephone switching circuits in Englewood, N.J.

---

> **Transistor**
>
> A three-terminal device whose output current, voltage, and/or power are controlled by its input.
>
> **Amplifier**
>
> A circuit used to increase the strength of an ac signal.

The main building block of modern electronic systems is the *transistor*. The **transistor** is a *three-terminal device whose output current, voltage, and/or power are controlled by its input*. In communications systems, it is used as the primary component in the **amplifier**, a circuit used to increase the strength of an ac signal. In digital computer systems, the transistor is used as a high-speed electronic switch capable of switching between two operating states (open and closed) at a rate of several billions of times per second.

There are two basic transistor types: the *bipolar junction transistor* (BJT), and the *field-effect transistor* (FET). As you will see, these two transistor types differ in their operating characteristics and their internal construction. You should note that the single term *transistor* is generally used in reference to the BJT. The field-effect transistor is simply referred to as a FET.

In this chapter we will take a look at the bipolar junction transistor and its basic operating principles. FETs are discussed in detail in Chapters 12 and 13.

---

## 6.1 Introduction to Bipolar Junction Transistors (BJTs)

**OBJECTIVE 1 ▶**

> **npn transistor**
>
> A BJT with *n*-type emitter and collector materials and a *p*-type base.
>
> **pnp transistor**
>
> A BJT with *p*-type emitter and collector materials and an *n*-type base.

The *bipolar junction transistor*, or *BJT*, is a *three-terminal* component. The three terminals are called the *emitter*, the *collector*, and the *base*. The emitter and collector are made up of the same type of semiconductor material (either *p*- or *n*-type), while the base is made of the other type of material. The structure of the BJT is illustrated in Figure 6.1.

As you can see, there are two types of BJTs. The first type, called the **npn transistor**, has *n*-type emitter and collector materials and a *p*-type base. The **pnp transistor** is constructed in the opposite manner; this transistor has *p*-type emitter and collector materials and an *n*-type base.

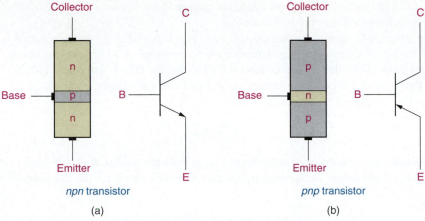

FIGURE 6.1   BJT construction and schematic symbols.

## 6.1.1    BJT Schematic Symbols

Figure 6.1 also shows the schematic symbols for the *npn* and *pnp* transistors. The arrow on the schematic symbol is important for three reasons:

1. It identifies the component terminals. *The arrow is always drawn on the emitter terminal*. The terminal opposite the emitter is the *collector*, and the center terminal is the *base*.
2. *The arrow always points toward the n-type material.* If the arrow points toward the *base*, the transistor is a *pnp* type. If it points toward the emitter, the transistor is an *npn* type.
3. *The arrow indicates the direction of the emitter current.* As with the *pn*-junction diode, electron flow is against the arrow. As you will see, knowing the direction of the emitter current tells you the directions of the other terminal currents.

## 6.1.2    Transistor Currents

The terminal currents of a transistor are illustrated in Figure 6.2. *The emitter, collector, and base currents* of the transistor are identified as $I_E$, $I_C$, and $I_B$, respectively. Under normal circumstances, $I_E$ has the greatest value of the three, followed by $I_C$. The base current ($I_B$) normally has a much lower value than either of the other currents. Note that the current directions for the *npn* transistor are the opposite of those for the *pnp* transistor.

The transistor is a *current-controlled device*; that is, the values of the collector and emitter currents are determined primarily by the value of the base *current*. Under normal circumstances, the values of $I_C$ and $I_E$ vary directly with the value of $I_B$. An increase or

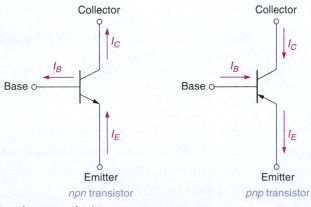

FIGURE 6.2   Transistor terminal currents.

decrease in the value of $I_B$ causes a similar change in the values of $I_C$ and $I_E$. This relationship is discussed in detail in Section 6.2.

The value of $I_C$ for a given transistor is normally some multiple of the value of $I_B$. The factor by which current increases from base to collector is referred to as the forward **current gain** of the device and is represented using the Greek letter *beta* ($\beta$). To determine the value of the collector current for a transistor, you simply multiply the base current by the component's beta rating as follows:

$$I_C = \beta I_B \tag{6.1}$$

**Current gain ($\beta$)**
The factor by which current increases from the base of a transistor to its collector.

For example, assume that the transistor in Figure 6.2a has values of $I_B = 50\ \mu\text{A}$ and $\beta = 120$. Using these values in equation (6.1), the collector current for the transistor is found as

$$I_C = \beta I_B = (120)(50\ \mu\text{A}) = \textbf{6 mA}$$

The significance of this relationship is demonstrated throughout this chapter.

### 6.1.3    Transistor Voltages

Several voltages are normally involved in any discussion of transistor operation. These voltages are described in Table 6.1, and most are identified in Figure 6.3. $V_{CC}$ and $V_{BB}$ (Figure 6.3a) are dc voltage sources that are used to bias the transistor. Some circuits also contain a dc biasing source in the emitter circuit labeled $V_{EE}$. $V_C$, $V_B$, and $V_E$ (Figure 6.3b) are transistor terminal voltages, and each is measured *from the identified transistor terminal to ground*. $V_{CE}$, $V_{CB}$, and $V_{BE}$ (Figure 6.3c) are measured across the identified transistor terminals. Be sure that you learn to distinguish quickly between the various voltages identified in Figure 6.3, as it will make future discussions easier to follow.

TABLE 6.1    Transistor Voltages

| Voltage Abbreviation | Definition |
|---|---|
| $V_{CC}$ | *Collector supply voltage.* This is a power supply voltage applied directly or indirectly to the collector of the transistor. |
| $V_{BB}$ | *Base supply voltage.* This is a dc voltage used to bias the base of the transistor. It may come directly from a dc voltage supply or be applied indirectly to the base by a resistive circuit. |
| $V_{EE}$ | *Emitter supply voltage.* This, again, is a dc biasing voltage. In many cases, $V_{EE}$ is simply a *ground* connection. |
| $V_C$ | This is the dc voltage *measured from the collector terminal of the component to ground*. |
| $V_B$ | This is the dc voltage *measured from the base terminal to ground*. |
| $V_E$ | This is the dc voltage *measured from the emitter terminal to ground*. |
| $V_{CE}$ | This is the dc voltage *measured between the collector and emitter terminals of the transistor*. |
| $V_{BE}$ | This is the dc voltage *measured between the base and emitter terminals of the transistor*. |
| $V_{CB}$ | This is the dc voltage *measured between the collector and base terminals of the transistor*. |

*A Memory Trick:*
There is a relatively simple way to remember the voltages listed here. When the voltage has a double subscript (such as *CC*, *BB*, or *EE*), it is a supply voltage. When two different subscripts are shown (such as *CE*, *BE*, or *CB*), the voltage is measured between the two terminals. When only one subscript is shown, the voltage is measured from that terminal to ground.

**Section Review ▶**

1. What is the primary application for the transistor in communications electronics?
2. What is the primary application for the transistor in digital electronics?
3. What are the two basic types of transistors?
4. What are the terminals of a BJT called?

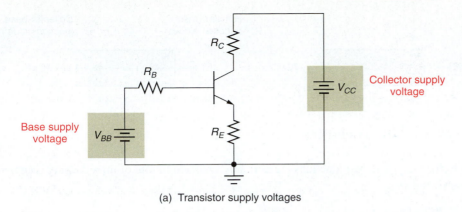

(a) Transistor supply voltages

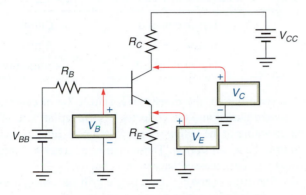

(b) Transistor terminal voltages to ground

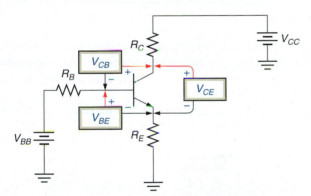

(c) The voltages measured across the transistor junctions

FIGURE 6.3   Transistor amplifier voltages.

5. What are the two types of BJTs? How do they differ from each other?
6. What does the arrow on the BJT schematic symbol indicate?
7. Draw and label the schematic symbols of an *npn* and a *pnp* transistor.
8. Identify the following transistor voltages: $V_{CC}$, $V_{BB}$, $V_{EE}$, $V_C$, $V_E$, $V_B$, $V_{CE}$, $V_{BE}$, and $V_{CB}$.

# 6.2   Transistor Construction and Operation

The transistor is made up of three separate semiconductor materials. The three materials are joined together so that they form two *pn* junctions, as shown in Figure 6.4.

The point at which the emitter and base are joined forms a single *pn* junction called the *base-emitter junction*. The *collector-base junction* is the point where the base and

◄ **OBJECTIVE 2**

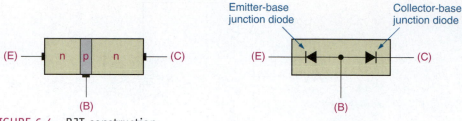

**FIGURE 6.4** BJT construction.

> The two junctions in a BJT are referred to as the *base-emitter junction* and the *collector-base junction*.

collector meet. The two junctions are normally operated in one of three biasing combinations as follows:

| Base-Emitter Junction | Collector-Base Junction | Operating Region |
|---|---|---|
| Reverse biased | Reverse biased | Cutoff |
| Forward biased | Reverse biased | Active |
| Forward biased | Forward biased | Saturation |

> The transistor operating regions are called *cutoff*, *active*, and *saturation*.

When both junctions are reverse biased, the transistor is said to be in *cutoff*. When the base-emitter junction is forward biased and the collector-base junction is reverse biased, the transistor is said to be operating in the *active* region. When both junctions are forward biased, the transistor is said to be in *saturation*. These operating "regions" refer to areas on a characteristic curve that will be discussed later in this chapter.

In our discussion of the transistor operating regions, we will be concentrating on the *npn* transistor. Note that all the principles covered apply equally to the *pnp* transistor.

### 6.2.1    Zero Bias

> **Zero bias**
>
> The biasing of the BJT at room temperature with no potentials applied.

Figure 6.5 shows the *npn* transistor at room temperature with no biasing potential applied. You may recall from Chapter 1 that an unbiased *pn* junction forms a depletion layer at room temperature due to recombination of free carriers produced by thermal energy. These depletion layers are shown in Figure 6.5.

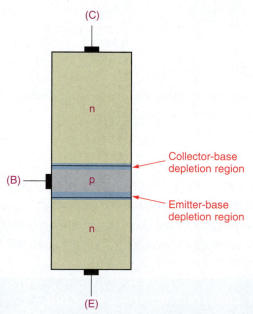

**FIGURE 6.5** Zero biasing.

Since both junctions have a depletion layer, they are both *reverse biased* at room temperature. Note that the depletion layers extend farther into the base region than either of the other two because the base region has a lower doping level. With a lower doping level, there are fewer free carriers for recombination, so the depletion layer extends farther into this region.

## 6.2.2    Cutoff

Figure 6.6 shows a transistor with two biasing potentials applied. As a result of these applied voltages, both transistor junctions are reverse biased.

◀ **OBJECTIVE 3**

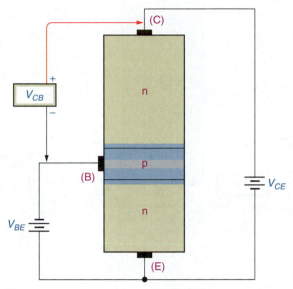

**FIGURE 6.6**   Cutoff.

With the polarities shown, the two depletion layers extend well into the emitter, base, and collector regions. With the larger depletion layers, only an extremely small amount of reverse current passes from the emitter to the collector, and the transistor is said to be in **cutoff**. For example, a 2N3904 transistor with a collector-emitter voltage ($V_{CE}$) of 40 V and a reverse base-emitter voltage ($V_{BE}$) as low as 3 V will allow only 50 nA of collector current ($I_C$). This is extremely small compared to the 200 mA value of $I_C$ that the component is capable of handling when the base-emitter junction is forward biased.

> **Cutoff**
> A BJT operating state where $I_C$ is nearly zero.

## 6.2.3    Saturation

The opposite of cutoff is **saturation**. Saturation is the condition where *further increases in $I_B$ do not result in further increases in $I_C$*. When a transistor is saturated, $I_C$ has reached its maximum possible value, as determined by the collector supply voltage ($V_{CC}$) and the total resistance in the collector-emitter circuit. This point can be explained using the circuit shown in Figure 6.7.

Assume for a moment that $V_{CE}$ for the transistor is 0 V (an *ideal* situation). If this is the case, $I_C$ will depend completely on the values of $V_{CC}$, $R_C$ (the collector resistor), and $R_E$ (the emitter resistor). According to Ohm's law, this maximum value of $I_C$ can be found as

> **Saturation**
> A BJT operating region where $I_C$ reaches its maximum value.

$$I_C = \frac{V_{CC}}{R_C + R_E}$$

Now, let's say that $I_B$ is increased to the point where $I_C$ reaches its maximum value and cannot increase any further. Once this point is reached, additional increases in $I_B$ do not increase the value of $I_C$, and the relationship $I_C = \beta I_B$ no longer holds true.

If $I_B$ is increased beyond the point where $I_C$ can increase, both of the transistor junctions become forward biased. This point is illustrated in Figure 6.7. $V_{CE}$ for the transistor is shown to be approximately 0.3 V, which is typical for a saturated transistor. With the 0.7 V value of $V_{BE}$, the collector-base junction is biased to the difference between the two,

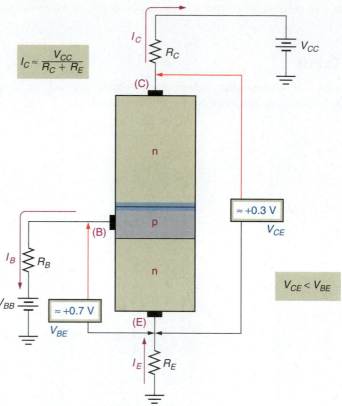

$$I_C \approx \frac{V_{CC}}{R_C + R_E}$$

$\approx +0.3$ V
$V_{CE}$

$V_{CE} < V_{BE}$

$\approx +0.7$ V
$V_{BE}$

**FIGURE 6.7**  Saturation circuit conditions.

0.4 V. Note that this voltage indicates that the collector-base junction of the transistor is *forward* biased (even though it isn't on *fully*).

### 6.2.4    Active Operation

**Active region**

The BJT operating region between saturation (maximum $I_C$) and cutoff (minimum $I_C$).

A transistor is said to be operating in the **active region** when the base-emitter junction is forward biased and the collector-base junction is reverse biased. Generally, the transistor is said to be in active operation when it is between cutoff and saturation. The biasing for active operation is illustrated in Figure 6.8.

The operation of the transistor in this region is easiest to understand by considering just the base-emitter voltage ($V_{BE}$). When $V_{BE}$ is sufficient to overcome the barrier potential of the junction, current is generated in the emitter and base regions.

If the combined base and emitter regions of the transistor acted as a normal diode, all emitter current would exit the component through the base. However, because the base region is very lightly doped, the resistance of the base material is greater than the resistance of the reverse-biased collector-base junction. Thus, the vast majority of the emitter current continues through the reverse-biased collector-base junction to the collector circuit.

The idea of current through a reverse-biased *pn* junction should not seem that strange to you. After all, the zener diode is designed to allow current through a reverse-biased junction. The collector-base junction of the transistor is also designed to allow a reverse current without doing damage to the junction.

### 6.2.5    The Bottom Line

When a transistor is in *cutoff*, both junctions are reverse biased, and the current through all three terminals is nearly zero. When *saturated*, $I_C$ is at its maximum possible value. In this case, both transistor junctions may be forward biased, depending on the value of $I_B$. In either case, $I_C$ is limited by $V_{CC}$ and the resistance in the collector-emitter circuit.

The region of operation between cutoff and saturation is called the *active region*. When a transistor is operating in this region, *its base-emitter junction is forward biased and its collector-base junction is reverse biased*. Because of the light doping of the base,

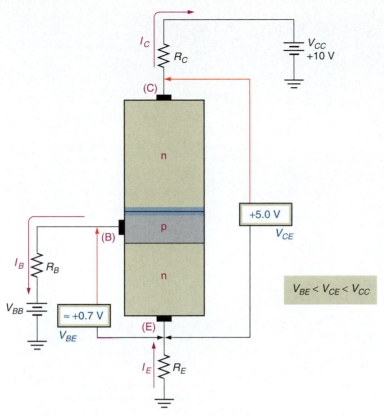

**FIGURE 6.8** Active operation.

very little recombination occurs in the base region, most of $I_E$ passes through to the collector circuit. The base-emitter voltage ($V_{BE}$) is approximately equal to 0.7 V (the barrier potential of the junction). The values of collector-base voltage ($V_{CB}$) and collector-emitter voltage ($V_{CE}$) depend on the amount of current through the transistor and on the values of the external circuit components.

As a reference, the characteristics of the three transistor operating regions are summarized in Figure 6.9.

---

1. How are the two junctions of a transistor biased when the component is in:
   a. Cutoff?
   b. The active region?
   c. Saturation?

2. What is the value of $I_C$ when a transistor is in cutoff?

3. What controls the value of $I_C$ when a transistor is saturated?

4. Describe the basic operation of a transistor biased for active-region operation.

5. How would you go about demonstrating that the depletion layers shown in Figure 6.5 exist when no bias is applied to a transistor?

6. When current passes through any resistance, heat is generated. Based on Figure 6.8, which part of the BJT generates the greatest amount of heat?

◀ **Section Review**

◀ **Critical Thinking**

## 6.3 Transistor Current and Voltage Ratings

There are several transistor current and voltage ratings. Some of these ratings are parameters, and some of them are typical electrical characteristics. In this section, we will take a look at several transistor current and voltage relationships, component ratings, and what they mean in everyday transistor applications.

# Transistor Operating Regions

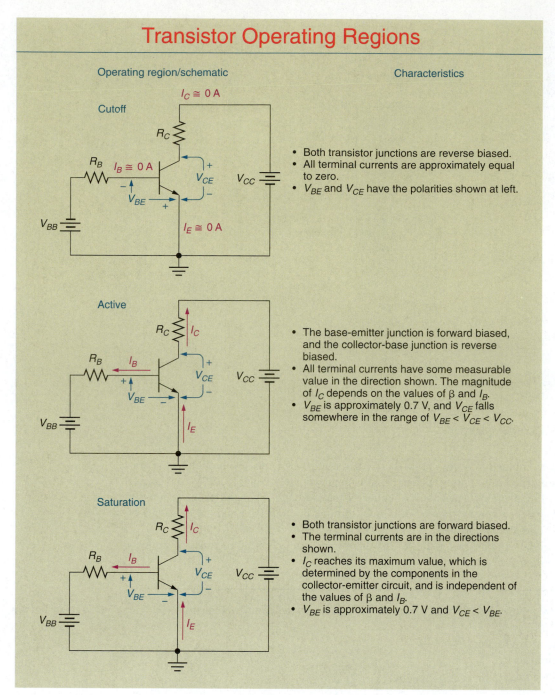

Operating region/schematic                      Characteristics

**Cutoff**

$I_C \cong 0\,A$
$I_B \cong 0\,A$
$I_E \cong 0\,A$

- Both transistor junctions are reverse biased.
- All terminal currents are approximately equal to zero.
- $V_{BE}$ and $V_{CE}$ have the polarities shown at left.

**Active**

- The base-emitter junction is forward biased, and the collector-base junction is reverse biased.
- All terminal currents have some measurable value in the direction shown. The magnitude of $I_C$ depends on the values of $\beta$ and $I_B$.
- $V_{BE}$ is approximately 0.7 V, and $V_{CE}$ falls somewhere in the range of $V_{BE} < V_{CE} < V_{CC}$.

**Saturation**

- Both transistor junctions are forward biased.
- The terminal currents are in the directions shown.
- $I_C$ reaches its maximum value, which is determined by the components in the collector-emitter circuit, and is independent of the values of $\beta$ and $I_B$.
- $V_{BE}$ is approximately 0.7 V and $V_{CE} < V_{BE}$.

**FIGURE 6.9**

## 6.3.1    Transistor Currents

*OBJECTIVE 4* ▶ As you know, the transistor is a current-controlled device. In many applications, the base current is varied to produce variations in $I_C$ and $I_E$. Because of the construction of the component, *a small change in $I_B$ results in a larger change in the other terminal currents*. This relationship is illustrated in Figure 6.10. As you can see, the small increase in base current (from 10 μA to 20 μA) produces a larger increase in $I_C$ and in $I_E$ (from 2 mA to 4 mA). The larger increase in the emitter and collector currents is due to the current gain of the transistor. This point is illustrated in Example 6.1.

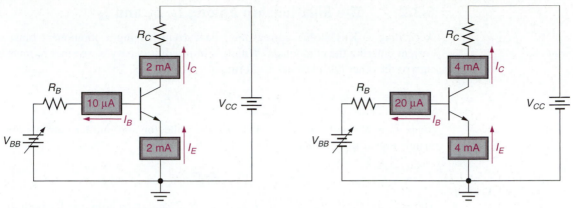

FIGURE 6.10   BJT current relationships.

## EXAMPLE 6.1

Determine the values of collector current for the values of base current shown in Figure 6.11.

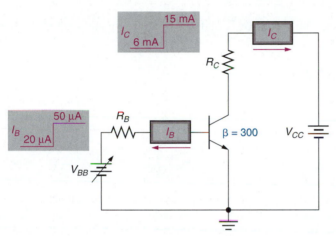

FIGURE 6.11

***Solution:***   The base current in Figure 6.11 has an initial value of 20 μA. The beta rating of the component is 300. Using these values, the initial value of collector current is found as

$$I_C = \beta I_B = (300)(20\ \mu A) = \textbf{6 mA}$$

When $I_B$ increases to 50 μA, the collector current also increases. At the new value of $I_B$, the collector current is found as

$$I_C = \beta I_B = (300)(50\ \mu A) = \textbf{15 mA}$$

Thus, a 30 μA change in base current causes a 9 mA change in collector current.

### PRACTICE PROBLEM 6.1

A transistor has values of $I_B = 50$ μA and $\beta = 350$. Determine the value of $I_C$ for the device.

The effect that a change in base current has on the output of a transistor will be demonstrated further when we discuss the ac operation of transistors.

### 6.3.2 The Relationship Among $I_E$, $I_C$, and $I_B$

According to Kirchhoff's current law, the current leaving a component must equal the current entering the component. With this in mind, it is easy to see that $I_E$ must equal the sum of the other two currents. By formula,

$$I_E = I_B + I_C \qquad (6.2)$$

Since $I_B$ is normally *much less* than $I_C$, the collector and emitter currents are approximately equal. By formula,

$$I_C \cong I_E \qquad (6.3)$$

The current relationship shown in equations (6.2) and (6.3) hold true for both the *npn* and *pnp* transistors.

The validity of equation (6.3) can be seen by looking at the results in Example 6.1. After the increase, $I_B$ was given as 50 µA and $I_C$ was determined to be 15 mA. According to equation (6.2), $I_E$ for the device can be found as

$$I_E = I_B + I_C = 50\ \mu A + 15\ mA = \textbf{15.05 mA}$$

As you can see, the values of $I_E$ and $I_C$ are approximately equal for the transistor in Example 6.1.

### 6.3.3 DC Beta

**OBJECTIVE 5** ▶

The **dc beta (β)** rating of a transistor is the *ratio of dc collector current to dc base current*. By formula,

$$\beta = \frac{I_C}{I_B} \qquad (6.4)$$

> **dc beta (β)**
> The ratio of dc collector current to dc base current.

This is an extremely important rating because *the most common transistor circuits have an input signal applied to the base and an output signal taken from the collector*. Thus, when the transistor is used in these circuits, the dc beta rating of the transistor represents *the overall dc current gain* of the transistor.

We can use equations (6.1) and (6.2) to define the other terminal currents. As given in equation (6.1),

$$I_C = \beta I_B$$

> **Lab Reference:** The dc beta of a transistor is determined using measured current values in Exercise 6.

If we combine this relationship with equation (6.2), we get

$$I_E = I_C + I_B = \beta I_B + I_B$$

or

$$I_E = I_B(\beta + 1) \qquad (6.5)$$

As Examples 6.2, 6.3, and 6.4 illustrate, you can use beta and any one terminal current to find the other two terminal currents.

#### EXAMPLE 6.2

Determine the values of $I_C$ and $I_E$ for the circuit shown in Figure 6.12.

*Solution:* The value of $I_C$ can be found as

$$I_C = \beta I_B = (200)(125\ \mu A) = \textbf{25 mA}$$

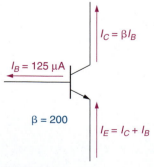

$I_C = \beta I_B$

$I_B = 125\ \mu A$

$\beta = 200$

$I_E = I_C + I_B$

**FIGURE 6.12**

and $I_E$ is found as

$$I_E = I_C + I_B = 25 \text{ mA} + 125 \text{ μA} = \textbf{25.125 mA}$$

**PRACTICE PROBLEM 6.2**
A transistor has values of $I_B = 50$ μA and $\beta = 400$. Determine the values of $I_C$ and $I_E$ for the device circuit.

## EXAMPLE 6.3

Determine the values of $I_C$ and $I_B$ for the circuit shown in Figure 6.13.

*Solution:*  Equation (6.5) can be rearranged to give us

$$I_B = \frac{I_E}{\beta + 1} = \frac{15 \text{ mA}}{200 + 1} = \textbf{74.6 μA}$$

Now, $I_C$ can be found as

$$I_C = I_E - I_B = 15 \text{ mA} - 74.6 \text{ μA} = \textbf{14.9 mA}$$

or

$$I_C = \beta I_B = (200)(74.6 \text{ μA}) = \textbf{14.9 mA}$$

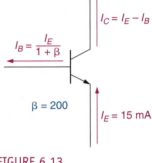

FIGURE 6.13

**PRACTICE PROBLEM 6.3**
A transistor has values of $I_E = 12$ mA and $\beta = 140$. Determine the values of $I_B$ and $I_C$ for the device.

## EXAMPLE 6.4

Determine the values of $I_B$ and $I_E$ for the circuit shown in Figure 6.14.

*Solution:*  The base current can be found as

$$I_B = \frac{I_C}{\beta} = \frac{50 \text{ mA}}{400} = \textbf{125 μA}$$

Now the emitter current can be found as

$$I_E = I_C + I_B = 50 \text{ mA} + 125 \text{ μA} = \textbf{50.125 mA}$$

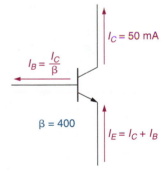

FIGURE 6.14

**PRACTICE PROBLEM 6.4**
A transistor has values of $I_C = 80$ mA and $\beta = 170$. Determine the values of $I_B$ and $I_E$ for the device.

Because the beta rating of a transistor is a *ratio* of current values, it has no unit of measure. Typical beta ratings can be as high as 300. This means that the typical transistor has a dc collector current that can be up to 300 times the value of the dc base current (when operated in the active region).

There is one point that should be made at this time: *Transistors have both dc beta ratings and ac beta ratings.* The ratio of *dc* collector current to *dc* base current is *dc beta*, while *ac beta* is the ratio of *ac* collector current to *ac* base current. We will discuss ac beta and its applications in Chapter 9.

Transistors have both dc and ac beta ratings.

### 6.3.4　DC Alpha

**OBJECTIVE 6 ▶**

The **dc alpha (α)** rating of a transistor is the *ratio of collector current to emitter current*. By formula,

**dc alpha (α)**
The ratio of dc collector current to dc emitter current. Also referred to as *collector current efficiency*.

$$\alpha = \frac{I_C}{I_E} \tag{6.6}$$

The alpha rating of a given transistor is *always less than unity* (1). The reason for this is illustrated in Figure 6.15.

Kirchhoff's current law states that the current leaving a point (or component) must equal the current entering the point (or component). The relationship among the three transistor terminal currents was stated in equation (6.2) as

$$I_E = I_B + I_C$$

Therefore,

$$I_C = I_E - I_B$$

Since $I_C$ is always less than $I_E$ (by an amount equal to $I_B$), the fraction $I_C/I_E$ must always work out to be less than 1.

The alpha rating of a transistor is usually 0.9 or higher. Note that alpha (like beta) has no units because it is a ratio between two current values. Equation (6.6) can be rearranged to give us the following useful relationships:

$$I_C = \alpha I_E \tag{6.7}$$

and

$$I_E = \frac{I_C}{\alpha} \tag{6.8}$$

Using these relationships, we can calculate base current ($I_B$) as

$$I_B = I_E - I_C = I_E - \alpha I_E$$

or

$$I_B = I_E(1 - \alpha) \tag{6.9}$$

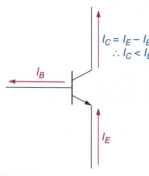

$I_C = I_E - I_B$
$\therefore I_C < I_E$

$I_B$

$I_E$

**FIGURE 6.15**　Why alpha is always less than unity.

### 6.3.5　The Relationship Between Alpha and Beta

As you will see later in this chapter, the spec sheet for a given transistor lists the value of beta for the device, but not the value of alpha. This is because beta is used far more commonly than alpha in transistor circuit calculations. (This fact will become evident when we cover the dc and ac analyses of transistor circuits.)

Since alpha is rarely listed on transistor spec sheets, you should be able to determine its value using the value of beta. You can determine the value of alpha using the rated value of beta with the following equation:

$$\alpha = \frac{\beta}{\beta + 1} \tag{6.10}$$

Example 6.5 illustrates the process of determining the value of alpha when the value of dc beta is known. It also demonstrates the validity of equation (6.10).

## EXAMPLE 6.5

Determine the value of alpha for the transistor shown in Figure 6.16. Then, determine the value of $I_C$ using both the alpha and the beta ratings of the transistor.

*Solution:* The beta rating is given as 300. Therefore, the alpha rating is found as

$$\alpha = \frac{\beta}{\beta + 1} = \frac{300}{301} = \mathbf{0.9967}$$

The value of $I_C$ can be found as

$$I_C = \alpha I_E = (0.9967)(30 \text{ mA}) = \mathbf{29.9 \text{ mA}}$$

or $I_C$ can be found as

$$I_C = \beta I_B = (300)(100 \text{ μA}) = \mathbf{30.0 \text{ mA}}$$

The two values of $I_C$ are close enough to each other to validate both methods of calculating $I_C$.

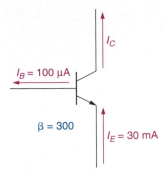

FIGURE 6.16

### PRACTICE PROBLEM 6.5

A transistor has the following values: $\beta = 349$, $I_E = 350$ mA, and $I_B = 1$ mA. Determine the value of $\alpha$ for the device. Then, calculate the value of $I_C$ using both $\alpha$ and $\beta$. How do the two values of $I_C$ compare?

## 6.3.6 Maximum Current Ratings

Most transistor spec sheets list maximum collector current ratings for both saturation and cutoff. When the transistor is saturated, the collector current can go as high as several hundred milliamperes. High-power transistors typically have current ratings as high as several amperes.

The maximum allowable base current for a given transistor can be found by dividing its maximum $I_C$ value by its *maximum* dc beta rating. By formula,

$$I_{B(max)} = \frac{I_{C(max)}}{\beta_{max}} \tag{6.11}$$

◄ **OBJECTIVE 7**

*A Practical Consideration:*
The beta rating of a transistor is usually listed as a *range* of values on the device spec sheet. Dealing with beta *ranges* is addressed later in this chapter.

Example 6.6 demonstrates the use of this equation.

## EXAMPLE 6.6

The transistor shown in Figure 6.17 has the following ratings: $I_{C(max)} = 500$ mA and $\beta_{max} = 300$. Determine the maximum allowable value of $I_B$ for the device.

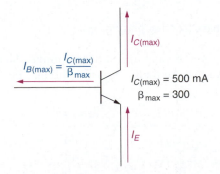

FIGURE 6.17

***Solution:*** Using the ratings given, the value of $I_{B(\text{max})}$ is found as

$$I_{B(\text{max})} = \frac{I_{C(\text{max})}}{\beta_{(\text{max})}} = \frac{500 \text{ mA}}{300} = \textbf{1.67 mA}$$

If the base current is allowed to exceed 1.67 mA for the transistor shown, the collector current may exceed its maximum rating of 500 mA, and the transistor will probably be destroyed.

**PRACTICE PROBLEM 6.6**
A transistor has ratings of $I_{C(\text{max})} = 1$ A and $\beta_{\text{max}} = 120$. Determine the value of $I_{B(\text{max})}$ for the device.

Transistors also have maximum *cutoff* current ratings. These ratings are usually in the low-nanoampere range and are specified for exact values of $V_{CE}$ and reverse $V_{BE}$. It was stated earlier that the 2N3904 has a maximum cutoff current rating of 50 nA when the reverse value of $V_{BE}$ is 3 V and the value of $V_{CE}$ is 40 V. These values are illustrated in Figure 6.18.

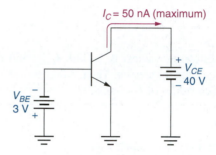

**FIGURE 6.18**

## 6.3.7    Transistor Voltage Ratings

**OBJECTIVE 8 ▶**

Most transistor spec sheets list a maximum value of *collector-base voltage* ($V_{CB}$). This rating indicates the maximum amount of reverse bias that can be applied to the collector-base junction without damaging the transistor. This rating is important because the collector-base junction is reverse biased for active region operation, as shown in Figure 6.19. In this circuit, $V_{CE}$ is 40 V and $V_{BE}$ is 0.75 V. The value of $V_{CB}$ is equal to the difference between the other two voltages: 39.25 V. If this voltage exceeds the $V_{CB}$ rating of the transistor, the component will probably be destroyed.

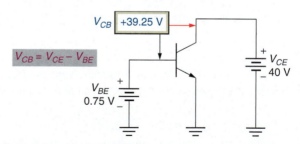

**FIGURE 6.19**    Collector-base junction biasing.

Every transistor has three breakdown voltage ratings. These ratings indicate the *maximum reverse voltages* that the transistor can withstand. For the 2N3904, these voltage ratings are as follows:

| Rating | Value ($V_{\text{dc}}$) |
|---|---|
| $V_{CBO}$ | 60 |
| $V_{CEO}$ | 40 |
| $V_{EBO}$ | 6 |

These voltages are illustrated in Figure 6.20. If any of these reverse voltage ratings are exceeded, the transistor may not survive the experience.

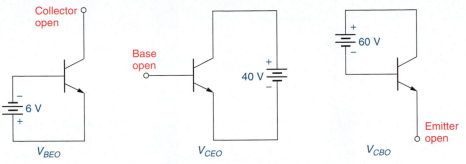

FIGURE 6.20   BJT breakdown voltage ratings.

A *Practical Consideration:* The "O" in the $V_{CEO}$, $V_{EBO}$, and $V_{CBO}$ ratings indicates that the third terminal is *open* when the rating is measured. For example, $V_{EBO}$ is measured with the collector terminal open. This ensures that the BJT is in cutoff when the parameter is measured.

There are many transistor ratings that have not been covered in this section. These ratings include *junction capacitance*, *maximum power dissipation*, *frequency limitations*, *operating temperature ranges*, and others. All these ratings will be covered as they are needed in future chapters.

◄ **Section Review**

1. What is meant by the term *current-controlled device?*
2. What is meant by the term *current gain?*
3. What symbol is commonly used to represent dc current gain?
4. Under normal circumstances, what is the relationship between:
   a. Base current and collector current?
   b. Collector current and emitter current?
5. What is *dc beta?*
6. Why doesn't beta have any units of measure?
7. What are the two types of beta ratings?
8. What is *dc alpha?*
9. What is the limit on the value of alpha?
10. Between beta and alpha, which rating is used more commonly?
11. Why do you need to be able to determine the value of alpha using the value of beta? How is this determination made?
12. What are the commonly used transistor voltage ratings?
13. What will happen if any of the reverse voltage ratings of a transistor are exceeded?
14. Equation (6.10) defines alpha ($\alpha$) in terms of beta ($\beta$). Why wouldn't we be interested in an equation defining beta ($\beta$) in terms of alpha ($\alpha$)?

◄ **Critical Thinking**

## 6.4    Transistor Characteristic Curves

In this section, we are going to take a look at three characteristic curves that illustrate the operation of the transistor. We will start by looking at the collector and base curves. (There is no need for an emitter curve since its current characteristics are the same as those of the collector.) We will also take a look at the beta curve, which shows the relationship among beta, $I_C$, and temperature.

◄ *OBJECTIVE 9*

### 6.4.1    Collector Curves

The **collector characteristic curve** illustrates the relationship among $I_C$, $I_B$, and $V_{CE}$. Each collector curve is derived for a specified value of $I_B$. Two such curves are shown in Figure 6.21. The upper curve represents the relationship between $I_C$ and $V_{CE}$ when $I_B = 100$ μA.

**Collector characteristic curve**
Relates the values of $I_C$, $I_B$, and $V_{CE}$.

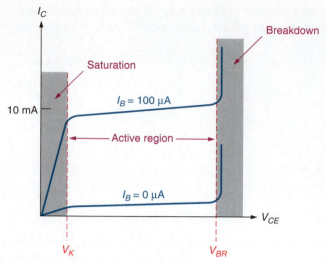

**FIGURE 6.21** A collector characteristic curve.

The lower curve represents the relationship between $I_C$ and $V_{CE}$ when $I_B = 0$ μA. Note that the $I_B = 0$ μA line represents the operation of the transistor when it is in *cutoff*.

As you can see, a collector curve is divided into three parts. The portion of the curve that lies below the knee voltage ($V_K$) represents the *saturation* characteristics of the device. The portion of the curve between $V_K$ and the breakdown voltage ($V_{BR}$) represents the characteristics of the device for its *active region* of operation. The portion of the curve to the right of $V_{BR}$ represents the characteristics of the device when driven into *breakdown*. To help you understand each portion of the characteristic curve better, we will discuss the operation of the circuit shown in Figure 6.22. (We will assume that the collector curve in Figure 6.21 applies to the transistor in this circuit.)

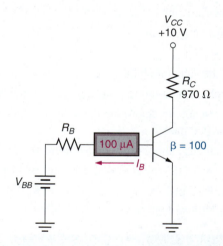

**FIGURE 6.22**

Earlier in the chapter, you were told that both junctions in a saturated transistor are forward biased. You were also told that the collector current is limited by the components external to the transistor. Now, consider the circuit in Figure 6.22. According to Kirchhoff's voltage law, the value of $V_{CE}$ for this circuit must equal the difference between $V_{CC}$ and the voltage across the collector resistor; that is,

$$V_{CE} = V_{CC} - V_{RC} \tag{6.12}$$

Substituting $I_C R_C$ for $V_{RC}$, we get

$$V_{CE} = V_{CC} - I_C R_C \tag{6.13}$$

The input to the base of the transistor is shown as $I_B = 100\ \mu A$, the value of $R_C$ is 970 $\Omega$, and beta is 100. These values give us the following output values:

$$I_C = \beta I_B = (100)(100\ \mu A) = \textbf{10 mA}$$

and

$$V_{CE} = V_{CC} - I_C R_C = 10\ V - (10\ mA)(970\ \Omega) = 10\ V - 9.7\ V = \textbf{0.3 V}$$

Now, let's relate these values to the transistor shown in Figure 6.23. As you can see, the value of $V_{CE} = 0.3$ V. If we assume that the transistor has a value of $V_{BE} = 0.75$ V, we get the voltages shown in the figure. The $-0.45$ V value of $V_{CB}$ equals the difference between the $V_{BE}$ and $V_{CE}$. In this case, the collector-base junction is forward biased, as is the base-emitter junction. These are characteristics of a transistor in saturation.

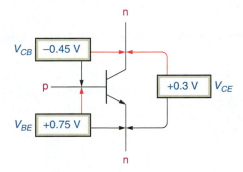

FIGURE 6.23

Most linear amplifiers make use of the *active region* of transistor operation. Figure 6.21 indicates that $I_C$ is determined by the values of $I_B$ and beta and is relatively independent of the value of $V_{CE}$ when a transistor is operating in its active region. There is little change in the value $I_C$ when $V_{CE}$ increases from $V_K$ to $V_{BR}$.

Changing the value of $R_C$ to 400 $\Omega$ in Figure 6.22 gives us the following conditions:

$$I_C = \beta I_B = (100)(100\ \mu A) = \textbf{10 mA}$$

and

$$V_{CE} = V_{CC} - I_C R_C = 10\ V - (10\ mA)(400\ \Omega) = \textbf{6 V}$$

As you can see, changing the value of $R_C$ has not affected the value of $I_C$. However, it *has* caused the value of $V_{CE}$ to change from 0.3 to 6 V. This demonstrates the fact that $I_C$ is not controlled by the value of $V_{CE}$ when a transistor is operated in its active region. The value of $I_C$ varies primarily as a result of changes in $I_B$ or beta.

Figure 6.24 shows what happens when the value of $I_B$ is increased to 150 $\mu A$. With beta held constant at 100, the value of $I_C$ increases proportionately with the increase in $I_B$. This change in $I_C$ is still relatively independent of changes in $V_{CE}$.

Transistor *breakdown* occurs when the value of $V_{CE}$ exceeds the breakdown voltage rating of the transistor. The value of $I_C$ increases dramatically until the transistor is destroyed by the excessive heat that results from the increase in current.

When several $I_B$ versus $I_C$ curves are plotted for a given transistor, a composite graph similar to the one in Figure 6.25 is created. The graph shows the collector currents produced by fixed values of $I_B$ for the transistor.

> *A Practical Consideration:*
> $I_C$ increases slightly when $V_{CE}$ increases over its entire range. In most cases, the change in $I_C$ is small enough to idealize to zero. However, collector currents in high-power transistors can increase by as much as 25% when $V_{CE}$ increases over its entire range.

> Most collector characteristic curves are composite curves.

## 6.4.2 Base Curves

The **base curve** of a transistor plots $I_B$ as a function of $V_{BE}$, as shown in Figure 6.26. Note that this curve closely resembles the forward operating curve of a typical *pn*-junction diode.

> **Base curve**
> A curve illustrating the relationship between $I_B$ and $V_{BE}$.

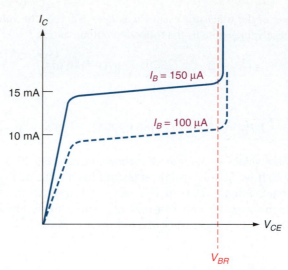

FIGURE 6.24   The effect of changing $I_B$ on a collector characteristic curve.

**Lab Reference:** A composite of collector curves is developed using measured values in Exercise 6.

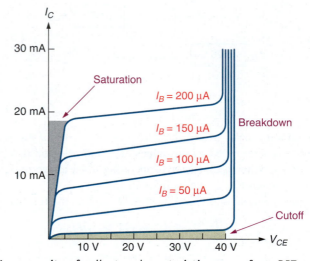

FIGURE 6.25   A composite of collector characteristic curves for a BJT.

### 6.4.3   Beta Curves

OBJECTIVE 10 ▶

**Beta curve**

A curve that shows the relationship between beta and temperature and/or collector current.

**Beta curves** show how the value of dc beta varies with both *temperature* and *dc collector current*. This point is illustrated in Figure 6.27. As you can see, the value of beta is greater at $T = 100°C$ than it is at 25°C. Also, beta increases (up to a point) for increases in the dc value of $I_C$. However, when $I_C$ increases beyond a certain value, beta starts to decrease.

The spec sheet for the 2N3904 transistor lists the following minimum beta values, each measured at the indicated value of $I_C$:

| Minimum Beta | dc Collector Current |
| --- | --- |
| 40 | $I_C = 0.1$ mA |
| 70 | $I_C = 1.0$ mA |
| 100 | $I_C = 10$ mA |
| 60 | $I_C = 50$ mA |
| 30 | $I_C = 100$ mA |

As you can see, the value of beta increases as $I_C$ is increased to 10 mA. However, as $I_C$ increases above 10 mA, the value of beta begins to decrease. This goes along with the curves shown in Figure 6.27.

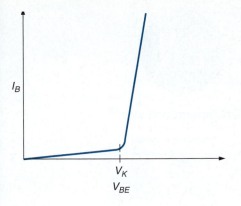

FIGURE 6.26   A base characteristic curve.

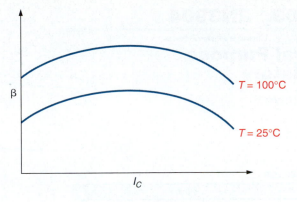

FIGURE 6.27   The relationship among beta, $I_C$, and temperature.

◀ Section Review

1. What does the saturation region represent on the transistor collector curve?

2. What does the active region represent on the transistor collector curve?

3. Does the value of $I_C$ depend on the value of $V_{CE}$? Explain your answer.

4. What happens when a transistor is operated in its breakdown region?

5. The base curve in Figure 6.26 indicates that the base-emitter junction of a transistor acts like what component? Explain your answer.

6. What is the relationship between beta and temperature?

7. What is the relationship between beta and dc collector current?

## 6.5    Transistor Specification Sheets

The spec sheet for a given transistor contains a wide variety of dc and ac operating characteristics. In this section, we will look at some of the commonly used *dc* maximum ratings and electrical characteristics. In Chapter 9, we will look at the commonly used *ac* characteristics and operating curves. For our discussion on the transistor *dc* ratings, we will use the spec sheet for the 2N3903/2N3904 transistors shown in Figure 6.28.

### 6.5.1    Maximum Ratings

You have already been introduced to many of the maximum ratings listed in Figure 6.28. The $V_{CEO}$, $V_{CBO}$, and $V_{EBO}$ ratings are the maximum reverse voltage ratings that we discussed in the last section. The $I_C$ rating is the maximum allowable continuous value of $I_C$, or 200 mA in this case.

The *total device dissipation* ($P_D$) rating of the transistor is the same type of rating that is used for the zener diode and *pn*-junction diode. As the spec sheet shows, the 2N3904 has a $P_D$ rating of 625 mW when the *ambient* (or *room*) temperature ($T_A$) is 25°C. If the *case* temperature ($T_C$) is held to 25°C, the device $P_D$ rating increases to 1.5 W. Note that the case temperature can be held to 25°C by fan cooling or through the use of a heat sink. As always, both ratings must be derated as temperature increases.

### 6.5.2    Thermal Characteristics

Transistor thermal ratings are used primarily in circuit development applications, so we will not discuss them in depth. However, you should be aware of these ratings and their implications.

◀ OBJECTIVE 11

Just as materials provide opposition to the flow of charge, they also provide opposition to the flow of heat (power dissipation). The opposition of any device to the flow of heat (power dissipation) is called its **thermal resistance ($R_\theta$)**.

> **Thermal resistance ($R_\theta$)**
> Any opposition to the flow of heat (power dissipation).

# 2N3903, 2N3904

2N3903 is a Preferred Device

## General Purpose Transistors

### NPN Silicon

**ON Semiconductor™**

http://onsemi.com

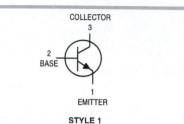

STYLE 1

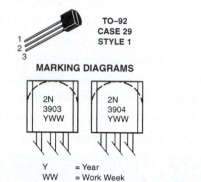

TO–92
CASE 29
STYLE 1

**MARKING DIAGRAMS**

| 2N 3903 YWW | 2N 3904 YWW |

Y = Year
WW = Work Week

### MAXIMUM RATINGS

| Rating | Symbol | Value | Unit |
|--------|--------|-------|------|
| Collector–Emitter Voltage | $V_{CEO}$ | 40 | Vdc |
| Collector–Base Voltage | $V_{CBO}$ | 60 | Vdc |
| Emitter–Base Voltage | $V_{EBO}$ | 6.0 | Vdc |
| Collector Current – Continuous | $I_C$ | 200 | mAdc |
| Total Device Dissipation @ $T_A$ = 25°C<br>Derate above 25°C | $P_D$ | 625<br>5.0 | mW<br>mW/°C |
| Total Device Dissipation @ $T_C$ = 25°C<br>Derate above 25°C | $P_D$ | 1.5<br>12 | Watts<br>mW/°C |
| Operating and Storage Junction Temperature Range | $T_J$, $T_{stg}$ | –55 to +150 | °C |

### THERMAL CHARACTERISTICS (Note 1.)

| Characteristic | Symbol | Max | Unit |
|----------------|--------|-----|------|
| Thermal Resistance, Junction to Ambient | $R_{\theta JA}$ | 200 | °C/W |
| Thermal Resistance, Junction to Case | $R_{\theta JC}$ | 83.3 | °C/W |

1. Indicates Data in addition to JEDEC Requirements.

### ELECTRICAL CHARACTERISTICS  ($T_A$ = 25°C unless otherwise noted)

| Characteristic | | Symbol | Min | Max | Unit |
|----------------|---|--------|-----|-----|------|
| **OFF CHARACTERISTICS** | | | | | |
| Collector–Emitter Breakdown Voltage (Note 2.) ($I_C$ = 1.0 mAdc, $I_B$ = 0) | | $V_{(BR)CEO}$ | 40 | – | Vdc |
| Collector–Base Breakdown Voltage ($I_C$ = 10 μAdc, $I_E$ = 0) | | $V_{(BR)CBO}$ | 60 | – | Vdc |
| Emitter–Base Breakdown Voltage ($I_E$ = 10 μAdc, $I_C$ = 0) | | $V_{(BR)EBO}$ | 6.0 | – | Vdc |
| Base Cutoff Current ($V_{CE}$ = 30 Vdc, $V_{EB}$ = 3.0 Vdc) | | $I_{BL}$ | – | 50 | nAdc |
| Collector Cutoff Current ($V_{CE}$ = 30 Vdc, $V_{EB}$ = 3.0 Vdc) | | $I_{CEX}$ | – | 50 | nAdc |
| **ON CHARACTERISTICS** | | | | | |
| DC Current Gain (Note 2.) | | $h_{FE}$ | | | – |
| ($I_C$ = 0.1 mAdc, $V_{CE}$ = 1.0 Vdc) | 2N3903 | | 20 | – | |
| | 2N3904 | | 40 | – | |
| ($I_C$ = 1.0 mAdc, $V_{CE}$ = 1.0 Vdc) | 2N3903 | | 35 | – | |
| | 2N3904 | | 70 | – | |
| ($I_C$ = 10 mAdc, $V_{CE}$ = 1.0 Vdc) | 2N3903 | | 50 | 150 | |
| | 2N3904 | | 100 | 300 | |
| ($I_C$ = 50 mAdc, $V_{CE}$ = 1.0 Vdc) | 2N3903 | | 30 | – | |
| | 2N3904 | | 60 | – | |
| ($I_C$ = 100 mAdc, $V_{CE}$ = 1.0 Vdc) | 2N3903 | | 15 | – | |
| | 2N3904 | | 30 | – | |
| Collector–Emitter Saturation Voltage (Note 2.) | | $V_{CE(sat)}$ | | | Vdc |
| ($I_C$ = 10 mAdc, $I_B$ = 1.0 mAdc) | | | – | 0.2 | |
| ($I_C$ = 50 mAdc, $I_B$ = 5.0 mAdc | | | – | 0.3 | |
| Base–Emitter Saturation Voltage (Note 2.) | | $V_{BE(sat)}$ | | | Vdc |
| ($I_C$ = 10 mAdc, $I_B$ = 1.0 mAdc) | | | 0.65 | 0.85 | |
| ($I_C$ = 50 mAdc, $I_B$ = 5.0 mAdc) | | | – | 0.95 | |

**FIGURE 6.28** The 2N3903/3904 specification sheet. (Copyright of Semiconductor Component Industries, LLC. Used by permission.)

The concept of thermal resistance can be illustrated as shown in Figure 6.29. Here, we have a heat source that is surrounded by a material. As the heat leaves the source, it must pass through the material to reach the air. If the material had no thermal resistance, the heat source and the outer surface of the material would be at the same temperature. In other words, *the temperature differential between the heat source and the outer surface of the material would be approximately 0°C*. However, since all materials provide some opposition to the flow of heat, *the temperature of the heat source will be greater than the temperature at the outer surface of the material*. The temperature differential between the heat source and the surrounding air depends on:

**1.** The thermal resistance of the material.
**2.** The amount of power being transferred.

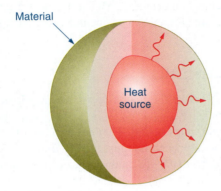

**Material**

**Heat source**

**FIGURE 6.29** The transfer of heat through a material.

Thermal resistance is measured in *degrees Celsius per watt* (°C/W), which is consistent with our description of thermal resistance. For example, the thermal resistance of an *ideal* conductor of heat would be 0°C/W, meaning that there would be no temperature differential across the material, regardless of the amount of power being dissipated. On the other hand, the thermal resistance of the 2N3904 (from junction to ambient) is 200°C/W (as shown in Figure 6.28). This means that *the temperature of the transistor collector-base junction is 200°C higher than the surrounding air when the device is dissipating 1 W*.

The temperature differential between two points can be found as the product of thermal resistance and power dissipation. For example, if a 2N3904 is dissipating 0.5 W, the difference in temperature between the collector-base junction and the surrounding air ($T_{JA}$) can be found as

$$\boldsymbol{T_{JA}} = R_{\theta JA}P_D = (200°C/W)(500 \text{ mW}) = \mathbf{100°C}$$

If the surrounding air is at 25°C, then the collector-base junction temperature of the device is found as

$$T_J = T_A + T_{JA} = 25°C + 100°C = \mathbf{125°C}$$

Note that the junction temperature equals the sum of the ambient temperature ($T_A$) and the temperature differential ($T_{JA}$).

The various temperature and power ratings of a transistor are related to each other because each is determined by the physical characteristics of the device. The relationship among the various temperature and power ratings can be demonstrated using the following 2N3904 ratings:

$$P_D = 625 \text{ mW (maximum)} \qquad R_{\theta JA} = 200°C/W \qquad T_J = 150°C \text{ (maximum)}$$

Note that the power dissipation rating assumes an ambient temperature of 25°C. Based on a maximum power dissipation of 625 mW, the maximum temperature differential between the collector-base junction and the surrounding air ($T_{JA}$) can be found as

$$T_{JA} = (R_{\theta JA})(P_D) = (200°C/W)(625 \text{ mW}) = \textbf{125°C}$$

Adding this value to the ambient temperature listed in the ratings (25°C), we get a maximum junction temperature of

$$T_J = T_A + T_{JA} = 25°C + 125°C = \textbf{150°C}$$

which is the rated maximum temperature for the device.

Thermal resistance is a part of any circuit design involving high component power dissipation and/or high operating temperatures. Most often, it is used in determining the type of component cooling required for a given device in a specified application.

### 6.5.3    Off Characteristics

Off characteristics
The characteristics of a transistor that is in *cutoff*.

Collector cutoff current ($I_{CEX}$)
The maximum value of $I_C$ through a cutoff transistor.

Base cutoff current ($I_{BL}$)
The maximum amount of current through a reverse-biased emitter-base junction.

The **off characteristics** describe the operation of the transistor when it is operated in *cutoff*. The first three ratings, $V_{(BR)CEO}$, $V_{(BR)CBO}$, and $V_{(BR)EBO}$, are the same ratings that appeared in the maximum ratings listing. They are repeated in this table for the convenience of the technician or engineer who must rapidly find the ratings.

The **collector cutoff current ($I_{CEX}$)** rating indicates the maximum value of $I_C$ when the device is in cutoff. For the 2N3904, $I_{CEX}$ is 50 nA.

The **base cutoff current ($I_{BL}$)** rating indicates the maximum amount of base current present when the emitter-base junction is in cutoff. For the 2N3904, $I_{BL}$ is a maximum of 50 nA.

As the $I_{BL}$ and $I_{CEX}$ ratings indicate, the terminal currents of a cutoff transistor are extremely low. In the case of the 2N3904, $I_B$ and $I_C$ each have a maximum value of 50 nA. The value of $I_E$ can be no greater than the sum of the two, or 100 nA.

### 6.5.4    On Characteristics

On characteristics
The characteristics of a transistor that is in either the active or saturation region of operation.

dc current gain ($h_{FE}$)
The label that is commonly used to represent dc beta.

Collector-emitter saturation voltage ($V_{CE(sat)}$)
The rated value of $V_{CE}$ when the transistor is in saturation.

Base-emitter saturation voltage ($V_{BE(sat)}$)
The rated value of $V_{BE}$ when the transistor is in saturation.

The **on characteristics** describe the dc operating characteristics for both the active and saturation regions of operation.

The **dc current gain ($h_{FE}$)** rating of the transistor is the value of *dc beta*. Note that the label $h_{FE}$ is normally used to represent dc beta, so we will be using this label in all our discussions from now on. The basis of this label is discussed in Chapter 9. For now, just remember that $h_{FE}$ is the label commonly used to represent dc beta.

As you can see, the values of $h_{FE}$ are measured at different values of $I_C$. This is consistent with the discussion we had earlier on *beta versus collector current*.

The **collector-emitter saturation voltage ($V_{CE(sat)}$)** rating is self-explanatory. It is the rated value of $V_{CE}$ when the transistor is operated in saturation. The 2N3904 spec sheet lists values of $V_{CE(sat)} = 0.2$ V (maximum) when $I_C = 10$ mA and of $V_{CE(sat)} = 0.3$ (maximum) when $I_C = 50$ mA. Other spec sheets may list *minimum* and *maximum* values of $V_{CE(sat)}$, providing a range of values.

The **base-emitter saturation voltage ($V_{BE(sat)}$)** rating is also self-explanatory. It is the rated value of $V_{BE}$ when the transistor is operated in saturation. For the 2N3904, the value of $V_{BE(sat)}$ has a range of 0.65 to 0.85 V when $I_B = 1$ mA and a maximum value of 0.95 V when $I_B = 5$ mA. As shown in the 2N3904 spec sheet, $V_{BE(sat)}$ is normally given as a range of values, or a *maximum* value.

There are many other commonly used transistor specifications. However, these specifications deal mainly with the *ac* operation of the device. We will cover these specs when we cover transistor ac operation in Chapter 9.

1. What do the *off characteristics* of a transistor indicate?
2. What is *base cutoff current*?
3. What is *collector cutoff current*?
4. What do the *on characteristics* of a transistor indicate?
5. What label is commonly used to represent dc beta?
6. What is *collector-emitter saturation voltage*?
7. What is *base-emitter saturation voltage*?

## 6.6 Transistor Testing

You may recall that a diode can be tested using an ohmmeter. The transistor can be checked in the same basic fashion; however, the BJT contains more than one junction that must be tested.

◀ *OBJECTIVE 12*

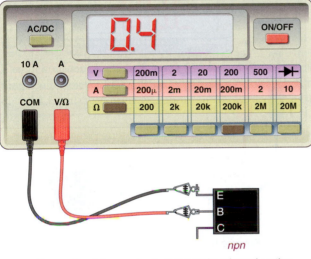

**Note:** Typical forward resistance readings for a good transistor junction are around 1 kΩ or lower.

**Lab Reference:** BJT testing with an ohmmeter is demonstrated in Exercise 6.

Forward resistance check of the emitter-base junction

(a)

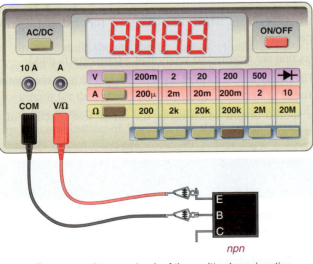

**Note:** Typical reverse resistance for a good transistor junction results in an "out of range" indication on the meter.

Reverse resistance check of the emitter-base junction

(b)

FIGURE 6.30 Transistor resistance checks.

**7.** Complete the following table.

| | Beta | $I_B$ | $I_C$ |
|---|---|---|---|
| a. | 150 | 25 μA | — |
| b. | — | 75 μA | 1.5 mA |
| c. | 240 | 100 μA | — |
| d. | 325 | — | 20 mA |

**8.** Complete the following table.

| | Beta | $I_B$ | $I_C$ |
|---|---|---|---|
| a. | — | 50 μA | 12 mA |
| b. | 440 | — | 35 mA |
| c. | 175 | 45 μA | — |
| d. | — | 120 μA | 84 mA |

**9.** Complete the following table.

| | $I_B$ | $I_C$ | $I_E$ |
|---|---|---|---|
| a. | 25 μA | 1 mA | — |
| b. | — | 1.8 mA | 1.98 mA |
| c. | 120 μA | — | 3 mA |
| d. | — | 7.5 mA | 8 mA |
| e. | 50 μA | — | 20 mA |
| f. | 175 μA | 9.825 mA | — |

**10.** A BJT has values of $I_B = 35$ μA and $\beta = 100$. Determine the values of $I_C$ and $I_E$ for the device.

**11.** A BJT has values of $I_B = 150$ μA and $\beta = 400$. Determine the values of $I_C$ and $I_E$ for the device.

**12.** A BJT has values of $I_B = 48$ μA and $\beta = 120$. Determine the values of $I_C$ and $I_E$ for the device.

**13.** A BJT has values of $I_C = 12$ mA and $\beta = 440$. Determine the values of $I_B$ and $I_E$ for the device.

**14.** A BJT has values of $I_C = 50$ mA and $\beta = 400$. Determine the values of $I_B$ and $I_E$ for the device.

**15.** A BJT has values of $I_E = 65$ mA and $\beta = 380$. Determine the values of $I_B$ and $I_C$ for the device.

**16.** A BJT has values of $I_E = 120$ mA and $\beta = 60$. Determine the values of $I_B$ and $I_C$ for the device.

**17.** A BJT has a value of $\beta = 426$. Determine the value of $\alpha$ for the device.

**18.** A BJT has a value of $\beta = 350$. Determine the value of $\alpha$ for the device.

**19.** A BJT has the following parameters: $I_{C(max)} = 120$ mA and $\beta = 50$ to 120. Determine the maximum allowable value of $I_B$ for the device.

**20.** A BJT has the following parameters: $I_{C(max)} = 250$ mA and $\beta = 35$ to 100. Determine the maximum allowable value of $I_B$ for the device.

### Section 6.4

**21.** Refer to Figure 6.39. What is the range of breakdown voltages of the transistor represented by the collector curves?

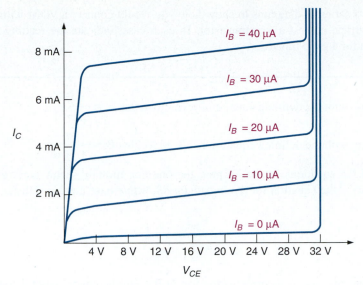

**FIGURE 6.39**

**22.** Refer to Figure 6.39. What is the approximate value of $I_C$ when $I_B$ is 40 µA?

**23.** Refer to Figure 6.39. What is the maximum value of $V_{CE}$ when the device is saturated?

### Section 6.7

**24.** For each circuit shown in Figure 6.40, identify the type of transistor and indicate the directions of the terminal currents.

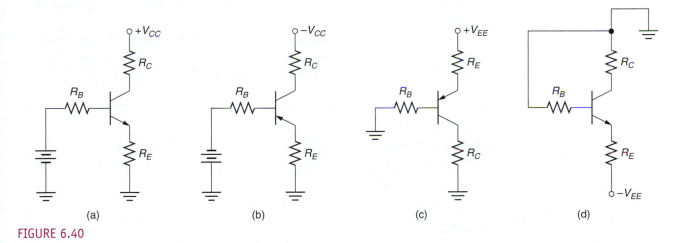

**FIGURE 6.40**

**TROUBLESHOOTING
PRACTICE PROBLEMS**

**25.** The following are the results of several transistor tests. In each case, determine whether or not the transistor is good.

| | Base-Emitter | | Collector-Base | | Emmiter-to-Collector Resistance |
|---|---|---|---|---|---|
| | *Forward* | *Reverse* | *Forward* | *Reverse* | |
| **a.** | 250 Ω | 300 kΩ | 150 kΩ | 140 kΩ | 1200 kΩ |
| **b.** | 800 Ω | 115 kΩ | 800 Ω | 115 kΩ | 14 kΩ |
| **c.** | 377 Ω | 152 kΩ | 900 Ω | 180 kΩ | 1500 kΩ |
| **d.** | 100 kΩ | 100 kΩ | 190 Ω | 144 kΩ | 3500 kΩ |

**26.** Draw a series of diagrams to show how you would connect a VOM with a negative ground lead to test a *pnp* transistor. In each case, indicate the reading you would obtain for a good transistor.

## PUSHING THE ENVELOPE

**27.** Starting with the defining equation for $\alpha$, prove that $\alpha = \dfrac{\beta}{\beta + 1}$.

**28.** Starting with the defining equation for $\beta$, prove that $\beta = \dfrac{\alpha}{1 - \alpha}$.

**29.** Of all the equations in this chapter, the one that initially seems incorrect (to most people) is the equation for the *maximum allowable* base current. This equation was given as

$$I_{B(\text{max})} = \frac{I_{C(\text{max})}}{\beta_{\text{max}}}$$

The part that seems incorrect is the use of $\beta_{\text{max}}$ in the denominator, because we are taught that a *minimum* value in the denominator of a fraction gives us a maximum result. Using the following transistor values:

$$I_{C(\text{max})} = 500 \text{ mA} \qquad \beta = 50 \text{ to } 250$$

demonstrate that the maximum allowable value of base current cannot be calculated using the value of $\beta_{\text{min}}$ in the denominator of the equation.

## SUGGESTED COMPUTER APPLICATIONS PROBLEMS

**30.** Write a program that will solve for $I_C$, $I_E$, and $\alpha$ given the values of $I_B$ and $h_{FE}$ (dc beta).

**31.** Write a program that will solve the table shown in Problem 9 given any two of the values.

**32.** Write a program that will accept information like that shown in Problem 25 and tell you whether a given transistor is good. Assume that a good transistor would have a minimum reverse-to-forward resistance ratio of 100:1.

## ANSWERS TO THE EXAMPLE PRACTICE PROBLEMS

**6.1** $I_C = 17.5$ mA

**6.2** $I_C = 20$ mA, $I_E = 20.05$ mA

**6.3** $I_B = 85.11$ μA, $I_C = 11.91$ mA

**6.4** $I_B = 471$ μA, $I_E = 80.47$ mA

**6.5** $\alpha = 0.997$, $I_C = 349$ mA (using $\beta$), $I_C = 348.95$ mA (using $\alpha$)

**6.6** $I_{B(\text{max})} = 8.33$ mA

## chapter 7

# DC Biasing Circuits

## Objectives

*After studying the material in this chapter, you should be able to:*

1. State the purpose of dc biasing circuits.
2. Plot the dc load line for an amplifier given the value of $V_{CC}$ and the total collector-emitter circuit resistance.
3. Describe the $Q$-point of an amplifier, and explain what it represents.
4. Describe and analyze the operation of a base-bias circuit.
5. Determine if a circuit is midpoint biased given the values of $I_C$, $V_{CE}$, and $V_{CC}$ for the circuit.
6. Describe and analyze the operation of a voltage-divider biasing circuit.
7. Estimate the value of $I_{CQ}$ for an amplifier without detailed calculation, and justify the use of this estimate.
8. Describe the troubleshooting procedure for a voltage-divider bias circuit.
9. Describe and analyze the operation of an emitter-bias circuit.
10. Describe and analyze the operation of a collector-feedback bias circuit.
11. Describe and analyze the operation of the emitter-feedback bias circuit.

## Outline

# The World of Silicon Miniature Components

Not long after the development of the transistor, the race was on to make miniature versions of almost every type of component and circuit. It was found that almost any type of circuit or component could be made using silicon. Resistors and capacitors, for example, could be produced on silicon. This made it possible for entire amplifiers (including the biasing components) to be made on a single silicon wafer.

Manufacturing technology has developed to the point where literally hundreds of thousands of components can be produced on a single silicon wafer that is much smaller than a penny. Yet the simplest of all electronic components, the *inductor*, took longer to develop in semiconductor form than any other component. Inductors were first fabricated in microminiature form in the late 1980s. Even now, these inductors are considered impractical because of the amount of chip space (or "real estate") that it takes to produce them.

**OBJECTIVE 1** ▶

The ac operation of a transistor amplifier depends on its initial dc values of $I_B$, $I_C$, and $V_{CE}$. As $I_B$ is varied from an initial value, $I_C$ and $V_{CE}$ are varied from their initial values. This operation is illustrated in Figure 7.1. *The function of dc biasing is to set the initial values of $I_B$, $I_C$, and $V_{CE}$.* Several bias methods are used to achieve different results, but each provides a means of setting the initial operating point of the transistor. In this chapter, we will discuss the operation and troubleshooting of the most commonly used dc biasing circuits.

> The purpose of the *dc biasing circuit* is to set up the initial dc values of $I_B$, $I_C$, and $V_{CE}$.

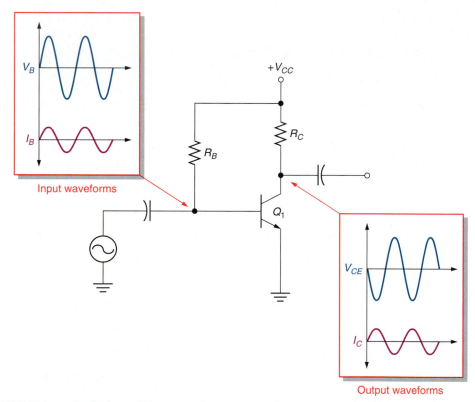

**FIGURE 7.1** Typical amplifier operation.

## 7.1 Introduction to DC Biasing: The DC Load Line

> **dc load line**
>
> A graph of all possible combinations of $V_{CE}$ and $I_C$ for an amplifier.

The **dc load line** is a graph that *represents all the possible combinations of $I_C$ and $V_{CE}$ for a given amplifier*. For every possible value of $I_C$, an amplifier has a corresponding value of $V_{CE}$. The dc load line represents all the $I_C/V_{CE}$ combinations for the circuit. A generic dc load line is shown in Figure 7.2.

The ends of the load line are labeled $I_{C(sat)}$ and $V_{CE(off)}$. The value of the $I_{C(sat)}$ point represents the ideal value of saturation current for the circuit. If we look at the saturated

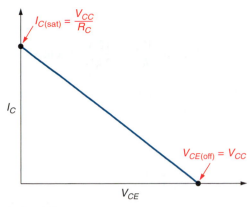

FIGURE 7.2    A generic dc load line.

transistor as being a short circuit from emitter to collector, then $V_{CE}$ is zero, and equation (6.12) becomes

Equation (6.12):
$$V_{CE} = V_{CC} - I_C R_C$$

$$V_{CC} = I_C R_C \quad \text{(saturation)} \tag{7.1}$$

Rearranging the formula for $I_C$, we get

$$I_{C(sat)} = \frac{V_{CC}}{R_C} \quad \text{(ideal)} \tag{7.2}$$

Conversely, when the transistor is in cutoff, we can look at the component as being an open circuit from emitter to collector. Because the transistor acts as an open, there is no collector current. Thus, $I_C R_C = 0$, and equation (6.12) becomes

$$V_{CE(off)} = V_{CC} \tag{7.3}$$

As the following example shows, equations (7.2) and (7.3) are used to plot the end points ◄ **OBJECTIVE 2** of the dc load line.

**EXAMPLE 7.1**

Plot the dc load line for the circuit shown in Figure 7.3a.

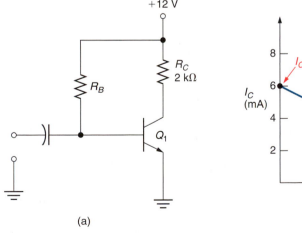

FIGURE 7.3

**Solution:**    With the circuit values shown,

$$V_{CE(off)} = V_{CC} = 12 \text{ V}$$

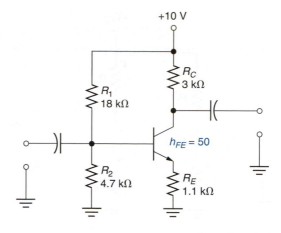

FIGURE 7.15

Because $I_{CQ} \cong I_E$, we can calculate the approximate the value of $I_{CQ}$ using

$$I_{CQ} \cong \frac{V_E}{R_E} = \frac{1.37 \text{ V}}{1.1 \text{ k}\Omega} = 1.25 \text{ mA}$$

Finally, $V_{CEQ}$ is found as

$$V_{CEQ} = V_{CC} - I_{CQ}(R_C + R_E) = 10 \text{ V} - (1.25 \text{ mA})(4.1 \text{ k}\Omega)$$
$$= 4.87 \text{ V}$$

**PRACTICE PROBLEM 7.7**

A circuit like the one in Figure 7.15 has the following values: $R_C = 620 \text{ }\Omega$, $R_E = 180 \text{ }\Omega$, $R_1 = 12 \text{ k}\Omega$, $R_2 = 2.7 \text{ k}\Omega$, and $V_{CC} = 10 \text{ V}$. Determine the values of $I_{CQ}$ and $V_{CEQ}$ for the circuit.

Equation (6.5): $I_E = I_B(h_{FE} + 1)$

Once the value of $I_E$ is known, the value of $I_B$ can be found simply by using equation (6.5), which can be rewritten as

$$I_B = \frac{I_E}{h_{FE} + 1} \tag{7.11}$$

**EXAMPLE 7.8**

Determine the value of $I_B$ for the circuit shown in Figure 7.16.

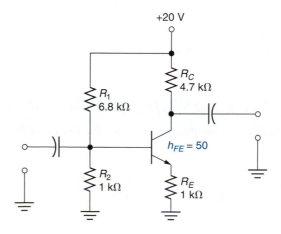

FIGURE 7.16

*Solution:*   First, $V_B$ is found as

$$V_B = V_{CC} \frac{R_2}{R_1 + R_2} = (20 \text{ V}) \frac{1 \text{ k}\Omega}{7.8 \text{ k}\Omega} = \textbf{2.56 V}$$

and $V_E$ is found as

$$V_E = V_B - 0.7 \text{ V} = 2.56 \text{ V} - 0.7 \text{ V} = \textbf{1.86 V}$$

Now, $I_E$ is found as

$$I_E = \frac{V_E}{R_E} = \frac{1.86 \text{ V}}{1 \text{ k}\Omega} = \textbf{1.86 mA}$$

Finally, $I_B$ is found as

$$I_B = \frac{I_E}{h_{FE} + 1} = \frac{1.86 \text{ mA}}{51} = \textbf{36.5 } \boldsymbol{\mu}\textbf{A}$$

**PRACTICE PROBLEM 7.8**
Determine the value of $I_B$ for the circuit described in Practice Problem 7.7. Assume that the value of $h_{FE}$ for the transistor is 200.

## 7.3.2   Which Value of $h_{FE}$ Do I Use?

When you are analyzing a voltage-divider bias circuit (or any other dc biasing circuit, for that matter), you will often have to refer to the spec sheet for the transistor to obtain its value of $h_{FE}$. The only problem is that the spec sheet may not list a single value of $h_{FE}$. Normally, a transistor spec sheet (like the one shown in Figure 7.17) will list any combination of the following values:

**1.** A *maximum* value of $h_{FE}$.
**2.** A *minimum* value of $h_{FE}$.
**3.** A *typical* value of $h_{FE}$.

When only one value of $h_{FE}$ is listed on the spec sheet, you must use that value in any circuit analysis.

When two or more values of $h_{FE}$ are listed, you should first look to see if one of them is the *typical* value of $h_{FE}$. If a *typical* value is listed, use that value.

When the spec sheet lists a *minimum* value and a *maximum* value of $h_{FE}$, you must use the *geometric average of the two values*. The geometric average of $h_{FE}$ is found as

$$h_{FE(ave)} = \sqrt{h_{FE(mm)} \times h_{FE(max)}} \qquad\qquad \textbf{(7.12)}$$

The following example illustrates the use of $h_{FE(ave)}$ in the analysis of a transistor amplifier.

ELECTRICAL CHARACTERISTICS ($T_A$ = 25°C unless otherwise noted)

| Characteristic | | Symbol | Min | Max | Unit |
|---|---|---|---|---|---|
| **OFF CHARACTERISTICS** | | | | | |
| Collector–Emitter Breakdown Voltage (Note 2.) ($I_C$ = 1.0 mAdc, $I_B$ = 0) | | $V_{(BR)CEO}$ | 40 | – | Vdc |
| Collector–Base Breakdown Voltage ($I_C$ = 10 µAdc, $I_E$ = 0) | | $V_{(BR)CBO}$ | 60 | – | Vdc |
| Emitter–Base Breakdown Voltage ($I_E$ = 10 µAdc, $I_C$ = 0) | | $V_{(BR)EBO}$ | 6.0 | – | Vdc |
| Base Cutoff Current ($V_{CE}$ = 30 Vdc, $V_{EB}$ = 3.0 Vdc) | | $I_{BL}$ | – | 50 | nAdc |
| Collector Cutoff Current ($V_{CE}$ = 30 Vdc, $V_{EB}$ = 3.0 Vdc) | | $I_{CEX}$ | – | 50 | nAdc |
| **ON CHARACTERISTICS** | | | | | |
| DC Current Gain (Note 2.) | | $h_{FE}$ | | | – |
| ($I_C$ = 0.1 mAdc, $V_{CE}$ = 1.0 Vdc) | 2N3903 | | 20 | – | |
| | 2N3904 | | 40 | – | |
| ($I_C$ = 1.0 mAdc, $V_{CE}$ = 1.0 Vdc) | 2N3903 | | 35 | – | |
| | 2N3904 | | 70 | – | |
| ($I_C$ = 10 mAdc, $V_{CE}$ = 1.0 Vdc) | 2N3903 | | 50 | 150 | |
| | 2N3904 | | 100 | 300 | |
| ($I_C$ = 50 mAdc, $V_{CE}$ = 1.0 Vdc) | 2N3903 | | 30 | – | |
| | 2N3904 | | 60 | – | |
| ($I_C$ = 100 mAdc, $V_{CE}$ = 1.0 Vdc) | 2N3903 | | 15 | – | |
| | 2N3904 | | 30 | – | |
| Collector–Emitter Saturation Voltage (Note 2.) | | $V_{CE(sat)}$ | | | Vdc |
| ($I_C$ = 10 mAdc, $I_B$ = 1.0 mAdc) | | | – | 0.2 | |
| ($I_C$ = 50 mAdc, $I_B$ = 5.0 mAdc | | | – | 0.3 | |
| Base–Emitter Saturation Voltage (Note 2.) | | $V_{BE(sat)}$ | | | Vdc |
| ($I_C$ = 10 mAdc, $I_B$ = 1.0 mAdc) | | | 0.65 | 0.85 | |
| ($I_C$ = 50 mAdc, $I_B$ = 5.0 mAdc) | | | – | 0.95 | |
| **SMALL-SIGNAL CHARACTERISTICS** | | | | | |
| Current–Gain – Bandwidth Product | | $f_T$ | | | MHz |
| ($I_C$ = 10 mAdc, $V_{CE}$ = 20 Vdc, f = 100 MHz) | 2N3903 | | 250 | – | |
| | 2N3904 | | 300 | – | |
| Output Capacitance ($V_{CB}$ = 5.0 Vdc, $I_E$ = 0, f = 1.0 MHz) | | $C_{obo}$ | – | 4.0 | pF |
| Input Capacitance ($V_{EB}$ = 0.5 Vdc, $I_C$ = 0, f = 1.0 MHz) | | $C_{ibo}$ | – | 8.0 | pF |
| Input Impedance | | $h_{ie}$ | | | k Ω |
| ($I_C$ = 1.0 mAdc, $V_{CE}$ = 10 Vdc, f = 1.0 kHz) | 2N3903 | | 1.0 | 8.0 | |
| | 2N3904 | | 1.0 | 10 | |
| Voltage Feedback Ratio | | $h_{re}$ | | | X $10^{-4}$ |
| ($I_C$ = 1.0 mAdc, $V_{CE}$ = 10 Vdc, f = 1.0 kHz) | 2N3903 | | 0.1 | 5.0 | |
| | 2N3904 | | 0.5 | 8.0 | |
| Small–Signal Current Gain | | $h_{fe}$ | | | – |
| ($I_C$ = 1.0 mAdc, $V_{CE}$ = 10 Vdc, f = 1.0 kHz) | 2N3903 | | 50 | 200 | |
| | 2N3904 | | 100 | 400 | |
| Output Admittance ($I_C$ = 1.0 mAdc, $V_{CE}$ = 10 Vdc, f = 1.0 kHz) | | $h_{oe}$ | 1.0 | 40 | µmhos |
| Noise Figure | | NF | | | dB |
| ($I_C$ = 100 µAdc, $V_{CE}$ = 5.0 Vdc, $R_S$ = 1.0 k Ω, f = 1.0 kHz) | 2N3903 | | – | 6.0 | |
| | 2N3904 | | – | 5.0 | |

**SWITCHING CHARACTERISTICS**

| | | | Symbol | Min | Max | Unit |
|---|---|---|---|---|---|---|
| Delay Time | ($V_{CC}$ = 3.0 Vdc, $V_{BE}$ = 0.5 Vdc, | | $t_d$ | – | 35 | ns |
| Rise Time | $I_C$ = 10 mAdc, $I_{B1}$ = 1.0 mAdc) | | $t_r$ | – | 35 | ns |
| Storage Time | ($V_{CC}$ = 3.0 Vdc, $I_C$ = 10 mAdc, | 2N3903 | $t_s$ | – | 175 | ns |
| | $I_{B1}$ = $I_{B2}$ = 1.0 mAdc) | 2N3904 | | – | 200 | |
| Fall Time | | | $t_f$ | – | 50 | ns |

2. Pulse Test: Pulse Width ≤ 300 µs; Duty Cycle ≤ 2%.

**FIGURE 7.17** Electrical characteristics listing for the 2N3904. (Copyright of Semiconductor Component Industries, LLC. Used by permission.)

## EXAMPLE 7.9

A voltage-divider bias circuit has the following values: $R_1 = 1.5 \text{ k}\Omega$, $R_2 = 680 \text{ }\Omega$, $R_C = 260 \text{ }\Omega$, $R_E = 240 \text{ }\Omega$, and $V_{CC} = +10 \text{ V}$. Assuming the transistor is a 2N3904, determine the value of $I_B$ for the circuit.

*Solution:*　First, $V_B$ is found as

$$V_B = V_{CC}\frac{R_2}{R_1 + R_2} = (10 \text{ V})\frac{680 \text{ }\Omega}{2180 \text{ }\Omega} = \textbf{3.12 V}$$

and $V_E$ is found as

$$V_E = V_B - 0.7 \text{ V} = 3.12 \text{ V} - 0.7 \text{ V} = \textbf{2.42 V}$$

Now, the value of $I_{CQ}$ is approximated as

$$I_{CQ} = \frac{V_E}{R_E} = \frac{2.42 \text{ V}}{240 \text{ }\Omega} \cong \textbf{10 mA}$$

Assuming that $I_{CQ} = 10 \text{ mA}$, we check the specification sheet shown in Figure 7.17. As shown, beta ($h_{FE}$) has a range of 100 to 300 when $I_C = 10 \text{ mA}$; thus $h_{FE(ave)}$ is found as

$$\boldsymbol{h_{FE(ave)}} = \sqrt{h_{FE(min)} \times h_{FE(max)}} = \sqrt{100 \times 300} = \textbf{173}$$

Finally, $I_B$ is found as

$$I_B = \frac{I_E}{h_{FE(ave)} + 1} = \frac{10 \text{ mA}}{174} = \textbf{57.5 }\boldsymbol{\mu}\textbf{A}$$

### PRACTICE PROBLEM 7.9

Determine the value of $I_B$ for the circuit in this example, if $Q_1$ is replaced with a 2N3903.

## 7.3.3　Bias Stability

Earlier, you were told that voltage-divider bias is the most commonly used biasing scheme. One reason is that it is a *beta-independent circuit*; that is, it provides values of $I_{CQ}$ and $V_{CEQ}$ that are highly stable against changes in beta ($h_{FE}$). This stability can be demonstrated using the values found in Example 7.9.

In Example 7.9, we calculated a value of $I_{CQ} \cong 10 \text{ mA}$. Let's assume that the transistor emitter current ($I_E$) is *exactly* 10 mA, and that the transistor has a range of $h_{FE} = 100$ to 300. At $h_{FE} = 100$,

$$I_B = \frac{I_E}{h_{FE} + 1} = \frac{10 \text{ mA}}{101} \cong 100 \text{ }\mu\text{A} \quad \text{and} \quad I_{CQ} = I_E - I_B \cong 9.90 \text{ mA}$$

At $h_{FE} = 300$,

$$I_B = \frac{I_E}{h_{FE} + 1} = \frac{10 \text{ mA}}{301} \cong 33 \text{ }\mu\text{A} \quad \text{and} \quad I_{CQ} = I_E - I_B \cong 9.97 \text{ mA}$$

As you can see, $I_{CQ}$ hardly changes over the entire range of $h_{FE}$ for the transistor. As a result, any changes in $h_{FE}$ have little effect on the Q-point of a voltage-divider bias circuit.

### 7.3.4    Saturation and Cutoff

The dc load line for the voltage-divider bias circuit is plotted using the end points of $I_{C(\text{sat})}$ and $V_{CE(\text{off})}$. When the transistor is saturated, $V_{CE}$ is approximately equal to 0 V. Thus, the collector current equals the supply voltage divided by the total resistance between $V_{CC}$ and ground. By formula,

$$I_{C(\text{sat})} = \frac{V_{CC}}{R_C + R_E} \qquad (7.13)$$

When the transistor is in cutoff, all the supply voltage is dropped across the transistor. Thus,

$$V_{CE(\text{off})} = V_{CC} \qquad (7.14)$$

Using these equations, the dc load line for the circuit in Example 7.9 is plotted as shown in Figure 7.18.

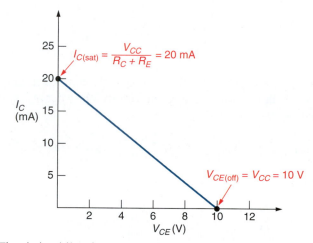

FIGURE 7.18    The dc load line for the circuit in Example 7.9.

### 7.3.5    Base Voltage: A Practical Consideration

According to equation (7.6), the value of $V_B$ for a voltage-divider bias circuit can be found as

$$V_B = V_{CC} \frac{R_2}{R_1 + R_2}$$

As you will see, this relationship is actually an approximation because it does not take into account the input resistance of the transistor base terminal.

#### BASE INPUT RESISTANCE

Figure 7.19 shows the base and emitter currents of a voltage-divider bias circuit. Equation (6.5) defines the relationship between $I_B$ and $I_E$ as follows:

$$I_E = I_B(h_{FE} + 1)$$

If we assume that $h_{FE} \gg 1$, then we can rewrite this relationship as

$$I_E \cong h_{FE}I_B$$

Just as current is magnified by a factor of $h_{FE}$ from base to emitter, $R_E$ is magnified by a factor of $h_{FE}$ from emitter to base. As a result, the transistor has a measurable amount of *base input resistance*, or $R_{IN(\text{base})}$. The value of this resistance can be found as

$$R_{IN(\text{base})} \cong h_{FE}R_E \qquad (7.15)$$

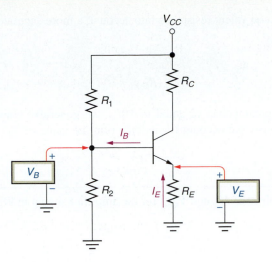

FIGURE 7.19

This relationship is derived in Appendix D and is demonstrated in Example 7.10.

### EXAMPLE 7.10

Determine the base input resistance of the transistor in Example 7.9.

*Solution:* Using the values of $h_{FE(ave)} = 173$ and $R_E = 240 \ \Omega$, the base input resistance of the transistor is found as

$$R_{IN(base)} \cong h_{FE}R_E = (173)(240 \ \Omega) = \textbf{41.5 k}\Omega$$

### PRACTICE PROBLEM 7.10

A voltage-divider bias circuit has values of $h_{FE} = 100$ to $450$ and $R_E = 1 \ \text{k}\Omega$. Calculate the base input resistance of the transistor.

### THE EFFECT OF $R_{IN(base)}$ ON THE BIASING CIRCUIT

The effect of $R_{IN(base)}$ on voltage-divider bias is illustrated in Figure 7.20. When we replace the base of the transistor with its resistive equivalent ($h_{FE}R_E$), we can see that this resistance is in parallel with $R_2$. Therefore, the resistance from the connection point to ground ($R_{EQ}$) can be found as

$$R_{EQ} = R_2 \parallel R_{IN(base)} \qquad\qquad (7.16)$$

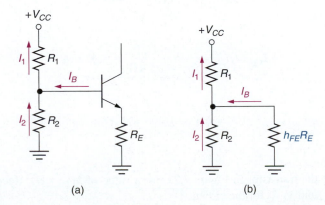

(a)                    (b)

FIGURE 7.20

Taking this parallel equivalent resistance into account, a more accurate value of $V_B$ can be found as

$$V_B = V_{CC} \frac{R_{EQ}}{R_1 + R_{EQ}} \mid R_{EQ} = R_2 \parallel h_{FE}R_E \qquad (7.17)$$

When $h_{FE}R_E$ is greater than or equal to $10R_2$, we generally ignore its effect on the value of $V_B$. Otherwise, we account for its effect on the value of $V_B$, as demonstrated in the following example.

### EXAMPLE 7.11

Determine the values of $I_{CQ}$ and $V_{CEQ}$ for the amplifier shown in Figure 7.21.

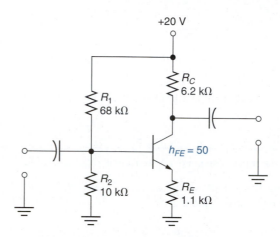

FIGURE 7.21

*Solution:*   The value of $h_{FE}R_E$ for the amplifier is found as

$$h_{FE}R_E = (50)(1.1 \text{ k}\Omega) = \mathbf{55 \text{ k}\Omega}$$

Since this value is less than 10 times the value of $R_2$, we use equations (7.16) and (7.17) to determine the value of $V_B$ as follows:

$$R_{EQ} = R_2 \parallel h_{FE}R_E = (10 \text{ k}\Omega) \parallel (55 \text{ k}\Omega) = \mathbf{8.46 \text{ k}\Omega}$$

and

$$V_B = V_{CC} \frac{R_{EQ}}{R_1 + R_{EQ}} = (20 \text{ V}) \frac{8.46 \text{ k}\Omega}{68 \text{ k}\Omega + 8.46 \text{ k}\Omega} = \mathbf{2.21 \text{ V}}$$

The rest of the problem is solved in the usual fashion as follows:

$$V_E = V_B - 0.7 \text{ V} = \mathbf{1.51 \text{ V}}$$

and

$$I_{CQ} \cong I_E = \frac{V_E}{R_E} = \frac{1.51}{1.1 \text{ k}\Omega} = \mathbf{1.37 \text{ mA}}$$

Finally,

$$V_{CEQ} = V_{CC} - I_{CQ}(R_C + R_E) = 20 \text{ V} - (1.37 \text{ mA})(7.3 \text{ k}\Omega) = \mathbf{9.99 \text{ V}}$$

### PRACTICE PROBLEM 7.11

Assume that the transistor in Figure 7.21 has a value of $h_{FE} = 80$. Recalculate the values of $I_{CQ}$ and $V_{CEQ}$ for the circuit.

## 7.3.6    Estimating the Value of $I_{CQ}$

◀ OBJECTIVE 7

You may have noticed that we have a bit of a problem. To determine whether $R_{IN(base)}$ must be considered in the calculation of $V_B$, we have to know the value of $h_{FE}$. However, we need to know the value of $I_{CQ}$ to find the value of $h_{FE}$ on any spec sheet. And we cannot determine the value of $I_{CQ}$ without having first determined the value of $V_B$.

To determine the value of $h_{FE}$ when the value of $V_B$ isn't certain, we need to determine the approximate value of $I_{CQ}$. We can approximate the value of $I_{CQ}$ by assuming that the circuit is midpoint biased. If the circuit is midpoint biased, then $I_{CQ}$ is approximately half the value of $I_{C(sat)}$. Therefore, to determine which value of $h_{FE}$ to use:

1. Calculate the value of $I_{C(sat)}$.
2. Approximate $I_{CQ}$ as being half the calculated value of $I_{C(sat)}$. Once you have approximated the value of $I_{CQ}$, you can use that approximated value to look up the value of $h_{FE}$ on the spec sheet of the transistor. Then, use the value of $h_{FE}$ from the spec sheet to proceed with your analysis.

The validity of this method can be seen by referring to Example 7.11. Using equation (7.13), the value of $I_{C(sat)}$ for the circuit is calculated to be 2.74 mA. Half of $I_{C(sat)}$ would then be 1.37 mA, which equals the calculated value of $I_{CQ}$ in the example.

You may be wondering why we can assume that every voltage-divider bias circuit is designed for midpoint bias. The reason is simple: Voltage-divider bias circuits are used primarily as linear amplifiers. As stated earlier in the chapter, these amplifiers are almost always designed for midpoint bias to provide for the largest possible output.

## 7.3.7    Circuit Troubleshooting

◀ OBJECTIVE 8

As stated earlier, voltage-divider bias circuits are typically used as linear amplifiers. As such, they tend to have terminal voltages that are predictable under normal circumstances. Typically, a working voltage-divider biased amplifier has the following values:

1. $V_C$ is slightly greater than $\dfrac{V_{CC}}{2}$.
2. $V_E$ is somewhere around $\dfrac{V_{CC}}{10}$.
3. $V_B$ is approximately 0.7 V greater than $V_E$.

Note that these guidelines are written for *npn* circuits, but they apply (in principle) to *pnp* circuits as well. In this section, we will discuss some problems that can arise in a voltage-divider biased amplifier and their symptoms.

### $R_1$ OPEN

When $R_1$ is open, there is no path for current in the base circuit. Therefore, no voltage is developed across $R_2$, and $V_B$ is zero. The loss of $V_B$ causes the transistor to be biased off. Thus, there is no emitter current, and $V_E$ equals 0 V.

Without any current in the collector circuit, $I_C R_C$ equals 0 V, and $V_C$ equals $V_{CC}$. Thus, the following circuit conditions exist when $R_1$ is open:

1. $V_B$ is 0 V.
2. $V_E$ is 0 V.
3. $V_C$ equals $V_{CC}$.

These circuit conditions are illustrated in Figure 7.22a.

### $R_2$ OPEN

If $R_2$ goes open circuit, the current through $R_1$ equals $I_B$. Since this current normally equals ($I_2 + I_B$), the current through $R_1$ *decreases* significantly. This causes the voltage drop across $R_1$ to decrease, and $V_B$ increases. The increase in $V_B$ causes the transistor to

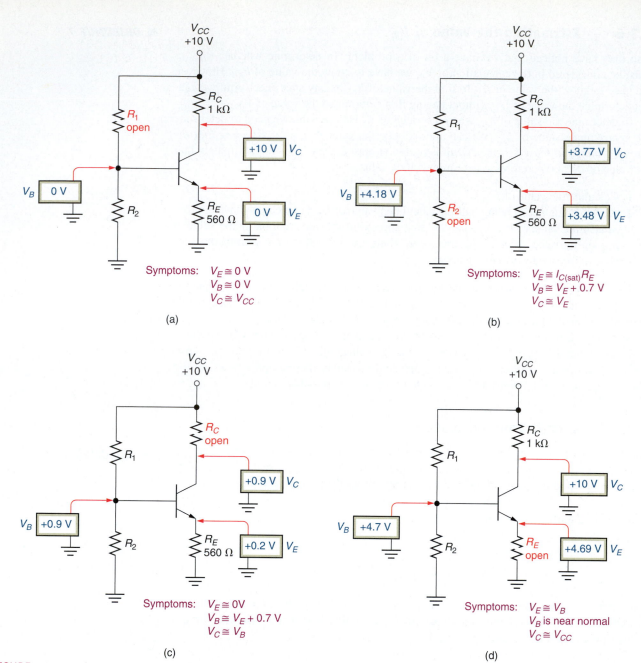

FIGURE 7.22  Voltage-divider bias fault system.

saturate. Thus, $V_C$ and $V_E$ are very close to equal in value, and $I_C$ equals $I_{C(\text{sat})}$. The existing circuit conditions are summarized as follows:

1. $V_B$ equals $V_E + 0.7$ V.
2. $V_E$ equals $R_E \times I_{C(\text{sat})}$.
3. $V_C$ is approximately equal to $V_E$.

These circuit conditions are illustrated in Figure 7.22b.

### $R_C$ OPEN

When $R_C$ opens, there is no path for collector current. In this case, $I_E$ equals $I_B$, and the value of $V_E$ drops significantly. The value of $V_B$ is still approximately 0.7 V greater than $V_E$. In summary:

1. $V_B$ is approximately 0.7 V greater than $V_E$.
2. $V_E$ is slightly greater than 0 V.
3. $V_C$ is approximately equal to $V_E$.

In other words, all three terminal voltages are *low*. These circuit conditions are illustrated in Figure 7.22c.

**Lab Reference:** Several voltage-divider bias circuit faults are simulated in Exercise 8.

## $R_E$ OPEN

When $R_E$ opens, there is no emitter current or collector current. The base current in the circuit drops to zero, but this has little effect on the value of $V_B$ (which is approximately equal to $I_2R_2$ under normal circumstances).

When you read the emitter voltage with a voltmeter, the meter completes the circuit, as shown in Figure 7.23. As a result, the meter creates a path for transistor emitter current through the meter resistance ($R_m$), causing you to obtain a reading that is somewhat higher than normal. The collector voltage equals $V_{CC}$ since there is no current in the collector circuit. In summary:

1. $V_B$ is normal.
2. The *meter reading* of $V_E$ may be slightly higher than normal.
3. $V_C$ equals $V_{CC}$.

These circuit conditions are illustrated in Figure 7.22d.

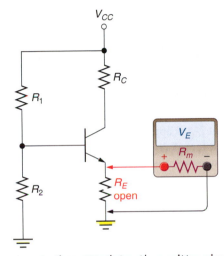

FIGURE 7.23   The meter connection completes the emitter circuit.

## $Q_1$ OPEN

This condition gives you the same output readings as having an open $R_1$; that is, the transistor will be in cutoff. A simple check of the transistor (as outlined in Chapter 6) or a resistance reading of $R_1$ tells you which of the two components is faulty.

**Leaky**
A term used to describe a component that is partially shorted.

## $Q_1$ LEAKY

The term **leaky** is used to describe a transistor that is partially shorted from emitter to collector. When a transistor is leaky, it may appear to be saturated, even when it should not be. The collector and emitter voltages may be nearly equal, and the 0.7 V drop from emitter to base will still be there. When a transistor is leaky, it must be replaced.

### 7.3.8   Practical Troubleshooting Considerations

When you are troubleshooting *any* dc biasing circuit, a few practical considerations should be kept in mind. These points, which can save you a great deal of time and trouble, are as follows:

1. *Standard carbon resistors do not short internally.* When a carbon-composition resistor appears to be short circuited, the short is caused by one of two conditions:
   a. Another component, wire, copper run (on a printed circuit board), or solder bridge has shorted the component. In this case, the problem is solved by removing the cause of the short.

What can cause a carbon resistor to appear to be shorted?

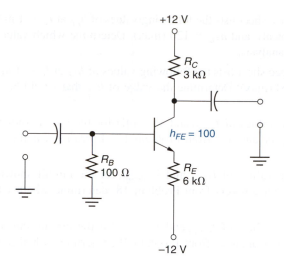

FIGURE 7.38

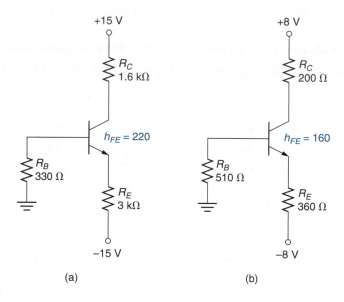

(a)                    (b)

FIGURE 7.39

**34.** Determine the values of $I_{CQ}$, $V_{CEQ}$, $I_{C(sat)}$, and $V_{CE\,(off)}$ for the circuit shown in Figure 7.39b.

**35.** Determine the value of $R_{IN(base)}$ for the circuit shown in Figure 7.39a.

**36.** Determine the value of $R_{IN(base)}$ for the circuit shown in Figure 7.39b.

**37.** Determine the values of $I_{CQ}$ and $V_{CEQ}$ for the circuit shown in Figure 7.40a.

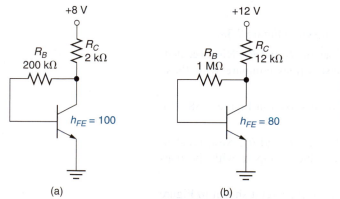

(a)                    (b)

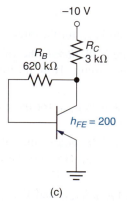

(c)

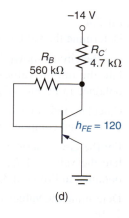

(d)

FIGURE 7.40

**38.** Determine the values of $I_{CQ}$ and $V_{CEQ}$ for the circuit shown in Figure 7.40b.

**39.** Determine the values of $I_{CQ}$ and $V_{CEQ}$ for the circuit shown in Figure 7.40c.

**40.** Determine the values of $I_{CQ}$ and $V_{CEQ}$ for the circuit shown in Figure 7.40d.

**41.** Determine the values of $I_{CQ}$ and $V_{CEQ}$ for the circuit shown in Figure 7.41a.

**42.** Determine the values of $I_{CQ}$ and $V_{CEQ}$ for the circuit shown in Figure 7.41b.

**43.** Determine the values of $I_{CQ}$ and $V_{CEQ}$ for the circuit shown in Figure 7.41c.

**44.** Determine the values of $I_{CQ}$ and $V_{CEQ}$ for the circuit shown in Figure 7.41d.

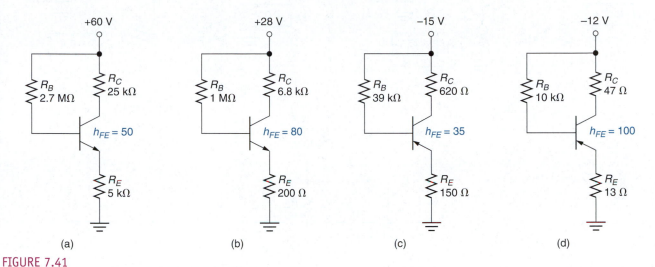

FIGURE 7.41

TROUBLESHOOTING PRACTICE PROBLEMS

**45.** What fault is indicated in Figure 7.42a? Explain your answer.

**46.** What fault is indicated in Figure 7.42b? Explain your answer.

**47.** What fault is indicated in Figure 7.42c? Explain your answer.

**48.** What fault is indicated in Figure 7.43?

PUSHING THE ENVELOPE

**49.** The transistor in Figure 7.44 has a value of $h_{FE} = 200$ for $I_C = 1$ to 8 mA. The value of $V_{CE(\text{sat})}$ for the component is 0.3 V. Derive the collector curves for the device for $I_C = 1$ to 4 mA and $V_{CE} = 0$ V to $V_{CE(\text{off})}$. Then, plot the dc load line for the circuit on the collector curves to determine the value of $I_B$ required for midpoint-bias operation.

**50.** The spec sheet for the transistor in Figure 7.45 is located in Appendix A. Determine the values of $I_{CQ}$, $V_{CEQ}$, $I_B$, $V_B$, $I_{C(\text{sat})}$, and $V_{CE\,(\text{off})}$.

**51.** *A collector-feedback bias circuit will be midpoint biased if $R_B = h_{FE}R_C$. Using equation (7.23), prove this statement to be true.* (*Hint:* Idealize the emitter-base junction diode in the equation.)

**52.** The 2N3904 transistor cannot be used in the circuit shown in Figure 7.46. Why not? (*Hint:* Consider the dc load line for the amplifier and the *maximum ratings* shown in the 2N3904 spec sheet in Section 6.5.)

SUGGESTED COMPUTER APPLICATIONS PROBLEMS

**53.** Write a program that will determine whether a given base-bias circuit is midpoint biased given the values of $R_C$, $R_B$, $V_{CC}$, and $h_{FE}$ for the transistor. Set up the program so that it will provide you with the values of $I_{CQ}$ and $V_{CEQ}$.

**54.** Write a program that will determine whether a given voltage-divider bias circuit is midpoint biased given the values of $R_1$, $R_2$, $R_C$, $R_E$, and $V_{CC}$. Set up the program so that it will provide you with the values of $I_{CQ}$ and $V_{CEQ}$.

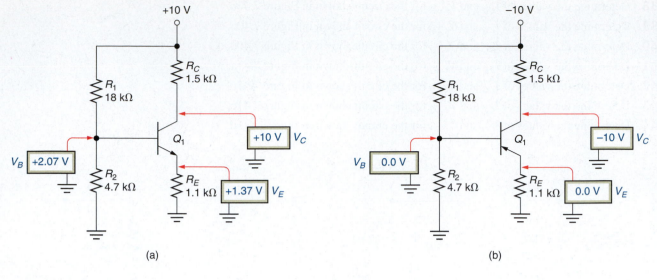

(a)

(b)

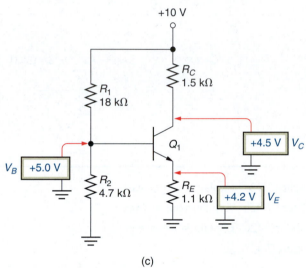

(c)

FIGURE 7.42

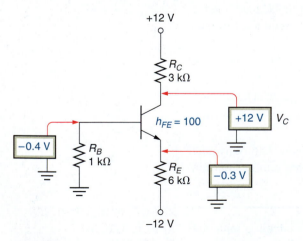

FIGURE 7.43

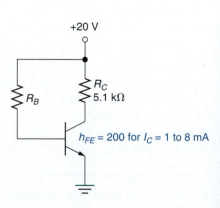

FIGURE 7.44

+8 V
$R_C$
390 Ω
2N4013
$R_B$
100 Ω
$R_E$
330 Ω
–4 V

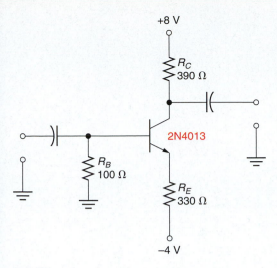

FIGURE 7.45

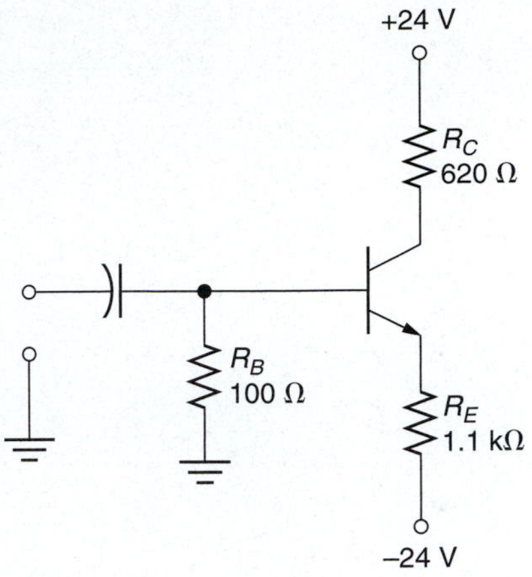

+24 V
$R_C$
620 Ω
$R_B$
100 Ω
$R_E$
1.1 kΩ
–24 V

FIGURE 7.46

## ANSWERS TO THE EXAMPLE PRACTICE PROBLEMS

**7.1** $V_{CE(\text{off})} = 8$ V, $I_{C(\text{sat})} = 7.27$ mA

**7.2** The end points of the load line are $V_{CE(\text{off})} = 16$ V and $I_{C(\text{sat})} = 8$ mA. $V_{CE} = 12$ V at $I_C = 2$ mA, $V_{CE} = 8$ V at $I_C = 4$ mA, and $V_{CE} = 4$ V at $I_C = 6$ mA

**7.3** $I_C = 9.85$ mA, $V_{CE} = 6.91$ V

**7.4** The circuit is *not* midpoint biased.

**7.5** The circuit *is* midpoint biased.

**7.6** $I_C = 18.72$ mA, $V_{CE} = 521.6$ mV

**7.7** $I_{CQ} = 6.33$ mA, $V_{CEQ} = 4.94$ V

**7.8** $I_B = 31.49$ μA

**7.9** $I_B = 114.2$ μA

**7.10** $R_{IN(\text{base})} = 212$ kΩ

**7.11** $I_{CQ} = 1.48$ mA, $V_{CEQ} = 9.2$ V

**7.12** $I_{CQ} = 4.77$ mA, $V_{CEQ} = 7.85$ V

**7.13** $I_{C(\text{sat})} = 6.67$ mA, $V_{CE(\text{off})} = 30$ V

**7.14** $I_{CQ} = 2.83$ mA, $V_{CEQ} = 6.34$ V

**7.15** $I_{CQ} = 2.72$ mA, $V_{CEQ} = 8.63$ V

## chapter 8

# Introduction to Amplifiers

## Objectives

*After studying the material in this chapter, you should be able to:*

1. List the three fundamental ac properties of amplifiers.
2. Discuss the concept of gain.
3. Describe the general model of a voltage amplifier.
4. Discuss the effects that amplifier input and output impedance have on the effective voltage gain of the circuit.
5. Describe the ideal voltage amplifier.
6. List, compare, and contrast the three BJT amplifier configurations.
7. Determine the configuration of any BJT amplifier.
8. Discuss the concept of amplifier efficiency.
9. List, compare, and contrast the various classes of amplifier operation.
10. Convert any value of power or voltage gain to and from decibel (dB) form.

## Outline

When you cannot hear the output from a stereo, you turn up the volume. When the picture on your television is too dark, you increase the brightness setting. In both of these cases, you are taking a relatively weak signal and making it stronger; that is, you are increasing its power level. The process of increasing the power of an ac signal is referred to as **amplification**. The circuits used to provide amplification are referred to as *amplifiers*. Several typical amplifiers are shown in Figure 8.1.

**Amplification**
The process of increasing the power of an ac signal.

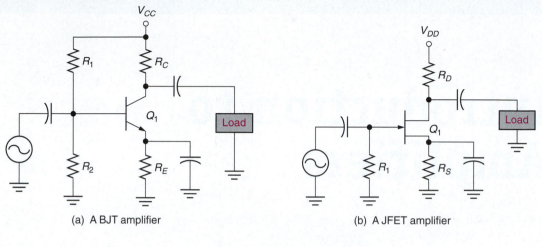

(a) A BJT amplifier

(b) A JFET amplifier

(c) An operational amplifier (op-amp) based circuit

**FIGURE 8.1** Typical amplifiers.

Amplifiers are some of the most widely used circuits that you will encounter. They are used extensively in audio, video, and telecommunications systems. They are also used in digital systems, biomedical systems, and so on. (In fact, you may find it difficult to come up with an electronic system that doesn't contain at least one amplifier.)

In the upcoming chapters, you will learn how to analyze and work with many different types of amplifiers. You'll be shown (among other things) how to calculate the values of several amplifier properties, such as *gain*, *input impedance*, and *output impedance*. In this chapter, we will discuss these properties and the roles they play in amplifier operation. We will also look at some other topics that relate to amplifiers in general.

## 8.1    Amplifier Properties

OBJECTIVE 1 ▶    All amplifiers have three fundamental properties: *gain*, *input impedance*, and *output impedance*. These properties can be combined to form a *general amplifier model*, like the one shown in Figure 8.2. Note that the diamond shape in the model is used (in this case) to represent the *gain* of the circuit. We'll modify this symbol slightly after we discuss the concept of gain.

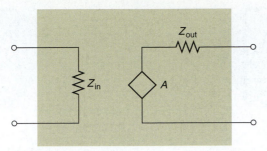

FIGURE 8.2 The general amplifier model.

## 8.1.1 Amplifier Gain

All amplifiers exhibit a property called **gain**. The gain of an amplifier is *a multiplier that exists between the circuit's input and output*. For example, if the gain of an amplifier is 100, then the output signal is 100 times as great as the input signal under normal operating conditions.

There are three types of gain: *voltage gain*, *current gain*, and *power gain*. Gain is represented using the letter $A$, as shown in Table 8.1. Note that the subscript in each symbol identifies the type of gain.

All amplifiers provide some degree of power gain. However, not all amplifiers are designed for this purpose. For example, a *voltage amplifier* is designed to provide a specific value of $A_v$. The fact that it also provides some value of power gain is usually a secondary consideration. The same can be said for a *current amplifier*, which is designed to provide a specific value of $A_i$. Only a *power amplifier* is designed to provide a specific value of $A_p$. The type of amplifier used in a given application depends on the type of gain desired.

In this chapter, we will focus primarily on the gain and impedance characteristics of voltage amplifiers. Current and power amplifiers are addressed in later chapters.

◀ OBJECTIVE 2

**Gain**
A multiplier that exists between the input and output of a circuit.

TABLE 8.1 Gain Symbols

| Type of Gain | Symbol |
| --- | --- |
| Voltage | $A_v$ |
| Current | $A_i$ |
| Power | $A_p$ |

## 8.1.2 Gain as a Ratio

Traditionally, *gain* is defined as *the ratio of an output value to its corresponding input value*. For example, *voltage gain* can be defined as *the ratio of ac output voltage to ac input voltage*. By formula,

$$A_v = \frac{v_{out}}{v_{in}} \tag{8.1}$$

where $v_{out}$ = the ac output voltage from the amplifier
$v_{in}$ = the ac input voltage to the amplifier

The calculation of voltage gain using input and output values is demonstrated in Example 8.1.

### EXAMPLE 8.1

The symbol shown in Figure 8.3 is a generic symbol for an amplifier. Calculate the voltage gain for the amplifier represented in the figure.

*Solution:* Using the $v_{out}$ and $v_{in}$ readings shown in the figure, the voltage gain of the amplifier is found as

$$A_v = \frac{v_{out}}{v_{in}} = \frac{250 \text{ mV}}{400 \text{ μV}} = 625$$

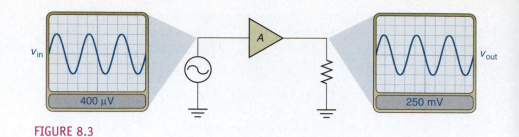

$v_{in}$   400 µV   A   250 mV   $v_{out}$

**FIGURE 8.3**

This result indicates that the ac output voltage for the circuit is, under normal operating circumstances, 625 times greater than the ac input voltage.

**PRACTICE PROBLEM 8.1**

An amplifier like the one represented in Figure 8.3 has the following values: $v_{out} = 72$ mV and $v_{in} = 300$ µV. Calculate the voltage gain of the circuit.

Does gain have a unit of measure?

Note that the value of voltage gain obtained in the example has no units. This is always the case. Since each type of gain can be defined as a ratio of an output value to an input value, gain has no unit of measure.

Equation (8.1) can be somewhat misleading because it implies that the value of $A_v$ is determined by the values of $v_{out}$ and $v_{in}$. In fact, the value of $A_v$ is determined by circuit component values.

Under normal circumstances, gain is a constant. Thus, a change in the input to an amplifier usually results in a change in the corresponding output value, not in a change in gain. For example, when the value of $v_{in}$ for an amplifier changes, the new value of $v_{out}$ is found as

$$v_{out} = A_v v_{in} \tag{8.2}$$

This relationship is demonstrated in Example 8.2.

**EXAMPLE 8.2**

The input of the circuit in Figure 8.3 changes to 240 µV. Calculate the new value of $v_{out}$.

*Solution:*   The gain of the circuit was found in Example 8.1 to be 625. Using this value of $A_v$, the new value of $v_{out}$ is found as

$$v_{out} = A_v v_{in} = (625)(240 \text{ µV}) = \textbf{150 mV}$$

**PRACTICE PROBLEM 8.2**

Refer to Practice Problem 8.1. Using the value of $A_v$ found in the problem, calculate the value of $v_{out}$ for $v_{in} = 360$ µV.

Like voltage gain, current and power gain can be defined as output-to-input ratios. These ratios are

$$A_i = \frac{i_{out}}{i_{in}} \tag{8.3}$$

and

$$A_p = \frac{P_{out}}{P_{in}} \tag{8.4}$$

Since they can be defined as ratios, current and power gain (like voltage gain) have no units. They are simply multipliers that exist between the input and output of an amplifier.

### 8.1.3   The General Voltage Amplifier Model

Figure 8.2 showed a general model for a voltage amplifier. Now that we have defined gain, we need to modify the amplifier model, as shown in Figure 8.4a. Note that the diamond shape is actually used to represent $v_{out}$ (as a *voltage source*). The value of this voltage source is given as $A_v v_{in}$ to show that the output voltage is a function of the input voltage and the voltage gain of the circuit.

◀ *OBJECTIVE 3*

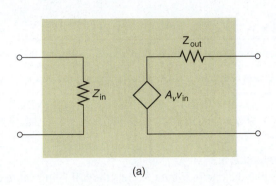

(a)

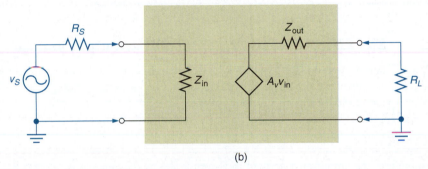

(b)

FIGURE 8.4   Voltage amplifier model.

*A Practical Consideration:* The $Z_{in}$ and $Z_{out}$ labels in Figure 8.4b (and throughout the chapter) are used because they represent transistor ratings that are listed on most spec sheets. Though listed as *impedances*, they can be combined algebraically with the resistances shown ($R_S$ and $R_L$). Remember . . . impedance can be purely resistive.

When we add a signal source and a load to the amplifier model in Figure 8.4a, we obtain the circuit shown in Figure 8.4b. The input circuit consists of $v_S$, $R_S$, and $Z_{in}$ (the amplifier input impedance). The output circuit consists of $v_{out}$, $Z_{out}$ (the amplifier output impedance), and $R_L$. As you can see, the circuits are nearly identical (in terms of the type of components they contain).

### 8.1.4   Amplifier Input Impedance ($Z_{in}$)

When an amplifier is connected to a signal source, the source sees the amplifier as a load. The **input impedance ($Z_{in}$)** of the amplifier is the value of this load. For example, the value of $Z_{in}$ for the amplifier in Figure 8.5 is shown to be 1.5 k$\Omega$. In this case, the amplifier acts as a 1.5 k$\Omega$ load that is in series with the source resistance ($R_S$).

If we assume that the input impedance of the amplifier in Figure 8.5 is purely resistive, the signal voltage at the amplifier input is found as

◀ *OBJECTIVE 4*

**Input impedance ($Z_{in}$)**
The load that an amplifier places on its source.

$$v_{in} = v_S \frac{Z_{in}}{R_S + Z_{in}}$$

(8.5)

Since $R_S$ and $Z_{in}$ form a voltage divider, the input voltage to the amplifier must be lower than the rated value of the source. This point is illustrated in the following example.

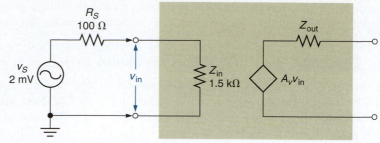

FIGURE 8.5    Amplifier input circuit.

### EXAMPLE 8.3

Calculate the value of $v_{in}$ for the circuit shown in Figure 8.5.

*Solution:*    Using the values shown in the figure, the value of $v_{in}$ is found as

$$v_{\text{in}} = v_S\frac{Z_{\text{in}}}{R_S + Z_{\text{in}}} = (2\ \text{mV})\frac{1.5\ \text{k}\Omega}{1.6\ \text{k}\Omega} = \mathbf{1.875\ mV}$$

### PRACTICE PROBLEM 8.3

An amplifier like the one in Figure 8.5 has the following values: $v_S = 800\ \mu V$, $R_S = 70\ \Omega$, and $Z_{in} = 750\ \Omega$. Calculate the value of $v_{in}$ for the circuit.

## 8.1.5    The Effect of $R_S$ and $Z_{in}$ on Amplifier Output Voltage

Example 8.3 demonstrated the effect that $R_S$ and $Z_{in}$ can have on the input to an amplifier. As the following example demonstrates, the reduction of the source voltage can cause a noticeable reduction in the circuit's output voltage.

### EXAMPLE 8.4

Assume that the amplifier described in Example 8.3 has a value of $A_v = 500$. Calculate the values of $v_{out}$ for $v_{in} = 2\ mV$ and $v_{in} = 1.875\ mV$.

*Solution:*    For $v_{in} = 1.2\ mV$, the value of $v_{out}$ is found as

$$v_{\text{out}} = A_v v_{\text{in}} = (500)(2\ \text{mV}) = \mathbf{1\ V}$$

For $v_{in} = 1.875\ mV$, the value of $v_{out}$ is found as

$$v_{\text{out}} = A_v v_{\text{in}} = (500)(1.875\ \text{mV}) = \mathbf{937.5\ mV}$$

### PRACTICE PROBLEM 8.4

Refer to Practice Problem 8.3. Using $A_v = 480$, calculate the values of $v_{out}$ for $v_{in} = v_S$ and for the value of $v_{in}$ found in the practice problem.

Here is what Example 8.4 has shown us: Had we ignored the effects of $R_S$ and $Z_{in}$ on the amplifier in Figure 8.5, we would have calculated an output voltage for the circuit of 1 V. However, because of the reduction in $v_S$, the actual value of $v_{out}$ is 937.5 mV. Thus, $R_S$ and $Z_{in}$ cause the output of the circuit to be 62.5 mV lower than it first appears to be. As you will see, the combination of $Z_{out}$ and $R_L$ causes an even greater reduction in the ideal output from an amplifier.

## 8.1.6 Amplifier Output Impedance ($Z_{out}$)

When a load is connected to an amplifier, the amplifier acts as the source for that load. As with any source, there is some measurable value of source impedance, in this case the **output impedance ($Z_{out}$)** of the amplifier. For example, consider the circuit shown in Figure 8.6. If we assume that the value of $Z_{out}$ for the amplifier is 200 Ω, the load sees the amplifier as a voltage source with an internal impedance of 200 Ω. If we assume that the output impedance of the amplifier in Figure 8.6 is purely resistive, the value of the load voltage can be found using the voltage-divider equation as follows:

> **Output impedance ($Z_{out}$)**
> The "source impedance" that an amplifier presents to its load.

$$v_L = v_{out} \frac{R_L}{Z_{out} + R_L} \qquad (8.6)$$

where $v_{out} = A_v v_{in}$

Example 8.5 demonstrates the effect that the combination of $Z_{out}$ and $R_L$ can have on the load voltage produced by an amplifier.

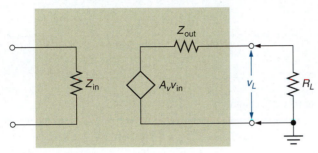

**FIGURE 8.6** Amplifier output circuit.

### EXAMPLE 8.5

Calculate the value of the load voltage ($v_L$) for the circuit shown in Figure 8.7.

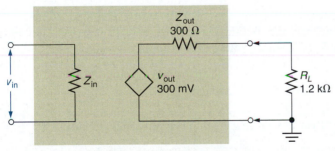

**FIGURE 8.7**

*Solution:* Using the values shown in the figure, the value of $v_L$ is found as

$$v_L = v_{out} \frac{R_L}{Z_{out} + R_L} = (300\ \text{mV}) \frac{1.2\ \text{k}\Omega}{1.5\ \text{k}\Omega} = \mathbf{240\ mV}$$

As you can see, $Z_{out}$ and $R_L$ have combined to cause a 60 mV reduction in the amplifier output voltage.

#### PRACTICE PROBLEM 8.5

A circuit like the one shown in Figure 8.7 has the following values: $v_{out} = 480$ mV, $Z_{out} = 240$ Ω, and $R_L = 1.5$ kΩ. Calculate the value of the load voltage for the circuit.

### 8.1.7    The Combined Effects of the Input and Output Circuits

The combination of the input and output circuits can cause a fairly significant reduction in the *effective* voltage gain of an amplifier. For example, consider the circuit shown in Figure 8.8. To see the combined effects of the input and output circuits, we need to calculate the value of $v_L$ for the circuit. The first step is to determine the value of $v_{in}$ as follows:

$$v_{in} = v_S \frac{Z_{in}}{R_S + Z_{in}} = (15 \text{ mV}) \frac{980 \text{ }\Omega}{1 \text{ k}\Omega} = \mathbf{14.7 \text{ mV}}$$

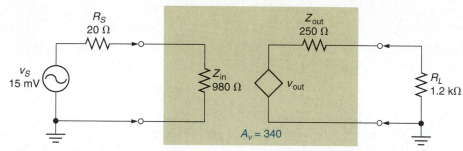

**FIGURE 8.8**

Now, the value of $v_{out}$ is found as

$$v_{out} = A_v v_{in} = (340)(14.7 \text{ mV}) = \mathbf{5 \text{ V}}$$

Finally, the load voltage is found as

$$v_L = v_{out} \frac{R_L}{Z_{out} + R_L} = (5 \text{ V}) \frac{1.2 \text{ k}\Omega}{1.45 \text{ k}\Omega} = \mathbf{4.14 \text{ V}}$$

According to these calculations, the amplifier has increased the 15 mV source voltage to a 4.14 V load voltage. We consider the *effective* voltage gain of the amplifier to equal the ratio of load voltage to source voltage. By formula,

$$A_{v(\text{eff})} = \frac{v_L}{v_S} \tag{8.7}$$

Therefore, the effective voltage gain of the circuit can be found as

$$A_{v(\text{eff})} = \frac{v_L}{v_S} = \frac{4.14 \text{ V}}{15 \text{ mV}} = \mathbf{276}$$

As you can see, the input and output circuits have reduced the voltage gain of this amplifier from 340 to an effective value of 276.

The effects that the input and output circuits have on the voltage gain of an amplifier can be significantly reduced by:

1. *Increasing* the value of $Z_{in}$.
2. *Decreasing* the value of $Z_{out}$.

For example, consider the circuit shown in Figure 8.9. This is essentially the same circuit as the one shown in Figure 8.8. However, for the sake of discussion, we have changed the values of $Z_{in}$ and $Z_{out}$. For this circuit,

$$v_{in} = v_S \frac{Z_{in}}{R_S + Z_{in}} = (15 \text{ mV}) \frac{8 \text{ k}\Omega}{8.02 \text{ k}\Omega} = \mathbf{14.96 \text{ mV}}$$

> How do you reduce the effects of the input and output circuits on amplifier voltage gain?

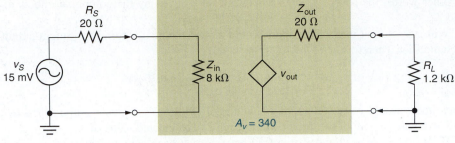

FIGURE 8.9

$$v_{out} = A_v v_{in} = (340)(14.96 \text{ mV}) = \mathbf{5.1 \text{ V}}$$

$$v_L = v_{out} \frac{R_L}{Z_{out} + R_L} = (5.1 \text{ V}) \frac{1.2 \text{ k}\Omega}{1.22 \text{ k}\Omega} = \mathbf{5 \text{ V}}$$

and the effective value of $A_v$ for the circuit can be found as

$$A_{v(eff)} = \frac{v_L}{v_S} = \frac{5 \text{ V}}{15 \text{ mV}} = \mathbf{333.3}$$

As you can see, the effective voltage gain of the circuit has increased significantly due to the changes made in the values of $Z_{in}$ and $Z_{out}$.

In most practical circuits, little can be done to change the resistance of a given signal source or load. However, the amplifier values of $Z_{in}$ and $Z_{out}$ are affected by the choice of the active components used as well as the type of biasing circuit and component values. This point will be demonstrated further in upcoming chapters.

## 8.1.8    The Ideal Voltage Amplifier

Now that you have been introduced to the basic amplifier properties, we will look at the characteristics of the *ideal voltage amplifier*. The ideal voltage amplifier, if it could be constructed, would have the following characteristics (among others):     ◀ **OBJECTIVE 5**

1. *Infinite* gain (if needed).
2. *Infinite* input impedance.
3. *Zero* output impedance.

The first of these characteristics needs little explanation. An ideal amplifier would be capable of providing any value of gain, no matter how high the value needed. In reality, values of $A_v$ are limited. The limit of $A_v$ depends in part on the type of active component(s) used in the circuit.

The impedance characteristics of the ideal voltage amplifier are illustrated in Figure 8.10. With infinite input impedance, there would be no current in the input circuit and, therefore, no voltage dropped across the source resistance ($R_S$). With no voltage dropped across $R_S$,

$$v_{in} = v_S \qquad \text{(for the ideal amplifier)}$$

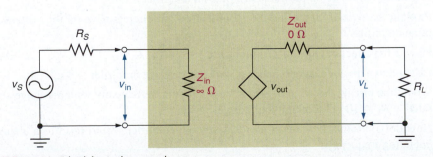

FIGURE 8.10    Ideal impedance values.

In other words, the source voltage would not be reduced by the combined effects of $R_S$ and $Z_{in}$.

With zero output impedance, there would be no voltage divider in the output circuit of the amplifier. Therefore,

$$v_L = v_{out} \qquad \text{(for the ideal amplifier)}$$

Since there would be no reduction of voltage by either the input or output circuits, the effective voltage gain of the circuit would equal the calculated value of $A_v$.

Values of $Z_{in} = \infty\ \Omega$ and $Z_{out} = 0\ \Omega$ have not yet been achieved in practical circuits. However, it is possible to "effectively" achieve them through proper circuit design. For example, consider the circuit shown in Figure 8.11. When compared to the value of $R_S$ (20 $\Omega$), the value of $Z_{in} = 100\ k\Omega$ is, for all practical purposes, infinite. When compared to the value of $R_L$ (1.2 k$\Omega$), the value of $Z_{out} = 3\ \Omega$ is, for all practical purposes, zero. As you can see, the values of $Z_{in}$ and $Z_{out}$ are a consideration in any circuit design since it is possible to minimize their effects on the voltage gain of the circuit.

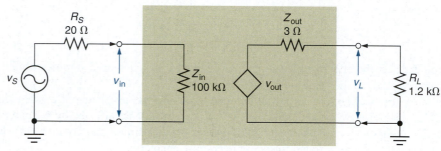

FIGURE 8.11

In upcoming chapters, you will see that the ideal amplifier has other characteristics that relate to power dissipation, frequency response, and signal reproduction. At this point, however, we have established the primary characteristics of the ideal voltage amplifier.

**Section Review ▶**

1. What is *amplification*?
2. What is *gain*? What are the three types of gain?
3. Define *voltage gain* as a ratio.
4. What determines the voltage gain of an amplifier?
5. What effect does the combination of source resistance ($R_S$) and amplifier input impedance ($Z_{in}$) have on the value of $v_{in}$ for an amplifier? What effect does it have on the value of $v_{out}$?
6. What effect does the combination of amplifier output impedance ($Z_{out}$) and load resistance ($R_L$) have on the value of $v_L$ for an amplifier?
7. Explain the overall effect that the input and output circuits have on the voltage gain of an amplifier.
8. How is the effect in Question 7 minimized?
9. List and explain the characteristics of the ideal voltage amplifier.

**Critical Thinking ▶**

10. Would the circuit shown in Figure 8.5 operate more effectively with a voltage source rated at $R_S = 10\ \Omega$? Explain your answer.
11. How do you think the gain of an ideal voltage amplifier would react to a change in operating frequency?

Now that we have established the gain, input impedance, and output impedance characteristics of the ideal voltage amplifier, we will take a brief look at several types of BJT amplifiers to see how they compare to the ideal amplifier. There are three BJT amplifier *configurations*, each having its unique combination of characteristics. These BJT amplifier configurations are shown in Figure 8.12.

◀ *OBJECTIVE 6*

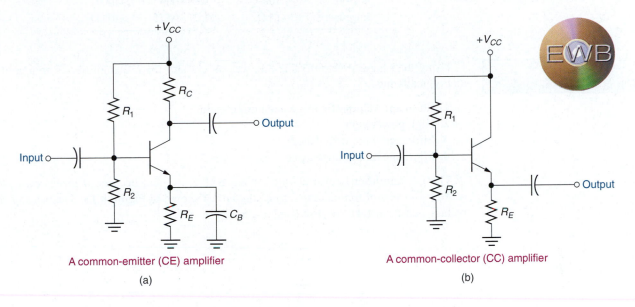

A common-emitter (CE) amplifier

(a)

A common-collector (CC) amplifier

(b)

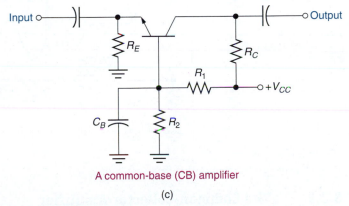

A common-base (CB) amplifier

(c)

FIGURE 8.12    BJT amplifier configurations.

### 8.2.1    The Common-Emitter Amplifier

The *common-emitter (CE) amplifier* is the most widely used BJT amplifier. A typical CE amplifier is shown in Figure 8.12a. As you can see, the input is applied to the *base* of the transistor, and the output is taken from the *collector*. The term *common-emitter* is used for two reasons:

1.  The emitter terminal of the transistor is *common* to both the input and output circuits.
2.  The emitter terminal of the transistor is normally returned to *ac ground* (or *ac common*). The ac ground is provided by the "bypass capacitor" ($C_B$) connected to the emitter terminal of the transistor. (The means by which this capacitor provides an ac ground will be discussed in Chapter 9.)

To discuss the characteristics of the CE amplifier (or any other), we need to establish some boundaries. For the sake of comparison, we'll classify gain and impedance values as being *low*, *midrange*, or *high*. These classifications are broken down as shown in Table 8.2. It should be noted that the ranges listed in the table are open to debate and should not be taken as standard values. They are provided merely as a basis for comparison.

**TABLE 8.2**  Property Ranges

| Property | Low | Midrange | High |
|---|---|---|---|
| Gain | <100 | 100–1000 | >1000 |
| Impedance | <1 kΩ | 1 kΩ–10 kΩ | >10 kΩ |

Typical CE amplifier characteristics.

Using the ranges given in Table 8.2, we can classify the CE amplifier as an amplifier that typically has:

1. Midrange values of voltage and current gain.
2. High power gain.
3. Midrange input impedance.
4. Midrange output impedance.

The CE amplifier is also unique among BJT amplifiers because *it produces a 180° voltage phase shift from its input to its output*, as shown in Figure 8.13. The basis of this voltage phase shift will be discussed in Chapter 9.

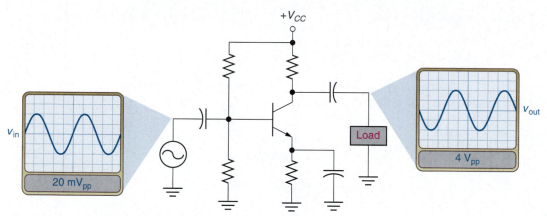

**FIGURE 8.13**  Common-emitter (CE) amplifier.

### 8.2.2  The Common-Collector Amplifier

The *common-collector (CC) amplifier* is another widely used BJT amplifier. A typical CC amplifier is shown in Figure 8.14. As you can see, the input is applied to the *base* of the transistor, and the output is taken from the *emitter*. In this case, it is the *collector* terminal of the transistor that is part of both the input and output circuits and provides the ac ground (or common).

Typical CC amplifier characteristics.

The CC amplifier typically has the following characteristics:

1. Midrange current gain.
2. Extremely low voltage gain (slightly less than 1).
3. High input impedance.
4. Low output impedance.

The most distinguishing characteristic here is the extremely *low voltage gain* of this configuration.

For reasons discussed in Chapter 10, the voltage gain of the CC amplifier is slightly less than *unity* (1). If we were to assume that $A_v = 1$ for a CC amplifier, the output waveform would be identical to the input waveform, as shown in Figure 8.14.

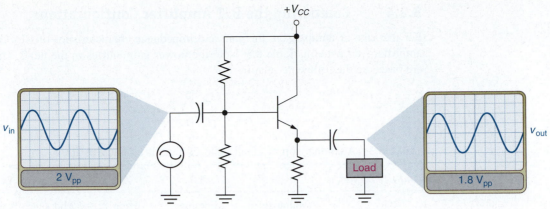

**FIGURE 8.14** Common-collector (CC) amplifier.

Since the ac signal at the emitter closely "follows" the ac voltage at the base, the CC amplifier is commonly referred to as an **emitter follower**. Note that this circuit is most commonly used for its current gain and impedance characteristics, as will be shown in Chapter 10.

**Emitter follower**

Another name for the common-collector amplifier.

### 8.2.3 The Common-Base Amplifier

The *common-base (CB) amplifier* is the least often used BJT amplifier configuration. A typical CB amplifier is shown in Figure 8.15. As you can see, the input is applied to the *emitter* of the transistor, and the output is taken from the *collector*. The "bypass capacitor" ($C_B$) in the base circuit provides the ac ground (or common) at that terminal.

The CB amplifier typically has the following characteristics:

Typical CB amplifier characteristics.

1. Midrange voltage gain.
2. Extremely low current gain (slightly less than 1).
3. Low input impedance.
4. High output impedance.

If you compare the CB amplifier to the ideal voltage amplifier, you'll see immediately one of the reasons why it is rarely used. The low input impedance and high output impedance of the circuit are exact opposites of the impedance characteristics of the ideal voltage amplifier. In other words, its effective voltage gain will be nowhere near its ideal value of $A_v$.

The extremely low current gain of the CB amplifier is because the input is applied to the emitter and the output is taken from the collector. Since collector current is always slightly less than emitter current, the value of $A_i$ for this circuit must be less than unity (1).

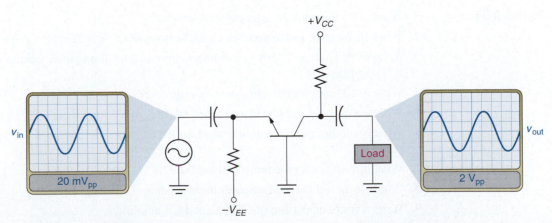

**FIGURE 8.15** Common-base (CB) amplifier.

### 8.2.4    Comparing the BJT Amplifier Configurations

For the sake of comparison, the gain and impedance characteristics of CE, CC, and CB amplifiers are listed in Table 8.3. Note the power gain entries in the table. These entries are based on the following relationship:

$$A_p = A_v A_i \tag{8.8}$$

TABLE 8.3   A Comparison of CE, CC, and CB Circuit Characteristics

| Amplifier Type | $A_v$ | $A_i$ | $A_p$ | $Z_{in}$ | $Z_{out}$ |
|---|---|---|---|---|---|
| CE | Midrange | Midrange | High | Midrange | Midrange |
| CC | <1 | Midrange | $\cong A_i$ | High | Low |
| CB | Midrange | <1 | $\cong A_v$ | Low | High |

Just as power equals the product of voltage and current, power gain equals the product of $A_v$ and $A_i$. In the case of the CC amplifier, if we assume that $A_v = 1$, then power gain must equal current gain. In the case of the CB amplifier, if we assume that $A_i = 1$, then power gain must equal voltage gain.

If you compare the values given in Table 8.3, you'll see that CC and CB amplifiers are nearly opposites in terms of their gain and impedance characteristics. At the same time, the CE amplifier is sort of a "middle-of-the-road" circuit.

### 8.2.5    Determining the Configuration of an Amplifier

**OBJECTIVE 7** ▶ There are many different ways to construct CE, CC, and CB amplifiers. The question is, when you see a BJT amplifier that you don't recognize, how can you tell which configuration you are dealing with?

The simplest way to determine the configuration of a given BJT amplifier is to use this technique: *Identify the input and output terminals. The third terminal is the common one.* For example, look at the circuit shown in Figure 8.15. The input is applied to the *emitter*, and the output is taken from the *collector*. The third terminal is the *base*. Therefore, the circuit is a *common-base amplifier*. The other two BJT amplifier configurations can be identified in the same fashion.

In Chapters 9 and 10, we will thoroughly analyze the amplifier configurations that have been introduced in this section. Among other things, you will be shown how to calculate the gain and impedance values that we have used to describe these circuits.

---

**Section Review** ▶
1. What is the basis for the term *common-emitter*?
2. What are the gain and impedance characteristics of CE amplifiers?
3. In terms of input and output ac voltages, how is the CE amplifier unique among the BJT amplifiers?
4. What is the basis for the term *common-collector*?
5. What are the gain and impedance characteristics of CC amplifiers?
6. What is another name commonly used for the CC amplifier? What is the basis of this name?
7. What is the basis for the term *common-base*?
8. What are the gain and impedance characteristics of CB amplifiers?
9. Write a brief comparison of CE, CC, and CB amplifiers.
10. How do you determine the configuration of a BJT amplifier that you don't recognize?

Some BJT amplifiers contain transistors that conduct during the entire cycle of the input signal. Others contain one or more transistors that conduct only during a portion of the input cycle. For example, consider the circuits shown in Figure 8.16. The transistor in Figure 8.16a conducts for the full 360° of the input cycle. When an amplifier contains a transistor that conducts during the entire cycle of the input, it is referred to as a **class A amplifier**. In contrast, the **class B amplifier** shown in Figure 8.16b contains two transistors that each conduct for approximately 180° of the input cycle. The **class C amplifier** shown in Figure 8.16c contains a single transistor that conducts for less than 180° of the input cycle.

   In this section, we will look at each of these types of amplifiers and the ways in which they differ. First, however, we will look at some of the factors that determine which class of amplifier is used in a given application.

> **Class A amplifier**
> An amplifier with a single transistor that conducts during the entire input cycle.
>
> **Class B amplifier**
> An amplifier with two transistors that each conduct for approximately half the input cycle.
>
> **Class C amplifier**
> An amplifier with one transistor that conducts for less than 180° of the input cycle.

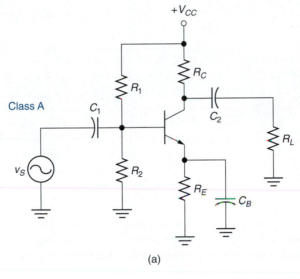

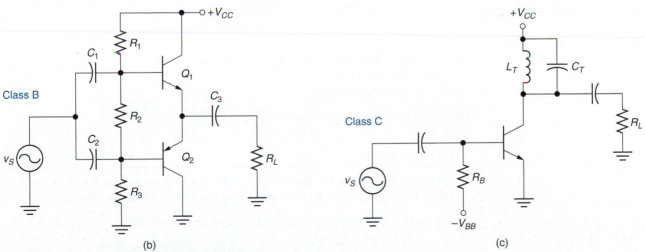

**FIGURE 8.16**   Different classes of amplifiers.

## 8.3.1    Amplifier Efficiency

Amplifiers actually increase the power level of an ac input *by transferring power from the dc power supply to the input signal.* For example, consider the circuit shown in Figure 8.17. The input signal is shown to have a power rating of 1.5 mW. With a value of $A_p = 300$, the load power is shown to be 450 mW. The difference between $P_{in}$ and $P_{out}$ is 448.5 mW. Where did this power come from? The power was actually transferred from the dc power supply to the load by the amplifier.

◄ *OBJECTIVE 8*

The ideal amplifier would deliver 100% of the power it draws from the dc power supply to the load. In practice, however, this does not occur because the components in the amplifier are all dissipating some amount of power. For example, consider the circuit shown in Figure 8.18. Assuming that the amplifier is operating normally, there is always some amount of current through each of the amplifier components. Since each component has a measurable value of current through it and a measurable voltage across it, each is dissipating power. This power dissipation, which is illustrated in the figure, reduces the amount of power available to be transferred to the load.

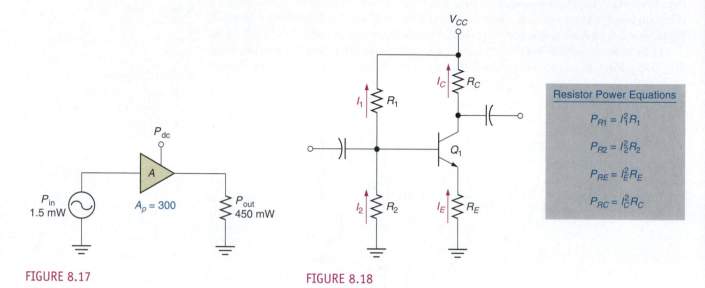

**Resistor Power Equations**

$$P_{R1} = I_1^2 R_1$$

$$P_{R2} = I_2^2 R_2$$

$$P_{RE} = I_E^2 R_E$$

$$P_{RC} = I_C^2 R_C$$

FIGURE 8.17                    FIGURE 8.18

**Efficiency (η)**
The percentage of the power drawn from the dc power supply that an amplifier actually delivers to its load.

A *figure of merit* for any amplifier is its **efficiency**. The efficiency of an amplifier is *the percentage of the power drawn from the supply that is actually delivered to the load.* Efficiency, which is represented by the Greek letter η (called *eta*), is found as

$$\eta = \frac{P_L}{P_{dc}} \times 100 \tag{8.9}$$

where η = the efficiency of the amplifier, written as a percentage (%)
  $P_L$ = the ac load power
  $P_{dc}$ = the dc input power

Example 8.6 further illustrates the concept of amplifier efficiency.

### EXAMPLE 8.6

An amplifier is continuously drawing 1.2 W from its dc power supply. If the ac load power for the circuit is 240 mW, what is the amplifier's efficiency rating?

*Solution:* The efficiency rating of the amplifier is found as

$$\eta = \frac{P_L}{P_{dc}} \times 100 = \frac{240 \text{ mV}}{1.2 \text{ W}} \times 100 = \mathbf{20\%}$$

Thus, only 20% of the power drawn from the supply is actually delivered to the load. The other 80% is used by the amplifier itself.

### PRACTICE PROBLEM 8.6
An amplifier is continuously drawing 3.3 W from its dc power supply. If the ac load power for the circuit is 450 mW, what is the efficiency rating of the circuit?

The higher the efficiency rating of an amplifier, the closer it comes to the ideal. As you will see, the maximum possible efficiency ratings for class A, class B, and class C amplifiers differ significantly.

### 8.3.2    Distortion

One of the goals in amplification is to produce an output waveform that has the exact same *shape* as the input waveform. Ideally, every linear amplifier should be capable of producing a duplicate of any input waveform.

**Distortion** is defined as *any undesired change in the shape of a waveform*. The waveforms shown in Figure 8.19 illustrate two different types of distortion that can be produced by amplifiers. As upcoming chapters will show, each of these types of distortion is characteristic of one or more of the amplifier classes.

**Distortion**

Any undesired change in the shape of a waveform.

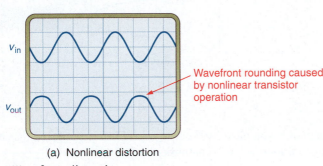

Wavefront rounding caused by nonlinear transistor operation

(a)  Nonlinear distortion

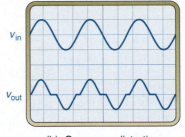

(b)  Crossover distortion

FIGURE 8.19    Waveform distortion.

### 8.3.3    Class A Amplifiers

Most of the amplifiers you will see in the field, whether they contain a BJT or some other active device, are class A amplifiers. Under normal operating conditions, a class A amplifier has:

◀ *OBJECTIVE 9*

1. An active device that conducts during the entire 360° of the input cycle.
2. An output that contains little or no distortion.
3. A maximum theoretical efficiency of 25%.

Class A operation is achieved in a BJT amplifier by midpoint biasing the transistor. As you may recall, midpoint bias allows the BJT output to vary widely around the *Q*-point without hitting saturation or cutoff, thus ensuring linear operation. This point is illustrated in Figure 8.20. As you can see, the output from the amplifier represented would be a near-perfect reproduction of the input. Biasing the BJT above or below midpoint would result in one of two situations:

Class A amplifier characteristics.

1. A reduction in the maximum possible amplifier output.
2. Distortion.

This point is discussed further in Chapter 11, along with the basis of the 25% maximum efficiency rating for the class A amplifier.

Because of their relatively poor efficiency ratings, class A amplifiers are generally used as *small-signal* (low-power) amplifiers. Small-signal amplifiers are primarily used to drive higher-power stages. One example of a higher-power stage is the audio output stage in a stereo system.

### 8.3.4    Class B Amplifiers

The class B amplifier typically contains *two* transistors that are connected as shown in Figure 8.21. Each transistor in this type of amplifier conducts during one alternation of the ac input cycle and is in cutoff during the other alternation. With one transistor conducting during the negative alternation of the ac input and the other conducting during

Class B amplifier characteristics.

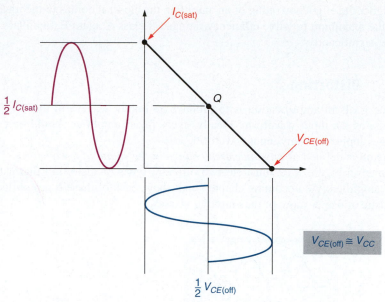

**FIGURE 8.20** Midpoint bias.

the positive alternation, a complete 360° output waveform is produced (as shown). Under normal operating conditions, a class B amplifier has:

1. Two transistors that are biased at cutoff (each conducts during one alternation of the ac input cycle).
2. An output that contains little or no distortion.
3. A maximum theoretical efficiency of approximately 78.5%.

The relatively high efficiency rating of the class B amplifier makes it very useful as a higher-power amplifier. The audio output stage mentioned earlier is typically a class B amplifier.

One variation of the class B amplifier is the **class AB amplifier** shown in Figure 8.22. In the class AB amplifier, which is also known as a *diode-biased amplifier*, each transistor conducts for *slightly more than 180° of the ac input cycle*. As you will see in Chapter 11, this circuit is used to prevent a specific type of distortion that can be produced by a standard class B amplifier.

> **Class AB amplifier**
> An amplifier with two transistors that each conduct for slightly more than 180° of the input cycle.

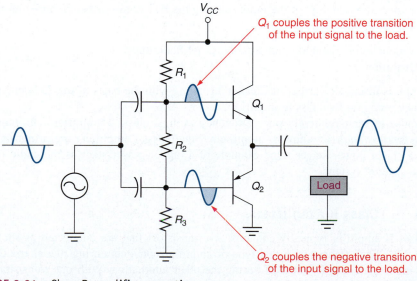

**FIGURE 8.21** Class B amplifier operation.

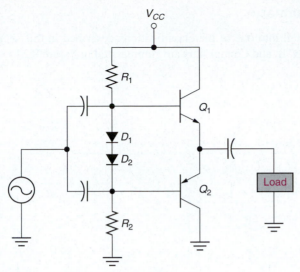

**FIGURE 8.22** The class AB (diode-bias) amplifier.

### 8.3.5 Class C Amplifiers

The class C amplifier contains a single transistor that conducts for less than 180° of the ac input cycle. A typical BJT class C amplifier is shown in Figure 8.23.

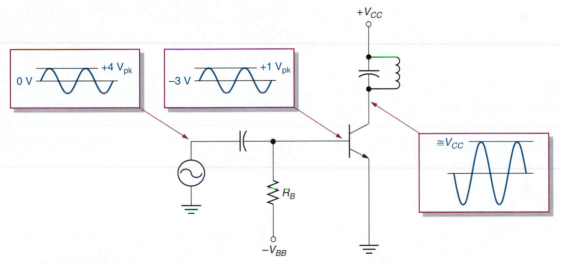

**FIGURE 8.23** Class C operation.

The BJT in the class C amplifier is biased deeply into cutoff. The ac input to the amplifier causes the transistor to conduct for a brief time during the input cycle. The rest of the output waveform shown in Figure 8.23 is produced by the LC tank in the collector circuit of the amplifier.

The class C amplifier, by its design, is a **tuned amplifier**. A tuned amplifier is one that produces a usable output for a specific range of input frequencies. Since the class C amplifier is a tuned amplifier, coverage of this type of circuit is reserved for Chapter 17. At this point, we are interested only in establishing the class C amplifier as one that typically has:

> **Tuned amplifier**
> An amplifier designed to have a specific value of gain over a specified range of frequencies.

1. A single transistor that conducts for less than 180° of the ac input cycle.
2. An output that may contain a significant amount of distortion.
3. A maximum theoretical efficiency rating of approximately 99%.

> Class C amplifier characteristics.

It would seem that the high efficiency rating of the class C amplifier would make it the ideal power amplifier. However, the distortion characteristics of the circuit limit its use.

### 8.3.6    Summary

Most amplifiers fall into one of the classifications described in this section. The characteristics of class A, B, and C amplifiers are summarized in Figure 8.24.

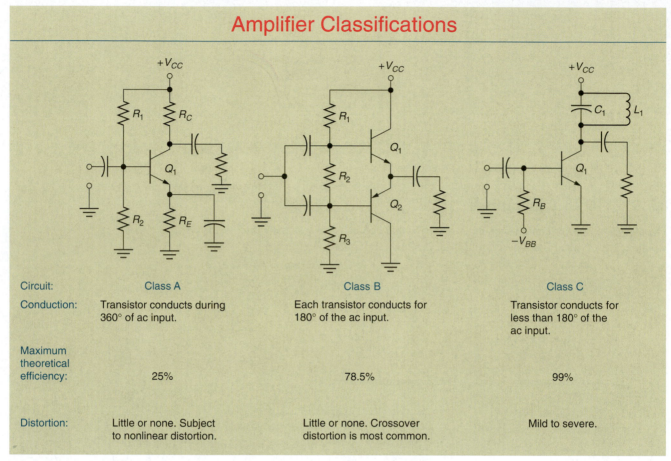

# Amplifier Classifications

| Circuit: | Class A | Class B | Class C |
|---|---|---|---|
| Conduction: | Transistor conducts during 360° of ac input. | Each transistor conducts for 180° of the ac input. | Transistor conducts for less than 180° of the ac input. |
| Maximum theoretical efficiency: | 25% | 78.5% | 99% |
| Distortion: | Little or none. Subject to nonlinear distortion. | Little or none. Crossover distortion is most common. | Mild to severe. |

FIGURE 8.24    Amplifier classes.

**Section Review ▶**

1. How does an amplifier increase the power level of an ac input signal?
2. What is the *efficiency* rating of an amplifier?
3. What would the efficiency rating be for an ideal amplifier? Why can't this rating be achieved in practice?
4. What is *distortion*? Why is it undesirable?
5. What are the typical characteristics of a *class A* amplifier?
6. In terms of biasing, how is class A operation achieved? Explain your answer.
7. Why are class A amplifiers typically used as small-signal amplifiers?
8. What are the typical characteristics of a *class B* amplifier?
9. What is a *class AB* amplifier? Why is it used?
10. What are the typical characteristics of a *class C* amplifier?
11. How is the output of a class C amplifier produced?
12. What factor limits the use of the class C amplifier?

Component and system specification sheets often list voltage gain and/or power gain values. These values, when provided, are often expressed in *decibel (dB)* form. A **decibel (dB)** is *a logarithmic unit used to express the ratio of one value to another*. Writing numbers in dB form allows us to easily represent very large gain values as relatively small numbers. For example, a power gain of 1,000,000 is equal to 60 dB.

In this section, you will be shown how to convert numbers back and forth between standard numeric form and dB form. You will also be shown the advantages and limitations of using dB gain values.

◄ *OBJECTIVE 10*

> **Decibel (dB)**
> A logarithmic unit used to express the ratio of one value to another.

### 8.4.1 dB Power Gain

The **dB power gain** of an amplifier is *the ratio of output power to input power, equal to 10 times the common log of that ratio*. By formula,

$$A_{p(\text{dB})} = 10 \log A_p \tag{8.10}$$

or

$$A_{p(\text{dB})} = 10 \log \frac{P_{\text{out}}}{P_{\text{in}}} \tag{8.11}$$

> **dB power gain**
> The ratio of circuit output power to input power, equal to 10 times the common log of that ratio.

Example 8.7 demonstrates the process for finding the decibel power gain of an amplifier.

#### EXAMPLE 8.7

A given amplifier has values of $P_{\text{in}} = 100\ \mu\text{W}$ and $P_{\text{out}} = 2\ \text{W}$. Calculate the dB power gain of the amplifier.

*Solution:* The dB power gain of the amplifier is found as

$$A_{p(\text{dB})} = 10 \log \frac{P_{\text{out}}}{P_{\text{in}}} = 10 \log \frac{2\ \text{W}}{100\ \mu\text{W}} = 10 \log (20{,}000) = \mathbf{43.01\ dB}$$

#### PRACTICE PROBLEM 8.7

An amplifier has values of $P_{\text{in}} = 420\ \mu\text{W}$ and $P_{\text{out}} = 6\ \text{W}$. What is the dB power gain of the amplifier?

When you need to convert a dB power gain value to standard numeric form, the following equation is used:

$$A_p = \log^{-1} \frac{A_{p(\text{dB})}}{10} \tag{8.12}$$

where $(\log^{-1})$ represents the *inverse log* of the fraction. Example 8.8 demonstrates the conversion of a dB power gain value to standard numeric form.

#### EXAMPLE 8.8

A given amplifier has a power gain of 3 dB. What is the ratio of output power to input power for the circuit?

*Solution:* The ratio of output power to input power is found as

$$A_p = \log^{-1} \frac{A_{p(\text{dB})}}{10} = \log^{-1} \frac{3\ \text{dB}}{10} = \log^{-1}(0.3) = \mathbf{1.995}$$

Thus, a gain of 3 dB represents a ratio of output power to input power that is approximately equal to 2; that is, the output power is approximately *twice* the value of the input power.

### PRACTICE PROBLEM 8.8

An amplifier has a dB power gain of 20 dB. What is the ratio of output power to input power for the circuit?

When the input power and dB power gain of an amplifier are known, the output power is found by first converting the dB power gain to standard numeric form. Then the output power is found as the product of $A_p$ and $P_{in}$. Example 8.9 demonstrates this process.

### EXAMPLE 8.9

A given amplifier has values of $P_{in}$ = 50 mW and $A_{p(dB)}$ = 3 dB. Calculate the amplifier output power.

*Solution:*  First, the value of 3 dB is converted to standard numeric form as follows:

$$A_p = \log^{-1}\frac{A_{p(dB)}}{10} = \log^{-1}\frac{3\text{ dB}}{10} = \log^{-1}(0.3) \cong 2$$

Now, using $A_p$ = 2, the value of the output power is found as

$$P_{out} = A_p P_{in} = (2)(50\text{ mW}) = 100\text{ mW}$$

### PRACTICE PROBLEM 8.9

An amplifier has values of $A_{p(dB)}$ = 5.2 dB and $P_{in}$ = 40 μW. Calculate the output power for the circuit.

The standard numeric form of 3 dB was rounded off to 2 in Example 8.9. When working with power gain values, it is standard practice to assume that 3 dB = 2.

When output power and dB power gain are known, the input power is found as shown in Example 8.10. Note the similarity between this procedure and the one shown in Example 8.9.

### EXAMPLE 8.10

A circuit has values of $A_{p(dB)}$ = −3 dB and $P_{out}$ = 50 mW. Calculate the value of the circuit input power.

*Solution:*  First, we must convert the value of −3 dB to standard numeric form as follows:

$$A_p = \log^{-1}\frac{A_{p(dB)}}{10} = \log^{-1}\frac{-3\text{ dB}}{10} = \log^{-1}(-0.3) \cong 0.5$$

Now, the value of the input power is found as

$$P_{in} = \frac{P_{out}}{A_p} = \frac{50\text{ mW}}{0.5} = 100\text{ mW}$$

### PRACTICE PROBLEM 8.10

An amplifier with a power gain of 12.4 dB has an output power of 2.2 W. Calculate the value of $P_{in}$ for the circuit.

If we take a moment to compare the results of Examples 8.9 and 8.10, we can come to some very important conclusions regarding the use of dB values. For convenience, the results for the two examples are listed here:

| Example | dB Power Gain | $P_{in}$ (mW) | $P_{out}$ (mW) |
|---------|---------------|---------------|----------------|
| 8.9 | 3 dB | 50 mW | 100 mW |
| 8.10 | −3 dB | 100 mW | 50 mW |

The conclusions that we can draw from these results are as follows:

1. *Positive* dB values represent a power *gain*, while *negative* dB values represent a power *loss*. When the gain was +3 dB, the output power was greater than the input power. When the gain was −3 dB, the output power was actually less than the input power.
2. Positive and negative decibels of equal magnitude represent reciprocal gains and losses. A +3 dB gain caused power to double, while a −3 dB gain caused power to be cut in half.

Positive versus negative dB values.

The relationship described in conclusion (2) can be stated mathematically as follows:

*If x dB represents a power gain of y, then*
*−x dB represents a power loss equal to 1/y.*

Table 8.4 lists some examples of this relationship.

**TABLE 8.4** Some Examples of Equal Positive and Negative dB Values

| dB Value | Gain | dB Value | Loss |
|----------|------|----------|------|
| 3 | 2 | −3 | ½ |
| 6 | 4 | −6 | ¼ |
| 12 | 16 | −12 | 1/16 |
| 20 | 100 | −20 | 1/100 |

The fact that equivalent (+) and (−) dB values represent gains and losses that are reciprocals is one of the reasons that dBs are commonly used. Granted, they can appear a bit intimidating at first. However, once you get used to seeing gains represented as dB values, you will have no problem using them.

Another advantage of using dB gain values is as follows: In multistage amplifiers, the total gain of the circuit is equal to the *sum* of the individual amplifier dB gains. For example, let's say that two amplifiers are connected as shown in Figure 8.25. The total power gain for the circuit is found as 20 dB + 6 dB = 26 dB. The basis of this relationship is discussed in Chapter 9.

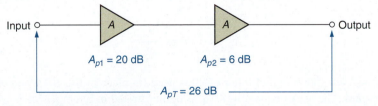

**FIGURE 8.25** dB gains are additive.

## 8.4.2   The dBm Reference

On some specification sheets, you will see power ratings that are listed as *dBm* values. For example, a stereo may be listed as having a maximum output power of 50 dBm. This rating tells you that the maximum output power from the stereo is *50 dB above 1 mW.* Power, measured in dBm, is found as

$$P_{dBm} = 10 \log \frac{P}{1 \text{ mW}}$$  (8.13)

Note that dBm values represent *actual power levels*. In contrast, dB values represent power *ratios*. This point is illustrated in Examples 8.11 and 8.12.

**EXAMPLE 8.11**

An amplifier has a rating of $A_p \times 50$ dB. Calculate the output power of the amplifier.

*Solution:* The problem cannot be solved as given. The value of 50 dB represents the *ratio* of output power to input power. Since no input power value was given, we cannot determine the value of the output power.

**EXAMPLE 8.12**

The output rating of an amplifier is given as 50 dBm. Calculate the output power for the circuit.

*Solution:* Since the circuit is rated in dBm, the actual output power can be found as follows:

$$A_p = \log^{-1} \frac{50 \text{ dB}}{10} = \log^{-1}(5.0) = \mathbf{1 \times 10^5}$$

and

$$P_{\text{out}} = A_p(1 \text{ mW}) = (1 \times 10^5)(1 \text{ mW}) = \mathbf{100 \text{ W}}$$

We have solved the problem by converting the dBm rating to a value of gain and then multiplying that gain by 1 mW. This is the standard approach to this problem.

**PRACTICE PROBLEM 8.12**

An amplifier has an output power rating of 32 dBm. Determine the actual output power from the circuit.

As you can see, dBm values can be used to determine actual power values, while dB values simply indicate the *ratio* of output power to input power.

### 8.4.3  dB Voltage Gain

The *dB voltage gain* of an amplifier is found as

$$A_{v(\text{dB})} = 20 \log A_v \tag{8.14}$$

or

$$A_{v(\text{dB})} = 20 \log \frac{v_{\text{out}}}{v_{\text{in}}} \tag{8.15}$$

As equation (8.15) shows, a multiplier of 20 is used (in place of the 10) when calculating dB voltage gain. The use of equation (8.15) is demonstrated in Example 8.13.

**EXAMPLE 8.13**

An amplifier has values of $v_{\text{in}} = 25$ mV and $v_{\text{out}} = 2$ V. Calculate the dB voltage gain of the circuit.

*Solution:* Using equation (8.14) and the values given in the problem, the dB voltage gain of the amplifier is found as

$$A_{v(\text{dB})} = 20 \log \frac{v_{\text{out}}}{v_{\text{in}}} = 20 \log \frac{2 \text{ V}}{25 \text{ mV}} = 20 \log(80) = \mathbf{38.1 \text{ dB}}$$

An amplifier with a 25 mV input has a 7.8 V output. Calculate the dB voltage gain of the circuit.

When you want to convert a dB voltage gain value to standard numeric form, the following equation is used:

$$A_v = \log^{-1} \frac{A_{v(\text{dB})}}{20}$$  **(8.16)**

Again, note the change in the denominator of the equation. Example 8.14 demonstrates the use of this equation.

### EXAMPLE 8.14

A given amplifier has a voltage gain of 6 dB. What is the ratio of output voltage to input voltage for the circuit?

*Solution:*   Using equation (8.16), the ratio of output voltage to input voltage is found as

$$A_v = \log^{-1} \frac{A_{v(\text{dB})}}{20} = \log^{-1} \frac{6 \text{ dB}}{20} = \log^{-1}(0.3) = \mathbf{2}$$

### PRACTICE PROBLEM 8.14

A given amplifier has a voltage gain of 12 dB. What is the ratio of output voltage to input voltage?

As is the case with dB power values, negative dB voltage values indicate a voltage *loss*. This point is illustrated in Example 8.15.

### EXAMPLE 8.15

A circuit has a dB voltage gain of −6 dB. What is the ratio of output voltage to input voltage for the circuit?

*Solution:*   Using equation (8.16), the ratio of output voltage to input voltage is found as

$$A_v = \log^{-1} \frac{A_{v(\text{dB})}}{20} = \log^{-1} \frac{-6 \text{ dB}}{20} = \log^{-1}(-0.3) = \mathbf{0.5}$$

### PRACTICE PROBLEM 8.15

A circuit has a voltage gain of −12 dB. What is the ratio of output voltage to input voltage for the circuit?

Do you see a pattern here? All the rules for dB power gains apply to dB voltage gains as well. The only difference is that dB voltage gain calculations use a constant of 20 instead of 10.

## 8.4.4   The Basis for Equation (8.15)

By definition, the dB power gain of an amplifier is found as

$$A_{p(\text{dB})} = 10 \log \frac{P_{\text{out}}}{P_{\text{in}}}$$

Using the basic power relationships, $P_{out}$ and $P_{in}$ can be found as

$$P_{out} = \frac{v_{out}^2}{R_{out}} \qquad \text{and} \qquad P_{in} = \frac{v_{in}^2}{R_{in}}$$

If we substitute these values into the equation for $A_{p(dB)}$, we get

$$A_{p(dB)} = 10 \log \frac{P_{out}}{P_{in}} = 10 \log \frac{v_{out}^2/R_{out}}{v_{in}^2/R_{in}} = 10 \log \frac{v_{out}^2}{v_{in}^2} + 10 \log \frac{R_{in}}{R_{out}}$$

or

$$A_{p(dB)} = 20 \log \frac{v_{out}}{v_{in}} + 10 \log \frac{R_{in}}{R_{out}}$$

This power gain equation contains dB voltage and resistance ratios. The voltage component of the equation is referred to as *dB voltage gain*. Stated mathematically,

$$A_{v(dB)} = 20 \log \frac{v_{out}}{v_{in}}$$

which is equation (8.15).

Something interesting happens when we have a circuit with equal input and output resistance values, that is, when $R_{in} = R_{out}$. In this case,

$$A_{p(dB)} = 20 \log \frac{v_{out}}{v_{in}} + 10 \log \frac{R_{in}}{R_{out}} = 20 \log \frac{v_{out}}{v_{in}} + 10 \log (1) = 20 \log \frac{v_{out}}{v_{in}} + 0 \text{ dB}$$

or

$$A_{p(dB)} = 20 \log \frac{v_{out}}{v_{in}} \quad \text{(when } R_{in} = R_{out}\text{)} \qquad \textbf{(8.17)}$$

As the above relationship indicates, dB power gain and dB voltage gain are equal when amplifier input and output resistance are equal.

### 8.4.5    Changes in dB Gain

When the voltage gain of an amplifier changes by a given number of decibels, the power gain of the amplifier changes by the same number of decibels. By formula,

$$\Delta A_p = \Delta A_v \qquad \textbf{(8.18)}$$

For example, if the voltage gain of an amplifier changes by $-3$ dB, the power gain of the amplifier also changes by $-3$ dB. The values of $A_p$ and $A_v$ for the circuit may or may not be equal, but they will *change* by the same number of decibels. This point will become important when we study the frequency response characteristics of amplifiers in Chapter 14.

### 8.4.6    One Final Note on Decibels

Experience has shown that most people take a while to get used to working with dB gain values. Although the whole concept of decibels may seem a bit strange at first, rest assured that everything will "click" for you eventually. In the meantime, the following list of summary points on decibels will help. Whenever you are not sure about the meaning of a given set of dB values, refer to this list:

Decibel (dB) characteristics.

1. Decibels are *logarithmic* representations of gain values.
2. Decibel power gain is found as $10 \log A_p$.

**3.** Decibel voltage gain is found as $20 \log A_v$.

**4.** When $A_v$ changes by a given number of decibels, $A_p$ changes by the same number of decibels.

**5.** You *cannot* use dB voltage and power gain values as multipliers. For example, if you want to determine $v_{out}$, given $v_{in}$ and $A_{v(dB)}$, you must convert $A_{v(dB)}$ to standard numeric form before multiplying to find $v_{out}$.

◀ **Section Review**

**1.** What is a *decibel*?

**2.** Why are power gain values often written in dB form?

**3.** What is the primary restriction on using dB power gain values?

**4.** What is the relationship between positive and negative dB values?

**5.** What is the dBm reference? How does it differ from dB power gain?

**6.** What is the primary difference between dB voltage calculations and dB power calculations?

**7.** What is the relationship between changes in dB voltage gain and dB power gain?

## CHAPTER SUMMARY

Here is a summary of the major points made in this chapter:

**1.** *Amplification* is the process of increasing the strength of an ac signal, that is, increasing its power level.

**2.** *Amplifiers* are circuits used to provide amplification.

**3.** All amplifiers have three fundamental properties:
  **a.** Gain.
  **b.** Input impedance.
  **c.** Output impedance.

**4.** The above properties can be used to form a *general amplifier model* (see Figure 8.1).

**5.** *Gain* is a multiplier that exists between a circuit's input and output.

**6.** There are three types of gain:
  **a.** Voltage gain ($A_v$).
  **b.** Current gain ($A_i$).
  **c.** Power gain ($A_p$).

**7.** Technically defined, gain is a ratio of an output value to its corresponding input value (see Example 8.1).
  **a.** Under normal circumstances, amplifier gain does not change significantly.
  **b.** A change to the input of an amplifier results in a change in the corresponding output value.

**8.** The general amplifier model can be used to demonstrate the input and output relationships of an amplifier (see Figure 8.4b).

**9.** The input impedance ($Z_{in}$) of an amplifier forms a voltage divider with the source resistance ($R_S$) (see Figure 8.5).

**10.** Amplifier output impedance ($Z_{out}$) forms a voltage divider with the load resistance ($R_L$) (see Figure 8.6).

**11.** The effect of $Z_{out}$ and $R_L$ on amplifier output voltage is demonstrated in Example 8.5.

**12.** The amplifier input and output circuits combine to reduce the *effective* voltage gain of an amplifier from its ideal value (see Section 8.1.7).

**13.** The reduction in voltage gain caused by the amplifier input and output circuits can be limited by:
  **a.** Increasing the value of $Z_{in}$.
  **b.** Decreasing the value of $Z_{out}$.

14. The *ideal voltage amplifier* has the following characteristics:
    a. Infinite gain (if needed).
    b. Infinite input impedance.
    c. Zero output impedance.
15. There are three BJT amplifier configurations:
    a. *Common-emitter* (CE) amplifier.
    b. *Common-collector* (CC) amplifier, or *emitter follower*.
    c. *Common-base* (CB) amplifier.
    All three are shown in Figure 8.12.
*16. The CE amplifier, which is the most widely used, has the following characteristics:
    a. Midrange values of voltage and current gain.
    b. High power gain.
    c. Midrange input impedance.
    d. Midrange output impedance.
*17. The CC amplifier has the following characteristics:
    a. Midrange current gain.
    b. Extremely low voltage gain (slightly less than 1).
    c. High input impedance.
    d. Low output impedance.
*18. The CB amplifier, which is the least often used, has the following characteristics:
    a. Midrange voltage gain.
    b. Extremely low current gain (slightly less than 1)
    c. Low input impedance.
    d. High output impedance.
19. The power gain of an amplifier equals the product of voltage gain and current gain.
    a. Since the voltage gain of a CC amplifier is slightly less than 1, $A_p \cong A_i$.
    b. Since the current gain of a CB amplifier is slightly less than 1, $A_p \cong A_v$.
20. The configuration of an amplifier can be identified by determining the input and output terminals. The third terminal is the common terminal.
21. A *class A amplifier* contains a single transistor that conducts during the entire input cycle (see Figure 8.16a).
22. A *class B amplifier* contains two transistors that each conduct for approximately 180° of the input cycle (see Figure 8.16b).
23. A *class C amplifier* contains a single transistor that conducts for less than 180° of the input cycle (see Figure 8.16c).
24. The *efficiency* of an amplifier indicates the percentage of the power drawn from the power supply that is actually delivered to the load.
    a. The maximum theoretical efficiency of a class A amplifier is 25%.
    b. The maximum theoretical efficiency of a class B amplifier is 78.5%.
    c. The maximum theoretical efficiency of a class C amplifier is 99%.
25. *Distortion* is any undesired change in the shape of a waveform.
26. *Decibel* (dB) power gain is a ratio of output power to input power, expressed as 10 times the common logarithm of that ratio.
27. The use of dB values allows us to represent very large gains using relatively small values; for example, 1,000,000 = 60 dB.
    a. The dB power gain of an amplifier is determined as shown in Example 8.7.
    b. The dB power gain of an amplifier is converted to standard numeric form as shown in Example 8.8.
28. Positive dB values represent power *gain*, while negative dB values represent power *loss*.

* See Table 8.2 for the limits of the ranges given.

**29.** Positive and negative dBs of equal magnitude represent reciprocal gains and losses.
  **a.** Example: 3 dB = 2, and −3 dB = ½.
  **b.** If $x$ dB represents a power gain of $y$, then $-x$ dB represents a power loss equal to $1/y$.
**30.** Spec sheet power ratings are commonly listed as *dBm* values. A dBm value references a power level to 1 mW. (See Example 8.12.)
**31.** dB voltage gain is a ratio of output voltage to input voltage, expressed as 20 times the common logarithm of that ratio.
**32.** The principles presented on dB power gain apply to dB voltage gain as well.
**33.** A change in dB voltage gain causes an identical change in dB power gain.

| Equation Number | Equation | Section Number |
|---|---|---|
| (8.1) | $A_v = \dfrac{v_{out}}{v_{in}}$ | 8.1 |
| (8.2) | $v_{out} = A_v v_{in}$ | 8.1 |
| (8.3) | $A_i = \dfrac{i_{out}}{i_{in}}$ | 8.1 |
| (8.4) | $A_p = \dfrac{P_{out}}{P_{in}}$ | 8.1 |
| (8.5) | $v_{in} = v_S \dfrac{Z_{in}}{R_S + Z_{in}}$ | 8.1 |
| (8.6) | $v_L = v_{out} \dfrac{R_L}{Z_{out} + R_L}$ | 8.1 |
| (8.7) | $A_{v(eff)} = \dfrac{v_L}{v_S}$ | 8.1 |
| (8.8) | $A_p = A_v A_i$ | 8.2 |
| (8.9) | $\eta = \dfrac{P_L}{P_{dc}} \times 100$ | 8.3 |
| (8.10) | $A_{p(dB)} = 10 \log A_p$ | 8.4 |
| (8.11) | $A_{p(dB)} = 10 \log \dfrac{P_{out}}{P_{in}}$ | 8.4 |
| (8.12) | $A_p = \log^{-1} \dfrac{A_{p(dB)}}{10}$ | 8.4 |
| (8.13) | $P_{dBm} = 10 \log \dfrac{P}{1\ mW}$ | 8.4 |
| (8.14) | $A_{v(dB)} = 20 \log A_v$ | 8.4 |
| (8.15) | $A_{v(dB)} = 20 \log \dfrac{v_{out}}{v_{in}}$ | 8.4 |
| (8.16) | $A_v = \log^{-1} \dfrac{A_{v(dB)}}{20}$ | 8.4 |
| (8.17) | $A_{p(dB)} = 20 \log \dfrac{v_{out}}{v_{in}}$  (when $R_{in} = R_{out}$) | 8.4 |
| (8.18) | $\Delta A_p = \Delta A_v$ | 8.4 |

amplification 284
amplifier 284
amplifier input
   impedance 287
amplifier output
   impedance 289
class A amplifier 297
class AB amplifier 300
class B amplifier 297
class C amplifier 297
common-base (CB)
   amplifier 295

common-collector (CC)
   amplifier 294
common-emitter (CE)
   amplifier 293
current gain ($A_i$) 285
dBm reference 305
dB power gain 303
dB voltage gain 306
decibel (dB) 303
distortion 299
efficiency ($\eta$) 298
emitter follower 295

gain 285
general amplifier
   model 284
ideal voltage
   amplifier 291
input impedance ($Z_{in}$) 287
output impedance
   ($Z_{out}$) 289
power gain ($A_p$) 285
tuned amplifier 301
voltage gain ($A_v$) 285

## PRACTICE PROBLEMS

### Section 8.1

**1.** Complete the table below.

| $v_{in}$ | $v_{out}$ | $A_v$ |
|---|---|---|
| 1.2 mV | 300 mV | _____ |
| 200 μV | 18 mV | _____ |
| 24 mV | 2.2 V | _____ |
| 800 μV | 140 mV | _____ |

**2.** Complete the table below.

| $v_{in}$ | $v_{out}$ | $A_v$ |
|---|---|---|
| 38 mV | 600 mV | _____ |
| 6 mV | 9.2 V | _____ |
| 500 μV | 88 mV | _____ |
| 48 mV | 48 mV | _____ |

**3.** For each combination of $v_{in}$ and $A_v$, calculate the value of $v_{out}$.

| $v_{in}$ | $A_v$ | $v_{out}$ |
|---|---|---|
| 240 μV | 540 | _____ |
| 1.4 mV | 300 | _____ |
| 24 mV | 440 | _____ |
| 800 μV | 720 | _____ |

**4.** Calculate the value of $v_{out}$ for each of the amplifiers shown in Figure 8.26.

**5.** An amplifier has values of $v_S = 600$ μV, $Z_{in} = 1.2$ kΩ, and $R_S = 180$ Ω. Calculate the value of $v_{in}$ for the circuit.

**6.** An amplifier has values of $v_S = 18$ mV, $Z_{in} = 720$ Ω, and $R_S = 60$ Ω. Calculate the value of $v_{in}$ for the circuit.

**7.** An amplifier has values of $v_{out} = 8.8$ V, $Z_{out} = 120$ Ω, and $R_L = 3$ kΩ. Calculate the value of $v_L$ for the circuit.

**8.** An amplifier has values of $v_{out} = 12$ V, $Z_{out} = 80$ Ω, and $R_L = 5.2$ kΩ. Calculate the value of $v_L$ for the circuit.

**9.** Calculate the *effective* voltage gain of the amplifier in Figure 8.27a.

**10.** Calculate the *effective* voltage gain of the amplifier in Figure 8.27b.

**11.** Calculate the *effective* voltage gain of the amplifier in Figure 8.28a.

**12.** Calculate the *effective* voltage gain of the amplifier in Figure 8.28b.

### Section 8.3

**13.** An amplifier draws 8.8 W from its dc power supply and delivers 1.2 W of power to its load. Calculate the efficiency of the circuit.

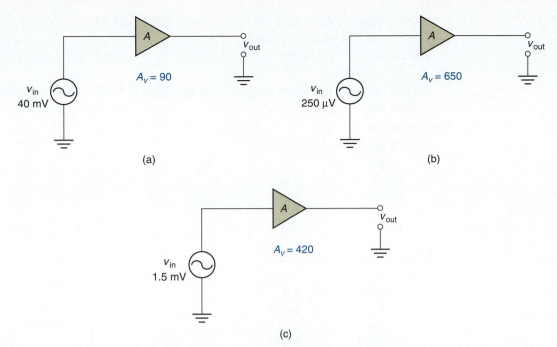

(a)

(b)

(c)

FIGURE 8.26

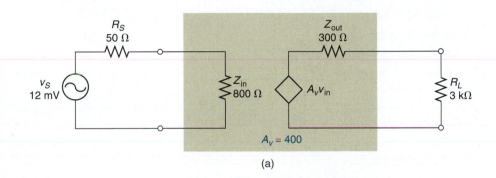

$A_v = 400$

(a)

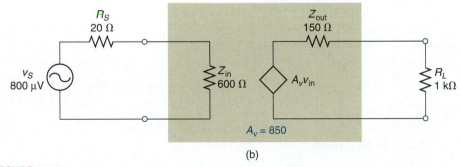

$A_v = 850$

(b)

FIGURE 8.27

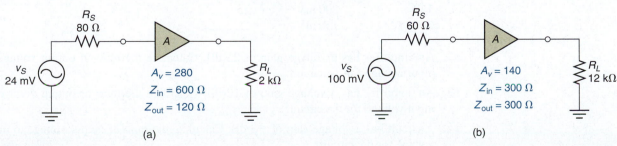

(a)

(b)

FIGURE 8.28

313

**14.** An amplifier draws 2 W from its dc power supply and delivers 300 mW of power to its load. Calculate the efficiency of the circuit.

**15.** An amplifier has values of $P_{dc} = 1.4$ W and $P_L = 200$ mW. What is the efficiency of the circuit?

**16.** An amplifier has values of $P_{dc} = 3$ W and $P_L = 115$ mW. What is the efficiency of the circuit?

**Section 8.4**

**17.** Complete the table below.

| $P_{in}$ | $P_{out}$ | $A_{p(dB)}$ |
|---|---|---|
| 10 mW | 5 W | _____ |
| 100 mW | 1 W | _____ |
| 1.25 mW | 4 W | _____ |
| 12 W | 6 W | _____ |

**18.** Complete the table below.

| $P_{in}$ | $P_{out}$ | $A_{p(dB)}$ |
|---|---|---|
| 500 μW | 1 W | _____ |
| 22 W | 1.4 W | _____ |
| 33 mW | 2.64 W | _____ |
| 120 mW | 2.2 W | _____ |

**19.** An amplifier has a power gain of 48 dB. What is the value of $A_p$ for the circuit?

**20.** An amplifier has a power gain of $-18$ dB. What is the value of $A_p$ for the circuit?

**21.** An amplifier has values of $A_{p(dB)} = 12$ dB and $P_{in} = 30$ mW. Calculate the output power for the circuit.

**22.** An amplifier has values of $A_{p(dB)} = 3$ dB and $P_{in} = 400$ mW. Calculate the output power for the circuit.

**23.** A circuit has values of $A_{p(dB)} = -12$ dB and $P_{in} = 800$ mW. Calculate the output power for the circuit.

**24.** A circuit has values of $A_{p(dB)} = -6$ dB and $P_{in} = 1.2$ mW. Calculate the output power for the circuit.

**25.** A circuit has values of $A_{p(dB)} = 5$ dB and $P_{out} = 2.2$ W. Calculate the value of $P_{in}$ for the circuit.

**26.** A circuit has values of $A_{p(dB)} = -12$ dB and $P_{out} = 600$ μW. Calculate the value of $P_{in}$ for the circuit.

**27.** The output of a circuit is rated at 40 dBm. Calculate the circuit output power.

**28.** The output of a circuit is rated at 22 dBm. Calculate the circuit output power.

**29.** The output of a circuit is rated at $-1.2$ dBm. Calculate the circuit output power.

**30.** The output of a circuit is rated at 60 dBm. Calculate the circuit output power.

**31.** Complete the table below.

| $v_{out}$ | $v_{in}$ | $A_{v(dB)}$ |
|---|---|---|
| 3.6 V | 120 mV | _____ |
| 800 mV | 50 μV | _____ |
| 14 V | 200 mV | _____ |
| 300 mV | 150 mV | _____ |

**32.** An amplifier has a voltage gain of 22 dB. Calculate the ratio of output voltage to input voltage for the circuit.

**33.** An amplifier has a voltage gain of 12 dB. Calculate the ratio of output voltage to input voltage for the circuit.

**34.** A circuit has a voltage gain of $-3$ dB. Calculate the ratio of output voltage to input voltage for the circuit.

**35.** A circuit has a voltage gain of $-14$ dB. Calculate the ratio of output voltage to input voltage for the circuit.

PUSHING THE ENVELOPE

**36.** Refer to the circuit in Figure 8.29. Calculate the *effective* dB voltage gain of the circuit and the output power in dBm.

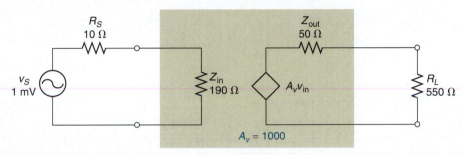

$R_S$
100 Ω

$v_S$
480 μV

$A_v = 30.8$ dB
$Z_{in} = 600$ Ω
$Z_{out} = 10$ Ω

$R_L$
120 Ω

FIGURE 8.29

**37.** Calculate the *effective* dB power gain of the circuit in Figure 8.30.

$R_S$
10 Ω

$v_S$
1 mV

$Z_{in}$
190 Ω

$A_v v_{in}$

$Z_{out}$
50 Ω

$R_L$
550 Ω

$A_v = 1000$

FIGURE 8.30

ANSWERS TO THE EXAMPLE PRACTICE PROBLEMS

| | |
|---|---|
| **8.1** | 240 |
| **8.2** | 86.4 mV |
| **8.3** | 731.7 μV |
| **8.4** | $v_{out} = 351$ mV when $v_{in} = 731.7$ μV |
| | $v_{out} = 384$ mV when $v_{in} = 800$ μV |
| **8.5** | 413.8 mV |
| **8.6** | 13.6% |
| **8.7** | 41.55 dB |
| **8.8** | 100 |
| **8.9** | 132.45 μW |
| **8.10** | 126.6 mW |
| **8.12** | 1.58 W |
| **8.13** | 49.9 dB |
| **8.14** | 3.98 |
| **8.15** | 0.25 |

## chapter 9

# Common-Emitter Amplifiers

## Objectives

*After studying the material in this chapter, you should be able to:*

1. Describe *gain*, and list the types of gain associated with each of the three transistor configurations.

2. Describe the input/output voltage and current phase relationships of the three transistor amplifier configurations.

3. Calculate the ac emitter resistance of a transistor.

4. List and discuss the two primary roles that capacitors serve in amplifiers.

5. Derive the ac equivalent circuit for a given amplifier.

6. Explain why the voltage gain of a common-emitter amplifier may be unstable.

7. Calculate the output power for a common-emitter amplifier given the amplifier values of $A_v$, $A_i$, and $P_{in}$.

8. Discuss the relationship between the load resistance and voltage gain of a common-emitter amplifier.

9. Calculate the values of $Z_{in(base)}$ and $Z_{in}$ for a common-emitter amplifier.

10. Discuss the effects of *swamping* on the ac characteristics of a common-emitter amplifier.

11. List and describe the four ac *h*-parameters.

12. Troubleshoot a multistage common-emitter amplifier to determine which amplifier stage is faulty.

## Outline

## Modern (?) Transistor Manufacturing

Here is a bit of history that you might enjoy. It has to do with the first automated transistor manufacturing machine, Mr. Meticulous. (Yes . . . that was its *real* name.)

The development of Mr. Meticulous is accredited to R. L. Wallace, Jr. Shortly after the debut run of this machine, the *Bell Laboratories Record* (March 1955) had this to say about Mr. Meticulous:

When fashioned by human hands over any extended period of time, some transistors are produced which are substandard and useless for research purposes. But the new machine, familiarly known as "Mr. Meticulous," never gets tired, never loses his precision or accuracy. His hand never shakes and his highly organized electronic "brain" rarely has mental lapses. The machine, originated by R. L. Wallace, Jr., may someday be a pilot model for industrial machines to be used in assembly-line transistor manufacture.

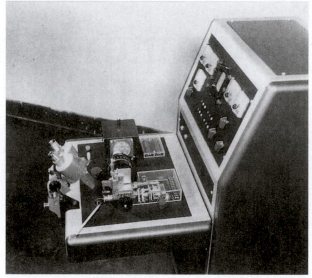

(Reprinted with the permission of Lucent Technologies, Inc./Bell Labs.)

In Chapter 8, you were introduced to the basic principles of amplifier ac operation. At this point, we will focus on the operation of common-emitter amplifiers. Then, in Chapter 10, we will look at both common-collector and common-base amplifiers.

### 9.1  AC Concepts

You have already been introduced to many amplifier properties, including *gain*, *amplifier input impedance*, and *amplifier output impedance*. In this section, we will look more closely at some ac concepts and how they relate specifically to the common-emitter amplifier.

#### 9.1.1  Amplifier Gain

**OBJECTIVE 1** ▶  As you know, there are three types of gain: *current gain* ($A_i$), *voltage gain* ($A_v$), and *power gain* ($A_p$). Under normal circumstances, the common-emitter amplifier exhibits all three types of gain. In contrast, the common-collector amplifier provides only current and power gain, and the common-base amplifier provides only voltage and power gain. The choice of amplifier configuration often depends on the type(s) of gain desired. Since the common-emitter amplifier can be used to provide any type of gain, it is the most often used of the BJT amplifiers.

#### 9.1.2  Input/Output Phase Relationships

**OBJECTIVE 2** ▶  In every BJT amplifier, *the input and output currents are in phase*. When the input current increases, the output current increases. When the input current decreases, the output current decreases.

The common-emitter amplifier is unique because *the input and output voltages are 180° out of phase*, even though the input and output currents are in phase. The cause of this input/output phase relationship is illustrated in Figure 9.1. The amplifier shown has a current gain ($A_i$) of 100. This means that the ac output current is 100 times the ac input current. The input current varies by 5 μA both above and below a 10 μA dc level. Multiplying the input current values by the $A_i$ of the amplifier gives us the output values shown. When the input current is at 10 μA, the output current is at (10 μA)(100) = 1 mA. When the input current is at 15 μA, the output current is at (15 μA)(100) = 1.5 mA, and so on. *Note that the input and output current waveforms are in phase.*

The common-emitter amplifier is the only BJT amplifier with a 180° voltage phase shift from input to output.

**Lab Reference:** The phase relationship between CE input and output signals is demonstrated in Exercise 11.

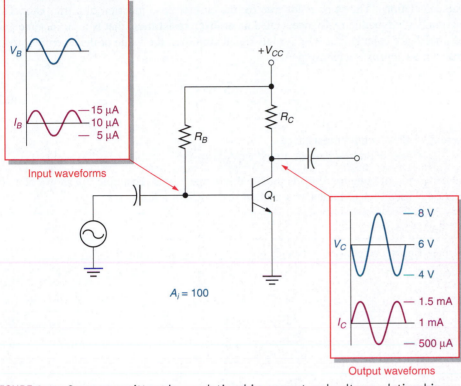

**FIGURE 9.1**   Common-emitter phase relationship: current and voltage relationship.

To understand the voltage relationship, two points need to be considered:

1. The output of the circuit is taken from the collector and is measured with respect to ground. Therefore,

$$V_{\text{out}} = V_C \qquad (9.1)$$

2. The collector voltage is equal to the supply voltage minus the drop across the collector resistor. This relationship has been expressed already as

$$V_C = V_{CC} - I_C R_C$$

If you calculate the value of $V_C$ for each of the values of $I_C$ indicated, you will see that $V_C$ *decreases as $I_C$ increases, and vice versa.* When $I_C$ increases from 1 to 1.5 mA, $V_C$ *decreases* from 6 to 4 V. As $I_C$ decreases from 1.5 mA to 500 μA, $V_C$ increases from 4 to 8 V. The phase relationship is due strictly to the fact that $I_C$ and $V_C$ vary inversely.

Now that you have seen the relationship between $I_C$ and $V_C$, the input/output voltage phase relationship can be explained as follows:

1. Input voltage and current are in phase.
2. Input current and output current are in phase. Therefore, input voltage and output current are in phase.

**3.** Output current is 180° out of phase with output voltage ($V_C$). Therefore, *input voltage and output voltage are 180° out of phase.*

Remember that the common-emitter amplifier is the only configuration that has this input/output voltage phase relationship. For both the common-base and common-collector amplifiers, the input and output voltages are in phase. For all three configurations, input and output currents are in phase.

### 9.1.3 AC Emitter Resistance

OBJECTIVE 3 ▶

**ac emitter resistance ($r_e'$)**
The dynamic resistance of the transistor base-emitter junction, used in voltage gain and input impedance calculations.

You may recall that the zener diode has a *dynamic* resistance value that is considered only in ac calculations. The same holds true for the emitter-base junction of a transistor. This junction has a dynamic resistance, called **ac emitter resistance**, that is used in some gain and impedance calculations. For a small-signal amplifier, the value of the ac emitter resistance can be *approximated* using

$$r_e' = \frac{25 \text{ mV}}{I_E} \qquad (9.2)$$

where $r_e'$ = the ac emitter resistance
$I_E$ = the dc emitter current, found as $V_E/R_E$

The derivation of equation (9.2) is provided in Appendix D. Example 9.1 shows how the equation is used to determine the ac emitter resistance of a transistor.

**EXAMPLE 9.1**

Determine the ac emitter resistance for the transistor in Figure 9.2.

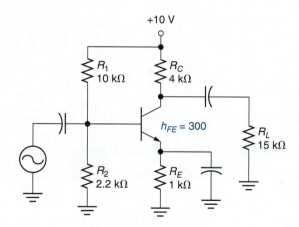

**FIGURE 9.2**

**Solution:** First, the value of $V_B$ is found as

$$V_B = V_{CC}\frac{R_2}{R_1 + R_2} = (10 \text{ V})\frac{2.2 \text{ k}\Omega}{12.2 \text{ k}\Omega} = \textbf{1.8 V}$$

$V_E$ is found as

$$V_E = V_B - 0.7 \text{ V} = \textbf{1.1 V}$$

and $I_E$ is found as

$$I_E = \frac{V_E}{R_E} = \frac{1.1 \text{ V}}{1 \text{ k}\Omega} = \textbf{1.1 mA}$$

Finally, the ac emitter resistance is found as

$$r_e' = \frac{25 \text{ mV}}{I_E} = \frac{25 \text{ mV}}{1.1 \text{ mA}} = 22.7 \text{ } \Omega$$

**PRACTICE PROBLEM 9.1**

A voltage-divider biased amplifier has values of $R_1 = 40$ k$\Omega$, $R_2 = 10$ k$\Omega$, $R_C = 6$ k$\Omega$, $R_E = 2$ k$\Omega$, $V_{CC} = +10$ V, and $h_{FE} = 80$. Determine the ac emitter resistance of the transistor.

Since $r_e'$ is a resistance, its value can also be calculated using Ohm's law and some measured circuit values. Figure 9.3 shows a plot of base-emitter voltage ($V_{BE}$) versus emitter current ($I_E$). When biased at the point labeled $Q_1$, the change in $V_{BE}$ results in the corresponding change in $I_E$ (labeled $\Delta V_{BE}$ and $\Delta I_E$, respectively). Using the values of $\Delta V_{BE}$ and $\Delta I_E$, the ac emitter resistance can be found as

$$r_e' = \frac{\Delta V_{BE}}{\Delta I_E} \tag{9.3}$$

Note that a lower $\Delta V_{BE}$ produces the same $\Delta I_E$ when the transistor is biased at the point labeled $Q_2$. Thus, the biasing point of the amplifier affects the value of $r_e'$.

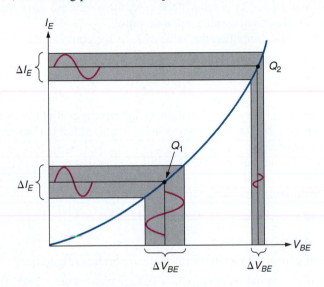

**FIGURE 9.3**

### 9.1.4    AC Beta

The ac current gain of a transistor is different from its dc current gain. This is because dc current gain is measured with $I_B$ and $I_C$ being constant, while ac current gain is measured with *changing* current values. Figure 9.4 helps to illustrate this point. Figure 9.4a shows the method by which dc beta is determined. A set value of $I_C$ is divided by a set value of $I_B$. Since the curve of $I_B$ versus $I_C$ is not linear, the value of dc beta changes as $I_C$ varies over the length of the curve.

Figure 9.4b shows the measurement of **ac beta**. $I_B$ is varied to cause a variation in $I_C$ *around the Q-point*. The change in $I_C$ is then divided by the change in $I_B$ to obtain the value of ac beta. By formula,

> **ac beta ($\beta_{ac}$ or $h_{fe}$)**
> The ratio of ac collector current to ac base current.

$$\beta_{ac} = \frac{\Delta I_C}{\Delta I_B} \tag{9.4}$$

or

$$\beta_{ac} = \frac{i_c}{i_b} \tag{9.5}$$

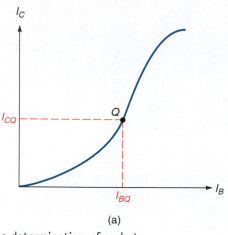

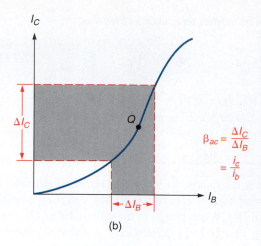

(a)                                         (b)

**FIGURE 9.4**   The determination of ac beta.

The measurement of $\beta_{ac}$ is illustrated in Figure 9.4b. Since the input and output currents for a transistor change at constant rates, you divide the change in output current by the change in input current to obtain the value of ac beta, as indicated in equation (9.5).

| $h_{FE}$ = dc beta |
| $h_{fe}$ = ac beta |

Transistor specification sheets list dc beta as $h_{FE}$. To distinguish the ac value of current gain, ac beta is listed as $h_{fe}$. As is usually the case, the lowercase subscript is used to tell you that the current gain is an ac value, not a dc value.

For a common-emitter amplifier, the value of $h_{fe}$ is the current gain ($A_i$) *for the transistor*. By formula,

$$A_i = h_{fe} \qquad (9.6)$$

It should be noted that equation (9.6) defines the current gain of the *transistor*, not the amplifier. The calculation of $A_i$ for a common-emitter amplifier is more involved, as will be shown later in this chapter.

The concepts and equations introduced in this section will be used constantly throughout our discussions on transistor ac operation. As you use them more and more, you will become more comfortable with them.

---

**Section Review ▶**

1. What are the three types of gain?

2. What types of gain are associated with each amplifier configuration?

3. Which amplifier type has an input/output voltage phase shift of 180°?

4. What is the phase relationship between the input and output voltage and current of each transistor amplifier configuration?

5. List, in order, the steps you need to take to determine the ac emitter resistance of a transistor.

6. What is the difference between $h_{FE}$ and $h_{fe}$?

7. Describe how $h_{FE}$ and $h_{fe}$ are measured.

## 9.2   The Roles of Capacitors in Amplifiers

**OBJECTIVE 4 ▶**

Capacitors play two primary roles in common-emitter amplifiers:

**Lab Reference:** The effects of bypass capacitors on amplifier operation are demonstrated in Exercise 11.

1. *Coupling capacitors* pass an ac signal from one amplifier to another, while providing dc isolation between the two.
2. *Bypass capacitors* are used to "short circuit" ac signals to ground while not affecting the dc operation of the circuit.

In this section, we will look at both of these capacitor applications.

## 9.2.1    Coupling Capacitors

In most practical applications, you will not see a single transistor amplifier. Rather, you will see a series of *cascaded* amplifiers. The term **cascaded** means *connected in series*.

When amplifiers are cascaded, the original signal is increased by the gain of each individual amplifier. As a result, the original signal is provided with more gain than could be provided by any single transistor amplifier. Each amplifier in the series string is referred to as a **stage**. The overall series of amplifier stages is referred to as a **multistage amplifier**. Multistage amplifiers are covered in detail later in this chapter. What we are interested in at this point is the fact that *capacitors are commonly used to connect one amplifier stage to another*. A capacitor used in this application is called a **coupling capacitor**.

Coupling capacitors are easy to spot in a schematic diagram. They are always placed in series between the stages of an amplifier. For example, in Figure 9.5, series capacitors are located between the signal source and $Q_1$, between $Q_1$ and $Q_2$, and between $Q_2$ and the load. When you see a capacitor positioned like one of these, it is a coupling capacitor. Coupling capacitors:

1. Pass the ac signal from one stage to the next with little or no distortion.
2. Provide dc isolation between the two stages.

These two functions are performed easily by the capacitor, which provides little opposition to an ac signal while completely blocking dc voltages and currents. This *isolation* between stages simply means that the dc collector voltage of one stage does not affect the biasing voltage at the base of the next stage, and vice versa. You may recall from your study of basic electronics that the reactance of a capacitor is found as

$$X_C = \frac{1}{2\pi f C} \tag{9.7}$$

where $X_C$ = the opposition to current provided by the capacitor
   $f$ = the frequency of the signal applied to the capacitor
   $C$ = the capacitance, in farads

**Cascaded**
A term meaning *connected in series*.

**Stage**
A single amplifier in a cascaded group of amplifiers.

**Multistage amplifier**
An amplifier with a series of stages.

**Coupling capacitor**
A capacitor connected between amplifier stages to provide dc isolation between the stages while allowing the ac signal to pass without distortion.

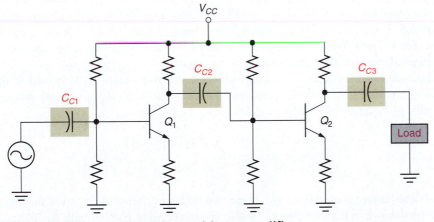

**FIGURE 9.5**    Coupling capacitors in a multistage amplifier.

This equation shows that *the opposition of a capacitor to current is inversely proportional to the frequency of the applied signal*. As frequency increases, the capacitor acts more and more like a short circuit. As frequency decreases, the capacitor acts more and more like an open circuit. For dc, which has a frequency of 0 Hz, the reactance of the capacitor is effectively *infinite*. As a result, the component blocks dc.

The effects of the coupling capacitor on cascaded amplifiers are illustrated in Figure 9.6. Figure 9.6a is an **ac equivalent circuit** for the circuit shown in Figure 9.5. Note that the capacitors have been replaced by wire connections. This represents the low opposition

**ac equivalent circuit**
A representation of a circuit that shows how the circuit appears to an ac source.

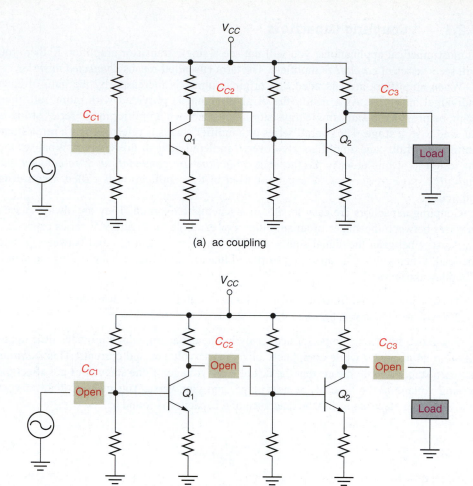

(a) ac coupling

(b) dc isolation

FIGURE 9.6    The characteristics of coupling capacitors.

that the coupling capacitors present to the ac signal. Figure 9.6b is the *dc equivalent* of Figure 9.5. In this circuit, the capacitors have been replaced by breaks in the connections between the stages. These breaks represent the near-infinite opposition the capacitors present to the dc voltage and current levels of the stages.

The need for dc isolation (blocking) between amplifier stages is illustrated in Figure 9.7. The capacitor in Figure 9.7a provides dc isolation between the circuits. In this circuit, the value of $V_B$ for the second stage is found as

$$V_B = V_{CC} \frac{R_4}{R_3 + R_4} = (10 \text{ V}) \frac{2.2 \text{ k}\Omega}{12.2 \text{ k}\Omega} = 1.8 \text{ V}$$

> The discussion here ignores the effects of $I_{CQ}$ for the first stage and the input resistance of $Q_2$ to keep things as simple as possible.

This is the biasing potential for $Q_2$. If the two stages of the amplifier were directly connected, as shown in Figure 9.7b, $R_{C1}$ would be connected in parallel with $R_3$. To see what this does to $V_B$ for the second stage, we start by determining the combined resistance of $R_{C1}$ and $R_3$. This resistance is found as

$$R_{EQ} = R_{C1} \parallel R_3 = (3.6 \text{ k}\Omega) \parallel (10 \text{ k}\Omega) = 2.65 \text{ k}\Omega$$

Now, $V_B$ for the second stage is found as

$$V_B = V_{CC} \frac{R_4}{R_{EQ} + R_4} = (10 \text{ V}) \frac{2.2 \text{ k}\Omega}{4.85 \text{ k}\Omega} = 4.54 \text{ V}$$

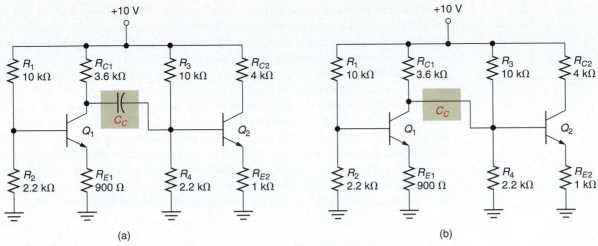

FIGURE 9.7    Capacitive vs. direct coupling.

As you can see, $V_B$ of stage 2 is significantly higher than it would be if the circuit was a "stand-alone" amplifier. By providing dc isolation between the amplifier stages, the biasing of each amplifier stage is independent of the other stages.

The use of the coupling capacitor allows each transistor amplifier stage to maintain its independent biasing characteristics while allowing the ac output from one stage to pass on to the next stage. *The exact value of the coupling capacitor is not critical, provided that its reactance is extremely low at the lowest operating frequency of the circuit.*

## 9.2.2    Bypass Capacitors

**Bypass capacitors** function in the same manner as coupling capacitors, but they are used for a different reason. Bypass capacitors are connected across the emitter resistor of an amplifier to keep the resistor from affecting the ac operation of the circuit. A bypass capacitor is shown in the circuit in Figure 9.8. Note that the capacitor effectively *bypasses* the emitter resistor, thus the name.

Figure 9.9 shows how the bypass capacitor affects the ac and dc operation of the amplifier. Figure 9.9a shows the *ac equivalent* of the bypass capacitor. Note that the capacitor has been replaced by a wire, just as it is in the coupling application. In this case, the capacitor establishes an *ac ground* at the emitter of the transistor. For ac purposes, $R_E$ is effectively shorted.

> **Bypass capacitor**
> A capacitor used to establish an ac ground at a specific point in a circuit.

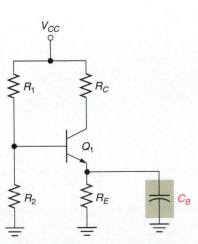

FIGURE 9.8    Bypass capacitor.

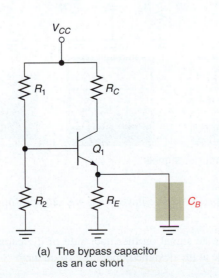

(a) The bypass capacitor as an ac short

(b) The bypass capacitor as a dc open

FIGURE 9.9    The characteristics of bypass capacitors.

What does this have to do with anything? As you will be shown later in this chapter, the voltage gain of a common-emitter amplifier depends on the *total ac emitter resistance*. The lower this ac resistance, the higher the voltage gain of the amplifier. By shorting out $R_E$, the bypass capacitor ensures that $R_E$ does not weigh into the voltage gain calculations for the amplifier. This increases the voltage gain of the circuit. We will examine this point more thoroughly later in this chapter.

Figure 9.9b shows the *dc equivalent* of the bypass capacitor. As before, the capacitor has been replaced with an open wire. This represents the near-infinite opposition that the capacitor presents to any dc voltage or current. Since the capacitor acts as an open to dc voltages and currents, it has no effect on the value of $I_E$ and, therefore, no effect on the value of $r'_e$.

Like a coupling capacitor, a bypass capacitor must act effectively as a short at the lowest operating frequency. The lower the value of $X_C$ (at the lowest frequency), the lower the ac resistance of the emitter circuit and the higher the circuit voltage gain.

### 9.2.3    Amplifier Signals

Capacitive coupling is not the only type of coupling used in amplifier circuits. Another type, called *transformer coupling*, is introduced in Chapter 11.

The effects that coupling and bypass capacitors have on amplifier voltages are illustrated in Figure 9.10. The signal at the collector of $Q_1$ is shown to be a sine wave riding on 5.6 $V_{dc}$, the quiescent value of $V_C$ for the first stage. The sine wave is applied to the coupling capacitor, $C_{C2}$. On the other side of the capacitor, you see the same sine wave. The only difference is that the sine wave is now riding on the dc value of $V_B$ for the second stage, 1.8 V. Thus, the ac signal has been passed from the first stage to the next, while the dc reference of the signal has been changed from $V_C$ of the first stage to $V_B$ of the second.

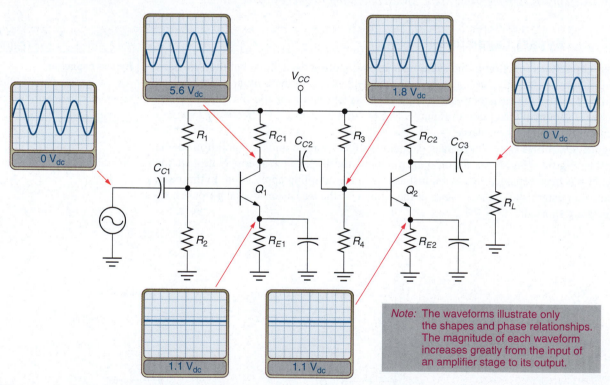

**FIGURE 9.10**   Typical common-emitter amplifier signals.

Based on what we have seen, we can sum up the effects of the coupling capacitor as follows:

**1.** The ac signal on the input side of the capacitor equals the ac signal on the output side at the designed operating frequency.

**2.** The dc voltage on the input side of the capacitor is determined by the source stage.

**3.** The dc voltage on the output side of the capacitor is determined by the load stage.

The emitters of both amplifiers show no change in $V_E$. This is caused by the bypass capacitors, which short the ac component of the emitter signal to ground. Thus, the signal present at each emitter is the pure dc value of $V_E$. Since $V_E$ remains constant, $I_E$ and $r'_e$ remain constant. This is important, since $r'_e$ weighs into all voltage gain calculations.

The effects of bypass capacitors on ac operation.

---

**1.** What two purposes are served by capacitors in multistage amplifiers?

**2.** What is meant by the term *coupling*?

**3.** Why is dc isolation between amplifier stages important?

**4.** What effect does a bypass capacitor have on the ac operation of an amplifier?

**5.** What effect does a coupling capacitor have on the ac signal that is coupled from one amplifier stage to the next?

**6.** Why don't you see an ac signal at the emitter terminal of an amplifier with a bypass capacitor?

◄ **Section Review**

## 9.3    The Common-Emitter AC Equivalent Circuit

At this point, we need to derive the ac equivalent circuit for the common-emitter amplifier. This circuit, which is used in many ac calculations, is easily derived using this two-step process:

◄ **OBJECTIVE 5**

**1.** *Short circuit all capacitors.*

**2.** *Replace all dc sources with a ground symbol.*

How is an ac equivalent circuit derived?

The basis for the first step is easy to understand. Since the capacitors act as short circuits to the ac signals, they are replaced with wires, as they were in Figures 9.6a and 9.9a.

The second step is based on the fact that *dc sources have extremely low internal resistance values*. This point is illustrated in Figure 9.11. In the circuit shown, $V_{CC}$ is represented as a simple dc battery. You may recall from your study of basic electronics that the internal resistance of a dc voltage source is *ideally* 0 Ω. Thus, we can replace the dc source with a simple wire, as shown in Figure 9.11b. Note that this wire is replaced by the ground (reference) symbol in Figure 9.11c.

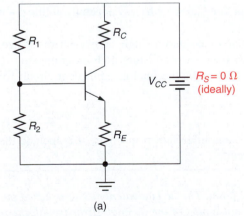
(a)

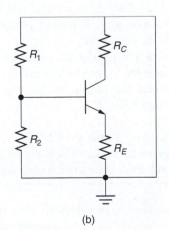
(b)

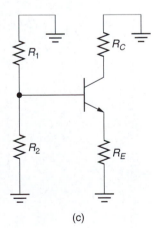
(c)

**FIGURE 9.11**   Deriving the common-emitter ac-equivalent circuit .

The process for finding the ac equivalent of a common-emitter amplifier is demonstrated in Example 9.2.

## EXAMPLE 9.2

Derive the ac equivalent circuit for the amplifier shown in Figure 9.12a.

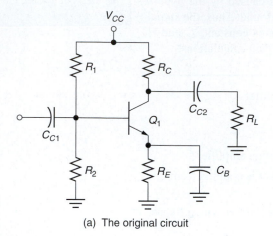

(a) The original circuit

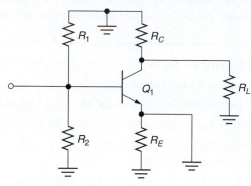

(b) The circuit with the capacitors shorted and the supply voltage replaced with a ground connection

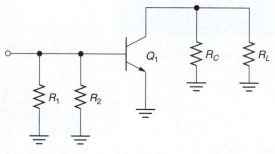

(c) The equivalent circuit in a more conventional form

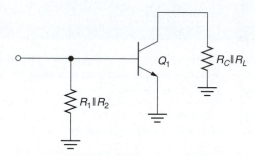

(d) The equivalent circuit with parallel components combined

**FIGURE 9.12**

**Solution:** The first step is to short circuit the capacitors ($C_C$ and $C_B$) and replace $V_{CC}$ with a ground symbol. This gives us the circuit shown in Figure 9.12b. As you can see, shorting the bypass capacitor leaves us with a straight wire across $R_E$. Thus, for the ac equivalent circuit, $R_E$ is shorted and is replaced by a wire. Figure 9.12c shows the equivalent circuit without $R_E$. Note also that $R_1$ and $R_C$ have been drawn as going to ground. (Be sure that you can see what is happening at this point before you go on.)

As a final step, $R_1$ and $R_2$ can be combined into a single equivalent resistance. Since the ac equivalent circuit shows these two resistors as both connected to ground, they are in parallel. Figure 9.12d shows the final ac equivalent circuit for Figure 9.12a.

### Practice Problem 9.2

Derive the ac equivalent circuit for the amplifier shown in Figure 9.2. Include the component values in the equivalent circuit.

You need to be absolutely clear on one point: *The ac equivalent circuit is based on the ac characteristics of the amplifier and is not a part of the dc analysis or troubleshooting of the amplifier.* The ac equivalent circuit is used for ac circuit analysis only.

**Section Review ▶**

1. What steps are taken to derive the ac equivalent circuit for a given amplifier?
2. What effects do the ac characteristics of an amplifier have on the dc characteristics of the circuit?

As you know, there are three types of gain: *voltage*, *current*, and *power* gain. In this section, we will take a closer look at all three. The main emphasis in this section is on the common-emitter circuit. The gain characteristics of the common-base and common-collector circuits are covered in detail in Chapter 10.

> *Remember:*
> Gain is a ratio, a multiplier that exists between the input and output of an amplifier.

### 9.4.1    Voltage Gain

As you have seen, the **voltage gain** ($A_v$) of an amplifier is the factor by which the ac signal voltage increases from the input of the amplifier to its output. The value of $A_v$ for a common-emitter amplifier can be found by dividing the ac output voltage by the ac input voltage as follows:

> **Voltage gain ($A_v$)**
> The factor by which ac signal voltage increases from the amplifier input to the amplifier output.

$$A_v = \frac{v_{out}}{v_{in}} \tag{9.8}$$

where $A_v$ = the voltage gain of the amplifier
   $v_{out}$ = the ac output voltage from the amplifier
   $v_{in}$ = the ac input voltage to the amplifier

Example 9.3 reviews the calculaton of amplifier voltage gain.

#### EXAMPLE 9.3

An amplifier has values of $v_{out}$ = 12 V and $v_{in}$ = 60 mV. What is the voltage gain of the circuit?

*Solution:*    The voltage gain for the circuit is found as follows:

$$A_v = \frac{v_{out}}{v_{in}} = \frac{12\ V}{60\ mV} = 200$$

Remember, voltage gain is a ratio and thus has no units, so the value of $A_v$ for the amplifier described above is simply 200.

> *A Practical Consideration:*
> The value of $A_v$ can be determined using peak, peak-to-peak, or rms values. As long as $v_{out}$ and $v_{in}$ are measured in the same fashion (that is, both peak, both rms, and so on), you will obtain the same value of $A_v$.

#### PRACTICE PROBLEM 9.3

An amplifier has measured values of $v_{out}$ = 14.2 V and $v_{in}$ = 120 mV. Determine the voltage gain of the circuit.

Equation (9.8) works well enough when the values of $v_{out}$ and $v_{in}$ are known, but there is a problem. The values of $v_{out}$ and $v_{in}$ are not usually shown on the schematic diagram for an amplifier. When this is the case, you may need to be able to determine the value of $A_v$ without knowing the values of $v_{out}$ and $v_{in}$.

The value of $A_v$ for a common-emitter amplifier equals *the ratio of total ac collector resistance to total ac emitter resistance*. For the circuit shown in Figure 9.12a, this ratio is found as

$$A_v = \frac{r_C}{r_e'} \tag{9.9}$$

where $r_C$ = the ac resistance in the collector circuit
   $r_e'$ = the ac emitter resistance of the transistor

As shown in Figure 9.12c, the load and collector resistors are in parallel. Thus, the ac resistance in the collector circuit is the parallel combination of $R_L$ and $R_C$. By formula,

$$r_C = R_C \parallel R_L \tag{9.10}$$

### 9.4.2 The Basis for Equation (9.9)

To understand the basis for equation (9.9), take a look at Figure 9.15. Figure 9.15a shows a common-emitter amplifier. The ac equivalent of this circuit is shown in Figure 9.15b. In Figure 9.15c, the transistor has been replaced with its own equivalent circuit. The equivalent circuit for the transistor contains three components:

1. A resistor ($r_e'$), which represents the ac emitter resistance of the transistor.
2. A diode, which represents the emitter-base junction of the transistor.
3. A current source, which represents the current supplied to $r_C$ by the collector of the transistor.

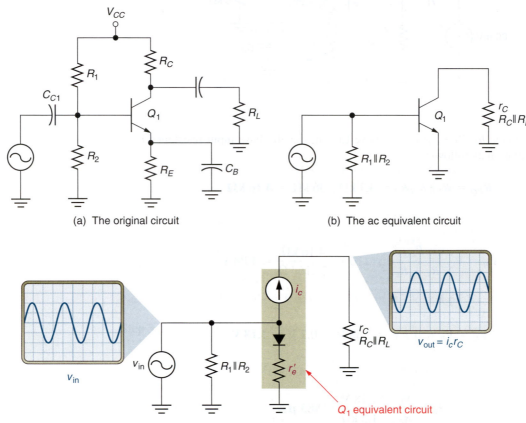

(a) The original circuit

(b) The ac equivalent circuit

(c) The ac equivalent circuit with the transistor ac model

FIGURE 9.15 An amplifier and its ac equivalents.

Note that the input voltage ($v_{in}$) is applied across the diode and $r_e'$. Assuming the diode to be *ideal*, the ac emitter current can be found as

$$i_e = \frac{v_{in}}{r_e'} \tag{9.11}$$

Thus,

$$v_{in} = i_e r_e' \tag{9.12}$$

and $v_{out}$ is found as

$$v_{out} = i_c r_C \tag{9.13}$$

Now, we can substitute equations (9.12) and (9.13) into equation (9.8) as follows:

$$A_v = \frac{v_{out}}{v_{in}} = \frac{i_c r_C}{i_e r_e'}$$

Since $i_e$ and $i_c$ are approximately equal, they can be dropped completely, leaving

$$A_v = \frac{r_C}{r'_e}$$

which is equation (9.9).

### 9.4.3 Voltage Gain Instability

◄ OBJECTIVE 6

One important consideration for the common-emitter amplifier is the *stability* of its voltage gain. An amplifier should have gain values that are stable so that the output from the circuit is predictable under all normal circumstances.

The voltage gain of a common-emitter amplifier can be somewhat unstable. As you know, the voltage gain of an amplifier is equal to the ratio of ac collector resistance to ac emitter resistance. Like beta, ac emitter resistance ($r'_e$) is somewhat dependent on temperature. Because of this, the voltage gain of a common-emitter amplifier, like the one shown in Figure 9.15, may vary with changes in temperature. To reduce the effects of temperature on voltage gain, the *gain stabilized* (or *swamped*) amplifier is used. Swamped amplifiers are discussed in detail in Section 9.6.

### 9.4.4 Calculating $v_{out}$

When you know the voltage gain of an amplifier, determining the ac output voltage is easy. Simply use a rewritten form of equation (9.8) as follows:

$$v_{out} = A_v v_{in} \qquad \text{(9.14)}$$

#### EXAMPLE 9.6

The circuit used in Example 9.5 has an ac input signal of 80 mV. Determine the ac output voltage from the circuit.

*Solution:* Using the value of $A_v$ from Example 9.5, we obtain

$$v_{out} = A_v v_{in} = (45.3)(80 \text{ mV}) = \textbf{3.62 V}$$

#### PRACTICE PROBLEM 9.6

The amplifier described in Practice Problem 9.5 has a value of $v_{in} = 20$ mV. Determine the value of $v_{out}$ for the amplifier.

### 9.4.5 Current Gain

**Current gain** ($A_i$) is *the factor by which ac current increases from the input of an amplifier to its output.* The value of $A_i$ for a common-emitter amplifier can be found by dividing the ac output current by the ac input current. By formula,

$$A_i = \frac{i_{out}}{i_{in}} \qquad \text{(9.15)}$$

| **Current gain ($A_i$)** |
| The factor by which ac current increases from the input of an amplifier to the output. |

where $A_i$ = the current gain of the amplifier
$\quad i_{out}$ = the ac output (load) current
$\quad i_{in}$ = the ac input (source) current

As you know, the current gain of the *transistor* in a common-emitter amplifier is given on its spec sheet as $h_{fe}$. However, the current gain for the *amplifier* is always *lower* than the value of $h_{fe}$. This is due to two factors:

1. The input signal current is divided between the transistor and the biasing network.
2. The output signal current (at the transistor collector) is divided between the collector resistor and the load.

## 9.5.2    Calculating Amplifier Input Impedance

The input impedance of an amplifier is determined using the ac equivalent circuit for the amplifier. As Figure 9.17 shows, the input impedance of an amplifier is the parallel impedance formed by $R_1$, $R_2$, and the base of the transistor. By formula,

$$Z_{in} = R_1 \parallel R_2 \parallel Z_{in(base)} \tag{9.19}$$

where  $Z_{in}$ = the input impedance to the amplifier
$Z_{in(base)}$ = the input impedance to the transistor base

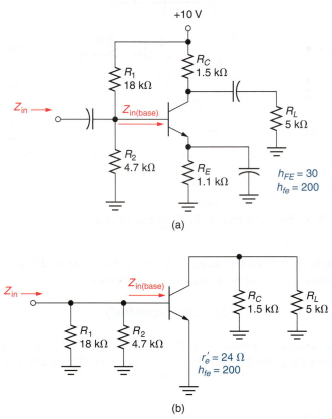

(a)

(b)

**FIGURE 9.17**   A common-emitter amplifier and its ac equivalent circuit.

Figure 9.17 shows a typical common-emitter amplifier and its ac equivalent circuit. Note that $Z_{in(base)}$ is the input impedance *to the base of the transistor*. The amplifier input impedance ($Z_{in}$) equals the parallel combination of $Z_{in(base)}$ and the biasing resistors.

You may recall that the dc input resistance of a transistor base is found as

$$R_{IN(base)} = h_{FE}R_E$$

This relationship was derived in Chapter 7. The input impedance of the base can be derived in the same manner. The only differences are as follows:

**1.** The ac emitter resistance is used in the calculation.
**2.** The value of ac beta ($h_{fe}$) is used in place of dc beta ($h_{FE}$).

Therefore, $Z_{in(base)}$ can be found as

$$Z_{in(base)} = h_{fe}r'_e \tag{9.20}$$

Once the input impedance of the transistor is found, it is considered to be in parallel with the base biasing resistors. This parallel circuit gives us the total input impedance to the amplifier, as illustrated in Example 9.9.

*A Practical Consideration:*
The value of $Z_{in(base)}$ for a transistor is given as $h_{ie}$ on the component's spec sheet. This rating is discussed later in this chapter.

## EXAMPLE 9.9

Determine the input impedance to the amplifier represented by the equivalent circuit in Figure 9.17b.

*Solution:* The value of $Z_{\text{in(base)}}$ can be found as

$$Z_{\text{in(base)}} = h_{fe}r'_e = (200)(24\ \Omega) = \textbf{4.8 k}\boldsymbol{\Omega}$$

$Z_{\text{in}}$ for the amplifier is found as

$$Z_{\text{in}} = R_1 \parallel R_2 \parallel Z_{\text{in(base)}} = 18\ \text{k}\Omega \parallel 4.7\ \text{k}\Omega \parallel 4.8\ \text{k}\Omega = \textbf{2.1 k}\boldsymbol{\Omega}$$

### PRACTICE PROBLEM 9.9

Determine the value of $Z_{\text{in}}$ for the amplifier described in Practice Problem 9.5. Assume a value of $h_{fe} = 200$ for the transistor.

## 9.5.3  Calculating the Value of $A_i$

Earlier in the chapter, you were told that the overall current gain ($A_i$) of a common-emitter amplifier is always *lower* than the current gain of the transistor ($h_{fe}$). We are now ready to take a closer look at this relationship.

As you know, the current gain of a common-emitter amplifier is found as

$$A_i = \frac{i_{\text{out}}}{i_{\text{in}}}$$

where $i_{\text{in}}$ is the *ac source current* and $i_{\text{out}}$ is the *ac load current.* You also know that the current gain of the *transistor* in a common-emitter amplifier is found as

$$h_{fe} = \frac{i_c}{i_b}$$

If you refer to Figure 9.17b, you'll see that the input of the common-emitter contains a *current divider.* This current divider is made up of $R_1$, $R_2$, and the base of the transistor. Since a portion of the ac source current passes through the biasing resistors, $i_b < i_{\text{in}}$. At the same time, the collector circuit of the amplifier forms another current divider. Since a portion of the ac collector current passes through the collector resistor, $i_{\text{out}} < i_c$. These factors combine to cause the overall current gain of the amplifier ($A_i$) to be significantly lower than the current gain of the transistor ($h_{fe}$).

In Appendix D, an equation is derived for amplifier current gain. This equation takes into account the reduction in gain caused by the transistor input and output circuitry as follows:

$$A_i = h_{fe}\left(\frac{Z_{\text{in}}r_C}{Z_{\text{in(base)}}R_L}\right) \tag{9.21}$$

where

$\quad A_i$ = the current gain of the common-emitter amplifier
$\quad h_{fe}$ = the current gain of the transistor
$(Z_{\text{in}}r_C)/(Z_{\text{in(base)}}R_L)$ = the reduction factor introduced by the biasing and output components

Example 9.10 demonstrates the relationship between $h_{fe}$ and $A_i$.

## EXAMPLE 9.10

Calculate the value of $A_i$ for the circuit shown in Figure 9.17b.

*Solution:* For the circuit shown,

$$Z_{\text{in(base)}} = h_{fe}r'_e = (200)(24\ \Omega) = \textbf{4.8 k}\boldsymbol{\Omega}$$

$$Z_{in} = R_1 \| R_2 \| Z_{in(base)} = \textbf{2.1 k}\boldsymbol{\Omega}$$

and

$$r_C = R_C \| R_L = \textbf{1.15 k}\boldsymbol{\Omega}$$

Now, these values are used (along with $h_{fe} = 200$) in equation (9.21) as follows:

$$A_i = h_{fe}\left(\frac{Z_{in}r_C}{Z_{in(base)}R_L}\right) = (200)\left[\frac{(2.1 \text{ k}\Omega)(1.15 \text{ k}\Omega)}{(4.8 \text{ k}\Omega)(5 \text{ k}\Omega)}\right] = (200)(0.101) = \textbf{20.2}$$

As you can see, the value of $A_i$ for this circuit is significantly lower than the value of $h_{fe}$.

**PRACTICE PROBLEM 9.10**

A common-emitter amplifier has the following values: $h_{fe} = 300$, $r_C = 2.15$ k$\Omega$, $R_L = 6.2$ k$\Omega$, $Z_{in(base)} = 8.2$ k$\Omega$, and $Z_{in} = 3.8$ k$\Omega$. Calculate the value of $A_i$ for the circuit.

### 9.5.4    Multistage Amplifier Gain Calculations

**Lab Reference:** The voltage gain characteristics of a multistage amplifier are demonstrated in Exercise 18.

When you want to determine the overall value of $A_v$, $A_i$, and/or $A_p$ for a multistage amplifier, *you must begin by determining the appropriate gain values for the individual stages.* Once the overall gain values for the individual stages are determined, you can determine the desired overall gain value by using any (or all) of the following equations:

$$A_{vT} = (A_{v1})(A_{v2})(A_{v3})\cdots \tag{9.22}$$

$$A_{iT} = (A_{i1})(A_{i2})(A_{i3})\cdots \tag{9.23}$$

$$A_{pT} = (A_{vT})(A_{iT}) \tag{9.24}$$

Equations (9.22) and (9.23) indicate that the overall value of $A_v$ or $A_i$ is simply the product of the individual stage gain values. Equation (9.24) indicates that the overall power gain is found as the product of the overall values of $A_v$ and $A_i$.

As you recall, the exact sequence of equations used to determine the overall value of $A_v$ for a given amplifier stage is as follows:

1. Perform a basic dc analysis of the amplifier to determine the value of $I_E$.
2. Using equation (9.2), determine the value of $r'_e$ for the amplifier.
3. Using equation (9.10), determine the value of $r_C$ for the amplifier.
4. Using the values of $r'_e$ and $r_C$ found in steps 2 and 3, determine the value of $A_v$ for the stage.

When you are dealing with a two-stage amplifier, the value of $Z_{in}$ for the second stage is used in place of $R_L$ for the $r_C$ calculation of the first stage. This point is illustrated in Example 9.11.

### EXAMPLE 9.11

Determine the voltage gain for the first stage of the circuit shown in Figure 9.18.

*Solution:*    Using the established procedure, $r'_e$ for the first stage is found to be 19.8 $\Omega$, and $r'_e$ for the second stage is found to be 17.4 $\Omega$. For the second stage, $h_{fe}$ is 200. Therefore,

$$Z_{in(base)} = h_{fe}r'_e = (200)(17.4 \text{ } \Omega) = \textbf{3.48 k}\boldsymbol{\Omega}$$

the value of $R_E$. The other emitter resistor, $r_E$, is part of the ac equivalent circuit, as shown in Figure 9.19b. Since the voltage gain of the amplifier is equal to the ratio of ac collector resistance to ac emitter resistance, the voltage gain for the amplifier is found as

$$A_v = \frac{r_C}{r_e' + r_E} \qquad (9.25)$$

The following example demonstrates the use of equation (9.25) in the calculation of $A_v$ for a swamped amplifier.

### EXAMPLE 9.13

Determine the value of $A_v$ for the amplifier in Figure 9.20.

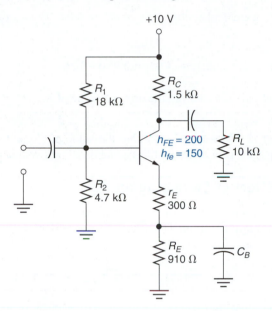

FIGURE 9.20

**Solution:** Using the established procedure, $V_E$ is found to equal 1.37 V. The total dc resistance in the emitter circuit equals $(R_E + r_E)$. Therefore, $I_E$ is found as

$$I_E = \frac{V_E}{R_E + r_E} = \frac{1.37\ \text{V}}{1.21\ \text{k}\Omega} = \textbf{1.13 mA}$$

Now, the value of $r_e'$ is found as

$$r_e' = \frac{25\ \text{mV}}{I_E} = \frac{25\ \text{mV}}{1.13\ \text{mA}} = \textbf{22.1}\ \boldsymbol{\Omega}$$

The value of $r_C$ is now found as

$$r_C = R_C \parallel R_L = \textbf{1.3 k}\boldsymbol{\Omega}$$

Finally, $A_v$ is found as

$$A_v = \frac{r_C}{r_e' + r_E} = \frac{1.3\ \text{k}\Omega}{322.1\ \Omega} = \textbf{4.04}$$

### PRACTICE PROBLEM 9.13
Assume that the amplifier in Figure 9.20 has values of $R_E = 820\ \Omega$ and $r_E = 330\ \Omega$. Determine the value of $A_v$ for the circuit.

Swamping improves the gain stability of an amplifier when $r_E \gg r_e'$. Since most of the ac emitter resistance is determined by the value of $r_E$, any change in $r_e'$ has little effect on the overall gain of the amplifier. This point is illustrated in Example 9.14.

### EXAMPLE 9.14

Determine the change in gain for the amplifier in Example 9.13 when $r_e'$ doubles in value.

*Solution:*   When $r_e'$ doubles in value, $A_v$ becomes

$$A_v = \frac{r_C}{r_e' + r_E} = \frac{1.3 \text{ k}\Omega}{344.2 \text{ }\Omega} = \mathbf{3.78}$$

In Example 9.13, we calculated a value of $A_v = 4.04$. Therefore, the change in gain is

$$\Delta A_v = 4.04 - 3.78 = \mathbf{0.26}$$

This is a change of only 6.44% from the original value of $A_v$.

When an amplifier is not swamped, doubling the value of $r_e'$ causes the value of $A_v$ to decrease by 50%. Thus, the gain of an amplifier is stabilized by swamping the emitter circuit.

### 9.6.1   The Effect of Swamping on $Z_{in}$

The input impedance of a transistor base is shown in equation (9.20) to equal $h_{fe}$ times the ac resistance of the emitter circuit. This ac resistance is equal to $(r_e' + r_E)$ for the swamped amplifier. Thus,

$$Z_{\text{in(base)}} = h_{fe}(r_e' + r_E) \tag{9.26}$$

The result is that the input impedance of the transistor base is increased by an amount equal to $h_{fe}r_E$, as shown in Example 9.15.

### EXAMPLE 9.15

Determine the value of $Z_{\text{in(base)}}$ for the circuits shown in Figure 9.21.

*Solution:*   The circuits have near-equal values of dc emitter resistance. Following the established procedure for finding $r_e'$ gives us a value of 25 $\Omega$ for both circuits. Now, for Figure 9.21a,

$$\mathbf{Z_{in(base)}} = h_{fe}r_e' = (200)(25 \text{ }\Omega) = \mathbf{5 \text{ k}\Omega}$$

For Figure 9.21b,

$$\mathbf{Z_{in(base)}} = h_{fe}(r_e' + r_E) = (200)(25 \text{ }\Omega + 200 \text{ }\Omega) = \mathbf{45 \text{ k}\Omega}$$

Increasing the value of $Z_{\text{in(base)}}$ *increases* the overall value of $Z_{in}$ for the amplifier, as shown in Example 9.16.

### EXAMPLE 9.16

Determine the value of $Z_{in}$ for the amplifiers in Figure 9.21.

*Solution:*   For Figure 9.21a,

$$\mathbf{Z_{in}} = R_1 \parallel R_2 \parallel Z_{\text{in(base)}} = 10 \text{ k}\Omega \parallel 2.2 \text{ k}\Omega \parallel 5 \text{ k}\Omega = \mathbf{1.33 \text{ k}\Omega}$$

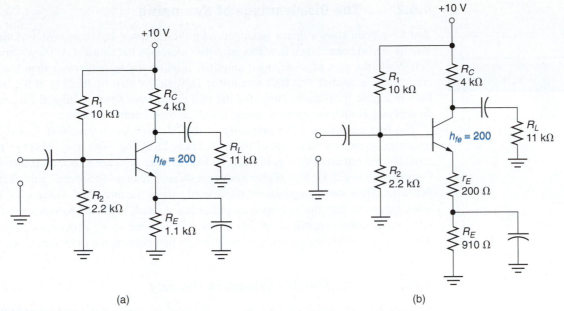

(a)                                   (b)

**FIGURE 9.21**

For Figure 9.21b,

$$\mathbf{Z_{in}} = R_1 \parallel R_2 \parallel Z_{in(base)} = 10 \text{ k}\Omega \parallel 2.2 \text{ k}\Omega \parallel 45 \text{ k}\Omega = \mathbf{1.73 \text{ k}\Omega}$$

As you can see, increasing the value of $Z_{in(base)}$ increases the overall value of $Z_{in}$ for an amplifier.

**PRACTICE PROBLEM 9.16**

Determine the value of $Z_{in}$ for the amplifier shown in Figure 9.20.

The increased $Z_{in}$ for Figure 9.21b means that this circuit will have less effect on the $A_v$ of a source amplifier. This point is illustrated in the following example.

**EXAMPLE 9.17**

The amplifiers in Figure 9.21 are both driven by a source amplifier with values of $r'_e = 25 \ \Omega$ and $R_C = 8 \text{ k}\Omega$. Determine the value of $A_v$ for the source amplifier when each circuit is connected as the load.

*Solution:*  When the circuit in Figure 9.21a is connected to the source amplifier, the voltage gain of the amplifier is found as

$$A_v = \frac{r_C}{r'_e} = \frac{8 \text{ k}\Omega \parallel 1.33 \text{ k}\Omega}{25 \ \Omega} = 45.6$$

> *Remember:*
> $$r_C = R_C \parallel Z_{in}$$

When the circuit shown in Figure 9.21b is connected to the source amplifier, the voltage gain of the amplifier is found as

$$A_v = \frac{r_C}{r'_e} = \frac{8 \text{ k}\Omega \parallel 1.73 \text{ k}\Omega}{25 \ \Omega} = 56.9$$

Thus, the reduced loading by the circuit in Figure 9.21b increased the gain of the source amplifier.

### 9.6.2    The Disadvantage of Swamping

Swamping improves stability but reduces $A_v$.

You have been shown that a swamped amplifier is more stable against variations in $r'_e$. You have also been shown how this amplifier increases the value of $A_v$ for a source amplifier. While the gain of a swamped amplifier is more stable, it is lower than the gain of a comparable amplifier that isn't swamped. This can be seen by looking at the two amplifiers in Figure 9.21 again. Note that the two circuits are nearly identical for purposes of dc analysis. Both have approximately 1.1 k$\Omega$ in their emitter circuits.

The two amplifiers differ primarily in their ac characteristics. Note that the total ac resistance in the emitter circuit of Figure 9.21a is 25 $\Omega$: the value of $r'_e$. This low resistance provides the circuit with a gain of 117. The total ac resistance in the emitter circuit of Figure 9.21b is 225 $\Omega$. This higher emitter resistance reduces the overall gain of the amplifier to 13. Thus, swamping reduces the overall gain of the amplifier. At the same time, you must remember that the swamped amplifier has a much more stable value of $A_v$ than the standard common-emitter amplifier. Also, the increase in $A_v$ experienced by a source amplifier that is driving a swamped amplifier partially offsets the reduction in gain.

### 9.6.3    The Emitter Circuit Is the Key

The key to distinguishing between a *swamped* (or *gain-stabilized*) common-emitter amplifier and a standard common-emitter amplifier is to look at the emitter circuit. If the emitter resistance is *completely bypassed*, you are dealing with a standard common-emitter amplifier. If the emitter resistance is only partially bypassed, you are dealing with a swamped, or gain-stabilized, amplifier. The circuit recognition features of these two types of common-emitter amplifiers, along with the different equations for the two, are summarized in Figure 9.22.

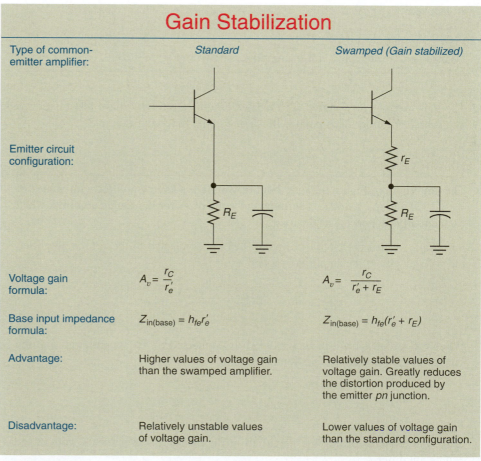

## Gain Stabilization

| Type of common-emitter amplifier: | Standard | Swamped (Gain stabilized) |
|---|---|---|
| Emitter circuit configuration: | | $r_E$ |
| Voltage gain formula: | $A_v = \dfrac{r_C}{r'_e}$ | $A_v = \dfrac{r_C}{r'_e + r_E}$ |
| Base input impedance formula: | $Z_{in(base)} = h_{fe}r'_e$ | $Z_{in(base)} = h_{fe}(r'_e + r_E)$ |
| Advantage: | Higher values of voltage gain than the swamped amplifier. | Relatively stable values of voltage gain. Greatly reduces the distortion produced by the emitter *pn* junction. |
| Disadvantage: | Relatively unstable values of voltage gain. | Lower values of voltage gain than the standard configuration. |

FIGURE 9.22

1. How does swamping reduce the effect of variations in the value of $r'_e$?

2. How does swamping affect the value of $A_v$ for a source amplifier? Explain your answer.

3. What is the primary disadvantage of using amplifier swamping? Explain your answer.

4. What effect (if any) would removing the bypass capacitor in Figure 9.20 have on the circuit's voltage gain? Explain.

## 9.7 h-Parameters

Hybrid parameters, or **h-parameters**, are transistor specifications that describe the component operating characteristics under specific circumstances. Each of the four h-parameters is measured under *no-load* or *full-load* conditions. These h-parameters are then used in circuit analysis applications.

The four h-parameters for a transistor in a common-emitter amplifier are as follows:

> $h_{ie}$ = the base input impedance
> $h_{fe}$ = the base-to-collector current gain
> $h_{oe}$ = the output admittance
> $h_{re}$ = the reverse voltage feedback ratio

**h-Parameters**
Transistor specifications that describe the operation of the device under full-load or no-load conditions.

It should be noted that all these values represent *ac characteristics* of the transistor as measured under specific circumstances. Before discussing the applications of h-parameters, let's take a look at the parameters and the method of measurement used for each.

### 9.7.1 Input Impedance ($h_{ie}$)

The **input impedance** ($h_{ie}$) is measured with the *output shorted*. A shorted output is a full load, so $h_{ie}$ *represents the input impedance to the transistor under full-load conditions*. The measurement of $h_{ie}$ is illustrated in Figure 9.23a. As you can see, a capacitor is connected between the emitter and collector terminals. This capacitor provides an *ac short* between the terminals when an ac signal is applied to the base-emitter junction of the transistor. With the input voltage applied, the base current is measured. Then, $h_{ie}$ is determined as

**Input impedance ($h_{ie}$)**
The input impedance of the transistor, measured under full-load conditions.

$$h_{ie} = \frac{v_{in}}{i_b} \quad \text{(output shorted)} \qquad (9.27)$$

Why short the output? You may recall that any resistance in the emitter circuit is reflected back to the base. This condition was described in the equation

**Why is $h_{ie}$ measured under full-load conditions?**

$$Z_{in(base)} = h_{fe}(r'_e + r_E)$$

By shorting the collector and emitter terminals, the measured value of $h_{ie}$ does not reflect any external resistance in the emitter circuit.

### 9.7.2 Current Gain ($h_{fe}$)

The base-to-collector **current gain** ($h_{fe}$) is also measured with an ac short connected across the emitter and collector terminals. Again, this represents a full ac load, so $h_{fe}$ *represents the current gain of the transistor under full-load conditions*. The measurement of

**Current gain ($h_{fe}$)**
The ac beta of the component, measured under full-load conditions.

The circuits shown in Figure 9.23 are simplified representations of *h*-parameter measurements. Actual *h*-parameter test circuits are more complex than shown here.

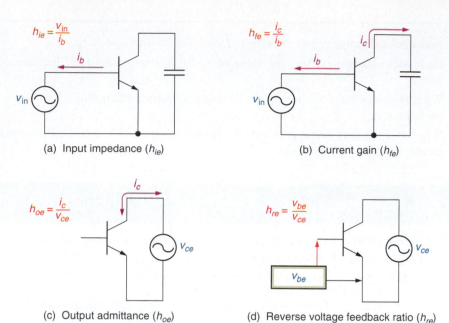

(a) Input impedance ($h_{ie}$)

(b) Current gain ($h_{fe}$)

(c) Output admittance ($h_{oe}$)

(d) Reverse voltage feedback ratio ($h_{re}$)

**FIGURE 9.23**    The measurement of *h*-parameters.

$h_{fe}$ is illustrated in Figure 9.23b. With a signal applied to the base and the output shorted, both the base and collector currents are measured. Then, $h_{fe}$ is determined as

$$h_{fe} = \frac{i_c}{i_b} \quad \text{(output shorted)} \tag{9.28}$$

Why is $h_{fe}$ measured under full-load conditions?

In this case, it is clear why the output is shorted. If the output were left open, $i_c$ would equal zero. Shorting the output gives us a measurable value of $i_c$ that can be reproduced in a practical test. In other words, anyone can achieve the same results simply by shorting the output terminals and applying the same signal to the transistor.

### 9.7.3    Output Admittance ($h_{oe}$)

**Output admittance ($h_{oe}$)**
The admittance of the collector-emitter circuit, measured under no-load conditions. Since admittance is the reciprocal of impedance, the unit of measure is often listed as *mhos* (although *siemens* is the preferred unit of measure.)

The **output admittance** ($h_{oe}$) is measured with the *input open*. The measurement of $h_{oe}$ is illustrated in Figure 9.23c. As you can see, a signal is applied across the collector-emitter terminals. Then, with this signal applied, the value of $i_c$ is measured. The value of $h_{oe}$ is then determined as

$$h_{oe} = \frac{i_c}{v_{ce}} \quad \text{(input open)} \tag{9.29}$$

Measuring $h_{oe}$ with the input open makes sense when you consider the effect of shorting the input. If the input to the transistor were shorted, there would be some base current ($i_b$). Since this current would come from the emitter, the value of $i_c$ would not be at its maximum value. In other words, by not allowing $i_b$ to be generated, $i_c$ is at its absolute maximum value. Therefore, $h_{oe}$ is an accurate measurement of the *maximum* output admittance.

**Reverse voltage feedback ratio ($h_{re}$)**
The ratio of $v_{be}$ to $v_{ce}$, measured under no-load conditions. This parameter is used mainly in amplifier design. Since $h_{re}$ is a ratio of two voltages, it has no unit of measure.

### 9.7.4    Reverse Voltage Feedback Ratio ($h_{re}$)

The **reverse voltage feedback ratio** ($h_{re}$) indicates *the amount of output voltage reflected back to the input*. This value is measured with the *input open*. The measurement of $h_{re}$ is illustrated in Figure 9.23d. A signal is applied to the collector-emitter terminals. Then, with the input open, the voltage that is fed back to the base is measured. The value of $h_{re}$ is then determined as

$$h_{re} = \frac{v_{be}}{v_{ce}} \quad \text{(input open)} \qquad (9.30)$$

Since the voltage at the base terminal is always less than the voltage across the emitter-collector terminals, $h_{re}$ is always less than 1. By measuring $h_{re}$ with the input open, you ensure that the voltage fed back to the base is always at its maximum possible value (since maximum voltage is always developed across an open circuit).

### 9.7.5    Circuit Calculations Involving $h$-Parameters

Circuit calculations involving $h$-parameters can be very simple or very complex, depending on the following:

1. What you are trying to determine.
2. How exact you want your calculations to be.

For our purposes, we are interested in only four $h$-parameter circuit equations. These equations are as follows:

$$A_i = h_{fe}\left(\frac{Z_{in}r_C}{h_{ie}R_L}\right) \qquad (9.21)$$

$$Z_{in(base)} = h_{ie} \qquad (9.31)$$

$$r'_e = \frac{h_{ie}}{h_{fe}} \qquad (9.32)$$

$$A_v = \frac{h_{fe}r_C}{h_{ie}} \qquad (9.33)$$

The $A_i$ equation (9.21) was introduced earlier in the chapter. The only change in the equation is the substitution of $h_{ie}$ for $Z_{in(base)}$, as given in equation (9.31). Equation (9.31) is relatively easy to understand if you refer to Figure 9.12d. Here, we see the ac equivalent circuit for the amplifier shown in Figure 9.12a. Looking at the ac equivalent circuit, it is easy to see that the only impedance between the base of the transistor and the emitter ground connection is the input impedance of the transistor. This input impedance is $h_{ie}$.

Equation (9.32) is derived using equations (9.20) and (9.31). You may recall that equation (9.20) defines $Z_{in(base)}$ as

$$Z_{in(base)} = h_{fe}r'_e$$

Substituting equation (9.31) for $Z_{in(base)}$, we obtain

$$h_{ie} = h_{fe}r'_e$$

Simply rearranging for $r'_e$ gives us equation (9.32).

You may be wondering why we would go to all this trouble to find $r'_e$ with $h$-parameters when we can simply use

$$r'_e = \frac{25 \text{ mV}}{I_E}$$

The fact of the matter is that the 25 mV/$I_E$ equation isn't always accurate. By using the $h$-parameter equation, we can obtain a much more accurate value of $r'_e$ and, thus, can calculate more closely the value of $A_v$ for a given common-emitter amplifier.

> *A Practical Consideration:*
> Depending on the transistor used, the given value of 25 mV in the $r'_e$ equation can actually be any value between 25 and 52 mV. This gives us
>
> $$\frac{25 \text{ mV}}{I_E} \leqslant r'_e \leqslant \frac{52 \text{ mV}}{I_E}$$
>
> This is why we use the more accurate $h$-parameter equations whenever possible.

Equation (9.33) is derived as follows:

$$A_v = \frac{r_C}{r_e'} = \frac{1}{r_e'} r_C = \frac{h_{fe}}{h_{ie}} r_C = \frac{h_{fe} r_C}{h_{ie}}$$

The equations in this section will be used throughout our discussion on ac amplifier analysis. It should be noted that the entire subject of *h*-parameters and their derivations is far more complex than represented here. The subject is covered more thoroughly in Appendix C for those readers who are interested. If you do not wish to get involved in *h*-parameter derivations, you may continue from this point without loss of continuity.

### 9.7.6 Determining *h*-Parameter Values

The spec sheet for a given transistor will list the values of the device's *h*-parameters in the *electrical characteristics* portion of the sheet. This is illustrated in Figure 9.24, which shows the electrical characteristics portion of the spec sheet for the 2N4400–4401 series transistors.

**2N4400, 2N4401**

| Characteristic | | Symbol | Min | Max | Unit |
|---|---|---|---|---|---|
| Emitter-Base Capacitance<br>($V_{BE} = 0.5$ Vdc, $I_C = 0$, f = 100 kHz) | | $C_{eb}$ | — | 30 | pF |
| Input Impedance<br>($I_C = 1.0$ mAdc, $V_{CE} = 10$ Vdc, f = 1.0 kHz) | 2N4400<br>2N4401 | $h_{ie}$ | 0.5<br>1.0 | 7.5<br>15 | kΩ |
| Voltage Feedback Ratio<br>($I_C = 1.0$ mAdc, $V_{CE} = 10$ Vdc, f = 1.0 kHz) | | $h_{re}$ | 0.1 | 8.0 | $\times 10^{-4}$ |
| Small-Signal Current Gain<br>($I_C = 1.0$ mAdc, $V_{CE} = 10$ Vdc, f = 1.0 kHz) | 2N4400<br>2N4401 | $h_{fe}$ | 20<br>40 | 250<br>500 | —<br>— |
| Output Admittance<br>($I_C = 1.0$ mAdc, $V_{CE} = 10$ Vdc, f = 1.0 kHz) | | $h_{oe}$ | 1.0 | 30 | μmhos |

**FIGURE 9.24** (Copyright of Semiconductor Component Industries, LLC. Used by permission.)

When *minimum* and *maximum* h-parameter values are given, we must determine the *geometric average* of the two values. Thus, the values of $h_{ie}$ and $h_{fe}$ that we would use in the analysis of a 2N4400 circuit would be found as

$$h_{ie} = \sqrt{h_{ie(min)} \times h_{ie(max)}} = \sqrt{(500\ \Omega)(7.5\ \text{k}\Omega)} = \mathbf{1.94\ k\Omega}$$

and

$$h_{fe} = \sqrt{h_{fe(min)} \times h_{fe(max)}} = \sqrt{(20)(250)} = \mathbf{71}$$

> *A Practical Consideration:*
> *h*-Parameter values are often listed under the heading *small-signal characteristics*.

Example 9.18 shows how the above values would be used in the ac analysis of a 2N4400 amplifier.

### EXAMPLE 9.18

Determine the values of $Z_{in}$ and $A_v$ for the circuit shown in Figure 9.25.

*Solution:* Using the established analysis procedure, the value of $I_C$ for the circuit is found to be approximately 1 mA. Since the values of $h_{ie}$ and $h_{fe}$ for the 2N4400 are listed at $I_C = 1$ mA (from the spec sheet in Figure 9.24), we can use the geometric averages of $h_{ie} = 1.94$ kΩ and $h_{fe} = 71$ that were determined earlier. Using these values,

$$Z_{in(base)} = h_{ie} = \mathbf{1.94\ k\Omega}$$

and

$$Z_{in} = R_1 \parallel R_2 \parallel Z_{in(base)} = \mathbf{1.36\ k\Omega}$$

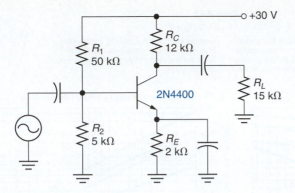

**FIGURE 9.25**

Now,

$$r_C = R_C \parallel R_L = \textbf{6.67 k}\Omega$$

and

$$A_v = \frac{h_{fe}r_C}{h_{ie}} = \frac{(71)(6.67 \text{ k}\Omega)}{1.94 \text{ k}\Omega} = \textbf{244}$$

**PRACTICE PROBLEM 9.18**

An amplifier like the one in Figure 9.25 has values of $R_C = 12$ k$\Omega$, $R_L = 4.7$ k$\Omega$, $R_1 = 33$ k$\Omega$, $R_2 = 4.7$ k$\Omega$, and $I_C = 1$ mA. At 1 mA, the transistor has $h$-parameter values of $h_{ie} = 1$ to 5 k$\Omega$ and $h_{fe} = 70$ to 350. Determine the values of $Z_{in}$ and $A_v$ for the circuit.

As Example 9.18 pointed out, the $h$-parameter values listed on the spec sheet of the 2N4400 were measured at $I_C = 1$ mA. As was the case with $h_{FE}$, the ac $h$-parameters all vary with the value of $I_C$. This point is illustrated in Figure 9.26, which shows the $h$-parameter curves for the 2N4400–4401 series transistors.

First, a word or two about the curves. Each $h$-parameter graph shows four curves. Two of the curves are identified as being for the 2N4400, and two of them are identified as being for the 2N4401. Just as we were given *minimum* and *maximum* $h$-parameter values at $I_{CQ} = 1$ mA, we have minimum (Unit 2) and maximum (Unit 1) curves for the transistors.

To determine the value of a given parameter at a given value of $I_{CQ}$, use the following procedure:

1. Using the graph, determine the minimum and maximum $h$-parameter values at the value of $I_{CQ}$.
2. Use the geometric average of the two values obtained in the analysis of the amplifier.

The use of this procedure is demonstrated in Example 9.19.

**EXAMPLE 9.19**

An amplifier that uses a 2N4400 has values of $I_{CQ} = 5$ mA and $r_C = 460$ $\Omega$. Determine the value of $A_v$ for the circuit.

**Solution:** From the $h_{fe}$ curve in Figure 9.26, we can approximate the limits of $h_{fe}$ to be 110 to 140 when $I_{CQ} = 5$ mA. The geometric average of these two values is found as

$$h_{fe} = \sqrt{h_{fe(min)} \times h_{fe(max)}} = \sqrt{(110)(140)} = \textbf{124}$$

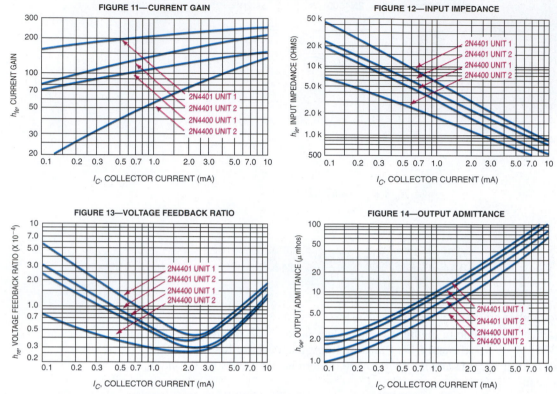

**h PARAMETERS**

$V_{CE} = 10$ Vdc, $f = 1.0$ kHz, $T_A = 25°C$

This group of graphs illustrates the relationship between hfe and other "*h*" parameters for this series of transistors. To obtain these curves, a high-gain and a low-gain unit were selected from both the 2N4400 and 2N4401 lines, and the same units were used to develop the correspondingly numbered curves on each graph.

**FIGURE 9.26**   (Copyright of Semiconductor Component Industries, LLC. Used by permission.)

The $h_{ie}$ graph shows the limits of $h_{ie}$ to be 600 and 800 $\Omega$ when $I_{CQ} = 5$ mA. The geometric average of these two values is found as

$$h_{ie} = \sqrt{h_{ie(\min)} \times h_{ie(\max)}} = \sqrt{(600 \ \Omega)(800 \ \Omega)} = \mathbf{693 \ \Omega}$$

Using $h_{fe} = 124$ and $h_{ie} = 693 \ \Omega$, the value of $A_v$ is determined to be

$$A_v = \frac{h_{fe}r_C}{h_{ie}} = \frac{(124)(460 \ \Omega)}{693 \ \Omega} = \mathbf{82.3}$$

**PRACTICE PROBLEM 9.19**

A 2N4401 is used in a circuit with values of $I_{CQ} = 2$ mA and $r_C = 1.64$ k$\Omega$. Determine the value of $A_v$ for the amplifier using the curves in Figure 9.26.

## 9.7.7    A Few More Points

In this section, we have concentrated on $h_{ie}$ and $h_{fe}$. That's because these two parameters are the ones most commonly used for circuit analysis. The other two *h*-parameters, $h_{oe}$ and $h_{re}$, are used primarily in circuit development applications.

Whenever you need to determine the voltage gain of an amplifier, the required *h*-parameter values can be obtained from the spec sheet of the transistor. If a range of values is given, you can actually follow either of two procedures:

1. If you want an *approximate* value of $A_v$, you can use the geometric average of the *h*-parameter values listed.

**2.** If you are interested in the *worst-case* values of $A_v$, you can analyze the amplifier using *both* the minimum and maximum values of $h_{fe}$ and $h_{ie}$. By doing this, you will obtain the *minimum* and *maximum* values of $A_v$. These two values represent the voltage gain limits for the amplifier.

---

**1.** What is $h_{ie}$? How is the value of $h_{ie}$ measured?

◀ **Section Review**

**2.** What is $h_{fe}$? How is the value of $h_{fe}$ measured?

**3.** What is $h_{oe}$? How is the value of $h_{oe}$ measured?

**4.** What is $h_{re}$? How is the value of $h_{re}$ measured?

**5.** Which of the four *h*-parameters are commonly used in circuit analysis?

**6.** Where are *h*-parameter values usually listed on a transistor spec sheet?

**7.** When minimum and maximum *h*-parameter values are listed, what value do you usually use for circuit analyses?

**8.** What is the procedure for determining the values of $h_{ie}$ and $h_{fe}$ at a specified value of $I_{CQ}$?

**9.** How is the circuit analysis procedure changed for determining the *worst-case* values of $A_v$?

**10.** Can replacing a transistor with one having the same part number cause a change in the voltage gain of a standard common-emitter amplifier? Explain your answer.

◀ **Critical Thinking**

## 9.8 Amplifier Troubleshooting

We have already discussed the dc troubleshooting procedure for an amplifier. However, that discussion assumed that you already knew there was a problem in the amplifier. How can you tell when one out of several amplifiers is the cause of a problem? For example, take a look at Figure 9.27. If the final output from this series of amplifiers is bad, how do you know which amplifier is the source of the trouble?

◀ *OBJECTIVE 12*

When you troubleshoot a series of amplifiers, *begin at the final-stage output.* When you verify that the output from the amplifier is faulty, check the amplifier input. If the input to the stage is good, then the problem exists in the final stage. If the input to the amplifier is faulty, disconnect the coupling component between the amplifier stages, and check the output from the source stage. For example, if the signal at test point (3) in Figure 9.27 appears faulty, disconnect either side of $C_4$ and check the signal at test point (4). The reason for doing this is to ensure that a faulty input is the result of a problem in the source stage. If the the output from a previous stage is good when the coupling component is disconnected, the fault lies in either the coupling component or the load stage.

### 9.8.1 Amplifier Input/Output Signals

Figure 9.28 shows the signals you should see at the inputs and outputs of each stage of Figure 9.27. Note that the standard common-emitter amplifiers (stages 2 and 3) show no ac signal at the emitter terminal of the transistors since the ac voltage is developed across $r_e'$ in the transistors themselves. The swamped amplifier (stage 1) has a small ac signal present at the emitter terminal but only a dc level at the bypass capacitor connection point. In all three cases, the collector signal is much larger in amplitude than the input signal at the base, and the two signals are 180° out of phase.

When troubleshooting the amplifier, locate the stage that has the faulty input/output combination (as described above). When you find that stage, troubleshoot the dc circuit as you were shown in Chapter 7.

### 9.8.2 Nonlinear Distortion

As you know, *distortion is an unwanted change in the shape of an ac signal.* One problem that commonly occurs in common-emitter amplifiers is called *nonlinear distortion.*

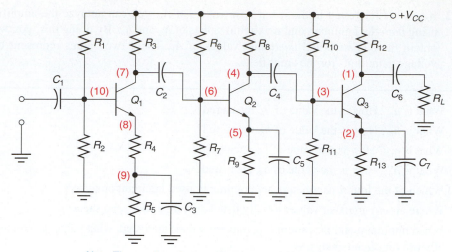

*Note*: The numbers in parentheses are used to identify test points.

FIGURE 9.27 A three-stage CE amplifier.

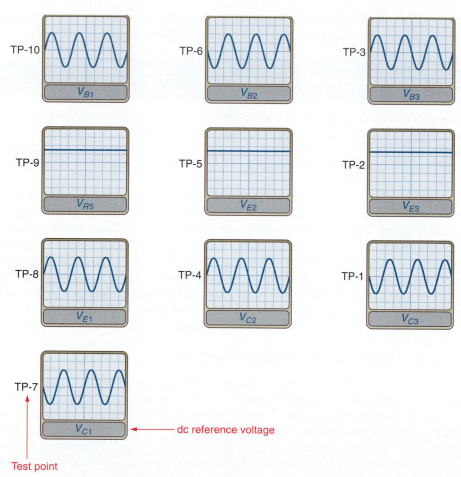

FIGURE 9.28 Signals for Figure 9.27.

The output waveform from an amplifier experiencing nonlinear distortion is illustrated in Figure 9.29a. Note the difference between the shape of the negative alternation of the signal (normal) and that of its positive alternation (distorted). The rounding of the positive alternation is a result of nonlinear distortion.

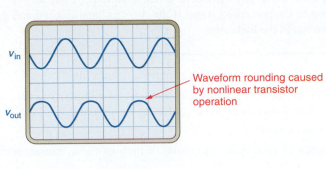

(a) Nonlinear distortion

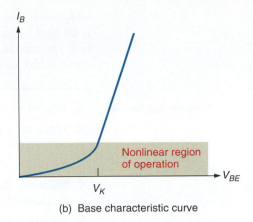

(b) Base characteristic curve

FIGURE 9.29 Nonlinear distortion.

**Nonlinear distortion** is *a type of distortion caused by driving the base-emitter junction of a transistor into its nonlinear operating region.* This point is illustrated with the help of Figure 9.29b, which shows the base characteristic curve of a transistor.

Normally, a transistor is operated so that the base-emitter junction stays in the linear region of operation. In this region, a change in $V_{BE}$ causes a *linear* (constant rate) change in $I_B$. If the transistor is operated in the nonlinear operating region, a change in $V_{BE}$ causes a change in $I_B$ that is not linear. For example, compare the slope of the curve at the points immediately above and below $V_K$. Since the slope of the curve changes, the rate of change in $I_B$ also changes.

The base-emitter junction of a transistor may be driven into its nonlinear operating region by either of two sets of circumstances:

> **Nonlinear distortion**
> A type of distortion caused by driving the base-emitter junction of a transistor into its nonlinear operating region.

1. A transistor is normally biased so that $I_B$ has a value well within the linear region of operation. If an amplifier is poorly biased (so that the $Q$-point value of $I_B$ is near the nonlinear region of operation), a relatively small ac input signal can drive the amplifier into the nonlinear region of operation, and nonlinear distortion will result.
2. The amplitude of the amplifier input signal may be sufficient to drive even a well-biased amplifier into nonlinear operation, causing nonlinear distortion.

If the first of these two problems (poor biasing) is present, the only solution is to redesign the amplifier for a higher value of $I_B$. When the second problem (overdriving the amplifier) is present, the solution is to reduce the amplitude of the amplifier input signal.

Nonlinear distortion can cause a variety of problems in communications electronics. In stereos, it may cause the audio to sound "grainy" or even cause the tones of various musical instruments to change. In television video circuits, it can cause distortion in the picture. To avoid these types of problems, amplifier design emphasizes avoiding nonlinear distortion. Since amplifiers are normally designed to avoid the problem, the most common cause of nonlinear distortion is overdriving an amplifier.

> *A Practical Consideration:*
> Swamping reduces nonlinear distortion because most of the ac input signal is developed across $r_E$, which is a linear device (as compared to the nonlinearity of the emitter *pn* junction). This linear operation allows us to use more of the load line and still have minimal distortion.

### 9.8.3    A Few Pointers

When you are troubleshooting common-emitter amplifiers, remember that the input signal for a given stage will have a much lower amplitude than the output. For this reason, you should reduce the vertical sensitivity (VOLTS/DIV) setting on your oscilloscope when going from output to input. For example, consider a circuit with a voltage gain of 100. *Under normal circumstances, this amplifier has an output signal that is 100 times as large as the input signal.* If you try to read these two signals using the same VOLTS/DIV setting on the oscilloscope, you will not be able to see the input signal.

Another point is this: Don't worry about seeing the 180° phase shift on your oscilloscope. If you want to, you can *external trigger* the oscilloscope on the amplifier input to view the phase shift. However, viewing the phase shift is not really critical. If the amplifier is working, the phase shift will be there. You will never run into a problem where the amplifier has the right signal but the phase shift is not there, so don't worry about it. Just look to see if the amplifier has a small-signal input and a large-signal output. If it does, it is all right (assuming that the output waveform is not distorted). If neither signal is normal, you need to return to the preceding stage. If the source signal is good and the output signal is bad, you have found the bad stage.

Section Review ▶

1. What is the procedure for troubleshooting a multistage amplifier?

2. When the signal at the input of an amplifier appears to be faulty, you disconnect the coupling between stages and check the output from the source stage. Why?

3. What is *nonlinear distortion*?

4. What are the causes of nonlinear distortion? Which of these causes occurs most frequently?

5. When troubleshooting an amplifier, why do you need to pay close attention to the VOLTS/DIV setting of the oscilloscope?

## CHAPTER SUMMARY

Here is a summary of the major points made in this chapter:

1. There are three types of gain: *current gain* ($A_i$), *voltage gain* ($A_v$), and *power gain* ($A_p$).
   a. Common-emitter amplifiers provide all three types of gain.
   b. Common-collector amplifiers provide only current gain and power gain.
   c. Common-base amplifiers provide only voltage gain and power gain.
   d. The choice of amplifier for a given application often depends on the types of gain desired.

2. The common-emitter amplifier is unique because the input and output signal voltages are 180° out of phase, even though the input and output currents are in phase (see Figure 9.1).
   a. Input voltage ($v_b$) and current ($i_b$) are in phase.
   b. Input and output currents are in phase.
   c. Output voltage ($v_c$) is 180° out of phase with output current ($i_c$). Therefore, input voltage ($v_b$) and output voltage ($v_c$) are 180° out of phase.

3. The *ac emitter resistance* ($r'_e$) of a transistor is the *dynamic* resistance of the base-emitter junction.
   a. The value of $r'_e$ is used in amplifier voltage gain and input impedance calculations.
   b. $r'_e$ equals the ratio of a change in $V_{BE}$ to the corresponding change in $I_E$.

4. The ratio in 3(b) varies with the location of the $Q$-point on the base curve of the transistor. Therefore, $r'_e$ varies with the bias point on the base curve. (see Figure 9.3.)

5. The ac current gain of a transistor is different from its dc current gain.
   a. The dc current gain ($\beta_{dc}$ or $h_{FE}$) is measured using *steady state* (unchanging) values of $I_C$ and $I_B$ (see Figure 9.4a).
   b. The ac current gain ($\beta_{ac}$ or $h_{fe}$) is measured using *changing* values of $I_C$ and $I_B$ (see Figure 9.4b).

6. The current gain ($A_i$) of the transistor in a common-emitter amplifier is listed as $h_{fe}$ on the component spec sheet.

7. Capacitors play two primary roles in common-emitter amplifiers.
   a. *Coupling capacitors* pass an ac signal from one amplifier stage to another while providing dc isolation between the two.
   b. *Bypass capacitors* are used to short circuit ac signals to ground while not affecting the dc operation of the circuit.

8. *Cascaded* amplifiers are connected in series.

9. Each amplifier in a group of cascaded amplifiers is called a *stage*.

10. A coupling capacitor is used to:
    a. Pass an ac signal from one amplifier stage to the next with little or no distortion.
    b. Provide dc isolation between amplifier stages.

11. An *ac equivalent circuit* shows how a circuit appears to an ac source.

12. The ac coupling and dc isolation characteristics of coupling capacitors are illustrated in Figure 9.6.

13. A *bypass capacitor* is used to establish an ac ground at a specific point in a circuit.
    a. In a common-emitter amplifier, a bypass capacitor is positioned as shown in Figure 9.8.
    b. The ac and dc characteristics of bypass capacitors are illustrated in Figure 9.9.

14. The effects of coupling and bypass capacitors on amplifier signal voltages are illustrated in Figure 9.10.

15. The ac equivalent of a common-emitter amplifier is derived by:
    a. Shorting all the capacitors in the circuit.
    b. Replacing all dc sources with a ground symbol.
    This procedure is illustrated in Example 9.2.

16. The *voltage gain* $(A_v)$ of an amplifier is the factor by which ac signal voltage increases from the input of an amplifier to its output.

17. The value of $A_v$ for a common-emitter amplifier equals the ratio of total ac collector resistance to total ac emitter resistance.

18. The value of $r'_e$ for a transistor can be affected by changes in temperature, which in turn can cause instability in the circuit $A_v$.

19. Current gain $(A_i)$ is the factor by which signal current increases from the input of an amplifier to its output.

20. The value of $A_i$ for a common-emitter amplifier is always lower than the transistor $h_{fe}$ rating for two reasons:
    a. The ac input is divided between the transistor and the biasing network.
    b. The ac output from the transistor collector is divided between the collector resistor and the load.

21. The *power gain* $(A_p)$ of an amplifier is the factor by which signal power increases from the input of an amplifier to its output.

22. Amplifier power gain equals the product of voltage gain and current gain.

23. For any amplifier, the lower the resistance of the load, the lower the voltage gain of the circuit.

24. Voltage gain reaches its maximum possible value under no-load conditions.

25. The base input impedance of a transistor is listed as $h_{ie}$ on the component's spec sheet.

26. In a cascaded amplifier, the input impedance of a load stage affects the voltage gain of the source stage (see Example 9.11).

27. A *swamped* amplifier uses a partially bypassed emitter resistance to increase its overall ac emitter resistance (see Figure 9.19). As a result of partially bypassing the emitter resistance:
    a. The voltage gain of the amplifier is made more stable against changes in $r'_e$. (For this reason, this type of amplifier is commonly referred to as *a gain-stabilized amplifier*.)
    b. Overall voltage gain is *decreased* drastically.
    c. Amplifier input impedance is *increased*.

28. *Hybrid parameters*, or *h-parameters*, are transistor specifications that describe the ac operation of the component under full-load or no-load conditions.
    a. $h_{ie}$ is the *base input impedance*.
    b. $h_{fe}$ is the *base-to-collector current gain*.
    c. $h_{oe}$ is the *output admittance*.
    d. $h_{re}$ is the *reverse voltage feedback ratio*.
    Each of these parameters is measured as shown in Figure 9.23.

29. When the spec sheet for a transistor lists *minimum* and *maximum* values, the *geometric average* of the two is used for circuit analysis problems (see Example 9.18).

30. Transistor spec sheets commonly provide graphs representing the *h*-parameter values under a variety of circuit conditions (see Figure 9.26). When graphs are used, specific *h*-parameter values are determined, as demonstrated in Example 9.19.

**31.** *Nonlinear distortion* is a type of distortion caused by driving the base-emitter junction of a transistor into its nonlinear operating region.

**32.** Nonlinear distortion can be caused by:

    **a.** Poorly biasing a transistor so that the Q-point value of $I_B$ is near the nonlinear region of the base curve.

    **b.** Overdriving the amplifier.

## EQUATION SUMMARY

| Equation Number | Equation | Section Number |
|---|:---:|---|
| **(9.1)** | $V_{\text{out}} = V_C$ | 9.1 |
| **(9.2)** | $r'_e = \dfrac{25 \text{ mV}}{I_E}$ | 9.1 |
| **(9.3)** | $r'_e = \dfrac{\Delta V_{BE}}{\Delta I_E}$ | 9.1 |
| **(9.4)** | $\beta_{\text{ac}} = \dfrac{\Delta I_C}{\Delta I_B}$ | 9.1 |
| **(9.5)** | $\beta_{\text{ac}} = \dfrac{i_c}{i_b}$ | 9.1 |
| **(9.6)** | $A_i = h_{fe}$   (for the transistor only) | 9.1 |
| **(9.7)** | $X_C = \dfrac{1}{2\pi f C}$ | 9.2 |
| **(9.8)** | $A_v = \dfrac{v_{\text{out}}}{v_{\text{in}}}$ | 9.4 |
| **(9.9)** | $A_v = \dfrac{r_C}{r'_e}$ | 9.4 |
| **(9.10)** | $r_C = R_C \parallel R_L$ | 9.4 |
| **(9.11)** | $i_e = \dfrac{v_{\text{in}}}{r'_e}$ | 9.4 |
| **(9.12)** | $v_{\text{in}} = i_e r'_e$ | 9.4 |
| **(9.13)** | $v_{\text{out}} = i_c r_C$ | 9.4 |
| **(9.14)** | $v_{\text{out}} = A_v v_{\text{in}}$ | 9.4 |
| **(9.15)** | $A_i = \dfrac{i_{\text{out}}}{i_{\text{in}}}$ | 9.4 |
| **(9.16)** | $A_p = A_i A_v$ | 9.4 |
| **(9.17)** | $P_{\text{out}} = A_p P_{\text{in}}$ | 9.4 |
| **(9.18)** | $r_C = R_C$   (open-load) | 9.5 |
| **(9.19)** | $Z_{\text{in}} = R_1 \parallel R_2 \parallel Z_{\text{in(base)}}$ | 9.5 |
| **(9.20)** | $Z_{\text{in(base)}} = h_{fe} r'_e$ | 9.5 |
| **(9.21)** | $A_i = h_{fe}\left(\dfrac{Z_{\text{in}} r_C}{Z_{\text{in(base)}} R_L}\right)$ | 9.5 |

| Equation Number | Equation | Section Number |
|---|---|---|
| **(9.22)** | $A_{vT} = (A_{v1})(A_{v2})(A_{v3}) \cdots$ | 9.5 |
| **(9.23)** | $A_{iT} = (A_{i1})(A_{i2})(A_{i3}) \cdots$ | 9.5 |
| **(9.24)** | $A_{pT} = (A_{vT})(A_{iT})$ | 9.5 |
| **(9.25)** | $A_v = \dfrac{r_C}{r'_e + r_E}$ | 9.6 |
| **(9.26)** | $Z_{\text{in(base)}} = h_{fe}(r'_e + r_E)$ | 9.6 |
| **(9.27)** | $h_{ie} = \dfrac{v_{\text{in}}}{i_b}$ (output shorted) | 9.7 |
| **(9.28)** | $h_{fe} = \dfrac{i_c}{i_b}$ (output shorted) | 9.7 |
| **(9.29)** | $h_{oe} = \dfrac{i_c}{v_{ce}}$ (input open) | 9.7 |
| **(9.30)** | $h_{re} = \dfrac{v_{be}}{v_{ce}}$ (input open) | 9.7 |
| **(9.31)** | $Z_{\text{in(base)}} = h_{ie}$ | 9.7 |
| **(9.32)** | $r'_e = \dfrac{h_{ie}}{h_{fe}}$ | 9.7 |
| **(9.33)** | $A_v = \dfrac{h_{fe} r_C}{h_{ie}}$ | 9.7 |

## KEY TERMS

ac beta ($\beta_{ac}$ or $h_{fe}$) 321
ac emitter resistance ($r'_e$) 320
ac equivalent circuit 323
bypass capacitor 325
cascaded 323
coupling capacitor 323
common-emitter amplifier 319
current gain ($A_i$) 333
current gain ($h_{fe}$) 345
gain-stabilized amplifier 340
$h$-parameters 345
input impedance ($h_{ie}$) 345
multistage amplifier 323
nonlinear distortion 352
output admittance ($h_{oe}$) 346
power gain ($A_p$) 334
reverse voltage feedback ratio ($h_{re}$) 346
stage 323
swamped amplifier 340
voltage gain ($A_v$) 329

## PRACTICE PROBLEMS

### Section 9.1

1. An amplifier has an emitter current of 12 mA. Determine the value of $r'_e$ for the circuit.
2. An amplifier has an emitter current of 10 mA. Determine the value of $r'_e$ for the circuit.
3. An amplifier has values of $V_E = 2.2$ V and $R_E = 910\ \Omega$. Determine the value of $r'_e$ for the circuit.
4. An amplifier has values of $V_E = 12$ V and $R_E = 4.7\ \text{k}\Omega$. Determine the value of $r'_e$ for the circuit.
5. An amplifier has values of $V_B = 3.2$ V and $R_E = 1.2\ \text{k}\Omega$. Determine the value of $r'_e$ for the circuit.
6. An amplifier has values of $V_B = 4.8$ V and $R_E = 3.9\ \text{k}\Omega$. Determine the value of $r'_e$ for the circuit.
7. Determine the value of $r'_e$ for the circuit shown in Figure 9.30.
8. Determine the value of $r'_e$ for the circuit shown in Figure 9.31.

**OBJECTIVE 2** ▶ Figure 10.5a shows a typical emitter follower. The key to the ac operation of the circuit lies in the fact that the load is capacitively coupled to the *emitter*. As a result, the amplifier has the ac equivalent shown in Figure 10.5b. Note that the resistor in the emitter circuit ($r_E$) represents the *ac resistance in the emitter circuit*. This resistance is found as

$$r_E = R_E \parallel R_L \qquad\qquad (10.7)$$

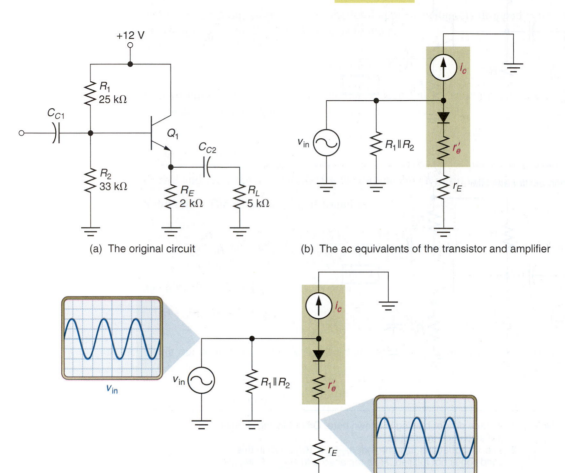

(a) The original circuit

(b) The ac equivalents of the transistor and amplifier

(c) The amplifier input and output signals

**FIGURE 10.5** A typical emitter follower and its ac equivalents.

**Lab Reference**: The gain characteristics of the emitter follower are demonstrated in Exercise 12.

The low voltage gain of the emitter follower is a result of the voltage divider in the output (emitter) circuit. Figure 10.5c shows the amplifier ac equivalent circuit with an input signal applied. Ignoring the effects of the base-emitter diode, the input signal is applied across the series combination of $r'_e$ and $r_E$, which form a voltage divider. Using the voltage-divider equation,

$$v_{out} = v_{in} \frac{r_E}{r'_e + r_E}$$

or

$$\frac{v_{out}}{v_{in}} = \frac{r_E}{r'_e + r_E}$$

Finally, since

$$A_v = \frac{v_{out}}{v_{in}}$$

we can calculate the voltage gain of the circuit as

$$A_v = \frac{r_E}{r'_e + r_E} \qquad \textbf{(10.8)}$$

In most practical applications, $r_E \gg r'_e$. When this is the case, the following approximation can be used for voltage gain:

$$A_v \cong 1 \qquad (\text{when } r_E \gg r'_e) \qquad \textbf{(10.9)}$$

The greater the difference between $r_E$ and $r'_e$, the closer the amplifier voltage gain comes to 1. In practice, the voltage gain of the emitter follower is usually between 0.8 and 0.999. This is illustrated in Example 10.3.

### EXAMPLE 10.3

Determine the value of $A_v$ for the circuit shown in Figure 10.5a.

*Solution:*　First, the value of $r_E$ is found as

$$r_E = R_E \parallel R_L = 2 \text{ k}\Omega \parallel 5 \text{ k}\Omega = \textbf{1.43 k}\Omega$$

We do not know the *h*-parameter values for the transistor, so we need to approximate the value of $r'_e$ using

$$r'_e = \frac{25 \text{ mV}}{I_E}$$

Using the established procedure, the value of $I_E$ for the circuit is found to be approximately 3.1 mA, and the value of $r'_e$ is found as

$$r'_e = \frac{25 \text{ mV}}{I_E} = \frac{25 \text{ mV}}{3.1 \text{ mV}} = \textbf{8.1 }\Omega$$

Using the value of $r'_e$ and the value of $r_E$, the voltage gain of the amplifier is found as

$$A_v = \frac{r_E}{r'_e + r_E} = \frac{1.43 \text{ k}\Omega}{8.1 \text{ }\Omega + 1.43 \text{ k}\Omega} = \textbf{0.9944}$$

### PRACTICE PROBLEM 10.3

Assume that the amplifier described in Practice Problem 10.1 has a 4 k$\Omega$ load. Determine the value of $A_v$ for the circuit.

## 10.2.1　Current Gain

In Chapter 9, you were shown that the current gain of a common-emitter amplifier ($A_i$) is significantly lower than the current gain of the transistor ($h_{fe}$). This relationship is due to the current divisions that occur in both the input and output circuits. The same principle holds true for the emitter follower. The current gain of an emitter follower is found as

$$A_i = h_{fc}\left(\frac{Z_{in}r_E}{Z_{in(base)}R_L}\right) \qquad \textbf{(10.10)}$$

*A Practical Consideration:*
If you look in Appendix C, you'll see that the exact equation for the value of $h_{fc}$ is

$$h_{fc} = h_{fe} + 1$$

Since $h_{fe}$ is normally much greater than 1, we normally assume that $h_{fc} \cong h_{fe}$.

where $h_{fc} \cong h_{fe}$. Note that the subscript $c$ indicates that the parameter applies to an emitter-follower (common-collector) amplifier rather than to a common-emitter amplifier. Although there is a more exact $h$-parameter derivation for $h_{fc}$, using $h_{fc} = h_{fe}$ in equation (10.10) will provide accurate results.

### 10.2.2 Power Gain

As with the common-emitter amplifier, the power gain ($A_p$) of an emitter follower is found as the product of current gain ($A_i$) and voltage gain ($A_v$). By formula,

$$A_p = A_i A_v \qquad (10.11)$$

Since the value of $A_v$ is always slightly less than 1, the power gain ($A_p$) of an emitter follower is always slightly less than the current gain ($A_i$) of the circuit. This point is illustrated in Example 10.4.

---

#### EXAMPLE 10.4

The amplifier shown in Figure 10.5 (Example 10.3) has a value of $A_i = 2.7$. Determine the power gain ($A_p$) of the amplifier.

*Solution:* In Example 10.3, we determined the voltage gain ($A_v$) of the amplifier to be 0.9944. With a current gain of $A_i = 2.7$, the power gain of the circuit is found as

$$A_p = A_i A_v = (2.7)(0.9944) = \textbf{2.68}$$

As you can see, the value of $A_p$ for the emitter follower is slightly less than the value of current gain, $A_i$.

#### PRACTICE PROBLEM 10.4

The amplifier described in Practice Problem 10.1 has a 4 k$\Omega$ load and a current gain of $A_i = 24$. Determine the power gain of the amplifier.

---

### 10.2.3 Input Impedance ($Z_{in}$)

Lab Reference: The input impedance of an emitter follower is measured in Exercise 12.

As with the common-emitter amplifier, the input impedance of the emitter follower is found as the parallel combination of the base resistors and the transistor input impedance. By formula,

$$Z_{in} = R_1 \parallel R_2 \parallel Z_{in(base)} \qquad (10.12)$$

The base input impedance is equal to the product of the device current gain and the total ac emitter resistance. By formula,

$$Z_{in(base)} = h_{fc}(r_e' + r_E) \qquad (10.13)$$

where $h_{fc}$ = the transistor current gain
$r_e'$ = the ac resistance of the transistor emitter
$r_E$ = the parallel combination of $R_E$ and $R_L$

Example 10.5 illustrates the procedure used to determine the input impedance of an emitter follower.

## EXAMPLE 10.5

Determine the input impedance of the amplifier shown in Figure 10.5a. Assume that the value of $h_{fe}$ for the transistor is listed on the spec sheet as 220.

**Solution:** In Example 10.3, we calculated values of $r_E = 1.43$ k$\Omega$ and $r'_e = 8.1$ $\Omega$. Assuming that $h_{fc}$ is approximately equal to $h_{fe}$, the impedance of the transistor is found as

$$Z_{in(base)} = h_{fc}(r'_e + r_E) = (220)(8.1\ \Omega + 1.43\ \text{k}\Omega) = \mathbf{316.4\ k\Omega}$$

Now, the input impedance of the amplifier is found as

$$Z_{in} = R_1 \parallel R_2 \parallel Z_{in(base)} = 25\ \text{k}\Omega \parallel 33\ \text{k}\Omega \parallel 316.4\ \text{k}\Omega = \mathbf{13.6\ k\Omega}$$

### PRACTICE PROBLEM 10.5

The amplifier described in Practice Problem 10.1 has a current gain of $h_{fe} = 240$ and a load resistance of $R_L = 2$ k$\Omega$. Determine the input impedance of the circuit.

When the value of $r_E$ is much greater than $r'_e$ (as was the case in Example 10.5), equation (10.13) can be simplified as follows:

$$Z_{in(base)} \cong h_{fc}r_E \qquad (\text{when } r_E \gg r'_e) \qquad \textbf{(10.14)}$$

The validity of this approximation can be seen by recalculating the values of $Z_{in(base)}$ and $Z_{in}$ for the circuit in Figure 10.5a. If you ignore the value of $r'_e$ in the calculations, you end up with the same value of $Z_{in}$ that was found in Example 10.5.

## 10.2.4    Output Impedance ($Z_{out}$)

As you may recall, **output impedance** is *the impedance that a circuit presents to its load.* When a load is connected to a circuit, the output impedance of the circuit acts as the source impedance for that load.

The output impedance of an emitter follower is found as

$$Z_{out} = R_E \parallel \left( r'_e + \frac{R'_{in}}{h_{fc}} \right) \qquad \textbf{(10.15)}$$

> **Output impedance**
> The impedance that a circuit presents to its load. The output impedance of a circuit is effectively the source impedance for its load.

where $Z_{out}$ = the output impedance of the amplifier
$\quad\quad R'_{in} = R_1 \parallel R_2 \parallel R_S$
$\quad\quad R_S$ = the output resistance of the input voltage source

Equation (10.15) is derived in Appendix D. Its use is demonstrated in Example 10.6.

### EXAMPLE 10.6

Determine the output impedance ($Z_{out}$) of the amplifier shown in Figure 10.6.

**Solution:** Using the $h$-parameter method of analysis, the value of $r'_e$ is found as

$$r'_e = \frac{h_{ic}}{h_{fc}} = \frac{4\ \text{k}\Omega}{200} = \mathbf{20\ \Omega}$$

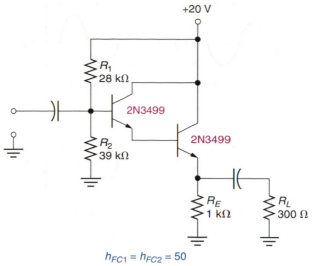

$h_{FC1} = h_{FC2} = 50$

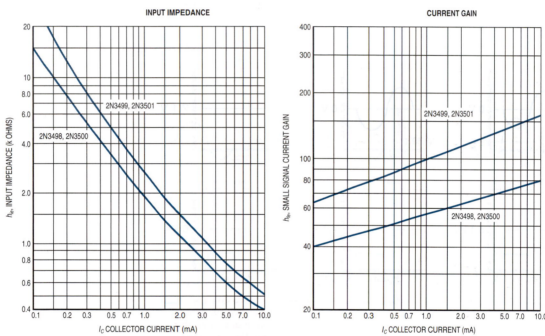

FIGURE 10.35   (Copyright of Semiconductor Component Industries, LLC. Used by permission.)

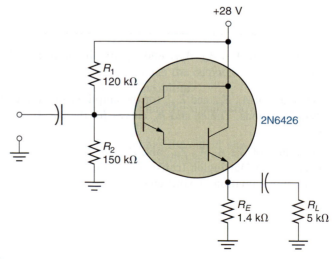

FIGURE 10.36

**38.** Write a program that will perform the dc analysis of a standard emitter follower or a Darlington emitter-follower given the required inputs.

**39.** Write a program that will perform the ac analysis of a standard emitter follower given the required input values.

**40.** Write a program that will perform the "quick" analysis of a Darlington emitter-follower given the required input values.

SUGGESTED COMPUTER
APPLICATIONS
PROBLEMS

ANSWERS TO THE
EXAMPLE PRACTICE
PROBLEMS

**10.1** $V_B = 9.92$ V, $V_E = 9.22$ V, $I_E = 10.13$ mA, $V_{CEQ} = 8.78$ V

**10.2** $V_{CE(off)} = 18$ V, $I_{C(sat)} = 19.78$ mA

**10.3** 0.9967

**10.4** 23.92

**10.5** 8.73 kΩ

**10.6** 15.54 Ω

**10.7** 83.8 kΩ, 3.57 kΩ

**10.8** 22.5 MΩ

**10.9** $A_i = 858.7$, $Z_{in} \cong 111$ kΩ, $Z_{out} = 25.3$ Ω

**10.10** $A_v = 47.71$, $A_i \cong 1$, $Z_{in} \cong 52.4$ Ω, $Z_{out} \cong 15$ kΩ

## chapter 11

# Power Amplifiers

## Objectives

*After studying the material in this chapter, you should be able to:*

1. Identify the $Q$-point location on the dc load line for class A, class B, class AB, and class C amplifiers.

2. Compare and contrast the ac load line and dc load line of a class A amplifier.

3. Explain the concept of amplifier compliance.

4. Calculate the values of dc input power and ac load power for an $RC$-coupled class A amplifier.

5. Explain why high amplifier efficiency ratings are desirable in power amplifiers.

6. Describe and analyze the operation of the transformer-coupled class A amplifier.

7. Describe and analyze the operation of class B amplifiers.

8. Perform all the calculations necessary to determine the values of source power and load power for a complementary-symmetry amplifier.

9. Define and describe class AB operation.

10. Explain how class AB operation eliminates crossover distortion and thermal runaway.

11. Perform the complete analysis of a class AB amplifier.

12. Calculate the transistor power dissipation requirements for a given class A, class B, or class AB amplifier.

## Outline

# Power Amplifiers: When Are They Used?

In this chapter, you will be studying a group of amplifiers referred to as *power amplifiers*. Many of the amplifier circuits and principles covered in this chapter are used in *communications electronics*. Communications electronics is the study of the circuits and systems used to transmit information between two or more locations. Television and radio transmitters and receivers, two-way transceivers, and radar systems are all examples of communications systems.

There are three basic types of power amplifiers: *class A*, *class B*, and *class C* power amplifiers. In this chapter, we concentrate on the first two (class A and class B) along with a special-case class B amplifier, called a *class AB* amplifier. These three circuits are commonly used as *audio amplifiers*. Audio amplifiers are circuits used to amplify signals whose frequencies lie in the range of approximately 20 Hz to 20 kHz.

Power amplifiers are used to deliver high values of power to low-resistance loads. The typical power amplifier will have an output power greater than 1 W with a load resistance that will range from 300 $\Omega$ (for transmission antennas) to 4 $\Omega$ (for audio speakers). Although these load values do not cover every possibility, they do illustrate that power amplifiers usually drive low-resistance loads.

The ideal power amplifier would deliver 100% of the power it draws from the supply to the load. In practice, however, this can never occur because the components in the amplifier all dissipate *some* of the power that is being drawn from the supply. As you were shown in Chapter 8, amplifier efficiency is calculated as follows:

> *Just a reminder:*
> $\eta$ is the Greek letter eta.

$$\eta = \frac{\text{ac output power}}{\text{dc input power}} \times 100$$

**OBJECTIVE 1** ▶ This equation indicates that lower dc input power results in higher amplifier efficiency. The dc input power to an amplifier varies according to the position of the $Q$-point on the load line. Typical $Q$-point locations for the four amplifier classes are shown in Figure 11.1.

> The lower the $Q$-point position on the dc load line, the lower the dc input power. Thus, the class A amplifier has the lowest maximum possible efficiency, and the class C amplifier has the greatest maximum possible efficiency.

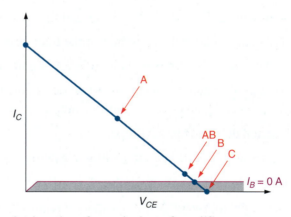

**FIGURE 11.1**   $Q$-point locations for each class of amplifier.

In this chapter, we will look at the operation of class A, class B, and class AB amplifiers. Class C amplifiers are *tuned circuits*, meaning that they are designed to operate within specific frequency ranges. Class C amplifiers, along with many other tuned circuits, are covered in Chapter 17.

## 11.1   The AC Load Line

In Chapter 7, you were introduced to the *dc load line*. This line is used to represent all possible combinations of $V_{CE}$ and $I_C$ for a given amplifier. The **ac load line** is used for the same basic purpose: It represents all possible ac combinations of $i_c$ and $v_{ce}$.

Up to this point, the ac load line has not been an important consideration. However, it serves as an effective tool for teaching many of the principles of class A and class B amplifier operation. For this reason, we will take a brief look at the ac load line before covering any of the common power amplifier circuits.

**ac load line**
A graph of all possible combinations of $i_c$ and $v_{ce}$.

◀ **OBJECTIVE 2**

The ac load line for a given amplifier does *not* follow the plot of the dc load line. The reason for this lies in the fact that the *dc load* of an amplifier is different from its *ac load*. For example, consider the circuits shown in Figure 11.2. In terms of dc operation, the load on the transistor is equal to $R_C$. The value of $R_L$ does not weigh into the picture because the coupling capacitor provides dc isolation between the transistor and $R_L$. The ac equivalent circuit (Figure 11.2b) shows the ac load to consist of $R_C$ in parallel with $R_L$. As you will see, this difference in the load resistance results in considerable differences between the dc and ac load lines.

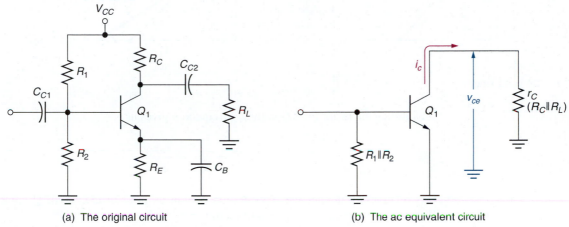

(a) The original circuit    (b) The ac equivalent circuit

**FIGURE 11.2**   A common-emitter amplifier and its ac equivalent circuit.

### 11.1.1    AC Saturation Current

If you take a close look at the ac equivalent circuit shown in Figure 11.2b, you will see that the transistor is the signal source for the ac load resistance ($r_c$). According to Ohm's law,

$$i_c = \frac{v_{ce}}{r_C} \tag{11.1}$$

This relationship can also be expressed as

$$\Delta I_C = \frac{\Delta V_{CE}}{r_C} \tag{11.2}$$

where $\Delta I_C$ = the *change in* $I_C$
$\Delta V_{CE}$ = the *change in* $V_{CE}$

As shown in Figure 11.3a, the ac saturation current can be found as

$$i_{c(sat)} = I_{CQ} + \Delta I_C$$

We know that the value of $V_{CE}$ is ideally 0 V when the transistor is in saturation. This means that the value of $\Delta I_C$ in Figure 11.3a can be found as

$$\Delta I_C = \frac{\Delta V_{CE}}{r_C} = \frac{V_{CEQ} - 0 \text{ V}}{r_C} = \frac{V_{CEQ}}{r_C}$$

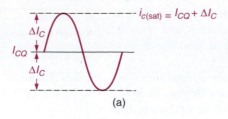

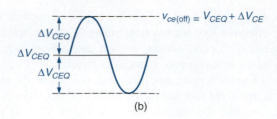

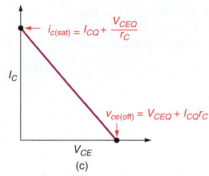

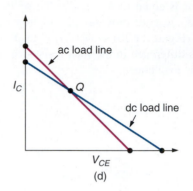

FIGURE 11.3   ac Load line.

Substituting the result for $\Delta I_C$ in the $i_{c(\text{sat})}$ equation gives us

$$i_{c(\text{sat})} = I_{CQ} + \frac{V_{CEQ}}{r_C} \tag{11.3}$$

This relationship defines the ac saturation current in terms of the circuit $Q$-point values. The point representing $i_{c(\text{sat})}$ is identified in Figure 11.3c.

### 11.1.2   AC Cutoff Voltage

As shown in Figure 11.3b, the ac cutoff voltage can be found as

$$v_{ce(\text{off})} = V_{CEQ} + \Delta V_{CE}$$

Ideally, $I_C = 0$ A when the transistor is in cutoff. Using a transposed version of equation (11.2), we can determine the value of $\Delta V_{CE}$ as follows:

$$\Delta V_{CE} = (\Delta I_C)(r_C) = (I_{CQ} - 0 \text{ A})(r_C) = I_{CQ} r_C$$

Substituting this result for $\Delta V_{CE}$ in the $v_{ce(\text{off})}$ equation gives us

$$v_{ce(\text{off})} = V_{CEQ} + I_{CQ} r_C \tag{11.4}$$

This equation defines the ac cutoff voltage in terms of the circuit $Q$-point values. The point representing $v_{ce(\text{off})}$ is identified in Figure 11.3c. Note that the ac load line crosses the dc load line at the $Q$-point, as is shown in Figure 11.3d. This is always the case for a given amplifier.

### 11.1.3   What Does the ac Load Line Tell You?

**OBJECTIVE 3** ▶    The ac load line of an amplifier is used to illustrate its *compliance*. The compliance of an amplifier is the limit that its output circuit places on the peak-to-peak output voltage. If the compliance of an amplifier is exceeded, *clipping* of the output waveform occurs. This point is discussed in detail later in this section.

Compliance
The limit that the output circuit
of an amplifier places on its
peak-to-peak output voltage.

   The **compliance** of an amplifier is found by determining the maximum possible transitions of $I_C$ and $V_{CE}$ from their respective values of $I_{CQ}$ and $V_{CEQ}$. The maximum possible transitions are illustrated in Figure 11.4. As Figure 11.4a shows, the maximum possible

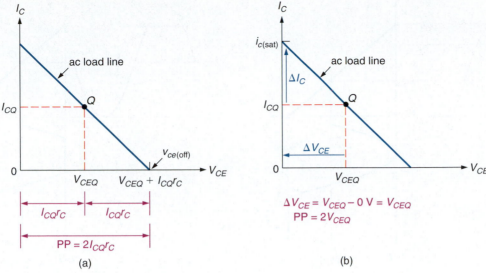

FIGURE 11.4  Amplifier compliance.

transition for $V_{CE}$ is equal to the difference between $v_{ce(off)}$ and $V_{CEQ}$. Since this transition is equal to $I_{CQ}r_C$, the maximum peak output voltage from the amplifier is equal to $I_{CQ}r_C$. Two times this value gives us the maximum peak-to-peak transition of the output voltage as follows:

$$PP = 2I_{CQ}r_C \qquad (11.5)$$

where PP = the output compliance, in peak-to-peak voltage
  $I_{CQ}$ = the Q-point value of $I_C$
  $r_C$ = the ac resistance in the collector circuit

In Figure 11.3a, you were shown that $i_{c(sat)} = I_{CQ} + \Delta I_C$. The transition from $I_{CQ}$ to $i_{c(sat)}$ is shown (as $\Delta I_C$) on the load line in Figure 11.4b. Note that $\Delta I_C$ corresponds to a change in collector-emitter voltage ($\Delta V_{CE}$). As you were shown earlier, $\Delta V_{CE} = V_{CEQ}$. Thus, the maximum peak-to-peak transition is equal to twice this value, as follows:

$$PP = 2V_{CEQ} \qquad (11.6)$$

where PP = the output compliance, in peak-to-peak voltage
  $V_{CEQ}$ = the Q-point value of $V_{CE}$

Equations (11.5) and (11.6) were derived using the relationships shown in Figure 11.4. When the Q-point of an amplifier lies at midpoint on its ac load line,

$$I_{CQ}r_C = V_{CEQ}$$

and its output can vary equally above and below the Q-point without clipping. When the Q-point lies *below midpoint* on the ac load line,

$$I_{CQ}r_C < V_{CEQ}$$

In this case, the output of the amplifier is limited by $I_{CQ}r_C$, and the amplifier can be driven into **cutoff clipping**, as shown in Figure 11.5a. When the Q-point lies *above midpoint* on the ac load line,

$$I_{CQ}r_C > V_{CEQ}$$

**Cutoff clipping**
A type of distortion caused by driving a transistor into cutoff.

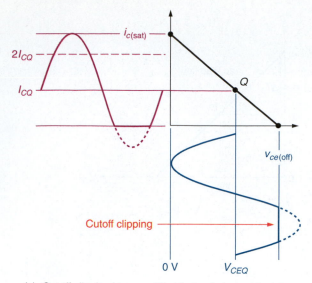

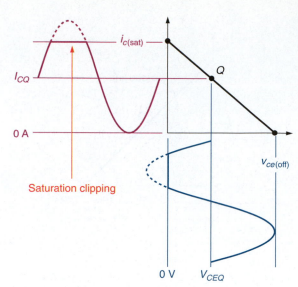

(a) Cutoff clipping is caused by biasing below midpoint on the load line.

(b) Saturation clipping is caused by biasing above midpoint on the load line.

FIGURE 11.5 Cutoff and saturation clipping.

---

**Saturation clipping**

A type of distortion caused by driving a transistor into saturation.

In this case, the output of the amplifier is limited by $V_{CEQ}$, and the amplifier can be driven into **saturation clipping**, as shown in Figure 11.5b.

To determine the compliance without using the ac load line, solve both equation (11.5) and (11.6). *The lower of the two results is the amplifier compliance.* If equation (11.5) yields the lower result, the circuit is biased as shown in Figure 11.5a and will experience cutoff clipping if overdriven. If equation (11.6) yields the lower result, the circuit is biased as shown in Figure 11.5b and will experience saturation clipping if overdriven. The calculation of compliance is demonstrated in Example 11.1.

### EXAMPLE 11.1

Determine the output compliance for the amplifier shown in Figure 11.6.

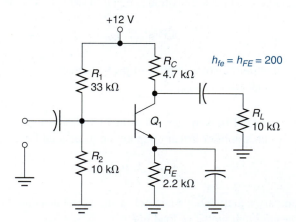

FIGURE 11.6

*Solution:* Using the established procedure, $I_{CQ}$ and $V_{CEQ}$ are found to be 950 μA and 5.45 V, respectively. (*Refer to Chapter 7 if you have trouble solving for these values.*) The value of $r_C$ is equal to the parallel combination of $R_C$ and $R_L$: 3.2 kΩ. Now, the amplifier is solved for *both* compliance values as follows:

$$PP = 2V_{CEQ} = 2(5.45 \text{ V}) = \mathbf{10.9 \ V_{PP}}$$

and

$$\mathbf{PP} = 2I_{CQ}r_C = 2(950 \ \mu A)(3.2 \ k\Omega) = \mathbf{6.08 \ V_{PP}}$$

Since the overall compliance equals the *lower* of the two values obtained, its value for this amplifier is equal to 6.08 $V_{PP}$. This means that the peak-to-peak output voltage can be no more than 6.08 V. If this value is exceeded, cutoff clipping will occur.

**PRACTICE PROBLEM 11.1**

A common-emitter amplifier like the one in Figure 11.6 has the following values: $V_{CC} = 20$ V, $R_1 = 10$ k$\Omega$, $R_2 = 1.8$ k$\Omega$, $R_C = 620$ $\Omega$, $R_E = 200$ $\Omega$, $R_L = 1.2$ k$\Omega$, and $h_{FE} = 180$. Determine the compliance (PP) of the amplifier.

The importance of the compliance values obtained in Example 11.1 is illustrated in Figure 11.7. Here, we see the dc and ac load lines along with two possible output signals. Waveform A shows the result of trying to drive the amplifier to PP = 10.9 $V_{PP}$. As you can see, the amplifier experiences *cutoff clipping* as a result of trying to exceed the *ac cutoff point* (8.49 V). On the other hand, waveform B is not clipped because its peak-to-peak value is limited to the compliance of the circuit, 6.08 $V_{PP}$.

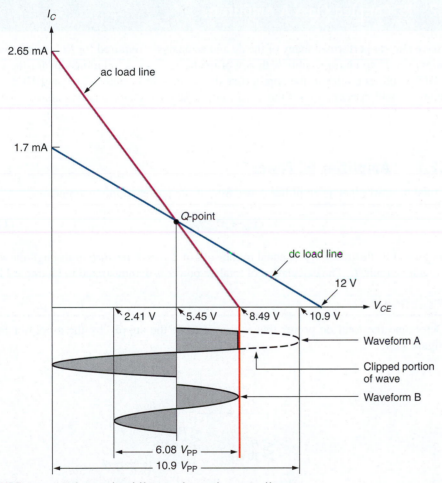

**FIGURE 11.7** Using ac load line to determine compliance.

## 11.1.4  A Practical Consideration

In Chapter 9, you were introduced to the concept of *nonlinear distortion*. In many cases, nonlinear distortion occurs long before cutoff clipping occurs. In other words, if the amplifier input amplitude is sufficient to cause cutoff clipping to occur, it is more than sufficient to generate nonlinear distortion in the amplifier output.

When the compliance of an amplifier is limited by cutoff clipping ($2I_{CQ}r_C$), you need to be aware that nonlinear distortion will occur before the compliance of the amplifier is reached. At the same time, if saturation clipping ($2V_{CEQ}$) is the limiting factor, nonlinear distortion should not be a problem.

Determining the compliance of an amplifier is not an everyday requirement for the average technician. At the same time, you should be aware that the *ac characteristics* of an amplifier often limit its output.

---

**Section Review ▶**

1. What is the *ac load line*?
2. Why does the ac load line of an amplifier differ from the dc load line?
3. What is *compliance*?
4. How is the compliance of an amplifier determined?

**Critical Thinking ▶**

5. Under what load condition would the ac and dc load lines for an amplifier be identical? Explain your answer.

---

## 11.2 *RC*-Coupled Class A Amplifiers

**OBJECTIVE 4 ▶**

We have already performed many of the dc and ac analyses required for *RC-coupled class A amplifiers*. If you have trouble with any of the basic dc or ac relationships for the various BJT amplifiers, refer to the appropriate discussions in Chapters 7, 9, and 10. In this section, we will concentrate on the power and efficiency characteristics of *RC*-coupled class A amplifiers.

### 11.2.1 Amplifier DC Power

The total dc power that an amplifier draws from its dc power supply is found as

$$P_S = V_{CC}I_{CC} \tag{11.7}$$

As Figure 11.8 illustrates, $I_{CC}$ is equal to the sum of $I_{CQ}$ and the current through the voltage divider circuit, $I_1$. The calculation of total dc power is demonstrated in Example 11.2.

#### EXAMPLE 11.2

Determine the total dc power that is drawn from the supply by the amplifier in Figure 11.9.

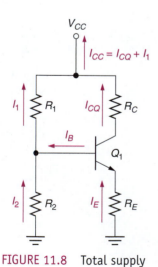

**FIGURE 11.8** Total supply current.

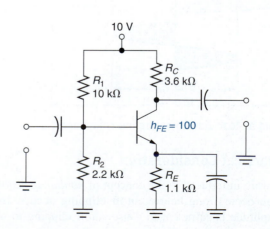

**FIGURE 11.9**

*Solution:* Assuming the base current to be negligible, the value of $I_1$ can be found as

$$I_1 = \frac{V_{CC}}{R_1 + R_2} = \frac{10 \text{ V}}{12.2 \text{ k}\Omega} = 820 \text{ }\mu\text{A}$$

Using the established procedure, the value of $I_{CQ}$ is found as

$$I_{CQ} \cong 1 \text{ mA}$$

Now, the value of $I_{CC}$ is found as

$$I_{CC} = I_1 + I_{CQ} = 1 \text{ mA} + 820 \text{ }\mu\text{A} = 1.82 \text{ mA}$$

Finally, $P_S$ is found as

$$P_S = V_{CC}I_{CC} = (10 \text{ V})(1.82 \text{ mA}) = 18.2 \text{ mW}$$

**PRACTICE PROBLEM 11.2**
Determine how much power is drawn from the dc power supply by the amplifier described in Practice Problem 11.1.

## 11.2.2  AC Load Power

AC load power can be calculated using the standard $V^2/R$ equation. Specifically,

$$P_L = \frac{V_L^2}{R_L}$$  **(11.8)**

*A Practical Consideration:* This equation is used when $V_L$ is measured with an ac voltmeter.

where $P_L$ = the ac load power
$V_L$ = the *rms* load voltage

The use of this equation is demonstrated in Example 11.3.

### EXAMPLE 11.3

Determine the ac load power for the circuit shown in Figure 11.10.

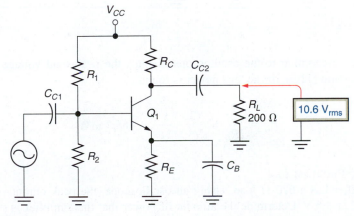

**FIGURE 11.10**

*Solution:* The voltmeter reads 10.6 V, as indicated. Since ac voltmeters provide rms measurements, the value shown can be plugged directly into equation (11.8). Thus,

$$P_L = \frac{V_L^2}{R_L} = \frac{(10.6 \text{ V})^2}{200 \text{ }\Omega} = 561.8 \text{ mW}$$

A Practical Consideration:
This equation can be used when $V_L$ is measured with an oscilloscope.

**PRACTICE PROBLEM 11.3**

An amplifier with a 1.2 k$\Omega$ load has a 4.62 $V_{rms}$ output. Determine the value of load power for the circuit.

A convenient form of equation (11.8) that can be used when you know the peak output voltage is

$$P_L = \frac{V_{pk}^2}{2R_L}$$ **(11.9)**

This equation is derived by substituting (0.707 $V_{pk}$) into the numerator of equation (11.8) and simplifying the fraction as follows:

$$P_L = \frac{V_L^2}{R_L} = \frac{(0.707\ V_{pk})^2}{R_L} = \frac{0.5\ V_{pk}^2}{R_L} = \frac{V_{pk}^2}{2R_L}$$

Example 11.4 demonstrates its application.

**EXAMPLE 11.4**

Determine the ac load power for the circuit shown in Figure 11.11.

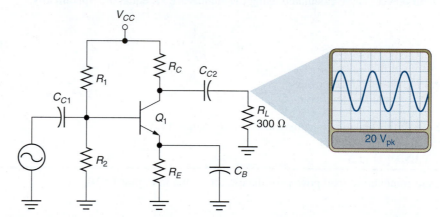

**FIGURE 11.11**

**Solution:** According to the oscilloscope display, the peak load voltage is 20 V. Using equation (11.9), the ac load power is found as

$$P_L = \frac{V_{pk}^2}{2R_L} = \frac{(20\ V)^2}{2(300\ \Omega)} = \textbf{666.7 mW}$$

**PRACTICE PROBLEM 11.4**

An amplifier has a 670 $\Omega$ load. Using an oscilloscope, the peak output voltage is measured at +8 V. Determine the amount of power that the amplifier is supplying to the load.

A Practical Consideration:
This equation can be used when $V_L$ is measured with an oscilloscope.

When the peak-to-peak load voltage is known, the ac load power can be found as

$$P_L = \frac{V_{PP}^2}{8R_L}$$ **(11.10)**

This equation is derived by substituting 0.5 $V_{PP}$ in place of $V_{pk}$ in equation (11.9) as follows:

$$P_L = \frac{V_{pk}^2}{2R_L} = \frac{(0.5\ V_{PP})^2}{2R_L} = \frac{0.25\ V_{PP}^2}{2R_L} = \frac{V_{PP}^2}{8R_L}$$

Equation (11.10) is important because it allows us to calculate the *maximum possible value of ac load power*. You may recall that the maximum possible output peak-to-peak voltage from an amplifier equals its *compliance*, PP. Therefore, the maximum possible ac load power can be found as

$$P_{L(max)} = \frac{PP^2}{8R_L} \qquad \textbf{(11.11)}$$

Example 11.5 demonstrates the use of this equation.

### EXAMPLE 11.5

A given amplifier has a compliance of 18 $V_{PP}$. If the load resistance is 100 Ω, what is the maximum possible value of ac load power?

*Solution:* Using the values given, the maximum possible load power is found as

$$P_{L(max)} = \frac{PP^2}{8R_L} = \frac{(18\ V_{PP})^2}{8(100\ \Omega)} = \textbf{405 mW}$$

### PRACTICE PROBLEM 11.5

An amplifier has a compliance of 20 $V_{PP}$ and a load resistance of 140 Ω. What is the maximum possible load power for the circuit?

## 11.2.3 Amplifier Efficiency

Once the values of $P_S$ and $P_L$ have been calculated for an amplifier, we can use these values to calculate the *efficiency* of the circuit. As you were told in Chapter 8, the efficiency of an amplifier is the percentage of the power drawn from the dc power supply that is actually delivered to the load.

The higher the efficiency of an amplifier, the better. Why? Because a high efficiency rating indicates that a low percentage of the power drawn from the supply is used by the amplifier itself. For example, an amplifier with a 90% efficiency rating uses only 10% of the power drawn from the dc power supply. The rest is delivered to the load. (*Remember:* The efficiency of an *ideal* amplifier is 100%.)

As you were shown in Chapter 8, the maximum theoretical efficiency of an *RC*-coupled class A amplifier is 25%. In practice, the efficiency of this type of amplifier is always much lower, as illustrated in Example 11.6.

◀ *OBJECTIVE 5*

*A Practical Consideration:* High efficiency helps extend the operating life of the components in an amplifier. Power used by an amplifier is dissipated in the form of heat. High efficiency means that relatively little power is dissipated by the components in the amplifier, which translates into longer component life.

### EXAMPLE 11.6

Determine the maximum efficiency of the amplifier described in Example 11.1.

*Solution:* In Example 11.1, we determined the compliance of the amplifier to be PP = 6.08 V. With a 10 kΩ load resistance (as shown in Figure 11.6), we can find the maximum value of load power as

$$P_{L(max)} = \frac{PP^2}{8R_L} = \frac{(6.08\ V_{PP})^2}{8(10\ k\Omega)} = \textbf{462 μW}$$

Using the established procedures, the following values are determined for the amplifier:

$$I_1 = 279.1 \ \mu A$$
$$I_{CQ} = 950.3 \ \mu A$$

and

$$I_{CC} = I_{CQ} + I_1 = \textbf{1.23 mA}$$

Now, $P_S$ is found as

$$P_S = V_{CC}I_{CC} = (12 \ V)(1.23 \ mA) = \textbf{14.76 mW}$$

Finally,

$$\eta = \frac{P_L}{P_S} \times 100 = \frac{462 \ \mu W}{14.76 \ mW} \times 100 = \textbf{3.13\%}$$

As you can see, even when this amplifier is driven to compliance, its efficiency rating is only 3.13%. This is considerably less than the maximum theoretical value of 25%.

### PRACTICE PROBLEM 11.6

The amplifier described in Practice Problem 11.1 is driven to compliance. Determine the maximum efficiency of the amplifier.

The maximum theoretical efficiency value of 25% for the *RC*-coupled class A amplifier is derived in Appendix D for those who wish to review the derivation.

The maximum theoretical efficiency of the transformer-coupled class A amplifier is 50%. The higher efficiency rating of this amplifier is due to its operating characteristics. We will take a look at this type of amplifier in the next section.

---

**Section Review ▶**

1. List, in order, the steps required to determine the value of $I_{CC}$ for an *RC*-coupled class A amplifier.

2. How do you calculate the value of $P_L$ when $V_L$ is measured with an ac voltmeter?

3. How do you calculate the value of $P_L$ when $V_L$ is measured with an oscilloscope?

4. When is load power ($P_L$) at its maximum value?

5. What is the maximum theoretical efficiency of an *RC*-coupled class A amplifier?

6. Why is a high efficiency rating desirable for a power amplifier?

**Critical Thinking ▶**

7. In this section, you were shown how to calculate load power using rms, peak, and peak-to-peak values. Why isn't there a $P_L$ equation for circuits like the one shown in Figure 11.11 that uses the dc average load voltage?

## 11.3 Transformer-Coupled Class A Amplifiers

**Transformer-coupled class A amplifier**
An amplifier that uses a transformer to couple the output signal to the load.

A **transformer-coupled class A amplifier** uses a transformer to couple its output signal to its load. A transformer-coupled amplifier is shown in Figure 11.12a. Because the ac characteristics of the transformer play an important role in the overall operation of the amplifier, we will briefly review some basic transformer characteristics.

## 11.3.1 Transformers

You may recall that the *turns ratio* of a transformer determines the relationship between primary and secondary values of voltage, current, and impedance. These relationships are summarized as follows:

$$\frac{N_P}{N_S} = \frac{V_P}{V_S} = \frac{I_S}{I_P} \qquad (11.12)$$

and

$$\left(\frac{N_P}{N_S}\right)^2 = \frac{Z_P}{Z_S} \qquad (11.13)$$

where $N_P, N_S$ = the number of turns in the primary and secondary, respectively
$V_P, V_S$ = the primary and secondary voltages
$I_P, I_S$ = the primary and secondary currents
$Z_P, Z_S$ = the primary and secondary impedances

Most transformers are classified as being either a *step-down transformer* or a *step-up transformer*. A **step-down transformer** is one with a secondary voltage that is less than the primary voltage. A step-down transformer will always have the following characteristics:

$$N_P > N_S$$
$$V_P > V_S$$
$$I_P < I_S$$
$$Z_P > Z_S$$

**Step-down transformer**
One with a secondary voltage that is less than the primary voltage.

A **step-up transformer** is one with a secondary voltage that is greater than the primary voltage. A step-up transformer will always have the following characteristics:

$$N_P < N_S$$
$$V_P < V_S$$
$$I_P > I_S$$
$$Z_P < Z_S$$

**Step-up transformer**
One with a secondary voltage greater than the primary voltage.

As you will see, these relationships play an important role in the ac operation of the transformer-coupled class A amplifier.

## 11.3.2 DC Operating Characteristics

The dc biasing for a transformer-coupled class A amplifier is very similar to that of any other class A amplifier with one important exception: *The circuit is designed for a value of $V_{CEQ}$ that is close to $V_{CC}$.*

◄ **OBJECTIVE 6**

$$V_{CEQ} \cong V_{CC}$$

You may recall that the value of $V_{CEQ}$ for an amplifier like the one in Figure 11.12a is found as

$$V_{CEQ} = V_{CC} - I_{CQ}(R_C + R_E)$$

Using the voltage-divider bias relationships established in Chapter 7, the circuit shown in Figure 11.12a is found to have a value of $I_{CQ} = 99.7$ mA. Using $R_W$ in place of $R_C$ in the $V_{CEQ}$ equation, we get

$$V_{CEQ} = V_{CC} - I_{CQ}(R_C + R_E) = 10\text{ V} - (99.7\text{ mA})(14\ \Omega) = 8.6\text{ V}$$

This value of $V_{CEQ}$ confirms that the circuit is designed for a value of $V_{CEQ}$ that is close to $V_{CC}$.

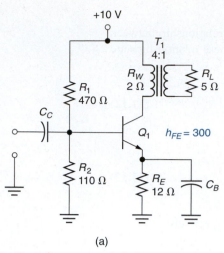

FIGURE 11.12  Transformer-coupled class A amplifier.

You should note that the value of $R_L$ is ignored in the dc analysis of the transformer-coupled class A amplifier, because transformers provide dc isolation between the primary and secondary. Since the load resistance is in the secondary of the transformer, it does not affect the dc analysis of the primary circuitry.

### 11.3.3  AC Operating Characteristics

Analyzing the ac operation of a transformer-coupled amplifier begins with plotting its ac load line. Earlier in the chapter, you were shown that the ac load line for a class A amplifier is plotted using

$$i_{c(sat)} = I_{CQ} + \frac{V_{CEQ}}{r_C} \quad \text{and} \quad v_{ce(off)} = V_{CEQ} + I_{CQ}r_C$$

We will use these relationships to plot the ac load line for the circuit shown in Figure 11.12a.

> How do you plot the ac load line?

Two of the values needed to plot the amplifier load are provided by the dc analysis of the circuit. For the circuit shown in Figure 11.12a, our dc analysis provided values of $I_{CQ} = 99.7$ mA and $V_{CEQ} = 8.6$ V. For the transformer-coupled amplifier, the value of $r_C$ in the load line equations equals the transformer primary impedance ($Z_P$). By formula,

$$r_C = Z_P \tag{11.14}$$

Using a transposed version of equation (11.13), the value of $Z_P$ is found as

$$Z_P = \left(\frac{N_P}{N_S}\right)^2 Z_S \tag{11.15}$$

For the circuit shown in Figure 11.12a,

$$Z_P = \left(\frac{N_P}{N_S}\right)^2 Z_S = (4)^2(5 \ \Omega) = \mathbf{80 \ \Omega}$$

Now, using a value of $r_C = Z_P = 80 \ \Omega$, the end points for the ac load line are found to have the following values:

$$i_{c(sat)} = I_{CQ} + \frac{V_{CEQ}}{r_C} = 99.7 \text{ mA} + \left(\frac{8.6 \text{ V}}{80 \ \Omega}\right) = 99.7 \text{ mA} + 107.5 \text{ mA} = \mathbf{207.2 \text{ mA}}$$

and

$$v_{ce(off)} = V_{CEQ} + I_{CQ}r_C = 8.6 \text{ V} + (99.7 \text{ mA})(80 \text{ }\Omega) = \textbf{16.6 V}$$

The ac load line is now drawn between points on the *x*- and *y*-axes that correspond (roughly) to 207.2 mA and 16.6 V. The completed load line is shown in Figure 11.12b. Note that the *Q*-point was plotted using the values of $V_{CEQ}$ and $I_{CQ}$ calculated earlier.

At this point, you might be wondering how the circuit can have a value of $v_{ce(off)}$ that is greater than $V_{CC}$. The high value of $v_{ce(off)}$ is made possible by the **counter emf** developed across the transformer primary. You may recall that *counter emf* is *a voltage developed across a coil as its magnetic field collapses. The polarity of the counter emf is the opposite of the voltage that originally generated the magnetic field*. This voltage and its effects are illustrated in Figure 11.13.

**Counter emf**

A voltage developed across a coil as its magnetic field collapses. The polarity of the voltage is the opposite of the voltage that originally generated the magnetic field.

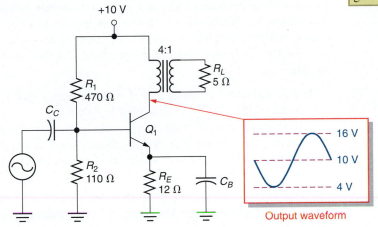

(a) The circuit from Figure 11.12a with an added output waveform

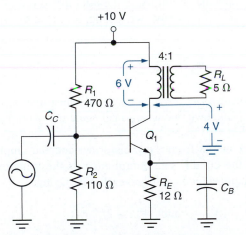

(b) The collector circuit voltages at the negative peak of the output waveform

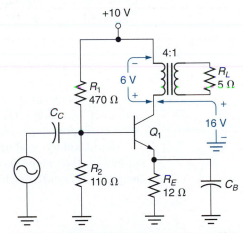

(c) The collector circuit voltages at the positive peak of the output waveform

FIGURE 11.13   Effects of counter emf.

Figure 11.13a shows the *ideal* circuit conditions for the amplifier in Figure 11.12a with an input signal applied. For ease of discussion, we will assume that the output is being driven back and forth between peak values of +16 and +4 V. Figure 11.13b shows the voltages in the collector circuit when the output signal is at its negative peak. Note that there is 6 V across the transformer primary with the polarity shown. When the output reaches its *positive* peak, the collector circuit has the voltages shown in Figure 11.3c. Note that there is still 6 V across the transformer primary, but also that *the polarity of the primary voltage has reversed*. This is the counter emf produced across the transformer primary as the collapsing magnetic field cuts through the transformer primary. As a result of this counter emf, the voltage at the collector of the transistor equals the sum of $V_{CC}$ and the voltage across the transformer primary, 16 V. As the magnetic field collapses into the transformer primary, the counter emf decreases in value until it reaches 0 V.

As you can see, a transformer-coupled amplifier has two very important characteristics:

1. $V_{CEQ}$ is very close to the value of $V_{CC}$.
2. The maximum output voltage is very close to $2V_{CEQ}$. Thus, it can approach the value of $2V_{CC}$.

As you will see, these characteristics are the basis for the higher efficiency rating for the transformer-coupled amplifier.

### 11.3.4  Amplifier Efficiency

The maximum theoretical efficiency of the transformer-coupled class A amplifier is 50%. (The derivation of this value is shown in Appendix D.) The actual efficiency rating of a transformer-coupled class A amplifier is generally less than 40%. There are several reasons for the difference between the practical and theoretical efficiency ratings for the amplifier:

1. The derivation of the $\eta = 50\%$ value assumes that $V_{CEQ} = V_{CC}$. In practice, the value of $V_{CEQ}$ is always somewhat less than $V_{CC}$.
2. The transformer is subject to various power losses (as you were taught in your study of ac electronics). Among these losses are *copper loss* and *hysteresis loss*. These transformer power losses are not considered in the derivation of the $\eta = 50\%$ value.

### 11.3.5  Calculating Maximum Load Power and Efficiency

To calculate the maximum load power for a transformer-coupled class A amplifier, start by determining the value of $I_{CQ}$ for the circuit using the established procedure for the dc biasing circuit. Once the value of $I_{CQ}$ is known, we can approximate the value of source power as

$$P_S \cong V_{CC}I_{CQ} \tag{11.16}$$

This approximation is valid because $I_{CQ} \gg I_1$ in a typical transformer-coupled class A amplifier. Once the value of $P_S$ is known, we have to calculate the maximum load power for the amplifier.

The maximum peak-to-peak output from the transformer-coupled amplifier is determined by the *compliance* of the circuit. The compliance of the transformer-coupled amplifier is found using the relationships established earlier in the text:

$$PP = 2V_{CEQ} \quad \text{and} \quad PP = 2I_{CQ}r_c$$

The lower of the two equation results is the compliance of the amplifier. Note that the compliance of the amplifier is the maximum possible peak-to-peak voltage *across the primary of the transformer*. Using this value and the turns ratio of the transformer, the maximum peak-to-peak load voltage is found as

$$V_{PP} = \frac{N_S}{N_P} PP \tag{11.17}$$

Once the value of peak-to-peak load voltage ($V_{PP}$) is known, the values of load power and efficiency are calculated as shown in Section 11.2. The entire process for calculating the maximum efficiency of a transformer-coupled class A amplifier is demonstrated in Example 11.7.

## EXAMPLE 11.7

Determine the maximum efficiency of the amplifier in Figure 11.12a.

*Solution:*   Using the voltage-divider bias relationships introduced in Chapter 7, we previously calculated values of $V_{CEQ} = 8.6$ V and $I_{CQ} = 99.7$ mA for the circuit. Ignoring the transistor base current, the current though $R_1$ (in the biasing circuit) can be found as

$$I_1 = \frac{V_{CC}}{R_1 + R_2} = \frac{10 \text{ V}}{580 \text{ }\Omega} = \textbf{17.2 mA}$$

The total current being drawn through the circuit is now found as

$$I_{CC} = I_1 + I_{CQ} \cong \textbf{117 mA}$$

and

$$P_S = V_{CC}I_{CC} = (10 \text{ V})(117 \text{ mA}) = \textbf{1.17 W}$$

The compliance of the amplifier is solved using

$$\textbf{PP} = 2V_{CEQ} = (2)(8.6 \text{ V}) = \textbf{17.2 V}$$

and

$$\textbf{PP} = 2I_{CQ}r_C = (2)(99.7 \text{ mA})(80 \text{ }\Omega) \cong \textbf{16 V}$$

The amplifier compliance is the lesser of the PP values. Using PP = 16 V, the maximum peak-to-peak load voltage is found as

$$V_{PP} = \frac{N_S}{N_P}\text{PP} = \frac{1}{4}(16 \text{ V}) = \textbf{4 V}$$

Now, the maximum load power is found as

$$P_{L(max)} = \frac{V_{PP}^2}{8R_L} = \frac{(4 \text{ V})^2}{40 \text{ }\Omega} = \textbf{400 mW}$$

Finally, the efficiency of the amplifier is found as

$$\eta = \frac{P_L}{P_S} \times 100 = \frac{400 \text{ mW}}{1.17 \text{ W}} \times 100 = \textbf{34.2\%}$$

Note that the actual efficiency of the amplifier is lower than the value calculated here, because the calculations are based on ideal values. They also ignore transformer power losses.

### PRACTICE PROBLEM 11.7

A transformer-coupled class A amplifier has the following values: $V_{CEQ} = 10$ V, $N_P = 5$, $N_S = 1$, $R_L = 4$ $\Omega$, $V_{CC} = 12$ V, and $I_{CQ} = 120$ mA. Calculate the maximum efficiency of the circuit.

### 11.3.6　One Final Note

The transformer-coupled class A amplifier has several advantages over the $RC$-coupled circuit. Three of these are:

1. The efficiency of the transformer-coupled circuit is higher.
2. The transformer can be used for impedance matching between the amplifier and its load.
3. The transformer-coupled circuit can be converted easily into a **tuned amplifier**, that is, an amplifier designed to provide gain over a specified range of frequencies.

**Tuned amplifier**

A circuit designed to have a specific value of gain over a specified range of frequencies.

Tuned amplifiers are discussed in Chapter 17.

In the next section, we will move on to the class B amplifier. A summary of the two class A amplifiers we have discussed is provided in Figure 11.14.

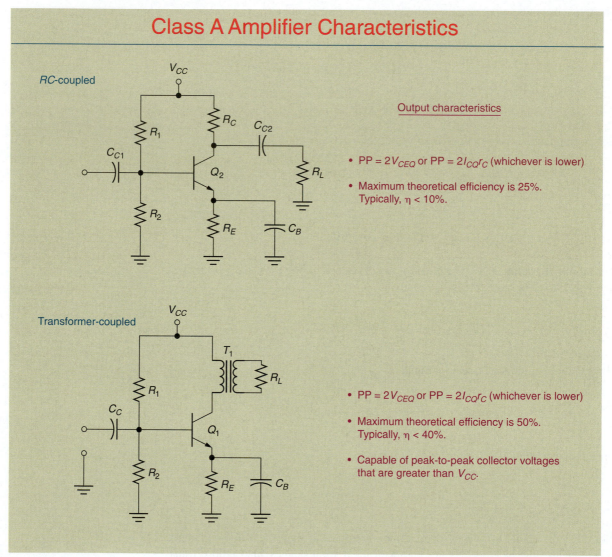

## Class A Amplifier Characteristics

*RC*-coupled

Output characteristics

- PP = $2V_{CEQ}$ or PP = $2I_{CQ}r_C$ (whichever is lower)
- Maximum theoretical efficiency is 25%. Typically, $\eta < 10\%$.

Transformer-coupled

- PP = $2V_{CEQ}$ or PP = $2I_{CQ}r_C$ (whichever is lower)
- Maximum theoretical efficiency is 50%. Typically, $\eta < 40\%$.
- Capable of peak-to-peak collector voltages that are greater than $V_{CC}$.

FIGURE 11.14

**Section Review ▶**

1. What is a *transformer-coupled class A amplifier*?
2. List the characteristics of the step-down transformer.
3. List the characteristics of the step-up transformer.
4. What is the process for plotting the ac load line of a transformer-coupled class A amplifier?

5. What is the maximum theoretical efficiency of the transformer-coupled class A amplifier? Why is the actual efficiency always less than this value?

6. List, in order, the steps required to calculate the maximum load power and efficiency rating of a transformer-coupled class A amplifier.

7. What are the advantages of using the transformer-coupled class A amplifier?

8. What is a tuned amplifier?

9. Using the circuit and graph in Figure 11.12, demonstrate that the dc load line for a transformer-coupled amplifier is nearly vertical. (*Hint*: What is $V_{CE\text{(off)}}$ for the amplifier?)

◀ **Critical Thinking**

## 11.4 Class B Amplifiers

The primary disadvantage of using class A power amplifiers is that their efficiency ratings are so low. As you have been shown, a majority of the power drawn from the dc power supply by a class A amplifier is used up by the amplifier itself. This goes against the primary purpose of a power amplifier, which is to transfer the power drawn from the dc power supply to a load.

The class B amplifier was developed to improve on the low efficiency rating of the class A amplifier. The maximum theoretical efficiency rating of a class B amplifier is approximately 78.5%. This means that up to 78.5% of the power drawn from the dc power supply can ideally be transferred to the load.

Unlike the class A amplifier, the class B amplifier consumes very little power when there is no input signal. This is a major improvement over the class A amplifier. However, because each transistor in a class B amplifier conducts for approximately 180°, each amplifier requires two transistors to reproduce an input waveform accurately.

Figure 11.15 shows the most commonly used type of class B configuration. This circuit configuration is referred to as a **complementary-symmetry amplifier**, or **push-pull emitter follower**. The circuit recognition feature is the use of *complementary transistors;* that is, one of the transistors is an *npn* and the other is a *pnp*. The biasing circuit components may change from one amplifier to another, but complementary-symmetry amplifiers always contain complementary transistors.

◀ *OBJECTIVE 7*

> **Complementary-symmetry amplifier (push-pull emitter follower)**
> A class B circuit configuration using *complementary transistors* (a pair, one *npn* and one *pnp*, with matched characteristics).

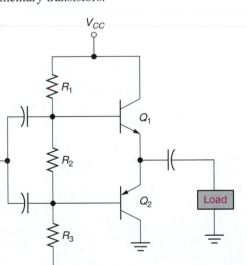

**FIGURE 11.15** Class B complementary-symmetry amplifier.

The **standard push-pull amplifier** contains two transistors of the same type with the emitters tied together. It uses a *center-tapped transformer*, or *a transistor phase splitter* on the input and a center-tapped transformer on the output. This amplifier type is shown in Figure 11.16. Note the transistor types and the transformer. This is the standard push-pull amplifier configuration.

> **Standard push-pull amplifier**
> A class B circuit that uses two identical transistors and a center-tapped transformer.

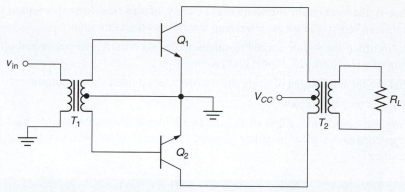

**FIGURE 11.16** Class B push-pull amplifier.

Why are complementary-symmetry amplifiers preferred?

Why is the complementary-symmetry configuration preferred over the standard push-pull? The center-tapped transformer makes the standard push-pull circuit much larger and more expensive to construct than the complementary-symmetry amplifier. Since the complementary-symmetry amplifier is by far the most commonly used, we will concentrate on this circuit configuration in our discussion on class B operation.

### 11.4.1   Class B Operation Overview

The term *push-pull* comes from the fact that the two transistors in a class B amplifier conduct on alternating half-cycles of the input signal. For example, consider the circuit shown in Figure 11.17. During the positive half-cycle of the input signal, $Q_1$ is biased *on* and $Q_2$ is biased *off*. During the negative half-cycle of the input signal, $Q_1$ is biased *off* and $Q_2$ is biased *on*. The fact that *both transistors are never fully conducting at the same time* is the key to the high efficiency rating of the amplifier. This point will be discussed in detail later in this section.

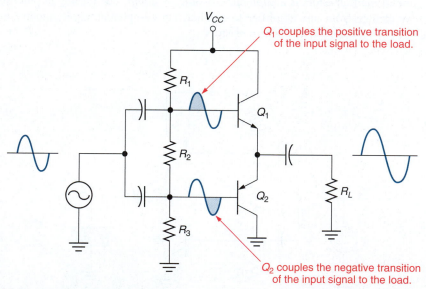

$Q_1$ couples the positive transition of the input signal to the load.

$Q_2$ couples the negative transition of the input signal to the load.

**FIGURE 11.17** Class B operation.

The biasing of the two transistors is the key to its operation. When the amplifier is in its *quiescent* state (no signal input), both transistors are biased at *cutoff*. When the input signal goes positive, $Q_1$ is biased above cutoff, and the transistor conducts. During this time, $Q_2$ is still biased at cutoff. When the input goes into its negative half-cycle, $Q_1$ is driven back into cutoff and $Q_2$ is biased above cutoff. As a result, conduction through $Q_2$ starts to increase while $Q_1$ remains off.

Because both transistors are biased in *cutoff*, class B amplifiers are subject to **crossover distortion**. Crossover distortion appears as flat lines between the output signal alternations, as shown in Figure 11.18. These flat lines are caused by both transistors being in cutoff for a short time between the positive and negative alternations of the input signal. Crossover distortion is eliminated by biasing both transistors at a level that is slightly *above* cutoff. This type of biasing, called *diode bias*, is discussed in the next section.

**Crossover distortion**
Distortion caused by class B transistor biasing. Crossover distortion occurs during the time neither of the transistors is fully conducting.

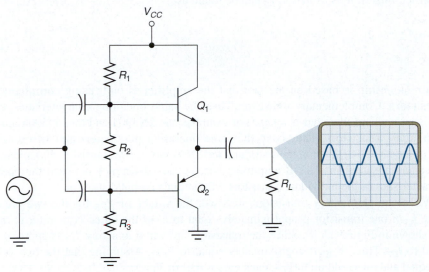

FIGURE 11.18   Crossover distortion.

## 11.4.2   DC Operating Characteristics

The class B amplifier has a vertical dc load line because there are no resistors in the emitter or collector circuits of the transistors. For example, consider the circuit shown in Figure 11.19a. Assuming that both transistors are biased right at the cutoff point, the Q-point is established at the $I_B = 0$ line on the collector characteristic curves. This is shown in Figure 11.19b. Now, assume that we could turn both transistors on at the same time. If they were both on, the following conditions would exist:

**1.** The voltage drops across the two transistors (from emitter to collector) would remain the same because the resistance *ratio* of the two components would not change.

**Lab Reference:** The dc characteristics of a class B amplifier are demonstrated in Exercise 14.

*A Practical Consideration:* Each transistor in Figure 11.19a would actually have a value of $I_C \cong I_{CO}$. $I_{CO}$ is the *collector cutoff current,* which is typically in the nanoampere (nA) range.

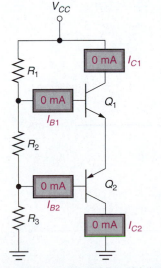

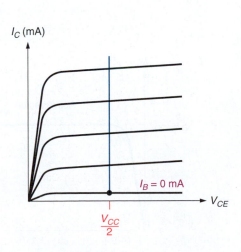

(a)  Approximate values of $I_B$ and $I_C$ in a quiescent class B amplifier

(b)  The dc load line for a class B amplifier

FIGURE 11.19   dc Load line for a class B amplifier.

**2.** The value of $I_C$ would be very high because no resistors are present in the emitter or collector circuits to restrict the current. Current would be limited only by the internal resistance of the conducting transistors.

Thus, the voltage across the transistors would remain fairly constant despite the increase in current through the collector and emitter circuits. This gives us a vertical dc load line.

The dc load line illustrates two other points about the dc operation of the class B amplifier. First, it tells us that $V_{CEQ}$ can be found as

$$V_{CEQ} = \frac{V_{CC}}{2}$$

(11.18)

> **Complementary transistors**
> Two BJTs, one *npn* and one *pnp*, with nearly identical electrical characteristics and ratings.

This relationship is based on the fact that the amplifier is built using **complementary transistors**. Complementary transistors have the same operating characteristics, except one is an *npn* and the other is a *pnp*. For example, the 2N3904 and the 2N3906 are complementary transistors. They have the same operating parameters and specifications, except the 2N3904 is an *npn* transistor and the 2N3906 is a *pnp* transistor. Complementary transistors are used in class B amplifiers because any difference between the operating characteristics of the individual transistors will produce output distortion.

When complementary transistors are used, the values of $V_{CE}$ for the two are equal when $I_C$ of one transistor is approximately equal to $I_C$ of the other. Now refer to the circuit shown in Figure 11.19. Since the transistors are wired in series, $I_{C1}$ is approximately equal to $I_{C2}$. Thus, $V_{CE1}$ is approximately equal to $V_{CE2}$. Assuming that the two voltages are equal and must add up to $V_{CC}$, each equals half of $V_{CC}$.

Another point to be made is shown in the following equation:

$$I_{CQ} \cong 0$$

(11.19)

This approximation is valid because both transistors are biased just inside the cutoff region. If they are both biased farther into the cutoff region, the value of $I_{CQ}$ approaches the ideal value, 0 A. To help you understand this point better, take a look at Figure 11.20. This is a "close-up" of the cutoff region of the characteristic curve for the transistors shown in Figure 11.19. For discussion purposes, we will refer to the two biasing points shown as *soft cutoff* and *hard cutoff*. When the transistors are biased at the soft-cutoff point, $I_{CQ}$ is at a higher level than that for the hard-cutoff point. As the transistor biasing is adjusted nearer the hard-cutoff point, the value of $I_{CQ}$ approaches zero. (It can never reach the ideal value of zero because there is always *some* amount of leakage current through the transistors.)

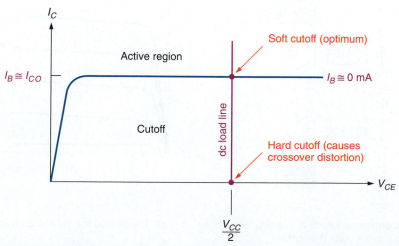

**FIGURE 11.20**   Magnified view of cutoff region.

It would seem that hard cutoff would be the ideal type of biasing to use on a class B amplifier. However, this is not the case. Biasing a class B amplifier at the hard-cutoff point causes crossover distortion. This is due to the transition time required for each transistor to come out of cutoff into the active region of operation. Biasing the transistors at the soft cutoff point reduces their transition time, thus eliminating crossover distortion.

What causes crossover distortion?

### 11.4.3    AC Operating Characteristics

The ac operating characteristics of the class B amplifier are significantly different from those of class A amplifiers. (This makes sense given the fact that there are virtually no physical similarities between the two types of circuits.) The ac characteristics of the class B amplifier are illustrated in Figure 11.21.

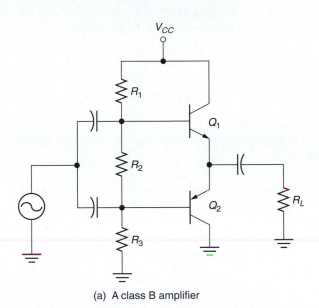

(a)  A class B amplifier

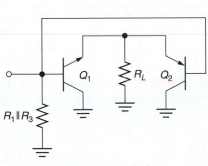

(b)  The ac equivalent

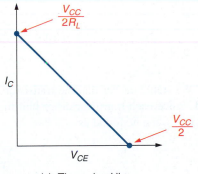

(c)  The ac load line

**FIGURE 11.21**    Class B ac characteristics.

Figure 11.21b shows the ac equivalent of the circuit in Figure 11.21a. You may recall that the ac equivalent is derived by grounding the dc voltage source ($V_{CC}$) and shorting the circuit capacitors. As shown in Figure 11.21b, the load is in parallel with the transistors. Therefore,

$$v_L = \Delta V_{CE} \tag{11.20}$$

and

$$i_L = \frac{\Delta V_{CE}}{R_L} \tag{11.21}$$

where $\Delta V_{CE}$ is the *change in* $V_{CE}$ that occurs as the result of an input signal to the amplifier.

**Lab Reference:** The ac operation of a class B amplifier is demonstrated in Exercise 14.

As you know, the value of $V_{CEQ}$ for the class B amplifier is found as

$$V_{CEQ} = \frac{V_{CC}}{2}$$

which was given earlier as equation (11.18). With this relationship in mind, look at the waveform shown in Figure 11.22. This waveform represents the maximum *ideal* output signal for the amplifier shown. The positive peak of the waveform is shown to equal $V_{CC}$. (This value assumes that $Q_1$ has a value of $V_{CE} = 0$ V when saturated.) Since $V_{CEQ}$ is half the value of $V_{CC}$, the maximum possible change in $V_{CE}$ is found as

$$\Delta V_{CE} = V_{CC} - \frac{V_{CC}}{2} = V_{CEQ}$$

or

$$\Delta V_{CE} = \frac{V_{CC}}{2} \qquad \textbf{(11.22)}$$

> Applying this approach to the negative alternation of the waveform gives us
>
> $$\Delta V_{CE} = V_{CEQ} - 0 \text{ V} = \frac{V_{CC}}{2}$$
>
> The fact that the calculations provide the same result supports the ac equivalent circuit, which shows the transistors to be wired in ac parallel. As such, their values of $\Delta V_{CE}$ *must* be equal; that is, the maximum change in voltage across either transistor equals the maximum change in voltage across the other.

Assuming that $Q_1$ is saturated at the positive peak of the output signal, the *ideal* value of $i_{c(\text{sat})}$ can be found as

$$i_{c(\text{sat})} = \frac{V_{CC}}{2R_L} \qquad \textbf{(11.23)}$$

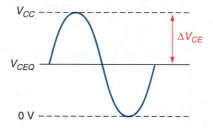

**FIGURE 11.22**

We stated earlier that the transistors in a class B amplifier are normally biased at cutoff. Since each transistor drops half the supply voltage when both are in cutoff, the value of $v_{ce(\text{off})}$ can be found as

$$v_{ce(\text{off})} = \frac{V_{CC}}{2} \qquad \textbf{(11.24)}$$

These two points are used to derive the ac load line of the class B amplifier. This load line is illustrated in Figure 11.21c.

### EXAMPLE 11.8

The amplifier in Figure 11.21a has values of $V_{CC} = 10$ V and $R_L = 10$ $\Omega$. Determine the end-point values for its ac load line.

*Solution:* The value of $i_{c(\text{sat})}$ is found as

$$i_{c(\text{sat})} = \frac{V_{CC}}{2R_L} = \frac{10 \text{ V}}{2(10 \ \Omega)} = \textbf{500 mA}$$

Now, the value of $v_{ce(off)}$ is found as

$$v_{ce(off)} = \frac{V_{CC}}{2} = \frac{10\text{ V}}{2} = 5\text{ V}$$

Thus, the ac load line for the circuit, which is shown in Figure 11.21c, has end-point values of $i_{c(sat)} = 500$ mA and $v_{ce(off)} = 5$ V.

**PRACTICE PROBLEM 11.8**

A class B amplifier has values of $V_{CC} = +12$ V and $R_L = 2.2$ kΩ. Calculate the end-point values for the circuit's ac load line.

### 11.4.4   Amplifier Impedance

You may recall that the base input impedance of an emitter follower is found as

$$Z_{in(base)} = h_{fc}(r_e' + r_E)$$

If you take a look at the class B amplifier, you will notice that the load resistor is connected to the emitters of the two transistors. Since the load resistor is not bypassed, its value must be considered in the calculation of $Z_{in(base)}$ for the amplifier. The transistor input impedance is therefore found as

$$Z_{in(base)} = h_{fc}(r_e' + R_L) \tag{11.25}$$

> *Don't forget:*
> $r_e'$ is the ac resistance of the transistor emitter, found as
> $$r_e' = \frac{h_{ic}}{h_{fc}}$$

The output of the class B amplifier is taken from the emitters of the transistors, so the amplifier output impedance is equal to the ac resistance of the emitter circuit. As you may recall from Chapter 10, this impedance is found as

$$Z_{out} = r_e' + \frac{R_{in}'}{h_{fc}} \tag{11.26}$$

where

$$R_{in}' = R_1 \parallel R_3 \parallel R_S$$

Note that the value of $R_2$ in the class B amplifier is not used in equation (11.26). This is because it is shorted by the input coupling capacitors in the ac equivalent circuit of the amplifier.

### 11.4.5   Amplifier Gain

Since the complementary-symmetry amplifier is basically an *emitter follower*, the current gain is found as with any emitter follower. By formula,

$$A_i = h_{fc}\left(\frac{Z_{in}r_E}{Z_{in(base)}R_L}\right)$$

Since $r_E = R_L$ for the class B amplifier, the equation for $A_i$ can be simplified to

$$A_i = h_{fc}\left(\frac{Z_{in}}{Z_{in(base)}}\right) \tag{11.27}$$

The voltage gain of the class B amplifier is found in the same manner as a standard emitter follower, as follows:

$$A_v = \frac{R_L}{R_L + r'_e}$$

As with any amplifier, the power gain is the product of $A_v$ and $A_i$. By formula,

$$A_p = A_v A_i$$

### 11.4.6    Power Calculations

OBJECTIVE 8 ▶ The output power for the class B amplifier is found in the same manner as it is for as the class A amplifier. By formula,

$$P_L = \frac{V_{PP}^2}{8R_L}$$

The maximum load power is also found in the same manner as it is for the class A amplifier. By formula,

$$P_{L(max)} = \frac{PP^2}{8R_L}$$

To calculate the maximum possible load power for a class B amplifier, we need to be able to determine its *compliance*. The compliance of a class B amplifier is found as

$$PP = 2V_{CEQ} \qquad\qquad (11.28)$$

Since $V_{CEQ} \cong V_{CC}/2$, equation (11.28) can be rewritten as

$$PP \cong V_{CC} \qquad\qquad (11.29)$$

The maximum load power for a class B amplifier is calculated as shown in Example 11.9.

### EXAMPLE 11.9

Determine the maximum load power for the circuit shown in Figure 11.23.

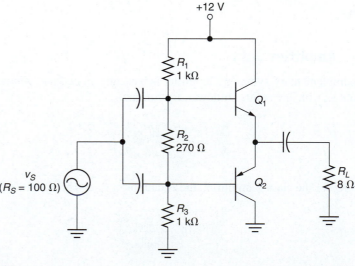

FIGURE 11.23

*Solution:* The compliance of the amplifier is found as

$$PP \cong V_{CC} = 12 \text{ V}$$

Now, the maximum load power is found as

$$P_{L(\max)} = \frac{PP^2}{8R_L} = \frac{(12 \text{ V}_{PP})^2}{8(8 \text{ }\Omega)} = \frac{144 \text{ V}^2}{64 \text{ }\Omega} = 2.25 \text{ W}$$

### PRACTICE PROBLEM 11.9
Determine the maximum load power for the circuit described in Practice Problem 11.8.

---

The total power that the amplifier draws from its dc power supply is found as

$$P_S = V_{CC}I_{CC}$$

where

$$I_{CC} = I_{C1(\text{ave})} + I_1 \qquad\qquad (11.30)$$

The equation for $P_S$ is the same one that is used for the class A amplifier. However, equation (11.30) needs some explaining.

The total current drawn from the supply is the sum of the *average* $Q_1$ collector current and the current through the amplifier base circuit, as shown in Figure 11.24. The average value of the current through the collector of $Q_1$ is given as

$$I_{C(\text{ave})} = \frac{I_{\text{pk}}}{\pi}$$

or

$$I_{C(\text{ave})} \cong 0.318 I_{\text{pk}}$$

where $I_{\text{pk}}$ is the peak current through the transistor. Note that this is the standard $I_{\text{ave}}$ equation for the half-wave rectifier. (Since the transistor is on for alternating half cycles, it effectively acts as a half-wave rectifier.)

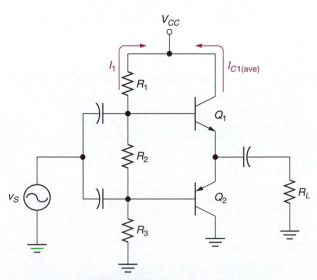

**FIGURE 11.24** Class B amplifier supply current.

According to equation (11.23) and Figure 11.21c, the maximum current through either transistor in the class B amplifier is found as

$$i_{c(\text{sat})} = \frac{V_{CC}}{2R_L}$$

Substituting this value into the first $I_{C(\text{ave})}$ equation (above), we get

$$I_{C1(\text{ave})} = \frac{V_{CC}}{2\pi R_L} \qquad \textbf{(11.31)}$$

Equation (11.31) uses $V_{CC}$ for the output voltage of the amplifier because it is assumed that the amplifier is driven to compliance. If the output of the amplifier is not at compliance, $V_{PP(\text{out})}$ must be substituted for $V_{CC}$. This equation for calculating $I_{C1(\text{ave})}$ is

$$I_{C1(\text{ave})} = \frac{V_{PP(\text{out})}}{2\pi R_L} \qquad \textbf{(11.32)}$$

Examples 11.10 and 11.11 demonstrate the process of determining the total power drawn from the supply and total load power.

### EXAMPLE 11.10

Determine the value of $P_S$ for the circuit shown in Figure 11.25.

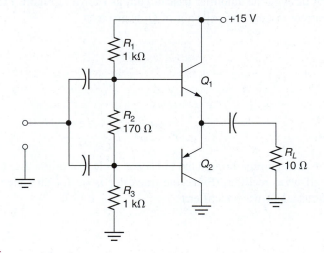

FIGURE 11.25

***Solution:*** Neglecting the base current from the two transistors, $I_1$ is found as

$$I_1 = \frac{V_{CC}}{R_1 + R_2 + R_3} = \frac{15\ \text{V}}{2.17\ \text{k}\Omega} = \textbf{6.91 mA}$$

Now, $I_{C(\text{ave})}$ is found as

$$I_{C(\text{ave})} = \frac{V_{CC}}{2\pi R_L} = \frac{15\ \text{V}}{2\pi(10\ \Omega)} = \textbf{238.7 mA}$$

Using these two values, $I_{CC}$ is found as

$$I_{CC} = I_{C1(\text{ave})} + I_1 = \textbf{245.6 mA}$$

Finally, the total power demand on the supply is determined as

$$P_S = V_{CC}I_{CC} = (15 \text{ V})(245.6 \text{ mA}) = \textbf{3.68 W}$$

**PRACTICE PROBLEM 11.10**

Refer to Figure 11.23. Determine the value of $P_S$ for the circuit.

---

**EXAMPLE 11.11**

Determine the maximum load power for the circuit in Figure 11.25.

*Solution:* The compliance of the amplifier is equal to $V_{CC}$, or 15 V. Using this value, the maximum load power is calculated as

$$P_{L(\text{max})} = \frac{PP^2}{8R_L} = \frac{(15 \text{ V}_{PP})^2}{8(10 \text{ }\Omega)} = \textbf{2.81 W}$$

**PRACTICE PROBLEM 11.11**

Refer to Figure 11.23. Determine the maximum value of $P_L$ for the circuit.

---

## 11.4.7 Class B Amplifier Efficiency

It was stated earlier in this chapter that the maximum theoretical efficiency of a class B amplifier is 78.5%. As is the case with class A amplifiers, any practical efficiency rating is always less than the maximum theoretical value.

The derivation of the $\eta = 78.5\%$ value is shown in Appendix D. If you take a look at the derivation, you will see it assumes that the compliance of the class B amplifier is equal to $V_{CC}$. Since both transistors in the class B amplifier still have a slight value of $V_{CE}$ when saturated, the actual compliance of the amplifier is slightly less than the value of $V_{CC}$. Thus, the class B amplifier efficiency rating never reaches the value of 78.5%.

Once the values of $P_S$ and $P_L$ for a given class B are known, the efficiency of the circuit is calculated in the same manner as it is for the class A amplifiers. This point is illustrated in Example 11.12.

**EXAMPLE 11.12**

Determine the efficiency of the amplifier used in Examples 11.10 and 11.11 (Figure 11.25).

*Solution:* The values of load power and dc supply power were calculated in Examples 11.10 and 11.11 as

$$P_L = 2.81 \text{ W} \quad \text{and} \quad P_S = 368 \text{ W}$$

Using these two values, the maximum efficiency of the amplifier is found as

$$\eta = \frac{P_L}{P_S} \times 100 = \frac{2.81 \text{ W}}{3.68 \text{ W}} \times 100 = \textbf{76.36\%}$$

**PRACTICE PROBLEM 11.12**

Determine the efficiency of the amplifier described in Practice Problems 11.10 and 11.11.

---

Example 11.13 shows how the efficiency of the class B amplifier changes when the output signal is reduced.

### EXAMPLE 11.13

Determine the efficiency of the amplifier used in Example 11.10 (Figure 11.25) if the load voltage is reduced by 50% to 7.5 $V_{PP}$.

*Solution:* The value of $I_1$ remains at 6.91 mA (as calculated in Example 11.10) because the biasing network has not changed. Since the amplifier is no longer driven to compliance, the value of $I_{C(ave)}$ is calculated as follows:

$$I_{C(ave)} = \frac{V_{PP(out)}}{2\pi R_L} = \frac{7.5 \text{ V}}{2\pi(10 \text{ }\Omega)} = \textbf{119.4 mA}$$

Thus, the circuit has a total current of

$$I_{CC} = I_1 + I_{C(ave)} = 6.91 \text{ mA} + 119.4 \text{ mA} = \textbf{126.31 mA}$$

Next, the total power drawn from the power supply is calculated. $V_{CC}$ is used for this calculation because it stays constant, despite the change in output voltage. Therefore,

$$P_S = V_{CC}I_{CC} = (15 \text{ V})(126.31 \text{ mA}) = \textbf{1.89 W}$$

Next, the maximum load power is found as

$$P_{L(max)} = \frac{V_{PP}^2}{8R_L} = \frac{(7.5 \text{ V})^2}{8(10 \text{ }\Omega)} = \textbf{703 mW}$$

Finally, the amplifier efficiency is found as

$$\eta = \frac{P_L}{P_S} \times 100 = \frac{703 \text{ mW}}{1.89 \text{ W}} \times 100 = \textbf{37.2\%}$$

### PRACTICE PROBLEM 11.13

Determine the efficiency of the amplifier described in Example 11.13 if the load voltage is 11 $V_{PP}$.

## 11.4.8   Summary

The class B amplifier is a two-transistor circuit that has a higher maximum efficiency rating than common class A amplifiers. There are two types of class B amplifiers: the *push-pull amplifier* and the *complementary-symmetry amplifier*. Of the two, the complementary-symmetry amplifier is more commonly used for two reasons:

1. The complementary-symmetry amplifier does not require the use of transformers and, thus, is cheaper to produce.
2. Since the complementary-symmetry amplifier doesn't have a transformer, it is not subject to transformer losses. Thus, it has higher efficiency than a comparable push-pull amplifier.

The two transistors in a class B amplifier are biased at cutoff. When a signal is applied to the amplifier, the positive alternation of the signal turns one transistor on, and the negative alternation turns the other transistor on. The resulting amplifier output is a signal that has a peak-to-peak value approximately equal to $V_{CC}$ (when driven to compliance). The dc and ac characteristics of the complementary-symmetry amplifier are summarized in Figure 11.26.

*A Practical Consideration:* Transformer inductive reactance and power losses both increase as frequency increases. For this reason, complementary-symmetry amplifiers can be operated at higher frequencies than comparable push-pull circuits.

# Class B Complementary-Symmetry Amplifiers

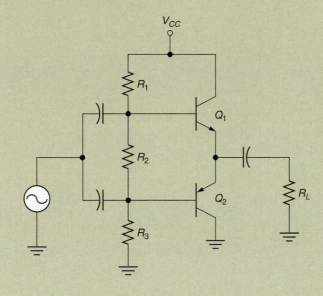

**Primary dc relationships**

$$V_{CEQ} = \frac{V_{CC}}{2}$$

$$I_{CQ} \cong 0 \text{ A}$$

$$I_{C(ave)} = \frac{V_{PP}}{2\pi R_L}$$

$$I_{CC} = I_{C(ave)} + I_1$$

**Primary power relationships**

$$P_L = \frac{V_{PP}^2}{8R_L}$$

$$PP \cong V_{CC}$$

$$P_{L(max)} = \frac{PP^2}{8R_L}$$

$$P_S = V_{CC}I_{CC}$$

$$\eta = \frac{P_L}{P_S} \times 100$$

Advantages:     Higher efficiency ratings than class A amplifiers.

Disadvantages:    Subject to crossover distortion.
                       Requires two transistors to operate.

**FIGURE 11.26**

---

◀ **Section Review**

1. What advantage does the class B amplifier have over the class A amplifier?
2. What are the two common types of class B amplifiers? What are the advantages and disadvantages of each?
3. What are *complementary transistors*? What kind of amplifier requires the use of complementary transistors?
4. Describe the operating characteristics of the class B amplifier.
5. What are the typical values of $I_{CQ}$ and $V_{CEQ}$ for a class B amplifier?
6. What is the typical compliance of a class B amplifier?
7. Why is the practical efficiency rating of a class B amplifier less than the maximum theoretical value of 78.5%?

## 11.5   Class AB Amplifiers (Diode Bias)

Up to this point, we have used voltage-divider bias for all our class B amplifiers. Problems can develop with the class B amplifier when voltage-divider bias is used:

1. Crossover distortion can occur.
2. Thermal runaway can occur.

**Diode bias**

A biasing circuit that uses two diodes in place of the resistor(s) between the bases of the two transistors.

**Compensating diodes**

Diodes used in the bias circuit of a class AB amplifier with forward voltage ratings that match the transistor $V_{BE}$ ratings.

A biasing circuit often used to eliminate the problems of crossover distortion and thermal runaway is shown in Figure 11.27. The circuit shown, called **diode bias**, uses two diodes in place of the resistor(s) between the transistor bases. The diodes in the bias circuit, called **compensating diodes**, are chosen to match the characteristic values of $V_{BE}$ for the two transistors. As you will see, the diodes eliminate both crossover distortion and thermal runaway when they are properly matched to the amplifier transistors.

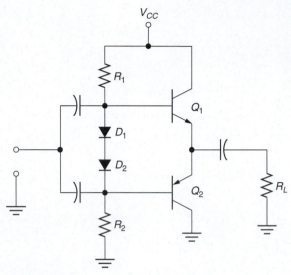

**FIGURE 11.27** Diode biasing.

When diode bias is used, an amplifier like the one in Figure 11.27 is referred to as a *class AB amplifier*. This type of operation is defined and discussed in detail later in this section.

### 11.5.1   Diode Bias DC Characteristics

You may recall that the transistors in a class B amplifier are biased at cutoff, causing the value of $I_{CQ}$ for the amplifier to be approximately equal to zero. When diode bias is used, the transistors are actually biased just above cutoff; that is, there is some measurable value of $I_{CQ}$ when diode bias is used. This point is illustrated with the help of Figure 11.28.

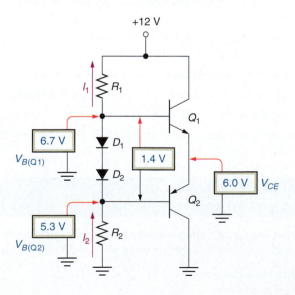

**FIGURE 11.28**

To understand the operation of the circuit shown, we must start with a few assumptions:

1. $V_{CEQ}$ is approximately half the value of $V_{CC}$, as is shown in the figure.
2. The current through $R_2$ causes 5.3 V to be developed across this resistor.

Assuming the above conditions are met, the base of $Q_2$ is at 5.3 V and $V_E$ of $Q_2$ equals 6 V. Since $V_B$ for this *pnp* transistor is 0.7 V more negative than $V_E$, $Q_2$ conducts.

With 1.4 V developed across the biasing diodes, $V_B$ of $Q_1$ is at 5.3 V + 1.4 V = 6.7 V. The value of $V_E$ for $Q_1$ is 6 V, as is shown in the figure. Since $V_B$ for this *npn* transistor is 0.7 V more positive than $V_E$, $Q_1$ also conducts. Thus, both transistors in the diode-biased amplifier conduct, and some measurable value of $I_C$ is present, as shown in Figure 11.29.

**Lab Reference:** The dc operation of a class AB amplifier is demonstrated in Exercise 14.

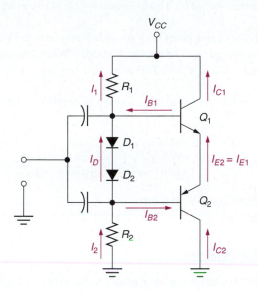

**FIGURE 11.29**

Now, let's take a look at the assumptions we've made and where they came from. Even though we have changed the transistor biasing circuit, the output transistors are in the same configuration as they were in the standard class B amplifier. Thus, the dc load line characteristics of the amplifier haven't changed, and $V_{CEQ}$ is always approximately half of $V_{CC}$.

To understand where the assumed value of $V_{R2}$ = 5.3 V came from, we have to look at the operation of the biasing circuit. The total voltage across the two resistors in the biasing circuit equals the difference between $V_{CC}$ and the 1.4 V drop across the diodes. By formula,

$$V_{R1} + V_{R2} = V_{CC} - 1.4 \text{ V}$$

Using the voltage-divider equation, we can find the value of $V_{R2}$ as

$$V_{R2} = \frac{R_2}{R_1 + R_2} (V_{CC} - 1.4 \text{ V})$$

Since $V_{B(Q2)} = V_{R2}$, the equation above can be rewritten as

$$V_{B(Q2)} = \frac{R_2}{R_1 + R_2} (V_{CC} - 1.4 \text{ V}) \tag{11.33}$$

When $R_1 = R_2$ (as was the case with the circuit in Figure 11.28), the fraction in equation (11.33) equals ½, and the equation can be rewritten as

$$V_{B(Q2)} = \frac{V_{CC}}{2} - 0.7 \text{ V}$$

*Don't Forget:*

$$V_{CEQ} \cong \frac{V_{CC}}{2}$$

or

$$V_{B(Q2)} = V_{CEQ} - 0.7 \text{ V} \quad (\text{when } R_1 = R_2) \qquad \textbf{(11.34)}$$

*A Practical Consideration:*
The values of the biasing resistors must be low enough to allow the diodes and the transistor base-emitter junctions to conduct. Normally, the biasing resistors will be in the low-kΩ range.

Equation (11.34) is important for several reasons: First, it points out that $Q_2$ (and, thus, $Q_1$) is automatically biased properly when $R_1 = R_2$. The exact values of the resistors are relatively unimportant as long as they are equal. When $R_1 = R_2$, you can assume that the base voltages are equal to $V_{CEQ} \pm 0.7$ V. When $R_1 \neq R_2$, you must use equation (11.30) to find the value of $V_{B(Q2)}$ and then add 1.4 V to this value to find $V_{B(Q1)}$. The second reason that equation (11.34) is so important is because almost all diode-biased circuits are designed with equal-value resistors. Thus, equation (11.34) is used far more often than equation (11.33).

You may be wondering at this point why we didn't consider the values of $I_{B1}$ and $I_{B2}$ in our analysis of the biasing circuit. The reason is simple: When the transistors are properly matched, $I_{B1}$ and $I_{B2}$ are equal in value. Therefore, the resistor currents ($I_1$ and $I_2$) are also equal in value.

When diode bias is used, the value of $I_1$ (which is needed in the calculation of $I_{CC}$) is found as

$$I_1 = \frac{V_{CC} - 1.4 \text{ V}}{R_1 + R_2} \qquad \textbf{(11.35)}$$

Again, we do not need to consider the effects of $I_{B1}$ and $I_{B2}$ since they do not affect the resistor current values.

### 11.5.2   Class AB Operation

OBJECTIVE 9 ▶ You have been told that $I_{CQ}$ has some measurable value when diode bias is used. For this reason, the amplifier cannot technically be classified as a class B amplifier. This point is illustrated with the help of Figure 11.30, which shows a typical input waveform for the circuit in Figure 11.29.

The alternations shown in Figure 11.30 are offset to illustrate a point. In reality, the waveform is identical to any other sinusoidal waveform.

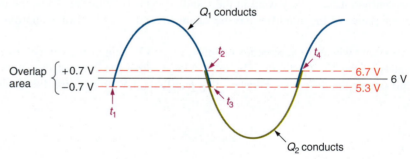

**FIGURE 11.30**

To simplify our discussion, we are going to make an assumption: A transistor conducts until its base and emitter voltages are equal, at which time it turns off. With this in mind, let's look at the circuit's response to the waveform shown in Figure 11.30.

$Q_1$ (in Figure 11.29) conducts as long as its base voltage is more positive than 6 V. The value of $V_B$ drops to 6 V at $t_1$ and $t_3$ because the $-0.7$ V value of $v_{in}$ subtracts from the 6.7 V value of $V_{B(Q1)}$. Thus, we can assume that $Q_1$ conducts for the entire time between $t_1$ and $t_3$. The same principle applies to $Q_2$. At $t_2$ and $t_4$, $Q_2$ turns off, because the $+0.7$ V value of $v_{in}$ adds to the value of $V_{B(Q2)}$, causing $V_B$ and $V_E$ to be equal. Thus, $Q_2$ conducts for the entire time between $t_2$ and $t_4$.

**Class AB amplifier**
An amplifier that typically contains two (or more) transistors, each conducting for more than 180° and less than 360° of the input cycle.

As you can see, the transistors in the diode bias circuit each conduct for *slightly more than* 180°. Because of this, the circuit is classified as a **class AB amplifier**. In class AB operation, the transistors conduct for a portion of the input cycle that is greater than 180° and less than 360°.

While the diode bias circuit is technically classified as a class AB amplifier, most technicians simply refer to it as a class B amplifier. This is because of the strong similarities in operation between the two types of amplifiers.

## 11.5.3 Eliminating Crossover Distortion

◀ *OBJECTIVE 10*

Since both transistors in the class AB amplifier are conducting when the input signal is at 0 V, the amplifier does not have the crossover distortion problems that the class B amplifier may have. Crossover distortion occurs only when *both* transistors are in cutoff, so the problem does not normally occur in class AB amplifiers.

## 11.5.4 Eliminating Thermal Runaway

For diode bias to eliminate the problem of thermal runaway, two conditions must be met:

1. The diodes and the transistor base-emitter junctions must be very nearly *perfectly* matched.
2. The diodes and the transistors must be in **thermal contact**. This means that they must be in physical contact with each other (or a common surface) so that the operating temperatures of the diodes always equal the operating temperatures of the transistors.

**Thermal contact**
Placing two or more components in physical contact with each other (or a common surface) so that their operating temperatures are equal.

When the diodes and transistors are matched, the forward voltage drops across the diodes ($V_{F1}$ and $V_{F2}$) are approximately equal to the $V_{BE}$ drops of the transistors (all other factors being equal). In other words, as long as their operating temperatures are the same,

$$V_{F1} = V_{BE(Q1)}$$

and

$$V_{F2} = V_{BE(Q2)}$$

When the transistors are in thermal contact with the diodes, all the components experience the same operating temperature. Thus, if the operating temperatures of the transistors increase by a given amount, the operating temperatures of the diodes increase by the same amount, and the values of $V_F$ and $V_{BE}$ remain approximately equal. With this in mind, let's take a look at the circuit response to an increase in temperature.

You may recall from our discussion on *temperature versus diode conduction* that an increase in diode temperature causes a slight increase in diode current and a slight decrease in the value of $V_F$. Assuming that the increases in transistor and diode temperature are equal, here's what happens:

1. The increase in diode temperature causes the 1.4 V drop across $D_1$ and $D_2$ to *decrease*. At the same time, $I_D$ *increases* (see Figure 11.29).
2. With the decrease in diode voltage, $I_1$ and $I_2$ *increase*. This causes the voltages across $R_1$ and $R_2$ to *increase*.
3. With the increases in the resistor voltages, the value of $V_{B(Q1)}$ *decreases* [$V_{B(Q1)} = V_{CC} - V_{R1}$], and the value of $V_{B(Q2)}$ *increases*.
4. The changes in base voltage bring the values of $V_B$ closer to the value of $V_E$ for the two transistors, reducing the forward bias on the transistors.
5. The reduction in forward bias decreases $I_B$, causing a decrease in $I_C$. The decrease in $I_C$ prevents thermal runaway from occurring.

As you can see, diode bias reduces the chance of thermal runaway. The diodes can be placed in thermal contact with the power transistors in one of several ways:

1. Attached to the heat-sink tab of a power transistor.
2. Attached to the heat sink on which a transistor is mounted.

When replacing power transistors in a class AB amplifier, you must reattach the compensating diodes to their original location for continued thermal protection.

Some class B and AB amplifiers have two additional resistors in the emitter output circuit. These resistors, shown in Figure 11.31, act as *swamping resistors*, reducing the effects of minor characteristic differences between the matched transistor pair. The resistors typically have low values, in the range of 0.1 to 0.47 Ω. Because the swamping resistors form voltage dividers with the load, any replacement resistor must have the same value as the original, or the result will be a distorted output.

*A Practical Consideration:* Swamping resistors like those shown in Figure 11.31 are available in 3-pin, 2-resistor ICs. These resistors, which are called *cement metal plate resistors*, are typically rated at 2 W or 5 W.

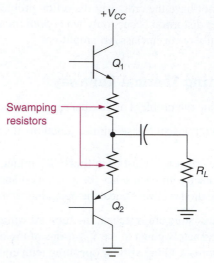

**FIGURE 11.31** Emitter output circuit.

### 11.5.5 Class AB Amplifier Analysis

OBJECTIVE 11 ▶ We can now use the information on dc biasing to approach the process for analyzing a class AB amplifier. For this discussion, we will use the example circuit shown in Figure 11.32a.

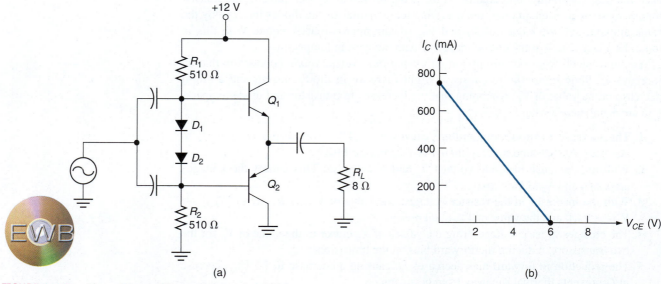

(a)                                           (b)

**FIGURE 11.32** Class AB amplifier and its ac load line.

**Lab Reference:** The ac operation of the class AB amplifier is demonstrated in Exercise 14.

We will start our analysis by determining the values of $i_{c(\text{sat})}$ and $v_{ce(\text{off})}$. For this circuit, $i_{c(\text{sat})}$ is found as

$$i_{c(\text{sat})} = \frac{V_{CC}}{2R_L} = \frac{12\ \text{V}}{16\ \Omega} = \textbf{750 mA}$$

and $v_{ce(off)}$ is found as

$$v_{ce(off)} = \frac{V_{CC}}{2} = 6 \text{ V}$$

Using these two values, the ac load line in Figure 11.32b is plotted.

Next, we determine the value of $I_1$. The value of $I_1$ is found as

$$I_1 = \frac{V_{CC} - 1.4 \text{ V}}{R_1 + R_2} = \frac{12 \text{ V} - 1.4 \text{ V}}{1020 \text{ }\Omega} = 10.4 \text{ mA}$$

The average current in the collector circuit of the amplifier is found as

$$I_{C1(ave)} = \frac{V_{CC}}{2\pi R_L} = \frac{12 \text{ V}}{2\pi(8 \text{ }\Omega)} = 238.7 \text{ mA}$$

Using this value and the value of $I_1$ calculated earlier, we can find the value of $I_{CC}$ as

$$I_{CC} = I_{C1(ave)} + I_1 = 238.7 \text{ mA} + 10.4 \text{ mA} = 249.1 \text{ mA}$$

Now, we can calculate the total power drawn from the supply as

$$P_S = V_{CC}I_{CC} = (12 \text{ V})(249.1 \text{ mA}) = 2.99 \text{ W}$$

Assuming the amplifier is being driven to compliance ($V_{PP} = V_{CC}$), we can calculate the maximum load power as

$$P_L \cong \frac{V_{PP}^2}{8R_L} = \frac{(12 \text{ V})^2}{8(8 \text{ }\Omega)} = 2.25 \text{ W}$$

Finally, the maximum efficiency of the amplifier is found as

$$\eta = \frac{P_L}{P_S} \times 100 = \frac{2.25 \text{ W}}{2.99 \text{ W}} \times 100 = 75.25\%$$

As you can see, the basic analysis of a class AB amplifier is actually fairly simple. In fact, with the exception of the $I_1$ calculation, it is identical to the analysis of the standard class B amplifier. When you deal with a standard class B amplifier, $I_1$ is found by dividing $V_{CC}$ by the total series base resistance. When dealing with the class AB amplifier, you must take the diode voltage drops into account.

The class AB amplifier is used far more commonly than the standard class B amplifier. For this reason, we will concentrate on the class AB amplifier from this point on. Just remember, except for the biasing circuit and the value of $I_{CQ}$, the class AB and the class B amplifiers are nearly identical. In fact, the term *class B* is commonly used to describe both amplifier types.

---

1. What are the two primary problems that can occur in standard class B amplifiers? What type of circuit is used to eliminate these problems? ◀ **Section Review**

2. Explain how the diode bias circuit develops the proper values of $V_B$ for the amplifier transistors.

3. How do you find the values of $V_B$ in a diode bias circuit when $R_1 = R_2$?

4. How do you find the values of $V_B$ in a diode bias circuit when $R_1 \neq R_2$?

5. What is *class AB operation*?

6. How does the class AB amplifier eliminate crossover distortion?

**7.** How does the class AB amplifier eliminate thermal runaway?

**8.** List, in order, the steps you would take to perform the complete analysis of a class AB amplifier.

**9.** Why is the class AB amplifier often referred to as a class B amplifier?

## 11.6 Class AB Amplifiers: Troubleshooting and Circuit Configurations

In this section, we will concentrate on some of the faults that may develop in the diode-biased complementary-symmetry amplifier. We also discuss some techniques you can use to isolate these faults. At the end of the section, you will be introduced to some other biasing circuits for class AB amplifiers. Although they are a bit more complex than the basic diode bias circuit, they each serve their own purpose. Do not let the more complex circuits fool you. They may contain more components than the amplifiers discussed previously, but they work in the same basic fashion.

### 11.6.1 Troubleshooting

Class AB amplifiers can be more difficult to troubleshoot than class A amplifiers because of the dual-transistor configuration. Any of the standard transistor faults can develop in either of the two transistors. However, there are some techniques you can use that will make troubleshooting these amplifiers relatively simple.

The first step in troubleshooting a class AB amplifier is the same as with any amplifier: *You must make sure that the amplifier is the source of the trouble.* For example, assume that the two-stage amplifier shown in Figure 11.33 is not working (there is no output signal). For this circuit, *three* checks must be made before you can assume there is a problem with the class AB amplifier:

**1.** You must verify that the $V_{CC}$ and ground connections in the class AB amplifier are good.

**2.** You must disconnect the load to make sure that it is not loading down the amplifier and preventing it from working.

**3.** You must be sure that the class AB amplifier is getting the proper input from the first stage.

**Lab Reference:** The operation of a two-stage amplifier similar to the one in Figure 11.33 is demonstrated in Exercise 15.

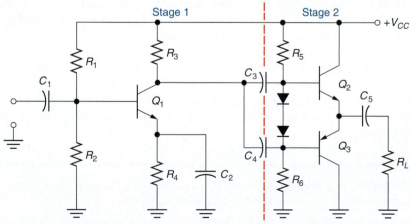

**FIGURE 11.33**

Picture this: You spend an hour troubleshooting the class AB amplifier, only to discover that the Stage 1 amplifier was the source of the fault! (This may sound a bit far-fetched, but 99 out of 100 technicians would admit to having done something like this at one time or another).

Many times, an amplifier malfunction is caused by its supporting circuitry. By checking each of the conditions noted, the process of troubleshooting can be greatly simplified. The first two places to check when troubleshooting an amplifier are $V_{CC}$ and ground. By verifying that you have the proper $V_{CC}$ and ground connections, you have eliminated any problem that could have been caused by blown fuses, disconnected plugs, or a faulty power supply.

Next, make sure that the load is not the source of the problem. A shorted load will always prevent an amplifier from having an output. If you disconnect the load and the amplifier starts to work properly, then the load is the cause of the problem. Once the load has been verified as being good, check the output from the *source amplifier*. As stated earlier, the source amplifier must be isolated when checking its output signal. Isolating the source amplifier can normally be accomplished by removing the coupling component(s) between the source amplifier and the class AB amplifier. If the output from the isolated source amplifier is good, then the class AB amplifier is the source of the problem.

When a class AB amplifier has been determined to be faulty, some relatively simple voltage checks can be performed *while the circuit is still isolated from its signal source*. Figure 11.34 shows the types of readings you should get in a class AB amplifier when it has been isolated and all the components are working properly. (The meter readings are specific to the circuit, but the relationships shown by the meters apply to class AB amplifiers in general.)

> After disconnecting the signal source, perform the dc voltage checks.

With most class A amplifiers, the troubleshooting procedure begins with analyzing the emitter and collector voltages of the transistor and then working your way back to the input. The procedure is essentially the same for the class AB amplifier. We check the voltage at the point where the two emitters are connected, and then we check the base voltages. Table 11.1 summarizes some common faults and their symptoms. The test points referenced in the table are shown in Figure 11.35.

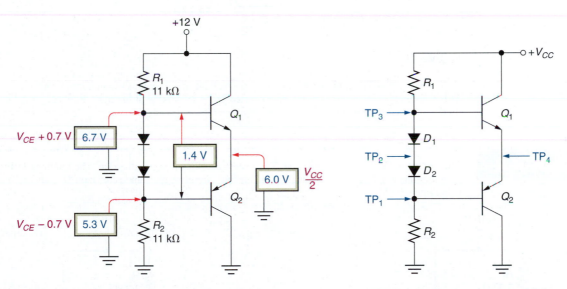

FIGURE 11.34   Typical class AB amplifier dc voltages.

FIGURE 11.35   Commonly used dc test points.

TABLE 11.1   Open-Component Troubleshooting

| Open Component | Symptoms |
| --- | --- |
| $R_2$ | All test point voltages will be higher than normal. |
| $D_1$ or $D_2$ | The voltage from $TP_3$ to $TP_1$ will be higher than normal. |
| | The voltage across the *good* diode will be near 0 V. |
| $R_1$ | All test point voltages will be close to 0 V. |
| $Q_1$ | All base circuit voltages will be normal. The voltage at $TP_4$ will be very low. |
| $Q_2$ | All base circuit voltages will be normal. The voltage at $TP_4$ will be slightly higher than normal |

What should you do if you
suspect the transistors are
faulty?

If the dc voltages in the circuit appear to be within tolerance and the amplifier does not work, simply replace both transistors. This saves the time required for testing to see which transistor is faulty. Besides, if one of the transistors has gone bad, odds are that it has damaged the other in the process. Therefore both transistors should be replaced. To help you get more comfortable with the information in this section, we will go through some troubleshooting applications.

## Application 11.1

The circuit shown in Figure 11.36 has no output signal. After checking the load, the $V_{CC}$ connection, and the ground connections, the source is isolated from the amplifier. When checked, the output signal from the source amplifier appears to be good, indicating that the class AB amplifier is the source of the fault. While isolated from the source, a voltmeter is used to obtain the readings shown.

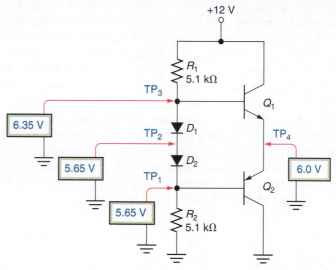

FIGURE 11.36

The voltage at $TP_1$ is higher than normal, the voltage at $TP_3$ is lower than normal, and $V_{D2} = 0$ V. This combination of readings points to a shorted $D_2$ as the source of the fault. Replacing the biasing diodes solves the problem. ∎

## Application 11.2

The circuit in Figure 11.37a is tested. The results are shown in the diagram. The readings in Figure 11.37a indicate that an open diode is in the biasing circuit. A resistance check on $D_2$ indicates that it is the open diode.

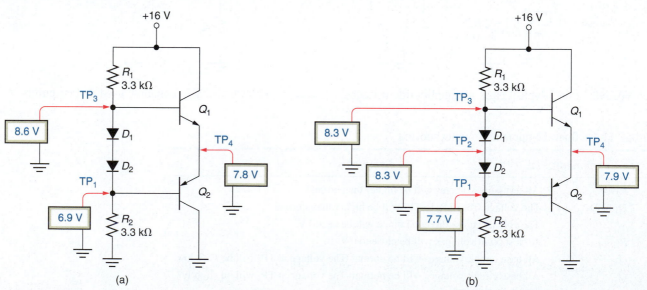

FIGURE 11.37

When $D_2$ is replaced, the circuit readings change to those shown in Figure 11.37b. These readings indicate that $D_1$ has shorted. Replacing $D_1$ causes the circuit to operate properly again. ■

What problem is indicated by the readings in Figure 11.38? What would you do to repair the problem? (Assume that the amplifier inputs all check out and that the load is not the problem.) The answer can be found after Application 11.4.

Application 11.3

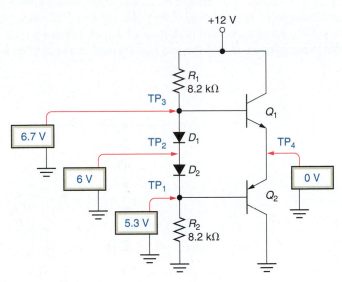

FIGURE 11.38

Figure 11.39a shows the initial readings in a faulty amplifier. All outside factors have been eliminated as being the cause of the problem. The readings indicate that $R_2$ is open. After replacing $R_2$, the circuit test points are checked again, giving the readings shown in Figure 11.39b. These readings indicate that $D_2$ is shorted. Replacing this component (and its matching diode) restores normal operation to the circuit. ■

Application 11.4

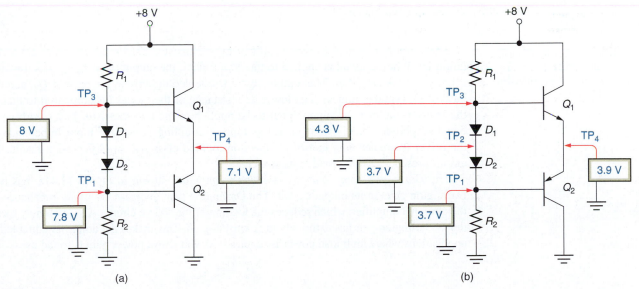

FIGURE 11.39

In Application 11.3, all the biasing voltages are correct. The only logical assumption at this point would be that there is a problem with one of the transistors. Therefore, the next step would be to replace both transistors.

## 11.6.2 Other Class AB Amplifiers

Now, we will look at a few other types of class AB amplifiers and biasing circuits. The purpose of this discussion is not to teach you *everything* about these circuits but, rather, to introduce you to some alternative circuit configurations.

In each circuit you've seen so far, the input signal has been coupled to the class AB amplifier via a pair of capacitors connected to the transistor bases. Two more commonly used input-coupling methods are shown in Figure 11.40. The circuit in Figure 11.40a uses a single coupling capacitor connected between the diodes in the biasing circuit. This coupling circuit has the advantage of requiring only one input capacitor. However, it has the disadvantage of a slightly lower maximum peak-to-peak output. In other words, the amplifier cannot be driven to compliance, though the loss in maximum output amplitude is slight.

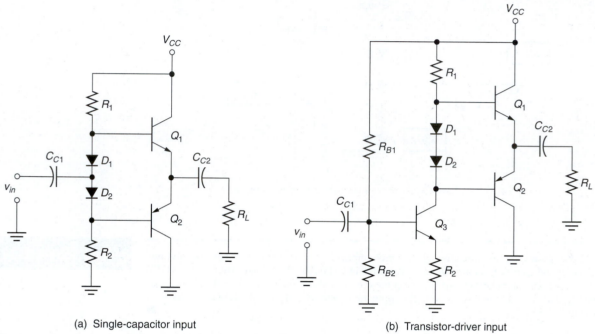

(a) Single-capacitor input     (b) Transistor-driver input

**FIGURE 11.40** More commonly used input coupling circuits.

**Lab Reference:** The operation of a transistor driver (like the one in Figure 11.40b) is demonstrated in Exercise 15.

The circuit shown in Figure 11.40b uses a transistor *driver* ($Q_3$) for the class AB amplifier. When a signal is applied to the driver input, the output ($v_{ce}$) changes as it would for any class A amplifier. This output signal is coupled directly to the base of $Q_2$, and the base of $Q_1$ (via the diodes). The lower ac resistance of the diodes (as compared to that of the resistors) allows the input signal to be applied to the two transistor bases with near-equal amplitude. One advantage of this type of coupling is that the input impedance of the driver is greater than that of a capacitor-coupled class AB amplifier. As a result, the circuit presents less of a load to its source.

The *Darlington complementary-symmetry amplifier*, shown in Figure 11.41a, has two Darlington pairs in its output circuit. The Darlington pairs increase the input impedance of the class AB amplifier, which reduces the loading of the source circuit. Also, you may recall that a Darlington pair has extremely high current gain. This makes the amplifier suitable for applications where high load power is required. As you know, power gain is found as

$$A_P = A_v A_i$$

Since the value of $A_i$ is much higher for a Darlington pair than it is for a single transistor, the power gain of the Darlington class AB amplifier is much greater than that of a standard push-pull.

Note the four biasing diodes between the bases of $Q_1$ and $Q_4$. Four diodes are needed to compensate for the 1.4 V value of $V_{BE}$ for *each* Darlington pair. Since each pair has a

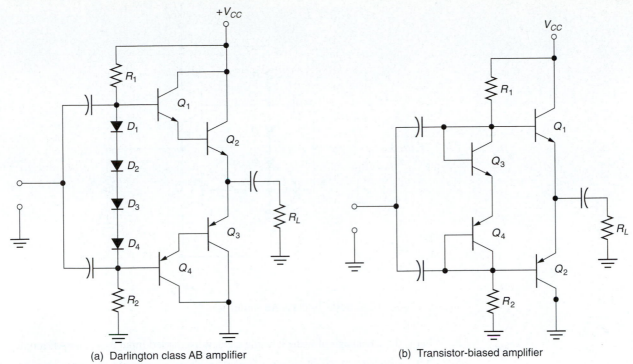

(a) Darlington class AB amplifier      (b) Transistor-biased amplifier

FIGURE 11.41   Darlington and transistor-biased amplifiers.

$V_{BE}$ of 1.4 V, there is a 2.8 V difference of potential between the bases of $Q_1$ and $Q_4$. The four diodes, when properly matched, will maintain the 2.8 V difference.

The *transistor-biased complementary-symmetry amplifier* uses transistors (rather than diodes) in the biasing circuit. This amplifier is shown in Figure 11.41b. The amplifier is used primarily in circuits that contain integrated (IC) transistors. Note that the biasing transistors are wired so that they act as diodes. The collector-base junction of each biasing transistor is shorted, leaving only an emitter-base junction in the circuit. This junction acts as a simple diode. Why go to the trouble of using transistors in the biasing circuit? Most transistor ICs contain four transistors. By using the extra transistors in an IC as the biasing diodes, you ensure that the diodes are perfectly matched to the circuit transistors ($Q_1$ and $Q_2$). At the same time, the fact that the transistors are all housed in a single IC limits its use to low-power applications.

When the output from a class AB amplifier must be centered around 0 V (instead of around the value of $V_{CC}/2$), the *split-supply class AB amplifier* may be used. This amplifier is shown in Figure 11.42. The two power supply connections for this circuit will be equal and opposite in polarity. For example, voltage supplies of $\pm10$ V may be used, but voltage supplies of $\pm10$ and $-5$ V would not be used. With matched power supplies, each transistor will drop its own supply voltage, and the output signal will be centered at 0 V. This allows the circuit to be directly coupled to the load, as shown in Figure 11.42.

There are many different biasing configurations for class AB amplifiers. When you come up against a biasing circuit you have never seen before, just remember the basic principles of class B and class AB operation. Within reason, the operation of any class B or class AB amplifier will follow these principles.

1. What is the first step in troubleshooting a class AB amplifier (or any other amplifier, for that matter)?   ◄ **Section Review**

2. Why are class AB amplifiers more difficult to troubleshoot than class A amplifiers?

3. What is the general approach to troubleshooting a class AB amplifier?

4. You are troubleshooting a class AB amplifier and come to the conclusion that the trouble is one of the two transistors. Briefly discuss the reasons for replacing both transistors rather than only one of them.

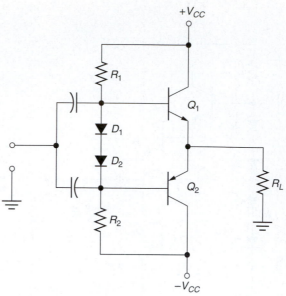

FIGURE 11.42   Dual-polarity class AB amplifier.

**5.** What is the advantage of using a single capacitor-coupled input to a class AB amplifier?

**6.** What is the advantage of using a transistor driver to couple an input signal to the class AB amplifier?

**7.** When is the *Darlington complementary-symmetry amplifier* used?

**8.** Why are four biasing diodes needed in the Darlington complementary-symmetry amplifier?

**9.** When is transistor biasing used in place of diode biasing?

**10.** When is a *split-supply class AB amplifier* used?

**Critical Thinking** ▶ **11.** Could a transistor driver like the one in Figure 11.40b be used in a class B amplifier? Explain your answer.

## 11.7   Related Topics

In this section, we look at a few topics that relate to power amplifiers, specifically maximum power ratings and calculations and the use of heat sinks.

### 11.7.1   Maximum Power Ratings

*OBJECTIVE 12* ▶  You may recall that transistors have a *maximum power dissipation rating*. When considering a transistor for a specific power amplifier application, you must make sure that the power dissipation rating of the transistor is sufficient for that application. For example, the specification sheet for the 2N3904 (Figure 6.28) shows a $P_{D(max)}$ rating of 625 mW. You cannot use the 2N3904 in any circuit that requires its transistor to dissipate more than 625 mW.

How do you determine the amount of power a transistor will have to handle in a specific circuit? For class A amplifiers, use the equation

$$P_D = V_{CEQ}I_{CQ}$$

(11.36)

For class B and class AB amplifiers, use the equation

$$P_D = \frac{V_{PP}^2}{40R_L}$$

(11.37)

to find the $P_D$ of the individual transistor. The derivations of these equations are fairly involved and are provided in Appendix D. Examples 11.14 and 11.15 show how the equations are used to determine the transistor power requirements of class A and class AB amplifiers.

### EXAMPLE 11.14

What is the value of $P_D$ for the transistor in Figure 11.43?

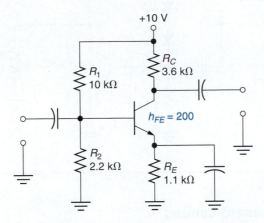

FIGURE 11.43

*Solution:* Using the established procedure, $I_{CQ}$ and $V_{CEQ}$ are found to be

$$I_{CQ} = 1 \text{ mA} \quad \text{and} \quad V_{CEQ} = 5.3 \text{ V}$$

Using these two values, $P_D$ is found as

$$P_D = V_{CEQ}I_{CQ} = (5.3 \text{ V})(1 \text{ mA})$$
$$= 5.3 \text{ mW}$$

### PRACTICE PROBLEM 11.14

A class A amplifier has the following values: $R_E = 1.2 \text{ k}\Omega$, $R_C = 2.7 \text{ k}\Omega$, $V_{CC} = 16 \text{ V}$, and $V_E = 2.4 \text{ V}$. Determine the required value of $P_D$ for the transistor in the amplifier.

### EXAMPLE 11.15

What is the value of $P_D$ for each transistor in Figure 11.44? (Assume that the output is equal to the compliance of the amplifier.)

*Solution:* The compliance of the amplifier is 12 V. Thus, $V_{PP} = 12 \text{ V}$, and

$$P_D = \frac{V_{PP}^2}{40R_L}$$
$$= \frac{(12 \text{ V})^2}{(40)(8 \text{ }\Omega)}$$
$$= \frac{144 \text{ V}^2}{320 \text{ }\Omega}$$
$$= 450 \text{ mW}$$

### PRACTICE PROBLEM 11.15

A class AB amplifier with values of $V_{CC} = 15 \text{ V}$ and $R_L = 12 \text{ }\Omega$ is driven to compliance. Determine the required value of $P_D$ for each transistor in the circuit.

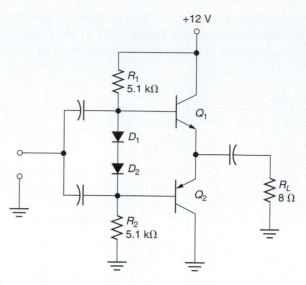

**FIGURE 11.44**

## 11.7.2    Component Cooling

When a large number of components are used in an enclosed system (such as a computer or stereo), the heat generated by those components causes the system temperature to rise. As the temperature rises, component operating temperature limits can become an important consideration. In extreme cases, high temperatures may exceed the maximum allowable junction temperatures for the semiconductors if some method of component cooling is not used.

You may recall that the power dissipation capability of a component is *derated* when the temperature rises above a certain value. For example, the spec sheet for the 2N3771 power transistor (Figure 6.37) lists the following values:

$$P_D = 150 \text{ W (maximum) at } T_C = 25°C$$
$$\text{Derating factor} = 0.855 \text{ W/°C (above } T_C = 25°C)$$

where $T_C$ is the *case temperature* of the component. These ratings indicate that the maximum power dissipation (150 W) must be decreased by 8.55 mW for each degree the case temperature rises above 25°C. The spec sheet also shows that the 2N3771 has a maximum allowable junction temperature of 150°C. If the component is used in an enclosed system, component cooling may be required to ensure that:

1. The required power dissipation at the given temperature does not exceed the derated value of $P_D$ for the device.
2. The junction temperature *never* exceeds 150°C.

There are several ways to provide cooling for the semiconductors within an enclosed system. One common method involves the use of one or more fans. Most computer systems, for example, are fan cooled. Typically, the fan draws in cooler outside air, helping to maintain relatively low temperatures within the system.

Another method used to avoid heat-related problems is to mount high-power transistors on structures called *heat sinks*. A **heat sink** is a metallic object that helps to cool components by increasing their *effective* surface area. Several heat sinks are shown in Figure 11.45.

When a high-power transistor is connected *properly* to a heat sink, heat flows from the component to the heat sink. Because the heat sink has more surface area, it can dissipate heat more rapidly than the transistor. As a result, the transistor's case temperature is held at a lower value, and the $P_D$ derating *value* is reduced.

When a power transistor is connected to a heat sink, **heat-sink compound** is used to aid the transfer of heat from the transistor to the sink. This compound can be purchased at

**Heat sink**
A large metallic object that helps to cool components by increasing their effective surface area.

**Heat-sink compound**
A compound used to aid in the transfer of heat from a component to a heat sink.

**FIGURE 11.45**   Heat sinks.

any electronics parts store. The heat-sink compound is applied to the transistor casing, and then the transistor is connected (attached) to the heat sink itself. The transistor must be attached to the heat sink mechanically. (Heat-sink compound is *not* glue. It does not act as an adhesive between the transistor and the heat sink.)

A power transistor case may not be at the same electrical potential as the heat sink. In this situation, a thin insulator is inserted between the two. When used with heat-sink compound, the spacer acts as an *electrical insulator* and a *thermal conductor*. Many times, these spacers will stick to an old transistor as it is removed from the circuit. If the spacer is not replaced correctly when the new transistor is installed, the replacement transistor may be damaged when power is restored to the circuit. Care must also be taken to ensure that the leads do not come in contact with the heat sink. The following procedure should be followed for replacing transistors that are mounted to a heat sink:

1. Remove the bad transistor, and wipe the old heat-sink compound off the heat sink.
2. *Lightly* coat the new transistor with heat-sink compound. Do not use more than is necessary to build a thin coat on the component.
3. Replace the insulator (spacer) if necessary.
4. Connect the transistor to the heat sink. If there were any mechanical connectors between the old transistor and the heat sink (such as screws), be sure to replace them. Any insulating sleeves on the mounting screws must also be replaced.
5. *Be sure that the transistor leads are not touching the heat sink.*

◄ **Section Review**

1. What is a *heat sink*?
2. What is *heat-sink compound*?
3. What is the proper procedure for replacing a transistor connected to a heat sink?
4. What precautions must be taken when performing the procedure in Question 3?

**CHAPTER SUMMARY**

Here is a summary of the major points made in this chapter:

1. Power amplifiers are used to deliver power to low-resistance loads.
   a. The output from a power amplifier is typically greater than 1 W.
   b. Typical power amplifier loads are in the range of $R_L \leq 300 \ \Omega$.
2. The *ideal* power amplifier would deliver 100% of its dc (power supply) input power to its load.
3. There are three basic types of power amplifiers: *class A*, *class B*, and *class C*.
4. Class C amplifiers are *tuned circuits*, meaning that they are designed to operate within specific frequency ranges.
5. The *ac load line* of an amplifier represents all possible combinations of $i_c$ and $v_{ce}$.
   a. The ac load line of an amplifier does not normally follow the plot of the dc load line.

**b.** The ac and dc load lines differ because the dc load on an amplifier is different from its ac load.

**c.** An ac load line is illustrated in Figure 11.3.

6. The ac load line of an amplifier is used to determine the amplifier's *compliance*.

   **a.** The compliance of an amplifier is the limit that the output circuit places on the peak-to-peak output voltage.

   **b.** If the compliance of an amplifier is exceeded, *clipping* occurs.

   **c.** Amplifier compliance is illustrated in Figure 11.4. Clipping is illustrated in Figure 11.5.

7. The higher the efficiency of an amplifier:

   **a.** The lower the percentage of dc power used by the amplifier itself.

   **b.** The longer the operating life of the components in the amplifier.

8. A *transformer-coupled class A amplifier* uses a transformer in place of a collector resistor to improve amplifier efficiency.

9. The dc biasing circuit of a transformer-coupled class A amplifier is designed so that $V_{CEQ}$ is as close as possible to the value of $V_{CC}$.

10. The dc load line of a transformer-coupled class A amplifier is nearly vertical, indicating that the value of $V_{CE}$ remains relatively constant over a wide range of $I_C$ values.

11. The procedure used to plot the ac load line of a transformer-coupled class A amplifier is described in Section 11.3.3.

12. The maximum output voltage from a transformer-coupled class A amplifier is very close to twice the value of $V_{CC}$.

13. The maximum theoretical efficiency of a transformer-coupled class A amplifier is approximately 50%. In practice, the efficiency of this type of amplifier is typically less than 40%.

14. The transformer-coupled class A amplifier has several advantages over the *RC*-coupled circuit:

   **a.** The efficiency of the transformer-coupled circuit is higher.

   **b.** The transformer can be used to match the amplifier output resistance to the load resistance.

   **c.** The transformer-coupled circuit can be converted into a *tuned amplifier*.

   A comparison of the transformer-coupled and *RC*-coupled circuits is provided in Figure 11.14.

15. There are two basic class B amplifiers.

   **a.** The *complementary-symmetry amplifier* is built using complementary transistors, that is, a *pnp* transistor and an *npn* transistor that have matching characteristics.

   **b.** A *push-pull amplifier* uses two identical transistors and a center-tapped transformer.

16. The complementary-symmetry amplifier is the more commonly used circuit because of the size and cost of the center-tapped transformer.

17. The overall operation of a complementary-symmetry amplifier is illustrated in Figure 11.17. As shown, each transistor conducts during one alternation of the input signal.

18. *Crossover distortion* (shown in Figure 11.18) occurs during the time that neither transistor in a class B amplifier is conducting.

   **a.** Crossover distortion appears as flat lines between the alternations of the amplifier output signal.

   **b.** Crossover distortion is eliminated by biasing both transistors at a level slightly *above* cutoff.

19. The class B amplifier has a vertical dc load line.

20. The class B amplifier is an emitter follower. As such, it has:

   **a.** Current gain that is greater than 1.

   **b.** Voltage gain that is slightly less than 1.

   **c.** Power gain that is slightly lower than the circuit current gain.

21. The compliance of a class B amplifier is approximately equal to its dc supply voltage.

22. The total current drawn from the dc power supply by a class B amplifier is approximately equal to the sum of the current through the biasing network and the *average* value of $I_{C1}$ (see Figure 11.24).

23. The maximum theoretical efficiency of a class B amplifier is 78.5%.
24. The complementary-symmetry amplifier is more efficient than a comparable push-pull amplifier because it is not subject to transformer power losses.
25. When voltage-divider biased, a class B amplifier is susceptible to crossover distortion and thermal runaway. Both of these problems can be eliminated through the use of *diode bias*. (See Figure 11.27).
26. When diode bias is used, each transistor conducts for *slightly more than* 180°. For this reason, the diode-biased amplifier is referred to as a *class AB amplifier*.
27. The diodes in a class AB amplifier are referred to as *compensating diodes*.
28. The method by which diode bias eliminates crossover distortion is illustrated in Figure 11.30.
29. Thermal runaway is prevented in a class AB amplifier by:
    a. Matching the compensating diodes to the base-emitter characteristics of the transistors.
    b. Placing the diodes in *thermal contact* with the transistors (so that both components operate at the same temperature).
30. The dual-transistor configuration of the class AB amplifier can make the circuit more difficult to troubleshoot than a class A amplifier.
31. A list of common class AB amplifier fault symptoms is provided in Table 11.1.
32. A *Darlington* complementary-symmetry amplifier (Figure 11.41a) can be used in applications requiring higher-than-normal amplifier input impedance and power gain.
33. The *transistor-biased* complementary-symmetry amplifier (Figure 11.41b) uses transistors in place of the biasing diodes. This configuration is commonly used in circuits containing integrated transistors.
34. A *dual-polarity* complementary-symmetry amplifier (Figure 11.42) is used in systems containing dual-polarity dc power supplies.
35. When power transistors are used in an enclosed system, additional means of component cooling may be required to ensure that:
    a. The required power dissipation at a given temperature does not exceed the *derated value* of $P_D$ of the device.
    b. The maximum allowable junction temperature is never exceeded.
36. Component cooling often involves the use of *fans* and/or *heat sinks*.
37. A heat sink is a large, metallic structure that helps to cool components by increasing their effective surface area.
38. *Heat-sink compound* is used to aid in the transfer of heat from a component to a heat sink.
39. The steps used to connect a component properly to a heat sink are listed in Section 11.7.2.

## EQUATION SUMMARY

| Equation Number | Equation | Section Number |
|---|---|---|
| (11.1) | $i_c = \dfrac{v_{ce}}{r_C}$ | 11.1 |
| (11.2) | $\Delta I_C = \dfrac{\Delta V_{CE}}{r_C}$ | 11.1 |
| (11.3) | $i_{c(\text{sat})} = I_{CQ} + \dfrac{V_{CEQ}}{r_C}$ | 11.1 |
| (11.4) | $v_{ce(\text{off})} = V_{CEQ} + I_{CQ}r_C$ | 11.1 |
| (11.5) | $PP = 2I_{CQ}r_C$ | 11.1 |
| (11.6) | $PP = 2V_{CEQ}$ | 11.1 |

| Equation Number | Equation | Section Number |
|---|---|---|
| **(11.7)** | $P_S = V_{CC}I_{CC}$ | 11.2 |
| **(11.8)** | $P_L = \dfrac{V_L^2}{R_L}$ | 11.2 |
| **(11.9)** | $P_L = \dfrac{V_{\text{pk}}^2}{2R_L}$ | 11.2 |
| **(11.10)** | $P_L = \dfrac{V_{\text{PP}}^2}{8R_L}$ | 11.2 |
| **(11.11)** | $P_{L(\text{max})} = \dfrac{\text{PP}^2}{8R_L}$ | 11.2 |
| **(11.12)** | $\dfrac{N_P}{N_S} = \dfrac{V_P}{V_S} = \dfrac{I_S}{I_P}$ | 11.3 |
| **(11.13)** | $\left(\dfrac{N_P}{N_S}\right)^2 = \dfrac{Z_P}{Z_S}$ | 11.3 |
| **(11.14)** | $r_C = Z_p$ | 11.3 |
| **(11.15)** | $Z_P = \left(\dfrac{N_P}{N_S}\right)^2 Z_S$ | 11.3 |
| **(11.16)** | $P_S \cong V_{CC}I_{CQ}$ | 11.3 |
| **(11.17)** | $V_{\text{PP}} = \dfrac{N_S}{N_P}\,\text{PP}$ | 11.3 |
| **(11.18)** | $V_{CEQ} = \dfrac{V_{CC}}{2}$ | 11.4 |
| **(11.19)** | $I_{CQ} \cong 0$ | 11.4 |
| **(11.20)** | $v_L = \Delta V_{CE}$ | 11.4 |
| **(11.21)** | $i_L = \dfrac{\Delta V_{CE}}{R_L}$ | 11.4 |
| **(11.22)** | $\Delta V_{CE} = \dfrac{V_{CC}}{2}$ | 11.4 |
| **(11.23)** | $i_{c(\text{sat})} = \dfrac{V_{CC}}{2R_L}$ | 11.4 |
| **(11.24)** | $v_{ce(\text{off})} = \dfrac{V_{CC}}{2}$ | 11.4 |
| **(11.25)** | $Z_{\text{in(base)}} = h_{fc}(r_e' + R_L)$ | 11.4 |
| **(11.26)** | $Z_{\text{out}} = r_e' + \dfrac{R_{\text{in}}'}{h_{fc}}$ | 11.4 |
| **(11.27)** | $A_i = h_{fc}\left(\dfrac{Z_{\text{in}}}{Z_{\text{in(base)}}}\right)$ | 11.4 |

| Equation Number | Equation | Section Number |
|:---:|:---:|:---:|
| **(11.28)** | $PP = 2V_{CEQ}$ | 11.4 |
| **(11.29)** | $PP \cong V_{CC}$ | 11.4 |
| **(11.30)** | $I_{CC} = I_{C1(\text{ave})} + I_1$ | 11.4 |
| **(11.31)** | $I_{C1(\text{ave})} = \dfrac{V_{CC}}{2\pi R_L}$ | 11.4 |
| **(11.32)** | $I_{C1(\text{ave})} = \dfrac{V_{PP(\text{out})}}{2\pi R_L}$ | 11.4 |
| **(11.33)** | $V_{B(Q2)} = \dfrac{R_2}{R_1 + R_2}(V_{CC} - 1.4 \text{ V})$ | 11.5 |
| **(11.34)** | $V_{B(Q2)} = V_{CEQ} - 0.7 \text{ V} \quad$ (when $R_1 = R_2$) | 11.5 |
| **(11.35)** | $I_1 = \dfrac{V_{CC} - 1.4 \text{ V}}{R_1 + R_2}$ | 11.5 |
| **(11.36)** | $P_D = V_{CEQ}I_{CQ}$ | 11.7 |
| **(11.37)** | $P_D = \dfrac{V_{PP}^2}{40R_L}$ | 11.7 |

## KEY TERMS

ac load line  404
class AB amplifier  436
compensating diodes  434
complementary-symmetry
   amplifier  421
complementary
   transistors  424
compliance  406
counter emf  417

crossover distortion  423
cutoff clipping  407
diode bias  434
heat sink  448
heat-sink compound  448
push-pull emitter follower
   421
RC-coupled class A
   amplifier  410

saturation clipping  408
standard push-pull
   amplifier  421
step-down transformer  415
step-up transformer  415
thermal contact  437
transformer-coupled
   class A amplifier  414
tuned amplifier  420

## PRACTICE PROBLEMS

### Section 11.1

**1.** Calculate the compliance of the amplifier in Figure 11.46.

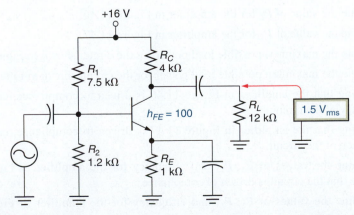

**FIGURE 11.46**

**2.** Calculate the compliance of the amplifier in Figure 11.47.

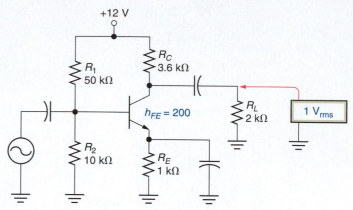

FIGURE 11.47

**3.** Calculate the compliance of the amplifier in Figure 11.48. Determine the type of clipping that the circuit would be most likely to experience.

**4.** Calculate the compliance of the amplifier in Figure 11.49. Determine the type of clipping that the circuit would be most likely to experience.

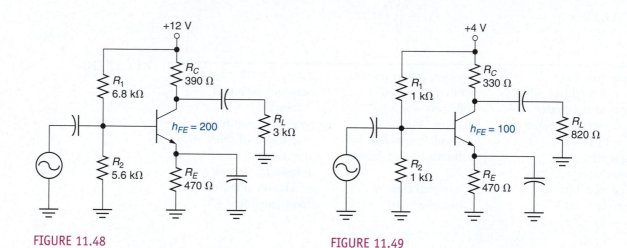

FIGURE 11.48

FIGURE 11.49

**Section 11.2**

**5.** Calculate the value of $P_S$ for the amplifier in Figure 11.46.

**6.** Calculate the value of $P_S$ for the amplifier in Figure 11.47.

**7.** Calculate the value of $P_L$ for the amplifier in Figure 11.46.

**8.** Calculate the value of $P_L$ for the amplifier in Figure 11.47.

**9.** Calculate the maximum possible load power for the circuit shown in Figure 11.46.

**10.** Calculate the maximum possible load power for the circuit shown in Figure 11.47.

**11.** Assuming that the amplifier in Figure 11.46 is driven to compliance, calculate the efficiency of the circuit.

**12.** Assuming that the amplifier in Figure 11.47 is driven to compliance, calculate the efficiency of the circuit.

**13.** Determine the values of $P_S$, $P_L$, and efficiency for the amplifier in Figure 11.48. Assume that the circuit is driven to compliance.

**14.** Determine the values of $P_S$, $P_L$, and efficiency for the amplifier in Figure 11.49. Assume that the circuit is driven to compliance.

**15.** Calculate the maximum efficiency for the amplifier in Figure 11.50.

**16.** Calculate the maximum efficiency for the amplifier in Figure 11.51.

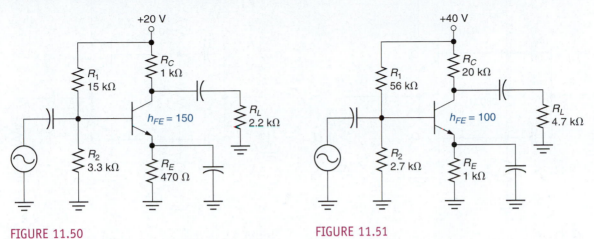

FIGURE 11.50    FIGURE 11.51

## Section 11.3

**17.** Derive the ac load line for the circuit shown in Figure 11.52.

**18.** Derive the ac load line for the circuit shown in Figure 11.53.

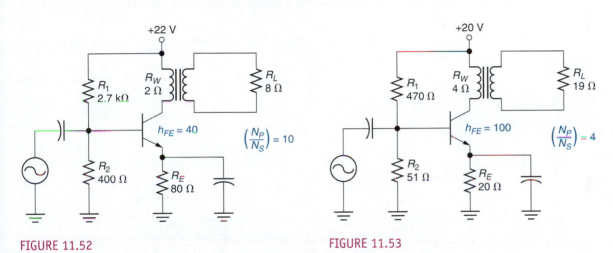

FIGURE 11.52    FIGURE 11.53

**19.** Determine the maximum load power for the circuit in Figure 11.52.

**20.** Determine the maximum load power for the circuit in Figure 11.53.

**21.** Calculate the maximum efficiency of the amplifier in Figure 11.52.

**22.** Calculate the maximum efficiency of the amplifier in Figure 11.53.

## Section 11.4

**23.** A class B amplifier has values of $V_{CC} = +18$ V and $R_L = 3$ kΩ. Plot the ac and dc load lines for the circuit.

**24.** A class B amplifier has values of $V_{CC} = +24$ V and $R_L = 200$ Ω. Plot the ac and dc load lines for the circuit.

**25.** Calculate the maximum load power for the amplifier described in Problem 23.

**26.** Calculate the maximum load power for the amplifier described in Problem 24.

**27.** Calculate the value of $P_S$ for the amplifier in Figure 11.54.

**28.** Calculate the value of $P_S$ for the amplifier in Figure 11.55.

**29.** Calculate the value of $P_{L(max)}$ for the amplifier in Figure 11.54.

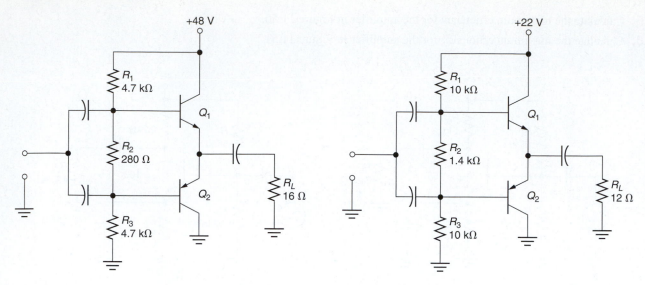

FIGURE 11.54

FIGURE 11.55

**30.** Calculate the value of $P_{L(max)}$ for the amplifier in Figure 11.55.

**31.** Calculate the maximum efficiency for the amplifier in Figure 11.54.

**32.** Calculate the maximum efficiency for the amplifier in Figure 11.55.

### Section 11.5

**33.** Determine the values of $V_{CEQ}$, $V_{B(Q1)}$, and $V_{B(Q2)}$ for the class AB amplifier in Figure 11.56.

**34.** Determine the values of $V_{CEQ}$, $V_{B(Q1)}$, and $V_{B(Q2)}$ for the class AB amplifier in Figure 11.57.

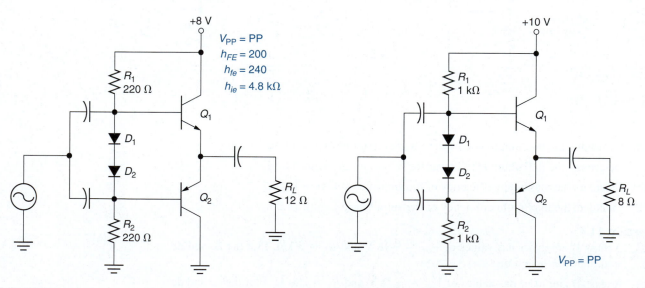

FIGURE 11.56

FIGURE 11.57

**35.** Calculate the maximum efficiency of the class AB amplifier in Figure 11.56.

**36.** Calculate the maximum efficiency of the class AB amplifier in Figure 11.57.

**37.** Calculate the maximum efficiency of the class AB amplifier in Figure 11.58.

**38.** Calculate the maximum efficiency of the class AB amplifier in Figure 11.59.

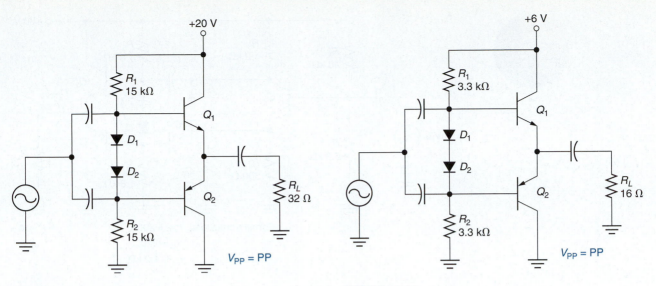

FIGURE 11.58                                    FIGURE 11.59

## Section 11.7

**39.** Calculate the value of $P_D$ for the transistor in Figure 11.48.

**40.** Calculate the value of $P_D$ for the transistor in Figure 11.49.

**41.** Calculate the value of $P_D$ for the transistor in Figure 11.52.

**42.** Determine the fault(s) that would cause the readings shown in Figure 11.60a.

**43.** Determine the fault(s) that would cause the readings shown in Figure 11.60b.

**TROUBLESHOOTING
PRACTICE PROBLEMS**

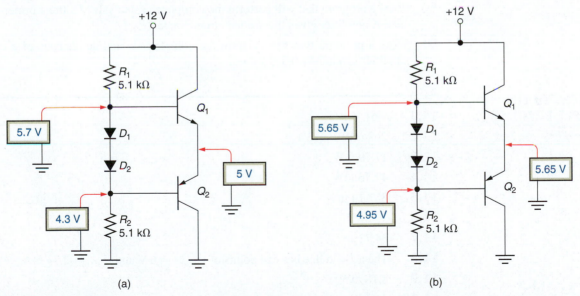

(a)                                             (b)

FIGURE 11.60

**PUSHING THE ENVELOPE**

**44.** Calculate the maximum allowable input power for the amplifier in Figure 11.56.

**45.** Calculate the values of $A_{vT}$, $A_{iT}$, and $A_{pT}$ for the two-stage amplifier in Figure 11.61.

**46.** Answer the following questions for the circuit in Figure 11.61. Use circuit calculations to explain your answers.
   **a.** Is the amplifier driven to compliance?
   **b.** What type of clipping would the amplifier be most likely to experience?

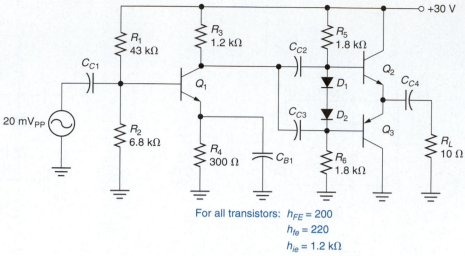

For all transistors: $h_{FE} = 200$
$h_{fe} = 220$
$h_{ie} = 1.2\ k\Omega$

FIGURE 11.61

**47.** Transistor $Q_2$ in Figure 11.61 is faulty. Can the 2N3904 be used in place of the transistor? Explain your answer using circuit calculations. The spec sheet for the 2N3904 is shown in Figure 6.28. (*Hint:* The power-handling requirement of the circuit is the primary consideration in this problem.)

**SUGGESTED COMPUTER APPLICATIONS PROBLEMS**

**48.** Write a program that will determine the efficiency of a class AB amplifier when it is provided with the needed input values.

**49.** Write a program that will perform the complete dc analysis of a transformer-coupled class A amplifier given the needed circuit values.

**50.** Write a program that will perform the complete ac analysis of a transformer-coupled class A amplifier given the needed circuit values.

**51.** Write a program that will perform the complete dc and ac analysis of a class AB amplifier given the needed values.

**ANSWERS TO THE EXAMPLE PRACTICE PROBLEMS**

| | |
|---|---|
| **11.1** | 9.61 $V_{PP}$ |
| **11.2** | 269 mW |
| **11.3** | 17.79 mW |
| **11.4** | 47.76 mW |
| **11.5** | 357.14 mW |
| **11.6** | 3.58% |
| **11.7** | 34.72% |
| **11.8** | The ac load line has end points of $v_{ce(off)} = 6$ V and $i_{c(sat)} = 2.73$ mA. |
| **11.9** | 8.18 mW |
| **11.10** | 2.93 W |
| **11.11** | 2.25 W |
| **11.12** | 76.79% |
| **11.13** | 55.3% |
| **11.14** | 16.4 mW |
| **11.15** | 468.75 mW |

# Field-Effect Transistors

## Objectives

*After studying the material in this chapter, you should be able to:*

1. List the types of field-effect transistors (FETs).

2. Explain the relationship between JFET channel width and drain current ($I_D$).

3. State the relationship between gate-source voltage ($V_{GS}$) and drain current ($I_D$).

4. Describe the gate input impedance characteristics of the JFET.

5. Determine the range of $Q$-point values for a given JFET biasing circuit.

6. List and explain the primary advantages and disadvantages of each of the three types of JFET biasing configurations.

7. Describe and analyze the ac operation of the common-source amplifier.

8. State the purpose of swamping a JFET amplifier.

9. Describe the relationship between the input impedance of a JFET amplifier and that of a comparable BJT amplifier.

10. Describe and analyze the ac operation of common-drain and common-gate amplifiers.

11. Describe the procedure used to troubleshoot a JFET amplifier.

12. List and define the commonly used JFET parameters and ratings.

13. Discuss the use of a JFET amplifier as a *buffer* or an *RF amplifier*.

## Outline

## That "Other" Development

Schockley, Brattain, and Bardeen have received their share of the limelight for the development of the transistor in 1948. However, there was another team of scientists that made a major contribution to the field of solid-state electronics in 1955.

In the May 1955 volume of the *Bell Laboratories Record*, an announcement was made of the development of another type of transistor. This transistor was developed by I. M. Ross and G. C. Dacey. The *Bell Laboratories Record* said of this development:

It would appear that it would find its main applications where considerations of size, weight, and power consumption dictate the use of a transistor, and where the required frequency response is higher than could be achieved with a simple junction transistor.

The component developed by Ross and Dacey has had a major impact over the years, especially in the area of integrated-circuit technology. What was this "other" component? It was the *field-effect transistor*.

---

> **Field-effect transistor (FET)**
> A three-terminal voltage-controlled device used in amplification and switching applications.

**OBJECTIVE 1 ▶**

You may recall that the *bipolar junction transistor* is a *current-controlled device*; that is, the output characteristics of the device are controlled by the base *current*. Another type of transistor, called the **field-effect transistor**, or **FET**, is a *voltage-controlled device*. The output characteristics of the **FET** are controlled by the input voltage, not by the input current.

There are two basic types of FETs: the *junction field-effect transistor*, or JFET, and the *metal-oxide-semiconductor FET*, or MOSFET. As you will be shown, the operating principles of these two components vary, as do their applications and limitations. In this chapter, we will discuss JFETs and their circuits. MOSFETs and their circuits are covered in Chapter 13.

## 12.1 Introduction to JFETs

The physical construction of the JFET is significantly different from that of the bipolar junction transistor, or BJT. You may recall that the bipolar transistor has three separate materials: either two *n*-type materials and a single *p*-type material or two *p*-type materials and a single *n*-type material. The JFET has only *two* materials: a single *n*-type material and a single *p*-type material. The construction of the JFET is illustrated in Figure 12.1.

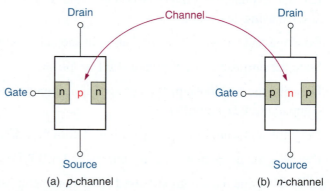

(a) *p*-channel  (b) *n*-channel

**FIGURE 12.1**  JFET construction.

---

> **Source**
> The JFET counterpart of the BJT emitter.
>
> **Drain**
> The JFET counterpart of the BJT collector.
>
> **Gate**
> The JFET counterpart of the BJT base.
>
> **Channel**
> The material that connects the source and drain.

As you can see, the device has three terminals, just like the bipolar junction transistor. However, the terminals of the JFET are labeled **source**, **drain**, and **gate**. The *source* can be viewed as the counterpart of the BJT's *emitter*, the *drain* as the counterpart of the *collector*, and the *gate* as the counterpart of the *base*.

The material that connects the source to the drain is referred to as the **channel**. When this material is an *n*-type, the JFET is referred to as an *n-channel JFET*. Obviously, an FET with a *p*-type channel would be referred to as a *p-channel JFET*. Both JFET types are shown in Figure 12.1. Note that the gate material *surrounds* the channel in the same way a belt surrounds your waist.

The schematic symbols for the *p*-channel and *n*-channel JFETs are shown in Figure 12.2. Note that the arrow points "in" or "out" from the component. Just as with the BJT

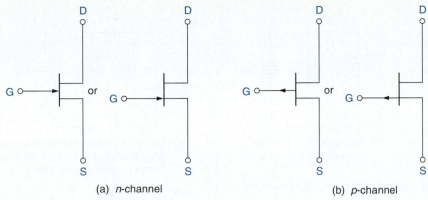

FIGURE 12.2  JFET schematic symbols.

schematic symbol, the arrow points toward the *n*-type material. When the arrow is pointing *in*, it is pointing toward the channel. Thus, Figure 12.2a is the schematic symbol for
the *n*-channel JFET. When the arrow is pointing *out*, it can be viewed as pointing toward
the gate. If the gate is an *n*-type material, the channel must be a *p*-type material. Thus,
Figure 12.2b is the schematic symbol for the *p*-channel JFET. If you have trouble distinguishing the two symbols, just think "*n* . . . in." This will remind you that the *n*-channel
JFET has the arrow pointing *in*.

You know that *npn* transistors normally require *positive* supply voltages, while *pnp* transistors normally require *negative* supply voltages. The same relationship holds true for
JFETs. *n-Channel* JFETs normally require *positive* supply voltages, while *p-channel* JFETs
normally require *negative* supply voltages. These supply voltages are shown in Figure 12.3.
Note that the *n*-channel JFET circuit (Figure 12.3a) has a *positive* drain supply voltage
($V_{DD}$), while the *p*-channel JFET circuit (Figure 12.3b) has a *negative* supply voltage.

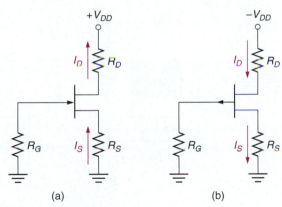

FIGURE 12.3  JFET supply voltages.

## 12.1.1  Operation Overview

In our discussions of BJTs, we concentrated on the *npn* transistor. At the same time, it
was stated that all the relationships discussed also held true for the *pnp* transistor. The
only differences were the voltage polarities and the current directions. In the same manner, our discussions on JFETs will concentrate on the *n-channel JFET*. Again, all the relationships covered also apply to the *p*-channel JFET. The only differences are the voltage
polarities and current directions.

The overall operation of the JFET is based on *varying the width of the channel to control the drain current*. As illustrated in Figure 12.4a, decreasing the width (diameter) of a
conductor increases its resistance. In Figure 12.4b, the drain-source voltage ($V_{DS}$) is generating current through a JFET. Assuming that $V_{DS}$ is fixed, we can decrease the JFET
drain current by increasing the resistance of the channel. Increasing the resistance of the
channel can be accomplished by decreasing its width.

◀ OBJECTIVE 2

Conductor diameters

The smaller the diameter of a conductor, the more it restricts current. Therefore, $R_X < R_Y < R_Z$

(a) Conductor width and resistance

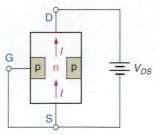

(b) $V_{DS}$ generates a current through the JFET channel.

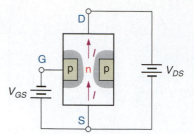

(c) The negative gate-source voltage ($V_{GS}$) reduces the width of the JFET channel, increasing its resistance.

FIGURE 12.4 The relationship between channel width and drain current.

**OBJECTIVE 3** ▶

Decreasing the width of the channel can be accomplished by effectively *increasing* the width of the gate. The width of the gate can be increased by applying a *reverse voltage* ($V_{GS}$) to the gate source-junction, as shown in Figure 12.4c. Note that the negative terminal of $V_{GS}$ is connected to the $p$-type gate and that the positive terminal of $V_{GS}$ is connected to the $n$-type source. This connection reverse biases the gate-source junction, causing a depletion layer to form. As you know, a depletion layer acts as an insulator. Thus, the cross-sectional area of the channel is effectively *decreased* when a depletion layer forms, causing the drain current to *decrease*.

> What is the relationship between $V_{GS}$ and $I_D$?

There are two ways to control channel width. First, by varying the value of $V_{GS}$, we can vary the width of the channel and, in turn, vary the amount of drain current. This point is illustrated in Figure 12.5. Note that as $V_{GS}$ increases, so does the size of the depletion layer. *As the size of the depletion layer increases, the effective width of the channel decreases*. As the effective width of the channel decreases, so does the amount of drain current. The relationship between $V_{GS}$ and $I_D$ can be summarized as follows: *As $V_{GS}$ increases (becomes more negative), $I_D$ decreases.*

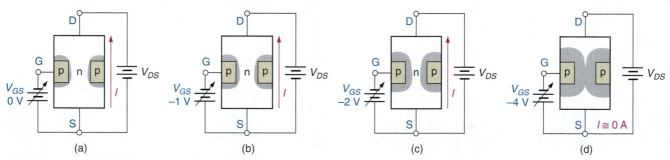

(a)  (b)  (c)  (d)

FIGURE 12.5 The relationship between $V_{GS}$ and $I_D$.

> What is the relationship between $V_{DS}$ and $I_D$?

Figure 12.5 serves to illustrate another important point. In Figure 12.5a, $V_{GS}$ is shown to be 0 V. At the same time, a small depletion layer is shown to be surrounding the gate. This small depletion layer is a result of the relationship between $V_{GS}$ and $V_{DS}$. Since the $p$-type gate is more negative than the $n$-type drain, a small depletion layer forms around the gate. The greater the value of $V_{DS}$, the greater the difference between the source and drain voltages and the larger the depletion layer. Thus, the second method of increasing the size of the depletion layer is to hold $V_{GS}$ constant while increasing the value of $V_{DS}$, as shown in Figure 12.6. Note that increasing $V_{DS}$ also causes $I_D$ to increase.

At this point, there seems to be a contradiction in the theory of operation. It would seem that increasing the size of the depletion layer would increase the resistance of the channel, preventing $I_D$ from increasing. How is it that $I_D$ is increasing when the resistance of the JFET channel is also increasing? While the resistance of the channel *is* increasing, it is doing so at a lower rate than the increase in $V_{DS}$. In other words, $V_{DS}$ is increasing in value faster than the resistance of the channel. As a result, the value of $I_D$ increases as $V_{DS}$ increases. However, a point is reached where further increases in $V_{DS}$ are offset by pro-

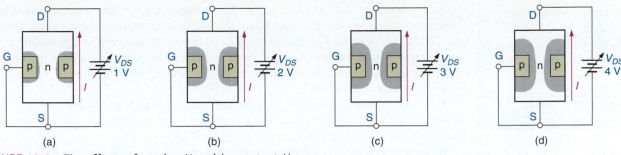

**FIGURE 12.6** The effects of varying $V_{DS}$ with constant $V_{GS}$.

portional increases in the resistance of the channel. The value of $V_{DS}$ at which this occurs is called the **pinch-off voltage ($V_P$)**. As $V_{DS}$ increases beyond the value of $V_P$, the value of $I_D$ levels off (becomes constant). This relationship is illustrated in the JFET drain curve shown in Figure 12.7a.

**Pinch-off voltage ($V_P$)**
The value of drain-source voltage ($V_{DS}$) that allows maximum JFET current ($I_D$) meaured at $V_{GS} = 0$ V.

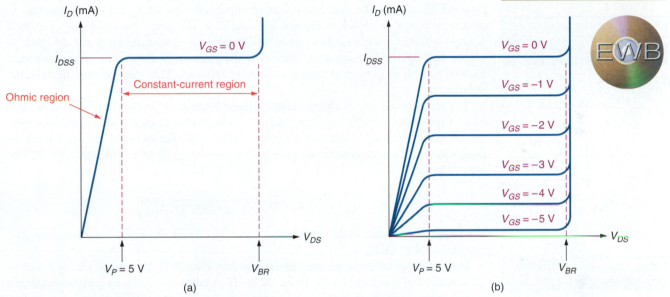

**FIGURE 12.7** JFET drain curves.

The portion of the curve to the left of $V_P$ is referred to as the **ohmic region**. As $V_{DS}$ increases from 0 V to $V_P$, drain current increases. As $V_{DS}$ reaches the value of $V_P$, drain current levels off and remains constant for all values of $V_{DS}$ between $V_P$ and the break-down voltage ($V_{BR}$). Note that the region of operation between $V_P$ and $V_{BR}$ (breakdown voltage) is called the **constant-current region**. As long as $V_{DS}$ is kept within this range, $I_D$ remains constant for a constant value of $V_{GS}$.

When $V_{GS} = 0$ V, you essentially have the gate and source terminals *shorted* together, and drain current reaches its maximum value, $I_{DSS}$. The **shorted-gate drain current ($I_{DSS}$)** is the maximum possible value of $I_D$. This value, which is listed on the spec sheet for a given JFET, is measured under the following conditions:

$$V_{GS} = 0 \text{ V} \quad \text{and} \quad V_{DS} = V_P$$

Any value of JFET drain current must be less than or equal to $I_{DSS}$. In this respect, $I_{DSS}$ can be viewed as the equivalent of $I_{C(\text{sat})}$ for a given BJT circuit. The relationship among $V_{GS}$, $V_{DS}$, and $I_{DSS}$ is illustrated in Figure 12.7b. When $V_{GS} = 0$ V, $I_D = I_{DSS}$. As $V_{GS}$ is made more negative than 0 V:

- The JFET drain pinches off at a voltage that is less than $V_P$.

- $I_D$ decreases from the value of $I_{DSS}$.

**Ohmic region**
The portion of the JFET operating curve that lies below $V_P$.

**Constant-current region**
The region of the JFET operating curve (between $V_P$ and $V_{BR}$) where drain current remains constant for fixed values of $V_{GS}$.

**Shorted-gate drain current ($I_{DSS}$)**
The maximum possible value of $I_D$.

**Lab Reference**: The relationship between $V_{DS}$ and $I_D$ is observed and graphed in Exercise 16.

**Gate-source cutoff voltage**
$(V_{GS(\text{off})})$
The value of $V_{GS}$ that reduces $I_D$ to approximately zero.

What is the relationship between $V_P$ and $V_{GS(\text{off})}$?

As $V_{GS}$ becomes more negative, a point is eventually reached where the depletion layer effectively closes off the channel. When this occurs, minimum current passes through the JFET. The value of $V_{GS}$ that reduces $I_D$ to approximately *zero* is referred to as the **gate-source cutoff voltage ($V_{GS(\text{off})}$)**. Note that the value of $V_{GS(\text{off})}$ defines the limit on the value of $V_{GS}$. For conduction to occur through the device, the value of $V_{GS}$ must be somewhere between 0 V and $V_{GS(\text{off})}$.

Here's an interesting point: $V_{GS(\text{off})}$ has the same magnitude as $V_P$. For example, if $V_P$ is 8 V, then $V_{GS(\text{off})}$ is $-8$ V. Since these two values are always equal and opposite, only one is usually listed on the spec sheet for a given JFET.

Remember that there is a definite difference between $V_P$ and $V_{GS(\text{off})}$. $V_P$ is the value of $V_{DS}$ that causes the JFET to become a constant-current component. It is measured at $V_{GS} = 0$ V and has a constant drain current of $I_D = I_{DSS}$. $V_{GS(\text{off})}$, on the other hand, is the value of $V_{GS}$ that reduces $I_D$ to approximately zero.

## 12.1.2 JFET Biasing

OBJECTIVE 4 ▶

Why shouldn't the gate-source junction of a JFET be forward biased?

The gate-source junction of a JFET is *always* reverse biased (under normal circumstances). Even when the gate and source terminals are shorted, $I_D$ causes the junction to be reverse biased, as shown in Figures 12.5 and 12.6. *The gate-source junction of a JFET is never allowed to become forward biased* because the gate material is not designed to handle any significant amount of current. If the junction is allowed to become forward biased, current is generated through the gate material. This current may destroy the component.

The fact that the gate is always reverse biased leads us to another important characteristic of the device: *JFETs have extremely high characteristic gate input impedance*. This impedance is typically in the high megohm (MΩ) range. For example, the spec sheet for the 2N5457 JFET lists a maximum *gate reverse current ($I_{GSS}$)* of 1.0 nA under the following conditions:

$$T = 25°C \quad V_{DS} = 0\text{ V} \quad V_{GS} = -15\text{ V}$$

If we apply Ohm's law to the values of $V_{GS}$ and $I_{GSS}$ listed above, we can calculate a gate impedance of 15 GΩ!

What is the primary advantage that FETs have over BJTs?

The advantage of this extremely high input impedance can be seen by taking a look at the circuit shown in Figure 12.8. Here, the JFET amplifier is driven by a high-impedance source. Because of the extremely high input impedance of the JFET gate, it draws no current from the source. It is as if there were no load on the source at all.

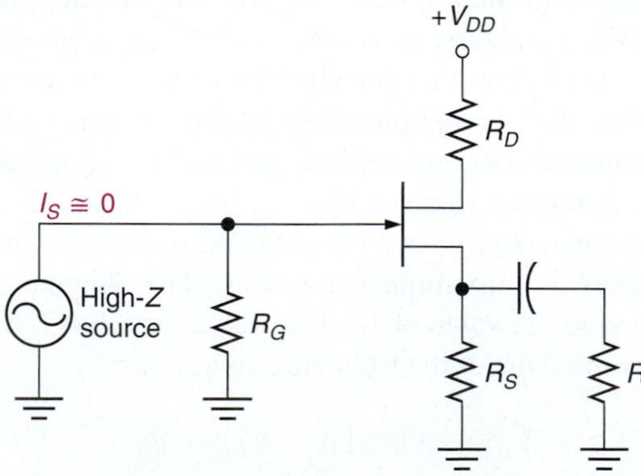

FIGURE 12.8

The high input impedance of the JFET has led to its extensive use in integrated circuits. The low current requirements of the component make it perfect for use in ICs, where thousands of transistors must be etched onto a single piece of silicon. The low current draw helps the IC to remain relatively cool, thus allowing more components to be placed in a smaller physical area.

## 12.1.3    Component Control

It was stated earlier that the JFET is a *voltage-controlled* device, while the BJT is a *current-controlled* device. We have now established the foundation necessary to look at this point in detail. We will start by reviewing the BJT as a current-controlled component. Consider the circuits shown in Figure 12.9. Using the formulas listed, the circuit values for Figure 12.9a can be calculated as

$$I_B = 16.3 \ \mu A \quad I_C = 1.63 \ mA \quad V_{CE} = 3.38 \ V$$

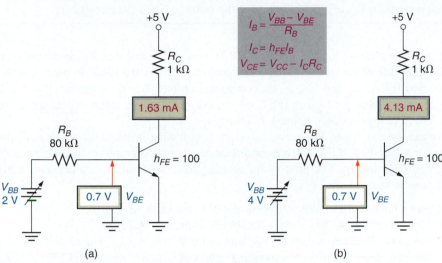

FIGURE 12.9    The BJT is a current-controlled device.

Using the same formulas on the circuit shown in Figure 12.9b gives the following results:

$$I_B = 41.3 \ \mu A \quad I_C = 4.13 \ mA \quad V_{CE} = 0.88 \ V$$

Note that $V_B = V_{BE}$ for each of the circuits shown. Thus, from one circuit to another, the value of base voltage did not change significantly. It stayed at approximately 0.7 V. Yet, there were drastic changes in the output characteristics of the two amplifiers. These changes were caused by the change in $I_B$, which demonstrates that the BJT is a *current-controlled* device. The output characteristics are determined by the input current, not by the input voltage.

In contrast, the JFET has *no* gate current. It has been shown that the size of the channel is controlled by the amount of reverse bias applied to the gate-source junction. Since $V_{GS}$ controls the JFET, it is *not* a current-controlled device but, rather, a voltage-controlled device.

Since the JFET has no gate current, there is no beta rating for the device. However, the output current ($I_D$) can be defined in terms of the circuit input voltage as follows:

$$I_D = I_{DSS} \left( 1 - \frac{V_{GS}}{V_{GS(off)}} \right)^2 \qquad (12.1)$$

where $I_{DSS}$ = the shorted-gate drain current rating of the device
   $V_{GS}$ = the gate-source voltage
$V_{GS(off)}$ = the gate-source cutoff voltage

Example 12.1 demonstrates the use of equation (12.1).

*A Practical Consideration:*
Equation (12.1) works only when

$$|V_{GS}| \leq |V_{GS(off)}|$$

If the above condition is not fulfilled, the equation will give you a false value of $I_D$. Just remember, if $V_{GS}$ is more negative than (or equal to) $V_{GS(off)}$, $I_D$ will be zero.

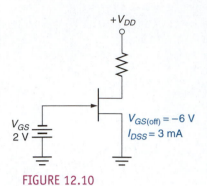

FIGURE 12.10

## EXAMPLE 12.1

Determine the value of drain current for the circuit shown in Figure 12.10.

*Solution:* The drain current for the circuit is found as

$$I_D = I_{DSS}\left(1 - \frac{V_{GS}}{V_{GS(off)}}\right)^2 = (3 \text{ mA})\left(1 - \frac{-2 \text{ V}}{-6 \text{ V}}\right)^2 = (3 \text{ mA})(0.444) = \textbf{1.33 mA}$$

### PRACTICE PROBLEM 12.1

A JFET with parameters of $I_{DSS} = 12$ mA and $V_{GS(off)} = -6$ V is used in a circuit that provides a $V_{GS}$ of $-3$ V. Determine the value of $I_D$ for the circuit.

If you take a closer look at equation (12.1), you will notice that $V_{GS}$ is the only value on the right side of the equation that will change for a specified JFET. In other words, for a specified JFET, $I_{DSS}$ and $V_{GS(off)}$ are component ratings. Based on the fact that $I_{DSS}$ and $V_{GS(off)}$ are constants for a given JFET, we can assume that $I_D$ is a function of the value of $V_{GS}$. As $V_{GS}$ changes (which it does), $I_D$ changes.

We can use a series of $V_{GS}$ versus $I_D$ values to plot what is called the *transconductance curve* for a specified JFET. A **transconductance curve** is a graph of all possible combinations of $V_{GS}$ and $I_D$. The process for plotting the transconductance curve of a given JFET is as follows:

> **Transconductance curve**
> A plot of all possible combinations of $V_{GS}$ and $I_D$.

> How is a transconductance curve plotted?

1. Plot a point on the *x*-axis that corresponds to the value of $V_{GS(off)}$.
2. Plot a point on the *y*-axis that corresponds to the value of $I_{DSS}$.
3. Select two or three values of $V_{GS}$ between 0 V and $V_{GS(off)}$. For each value of $V_{GS}$ selected, determine the corresponding values of $I_D$ using equation (12.1).
4. Plot the points from step 3, and connect all the plotted points with a smooth curve.

Step 1 is based on the fact that $I_D = 0$ when $V_{GS} = V_{GS(off)}$. Thus, the point for $V_{GS(off)}$ versus $I_D$ always falls on the *x*-axis of the graph. Step 2 is based on the fact that $I_D = I_{DSS}$ when $V_{GS} = 0$ V. Thus, the point for $V_{GS}$ versus $I_{DSS}$ always falls on the *y*-axis of the graph. Example 12.2 demonstrates the process given for plotting the transconductance curve of a given JFET.

## EXAMPLE 12.2

Plot the transconductance curve for a JFET having values of $V_{GS(off)} = -6$ V and $I_{DSS} = 3$ mA.

*Solution:* With the values given, we know that our end points for the curve are $(-6 \text{ V}, 0 \text{ mA})$ and $(0 \text{ V}, 3 \text{ mA})$. We now use three values of $V_{GS}$ ($-1$, $-3$, and $-5$ V) and calculate the corresponding value of $I_D$ for each.

At $V_{GS} = -1$ V,

$$I_D = (3 \text{ mA})\left(1 - \frac{-1 \text{ V}}{-6 \text{ V}}\right)^2 = \textbf{2.08 mA}$$

At $V_{GS} = -3$ V,

$$I_D = (3 \text{ mA})\left(1 - \frac{-3 \text{ V}}{-6 \text{ V}}\right)^2 = \textbf{0.75 mA} \quad (750 \text{ μA})$$

At $V_{GS} = -5$ V,

$$I_D = (3 \text{ mA})\left(1 - \frac{-5 \text{ V}}{-6 \text{ V}}\right)^2 = \textbf{0.083 mA} \quad (83 \text{ μA})$$

We now have the following combinations of $V_{GS}$ and $I_D$:

| $V_{GS}$ (V) | $I_D$ (mA) |
|:---:|:---:|
| −6 | 0 |
| −5 | 0.083 |
| −3 | 0.75 |
| −1 | 2.08 |
| 0 | 3 |

The points representing these combinations are now plotted and connected to form the curve shown in Figure 12.11.

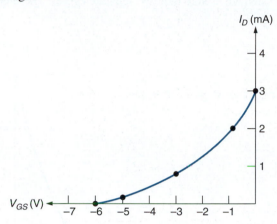

FIGURE 12.11

**Lab Reference**: A curve similar to the one in Figure 12.11 is plotted using measured values in Exercise 16.

### PRACTICE PROBLEM 12.2

A JFET has parameters of $V_{GS(off)} = -20$ V and $I_{DSS} = 12$ mA. Plot the transconductance curve for the device using $V_{GS}$ values of 0, −5, −10, −15, and −20 V.

The transconductance curve for a given JFET will be used in both the dc and ac analyses of any amplifier using that JFET. For this reason, it is important that you be able to plot the transconductance curve for any given JFET. When you need to plot a transconductance curve, simply obtain the values of $V_{GS(off)}$ and $I_{DSS}$ from the spec sheet for the device, and then follow the procedure outlined in this section.

One important point needs to be made at this time: Most JFET spec sheets list more than one value of $V_{GS(off)}$ and $I_{DSS}$. For example, the spec sheet for the 2N5457 lists the following:

$$V_{GS(off)} = -0.5 \text{ V(minimum)} \qquad I_{DSS} = 1 \text{ mA(minimum)}$$

$$V_{GS(off)} = -6 \text{ V(maximum)} \qquad I_{DSS} = 5 \text{ mA(maximum)}$$

When a range of values is given, you must use the two *minimum* values to plot one curve and the two *maximum* values to plot another curve on the same graph. This procedure is demonstrated in Example 12.3.

### EXAMPLE 12.3

Using the minimum and maximum values of $V_{GS(off)}$ and $I_{DSS}$ for the 2N5457, plot the two transconductance curves for the device.

*Solution:* The maximum values of $V_{GS(off)}$ and $I_{DSS}$ are −6 V and 5 mA, respectively. Using these values and equation (12.1), we can solve for the following points:

| $V_{GS}$ (V) | $I_D$ (mA) | |
|:---:|:---:|:---:|
| −6 | 0 | |
| −4 | 0.556 | (556 μA) |
| −2 | 2.222 | |
| 0 | 5 | |

Using the minimum values of $V_{GS(\text{off})} = -0.5$ V and $I_{DSS} = 1$ mA, we can solve for the following points with equation (12.1):

| $V_{GS}$ (V) | $I_D$ (mA) | |
|---|---|---|
| −0.5 | 0 | |
| −0.4 | 0.04 | (40 μA) |
| −0.2 | 0.36 | (360 μA) |
| 0 | 1 | |

The point sets listed are used to plot two separate curves. These curves are shown in Figure 12.12.

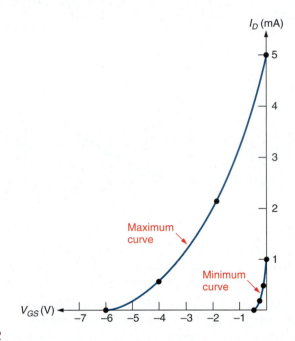

FIGURE 12.12

### PRACTICE PROBLEM 12.3

The 2N5486 JFET has values of $V_{GS(\text{off})} = -2$ to $-6$ V and $I_{DSS} = 8$ to 20 mA. Plot the minimum and maximum transconductance curves for the device.

A few more relationships that involve the principle of transconductance must be discussed. However, these principles relate to the ac operation of the JFET, so we will cover them in our discussion on JFET ac characteristics. At this point, we have the foundation needed to cover JFET biasing circuits.

## 12.1.4   One Final Note

In this section, we have used *n*-channel JFETs in all the examples because they are used far more commonly than *p*-channel JFETs. Just remember, all the *n*-channel JFET principles apply equally to *p*-channel JFETs. The only differences are the voltage polarities and the direction of the drain and source currents. The *n*-channel and *p*-channel JFETs are contrasted in Figure 12.13.

Section Review ▶

1. What is a *field-effect transistor*? What are the two types of FETs?
2. What are the three terminals of a JFET?
3. Describe the physical relationship between the JFET gate and channel.

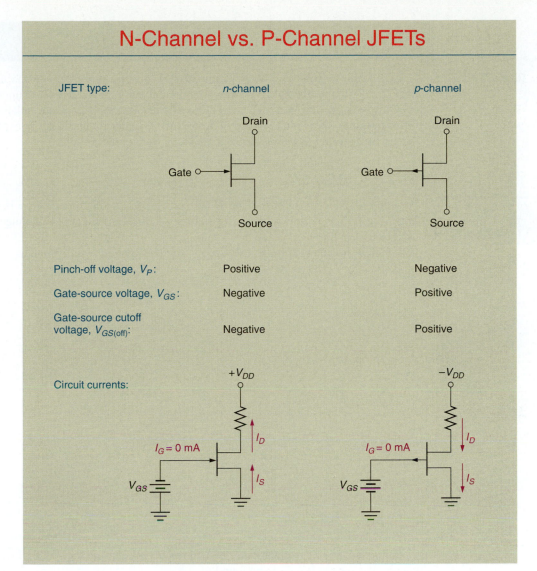

# N-Channel vs. P-Channel JFETs

| JFET type: | n-channel | p-channel |
|---|---|---|
| Pinch-off voltage, $V_P$: | Positive | Negative |
| Gate-source voltage, $V_{GS}$: | Negative | Positive |
| Gate-source cutoff voltage, $V_{GS(off)}$: | Negative | Positive |
| Circuit currents: | | |

**FIGURE 12.13**

4. What are the two types of JFETs?
5. Draw the schematic symbol for each type of JFET.
6. What supply voltage polarity is typically used for each type of JFET?
7. What is the relationship between channel width and drain current?
8. What is the relationship between $V_{GS}$ and channel width?
9. What is the relationship between $V_{GS}$ and drain current?
10. What is *pinch-off voltage*?
11. What effect does an increase in $V_{DS}$ have on $I_D$ when $V_{DS} < V_P$?
12. What effect does an increase in $V_{DS}$ have on $I_D$ when $V_{DS} > V_P$?
13. What is the operating region above $V_P$ called?
14. What is $I_{DSS}$?
15. What is the relationship between $I_D$ and $I_{DSS}$?
16. What is $V_{GS(off)}$?
17. What is the primary restriction on the value of $V_{GS}$?
18. Why do JFETs typically have extremely high gate input impedance?

**OBJECTIVE 5** ▶ JFET biasing circuits are very similar to BJT biasing circuits. The main difference between JFET circuits and BJT circuits is the operation of the active components themselves. In this section, we will cover those dc operating principles and relationships that differ from the BJT circuits already covered. We will not go into lengthy explanations of the circuit operation because these explanations were made earlier.

### 12.2.1 Gate Bias

| Gate bias |
|---|
| The JFET counterpart of BJT base bias. |

Do you remember *base bias*? The JFET counterpart of this circuit type is called **gate bias**. A gate-bias circuit is shown in Figure 12.14. The gate supply voltage ($-V_{GG}$) is used to ensure that the gate-source junction is reverse biased. Since there is no gate current, there is no voltage dropped across $R_G$, and the value of $V_{GS}$ is found as

$$V_{GS} = -V_{GG} \tag{12.2}$$

Lab Reference: The operation of gate bias is demonstrated in Exercise 17.

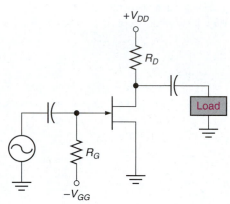

**FIGURE 12.14** Gate bias.

Using $V_{GG}$ as $V_{GS}$ in equation (12.1) allows us to calculate the value of $I_D$. Once $I_D$ is known, $V_{DS}$ for the JFET can be found as

$$V_{DS} = V_{DD} - I_D R_D \tag{12.3}$$

Example 12.4 demonstrates the complete dc analysis of a simple gate-bias circuit.

### EXAMPLE 12.4

The JFET in Figure 12.15 has values of $V_{GS(\text{off})} = -8$ V and $I_{DSS} = 16$ mA. Determine the values of $V_{GS}$, $I_D$, and $V_{DS}$ for the circuit.

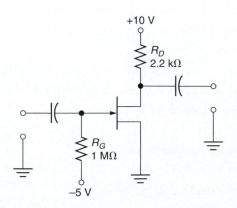

**FIGURE 12.15**

**Solution:** Since none of $V_{GG}$ is dropped across the gate resistor, $V_{GS}$ is found as

$$V_{GS} = V_{GG} = -5 \text{ V}$$

Using this value of $V_{GS}$ and the parameters listed above, the value of $I_D$ is found as

$$I_D = I_{DSS}\left(1 - \frac{V_{GS}}{V_{GS(\text{off})}}\right)^2 = (16 \text{ mA})\left(1 - \frac{-5 \text{ V}}{-8 \text{ V}}\right)^2 = 2.25 \text{ mA}$$

Now, the value of $V_{DS}$ is found as

$$V_{DS} = V_{DD} - I_D R_D = 10 \text{ V} - (2.25 \text{ mA})(2.2 \text{ k}\Omega) = 5.05 \text{ V}$$

### PRACTICE PROBLEM 12.4
Assume that the JFET in Figure 12.15 has values of $V_{GS(\text{off})} = -10$ V and $I_{DSS} = 12$ mA. Determine the values of $V_{GS}$, $I_D$, and $V_{DS}$ for the circuit.

Example 12.4 illustrated the process for analyzing a gate-bias circuit. However, it assumed that $V_{GS(\text{off})}$ and $I_{DSS}$ have one specified value each. As you know, this is not usually the case.

The fact that a given type of JFET can have a range of values for $V_{GS(\text{off})}$ and $I_{DSS}$ leads us to a major problem with the gate-bias circuit: *Gate bias does not provide a stable Q-point from one JFET to another*. This problem is illustrated in Example 12.5.

As Example 12.5 shows, the circuit does not provide a stable Q-point. You may place one 2N5458 in the circuit and get an $I_D$ of 5 mA. Another 2N5458 may give you an $I_D$ of 2 mA, and so on. Because of the instability of the circuit, gate bias is rarely used for anything other than switching applications.

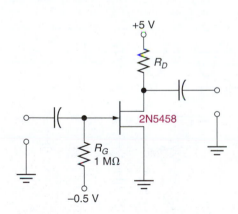

**FIGURE 12.16**

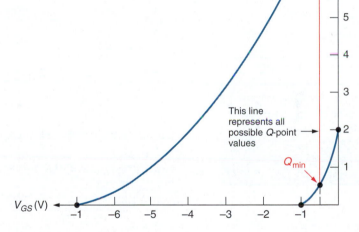

**FIGURE 12.17**

### EXAMPLE 12.5

Determine the *range* of Q-point values for the circuit shown in Figure 12.16. Assume that the JFET has ranges of $V_{GS(\text{off})} = -1$ to $-7$ V and $I_{DSS} = 2$ to $9$ mA.

**Solution:** Using the established procedure, the two transconductance curves for the 2N5458 are plotted. These curves are shown in Figure 12.17. Since $V_{GS}$ is set to $-0.5$ V, a vertical line called a *bias line* is drawn from this point through the two transconductance curves, as shown in Figure 12.17. The point where the line intersects the maximum curve is the value of $Q_{\text{max}}$. The point where the line intersects the minimum curve is the value of $Q_{\text{min}}$. The line between the two points represents all the possible Q-points for a 2N5458 used in the circuit shown in Figure 12.16.

**Lab Reference**: Gate bias instability is demonstrated in Exercise 17.

*A Practical Consideration:* When only a maximum curve can be plotted, the Q-point can fall anywhere between the points where the bias line intersects the curve and either axis of the graph.

You may be wondering why the gate-bias circuit contains a gate resistor when there is no gate current. Since resistors are usually used to develop a voltage, this resistor would seem to be useless. After all, there is no voltage drop across the component and no current for it to limit. Actually, $R_G$ is used for ac operation purposes. Take a look at the circuits shown in Figure 12.18. Figure 12.18a shows the results of eliminating $R_G$. The ac signal produced by the first stage is coupled to the second, only to be shorted to the dc supply, $-V_{GG}$. The resistor is needed to prevent this from happening. As Figure 12.18b shows, placing a gate resistor in the second stage restores the ac signal. Therefore, while the gate resistor has little to do with the dc operation of the circuit, it is a vital part of the ac operation of the circuit. As you will see, *self-bias* uses a gate resistor for the same reason.

The waveform shown at the drain of $Q_1$ in Figure 12.18a is included to illustrate a point. In practice, this waveform would be shorted to $-V_{GG}$ through the low reactance of the capacitor.

*A Practical Consideration:* The value of $R_G$ does not impact the dc biasing because there is no current in the gate circuit. The value of $R_G$ is normally determined by the input impedance requirements of the circuit.

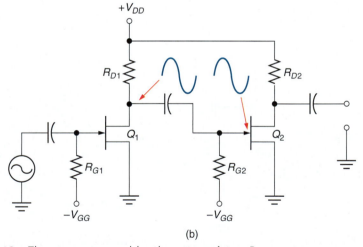

FIGURE 12.18  The purpose served by the gate resistor, $R_G$.

### 12.2.2   Self-Bias

**Self-bias**

A JFET biasing circuit that uses a source resistor to help establish a negative $V_{GS}$.

**Self-bias** is a more viable type of JFET biasing. The self-bias circuit replaces the gate supply $(-V_{GG})$ with a source resistor $(R_S)$. The self-bias circuit is shown in Figure 12.19. Note that the gate is returned to ground via $R_G$, and a resistor has been added in the source circuit. This resistor helps to produce the $-V_{GS}$ needed for the operation of the JFET. Figure 12.20 helps to illustrate this point.

In any JFET circuit, *all* the source current passes through the device to the drain circuit. This is due to the fact that there is no significant gate current. Therefore, we can define source current as

$$I_S = I_D \qquad\qquad (12.4)$$

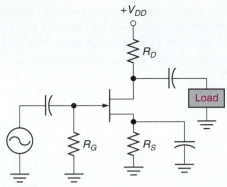

FIGURE 12.19   Self-bias.

Lab Reference: Self-bias operation is demonstrated in Exercise 17.

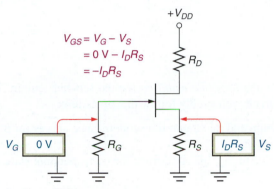

FIGURE 12.20   $V_{GS}$ for the self-bias current.

With all of the drain current passing through the source resistor, the voltage across the resistor is found as

$$V_S = I_D R_S \qquad (12.5)$$

Since there is no significant current in the gate circuit, no voltage is developed across the gate resistor. Thus, the gate voltage (with respect to ground) is given as

$$V_G = 0 \text{ V} \qquad (12.6)$$

Now, we can use the relationships shown in equations (12.5) and (12.6) to show how the source resistor is used to reverse bias the gate-source junction. The voltage *from gate to source* can be expressed as

$$V_{GS} = V_G - V_S \qquad (12.7)$$

Substituting the values in equations (12.5) and (12.6) in place of $V_G$ and $V_S$ gives us

$$V_{GS} = 0 \text{ V} - I_D R_S$$

which can be written as

$$V_{GS} = -I_D R_S \qquad (12.8)$$

Thus, a negative $V_{GS}$ is developed by the current through the source resistor.

If equation (12.8) is confusing, just remember that $V_{GS}$ is the voltage measured *from the gate terminal to the source terminal*. There is no rule stating that the voltage at the gate terminal must be negative, only that it must be *more negative* than the voltage at the source terminal. If the source terminal is at some positive voltage and the gate is grounded, the gate terminal is *more negative* (less positive) than the source terminal. Figure 12.21 helps to illustrate this point. In the first circuit, a value of $V_S = +2$ V results in a value of $V_{GS} = -2$ V. A value of $V_S = +4$ V in the second circuit results in a value of $V_{GS} = -4$ V. In either case, $V_{GS}$ is the *negative* equivalent of $V_S$.

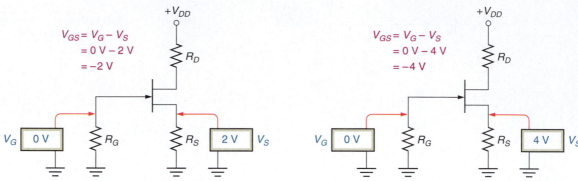

**FIGURE 12.21**   How self-bias produces a negative $V_{GS}$.

Equation (12.8) is the *bias line equation* for the self-bias circuit. To plot the dc bias line for a self-bias circuit, you simply follow this procedure:

How is the dc bias line for a self-bias circuit plotted?

1. Plot the minimum and maximum transconductance curves for the JFET used in the circuit.
2. Choose any value of $V_{GS}$, and determine the corresponding value of $I_D$ using

$$I_D = \frac{-V_{GS}}{R_S}$$

(12.9)

3. Plot the point determined by equation (12.9), and draw a line from this point to the graph *origin* (the [0, 0] point).
4. The points where the line crosses the two transconductance curves define the limits of the $Q$-point operation of the circuit.

This procedure is demonstrated in Example 12.6.

### EXAMPLE 12.6

Determine the range of $Q$-point values for the circuit shown in Figure 12.22.

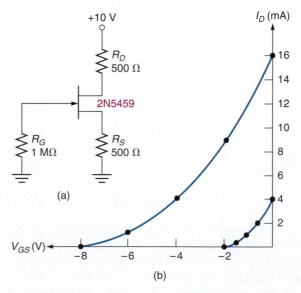

(a)

(b)

**FIGURE 12.22**

***Solution:*** The 2N5459 spec sheet lists the following values for $V_{GS(off)}$ and $I_{DSS}$:

$$V_{GS(off)} = -2 \text{ to } -8 \text{ V}$$

$$I_{DSS} = 4 \text{ to } 16 \text{ mA}$$

Using these two sets of values and equation (12.1), the transconductance curves are plotted as shown in Figure 12.22b.

The value of $V_{GS} = -4$ V is chosen at random to calculate $I_D$. Using equation (12.9), $I_D$ at $V_{GS} = -4$ V is found as

$$I_D = \frac{-V_{GS}}{R_S} = \frac{4 \text{ V}}{500 \text{ }\Omega} = 8 \text{ mA}$$

This point $(-4, 8)$ is plotted on the graph, and a line is drawn from the point to the graph origin. The plot of the line is shown in Figure 12.23. Note that the points where the bias line intersects the transconductance curves are labeled $Q_{max}$ and $Q_{min}$. Using the graph coordinates of these two points, we obtain the maximum and minimum values of $V_{GS}$ and $I_D$ shown in Figure 12.23.

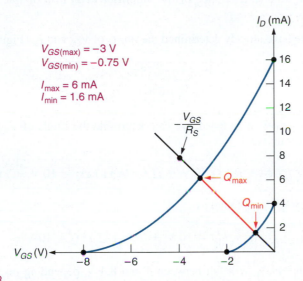

**FIGURE 12.23**

**PRACTICE PROBLEM 12.6**

A JFET with parameters of $V_{GS(off)} = -5$ to $-10$ V and $I_{DSS} = 5$ to $10$ mA is used in a self-bias circuit with a value of $R_S = 2$ kΩ. Plot the transconductance curves and bias line for the circuit.

As Example 12.6 demonstrates, there is still some instability in the $Q$-point. In other words, we have not succeeded in obtaining a completely stable $Q$-point value by switching to self-bias. However, as Figure 12.24 indicates, the $Q$-point of a self-bias circuit is far more stable for a given JFET than the gate-bias circuit. In Figure 12.24, the bias line from Example 12.6 is shown as plotted in the example. A dashed line is plotted showing the same transistor in a gate-bias circuit. From these two dc bias lines, a couple of observations can be made:

1. The gate-bias circuit is not subject to any change in $V_{GS}$ and, thus, is more stable in this respect.
2. The possible change in $I_D$ from one 2N5459 to another is much greater for the gate-bias circuit than for the self-bias circuit.

Thus, the gate-bias circuit has the more stable value of $V_{GS}$, while the self-bias circuit has the more stable value of $I_D$. Which is more important? This question is best answered by

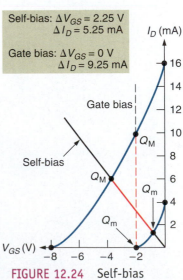

**FIGURE 12.24** Self-bias stability versus gate-bias stability.

considering the ac operation of amplifiers in general. The ac characteristics like compliance, amplifier power dissipation, and load power all depend on the output voltage and current characteristics of the amplifier. You may recall that PP, $P_L$, and $P_S$ for the BJT amplifier are all defined in terms of $V_{CEQ}$ and $I_{CQ}$. For the JFET amplifier, the ac output characteristics are defined in terms of $V_{DS}$ and $I_D$. The relationship between $V_{DS}$ and $I_D$ is given as

$$V_{DS} = V_{DD} - I_D(R_D + R_S) \qquad \text{(12.10)}$$

If the ac output characteristics of the JFET amplifier are to be stable, $I_D$ must be as stable as possible. Since self-bias provides a more stable value of $I_D$ than gate bias, it is the preferred biasing method of the two.

The complete dc analysis of a self-bias circuit is actually pretty simple. The process starts with plotting the transconductance curves and the dc bias line, as was done in Example 12.6. Then, using the $Q$-point limits, the minimum and maximum values of $V_{GS}$ and $I_D$ are determined. The value of $I_D$ can then be used to determine the minimum and maximum values of $V_{DS}$. This analysis process is demonstrated in Example 12.7.

### EXAMPLE 12.7

Determine the dc characteristics of the amplifier used in Example 12.6 (Figure 12.22).

*Solution:*   We have already determined the limits of $V_{GS}$ and $I_D$ (Figure 12.23) as

$$V_{GS} = -0.75 \text{ to } -3 \text{ V}$$
$$I_D = 1.6 \text{ to } 6 \text{ mA}$$

Using the two values of $I_D$ listed, we can determine the limits of $V_{DS}$. When $I_D = 1.6$ mA, $V_{DS}$ is found as

$$V_{DS} = V_{DD} - I_D(R_D + R_S) = 10 \text{ V} - (1.6 \text{ mA})(1 \text{ k}\Omega) = 10 \text{ V} - 1.6 \text{ V} = \textbf{8.4 V}$$

When $I_D = 6$ mA, $V_{DS}$ is found as

$$V_{DS} = V_{DD} - I_D(R_D + R_S) = 10 \text{ V} - (6 \text{ mA})(1 \text{ k}\Omega) = 10 \text{ V} - 6 \text{ V} = \textbf{4 V}$$

Thus, the value of $V_{DS}$ will fall between 4 and 8.4 V, depending on the particular 2N5459 used in the circuit.

### PRACTICE PROBLEM 12.7
The circuit described in Practice Problem 12.6 has a 1 k$\Omega$ drain resistor and a 9 V source. Determine the range of $V_{DS}$ values for the circuit.

Example 12.7 illustrates one important point: Even though the self-bias circuit is more stable than gate bias, it still leaves a lot to be desired. A range of $V_{DS} = 4$ to 8.4 V would hardly be stable enough for most linear applications. We must therefore look for a circuit to provide a much greater amount of stability. *Voltage-divider bias* does the job very well.

### 12.2.3    Voltage-Divider Bias

The voltage-divider biased JFET amplifier is very similar to its BJT counterpart. This biasing circuit is shown in Figure 12.25. The gate voltage for this amplifier is found in the same manner as $V_B$ in the BJT circuit. By formula,

$$V_G = V_{DD} \frac{R_2}{R_1 + R_2} \qquad \text{(12.11)}$$

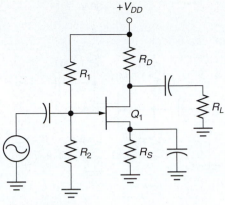

**FIGURE 12.25** Voltage-divider bias.

Also, the difference between $V_G$ and $V_S$ is equal to $V_{GS}$. According to Ohm's law, $I_D$ can be found as

$$I_D = \frac{V_S}{R_S}$$

Since $V_S + V_{GS} = V_G$, we can rewrite this equation as

$$I_D = \frac{V_G - V_{GS}}{R_S} \qquad \textbf{(12.12)}$$

Now, we have a problem. If you refer to Examples 12.6 and 12.7 on the self-bias circuit, you will see that $V_{GS}$ changes from one transistor to another. The same holds true for the voltage-divider biased circuit. This can make it difficult to determine the exact value of $I_D$. However, plotting the dc bias line for this circuit will show you that the problem is not a major one. The method used to plot the dc bias line for a voltage-divider bias circuit is as follows:

> How is the dc bias line for voltage-divider bias plotted?

1. As usual, plot the transconductance curves for the specific JFET.
2. Calculate the value of $V_G$ using equation (12.11).
3. *On the positive x-axis of the graph*, plot a point at the value of $V_G$.
4. Solve for $I_D$ using

$$I_D = \frac{V_G}{R_S} \qquad \textbf{(12.13)}$$

5. Locate the point *on the y-axis* that corresponds to the value found in step 4.
6. Draw a line from the $V_G$ point through the point found with equation (12.13), and continue the line to intersect *both* transconductance curves. The intersection points represent the $Q_{max}$ and $Q_{min}$ points.

Example 12.8 demonstrates this procedure.

### EXAMPLE 12.8

Plot the dc bias line for the circuit shown in Figure 12.26.

*Solution:* Figure 12.26 contains the 2N5459, the JFET whose transconductance curves were plotted in Example 12.6. Those transconductance curves can be seen in Figure 12.27. The value of $V_G$ for the circuit is calculated as

$$V_G = V_{DD}\frac{R_2}{R_1 + R_2} = (30\ \text{V})\frac{1.5\ \text{M}\Omega}{3\ \text{M}\Omega} = \textbf{15 V}$$

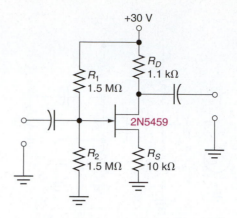

**FIGURE 12.26**

This point is now plotted on the *positive x*-axis, as shown in Figure 12.27. The value of $V_G$ is also used to find the *y*-axis intersect point as follows:

$$I_D = \frac{V_G}{R_S} = \frac{15\,V}{10\,k\Omega} = \mathbf{1.5\ mA}$$

Note that this point is marked on the *y*-axis of the graph. A line is then drawn through the two points and on through the two curves to establish the $Q_{max}$ and $Q_{min}$ points.

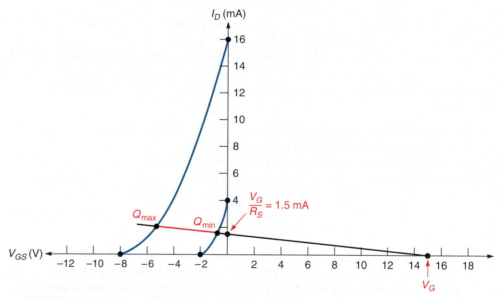

**FIGURE 12.27** The voltage-divider bias dc bias line.

### PRACTICE PROBLEM 12.8

In Practice Problem 12.3, you plotted the two transconductance curves for the 2N5486. This component is used in a voltage-divider biased amplifier with values of $V_{DD} = 36\,V$, $R_1 = 10\,M\Omega$, $R_2 = 3.3\,M\Omega$, $R_D = 1.8\,k\Omega$, and $R_S = 3\,k\Omega$. Plot the dc bias line for the amplifier.

Figure 12.27 shows that the stability of $I_D$ has improved a great deal. Figure 12.28 shows a close-up of the lower portion of Figure 12.27 to help illustrate this point. While $V_{GS}$ is varying between approximately $-5$ and $-0.5$ V, $I_D$ varies only from 2 mA to approximately 1.5 mA. Thus, this amplifier provides a much more stable value of $I_D$ than

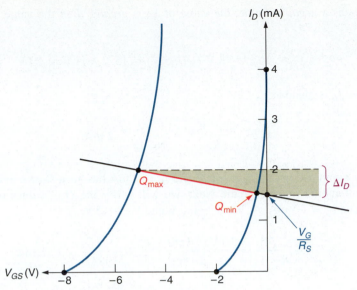

FIGURE 12.28 A close-up of Figure 12.27.

either gate bias or self-bias. For comparison, the results from Figures 12.24 and 12.28 are summarized as follows:

| Circuit Type | Change in 2N5459 Drain Current Values (mA) |
|---|---|
| Gate bias | 9.25 |
| Self-bias | 5.25 |
| Voltage-divider bias | 0.5 |

As you can see, the voltage-divider bias circuit is by far the most stable of the three.

It would seem that this stability of $I_D$ would be impossible given the large possible variations in $V_{GS}$ for the amplifier in Example 12.8. Since $I_D$ is a function of $V_{GS}$, as was stated in equation (12.1), it would seem that differences in $V_{GS}$ would result in significant differences in $I_D$. However, an analysis of the circuit shows the Q-point values obtained to be valid. This analysis is performed in Example 12.9.

### EXAMPLE 12.9

Example 12.8 gave minimum and maximum Q-point values for Figure 12.26 as

$$Q_{max} = -5 \text{ V at 2 mA}$$
$$Q_{min} = -0.5 \text{ V at 1.5 mA} \qquad \text{(approximate)}$$

Verify these two Q-point values as being possible operating combinations of $V_{GS}$ and $I_D$.

**Solution:** We will tackle the $Q_{max}$ values of $V_{GS} = -5$ V and $I_D = 2$ mA first. Just as before, we start by determining the value of $V_G$ as follows:

$$V_G = V_{DD} \frac{R_2}{R_1 + R_2} = \textbf{15 V}$$

Now, recall that $V_S$ is equal to the difference between $V_G$ and $V_{GS}$. By formula,

$$V_S = V_G - V_{GS}$$

Since $V_{GS}$ is a *negative* value, the value of $V_S$ is *greater than* the value of $V_G$ as follows:

$$V_S = V_G - V_{GS} = 15 \text{ V} - (-5 \text{ V}) = 15 \text{ V} + 5 \text{ V} = \mathbf{20 \text{ V}}$$

Therefore, $I_D$ is found as

$$I_D = \frac{V_G - V_{GS}}{R_S} = \frac{20 \text{ V}}{10 \text{ k}\Omega} = \mathbf{2 \text{ mA}}$$

Thus, the combination of $V_{GS} = -5$ V and $I_D = 2$ mA has been shown to be a possible combination for the circuit. Next, we will verify the $Q_{min}$ values of $V_{GS} = -0.5$ V and $I_D \cong 1.5$ mA. At $V_{GS} = -0.5$ V and $V_G = 15$ V,

$$I_D = \frac{V_G - V_{GS}}{R_S} = \frac{15 \text{ V} - (-0.5 \text{ V})}{10 \text{ k}\Omega} = \frac{15.5 \text{ V}}{10 \text{ k}\Omega} = \mathbf{1.55 \text{ mA}}$$

Again, the results obtained from plotting the bias line of Figure 12.24 have been shown to be accurate.

### PRACTICE PROBLEM 12.9

Using the technique shown in Example 12.9, verify your $Q$-point values for Practice Problem 12.8.

When you are analyzing a voltage-divider biased JFET amplifier, the procedure is the same as the one that we followed in plotting the transconductance curves and verifying the results. The only other value that must be determined is $V_{DS}$. $V_{DS}$ is determined for this circuit as it is with all the others—using equation (12.10). In Example 12.10, we calculate the range of possible $V_{DS}$ values for the circuit we have covered in this section.

### EXAMPLE 12.10

Determine the minimum and maximum values of $V_{DS}$ for the circuit shown in Figure 12.26.

*Solution:* When $I_D = 2$ mA,

$$V_{DS} = V_{DD} - I_D(R_D + R_S) = 30 \text{ V} - (2 \text{ mA})(11.1 \text{ k}\Omega) = \mathbf{7.8 \text{ V}}$$

When $I_D = 1.55$ mA,

$$V_{DS} = V_{DD} - I_D(R_D + R_S) = 30 \text{ V} - (1.55 \text{ mA})(11.1 \text{ k}\Omega) = \mathbf{12.8 \text{ V}}$$

### PRACTICE PROBLEM 12.10

Determine the minimum and maximum values of $V_{DS}$ for the circuit described in Practice Problem 12.8. Use the calculated values of $I_D$ from Practice Problem 12.9.

We now have a situation where the value of $I_D$ is much more stable than it has been with previous circuits. However, the smaller variation in $I_D$ *still* allows a rather drastic change in $V_{DS}$.

### 12.2.4    Current-Source Bias

**Current-source bias** provides high $Q$-point stability by making the value of $I_D$ independent of the JFET. Two current-source bias circuits are shown in Figure 12.29. For both circuits, the JFET drain current is equal to the *collector* current of the BJT. By formula,

**Current-source bias**
A JFET biasing circuit that uses a BJT to maintain a constant value of drain current ($I_D$).

$$I_D = I_C \qquad\qquad (12.14)$$

Since the value of $I_C$ is independent of the variations in JFET parameters, so is the value of $I_D$. This assumes, of course, that the value of $I_C$ is less than the *lowest* value of $I_{DSS}$ for the JFET. For example, let's assume that the JFET in Figure 12.29a has a value of $I_{DSS} =$ 5 to 12 mA. As long as $I_C$ is less than 5 mA, the value of $I_D$ is independent of the JFET itself. Thus, a stable $Q$-point is obtained by designing the circuit so that $I_C$ equals the desired value of $I_D$.

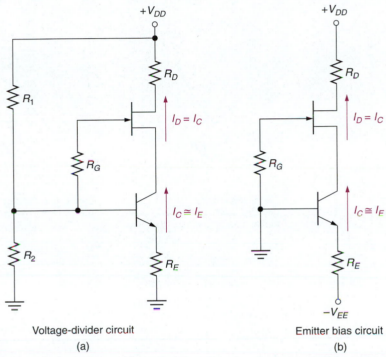

**FIGURE 12.29**    Current-source bias.

In Figure 12.29a, the value of $I_C$ is found as it would be for any voltage-divider biased BJT amplifier. In Figure 12.29b, the value of $I_C$ is found as it would be for any emitter bias circuit.

Even though current-source bias provides the most stable $Q$-point value of $I_D$, the complexity of the circuit makes it undesirable for most applications; that is, the improved stability (over voltage-divider bias) is usually insufficient to warrant the extra circuitry.

### 12.2.5    Summary

Gate bias, self-bias, and voltage-divider bias are the three most commonly used JFET biasing circuits. Each of these circuits has its own type of dc bias line as well as advantages and disadvantages. The characteristics of these three dc biasing circuits are summarized in Figure 12.30.

◀ *OBJECTIVE 6*

# Common JFET Biasing Circuits

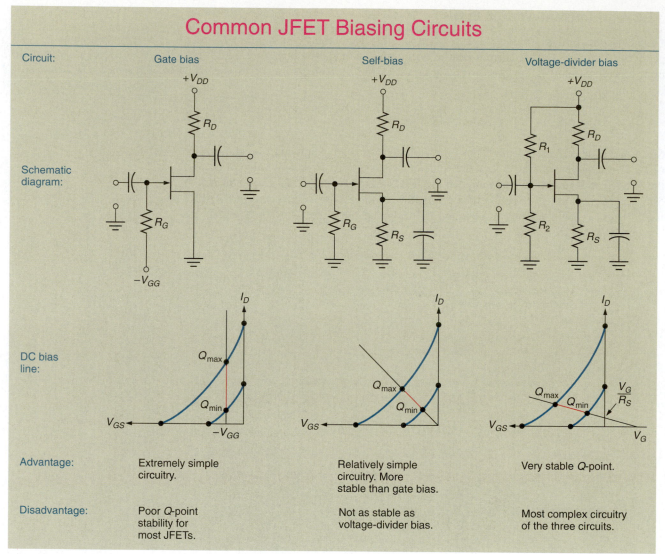

| | Gate bias | Self-bias | Voltage-divider bias |
|---|---|---|---|
| **Circuit:** Schematic diagram: | | | |
| **DC bias line:** | | | |
| **Advantage:** | Extremely simple circuitry. | Relatively simple circuitry. More stable than gate bias. | Very stable $Q$-point. |
| **Disadvantage:** | Poor $Q$-point stability for most JFETs. | Not as stable as voltage-divider bias. | Most complex circuitry of the three circuits. |

FIGURE 12.30

---

**Section Review ▶**

1. What is the primary disadvantage of gate bias?
2. How does self-bias produce a negative value of $V_{GS}$?
3. What is the process used to plot the dc bias line for the self-bias circuit?
4. What is the process used to plot the dc bias line for the voltage-divider bias circuit?

**Critical Thinking ▶**

5. Designing for $I_D \cong 0$ A is common for switching applications. What change would you make to the circuit shown in Figure 12.16 to ensure that $I_D \cong 0$ A for any 2N5458 used in the circuit?
6. What biasing circuit could be viewed as the BJT counterpart of self-bias? Explain.
7. Refer to Figure 12.29a. If $R_2$ opens, what effect (if any) does it have on the JFET drain voltage ($V_D$)?

## 12.3    AC Operating Characteristics: The Common-Source Amplifier

In many ways, the ac operation of a JFET amplifier is similar to that of the BJT amplifier. However, some differences need to be examined in detail, such as the effects of changes in JFET transconductance.

In the next two sections, we will examine the ac operation of JFET amplifiers. We will start here with the *common-source* amplifier simply because it is the most widely used of the JFET amplifiers. Then, in the next section, we will look at the operation of the *common-gate* and *source-follower* (common-drain) amplifiers.

### 12.3.1    Operation Overview

◄ *OBJECTIVE 7*

There are many similarities between BJT amplifiers and JFET amplifiers. At the same time, there are differences that make JFET amplifiers more desirable in some applications and that limit their use in others. One of the biggest differences is that JFETs are voltage-controlled components and BJTs are current-controlled components. Another major difference is the input impedance of the JFET, as opposed to the input impedance of the BJT. As you will see, these differences play a major role in which transistor type is selected for a specific application.

In general terms, the JFET amplifier responds to an input signal in much the same way as a BJT amplifier. Figure 12.31 shows a **common-source amplifier**, which is the JFET counterpart of the *common-emitter* amplifier. Note that the input and output signals for the common-source amplifier closely resemble those of the common-emitter amplifier. Both amplifiers have a 180° signal voltage phase shift from input to output. Both amplifiers provide voltage gain for the input signal.

> **Common-source amplifier**
> The JFET counterpart of a common-emitter amplifier.

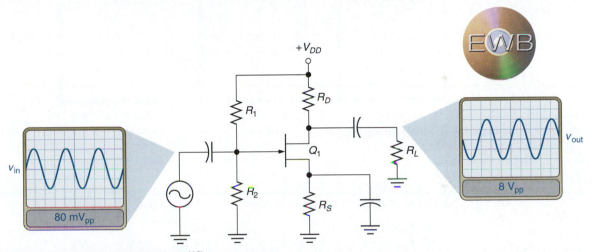

**FIGURE 12.31**   The common-source amplifier.

Even though the common-source and common-emitter amplifiers serve the same basic purpose, the means by which they operate are quite different. The ac operation of the common-source amplifier is better explained with the help of Figure 12.32. The initial dc operating values are represented as straight-line values in the blocks. These voltage values are assumed to be as follows:

$$V_G = +8 \text{ V}$$
$$V_S = +12 \text{ V}$$
$$V_D = +24 \text{ V}$$
$$V_{GS} = V_G - V_S = -4 \text{ V}$$

Here's what happens when $V_G$ is caused to increase by the input signal:

> How does the common-source amplifier respond to an input signal?

1. $V_G$ increases to $+10$ V.
2. $V_{GS}$ decreases to $-2$ V ($V_{GS} = V_G - V_S = 10 \text{ V} - 12 \text{ V} = -2 \text{ V}$).
3. The decrease in $V_{GS}$ causes an *increase* in $I_D$, as shown in the transconductance curve.
4. Increasing $I_D$ causes $V_D$ to decrease ($V_D = V_{DD} - I_D R_D$).

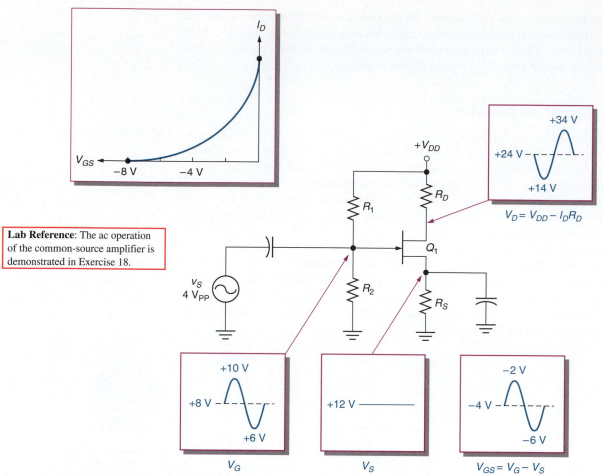

**Lab Reference**: The ac operation of the common-source amplifier is demonstrated in Exercise 18.

**FIGURE 12.32** Common-source amplifier ac operation.

Thus, when the input voltage increases, the output voltage decreases. Here is what happens when $V_G$ is caused to decrease by the input signal:

1. $V_G$ decreases to $+6$ V.
2. $V_{GS}$ increases to $-6$ V ($V_{GS} = V_G - V_S = 6\text{ V} - 12\text{ V} = -6\text{ V}$).
3. The *increase* in $V_{GS}$ causes a *decrease* in $I_D$, as shown in the transconductance curve.
4. Decreasing $I_D$ causes $V_D$ to increase ($V_D = V_{DD} - I_D R_D$).

We have seen how *a change in $V_{GS}$ causes a change in $I_D$*. When you start dealing with changing quantities of $V_{GS}$ and $I_D$, you get into the area of transconductance. Next, we will take a look at transconductance and its effects on ac operation.

### 12.3.2 Transconductance

We have been plotting transconductance curves throughout this chapter as graphs of $V_{GS}$ versus $I_D$. **Transconductance** is *a ratio of a change in drain current to a change in gate-source voltage*. By formula,

**Transconductance ($g_m$)**
A ratio of a change in drain current ($I_D$) to a change in $V_{GS}$, measured in microsiemens ($\mu$S) or micromhos ($\mu$mhos).

$$g_m = \frac{\Delta I_D}{\Delta V_{GS}} \qquad (12.15)$$

where $g_m =$ the transconductance of the JFET at a given value of $V_{GS}$
$\Delta I_D =$ the change in $I_D$
$\Delta V_{GS} =$ the change in the value of $V_{GS}$

As a rating, transconductance is measured in *microsiemens* (μS) or *micromhos* (μmhos). You may recall that conductance is measured in *siemens* or *mhos*.

We have plotted several transconductance curves throughout the chapter. The maximum-value transconductance curve for the 2N5459 (shown in Figure 12.33) will help you to see one of the problems with using transconductance in ac calculations. Note that two sets of points were selected on the curve, each marking a $\Delta V_{GS}$ of 1 V. Points $A_1$ and $A_2$ correspond to a change in $V_{GS}$ of 1 V, as do points $B_1$ and $B_2$. For each set of points, the corresponding change in $I_D$ is shown.

*A Practical Consideration:* Although *siemens* is the preferred unit of measure, some spec sheets still rate transconductance in *mhos*. This is why the unit is identified here.

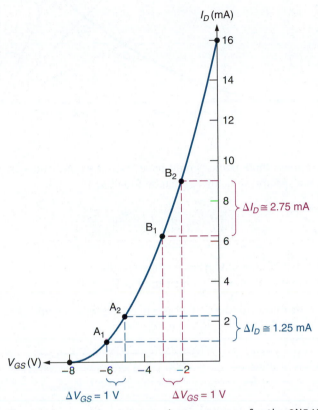

FIGURE 12.33   The maximum-value transconductance curve for the 2N5459.

The values of $\Delta V_{GS}$ and their corresponding values of $\Delta I_D$ are summarized as follows:

| $\Delta V_{GS}$ (V) | $\Delta I_D$ (mA) |
| --- | --- |
| −6 to −5 = 1 | 1.25 |
| −3 to −2 = 1 | 2.75 |

Using equation (12.15), we can calculate the values of $g_m$ that correspond to the changes above as

$$g_{m(A)} = \frac{\Delta I_D}{\Delta V_{GS}} = \frac{1.25 \text{ mA}}{1 \text{ V}} = 1250 \text{ μS}$$

and

$$g_{m(B)} = \frac{\Delta I_D}{\Delta V_{GS}} = \frac{2.75 \text{ mA}}{1 \text{ V}} = 2750 \text{ μS}$$

As you can see, the value of $g_m$ is not constant across the transconductance curve. Figure 12.34 shows a single transconductance curve with three possible dc bias lines. A single

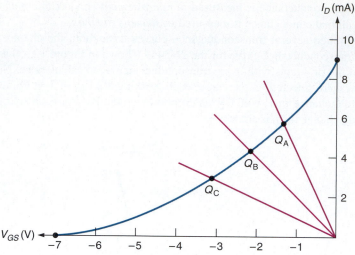

**FIGURE 12.34**

JFET operated at these three points would have three separate values of $g_m$. Fortunately, the value of $g_m$ at a specific value of $V_{GS}$ can be found as

$$g_m = g_{m0}\left(1 - \frac{V_{GS}}{V_{GS(\text{off})}}\right) \qquad (12.16)$$

where $g_m$ = the value of transconductance at the specific value of $V_{GS}$
$g_{m0}$ = the maximum value of $g_m$, measured at $V_{GS} = 0$ V

Example 12.11 shows how the value of $g_m$ is calculated for a specific value of $V_{GS}$.

### EXAMPLE 12.11

The 2N5459 has a rating of $g_{m0} = 4000$ μS. Determine the values of $g_m$ at $V_{GS} = -3$ V and $V_{GS} = -5$ V for this device. For the 2N5459, $V_{GS(\text{off})} = -8$ V.

*Solution:* At $V_{GS} = -3$ V,

$$g_m = g_{m0}\left(1 - \frac{V_{GS}}{V_{GS(\text{off})}}\right) = (4000 \text{ μS})\left(1 - \frac{-3 \text{ V}}{-8 \text{ V}}\right) = (4000 \text{ μS})(0.625) = \textbf{2500 μS}$$

At $V_{GS} = -5$ V,

$$g_m = (4000 \text{ μS})\left(1 - \frac{-5 \text{ V}}{-8 \text{ V}}\right) = (4000 \text{ μS})(0.375) = \textbf{1500 μS}$$

### PRACTICE PROBLEM 12.11

The 2N5486 has maximum values of $g_{m0} = 8000$ μS and $V_{GS(\text{off})} = -6$ V. Using these values, determine the values of $g_m$ at $V_{GS} = -2$ V and $V_{GS} = -4$ V.

On many specification sheets, the value of $g_{m0}$ is labeled $y_{fs}$ or $g_{fs}$. When no value of maximum transconductance is given, you can approximate the value of $g_{m0}$ as

$$g_{m0} \cong \frac{2I_{DSS}}{V_{GS(\text{off})}} \qquad (12.17)$$

While equation (12.17) does not give you an exact value of $g_{m0}$, the value obtained is usually close enough to serve as a valid approximation.

## 12.3.3　Amplifier Voltage Gain

The ac operation of a JFET amplifier is closely related to the value of $g_m$ at the dc value of $V_{GS}$. Consider the common-source amplifier shown in Figure 12.35a and its ac equivalent. Note that the biasing resistors ($R_1$ and $R_2$) have been replaced in Figure 12.35b with a single parallel equivalent resistor, $r_G$. Also, the drain and load resistances ($R_D$ and $R_L$) have been replaced by the single parallel equivalent resistor, $r_D$. This is the standard ac equivalent circuit, as developed in our discussions on BJT amplifiers.

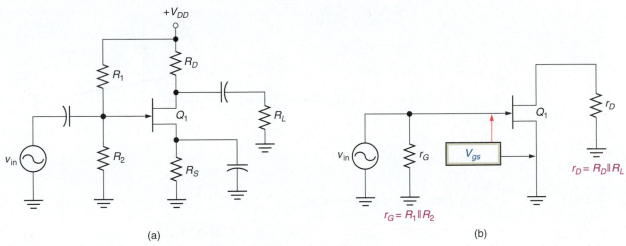

(a)                                                                 (b)

FIGURE 12.35　The common-source amplifier and its ac equivalent circuit.

For Figure 12.35b, the voltage gain is found as

$$A_v = g_m r_D \qquad\qquad\qquad \textbf{(12.18)}$$

Equation (12.18) points out another problem with the JFET amplifier. As you can see, the value of $A_v$ depends on the value of $g_m$. You have also been shown the following:

1. The value of $g_m$ depends on the value of $V_{GS}$.
2. The value of $V_{GS}$ can vary from one JFET to another.

This leads to the conclusion that *the voltage gain of an amplifier can vary significantly from one JFET to another*. This point is illustrated in Example 12.12.

### EXAMPLE 12.12

Determine the maximum and minimum values of $A_v$ for the amplifier shown in Figure 12.36. Assume that the 2N5459 has values of $g_{m0} = 6000\ \mu\text{S}$ at $V_{GS(\text{off})} = -8$ V and $g_{m0} = 2000\ \mu\text{S}$ at $V_{GS(\text{off})} = -2$ V.

*Solution:*　The transconductance curves and dc bias line for the amplifier are shown in Figure 12.36b. As shown, the maximum and minimum values of $V_{GS}$ are $-5$ V and $-1$ V, respectively. Using the maximum values of $V_{GS(\text{off})}$ and $V_{GS}$, $g_m$ is found as

$$g_m = g_{m0}\left(1 - \frac{V_{GS}}{V_{GS(\text{off})}}\right) = (6000\ \mu\text{S})\left(1 - \frac{-5\text{ V}}{-8\text{ V}}\right) = 2250\ \mu\text{S}$$

Using the minimum values of $V_{GS(\text{off})}$ and $V_{GS}$, $g_m$ is found as

$$g_m = (2000\ \mu\text{S})\left(1 - \frac{-0.75\text{ V}}{-2\text{ V}}\right) = 1250\ \mu\text{S}$$

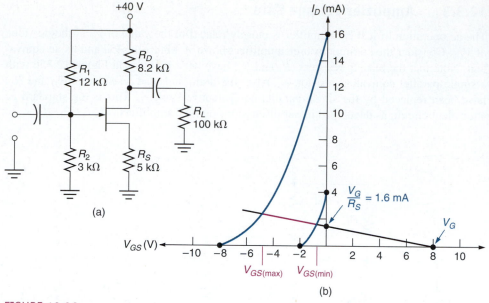

FIGURE 12.36

The value of $r_D$ is found as

$$r_D = R_D \parallel R_L = \mathbf{7.58 \ k\Omega}$$

The maximum value of $A_v$ can now be found as

$$A_v = r_D g_{m(max)} = (7.58 \ k\Omega)(2250 \ \mu S) = \mathbf{17.06}$$

The minimum value of $A_v$ is found as

$$A_v = r_D g_{m(min)} = (7.58 \ k\Omega)(1250 \ \mu S) = \mathbf{9.48}$$

**PRACTICE PROBLEM 12.12**
The amplifier described in Practice Problem 12.8 has a load resistance of 4.7 k$\Omega$. The 2N5486 has a $g_{m0}$ range of 4000 to 8000 $\mu$S. Using the values of $V_{GS}$ from Practice Problem 12.9, calculate the range of $A_v$ values for the amplifier.

### 12.3.4 The Basis for Equation (12.18)

The ac equivalent circuit shown in Figure 12.35b has an output voltage found as

$$v_{\text{out}} = i_d r_D \qquad (12.19)$$

where $v_{\text{out}}$ = the output signal voltage
$i_d$ = the change $I_D$ in caused by the amplifier input signal

Now, recall that $g_m$ is defined in equation (12.15) as

$$g_m = \frac{\Delta I_D}{\Delta V_{GS}}$$

Another way of writing this equation is

$$g_m = \frac{i_d}{v_{gs}} \qquad (12.20)$$

where $v_{gs}$ is the change in $V_{GS}$ caused by the amplifier input signal, equal to that input signal. Equation (12.20) can be rewritten as

$$i_d = g_m v_{gs}$$

Substituting this equation into equation (12.19) gives us

$$v_{out} = g_m v_{gs} r_D \qquad \textbf{(12.21)}$$

Now, recall that voltage gain ($A_v$) is the ratio of output voltage to input voltage. By formula,

$$A_v = \frac{v_{out}}{v_{in}} \quad (v_{in} = v_{gs})$$

We can rearrange equation (12.21) to obtain a valid voltage-gain formula for the common-source JFET amplifier:

$$\frac{v_{out}}{v_{in}} = g_m r_D$$

or

$$A_v = g_m r_D$$

This is equation (12.18).

### 12.3.5    JFET Swamping

Just as the BJT amplifier can be swamped to reduce the effects of variations in $r'_e$, we can swamp the JFET amplifier to reduce the effects of variations in $g_m$. The swamped JFET amplifier closely resembles the swamped BJT amplifier shown in Figure 12.37. As you can see, the source resistance is only partially bypassed, just as the emitter resistance is partially bypassed in the swamped BJT amplifier.    ◄ **OBJECTIVE 8**

The voltage gain of the swamped JFET amplifier is found as

$$A_v = \frac{r_D}{r_S + (1/g_m)} \qquad \textbf{(12.22)}$$

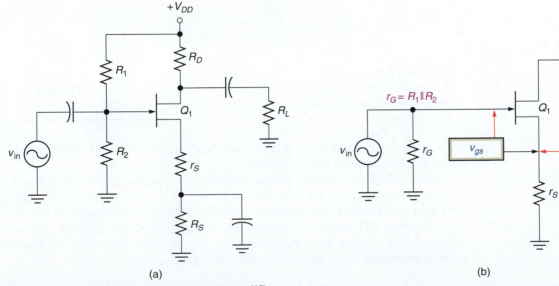

(a)                                          (b)

FIGURE 12.37    A swamped common-source amplifier.

If the value of $r_S \gg 1/g_m$, the resistor will swamp out the effects of variations in $g_m$. This point is illustrated in Example 12.13.

**EXAMPLE 12.13**

Determine the maximum and minimum gain values for the circuit shown in Figure 12.38.

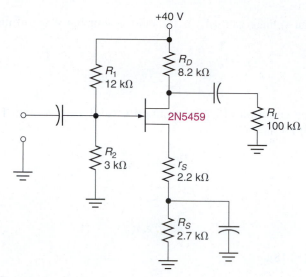

**FIGURE 12.38**

**Solution:** Basically, the circuit shown in Figure 12.38 is the same as the one shown in Figure 12.36. The difference of 100 Ω in the source circuit is not enough to change significantly the values obtained in Example 12.12, so we will use those values here.

Using the values from Figure 12.38 and Example 12.12, the maximum value of $A_v$ is found as

$$A_v = \frac{r_D}{r_S + (1/g_m)} = \frac{7.58 \text{ k}\Omega}{2.2 \text{ k}\Omega + 444 \text{ }\Omega} = \mathbf{2.87}$$

The minimum value of $A_v$ can be calculated as

$$A_v = \frac{r_D}{r_S + (1/g_m)} = \frac{7.58 \text{ k}\Omega}{2.2 \text{ k}\Omega + 1 \text{ k}\Omega} = \mathbf{2.37}$$

**PRACTICE PROBLEM 12.13**

A swamped JFET amplifier has values of $r_D = 6.6$ kΩ, $r_S = 1.5$ kΩ, and $g_m = 2000$ to 4000 μS. Determine the range of $A_v$ values for the circuit.

As you can see, the stability of $A_v$ is much better for the swamped amplifier than it is for the standard common-source amplifier.

Example 12.13 demonstrates the one drawback to swamping a JFET amplifier. If you compare the values of $A_v$ for Examples 12.12 and 12.13, you will notice that you lose quite a bit of gain when you swamp the amplifier. This is the same problem that we had with swamping BJT amplifiers. *Anytime you swamp an amplifier, you improve stability but you decrease voltage gain.* Your value of $A_v$ may be much more stable, but it will also be much lower than it was without the added swamping resistor.

## 12.3.6 The Basis for Equation (12.22)

The input voltage to the circuit in Figure 12.37b is in parallel with the gate-source circuit. Because of this, $v_{in}$ must equal the sum of $v_{gs}$ and the voltage developed across $r_S$. By formula,

$$v_{in} = v_{gs} + i_d r_S$$

Since $i_d = v_{gs} g_m$, the equation above can be rewritten as

$$v_{in} = v_{gs} + v_{gs} g_m r_S$$

or

$$v_{in} = v_{gs}(1 + g_m r_S)$$

The voltage gain of the amplifier is found as

$$A_v = \frac{v_{out}}{v_{in}} = \frac{g_m v_{gs} r_D}{v_{gs}(1 + g_m r_S)}$$

Eliminating $v_{gs}$ from the equation above gives us

$$A_v = \frac{g_m r_D}{1 + g_m r_S}$$

Finally, we can multiply the fraction above by 1 in the form of

$$\frac{1/g_m}{1/g_m}$$

to obtain

$$A_v = \frac{r_D}{r_S + (1/g_m)}$$

This is equation (12.22).

## 12.3.7 Amplifier Input Impedance

Due to the extremely high input impedance of the JFET, the overall input impedance to a JFET amplifier is higher than the input impedance to a similar BJT amplifier. For the *gate-bias* and *self-bias* circuits, the amplifier input impedance is found as

◄ **OBJECTIVE 9**

$$Z_{in} \cong R_G \tag{12.23}$$

For the *voltage-divider biased* amplifier, $Z_{in}$ is found as

$$Z_{in} \cong R_1 \parallel R_2 \tag{12.24}$$

Example 12.14 serves to illustrate the fact that a JFET amplifier has a higher value of $Z_{in}$ than a similar BJT amplifier.

## EXAMPLE 12.14

Determine the values of $Z_{in}$ for the amplifiers shown in Figure 12.39.

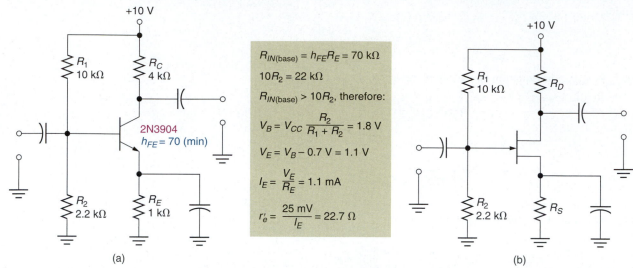

FIGURE 12.39

*Solution:* Figure 12.39 shows the calculations for the BJT amplifier up to the determination of $r_e'$. The geometric average of $h_{fe}$ for the 2N3904 is 200 at $I_C = 1$ mA. Using the values shown, the input impedance to the BJT amplifier is found as

$$Z_{in} = R_1 \| R_2 \| h_{fe} r_e' = 1.29 \text{ k}\Omega$$

For the JFET amplifier, the input impedance is found as

$$Z_{in} \cong R_1 \| R_2 = 1.8 \text{ k}\Omega$$

Thus, using the same biasing resistors, the JFET amplifier has a higher overall value of $Z_{in}$ than that of the BJT amplifier.

Because of the higher input impedance of the JFET amplifier, the circuit presents less of a load to its source. Thus, in any multistage amplifier, the overall gain of a source amplifier is higher when its load is a JFET amplifier rather than an equivalent BJT amplifier. Example 12.15 helps to illustrate this point.

## EXAMPLE 12.15

The amplifier in Example 12.12 is used to drive the amplifiers described in Example 12.14. Assume that $g_m$ for the driving amplifier is fixed at 2000 µS. Determine the value of $A_v$ for the driving amplifier when each load amplifier is connected to the output.

*Solution:* When the BJT amplifier is the load, the value of $r_D$ for Figure 12.36 is found as

$$r_D = R_D \| Z_{in} = 8.2 \text{ k}\Omega \| 1.29 \text{ k}\Omega = 1.11 \text{ k}\Omega$$

With this load, the voltage gain of the amplifier is

$$A_v = g_m r_D = (2000 \text{ µS})(1.11 \text{ k}\Omega) = 2.22$$

When the JFET amplifier is the load, the value of $r_D$ for Figure 12.36 is found as

$$r_D = R_D \parallel Z_{in} = 8.2 \text{ k}\Omega \parallel 1.8 \text{ k}\Omega = \mathbf{1.476 \text{ k}\Omega}$$

With this load, the voltage gain of the amplifier is

$$A_v = g_m r_D = (2000 \text{ } \mu\text{S})(1.476 \text{ k}\Omega) = \mathbf{2.952}$$

Thus, the gain of the circuit in Figure 12.36 is higher when driving the JFET amplifier than it is when driving the BJT amplifier.

### 12.3.8 One Final Note

You may have noticed that the material in this section was discussed in terms of the *voltage-divider biased* JFET amplifier. If you derive the ac equivalent circuits for the gate-bias and self-bias circuits, you will see that these ac equivalent circuits are the same as the voltage-divider bias equivalent circuit. For this reason, these other biasing circuits were not discussed.

In this section, you were shown that the JFET common-source amplifier has high input impedance when compared to the common-emitter amplifier. You were also shown that the voltage gain of the JFET amplifier (which tends to be relatively low) changes when the value of transconductance for the JFET changes. By swamping the JFET amplifier, we can reduce the variations in voltage gain that occur when transconductance varies. However, the increased stability in voltage gain is obtained at the cost of even lower voltage gain.

◄ Section Review

1. Explain the common-source amplifier circuit response to an ac input signal.
2. What is *transconductance*?
3. What are the units of transconductance?
4. How does the value of $g_m$ relate to the position of a circuit $Q$-point on the transconductance curve?
5. What is the relationship between JFET amplifier voltage gain and the value of $V_{GS}$ for the circuit?
6. What purpose is served by swamping a JFET amplifier?
7. How does JFET amplifier input impedance compare to that of a comparable BJT amplifier? Explain your answer.

## 12.4 AC Operating Characteristics: Common-Drain and Common-Gate Amplifiers

The *common-drain* and *common-gate* configurations are the JFET counterparts of the *common-collector* and *common-base* BJT amplifiers. In this section, we will take a brief look at each of these amplifiers.

◄ OBJECTIVE 10

### 12.4.1 The Common-Drain Amplifier (Source Follower)

The **source follower** accepts an input signal at its gate and provides an output signal at its source terminal. The input and output signals for this amplifier are *in phase*, thus the name *source follower*. The basic source follower is shown in Figure 12.40. The characteristics of the source follower are summarized as follows:

> **Source follower (common-drain amplifier)**
> The JFET counterpart of the emitter follower.

Input impedance:    High
Output impedance:   Low
Gain:                         $A_v < 1$

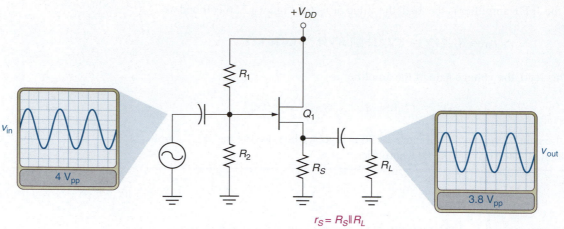

FIGURE 12.40 The source follower.

Because of its input and output impedance characteristics, the source follower is used primarily as a *buffer*; that is, it is used to couple a high-impedance source to a low-impedance load. You may recall from Chapter 10 that the emitter follower is also used for this purpose.

The voltage gain ($A_v$) of the source follower is found using the following equation:

$$A_v = \frac{r_S}{r_S + (1/g_m)} \tag{12.25}$$

> The voltage gain of the swamped common-source amplifier is found as
> $$A_v = \frac{r_D}{r_S + (1/g_m)}$$

where $r_S = R_S \parallel R_L$. The basis for equation (12.25) is easy to see if you compare the source follower in Figure 12.40 with the swamped common-source amplifier shown in Figure 12.37. As you can see, both circuits have some value of unbypassed source resistance. Thus, the voltage gains of the two circuits are calculated in the same basic fashion. The only differences are as follows:

1. The unbypassed source resistance in the source follower is the parallel combination of the source resistor ($R_S$) and the circuit load resistance ($R_L$).
2. The output voltage from the source follower appears across $r_S$. Therefore, $r_S$ is used in place of $r_D$ in the swamped amplifier equation.

Since $r_S$ appears in both the numerator and the denominator of equation (12.25), it should be apparent that the value of $A_v$ for the source follower will always be less than 1.

The high input impedance of the amplifier is caused, again, by the high $Z_{in}$ of the JFET. As with the common-source amplifier, the total input impedance to the amplifier equals the parallel equivalent resistance of the input circuit. When voltage-divider bias is used, $Z_{in}$ is found as

$$Z_{in} = R_1 \parallel R_2 \tag{12.26}$$

When gate bias or self-bias is used,

$$Z_{in} = R_G \tag{12.27}$$

Again, the high gate input impedance eliminates it from any $Z_{in}$ calculations.

The output impedance of the amplifier is explained using the ac equivalent circuit shown in Figure 12.41. As shown, the load "sees" the circuit as a source resistor in parallel with the input resistance of the JFET source ($1/g_m$). Thus, the output impedance of the amplifier is found as

> *Don't Forget:*
> Output impedance is measured with the load disconnected. Thus, $r_S$ is replaced by $R_S$ in Figure 12.41 and equation (12.28).

$$Z_{out} = R_S \parallel \frac{1}{g_m} \tag{12.28}$$

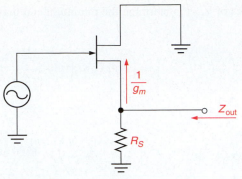

**FIGURE 12.41** Source-follower output impedance.

Now, how does equation (12.28) show that the value of $Z_{out}$ is always low? The normal range of $g_m$ is from 1000 μS up. For $g_m = 1000$ μS,

$$\frac{1}{g_m} = \frac{1}{0.001} = 1 \text{ k}\Omega$$

Since we used the lowest typical value of $g_m$ in the calculation above, it is safe to say that $1/g_m$ is *typically lower than* 1 kΩ. Since the total resistance of a parallel circuit must be lower than any individual resistance value, it is safe to say that $Z_{out}$ *is typically less than* 1 kΩ.

Example 12.16 demonstrates the process of analyzing a source-follower circuit. It also illustrates the fact that the source follower has a high $Z_{in}$, a low $Z_{out}$, and an overall voltage gain that is less than unity.

### EXAMPLE 12.16

Determine the maximum and minimum values of $A_v$ and $Z_{out}$ for the circuit shown in Figure 12.42. Also, determine the value of $Z_{in}$ for the circuit.

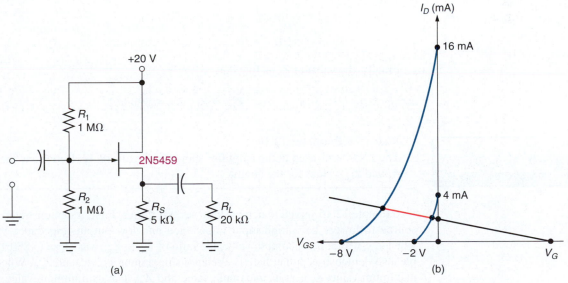

**FIGURE 12.42**

*Solution:* Using the dc bias line shown in Figure 12.42b, we approximate the minimum and maximum values of $V_{GS}$ to be −0.5 V and −5 V, respectively. The 2N5459 has values stated on the spec sheet of $g_{m0} = 6000$ μS at $V_{GS(off)} = -8$ V and $g_{m0} = 2000$ μS at $V_{GS(off)} = -2$ V. These values are used with our minimum and maximum values of $V_{GS}$ to determine the minimum and maximum values of $g_m$ as follows:

$$g_{m(\text{max})} = g_{m0}\left(1 - \frac{V_{GS}}{V_{GS(\text{off})}}\right) = \left(6000\ \mu S\right)\left(1 - \frac{-5\ V}{-8\ V}\right) = \mathbf{2250\ \mu S}$$

and

$$g_{m(\text{min})} = (2000\ \mu S)\left(1 - \frac{-0.5\ V}{-2\ V}\right) = \mathbf{1500\ \mu S}$$

The value of $r_S$ is now found as

$$r_S = R_S \parallel R_L = 5\ k\Omega \parallel 20\ k\Omega = \mathbf{4\ k\Omega}$$

Using the maximum value of $g_m$, the maximum value of $A_v$ is found as

$$A_v = \frac{r_S}{r_S + (1/g_m)} = \frac{4\ k\Omega}{4\ k\Omega + 444\ \Omega} = \mathbf{0.9001} \qquad \text{(maximum)}$$

Using the minimum value of $g_m$, the minimum value of $A_v$ is found as

$$A_v = \frac{r_S}{r_S + (1/g_m)} = \frac{4\ k\Omega}{4\ k\Omega + 667\ \Omega} = \mathbf{0.8571} \qquad \text{(minimum)}$$

Using the maximum and minimum values of $g_m$, the corresponding values of $Z_{\text{out}}$ are found as

$$Z_{\text{out}} = R_S \parallel \frac{1}{g_m} = 5\ k\Omega \parallel 667\ \Omega = \mathbf{588\ \Omega} \qquad \text{(maximum)}$$

and

$$Z_{\text{out}} = R_S \parallel \frac{1}{g_m} = 5\ k\Omega \parallel 444\ \Omega = \mathbf{408\ \Omega} \qquad \text{(minimum)}$$

Finally, the value of $Z_{\text{in}}$ is found as

$$Z_{\text{in}} = R_1 \parallel R_2 = \mathbf{500\ k\Omega}$$

Reference:
The data sheet for the 2N5486 JFET is shown in Figure 12.50.

**PRACTICE PROBLEM 12.16**
The 2N5486 is used in the amplifier shown in Figure 12.42. Determine the range of $A_v$ and $Z_{\text{out}}$ values for the circuit.

Example 12.16 served to demonstrate several things. First, it demonstrated the fact that source followers have high input impedance and low output impedance values. Even when $Z_{\text{out}}$ was at its maximum value, the ratio of $Z_{\text{in}}$ to $Z_{\text{out}}$ was close to 850:1. The example also demonstrated that there is a relationship among $g_m$, $A_v$, and $Z_{\text{out}}$. When $g_m$ is at its maximum value, $A_v$ is at its maximum value and $Z_{\text{out}}$ is at its minimum value. The reverse also holds true. Based on these observations, we can state the following relationships for the source follower:

**1.** $A_v$ *is directly proportional to* $g_m$.
**2.** $Z_{\text{out}}$ *is inversely proportional to* $g_m$.

As always, the value of $Z_{\text{in}}$ is not related to the value of $g_m$.

You have been shown that the source follower has high input impedance, low output impedance, and a voltage gain that is less than 1. At this point, we will take a look at the common-gate amplifier.

## 12.4.2  The Common-Gate Amplifier

The **common-gate amplifier** accepts an input signal at its source terminal and provides an output signal at its drain terminal. The basic common-gate amplifier is shown in Figure 12.43. The characteristics of the common-gate circuit are summarized as follows:

**Common-gate amplifier**
The JFET counterpart of the common-base amplifier.

| | |
|---|---|
| Input impedance: | Low |
| Output impedance: | High (compared to $Z_{in}$) |
| Gain: | $A_v > 1$ |

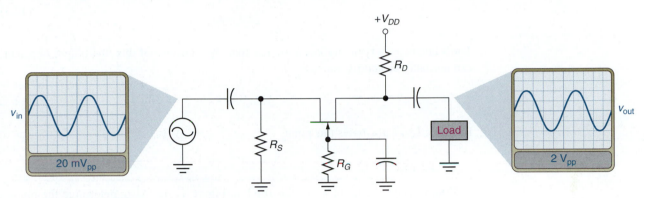

FIGURE 12.43   The common-gate amplifier.

As with the common-base circuit, the common-gate amplifier can be used to couple a low-impedance source to a high-impedance load. In this respect, the common-gate circuit (like the source follower) is used as an *impedance-matching* circuit. Also, like the common-base amplifier, the common-gate amplifier has a value of $A_v$ greater than 1.

Like the common-source amplifier, the voltage gain of the common-gate amplifier equals the product of its transconductance and ac drain resistance. By formula,

$$A_v = g_m r_D$$

You saw the derivation of this equation in Section 12.3.

The input signal to a common-gate amplifier is applied to its source. Figure 12.41 showed that the impedance of the source is found as $R_S$ in parallel with $1/g_m$. Thus, the input impedance of the common-gate amplifier is found as

$$Z_{in} = R_S \parallel \frac{1}{g_m} \tag{12.29}$$

As you can see, this is identical to the $Z_{out}$ equation (12.28) for the source follower. This makes sense because the input impedance of the common-gate amplifier is measured at the same terminal (the source) as the output impedance of the source follower.

Figure 12.44 shows that the output impedance is made up of the parallel combination of $R_D$ and the resistance of the JFET drain. By formula,

$$Z_{out} = R_D \parallel r_d \tag{12.30}$$

where $r_d$ = the resistance of the JFET drain terminal. The resistance of the JFET drain can be calculated using another JFET parameter: **output admittance** ($y_{os}$). In your study of basic electronics, you were taught that admittance is the reciprocal of impedance. We

**Output admittance ($y_{os}$)**
The admittance of the JFET drain, given on the component's spec sheet.

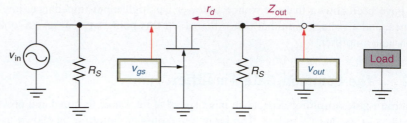

**FIGURE 12.44** Common-gate ac equivalent circuit.

can therefore determine the value of the JFET drain resistance by taking the reciprocal of $y_{os}$. By formula,

$$r_d = \frac{1}{y_{os}}$$ 

(12.31)

The value of $r_d$ is typically *much greater* than $R_D$. Because of this, the output impedance can normally be approximated as

$$Z_{out} \cong R_D$$

(12.32)

Example 12.17 illustrates this point.

### EXAMPLE 12.17

Determine the value of $r_d$ for the JFET in Figure 12.45. Also, determine the minimum and approximated values of $Z_{out}$.

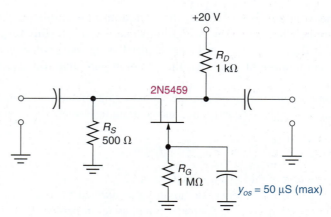

**FIGURE 12.45**

*Solution:* The spec sheet for the 2N5459 lists a maximum value of $y_{os} = 50\ \mu S$. Using this value, the *minimum* value of $r_d$ is found as

$$r_d = \frac{1}{y_{os}} = \frac{1}{50\ \mu S} = \mathbf{20\ k\Omega}$$

Using this value of $r_d$, the minimum value of $Z_{out}$ is found as

$$Z_{out} = R_D \parallel r_d = 1\ k\Omega \parallel 20\ k\Omega = \mathbf{952\ \Omega}$$

The approximated value of $Z_{out}$ is found as

$$Z_{out} \cong R_D = \mathbf{1\ k\Omega}$$

In Example 12.17, the *maximum* value of $y_{os}$ was used to find the *minimum* value of $r_d$. This was done to illustrate a point. When the minimum value of $r_d$ was used, the value of $Z_{out}$ was still within 10% of the approximated value. If you were to use any higher value of $r_d$, the minimum and approximated values of $Z_{out}$ would be even closer to each other. For example, the spec sheet for the 2N5459 lists a *typical* value of $y_{os} = 10$ μS. Using this value, $r_d = 100$ kΩ. Using this value for $r_d$, $Z_{out}$ is calculated as 990 Ω. This is even closer to the approximated value of $Z_{out}$ than the value obtained in Example 12.17. We can therefore assume that equation (12.32) is valid in all cases.

At this point, we have covered the most basic JFET circuits. Other circuits that use JFETs will appear off and on throughout the remainder of this book and will be discussed at the appropriate points.

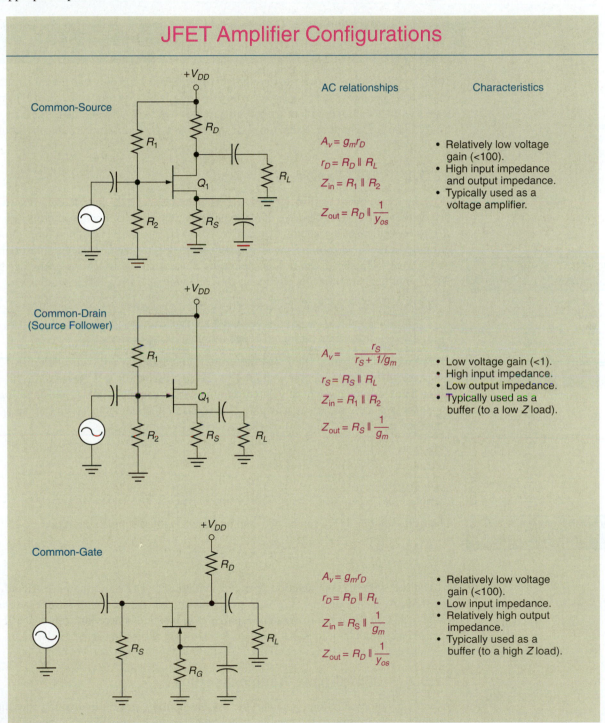

# JFET Amplifier Configurations

**Common-Source**

AC relationships

$A_v = g_m r_D$

$r_D = R_D \parallel R_L$

$Z_{in} = R_1 \parallel R_2$

$Z_{out} = R_D \parallel \dfrac{1}{y_{os}}$

Characteristics

- Relatively low voltage gain (<100).
- High input impedance and output impedance.
- Typically used as a voltage amplifier.

**Common-Drain (Source Follower)**

$A_v = \dfrac{r_S}{r_S + 1/g_m}$

$r_S = R_S \parallel R_L$

$Z_{in} = R_1 \parallel R_2$

$Z_{out} = R_S \parallel \dfrac{1}{g_m}$

- Low voltage gain (<1).
- High input impedance.
- Low output impedance.
- Typically used as a buffer (to a low Z load).

**Common-Gate**

$A_v = g_m r_D$

$r_D = R_D \parallel R_L$

$Z_{in} = R_S \parallel \dfrac{1}{g_m}$

$Z_{out} = R_D \parallel \dfrac{1}{y_{os}}$

- Relatively low voltage gain (<100).
- Low input impedance.
- Relatively high output impedance.
- Typically used as a buffer (to a high Z load).

FIGURE 12.46

### 12.4.3    Summary

You have now been shown the gain and impedance characteristics of the common-source, source-follower, and common-gate amplifiers. These three amplifiers and their characteristics are summarized in Figure 12.46.

**Section Review ▶**

1. What are the gain and impedance characteristics of the source follower?
2. List, in order, the steps taken to determine the range of $A_v$ values for a source follower.
3. What are the gain and impedance characteristics of the common-gate amplifier?

## 12.5    Troubleshooting JFET Circuits

JFET faults are relatively easy to diagnose because JFETs have only one junction (although shorts and opens may develop between any two terminals). As you will see, JFETs can be tested without removing them from their circuits. You will also see that JFET amplifier troubleshooting is approached in the same manner as is BJT amplifier troubleshooting.

### 12.5.1    JFET Faults

**OBJECTIVE 11 ▶**

JFET faults are about as difficult to detect as a flat tire because there is only *one* junction that can develop a fault. To make things even easier, the shorted-junction and open-junction JFETs have distinct symptoms that can be observed in the circuit. Therefore, minimum effort is required to determine whether or not the JFET is the cause of an amplifier failure.

| What are the normal JFET operating characteristics? |
|---|

Figure 12.47a shows the normal operating characteristics in a voltage-divider biased JFET amplifier. Note that the circuit has the following operating characteristics:

1. The current through $R_1$ is equal to the current through $R_2$.
2. The source and drain terminal currents are equal.
3. There is no current through the gate terminal of the device.

| What are the symptoms of a *shorted* gate? |
|---|

The *shorted-junction* JFET has the current characteristics shown in Figure 12.47b. Since the gate-source junction is shorted, $V_{GS}$ is approximately 0 V. As a result, $I_D \cong I_{DSS}$. Since there is some amount of gate current (as shown in the figure), $I_S$ and $I_D$ will not be equal (as is usually the case). The gate current also affects the biasing circuit, so $I_1 \neq I_2$. These are the classic symptoms of a shorted gate. In summary:

1. $I_1 \neq I_2$
2. $I_D \cong I_{DSS}$
3. $I_S \neq I_D$
4. $I_G > 0$

| What are the symptoms of an *open* gate? |
|---|

The *open-junction* JFET can be detected by observing two simple conditions:

1. $V_{GS} \neq 0$ V
2. $I_D \cong I_{DSS}$

You see, with the gate junction open, the depletion layer that normally restricts $I_D$ cannot form. At the same time, $V_{GS}$ is at some measurable value. If you have a measurable value of $V_{GS}$ and $I_D$ is approximately equal to $I_{DSS}$, the gate-source junction of the JFET is open. The circuit conditions that relate to this fault are shown in Figure 12.47c.

### 12.5.2    JFET Circuit Troubleshooting

As with any other amplifier, you start the troubleshooting process by making sure that the amplifier is the source of the trouble. (This point cannot be stressed often enough.) When you find a JFET amplifier with an abnormal output, check to make sure that the amplifier

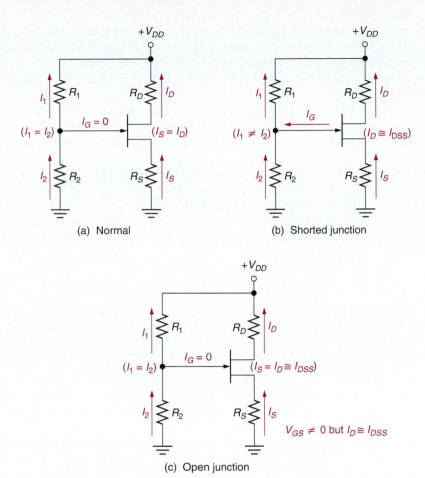

(a) Normal

(b) Shorted junction

(c) Open junction

$V_{GS} \neq 0$ but $I_D \cong I_{DSS}$

FIGURE 12.47   JFET fault symptoms.

source has a good output signal, and check the $V_{DD}$ and ground connections to the circuit. If all these tests have the proper results, you can assume that the amplifier is faulty.

We will restrict our discussion of amplifier faults to *open* components. As you know, most resistors do not simply develop a *shorted* condition. If a resistor is shorted, the short is usually caused by another component or a solder bridge. A visual inspection will eliminate these possibilities.

For the voltage-divider biased circuit, the common faults and their symptoms are listed in Table 12.1. Refer to Figure 12.47a to identify the components.

TABLE 12.1   JFET Troubleshooting

| Fault | Symptoms |
|---|---|
| $R_1$ open | $V_{GS}$ increases; $I_2 = 0$; $V_D$ increases; the circuit acts as a self-bias circuit. |
| $R_2$ open | JFET is destroyed by forward bias; if the JFET is replaced, the new component is also destroyed; $V_2$ increases. |
| $R_D$ open | $V_D = 0$ V; all terminal currents drop to 0. |
| $R_S$ open | $V_D = V_{DD}$; $V_{RD} = 0$ V. |

Of the problems listed, $R_1$ going open is the most difficult to detect. If $R_1$ opens, it is like removing the component from the circuit. If you do that, you simply have a self-bias circuit. Because of this, $R_1$ could open and the unit containing the amplifier could take a long time to fail. The main effect of $R_1$ opening on circuit operation is a shift of the $Q$-point. If the amplifier develops a problem with output signal clipping, check the voltage across $R_1$. If this voltage is equal to $V_{DD}$, the component is open.

If $R_2$ opens, the component is effectively removed from the gate circuit, leaving $V_{DD}$ and $R_1$. At the instant $R_2$ opens, the current path through $R_1$ is interrupted, dropping the voltage across $R_1$ to 0 V. This places $V_{DD}$ at the gate of the JFET, which will forward bias (and possibly destroy) the component. If the JFET is replaced, the new JFET may also be damaged or destroyed.

If $R_D$ opens, it effectively removes $V_{DD}$ from the drain-source circuit. All the supply voltage is dropped across the open component, so the voltage from the drain to ground ($V_D$) equals 0 V. Also, the lack of supply voltage causes the drain and source currents to drop to zero.

If $R_S$ opens, the current through the source terminal of the JFET drops to zero. Since $R_S$ is open, $V_S$ equals the supply voltage ($V_{DD}$). Since $V_S = V_{DD}$, $V_{GS}$ is more negative than normal, possibly more negative than the JFET's $V_{GS(off)}$ rating. At the same time, the gate circuit is unaffected, meaning that all voltages in the gate circuit are normal.

Now, let's put all this into perspective by going through a couple of troubleshooting applications.

---

**Application 12.1**

The circuit shown in Figure 12.48 is verified as being faulty when the amplifier signals and supply connections are checked. After disconnecting the input, the voltages shown are measured. The supply voltage ($V_{DD}$) is dropped across one of the biasing resistors ($R_2$), indicating that the resistor is open. But where do the other readings come from? Is there another problem in the circuit? The answer is *yes*.

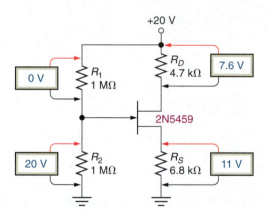

**FIGURE 12.48**

When $R_2$ opened, the gate-source junction of the JFET became forward biased by the +20 V applied to its gate. The current drawn through the gate by this voltage caused the gate to *open*. When the gate-source junction of the JFET opened, there was no depletion layer to restrict the current through its channel. Thus, the current through the drain and source circuits increased as much as $R_D$, $R_S$, and the channel resistance would allow—in this case, to approximately 1.62 mA. The 1.62 mA through the source circuit developed 11 V across $R_S$. The 9 V difference between $V_G$ and $V_S$ is measured across the open gate-source junction of the JFET. *Note the polarity of $V_{GS}$. The positive value of $V_{GS}$ is a good indication that the JFET gate-source junction has opened.* In this case, both $R_2$ and the JFET must be replaced. ■

---

**Application 12.2**

The circuit shown in Figure 12.49 has been verified as being faulty. After disconnecting the input, the voltages shown are measured. This problem is a simple one. The fact that $V_{DD}$ is measured across the source resistor indicates that $R_S$ is open. Note that the lack of $I_D$ causes the reading shown across the drain resistor. Also, the gate circuit is not affected by the fault, so the voltage readings across $R_1$ and $R_2$ are normal. ■

We have touched on only two problems that can develop in a JFET amplifier. However, using a solid understanding of the component and a little common sense, you should have no difficulty in diagnosing JFET amplifier faults.

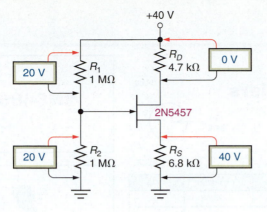

**FIGURE 12.49**

---

◄ **Section Review**

1. What is the primary difference between troubleshooting JFET amplifiers and BJT amplifiers?
2. List the common symptoms of a shorted-gate junction.
3. List the common symptoms of an open-gate junction.
4. Describe the procedure used to troubleshoot a JFET amplifier.

## 12.6 JFET Specification Sheets and Applications

Up to this point, we have based our calculations on several JFET ratings. Now, we will look at the common JFET ratings, what they mean, and when they are used. In this section, we also take a look at some practical JFET applications.

### 12.6.1 JFET Specification Sheets

We have used the parameters of the 2N5486 in many of the example practice problems. In this section, we use the spec sheet for this device, shown in Figure 12.50.

◄ *OBJECTIVE 12*

### 12.6.2 Maximum Ratings

Most of the maximum ratings for these devices are self-explanatory at this point. In our BJT spec sheet discussions, we covered all the commonly used maximum voltage, current, and power ratings. The JFET maximum ratings are no different.

One maximum rating may surprise you: the **forward gate current** $(I_{G(f)})$. In the section of maximum ratings in Figure 12.50, you will see the following:

| Rating | Symbol | Value | Unit |
|---|---|---|---|
| Forward gate current | $I_{G(f)}$ | 10 | mA |

> **Forward gate current $(I_{G(f)})$**
> The maximum amount of current that can be drawn through the JFET gate without damaging the device.

This rating indicates that the gate can handle a maximum forward current of 10 mA. Does this mean that the device can be operated in the forward region? No. The control of drain current depends on the amount of *reverse* bias on the JFET gate. The maximum drain current $(I_{DSS})$ is reached when the reverse bias reaches 0 V. Any increase in $V_{GS}$ above 0 V does not cause an increase in $I_D$. However, should the gate *accidentally* become forward biased, $I_G$ must be greater than 10 mA for the device to be destroyed. As long as $I_G < 10$ mA, the device will be safe.

# JFET VHF/UHF Amplifiers
## N–Channel–Depletion

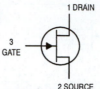

1 DRAIN

3 GATE

2 SOURCE

**2N5486**

**CASE 29–11, STYLE 5
TO–92 (TO–226AA)**

## MAXIMUM RATINGS

| Rating | Symbol | Value | Unit |
|---|---|---|---|
| Drain–Gate Voltage | $V_{DG}$ | 25 | Vdc |
| Reverse Gate–Source Voltage | $V_{GSR}$ | 25 | Vdc |
| Drain Current | $I_D$ | 30 | mAdc |
| Forward Gate Current | $I_{G(f)}$ | 10 | mAdc |
| Total Device Dissipation @ $T_C$ = 25°C     Derate above 25°C | $P_D$ | 350   2.8 | mW   mW/°C |
| Operating and Storage Junction     Temperature Range | $T_J$, $T_{stg}$ | −65 to +150 | °C |

## ELECTRICAL CHARACTERISTICS    ($T_A$ = 25°C unless otherwise noted)

| Characteristic | Symbol | Min | Typ | Max | Unit |
|---|---|---|---|---|---|
| **OFF CHARACTERISTICS** | | | | | |
| Gate–Source Breakdown Voltage     ($I_G$ = −1.0 μAdc, $V_{DS}$ = 0) | $V_{(BR)GSS}$ | −25 | — | — | Vdc |
| Gate Reverse Current     ($V_{GS}$ = −20 Vdc, $V_{DS}$ = 0)     ($V_{GS}$ = −20 Vdc, $V_{DS}$ = 0, $T_A$ = 100°C) | $I_{GSS}$ | —   — | —   — | −1.0   −0.2 | nAdc   μAdc |
| Gate Source Cutoff Voltage     ($V_{DS}$ = 15 Vdc, $I_D$ = 10 nAdc) | $V_{GS(off)}$ | −2.0 | — | −6.0 | Vdc |
| **ON CHARACTERISTICS** | | | | | |
| Zero–Gate–Voltage Drain Current     ($V_{DS}$ = 15 Vdc, $V_{GS}$ = 0) | $I_{DSS}$ | 8.0 | — | 20 | mAdc |
| **SMALL–SIGNAL CHARACTERISTICS** | | | | | |
| Forward Transfer Admittance     ($V_{DS}$ = 15 Vdc, $V_{GS}$ = 0, f = 1.0 kHz) | $|y_{fs}|$ | 4000 | — | 8000 | μmhos |
| Input Admittance     ($V_{DS}$ = 15 Vdc, $V_{GS}$ = 0, f = 400 MHz) | $Re(y_{is})$ | — | — | 1000 | μmhos |
| Output Admittance     ($V_{DS}$ = 15 Vdc, $V_{GS}$ = 0, f = 1.0 kHz) | $|y_{os}|$ | — | — | 75 | μmhos |
| Output Conductance     ($V_{DS}$ = 15 Vdc, $V_{GS}$ = 0, f = 400 MHz) | $Re(y_{os})$ | — | — | 100 | μmhos |
| Forward Transconductance     ($V_{DS}$ = 15 Vdc, $V_{GS}$ = 0, f = 400 MHz) | $Re(y_{fs})$ | 3500 | — | — | μmhos |

**FIGURE 12.50**    (Copyright of Semiconductor Component Industries, LLC. Used by permission.)

---

**Gate-source breakdown voltage ($V_{(BR)GSS}$)**
The value of $V_{GS}$ that can cause the gate-source junction to break down.

**Gate reverse current ($I_{GSS}$)**
The maximum amount of gate current that will be generated through a reverse-biased gate-source junction.

### 12.6.3    Off Characteristics

The **gate-source breakdown voltage** ($V_{(BR)GSS}$) defines the limit on $V_{GS}$. If $V_{GS}$ is allowed to reach this value, the junction may break down and the JFET may have to be replaced. For the 2N5486 JFET, the gate-source breakdown voltage is shown to be −25 V.

The **gate reverse current** ($I_{GSS}$) rating indicates the maximum value of gate current when the gate-source junction is reverse biased. For the 2N5486 JFET, this rating is −1 nA when the ambient temperature ($T_A$) is 25°C. Note that the negative current value is used to indicate the direction of the gate current. $I_{GSS}$ is a *temperature-dependent* rating. As the spec sheet shows, the value of $I_{GSS}$ increases as temperature increases. (The relationship between junction reverse current and temperature was first discussed in Chapter 2.)

Since we have already discussed $V_{GS(off)}$ and $I_{DSS}$ (which is listed under "On Characteristics" in Figure 12.50) in detail, we will not discuss them further here.

## 12.6.4　Small-Signal Characteristics

At first glance, this section of the spec sheet can be somewhat confusing because there are several $y_{fs}$ and $y_{os}$ parameters listed. Which do we use? To answer this question, we have to review briefly some basic ac principles.

*Conductance is the reciprocal of resistance. Susceptance is the reciprocal of reactance.* Combined, they make up *admittance*, which is *the reciprocal of impedance* (the combination of resistance and reactance). While resistance, reactance, and impedance are all measures of *opposition to current*, conductance, susceptance, and admittance are all measures of *the relative ease with which current can be generated through a component.*

The *forward transfer admittance*, *input admittance*, and *output admittance* ratings all take the effects of susceptance into account. You see, the JFET has measurable input and output *capacitances*. These capacitances all have some amount of reactance and, thus, some amount of susceptance. The admittance ratings include these values of susceptance. The *output conductance* and *forward transconductance* ratings, on the other hand, take only the component resistance into account.

The bottom line is this: Admittance values are based on both component resistance and component reactance. Conductance values are based solely on component resistance values. The question now is: Which ratings should be used for circuit analysis?

If you look closely at the 2N5486 spec sheet, you'll see that the following ratings were measured at $f = 400$ MHz:

- Input admittance
- Output conductance
- Forward transconductance

If you are analyzing a circuit that is operated at or near this frequency range, you should use these ratings in your analysis.

The following ratings were all measured at $f = 1$ kHz:

- Forward transfer admittance
- Output admittance

If you are analyzing a circuit that is operated at or near this frequency, you should use these ratings in your analysis.

The problem we have encountered in this section of the JFET spec sheet brings up a very interesting point. On many spec sheets, you will find multiple ratings that use the same unit of measure. When this occurs, you will always find that one or more of the conditions under which the ratings were measured differ between the two ratings. (In this case, the frequency of operation differed among the various ratings.) When this occurs, you should check to see which ratings were measured under the conditions that most closely resemble those of the circuit you are analyzing, then use those ratings in your circuit analysis. For example, if we were analyzing a circuit that has an operating frequency of 10 kHz, we should use the *forward transfer admittance* rating for our $g_{m0}$ values and the *output admittance* to find the value of $r_d$. If we were analyzing a circuit that has an operating frequency of 70 MHz, we would use the *forward transconductance* rating to obtain our $g_{m0}$ value and the *output conductance* rating to obtain our value of $r_d$.

## 12.6.5　JFET Capacitance Ratings

JFETs have three capacitance ratings that affect the high-frequency operation of the components. These ratings are discussed in detail in Chapter 14. At this point, we will briefly discuss several JFET circuit applications.

> The discussion here on conductance, susceptance, and admittance is very brief and is intended only as a review. You can review these principles further in the appropriate sections of your ac circuits text.

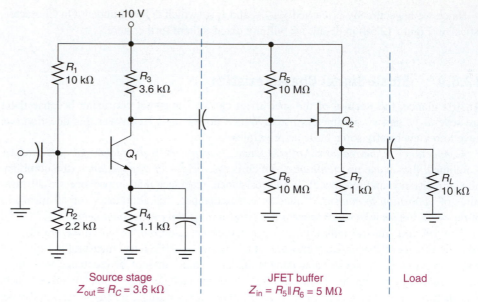

+10 V

Source stage
$Z_{out} \cong R_C = 3.6 \text{ k}\Omega$

JFET buffer
$Z_{in} = R_5 \| R_6 = 5 \text{ M}\Omega$

Load

**FIGURE 12.51** The JFET buffer.

### 12.6.6 JFET Applications

The high input impedance of the JFET amplifier makes it ideal for use in situations where source loading is a critical consideration. For example, consider the circuit shown in Figure 12.51. The JFET amplifier is used as a *buffer* between the source amplifier and the load. Because of the high input impedance of the JFET amplifier (5 MΩ), the amplifier presents virtually no load to the source amplifier. This has the effect of increasing the voltage gain of the source stage. Recall that the gain of a BJT common-emitter amplifier is found as

$$A_v = \frac{R_C \| R_L}{r'_e}$$

If the load resistor had been connected directly to the source stage, the gain of the source stage would have been found as

$$A_v = \frac{2.65 \text{ k}\Omega}{r'_e}$$

where 2.65 kΩ is the parallel combination of $R_C$ and $R_L$. With the JFET buffer, the gain of the first stage is found as

$$A_v = \frac{3.59 \text{ k}\Omega}{r'_e}$$

Regardless of the value of $r'_e$ for the first stage, the gain of that stage is increased due to the use of the buffer amplifier. The low $Z_{out}$ of the buffer would also increase the signal voltage at the load. Thus, the overall circuit works much more efficiently with the buffer amplifier than without it.

In communications electronics, the JFET is used for a variety of functions. One of these is as the active component in an **RF amplifier** (radio frequency amplifier). The RF amplifier is the first circuit to which a receiver input signal is applied. A JFET RF amplifier is shown in Figure 12.52. The gate and drain circuits of the amplifier are *tuned* circuits, meaning that they are designed to operate at or around specified frequencies. We will cover tuned amplifiers later in the text. The advantage of using the JFET in this circuit is that JFETs are *low-noise* components. As you know, *noise* is any unwanted interference with a transmitted signal. Noise can be generated by either the transmitter or

**RF amplifier**
Radio frequency amplifier; the input circuit of a communications receiver.

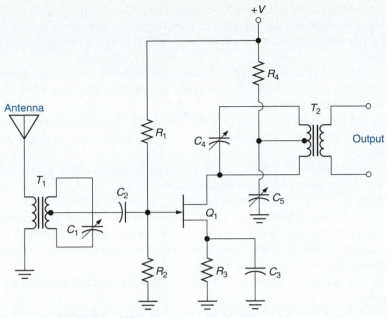

FIGURE 12.52   A JFET RF amplifier.

receiver or by factors outside the communications system. The JFET will not generate any significant amount of noise and, thus, is useful as an RF amplifier. Another advantage of using the JFET RF amplifier is that it requires no input current to operate. The antenna receives a weak signal that generates very little current. Since the JFET is a voltage-controlled component, it is well suited to respond to the low current signal provided by the antenna.

There are many other applications for JFET amplifiers. In this section, we have looked at only a few of them. The main thing to remember is that a JFET amplifier is used to serve the same basic purpose as a BJT amplifier. In applications where it is advantageous to have a high $Z_{in}$ amplifier, the JFET amplifier is preferred over the BJT amplifier.

◄ Section Review

1. Define each of the following JFET ratings: gate current, gate-source breakdown voltage, gate reverse current, gate-source cutoff voltage, and zero gate voltage drain current.

2. What is the advantage of using a JFET buffer over a BJT buffer?

3. What is an *RF amplifier*?

4. What advantages does the JFET RF amplifier have over the BJT RF amplifier?

## CHAPTER SUMMARY

Here is a summary of the major points made in this chapter:

1. A *field-effect transistor* (FET) is a voltage-controlled device.
2. There are two basic types of FETs:
   a. The *junction* FET (or JFET)
   b. The *metal-oxide-semiconductor* FET (or MOSFET)
3. A JFET has only a single *n*-type material and a single *p*-type material (see Figure 12.1).
4. The terminals of a JFET are called the *source*, *drain*, and *gate*.
   a. The *source* is the JFET counterpart of the BJT *emitter*.
   b. The *drain* is the JFET counterpart of the BJT *collector*.
   c. The *gate* is the JFET counterpart of the BJT *base*.
5. There are two types of JFETs:
   a. The *n*-channel JFET has an *n-type channel* surrounded by a *p-type gate*.
   b. The *p*-channel JFET has a *p-type channel* surrounded by an *n-type gate*.
   The schematic symbols for these components are shown in Figure 12.2.

6. In terms of circuit operation, the $n$-channel and $p$-channel JFETs differ only in their current directions and the polarities of their supply voltages (see Figure 12.3).
7. JFET operation is based on varying the width of the channel to control drain current.
    a. Drain current varies inversely with the cross-sectional area of the channel.
    b. The cross-sectional area of the channel can be decreased by effectively increasing the width of the gate.
8. When the gate-source junction of a JFET is reverse biased, a depletion layer forms around the junction. This depletion layer reduces the cross-sectional area of the channel.
9. The channel width of a JFET can be decreased by maintaining a constant $V_{DS}$ and increasing the value of $V_{GS}$ (see Figure 12.5).
    a. When $V_{GS} = 0$ V, drain current reaches its maximum value, $I_{DSS}$.
    b. As $V_{GS}$ becomes more negative, the channel narrows. When $V_{GS}$ reaches a specified value, $V_{GS(off)}$, the channel is closed off and drain current drops to near 0 A.
10. The channel width of a JFET can be decreased by maintaining a constant $V_{GS}$ and increasing the value of $V_{DS}$ (see Figure 12.6).
    a. As $V_{DS}$ increases, channel width decreases and drain current increases (to a point).
    b. Once $V_{DS}$ reaches a specified value, called the *pinch-off voltage* ($V_P$), further increases in $V_{DS}$ do not cause increases in $I_D$.
11. The region of JFET operation defined by $V_P < V_{DS} < V_{BR}$ is called the *constant-current region* (see Figure 12.7).
12. $V_{GS(off)}$ and $V_P$ always have the same magnitude; that is, $V_{GS(off)} = -V_P$. Since they have the same magnitude, only one of the two values is usually listed on a JFET spec sheet.
13. Since the gate-source junction of a JFET is reverse biased under normal circumstances, the gate has extremely high input impedance, typically in the high M$\Omega$ range.
14. The gate-source junction of a JFET is not designed to handle any significant amount of current. As such, the component may be destroyed if the junction is forward biased.
15. The *transconductance curve* for a JFET represents all the possible combinations of $I_D$ and $V_{GS}$ for the device.
16. The $I_{DSS}$ and $V_{GS(off)}$ ratings for a JFET always form the end points of its transconductance curve.
17. Most JFET spec sheets list *minimum* and *maximum* values of $I_{DSS}$ and $V_{GS(off)}$. When this is the case, both curves are plotted and used in the analysis of the circuit. (See Example 12.3.)
18. *Gate bias* is the JFET counterpart of base bias.
    a. A gate-bias circuit uses a negative gate supply voltage ($-V_{GG}$) to establish a negative $V_{GS}$ (see Figure 12.14).
    b. The primary drawback of gate bias is that it does not provide a stable $Q$-point from one JFET to another (see Figure 12.17 in Example 12.5).
19. The line that represents all possible $Q$-point combinations of $I_D$ and $V_{GS}$ for a JFET biasing circuit is referred to as the *dc bias line*.
20. *Self-bias* is a JFET biasing circuit that uses a source resistor to establish a negative $V_{GS}$.
    a. The gate terminal is returned to ground via a gate resistor (see Figure 12.20).
    b. Since there is no current in the gate circuit, the JFET gate is at 0 V.
    c. The current through the source resistor ($R_S$) establishes a positive voltage at the JFET's source terminal.
    d. Since the gate potential (0 V) is more negative than the source potential ($I_D R_S$), the gate-source junction of the JFET is reverse biased.
21. Self-bias produces a more stable $Q$-point value of $I_D$ than gate bias (see Figure 12.24).
22. Voltage-divider bias can be used to provide even greater $Q$-point stability than self-bias. The $Q$-point stability for voltage-divider bias is illustrated in Figure 12.28.

23. *Current-source bias* provides high $Q$-point stability by making the value of $I_D$ independent of the JFET.
   a. The value of $I_D$ for current-source bias is established using a BJT (see Figure 12.29).
   b. Because of the complexity of the circuit, current-source bias is rarely used. (In most applications, voltage-divider bias provides sufficient $Q$-point stability.)
24. The *common-source* amplifier is the JFET counterpart of the *common-emitter* amplifier.
25. Transconductance ($g_m$) is a ratio of a change in drain current ($\Delta I_D$) to a change in gate-source voltage ($\Delta V_{GS}$).
   a. Transconductance is typically measured in *microsiemens* ($\mu$S) or *micromhos* ($\mu$mhos).
   b. The value of $g_m$ for a JFET varies from one value of $V_{GS}$ to another (see Figure 12.33).
26. The voltage gain of the common-source amplifier is directly proportional to the value of $g_m$.
27. Since $g_m$ depends on the value of $V_{GS}$ (which can vary from one JFET to another), the value of $A_v$ can vary significantly from one JFET to another.
28. A swamping resistor can be added to the source circuit of a common-source amplifier to reduce the effects of variations in $g_m$ (see Figure 12.37). The addition of a swamping resistor results in lower but more stable voltage gain.
29. The overall input impedance of a JFET amplifier is greater than that of a comparable BJT amplifier.
30. The *common-drain* amplifier (or *source follower*) is the JFET counterpart of the emitter follower. The source follower has:
   a. High input impedance
   b. Low output impedance
   c. Voltage gain that is less than 1
   d. Input and output signal voltages that are in phase
31. The *common-gate* amplifier is the counterpart of the *common-base* amplifier.
32. The common gate amplifier has:
   a. Low input impedance
   b. High output impedance (compared to $Z_{in}$)
   c. Voltage gain that is greater than 1
   d. Input and output signal voltages that are in phase
33. JFET faults are relatively easy to detect because the component contains only one *pn* junction.
   a. The primary symptoms of a *shorted* gate-source junction are lower than normal $V_{GS}$, $I_G > 0$ A, and $I_D \cong I_{DSS}$.
   b. The primary symptoms of an *open* gate-source junction are $I_D \cong I_{DSS}$ while $V_{GS} \neq 0$ V.
34. The *gate current* ($I_G$) rating of a JFET indicates the maximum amount of current that can be drawn through the JFET gate without damaging the device.
35. The *gate-source breakdown voltage* ($V_{(BR)GSS}$) is the maximum allowable value of $V_{GS}$.
36. JFET spec sheets commonly identify maximum transconductance ($g_{m0}$) as either $g_{fs}$ or $y_{fs}$.

**EQUATION SUMMARY**

| Equation Number | Equation | Section Number |
|---|---|---|
| (12.1) | $I_D = I_{DSS}\left(1 - \dfrac{V_{GS}}{V_{GS(\text{off})}}\right)^2$ | 12.1 |
| (12.2) | $V_{GS} = -V_{GG}$ | 12.2 |
| (12.3) | $V_{DS} = V_{DD} - I_D R_D$ | 12.2 |
| (12.4) | $I_S = I_D$ | 12.2 |

| | | |
|---|---|---|
| **(12.5)** | $V_S = I_D R_S$ | 12.2 |
| **(12.6)** | $V_G = 0 \text{ V}$ | 12.2 |
| **(12.7)** | $V_{GS} = V_G - V_S$ | 12.2 |
| **(12.8)** | $V_{GS} = -I_D R_S$ | 12.2 |
| **(12.9)** | $I_D = \dfrac{-V_{GS}}{R_S}$ | 12.2 |
| **(12.10)** | $V_{DS} = V_{DD} - I_D(R_D + R_S)$ | 12.2 |
| **(12.11)** | $V_G = V_{DD}\dfrac{R_2}{R_1 + R_2}$ | 12.2 |
| **(12.12)** | $I_D = \dfrac{V_G - V_{GS}}{R_S}$ | 12.2 |
| **(12.13)** | $I_D = \dfrac{V_G}{R_S}$ | 12.2 |
| **(12.14)** | $I_D = I_C$ | 12.2 |
| **(12.15)** | $g_m = \dfrac{\Delta I_D}{\Delta V_{GS}}$ | 12.3 |
| **(12.16)** | $g_m = g_{m0}\left(1 - \dfrac{V_{GS}}{V_{GS(\text{off})}}\right)$ | 12.3 |
| **(12.17)** | $g_{m0} \cong \dfrac{2I_{DSS}}{V_{GS(\text{off})}}$ | 12.3 |
| **(12.18)** | $A_v = g_m r_D$ | 12.3 |
| **(12.19)** | $v_{\text{out}} = i_d r_D$ | 12.3 |
| **(12.20)** | $g_m = \dfrac{i_d}{v_{gs}}$ | 12.3 |
| **(12.21)** | $v_{\text{out}} = g_m v_{gs} r_D$ | 12.3 |
| **(12.22)** | $A_v = \dfrac{r_D}{r_S + (1/g_m)}$ | 12.3 |
| **(12.23)** | $Z_{\text{in}} \cong R_G$ | 12.3 |
| **(12.24)** | $Z_{\text{in}} \cong R_1 \parallel R_2$ | 12.3 |
| **(12.25)** | $A_v = \dfrac{r_S}{r_S + (1/g_m)}$ | 12.4 |
| **(12.26)** | $Z_{\text{in}} = R_1 \parallel R_2$ | 12.4 |
| **(12.27)** | $Z_{\text{in}} = R_G$ | 12.4 |
| **(12.28)** | $Z_{\text{out}} = R_S \parallel \dfrac{1}{g_m}$ | 12.4 |

| Equation Number | Equation | Section Number |
|---|---|---|
| (12.29) | $Z_{\text{in}} = R_S \parallel \dfrac{1}{g_m}$ | 12.4 |
| (12.30) | $Z_{\text{out}} = R_D \parallel r_d$ | 12.4 |
| (12.31) | $r_d = \dfrac{1}{y_{os}}$ | 12.4 |
| (12.32) | $Z_{\text{out}} \cong R_D$ | 12.4 |

channel 460
common-drain
    amplifier 493
common-gate
    amplifier 497
common-source
    amplifier 483
constant-current
    region 463
current-source bias 481
drain 460
field-effect transistor
    (FET) 460

forward gate current
    ($I_{G(f)}$) 503
gate 460
gate bias 470
gate reverse current
    ($I_{GSS}$) 504
gate-source breakdown
    voltage ($V_{(BR)GSS}$) 504
gate-source cutoff voltage
    ($V_{GS(off)}$) 464
ohmic region 463
output admittance ($y_{os}$) 497
pinch-off voltage ($V_P$) 463

RF amplifier 506
self-bias 472
shorted-gate drain current
    ($I_{DSS}$) 463
source 460
source follower 493
transconductance ($g_m$, $g_{fs}$,
    or $y_{fs}$) 484
transconductance
    curve 466

## PRACTICE PROBLEMS

### Section 12.1

1. A JFET amplifier has values of $I_{DSS} = 8$ mA, $V_{GS(off)} = -12$ V, and $V_{GS} = -6$ V. Determine the value of $I_D$ for the circuit.

2. A JFET amplifier has values of $I_{DSS} = 16$ mA, $V_{GS(off)} = -5$ V, and $V_{GS} = -4$ V. Determine the value of $I_D$ for the amplifier.

3. A JFET amplifier has values of $I_{DSS} = 14$ mA, $V_{GS(off)} = -6$ V, and $V_{GS} = 0$ V. Determine the value of $I_D$ for the circuit.

4. A JFET amplifier has values of $I_{DSS} = 10$ mA, $V_{GS(off)} = -8$ V, and $V_{GS} = -5$ V. Determine the value of $I_D$ for the circuit.

5. A JFET amplifier has values of $I_{DSS} = 12$ mA, $V_{GS(off)} = -10$ V, and $V_{GS} = -4$ V. Determine the value of $I_D$ for the circuit.

6. The JFET amplifier described in Problem 5 has a value of $V_{GS} = -14$ V. Determine the value of $I_D$ for the circuit.

7. The 2N5484 has values of $V_{GS(off)} = -0.3$ to $-3$ V and $I_{DSS} = 1.0$ to 5.0 mA. Plot the minimum and maximum transconductance curves for the device.

8. The 2N5485 has values of $V_{GS(off)} = -0.5$ to $-4.0$ V and $I_{DSS} = 4.0$ to 10 mA. Plot the minimum and maximum transconductance curves for the device.

9. The 2N5457 has values of $V_{GS(off)} = -0.5$ to $-6.0$ V and $I_{DSS} = 1.0$ to 5.0 mA. Plot the minimum and maximum transconductance curves for the device.

10. The 2N5458 has values of $V_{GS(off)} = -1.0$ to $-7.0$ V and $I_{DSS} = 2.0$ to 9.0 mA. Plot the minimum and maximum transconductance curves for the device.

11. The 2N3437 has parameters of $V_{GS(off)} = -5.0$ V (maximum) and $I_{DSS} = 0.8$ to 4.0 mA. Plot the maximum transconductance curve for the device.

12. The 2N3438 has parameters of $V_{GS(off)} = -2.5$ V (maximum) and $I_{DSS} = 0.2$ to 1.0 mA. Plot the maximum transconductance curve for the device.

CHAPTER 12   Practice Problems   511

**13.** Determine the values of $I_D$ and $V_{DS}$ for the amplifier in Figure 12.53a.

**14.** Determine the values of $I_D$ and $V_{DS}$ for the amplifier in Figure 12.53b.

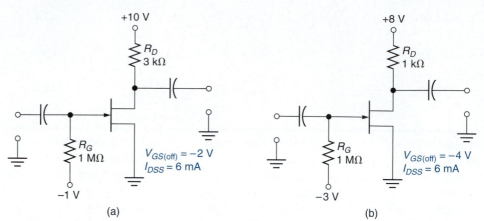

FIGURE 12.53

**15.** The 2N5486 (described in Section 12.1, Practice Problem 12.3) is used in the circuit in Figure 12.53a. Determine the range of $I_D$ values for the circuit.

**16.** A JFET with values of $V_{GS(off)} = -5$ to $-10$ V and $I_{DSS} = 4$ to 8 mA is used in the circuit in Figure 12.53b. Determine the range of $I_D$ values for the circuit.

**17.** The 2N3437 (described in Problem 11) is used in the circuit shown in Figure 12.54. Determine the range of $I_D$ values for the circuit. (*Hint:* See the margin note under *A Practical Consideration* on page 471 of the text.)

**18.** The 2N3438 (described in Problem 12) is used in the circuit shown in Figure 12.55. Determine the range of $I_D$ values for the circuit. (*Hint:* See the margin note under *A Practical Consideration* on page 471 of the text.)

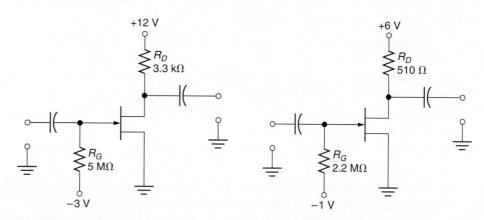

FIGURE 12.54                    FIGURE 12.55

**19.** Determine the ranges of $V_{GS}$, $I_D$, and $V_{DS}$ for the circuit shown in Figure 12.56. The 2N5484 is described in Problem 7.

**20.** Determine the ranges of $V_{GS}$, $I_D$, and $V_{DS}$ for the circuit shown in Figure 12.57. Assume the JFET is a 2N5485 (which is described in Problem 8).

**21.** Determine the ranges of $V_{GS}$, $I_D$, and $V_{DS}$ for the circuit shown in Figure 12.58. Assume the JFET is a 2N5458 (which is described in Problem 10).

**22.** Determine the ranges of $V_{GS}$, $I_D$, and $V_{DS}$ for the circuit shown in Figure 12.59. Assume the JFET is a 2N3437 (which is described in Problem 11).

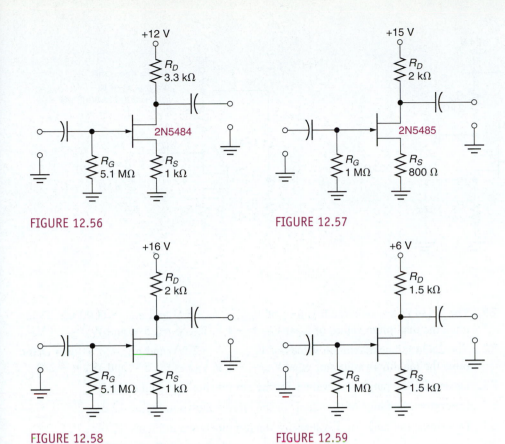

FIGURE 12.56

FIGURE 12.57

FIGURE 12.58

FIGURE 12.59

**23.** Determine the ranges of $I_D$ and $V_{DS}$ for the circuit shown in Figure 12.60.

**24.** Determine the ranges of $I_D$ and $V_{DS}$ for the circuit shown in Figure 12.61.

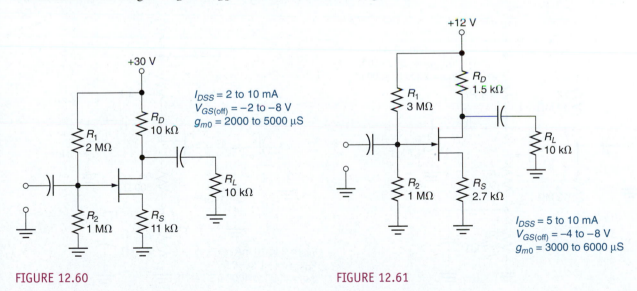

FIGURE 12.60

FIGURE 12.61

**25.** Determine the ranges of $I_D$ and $V_{DS}$ for the circuit shown in Figure 12.62.

**26.** Determine the ranges of $I_D$ and $V_{DS}$ for the circuit shown in Figure 12.63.

### Section 12.3

**27.** The 2N5484 has values of $g_{m0} = 3000$ to $6000$ μS and $V_{GS(off)} = -0.3$ to $-3.0$ V. Determine the range of $g_m$ when $V_{GS} = -0.2$ V.

**28.** The 2N5485 has values of $g_{m0} = 3500$ to $7000$ μS and $V_{GS(off)} = -0.5$ to $-4.0$ V. Determine the range of $g_m$ when $V_{GS} = -0.4$ V.

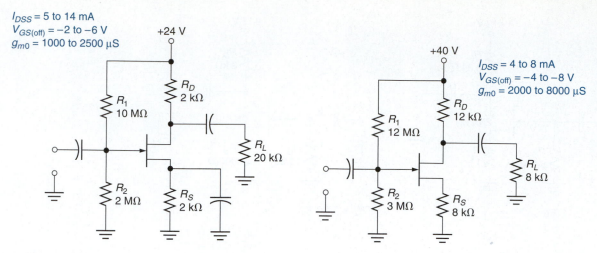

FIGURE 12.62

FIGURE 12.63

**29.** The 2N3437 has maximum values of $V_{GS(off)} = -5$ V and $g_{m0} = 6000$ μS. Determine the maximum values of $g_m$ at $V_{GS} = -1$ V, $V_{GS} = -2.5$ V, and $V_{GS} = -4$ V.

**30.** The 2N3438 has maximum values of $V_{GS(off)} = -2.5$ V and $g_{m0} = 4500$ μS. Determine the maximum values of $g_m$ at $V_{GS} = -1$ V, $V_{GS} = -1.5$ V, and $V_{GS} = -2$ V.

**31.** Determine the range of $A_v$ values for the circuit shown in Figure 12.60.

**32.** Determine the range of $A_v$ values for the circuit shown in Figure 12.61.

**33.** Determine the range of $A_v$ values for the circuit shown in Figure 12.62.

**34.** Determine the range of $A_v$ values for the circuit shown in Figure 12.63.

**35.** Determine the range of $A_v$ values for the swamped amplifier shown in Figure 12.64.

**36.** Determine the range of $A_v$ values for the swamped amplifier shown in Figure 12.65.

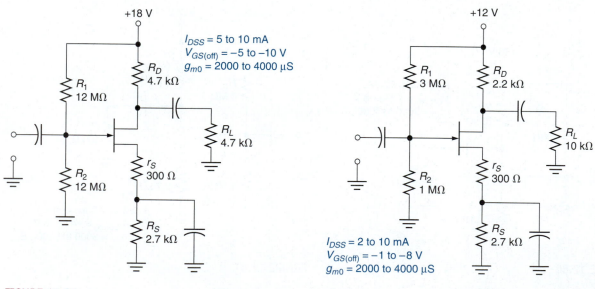

FIGURE 12.64

FIGURE 12.65

**37.** Determine the value of $Z_{in}$ for the amplifier in Figure 12.60.

**38.** Determine the value of $Z_{in}$ for the amplifier in Figure 12.61.

### Section 12.4

**39.** Determine the values of $A_v$, $Z_{in}$, and $Z_{out}$ for the amplifier in Figure 12.66.

**40.** Determine the values of $A_v$, $Z_{in}$, and $Z_{out}$ for the amplifier in Figure 12.67.

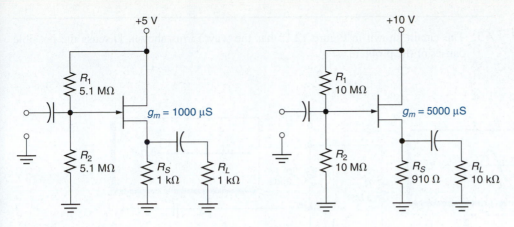

FIGURE 12.66

FIGURE 12.67

**41.** Determine the ranges of $A_v$, $Z_{in}$, and $Z_{out}$ for the amplifier in Figure 12.68.

**42.** Determine the ranges of $A_v$, $Z_{in}$, and $Z_{out}$ for the amplifier in Figure 12.69.

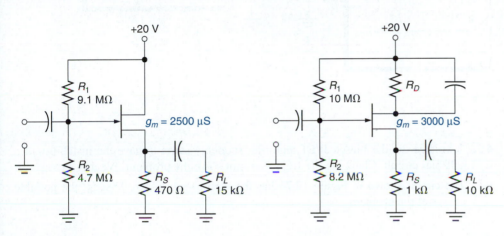

FIGURE 12.68

FIGURE 12.69

**43.** Determine the ranges of $A_v$, $Z_{in}$, and $Z_{out}$ for the circuit shown in Figure 12.70.

**44.** Determine the ranges of $A_v$, $Z_{in}$, and $Z_{out}$ for the circuit shown in Figure 12.71.

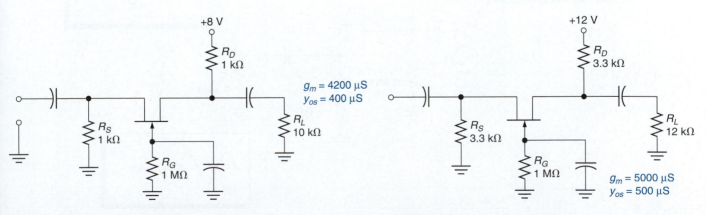

FIGURE 12.70

FIGURE 12.71

**45.** The circuit shown in Figure 12.72 has the waveforms shown. Discuss the possible causes of the problem.

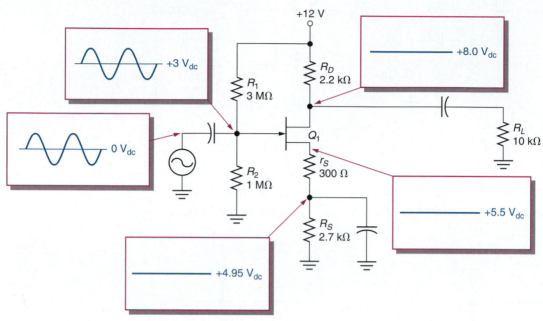

FIGURE 12.72

**46.** A voltage-divider biased JFET amplifier suddenly starts to have the instability of a self-bias circuit. Changing the JFET does not solve the problem. What is wrong?

**47.** The circuit shown in Figure 12.73 has the waveforms shown. Discuss the possible causes of the problem.

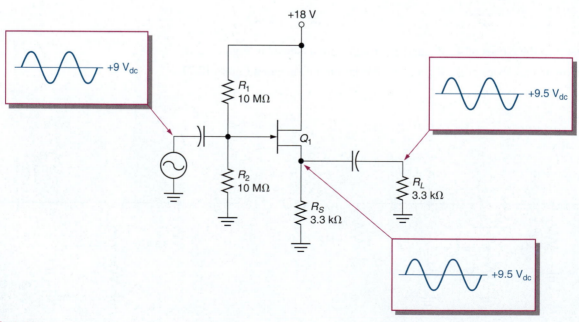

FIGURE 12.73

**48.** The circuit shown in Figure 12.74 has the dc voltages shown. Discuss the possible causes of the problem.

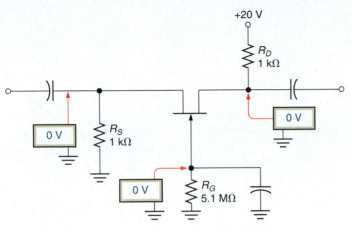

FIGURE 12.74

PUSHING THE ENVELOPE

**49.** The biasing circuit shown in Figure 12.75 is a *current-source bias* circuit. Determine how this circuit obtains a stable value of $I_D$.

**50.** Can the 2N5486 be used safely in the circuit shown in Figure 12.76? Explain your answer using circuit calculations.

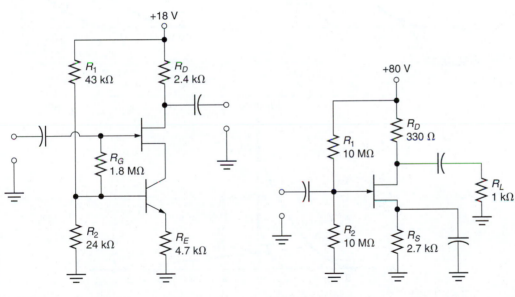

FIGURE 12.75                    FIGURE 12.76

**51.** The 2N5486 cannot be substituted for the JFET in Figure 12.77. Why not?

ANSWERS TO THE EXAMPLE PRACTICE PROBLEMS

**12.1**  3 mA

**12.2**  See Figure 12.78.

**12.3**  See Figure 12.79.

**12.4**  $V_{GS} = -5$ V, $I_D = 3$ mA, $V_{DS} = 3.4$ V

**12.6**  See Figure 12.80.

**12.7**  1.5 to 5.25 V

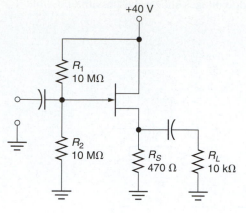

FIGURE 12.77

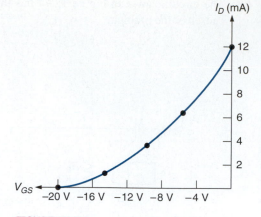

FIGURE 12.78

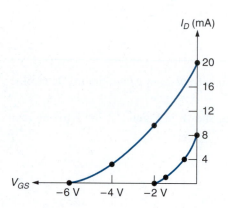

FIGURE 12.79

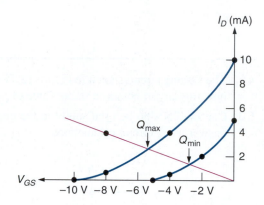

FIGURE 12.80

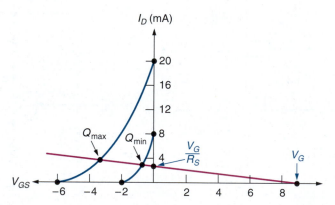

FIGURE 12.81

**12.8** See Figure 12.81.

**12.9** Using values of $V_{GS} = -750$ mV to $-3.5$ V: $I_D = 3.23$ to $4.04$ mA.

**12.10** 16.61 V (minimum), 20.5 V (maximum)

**12.11** $g_m = 5333$ μS at $V_{GS} = -2$ V, $g_m = 2667$ μS at $V_{GS} = -4$ V

**12.12** $A_v = 3.25$ to $6.24$

**12.13** $A_v = 3.3$ to $3.77$

**12.16** $A_v = 0.8889$ to $0.9143$, $Z_{out} = 349$ to $455$ Ω

## chapter 13

# MOSFETs

## Objectives

*After studying the material in this chapter, you should be able to:*

1. Identify the two types of MOSFETs, and describe the construction of each.
2. Describe the devices and precautions used to protect MOSFETs.
3. Describe *depletion-mode* operation.
4. Describe *enhancement-mode* operation.
5. Describe and analyze zero-bias operation.
6. Define *threshold voltage*, $V_{GS(\text{th})}$, and discuss its significance.
7. List the FET biasing circuits that can and cannot be used with E-MOSFETs.
8. Describe the physical construction and capacitance characteristics of the dual-gate MOSFET.
9. Describe the physical construction and current-handling capabilities of VMOS devices.
10. List the advantages of VMOS over other devices.
11. List the advantages that CMOS logic circuits have over typical BJT logic circuits.
12. Describe the purpose served by power MOSFET drivers in digital communications.

## Outline

13.1 MOSFET Construction and Handling
13.2 D-MOSFETs
13.3 E-MOSFETs
13.4 Dual-Gate MOSFETs
13.5 Power MOSFETs
13.6 Complementary MOSFETs (CMOS): A MOSFET Application
13.7 Other MOSFET Applications
Chapter Summary

# The MOSFET: A Major Component in Integrated Circuit Technology

Odds are that you have never heard of MOS (metal-oxide-semiconductor) technology before. At the same time, you probably have one or more items with you that contain MOSFET circuitry.

Most calculators, digital watches, and desktop computers are made using VLSI (very large scale integrated) circuits. These circuits contain up to hundreds of thousands of components that are all etched on a single piece of semiconductor material. The development of VLSI technology has made many items, such as those listed above, possible.

All electronic components dissipate heat. All electronic components must have some measurable size. These are two of the problems that have always faced the developers of integrated circuits, or ICs. The goal of IC technology has always been to produce the maximum number of components in the smallest amount of space possible. However, the need for heat-dissipation space and the actual space taken up by the components have limited the number of components that could be etched into a single piece of silicon.

The development of MOS technology has provided one solution to the problem of component spacing. Since MOSFET circuits are extremely low input current circuits, the heat-dissipation problem has been greatly reduced in these circuits. At the same time, MOSFET circuits can be made much smaller than their BJT counterparts. For these reasons, MOSFET technology has come to dominate most of the VLSI circuit market.

---

**Depletion-mode operation**
Using an input voltage to effectively decrease the channel size of an FET.

**MOSFET**
An FET that can be operated in the enhancement mode.

**Enhancement-mode operation**
Using an input voltage to effectively increase the channel size of an FET.

The main drawback to JFET operation is that the JFET gate *must* be reverse biased for the device to operate properly. Recall that a reverse gate-source voltage is varied to deplete the channel of free carriers, thus controlling the effective size of the channel. This type of operation is referred to as **depletion-mode operation**. A depletion-type device would be one that uses an input voltage to *reduce* the channel size from its *zero-bias* size.

The *metal-oxide-semiconductor FET*, or **MOSFET**, is a device that can be operated in the **enhancement mode**. This means that the input signal can be used to *increase* the effective size of the channel. This also means that the device is not restricted to operating with its gate reverse biased. This is an improvement over the standard JFET. In this chapter, we will look at MOSFETs, their construction and operation, as well as their applications and fault symptoms.

---

## 13.1 MOSFET Construction and Handling

OBJECTIVE 1 ▶

**D-MOSFET**
A MOSFET that can be operated in both the depletion mode and the enhancement mode.

**E-MOSFET**
A MOSFET that is restricted to enhancement-mode operation.

There are two basic types of MOSFETs: *depletion-type MOSFETs*, or *D-MOSFETs*, and *enhancement-type MOSFETs*, or *E-MOSFETs*. **D-MOSFETs** can be operated in both the depletion mode and the enhancement mode, whereas **E-MOSFETs** are restricted to operating in the enhancement mode. The primary difference between D-MOSFETs and E-MOSFETs is their physical construction. The construction difference between the two is illustrated in Figure 13.1. As you can see, the D-MOSFET has a physical channel (shaded area) between the source and drain terminals. The E-MOSFET, on the other hand, has no such channel. The E-MOSFET depends on the gate voltage to *form* a channel between the source and drain terminals. This point will be discussed in detail later in this chapter.

Both MOSFETs in Figure 13.1 show an *insulating layer* between the gate and the rest of the component. This insulating layer is made up of *silicon dioxide* ($SiO_2$), a glass-like insulating material. The gate terminal is made of a metal conductor. Thus, going from gate to substrate, you have *metal-oxide-semiconductor*, which is where the term *MOSFET* comes from. Since the gate is insulated from the rest of the component, the MOSFET is sometimes referred to as an *insulated-gate FET*, or *IGFET*.

**Substrate**
The foundation of a MOSFET.

The foundation of the MOSFET is called the **substrate**. This material is represented in the schematic symbol by the center line that is connected to the source terminal. Note that an *n*-channel MOSFET has a *p*-material substrate, and a *p*-channel MOSFET has an *n*-material substrate.

In the schematic symbol for the MOSFET, the arrow is placed on the substrate. As with the JFET, an arrow pointing in represents an *n-channel* device, while an arrow pointing *out* represents a *p-channel* device. Figure 13.2 shows the *p*-channel *depletion-type* and *enhancement-type* MOSFET symbols.

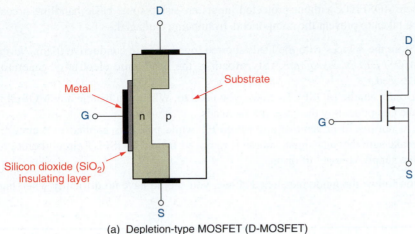

(a) Depletion-type MOSFET (D-MOSFET)

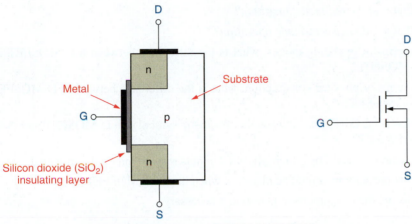

(b) Enhancement-type MOSFET (E-MOSFET)

FIGURE 13.1   *n*-Channel MOSFETs.

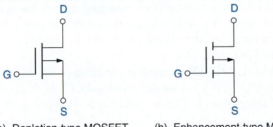

(a) Depletion-type MOSFET        (b) Enhancement-type MOSFET

FIGURE 13.2   *p*-Channel MOSFET symbols.

## 13.1.1   Component Handling

The layer of $SiO_2$ that insulates the gate from the channel is extremely thin and can be destroyed easily by *static electricity*. The static electricity generated by the human body can be sufficient to ruin the $SiO_2$ layer, so some precautions must be taken to protect the component.

Many MOSFETs are now manufactured with protective diodes etched between the gate and the source, as shown in Figure 13.3. The diodes are configured so that they will conduct in either direction provided that a predetermined voltage is reached. This voltage will be higher than any working voltage normally applied to the MOSFET. For example, if a MOSFET with a protected input is rated for a maximum $V_{GS}$ of $\pm 30$ V, the zener diodes will be designed to conduct at any voltage outside the $\pm 30$ V range. This protects the device from excessive static buildup as well as abnormally high values of $V_{GS}$.

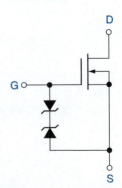

FIGURE 13.3   MOSFET static protection.

◀ *OBJECTIVE 2*

When MOSFETs without protected inputs are used, some basic handling precautions must be taken to prevent the components from being damaged:

**Lab Reference:** You will need to keep these precautions in mind while performing Exercise 19.

1. Store the devices with the leads shorted together or in conductive foam. *Never store MOSFETs in Styrofoam.* (Styrofoam is the best static electricity generator ever devised.)
2. Do not handle MOSFETs unless you need to. When you handle any MOSFET, hold the component by the case, not the leads.
3. Do not install or remove any MOSFET while power is applied to a circuit. Also, make sure that any signal source is removed from a MOSFET circuit before turning the supply voltage off or on.

If you follow the guidelines listed above, you should have no difficulty when handling MOSFETs.

**Section Review ▶**

1. What is a *MOSFET*?
2. What is *depletion-mode operation*?
3. What is *enhancement-mode operation*?
4. In terms of operating modes, what is the difference between a D-MOSFET and an E-MOSFET?
5. In terms of physical construction, what is the difference between a D-MOSFET and an E-MOSFET?
6. What is the difference between the schematic symbol of a D-MOSFET and that of an E-MOSFET?
7. Discuss the means by which MOSFET inputs may be internally protected.
8. What precautions should be observed when handling MOSFETs?
9. Why are the precautions in Question 8 necessary?

## 13.2 D-MOSFETs

As stated earlier, the D-MOSFET is capable of operating in both the depletion mode and the enhancement mode. When it is operating in the depletion mode, the characteristics of the D-MOSFET are very similar to those of the JFET. The overall operation of the D-MOSFET is illustrated in Figure 13.4.

Figure 13.4a shows the D-MOSFET operating conditions with the gate and source terminals shorted together ($V_{GS} = 0$ V). As stated in Chapter 12, $I_D = I_{DSS}$ when $V_{GS} = 0$ V. Thus, the D-MOSFET represented in Figure 13.4a has a value of $I_{DSS} = 5$ mA (the value indicated by the meter).

**OBJECTIVE 3 ▶**

When $V_{GS}$ is negative, the biasing voltage depletes the channel of free carriers. This *effectively* reduces the width of the channel, increasing its resistance. This *depletion-mode* operation is illustrated in Figure 13.4b. Note that the negative $V_{GS}$ has the same effect on the MOSFET as it has on the JFET. In Figure 13.4b, the depletion layer generated by $V_{GS}$ (represented by the white space between the insulating material and the channel) cuts into the channel, reducing its width. As a result, $I_D < I_{DSS}$. Though Figure 13.4b shows a value of $I_D = 0.8$ mA, the actual value of $I_D$ depends on the values of $I_{DSS}$, $V_{GS(off)}$, and $V_{GS}$.

**OBJECTIVE 4 ▶**

Figure 13.4c illustrates the *enhancement-mode* operation of the D-MOSFET. This operating mode is a result of applying a *positive* $V_{GS}$ to the component. When $V_{GS}$ is positive, the channel is *effectively* widened. This reduces the resistance of the channel, allowing $I_D$ to exceed the value of $I_{DSS}$. For example, Figure 13.4c shows a value of $I_D = 7.4$ mA, which is greater than the value of $I_{DSS}$ shown in Figure 13.4a. Again, the actual value of $I_D$ depends on the values of $I_{DSS}$, $V_{GS(off)}$, and $V_{GS}$.

How does enhancement-mode operation work?

How does a positive $V_{GS}$ effectively *widen* the channel? You must remember that the majority carriers in a *p*-type material are valence band *holes*. The holes in the *p*-type substrate are *repelled* by the positive gate voltage. Left behind is an area that is depleted of

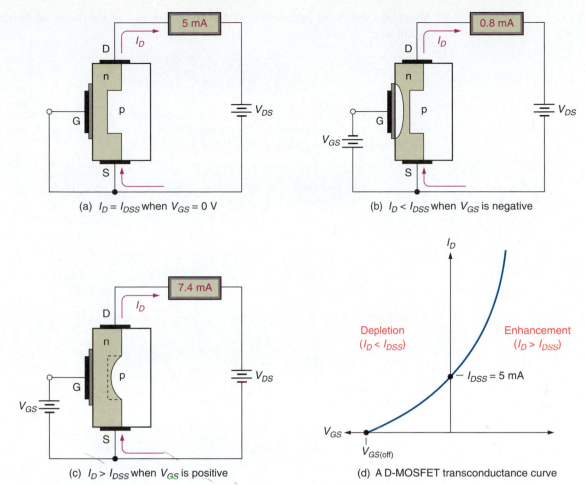

(a) $I_D = I_{DSS}$ when $V_{GS} = 0$ V

(b) $I_D < I_{DSS}$ when $V_{GS}$ is negative

(c) $I_D > I_{DSS}$ when $V_{GS}$ is positive

(d) A D-MOSFET transconductance curve

FIGURE 13.4  D-MOSFET operation.

valence band holes. At the same time, the conduction band electrons (minority carriers) in the *p*-type material are *attracted* toward the channel by the positive gate voltage. With the buildup of electrons near the channel, the area to the right of the physical channel *effectively* becomes an *n*-type material. The extended *n*-type channel now allows more current, and $I_D > I_{DSS}$.

The combination of these three operating states is represented by the D-MOSFET transconductance curve in Figure 13.4d. Note that $I_D = I_{DSS} = 5$ mA when $V_{GS} = 0$ V. When $V_{GS}$ is *negative*, $I_D < I_{DSS}$. When $V_{GS} = V_{GS(\text{off})}$, $I_D$ is reduced to approximately 0 mA. When $V_{GS}$ is *positive*, $I_D > I_{DSS}$. Obviously, $I_{DSS}$ is *not* the maximum possible value of $I_D$ for a MOSFET. The maximum allowable value of $I_D$ for a given D-MOSFET is provided on the component's spec sheet.

You may have noticed that the transconductance curve for the D-MOSFET is very similar to the curve for a JFET. Because of this similarity, the JFET and the D-MOSFET have the same transconductance equation. This equation was given in Chapter 12 as

$$I_D = I_{DSS}\left(1 - \frac{V_{GS}}{V_{GS(\text{off})}}\right)^2 \qquad (13.1)$$

Example 13.1 demonstrates the use of equation (13.1) in plotting the transconductance curve for a D-MOSFET.

## EXAMPLE 13.1

A D-MOSFET has parameters of $V_{GS(\text{off})} = -6$ V and $I_{DSS} = 1$ mA. Plot the transconductance curve for the device.

*Solution:* From the parameters, we can determine two of the points on the curve as follows:

$$V_{GS} = -6 \text{ V} \quad \text{and} \quad I_D = 0 \text{ mA}$$
$$V_{GS} = 0 \text{ V} \quad \text{and} \quad I_D = 1 \text{ mA}$$

These points define the conditions at $V_{GS} = V_{GS(off)}$ and at $I_D = I_{DSS}$. Now, we will use equation (13.1) to determine the coordinates of several more points. Remember that the only value we change is $V_{GS}$. When $V_{GS} = -3$ V,

$$I_D = (1 \text{ mA})\left(1 - \frac{-3 \text{ V}}{-6 \text{ V}}\right)^2 = 0.25 \text{ mA} \qquad (250 \text{ }\mu\text{A})$$

When $V_{GS} = -1$ V,

$$I_D = (1 \text{ mA})\left(1 - \frac{-1 \text{ V}}{-6 \text{ V}}\right)^2 = 0.694 \text{ mA} \qquad (694 \text{ }\mu\text{A})$$

When $V_{GS} = +1$ V,

$$I_D = (1 \text{ mA})\left(1 - \frac{+1 \text{ V}}{-6 \text{ V}}\right)^2 = 1.36 \text{ mA}$$

When $V_{GS} = +3$ V,

$$I_D = (1 \text{ mA})\left(1 - \frac{+3 \text{ V}}{-6 \text{ V}}\right)^2 = 2.25 \text{ mA}$$

We now have the following combinations of $V_{GS}$ and $I_D$:

| $V_{GS}$ (V) | $I_D$ (mA) |
|---|---|
| −6 | 0 |
| −3 | 0.25 |
| −1 | 0.69 |
| 0 | 1.00 |
| +1 | 1.36 |
| +3 | 2.25 |

Using the values listed, we can plot the transconductance curve shown in Figure 13.5.

**PRACTICE PROBLEM 13.1**

A D-MOSFET has values of $V_{GS(off)} = -4$ to $-6$ V and $I_{DSS} = 8$ to 12 mA. Plot the minimum and maximum transconductance curves for the device. Use $V_{GS}$ values of −4, −2, 0, 2, and 4 V.

As you can see, the process for plotting a D-MOSFET transconductance curve is almost the same as plotting one for a JFET. The only differences are that you must use positive values of $V_{GS}$ (as well as negative) and that you will get values of $I_D$ that are greater than $I_{DSS}$.

## 13.2.1 D-MOSFET Drain Curves

The drain curves of a D-MOSFET can be used to plot the dc load line for a given circuit, just as collector curves can be used to plot the load line for a given BJT. The drain curves for a D-MOSFET are plotted using corresponding values of $V_{GS}$ and $I_D$. A composite of drain curves is shown in Figure 13.6. The curves shown were plotted using the values from

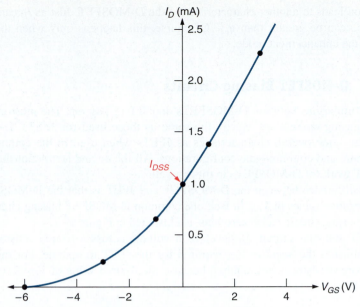

**FIGURE 13.5**

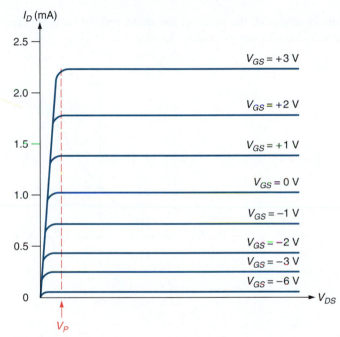

**FIGURE 13.6** MOSFET drain curves.

Example 13.1. Also, the curves for $V_{GS} = -2$ V and $V_{GS} = 2$ V were added. Note that above $V_P$, the value of $V_{DS}$ becomes constant for each combination of $V_{GS}$ and $I_D$. Also, note that the value of $V_P$ is relative to the gate voltage and, thus, changes from one curve to another. The value of $V_P$ for $V_{GS} = 0$ V is highlighted in Figure 13.6 to illustrate this point. At lower values of $V_{GS}$, $V_P$ is lower. Increasing the value of $V_{GS}$ increases the value of $V_P$.*

## 13.2.2 Transconductance

The value of $g_m$ for a D-MOSFET is found the same way that it is for a JFET. By formula,

$$g_m = g_{m0}\left(1 - \frac{V_{GS}}{V_{GS(\text{off})}}\right) \qquad \textbf{(13.2)}$$

* Note that the $V_P$ *rating* of a MOSFET is measured at $V_{GS} = 0$ V.

This equation leads to another characteristic of the D-MOSFET. Just as $I_D$ can be greater than $I_{DSS}$, $g_m$ can be greater than $g_{m0}$. Of course, this happens only when the device is operated in the enhancement mode.

### 13.2.3　D-MOSFET Biasing Circuits

Again, the similarities between D-MOSFETs and JFETs pay off. The most common D-MOSFET biasing circuits are exactly the same as those used for JFETs. D-MOSFETs also have the same overall characteristics as JFETs when used in the common-source, common-drain, and common-gate configurations. All the dc and ac relationships covered for the JFET work for D-MOSFET circuits as well.

**OBJECTIVE 5** ▶

A major difference between the D-MOSFET and JFET is that the D-MOSFET is not limited to negative values of $V_{GS}$. In fact, one common D-MOSFET biasing circuit sets $V_{GS}$ to 0 V. This biasing circuit, called **zero bias**, is illustrated in Figure 13.7.

In a JFET self-bias circuit, $I_D$ develops a voltage across a source resistor, $R_S$. This voltage establishes the *negative* $V_{GS}$ required for the JFET to operate. For the *zero-bias* circuit, a source resistor is not required because the desired value of $V_{GS}$ is 0 V. With $V_S$ and $V_G$ at 0 V:

> **Zero bias**
> A D-MOSFET biasing circuit that has quiescent values of $V_{GS} = 0$ V and $I_D = I_{DSS}$.

$$V_{GS} = 0\text{ V} \quad \text{and} \quad I_D = I_{DSS}$$

When the circuit is designed, the value of the drain resistor ($R_D$) is selected so that it drops approximately half the supply voltage, leaving $V_{DS} \cong \frac{1}{2}V_{DD}$.

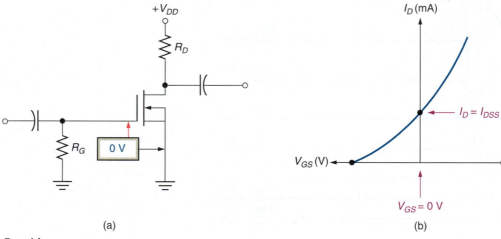

(a)　　　　　　　　　　　　　　　　　(b)

FIGURE 13.7　Zero bias.

### 13.2.4　D-MOSFET Input Impedance

The gate impedance of a D-MOSFET is extremely high. For example, one MOSFET has a maximum gate current of 10 pA when $V_{GS}$ is 35 V. This calculates to a gate impedance of $3.5 \times 10^{12}$ $\Omega$! With an input impedance in this range, the MOSFET presents virtually no load to its source circuit.

### 13.2.5　D-MOSFETs Versus JFETs

Figure 13.8 summarizes many of the characteristics of JFETs and D-MOSFETs. As you can see, the components are very similar in a number of respects. The D-MOSFET has the advantages of higher input impedance and the ability to operate in the enhancement mode. It can also use zero bias, while the JFET cannot. At the same time, D-MOSFETs are more sensitive to increases in temperature, and precautions must be taken when they are handled.

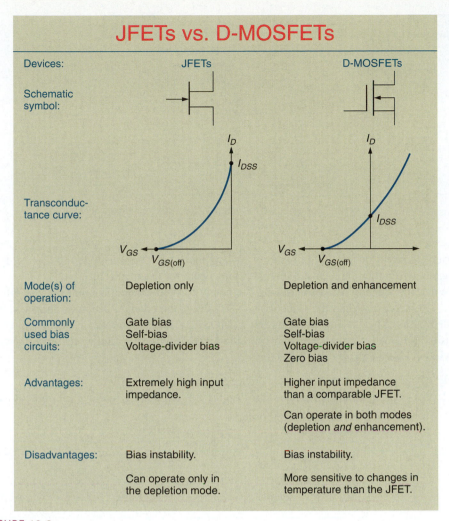

## JFETs vs. D-MOSFETs

| Devices: | JFETs | D-MOSFETs |
|---|---|---|
| Schematic symbol: | | |
| Transconductance curve: | | |
| Mode(s) of operation: | Depletion only | Depletion and enhancement |
| Commonly used bias circuits: | Gate bias<br>Self-bias<br>Voltage-divider bias | Gate bias<br>Self-bias<br>Voltage-divider bias<br>Zero bias |
| Advantages: | Extremely high input impedance. | Higher input impedance than a comparable JFET.<br><br>Can operate in both modes (depletion *and* enhancement). |
| Disadvantages: | Bias instability.<br><br>Can operate only in the depletion mode. | Bias instability.<br><br>More sensitive to changes in temperature than the JFET. |

FIGURE 13.8

1. Describe the depletion-mode relationship between $V_{GS}$ and $I_D$ for a D-MOSFET.  ◀ **Section Review**

2. Describe the enhancement-mode relationship between $V_{GS}$ and $I_D$ for a D-MOSFET.

3. Describe the relationship between $I_D$ and $I_{DSS}$ for a D-MOSFET.

4. Describe the relationship between $g_m$ and $g_{m0}$ for a D-MOSFET.

5. Describe the quiescent conditions of the *zero-bias* circuit.

6. List the similarities and differences between JFETs and D-MOSFETs.

7. Refer to Figure 13.7a. What purpose is served by the gate resistor in the zero-bias circuit?  ◀ **Critical Thinking**

## 13.3   E-MOSFETs

It was stated earlier that an E-MOSFET is capable of operating only in the enhancement mode. In other words, the gate potential must be *positive* with respect to the source. The reason for this can be seen in Figure 13.9. When the value of $V_{GS}$ is 0 V, there is no channel connecting the source and drain materials (as shown in Figure 13.9a). As a result, there can be no significant amount of drain current.

When a positive potential is applied to the gate, the *n*-channel E-MOSFET responds in the same manner as a D-MOSFET. As Figure 13.9b shows, the positive gate voltage forms a channel between the source and drain by depleting the area of valence band holes. As the holes are repelled by the positive gate voltage, that voltage is also attracting the minority

> *A Practical Consideration:*
> In this section, our discussion is limited to the *n*-channel E-MOSFET. Its *p*-channel counterpart operates according to the same principles. The only operating differences (as usual) are the current directions and the voltage polarities.

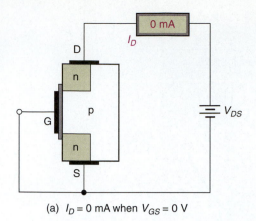

(a) $I_D = 0$ mA when $V_{GS} = 0$ V

(b) $I_D > 0$ mA when $V_{GS}$ is sufficient to form a channel.

**FIGURE 13.9**   E-MOSFET operation.

carrier electrons in the *p*-type material toward the gate. This forms an *effective n*-type bridge between the source and drain, providing a path for drain current.

**OBJECTIVE 6** ▶
When the value of $V_{GS}$ is *increased*, the newly formed channel becomes wider, allowing $I_D$ to increase. When the value of $V_{GS}$ *decreases*, the channel becomes narrower, and $I_D$ decreases. This is illustrated by the E-MOSFET transconductance curve shown in Figure 13.10. As you can see, this transconductance curve is similar to those covered previously. The primary differences are:

> **Threshold voltage ($V_{GS(th)}$)**
> The value of $V_{GS}$ that turns the E-MOSFET *on*.

1. All values of $V_{GS}$ that cause the device to conduct are *positive*.
2. The point at which the device turns on (or off, depending on how you look at it) is called the **threshold voltage ($V_{GS(th)}$)**.

> **Lab reference:** The curve for an E-MOSFET is plotted using measured values of $V_{GS}$ and $I_D$ in Exercise 19.

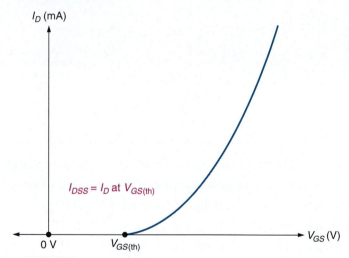

**FIGURE 13.10**   E-MOSFET transconductance curve.

Note that the value of $I_{DSS}$ for an E-MOSFET is approximately 0 A. For example, the 3N169 E-MOSFET spec sheet lists an $I_{DSS}$ of 10 nA when the ambient temperature is 25°C. For all practical purposes, this value of $I_D$ is zero. Obviously, since the value of $I_{DSS}$ for the E-MOSFET is near zero, the standard transconductance formula will not work for the E-MOSFET. To determine the value of $I_D$ at a given value of $V_{GS}$, you must use the following relationship:

$$I_D = k[V_{GS} - V_{GS(th)}]^2 \qquad \text{(13.3)}$$

where *k* is a *constant* for the MOSFET, found as

$$k = \frac{I_{D(on)}}{[V_{GS(on)} - V_{GS(th)}]^2} \qquad \text{(13.4)}$$

The values used in equation (13.4) to determine the value of $k$ are obtained from the spec sheet of the E-MOSFET being used. For example, consider the specifications listed in Figure 13.11. According to the charts shown, the 2N7000 has the following ratings:

$$I_{D(on)} = 75 \text{ mA (minimum)} \quad \text{and} \quad V_{GS(th)} = 0.8 \text{ V (minimum)}$$

# 2N7000

**Preferred Device**

## Small Signal MOSFET
## 200 mAmps, 60 Volts

### N–Channel TO–92

**ON Semiconductor™**

http://onsemi.com

**200 mAMPS**
**60 VOLTS**
$R_{DS(on)} = 5 \ \Omega$

N–Channel

### MAXIMUM RATINGS

| Rating | Symbol | Value | Unit |
|---|---|---|---|
| Drain Source Voltage | $V_{DSS}$ | 60 | Vdc |
| Drain–Gate Voltage ($R_{GS} = 1.0$ M$\Omega$) | $V_{DGR}$ | 60 | Vdc |
| Gate–Source Voltage<br>– Continuous<br>– Non–repetitive ($t_p \leq 50$ μs) | $V_{GS}$<br>$V_{GSM}$ | ±20<br>±40 | Vdc<br>Vpk |
| Drain Current<br>– Continuous<br>– Pulsed | $I_D$<br>$I_{DM}$ | 200<br>500 | mAdc |
| Total Power Dissipation @ $T_C = 25°C$<br>Derate above 25°C | $P_D$ | 350<br>2.8 | mW<br>mW/°C |
| Operating and Storage Temperature Range | $T_J$, $T_{stg}$ | −55 to<br>+150 | °C |

### ELECTRICAL CHARACTERISTICS ($T_C = 25°C$ unless otherwise noted)

| Characteristic | Symbol | Min | Max | Unit |
|---|---|---|---|---|
| **OFF CHARACTERISTICS** | | | | |
| Drain–Source Breakdown Voltage<br>($V_{GS} = 0$, $I_D = 10$ μAdc) | $V_{(BR)DSS}$ | 60 | – | Vdc |
| Zero Gate Voltage Drain Current<br>($V_{DS} = 48$ Vdc, $V_{GS} = 0$)<br>($V_{DS} = 48$ Vdc, $V_{GS} = 0$, $T_J = 125°C$) | $I_{DSS}$ | –<br>– | 1.0<br>1.0 | μAdc<br>mAdc |
| Gate–Body Leakage Current, Forward<br>($V_{GSF} = 15$ Vdc, $V_{DS} = 0$) | $I_{GSSF}$ | – | −10 | nAdc |
| **ON CHARACTERISTICS** | | | | |
| Gate Threshold Voltage<br>($V_{DS} = V_{GS}$, $I_D = 1.0$ mAdc) | $V_{GS(th)}$ | 0.8 | 3.0 | Vdc |
| Static Drain–Source On–Resistance<br>($V_{GS} = 10$ Vdc, $I_D = 0.5$ Adc)<br>($V_{GS} = 4.5$ Vdc, $I_D = 75$ mAdc) | $r_{DS(on)}$ | –<br>– | 5.0<br>6.0 | Ohm |
| Drain–Source On–Voltage<br>($V_{GS} = 10$ Vdc, $I_D = 0.5$ Adc)<br>($V_{GS} = 4.5$ Vdc, $I_D = 75$ mAdc) | $V_{DS(on)}$ | –<br>– | 2.5<br>0.45 | Vdc |
| On–State Drain Current<br>($V_{GS} = 4.5$ Vdc, $V_{DS} = 10$ Vdc) | $I_{d(on)}$ | 75 | – | mAdc |
| Forward Transconductance<br>($V_{DS} = 10$ Vdc, $I_D = 200$ mAdc) | $g_{fs}$ | 100 | – | μmhos |

### DYNAMIC CHARACTERISTICS

| Characteristic | | Symbol | Min | Max | Unit |
|---|---|---|---|---|---|
| Input Capacitance | | $C_{iss}$ | – | 60 | pF |
| Output Capacitance | ($V_{DS} = 25$ V, $V_{GS} = 0$, | $C_{oss}$ | – | 25 | |
| Reverse Transfer Capacitance | f = 1.0 MHz) | $C_{rss}$ | – | 5.0 | |

### SWITCHING CHARACTERISTICS (Note 1.)

| Characteristic | | Symbol | Min | Max | Unit |
|---|---|---|---|---|---|
| Turn–On Delay Time | ($V_{DD} = 15$ V, $I_D = 500$ mA, | $t_{on}$ | – | 10 | ns |
| Turn–Off Delay Time | $R_G = 25$ $\Omega$, $R_L = 30$ $\Omega$, $V_{gen} = 10$ V) | $t_{off}$ | – | 10 | |

1. Pulse Test: Pulse Width ≤ 300 μs, Duty Cycle ≤ 2.0%.

**FIGURE 13.11** The 2N7000 specifications. (Copyright of Semiconductor Component Industries, LLC. Used by permission.)

If you look closely at the test conditions for the $I_{D(on)}$ rating, you'll see that this current was measured at $V_{GS} = 4.5$ V. Using this value with the other ratings shown on the spec sheet, $k$ can be found as

$$k = \frac{I_{D(on)}}{[V_{GS} - V_{GS(th)}]^2} = \frac{75 \text{ mA}}{[4.5 \text{ V} - 0.8 \text{ V}]^2} = 5.48 \times 10^{-3} \text{ mA/V}^2$$

This value of $k$ would be used in equation (13.3) to determine the value of $I_D$ at a given value of $V_{GS}$. The entire process is demonstrated in Example 13.2.

### EXAMPLE 13.2

The spec sheet for the 3N171 lists the following minimum values:

$$I_{D(on)} = 10 \text{ mA at } V_{GS} = 10 \text{ V} \quad \text{and} \quad V_{GS(th)} = 1.5 \text{ V}$$

Using these ratings, determine the value of $I_D$ for the circuit shown in Figure 13.12.

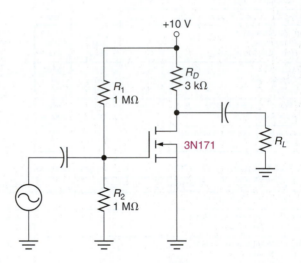

FIGURE 13.12

*Solution:* First, the value of $k$ is found as

$$k = \frac{I_{D(on)}}{[V_{GS(on)} - V_{GS(th)}]^2} = \frac{10 \text{ mA}}{(10 \text{ V} - 1.5 \text{ V})^2} = \mathbf{1.38 \times 10^{-4} \text{ mA/V}^2}$$

The circuit does not contain a source resistor, so $V_{GS} = V_G$. The value of $V_G$ is found as

$$V_G = V_{DD} \frac{R_2}{R_1 + R_2} = (10 \text{ V}) \frac{1 \text{ M}\Omega}{2 \text{ M}\Omega} = \mathbf{5 \text{ V}}$$

Using the value of $V_{GS} = 5$ V in equation (13.3), the value of $I_D$ for the circuit is found as

$$I_D = k[V_{GS} - V_{GS(th)}]^2 = (1.38 \times 10^{-4} \text{ mA/V}^2)(5 \text{ V} - 1.5 \text{ V})^2 = \mathbf{1.69 \text{ mA}}$$

### PRACTICE PROBLEM 13.2

An E-MOSFET has values of $I_{D(on)} = 14$ mA at $V_{GS} = 12$ V and $V_{GS(th)} = 2$ V. Determine the value of $I_D$ at $V_{GS} = 16$ V.

## 13.3.1 E-MOSFET Biasing Circuits

Several of the biasing circuits used for JFETs and D-MOSFETs cannot be used to bias E-MOSFETs because enhancement-mode operation requires a *positive* value of $V_{GS}$. For example, zero bias and self-bias produce values of $V_{GS}$ that are negative or equal to 0 V. Therefore, neither of these bias circuits can be used with an E-MOSFET. However, *voltage-divider bias* can be used, as can *gate-bias* and *drain-feedback bias*.

In our study of BJTs, we discussed a biasing circuit called *collector-feedback bias*. **Drain-feedback bias** is the MOSFET counterpart of this biasing method. Figure 13.13a shows a drain-feedback bias circuit.

The analysis of a drain-feedback bias circuit is often very simple. First, note that the gate resistor ($R_G$) in Figure 13.13a is connected between the drain and the gate. With the superhigh resistance of the MOSFET gate, there is no current through the drain circuit and, therefore, no voltage developed across $R_G$. As such, the gate voltage equals the drain voltage. By formula,

◀ OBJECTIVE 7

> **Drain-feedback bias**
> The MOSFET counterpart of collector-feedback bias.

$$V_{GS} = V_{DS} \qquad (13.5)$$

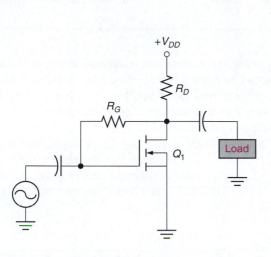

(a) Drain-feedback bias

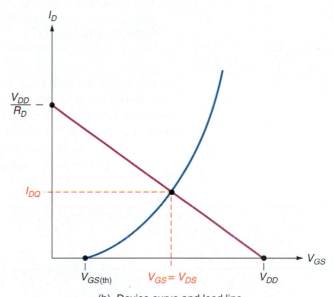

(b) Device curve and load line

**FIGURE 13.13**  Drain-feedback bias.

In most cases, the value of $I_{D(on)}$ for an E-MOSFET is measured under the condition described in equation (13.5). For example, the spec sheet for the 3N170 shows that the rated value of $I_{D(on)}$ is measured at values of $V_{GS} = 10$ V and $V_{DS} = 10$ V. Thus, when the 3N170 is drain-feedback biased, $I_D = I_{D(on)}$, and

$$V_{DS} = V_{DD} - R_D I_{D(on)} \qquad (13.6)$$

Note that equation (13.6) is valid for any E-MOSFET with an $I_{D(on)}$ rating that is measured at $V_{GS} = V_{DS}$. Example 13.3 demonstrates this approach to analyzing a drain-feedback bias circuit.

### EXAMPLE 13.3

Determine the values of $I_D$ and $V_{DS}$ for the circuit shown in Figure 13.14.

*Solution:*   The spec sheet for the 3N170 lists an $I_{D(on)}$ of 10 mA when $V_{GS} = V_{DS}$. Since these voltages *are* equal in the drain-feedback bias circuit, $I_D = I_{D(on)} = 10$ mA. Using this value, $V_{DS}$ (and, thus, $V_{GS}$) is found as

$$V_{DS} = V_{DD} - R_D I_{D(on)} = 20 \text{ V} - (1 \text{ k}\Omega)(10 \text{ mA}) = \mathbf{10 \text{ V}}$$

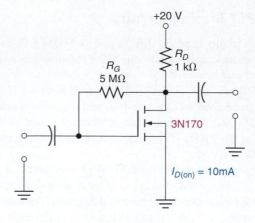

FIGURE 13.14

**PRACTICE PROBLEM 13.3**

The E-MOSFET described in Practice Problem 13.2 is used in a drain-feedback bias circuit with values of $V_{DD} = 14$ V and $R_D = 510 \, \Omega$. Determine the values of $I_D$ and $V_{DS}$ for the circuit.

When an E-MOSFET has an $I_{D(on)}$ rating that is measured at unequal values of $V_{GS}$ and $V_{DS}$, the circuit is analyzed as illustrated in Figure 13.13b. First, the transconductance curve for the devices is plotted (or obtained from the component spec sheet). Then, the load line for the circuit is plotted using the relationships shown. The point where the curve and load line intersect gives you the circuit values of $I_D$, $V_{GS}$, and $V_{DS}$.

That is all there is to analyzing a drain-feedback bias circuit. Don't forget: The drain-feedback bias circuit provides a *positive* value of $V_{GS}$. Because of this, it cannot be used with JFETs or D-MOSFETs. (D-MOSFETs can be operated in the enhancement mode, but they are never biased for linear applications with a value of $V_{GS}$ that is more positive than 0 V.)

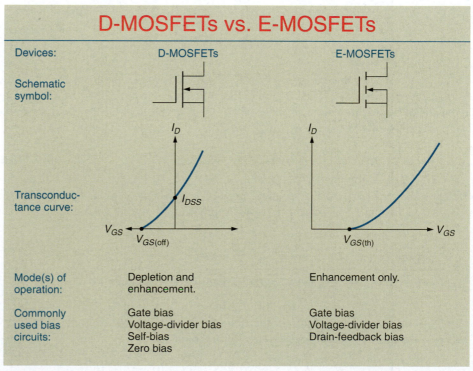

FIGURE 13.15

## 13.3.2 Summary

A comparison of D-MOSFETs and E-MOSFETs is provided in Figure 13.15, which summarizes many of the important characteristics of these devices.

◄ Section Review

1. How does a positive value of $V_{GS}$ develop a channel through an E-MOSFET?
2. What is *threshold voltage* $(V_{GS(th)})$?
3. When $V_{GS} = V_{GS(th)}$, what is the value of $I_D$ for an E-MOSFET?
4. List, in order, the steps required to determine the value of $I_D$ for an E-MOSFET at a given value of $V_{GS}$.
5. Which D-MOSFET biasing circuits can be used with E-MOSFETs?
6. Which D-MOSFET biasing circuits cannot be used with E-MOSFETs?
7. Describe the quiescent conditions for the *drain-feedback bias* circuit.
8. What change(s) would you make to the circuit shown in Figure 13.14 to increase the Q-point value of $I_D$?

◄ Critical Thinking

## 13.4 Dual-Gate MOSFETs

The operation of most MOSFETs is limited at high frequencies because of their high gate-to-channel capacitance. The source of this high capacitance is illustrated in Figure 13.16. The metal plate used for the gate is a conductor. The silicon dioxide between the gate and channel is an insulator. The channel itself can be viewed as a conductor (when compared to the silicon dioxide layer). Thus, the combination of the three forms a capacitor.

◄ OBJECTIVE 8

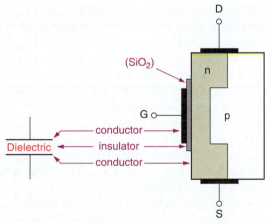

FIGURE 13.16   MOSFET input capacitance.

A **dual-gate MOSFET** uses two gate terminals to reduce the overall capacitance of the component. The physical construction and schematic symbols for the dual-gate MOSFET are shown in Figure 13.17. The reduced capacitance of the dual-gate MOSFET is a result of the way in which the component is used. Normally, the component is used so it acts as two series-connected MOSFETs. This is demonstrated in Section 13.7. When the dual-gate MOSFET is used as two series MOSFETs, the effect is similar to connecting two capacitors in series. You should recall that the total capacitance in a series connection is lower than either individual component value. Thus, by connecting the two gates in a series configuration, the overall capacitance is reduced.

**Dual-gate MOSFET**
A MOSFET constructed with two gates to reduce gate input capacitance.

◄ Section Review

1. Why does the MOSFET have high input capacitance?
2. How does the dual-gate MOSFET provide reduced input capacitance?

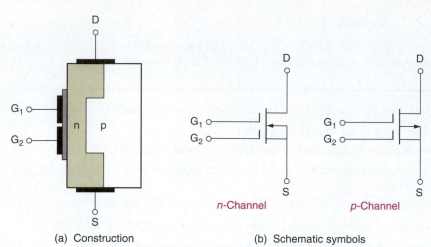

(a) Construction      (b) Schematic symbols

**FIGURE 13.17**    Dual-gate MOSFET construction and schematic symbols.

## 13.5    Power MOSFETs

Advances in engineering have produced a variety of MOSFETs that are designed specifically for high-power applications. The development of these components gives MOSFETs the ability to compete with high-power BJTs. In this section, we will take a brief look at two power MOSFETs. Some applications for power MOSFETs are introduced in Section 13.7.

### 13.5.1    VMOS Devices

◀ **OBJECTIVE 9** ▶

> **V-MOSFET (VMOS)**
> An E-MOSFET designed to handle high values of drain current.

A **V-MOSFET**, or **VMOS**, is an enhancement MOSFET that can handle much higher drain currents than a standard E-MOSFET. The current capability of a VMOS device results from its physical construction, which is illustrated in Figure 13.18. As you can see, the component has a V-shaped gate and materials that are labeled $p$, $n^+$, and $n^-$. The $n$-material labels indicate differences in doping levels, $n^+$ being *lightly doped* and $n^-$ being *heavily doped*. Also, there is no physical channel connecting the source and drain terminals. Thus, the device is an *enhancement-type* MOSFET.

**OBJECTIVE 10** ▶

When a positive gate voltage is applied to the device, an $n$-type channel forms around the V-shaped gate, as shown in Figure 13.19. The shape of the gate results in a wider channel than is generated in a standard E-MOSFET. The wider channel allows a higher drain current to be generated at a given value of $V_{GS}$. This is the primary advantage that VMOS devices have over other E-MOSFETs.

> VMOS is not susceptible to thermal runaway.

Another advantage of using VMOS devices is that they are not susceptible to thermal runaway. A VMOS device has a *positive temperature coefficient*, which means that the channel resistance of the device increases when temperature increases. This also means that drain current *decreases* when temperature increases. The positive temperature coeffi-

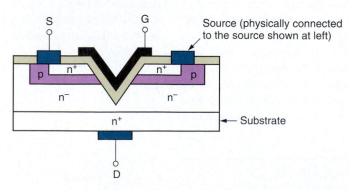

**FIGURE 13.18**    VMOS construction.

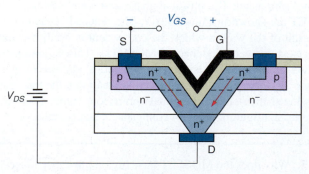

**FIGURE 13.19**    VMOS operation.

cient also makes it possible for several VMOS devices to be connected *in parallel* to increase their overall power-handling capability.

Two (or more) VMOS devices connected in parallel automatically regulate current. Here's how it works: If the current through one of the transistors ($I_D$) begins to increase, that component experiences an increase in temperature. The increase in temperature causes the channel resistance to increase, reducing the value of $I_D$. This guarantees that the two VMOS transistors have approximately equal values of drain current.

Because of its higher drain current ratings and positive temperature coefficients, the VMOS device can be used in several applications where the standard MOSFET cannot be used.

## 13.5.2 LDMOS

Another type of power MOSFET is the **lateral double-diffused MOSFET**, or **LDMOS**. This type of E-MOSFET uses a very short channel length and a heavily doped *n*-type substrate ($n^-$) to obtain high drain current and low channel resistance ($r_{d(on)}$). The basic construction of this enhancement-type MOSFET is shown in Figure 13.20.

> **Lateral double-diffused MOSFET (LDMOS)**
> A high-power MOSFET that uses a narrow channel and a heavily doped *n*-type region to obtain high $I_D$ and low $r_{d(on)}$.

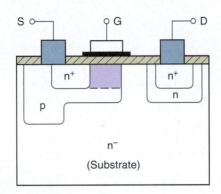

FIGURE 13.20   LDMOS.

The short channel (shaded area) is made up of the *p*-type material that lies between the $n^-$ substrate and the $n^+$ (lightly doped) source material. Since only *n*-type materials lie between the channel and the other two terminals, the effective length of the channel is extremely short. This, coupled with the low resistance of the ($n^-$) material in the channel-to-drain path, provides an extremely low typical value of $r_{d(on)}$. With a low channel resistance, an LDMOS device can handle very high currents without generating any damaging amount of heat (power dissipation).

LDMOS devices have typical values of $r_{d(on)}$ that are in the range of 2 Ω or less. With this low channel resistance, it is typically capable of handling currents as high as 20 A.

> *A Practical Consideration:*
> VMOS and LDMOS devices have $g_{m0}$ values in the mhos (or siemens) range. This means that they are capable of very high voltage gain values when used in common-source amplifiers.

◀ **Section Review**

1. Describe the physical construction of the VMOS.

2. How does the physical construction of the VMOS allow for a high value of $I_D$?

3. Describe the physical construction of the LDMOS.

4. How does the physical construction of the LDMOS allow for high values of $I_D$?

## 13.6   Complementary MOSFETs (CMOS): A MOSFET Application

The main contribution to electronics made by MOSFETs can be found in the area of *digital* (computer) electronics. In digital circuits, signals are made up of rapidly switching dc levels. An example of a digital signal is shown in Figure 13.21. This type of signal, referred to as a *rectangular wave*, is made up of two dc levels, called *logic levels*. For the waveform shown in Figure 13.21, the logic levels are 0 and +5 V.

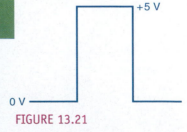

FIGURE 13.21

**Logic families**

Groups of digital circuits with nearly identical characteristics.

**Complementary MOS (CMOS)**

A logic family made up of MOSFETs.

**Inverter**

A logic-level converter.

A group of circuits that have similar operating characteristics is referred to as a **logic family**. All the circuits in a given logic family respond to the same logic levels, have similar speed and power dissipation characteristics, and can be directly connected together. One such logic family is **complementary MOS**, or **CMOS**, logic. This logic family is made up entirely of MOSFETs. The basic CMOS *inverter* is shown in Figure 13.22. An **inverter** is a digital circuit that converts one logic level to the other. When the input is at one logic level, the output is at the other. For the circuit in Figure 13.22, a 0 V input produces a +5 V output, and a +5 V input produces a 0 V output. The purpose served by such a circuit is beyond the scope of this book, but we will take a brief look at its operation.

The operation of the CMOS inverter is easy to understand if we look at each of the transistors as an individual *switch*. Figure 13.23 shows each of the MOSFETs as individual circuits. The *p*-channel MOSFET (Figure 13.23a) is an *enhancement*-type MOSFET, meaning that there is no physical channel from source to drain. When $Q_1$ has a value of $V_{GS} = 0$ V, no channel is formed between the source and drain. As a result, current cannot pass through the component. When $Q_1$ has a value of $V_{GS} = -5$ V, a channel is formed and the component can conduct.

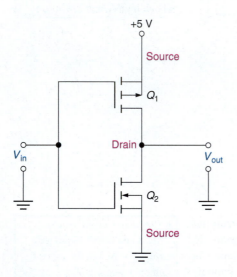

FIGURE 13.22   CMOS inverter.

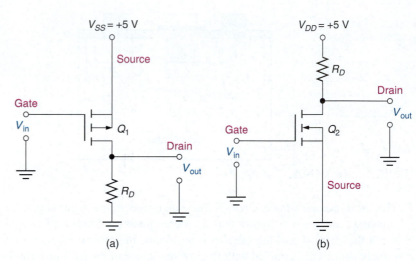

(a)                    (b)

FIGURE 13.23

The source terminal of $Q_1$ is connected to a +5 V supply ($V_{SS}$). Assuming that the value of $V_{in}$ for the circuit can be only 0 or +5 V, the circuit has the following relationships:

| $V_{in}$ | $V_{SS}$ | $V_{GS} = V_{in} - V_{SS}$ | $Q_1$ Operating State |
| --- | --- | --- | --- |
| +5 V | +5 V | 0 V | Not conducting (*off*) |
| 0 V | +5 V | −5 V | Conducting (*on*) |

As the chart shows, $Q_1$ conducts when $V_{in} = 0$ V and does not conduct when $V_{in} = +5$ V. Note that the MOSFET is said to be *on* when conducting and *off* when not conducting.

Now, consider the *n*-channel MOSFET ($Q_2$) in Figure 13.23b. This device is also an enhancement-type MOSFET. However, because it is an *n*-channel device, it has the opposite characteristics of $Q_1$. That is, it conducts when $V_{GS} = +5$ V and does not conduct when $V_{GS} = 0$ V. The circuit relationships for $Q_2$ are summarized as follows:

| $V_{in}$ | $V_{GS} = V_{in}$ | $Q_1$ Operating State |
| --- | --- | --- |
| +5 V | +5 V | Conducting (*on*) |
| 0 V | 0 V | Not conducting (*off*) |

Note that the source of $Q_2$ is returned to ground. Thus, $V_{GS} = V_{in}$. As the chart shows, $Q_2$ conducts when $V_{in} = +5$ V and does not conduct when $V_{in} = 0$ V. Now, let's put the two circuits together (as shown in Figure 13.22). Note that $Q_1$ serves as the drain resistor for $Q_2$, while $Q_2$ serves as the drain resistor for $Q_1$. When the input to the circuit is $+5$ V, $Q_1$ is off and $Q_2$ is on. Thus, the conduction path is between the output and ground through $Q_2$. When the input is at 0 V, $Q_1$ is on and $Q_2$ is off. Thus, the conduction path is between the output and $V_{SS}$ through $Q_1$. This operation is summarized as follows:

| $V_{in}$ (V) | $Q_1$ | $Q_2$ | $V_{out}$ (V) |
|:---:|:---:|:---:|:---:|
| 0 | On | Off | +5 |
| +5 | Off | On | 0 |

The relationship between $V_{in}$ and $V_{out}$ is as it should be for an inverter.

So, why is the CMOS logic circuit so popular? This question is best answered by taking a quick look at the BJT counterpart of the CMOS inverter.

◀ OBJECTIVE 11

The basic TTL (a bipolar transistor logic family) inverter is shown in Figure 13.24. It is obvious that the CMOS circuit is far less complex than its TTL counterpart. This means that many more CMOS circuits can be located in a given amount of space on an integrated circuit; that is, the CMOS circuits have a greater *packing density* than TTL circuits. In addition to the improved density, CMOS circuits have the following advantages over TTL circuits:

> Why is CMOS logic so popular?

1. They draw little current from the supply, so they consume very little power.
2. The low input current requirement of CMOS circuitry allows one CMOS output to drive an unlimited number of parallel CMOS loads; that is, you could connect any number of CMOS circuits in parallel and drive them with a single source. In contrast, the TTL circuit is usually restricted to driving no more than 10 other TTL inputs.

> The number of inputs that can be driven by a single output is called the **fanout** of a logic circuit.

These advantages will become clearer when you study the area of digital electronics. At this point, however, you should have a pretty good idea of the benefits of using CMOS logic.

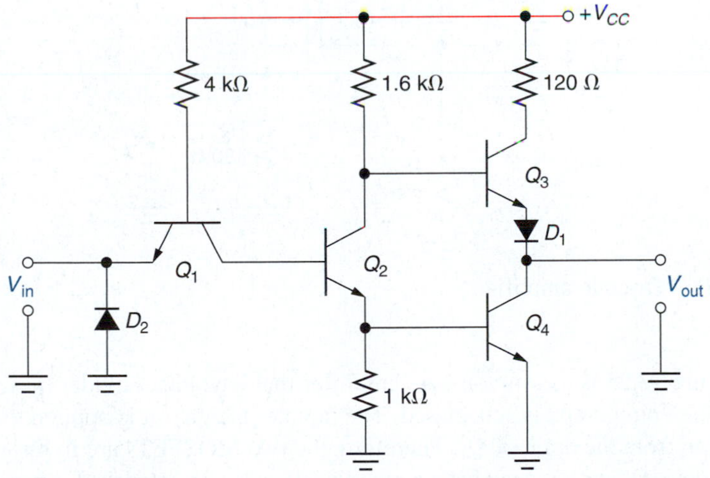

FIGURE 13.24   TTL inverter.

◀ Section Review

1. What are *logic levels*?
2. What is a *logic family*?
3. What is an *inverter*?
4. Describe the physical construction of the CMOS inverter.
5. Describe the operation of the CMOS inverter.
6. List the advantages that CMOS logic has over TTL logic.

While CMOS logic is one of the primary MOSFET applications, there are many other applications as well. In this section, we will look at a representative sample of these MOSFET applications.

It is important to note that the D-MOSFET can be used in most applications where the JFET can be used. Thus, the full range of MOSFET applications goes well beyond those discussed in this section.

### 13.7.1   Cascode Amplifiers

**Cascode amplifier**

A low-$C_{in}$ amplifier used in high-frequency applications.

Both JFETs and MOSFETs have relatively high input capacitances, as stated earlier. These input capacitances can adversely affect the high-frequency operation of the components. To overcome the effects of high input capacitance, the **cascode amplifier** was developed. The cascode amplifier consists of a common-source amplifier in series with a common-gate amplifier, as shown in Figure 13.25.

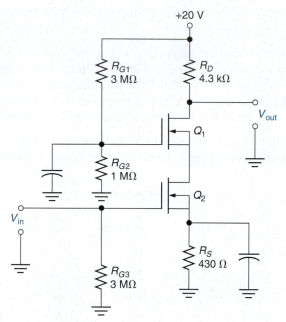

FIGURE 13.25   Cascode amplifier.

$Q_1$ in Figure 13.25 is a common-gate amplifier that is voltage-divider biased. $Q_2$ is the common-source circuit and is self-biased. The input to the circuit is applied to $Q_2$, and the output is taken from the drain of $Q_1$. Therefore, the two MOSFETs are in series.

As you know, $Q_1$ and $Q_2$ each have a measurable value of gate input capacitance. Connecting the two components in series has the same effect on this input capacitance as connecting any two capacitances in series: *The total series capacitance is lower than either individual capacitance.* Therefore, the cascode amplifier has lower input capacitance than a standard common-source MOSFET amplifier. This reduced capacitance makes the circuit better suited for high-frequency applications, where values of capacitive reactance can become critical.

A simple cascode amplifier can be constructed using a dual-gate MOSFET. The dual-gate MOSFET equivalent of Figure 13.25 is shown in Figure 13.26. Gate 1 of the MOSFET is connected to act as a common-gate amplifier, while gate 2 is connected to act as the common-source amplifier. This amplifier has the same capacitance characteristics and high-frequency capabilities as those of the amplifier in Figure 13.25.

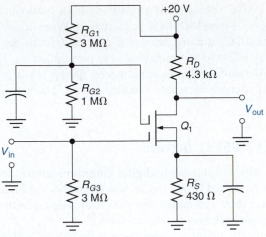

FIGURE 13.26  Dual-gate MOSFET cascode amplifier.

## 13.7.2    RF Amplifier

The dual-gate MOSFET can also be used in an RF amplifier. Recall that the RF amplifier is the "front-end" amplifier in an AM receiver. A dual-gate MOSFET RF amplifier is shown in Figure 13.27. Note the zener diodes drawn inside the dual-gate MOSFET. These diodes are built into the MOSFET to protect the component from static electricity. As you know, the MOSFET is extremely sensitive to static electricity. The zener diodes protect the component inputs. Except for that, they have no effect on the operation of the circuit.

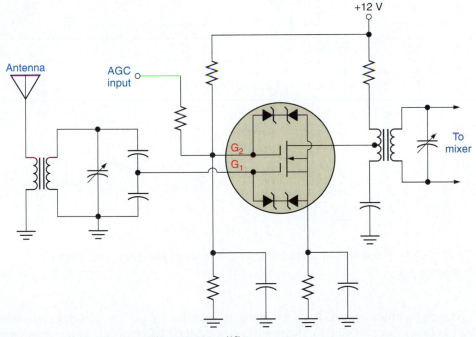

FIGURE 13.27    Dual-gate MOSFET RF amplifier.

The RF amplifier shown is similar to the one discussed in Chapter 12. The antenna is connected to gate 1 by a tuned circuit, and the output of the amplifier is coupled to the next circuit (the mixer) by another tuned circuit. When the dual-gate MOSFET is used, almost no current is required from the antenna for the circuit to operate properly.

The advantage of using the dual-gate MOSFET is that it allows for the use of an *automatic gain control* (AGC) in the RF amplifier. A dc voltage is applied to gate 2 from the AGC circuit. This voltage is directly proportional to the strength of the signal

> The AGC input shown in Figure 13.27 is a *feedback path* from a later stage in the receiver. The dc level at this input is determined by the amplitude of the signal in the audio circuitry.

received. When this AGC voltage is at a relatively high level (indicating a strong signal), the gain of the dual-gate MOSFET is decreased. When a weak signal is received, the voltage from the AGC circuitry decreases, and the gain of the dual-gate MOSFET increases. The overall effect of this operation is to prevent *fading*. If you have ever been in a car with an AM radio, you have experienced fading when going under a bridge. When the AM signal strength decreases, the audio fades. The use of AGC circuitry minimizes this fading.

### 13.7.3 Power MOSFET Drivers

**Digital communications**
A method of transmitting and receiving information in digital form.

**◄ OBJECTIVE 12**

At some point, you will probably study **digital communications**. In digital communications, information is converted (using one of several methods) into a series of digital signals. Those signals are then transmitted and received in digital form. Finally, the receiver converts the information back into its original form.

Many digital communications systems require the use of power amplifiers that can produce high-speed, high-current digital outputs. These signals can be produced using a power MOSFET driver circuit like the one shown in Figure 13.28.

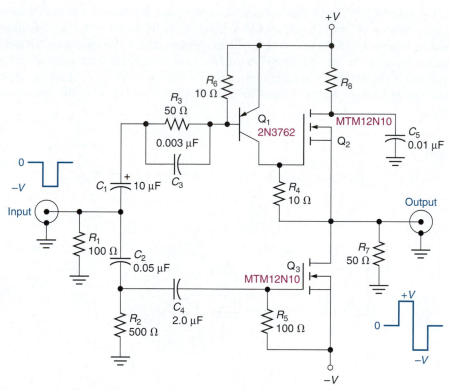

**FIGURE 13.28** (Copyright of Semiconductor Component Industries, LLC. Used by permission.)

The high-current capability of the driver is provided by the low channel resistance ($r_{d(on)}$) of the power MOSFETs. (Remember, power MOSFETs typically have $r_{d(on)}$ values of 2 Ω or less.) Thus, depending on the values of $+V$, $-V$, and $R_8$, the MOSFET pair can output currents as high as 20 to 35 A.

The high-speed quality of the circuit is provided by a number of factors. First, the resistance values in the driver circuit are extremely low. This allows the capacitances in the circuit to charge and discharge extremely rapidly. Also, note the parallel circuit made up of $R_3$ and $C_3$ in the base circuit of $Q_1$. The capacitor ($C_3$) is called a **speed-up capacitor**. This capacitor is used to improve the switching time (the time taken to go back and forth between saturation and cutoff) of $Q_1$. Speed-up capacitor operation is discussed in detail in Chapter 19.

**Speed-up capacitor**
A capacitor used in the base circuit of a BJT to help the device switch rapidly between saturation and cutoff.

## 13.7.4   VMOS Applications

You may remember that high-current BJTs are used in applications where the standard BJT will not handle the current requirements. The same relationship exists between VMOS devices and standard MOSFETs. A VMOS device may be used in almost any enhancement-type MOSFET circuit. However, it is usually used only when its high-current capabilities are required.

◀ Section Review

1. What is a *cascode amplifier*?
2. How does the cascode amplifier reduce input capacitance?
3. Describe the dual-gate MOSFET cascode amplifier.
4. Describe the dual-gate MOSFET RF amplifier.
5. What is *automatic gain control* (AGC)? How is it achieved in the dual-gate MOSFET RF amplifier?
6. What is *fading*?
7. What is *digital communications*?
8. What purpose does the power MOSFET driver serve in digital communications?
9. When are VMOS devices typically used?

## CHAPTER SUMMARY

Here is a summary of the major points made in this chapter:

1. A *metal-oxide-semiconductor* FET, or *MOSFET*, is a device that is capable of enhancement-mode operation. (In contrast, JFETs are capable of depletion-mode operation only.)
   a. Depletion MOSFETs, or D-MOSFETs, are capable of depletion-mode and enhancement-mode operation.
   b. Enhancement MOSFETs, or E-MOSFETs, are capable of enhancement-mode operation only.
2. The D-MOSFET has a physical channel between its source and drain terminals (see Figure 13.1a). $V_{GS}$ is used to reduce or increase the width of this channel.
3. The E-MOSFET has no physical channel between its source and drain terminals (see Figure 13.1b). $V_{GS}$ is used to form a channel between the source and drain terminals.
4. The material that forms the physical base of a MOSFET is called the *substrate*.
5. The silicon dioxide ($SiO_2$) layer between the gate and the rest of the component is highly sensitive to static electricity.
   a. Many MOSFETs contain internal static protection circuitry (see Figure 13.3).
   b. MOSFETs should never be stored in Styrofoam (which generates static electricity).
   c. MOSFETs should not be held by their leads; they should be held by their cases only.
   d. MOSFETs should not be installed or removed while power is applied.
6. The operation of an *n*-channel D-MOSFET is illustrated in Figure 13.4.
   a. When $V_{GS} = 0$ V, $I_D = I_{DSS}$ (see Figure 13.4a).
   b. When $V_{GS}$ is negative, $I_D < I_{DSS}$ (see Figure 13.4b).
   c. When $V_{GS}$ is positive, $I_D > I_{DSS}$ (see Figure 13.4c).
   These operating states are all represented on the component transconductance curve (see Figure 13.4d).
7. D-MOSFETs can use any of the JFET biasing circuits described in Chapter 12.
8. D-MOSFETs can be *zero biased*. When zero biased, $I_D = I_{DSS}$ (see Figure 13.7).
9. D-MOSFETs are compared to JFETs in Figure 13.8.
10. The operation of an n-channel E-MOSFET is illustrated in Figure 13.9.
    a. When $V_{GS} = 0$ V, there is no channel between the source and drain terminals, and $I_D = I_{DSS} \cong 0$ A.

**b.** When $V_{GS}$ reaches a specified *positive* value called the *threshold voltage* ($V_{GS(\text{th})}$), a channel forms between the source and drain terminals, and $I_D > 0$ A. The transconductance curve for an *n*-channel E-MOSFET is shown in Figure 13.10.

11. Zero bias and self-bias cannot be used to bias E-MOSFETs because they are enhancement-only devices.

12. E-MOSFETs are typically biased using voltage-divider bias, gate bias (with a positive $V_{GS}$), and drain-feedback bias.

  **a.** Drain-feedback bias is similar (in appearance) to collector-feedback bias (see Figure 13.13).

  **b.** Drain-feedback bias automatically establishes a value of $I_D = I_{D(\text{on})}$. $I_{D(\text{on})}$ is the rated value of E-MOSFET current when $V_{GS} = V_{DS}$.

13. MOSFET operation is limited at high frequencies because they have high gate capacitance (see Figure 13.16).

14. A *dual-gate* MOSFET uses two gate terminals to reduce the overall gate capacitance of the component.

  **a.** Normally, a dual-gate MOSFET is wired to operate as two series MOSFETs.

  **b.** Dual-gate MOSFET construction is illustrated in Figure 13.17.

15. A *V-MOSFET* (VMOS) is designed to handle much higher drain currents than a standard MOSFET.

  **a.** VMOS devices are enhancement-mode devices.

  **b.** VMOS device construction is illustrated in Figure 13.18. The shape of the gate results in a wider channel than that in a standard MOSFET.

16. When a VMOS is biased as shown in Figure 13.19, a channel is formed that allows conduction.

17. VMOS devices have *positive* temperature coefficients.

  **a.** A positive temperature coefficient means that device resistance increases when temperature increases.

  **b.** Their positive temperature coefficients allow several VMOS devices to be wired in parallel for increased maximum current. Parallel wiring is not possible for BJTs.

18. The *lateral double-diffused* MOSFET, or LDMOS, uses a short channel and a heavily doped *n*-type region to obtain low channel resistance and, therefore, high drain current. LDMOS construction is illustrated in Figure 13.20.

19. A group of digital circuits that have similar operating characteristics is referred to as a *logic family*.

20. *Complementary MOS* (CMOS) logic is made up entirely of MOSFETs.

  **a.** An *inverter* is a logic-level converter; that is, it converts one dc voltage to another, and vice versa.

  **b.** A CMOS inverter is shown in Figure 13.22.

21. CMOS circuits have the following advantages over TTL (a logic family based on BJTs):

  **a.** They draw little current from the supply, so they consume very little power.

  **b.** One CMOS output can drive an unlimited number of CMOS inputs connected in parallel.

22. A *cascode amplifier* is a low-$C_{\text{in}}$ MOSFET amplifier that is used in high-frequency applications.

  **a.** A cascode amplifier has a common-source circuit directly wired to a common-gate circuit (see Figure 13.25).

  **b.** The series connection of MOSFETs in a cascode amplifier causes the circuit to have lower input capacitance than either individual MOSFET.

  **c.** A dual-gate MOSFET can be wired as a cascode amplifier (see Figure 13.26).

23. An RF amplifier that uses a dual-gate MOSFET is shown in Figure 13.27. One gate is used as a signal input, and the other is used for an *automatic gain control* (AGC) input.

24. MOSFET power drivers are commonly used in digital communications circuitry.

| Equation Number | Equation | Section Number |
|:---:|:---:|:---:|
| **(13.1)** | $I_D = I_{DSS}\left(1 - \dfrac{V_{GS}}{V_{GS(\text{off})}}\right)^2$ | 13.2 |
| **(13.2)** | $g_m = g_{m0}\left(1 - \dfrac{V_{GS}}{V_{GS(\text{off})}}\right)$ | 13.2 |
| **(13.3)** | $I_D = k[V_{GS} - V_{GS(\text{th})}]^2$ | 13.3 |
| **(13.4)** | $k = \dfrac{I_{D(\text{on})}}{[V_{GS(\text{on})} - V_{GS(\text{th})}]^2}$ | 13.3 |
| **(13.5)** | $V_{GS} = V_{DS}$ | 13.3 |
| **(13.6)** | $V_{DS} = V_{DD} - R_D I_{D(\text{on})}$ | 13.3 |

cascode amplifier  538
complementary MOS
   (CMOS)  536
depletion-mode
   operation  520
digital communications
   540
D-MOSFET  520
drain-feedback bias  531

dual-gate MOSFET  533
E-MOSFET  520
enhancement-mode
   operation  520
fanout  537
inverter  536
lateral double-diffused
   MOSFET (LDMOS)
   535

logic families  536
logic levels  535
MOSFET  520
speed-up capacitor  540
substrate  520
threshold voltage
   ($V_{GS(\text{th})}$)  528
V-MOSFET (VMOS)  534
zero bias  526

## Section 13.2

**1.** A D-MOSFET has values of $V_{GS(\text{off})} = -4$ V and $I_{DSS} = 8$ mA. Plot the transconductance curve for the device.

**2.** A D-MOSFET has values of $V_{GS(\text{off})} = -8$ V and $I_{DSS} = 12$ mA. Plot the transconductance curve for the device.

**3.** A D-MOSFET has values of $V_{GS(\text{off})} = -4$ to $-6$ V and $I_{DSS} = 6$ to $9$ mA. Plot the minimum and maximum transconductance curves for the device.

**4.** A D-MOSFET has values of $V_{GS(\text{off})} = -4$ to $-10$ V and $I_{DSS} = 5$ to $10$ mA. Plot the minimum and maximum transconductance curves for the device.

**5.** The D-MOSFET described in Problem 1 has a value of $g_{m0} = 2000$ μS. Determine the values of $g_m$ at $V_{GS} = -2$ V, $V_{GS} = 0$ V, and $V_{GS} = +2$ V.

**6.** The D-MOSFET described in Problem 2 has a value of $g_{m0} = 3000$ μS. Determine the values of $g_m$ at $V_{GS} = -5$ V, $V_{GS} = 0$ V, and $V_{GS} = +5$ V.

## Section 13.3

**7.** An E-MOSFET has ratings of $V_{GS(\text{th})} = 4$ V and $I_{D(\text{on})} = 12$ mA at $V_{GS} = 8$ V. Determine the value of $I_D$ for the device when $V_{GS} = 6$ V.

**8.** An E-MOSFET has ratings of $V_{GS(\text{th})} = 2$ V and $I_{D(\text{on})} = 10$ mA at $V_{GS} = 8$ V. Determine the value of $I_D$ for the device when $V_{GS} = +12$ V.

**9.** An E-MOSFET has ratings of $V_{GS(\text{th})} = 1$ V and $I_{D(\text{on})} = 8$ mA at $V_{GS} = 4$ V. Determine the values of $I_D$ for the device when $V_{GS} = 1$ V, $V_{GS} = 4$ V, and $V_{GS} = 5$ V.

**10.** An E-MOSFET has values of $V_{GS(\text{th})} = 3$ V and $I_{D(\text{on})} = 2$ mA at $V_{GS} = 5$ V. Determine the values of $I_D$ for the device when $V_{GS} = 3$ V, $V_{GS} = 5$ V, and $V_{GS} = 8$ V.

**11.** Calculate the values of $I_D$ and $V_{DS}$ for the amplifier in Figure 13.29.

**12.** Calculate the values of $I_D$ and $V_{DS}$ for the amplifier in Figure 13.30.

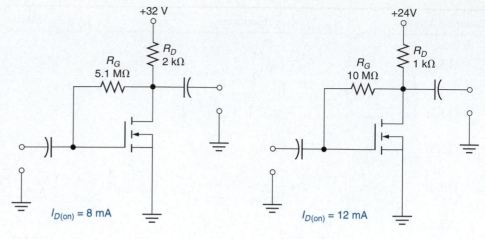

Assume that the rated values of $I_{D(on)}$ in Figures 13.29 through 13.34 were measured at $V_{GS} = V_{DS}$.

**FIGURE 13.29**

**FIGURE 13.30**

13. Calculate the values of $I_D$ and $V_{DS}$ for the amplifier in Figure 13.31.
14. Calculate the values of $I_D$ and $V_{DS}$ for the amplifier in Figure 13.32.
15. Calculate the values of $I_D$ and $V_{DS}$ for the amplifier in Figure 13.33.
16. Calculate the values of $I_D$ and $V_{DS}$ for the amplifier in Figure 13.34.

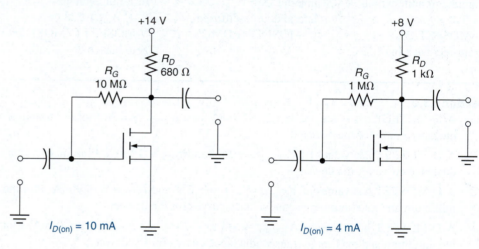

**FIGURE 13.31**

**FIGURE 13.32**

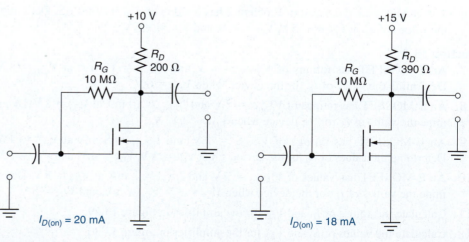

**FIGURE 13.33**

**FIGURE 13.34**

**17.** The circuit shown in Figure 13.35 has the waveforms shown. Discuss the possible cause(s) of the problem.

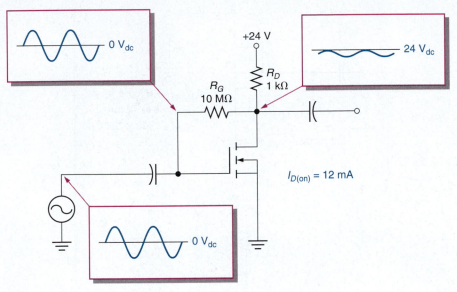

**FIGURE 13.35**

**18.** The circuit shown in Figure 13.36 has the dc voltages indicated. Discuss the possible cause(s) of the problem.

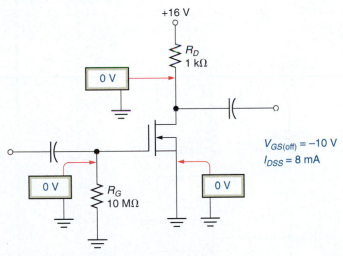

**FIGURE 13.36**

**19.** Determine the range of $A_v$ values for the circuit shown in Figure 13.37.

**20.** Determine the range of $A_v$ values for the circuit shown in Figure 13.38.

**21.** Determine the range of $I_D$ values for the circuit shown in Figure 13.39.

**22.** The constant-current bias circuit shown in Figure 13.40 won't work. Why?

**23.** The MOSFET shown in Figure 13.41 has a maximum gate current of 800 pA when $V_{GS} = 32$ V. Determine the input impedance of the device and the circuit. Then, determine the input impedance of the circuit, assuming that the MOSFET has infinite input impedance. What is the difference between the two values of $Z_{in}$?

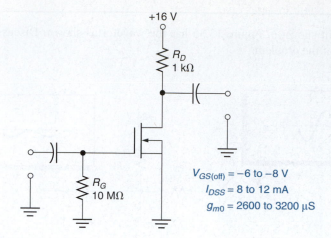

FIGURE 13.37

$V_{GS(off)} = -6$ to $-8$ V
$I_{DSS} = 8$ to 12 mA
$g_{m0} = 2600$ to 3200 µS

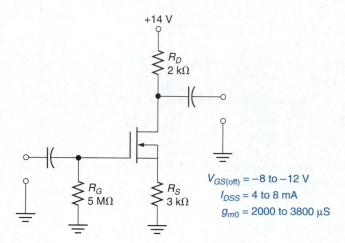

FIGURE 13.38

$V_{GS(off)} = -8$ to $-12$ V
$I_{DSS} = 4$ to 8 mA
$g_{m0} = 2000$ to 3800 µS

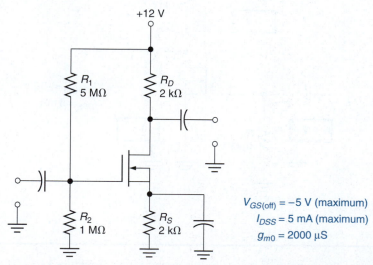

FIGURE 13.39

$V_{GS(off)} = -5$ V (maximum)
$I_{DSS} = 5$ mA (maximum)
$g_{m0} = 2000$ µS

**SUGGESTED COMPUTER APPLICATIONS PROBLEMS**

**24.** Write a program that will perform the complete dc analysis of a zero-bias circuit when provided with the proper information.

**25.** Write a program that will determine the value of $I_D$ for an E-MOSFET at any given value of $V_{GS}$. The program should include the determination of the value of $k$ for the device.

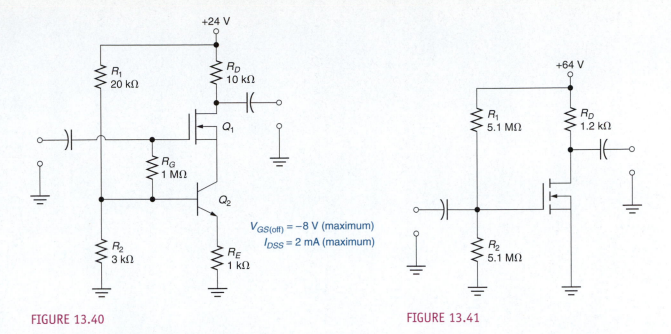

FIGURE 13.40

FIGURE 13.41

**26.** Write a program that will perform the complete dc analysis of a drain-feedback bias circuit when provided with the required information.

---

## ANSWERS TO THE EXAMPLE PRACTICE PROBLEMS

**13.1** See Figure 13.42.

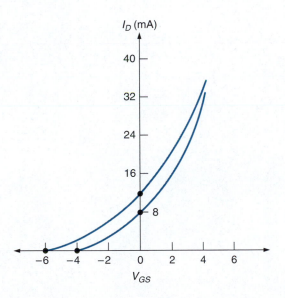

FIGURE 13.42

**13.2** 27.44 mA

**13.3** $I_D$ = 14 mA, $V_{DS}$ = 6.86 V

---

## chapter 14

# Amplifier Frequency Response

## Objectives

*After studying the material in this chapter, you should be able to:*

1. Define *bandwidth*, *cutoff frequency*, and *geometric center frequency*, and identify each on a frequency-response curve.

2. Calculate any two of the following values given the other two: $f_{C1}, f_{C2}$, geometric center frequency ($f_0$), or bandwidth.

3. Describe the *decade* and *octave* frequency multipliers.

4. Compare and contrast the *Bode plot* with the frequency-response curve.

5. Perform a complete low-frequency analysis of a BJT amplifier.

6. Discuss the concept of gain roll-off, and calculate its effect on voltage gain at a given operating frequency.

7. Explain why BJT internal capacitances are not considered in low-frequency analyses.

8. Calculate the Miller input and output capacitance values for a BJT amplifier.

9. Perform a complete high-frequency analysis of a BJT amplifier.

10. Compare high-frequency roll-off rates to low-frequency roll-off rates.

11. Perform the low-frequency-response analysis of an FET amplifier.

12. Perform the high-frequency-response analysis of an FET amplifier.

13. Describe and analyze the frequency response of a multistage amplifier.

## Outline

In this chapter, we will take a look at the effects of operating frequency on BJT and FET circuit gain. As we progress through our discussions, you will see that several principles apply to all types of amplifiers. In this section, we will discuss each of these principles.

### 14.1.1 Bandwidth

OBJECTIVE 1 ▶

**Bandwidth (BW)**
The range of frequencies over which gain is relatively constant.

Most amplifiers have relatively constant gain over a certain range, or *band*, of frequencies. This band of frequencies is called the **bandwidth (BW)** of the amplifier. The bandwidth for a given amplifier depends on the circuit component values and the type of active component(s) used. Later in this chapter, you will be shown how to calculate the bandwidth for several BJT and FET amplifiers.

When an amplifier is operated within its bandwidth, the current, voltage, and power gain values for the amplifier are calculated as shown earlier in the text. For clarity, these values of $A_i$, $A_v$, and $A_p$ are referred to as *midband gain* values. For example, the *midband power gain*, $A_{p(\text{mid})}$, of an amplifier is the power gain of the circuit when it is operated within its bandwidth. Again, the exact value of $A_{p(\text{mid})}$ (as well as the other gain values) are calculated as shown earlier in the text and vary from one circuit to another.

**Frequency-response curve**
A curve showing the relationship between gain and operating frequency.

A graphic representation of the relationship between amplifier gain and operating frequency is called a **frequency-response curve**. A simplified frequency-response curve is shown in Figure 14.1.

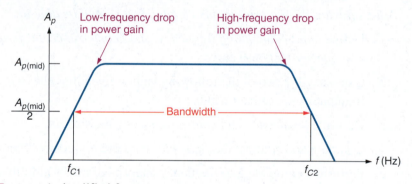

**FIGURE 14.1** A simplified frequency response curve.

As the frequency-response curve shows, the power gain of an amplifier remains relatively constant across a band of frequencies. When the operating frequency starts to go outside this frequency range, the gain begins to drop off. The greater the increase or decrease in operating frequency (outside the constant-gain band), the greater the decrease in gain.

Two frequencies of interest, $f_{C1}$ and $f_{C2}$, are identified on the frequency-response curve. These are the frequencies at which power gain decreases to approximately 50% of $A_{p(\text{mid})}$. For example, let's assume that the amplifier represented by the curve in Figure 14.1 has the following values: $f_{C1} = 10$ kHz, $f_{C2} = 100$ kHz, and $A_{p(\text{mid})} = 500$. When operated at a point between its cutoff frequencies, the amplifier has a power gain of 500. Decreasing the operating frequency toward $f_{C1}$ causes a decrease in power gain. When the operating frequency reaches 10 kHz ($f_{C1}$), the amplifier power gain has decreased to 250. Likewise, *increasing* the operating frequency toward $f_{C2}$ causes a decrease in power gain. When the operating frequency reaches 100 kHz, the amplifier power gain has decreased to 250. If the circuit operating frequency varies far enough outside the limits set by $f_{C1}$ and $f_{C2}$, the amplifier eventually reaches *unity gain* ($A_p = 1$). At this point, the amplifier no longer serves any purpose.

*A Practical Consideration:*
The gain of an amplifier will start to drop off well before the operating frequency reaches $f_{C1}$ or $f_{C2}$. However, by the time $f_{C1}$ or $f_{C2}$ is reached, the value of $A_p$ will have decreased by approximately 50%.

The frequencies labeled $f_{C1}$ and $f_{C2}$ are the amplifier **cutoff frequencies**. The cutoff frequencies frequencies for a given amplifier define its bandwidth limits, as shown in Figure 14.1. Note that the bandwidth of an amplifier is found as the *difference between* $f_{C1}$ and $f_{C2}$ as follows:

◀ **OBJECTIVE 2**

$$BW = f_{C2} - f_{C1} \qquad (14.1)$$

> **Cutoff frequencies**
> The frequencies at which $A_p$ decreases to 50% of $A_{p(mid)}$.

where BW = the amplifier bandwidth, in hertz
  $f_{C2}$ = the *upper* cutoff frequency
  $f_{C1}$ = the *lower* cutoff frequency

The following chart shows some examples of $f_{C1}, f_{C2}$, and BW value combinations.

| $f_{C1}$ (kHz) | $f_{C2}$ (kHz) | BW (kHz) |
|---|---|---|
| 10 | 100 | 90 |
| 25 | 250 | 225 |
| 0 (dc) | 400 | 400 |

Note that the bandwidth in each case equals the *difference* between $f_{C1}$ and $f_{C2}$. This relationship is illustrated further in Example 14.1.

## EXAMPLE 14.1

The frequency-response curve in Figure 14.2 represents the operation of an amplifier. Determine the bandwidth of the circuit.

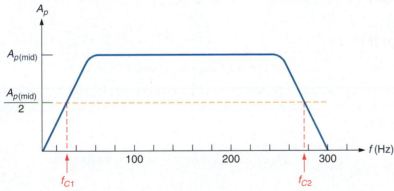

FIGURE 14.2

*Solution:* The cutoff frequencies correspond to the *half-power points* on the curve, as shown. For this particular circuit, these values are approximately

$$f_{C1} = 30 \text{ kHz} \quad \text{and} \quad f_{C2} = 275 \text{ kHz}$$

The bandwidth is found as

$$BW = f_{C2} - f_{C1} = 275 \text{ kHz} - 30 \text{ kHz} = \textbf{245 kHz}$$

### PRACTICE PROBLEM 14.1

Determine the bandwidth for the amplifier represented by the frequency-response curve in Figure 14.3.

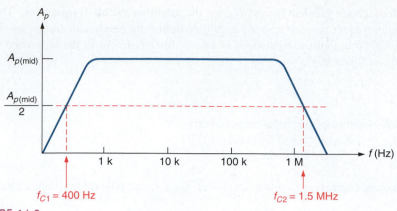

FIGURE 14.3

**Geometric center frequency ($f_0$)**
The frequency that equals the geometric average of $f_{C1}$ and $f_{C2}$.

What is the relationship between geometric center frequency ($f_0$) and power gain?

There are cases when it is desirable to know the **geometric center frequency ($f_0$)** of a given amplifier. The geometric center frequency is *the geometric average of the cutoff frequencies*. By formula,

$$f_0 = \sqrt{f_{C1}f_{C2}} \qquad (14.2)$$

When the operating frequency of an amplifier is equal to the value of $f_0$, the power gain of the amplifier is at its maximum value. Then, as the operating frequency varies toward $f_{C1}$ or $f_{C2}$, the power gain begins to drop off. By the time the operating frequency reaches either of the cutoff frequencies, power gain has dropped to 50% of $A_{p(mid)}$. Example 14.2 demonstrates the calculation of $f_0$ for an amplifier.

### EXAMPLE 14.2

Determine the value of $f_0$ for the amplifier represented by Figure 14.2.

*Solution:* The values of $f_{C1}$ and $f_{C2}$ were found to be 30 kHz and 275 kHz, respectively. Using these two values in equation (14.2), the value of the center frequency is found as

$$f_0 = \sqrt{f_{C1}f_{C2}} = \sqrt{(30 \text{ kHz})(275 \text{ kHz})} = \textbf{90.83 kHz}$$

### PRACTICE PROBLEM 14.2

Determine the value of $f_0$ for the amplifier represented by the curve in Figure 14.3.

Since $f_0$ is the *geometric average* of the cutoff frequencies, the ratio of $f_0$ to $f_{C1}$ equals the ratio of $f_{C2}$ to $f_0$. By formula,

$$\frac{f_0}{f_{C1}} = \frac{f_{C2}}{f_0} \qquad (14.3)$$

Example 14.3 illustrates the fact that these ratios are equal.

### EXAMPLE 14.3

Show that the ratio of $f_0$ to $f_{C1}$ is equal to the ratio of $f_{C2}$ to $f_0$ for the amplifier described in Examples 14.1 and 14.2.

*Solution:*  The following values were obtained in the two examples:

$$f_{C1} = 30 \text{ kHz}$$
$$f_{C2} = 275 \text{ kHz}$$
$$f_0 = 90.83 \text{ kHz}$$

The ratio of $f_0$ to $f_{C1}$ is found as

$$\frac{f_0}{f_{C1}} = \frac{90.83 \text{ kHz}}{30 \text{ kHz}} = \textbf{3.03}$$

The ratio of $f_{C2}$ to $f_0$ is found as

$$\frac{f_{C2}}{f_0} = \frac{275 \text{ kHz}}{90.83 \text{ kHz}} = \textbf{3.03}$$

As you can see, the frequency ratios are equal.

### PRACTICE PROBLEM 14.3
Show that equation (14.3) holds true for the circuit described in Practice Problems 14.1 and 14.2.

Equation (14.3) is important because it provides us with some insight into the relationship between $f_0$, $f_{C1}$, and $f_{C2}$. When an amplifier is operated at its geometric center frequency, the input frequency can vary by the same *factor* in each direction before the power gain of the amplifier drops to 50% of $A_{p(\text{mid})}$. For example, let's say that an amplifier has the following values:

$$f_{C1} = 6 \text{ kHz}$$
$$f_{C2} = 150 \text{ kHz}$$
$$f_0 = 30 \text{ kHz}$$

As you can see, $f_0$ is five times the value of $f_{C1}$. At the same time, $f_{C2}$ is five times the value of $f_0$. Thus, if the amplifier is operated at $f_0$, the input frequency can decrease by a factor of five before hitting the lower cutoff frequency ($f_{C1}$) or can increase by a factor of five before hitting the upper cutoff frequency ($f_{C2}$). This is where the term *geometric center frequency* comes from.

Equation (14.3) can be transposed to provide us with two other useful equations. Each can be used to predict the value of $f_{C1}$ or $f_{C2}$ when the other cutoff frequency and the center frequency are known, as follows:

$$f_{C1} = \frac{f_0^2}{f_{C2}} \qquad\qquad\qquad \textbf{(14.4)}$$

and

$$f_{C2} = \frac{f_0^2}{f_{C1}} \qquad\qquad\qquad \textbf{(14.5)}$$

Once the unknown cutoff frequency has been determined, we can calculate the bandwidth of the amplifier, as Examples 14.4 and 14.5 demonstrate.

## EXAMPLE 14.4

The values of $f_0$ and $f_{C2}$ for an amplifier are measured at 60 kHz and 300 kHz, respectively. Determine the values of $f_{C1}$ and BW for the amplifier.

*Solution:*    The value of $f_{C1}$ is found using equation (14.4) as follows:

$$f_{C1} = \frac{f_0^2}{f_{C2}} = \frac{(60 \text{ kHz})^2}{300 \text{ kHz}} = \mathbf{12 \text{ kHz}}$$

The value of BW is now found as

$$\mathbf{BW} = f_{C2} - f_{C1} = 300 \text{ kHz} - 12 \text{ kHz} = \mathbf{288 \text{ kHz}}$$

### PRACTICE PROBLEM 14.4

An amplifier has values of $f_{C2} = 300$ kHz and $f_0 = 30$ kHz. Determine the bandwidth of the amplifier.

## EXAMPLE 14.5

The values of $f_0$ and $f_{C1}$ for an amplifier are measured at 40 kHz and 8 kHz, respectively. Determine the values of $f_{C2}$ and BW for the amplifier.

*Solution:*    The value of $f_{C2}$ is found using equation (14.5) as follows:

$$f_{C2} = \frac{f_0^2}{f_{C1}} = \frac{(40 \text{ kHz})^2}{8 \text{ kHz}} = \mathbf{200 \text{ kHz}}$$

The value of BW is now found as

$$\mathbf{BW} = f_{C2} - f_{C1} = 200 \text{ kHz} - 8 \text{ kHz} = \mathbf{192 \text{ kHz}}$$

### PRACTICE PROBLEM 14.5

An amplifier has values of $f_0 = 25$ kHz and $f_{C1} = 5$ kHz. Determine the bandwidth of the amplifier.

You have been shown the relationship among the cutoff frequencies ($f_{C1}$ and $f_{C2}$), the center frequency ($f_0$), and the bandwidth (BW) of an amplifier. In later sections, you will be shown the circuit analysis equations used to determine all these values.

### 14.1.2    Measuring $f_{C1}$ and $f_{C2}$

When you are working with an amplifier, its cutoff frequencies can be measured easily using an oscilloscope. The procedure is as follows:

1. Set up the amplifier for the maximum undistorted output signal. This is done by varying the *amplitude* of the amplifier input signal. As a first step, set the input frequency to approximately 100 kHz.
2. Establish that you are operating in the midband of the amplifier by varying the *frequency* of the input signal several kilohertz in both directions. If you are in the midband range of the amplifier, the *amplitude* of the amplifier output will *not* change significantly as you vary the input frequency.
3. If you are not in the midband frequency range, adjust $f_{in}$ until you are.
4. Adjust the *volts-div* calibration control until the waveform fills exactly *seven* major divisions (peak-to-peak).

> *A Practical Consideration:*
> Most ac voltmeters do not have the frequency-handling capability needed to measure bandwidth. For this reason, the method covered here emphasizes the use of the oscilloscope. When using the oscilloscope to measure bandwidth, you should use a $\times 10$ probe to reduce the effects of the scope's input capacitance on your measurements.

**5.** To determine the value of $f_{C1}$, decrease $f_{in}$ until the output waveform fills only *five* major divisions. At this frequency, the amplitude of the output has changed by a ratio of $5/7 \cong 0.707$. This ratio indicates that we are operating the circuit at its cutoff frequency.

**6.** Increase the frequency until the same thing happens on the high end of operation. Measure $f_{C2}$ at this frequency.

**Lab Reference:** This technique for measuring amplifier cutoff frequencies can be used in Exercise 20.

In basic electronics, you learned that *power varies with the square of voltage*. The same principle holds true for $A_p$ and $A_v$. Thus, when $A_p$ drops to 0.5 of $A_{p(mid)}$, the voltage gain will drop by the *square root* of the *0.5* factor, 0.707. As you will see, the fact that $A_v$ equals $0.707A_{v(mid)}$ at the cutoff frequencies is important when deriving the equations for $f_{C1}$ and $f_{C2}$.

### 14.1.3 Gain and Frequency Measurements

Up to this point, we have used simplified frequency-response curves in the various illustrations. A more practical frequency-response curve is shown in Figure 14.4. As you can see, this curve is a bit more complicated than any you have seen thus far. The primary differences are as follows:

**1.** The *y*-axis of the graph represents *changes in power gain* rather than specific power gain values.
**2.** The changes in power gain are expressed in *decibels* (dB).
**3.** The frequency units are based on a *logarithmic* scale; that is, each major division is a whole-number multiple of the previous major division.

We will look at all of these differences so that you will have no problem with reading an actual frequency-response curve.

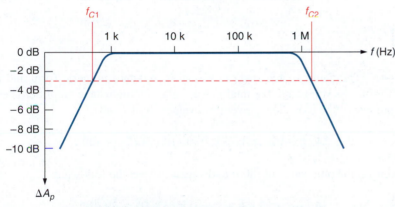

**FIGURE 14.4** A more practical frequency-response curve.

### 14.1.4 Decibel Power Gain

A decibel (dB) is a logarithmic representation of a ratio. One advantage of using decibels is that they allow us to represent very large ratios as relatively small numbers. In Chapter 8, we discussed the concept of dB gain in detail. The material presented here is intended only as a brief review.

Frequency-response curves and specification sheets often list gain values that are measured in decibels. As shown in Chapter 8, the decibel power gain of an amplifier is found as

$$A_{p(dB)} = 10 \log A_p \qquad \textbf{(14.6)}$$

or

$$A_{p(dB)} = 10 \log \frac{P_{out}}{P_{in}} \qquad \textbf{(14.7)}$$

Examples 14.6 and 14.7 review the process for finding the power gain of an amplifier in decibels.

**EXAMPLE 14.6**

An amplifier has values of $P_{in} = 100\ \mu W$ and $P_{out} = 2\ W$. What is the dB power gain of the amplifier?

*Solution:* The dB power gain of the amplifier is found as

$$A_{p(dB)} = 10 \log \frac{P_{out}}{P_{in}} = 10 \log \frac{2\ W}{100\ \mu W} = 10 \log (20{,}000) = \textbf{43.01 dB}$$

**PRACTICE PROBLEM 14.6**

An amplifier has values of $P_{in} = 420\ \mu W$ and $P_{out} = 6\ W$. What is the dB power gain of the amplifier?

For multistage amplifiers, the total dB gain of the circuit equals the *sum* of the dB gains of the stages.

**EXAMPLE 14.7**

For the multistage amplifier represented in Figure 14.5, prove that the total power gain is equal to $A_{p1(dB)} + A_{p2(dB)} + A_{p3(dB)}$.

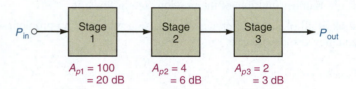

$P_{in}$ ○→ **Stage 1** → **Stage 2** → **Stage 3** → $P_{out}$

$A_{p1} = 100$    $A_{p2} = 4$    $A_{p3} = 2$
$= 20\ dB$    $= 6\ dB$    $= 3\ dB$

**FIGURE 14.5**

*Solution:* The gain of each stage is shown in standard numeric form and as a decibel value. As you recall, the total power gain of a multistage amplifier is equal to the product of the individual power ratio values. By formula,

$$A_{p(T)} = (A_{p1})(A_{p2})(A_{p3}) = (100)(4)(2) = \textbf{800}$$

If we convert the ratio value of (800) to decibels, we get the following:

$$A_{p(dB)} = 10 \log A_p = 10 \log (800) = \textbf{29 dB}$$

Now, if we simply add the decibel gains of the three stages, we get the following:

$$A_{p(T)} = A_{p1(dB)} + A_{p2(dB)} + A_{p3(dB)} = 20\ dB + 6\ dB + 3\ dB = \textbf{29 dB}$$

Since both methods of determining total dB power gain yield the same result, we have shown that you can determine total multistage gain by simply adding the individual dB power gain values.

**PRACTICE PROBLEM 14.7**

A three-stage amplifier has gain values of 100, 86, and 45. Show that the decibel gain values can be added to determine the total amplifier gain.

Now, let's relate what you have learned about dB gain values to the frequency-response curve in Figure 14.4. Note that the *y*-axis of the graph represents *a change in power gain*, expressed in dB. When the circuit is operated in its midband frequency range,

$$A_p = A_{p(mid)}$$

and

$$\frac{A_p}{A_{p(mid)}} = 1$$

If we convert this power gain ratio to dB form, we get

$$A_{p(mid)} = 10 \log (1) = 0 \text{ dB}$$

This is why midband power gain is labeled as 0 dB in Figure 14.4.

When $\Delta A_p = -3$ dB, the power gain of the amplifier is half its midband value. That is,

$$A_p = \frac{A_{p(mid)}}{2} \quad (\text{when } \Delta A_p = -3 \text{ dB})$$

Based on this relationship, we can redefine $f_{C1}$ and $f_{C2}$ (the cutoff frequencies) as follows: *The cutoff frequencies of an amplifier are those frequencies at which the power gain of the amplifier is 3 dB lower than its midband value.* For this reason, the cutoff frequencies of an amplifier are often called the *upper and lower 3 dB frequencies.*

### 14.1.5  Frequency Units and Terminology

The frequency-response curve in Figure 14.4 utilizes a *logarithmic scale* to measure frequency values. A logarithmic scale is one that uses a *geometric* progression of units. In a geometric progression, the value of each division is *a whole number multiple of the value for the previous division.* For example, in Figure 14.4, the value of each division on the horizontal scale is 10 times the value of the previous division.

When the multiplier rate for a frequency scale is 10, the scale is referred to as a *decade scale.* Note that a **decade** is a frequency multiplier equal to 10. The scale in Figure 14.4 is a decade scale.

Another commonly used frequency multiplier is the **octave**. One octave is a frequency multiplier of *two.* Thus, when you increase frequency by one octave, you *double* the original frequency. The following table shows a series of *fundamental* (starting) frequencies and the decade and octave frequencies for each:

◀ *OBJECTIVE 3*

**Decade**
A frequency multiplier equal to 10.

**Octave**
A frequency multiplier equal to 2.

| Fundamental | Octave | Decade |
|---|---|---|
| 100 Hz | 200 Hz | 1 kHz |
| 500 Hz | 1 kHz | 5 kHz |
| 1.5 MHz | 3 MHz | 15 MHz |
| $f_x$ | $2f_x$ | $10f_x$ |

The terms *octave* and *decade* are used extensively in describing the rate at which gain varies with frequency. For example, an amplifier may be said to have a *roll-off rate of 20 dB/decade.* This means that the gain of the amplifier decreases at a rate of 20 dB for each decade that the frequency goes beyond the limit of either $f_{C1}$ or $f_{C2}$. Roll-off rates are discussed in more detail later in this chapter.

### 14.1.6  Bode Plots

The **Bode plot** (pronounced "bō-dē") is a variation on the basic frequency-response curve. A Bode plot is shown in Figure 14.6. As you can see, the Bode plot differs from the frequency-response curves you have seen so far in one respect: The value of $\Delta A_{p(mid)}$ is assumed to be zero until the cutoff frequencies are reached. Then, the power gain of the amplifier is assumed to drop at a set rate of 20 dB/decade. For comparison, the standard frequency-response curve (labeled FRC) is included in the figure. Note that the power gain of the amplifier is shown to be down by 3 dB at each of the cutoff frequencies on the FRC, while the power gain is still shown to be at its midband value on the Bode plot.

◀ *OBJECTIVE 4*

**Bode plot**
A frequency-response curve that assumes $\Delta A_{p(mid)}$ is zero until the cutoff frequency is reached.

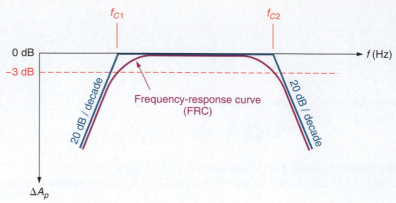

FIGURE 14.6   Bode plot.

Bode plots are typically used in place of other types of frequency-response curves because they are easier to read. Since Bode plots are the most commonly used frequency-response graphs, we will use them throughout the remainder of this chapter. Just remember, while the Bode plot shows $\Delta A_p$ to be zero at the cutoff frequencies, the actual value is $-3$ dB.

At this point, we have looked at the overall relationship between gain and frequency. We have also discussed the basic units used to measure gain and frequency. Now, we will move on to specific types of circuits, their respective calculations, and the reasons for their response characteristics.

---

**Section Review** ▶

1. What is *bandwidth*?

2. Draw a frequency-response curve, and identify the following: $f_{C1}$, $f_{C2}$, bandwidth, and geometric center frequency.

3. What is the numeric relationship among the values of $f_{C1}$, $f_0$, and $f_{C2}$ for a given amplifier?

4. What is the procedure for measuring $f_{C1}$ and $f_{C2}$ with an oscilloscope?

5. Which oscilloscope probe should be used for high-frequency measurements? Why?

6. In what form are gain values typically written?

7. What are the advantages of using decibel gain values?

8. Why are the cutoff frequencies referred to as the *upper* and *lower 3 dB points*?

9. What type of frequency scale is normally used in frequency-response curves?

10. What is a *decade*? What is an *octave*?

11. Why are decade and octave frequency scales used?

12. Compare and contrast the *Bode plot* with a frequency-response curve.

## 14.2   BJT Amplifier Frequency Response

The cutoff frequencies of an *RC*-coupled amplifier are determined by the combination of amplifier capacitance and resistance values. In this section, you will be shown how the capacitance and resistance values in an *RC*-coupled BJT amplifier determine the frequency-response characteristics of the circuit.

### 14.2.1   Low-Frequency Response

**OBJECTIVE 5** ▶

Our discussion on BJT amplifier low-frequency response begins with the base circuit. For reference, this circuit is shown in Figure 14.7. The resistor shown in the base circuit ($R_1 \parallel R_2$) represents the ac equivalent of a voltage-divider bias network. This resistance is in parallel with the input impedance of the transistor ($h_{ie}$). As shown in the modified circuit on

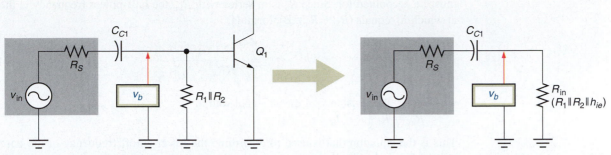

FIGURE 14.7   The base circuit of a BJT amplifier.

the right, $R_1$, $R_2$, and $h_{ie}$ can be combined into a single amplifier input resistance ($R_{in}$). Assuming that the signal source is ideal ($R_S = 0\ \Omega$), the amplifier input resistance forms a voltage divider with the reactance of the coupling capacitor. For this divider circuit, the value of $v_b$ is found as

$$v_b = v_{in}\frac{R_{in}}{\sqrt{X_C^2 + R_{in}^2}} \qquad\qquad (14.8)$$

Now, let's take a look at what happens to the fraction in equation (14.8) when $X_C = R_{in}$. When this happens, the fraction simplifies as follows:

$$\frac{R_{in}}{\sqrt{X_C^2 + R_{in}^2}} = \frac{R_{in}}{\sqrt{R_{in}^2 + R_{in}^2}} = \frac{R_{in}}{\sqrt{2R_{in}^2}} = \frac{R_{in}}{\sqrt{2R_{in}^2}} = \frac{1}{\sqrt{2}} \cong 0.707$$

If we replace the fraction in equation (14.8) with the 0.707 value obtained, we get the following relationship:

$$v_b = 0.707v_{in} \quad (\text{when } X_C = R_{in}) \qquad\qquad (14.9)$$

Thus, the circuit shown in Figure 14.7 has a value of $v_b = 0.707v_{in}$ when $X_C = R_{in}$. (This assumes that the signal source is ideal; that is, it assumes that $R_S = 0\ \Omega$.) When $v_b = 0.707v_{in}$, the voltage gain of the circuit is *effectively* reduced to $0.707A_{v(mid)}$. As you may recall, $A_v = 0.707A_{v(mid)}$ at the half-power points. Thus, the lower half-power point ($f_{C1}$) for Figure 14.7 occurs at the frequency that causes $X_C$ to equal $R_{in}$.

At what frequency does this occur? If we swap the values of $f$ and $X_C$ in the basic capacitive reactance equation, we obtain the following equation:

$$f = \frac{1}{2\pi X_C C} \qquad\qquad (14.10)$$

Since $X_C$ is equal to the total ac resistance in the base circuit ($R$) at $f_{C1}$, we can rewrite this equation as

$$f_{1B} = \frac{1}{2\pi RC} \qquad\qquad (14.11)$$

where $f_{1B}$ = the output lower cutoff frequency of the base circuit
$\quad\ R$ = the *total* ac resistance in the base circuit
$\quad\ C$ = the value of the base coupling capacitor

The value of $R$ in equation (14.11) represents the *total* ac resistance in the base circuit. Though we assumed a value of $R_S = 0\ \Omega$ earlier, the source resistance in Figure 14.7

> When referring to the overall circuit response of an amplifier, we designate the cutoff frequencies as $f_{C1}$ and $f_{C2}$. To distinguish the cutoff frequencies of the individual transistor terminal circuits, we will designate them as $f_{1B}$, $f_{1E}$, $f_{1C}$, and so on.

must be accounted for. Since $R_S$ is in series with $R_{in}$, the half-power frequency is the one at which $X_C$ equals $(R_S + R_{in})$. By formula,

$$f_{1B} = \frac{1}{2\pi(R_S + R_{in})C_{C1}}$$

(14.12)

where $R_S$ = the resistance of the ac signal source
$R_{in} = R_1 \| R_2 \| h_{ie}$

This is the equation that is used to determine the lower cutoff frequency of the base circuit, as Example 14.8 demonstrates.

### EXAMPLE 14.8

Determine the value of $f_{1B}$ for the circuit shown in Figure 14.8.

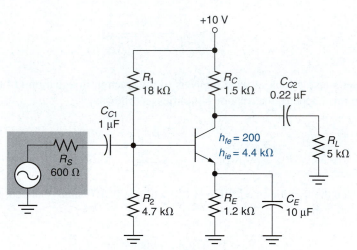

FIGURE 14.8

*Solution:*   First, the value of $R_{in}$ is found as

$$R_{in} = R_1 \| R_2 \| h_{ie} = 18 \text{ k}\Omega \| 4.7 \text{ k}\Omega \| 4.4 \text{ k}\Omega = 2018 \ \Omega$$

Now, using this value in equation (14.12), we obtain the value of $f_{1B}$ as follows:

$$f_{1B} = \frac{1}{2\pi(R_S + R_{in})C} = \frac{1}{2\pi(2618 \ \Omega)(1 \ \mu F)} = \frac{1}{16.45 \times 10^{-3}} \text{ Hz} = 60.79 \text{ Hz}$$

Thus, at 60.79 Hz, the base circuit of Figure 14.8 reduces the power gain of the amplifier by 3 dB.

### PRACTICE PROBLEM 14.8

A BJT amplifier has values of $R_1 = 56$ k$\Omega$, $R_2 = 5.6$ k$\Omega$, $R_S = 1$ k$\Omega$, $C_{C1} = 1$ $\mu$F, and $h_{ie} = 5$ k$\Omega$. Determine the value of $f_{1B}$ for the circuit.

The collector circuit of the BJT amplifier works according to the same principle as the base circuit. When the reactance of the output coupling capacitor is equal to the total series resistance in the collector circuit $(R_C + R_L)$, the output voltage is reduced to 0.707 times its midband value. Again, the value of $A_v$ has then been effectively reduced to $0.707A_{v(mid)}$.

It should come as no surprise that the equation for the cutoff frequency of the collector circuit is very similar to that for the base circuit. The actual equation is as follows:

$$f_{1C} = \frac{1}{2\pi(R_C + R_L)C}$$   (14.13)

where   $f_{1C}$ = the lower cutoff frequency of the collector circuit
   $(R_C + R_L)$ = the sum of the resistances in the collector circuit
   $C$ = the value of the output coupling capacitor

Example 14.9 demonstrates the use of this equation.

### EXAMPLE 14.9

Determine the value of $f_{1C}$ for the circuit shown in Figure 14.8.

*Solution:*   Using the values shown in the figure, the value of $f_{1C}$ is found as

$$f_{1C} = \frac{1}{2\pi(R_C + R_L)C} = \frac{1}{2\pi(6.5 \text{ k}\Omega)(0.22 \text{ }\mu\text{F})} = \frac{1}{8.984 \times 10^{-3}} \text{ Hz} = \textbf{111.3 Hz}$$

### PRACTICE PROBLEM 14.9

The circuit described in Practice Problem 14.8 has values of $R_C$ = 3.6 k$\Omega$, $R_L$ = 1.5 k$\Omega$, and $C_{C2}$ = 2.2 $\mu$F. Determine the value of $f_{1C}$ for the circuit.

In Examples 14.8 and 14.9, we got values of $f_{1B}$ = 60.79 Hz and $f_{1C}$ = 111.3 Hz for the amplifier in Figure 14.8. Based on these two values, what is the cutoff frequency for the amplifier? If you said 111.3 Hz, you were right. An amplifier cuts off at the frequency closest to $f_0$. Thus, for low-frequency response, the amplifier cuts off at the *highest* calculated value of $f_{C1}$.

Before deriving the equation for the emitter cutoff frequency, we need to refer to some relationships established in Chapter 10. These were

$$R_{out} = R_E \left\| \left(r'_e + \frac{R'_{in}}{h_{fc}}\right)\right.$$   (14.14)

Remember:
   $h_{fc} = h_{fe} + 1$

where

$$R'_{in} = R_1 \| R_2 \| R_S$$

In equation (14.14), the output resistance of the amplifier is considered to be the parallel combination of $R_E$ and $(r'_e + R'_{in}/h_{fc})$. Under normal circumstances, the value of $R_E$ is much greater than the value of $(r'_e + R'_{in}/h_{fc})$. Therefore, we can approximate the total ac resistance in the emitter circuit as

$$R_{out} \cong r'_e + \frac{R'_{in}}{h_{fc}}$$   (14.15)

Using this value, the value of $f_{1E}$ is found as

$$f_{1E} = \frac{1}{2\pi R_{out}C_E}$$   (14.16)

The following example demonstrates the use of this equation.

## EXAMPLE 14.10

Determine the value of $f_{1E}$ for the circuit shown in Figure 14.8.

**Solution:** First, the value of $R_{out}$ is found as

$$R_{out} \cong r_e' + \frac{R_{in}'}{h_{fc}} = 22\,\Omega + \frac{4.7\,\text{k}\Omega \parallel 18\,\text{k}\Omega \parallel 600\,\Omega}{201} = 22\,\Omega + 2.57\,\Omega = \textbf{24.57}\,\boldsymbol{\Omega}$$

Now, the value of $f_{1E}$ can be found as

$$f_{1E} = \frac{1}{2\pi R_{out}C_E} = \frac{1}{2\pi(24.57\,\Omega)(10\,\mu\text{F})} = \frac{1}{1.544 \times 10^{-3}}\,\text{Hz} = \textbf{647.8 Hz}$$

### PRACTICE PROBLEM 14.10

The amplifier described in Practice Problems 14.8 and 14.9 has values of $h_{fe} = 200$, $R_E = 1.3\,\text{k}\Omega$, and $C_E = 100\,\mu\text{F}$. Determine the value of $f_{1E}$ for the circuit.

At this point, what would you say is the value of $f_{C1}$ for the circuit shown in Figure 14.8? Would you say 647.8 Hz? Good!

## 14.2.2 Gain Roll-Off

**OBJECTIVE 6** ▶

The term **roll-off rate** is used to describe the rate at which the voltage gain of an amplifier (in dB) drops off after $f_{C1}$ or $f_{C2}$ has been passed. The *low-frequency* roll-off for the common-emitter amplifier is calculated using the following equation:

**Roll-off rate**

The rate of gain reduction for a circuit when operated beyond its cutoff frequencies.

$$\Delta A_v = 20 \log \frac{1}{\sqrt{1+(f_{C1}/f)^2}} \tag{14.17}$$

where $f$ = the frequency of operation
$f_{C1}$ = the lower cutoff frequency of the amplifier

The derivation of equation (14.17) is shown in Appendix D.

Before we get involved in any calculations, one point needs to be made about equation (14.17): *Equation (14.17) is used to determine the change in voltage gain.* For example, if equation (14.17) yields a result of $-3$ dB at a given frequency ($f$), this means that $A_{v(dB)}$ is 3 dB lower than its midband value at that frequency. The following example serves to illustrate this point.

### EXAMPLE 14.11

An amplifier has values of $A_{v(mid)} = 45$ dB and $f_{C1} = 2$ kHz. What is the gain of the amplifier when it is operated at 500 Hz?

**Solution:** The *change in* $A_v$ is found using equation (14.17) as follows:

$$\Delta A_v = 20 \log \frac{1}{\sqrt{1 + (f_{C1}/f)^2}} = 20 \log \frac{1}{\sqrt{1 + (2000/500)^2}} = 20 \log \frac{1}{\sqrt{1 + 16}}$$
$$= 20 \log (0.2425) = \textbf{-12.3 dB}$$

Now, the value of $A_{v(dB)}$ at 500 Hz is found as

$$A_{v(\textbf{dB})} = A_{v(mid)} + \Delta A_v = 45\,\text{dB} + (-12.3\,\text{dB}) = \textbf{32.7 dB}$$

### PRACTICE PROBLEM 14.11

An amplifier with a gain of 23 dB has a lower cutoff frequency of 8 kHz. If the circuit is operated at 4 kHz, what is the gain of the amplifier?

Equation (14.17) tells us one very important fact about the low-frequency roll-off of an amplifier: *The values of R and C in the amplifier have nothing to do with the roll-off rate of a given terminal circuit.* For each terminal circuit (base versus emitter versus collector), the roll-off rate is strictly a function of the difference between $f$ and $f_{C1}$. The *RC* circuit determines the frequency at which the roll-off begins. However, once it has begun, the values of *R* and *C* are no longer significant. We can therefore conclude that *all RC circuits roll off at the same rate.*

How do the values of *R* and *C* affect gain roll-off?

What *is* the low-frequency roll-off rate? We can determine this by setting $f$ to several multiples of $f_{C1}$ and solving the equation. For example, when $f = f_{C1}$, $\Delta A_v$ is found as

$$\Delta A_v = 20 \log \frac{1}{\sqrt{1 + 1^2}} = -3 \text{ db} \left| \frac{f_{C1}}{f} = 1 \right.$$

When $f = 0.1 f_{C1}$,

$$\Delta A_v = 20 \log \frac{1}{\sqrt{1 + 10^2}} = -20 \text{ dB} \left| \frac{f_{C1}}{f} = 10 \right.$$

When $f = 0.01 f_{C1}$,

$$\Delta A_v = 20 \log \frac{1}{\sqrt{1 + 100^2}} = -40 \text{ dB} \left| \frac{f_{C1}}{f} = 100 \right.$$

When $f = 0.001 f_{C1}$,

$$\Delta A_v = 20 \log \frac{1}{\sqrt{1 + 1000^2}} = -60 \text{ dB} \left| \frac{f_{C1}}{f} = 1000 \right.$$

Do you see a pattern here? Every time we decrease the value of $f$ by a factor of 10 (a *decade*), the gain of the amplifier drops another 20 dB. Therefore, we can state that the low-frequency roll-off for voltage gain is *20 dB/decade*. The Bode plot that corresponds to this roll-off rate is shown in Figure 14.9.

What is the roll-off rate of an *RC* circuit?

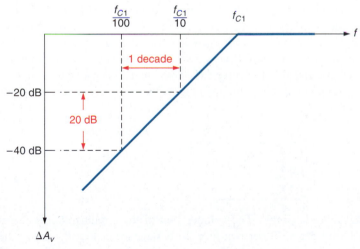

FIGURE 14.9

The Bode plot in Figure 14.9 is best explained in a series of steps:

1. At $f_{C1}$, the change in voltage gain is shown to be 0 dB.
2. At $f = 0.1 f_{C1}$ (one decade down), the voltage gain of the amplifier is reduced by 20 dB.
3. At $f = 0.01 f_{C1}$ (next decade down), the voltage gain of the amplifier is reduced by another 20 dB. The total value of $\Delta A_v$ is $-40$ dB.

Lab Reference: Gain roll-off is demonstrated in Exercise 20.

If the process were to continue, the next several decades would have total $\Delta A_v$ values of $-60$ dB, $-80$ dB, and so on. At each decade interval, the voltage gain of the circuit would be 20 dB lower than the previous decade interval.

Let's relate this to the circuit shown in Figure 14.8. We determined that the cutoff frequency for the emitter circuit ($f_{1E}$) was 647.5 Hz. For the sake of discussion, we will make two assumptions:

1. The cutoff frequency is 600 Hz. (This will simplify our frequency scale.)
2. The emitter circuit is the *only* circuit that will cut off. In other words, we will neglect the effects of the base and collector circuits. (You will be shown very shortly what happens when these circuits are brought into the picture.)

Figure 14.10 shows the Bode plot for the emitter circuit shown in Figure 14.8. The cutoff frequency is 600 Hz. One decade down from this frequency is 60 Hz. At this frequency, the voltage gain of the amplifier is reduced by 20 dB, giving us a total $\Delta A_v$ of $-20$ dB, and so on.

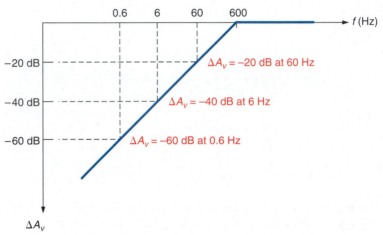

**FIGURE 14.10** The Bode plot for the emitter circuit of Figure 14.8.

If we were to plot the gain roll-off for the base circuit in Figure 14.8, the roll-off rate of 20 dB per decade would not be very useful. Since $f_{1B}$ is about 60 Hz, one or two decades down does not leave us with a whole lot of frequency! It helps, then, to be able to determine the voltage gain roll-off at *octave* frequency rates. The following table shows the $\Delta A_v$ values associated with octave frequency intervals. The values shown were determined by substituting the value of ($f_{C1}/f$) into equation (14.17) and solving for $\Delta A_v$.

| $f$ | $f_{C1}/f$ | $\Delta A_v$ (approximate) |
|---|---|---|
| $1/2\, f_{C1}$ | 2 | $-6$ |
| $1/4\, f_{C1}$ | 4 | $-12$ |
| $1/8\, f_{C1}$ | 8 | $-18$ |

Can roll-off rates be measured at *octave* intervals?

The pattern here is simple. The voltage gain of the circuit rolls off at a rate of *6 dB per octave*. Using the 6 dB per octave roll-off rate, we can easily derive the Bode plot for the base circuit of Figure 14.8. This plot is shown in Figure 14.11. Note that the $\Delta A_v$ intervals are now equal to 6 dB and that the frequency is measured in octaves instead of decades. Other than these two points, the plot shown is read in the same manner as the one in Figure 14.10.

Before we go on to look at the effect of having three cutoff frequencies in an amplifier ($f_{1B}, f_{1C},$ and $f_{1E}$), let's summarize the points that have been made in this section:

Summary of *RC* roll-off rates.

1. The low-frequency roll-off rate for an *RC* circuit is *20 dB per decade*. This is equal to a rate of *6 dB per octave*.

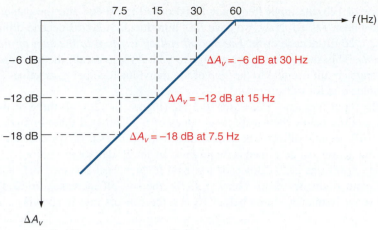

FIGURE 14.11  Roll-off measured in decibels per octave.

2. The roll-off rate is independent of the values of $R$ and $C$ in the circuit. Thus, the rates given in statement 1 apply to all $RC$ circuits.
3. Equation (14.17) can be used to determine the value of $A_v$ *relative to* $A_{v(\text{mid})}$ at a given frequency, as demonstrated in Example 14.11.

### 14.2.3  The Combined Effects of $f_{1B}$, $f_{1C}$, and $f_{1E}$

You have seen that a circuit like the one shown in Figure 14.8 has *three RC* terminal circuits, each with its own cutoff frequency. Next, we will look at what happens when more than one cutoff frequency is passed. The overall concept is easy to understand with the help of Figure 14.12. Here, we have three separate $RC$ circuits, each having the indicated value of $f_{C1}$. The cutoff frequencies were chosen at one-decade intervals, which simplifies a complex analysis.

In Figure 14.12a, the input frequency is shown to be 500 Hz. Since this frequency is below the value of $f_{C1}$ for $RC_1$, the circuit is causing a voltage gain roll-off of 20 dB/decade. At this point, the input frequency is still above $RC_2$ and $RC_3$, so these two circuits are still operating at their respective values of $A_{v(\text{mid})}$. Thus, at this point, the overall gain is dropping by 20 dB/decade, the roll-off rate of $RC_1$.

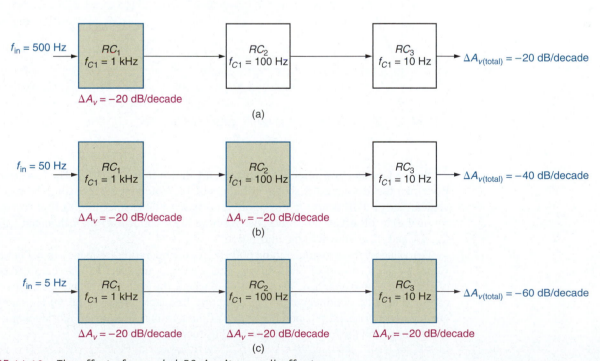

FIGURE 14.12  The effect of cascaded $RC$ circuits on roll-off rates.

In Figure 14.12b, the input frequency has dropped to 50 Hz, and the value of $f_{C1}$ for $RC_2$ has also been passed. Now, we have $RC_1$ introducing a 20 dB/decade drop and $RC_2$ introducing a 20 dB/decade drop. *The overall roll-off is equal to the sum of these drops, 40 dB/decade.* This should not seem strange to you. As you recall, the total decibel gain of a multistage circuit is equal to the *sum* of the individual decibel gain values. The same principle holds true for roll-off rates.

When the input frequency decreases to 5 Hz (Figure 14.12c), all three of the $RC$ circuits are operating below their respective values of $f_{C1}$. Thus, all three circuits are introducing a 20 dB/decade roll-off. The sum of these roll-off rates is 60 dB/decade.

The circuit action just described is represented by the Bode plot in Figure 14.13. As you can see, the roll-off rate is 20 dB/decade for input frequencies between 1 kHz and 100 Hz. When the input frequency drops below 100 Hz, the roll-off increases to 40 dB/decade. When the input frequency drops below 10 Hz, the roll-off rate is shown to be 60 dB/decade.

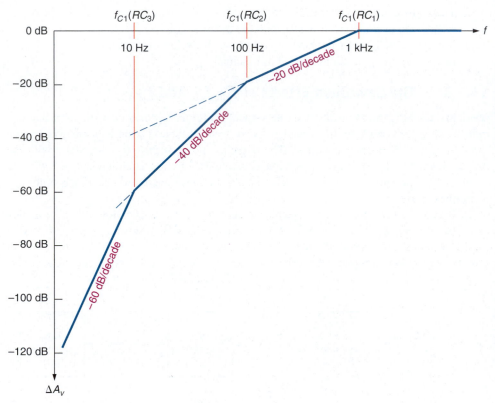

**FIGURE 14.13** The effects of multiple cutoff frequencies.

Now, let's relate this principle to the circuit shown in Figure 14.8. The low-frequency Bode plot for this circuit is shown in Figure 14.14. Note that we are using the more convenient rate of *6 dB/octave* in this figure. This is acceptable since a roll-off of 6 dB/octave is equal to a roll-off of 20 dB/decade. We have also rounded off the calculated values of $f_{1B}, f_{1C}$, and $f_{1E}$ as shown in the illustration.

The value of $f_{1E}$ is assumed to be 600 Hz. As the operating frequency drops below this value, we get a roll-off of 6 dB/octave. This continues until the 120 Hz cutoff frequency of the collector is reached. Once this is reached, the roll-off rate increases to 12 dB/octave. This rate continues until the operating frequency reaches the 60 Hz cutoff frequency of the base. At this point, all three of the terminal circuits are introducing a 6 dB/octave drop, and the total roll-off rate is 18 dB/octave.

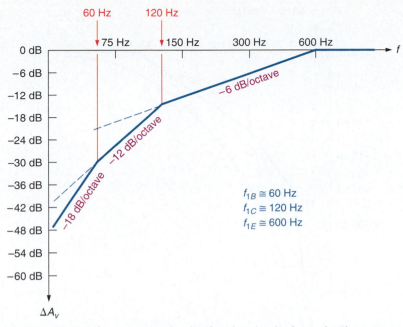

FIGURE 14.14   The low-frequency Bode plot for the circuit shown in Figure 14.8.

### 14.2.4   A Practical Consideration

The Bode plot represents the *ideal* amplifier response to a change in frequency. For example, the Bode plot in Figure 14.14 idealizes the low-frequency response of the circuit in Figure 14.8 as follows:

1. It assumes that all cutoff frequencies are exact. In other words, there are no resistor or capacitor tolerances that affect the actual cutoff frequency values.
2. It assumes that each transistor terminal circuit has a 0 dB value of $\Delta A_v$ until the cutoff frequency is reached.
3. It assumes that roll-off is constant at the given roll-off rate.

In practice, none of these assumptions is entirely accurate. Circuit component tolerances and stray capacitance values *do* affect the actual cutoff frequencies. You will find that, in practice, measured cutoff frequencies vary from the calculated cutoff frequencies.

> Component tolerances affect actual cutoff frequencies.

A more practical description of the circuit frequency response is as follows:

1. Each *RC* circuit noticeably begins to reduce the gain of an amplifier at approximately *twice* its value of $f_1$. The gain of the amplifier is reduced by approximately 3 dB by the time the lower cutoff frequency is reached.
2. The roll-off rate for a given *RC* circuit is 20 dB/decade (6 dB/octave).
3. When multiple *RC* circuits are involved (as in the BJT amplifier), the effects of each are felt well before their actual cutoff frequencies are reached.

> Practical frequency-response characteristics.

How much will these considerations affect the frequency-response curve of a given amplifier? Figure 14.15 shows the original Bode plot for the amplifier in Figure 14.8 and the more realistic frequency-response curve. As you can see, the curve resembles the Bode plot. However, the combined effects of the *RC* circuits cause the actual curve to have more rounded turning points that occur at higher frequencies than those calculated.

The Bode plot provides a means of predicting the response of a circuit to a change in frequency. However, when we are making these predictions, we need to keep in mind that they are *approximations* based on assuming a number of ideal conditions. Therefore, we should not be surprised when the percentage of error between our predicted and measured frequency-response values approaches 10% or 20%.

> When using Bode plots, keep in mind that they provide approximate values. This is because they are plotted assuming a number of ideal conditions.

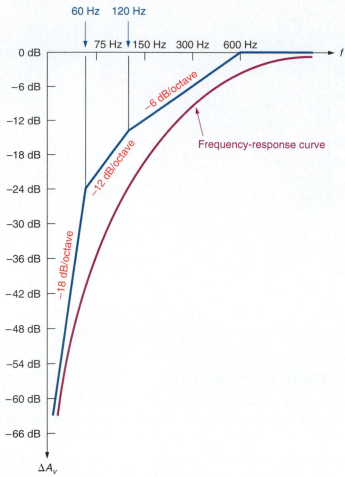

**FIGURE 14.15**

One point of interest: The greater the separation between cutoff frequencies, the closer the actual frequency-response curve comes to the Bode plot for an amplifier. For example, an amplifier with cutoff frequencies of 10 kHz, 1 kHz, and 10 Hz has a frequency-response curve that is closer to the ideal (the Bode plot) than an amplifier with cutoff frequencies of 10, 5, and 1 kHz. A demonstration of this concept is included as a "Pushing the Envelope" problem at the end of the chapter.

### 14.2.5   BJT Amplifier High-Frequency Response

In many ways, the high-frequency operation of the BJT amplifier is similar to its low-frequency operation. You have a cutoff frequency for each terminal circuit. Once a given value of $f_{C2}$ is passed, the circuit cuts off and introduces a roll-off rate that is approximately equal to 20 dB/decade (6 dB/octave).

The primary differences between low-frequency and high-frequency responses are as follows:

> How do $f_{C2}$ calculations differ from $f_{C1}$ calculations?

1. The values of capacitance used in $f_{C2}$ calculations.
2. The methods used to determine the total resistance in each terminal circuit.

These differences will now be discussed in detail.

### 14.2.6   BJT Internal Capacitance

You may recall from Chapter 2 that a *pn* junction has some measurable amount of capacitance. Since the BJT has two internal *pn* junctions, it has two internal capacitances, as shown in Figure 14.16a. As you will be shown in this section, these internal capacitances control the high-frequency response of the device.

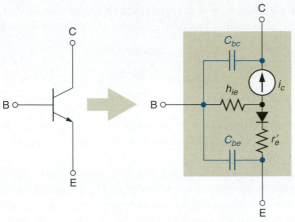

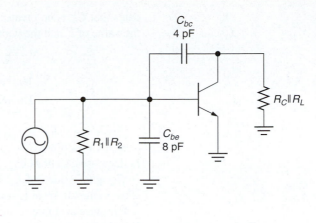

(a) A BJT and its ac equivalent circuit

(b) Representing BJT capacitance in the amplifier equivalent circuit

FIGURE 14.16   BJT internal capacitance.

The capacitance of the collector-base junction ($C_{bc}$) is usually listed on the spec sheet for a given transistor. Note that it may be identified as *collector-base capacitance* or *output capacitance* and may be labeled as $C_{bc}$, $C_{out}$, $C_{ob}$, or $C_{obo}$. For example, the 2N3904 spec sheet (Figure 7.17) lists an *output capacitance* rating of $C_{obo} = 4$ pF.

The value of $C_{bc}$ may or may not be listed on a transistor's spec sheet. When it is listed, it is generally identified as *base-emitter junction capacitance, base input capacitance,* or *input capacitance* and may be labeled as $C_{be}$, $C_{in}$, $C_b$, $C_{ib}$, or $C_{ibo}$. For example, the 2N3904 spec sheet (Figure 7.17) lists an *input capacitance* rating of $C_{ibo} = 8$ pF. When the value of $C_{be}$ is not listed on a transistor's spec sheet, its value can be approximated as

$$C_{be} \cong \frac{1}{2\pi f_T r'_e} \qquad (14.18)$$

where $f_T$ is the **current gain–bandwidth product** of the transistor, as listed on the spec sheet. The current gain–bandwidth product of a transistor is the frequency at which $C_{be}$ has a low enough reactance to cause the transistor current gain to drop to *unity*. Example 14.12 demonstrates the use of equation (14.18) in determining the value of $C_{be}$.

> **Current gain–bandwidth product**
> The frequency at which BJT current gain drops to unity.

### EXAMPLE 14.12

The spec sheet for the 2N3904 lists a value of $f_T = 300$ MHz at $I_C = 10$ mA. Determine the value of $C_{be}$ at this current.

*Solution:*   Using values obtained from the 2N3904 operating curves, we can determine the value of $r'_e$ to equal 2.9 Ω. Using this value in equation (14.18), we can approximate the value of $C_{be}$ as

$$C_{be} \cong \frac{1}{2\pi f_T r'_e} = \frac{1}{2\pi (300 \text{ MHz})(2.9 \text{ Ω})} = \textbf{183 pF}$$

> The value of $r'_e$ used in this example was found using
> $$r'_e = \frac{h_{ie}}{h_{fe}}$$
> The operating curves for the 2N3904 (not shown) provide the following values at $I_C = 10$ mA:
> $$h_{ie} = 500 \text{ Ω}$$
> $$h_{fe} = 170$$
> Using these values, $r'_e$ was found to be 2.9 Ω

### PRACTICE PROBLEM 14.12

A transistor has a current gain–bandwidth product of $f_T = 200$ MHz, $h_{ie} = 650$ Ω, and $h_{fe} = 250$ at $I_C = 10$ mA. Determine the value of $C_{be}$ for the device.

As you can see, the value of $C_{be} = 183$ pF in the example is significantly different from the spec sheet rating of 8 pF. The difference between the two values is due to the fact that they were determined for two different values of $I_E$. The spec sheet rating indicates that $C_{be}$ is no greater than 8 pF when $I_E = 0$ mA. However, as $I_E$ changes, so do both the value of $r_e'$ and the width of the base-emitter junction, as shown in the following table.

| $I_E$ | $r_e'$ | Junction Width |
|---|---|---|
| Increase | Decrease | Decrease |
| Decrease | Increase | Increase |

Since the value of $C_{be}$ is *inversely* proportional to the width of the emitter-base junction, it varies directly with variations in $I_E$. In other words, as $I_E$ increases, so does $C_{be}$. As $I_E$ decreases, so does $C_{be}$.

So which value of $C_{be}$ do you use for a given circuit, the calculated value or the spec sheet value? In most cases, you should use the *calculated* value of $C_{be}$ because it provides much more accurate results in later calculations. However, remember that even the calculated value of $C_{be}$ is only an approximate figure. Since the calculation of $r_e'$ is only an approximation, the value obtained by using equation (14.18) is an approximation. However, in most cases, this approximated value of $C_{be}$ is much closer to the actual emitter-base capacitance than the spec sheet rating.

**OBJECTIVE 7** ▶ Why have we ignored these capacitance values until now? As you can see, $C_{be}$ and $C_{bc}$ have very low values. Because of this, their low-frequency reactance is extremely high. For example, the reactance of $C_{bc}$ (in Figure 14.16b) is approximately 3.98 MΩ at $f = 10$ kHz. This reactance is so high that it can be considered to be open for all practical purposes, but what happens when the operating frequency of the transistor increases? Since reactance is inversely proportional to frequency for a given capacitor, the value of $X_C$ for $C_{bc}$ decreases. At 100 MHz, the reactance of $C_{bc}$ is 398 Ω. This reactance *cannot* be ignored because it will have a significant impact on the operation of the transistor.

### 14.2.7    Miller's Theorem

**OBJECTIVE 8** ▶ Before we can get on to some actual circuit calculations, we have to solve one more problem that involves the value of $C_{bc}$. As you saw in Figure 14.16, $C_{bc}$ is the capacitance between the collector and base terminals. We can represent this capacitance as shown in Figure 14.17.

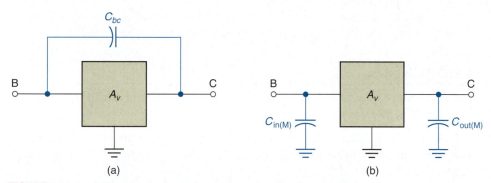

**FIGURE 14.17**    Miller equivalent circuit for a feedback capacitor.

In Figure 14.17a, the transistor has been shown as a block having a certain value of voltage gain, $A_v$. Since $C_{bc}$ exists between the collector and base terminals, it is shown as an external capacitor. The question now is this: Is $C_{bc}$ part of the input (base) circuit or part of the output (collector) circuit? The answer is: *both*.

**Miller's theorem** allows $C_{bc}$ to be represented as two capacitors, one in the base circuit and one in the collector circuit. These two capacitors are shown in Figure 14.17b. For this circuit, the *Miller input capacitance* is given as

**Miller's theorem**
Allows a feedback capacitor to be represented as separate input and output capacitances.

$$C_{in(M)} = C_{bc}(A_v + 1) \qquad \text{(14.19)}$$

and the *Miller output capacitance* is given as

$$C_{out(M)} = C_{bc} \frac{A_v + 1}{A_v} \qquad (14.20)$$

*A Practical Consideration:*
Miller's theorem applies only
to inverting amplifiers (those
with a 180° voltage phase
shift).

In most cases, $A_v > 10$. When this is the case, the above equations can be simplified as

$$C_{in(M)} \cong A_v C_{bc} \qquad (14.21)$$

and

$$C_{out(M)} \cong C_{bc} \qquad (14.22)$$

Note that the above equations can be used whenever the value of $A_v$ is greater than 10.

So, why use Miller's theorem? The *feedback* capacitor shown in Figure 14.17a can make circuit calculations extremely complex. A much simpler approach is to represent $C_{bc}$ as two separate capacitors. This way, one of the capacitors is used strictly for input (base) circuit calculations, and the other is used strictly for output (collector) circuit calculations. Example 14.13 demonstrates the method for determining the Miller input and output capacitance values for a given amplifier. After the Miller capacitance values are determined, we are ready to determine its upper cutoff frequencies.

Why do we use Miller's theorem?

### EXAMPLE 14.13

Determine the Miller input and output capacitance values for the circuit shown in Figure 14.18a.

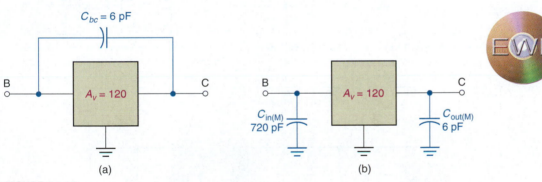

(a)                    (b)

FIGURE 14.18

*Solution:*    The Miller input capacitance is determined as

$$C_{in(M)} \cong A_v C_{bc} = (120)(6 \text{ pF}) = \mathbf{720\ pF}$$

The Miller output capacitance is determined as

$$C_{out(M)} \cong C_{bc} = \mathbf{6\ pF}$$

Using the calculated values of $C_{in(M)}$ and $C_{out(M)}$, we can redraw the circuit as shown in Figure 14.18b.

#### PRACTICE PROBLEM 14.13

A BJT with a value of $C_{bc} = 4$ pF is used in an amplifier that has a value of $A_v = 240$. Determine the Miller input and output equivalent capacitances for $C_{bc}$.

## 14.2.8 High-Frequency Response

When we discussed the low-frequency response of the common-emitter amplifier, you were shown that the value of $f_{C1}$ depends on the various values of circuit resistance and capacitance. The same holds true for the value of $f_{C2}$.

The high-frequency equivalent circuit for the common-emitter amplifier is shown in Figure 14.19. We will reduce this circuit to its simplest $RC$ equivalent so that you will see *why* the values of $R$ and $C$ are combined as they are. First, there is one value that requires explanation: $C_L$ is the *load capacitance*. If the amplifier is driving another amplifier stage, $C_L$ equals the *high-frequency* input capacitance of the load stage. This point is discussed further in our coverage of multistage circuits. (At this point, we will simply assume a value of $C_L$ for any example problems.)

*A Practical Consideration:*
The value of $C_L$ in the collector circuit of Figure 14.19 assumes that the amplifier load has some amount of capacitance. If the load is purely resistive, $C_L$ is zero, and the total capacitance in the output circuit can be assumed to be equal to $C_{out(M)}$ for the transistor.

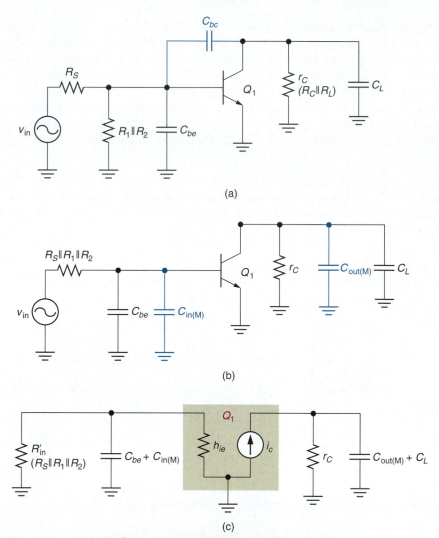

FIGURE 14.19   BJT high-frequency ac equivalent circuit.

The equivalent circuit in Figure 14.19b was derived by performing *two* steps:

1. The collector-base capacitor, $C_{bc}$, was replaced by its equivalent Miller input and output capacitors.
2. The source resistance ($R_S$) and the base biasing network were combined into a single *parallel* equivalent resistance, $R'_{in}$.

Now, we will simplify the circuit further by taking three additional steps. As Figure 14.19c shows:

3. The parallel capacitors in the base circuit, $C_{be}$ and $C_{in(M)}$, are combined into a single input capacitance value.

**4.** The two parallel capacitors in the collector circuit, $C_L$ and $C_{out(M)}$, are combined into a single output capacitance value.

**5.** The transistor has been replaced to show it as a resistance equal to $h_{ie}$, and a current source.

We can now use this equivalent circuit to analyze the high-frequency response of the orig-    ◄ **OBJECTIVE 9**
inal amplifier.

The upper cutoff frequency ($f_{C2}$) is determined using the same basic equation we used to determine $f_{C1}$:

$$f_{C2} = \frac{1}{2\pi RC}$$

For the base circuit, $R$ is the parallel combination of $R'_{in}$ and $h_{ie}$. Also, $C$ is the combination of $C_{be}$ and $C_{in(M)}$. Thus, for the base circuit,

$$f_{2B} = \frac{1}{2\pi(R'_{in} \parallel h_{ie})(C_{be} + C_{in(M)})} \qquad \textbf{(14.23)}$$

where $R'_{in} = R_S \parallel R_1 \parallel R_2$
$h_{ie}$ = the base input impedance to the transistor
$C_{be}$ = the base-emitter junction capacitance
$C_{in(M)}$ = the Miller input capacitance

Example 14.14 demonstrates the process for determining the upper cutoff frequency for the base circuit.

### EXAMPLE 14.14

Determine the value of $f_{2B}$ for the amplifier shown in Figure 14.20.

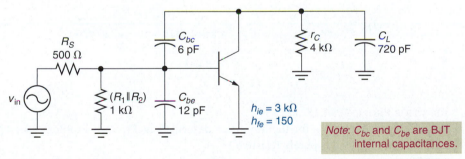

**FIGURE 14.20**

***Solution:*** The first step is to determine the value of $A_v$ for the amplifier as follows:

$$A_v = \frac{h_{fe}r_C}{h_{ie}} = \frac{(150)(4\text{ k}\Omega)}{3\text{ k}\Omega} = \textbf{200}$$

The value of $A_v$ is now used (along with $C_{bc}$) to determine the value of $C_{in(M)}$:

$$C_{in(M)} = A_v C_{bc} = (200)(6\text{ pF}) = \textbf{1.2 nF}$$

The value of $R'_{in}$ is found as

$$R'_{in} = R_S \parallel R_1 \parallel R_2 = \textbf{333 }\Omega$$

Now that we have all the needed values of resistance and capacitance, we can find the value of $f_{2B}$ for the circuit as follows:

$$f_{2B} = \frac{1}{2\pi(R'_{in} \parallel h_{ie})(C_{be} + C_{in(M)})} = \frac{1}{2\pi(333\ \Omega \parallel 3\ k\Omega)(12\ pF + 1.2\ nF)}$$

$$= \frac{1}{2\pi(300\ \Omega)(1.212\ nF)} = \frac{1}{2.285 \times 10^{-6}}\ Hz = \mathbf{437.72\ kHz}$$

**PRACTICE PROBLEM 14.14**

An amplifier has values of $R_1 = 56\ k\Omega$, $R_2 = 5.6\ k\Omega$, $R_S = 1\ k\Omega$, $R_C = 3.6\ k\Omega$, $R_L = 1.5\ k\Omega$, $h_{fe} = 200$, $h_{ie} = 5\ k\Omega$, $f_T = 300\ MHz$ at $I_C = 10\ mA$, and $C_{bc} = 4\ pF$. Determine the value of $f_{2B}$ for the circuit.

For the collector circuit, the value of $R$ is equal to $r_C$, and $C$ is the combination of $C_L$ and $C_{out(M)}$. Thus, for the collector circuit,

$$f_{2C} = \frac{1}{2\pi r_C(C_{out(M)} + C_L)} \tag{14.24}$$

The following example demonstrates the use of this equation.

**EXAMPLE 14.15**

Determine the value of $f_{2C}$ for the circuit shown in Figure 14.20.

*Solution:* Since the value of $A_v \gg 10$, the value of $C_{out(M)}$ is approximately equal to $C_{bc}$. Thus,

$$C_{out(M)} = 6\ pF$$

Since the other circuit values are already known, we can proceed to find $f_{2C}$ as follows:

$$f_{2C} = \frac{1}{2\pi r_C(C_{out(M)} + C_L)} = \frac{1}{2\pi(4\ k\Omega)(6\ pF + 720\ pF)}$$

$$= \frac{1}{1.825 \times 10^{-5}}\ Hz = \mathbf{54.8\ kHz}$$

**PRACTICE PROBLEM 14.15**

What is the value of $f_{2C}$ for the amplifier described in Practice Problem 14.14? (Assume that the load is purely resistive.)

### 14.2.9   Gain Roll-Off

OBJECTIVE 10 ▶ The *high-frequency* gain roll-off is found in the same manner as the low-frequency gain roll-off. The exact equation is

$$\Delta A_v = 20 \log \frac{1}{\sqrt{1 + (f/f_{C2})^2}} \tag{14.25}$$

As you can see from equation (14.25), the gain roll-off for the high-frequency circuit is also independent of the values of $R$ and $C$. We can use equation (14.25) to determine the high-frequency roll-off in *dB per decade* as follows. When $f = f_{C2}$,

$$\Delta A_v = 20 \log \frac{1}{\sqrt{1 + 1^2}} = \mathbf{-3\ dB}$$

When $f = 10f_{C2}$,

$$\Delta A_v = 20 \log \frac{1}{\sqrt{1 + 10^2}} = -20 \text{ dB}$$

When $f = 100f_{C2}$,

$$\Delta A_v = 20 \log \frac{1}{\sqrt{1 + 100^2}} = -40 \text{ dB}$$

The high-frequency roll-off of the common-emitter amplifier is equal to the low-frequency roll-off, *20 dB/decade*. Similarly, the high-frequency roll-off can also be computed as being equal to 6 dB/octave.

The high-frequency operation of a given common-emitter amplifier is exactly the same as the low-frequency operation. The only differences are the methods used to determine $f_{C2}$.

### 14.2.10 Theory Versus Practice

A wide variety of factors can affect the frequency response of a given amplifier. Component tolerances, estimations of BJT internal capacitances, and the effects of test equipment input capacitance can cause the measured values of $f_{C1}$ and $f_{C2}$ to vary significantly from their predicted (calculated) values.

When measuring the values of $f_{C1}$ and $f_{C2}$ for a BJT amplifier, you will find that the largest percentage of error will occur in the $f_{C2}$ measurements. There are two reasons for this:

1. The BJT internal capacitance values are estimated.
2. The capacitance values used in $f_{C2}$ calculations are in the picofarad (pF) range, as are the input capacitance ratings of most pieces of test equipment. Thus, by connecting the test equipment to the circuit, you are significantly altering the total capacitance that determines the value of $f_{C2}$.

When you are predicting the values of $f_{C1}$ and $f_{C2}$ for your standard BJT amplifier, you need to keep in mind that you are dealing with approximated values and, therefore, will get results that are approximations. While this may seem frustrating, it shouldn't be. You see, the purpose of performing the frequency analysis of an amplifier is to ensure that the circuit gain will not be affected by the input frequency. For example, if you want to operate a given amplifier at a frequency of 50 kHz, you want to know that an input frequency of 50 kHz will not cause the amplifier gain to be reduced. Beyond this application, the exact values of $f_{C1}$ and $f_{C2}$ are rarely of any consequence for standard *RC*-coupled BJT amplifiers.

There is one common application where the exact values of $f_{C1}$ and $f_{C2}$ are important. This is the case where you are dealing with *tuned amplifiers*. You may recall that tuned amplifiers are designed for specific bandwidths. Since they are designed for specific values of $f_{C1}$ and $f_{C2}$, these values are important when you are dealing with tuned amplifiers. This point is discussed further in Chapter 17.

*A Practical Consideration:* When you connect an oscilloscope or frequency counter to the output of an amplifier, the value of $f_{2C}$ will decrease. The reason is the added input capacitance of the particular piece of test equipment. For example, if an oscilloscope has an input capacitance of 30 pF, you are adding 30 pF when you connect the oscilloscope to the output of the circuit. This added 30 pF may significantly reduce the value of $f_{2C}$ for the circuit.

One way of reducing the percentage of error introduced by an oscilloscope is to use a ×10 *probe* when making high-frequency measurements. A ×10 probe decreases the input capacitance of the oscilloscope and, thus, reduces the error introduced by the device.

◀ **Section Review**

1. Explain why the power gain of a given amplifier drops by 50% when the reactance of the input coupling capacitor equals the input resistance of the amplifier.

2. Define the term *roll-off rate*.

3. What is the relationship between the roll-off rate for a given *RC* circuit and the circuit values of *R* and *C*? Explain your answer.

4. What are the standard low-frequency roll-off rates?

5. Describe the BJT amplifier low-frequency Bode plot.

6. Contrast the Bode plot with the actual low-frequency response curve of a BJT amplifier.

7. What are the primary differences between BJT $f_{C1}$ and $f_{C2}$ calculations?

8. Why aren't the BJT internal capacitances considered in the low-frequency analysis of a given amplifier?

9. What does Miller's theorem state? Why is it used?

10. List, in order, the steps taken to perform the high-frequency analysis of a BJT amplifier.

11. Compare high-frequency roll-off rates to low-frequency roll-off rates.

12. Explain why $f_{C2}$ measurements tend to have larger percentage of error values than $f_{C1}$ measurements.

## 14.3 FET Amplifier Frequency Response

The transition from BJT circuits to FET circuits is actually very simple. All we have to do is come up with the equations for $f_{C1}$ and $f_{C2}$ for a given FET amplifier. Everything else is identical to what we have been doing up to this point.

In this section, we will concentrate on the voltage-divider biased common-source amplifier. This circuit, along with its low-frequency ac equivalent, is shown in Figure 14.21.

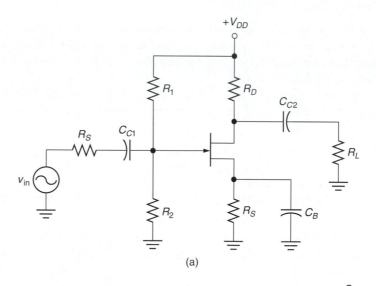

(a)

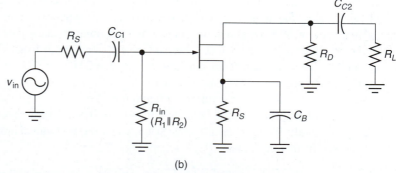

(b)

**FIGURE 14.21** FET amplifier low-frequency ac equivalent circuit.

An Important Point:
The value of $R_S$ in equation (14.26) represents the resistance of the input voltage source, not the JFET source resistor.

### 14.3.1 Low-Frequency Response

**OBJECTIVE 11 ▶** As you can see, there is very little difference between the FET amplifier and the BJT amplifier as far as the low-frequency equivalent circuit is concerned. In fact, the gate and drain cutoff frequency equations are nearly identical to those used for the BJT amplifier. Thus,

$$f_{1G} = \frac{1}{2\pi(R_S + R_{in})C_{C1}}$$

(14.26)

and

$$f_{1D} = \frac{1}{2\pi(R_D + R_L)C_{C2}}$$  **(14.27)**

Because of the extremely high input impedance of the FET, the value of $R_{in}$ for the amplifier is found as

$$R_{in} = R_1 \| R_2$$  **(14.28)**

As Example 14.16 shows, the high input impedance of the FET means that the FET amplifier normally has a much lower input cutoff frequency than that of a similar BJT amplifier.

## EXAMPLE 14.16

Determine which of the two circuits shown in Figure 14.22 has the lower input circuit cutoff frequency.

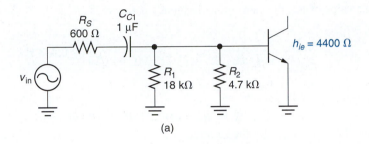

(a)

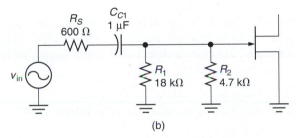

(b)

**FIGURE 14.22**

**Solution:** Figure 14.22a is the ac circuit that we analyzed in Example 14.8. At that point, we determined the value of $f_{1B}$ for the circuit to be 60.79 Hz (review Example 14.8 if necessary).

The circuit in Figure 14.22b is identical to the one in Figure 14.22a, except that the BJT has been replaced by an FET. Because of the high input impedance of the FET,

$$R_{in} = R_1 \| R_2 = 18 \text{ k}\Omega \| 4.7 \text{ k}\Omega = 3727 \ \Omega$$

Using this value for $R_{in}$, the value of $f_{1G}$ is found as

$$f_{1G} = \frac{1}{2\pi(R_S + R_{in})C_{C1}} = \frac{1}{2\pi(4327 \ \Omega)(1 \ \mu\text{F})} = 36.78 \text{ Hz}$$

Thus, by replacing the BJT with an FET, we have decreased the value of $f_1$ for the input circuit by nearly 50%.

In practice, FET amplifiers tend to use higher-value biasing resistors than do BJT amplifiers. As Example 14.17 demonstrates, this has the effect of lowering the value of $f_{1G}$ even more.

## EXAMPLE 14.17

Determine the value of $f_{1G}$ for the amplifier shown in Figure 14.23.

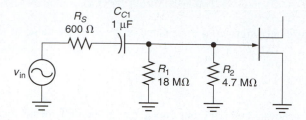

**FIGURE 14.23**

**Solution:** For this circuit,

$$R_{in} = R_1 \| R_2 = 18 \text{ M}\Omega \| 4.7 \text{ M}\Omega = \mathbf{3.727 \text{ M}\Omega}$$

Using this value of $R_{in}$, the gate cutoff frequency is found as

$$f_{1G} = \frac{1}{2\pi (R_S + R_{in})C_{C1}} = \frac{1}{2\pi (3.727 \text{ M}\Omega)(1 \ \mu\text{F})} = \mathbf{0.043 \text{ Hz}}$$

### PRACTICE PROBLEM 14.17

An FET amplifier has values of $R_1 = R_2 = 10 \text{ M}\Omega$, $R_S = 100 \ \Omega$, and $C_{C1} = 3.3 \ \mu\text{F}$. Determine the value of $f_{1G}$ for the circuit.

The low-frequency analysis of the drain circuit is just like that for the collector circuit of a BJT amplifier. For this reason, it needs no further explanation. The entire low-frequency analysis of a basic FET amplifier is demonstrated in Example 14.18.

### EXAMPLE 14.18

Determine the overall value of $f_{C1}$ for the circuit shown in Figure 14.24.

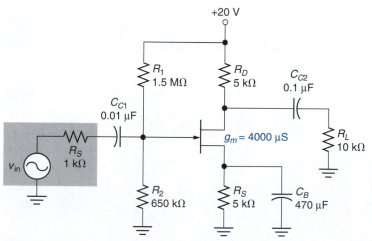

**FIGURE 14.24**

**Solution:** The input resistance to the amplifier is found as

$$R_{in} = R_1 \| R_2 = 1.5 \text{ M}\Omega \| 650 \text{ k}\Omega = \mathbf{453.5 \text{ k}\Omega}$$

Now, $f_{1G}$ is found as

$$f_{1G} = \frac{1}{2\pi(R_S + R_{in})C_{C1}} = \frac{1}{2\pi(454.5 \text{ k}\Omega)(0.01 \text{ }\mu\text{F})} = 35 \text{ Hz}$$

The value of $f_{1D}$ is found as

$$f_{1D} = \frac{1}{2\pi(R_D + R_L)C_{C2}} = \frac{1}{2\pi(15 \text{ k}\Omega)(0.1 \text{ }\mu\text{F})} = 106 \text{ Hz}$$

**PRACTICE PROBLEM 14.18**

An FET amplifier has values of $R_1 = 10 \text{ M}\Omega$, $R_2 = 1 \text{ M}\Omega$, $R_D = 1.1 \text{ k}\Omega$, $R_S = 820 \text{ }\Omega$, $R_L = 5 \text{ k}\Omega$, $C_{C1} = 0.01 \text{ }\mu\text{F}$, $C_{C2} = 0.1 \text{ }\mu\text{F}$, and $g_m = 2200 \text{ }\mu\text{S}$. Determine the overall value of $f_{C1}$ for the amplifier.

The value of $f_{C1}$ for the amplifier is assumed to be 106 Hz, the higher of the two cutoff frequencies. Just as with the BJT amplifier, the gain is reduced by 3 dB as the operating frequency decreases to 106 Hz. At this point, the gain continues to drop (with a continual decrease in frequency) at a rate of *6 dB/octave* until 35 Hz is reached. Then, the roll-off rate increases to *12 dB/octave*, exactly the same as with the BJT amplifier.

You may be wondering why we aren't considering the cutoff frequency of the source circuit in our analysis. There are two reasons:

1. Determining the cutoff frequency of the source circuit is an extremely complex procedure.
2. The lower cutoff frequency of the source circuit is much lower than the values of $f_{1G}$ and $f_{1D}$ under normal circumstances. Therefore, its value has little impact on the low-frequency analysis of the amplifier.

Since the lower cutoff frequency of the source circuit does not affect the value of $f_{C1}$ for the amplifier, we will consider our analysis complete with finding the values of $f_{1G}$ and $f_{1D}$.

## 14.3.2   High-Frequency Response

The high-frequency response of the FET is limited by values of *internal* capacitance, just as for the BJT. These capacitances are shown in Figure 14.25. As you can see, we again have a situation very similar to that of the BJT. There is a measurable amount of capacitance between each terminal *pair* of the FET. These capacitances each have a reactance that decreases as frequency increases. As the reactance of a given terminal capacitance decreases, more and more of the signal at the terminal is shorted through the component.

◀ *OBJECTIVE 12*

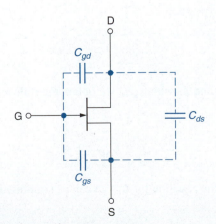

**FIGURE 14.25**   FET internal capacitances.

The high-frequency equivalent circuit of the basic FET amplifier is analyzed in the same fashion as the BJT amplifier. The high-frequency equivalent for the amplifier in Figure 14.21a is shown in Figure 14.26. As you can see, all the terminal capacitance values are included, with the exception of $C_{gd}$. This capacitor has been replaced with the Miller equivalent input and output capacitance values. (This is the same thing that we did with $C_{bc}$ in the BJT amplifier.)

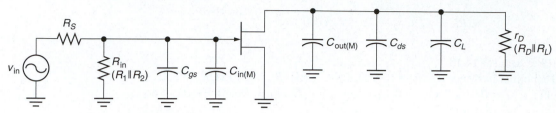

**FIGURE 14.26** FET amplifier high-frequency ac equivalent circuit.

The value of $C_{in(M)}$ is found using a form of equation (14.19) as follows:

$$C_{in(M)} = C_{gd}(A_v + 1)$$

Since the value of $A_v$ for the FET amplifier is equal to $g_m r_D$, the equation above can be rewritten as

$$C_{in(M)} = C_{gd}(g_m r_D + 1) \qquad (14.29)$$

*A Practical Consideration:*
When the value of $g_m r_D$ for the amplifier is greater than (or equal to) 10, equation (14.29) can be simplified to:

$$C_{in(M)} \cong C_{gd} g_m r_D$$

The calculation of $C_{in(M)}$ is demonstrated in Example 14.19.

### EXAMPLE 14.19

An FET amplifier has values of $C_{gd} = 4$ pF, $g_m = 2500$ μS, and $r_D = 5.6$ kΩ. Determine the value of $C_{in(M)}$ for the amplifier.

*Solution:* The value of $C_{in(M)}$ is found as

$$C_{in(M)} \cong C_{gd}(g_m r_D + 1) = (4 \text{ pF})[(2500 \text{ μS})(5.6 \text{ kΩ}) + 1]$$
$$= (4 \text{ pF})(15) = \mathbf{60 \text{ pF}}$$

### PRACTICE PROBLEM 14.19

A given FET amplifier has values of $C_{gd} = 3$ pF, $g_m = 3200$ μS, and $r_D = 1.8$ kΩ. Determine the value of $C_{in(M)}$ for the circuit.

Equation (14.20) defined $C_{out(M)}$ as being

$$C_{out(M)} = C\frac{A_v + 1}{A_v}$$

If we replace $C$ with $C_{gd}$, and $A_v$ with $g_m r_D$, we obtain the following equation for the FET amplifier value of $C_{out(M)}$:

$$C_{out(M)} = C_{gd}\frac{g_m r_D + 1}{g_m r_D} \qquad (14.30)$$

Again, when $g_m r_D > 10$, the equation for $C_{out(M)}$ can be simplified to

$$C_{out(M)} \cong C_{gd} \qquad (14.31)$$

From the circuit shown in Figure 14.26, you can see where the following equations come from:

$$C_G = C_{gs} + C_{\text{in(M)}} \qquad \text{(14.32)}$$

and

$$C_D = C_{\text{out(M)}} + C_{ds} + C_L \qquad \text{(14.33)}$$

where $C_L$ is the input capacitance of the load. We can now use these values to define the values of $f_{C2}$ for the gate and drain circuits. These equations, which will look very familiar at this point, are as follows:

$$f_{2G} = \frac{1}{2\pi R'_{\text{in}} C_G} \qquad \text{(14.34)}$$

where

$$R'_{\text{in}} = R_S \parallel R_{\text{in}}$$

and

$$f_{2D} = \frac{1}{2\pi r_D C_D} \qquad \text{(14.35)}$$

Example 14.20 demonstrates the process for determining the upper cutoff frequency for an FET amplifier.

### EXAMPLE 14.20

Determine the values of $f_{2G}$ and $f_{2D}$ for the amplifier in Figure 14.24. Assume that the FET has values of $C_{gd} = 4$ pF, $C_{gs} = 5$ pF, and $C_{ds} = 2$ pF. Also assume that the load capacitance is 1 pF.

*Solution:* The first step is to draw the high-frequency equivalent of the circuit. This equivalent circuit is shown in Figure 14.27.

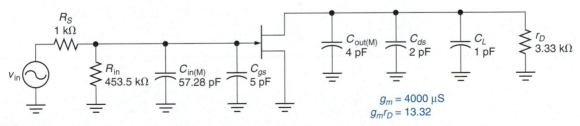

FIGURE 14.27

The value of $R_{\text{in}}$ was found by taking $R_1$ in parallel with $R_2$. Combining this value with $R_S$, we obtain the total ac resistance in the gate circuit as follows:

$$R'_{\text{in}} = R_S \parallel R_{\text{in}} = 998 \ \Omega$$

Now, the Miller input capacitance is found as

$$C_{\text{in(M)}} \cong C_{gd}(g_m r_D + 1) = (4 \text{ pF})(14.32) = 57.28 \text{ pF}$$

Using this value of $C_{in(M)}$ and the value of $C_{gs}$, the total gate circuit capacitance is found as

$$C_G = C_{gs} + C_{in(M)} = 5 \text{ pF} + 57.28 \text{ pF} = \mathbf{62.28 \text{ pF}}$$

We can now use this value along with the value of $R'_{in}$ to find $f_{2G}$ as follows:

$$f_{2G} = \frac{1}{2\pi R'_{in} C_G} = \frac{1}{2\pi (998 \ \Omega)(62.28 \text{ pF})} = \mathbf{2.56 \text{ MHz}}$$

Since $g_m r_D > 10$ for this circuit, the Miller output capacitance is equal to $C_{gd}$, 4 pF. Combining this value with $C_{ds}$ and $C_L$, we get the total capacitance in the drain circuit as follows:

$$C_D = C_{out(M)} + C_{ds} + C_L = 4 \text{ pF} + 2 \text{ pF} + 1 \text{ pF} = \mathbf{7 \text{ pF}}$$

Using the values of $C_D$ and $r_D$, we can find the upper cutoff frequency of the drain circuit as follows:

$$f_{2D} = \frac{1}{2\pi r_D C_D} = \frac{1}{2\pi (3.33 \text{ k}\Omega)(7 \text{ pF})} = \mathbf{6.83 \text{ MHz}}$$

### PRACTICE PROBLEM 14.20

The amplifier described in Practice Problem 14.18 has values of $C_{gd} = 4$ pF, $C_{gs} = 5$ pF, and $C_{ds} = 2$ pF. Determine the values of $f_{2G}$ and $f_{2D}$ for the circuit. Assume a load capacitance of 0 F.

Since the value of $f_{2G}$ is lower than $f_{2D}$, the gate circuit determines the overall value of $f_{C2}$ for the amplifier. Thus, the upper cutoff frequency is 2.56 MHz. If the input frequency reaches this value, the value of $A_v$ will be 3 dB lower than $A_{v(mid)}$. Also, further increases in input frequency cause the value of $A_v$ to continue to drop at a rate of 6 dB/octave until 6.83 MHz is reached. If the input frequency continues to increase, the value of $A_v$ drops at a rate of 12 dB/octave. The Bode plot representing this frequency response is shown in Figure 14.28.

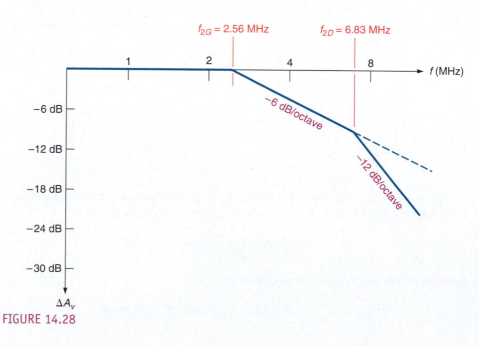

FIGURE 14.28

## 14.3.3    Capacitance Specifications

The high-frequency analysis of a JFET circuit is complicated slightly by the fact that $C_{ds}$, $C_{gd}$, and $C_{gs}$ are not commonly listed on JFET spec sheets. Rather, JFET spec sheets typically contain the capacitance ratings shown in Figure 14.29. The *input capacitance* ($C_{iss}$) rating equals the sum of $C_{gs}$ and $C_{gd}$. By formula,

$$C_{iss} = C_{gs} + C_{gd} \qquad\qquad (14.36)$$

As shown in Figure 14.29, the 2N5486 has an input capacitance rating of $C_{iss} = 5$ pF (maximum).

### 2N5486

**ELECTRICAL CHARACTERISTICS**    ($T_A = 25°C$ unless otherwise noted) (Continued)

| Characteristic | Symbol | Min | Typ | Max | Unit |
|---|---|---|---|---|---|
| **SMALL–SIGNAL CHARACTERISTICS (continued)** | | | | | |
| Input Capacitance ($V_{DS} = 15$ Vdc, $V_{GS} = 0$, $f = 1.0$ MHz) | $C_{iss}$ | — | — | 5.0 | pF |
| Reverse Transfer Capacitance ($V_{DS} = 15$ Vdc, $V_{GS} = 0$, $f = 1.0$ MHz) | $C_{rss}$ | — | — | 1.0 | pF |
| Output Capacitance ($V_{DS} = 15$ Vdc, $V_{GS} = 0$, $f = 1.0$ MHz) | $C_{oss}$ | — | — | 2.0 | pF |

FIGURE 14.29    2N5486 capacitance ratings. (Copyright of Semiconductor Component Industries, LLC. Used by permission.)

The *output capacitance* ($C_{oss}$) rating equals the sum of $C_{ds}$ and $C_{gd}$. By formula,

$$C_{oss} = C_{ds} + C_{gd} \qquad\qquad (14.37)$$

As shown in Figure 14.29, the 2N5486 has an output capacitance rating of $C_{oss} = 2$ pF (maximum). Finally, the *reverse transfer capacitance* ($C_{rss}$) rating equals $C_{gd}$. By formula,

$$C_{rss} = C_{gd} \qquad\qquad (14.38)$$

As shown in Figure 14.29, the 2N5486 has an reverse transfer capacitance rating of $C_{rss} = 1$ pF (maximum). Equations (14.36), (14.37), and (14.38) can be combined to derive the following useful relationships:

$$C_{gd} = C_{rss} \qquad\qquad (14.39)$$

$$C_{ds} = C_{oss} - C_{rss} \qquad\qquad (14.40)$$

and

$$C_{gs} = C_{iss} - C_{rss} \qquad\qquad (14.41)$$

Example 14.21 demonstrates the use of these equations.

### EXAMPLE 14.21

Using the ratings shown in Figure 14.29, calculate the values of $C_{ds}$, $C_{gd}$, and $C_{gs}$ for the 2N5486.

*Solution:*    The gate-drain capacitance is found as

$$C_{gd} = C_{rss} = 1 \text{ pF}$$

The gate-source capacitance is found as

$$C_{gs} = C_{iss} - C_{rss} = 5 \text{ pF} - 1 \text{ pF} = \mathbf{4 \text{ pF}}$$

Finally, the drain-source capacitance is found as

$$C_{ds} = C_{oss} - C_{rss} = 2 \text{ pF} - 1 \text{ pF} = \mathbf{1 \text{ pF}}$$

**PRACTICE PROBLEM 14.21**
An FET has values of $C_{oss} = 4$ pF, $C_{rss} = 3$ pF, and $C_{iss} = 12$ pF. Determine the values of $C_{gs}$, $C_{ds}$, and $C_{gd}$.

### 14.3.4    Theory Versus Practice

In our discussion on BJT amplifiers, you were shown that there tends to be a high percentage of error in the circuit $f_{C2}$ calculations. For the FET amplifier, the largest percentage of error tends to show up in the $f_{2G}$ calculation. The reason for this is that $g_m$ is involved in the calculation of $C_{in(M)}$.

In Chapter 12, you were shown that a JFET has two transconductance curves and, therefore, a range in values of $g_m$. The large possible range in $g_m$ values can cause a wide range in the value of $A_v$ for the amplifier. The range of $A_v$ values for an FET amplifier can affect the calculation of $C_{in(M)}$ and, therefore, the calculation of $f_{2G}$.

With the wide possible variation in the value of $f_{2G}$, you should not be surprised when the expected value of $f_{2G}$ is significantly lower (or higher) than the measured value. Again, the exact value of $f_{C2}$ for an amplifier is critical only when you are dealing with tuned amplifiers. In any standard FET amplifier, you are concerned only with the approximate value of $f_{C2}$ for the circuit.

**Section Review ▶**

1. List, in order, the steps taken to perform the low-frequency analysis of an FET amplifier.

2. List, in order, the steps taken to perform the high-frequency analysis of an FET amplifier.

3. List the typical FET capacitance ratings and the equations used to convert them into usable terminal capacitances.

## 14.4    Multistage Amplifiers

**OBJECTIVE 13 ▶**    When you cascade amplifier circuits with identical values of $f_{C2}$, the overall value of $f_{C2}$ is found as

$$f_{C2(T)} = f_{C2}\sqrt{2^{1/n} - 1} \qquad \textbf{(14.42)}$$

where $f_{C2(T)}$ = the overall value of $f_{C2}$ for the circuit
$\qquad f_{C2}$ = the upper cutoff frequency for one stage
$\qquad n$ = the total number of stages

The derivation of equation (14.42) is included in Appendix D.

When you cascade several amplifier stages, the overall value of $f_{C2}$ is *lower* than the value of $f_{C2}$ for a single stage. This point is illustrated in Example 14.22.

## EXAMPLE 14.22

Two amplifier circuits, each having a value of $f_{C2} = 500$ kHz, are cascaded. Determine the overall value of $f_{C2}$ for the two-stage amplifier.

**Solution:** The overall value of $f_{C2}$ is found as

$$f_{C2(T)} = f_{C2}\sqrt{2^{1/n} - 1} = (500 \text{ kHz})\sqrt{2^{1/2} - 1} = (500 \text{ kHz})(0.643) = \mathbf{321.8 \text{ kHz}}$$

### PRACTICE PROBLEM 14.22

Four amplifier stages, each having an upper cutoff frequency of $f_{C2} = 800$ kHz, are cascaded. Determine the overall value of $f_{C2}$ for the amplifier.

Most BJT and FET amplifiers have values of $f_{C1}$, as you have seen in this chapter. When a value of $f_{C1}$ exists for identical cascaded amplifiers, the overall value of $f_{C1}$ changes as follows:

$$f_{C1(T)} = \frac{f_{C1}}{\sqrt{2^{1/n} - 1}} \tag{14.43}$$

This equation is also derived in Appendix D.

When you cascade several amplifiers with values of $f_{C1}$, the overall value of $f_{C1}$ is higher than the value of $f_{C1}$ for a single stage. This point is illustrated in Example 14.23.

## EXAMPLE 14.23

Two amplifiers with values of $f_{C1} = 5$ kHz are cascaded. Determine the overall value of $f_{C1}$ for the two-stage amplifier.

**Solution:** The overall value of $f_{C1}$ is found as

$$f_{C1(T)} = \frac{f_{C1}}{\sqrt{2^{1/n} - 1}} = \frac{5 \text{ kHz}}{\sqrt{2^{1/2} - 1}} = \frac{5 \text{ kHz}}{0.0644} = \mathbf{7.76 \text{ kHz}}$$

### PRACTICE PROBLEM 14.23

Three amplifier stages, each having a value of $f_{C1} = 800$ Hz, are cascaded. Determine the overall value of $f_{C1}$ for the three-stage amplifier.

For any multistage amplifier, the total bandwidth is found as

$$BW_T = f_{C2(T)} - f_{C1(T)} \tag{14.44}$$

When you cascade two or more amplifiers that are *not* identical, the overall circuit is analyzed in the same fashion as we used to analyze the combined effects of $f_{1B}$, $f_{1C}$, and $f_{1E}$ for the BJT amplifier. For example, consider the Bode plot shown in Figure 14.30. This plot represents the overall response of a two-stage amplifier with the following characteristics:

| Stage | $f_{C1}$ | $f_{C2}$ |
|-------|----------|----------|
| 1 | 1 kHz | 100 kHz |
| 2 | 10 kHz | 1 MHz |

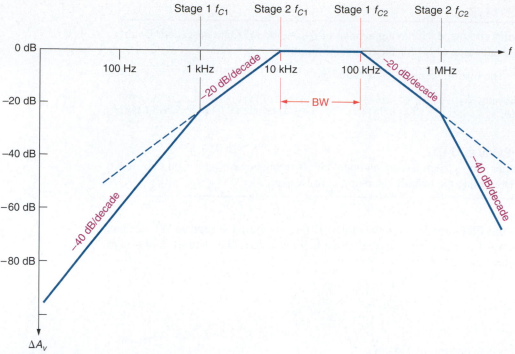

FIGURE 14.30   Overall amplifier frequency response.

> This discussion assumes (for simplicity) that each amplifier stage has only one lower and one upper cutoff frequency. In practice, each may have more (as demonstrated in earlier examples).

On the low end, the two-stage amplifier cutoff frequency is at 10 kHz (the value of $f_{C1}$ for stage 2). The gain of the amplifier decreases at a *20 dB/decade* rate until the value of $f_{C1}$ for stage 1 is reached (1 kHz). At this point, the overall gain decreases at a rate of *40 dB/decade*. The same principles apply to the high-frequency operation. As you can see, the two-stage gain is reduced by 3 dB at the cutoff frequency of stage 1 (100 kHz). The *20 dB/decade* drop continues until $f_{C2}$ of stage 2 (1 MHz) is reached. At that point, the roll-off rate increases to *40 dB/decade*.

Note that the bandwidth of the overall amplifier is determined by the *highest* $f_{C1}$ value and the lowest $f_{C2}$ value. Also, these principles apply to circuits with more than two stages. As each amplifier cuts off, the reduction in gain increases by another 20 dB/decade.

**Section Review ▶**

1. As you cascade identical amplifier stages, what happens to the overall value of $f_{C2}$?
2. As you cascade identical amplifier stages, what happens to the overall value of $f_{C1}$?
3. As you cascade identical amplifier stages, what happens to the overall value of BW?
4. How do you plot the frequency response of a cascaded amplifier that does not contain identical stages?

**CHAPTER SUMMARY**

Here is a summary of the major points made in this chapter:

1. The *bandwidth* of an amplifier is the range of frequencies over which gain is relatively constant.
2. A *frequency-response curve* shows the relationship between amplifier operating frequency and gain (see Figure 14.1).
   a. The limits of the bandwidth are defined by the *cutoff frequencies*, $f_{C1}$ and $f_{C2}$.
   b. The cutoff frequencies are the frequencies at which amplifier power gain is 50% of its midband value.
3. The numeric value of amplifier bandwidth is equal to the difference between its cutoff frequencies.
4. The cutoff frequencies of an amplifier are sometimes referred to as the *half-power frequencies*.

5. The *geometric center frequency* ($f_0$) of an amplifier equals the geometric average of $f_{C1}$ and $f_{C2}$.
   a. Power gain is maximum when an amplifier is operated at $f_0$.
   b. The ratio of $f_0 : f_{C1}$ equals the ratio of $f_{C2} : f_0$.
6. The cutoff frequencies for an amplifier can be measured using the technique outlined in Section 14.1.2.
7. Frequency response curves normally represent gain and frequency as follows:
   a. Power (or voltage) gain is measured as a *ratio*, expressed in dB.
   b. Frequency is measured using a *logarithmic scale*; that is, each division equals a whole number multiple (which is a constant) of the previous division.
   See Figure 14.4.
8. A *decade* is a frequency multiplier of 10. Each division on a *decade scale* has 10 times the value of the previous division.
9. An *octave* is a frequency multiplier of two. Each division on an *octave scale* has twice the value of the previous division.
10. Decade and octave intervals are commonly used to describe the rate at which gain varies with frequency. Two common examples are *20 dB per decade* and *6 dB per octave*.
11. A *Bode plot* is a frequency response curve that assumes $\Delta A_{p(\text{mid})}$ is constant until the cutoff frequencies are reached (see Figure 14.6).
12. The *low-frequency* response of a BJT amplifier is determined by:
   a. The resistance values in the circuit.
   b. The capacitance values in the circuit that are external to the BJT.
13. The term *roll-off rate* is used to describe the rate at which the gain of an amplifier is reduced (in dB) when operated outside its bandwidth.
14. Every *RC* circuit rolls off at a rate of *20 dB per decade*, which is the same as a roll-off rate of *6 dB per octave*. (See Figures 14.10 and 14.11.)
15. The roll-off rate of an *RC* circuit is independent of the values of *R* and *C*.
16. The overall roll-off rate for a circuit containing multiple *RC* circuits equals the sum of the individual roll-off rates (see Figure 14.12).
17. The Bode plot for a circuit represents the *ideal* amplifier response to a change in frequency.
   a. It assumes that all cutoff frequencies are exactly equal to their predicted values.
   b. It assumes that each *RC* circuit has a roll-off rate of 0 dB until the cutoff frequency is reached.
   c. It assumes that roll-off is constant at the given roll-off rate.
18. An *RC* circuit begins to reduce amplifier gain at approximately *twice* the value of $f_{C1}$.
19. The effects of multiple cutoff frequencies are illustrated in Figure 14.13.
20. The difference between a Bode plot and a typical frequency-response curve is illustrated in Figure 14.15.
21. The *high-frequency* response of a BJT amplifier is determined by:
   a. The circuit resistance values.
   b. The BJT internal capacitance values.
   See Figure 14.16.
22. The *current gain–bandwidth product* of a BJT is the frequency at which its current gain drops to *unity* (1).
23. *Miller's theorem* allows a feedback capacitor (across inverting terminals) to be represented as separate input and output capacitors.
24. The high-frequency roll-off characteristics of an amplifier are identical to the low-frequency roll-off characteristics.
25. Predicted values of $f_{C2}$ for a BJT amplifier tend to have high percentages of error because:
   a. The BJT internal capacitance values are estimated.
   b. The BJT internal capacitances are in the pF range, as are the input capacitances of many pieces of test equipment. Therefore, connecting test equipment to the circuit can have a profound effect on the value of $f_{C2}$.
26. FET amplifier frequency response is nearly identical to that of BJT amplifiers.

**27.** FET amplifiers tend to have lower values of $f_{C1}$ than do BJT amplifiers because:

    **a.** The input impedance of an FET is extremely high.

    **b.** FET amplifiers tend to use higher-value biasing resistors than do BJT amplifiers.

**28.** FET amplifier high-frequency response is determined by:

    **a.** The circuit resistances.

    **b.** The *internal* FET capacitances.

**29.** When identical amplifier stages are cascaded:

    **a.** The overall value of $f_{C2}$ is *lower than* that of an individual stage.

    **b.** The overall value of $f_{C1}$ is *higher than* that of an individual stage.

    **c.** The overall value of BW is *lower than* that of an individual stage.

## EQUATION SUMMARY

| Equation Number | Equation | Section Number |
|:---:|:---:|:---:|
| (14.1) | $BW = f_{C2} - f_{C1}$ | 14.1 |
| (14.2) | $f_0 = \sqrt{f_{C1}f_{C2}}$ | 14.1 |
| (14.3) | $\dfrac{f_0}{f_{C1}} = \dfrac{f_{C2}}{f_0}$ | 14.1 |
| (14.4) | $f_{C1} = \dfrac{f_0^2}{f_{C2}}$ | 14.1 |
| (14.5) | $f_{C2} = \dfrac{f_0^2}{f_{C1}}$ | 14.1 |
| (14.6) | $A_{p(dB)} = 10 \log A_p$ | 14.1 |
| (14.7) | $A_{p(dB)} = 10 \log \dfrac{P_{out}}{P_{in}}$ | 14.1 |
| (14.8) | $v_b = v_{in} \dfrac{R_{in}}{\sqrt{X_C^2 + R_{in}^2}}$ | 14.2 |
| (14.9) | $v_b = 0.707 v_{in}$    (when $X_C = R_{in}$) | 14.2 |
| (14.10) | $f = \dfrac{1}{2\pi X_C C}$ | 14.2 |
| (14.11) | $f_{1B} = \dfrac{1}{2\pi RC}$ | 14.2 |
| (14.12) | $f_{1B} = \dfrac{1}{2\pi (R_S + R_{in})C_{C1}}$ | 14.2 |
| (14.13) | $f_{1C} = \dfrac{1}{2\pi (R_C + R_L)C}$ | 14.2 |
| (14.14) | $R_{out} = R_E \parallel \left( r_e' + \dfrac{R_{in}'}{h_{fc}} \right)$ | 14.2 |
| (14.15) | $R_{out} \cong r_e' + \dfrac{R_{in}'}{h_{fc}}$ | 14.2 |
| (14.16) | $f_{1E} = \dfrac{1}{2\pi R_{out} C_E}$ | 14.2 |

| Equation Number | Equation | Section Number |
|---|---|---|
| **(14.17)** | $\Delta A_v = 20 \log \dfrac{1}{\sqrt{1 + (f_{C1}/f)^2}}$ | 14.2 |
| **(14.18)** | $C_{be} \cong \dfrac{1}{2\pi f_T r_e'}$ | 14.2 |
| **(14.19)** | $C_{in(M)} = C_{bc}(A_v + 1)$ | 14.2 |
| **(14.20)** | $C_{out(M)} = C_{bc} \dfrac{A_v + 1}{A_v}$ | 14.2 |
| **(14.21)** | $C_{in(M)} \cong A_v C_{bc}$ | 14.2 |
| **(14.22)** | $C_{out(M)} \cong C_{bc}$ | 14.2 |
| **(14.23)** | $f_{2B} = \dfrac{1}{2\pi(R_{in}' \parallel h_{ie})(C_{be} + C_{in(M)})}$ | 14.2 |
| **(14.24)** | $f_{2C} = \dfrac{1}{2\pi r_C(C_{out(M)} + C_L)}$ | 14.2 |
| **(14.25)** | $\Delta A_v = 20 \log \dfrac{1}{\sqrt{1 + (f/f_{C2})^2}}$ | 14.2 |
| **(14.26)** | $f_{1G} = \dfrac{1}{2\pi(R_S + R_{in})C_{C1}}$ | 14.3 |
| **(14.27)** | $f_{1D} = \dfrac{1}{2\pi(R_D + R_L)C_{C2}}$ | 14.3 |
| **(14.28)** | $R_{in} = R_1 \parallel R_2$ | 14.3 |
| **(14.29)** | $C_{in(M)} = C_{gd}(g_m r_D + 1)$ | 14.3 |
| **(14.30)** | $C_{out(M)} = C_{gd} \dfrac{g_m r_D + 1}{g_m r_D}$ | 14.3 |
| **(14.31)** | $C_{out(M)} \cong C_{gd}$ | 14.3 |
| **(14.32)** | $C_G = C_{gs} + C_{in(M)}$ | 14.3 |
| **(14.33)** | $C_D = C_{out(M)} + C_{ds} + C_L$ | 14.3 |
| **(14.34)** | $f_{2G} = \dfrac{1}{2\pi R_{in}' C_G}$ | 14.3 |
| **(14.35)** | $f_{2D} = \dfrac{1}{2\pi r_D C_D}$ | 14.3 |
| **(14.36)** | $C_{iss} = C_{gs} + C_{gd}$ | 14.3 |
| **(14.37)** | $C_{oss} = C_{ds} + C_{gd}$ | 14.3 |
| **(14.38)** | $C_{rss} = C_{gd}$ | 14.3 |
| **(14.39)** | $C_{gd} = C_{rss}$ | 14.3 |
| **(14.40)** | $C_{ds} = C_{oss} - C_{rss}$ | 14.3 |

| Equation Number | Equation | Section Number |
|---|---|---|
| **(14.41)** | $C_{gs} = C_{iss} - C_{rss}$ | 14.3 |
| **(14.42)** | $f_{C2(T)} = f_{C2}\sqrt{2^{1/n} - 1}$ | 14.4 |
| **(14.43)** | $f_{C1(T)} = \dfrac{f_{C1}}{\sqrt{2^{1/n} - 1}}$ | 14.4 |
| **(14.44)** | $\mathrm{BW}_T = f_{C2(T)} - f_{C1(T)}$ | 14.4 |

## KEY TERMS

bandwidth (BW)  550
Bode plot  557
current gain–bandwidth
   product  569
cutoff frequencies  551

decade  557
frequency-response
   curve  550
geometric center
   frequency ($f_0$)  552

Miller's theorem  570
octave  557
roll-off rate  562

## PRACTICE PROBLEMS

### Section 14.1

**1.** An amplifier has cutoff frequencies of 1.2 kHz and 640 kHz. Calculate the bandwidth and center frequency for the circuit.

**2.** An amplifier has cutoff frequencies of 3.4 kHz and 748 kHz. Calculate the bandwidth and center frequency for the circuit.

**3.** An amplifier has cutoff frequencies of 5.2 kHz and 489.6 kHz. Calculate the bandwidth and center frequency for the circuit.

**4.** An amplifier has cutoff frequencies of 1.4 kHz and 822.7 kHz. Calculate the bandwidth and center frequency for the circuit.

**5.** An amplifier has cutoff frequencies of 2.6 kHz and 483.6 kHz. Show that the ratio of $f_0$ to $f_{C1}$ is equal to the ratio of $f_{C2}$ to $f_0$ for the circuit.

**6.** An amplifier has cutoff frequencies of 1 kHz and 345 kHz. Show that the ratio of $f_0$ to $f_{C1}$ is equal to the ratio of $f_{C2}$ to $f_0$ for the circuit.

**7.** The values of $f_0$ and $f_{C2}$ for an amplifier are measured at 72 kHz and 548 kHz, respectively. Calculate the values of $f_{C1}$ and bandwidth for the amplifier.

**8.** The values of $f_0$ and $f_{C2}$ for an amplifier are measured at 22.8 kHz and 321 kHz, respectively. Calculate the values of $f_{C1}$ and bandwidth for the circuit.

**9.** The values of $f_0$ and $f_{C1}$ for an amplifier are measured at 48 kHz and 2 kHz, respectively. Calculate the values of $f_{C2}$ and bandwidth for the circuit.

**10.** The values of $f_0$ and $f_{C1}$ for an amplifier are measured at 36 kHz and 4.3 kHz, respectively. Calculate the values of $f_{C2}$ and bandwidth for the circuit.

### Section 14.2

**11.** Calculate the value of $f_{1B}$ for the amplifier in Figure 14.31.

**12.** Calculate the value of $f_{1B}$ for the amplifier in Figure 14.32.

**13.** Calculate the value of $f_{1C}$ for the amplifier in Figure 14.31.

**14.** Calculate the value of $f_{1C}$ for the amplifier in Figure 14.32.

**15.** Calculate the value of $f_{1E}$ for the amplifier in Figure 14.31.

**16.** Calculate the value of $f_{1E}$ for the amplifier in Figure 14.32.

**17.** A given amplifier has values of $A_{v(\mathrm{mid})} = 32$ dB and $f_{C1} = 8$ kHz. Calculate the dB voltage gain of the amplifier at operating frequencies of 7, 5, 4, and 1 kHz.

**18.** A given amplifier has values of $A_{v(\mathrm{mid})} = 16$ dB and $f_{C1} = 12$ kHz. Calculate the dB voltage gain of the amplifier at operating frequencies of 18, 12, 10, and 2 kHz.

**19.** Compare the results of Problems 11, 13, and 15. What is the approximate value of $f_{C1}$ for the amplifier described in these three problems?

**20.** Compare the results of Problems 12, 14, and 16. What is the approximate value of $f_{C1}$ for the amplifier described in these three problems?

**21.** A transistor has a value of $f_T = 200$ MHz at $I_C = 10$ mA. Determine the value of $C_{be}$ at this current.

**22.** A transistor has a value of $f_T = 400$ MHz at $I_C = 1$ mA. Determine the value of $C_{be}$ at this current.

**23.** An inverting amplifier has values of $C_{bc} = 7$ pF and $A_v = 100$. Determine the Miller input and output capacitance values for the circuit.

**24.** An inverting amplifier has values of $C_{bc} = 3$ pF and $A_v = 4$. Determine the Miller input and output capacitance values for the circuit.

**25.** Determine the value of $f_{2B}$ for the amplifier in Figure 14.31.

**26.** Determine the value of $f_{2B}$ for the amplifier in Figure 14.32.

**27.** Determine the value of $f_{2C}$ for the amplifier in Figure 14.31.

**28.** Determine the value of $f_{2C}$ for the amplifier in Figure 14.32.

**29.** Calculate the overall values of $f_{C1}$ and $f_{C2}$ for the amplifier in Figure 14.33.

**30.** Calculate the overall values of $f_{C1}$ and $f_{C2}$ for the amplifier in Figure 14.34.

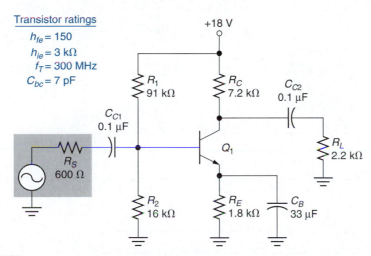

FIGURE 14.31

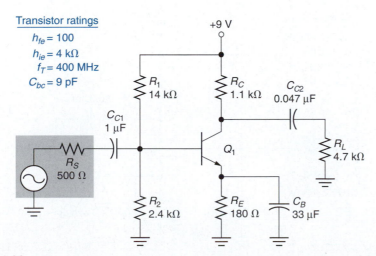

FIGURE 14.32

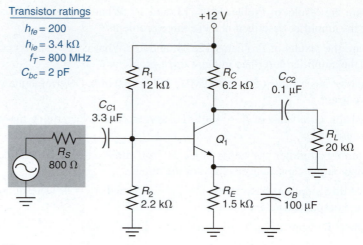

Transistor ratings
$h_{fe} = 200$
$h_{ie} = 3.4 \text{ k}\Omega$
$f_T = 800 \text{ MHz}$
$C_{bc} = 2 \text{ pF}$

+12 V

$R_1$ 12 kΩ
$R_C$ 6.2 kΩ
$C_{C2}$ 0.1 µF
$C_{C1}$ 3.3 µF
$R_S$ 800 Ω
$Q_1$
$R_L$ 20 kΩ
$R_2$ 2.2 kΩ
$R_E$ 1.5 kΩ
$C_B$ 100 µF

FIGURE 14.33

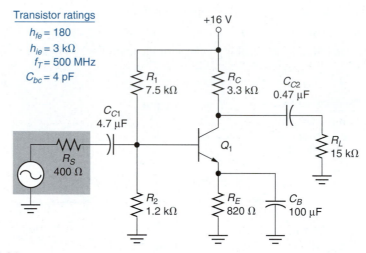

Transistor ratings
$h_{fe} = 180$
$h_{ie} = 3 \text{ k}\Omega$
$f_T = 500 \text{ MHz}$
$C_{bc} = 4 \text{ pF}$

+16 V

$R_1$ 7.5 kΩ
$R_C$ 3.3 kΩ
$C_{C2}$ 0.47 µF
$C_{C1}$ 4.7 µF
$R_S$ 400 Ω
$Q_1$
$R_L$ 15 kΩ
$R_2$ 1.2 kΩ
$R_E$ 820 Ω
$C_B$ 100 µF

FIGURE 14.34

## Section 14.3

**31.** Calculate the value of $f_{1G}$ for the amplifier in Figure 14.35.

**32.** Calculate the value of $f_{1G}$ for the amplifier in Figure 14.36.

**33.** Calculate the value of $f_{1D}$ for the amplifier in Figure 14.35.

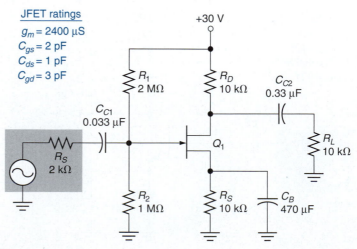

JFET ratings
$g_m = 2400 \text{ µS}$
$C_{gs} = 2 \text{ pF}$
$C_{ds} = 1 \text{ pF}$
$C_{gd} = 3 \text{ pF}$

+30 V

$R_1$ 2 MΩ
$R_D$ 10 kΩ
$C_{C2}$ 0.33 µF
$C_{C1}$ 0.033 µF
$R_S$ 2 kΩ
$Q_1$
$R_L$ 10 kΩ
$R_2$ 1 MΩ
$R_S$ 10 kΩ
$C_B$ 470 µF

FIGURE 14.35

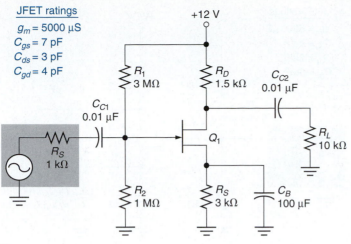

FIGURE 14.36

34. Calculate the value of $f_{1D}$ for the amplifier in Figure 14.36.

35. Calculate the values of $f_{2G}$ and $f_{2D}$ for the amplifier in Figure 14.35.

36. Calculate the values of $f_{2G}$ and $f_{2D}$ for the amplifier in Figure 14.36.

37. An FET has ratings of $C_{iss} = 14$ pF, $C_{rss} = 9$ pF, and $C_{oss} = 10$ pF. Determine the values of $C_{gd}$, $C_{gs}$, and $C_{ds}$ for the device.

38. An FET has ratings of $C_{oss} = 16$ pF, $C_{rss} = 9$ pF, and $C_{iss} = 12$ pF. Determine the values of $C_{gd}$, $C_{gs}$, and $C_{ds}$ for the device.

39. Calculate the overall values of $f_{C1}$ and $f_{C2}$ for the amplifier in Figure 14.37.

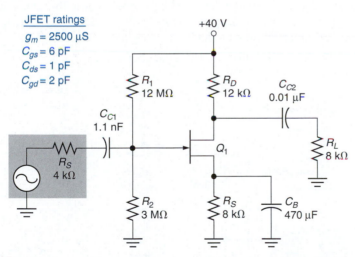

FIGURE 14.37

40. Calculate the overall values of $f_{C1}$ and $f_{C2}$ for the amplifier in Figure 14.38.

41. Two amplifiers, each having a value of $f_{C2} = 120$ kHz, are cascaded. Determine the overall value of $f_{C2}$ for the circuit.

42. Two amplifiers, each having a value of $f_{C1} = 3$ kHz, are cascaded. Determine the overall value of $f_{C1}$ for the circuit.

43. Two amplifiers, each having values of $f_{C1} = 2$ kHz and $f_{C2} = 840$ kHz, are cascaded. Determine the bandwidth of the circuit.

44. Four amplifiers, each having values of $f_{C1} = 1.5$ kHz and $f_{C2} = 620$ kHz, are cascaded. Determine the bandwidth of the circuit.

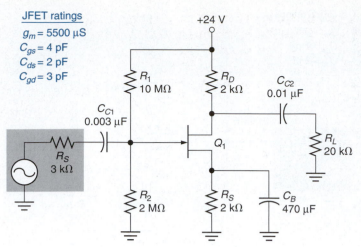

FIGURE 14.38

**45.** We have two amplifiers. The first amplifier has values of $f_{1B} = 2$ kHz, $f_{1C} = 16$ kHz, and $f_{1E} = 32$ kHz. The second amplifier has values of $f_{1B} = 8$ kHz, $f_{1C} = 12$ kHz, and $f_{1E} = 4$ kHz. For each of these amplifiers:

   **a.** Draw the Bode plot, including the appropriate roll-off rates. The Bode plots should have an octave frequency scale that starts at 2 kHz and continues up to 32 kHz.

   **b.** Plot the frequency-response curve on the same graph as the Bode plot. The frequency-response curve must take into account the combined effects of $f_{1B}, f_{1C}$, and $f_{1E}$ at each major division on the graph. In other words, at each major division, calculate the values of $\Delta A_v$ for each of the terminal circuits, determine the total value of $\Delta A_v$, and plot the point that corresponds to the total value of $\Delta A_v$.

After completing the two graphs, compare the results to see which frequency-response curve more closely resembles its Bode plot.

**46.** Calculate the change in $f_0$ that occurs in the circuit in Figure 14.39 if the load resistance opens.

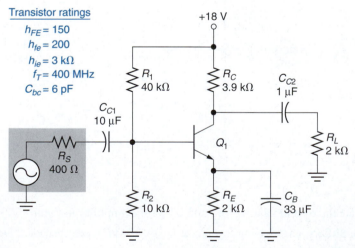

FIGURE 14.39

**47.** Write a program that will determine the lower cutoff frequencies of a BJT amplifier given the proper input values.

**48.** Write a program that will determine the upper cutoff frequencies of a BJT amplifier, given the proper input values.

**49.** Combine the programs from Problems 47 and 48. Then, add program steps to determine the overall values of $f_{C1}, f_{C2}, f_0$, and bandwidth for the circuit.

**50.** Repeat Problems 47 through 49 for a JFET amplifier.

**51.** Repeat Problem 50 for an inverting amplifier.

**14.1**   1.4996 MHz

**14.2**   24.49 kHz

**14.3**   Both ratios equal 61.23.

**14.4**   297 kHz

**14.5**   120 kHz

**14.6**   41.55 dB

**14.7**   $A_{pT} = 55.88$ dB (when solved using either approach)

**14.8**   45.2 Hz

**14.9**   14.2 Hz

**14.10** 54.5 Hz

**14.11** 16.01 dB

**14.12** 306 pF

**14.13** 960 pF, 4 pF

**14.14** 1.08 MHz

**14.15** 37.54 MHz

**14.17** 0.01 Hz

**14.18** 260.9 Hz

**14.19** 20.28 pF

**14.20** $f_{2G} = 11.47$ MHz, $f_{2D} = 22.01$ MHz

**14.21** $C_{gs} = 9$ pF, $C_{ds} = 1$ pF, $C_{gd} = 3$ pF

**14.22** 348 kHz

**14.23** 1.57 kHz

## chapter 15

# Operational Amplifiers

## Objectives

*After studying the material in this chapter, you should be able to:*

1. Describe the *operational amplifier*, or *op-amp*.
2. State the purpose served by a *differential amplifier*.
3. Describe open-loop voltage gain, and state its typical range of values.
4. Describe the operation of a discrete differential amplifier.
5. Describe the various voltage and current ratings for an op-amp.
6. Discuss the effects of common-mode rejection ratio on op-amp operation.
7. Discuss the relationship between slew-rate and op-amp operating frequency.
8. Discuss slew-rate distortion and the means by which it can be reduced.
9. Describe and analyze the operation of the inverting amplifier.
10. Describe and analyze the operation of the noninverting amplifier.
11. Compare and contrast the operating characteristics of the inverting and noninverting amplifiers.
12. List the common op-amp faults and the symptoms of each.
13. Define gain-bandwidth product, and explain its significance.
14. Describe the various types of feedback.
15. Describe the effects of negative feedback on inverting amplifier gain and bandwidth.
16. Calculate the *attenuation factor* and *feedback factor* for a given feedback amplifier.
17. Calculate the input and output impedance values for inverting and noninverting amplifiers.

## Outline

## Shrinking Circuits

The impact that integrated-circuit technology has had on the field of electronics can easily be understood if you consider the history of the operational amplifier, or op-amp. At one time, op-amps were made using vacuum tubes. A typical vacuum tube op-amp contained approximately six vacuum tubes that were housed in a single module. The typical module measured $3 \times 5 \times 12$ in. (approximately $8 \times 13 \times 30$ cm).

In the early 1960s, the semiconductor op-amp became available. This component (whose *equivalent* circuit is shown in Figure 15.1) typically contains close to 200 internal active devices in a case that is approximately $0.4 \times 0.26$ in. ($10.2 \times 6.6$ mm)!

---

**Discrete components**
Components housed in individual packages; that is, one package—one component.

**Integrated circuit (IC)**
A single package that contains any number of active and/or passive components, all constructed on a single piece of semiconductor material.

Individual components, such as the 2N3904 BJT and the 2N5459 FET, are classified as **discrete components**. The term *discrete* indicates that each physical package contains only one component. For example, when you purchase a 2N3904, you are buying a single component housed in its own casing.

Over the years, advances in manufacturing technology have made it possible to produce entire *circuits* on a single piece of semiconductor material. This type of circuit, which is housed in a single casing, is referred to as an **integrated circuit**, or **IC**. ICs range in complexity from simple circuits containing a few active and/or passive components to complex circuits containing hundreds of thousands of components. The more complex the internal circuitry of an IC, the more complex its function.

The major impacts of ICs all relate to their internal operation and relatively low manufacturing cost. Circuit operations that once took hundreds of discrete components to perform can now be accomplished with a single IC. This has made circuits easier to design and troubleshoot. At the same time, the cost of an IC is generally lower than the cost of a comparable discrete-component circuit. This has made electronic systems less expensive to manufacture.

It would be impossible for a single book to cover *every* type of IC available. As your study of electronics continues, you will be introduced to more and more types of ICs. In this book, we will concentrate on the most commonly used *linear* IC, the *operational amplifier*, or *op-amp*. We will discuss the operating principles and basic amplifier applications of op-amps, including op-amp circuit troubleshooting.

## 15.1   Op-Amps: An Overview

OBJECTIVE 1 ▶ The op-amp is a *high-gain* dc amplifier that has *high input impedance* and *low output impedance*. The internal circuitry, schematic symbol, and pin diagram for the 741 general-purpose **operational amplifier (op-amp)** are shown in Figure 15.1.

Check out that circuit! How would you like to troubleshoot *that* on your average Monday morning? Fortunately, all the circuitry of the 741 is contained in a single component.

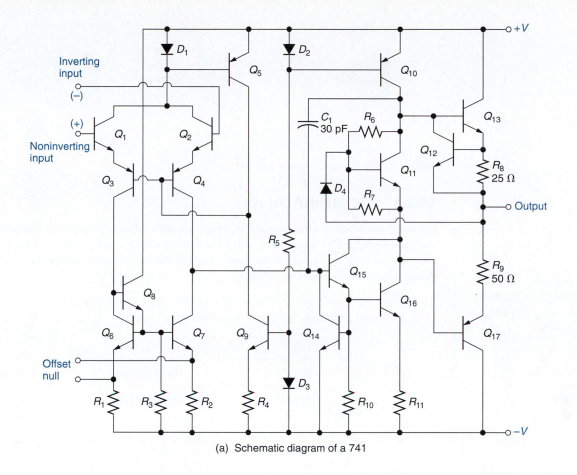

(a) Schematic diagram of a 741

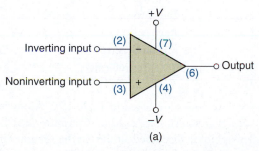

(b) The schematic symbol for an op-amp

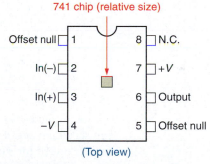

(Top view)

(c) The 741 chip packaged in an 8-pin DIP (dual-in-line package)

FIGURE 15.1   A 741 operational amplifier (op-amp).

Since we're dealing with a single component, all we need to concern ourselves with are the input/output relationships and characteristics of the component. You cannot get into a 741 to repair the internal circuitry, so its complexity is of no consequence.

The signal inputs to the op-amp are labeled *inverting* and *noninverting*. Normally, an input signal is applied to either of these two inputs. The other input is usually wired to control the operating characteristics of the component. The particular application determines which input pin is used as the active input.

The op-amp has *two* dc power supply inputs, labeled $+V$ and $-V$. These dc supply pins are normally connected in one of two ways, as illustrated in Figure 15.2. One way is to have $+V$ and $-V$ set to equal voltages that are of opposite polarity, as shown in Figure 15.2a. The other way is to provide a single supply voltage to one of the supply pins while grounding the other. This connection is shown in Figures 15.2b and c. Again, the specific wiring of these two pins depends on the particular application. The *offset null* pins (Figure 15.1) are discussed later in this chapter.

> **Operational amplifier (op-amp)**
> A high-gain dc amplifier that has high input impedance and low output impedance.

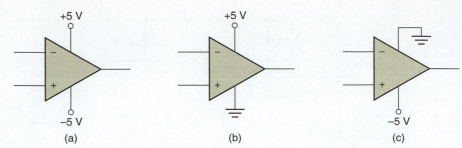

FIGURE 15.2 Op-amp supply voltages.

### 15.1.1 IC Identification

Op-amps are produced by a variety of manufacturers. Many op-amps are identified using a seven-character ID code like the one shown in Figure 15.3. The *prefix* is used to identify the manufacturer. A listing of the most common prefixes is provided in Table 15.1. The **designator code** indicates two things:

> **Designator code**
> An IC code that indicates the type of circuit and its operating temperature range.

1. The three-digit number indicates the specific type of op-amp.
2. The letter indicates the operating temperature range.

| Prefix | Designator | Suffix |
|--------|-----------|--------|
| MC | 741C | N |

FIGURE 15.3 Op-amp ID code.

TABLE 15.1 Manufacturers' Prefixes

| Prefix | Manufacturer |
|--------|-------------|
| AD/OP | Analog Devices |
| CA/HA | Harris |
| KA | Fairchild |
| LM | National Semiconductor |
| MC | ON Semiconductor |
| NE/SE | Signetics |
| OPA | Burr-Brown |
| RC/RM | Raytheon |
| SG | Silicon General |
| TI | Texas Instruments |

Some commonly used temperature codes are listed in Table 15.2. Along with identifying specific types of op-amps, designator codes are used to determine which op-amps can be substituted for each other. This point is discussed further in the section on op-amp circuit troubleshooting. The *suffix* identifies the op-amp package. The commonly used suffix codes are listed in Table 15.3.

TABLE 15.2 Temperature Codes

| Code | Application | Temperature Range (°C) |
|------|-------------|------------------------|
| C | Commercial | 0 to 70 |
| I | Industrial | −25 to 85 |
| M | Military | −55 to 125 |

TABLE 15.3 Suffix Codes

| Code | Package Type |
|------|-------------|
| D, VD | Surface-mount package (SMP) |
| J | Ceramic dual-in-line (DIP) |
| N, P, VP | Plastic DIP |
| DM | Micro SMP |

## 15.1.2 Op-Amp Packages

Op-amps are available in metal cans, *dual-in-line packages* (DIPs), and *surface-mount packages* (SMPs). Typical pin configurations for these packages are illustrated in Figure 15.4. Note that the type of package is extremely important when considering whether one component may be substituted for another.

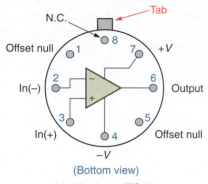

N.C. stands for *not connected.* The pins labeled N.C. are not used.

(a) Metal can (TO-5)          (b) 8-pin DIP or SMP

FIGURE 15.4   Op-amp packages.

---

◀ Section Review

1. What is a *discrete component*?
2. What is an *integrated circuit*?
3. What is an *operational amplifier*, or *op-amp*?
4. Draw the schematic symbol for an op-amp, and identify the signal and supply voltage inputs.
5. What are the three parts of the op-amp identification code? What does each part of the code tell you?
6. What are the three types of op-amp packages?

## 15.2   Operation Overview

The input stage of the op-amp is a **differential amplifier**. This type of circuit amplifies the *difference between* two input voltages, $V_1$ and $V_2$. Any difference between the values of $V_1$ and $V_2$ appears as a difference of potential ($V_{diff}$) across the input terminals, as shown in Figure 15.5. By formula,

◀ OBJECTIVE 2

$$V_{diff} = V_2 - V_1 \qquad (15.1)$$

**Differential amplifier**
A circuit that amplifies the difference between two input voltages.

where $V_{diff}$ = the voltage that will be amplified
$V_1$ = the voltage applied to the *inverting* input
$V_2$ = the voltage applied to the *noninverting* input

*It is important for you to remember that the op-amp is amplifying the difference between the input terminal voltages.*

The output from the amplifier for a given pair of input voltages depends on several factors:

1. The *gain* of the amplifier.
2. The polarity relationship between $V_1$ and $V_2$.
3. The values of the supply voltages, $+V$ and $-V$.
4. The load resistance.

We will now look at these four factors in detail.

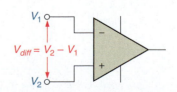

FIGURE 15.5   Op-amp signal inputs.

## 15.2.1   Op-Amp Gain

**OBJECTIVE 3 ▶**

**Open-loop voltage gain ($A_{OL}$)**
The maximum possible gain of a given op-amp. The voltage gain of an op-amp with no feedback path.

The *maximum* possible gain provided by a given op-amp is referred to as its **open-loop voltage gain ($A_{OL}$)**. The value of $A_{OL}$ is generally greater than 10,000. For example, the Fairchild KA741 op-amp has an open-loop voltage gain of 200,000 (typical).

The term *open-loop* indicates a circuit condition where *there is no feedback path from the output to the op-amp input.* You may recall that a feedback path is a connection used to "feed" a portion of the output signal back to the input. Op-amp circuits usually contain one or more feedback paths. In Figure 15.6, the circuit containing $R_f$ is the feedback path. When part of the output signal is fed back to the input, the effective gain of the op-amp is reduced. As we cover specific circuits, you will be shown how to determine the gain of each. In the meantime, you should remember:

1. The *open-loop voltage gain ($A_{OL}$)* of an op-amp is its *maximum* voltage gain, typically 10,000 or greater.
2. The effective voltage gain of an op-amp is reduced when a feedback path is added between the component output and input.

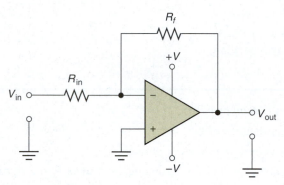

FIGURE 15.6   Op-amp feedback path.

The high gain of the op-amp is another advantage that this component has over BJTs and FETs. Values of $A_v$ near the 200,000 mark (which are common for op-amps) are virtually impossible for single-stage discrete amplifiers.

## 15.2.2   Input/Output Polarity

The polarity relationship between $V_1$ and $V_2$ determines whether the op-amp output voltage swings toward $+V$ or $-V$. When $V_1$ is *more negative* than $V_2$, the op-amp output voltage swings toward $+V$, as shown in Figure 15.7a. When $V_1$ is *more positive* than $V_2$, the op-amp output voltage swings toward $-V$, as shown in Figure 15.7b. Note the relationships between the input voltage polarities and the input signs in the schematic symbol. In Figure 15.7a, the relative polarities of $V_1$ and $V_2$ match the polarity signs in the schematic symbol, and the output is positive. In Figure 15.7b, the relative polarities for $V_1$ and $V_2$ do not match the polarity signs in the schematic symbol, and the output is negative. These input/output relationships can be summarized as follows:

What determines op-amp output polarity?

*When the input voltage polarities, with respect to each other, match the polarity signs in the schematic symbol, the output voltage is positive. When they do not, the output is negative.*

This relationship is illustrated further in Figure 15.8.

It is important to note that the output polarity is determined by the *relationship between* the polarities of $V_1$ and $V_2$, not by their polarities with respect to ground. For example, look at circuits (a) and (c) in Figure 15.8. In both of these cases, $V_1$ is *more negative* than $V_2$ (the relative polarities match the polarity symbols), and the output is *positive*. It did not matter whether both inputs were positive or negative, only that $V_1$ was *negative with respect to* $V_2$. If you take a moment to look at circuits (b) and (d), you'll see that $V_1$ is *more positive* than $V_2$ (the relative polarities do not match the polarity signs),

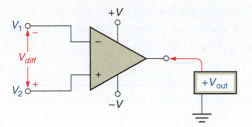 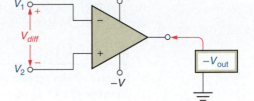

(a) $V_{out}$ is positive when $V_2$ is more positive than $V_1$.  (b) $V_{out}$ is negative when $V_2$ is more negative than $V_1$.

FIGURE 15.7  Op-amp input polarities.

and the output is *negative*. Again, the *relationship between $V_1$ and $V_2$* has determined the output polarity.

Now take a look at circuit (e). Note that the noninverting (+) input is *grounded*. When the input signal goes more positive than ground, the relative input polarities do not match the polarity signs, and the output goes negative. When the input signal goes more negative than ground, the relative input polarities match the polarity signs, and the output goes positive. Note the 180° phase shift between the op-amp input and output signals. This is where the term **inverting input** comes from. If you apply the same reasoning to circuit (f), you will see why the (+) input is referred to as the **noninverting input**.

There is another method you can use to determine the polarity of the op-amp output voltage for a given set of input voltages. You may remember that equation (15.1) defines the differential input as

$$V_{diff} = V_2 - V_1$$

> **Inverting input**
> The op-amp input that produces a 180° voltage phase shift (from input to output) when used as a signal input.
>
> **Noninverting input**
> The op-amp input that does not produce a voltage phase shift (from input to output) when used as a signal input.

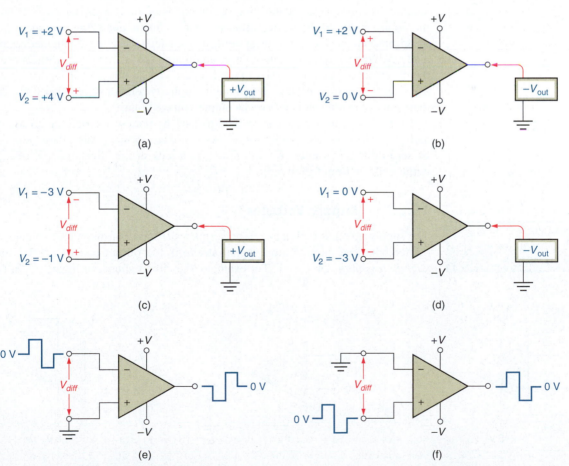

FIGURE 15.8  Input/output polarity relationships.

When the result of this equation is *positive*, the op-amp output voltage is *positive*. When the result of this equation is *negative*, the output voltage is *negative*. These relationships are illustrated in Example 15.1.

### EXAMPLE 15.1

Determine the output voltage polarity for circuits (a) and (b) in Figure 15.8 using equation (15.1).

***Solution:*** For circuit (a),

$$V_{diff} = V_2 - V_1 = 4\,V - 2\,V = 2\,V$$

Since $V_{diff}$ is *positive*, the op-amp output voltage is *positive*. For circuit (b),

$$V_{diff} = V_2 - V_1 = 0\,V - 2\,V = -2\,V$$

Since $V_{diff}$ is *negative*, the op-amp output voltage is *negative*.

### PRACTICE PROBLEM 15.1

Using equation (15.1), determine the output voltage polarities for circuits (c) and (d) in Figure 15.8.

If you compare the polarities obtained in the example with those determined in the discussion, you will see that we reached the same conclusions in both cases.

You may be wondering why we went to all the trouble of developing an *observation method* of analysis (comparing the voltage polarities to the polarity signs in the schematic symbol) when the *mathematical method* of analysis used in Example 15.1 seems so much simpler. The reason for establishing two methods of analysis is that each method is simpler to use under different circumstances. When you are analyzing the schematic diagram of an op amp–based circuit or system, it is easier to use the mathematical method of analysis. When you are troubleshooting an op-amp circuit or system with an oscilloscope, the observation method is easier to use because you do not need to determine exact voltage values. You only need to determine the polarity relationship between the inputs, and then you can predict and observe the output voltage polarity.

Up to this point, we have simplified matters by not considering the effects of the $+V$ and $-V$ values on the op-amp output. As you were shown earlier, these supply pins can be set to different values. We will now take a look at the effects of $+V$ and $-V$ on the output voltage from the op-amp.

### 15.2.3 Supply Voltages

> How do the supply voltages affect the op-amp output voltage?

The supply voltages ($+V$ and $-V$) determine the *limits* of the output voltage swing. No matter what the gain or input signal strength, the output cannot exceed some value *slightly less than* $+V$ or $-V$. For example, the circuit shown in Figure 15.9a has supply

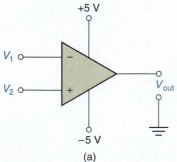

(a)

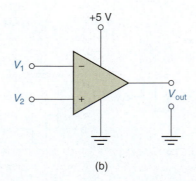

(b)

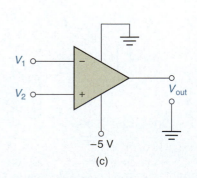

(c)

FIGURE 15.9   Op-amp supply voltages.

voltages of $\pm 5$ V. Assuming that the output can make the full transition between the supply voltages, it cannot exceed peak values of $+5$ V and $-5$ V. For the circuit shown in Figure 15.9b, the output would be limited to $+5$ V on the positive transition and 0 V on the negative transition. Finally, the output from the circuit shown in Figure 15.9c would have positive and negative limits of 0 and $-5$ V, respectively.

In practice, the peak output voltage cannot reach either $+V$ or $-V$. The reason for this lies in the fact that some voltage is dropped across the components in the op-amp output circuit (refer to Figure 15.1a). The actual limits on $V_{\text{out}}$ depend on the op-amp being used and the value of the load resistance. For example, the spec sheet for the KA741 op-amp lists the following parameters:

| Parameter | Condition | Typical Value[a] |
|---|---|---|
| Output voltage swing | $R_L \geq 10$ k$\Omega$ | $\pm 14$ V |
| | $R_L \geq 2$ k$\Omega$ | $\pm 13$ V |

[a]$V_S = \pm 15$ V, $T_A = 25°$C.

If the load resistance is less than 2 k$\Omega$, a graph like the one in Figure 15.10 can be used to determine the limits on the output voltage. Note that the output voltages in the curve are given as *peak-to-peak* values. For example, the curve shows that a 200 $\Omega$ load resistance limits the output voltage swing to 10 $V_{\text{PP}}$, or $\pm 5$ V.

The curve in Figure 15.10 assumes power supply values of $\pm 15$ V. If the power supply is set to values other than $\pm 15$ V, a graph like the one in Figure 15.11 can be used to determine the limits on the output voltage. Generally, the following guidelines apply to any 741 op-amp:

| Value of Load Resistance (k$\Omega$) | Max. (+) Output | Max. (−) Output |
|---|---|---|
| Greater than 10 | $(+V) - 1$ V | $(-V) + 1$ V |
| 2 to 10 | $(+V) - 2$ V | $(-V) + 2$ V |

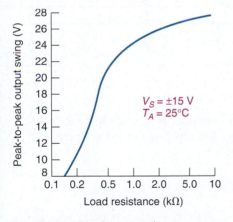

FIGURE 15.10 Output voltage as a function of load resistance.

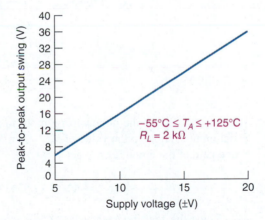

FIGURE 15.11 Output voltage as a function of supply voltage.

For example, consider the circuit shown in Figure 15.12. The supply voltages are $+10$ V and ground. If the value of $R_L$ is 10 k$\Omega$ or more, the maximum output transition is from $+9$ to $+1$ V. If the load resistance is between 2 and 10 k$\Omega$, the maximum output voltage transition is from $+8$ to $+2$ V.

## 15.2.4 Putting It All Together

We have considered the factors of gain, input polarity relationships, load resistance, and supply voltages on the operation of an op-amp. In this section, we will go through a series of examples that put all these factors together. In each example, a voltage gain value is

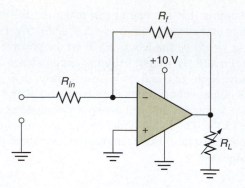

**FIGURE 15.12**

assumed to simplify the problem. Later in the chapter, you will be shown how to determine the gain for each of these circuits.

### EXAMPLE 15.2

Determine the peak-to-peak output voltage for the circuit shown in Figure 15.13. Also, determine the maximum possible output values for the amplifier. Assume that the voltage gain of the circuit is 150.

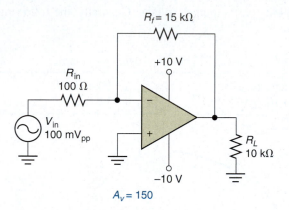

**FIGURE 15.13**

***Solution:*** Since the noninverting input to the amplifier is grounded, the differential input is equal to the value of $V_{in}$, 100 mV$_{PP}$. Multilplying this value by the voltage gain of the amplifier, we get

$$V_{out} = A_v V_{in} = (150)(100\,\text{mV}_{PP}) = \textbf{15 V}_{\textbf{pp}}$$

Since the load resistance is 10 kΩ, we can determine the maximum possible peak output voltages using the general guidelines established for the 741 earlier in the chapter. Thus, the maximum positive and negative peak values are found as

$$V_{pk(+)} = (+V) - 1\,\text{V} = \textbf{+9 V} \quad (\text{maximum})$$

and

$$V_{pk(-)} = (-V) + 1\,\text{V} = \textbf{-9 V} \quad (\text{maximum})$$

### PRACTICE PROBLEM 15.2

An op-amp circuit has the following values: $+V = 12$ V, $-V = -12$ V, $V_{in} = 40\,\text{mV}_{PP}$, $V_2 = 0$ V (ground), and $A_v = 140$. Determine the value of $V_{out}$ for the circuit.

## EXAMPLE 15.3

Determine the maximum allowable value of $V_{in}$ for the circuit shown in Figure 15.14. Assume that the voltage gain of the amplifier is 200.

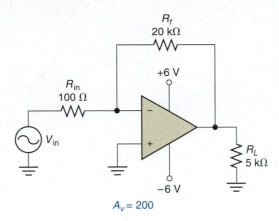

**FIGURE 15.14**

**Solution:** The first step is to determine the maximum allowable peak output voltage values. Since the load resistance is between 2 and 10 k$\Omega$, the peak output values are found as

$$V_{pk(+)} = (+V) - 2\text{ V} = +4\text{ V} \quad \text{(maximum)}$$

and

$$V_{pk(-)} = (-V) - 2\text{ V} = -4\text{ V} \quad \text{(maximum)}$$

The maximum peak-to-peak output voltage equals the difference between these two values, 8 V. Dividing this value by the gain of the amplifier gives us the maximum allowable peak-to-peak value of $V_{in}$ as follows:

$$V_{in} = \frac{V_{out}}{A_v} = \frac{8\text{ V}_{PP}}{200} = 40\text{ mV}_{PP}$$

### PRACTICE PROBLEM 15.3

Determine the maximum allowable peak-to-peak input voltage for the amplifier described in Practice Problem 15.2. Assume that the load resistance for the circuit is 20 k$\Omega$.

## EXAMPLE 15.4

Determine the maximum input and output voltage values for the circuit in Figure 15.15. Assume that the op-amp has the output limits shown in Figure 15.10.

**Solution:** Since the load resistance is less than 2 k$\Omega$, we must make sure that the output voltage swing is not limited by the load. Checking the graph in Figure 15.10, we see that the maximum possible output swing with a load of 1.5 k$\Omega$ is 26 $V_{PP}$, or $\pm 13$ $V_{pk}$. Since the supply voltages are well below this value, we can safely assume that the load will not significantly restrict the output voltage swing. With this in mind, we can determine the maximum output voltages using the guidelines established for loads *under* 10 k$\Omega$. Thus,

$$V_{pk(+)} = (+V) - 2\text{ V} = +2\text{ V} \quad \text{(maximum)}$$

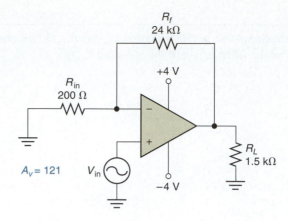

FIGURE 15.15

and

$$V_{pk(-)} = (-V) + 2\,V = -2\,V \quad \text{(maximum)}$$

and our maximum output from the circuit is 4 $V_{PP}$. Using this value, the maximum peak-to-peak value of the input is found as

$$V_{in} = \frac{V_{out}}{A_v} = \frac{4\,V_{PP}}{121} = 33.06\,mV_{PP}$$

**PRACTICE PROBLEM 15.4**

Determine the maximum allowable peak-to-peak input voltage for the amplifier described in Practice Problem 15.2. Assume that the load resistance for the circuit is 1 k$\Omega$. Use the curve in Figure 15.10 to determine the effect of the load resistance on the op-amp output voltage limits.

Incidentally, the circuit shown in Figure 15.15 has an input applied to the *noninverting* terminal, while the circuits shown in Figures 15.13 and 15.14 have inputs applied to the *inverting* terminals. The main difference between the circuit shown in Figure 15.15 and the other two is that there is no voltage phase shift from input to output. The circuits shown in Figures 15.13 and 15.14 both have a 180° voltage phase shift from input to output.

Another point can be made at this time: The circuit configuration shown in Figures 15.13 and 15.14 are the op-amp counterparts of the *common-emitter* amplifier. If you compare these circuits to the voltage-divider biased amplifier, you will see that the op-amp circuits contain far fewer components. This is another advantage of using op-amp circuits in place of discrete component circuits. Op-amp circuits require fewer external components to establish the desired operation.

**Section Review ▶**

1. What is a *differential amplifier*?
2. What determines the polarity of an op-amp output?
3. What is *open-loop voltage gain* ($A_{OL}$)?
4. What is the typical range of $A_{OL}$?
5. Describe the *observation method* for determining the output voltage polarity for an op-amp.
6. Describe the *mathematical method* for determining the output voltage polarity for an op-amp.

7. What effect do the supply voltages of an op-amp have on the maximum peak-to-peak output voltage?

8. What effect does the load resistance of an op-amp have on the maximum peak-to-peak output voltage?

9. Refer to Figure 15.1a. Using the schematic shown, explain why op-amp output voltage *decreases* as load resistance decreases.

◄ **Critical Thinking**

## 15.3 Differential Amplifiers and Op-Amp Specifications

Diodes, BJTs, and FETs all have parameters and electrical characteristics that affect their operation. The op-amp is no different. Up to this point, we have ignored the op-amp parameters and electrical characteristics to give you a chance to grasp the fundamental concepts of component operation without being bogged down with detail. Now, it is time to bring the op-amp parameters and electrical characteristics into the picture.

Earlier in this chapter, you were told that the input circuit to an op-amp is a *differential amplifier* that is driven by the inverting and noninverting inputs. As you will see, this circuit determines many op-amp electrical characteristics.

### 15.3.1 The Basic Differential Amplifier

A *differential amplifier* is a circuit that accepts two inputs and produces an output that is proportional to the *difference between* those inputs. A basic differential amplifier is shown in Figure 15.16. Note that the input to $Q_2$ is identified as the *noninverting input* (*NI*) and the input to $Q_1$ as the *inverting input* (*I*).

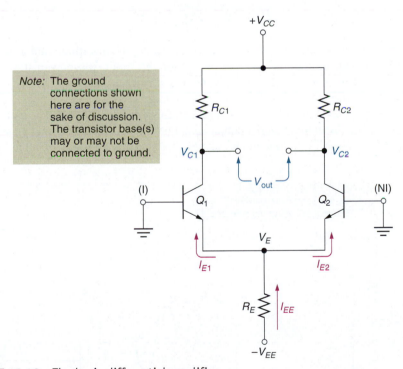

Note: The ground connections shown here are for the sake of discussion. The transistor base(s) may or may not be connected to ground.

**FIGURE 15.16** The basic differential amplifier.

Ideally, the $Q_1$ and $Q_2$ circuits have identical characteristics. Since they are etched on a single piece of silicon, the transistors are identical. By design, the values of the two collector resistors are also equal. With identical characteristics, the quiescent values of $I_E$ for the transistors are equal. When the two inputs are grounded as shown in Figure 15.16:

◄ **OBJECTIVE 4**

$$I_{E1} = I_{E2} \qquad\qquad \textbf{(15.2)}$$

Since both emitter currents pass through $R_E$,

$$I_{E1} = I_{E2} = \frac{I_{EE}}{2} \qquad (15.3)$$

where

$$I_{EE} = \frac{V_{RE}}{R_E} \qquad (15.4)$$

Assuming the base currents are negligible,

$$I_C \cong I_E$$

and

$$I_{C1} = I_{C2} \cong \frac{I_{EE}}{2}$$

When both collector currents and both collector resistors are equal, then

$$V_{C1} = V_{C2} = V_{CC} - I_C R_C \qquad (15.5)$$

and

$$V_{out} = V_{C1} - V_{C2} = 0 \text{ V}$$

Now, consider what happens when we apply a signal to the inverting input as shown in Figure 15.17a. During the positive alternation of the input, the current through $Q_1$ increases. Assuming that the value of $I_{EE}$ is relatively constant, the increase in $I_{E1}$ causes $I_{E2}$ to *decrease*. The increase in $I_{E1}$ causes the voltage across $R_{C1}$ to increase, and $V_{C1}$ *decreases*. At the same time, the decrease in $I_{E2}$ causes the voltage across $R_{C2}$ to decrease, and $V_{C2}$ *increases*. Assuming that the output signal ($V_{out}$) is taken from output 1 (with output 2 being the reference), we obtain the negative output alternation shown in the figure.

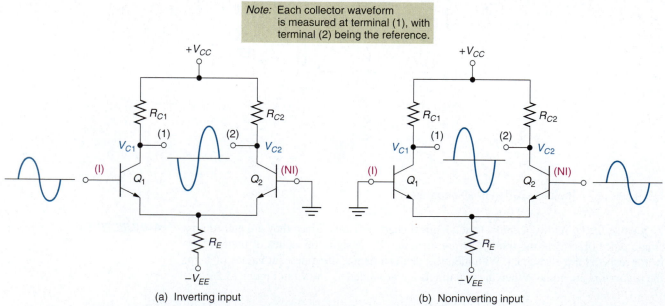

*Note:* Each collector waveform is measured at terminal (1), with terminal (2) being the reference.

(a) Inverting input          (b) Noninverting input

FIGURE 15.17    Differential amplifier input/output relationships.

When the input in Figure 15.17a goes *negative*, $I_{E1}$ decreases and $I_{E2}$ increases. This time, the current changes cause $V_{C1}$ to increase and $V_{C2}$ to decrease. These changes cause $V_{out}$ to increase, as shown in the figure. Thus, the output voltage is 180° out of phase with the input voltage.

In Figure 15.17b, a positive input causes $I_{E2}$ to increase and $I_{E1}$ to decrease. As you have been shown, these changes cause a positive alternation to be produced at output 1. A negative-going input causes $I_{E2}$ to decrease and $I_{E1}$ to increase. These changes cause a negative alternation to be produced at output 1. Thus, the input and output voltages (in this case) are in phase.

## 15.3.2 Modes of Operation

There are three basic modes of operation for a differential amplifier:

1. *Single-ended mode.* A differential amplifier is operating in a single-ended mode when an active signal is applied to only one input, as shown in Figure 15.17. The inactive input is normally connected directly (or via a resistor) to ground. Depending on which input is active, the amplifier is classified as either an *inverting amplifier* (Figure 15.17a) or a *noninverting amplifier* (Figure 15.17b).
2. *Differential mode.* For differential operation, two active signals are applied to the amplifier inputs. The output magnitude and polarity reflect the relationship between the two inputs, as described earlier in this chapter.
3. *Common mode.* This mode of operation occurs when identical signals are applied simultaneously to the inverting and noninverting inputs. Since the transistors have identical inputs, they provide identical ouputs. As a result, the collector voltages are equal throughout the entire cycle of the input waveform, and the output from the differential amplifier is 0 V. The benefit of this mode of operation is that any noise or undesired signal that appears at the two inputs simultaneously does not generate an output from the circuit (ideally). This point is discussed in detail later in this section.

## 15.3.3 Output Offset Voltage

Even though the transistors in the differential amplifier are very closely matched, there are some differences in their electrical characteristics. One of these differences is found in the values of $V_{BE}$ for the two transistors. When $V_{BE1} \neq V_{BE2}$, an imbalance is created in the differential amplifier that may result in an **output offset voltage** ($V_{out(offset)}$). This condition is illustrated in Figure 15.18. Note that with the op-amp inputs grounded, the output shows a measurable voltage. This voltage (which can be positive or negative) is a result of the imbalance in the differential amplifier, which causes one of the transistors to conduct harder than the other.

◀ *OBJECTIVE 5*

> **Output offset voltage**
> ($V_{out(offset)}$)
> A voltage that may appear at the output of an op-amp; caused by an imbalance in the differential amplifier.

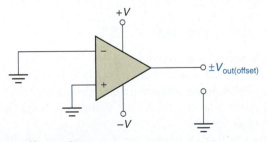

FIGURE 15.18   Output offset voltage.

The difference between the base-emitter voltages in the differential amplifier is referred to as **input offset voltage** ($V_{io}$) and is normally represented as shown in Figure 15.19. Note that the battery ($V_{io}$) represents the difference between $V_{BE1}$ and $V_{BE2}$. The value of $V_{io}$ can be found as

> **Input offset voltage ($V_{io}$)**
> The difference between the base-emitter voltages in a differential amplifier that produces an offset output voltage when the signal inputs are grounded.

$$V_{io} = \frac{V_{out(offset)}}{A_v} \qquad (15.6)$$

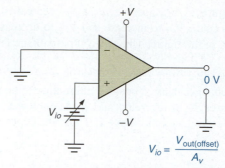

$$V_{io} = \frac{V_{out(offset)}}{A_v}$$

**FIGURE 15.19** A battery representing the input offset voltage.

**Offset null**

Input pins on an op-amp that are used to eliminate an output offset voltage.

One method of eliminating output offset voltage is to connect the op-amp's **offset null** pins as shown in Figure 15.20a. When the offset null is used, power is applied to the circuit and the potentiometer is adjusted to eliminate the output offset.

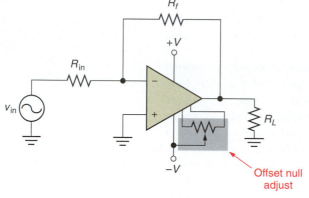

(a) Using the offset null input pins

(b) Using a compensating resistor

**FIGURE 15.20** Reducing output offset voltage.

Figure 15.1 shows that the offset null pins are connected (indirectly) to the input differential amplifier. When properly adjusted, the offset null circuitry corrects the imbalance in the differential amplifier, causing the output of the op-amp to go to 0 V. It should be noted that the offset null pins are rarely used. In some applications, critical operation of the op-amp is not required. In other applications, simpler methods can be used to eliminate any output offset voltage.

### 15.3.4 Input Offset Current

**Input offset current**

A slight difference in op-amp input currents, caused by differences in the transistor beta ratings.

**Compensating resistor**

A resistor connected to the noninverting input to an op-amp to compensate for any difference in the input currents.

When an op-amp is connected as shown in Figure 15.18, there may also be a slight difference between the inverting and noninverting input currents ($I_1$ and $I_2$). This slight difference in input currents, called **input offset current**, is caused by a beta mismatch between the transistors in the differential amplifier. Left unchecked, input offset current can contribute to any output offset voltage.

One commonly used method of minimizing the effects of any difference between the signal input currents is to place a **compensating resistor** between the noninverting input and ground. For example, Figure 15.20b shows an op-amp circuit with a compensating resistor ($R_C$). The value of $R_C$ equals the parallel combination of $R_{in}$ and $R_f$. The reason for this is simple: The circuit connected to the inverting input *as seen by that input* is made up of $R_{in} \parallel R_f$. By setting $R_C = R_{in} \parallel R_f$, the two inputs "see" the same resistance.

When the proper value of $R_C$ is used, any difference between the input currents generates a proportional differential input voltage. This differential input helps to offset the imbalance caused by the difference in the input currents, which helps to hold any resulting output offset voltage to a minimum.

## 15.3.5 Input Bias Current

The inputs to an op-amp require some amount of dc biasing current for the BJTs in the differential amplifier. The average quiescent value of dc biasing current drawn by the signal inputs of the op-amp is the **input bias current** rating of the amplifier. For the KA741, this rating is between 80 nA (typical) and 500 nA (maximum). This means that the op-amp signal inputs draw between 80 and 500 nA from the external circuitry when no active signal is applied to the device.

The fact that both transistors in the differential amplifier require an input biasing current leads to the following operating restriction: *An op-amp will not produce the expected output if either of its inputs is open.* For example, look at the circuit shown in Figure 15.21. The noninverting input is shown to have an open between the op-amp and ground. This open circuit does not allow the dc biasing current required for the operation of the differential amplifier. (*The transistor associated with the inverting input will work, but not the one associated with the noninverting input.*) Since the differential amplifier will not work, the overall op-amp circuit will not work. Thus, an input bias current path must always be provided for *both* op-amp inputs.

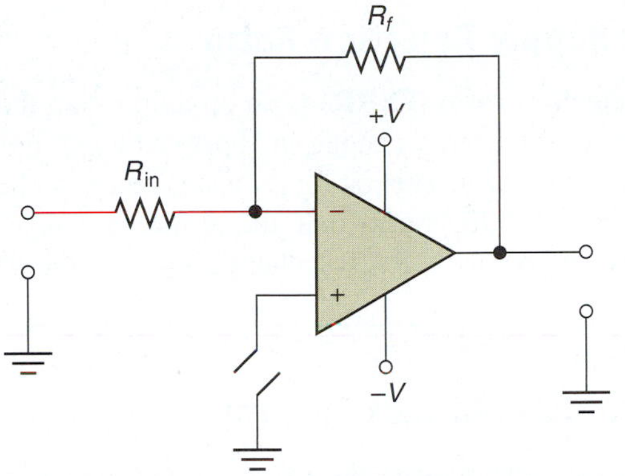

**Input bias current**
The average value of quiescent dc biasing current drawn by the signal inputs of an op-amp.

**FIGURE 15.21** An open input prevents the op-amp from working.

## 15.3.6 Common-Mode Rejection Ratio (CMRR)

**◄ OBJECTIVE 6**

**Common-mode signals** are *identical signals that are applied simultaneously to the two inputs of an op-amp*. For example, the two signals shown in Figure 15.22 are common-mode signals.

**Common-mode signals**
Identical signals that are applied simultaneously to the two inputs of an op-amp.

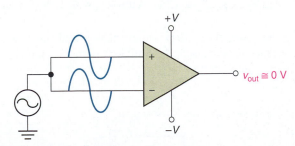

**FIGURE 15.22** Common-mode signals.

If the two signals shown *occur at the same time and have the same amplitude*, the *ideal* op-amp does not respond to their presence. Remember that the op-amp is designed to respond to *the difference between* two input signals. If there is no difference between the two inputs, the ideal op-amp doesn't produce an output.

A measure of the ability of an op-amp to "ignore" common-mode signals is its **common-mode rejection ratio (CMRR)**. Technically, it is *the ratio of differential voltage gain to common-mode voltage gain*. For example, let's assume for a moment that the op-amp in

**Common-mode rejection ratio (CMRR)**
The ratio of differential gain to common-mode gain.

Figure 15.22 has a differential voltage gain of 1500. Let's also assume that the op-amp has a common-mode voltage gain of 0.01. The common-mode rejection ratio for the op-amp would be found as

$$\text{CMRR} = \frac{A_v \text{ (differential)}}{A_v \text{ (common mode)}} = \frac{1500}{0.01} = \textbf{150,000} \quad \textbf{(103.5 dB)}$$

This result indicates that the gain provided to a differential input ($V_{diff}$) is 150,000 times as great as the gain provided to two common-mode signals.

The ability of the op-amp to reject common-mode signals is one of its strengths. Common-mode signals are usually *undesired* signals, caused by external interference. For example, any RF noise picked up by the op-amp inputs is considered undesirable. The common-mode rejection ratio indicates the op-amp's ability to reject such unwanted signals.

> Common-mode signals are usually undesired signals.

The common-mode rejection ratio for a given op-amp is usually measured in *decibels* (dB). For example, the KA741 op-amp has a minimum common-mode rejection ratio of 70 dB. This means that differential input gain is at least 3162 times as great as common-mode signal gain.

### 15.3.7    Power Supply Rejection Ratio

> **Power supply rejection ratio (PSRR)**
>
> The ratio of a change in op-amp output voltage to a change in supply voltage.

The **power supply rejection ratio (PSRR)** is an op-amp rating that indicates the change in output voltage that results from a change in supply voltage. For example, the KA741 has a PSRR of 77 dB (minimum). This rating indicates that any change in the power supply voltage will be at least 77 dB greater than the resulting change in output voltage. For example, if the supply voltage for a KA741 changes by 1 V, the output will change by a *maximum* of 141 μV.

### 15.3.8    Output Short-Circuit Current

> **Output short-circuit current**
>
> The maximum output current for an op-amp, measured under shorted load conditions.

Op-amps are protected internally from the excessive current demand of a shorted load. The **output short-circuit current** rating of an op-amp is the maximum value of output current, measured with the load shorted. For the KA741, this rating is 25 mA. Thus, with a load of 0 Ω, the output current from this op-amp will be no more than 25 mA.

The short-circuit current rating helps to explain why the output voltage from an op-amp drops when the load resistance decreases. For example, consider the circuit shown in Figure 15.23. The load resistance is shown to be 50 Ω. Assuming that the op-amp is producing the maximum output current of 25 mA, the maximum load voltage is found as

$$V_L = I_{out}R_L = (25 \text{ mA})(50 \text{ Ω}) = \textbf{1.25 V}$$

This is considerably less than the ±10 V supply that would otherwise limit the output voltage. As load resistance is decreased, the limit on op-amp output current can cause the load voltage to decrease significantly.

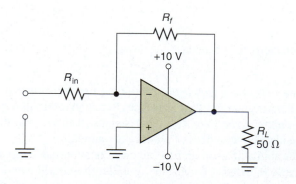

**FIGURE 15.23**

## 15.3.9 Slew Rate

The **slew rate** of an op-amp is a measure of *how fast the output voltage can change in response to a change at either signal input*. The slew rate of the KA741 is 0.5 V/μs (typical). This means that the output from this amplifier can change by 0.5 V every microsecond. Since frequency is related to time, the *slew rate can be used to determine the maximum operating frequency of the op-amp* as follows:

◄ **OBJECTIVE 7**

**Slew rate**
The maximum rate at which op-amp output voltage can change.

$$f_{max} = \frac{\text{slew rate}}{2\pi V_{pk}} \qquad (15.7)$$

**Lab Reference:** The slew rate of an op-amp is measured in Exercise 21.

where $V_{pk}$ = the peak output voltage from the op-amp. Example 15.5 demonstrates the use of this relationship.

### EXAMPLE 15.5

Determine the maximum operating frequency for the circuit shown in Figure 15.24. (Assume that $2\ k\Omega < R_L < 10\ k\Omega$.)

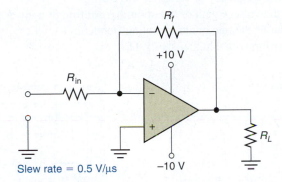

$R_f$

+10 V

$R_{in}$

$R_L$

Slew rate = 0.5 V/μs

−10 V

**FIGURE 15.24**

*Solution:* The maximum peak output voltage for this circuit is approximately 8 V. Using this value as $V_{pk}$ in equation (15.7), the maximum operating frequency for the amplifier is determined as

$$f_{max} = \frac{\text{slew rate}}{2\pi V_{pk}} = \frac{0.5\ \text{V/μs}}{2\pi(8\ \text{V})} = \frac{500\ \text{kHz}}{50.27} = \textbf{9.95 kHz}$$

### PRACTICE PROBLEM 15.5

An op-amp with a slew rate of 0.4 V/μs has a 10 $V_{pk}$ output. Determine the maximum operating frequency for the component.

Note that the slew rate in Example 15.5 was converted to a frequency as follows:

$$\frac{0.5}{1\ \text{μs}} = \frac{5 \times 10^{-1}}{1 \times 10^{-6}\ \text{s}} = 500\ \text{kHz}$$

When the maximum required output from the op-amp *decreases*, the maximum possible operating frequency *increases*. This relationship is demonstrated in Example 15.6.

### EXAMPLE 15.6

The amplifier in Figure 15.24 is used to amplify an input signal to a peak output voltage of 100 mV. What is the maximum operating frequency of the amplifier?

**Solution:** Using 100 mV as the peak output voltage, $f_{max}$ is found as

$$f_{\text{max}} = \frac{\text{slew rate}}{2\pi V_{\text{pk}}} = \frac{0.5 \text{ V/}\mu\text{s}}{2\pi(100 \text{ mV})} = \frac{500 \text{ kHz}}{0.628} = \textbf{796 kHz}$$

**PRACTICE PROBLEM 15.6**

The output peak voltage for the op-amp in Practice Problem 15.5 is reduced to 2 $V_{pk}$. Determine the new maximum operating frequency for the device.

Examples 15.5 and 15.6 show that the op-amp can be operated at a much higher frequency when it is used as a small-signal amplifier than when it is used as a large-signal amplifier.

**◀ OBJECTIVE 8 ▶**

> The phase relationships shown in Figure 15.25 will vary with the op-amp configuration (inverting or noninverting) and the operating frequency.

The effects of operating an op-amp beyond its frequency limit ($f_{max}$) are illustrated in Figure 15.25. This figure shows two input waveforms and the distorted output waveforms. In each case, the input waveform is changing at a rate that is greater than the op-amp can handle. When this happens, the output from the op-amp changes more slowly than the input, resulting in the distortion shown. This problem can be corrected by:

1. Decreasing the frequency of the input waveform.
2. Decreasing the voltage gain of the op-amp, resulting in a lower peak output voltage (which increases the maximum operating frequency).
3. Using an op-amp with a higher slew rate.

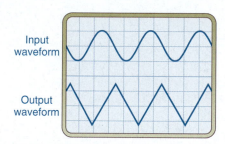

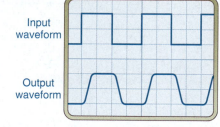

(a) The effect of slew rate distortion on a sine wave   (b) The effect of slew rate distortion on a square wave

**FIGURE 15.25**   Slew-rate distortion.

### 15.3.10   Input/Output Resistance

As stated earlier in the chapter, op-amps typically have high input resistance and low output resistance ratings. For the KA741, these ratings are 2 M$\Omega$ and 75 $\Omega$, respectively. Note that the high input resistance and low output resistance of the op-amp closely resemble the characteristics of the *ideal* voltage amplifier, which has infinite input resistance and zero output resistance.

### 15.3.11   Putting It All Together

You have been exposed to many op-amp operating characteristics in this section. Now, we will take a look at how they relate to each other. As the graphs in Figure 15.26 illustrate, the various op-amp characteristics are not independent of each other. Table 15.4 has been derived using the graphs in Figure 15.26. As you read the comments on each characteristic, refer to the indicated graph to verify the comments for yourself. Graph references are given in parentheses after each statement.

The following series of statements provide *cause-and-effect* relationships from the 741 operating curves:

1. As you *increase supply voltage*, you also:
   a. Increase the open-loop voltage gain.
   b. Increase the possible output voltage swing.

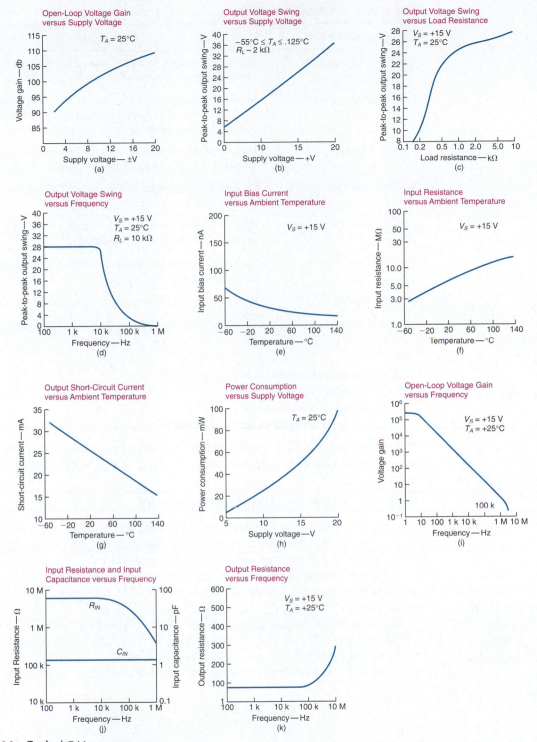

**FIGURE 15.26** Typical 741 op-amp curves.

   **c.** Increase the component power consumption.

   **d.** Vary the input offset current very slightly.

**2.** As you *increase the operating frequency*, you also:

   **a.** Decrease the maximum output voltage swing ($f > 10$ kHz).

   **b.** Decrease the open-loop voltage gain.

   **c.** Decrease the input resistance ($f > 10$ kHz).

   **d.** Increase the output resistance ($f > 100$ kHz).

**TABLE 15.4** Op-Amp Characteristics

| *Characteristic* | *Comments* |
|---|---|
| Open-loop voltage gain | Varies with changes in supply voltage (a); decreases at a linear rate as frequency increases from 10 Hz to 1 MHz (i). |
| Output voltage swing | Varies with changes in $R_L$; is severely limited by values of $R_L < 500\ \Omega$ (c); directly proportional to supply voltage (b); constant for frequencies up to 10 kHz, then drops off rapidly (d). |
| Input resistance | Directly proportional to ambient temperature (f); constant for frequencies up to 10 kHz, then starts to drop off (j). |
| Input bias current | Decreases as temperature increases (e). This is consistent with the increase in input resistance. |
| Output short-circuit current | Rated at a temperature of 25°C; increases below this temperature, and decreases above this temperature (g). |
| Power consumption | The power used by the device is directly proportional to the supply voltage (h). |
| Output resistance | Remains constant for frequencies up to 100 kHz, then increases rapidly (k). |

**3.** As the *ambient temperature increases*:
- **a.** Input bias current decreases.
- **b.** Input resistance increases.
- **c.** Output short-circuit current decreases.

As you will see, these cause-and-effect relationships can come in handy when you are trying to build a circuit in the lab or are troubleshooting a given op-amp circuit.

### 15.3.12    Other Op-Amp Characteristics

The *electrical characteristics* table from the KA741 spec sheet is shown in Figure 15.27. As you can see, the spec sheet contains all the characteristics discussed in this section. At this point, we take a brief look at four additional ratings listed on the spec sheet.

The **input voltage range** indicates the *maximum differential input to the op-amp*. As shown in the figure, the KA741 can withstand a $\pm 12$ V difference of potential across its signal inputs.

The **large-signal voltage gain** is the *open-loop voltage gain* rating for the component. The spec sheet in Figure 15.27 shows a rating of $G_V = 200$ V/mV (typical). This value indicates that the component can provide an output of 200 V (maximum) with a differential input of 1 mV. The above rating can be converted to standard form as follows:

$$A_{\text{OL}} = \frac{200\ \text{V}}{1\ \text{mV}} = \frac{200\ \text{V}}{1 \times 10^{-3}\ \text{V}} = 200{,}000$$

> **Input voltage range**
> The maximum differential input that an op-amp can accept without risking damage to its input differential amplifier.
>
> **Large-signal voltage gain**
> The *open-loop voltage gain* of an op-amp.

The **supply current** rating indicates *the quiescent current that is drawn from the power supply*. When the KA741 does not have an input signal, it will draw a maximum of 2.8 mA from the $+V$ and $-V$ power supplies.

The **power consumption** rating indicates *the amount of power that the op-amp will dissipate when it is in its quiescent state*. As you can see, the power dissipation rating of the KA741 has a maximum value of 85 mW.

There are two more ratings that we have not covered. These are *bandwidth* and *transient response*. Bandwidth is discussed later in this chapter. Transient response is covered in Chapter 19. In the upcoming sections of this chapter, we will discuss the operation of the two most basic op-amp circuits, the *inverting amplifier* and the *noninverting amplifier*.

> **Supply current**
> A rating that indicates the value of quiescent (inactive) current that an op-amp draws from its power supply.
>
> **Power consumption**
> A rating that indicates the amount of power dissipated by an op-amp when it is operated in its quiescent state.

## KA741/KA741E

### Single Operational Amplifier

# Electrical Characteristics

($V_{CC}$ = 15V, $V_{EE}$ = –15V. $T_A$ = 25°C, unless otherwise specified)

| Parameter | | Symbol | Conditions | | KA741E | | | KA741 | | | Unit |
|---|---|---|---|---|---|---|---|---|---|---|---|
| | | | | | Min. | Typ. | Max. | Min. | Typ. | Max. | |
| Input Offset Voltage | | $V_{IO}$ | $R_S{\le}10k\Omega$ | | – | – | – | – | 2.0 | 6.0 | mV |
| | | | $R_S{\le}50\Omega$ | | | 0.8 | 3.0 | – | – | – | |
| Input Offset Voltage Adjustment Range | | $V_{IO(R)}$ | $V_{CC} = \pm20V$ | | ±10 | – | – | – | ±15 | – | mV |
| Input Offset Current | | $I_{IO}$ | – | | – | 3.0 | 30 | – | 20 | 200 | nA |
| Input Bias Current | | $I_{BIAS}$ | – | | – | 30 | 80 | – | 80 | 500 | nA |
| Input Resistance | | $R_I$ | $V_{CC} = \pm20V$ | | 1.0 | 6.0 | – | 0.3 | 2.0 | – | M$\Omega$ |
| Input Voltage Range | | $V_{I(R)}$ | – | | ±12 | ±13 | – | ±12 | ±13 | – | V |
| Large Signal Voltage Gain | | $G_V$ | $R_L \ge 2K\Omega$ | $V_{CC} = \pm20V$, $V_{O(P-P)} = \pm15V$ | 50 | – | – | – | – | – | V/mV |
| | | | | $V_{CC} = \pm15V$, $V_{O(P-P)} = \pm10V$ | – | – | – | 20 | 200 | – | |
| Output Short Circuit Current | | $I_{SC}$ | – | | 10 | 25 | 35 | – | 25 | – | mA |
| Output Voltage Swing | | $V_{O(P-P)}$ | $V_{CC} = \pm20V$ | $R_L \ge 10K\Omega$ | ±16 | – | – | – | – | – | V |
| | | | | $R_L \ge 2K\Omega$ | ±15 | – | – | – | – | – | |
| | | | $V_{CC} = \pm15V$ | $R_L \ge 10K\Omega$ | – | – | – | ±12 | ±14 | – | |
| | | | | $R_L \ge 2K\Omega$ | – | – | – | ±10 | ±13 | – | |
| Common Mode Rejection Ratio | | CMRR | $R_S \le 10K\Omega$. $V_{CM} = \pm12V$ | | – | – | – | 70 | 90 | – | dB |
| | | | $R_S \le 50K\Omega$. $V_{CM} = \pm12V$ | | 80 | 95 | – | – | – | – | |
| Power Supply Rejection Ratio | | PSRR | $V_{CC} = \pm15V$ to $V_{CC} = \pm15V$ $R_S{\le}50\Omega$ | | 86 | 96 | – | – | – | – | dB |
| | | | $V_{CC} = \pm15V$ to $V_{CC} = \pm15V$ $R_S{\le}10K\Omega$ | | – | – | – | 77 | 96 | – | |
| Transient Response | Rise Time | $I_R$ | Unity Gain | | – | 0.25 | 0.8 | – | 0.3 | – | µs |
| | Overshoot | OS | | | – | 6.0 | 20 | – | 10 | – | % |
| Bandwidth | | BW | | | 0.43 | 1.5 | – | – | – | – | MHz |
| Slew Rate | | SR | Unity Gain | | 0.3 | 0.7 | – | – | 0.5 | – | V/µs |
| Supply Current | | $I_{CC}$ | $R_L = \infty\Omega$ | | – | – | – | – | 1.5 | 2.8 | mA |
| Power Consumption | | $P_C$ | $V_{CC} = \pm20V$ | | – | 80 | 150 | – | – | – | mW |
| | | | $V_{CC} = \pm15V$ | | – | – | – | – | 50 | 85 | |

**LIFE SUPPORT POLICY**

FAIRCHILD'S PRODUCTS ARE NOT AUTHORIZED FOR USE AS CRITICAL COMPONENTS IN LIFE SUPPORT DEVICES OR SYSTEMS WITHOUT THE EXPRESS WRITTEN APPROVAL OF THE PRESIDENT OF FAIRCHILD SEMICONDUCTOR INTERNATIONAL. As used herein:

1. Life support devices or systems are devices or systems which, (a) are intended for surgical implant into the body, or (b) support or sustain life, and (c) whose failure to perform when properly used in accordance with instructions, for use provided in the labeling, can be reasonably expected to result in a significant injury of the user.

2. A critical component in any component of a life support device or system whose failure to perform can be reasonably expected to cause the failure of the life support device or system, or to affect its safety or effectiveness.

**FIGURE 15.27** KA741 specification sheet. (Courtesy of Fairchild Semiconductor. Used by permission.)

## 15.3.13 Op-Amp Classifications

The KA741 is classified as a *general-purpose* op-amp, meaning that it is well suited for many of the most common op-amp applications. While most op-amps fall into this category, some op-amps are designed for specific applications.

*High-current op-amps* are designed to produce high output currents. For example, the TCA0372 can provide a 1 A output. High-current op-amps tend to have higher slew rates than general-purpose components, so they can be operated at higher frequencies. They are also capable of providing relatively high voltage gain even at low values of load resistance.

*High-speed op-amps* are designed to have high slew rates. For example, the MC33071 has a slew rate of 13 V/μs. When you compare this value to the slew rate listed in Figure 15.27, you can see that the MC33071 is approximately 26 times as fast as the KA741 (at a given peak output voltage). Note that high-speed op-amps tend to have lower $A_{OL}$ ratings than general-purpose components.

*Low-power op-amps* are designed to draw little power supply current when no active signal is applied. For example, the MC33171 draws only 180 μA from the dc power supply in its quiescent state. In contrast, Figure 15.27 shows that the KA741 typically draws 1.5 mA from its dc power supply. As such, low-power op-amps are better suited than general-purpose components for use in battery-powered circuits and systems. Note that low-power op-amps have lower output current ratings than general-purpose components. As a result, they are more severely affected by changes in load current demand.

The input voltage range and maximum output for most op-amps are limited to values lower than the power supply voltages. For example, the spec sheet in Figure 15.27 shows the following limits when the supply voltages are ±15 V:

$$Input\ voltage\ range: \quad V_{I(R)} = \pm13\ V$$
$$Output\ voltage\ swing: \quad V_{O(P\text{-}P)} = \pm14\ V$$

As these values show, the input and output voltage limits are lower than the supply voltages (±15 V). A *rail-to-rail op-amp* is designed to allow *rail-to-rail operation*. The term **rail** is used in reference to the op-amp supply voltages. As such, *rail-to-rail operation* means the input and output voltages can come much closer to the values of $+V$ and $-V$. For example, the MC33204 has the following capabilities:

| **Rail** |
|---|
| A term used to describe either op-amp supply voltage. |

1. The inputs can be driven as high as 200 mV beyond each rail.
2. The output swings can come within 50 mV of the rails.

As such, the component is capable of making full use of the voltage range between $+V$ and $-V$.

---

**Section Review ▶**

1. Describe the differential amplifier's response to an input signal at its inverting input.
2. Describe the differential amplifier's response to an input signal at its noninverting input.
3. What is *input offset voltage*?
4. What is *output offset voltage*?
5. What effect does the *offset null* control have on the input offset voltage and output offset voltage? Explain your answer.
6. What is *input offset current*?
7. What is *input bias current*? What restriction does input bias current place on the wiring of an op-amp?
8. What are *common-mode signals*?
9. What is the *common-mode rejection ratio (CMRR)*? Why is a high CMRR considered desirable?
10. What is the *power supply rejection ratio*?
11. What is the *output short-circuit current* rating?
12. Discuss the relationship between load resistance and maximum output voltage swing in terms of the output short-circuit current rating.
13. What is *slew rate*? What circuit parameter is limited by the op-amp slew rate?
14. What is the result of trying to operate an op-amp beyond its slew rate limit?
15. Define each of the following op-amp characteristics:
    a. Input voltage range
    b. Large-signal voltage range

**c.** Supply current rating

**d.** Power consumption rating

**16.** Briefly describe each type of op-amp listed:

    **a.** High-current

    **b.** High-speed

    **c.** Low-power

    **d.** Rail-to-rail

## 15.4 Inverting Amplifiers

In most of our examples, we have used an op-amp circuit that contains a single input resistor ($R_{in}$) and a single feedback resistor ($R_f$). This circuit is the **inverting amplifier**, which is the op-amp counterpart of the common-emitter and common-source circuits. The operation of the inverting amplifier is illustrated in Figure 15.28.

The key to the operation of the inverting amplifier lies in the differential input circuit of the op-amp. Assuming that the op-amp input transistors are perfectly matched, $V_1$ and $V_2$ (Figure 15.28a) are always *approximately* equal to each other. They may differ by a few millivolts, but that is all.

Now, let's idealize the op-amp input circuit for a moment. For the sake of discussion, we will assume that $V_1$ and $V_2$ are *exactly* equal. With this in mind, look at the circuit shown in Figure 15.28b. If the voltages at the two inputs are equal and the noninverting input is grounded, the inverting input to the op-amp is also at ground. Remember, this *virtual ground* is caused by the fact that $V_1$ and $V_2$ are approximately equal and the noninverting input is grounded.

◄ **OBJECTIVE 9**

**Inverting amplifier**
A basic op-amp circuit that produces a 180° signal phase shift. The op-amp counterpart of the common-emitter and common-source circuits.

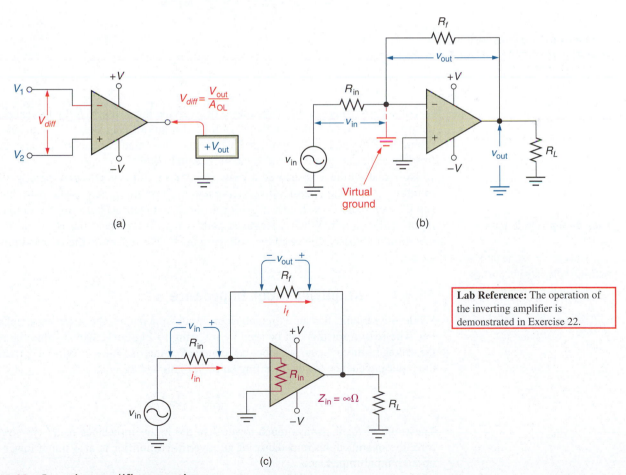

**Lab Reference:** The operation of the inverting amplifier is demonstrated in Exercise 22.

**FIGURE 15.28** Inverting amplifier operation.

With the inverting input of the op-amp at virtual ground, the input voltage can be measured across $R_{in}$. Also, the output voltage can be measured across $R_f$. These relationships are illustrated in Figure 15.28b.

Since $v_{out}$ appears across the feedback resistor, its value can be found as

$$v_{out} = i_f R_f \qquad \textbf{(15.8)}$$

Since $v_{in}$ appears across the input resistor, its value can be found as

$$v_{in} = i_{in} R_{in} \qquad \textbf{(15.9)}$$

Since an op-amp has extremely high input resistance (as shown in Figure 15.28c), it has almost zero input current. Therefore, $i_{in} \cong i_f$, and equation (15.8) can be rewritten as

$$v_{out} = i_{in} R_f \qquad \textbf{(15.10)}$$

The value of $A_v$ for an inverting amplifier is often written as a negative value to indicate the $180°$ phase shift.

As you know, voltage gain ($A_v$) is found as the ratio of output voltage to input voltage. By formula,

$$A_v = \frac{v_{out}}{v_{in}}$$

Since $v_{out} = i_{in} R_f$ and $v_{in} = i_{in} R_{in}$, the above equation can be rewritten as

$$A_v = \frac{i_{in} R_f}{i_{in} R_{in}}$$

**Lab Reference:** The relationship between the resistor values and voltage gain is demonstrated in Exercise 22.

or

$$A_v = \frac{R_f}{R_{in}} \qquad \textbf{(15.11)}$$

For the inverting amplifier, you need only divide the value of $R_f$ by the value of $R_{in}$ to determine the approximate voltage gain of the amplifier. In fact, if you look back at Examples 15.2 and 15.3, you'll see where the "assumed" values of voltage gain came from. In each case, the assumed value of $A_v$ equals $R_f/R_{in}$.

Now that we have discussed a basic op-amp circuit, we can give practical meaning to a few terms. The *open-loop voltage gain* ($A_{OL}$) of an op-amp is the gain that is measured when there is no *feedback path* (physical connection) between the output and the input of the circuit. When a feedback path is present (such as the $R_f$ connection in the inverting amplifier), the resulting voltage gain is referred to as the **closed-loop voltage gain** ($A_{CL}$).

**Closed-loop voltage gain ($A_{CL}$)**
The voltage gain of an op-amp with a feedback path; always lower than the value of $A_{OL}$.

### 15.4.1    Amplifier Input Impedance

While the op-amp has an extremely high input impedance, the inverting amplifier does not. The reason for this can be seen by referring to Figure 15.28b. As this figure shows, the voltage source "sees" an input resistance ($R_{in}$) that is going to (virtual) ground. Thus, the input impedance for the inverting amplifier is found as

$$Z_{in} \cong R_{in} \qquad \textbf{(15.12)}$$

The value of $R_{in}$ is always much lower than the input impedance of the op-amp. Therefore, the overall input impedance of an inverting amplifier is also much lower than the op-amp input impedance.

## 15.4.2 Amplifier Output Impedance

If you look at the circuit shown in Figure 15.29, you can see that the output impedance of the inverting amplifier is the parallel combination of $R_f$ and the output impedance of the op-amp itself. Since $R_f$ is normally much greater than the value of $Z_{out}$ for the op-amp, the output impedance of the circuit is usually assumed to equal the $Z_{out}$ rating of the op-amp.

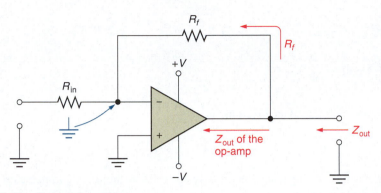

FIGURE 15.29   Output impedance.

## 15.4.3   Amplifier CMRR

In the last section, the *common-mode rejection ratio* (CMRR) of an op-amp was defined as the ratio of differential gain to **common-mode gain** ($A_{CM}$). Since the differential gain of the inverting amplifier ($A_{CL}$) is much lower than the differential gain of the op-amp ($A_{OL}$), the CMRR of the inverting amplifier is typically much lower than that of the op-amp.

The CMRR of the inverting amplifier is found as the ratio of closed-loop gain to common-mode gain. By formula,

$$\text{CMRR} = \frac{A_{CL}}{A_{CM}}$$

(15.13)

where $A_{CL}$ = the *closed-loop* voltage gain of the inverting amplifier
$A_{CM}$ = the common-mode gain of the op-amp

Even though the CMRR of an inverting amplifier is always lower than that of the op-amp, it is still extremely high in most cases. This point is illustrated in the upcoming example.

## 15.4.4   Inverting Amplifier Analysis

The complete analysis of an inverting amplifier involves determining the circuit values of $A_{CL}$, $Z_{in}$, $Z_{out}$, CMRR, and $f_{max}$. The complete analysis of an inverting amplifier is demonstrated in Example 15.7.

> **Common-mode gain ($A_{CM}$)**
> The gain a differential amplifier provides to common-mode signals.

> *A Practical Consideration:*
> Most op-amp spec sheets do not provide a value $A_{CM}$. Rather, they list values of $A_{OL}$ and CMRR. When this is the case, the value of $A_{CM}$ can be obtained using
>
> $$A_{CM} = \frac{A_{OL}}{\text{CMRR}}$$
>
> If the CMRR is listed in decibel (dB) form, it must be converted to standard numeric form before using the above equation:
>
> $$\text{CMRR} = \log^{-1}\left(\frac{\text{CMRR}(dB)}{20}\right)$$

### EXAMPLE 15.7

Perform a complete analysis of the circuit shown in Figure 15.30.

*Solution:*   First, the closed-loop voltage gain ($A_{CL}$) of the circuit is found as

$$A_{CL} = \frac{R_f}{R_{in}} = \frac{100 \text{ k}\Omega}{10 \text{ k}\Omega} = 10$$

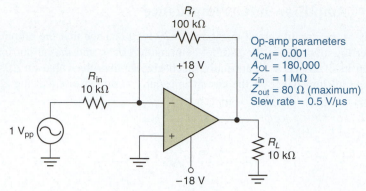

FIGURE 15.30

The circuit input impedance is found as

$$Z_{in} \cong R_{in} = 10 \text{ k}\Omega$$

The circuit output impedance is slightly lower than the output impedance of the op-amp, 80 Ω (maximum).

The CMRR of the circuit is found as

$$\text{CMRR} = \frac{A_{CL}}{A_{CM}} = \frac{10}{0.001} = 10{,}000$$

To calculate the maximum operating frequency for the inverting amplifier, we need to determine its peak output voltage. With values of $v_{in} = 1 \text{ V}_{PP}$ and $A_{CL} = 10$, the peak-to-peak output voltage is found as

$$v_{out} = A_{CL}v_{in} = (10)(1 \text{ V}_{PP}) = 10 \text{ V}_{PP}$$

The peak output voltage is half the peak-to-peak value, 5 $V_{pk}$. Using this peak output value, the maximum operating frequency is found as

$$f_{max} = \frac{\text{slew rate}}{2\pi V_{pk}} = \frac{0.5 \text{ V/}\mu\text{s}}{31.42 \text{ V}} = \frac{500 \text{ kHz}}{31.42} = 15.9 \text{ kHz}$$

**PRACTICE PROBLEM 15.7**

An op-amp has the following parameters: $A_{CM} = 0.02$, $A_{OL} = 150{,}000$, $Z_{in} = 1.5 \text{ M}\Omega$, $Z_{out} = 50 \text{ }\Omega$ (maximum), and slew rate = 0.75 V/μs. The op-amp is used in an inverting amplifier with ±12 V supply voltages and values of $v_{in} = 50 \text{ mV}_{PP}$, $R_f = 250 \text{ k}\Omega$, and $R_{in} = 1 \text{ k}\Omega$. Perform the complete analysis of the circuit.

## 15.4.5 Other Inverting Amplifier Configurations

In this section, we have covered only the simplest of the inverting amplifiers. While the calculations of gain, input impedance, and output impedance may vary with different configurations, the calculations of $f_{max}$ and CMRR do not. You will also see that the various inverting amplifiers all have the characteristic 180° voltage phase shift from input to output.

**Section Review ▶**

1. Explain why the voltage gain of an inverting amplifier ($A_{CL}$) is equal to the ratio of $R_f$ to $R_{in}$.

2. Why is the input impedance of an inverting amplifier lower than the input impedance of the op-amp?

**3.** Why isn't the value of $R_f$ normally considered in the output impedance calculation for an inverting amplifier?

**4.** Why is the CMRR ratio of an inverting amplifier lower than that of its op-amp?

**5.** List, in order, the steps involved in performing the complete analysis of an inverting amplifier.

**6.** Keeping equation (15.12) in mind, how would you go about increasing the voltage gain of an inverting amplifier? Explain your answer.

◀ **Critical Thinking**

**7.** Refer to Example 15.7. What effect would increasing the circuit value of $R_{in}$ have on each of the values calculated in the example?

## 15.5 Noninverting Amplifiers

The **noninverting amplifier** has most of the characteristics of the inverting amplifier, with two exceptions:

**1.** The noninverting amplifier has much higher circuit input impedance.
**2.** The noninverting amplifier does not produce a 180° voltage phase shift from input to output. Thus, the input and output signals are in phase.

The basic noninverting amplifier is shown in Figure 15.31. Note that the input is applied to the noninverting op-amp input, and the input resistor ($R_{in}$) is returned to ground.

◀ *OBJECTIVE 10*

> **Noninverting amplifier**
> An op-amp cicuit with no signal phase shift.

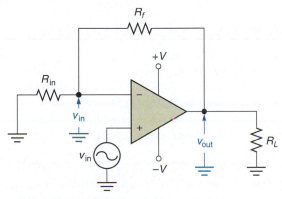

**FIGURE 15.31** Noninverting amplifier.

> **Lab Reference:** The operation of the noninverting amplifier is demonstrated in Exercise 23.

Because the input signal is applied to the noninverting terminal, we must calculate the gain of this circuit differently. We will start our discussion of circuit gain by again assuming that $v_1 = v_2$ and that the currents through $R_{in}$ and $R_f$ are equal. Since $v_1 = v_2$ and $v_2 = v_{in}$, we can state the following relationship:

> *Don't Forget:*
> $v_1$ is the voltage at the inverting input of the op-amp, and $v_2$ is the voltage at the noninverting input.

$$v_1 = v_{in} \qquad (15.14)$$

Note that this is different from the relationship of $v_1 = 0$ V stated for the inverting amplifier with a grounded (+) input. Once again, we have $v_{in}$ measured across $R_{in}$, and $i_{in}$ can be found as

$$i_{in} = \frac{v_{in}}{R_{in}}$$

or

$$v_{in} = i_{in}R_{in}$$

Since the voltage across $R_f$ is equal to the difference between $v_{in}$ and $v_{out}$ and $i_f = i_{in}$, we can state that the current through the resistor is found as

$$i_f = \frac{v_{out} - v_{in}}{R_f} \qquad (15.15)$$

and

$$v_{out} = i_{in}R_f + v_{in}$$

Now, we can use the established equations to define $A_{CL}$ as

$$A_{CL} = \frac{v_{out}}{v_{in}} = \frac{i_{in}R_f + i_{in}R_{in}}{i_{in}R_{in}} = \frac{i_{in}R_f}{i_{in}R_{in}} + \frac{i_{in}R_{in}}{i_{in}R_{in}}$$

or

$$A_{CL} = \frac{R_f}{R_{in}} + 1 \qquad (15.16)$$

where $A_{CL}$ is the *closed-loop* voltage gain of the amplifier. Thus, the voltage gain of a noninverting amplifier is always greater than the gain of a similar inverting amplifier by a value of 1. If an inverting amplifier has a voltage gain of 150, a similar noninverting amplifier has a voltage gain of 151.

### 15.5.1 Amplifier Input and Output Impedance

Since the input signal is applied directly to the op-amp, the noninverting amplifier has extremely high input impedance. In fact, the presence of the feedback network causes the amplifier input impedance to be even greater than the input impedance of the op-amp in most cases. This point is discussed in detail later in this chapter.

The output impedance of the noninverting amplifier is always less than (or approximately equal to) the output impedance of the op-amp, as is the case with the inverting amplifier.

The extremely high input impedance and extremely low output impedance of the noninverting amplifier makes the circuit very useful as a *buffer*. You may recall that buffers like the *emitter follower* and the *source follower* can be used to match a voltage source to a low-impedance load. The noninverting amplifier can be used for the same purpose. The primary difference is that the noninverting amplifier is capable of high voltage gain values, while the emitter follower and the source follower have $A_v$ values that are less than 1.

### 15.5.2 Noninverting Amplifier Analysis

The complete analysis of the noninverting amplifier is nearly identical to that of the inverting amplifier. The values of $Z_{out}$, CMRR, and $f_{max}$ are found using the same equations we used for the inverting amplifier. The values of $A_{CL}$ and $Z_{in}$ are found using the equations and principles established in this section. The complete analysis of the noninverting amplifier is demonstrated in Example 15.8.

#### EXAMPLE 15.8

Perform the complete analysis of the noninverting amplifier shown in Figure 15.32.

*Solution:* The closed-loop voltage gain ($A_{CL}$) of the circuit is found as

$$A_{CL} = \frac{R_f}{R_{in}} + 1 = \frac{100\ k\Omega}{10\ k\Omega} + 1 = 11$$

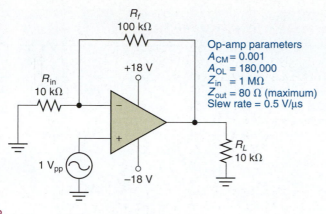

**FIGURE 15.32**

The circuit values of $Z_{in}$ and $Z_{out}$ are determined by the op-amp ratings. Using the ratings shown in the figure, we can assume the amplifier has values of $Z_{in} \geq$ 1 MΩ and $Z_{out} < 80$ Ω.

The CMRR for the circuit is found as

$$\text{CMRR} = \frac{A_{CL}}{A_{CM}} = \frac{11}{0.001} = \textbf{11,000}$$

To determine the value of $f_{max}$, we need to calculate the peak output voltage for the amplifier. The peak-to-peak output voltage is found as

$$v_{out} = A_{CL}v_{in} = (11)(1 \text{ V}_{PP}) = \textbf{11 V}_{PP}$$

The peak output voltage is one-half of this value, 5.5 $V_{pk}$. Using this value and the slew rate shown in the figure, the value of $f_{max}$ is found as

$$f_{max} = \frac{\text{slew rate}}{2\pi V_{pk}} = \frac{0.5 \text{ V/}\mu\text{s}}{34.56 \text{ V}} = \frac{500 \text{ kHz}}{34.56} = \textbf{14.47 kHz}$$

**PRACTICE PROBLEM 15.8**
The circuit described in Practice Problem 15.7 is wired as a noninverting amplifier. Perform a complete analysis on the new circuit.

The circuit values and op-amp parameters used in Example 15.8 are identical to those used for the inverting amplifier in Example 15.7. The results found in the two examples are summarized below.    ◀ **OBJECTIVE 11**

| Value | Inverting Amplifier | Noninverting Amplifier |
|---|---|---|
| $A_{CL}$ | 10 | 11 |
| $Z_{in}$ | 1 kΩ | 1 MΩ (minimum) |
| $Z_{out}$ | 80 Ω (maximum) | 80 Ω (maximum) |
| CMRR | 10,000 | 11,000 |
| $f_{max}$ | 15.9 kHz | 14.47 kHz |

As you can see, the noninverting amplifier has slightly greater values of $A_{CL}$ and CMRR. It also has a much greater value of input impedance. At the same time, the inverting amplifier has a slightly greater maximum operating frequency. This is due to the higher peak output voltage of the comparable noninverting amplifier. Figure 15.33 summarizes the operation of the noninverting and inverting amplifiers.

## Inverting and Noninverting Amplifiers

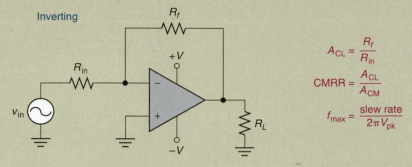

Inverting

$$A_{CL} = \frac{R_f}{R_{in}}$$

$$CMRR = \frac{A_{CL}}{A_{CM}}$$

$$f_{max} = \frac{\text{slew rate}}{2\pi V_{pk}}$$

Circuit characteristics:

- A 180° voltage phase shift is produced by the circuit.

- Input impedance ($Z_{in}$) is approximately equal to the value of $R_{in}$. Output impedance ($Z_{out}$) is lower than the $Z_{out}$ rating of the op-amp.

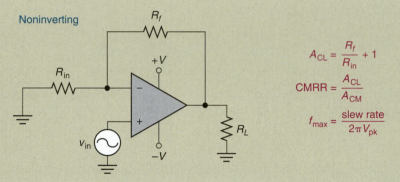

Noninverting

$$A_{CL} = \frac{R_f}{R_{in}} + 1$$

$$CMRR = \frac{A_{CL}}{A_{CM}}$$

$$f_{max} = \frac{\text{slew rate}}{2\pi V_{pk}}$$

Circuit characteristics:

- The input and output signals are in phase.

- Input impedance ($Z_{in}$) is greater than (or equal to) the $Z_{in}$ rating of the op-amp. Output impedance ($Z_{out}$) is lower than the $Z_{out}$ rating of the op-amp.

FIGURE 15.33

### 15.5.3   The Voltage Follower

**Voltage follower**

The op-amp counterpart of the emitter follower and the source follower.

If we remove $R_{in}$ and $R_f$ from the noninverting amplifier and short the output of the amplifier to the inverting input, we have a **voltage follower**. This circuit, which is the op-amp counterpart of the emitter follower and source follower, is shown in Figure 15.34. You may recall that the characteristics of the emitter and source followers are as follows:

1. High $Z_{in}$ and low $Z_{out}$
2. $A_v$ that is less than 1
3. Input and output signals that are in phase

Characteristics 1 and 3 are accomplished by using an op-amp in a noninverting circuit configuration. The voltage gain for the voltage follower is calculated as

$$A_{CL} = \frac{R_f}{R_{in}} + 1 = \frac{0\,\Omega}{R_{in}} + 1 = 1$$

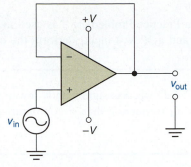

**Lab Reference:** The operation of the voltage follower is demonstrated in Exercise 23.

FIGURE 15.34   Voltage follower.

This result is based on the fact that $R_f = 0\ \Omega$. Simple enough? (*If you think that calculating $A_{CL}$ for this circuit is easy, wait until you try troubleshooting it!*)

The values of $Z_{in}$, $Z_{out}$, and $f_{max}$ for the voltage follower are calculated using the same equations we used for the basic noninverting amplifier. Since $A_{CL} = 1$ for the voltage follower, the circuit CMRR is found using

$$\text{CMRR} = \frac{1}{A_{CM}} \qquad\qquad (15.17)$$

As Example 15.9 demonstrates, the voltage follower is by far the easiest op-amp circuit to analyze.

### EXAMPLE 15.9

Perform the complete analysis of the voltage follower in Figure 15.35.

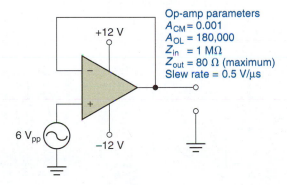

Op-amp parameters
$A_{CM} = 0.001$
$A_{OL} = 180,000$
$Z_{in} = 1\ \text{M}\Omega$
$Z_{out} = 80\ \Omega$ (maximum)
Slew rate = 0.5 V/μs

FIGURE 15.35

**Solution:**   For the voltage follower,

$$A_{CL} = 1$$

The values of $Z_{in}$ and $Z_{out}$ are equal to the rated values for the op-amp, 1 M$\Omega$ and 80 $\Omega$ (maximum), respectively. The CMRR of the circuit is found as

$$\text{CMRR} = \frac{1}{A_{CM}} = \frac{1}{0.001} = 1000$$

Since $A_{CL} = 1$ for the circuit, $v_{out} = v_{in}$. Thus, the peak output voltage is half of 6 $V_{PP}$, or 3 $V_{pk}$, and $f_{max}$ is found as

$$f_{max} = \frac{\text{slew rate}}{2\pi V_{pk}} = \frac{0.5\ \text{V/μs}}{18.85\ \text{V}} = \frac{500\ \text{kHz}}{18.85} = 26.53\ \text{kHz}$$

**PRACTICE PROBLEM 15.9**

The op-amp described in Practice Problem 15.7 is used in a voltage follower with $\pm 14$ V supply voltages and a 12 $V_{PP}$ input. Perform the complete analysis of the amplifier.

---

As you can see, a voltage follower has significantly lower $A_{CL}$ and CMRR values and a higher maximum operating frequency than a comparable "standard" noninverting amplifier.

---

**Section Review ▶**

1. How do you calculate the value of $A_{CL}$ for a noninverting amplifier?
2. How do you determine the values of $Z_{in}$ and $Z_{out}$ for a noninverting amplifier?
3. How do you determine the values of CMRR and $f_{max}$ for a noninverting amplifier?
4. Compare and contrast the inverting amplifier with the noninverting amplifier.
5. Describe the gain and impedance characteristics of the voltage follower.

## 15.6   Troubleshooting Basic Op-Amp Circuits

From the technician's viewpoint, op-amp circuits are a dream come true. Basic op-amp circuits have up to three components that can become faulty, and each fault has distinct symptoms. For example, consider the circuit shown in Figure 15.36. Assuming that the load, source, and dc power supplies are goood, only three components could cause a given fault: $R_{in}$, $R_f$, and the op-amp itself. Let's take a look at what happens when one of the resistors goes bad.

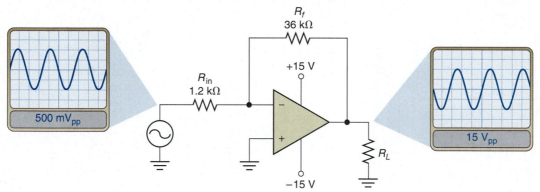

**FIGURE 15.36**

### 15.6.1   $R_f$ Open

**OBJECTIVE 12 ▶**

When the feedback resistor opens, the entire feedback network is effectively removed from the circuit. This causes the gain of the amplifier to increase from $A_{CL}$ to the open-loop voltage gain of the op-amp ($A_{OL}$). If we assume that the op-amp in Figure 15.36 has a value of $A_{OL} = 200,000$, an open feedback resistor would cause the gain of the circuit to increase from 30 to 200,000. The result of this increase in gain is shown in Figure 15.37.

Figure 15.37a shows the normal output from the circuit shown in Figure 15.36. As you can see, the signal is a clear, unclipped sine wave. The waveform shown in Figure 15.37b is the result of $R_f$ opening. The gain of the amplifier has become so great that the output waveform is clipped on both the positive and negative alternations. This output waveform is classic for an open feedback loop.

**Lab Reference:** Typical fault symptoms for the inverting and noninverting amplifiers are simulated in Exercises 22 and 23.

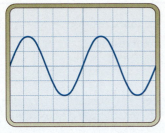

(a) Normal output waveform for the circuit in Figure 15.36

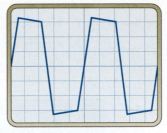

(b) Output waveform for the circuit in Figure 15.36 with $R_1$ open

FIGURE 15.37   Effect of an open $R_f$ on the inverting amplifier output.

### 15.6.2   $R_{in}$ Open

This fault is an interesting one. You would think that an open $R_{in}$ would simply block the input signal, and that the output would therefore go to 0 V. However, this may not be the case. Let's take a look at the circuit you have when $R_{in}$ opens. This circuit is shown in Figure 15.38.

For now, we will assume that the output from this circuit is equal to $+V$ at the moment when $R_i$ opens. Here's what can happen:

1. A *positive* signal is fed back to the inverting input from the output via $R_f$.
2. The positive inverting input causes the output from the amplifier to go negative toward $-V$.
3. A *negative* signal is now fed back to the inverting input from the output.
4. As the inverting input goes negative, the output again goes positive. This takes the amplifier back to step 1.

The process above repeats over and over, *causing the amplifier to produce an output signal with no input signal*. This signal will be in the low-millivolt range. Incidentally, there are circuits that are *designed* to work in the fashion described above. These circuits, called *oscillators*, are discussed in Chapter 18.

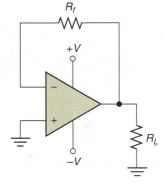

FIGURE 15.38   Equivalent circuit for an inverting amplifier with open $R_{in}$.

### 15.6.3   What Happens if the Op-Amp Is Bad?

The answer to this question depends on what goes wrong with the op-amp. If you refer to Figure 15.1, you will see that there are a lot of components in the op-amp that could go bad.

The best way to determine that an op-amp is faulty is to eliminate all other possible causes of a circuit failure. A faulty op-amp is most likely the cause of a circuit failure when:

1. All the resistors in the amplifier are good.
2. All the supply voltages are at their normal levels.
3. There are no problems with the IC socket.

### 15.6.4   IC Sockets

ICs are sometimes placed into circuits using IC sockets. These sockets are used to allow you to remove and replace the ICs easily.

When IC sockets are used, a problem can develop. Sometimes an IC develops an *oxide layer* on its pins. If this oxidation becomes severe enough, it can cause the circuit to have an erratic output. The circuit may work one minute and not the next. When this happens, remove the IC and clean the pins and the socket with contact cleaner. This should eliminate the problem.

One more point: When you replace an IC, make sure that you put it into the circuit properly. It is very simple to put an IC in backward if you are not paying attention. If you *do* place an IC in a socket backward and apply power to the circuit, odds are that you will have to replace that IC with a new one.

*Component Substitution:*
Most common op-amps are available at any electronics parts store. If an op-amp is bad, you can generally replace it with an equivalent from any manufacturer. Equivalent op-amps have the same package, designator code, and suffix code (see Figure 15.3).

1. What is the primary symptom of an open feedback resistor?
2. What is the primary symptom of an open input resistor?
3. How can you tell when an op-amp is faulty?

## 15.7 Op-Amp Frequency Response

You have already been introduced to some of the frequency considerations involved in dealing with op-amps. As a review, here are some of the major points that were made regarding the frequency response of an op-amp:

1. The *slew rate* of an op-amp is a measure of *how fast its output voltage can change*, measured in volts per microsecond (V/μs).
2. The maximum operating frequency of an op-amp is found as

$$ f_{\text{max}} = \frac{\text{slew rate}}{2\pi V_{\text{pk}}} $$

Thus, the *peak output voltage limits the maximum operating frequency.*
3. When the maximum output frequency of an op-amp is exceeded, the result is a *distorted* output waveform.
4. Increasing the operating frequency of an op-amp beyond a certain point will:
   a. Decrease the maximum output voltage swing.
   b. Decrease the open-loop voltage gain.

In this section, we will take a closer look at the effect that frequency has on the operation of an op-amp.

### 15.7.1 Frequency Versus Gain

dc amplifier

An amplifier that exhibits midband gain when operated at 0 Hz.

The op-amp is classified as a **dc amplifier** because it has no lower cutoff frequency. This means that the gain of the component remains relatively constant down to 0 Hz, as shown in Figure 15.39. As frequency increases from 0 Hz, a point is reached where the voltage gain starts to decrease. At its upper cutoff frequency ($f_{C2}$), the voltage gain of an op-amp is 3 dB lower than its *midband* (maximum) value. Note that the voltage gain of the op-amp continues to decrease at the standard rate of 20 dB per decade above $f_{C2}$. Thus, *increasing the operating frequency decreases the component voltage gain.*

*Don't Forget:*
A Bode plot doesn't show the 3 dB drop at $f_{C2}$.

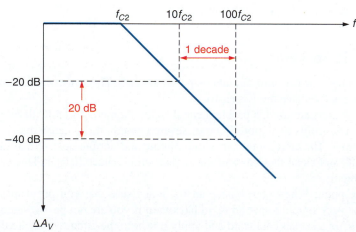

FIGURE 15.39  Op-amp frequency response.

There is another way we can look at this frequency versus gain relationship: *Decreasing the voltage gain of an op-amp increases its maximum operating frequency*. This relationship can be demonstrated using the op-amp frequency-response curve shown in Figure 15.40.

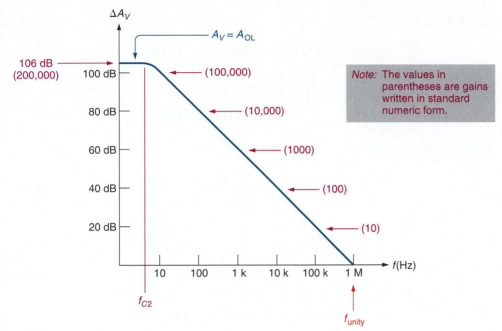

**FIGURE 15.40**   Op-amp frequency response curve.

The maximum voltage gain on the curve is the open-loop voltage gain of the component ($A_{OL}$), approximately 106 dB (200,000). If you want to operate the component so that $A_v = A_{OL(max)}$, you are limited to a maximum operating frequency of approximately 5 Hz. Above this frequency, $A_{OL}$ begins to decrease.

Now, what if we use a feedback path that reduces the closed-loop gain of the amplifier to 60 dB? Figure 15.40 shows that $f_{C2} = 1$ kHz at $A_{CL} = 60$ dB. As you can see, decreasing the gain of the amplifier increases the value of $f_{C2}$ of the device. If fact, each time we decrease $A_{CL}$ by 20 dB, we increase the cutoff frequency by one decade. The maximum operating frequency of an op-amp is measured at $A_{CL} = 0$ dB (unity). This frequency is called the **unity-gain frequency** ($f_{unity}$). For the op-amp represented by the curve in Figure 15.40, $f_{unity}$ is approximately equal to 1 MHz.

Since bandwidth increases as voltage gain decreases, we can make the following statements:

1. The higher the gain of an op-amp, the narrower its bandwidth.
2. The lower the gain of an op-amp, the wider its bandwidth.

Thus, we have a *gain-bandwidth trade-off*. If you want a wide bandwidth, you have to settle for lower gain. If you want high gain, you have to settle for a narrower bandwidth.

### 15.7.2   Gain-Bandwidth Product

The **gain-bandwith product** of an amplifier is *a constant that always equals the unity-gain frequency of its op-amp*. The gain-bandwith product can be used to determine:

1. The maximum possible value of $A_{CL}$ at a specified upper cutoff frequency ($f_{C2}$).
2. The upper cutoff frequency ($f_{C2}$) at a specified value of $A_{CL}$.

The relationship between $A_{CL}, f_{C2}$, and $f_{unity}$ for an amplifier is given as:

$$A_{CL} f_{C2} = f_{unity}$$  (15.18)

**Unity-gain frequency ($f_{unity}$)**
The maximum possible operating frequency for an op-amp, measured at $A_{OL} = 0$ dB.

◀ *OBJECTIVE 13*

**Gain-bandwidth product**
A constant equal to the unity-gain frequency of an op-amp. The product of $A_{CL}$ and bandwidth is always approximately equal to this constant.

At any frequency, the product of $A_{CL}$ and $f_{C2}$ must equal the unity-gain frequency of the op-amp. For example, refer to Figure 15.40. As you were shown, $f_{unity} = 1$ MHz for the op-amp represented by the curve. At 10 Hz,

The values used in these calculations were obtained from the curve in Figure 15.40.

$$A_{CL}f_{C2} = (100{,}000)(10 \text{ Hz}) = 1 \text{ MHz}$$

At 100 Hz,

$$A_{CL}f_{C2} = (10{,}000)(100 \text{ Hz}) = 1 \text{ MHz}$$

At 1 kHz,

$$A_{CL}f_{C2} = (1000)(1 \text{ kHz}) = 1 \text{ MHz}$$

and so on. By rearranging equation (15.18), we can derive the following equations:

$$A_{CL} = \frac{f_{unity}}{f_{C2}} \qquad (15.19)$$

and

$$f_{C2} = \frac{f_{unity}}{A_{CL}} \qquad (15.20)$$

Examples 15.10 and 15.11 demonstrate the use of these equations.

### EXAMPLE 15.10

The LM318 op-amp has a unity-gain frequency of 15 MHz. Determine the bandwidth of the LM318 when $A_{CL} = 500$, and the maximum value of $A_{CL}$ when $f_{C2} = 200$ kHz.

*Solution:* When $A_{CL} = 500$, the value of $f_{C2}$ is found as

$$f_{C2} = \frac{f_{unity}}{A_{CL}} = \frac{15 \text{ MHz}}{5000} = 30 \text{ kHz}$$

Since the op-amp is capable of operating as a dc amplifier,

$$BW = f_{C2} = 30 \text{ kHz}$$

When $f_{C2} = 200$ kHz, the maximum value of $A_{CL}$ is found as

$$A_{CL} = \frac{f_{unity}}{f_{C2}} = \frac{15 \text{ MHz}}{200 \text{ kHz}} = 75$$

### PRACTICE PROBLEM 15.10

An op-amp has a unity-gain frequency of 25 MHz. What is the bandwidth of the device when $A_{CL} = 200$?

The greatest thing about using the gain-bandwidth product is that it allows us to solve various gain-bandwidth problems without the use of a Bode plot. Here's a *component substitution* problem that can be solved using the gain-bandwidth product.

### EXAMPLE 15.11

We need to replace the op-amp in an amplifier that has values of $A_{CL} = 500$ and BW = 80 kHz. Can the LM318 op-amp be used in this circuit?

**Solution:** The LM318 has a value of $f_{unity}$ = 15 MHz. Therefore, any product of $A_{CL}f_{C2}$ must be *less than or equal to* this value for the op-amp to be used. For our application,

$$A_{CL}f_{C2} = (500)(80 \text{ kHz}) = \textbf{40 MHz}$$

Since this value of $A_{CL}f_{C2}$ is greater than the LM318 $f_{unity}$ rating, the LM318 *cannot* be used as a substitute for the original component.

**PRACTICE PROBLEM 15.11**
We need to construct an amplifier with values of $A_{CL}$ = 52 dB and BW = 10 kHz. We have an op-amp with a gain-bandwidth product of 5 MHz. Determine whether the op-amp can be used in this application.

### 15.7.3   Op-Amp Internal Capacitance

If you refer to the internal circuitry of the 741 op-amp shown in Figure 15.1, you'll see that the circuit contains an internal capacitor, $C_1$. This capacitor limits the high-frequency operation of the component.

As frequency increases, the reactance of $C_1$ decreases. As this reactance decreases, the capacitor acts more and more like a short circuit. Eventually, a point is reached where a portion of the op-amp internal circuitry is short circuited, effectively reducing the gain of the amplifier to unity. The frequency at which this occurs is the unity-gain frequency of the device.

### 15.7.4   Determining the Value of $f_{unity}$

The unity-gain frequency of an op-amp can be determined in a number of ways. Some op-amp spec sheets list an $f_{unity}$ rating. Others will simply list a *bandwidth* rating. For example, the KA741 spec sheet (Figure 15.27) lists a bandwidth rating of 1.5 MHz. This is the unity-gain frequency of the device.

When the operating curves for an op-amp are available, the *voltage gain versus operating frequency* curve can be used to determine the value of $f_{unity}$. For example, refer to the frequency-response curve in Figure 15.40. The value of $f_{unity}$ for the op-amp equals the gain-bandwidth product at any point on the curve. Using the values at $f$ = 100 kHz,

$$f_{unity} = (100 \text{ kHz})(10) = \textbf{1 MHz}$$

Note that the value of $A_{OL}$ = 10 (for $f$ = 100 kHz) is indicated on the curve.

The value of $f_{unity}$ for an op-amp can be measured using the following procedure:

1. Set up an inverting amplifier with a closed-loop gain of 100. (We use this value because it is easy to construct with resistor values of $R_f$ = 100 k$\Omega$ and $R_{in}$ = 1 k$\Omega$.)
2. Apply an input signal to the amplifier, and increase the operating frequency until $f_{C2}$ is reached, that is, until the peak-to-peak output voltage drops to 0.707 times the midband value.
3. Take the measured value of $f_{C2}$ and plug it into the following equation:

$$f_{unity} = 100f_{C2}$$

Note that this equation is simply a form of equation (15.18) with an assumed voltage gain of 100.

### 15.7.5   One Final Point

As you can see, the bandwidth calculations for the op-amp are much simpler than those for the BJT and FET amplifiers. This is another of the many advantages that op-amp circuits have over discrete amplifiers.

1. What is the relationship between slew rate and maximum operating frequency?

2. What is the relationship between peak output voltage and maximum operating frequency?

3. What is the relationship between op-amp operating frequency and voltage gain?

4. What is *gain-bandwidth product*?

5. How can the *voltage gain versus operating frequency* curve for an op-amp be used to determine the value of $f_{unity}$ for the device?

6. How can the value of $f_{unity}$ for an op-amp be measured?

## 15.8 Negative Feedback

As you know, *feedback* is a term that describes the process of providing a signal path from the output of a circuit back to its input. For example, take a look at the op-amp circuit shown in Figure 15.41. The *feedback resistor* ($R_f$) used in the amplifier provides a signal path from the output of the op-amp back to its input. The effect that feedback has on the operation of the circuit depends on a number of factors, as you will be shown in this section.

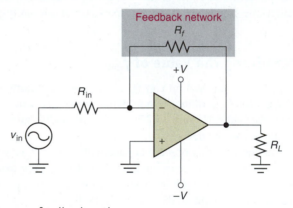

**FIGURE 15.41** Op-amp feedback path.

### 15.8.1 Negative Versus Positive Feedback

**OBJECTIVE 14 ▶**

**Negative feedback**
A type of feedback in which the feedback signal is 180° out of phase with the input signal.

**Positive feedback**
A type of feedback in which the feedback signal is in phase with the input signal.

**Oscillator**
A circuit that converts dc to a sinusoidal (or other) waveform.

Feedback is generally classified as either *negative* or *positive*. **Negative feedback** provides a feedback signal that is 180° out of phase with the input signal. One method of obtaining negative feedback is illustrated in Figure 15.42. In the circuit shown, the amplifier is providing a 180° voltage phase shift, but the feedback network is not. The result is that that total voltage phase shift around the loop is 180°, and the feedback signal is out of phase with the input signal. The same result can be achieved by using an amplifier with a 0° phase shift and a feedback network with a 180° phase shift.

**Positive feedback** provides a feedback signal that is in phase with the circuit input. One method of obtaining positive feedback is represented in Figure 15.43. In this case, the amplifier and the feedback network *each* introduce a 180° voltage phase shift into the loop. This results in a total voltage phase shift of 360° (or 0°), and the feedback signal is in phase with the circuit input. The same result can be achieved using an amplifier and feedback network that each introduce a 0° phase shift. In either case, the feedback signal and input signals are in phase.

Positive feedback is used in a special type of amplifier called an **oscillator**. An oscillator is a circuit that converts dc to a sinusoidal (or other) output waveform. We will discuss the operation and applications of oscillators in Chapter 18. In this section, we concentrate on the effects of negative feedback on amplifier operation.

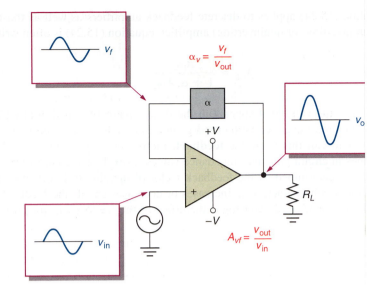

**FIGURE 15.46**

Written for an

Equation (1
for an inverting
tions establishe

Once the value
circuit. The fee
be shown later
the value of the

**EXAMPLE**

Calculate th
Figure 15.4

**Solution:**

loop voltag

Now, using
found as

Finally, the

**PRACTICE**

An *inverting*
amp has a r

We will treat the feedback amplifier as a combination of two distinct circu
and a feedback network. Each of these circuits serves a specific purpose. Th
vides voltage gain, defined as

$$A_{OL} = \frac{v_{out}}{v_{diff}}$$

where $v_{diff}$ = the difference between the voltages at the inverting and noni
to the op-amp.

Remember that $A_{OL}$ is an op-amp parameter and does not necessarily eq
gain of the feedback amplifier.

The feedback network is represented as an *attenuation*. **Attenuation**
*tion* in the amplitude of a signal. Thus, the feedback network can be view
value of $A_v < 1$. Note that the Greek letter $\alpha$ is commonly used to repre
and/or power *loss*. (Don't confuse this use of $\alpha$ with the BJT current rat
Chapter 6.)

The **attenuation factor ($\alpha_v$)** of the feedback network is found as

$$\alpha_v = \frac{v_f}{v_{out}}$$

Note that $v_f$ is always less than $v_{out}$. For this reason, $\alpha_v$ is always less than 1
As with any amplifier, the *effective* voltage gain of a feedback amplifier

$$A_{vf} = \frac{v_{out}}{v_{in}}$$

In Appendix D, we use equations (15.21), (15.22), and (15.23) to deriv
relationship:

$$A_{vf} = \frac{A_v}{1 + \alpha_v A_v}$$

where $A_{vf}$ = the *effective* gain of the voltage feedback amplifier
$A_v$ = the *open-loop* voltage gain of the amplifier (that is, the voltag
amplifier would exhibit if there were no feedback path presen

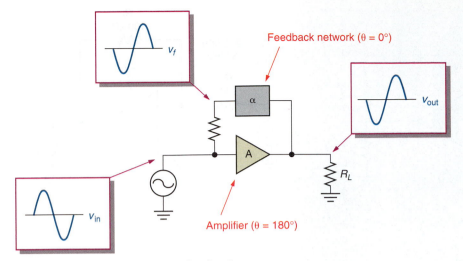

**FIGURE 15.42** Obtaining negative feedback.

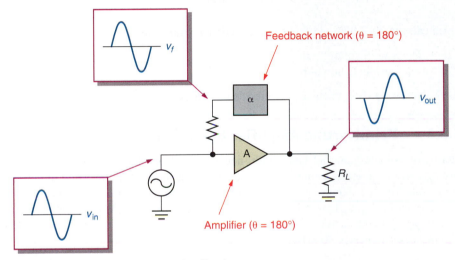

**FIGURE 15.43** Obtaining positive feedback.

### 15.8.2 Inverting Amplifier Operation

The feedback resistor ($R_f$) in an inverting amplifier forms a *negative feedback* network. When
a negative feedback network is connected to an op-amp, it results in a *decrease in voltage
gain* and an *increase in operating bandwidth*. This point is illustrated in Figure 15.44.

Figure 15.44a shows an inverting amplifier with no feedback path. Without the feedback path, the voltage gain of the circuit is equal to the open-loop voltage gain ($A_{OL}$) of
the op-amp. In this case, $A_V = A_{OL} = 200,000$. With a unity-gain frequency rating of
3 MHz, the bandwidth of the circuit is found as

$$BW = \frac{f_{unity}}{A_{OL}} = \frac{3 \text{ MHz}}{200,000} = 15 \text{ Hz}$$

When a negative feedback path is added to the op-amp (as shown in Figure 15.44b),
the voltage gain and bandwidth of the circuit are found as

$$A_{CL} = \frac{R_f}{R_{in}} = \frac{100 \text{ k}\Omega}{1 \text{ k}\Omega} = 100 \quad (40 \text{ dB})$$

◀ **OBJECTIVE 15**

The discussion here is
consistent with the earlier
discussion on op-amp
frequency response. It is
intended to provide an alternate
approach to the topic.

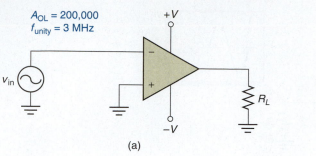

$A_{OL} = 200,000$
$f_{unity} = 3\text{ MHz}$

**FIGURE 15.44**

and

**BW** =

As you can see, the addition of a r

1. A *decrease* in voltage gain.
2. An *increase* in bandwidth.

This is always the case when nega

### 15.8.3 Noninverting A

In terms of voltage gain and ban
characteristics of a noninverting a
demonstrated by the circuit calcu
circuit shown in Figure 15.45a ha
feedback path is added, the circui
width increases to nearly 30 kHz.

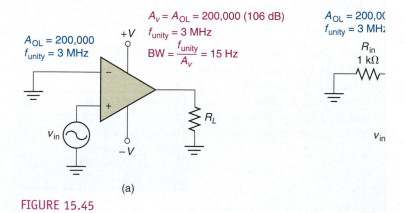

$A_{OL} = 200,000$
$f_{unity} = 3\text{ MHz}$

$A_v = A_{OL} = 200,000\ (106\text{ dB})$
$f_{unity} = 3\text{ MHz}$
$BW = \dfrac{f_{unity}}{A_v} = 15\text{ Hz}$

$A_{OL} = 200,0($
$f_{unity} = 3\text{ MH}$

$R_{in}$
$1\text{ k}\Omega$

$v_{in}$

(a)

**FIGURE 15.45**

**OBJECTIVE 16 ▶**

### 15.8.4 Mathematical A

Now that you have seen how neg
bandwidth of an op-amp, it is time
sion, we will refer to the circuit sho

---

**Feedb**
A valu
impeda
given f

---

## 15.8.5 The Effects of Negative Feedback on Circuit Impedance Values

**OBJECTIVE 17 ▶**  The input and output impedance values for inverting and noninverting amplifiers are calculated as shown in Figure 15.49. For the inverting amplifier, the presence of the *virtual ground* at the inverting input results in the amplifier input impedance being approximately equal to the value of the input resistor ($R_{in}$). This point was discussed earlier in the chapter.

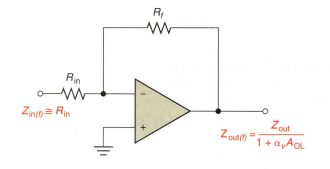

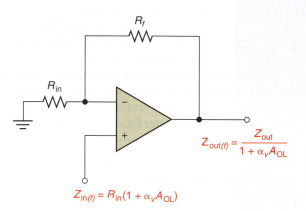

**FIGURE 15.49**  Amplifier input and output impedance.

For the noninverting amplifier, the presence of the feedback signal at the inverting input reduces the input differential voltage ($v_{diff}$) and, therefore, the amount of current that the amplifier draws from the source. Since

$$Z_{in} = \frac{v_s}{i_s}$$

the reduction in source current causes an effective *increase* in amplifier input impedance. The amount by which the input impedance is increased is equal to the feedback factor of the circuit. By formula,

$$Z_{in(f)} = Z_{in}(1 + \alpha_v A_{OL}) \qquad (15.29)$$

where $Z_{in(f)}$ = the input impedance to the noninverting amplifier
$Z_{in}$ = the op-amp input impedance

As the following example demonstrates, the input impedance to a noninverting amplifier can be significantly higher than that of its op-amp.

#### EXAMPLE 15.14

Refer to the noninverting amplifier in Figure 15.49b. Assume that the op-amp has ratings of $Z_{in} = 5\text{ M}\Omega$ and $A_{OL} = 180,000$. If $R_{in} = 1.2\text{ k}\Omega$ and $R_f = 180\text{ k}\Omega$, what is the value of the amplifier input impedance?

*Solution:*  We have to start our calculations by finding the value of the attenuation factor. Using the method established earlier in the text, the closed loop voltage gain of the circuit is found to be 151. Thus,

$$\alpha_v = \frac{1}{A_{CL}} = \frac{1}{151} = 0.0066$$

Now, using $\alpha_v = 0.0066$ and $A_{OL} = 180{,}000$, the value of the feedback factor is found as

$$1 + \alpha_v A_{OL} = 1 + (0.0066)(180{,}000) = \mathbf{1189}$$

Finally, the amplifier input impedance is found as

$$Z_{in(f)} = Z_{in}(1 + \alpha_v A_{OL}) = (5 \text{ M}\Omega)(1189) = \mathbf{5.95 \text{ G}\Omega}$$

### PRACTICE PROBLEM 15.14

An op-amp has the following ratings: $Z_{in} = 2 \text{ M}\Omega$ and $A_{OL} = 200{,}000$. The op-amp is used in a noninverting amplifier with values of $R_f = 220 \text{ k}\Omega$ and $R_{in} = 1 \text{ k}\Omega$. Calculate the value of the amplifier input impedance.

As you can see, the addition of a feedback network causes the noninverting amplifier to have extremely high input impedance. As a result, the amplifier input impedance presents almost no load on its source circuit. This is one of the benefits of using negative feedback.

Just as negative feedback effectively increases the input impedance of an op-amp, *it also decreases the op-amp's output impedance*. The amount by which $Z_{out}$ is reduced is also determined by the feedback factor of the circuit. By formula,

$$Z_{out(f)} = \frac{Z_{out}}{1 + \alpha_v A_{OL}} \tag{15.30}$$

where $Z_{out(f)}$ = the outut impedance of the amplifier
$\quad\quad Z_{out}$ = the op-amp output impedance

The following example demonstrates the effect of negative feedback on the output impedance of a noninverting amplifier.

### EXAMPLE 15.15

Refer to Example 15.14. If the op-amp has a rating of $Z_{out} = 80 \ \Omega$, what is the value of the amplifier output impedance?

*Solution:* In Example 15.14, the feedback factor was found to have a value of 1189. Using this value and the rated output impedance of the op-amp, the output impedance of the noninverting amplifier is found as

$$Z_{out(f)} = \frac{Z_{out}}{1 + \alpha_v A_{OL}} = \frac{80 \ \Omega}{1189} = \mathbf{67.3 \text{ m}\Omega}$$

### PRACTICE PROBLEM 15.15

Refer to Practice Problem 15.14. If the op-amp has a rating of $Z_{out} = 75 \ \Omega$, what is the output impedance of the amplifier?

As you can see, the feedback network has greatly reduced the effective output impedance of the op-amp. Again, this is an added benefit of using negative feedback. With the lower output impedance, the circuit is much better suited to driving low-impedance loads. If you refer to Figure 15.49, you'll see that the use of negative feedback has the same effect on the output impedance of an *inverting* amplifier as it has on the output impedance of a noninverting amplifier.

1. How does negative voltage feedback reduce the effective voltage gain of an amplifier?
2. List and describe the various types of commonly used feedback.
3. What is *attenuation*?
4. What is the *attenuation factor* of a feedback network?
5. What is the *feedback factor* of a negative feedback amplifier?
6. What effect does negative feedback have on the input impedance of an inverting amplifier? A noninverting amplifier?
7. What effect does negative feedback have on the output impedance of an inverting or a noninverting amplifier?

## CHAPTER SUMMARY

Here is a summary of the major points made in this chapter:

1. *Discrete components* are devices housed in individual packages; that is, each package holds one component.
2. An *integrated circuit* (IC) is a single package that contains any number of active and/or passive components, all constructed on a single piece of semiconductor material.
   a. ICs range in complexity from simple circuits containing a few active and/or passive components to complex circuits containing hundreds of thousands of components.
   b. The more complex the internal circuitry of an IC, the more complex its function.
3. The *operational amplifier*, or *op-amp*, is a high-gain dc amplifier that has high input impedance and low output impedance.
4. The internal circuitry, schematic diagram, and pin layout for the 741 op-amp (a common component) are shown in Figure 15.1. The component has:
   a. Two signal inputs: the *inverting* input and the *noninverting* input.
   b. Two power supply inputs: the $+V$ input and $-V$ input.
   c. Two *offset null* input pins.
   d. One output pin.
5. Op-amps are identified using a seven-character code printed on the case (see Figure 15.3).
   a. The prefix identifies the manufacturer (see Table 15.1).
   b. The designator identifies the component type and operating temperature range (see Table 15.2).
   c. The suffix identifies the package type (see Table 15.3).
   Several op-amp packages are illustrated in Figure 15.4.
6. The input stage of the op-amp is a *differential amplifier*.
   a. This type of circuit amplifies the *difference between* two input voltages, $V_1$ and $V_2$.
   b. Any difference between $V_1$ and $V_2$ shows up as a *difference of potential* ($V_{diff}$) across the input terminals. This voltage ($V_{diff}$) is amplified by the circuit.
7. The maximum possible voltage gain provided by an op-amp is referred to as its *open-loop voltage gain* ($A_{OL}$).
   a. The term *open loop* indicates that there is no feedback path from the output to the op-amp input.
   b. $A_{OL}$ typically has a value of 10,000 or more.
   c. Connecting a feedback path to an op-amp reduces its voltage gain.
8. The polarity relationship between $V_1$ and $V_2$ determines whether the op-amp output swings toward $+V$ or $-V$ (see Figure 15.7).
9. The *inverting input* to an op-amp produces a 180° output voltage phase shift when used as a signal input (see Figure 15.8e).
10. The *noninverting input* to an op-amp does not produce an output voltage phase shift (see Figure 15.8f).

11. The supply voltages ($+V$ and $-V$) determine the limits of the output voltage swing.
    a. Ideally, the limits to any output voltage swing equal the values of $+V$ and $-V$.
    b. In practice, the output cannot reach either $+V$ or $-V$ because some voltage is dropped across internal components in the op-amp output circuitry (see Figure 15.1).
    c. The actual limits on the output voltage swing depend on the value of the load resistance and the supply voltages (see Figures 15.10 and 15.11).
12. Many op-amp electrical characteristics describe the operation of its input circuit: the differential amplifier.
    a. The simplest differential amplifier is a two-transistor, two-input circuit (see Figure 15.16).
    b. The responses of the differential amplifier to inputs at its inverting and noninverting inputs are illustrated in Figure 15.17.
13. A differential amplifier has three modes of operation:
    a. *Single-ended* operation involves applying an active signal to one input only. The other input is normally connected directly (or indirectly) to ground. Depending on the input used, the differential amplifier acts as an inverting or noninverting amplifier.
    b. *Differential* operation involves applying two active inputs to the amplifier. The output (in this case) depends on the characteristics of the two waveforms.
    c. *Common-mode* operation occurs when two signals of equal magnitude, frequency, and phase are applied to the amplifier. Ideally, the output from a differential amplifier is 0 V when common-mode signals are applied to the inputs.
14. The *output offset voltage* rating of an op-amp is a voltage (usually in the mV range) that may appear at the output, and it is produced by an imbalance in the differential amplifier. Offset output voltage can be eliminated by:
    a. Using a potentiometer to set a potential across the *offset null* pins (see Figure 15.20a).
    b. Connecting a *compensating resistor* to the noninverting input (see Figure 15.20b).
15. The *input bias current* rating of an op-amp is the average value of quiescent dc biasing current drawn by its signal inputs.
16. Since both signal inputs draw some amount of biasing current, an op-amp will not work if either signal input is left open.
17. *Common-mode signals* are identical signals that appear simultaneously at the two signal inputs of an op-amp.
    a. Common-mode signals are usually undesirable. For example, noise may appear as common-mode signals.
    b. Ideally, an op-amp does not respond to common-mode signals.
18. The *common-mode rejection ratio* (CMRR) of an op-amp is a measure of its ability to "ignore" common-mode signals.
    a. Technically, CMRR is the ratio of differential gain to common-mode gain.
    b. CMRR is normally expressed in dB.
19. The *power supply rejection ratio* of an op-amp indicates how much the output changes as a result of changes in supply voltage.
20. The *output short-circuit current* rating of an op-amp is the maximum possible value of output current measured under shorted load conditions.
21. The *slew rate* of an op-amp is a measure of how fast the output voltage can change in response to a change at either signal input.
    a. Slew rates are typically measured in volts per microsecond (V/$\mu$s).
    b. The slew rate of an op-amp affects its maximum possible operating frequency (see Example 15.5).
22. The operating frequency of an op-amp can be increased (within limits) by reducing its output voltage swing (see Example 15.6).
23. When operated beyond its frequency limit, an op-amp distorts its output waveform (see Figure 15.25).
24. Op-amps typically have high input resistance and low output resistance.

25. The *input voltage range* rating for an op-amp is the maximum differential input that can be applied to the component without risking damage to the differential amplifier.

26. An *inverting amplifier* is an op-amp circuit that produces a 180° voltage phase shift from input to output.

27. The inverting amplifier is the op-amp counterpart of the common-emitter and common-source amplifiers.

28. The *closed-loop voltage gain* ($A_{CL}$) of an inverting amplifier is always lower than the open-loop voltage gain ($A_{OL}$) of the op-amp.

29. The input impedance of an inverting amplifier is approximately equal to the value of its input resistor ($R_{in}$).

30. The output impedance of an inverting amplifier is always lower than the output impedance of the op-amp, because the feedback path is in parallel with the op-amp output.

31. The common-mode rejection ratio (CMRR) of an inverting amplifier is lower than the CMRR of the op-amp.
    a. The CMRR of an op-amp is the ratio of $A_{OL}$ to common-mode gain.
    b. The CMRR of an inverting amplifier is the ratio of $A_{CL}$ to common-mode gain.

32. The primary differences between noninverting and inverting amplifiers are:
    a. The noninverting amplifier has much higher circuit input impedance.
    b. The noninverting amplifier does not produce a 180° voltage phase shift from the amplifier input to its output.

33. The closed-loop voltage gain ($A_{CL}$) of a noninverting amplifier is slightly higher than the value of $A_{CL}$ for a comparable inverting amplifier.
    a. For the inverting amplifier, $A_{CL} = R_f/R_{in}$.
    b. For the noninverting amplifier, $A_{CL} = (R_f/R_{in}) + 1$.

34. The input impedance of the noninverting amplifier is *greater than* the input impedance of the op-amp.

35. The output impedance of a noninverting amplifier is lower than the output impedance of the op-amp.

36. The characteristics of the noninverting amplifier make the circuit useful as a *buffer*.

37. The *voltage follower* is the op-amp counterpart of the emitter follower and source follower.
    a. The voltage follower is a noninverting amplifier that has the output tied directly back to the inverting input (see Figure 15.34).
    b. The voltage follower has a value of $A_{CL} = 1$.

38. The best way to determine that an op-amp is faulty is to eliminate all other possible causes of a circuit failure. A faulty op-amp is the most likely cause of a circuit failure when:
    a. All the resistors in the amplifier are good.
    b. All the supply voltages are at their normal levels.
    c. There are no problems with the IC socket used to hold the op-amp.

39. A *dc amplifier* exhibits midband voltage gain at 0 Hz.

40. As the operating frequency of an op-amp increases beyond its upper cutoff frequency ($f_{C2}$), the $A_{OL}$ of the component decreases at the standard *20 dB per decade* rate.

41. Decreasing the voltage gain of an op-amp (through the use of feedback) increases its maximum operating frequency (see Figure 15.40).

42. The *unity-gain frequency* ($f_{unity}$) of an op-amp is the frequency that corresponds to a voltage gain of 0 dB.
    a. The product of $A_{CL}f_{C2}$ always equals the unity-gain frequency.
    b. The value of $A_{CL}f_{C2}$ is referred to as the *gain-bandwidth product*. The relationship among $A_{CL}, f_{C2}$, and $f_{unity}$ is demonstrated in Examples 15.10 and 15.11.

43. The high-frequency roll-off of an op-amp is caused by the internal capacitance of the component.

44. *Negative feedback* provides a feedback signal that is 180° out of phase with the amplifier input signal (see Figure 15.42).

45. Adding a negative feedback path to an op-amp:
    a. Decreases voltage gain.
    b. Increases amplifier bandwidth.

**46.** *Attenuation* is a reduction in the amplitude of a signal.

**47.** The *attenuation factor* ($\alpha_v$) of a feedback path is the ratio of feedback voltage to amplifier output voltage. The value of $\alpha_v$ is always less than 1.

**48.** The *feedback factor* ($1 + \alpha_v A_{OL}$) of an op-amp circuit is used in most circuit impedance calculations.

**49.** The input impedance of a noninverting amplifier equals the product of the op-amp input impedance and the circuit feedback factor (see Example 15.14).

**50.** The output impedance of an inverting (or noninverting) amplifier equals the op-amp output impedance divided by the feedback factor (see Example 15.15).

**EQUATION SUMMARY**

| Equation Number | Equation | Section Number |
|---|---|---|
| (15.1) | $V_{diff} = V_2 - V_1$ | 15.2 |
| (15.2) | $I_{E1} = I_{E2}$ | 15.3 |
| (15.3) | $I_{E1} = I_{E2} = \dfrac{I_{EE}}{2}$ | 15.3 |
| (15.4) | $I_{EE} = \dfrac{V_{RE}}{R_E}$ | 15.3 |
| (15.5) | $V_{C1} = V_{C2} = V_{CC} - I_C R_C$ | 15.3 |
| (15.6) | $V_{io} = \dfrac{V_{out(offset)}}{A_v}$ | 15.3 |
| (15.7) | $f_{max} = \dfrac{\text{slew rate}}{2\pi V_{pk}}$ | 15.3 |
| (15.8) | $v_{out} = i_f R_f$ | 15.4 |
| (15.9) | $v_{in} = i_{in} R_{in}$ | 15.4 |
| (15.10) | $v_{out} = i_{in} R_f$ | 15.4 |
| (15.11) | $A_v = \dfrac{R_f}{R_{in}}$ | 15.4 |
| (15.12) | $Z_{in} \cong R_{in}$ | 15.4 |
| (15.13) | $\text{CMRR} = \dfrac{A_{CL}}{A_{CM}}$ | 15.4 |
| (15.14) | $v_1 = v_{in}$ | 15.5 |
| (15.15) | $i_f = \dfrac{v_{out} - v_{in}}{R_f}$ | 15.5 |
| (15.16) | $A_{CL} = \dfrac{R_f}{R_{in}} + 1$ | 15.5 |
| (15.17) | $\text{CMRR} = \dfrac{1}{A_{CM}}$ | 15.5 |
| (15.18) | $A_{CL} f_{C2} = f_{unity}$ | 15.7 |
| (15.19) | $A_{CL} = \dfrac{f_{unity}}{f_{C2}}$ | 15.7 |

## KEY TERMS

attenuation ($\alpha$) 639
attenuation factor ($\alpha_v$) 639
closed-loop voltage gain
 ($A_{\text{CL}}$) 622
common-mode gain
 ($A_{\text{CM}}$) 623
common-mode rejection
 ratio (CMRR) 613
common-mode
 signals 613
compensating resistor 612
current feedback 602
dc amplifier 632
designator code 600
differential amplifier 601
discrete components 598
feedback factor
 ($1 + \alpha_v A_{\text{OL}}$) 640

gain-bandwidth
 product 633
input bias current 613
input offset current 612
input offset voltage
 ($V_{io}$) 611
input voltage range 618
integrated circuit (IC) 598
inverting amplifier 621
inverting input 603
large-signal voltage
 gain 618
negative feedback 636
noninverting amplifier 625
noninverting input 603
offset null 612
open-loop voltage gain
 ($A_{\text{OL}}$) 602

operational amplifier
 (op-amp) 599
oscillator 636
output offset voltage
 ($V_{\text{out(offset)}}$) 611
output short-circuit
 current 614
positive feedback 636
power consumption 618
power supply rejection
 ratio (PSRR) 614
rail 620
slew rate 615
supply current 618
unity-gain frequency
 ($f_{\text{unity}}$) 633
voltage follower 628

## PRACTICE PROBLEMS

### Section 15.2

**1.** Determine the output polarity for each of the op-amps in Figure 15.50.

**2.** Determine the output polarity for each of the op-amps in Figure 15.51.

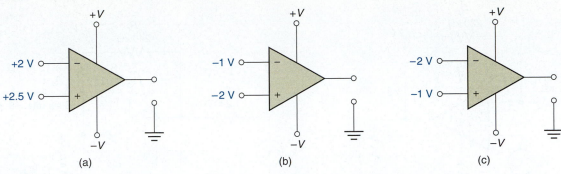

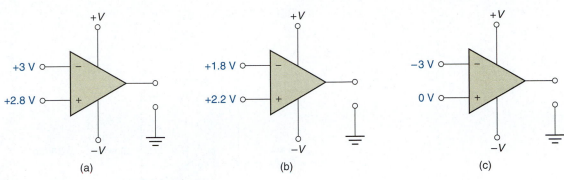

**FIGURE 15.50**

**FIGURE 15.51**

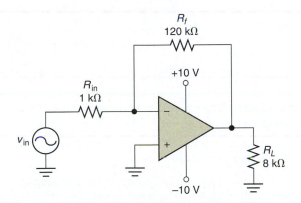

**FIGURE 15.52**

3. Determine the maximum peak-to-peak output voltage for the amplifier in Figure 15.52.

4. Determine the maximum peak-to-peak output voltage for the amplifier in Figure 15.53.

5. The amplifier in Problem 3 has a voltage gain of 120. Determine the maximum allowable peak-to-peak input voltage for the circuit.

6. The amplifier in Problem 4 has a voltage gain of 220. Determine the maximum allowable peak-to-peak input voltage for the circuit.

7. Determine the maximum allowable peak-to-peak input voltage for the amplifier in Figure 15.54.

8. Determine the maximum allowable peak-to-peak input voltage for the amplifier in Figure 15.55.

**Section 15.3**

9. The amplifier in Figure 15.54 has an output offset voltage of 2.4 V. Determine the input offset voltage for the circuit.

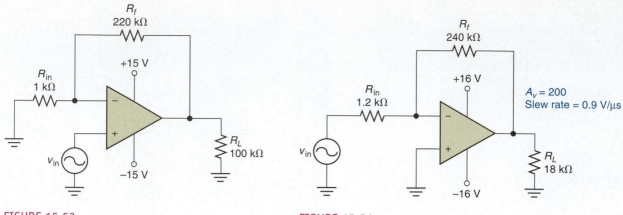

FIGURE 15.53

FIGURE 15.54

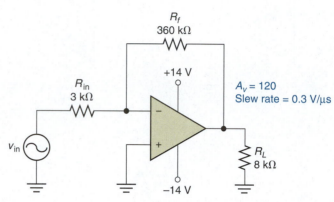

FIGURE 15.55

**10.** The amplifier in Figure 15.55 has an output offset voltage of 960 mV. Determine the input offset voltage for the circuit.

**11.** Determine the maximum operating frequency for the amplifier in Figure 15.54. Assume that the circuit has a 10 mV$_{PP}$ input signal.

**12.** Determine the maximum operating frequency for the amplifier in Figure 15.55. Assume that the circuit has a 40 mV$_{PP}$ input signal.

**13.** Determine the maximum operating frequency for the amplifier in Figure 15.56a.

**14.** Determine the maximum operating frequency for the amplifier in Figure 15.56b.

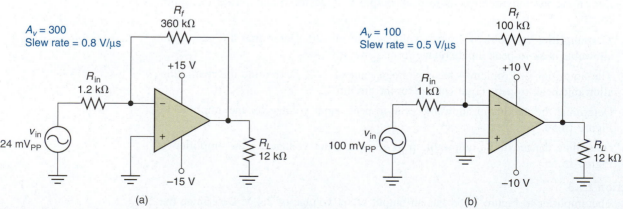

FIGURE 15.56

## Section 15.4

**15.** Perform a complete analysis of the amplifier in Figure 15.57.

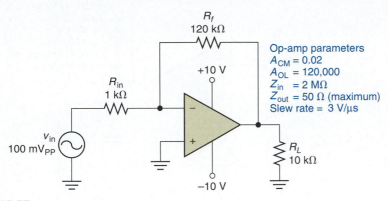

Op-amp parameters
$A_{CM} = 0.02$
$A_{OL} = 120,000$
$Z_{in} = 2\ M\Omega$
$Z_{out} = 50\ \Omega$ (maximum)
Slew rate $= 3\ V/\mu s$

FIGURE 15.57

**16.** Perform a complete analysis of the amplifier in Figure 15.58.

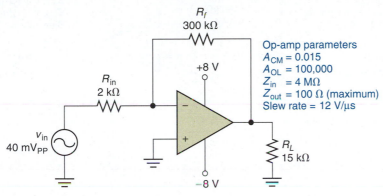

Op-amp parameters
$A_{CM} = 0.015$
$A_{OL} = 100,000$
$Z_{in} = 4\ M\Omega$
$Z_{out} = 100\ \Omega$ (maximum)
Slew rate $= 12\ V/\mu s$

FIGURE 15.58

**17.** Perform a complete analysis of the amplifier in Figure 15.59.

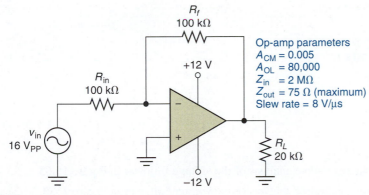

Op-amp parameters
$A_{CM} = 0.005$
$A_{OL} = 80,000$
$Z_{in} = 2\ M\Omega$
$Z_{out} = 75\ \Omega$ (maximum)
Slew rate $= 8\ V/\mu s$

FIGURE 15.59

**18.** Perform a complete analysis of the amplifier in Figure 15.60.
**19.** Perform a complete analysis of the amplifier in Figure 15.61.
**20.** Perform a complete analysis of the amplifier in Figure 15.62.
**21.** Perform a complete analysis of the amplifier in Figure 15.63.
**22.** Perform a complete analysis of the amplifier in Figure 15.64.

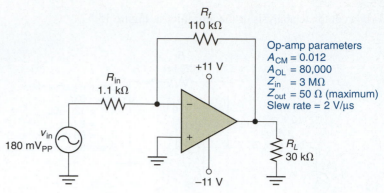

FIGURE 15.60

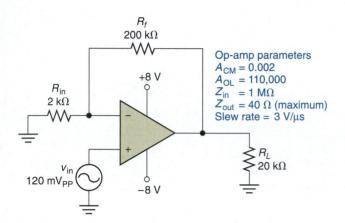

FIGURE 15.61

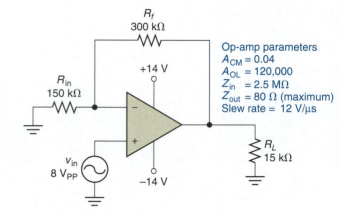

FIGURE 15.62

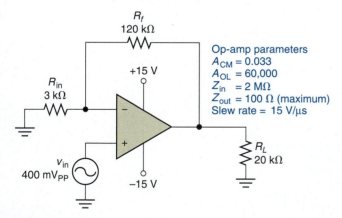

FIGURE 15.63

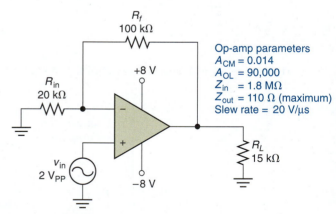

FIGURE 15.64

**23.** Perform a complete analysis of the voltage follower in Figure 15.65.

**24.** Perform a complete analysis of the voltage follower in Figure 15.66.

**Section 15.6**

**25.** An op-amp has a gain-bandwidth product of 12 MHz. Determine the bandwidth of the device when $A_{CL} = 400$.

**26.** An op-amp has a gain-bandwidth product of 14 MHz. Determine the bandwidth of the device when $A_{CL} = 320$.

**27.** An op-amp has a gain-bandwidth product of 25 MHz. Determine the bandwidth of the device when $A_{CL} = 42$ dB.

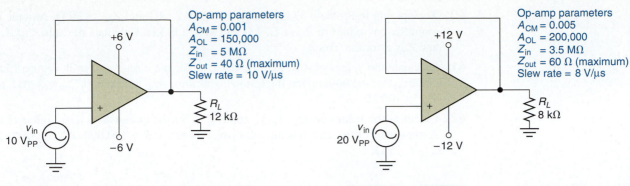

**Op-amp parameters**
$A_{CM} = 0.001$
$A_{OL} = 150,000$
$Z_{in} = 5\ M\Omega$
$Z_{out} = 40\ \Omega$ (maximum)
Slew rate = 10 V/μs

+6 V
$V_{in}$
10 V$_{PP}$
−6 V
$R_L$
12 kΩ

FIGURE 15.65

**Op-amp parameters**
$A_{CM} = 0.005$
$A_{OL} = 200,000$
$Z_{in} = 3.5\ M\Omega$
$Z_{out} = 60\ \Omega$ (maximum)
Slew rate = 8 V/μs

+12 V
$V_{in}$
20 V$_{PP}$
−12 V
$R_L$
8 kΩ

FIGURE 15.66

**28.** An op-amp has a gain-bandwidth product of 1 MHz. Determine the bandwidth of the device when $A_{CL} = 20$ dB.

**29.** We need to construct an amplifier that has values of $A_{CL} = 200$ and $f_{C2} = 120$ kHz. Can we use an op-amp with $f_{unity} = 28$ MHz for this application?

**30.** We need to construct an amplifier that has values of $A_{CL} = 24$ dB and $f_{C2} = 40$ kHz. Can we use an op-amp with $f_{unity} = 1$ MHz for this application?

**31.** An op-amp circuit with $A_{CL} = 120$ has a measured value of $f_{C2} = 100$ kHz. What is the gain-bandwidth product of the op-amp?

**32.** An op-amp circuit with $A_{CL} = 300$ has a measured value of $f_{C2} = 88$ kHz. What is the gain-bandwidth product of the op-amp?

**33.** The circuit shown in Figure 15.67 has a measured $f_{C2}$ of 250 kHz. What is the value of $f_{unity}$ for the op-amp?

**34.** The circuit shown in Figure 15.68 has a measured $f_{C2}$ of 100 kHz. What is the value of $f_{unity}$ for the op-amp?

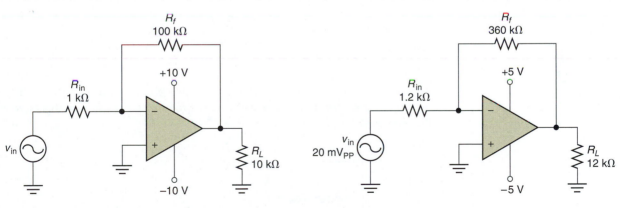

$R_f$
100 kΩ
+10 V
$R_{in}$
1 kΩ
$V_{in}$
$R_L$
10 kΩ
−10 V

FIGURE 15.67

$R_f$
360 kΩ
+5 V
$R_{in}$
1.2 kΩ
$V_{in}$
20 mV$_{PP}$
$R_L$
12 kΩ
−5 V

FIGURE 15.68

**35.** An inverting amplifier has values of $A_{OL} = 1000$ and $\alpha_v = 0.22$. Determine the value of $A_{CL}$ for the circuit.

**36.** An inverting amplifier has values of $A_{OL} = 588$ and $\alpha_v = 0.092$. Determine the value of $A_{CL}$ for the circuit.

**37.** An inverting amplifier has values of $A_{OL} = 150,000$ and $\alpha_v = 0.008$. Determine the value of $A_{CL}$ for the circuit.

**38.** An inverting amplifier has values of $A_{OL} = 200,000$ and $\alpha_v = 0.0015$. Determine the value of $A_{CL}$ for the circuit.

**39.** The amplifier in Problem 35 has values of $Z_{in} = 48$ kΩ and $Z_{out} = 220$ Ω. Calculate the values of $Z_{in(f)}$ and $Z_{out(f)}$ for the circuit.

**40.** The op-amp in Problem 37 has values of $Z_{in} = 2$ MΩ and $Z_{out} = 80$ Ω. Assume the amplifier has values of $R_{in} = 1.2$ kΩ and $R_f = 150$ kΩ. Calculate the values of $Z_{in(f)}$ and $Z_{out(f)}$ for the circuit.

**41.** Calculate the values of $A_{CL}$, $Z_{in(f)}$, and $Z_{out(f)}$ for the circuit shown in Figure 15.67. Assume that the op-amp has the following ratings: $A_{OL} = 200,000$, $Z_{in} = 5$ MΩ, and $Z_{out} = 100$ Ω.

**42.** Calculate the values of $A_{CL}$, $Z_{in(f)}$, and $Z_{out(f)}$ for the circuit shown in Figure 15.68. Assume that the op-amp has the following ratings: $A_{OL} = 180,000$, $Z_{in} = 3$ MΩ, and $Z_{out} = 75$ Ω.

**TROUBLESHOOTING PRACTICE PROBLEMS**

**43.** The circuit shown in Figure 15.69 has the waveforms shown. Discuss the possible cause(s) of the problem.

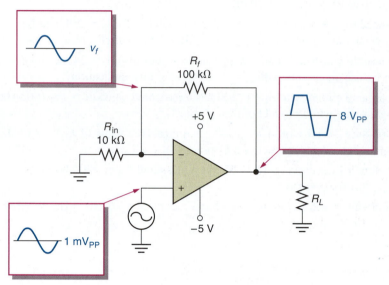

**FIGURE 15.69**

**44.** The circuit shown in Figure 15.70 has the waveforms shown. Determine the cause of the problem. (*Hint*: The measured value of $R_f$ is exactly as shown; that is, $R_f$ is not the problem.)

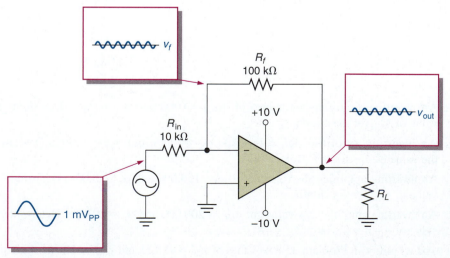

**FIGURE 15.70**

**45.** The circuit shown in Figure 15.71 has the following readings: TP-1 = +10 mV, TP-2 = +20 mV, TP-3 = +120 mV, TP-4 = 0 V, and TP-5 = −240 mV. Determine the possible cause(s) of the problem.

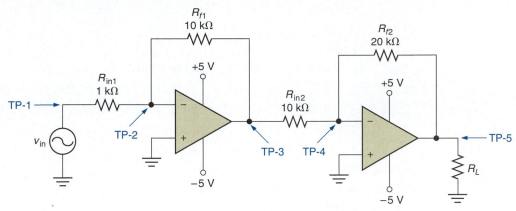

**FIGURE 15.71**

**46.** The circuit in Figure 15.71 has the following readings: TP-1 = −5 mV, TP-2 = 0 V, TP-3 = +4 V, TP-4 = 0 V, and TP-5 = −4 V. Determine the possible cause(s) of the problem.

**47.** The circuit in Figure 15.71 has the following readings: TP-1 = +2 mV, TP-2 = +1.8 mV, TP-3 = +4 V, TP-4 = 0 V, and TP-5 = 0 V. Determine the possible cause(s) of the problem.

**48.** The circuit in Figure 15.71 has the following readings: TP-1 = −1 V, TP-2 = 0 V, TP-3 = 0 V, TP-4 = 0 V, and TP-5 = 0 V. Determine the possible cause(s) of the problem.

**PUSHING THE ENVELOPE**

**49.** The op-amp in Figure 15.72 has the input and output voltages shown. Determine the CMRR of the device.

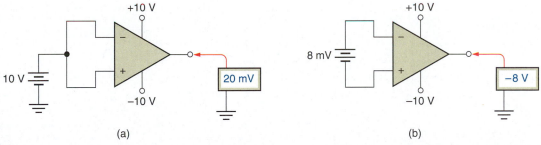

(a)                                   (b)

**FIGURE 15.72**

**50.** Using the KA741, design a noninverting amplifier that will deliver a 25 $V_{PP}$ output to a 20 kΩ load resistance with a 100 $mV_{pk}$ input signal. The available supply voltages are ±18 $V_{dc}$. Include the 8-pin DIP pin numbers in your schematic diagram.

**SUGGESTED COMPUTER APPLICATIONS PROBLEMS**

**51.** Write a program that will determine the maximum allowable input voltage for an inverting amplifier when you know the values of $R_f$, $R_i$, $R_1$, +V, and −V.

**52.** Write a program like the one described in Problem 51 for the noninverting amplifier.

**15.1**    15.8c: $V_{diff} = +2$ V, positive; and 15.8d: $V_{diff} = -3$ V, negative

**15.2**    $5.6\ V_{PP}$

**15.3**    $157\ mV_{PP}$

**15.4**    $142.9\ mV_{PP}$

**15.5**    6.37 kHz

**15.6**    31.8 kHz

**15.7**    $A_{CL} = 250$, $Z_{in} \cong 1$ k$\Omega$, $Z_{out} < 50\ \Omega$, CMRR = 12,500, $f_{max} = 19.1$ kHz

**15.8**    $A_{CL} = 251$, $Z_{in} \geq 1.5$ M$\Omega$, $Z_{out} \leq 50\ \Omega$, CMRR = 12,550, $f_{max} = 19.02$ kHz

**15.9**    $A_{CL} = 1$, $Z_{in} = 1.5$ M$\Omega$, $Z_{out} = 50\ \Omega$, CMRR = 50, $f_{max} = 19.9$ kHz

**15.10**    125 kHz

**15.11**    $A_{CL} f_{C2} = 3.98$ MHz. Yes, it can be used.

**15.12**    99.95

**15.13**    1639

**15.14**    1.8 G$\Omega$

**15.15**    83.2 m$\Omega$

## chapter 16

# Additional Op-Amp Applications

## Objectives

*After studying the material in this chapter, you should be able to:*

1. State the purpose served by a *comparator*.
2. Describe the operation of a basic comparator circuit.
3. List the typical comparator fault symptoms and the possible causes of each.
4. State the purpose served by an *integrator*, and describe its operation.
5. State the purpose served by a *differentiator*, and describe its operation.
6. State the purpose served by a *summing amplifier*, and describe its operation.
7. Describe the use of the summing amplifier in a *digital-to-analog converter*.
8. Describe the operation of the *averaging amplifier* and the *subtractor*.
9. List the typical applications of the *instrumentation amplifier*, and describe its operation.
10. Describe the use of an op-amp in an audio amplifier.
11. Describe the operation of a *voltage-controlled current source*.
12. Describe the construction and operation of a *precision rectifier*.

## Outline

# Looking to the Future

Through the course of this book, you have been provided with glimpses of the history of solid-state devices and their development. The next question is simple: *Where will it go from here?*

While there are varying opinions on the future of solid-state electronics, one thing seems certain: The future of discrete devices is rather bleak. Most IC amplifiers and switching circuits are faster, cheaper, and easier to work with than are discrete component circuits. At the same time, limits on the power-handling capabilities of most ICs limit their use in high-power circuits. This guarantees that discrete devices (which can generally handle more power than most ICs) will remain in use during the foreseeable future. In fact, some high-power systems still require the use of vacuum tubes for applications that exceed the power-handling capabilities of discrete components.

In this chapter, we look at some op-amp circuits that are used for a wide variety of applications. These circuits do not necessarily relate to each other, except for the fact that they all use one or more op-amps. Most of the chapter deals with five common circuits: *comparators*, *integrators*, *differentiators*, *summing amplifiers*, and *instrumentation amplifiers*. Each of these circuits is given extensive coverage. Then, we will take a brief look at several other typical op-amp circuits. Remember as you go through the chapter that, other than the use of one or more op-amps, the circuits covered are not necessarily related to each other. Each section should be approached as a separate entity.

## 16.1   Comparators

OBJECTIVE 1 ▶

**Comparator**
A circuit used to compare two voltages.

The **comparator** is a relatively simple circuit used to compare two voltages and provide an output indicating the relationship between those two voltages. Generally, comparators are used to compare either:

1. Two changing voltages to each other (as in comparing two sine waves).
2. A changing voltage to a set dc reference voltage.

We will look at the second type of comparator in this section. However, you may find it helpful if we take a moment to discuss the applications of comparators before going into any circuit detail.

### 16.1.1   Applications

**Digital circuit**
A circuit designed to respond to specific alternating dc voltage levels.

Comparators are most commonly used in *digital* applications. A **digital circuit** is one that responds to *alternating dc voltage levels* rather than sinusoidal waveforms. For example, the waveform shown in Figure 16.1 consists of "high" and "low" dc voltage levels and the transitions between them. You will spend a great deal of time studying digitial systems such as personal computers at some point. For now, we only need to establish that digital systems respond to alternating dc levels. These alternating dc levels almost always take the form of *rectangular waves* or *square waves*.

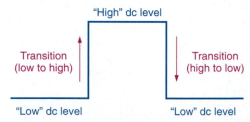

"High" dc level

Transition (low to high)

Transition (high to low)

"Low" dc level          "Low" dc level

**FIGURE 16.1**   Digital waveform characteristics.

Now, consider the circuit shown in Figure 16.2a. Let's assume that the variable voltage source is generating a sinusoidal output and that the digital circuit is intended to perform some function whenever that sine wave exceeds +10 V. (The nature of the function is not

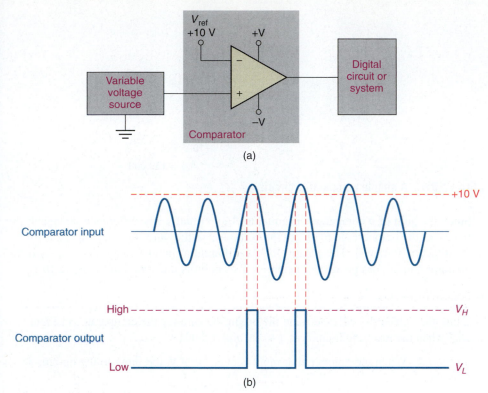

FIGURE 16.2  Basic comparator operation.

important at this point.) The comparator is used to couple the variable voltage source to the digital circuit. Here's how the circuit works:

- The inverting input of the comparator is connected to a $+10$ V *reference*.
- When the output from the variable voltage source is less than $+10$ V, the output from the comparator is *low*. The output from the comparator remains *low* until the sine wave exceeds the value of the reference voltage.
- When the output from the variable voltage source exceeds $+10$ V, the output from the comparator goes *high*. The output from the comparator remains *high* until the sine wave drops below the value of the reference voltage.

This circuit operation is illustrated by the voltage waveforms shown in Figure 16.2b. Since the digital circuit is designed to respond to high and low dc levels, we have successfully converted information about the sine wave into a form that the digital circuit can handle. This is the primary function of a comparator.

Several points can be made at this time:

1. The output from a comparator is normally a dc voltage that indicates the polarity (or magnitude) relationship between the two input voltages.
2. A comparator is not normally used to convert a sine wave to a square wave. This function is normally performed by a circuit called a *Schmitt trigger* (which is discussed in Chapter 19).
3. When a comparator is used to compare a signal amplitude to a fixed dc level (as was the case in Figure 16.2), the circuit is referred to as a **level detector**.

> **Level detector**
> Another name for a comparator used to compare an input voltage to a fixed dc reference voltage.

Now that you have an idea of what a comparator is used for, we will take a look at some common circuit configurations.

## 16.1.2  Comparator Circuits

The most noticeable circuit recognition feature of the comparator is *the lack of any feedback path* in the circuit. A typical comparator circuit is shown in Figure 16.3. Without a feedback path, the voltage gain of the circuit equals the open loop gain ($A_{OL}$) of the op-amp.

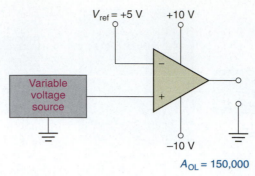

$V_{ref} = +5$ V    +10 V

Variable
voltage
source

−10 V

$A_{OL} = 150,000$

**FIGURE 16.3**   A simple comparator circuit.

OBJECTIVE 2 ▶ Since the gain of a comparator is equal to $A_{OL}$, virtually any difference voltage at the input causes the output to go to one of the voltage extremes and stay there until the difference voltage is removed. The polarity of the differential input voltage determines if the comparator output goes positive or negative. This point is illustrated in Example 16.1.

### EXAMPLE 16.1

Using the values shown, determine the output voltage for the comparator in Figure 16.3 when the inverting input is at $+4.999$ and $+5.001$ V.

*Solution:*   When the noninverting input is at $+4.999$ V, the input to the op-amp is found as

$$V_{diff} = V_{in} - V_{ref} = 4.999 \text{ V} - 5 \text{ V} = -1 \text{ mV}$$

and the op-amp output voltage is found as

$$V_{out} = A_{OL}V_{diff} = (150,000)(-1 \text{ mV}) = -150 \text{ V}$$

Since this is clearly beyond the limits of the output, $V_{out}$ is found as

$$V_{out} \cong -V + 1 \text{ V} = -9 \text{ V}$$

Note that this calculation assumes a value of $R_L \geq 10$ kΩ. When the noninverting input is at $+5.001$ V, the input to the op-amp is found as

$$V_{diff} = V_{in} - V_{ref} = 5.001 \text{ V} - 5 \text{ V} = 1 \text{ mV}$$

and the op-amp output voltage is found as

$$V_{out} = A_{OL}V_{diff} = (150,000)(1 \text{ mV}) = 150 \text{ V}$$

Since this is clearly beyond the limits of the output, $V_{out}$ is found as

$$V_{out} \cong +V - 1 \text{ V} = 9 \text{ V}$$

Again, we are assuming a value of $R_L \geq 10$ kΩ in the calculation of $V_{out}$.

### PRACTICE PROBLEM 16.1

A comparator like the one in Figure 16.3 has its inverting input connected to a $+2$ V reference. The op-amp is connected to $\pm 12$ V supplies and has a value of $A_{OL} = 70,000$. Determine the output voltage for input voltages of $+2.001$ and $+1.999$ V.

As Example 16.1 shows, the comparator output voltage goes to one of the extremes even when there is very little difference between the two inputs. In fact, if you divide the

9 V maximum output voltage by the value of $A_{OL}$, you will see that the circuit needs a difference voltage of only 60 μV to drive the output to either extreme (+9 or −9 V). The polarity of the input determines to which extreme the output is driven.

### 16.1.3   Setting the Reference Level

Most often, a voltage-divider circuit is used to set the reference voltage for a given level detector. Such a circuit is shown in Figure 16.4. For the circuit shown, the reference voltage would be found using the standard voltage-divider formula as follows:

$$V_{ref} = +V \frac{R_2}{R_1 + R_2} \qquad \textbf{(16.1)}$$

Example 16.2 demonstrates the use of this formula in a practical analysis situation.

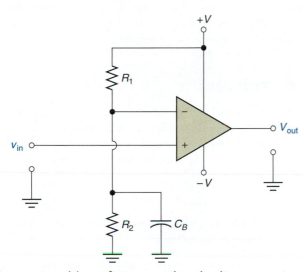

**Lab Reference:** The operation of a comparator like the one in Figure 16.4 is demonstrated in Exercise 24.

FIGURE 16.4   Comparator with a reference-setting circuit.

### EXAMPLE 16.2

Determine the reference voltage for the comparator in Figure 16.5.

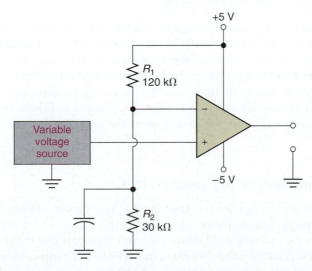

FIGURE 16.5

***Solution:*** The reference voltage is set by the voltage-divider circuit. Thus,

$$V_{ref} = +V\frac{R_2}{R_1 + R_2} = (+5\text{ V})\frac{30\text{ k}\Omega}{150\text{ k}\Omega} = +1\text{ V}$$

**PRACTICE PROBLEM 16.2**

A comparator like the one in Figure 16.5 has $\pm 8$ V supplies and resistor values of $R_1 = 3$ k$\Omega$ and $R_2 = 1.8$ k$\Omega$. Determine the input reference voltage for the circuit.

The circuits shown in Figures 16.4 and 16.5 both contain a *bypass capacitor* in the voltage-divider circuit. This bypass capacitor is included to prevent the variations in $v_{in}$ from being coupled to the voltage-divider circuit through the op-amp. The result is that the output of the circuit is much more reliable.

Now, we will take a look at some other comparator circuit configurations. As you will see, each comparator circuit configuration is used for a specific application. However, they all work according to the same basic principles.

### 16.1.4    Circuit Variations

Up to this point, we have assumed that the comparator is used to provide a *positive* output when the input voltage is more *positive* than some *positive* reference voltage. However, this is not always the case. In fact, you could substitute any combination of the words *positive* and *negative* into the following statement and you would be correct:

*A comparator is a circuit used to provide a _____ output when the input voltage is more _____ than some _____ reference voltage.*

For example, take a look at the circuits shown in Figure 16.6. The input/output relationships of the circuits shown are as follows:

> *A Practical Consideration:*
> It is easy to determine the input/output relationship for a comparator. If the input signal is applied to the inverting ($-$) terminal of the op-amp, the output is *negative* when $v_{in}$ is more positive than $V_{ref}$. If the input signal is applied to the noninverting ($+$) terminal of the op-amp, the output is *positive* when $v_{in}$ is more positive than $V_{ref}$.

- Figure 16.6a: The circuit has a *positive* output when the input voltage is more *negative* than the *positive* reference voltage. Otherwise, the output is negative.

- Figure 16.6b: The circuit has a *positive* output when the input voltage is more *positive* than the *negative* reference voltage. Otherwise, the output is negative.

- Figure 16.6c: The circuit has a *positive* output when the input voltage is *positive*. Otherwise, the output is negative.

- Figure 16.6d: The circuit has a *positive* output when the input voltage is *negative*. Otherwise, the output is negative.

When you apply your understanding of the relationship between the inputs to an op-amp and its output, the statements above are relatively easy to visualize.

> **Variable comparator**
> A comparator with an adjustable reference voltage.

Another circuit variation is the **variable comparator**. The variable comparator allows you to change the dc reference voltage. Such a circuit is shown in Figure 16.7. By adjusting the value of $R_2$ in the circuit, the reference voltage can be set to any value within the limits of $+V$ and ground. Note that any of the comparators in Figure 16.6 can be made variable in the same way (with the exception of the ground reference comparators, of course).

### 16.1.5    Troubleshooting Comparators

**OBJECTIVE 3** ▶    One of the really nice things about comparators is that they're extremely easy to troubleshoot. For example, refer to the circuit in Figure 16.7. There isn't a whole lot here that can go wrong. In fact, there are only three common problems that could develop. These problems and their possible causes are listed in Table 16.1. Example 16.3 shows how the table can be applied to a practical troubleshooting situation.

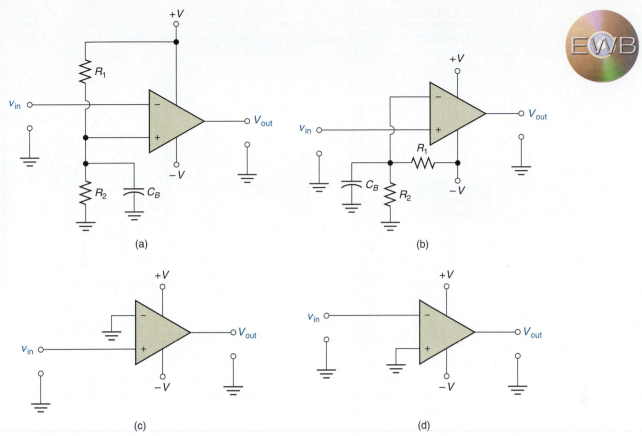

FIGURE 16.6   Comparator circuit variations.

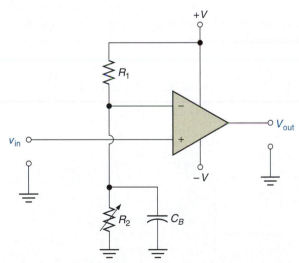

FIGURE 16.7   Comparator with a variable reference circuit.

TABLE 16.1   Comparator Troubleshooting

| Problem | Possible Causes |
|---|---|
| 1. No output. | No input signal<br>One or both supply voltages out<br>A bad op-amp |
| 2. The output changes at an input voltage that is too *high*. | $R_2$ is out of adjustment<br>$R_1$ is shorted (not likely) |
| 3. The output changes at an input voltage that is too *low*. | $R_1$ is open<br>$R_2$ is out of adjustment<br>$C_B$ is shorted (not likely) |

## EXAMPLE 16.3

Determine the possible cause(s) of the problem in the circuit shown in Figure 16.8.

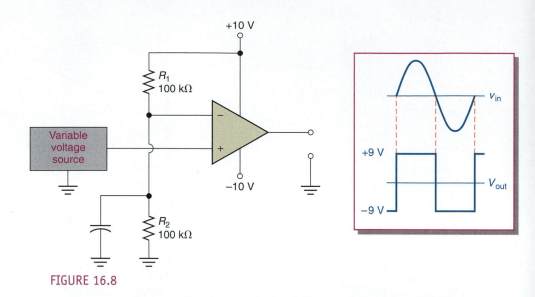

**FIGURE 16.8**

**Solution:** The first step, of course, is to determine whether any problem exists at all. If you look at the voltage-divider circuit, you will notice that $R_1 = R_2$. Therefore, the reference voltage should be half of $+V$, or $+5$ V. Since the output is changing at the point when the input is at ground rather than $+5$ V, there *is* a problem.

Table 16.1 indicates that this problem could be caused by $R_1$ open, $R_2$ shorted, or $C_B$ shorted. Since the problem of $R_1$ open is by far the most likely, the resistance of this component is checked and found to be too high to measure. Replacing the component solves the problem.

### PRACTICE PROBLEM 16.3

Determine the possible cause(s) of the problem (if any) in the circuit shown in Figure 16.9.

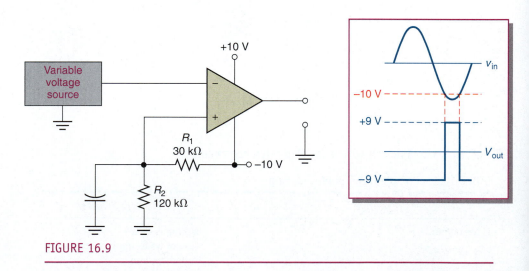

**FIGURE 16.9**

As you can see, the comparator is an easy circuit to analyze and troubleshoot. No matter which circuit configuration is used, you will have no problem dealing with the circuit if you remember the basic comparator operating principles. The characteristics of comparators are summarized in Figure 16.10.

## Comparator Characteristics

*Schematic diagram:

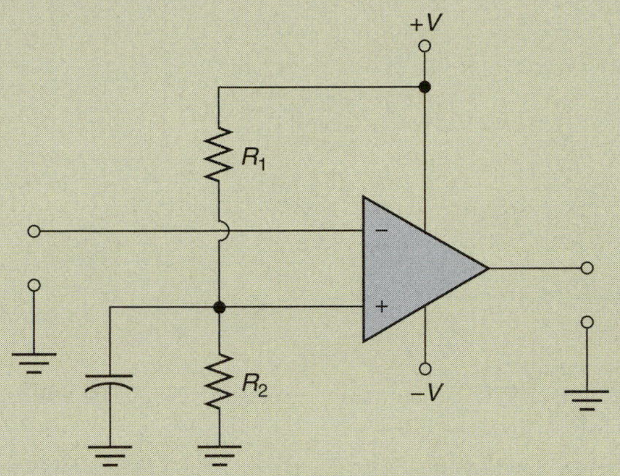

**Circuit recognition feature:** No feedback path to limit voltage gain.

**Primary application:** Voltage-level detector: Comparing the instantaneous value of an active signal to a fixed dc voltage.

**Input/output relationship:** When the active signal is applied to the inverting (−) input, the output goes *positive* if $V_{in}$ is more *negative* than $V_{ref}$. If the active signal is applied to the noninverting (+) input, the output goes *positive* if $V_{in}$ is more *positive* than $V_{ref}$.

**Reference voltage:** Determined by the voltage-divider circuit. Equal to 0 V if the inactive input of the op-amp is connected directly to ground.

*Many configurations are possible. The circuit recognition feature for all configurations is the lack of a feedback path.

FIGURE 16.10

1. What is a *comparator*?

2. Discuss the purpose served by comparators in digital systems.

3. What is the circuit recognition feature of a comparator?

4. What does the voltage gain of a comparator equal?

5. How do you determine the reference voltage and the input/output phase relationship of a comparator?

6. What is a *variable comparator*?

7. List the common comparator faults and their causes.

◄ Section Review

## 16.2 Integrators and Differentiators

In this section, we will look at the operation of op-amp *integrators* and *differentiators*. These two circuits are shown in Figure 16.11. As you can see, the two circuits are nearly identical. Each contains a single op-amp and an *RC* circuit. However, the resistor and capacitor positions in the two circuits are reversed. As a result, the circuits have opposite input/output relationships. As shown in Figure 16.11, the integrator converts a square wave into a triangular wave, while the differentiator converts a triangular wave into a square wave.

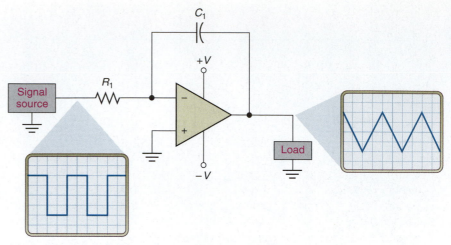

(a)  The integrator converts a rectangular wave to a triangular wave

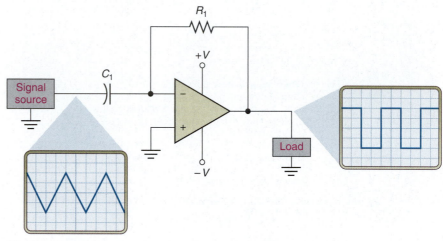

(b)  The differentiator converts a triangular wave to a rectangular wave

**FIGURE 16.11**   Op-amp integrator and differentiator.

## 16.2.1   Integrators

*OBJECTIVE 4* ▶

**Integrator**
A circuit whose output is proportional to the area of the input waveform.

The **integrator** provides an output that is proportional to the area of its input waveform. The concept of waveform area is illustrated in Figure 16.12a.

In geometry, we learn that the area of a rectangle equals the product of its length and its height. For the waveform in Figure 16.12a, the length is the *pulse width* (measured in

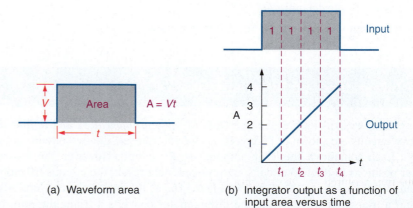

(a)  Waveform area

(b)  Integrator output as a function of input area versus time

**FIGURE 16.12**

units of time) and the height is its *amplitude* (measured in volts). Therefore, the area of the waveform is found as

$$A = Vt$$

where $A$ = the area of the waveform
$\quad V$ = the peak voltage of the waveform
$\quad t$ = the pulse width of the waveform

Figure 16.12b shows how the output of the integrator varies with the area of the input waveform. As you can see, the input waveform has been divided into four equal sections. For ease of discussion, each section is referred to as one *unit of area*. From the positive-going transition of the input to $t_1$, we have one unit of area. In response to this input condition, the output waveform of the integrator goes to 1. When $t_2$ is reached, the area of the waveform has increased to two units of area and the integrator output goes to 2, and so on. As you can see, the output of the integrator indicates the total number of area units at any point on the input waveform.

As the input cycles continue, the integrator produces the corresponding output waveform shown in Figure 16.11a. Note that the waveform is shifted 180° because the input signal to an integrator is applied to the *inverting* input of the op-amp. As you can see, the integrator is acting as a square wave–to–triangular waveform converter.

> The integrator can be viewed as a square wave–to–triangular wave converter.

## 16.2.2  Integrator Operation

The simplest integrator is an *RC* circuit like the one shown in Figure 16.13. With the square-wave input shown, the *RC* integrator would *ideally* have a linear triangular output. This ideal output waveform is shown in the figure. The only problem is that the capacitor does not charge/discharge at a linear rate but, rather, at an exponential rate. This produces the "actual" waveform shown in the figure.

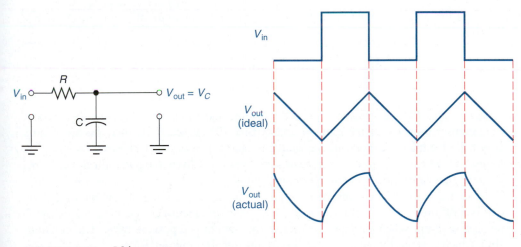

FIGURE 16.13  *RC* integrator.

In Chapter 3, you were shown that the capacitor in an *RC* circuit requires five *time constants* to reach its full potential charge, where a time constant is the time period found as

$$\tau = RC$$

Thus, the time required for the capacitor to reach full charge can be calculated as

$$T_C = 5RC$$

When the value of $T_C$ equals the pulse width of the input, the $RC$ integrator produces the nonlinear (actual) output shown in Figure 16.13.

Unfortunately, the nonlinear output of the $RC$ integrator is not nearly close enough to the ideal integrator output for most applications. However, by adding an op-amp to the circuit, we can obtain a linear output from the integrator that is much closer to the ideal.

How does an op-amp integrator produce a linear output?

The key to obtaining a linear output from the integrator is to provide a *constant-current* charge path for the capacitor. In other words, if we can make the capacitor charging current constant, the *rate* of charge for the component becomes constant and a linear output is produced. Keeping this in mind, let's take a look at the circuit in Figure 16.14. The *constant-current* characteristic of this circuit is based on two well-known points:

1. The inverting input to the op-amp is held at *virtual ground* by the differential amplifier in the component's input circuit.
2. The input impedance of the op-amp is so high that virtually all of $I_1$ passes through the feedback path.

These two points were established in Chapter 15.

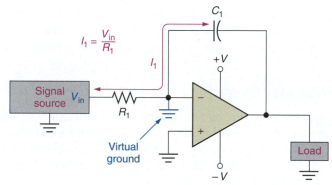

**FIGURE 16.14**  Op-amp integrator constant-current characteristics.

Since the inverting input is held at virtual ground, the value of input current ($I_1$) is found as

$$ I_1 = \frac{V_{in}}{R_1} $$

Assuming that $V_{in}$ is constant over a period of time and $R_1$ has a fixed value, the value of $I_1$ remains constant over the same period of time. Since nearly all of $I_1$ passes to (and from) the feedback network, the capacitor is charged by a constant-current source. As long as $V_{in}$ remains constant, the capacitor charges (and discharges) at a linear rate. This produces the **ramp** (ideal) output shown in Figure 16.13.

Since the input to the integrator is applied to the inverting input, the output of the circuit is 180° out of phase with the input. Thus, when the input goes positive, the output is a negative ramp. When the input is negative, the output is a positive ramp. This relationship, which was briefly mentioned earlier, is illustrated in Figure 16.15.

**Ramp**
Another name for a voltage that changes at a constant (or *linear*) rate.

**Lab Reference:** The phase relationship shown in Figure 16.15 is observed in Exercise 25.

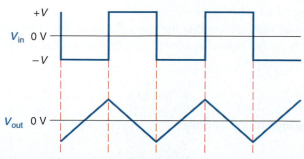

**FIGURE 16.15**  Op-amp integrator phase relationship.

The output waveform in Figure 16.15 is centered around 0 V. In practice, this may not be the case. If the op-amp has not been properly compensated, an output offset voltage may center $V_{out}$ around some value other than 0 V. One method for ensuring that the output is centered around 0 V is to add a *feedback resistor* ($R_f$) as shown in Figure 16.16.

The feedback resistor (Figure 16.16) eliminates any output offset voltage.

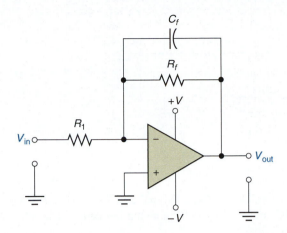

**FIGURE 16.16**

Adding a feedback resistor ensures that the output is centered around 0 V, but it also introduces a restriction on the circuit operation. Like any *RC* circuit, the *RC* feedback circuit has a *cutoff frequency*. In this case, it is a *lower* cutoff frequency that can affect the circuit operation. As frequency decreases, the reactance of the capacitor increases. At some point, the reactance of the capacitor exceeds the value of $R_f$, and the integrating action of the circuit is lost. As always, the cutoff frequency is found as

$$f_{C1} = \frac{1}{2\pi R_f C_f} \qquad (16.2)$$

If operating frequency decreases, an integrator *starts* to lose its linear output characteristics before the frequency found in equation (16.2) is reached. For an optimum output, the value of $X_C$ should always be less than $0.1R_f$. Therefore, the circuit should not be operated below the frequency found by

$$f_{min} = \frac{10}{2\pi R_f C_f} \qquad (16.3)$$

Example 16.4 illustrates the frequency limits of the integrator.

### EXAMPLE 16.4

Determine the cutoff frequency for the circuit shown in Figure 16.17. Also, determine the frequency at which the output starts to lose its linear characteristics.

*Solution:* The cutoff frequency for the circuit is found as

$$f_{C1} = \frac{1}{2\pi R_f C_f} = \frac{1}{2\pi(100 \text{ k}\Omega)(0.01 \text{ }\mu\text{F})} = \textbf{159 Hz}$$

If you compare equations (16.2) and (16.3), you will see that linearity starts to go when the operating frequency is 10 times the cutoff frequency. Therefore, we can save some time and trouble by using

$$f_{min} = 10f_{C1} = \textbf{1.59 kHz}$$

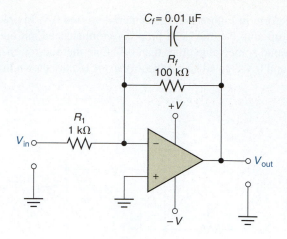

FIGURE 16.17

**PRACTICE PROBLEM 16.4**

An integrator has values of $C_f = 0.1$ μF and $R_f = 51$ kΩ. Determine the cutoff frequency for the circuit. Also, determine the frequency at which it will start to lose its linear characteristics.

As you may recall, Miller's theorem allows us to represent a feedback capacitor as two shunt capacitances, one at the input and one at the output, as shown in Figure 16.18. The shunt capacitances cause the circuit to have an *upper cutoff frequency* ($f_{C2}$). If the operating frequency increases, the capacitors begin shorting out the input and output signals. If the operating frequency continues to increase, the input and output signals are eventually attenuated to the point where the circuit gain drops to zero.

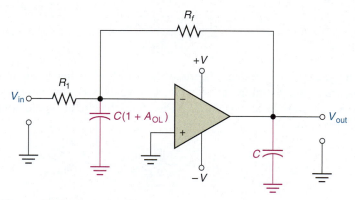

FIGURE 16.18  ac Equivalent circuit.

### 16.2.3  Integrator Troubleshooting

The integrator is another circuit that is relatively easy to troubleshoot. Some potential integrator problems are listed in Table 16.2, along with the symptoms of each. Circuit faults involving $R_1$, the IC socket, or the op-amp are approached in the manner described in Section 15.6.

### 16.2.4  The Differentiator

**OBJECTIVE 5 ▶**

**Differentiator**
A circuit whose output is proportional to the rate of change of its input signal.

If we reverse the integrator capacitor and resistor (as shown in Figure 16.11b), we have a **differentiator**. The differentiator provides an output that is proportional to the *rate of change* of its input signal. The input/output relationship of the differentiator is illustrated in Figure 16.19.

The input signal for the differentiator is shown to be a triangular waveform. Between times $t_0$ and $t_2$, the rate of change of the input is constant. Since the change is in a *positive* direction, the rate of change is a *positive constant*. As a result of the *inverting* action of its

## TABLE 16.2 Integrator Troubleshooting

| Fault | Symptoms |
|---|---|
| $R_f$ open | If $R_f$ opens, the resistor is effectively removed from the circuit. Since this resistor is used to keep the output referenced around 0 V, the major symptom is the loss of this output reference. Also, the circuit cutoff frequency drops significantly. |
| $R_f$ shorted | As you know, resistors do not usually short. However, if something shorts out $R_f$, integration is lost, as is the gain of the circuit. In other words, the circuit has little, if any, output signal. |
| $C_f$ open | As you know, a capacitor is an open circuit by nature. However, if the component opens so that it prevents coupling, the circuit starts to act as a common inverting amplifier. The gain of the circuit becomes $$A_{CL} = \frac{R_f}{R_{in}}$$ and the output changes to a square wave. |
| $C_f$ shorted | This has the same effect as $R_f$ shorting. In fact, if either of these components shorts, the result is that the other component is also shorted. Therefore, you must test $C_f$ and $R_f$ individually to see which component is actually shorted. |

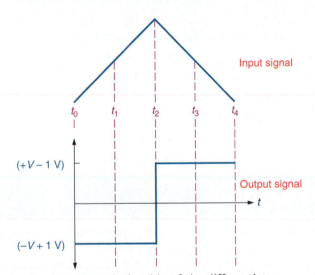

**Lab Reference:** The input/output relationship for the differentiator is demonstrated in Exercise 25.

**FIGURE 16.19** The input/output relationship of the differentiator.

op-amp, the differentiator responds to this input by producing a constant *negative* output voltage. Between times $t_2$ and $t_4$, the rate of change is a *negative constant*. As a result, the differentiator output switches to a constant *positive* voltage. Again, the polarity reversal is due to the inverting action of the op-amp. Thus, with a triangular input waveform, the differentiator produces a rectangular output waveform.

Equation (16.2) gave us a limit on the low-frequency operation of the integrator. The same equation can be used to provide a limit on the *high-frequency* operation of the differentiator. Once again, the operating characteristics of the circuits are opposites.

Just as the integrator starts to lose its operating characteristics at 10 times the lower cutoff frequency, the differentiator starts to lose its operating characteristics at one-tenth the value of $f_{C2}$. By formula,

$$f_{max} = \frac{1}{20\pi R_f C_1} \tag{16.4}$$

If the input frequency to the differentiator exceeds the limit determined by equation (16.4), the output square wave starts to become distorted.

## 16.2.5 Differentiator Troubleshooting

Like the integrator, the differentiator is relatively easy to troubleshoot. The primary faults that can develop (other than a bad op-amp) are listed in Table 16.3, along with the symptoms of each. If the differentiator fails to operate and none of the symptoms in Table 16.3 appears, the op-amp (or IC socket) is the most likely source of the fault.

**TABLE 16.3**  Differentiator Troubleshooting

| *Fault* | *Symptoms* |
|---------|-----------|
| $C_1$ open | The input is removed from the op-amp, and the output of the circuit remains at (or near) zero. |
| $C_1$ shorted | If something shorts out $C_1$, the gain of the circuit increases drastically and the differentiating action of the circuit is lost. |
| $R_f$ open | In this case, the gain of the circuit equals the value of $A_{OL}$ for the op-amp. Also, differentiating action is lost. |
| $R_f$ shorted | The gain of the circuit drops to zero, and there is little (if any) output signal. |

**Section Review ▶**

1. What is an *integrator*?
2. Describe the circuit operation of the op-amp integrator.
3. What is a *ramp*?
4. List the common faults that occur in op-amp integrators and the symptoms of each.
5. What is a *differentiator*?
6. Describe the circuit operation of the differentiator.
7. List the common differentiator faults and the symptoms of each.

## 16.3  Summing Amplifiers

**OBJECTIVE 6 ▶**

**Summing amplifier**
An op-amp circuit that produces an output proportional to the sum of its input voltages.

**Lab Reference:** Summing amplifier operation is demonstrated in Exercise 24.

The **summing amplifier** is an op-amp circuit that provides an output proportional to the *sum* of its inputs. A basic summing amplifier is shown in Figure 16.20. The key to understanding this circuit is to start by considering each input as an individual circuit. If $V_1$ were the only input, the output would be found as

$$V_{out} = -V_1\frac{R_f}{R_1}$$

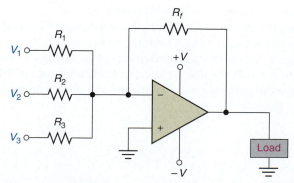

**FIGURE 16.20**  Summing amplifier.

Similarly, if $V_2$ were the only input, we would have

$$V_{\text{out}} = -V_2 \frac{R_f}{R_2}$$

And if $V_3$ were the only input, we would have

$$V_{\text{out}} = -V_3 \frac{R_f}{R_3}$$

If we have voltages at all three of the input terminals, the output is found using the sum of the equations for each as follows:

$$V_{\text{out}} = \frac{-V_1 R_f}{R_1} + \frac{-V_2 R_f}{R_2} + \frac{-V_3 R_f}{R_3}$$

Or, after factoring out the value of $-R_f$,

$$V_{\text{out}} = -R_f\left(\frac{V_1}{R_1} + \frac{V_2}{R_2} + \frac{V_3}{R_3}\right) \qquad (16.5)$$

Example 16.5 demonstrates the procedure for determining the output from a summing amplifier.

### EXAMPLE 16.5

Determine the output voltage from the summing amplifier in Figure 16.21.

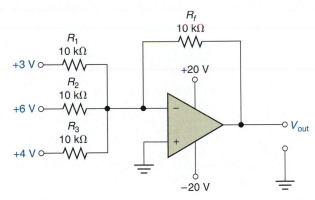

FIGURE 16.21

*Solution:* Using the values shown in the circuit and equation (16.5), the output voltage is found as

$$V_{\text{out}} = -R_f\left(\frac{V_1}{R_1} + \frac{V_2}{R_2} + \frac{V_3}{R_3}\right) = \left(-10\text{ k}\Omega\right)\left(\frac{+3\text{ V}}{10\text{ k}\Omega} + \frac{+6\text{ V}}{10\text{ k}\Omega} + \frac{+4\text{ V}}{10\text{ k}\Omega}\right) = -13\text{ V}$$

### PRACTICE PROBLEM 16.5

A summing amplifier like the one in Figure 16.21 has the following values: $R_f = 2$ k$\Omega$, $R_1 = R_2 = R_3 = 1$ k$\Omega$, $V_1 = 1$ V, $V_2 = 500$ mV, and $V_3 = 1.5$ V. The supply voltages for the circuit are $\pm 18$ V. Determine the value of $V_{\text{out}}$ for the circuit.

If you compare the result in Example 16.5 with the input voltages, you will see that the output magnitude equals the sum of the input voltages. Thus, the term *summing amplifier* is appropriate for the circuit.

In practice, it is not always possible to have an output that is equal to the sum of the input voltages. For example, what if the inputs to Figure 16.21 had been +10, +8, and +7 V? Clearly, the sum of these voltages is +25 V. However, the maximum possible output voltages from the circuit are ±19 V. Thus, an output of −25 V isn't possible. To solve this problem, the value of $R_f$ is usually chosen so that the output voltage is *proportional* to the sum of the inputs. This point is illustrated in Example 16.6.

### EXAMPLE 16.6

Determine the output voltage from the circuit shown in Figure 16.22.

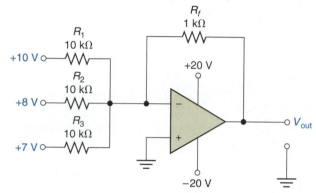

FIGURE 16.22

*Solution:* Using the values shown in the circuit and equation (16.5), the output voltage is found as

$$V_{out} = -R_f\left(\frac{V_1}{R_1} + \frac{V_2}{R_2} + \frac{V_3}{R_3}\right)$$
$$= (-1\ k\Omega)\left(\frac{+10\ V}{10\ k\Omega} + \frac{+8\ V}{10\ k\Omega} + \frac{+7\ V}{10\ k\Omega}\right) = -2.5\ V$$

### PRACTICE PROBLEM 16.6

A summing amplifier like the one in Figure 16.22 has the following values: $R_f$ = 1 kΩ, $R_1 = R_2 = R_3$ = 3 kΩ, $V_1$ = 2 V, $V_2$ = 1 V, and $V_3$ = 6 V. The supply voltages for the circuit are ±12 V. Determine the value of $V_{out}$ for the circuit.

In Example 16.6, we solved the problem of overdriving the amplifier with +10, +8, and +7 V inputs by *decreasing* the value of $R_f$. Now, the output magnitude is not equal to the sum of the inputs but, rather, is *proportional to* the sum of the inputs. In this case, it is equal to one-tenth of the input sum. Picking random values of $V_1$, $V_2$, and $V_3$ and using them in equation (16.5) will show that the circuit always provides an output that is one-tenth of the sum of the inputs. (This assumes, of course, that the value obtained by the equation does not exceed ±19 V.)

At this time, several points should be made regarding the summing amplifier:

1. There is no practical limit on the number of summing inputs. As long as the amplifier is capable of providing the correct output voltage, the circuit can have any number of inputs. In this case, equation (16.5) would be expanded (or reduced) to include all the amplifier inputs.

2. The input resistors of a summing amplifier do not always have equal values. In other words, the inputs can be *weighted* so that certain inputs have more of an effect on the output.

The second point will be discussed in detail when we look at a common application for the summing amplifier later in this section. First, however, we will look at how you can develop a general formula for any summing amplifier that will make the analysis of that circuit easier.

**Lab Reference:** The effect of weighting the inputs to a summing amplifier is demonstrated in Exercise 24.

### 16.3.1 Circuit Analysis

Often, you need to predict quickly the output from a summing amplifier for a variety of input combinations. This task is made easier if you derive a *general-class* output equation for the amplifier.

The general-class equation for a given summing amplifier is derived as follows:

1. Determine the $R_f/R$ ratio for each input branch.
2. Represent each branch as the product of its resistance ratio and its input voltage.
3. Add the products found in step 2.

Let's apply this process to the circuit shown in Figure 16.23. For the first branch in the circuit, the resistance ratio is

$$\frac{R_f}{R_1} = 1$$

and the branch would be represented as this ratio times $V_1$, or simply $V_1$. The resistance ratio of the second branch would be

$$\frac{R_f}{R_2} = 0.2$$

and the branch would be represented as $0.2V_2$. Performing this procedure on the third and fourth branches would yield results of 0.1 and 0.05, respectively. The output of the circuit can now be found as

$$-V_{\text{out}} = V_1 + 0.2V_2 + 0.1V_3 + 0.05V_4$$

This is the **general-class equation** for the circuit and can be used to determine the output voltage for any combination of input voltages, as shown in Example 16.7.

**General-class equation**
An equation derived for a summing amplifier that is used to predict the output from a circuit for any combination of inputs.

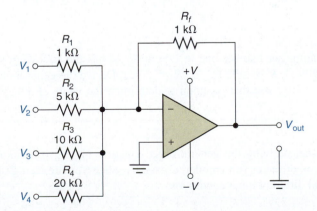

FIGURE 16.23

## EXAMPLE 16.7

Determine the output from the circuit shown in Figure 16.24 for each of the following input combinations:

| $V_1$ (V) | $V_2$ (V) | $V_3$ (V) |
|-----------|-----------|-----------|
| +10 | 0 | +10 |
| 0 | +10 | +10 |
| +10 | +10 | +10 |

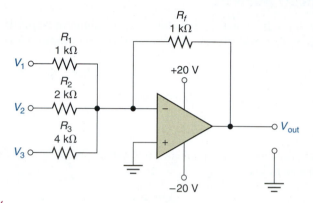

FIGURE 16.24

**Solution:** First, the circuit can be represented by its general-class equation. For this circuit, the equation is derived as follows:

$$-V_{out} = \frac{R_f}{R_1}V_1 + \frac{R_f}{R_2}V_2 + \frac{R_f}{R_3}V_3 = V_1 + 0.5V_2 + 0.25V_3$$

> The results shown here are values of $-V_{out}$. Therefore, the values of $V_{out}$ are $-12.5$, $-7.5$, and $-17.5$ V.

Now, using this equation, we can determine the output for the first set of inputs as

$$-V_{out} = 10\ V + 0.5(0\ V) + 0.25(10\ V) = \textbf{12.5 V}$$

For the second set of inputs,

$$-V_{out} = 0\ V + 0.5(10\ V) + 0.25(10\ V) = \textbf{7.5 V}$$

Finally, for the third set of inputs,

$$-V_{out} = 10\ V + 0.5(10\ V) + 0.25(10\ V) = \textbf{17.5 V}$$

### PRACTICE PROBLEM 16.7

A summing amplifier like the one in Figure 16.24 has the following values: $R_f = 2\ k\Omega$, $R_1 = 200\ \Omega$, $R_2 = 400\ \Omega$, and $R_3 = 2\ k\Omega$. Derive the general-class equation for the circuit. Then use that equation to determine the value of $-V_{out}$ when $V_1 = V_2 = V_3 = 1$ V.

As you can see, deriving the general-class equation for a given summing amplifier makes the analysis of the circuit simpler, especially when you are trying to determine the output for several different input conditions.

## 16.3.2  Summing Amplifier Applications

One summing amplifier application is as a **digital-to-analog (D/A) converter**. This application is illustrated in Figure 16.25.

◀ *OBJECTIVE 7*

**Digital-to-analog (D/A) converter**
A circuit that converts digital circuit outputs to equivalent analog voltages.

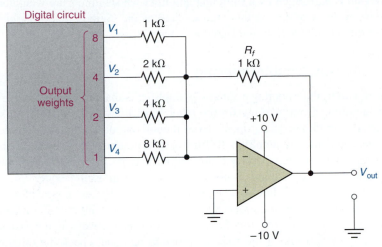

FIGURE 16.25  Simplified D/A converter.

First, a word or two about the *digital circuit*: This circuit has "weighted" outputs (labeled 8, 4, 2, and 1) that combine to represent any number between 0 and 15. For example, let's say that each of the weighted outputs from the digital circuit is always at either +5 or 0 V. The digital circuit represents the number 10 by providing the following combination of output voltages:

| $V_1(8)$ | $V_2(4)$ | $V_3(2)$ | $V_4(1)$ |
|----------|----------|----------|----------|
| +5 V | 0 V | +5 V | 0 V |

With +5 V outputs on lines 8 and 2, the output from the digital circuit represents a decimal value of $8 + 2 = 10$. To represent a decimal value of 5, the digital circuit provides an output combination of $V_1 = 0$ V, $V_2 = +5$ V, $V_3 = 0$ V, and $V_4 = +5$ V. (This combination for the decimal number 5 agrees with the outputs shown in the table below.)

The D/A converter is used to convert the outputs from the digital circuit into a single voltage that is proportional to the numeric output of the circuit. To see how this works, we start by deriving the general-class equation for the circuit. This equation is

$$-V_{out} = V_1 + 0.5V_2 + 0.25V_3 + 0.125V_4$$

Using this equation, the output values in the following table were derived:

| *Decimal Value* | $V_1(8)$ | $V_2(4)$ | $V_3(2)$ | $V_4(1)$ | $V_{out}$ |
|-----------------|----------|----------|----------|----------|-----------|
| 0 | 0 V | 0 V | 0 V | 0 V | −0 V |
| 1 | 0 V | 0 V | 0 V | +5 V | −0.625 V |
| 2 | 0 V | 0 V | +5 V | 0 V | −1.25 V |
| 3 | 0 V | 0 V | +5 V | +5 V | −1.875 V |
| 4 | 0 V | +5 V | 0 V | 0 V | −2.5 V |
| 5 | 0 V | +5 V | 0 V | +5 V | −3.125 V |
| . | | | | | |
| . | | | | | |
| . | | | | | |
| 15 | +5 V | +5 V | +5 V | +5 V | −9.375 V |

It is not important for you to understand the exact workings of the digital circuit output. It is hoped that you can see a pattern in the $+5$ V outputs from the circuit. The main point is this: Often, it is necessary to convert the voltages produced by a group of digital outputs into a single voltage. This is accomplished using a process called D/A conversion. This conversion is performed by a summing amplifier whose inputs are weighted proportionally to the weights of the digital outputs. Thus, as the digital circuit output increases in value, the output voltage from the D/A converter increases.

### 16.3.3    Circuit Troubleshooting

The problem with troubleshooting a summing amplifier is that an open input line does not prevent the amplifier from working for the other input lines. For example, let's say that the 2 kΩ resistor in Figure 16.25 opens. Even though output 4 from the digital circuit is now disabled, the summing amplifier still provides an output that is proportional to the sum of the remaining outputs. Granted, this output may be incorrect for the values produced by the digital circuit, but this is not easy to see with an oscilloscope.

If you isolate a system failure to a summing amplifier, the best approach is simply to measure the resistance of each component in the circuit. Since all the circuit components (other than the op-amp) are resistors, a series of resistance checks is the fastest way to locate the faulty component. If the resistors prove to be good, check the op-amp and its socket (if any).

How do you tell when a summing amplifier isn't working properly? This question can be answered using the block diagram in Figure 16.26. Let's say that circuit B is not working properly. A check of the input verifies that the circuit is not receiving the proper input signal. Testing circuit A verifies that it *is* working properly. Under these circumstances, the summing amplifier must be the source of the problem. In this case, *the summing amplifier is determined to be faulty by verifying that the problem isn't located anywhere else*. As inexact as this procedure may seem, it is often the best since summing amplifiers tend to be used in relatively complex digital circuits.

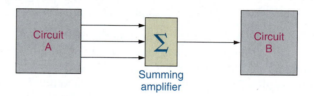

FIGURE 16.26

### 16.3.4    Circuit Variations

OBJECTIVE 8 ▶

By using the proper input and feedback resistor values, a summing amplifier can be designed to provide an output that equals the *average* of any number of input voltages. An example of this type of circuit, which is called an **averaging amplifier**, is shown in Figure 16.27.

> **Averaging amplifier**
> A summing amplifier that provides an output proportional to the *average* of the input voltage.

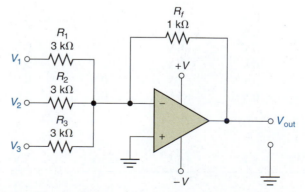

FIGURE 16.27    Averaging amplifier.

If we derive the general-class formula for the circuit shown in Figure 16.27, we obtain

$$-V_{\text{out}} = \frac{V_1}{3} + \frac{V_2}{3} + \frac{V_3}{3}$$

or

$$-V_{\text{out}} = \frac{V_1 + V_2 + V_3}{3}$$

This, by definition, is the average of the three input voltages.

A summing amplifier is wired as an averaging amplifier whenever *both* of the following conditions are met:

1. All input resistors ($R_1$, $R_2$, etc.) are *equal in value*.
2. The ratio of any input resistor to the feedback resistor is equal to the number of circuit inputs.

For example, in Figure 16.27, all input resistors are equal in value (3 kΩ). If we take the ratio of any input resistor to the feedback resistor, we get 3 kΩ/1 kΩ = 3. This is equal to the number of inputs to the circuit.

Another variation on the summing amplifier is shown in Figure 16.28. This circuit, called a **subtractor**, provides an output that is equal to *the difference between $V_1$ and $V_2$*.

**Subtractor**

A summing amplifier that provides an output proportional to the difference between two input voltages.

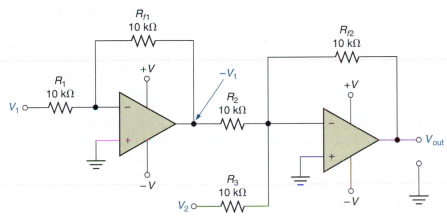

FIGURE 16.28   Subtractor.

$V_1$ is applied to a standard inverting amplifier that has unity gain. Because of this, the output from the inverting amplifier equals $-V_1$. This output is then applied to a summing amplifier (also having unity gain) along with $V_2$. The output from the second op-amp is found as

$$-V_{\text{out}} = V_2 + (-V_1) = V_2 - V_1$$

It should be noted that the gain of the second stage in the subtractor can be varied to provide an output that is proportional to (rather than equal to) the difference between the input voltages. However, if the circuit is to act as a subtractor, the input inverting amplifier *must* have unity gain. Otherwise, the output will not be proportional to the true difference between $V_1$ and $V_2$.

## 16.3.5   Summary

The summing amplifier can be designed to provide an output voltage that is equal to the sum of any number of input voltages. Two variations on the basic summing amplifier are the averaging amplifier and the subtractor. Summing amplifier characteristics are summarized in Figure 16.29.

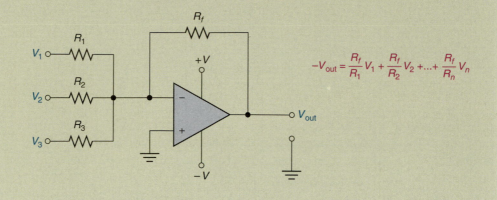

# Summing Amplifiers

$$-V_{out} = \frac{R_f}{R_1}V_1 + \frac{R_f}{R_2}V_2 + \dots + \frac{R_f}{R_n}V_n$$

| | |
|---|---|
| Primary application: | Digital-to-analog (D/A) conversion |
| Input restrictions: | There is no limit on the number of active inputs. However, the resistance ratios must be chosen to ensure that $V_{out}$ never tries to exceed the limits set by the op-amp supply voltages. |
| Circuit variations: | Averaging amplifier<br>Subtractor |
| Circuit recognition: | Multiple inputs connected to the inverting input of the op-amp. The inputs may come from a single source or a variety of sources. |

FIGURE 16.29

**Section Review ▶**

1. What is a *summing amplifier*?
2. Why is the output from a summing amplifier often proportional to (rather than equal to) the sum of its input voltages?
3. Is there a limit on the number of inputs to a summing amplifier? Explain your answer.
4. What is the *general-class equation* of a summing amplifier?
5. List the steps used to derive the general-class equation for a summing amplifier.
6. What is a *digital-to-analog (D/A) converter*?
7. Describe the use of a summing amplifier as a D/A converter.
8. Why are summing amplifiers difficult to troubleshoot?
9. How is a summing amplifier usually determined to be faulty?
10. What is an *averaging amplifier*?
11. What are the resistor requirements of an averaging amplifier?
12. What is a *subtractor*?

**Critical Thinking ▶**

13. Refer to the D/A converter output table in Section 16.3.2. How does a value of $-9.375$ V relate to a decimal value of 15?

## 16.4 Instrumentation Amplifiers

Up to this point, we have concentrated on op-amp circuits that are most commonly used in digital applications. Now, we will look at an op-amp circuit that is used primarily in process control and measurement applications: the **instrumentation amplifier**. The instrumentation amplifier is a *high-gain, high-CMRR circuit that is used to detect and amplify low-level signals.*

The inputs to an instrumentation amplifier are typically in the microvolt or low millivolt range. As such, they can be severely affected by any noise at the amplifier inputs. Noise generated at the amplifier inputs appears as common-mode signals, so the high CMRR of the instrumentation amplifier gives it the ability to ignore noise while amplifying any signal input.

An instrumentation amplifier is shown in Figure 16.30. Circuits A and B are *noninverting* amplifiers, and circuit C acts as a *differential amplifier.* We'll start our analysis of the circuit operation by defining the output from circuit A.

◄ *OBJECTIVE 9*

**Instrumentation amplifier**
A circuit used to amplify low-level signals in process control or measurement applications.

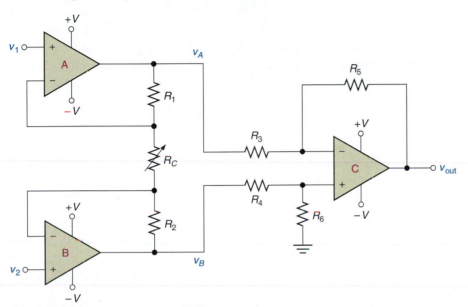

FIGURE 16.30   Instrumentation amplifier.

Since A is a noninverting amplifier, $v_1$ is amplified by a factor of $(1 + R_1/R_C)$. The signal at the inverting input to A is provided by the output of B. The amplitude of this signal at the inverting input can be shown to equal

$$\frac{R_2}{R_C} v_2$$

Since the output of A is equal to *the difference between its input voltages,* $v_{A(\text{out})}$ can be found as

$$v_{A(\text{out})} = \left(1 + \frac{R_1}{R_C}\right) v_1 - \frac{R_2}{R_C} v_2$$

Circuit B is identical to circuit A. Therefore, the output from op-amp B can be found as

$$v_{B(\text{out})} = \left(1 + \frac{R_2}{R_C}\right) v_2 - \frac{R_1}{R_C} v_1$$

Assuming that $R_3 = R_4$, the differential input to C is equal to the difference between the above outputs. This differential input is found as

$$v_{B(\text{out})} - v_{A(\text{out})} = \left[\left(1 + \frac{R}{R_C}\right)v_2 - \frac{R}{R_C}v_1\right] - \left[\left(1 + \frac{R}{R_C}\right)v_1 - \frac{R}{R_C}v_2\right]$$

Note that $R_1 = R_2$ (by design), so we don't distinguish between the two in the above equation. Simplifying this equation, we get

$$v_{B(\text{out})} - v_{A(\text{out})} = \left(1 + \frac{2R}{R_C}\right)(v_2 - v_1)$$

The differential amplifier is designed for unity gain. Therefore, the final circuit output is found as

$$v_{(\text{out})} = \left(1 + \frac{2R}{R_C}\right)(v_2 - v_1) \tag{16.6}$$

Since the output from the op-amp circuit is equal to the product of its closed-loop voltage gain and differential input, we can define the closed-loop voltage gain of the circuit as

$$A_{\text{CL}} = 1 + \frac{2R}{R_C} \tag{16.7}$$

If we were to assume that $v_2 = v_1$, the output from the amplifier, as found using equation (16.6), would equal 0 V. Therefore, the common-mode gain of the circuit equals *zero*. In other words, the CMRR of the circuit is (for all practical purposes) *infinite*.

### 16.4.1 Circuit Calibration

For the instrumentation amplifier to work properly, the gain values of the input amplifiers must be identical. The potentiometer ($R_C$) is included to provide a means of adjusting for any difference between the gain of A and the gain of B. Since the value of $R_C$ affects both $v_{A(\text{out})}$ and $v_{B(\text{out})}$, we can adjust it to compensate for any difference between the input gain values.

The simplest way to set $R_C$ to its proper value is to apply a common-mode signal to the amplifier inputs. Any difference in the input gain values will result in the circuit producing an output voltage. So with the common-mode signals applied, $R_C$ is varied until the output drops to 0 V. When this occurs, the circuit is calibrated properly.

---

**Section Review ▶**

1. What purpose is served by an instrumentation amplifier?
2. Why is it important for an instrumentation amplifier to have extremely high gain and CMRR values?
3. Describe the calibration process for a basic instrumentation amplifier.

## 16.5 Other Op-Amp Circuits

In this section, we will take a very brief look at some other op-amp circuits. These circuits are presented merely to give you an idea of the wide variety of applications for the op-amp. We will not go into the details of circuit analysis or troubleshooting for the circuits in this section; it is strictly an applications section.

## 16.5.1  Audio Amplifier

The final output stage of most communications receivers is an **audio amplifier**. The audio amplifier is the circuit that drives the system speakers. The ideal audio amplifier has the following characteristics:

1. High gain.
2. Minimum distortion in the audio-frequency range (approximately 20 Hz to 20 kHz).
3. High input impedance.
4. Extremely low output impedance (to provide maximum coupling to the speakers).

In a *low-power audio system*, an op-amp audio amplifier fulfills the requirements listed very nicely. An op-amp audio amplifier is shown in Figure 16.31. The first thing you will probably notice about this circuit is that the $-V$ input to the op-amp is connected to *ground*. This means that the op-amp output always has some *positive* value. Note that the coupling capacitor between the output of the op-amp and the speaker is used to reference the speaker voltage around 0 V; that is, it removes the positive dc reference from the op-amp output. Also, $C_5$ is included in the $V_{CC}$ line to prevent any transient current caused by the operation of the op-amp from being coupled back to $Q_1$ through the power supply.

◄ **OBJECTIVE 10**

**Audio amplifier**
The final audio stage in communications receivers; used to drive the speakers.

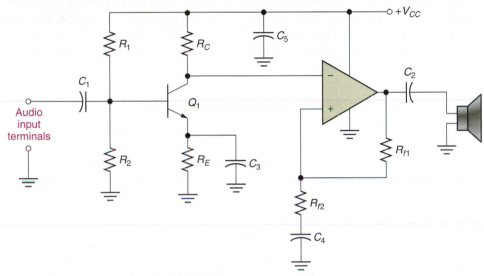

**FIGURE 16.31**  Low-power audio amplifier.

The high-gain requirement is accomplished by the combination of the two amplifier stages. The low $Z_{out}$ of the audio amplifier is accomplished by the op-amp itself, as is the low distortion characteristic.

## 16.5.2  Voltage-Controlled Current Source

As you know, a *current source* is a circuit that produces a constant-value output current. A **voltage-controlled current source** uses an input voltage to set the value of its output current. Such a circuit is shown in Figure 16.32.

To give you an idea of how the circuit works, the two sides of the zener diode have been labeled (A) and (B). Side (B) of the diode is common to the $+V$ side of $R_2$, so this point is also labeled as (B). The lower side of the zener and the input to the op-amp are both labeled (A) since these points are common to each other. The virtual-ground principle allows us to also label the inverting input to the op-amp as (A). Continuing the process, the lower side of $R_2$ is also labeled as (A).

Since the zener diode and $R_2$ have the same (A) and (B) labels, the voltages across the two components are always equal. The zener voltage is constant (and equal to the $V_Z$

◄ **OBJECTIVE 11**

**Voltage-controlled current source**
A circuit with a constant-current output controlled by the circuit input voltage.

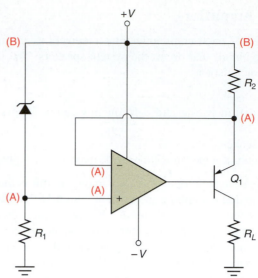

**FIGURE 16.32** Voltage-controlled current source.

rating of the component), so the voltage across $R_2$ is also constant. As a result, the current through $R_2$ has a fixed value. Because this is the emitter current of the *pnp* transistor, the $Q_1$ collector current is also constant, as is the load current. Note that the load current stays constant as long as $(I_C R_L)$ does not cause the transistor to saturate. If the transistor goes into saturation, current regulation is lost.

### 16.5.3 Precision Rectifier

**OBJECTIVE 12 ▶**

**Precision rectifier**

A clipper that consists of a diode and an op-amp. The circuit is characterized by the ability to clip extremely low-level input signals.

A **precision rectifier** is an op-amp/diode circuit characterized by its ability to conduct at extremely low diode forward voltages. This relatively simple circuit can be used in any clipper application. For example, consider the positive clipper shown in Figure 16.33a. You may recall that this circuit is used to eliminate the positive portion of its input signal. There is just one problem with the circuit: As shown, the input signal is clipped at approximately 700 mV. What if you want to clip a 100 mV$_{pk}$ input signal at 0 V? The standard diode clipper cannot be used in this application because of the forward voltage required to turn on the diode.

The precision rectifier circuit shown in Figure 16.33b can be used to clip a low-level signal. When $V_{in}$ to the circuit is more positive than 0 V by even a few millivolts, the op-amp output goes positive, cutting $D_1$ off. When the input goes negative, so does the output, which turns $D_1$ on. Since there is no voltage divider in the feedback path, the circuit acts as a voltage follower and the output is identical to the input. The 0.7 V drop across $D_1$ is offset by the op-amp.

Reversing the direction of $D_1$ produces a *negative* clipper. This circuit acts as described above. The only difference is that the *negative* alternation of the input is clipped.

### 16.5.4 One Final Note

Obviously, there is a nearly endless list of applications for the op-amp. We could not hope to cover them all in this chapter. In upcoming chapters, you will see more op-amp circuits, such as *active filters*, *Schmitt triggers*, and so on.

---

**Section Review ▶**

1. What is an *audio amplifier*?

2. What characteristics of the op-amp make it ideal for audio-amplifier applications?

3. What characteristics of the op-amp make it ideal for use as the input circuit for a voltmeter?

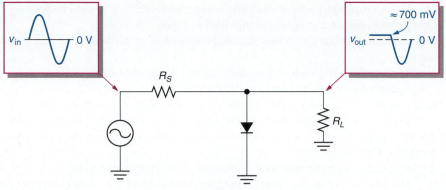

(a) Shunt rectifier input and output signals

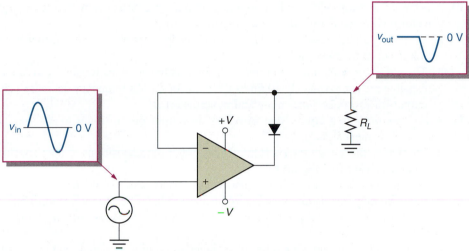

(b) Precision rectifier input and output signals

FIGURE 16.33    Precision rectifier circuit.

4. Describe the operation of the *voltage-controlled current source*.
5. Describe the operation of the *precision rectifier*.

Here is a summary of the major points made in this chapter:

1. A *comparator* is a circuit used to compare two voltages. Usually, a comparator is used to make either of the following comparisons:
   a. Two changing voltages to each other.
   b. One changing voltage to a set dc reference voltage. A comparator used for this purpose is often called a *level detector*.
2. The output from a comparator is normally a dc voltage that indicates the polarity (or magnitude) relationship between two voltages.
3. The circuit recognition feature of any comparator is the lack of a feedback path from the op-amp output to its input (see Figure 16.3).
4. Most comparators use a voltage-divider circuit to set the value of the reference voltage ($V_{ref}$) (see Figure 16.4).
5. Comparators come in a variety of configurations. Any combination of the words *positive* and *negative* in the following statement would be correct: "A comparator is a circuit used to provide a _____ output when the input voltage is more _____ than some _____ reference voltage."
   a. Several example comparator configurations are shown in Figure 16.6.
   b. A *variable comparator* is shown in Figure 16.7.

6. Due to their extremely simple circuitry, comparators are easy to troubleshoot.
7. Comparator characteristics are summarized in Figure 16.10.
8. An *integrator* provides an output that is proportional to the area of its input wave-form.
   a. The area of the rectangular waveform equals the product of its *pulse width* (measured in units of time) and its amplitude.
   b. An integrator is often used as a square wave–to–triangular wave converter.
   c. The circuit recognition feature of the integrator is a *capacitive feedback path*.
9. An *RC* integrator with a square-wave input provides an *exponential* output wave-form (see Figure 16.13).
   a. An op-amp integrator provides a linear, or *ramp*, output waveform.
   b. The linear output of the op-amp integrator is based on the op-amp input circuit, which acts as a *constant-current* circuit. This circuit allows the capacitor to charge (and discharge) at a linear rate.
10. A *feedback resistor* is often added to an op-amp integrator to ensure that the output waveform is centered around 0 V (see Figure 16.16).
   a. When a feedback resistor is added, an *RC* feedback path is produced that has a *lower* cutoff frequency ($f_{C1}$).
   b. The output from the integrator begins to lose its linear characteristics at frequencies that are approximately 10 times the value of $f_{C1}$. As frequency decreases, the output waveform becomes more and more distorted.
11. A *differentiator* is a circuit that provides an output proportional to *the rate of change* of its input signals.
   a. Physically, this circuit differs from the integrator in the positions of the capacitor and the resistor (see Figure 16.11).
   b. A differentiator acts as a triangular wave–to–square wave converter.
12. A *summing amplifier* is an op-amp circuit that produces an output proportional to the sum of its input voltages. (A basic summing amplifier is shown in Figure 16.20.)
13. In many cases, a summing amplifier is designed to provide an output that is *proportional to* (rather than equal to) the sum of its inputs. This prevents the inputs from overdriving the op-amp output.
14. The *general-class equation* for a summing amplifier is used to quickly determine the output for any set of input voltages. To derive this equation for a given summing amplifier:
   a. Determine the $R_f/R$ for each input branch.
   b. Represent each branch as the product of its resistance ratio and input voltage.
   c. Add the products found in step (b).
15. Summing amplifiers are often used as *digital-to-analog* (D/A) converters.
   a. A D/A converter is a circuit that converts a group of digital circuit outputs to an equivalent analog voltage.
   b. A simple D/A converter is shown in Figure 16.25.
16. An *averaging amplifier* is a circuit that provides an output proportional to the average of its input values (see Figure 16.27).
17. A summing amplifier is wired as an averaging amplifier when:
   a. All input resistors are equal in value.
   b. The ratio of any input resistor to the feedback resistor ($R/R_f$) equals the number of circuit inputs.
18. A *subtractor* is a variation on the summing amplifier that provides an output proportional to the difference between its input voltages (see Figure 16.28).
19. An *instrumentation amplifier* is a high-gain, high-CMRR amplifier used to detect and amplify low-level signals (see Figure 16.30).
20. The final output stage of most communications receivers is an *audio amplifier*. The ideal audio amplifier has:
   a. High gain.
   b. Minimum distortion in the audio-frequency range (approximately 20 Hz to 20 kHz).

c. High input impedance.

d. Extremely low output impedance.

21. A *voltage-controlled current source* is a circuit with a constant-current output that is controlled by the circuit input voltage (see Figure 16.32).

22. A *precision rectifier* is a clipper (made up of a diode and an op-amp) that can clip extremely low-level input signals (see Figure 16.33).

## EQUATION SUMMARY

| *Equation Number* | *Equation* | *Section Number* |
|---|---|---|
| **(16.1)** | $V_{ref} = +V\frac{R_2}{R_1 + R_2}$ | 16.1 |
| **(16.2)** | $f_{C1} = \frac{1}{2\pi R_f C_f}$ | 16.2 |
| **(16.3)** | $f_{min} = \frac{10}{2\pi R_f C_f}$ | 16.2 |
| **(16.4)** | $f_{max} = \frac{1}{20\pi R_f C_1}$ | 16.2 |
| **(16.5)** | $V_{out} = -R_f\left(\frac{V_1}{R_1} + \frac{V_2}{R_2} + \frac{V_3}{R_3}\right)$ | 16.3 |
| **(16.6)** | $v_{(out)} = \left(1 + \frac{2R}{R_C}\right)(v_2 - v_1)$ | 16.4 |
| **(16.7)** | $A_{CL} = 1 + \frac{2R}{R_C}$ | 16.4 |

## KEY TERMS

audio amplifier 683
averaging amplifier 678
comparator 658
differentiator 670
digital circuit 658
digital-to-analog (D/A) converter 677
general-class equation 675
instrumentation amplifier 681
integrator 666
level detector 659
precision rectifier 684
ramp 668
subtractor 679
summing amplifier 672
variable comparator 662
voltage-controlled current source 683

## PRACTICE PROBLEMS

### Section 16.1

1. Determine the value of $V_{ref}$ for the comparator in Figure 16.34.
2. Determine the value of $V_{ref}$ for the comparator in Figure 16.35.
3. Determine the value of $V_{ref}$ for the comparator in Figure 16.36.
4. Determine the value of $V_{ref}$ for the comparator in Figure 16.37.

### Section 16.2

5. Determine the cutoff frequency for the circuit shown in Figure 16.38.
6. Determine the frequency at which the circuit shown in Figure 16.38 will start to lose its linear output characteristics.
7. Determine the cutoff frequency of the circuit shown in Figure 16.39.
8. Determine the frequency at which the circuit shown in Figure 16.39 will start to lose its linear output characteristics.
9. Determine the cutoff frequency of the circuit shown in Figure 16.40.

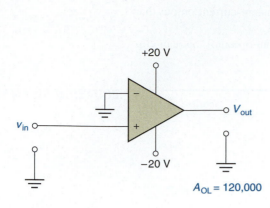

FIGURE 16.34

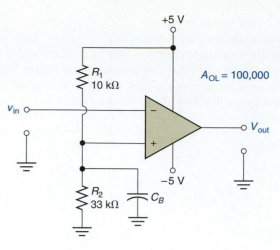

FIGURE 16.35

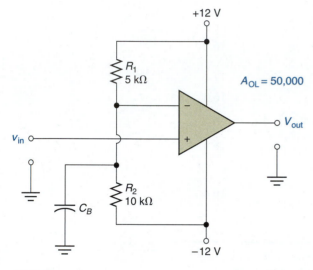

FIGURE 16.36

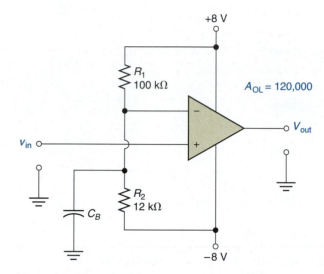

FIGURE 16.37

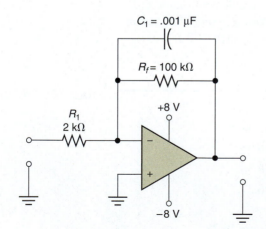

FIGURE 16.38

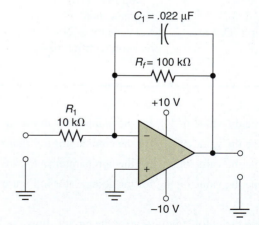

FIGURE 16.39

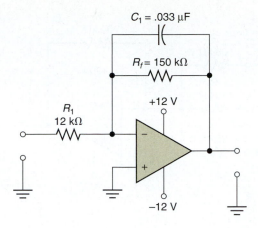

FIGURE 16.40

**10.** Determine the frequency at which the circuit shown in Figure 16.40 will start to lose its linear output characteristics.

**11.** A differentiator has values of $R_1 = 10 \text{ k}\Omega$ and $C_1 = 0.01 \ \mu\text{F}$. Determine its maximum linear operating frequency and its cutoff frequency.

**12.** A differentiator has values of $R_1 = 2.2 \text{ k}\Omega$ and $C_1 = 0.15 \ \mu\text{F}$. Determine its maximum linear operating frequency and its cutoff frequency.

**Section 16.3**

**13.** Derive the general-class equation for the summing amplifier in Figure 16.41a. Then, use the equation to determine the output voltage for the circuit.

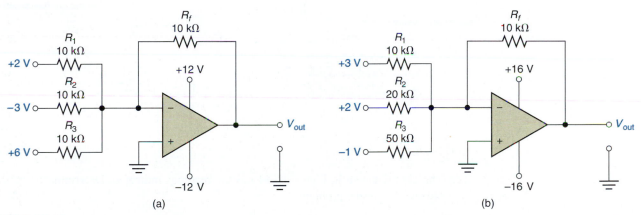

FIGURE 16.41

**14.** Derive the general-class equation for the summing amplifier in Figure 16.41b. Then, use the equation to determine the output voltage for the circuit.

**15.** Derive the general-class equation for the summing amplifier in Figure 16.42. Then, use the equation to determine the output voltage for the circuit.

**16.** The feedback resistor in Figure 16.42 is changed to 24 kΩ. Derive the new general-class equation for the circuit, and determine its new output voltage.

**17.** A five-input summing amplifier has values of $R_1 = R_2 = R_3 = R_4 = R_5 = 15 \text{ k}\Omega$. What value of feedback resistor is required to produce an averaging amplifier?

**18.** A four-input summing amplifier has values of $R_1 = R_2 = R_3 = R_4 = 20 \text{ k}\Omega$. What value of feedback resistor is required to produce an averaging amplifier?

**19.** Refer to Figure 16.28. Assume that $V_1 = 4$ V and $V_2 = 6$ V. Determine the output voltage from the circuit.

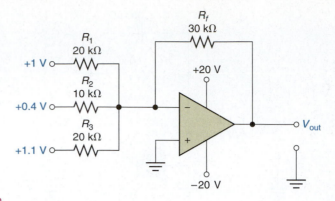

FIGURE 16.42

**20.** Refer to Figure 16.28. Assume that $R_{f2}$ in the circuit has been changed to 30 kΩ. Also, assume that the input voltages are $V_1 = 1$ V and $V_2 = 3$ V. Determine the output voltage from the circuit.

**TROUBLESHOOTING PRACTICE PROBLEMS**

**21.** The circuit shown in Figure 16.43 has the readings indicated. Determine the possible cause(s) of the problem.

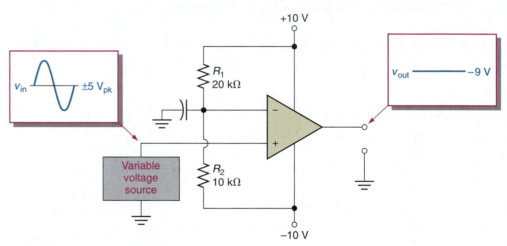

FIGURE 16.43

**22.** The circuit shown in Figure 16.44 has the readings indicated. Determine the possible cause(s) of the problem.

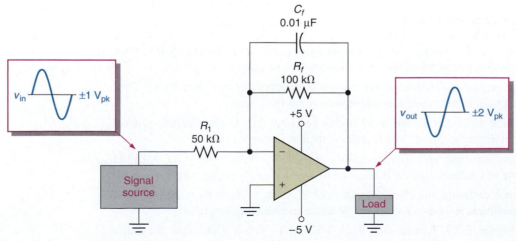

FIGURE 16.44

**23.** The circuit shown in Figure 16.45 has the readings indicated. Determine the possible cause(s) of the problem.

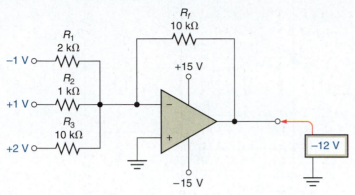

FIGURE 16.45

---

PUSHING THE ENVELOPE

**24.** Calculate the voltage gain of the audio amplifier in Figure 16.46.

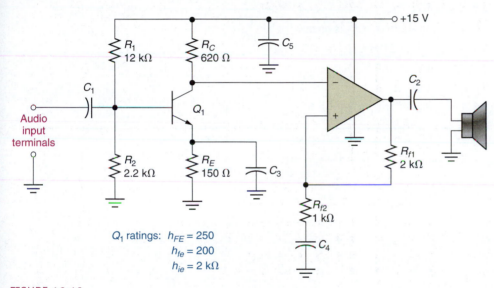

$Q_1$ ratings: $h_{FE} = 250$
$h_{fe} = 200$
$h_{ie} = 2$ k$\Omega$

FIGURE 16.46

**25.** The circuit in Figure 16.47 has the component values shown. Determine the value of $V_{CE}$ for $R_L = 1$ k$\Omega$ and $R_L = 3$ k$\Omega$.

**26.** Design a circuit to solve the following equation:

$$-V_{out} = \frac{V_1 - V_2}{2} + \frac{V_3}{3}$$

The circuit is to contain only two op-amps.

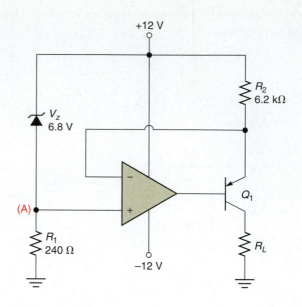

FIGURE 16.47

**ANSWERS TO THE EXAMPLE PRACTICE PROBLEMS**

**16.1** $V_{out} = \pm 11$ V

**16.2** $+3$ V

**16.3** $R_1$ is shorted or $R_2$ is open.

**16.4** $f_{C1} = 31.21$ Hz, $f_{min} = 312.1$ Hz

**16.5** $-6$ V

**16.6** $-3$ V

**16.7** $-V_{out} = 10V_1 + 5V_2 + V_3, V_{out} = -16$ V

chapter 17

# Tuned Amplifiers

## Objectives

*After studying the material in this chapter, you should be able to:*

1. Compare and contrast the frequency-response characteristics of ideal and practical tuned amplifiers.

2. Discuss the *quality (Q) factor* of a tuned amplifier, the factors that affect its value, and its relationship to amplifier bandwidth.

3. Describe the frequency-response curves of the *low-pass*, *high-pass*, *band-pass*, and *band-stop (notch)* filters.

4. Compare and contrast the frequency-response curves of Butterworth, Chebyshev, and Bessel filters.

5. Perform the gain and frequency analysis of *low-pass* and *high-pass* active filters.

6. Perform the gain and frequency analysis of the *two-stage band-pass*, *multiple-feedback band-pass*, and *notch* filters.

7. Describe active filter fault symptoms and troubleshooting.

8. Describe the frequency-response characteristics of discrete tuned amplifiers.

9. Perform the complete frequency analysis of a discrete tuned amplifier.

10. Discuss the tuning and troubleshooting of common discrete tuned amplifiers.

11. Describe the operation of the basic *class C* amplifier.

## Outline

**17.6** Discrete Tuned Amplifiers

**17.7** Discrete Tuned Amplifiers: Practical Considerations and Troubleshooting

**17.8** Class C Amplifiers

Chapter Summary

<table>
<tr><td>

**Tuned amplifier**

An amplifier designed for a specific bandwidth.

</td></tr>
</table>

The heart of any communications system is the **tuned amplifier**. A tuned amplifier is designed for a specific bandwidth. For example, a given tuned amplifier may be designed to amplify only those frequencies that are within $\pm 20$ kHz of 1000 kHz, that is, between 980 and 1020 kHz. As long as the input signal is within these frequencies, it is amplified. If it goes outside this frequency range, amplification is drastically reduced.

In Chapter 14, you were introduced to the concept of bandwidth. You were also shown how amplifier resistance and capacitance values limit the frequency response of a given amplifier. You may be wondering, then, what distinguishes a *tuned* amplifier from any other type of amplifier. After all, don't *all* amplifiers provide gain over a limited band of frequencies? The difference is that tuned amplifiers are designed to have *specific* bandwidths. For standard amplifiers, bandwith is viewed as a limitation. For tuned amplifiers, a specific bandwidth is a desired characteristic that is achieved through circuit design. In many cases, the bandwidth of a tuned amplifier is narrower than that of a standard amplifier having the same geometric center frequency ($f_0$). This point is illustrated in Figure 17.1

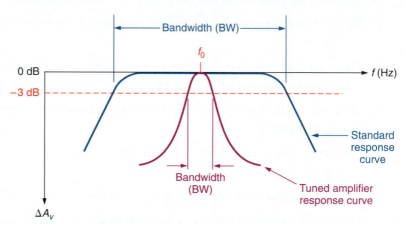

**FIGURE 17.1** Tuned amplifier frequency response.

In this chapter, we discuss the characteristics of tuned op-amp and discrete circuits. We will also take a brief look at some tuned circuit applications.

## 17.1 Tuned Amplifier Characteristics

**OBJECTIVE 1 ▶**
Before we get into any specific circuits, let's take a moment to discuss some of the bandwidth characteristics of tuned amplifiers. The *ideal* characteristics of such an amplifier are illustrated in Figure 17.2.

The ideal tuned amplifier would have zero ($-\infty$ dB) gain for all frequencies between 0 Hz and the lower cutoff frequency, ($f_{C1}$). Between $f_{C1}$ and the upper cutoff frequency ($f_{C2}$), the circuit gain would equal its design value, $A_{v(\text{mid})}$. If the operating frequency were to exceed the value of $f_{C2}$, the gain would again drop to zero. Thus, all frequencies within the bandwidth of the amplifier would be effectively *passed*, while all others would be effectively *stopped*. This is where the terms **pass band** and **stop band** come from.

In practice, the ideal characteristics of the tuned amplifier have not been achieved. Figure 17.3 compares the characteristics of the ideal tuned amplifier to those of a more practical circuit. Note that the gain roll-off of the practical circuit curve is not instantaneous.

<table>
<tr><td>

**Pass band**

The range of frequencies passed (amplified) by a tuned amplifier.

**Stop band**

The range of frequencies outside an amplifier's pass band.

</td></tr>
</table>

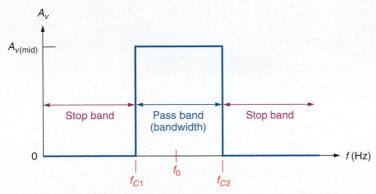

FIGURE 17.2   The ideal characteristics of a tuned amplifier.

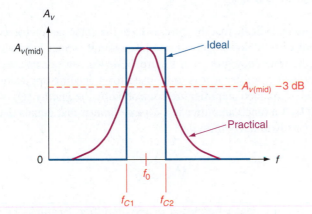

FIGURE 17.3   Ideal versus practical pass-band characteristics.

Rather, the gain rolls off from $A_{v(mid)}$ gradually as the frequency passes outside the pass band. As you will now be shown, the roll-off rate is an important factor in the overall operation of a tuned amplifier.

## 17.1.1   Roll-Off Rate Versus Bandwidth

We need to start this discussion by establishing two "bottom line" principles:

1. The closer the frequency-response curve of a tuned amplifier comes to that of the ideal, the better.
2. *In some applications*, the narrower the bandwidth of a tuned amplifier, the better.

The first statement almost goes without saying. As with any circuit, the closer a tuned cir-cuit comes to its ideal characteristics, the better. (This has been the case with every compo-nent and circuit we have discussed.) The second statement warrants discussion. In Chapter 14, we viewed the frequency response of an amplifier as being a *limitation*. In other words, we normally consider an *infinite* bandwidth to be ideal. This is not the case with tuned amplifiers. *With tuned amplifiers, we want a specific—usually narrow—bandwidth.* For example, take a look at Figure 17.4. Here we see a *pass band* (shaded) with two *stop bands*, A and B. If a tuned amplifier is designed to have a center frequency of 1000 kHz and a bandwidth of 40 kHz, that amplifier can pass all the frequencies within the pass band while rejecting all those in either of the stop bands. However, if we used a tuned amplifier with the same center frequency and a bandwidth of 100 kHz, that amplifier would pass a portion of the frequencies in both stop bands. This is not an acceptable situation. A tuned amplifier must pass all the frequencies within the pass band while stopping all others. Therefore, we want a given tuned amplifier to have a specific, usually narrow, bandwidth. With this in mind, let's take a look at the relationship between roll-off rate and bandwidth.

The lower the roll-off rate of an amplifier, the greater its bandwidth. This relationship is illustrated in Figure 17.5, which shows the frequency-response curves of two tuned

◀ *OBJECTIVE 2*

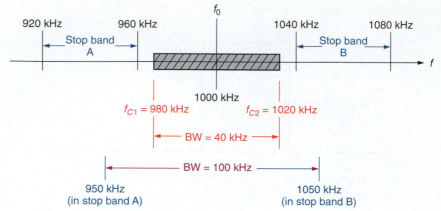

FIGURE 17.4   Frequency bands.

*An Example:*
The frequency ranges shown in Figure 17.4 could represent the frequency ranges of three AM radio stations. If the tuned amplifier has a bandwidth of 40 kHz, it will pick up only the radio station to which it is tuned (the center station). If the tuned amplifier has a bandwidth of 100 kHz, it will not only pick up the center station, it will pick up portions of the other two station signals as well.

amplifiers. The curves indicate that the circuits have the same geometric center frequency ($f_0$), yet the circuit with the lower roll-off rate has a much greater bandwidth. If we apply this principle to the frequency groups in Figure 17.4, we can see that we must limit the roll-off rate of a tuned amplifier if it is to be used for a specific application. The roll-off rate and bandwidth of a tuned amplifier are controlled by the *quality (Q)* of the circuit.

**Quality (Q)**
A figure of merit for a tuned amplifier that is equal to the ratio of center frequency to bandwidth.

The **quality (Q)** of a tuned amplifier is a *figure of merit* that equals the ratio of center frequency ($f_0$) to bandwidth. By formula,

$$Q = \frac{f_0}{BW} \qquad (17.1)$$

where $f_0$ is the geometric center frequency of the amplifier. Example 17.1 illustrates the relationship given in equation (17.1).

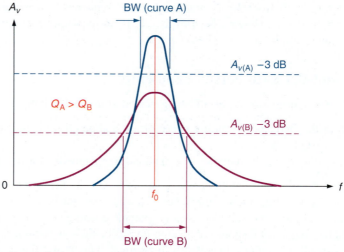

FIGURE 17.5   Bandwidth versus roll-off rate.

**EXAMPLE 17.1**

What is the required $Q$ of a tuned amplifier used for the application shown in Figure 17.4?

**Solution:**   The pass band in Figure 17.4 has a geometric center frequency of 1000 kHz and a bandwidth of 40 kHz. Using these two values, the required $Q$ is found as

$$Q = \frac{f_0}{BW} = \frac{1000 \text{ kHz}}{40 \text{ kHz}} = 25$$

Thus, the tuned amplifier used in this application would require a $Q$ of approximately 25.

**PRACTICE PROBLEM 17.1**
An amplifier with a geometric center frequency of 1200 kHz has a bandwidth of 20 kHz. Determine the $Q$ of the amplifier.

Equation (17.1) is somewhat misleading because it implies that the $Q$ of an amplifier depends on the circuit's geometric center frequency and bandwidth. The $Q$ of a tuned amplifier actually depends on component values within the circuit (as does the value of $f_0$). Once these values have been calculated, the bandwidth of a tuned amplifier is found using

> What determines the $Q$ of an amplifier?

$$BW = \frac{f_0}{Q} \qquad (17.2)$$

The following example further illustrates this point.

**EXAMPLE 17.2**

Using the proper circuit calculations, the $Q$ of a given amplifier is found to be 60. If the geometric center frequency for the amplifier is 1860 kHz, what is its bandwidth?

*Solution:* With a geometric center frequency of 1860 kHz and a $Q$ of 60, the amplifier bandwidth is found as

$$BW = \frac{f_0}{Q} = \frac{1860 \text{ kHz}}{60} = 31 \text{ kHz}$$

**PRACTICE PROBLEM 17.2**
The $Q$ of an amplifier is determined to be 25. If the geometric center frequency of the circuit is 1400 kHz, what is its bandwidth?

As we discuss various tuned amplifiers, you will be shown how to calculate the values of $f_0$ and $Q$ for each. Once these values are known, determining the bandwidth of a tuned amplifier is simple.

## 17.1.2 Geometric Center Frequency

As you may recall, $f_0$ is the geometric average of $f_{C1}$ and $f_{C2}$. By formula,

$$f_0 = \sqrt{f_{C1}f_{C2}} \qquad (17.3)$$

When the $Q$ of an amplifier is greater than (or equal to) 2, the value of $f_0$ approaches the *algebraic average* of $f_{C1}$ and $f_{C2}$ (designated $f_{\text{ave}}$); that is, $f_0$ approaches a value that is exactly halfway between $f_{C1}$ and $f_{C2}$. By formula,

$$f_0 \cong f_{\text{ave}} \quad (\text{when } Q \geqslant 2)$$

where $f_{\text{ave}}$ is the *algebraic* average of $f_{C1}$ and $f_{C2}$, which is found as

$$f_{\text{ave}} = \frac{f_{C1} + f_{C2}}{2} \qquad (17.4)$$

Example 17.3 demonstrates the relationship between $f_0$ and $f_{\text{ave}}$ when $Q \geqslant 2$.

## EXAMPLE 17.3

The cutoff frequencies for a tuned amplifier are measured as $f_{C1} = 960$ kHz and $f_{C2} = 1440$ kHz. Using equations (17.3) and (17.4), determine the values of $f_0$ and $f_{ave}$ for the circuit. Then, verify that the value of $Q$ for the circuit is greater than 2.

*Solution:*  Using equation (17.3), the value of $f_0$ is found as

$$f_0 = \sqrt{f_{C1}f_{C2}} = \sqrt{(960 \text{ kHz})(1440 \text{ kHz})} = \textbf{1176 kHz}$$

*A Practical Consideration:*
The cutoff frequencies of a tuned amplifier are measured using the technique described in Chapter 14 (Section 14.1.2)

Using equation (17.4), $f_{ave}$ is found as

$$f_{ave} = \frac{f_{C1} + f_{C2}}{2} = \frac{960 \text{ kHz} + 1440 \text{ kHz}}{2} = \textbf{1200 kHz}$$

The value of $Q$ for the circuit is now found as

$$Q = \frac{f_0}{\text{BW}} = \frac{1176 \text{ kHz}}{480 \text{ kHz}} = \textbf{2.45}$$

As you can see, the value of $f_0$ approaches the value of $f_{ave}$ when the $Q$ of an amplifier is greater than (or equal to) 2. (In this case, the difference between $f_0$ and $f_{ave}$ was approximately 2%.)

### PRACTICE PROBLEM 17.3

A tuned amplifier has measured cutoff frequencies of 980 and 1080 kHz. Show that $f_0 \cong f_{ave}$ for the circuit and that the circuit value of $Q$ is greater than 2.

You may be wondering how you can tell when the value of $Q$ for an amplifier is greater than 2. You can assume that the value of $Q$ is greater than (or equal to) 2 *when the value of $f_0$ for the circuit is at least two times its bandwidth.* As long as this condition is met, the $Q$ of the amplifier will be greater than (or equal to) 2, and $f_0$ will be approximately equal to $f_{ave}$. As you will learn, this relationship simplifies the frequency analysis of many tuned amplifiers.

**Section Review ▶**

1. What are the characteristics of the *ideal* tuned amplifier?
2. How does the frequency response of the practical tuned amplifier vary from that of the ideal tuned amplifier?
3. What is the relationship between the roll-off rate and the bandwidth of a tuned amplifier?
4. What is the *quality* ($Q$) rating of a tuned amplifier?
5. What is the relationship between the values of $Q$ and BW for a tuned amplifier?
6. What determines the value of $Q$ for a tuned amplifier?
7. What is the relationship between the values of $Q$ and $f_0$ for a tuned amplifier?

**Critical Thinking ▶**

8. Based on the curves in Figure 17.5, what is the relationship (if any) between amplifier $Q$ and voltage gain?

## 17.2  Active Filters: An Overview

**Active filter**

A tuned op-amp circuit.

Tuned op-amp circuits are generally referred to as **active filters**. There are four basic types of active filters, each with its own circuit configurations and frequency response curve.

## 17.2.1    Overview

The frequency-response curve for each type of filter is shown in Figure 17.6. The **low-pass filter** represented in Figure 17.6a passes all frequencies from 0 Hz (dc) up to an upper cutoff frequency ($f_C$). The **high-pass filter** represented in Figure 17.6b passes all frequencies above a lower cutoff frequency ($f_C$).

The curve in Figure 17.6b implies that a high-pass filter does not have an upper cutoff frequency. Since the op-amp has a unity-gain frequency, the circuit must have an upper cutoff frequency. However, this frequency is normally well beyond the operating frequency of the filter, so it is of no consequence.

◄ **OBJECTIVE 3**

**Low-pass filter**
A filter designed to pass all frequencies *below* a given cutoff frequency.

**High-pass filter**
A filter designed to pass all frequencies *above* a given cutoff frequency.

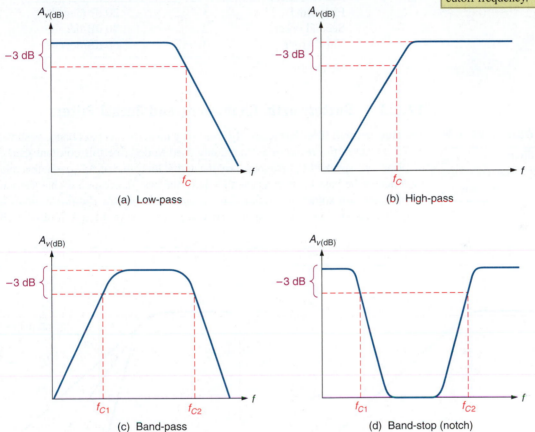

FIGURE 17.6    Active filter frequency-response curves.

The **band-pass filter** (represented in Figure 17.6c) passes all frequencies that fall between its values of $f_{C1}$ and $f_{C2}$. The **band-stop filter**, or **notch filter**, *blocks* all frequencies between its values of $f_{C1}$ and $f_{C2}$. In other words, it passes all frequencies *below* $f_{C1}$ and *above* $f_{C2}$ (as shown in Figure 17.6d). Note that band-pass and notch filters have opposite frequency-response characteristics.

One critical point needs to be made at this time: The concepts of $Q$, center frequency, and bandwidth are related primarily to the band-pass and notch filters. When we are dealing with band-pass and notch filters, we are generally concerned with the bandwidths of such amplifiers, along with their respective values of $Q$ and $f_0$. However, when we are dealing with low-pass and high-pass filters, we are concerned only with the circuit cutoff frequency. The concepts of $Q$ and center frequency generally are not applied to these circuits.

**Band-pass filter**
One designed to pass all frequencies that fall between its cutoff frequencies ($f_{C1}$ and $f_{C2}$).

**Band-stop (notch) filter**
One designed to block all frequencies that fall between its cutoff frequencies ($f_{C1}$ and $f_{C2}$).

## 17.2.2    General Terminology

Before we analyze the operation of any specific circuits, we need to establish some terms that are commonly used to describe active filters. The first of these is the term **pole**. A *pole* is simply an *RC* circuit. Thus, a one-pole filter contains one *RC* circuit, a two-pole

**Pole**
A single *RC* circuit.

filter contains two *RC* circuits, and so on. The term *order* is also used to identify the number of *RC* circuits in an active filter. For example, a *first-order* active filter contains one pole, a *second-order* active filter contains two poles, and so on.

Why do we emphasize the number of poles? As you learned in Chapter 14, gain roll-off rates are additive. *The more poles in an active filter, the higher the gain roll-off rate when the circuit is operated outside its pass band.* For example, one type of active filter (called a *Butterworth filter*) has a roll-off rate of 20 dB/decade per pole. The following table illustrates the relationship among order, poles, and gain roll-off for Butterworth filters:

| Filter Type | Number of Poles[a] | Total Gain Roll-Off |
|---|---|---|
| First order | 1 | 20 dB/decade |
| Second order | 2 | 40 dB/decade |
| Third order | 3 | 60 dB/decade |

[a]*RC* circuits.

### 17.2.3 Butterworth, Chebyshev, and Bessel Filters

OBJECTIVE 4 ▶

**Butterworth filter**

An active filter characterized by a flat pass-band response and 20 dB/decade roll-off rates.

The **Butterworth filter** has relatively constant gain across its pass band, as shown in Figure 17.7a. The term *flat response* is commonly used to describe this constant-gain characteristic. When we speak of flat response, we mean that the value of $A_{v(dB)}$ remains relatively constant across the pass band. Because they have the best flat-response characteristics, Butterworth filters are sometimes referred to as *maximally flat* or *flat-flat* filters. Figure 17.7b shows the differences in the response curves of first-, second-, and third-order Butterworth low-pass filters.

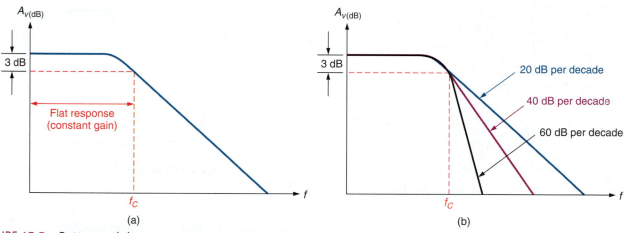

(a)    (b)

FIGURE 17.7    Butterworth low-pass response curves.

**Chebyshev filter**

An active filter characterized by an irregular pass-band response and high initial gain roll-off rates.

Another type of filter, the **Chebyshev filter**, has a higher *initial* roll-off rate (per pole) than a Butterworth filter. However, there are two inherent problems with the Chebyshev filter:

1. The gain of a Chebyshev filter is not constant across its pass band.
2. The Chebyshev filter has a greater roll-off rate *only for frequencies just outside the pass band.* As the operating frequency moves futher outside the pass band, the Chebyshev and Butterworth filters have equal roll-off rates.

**Ripple width**

The maximum variation in filter gain, measured in dB.

These two characteristics are illustrated in Figure 17.8. Figure 17.8a shows the response curve of a first-order Chebyshev filter. As you can see, the gain of the Chebyshev filter varies across the pass band. The **ripple width** of the filter is a maximum variation in filter gain, measured in dB. For the curve in Figure 17.8a, the ripple width is 3 dB. It should be noted that the ripple width of a Chebyshev filter can be reduced (by design)

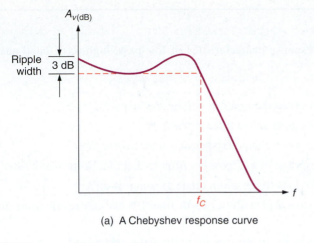

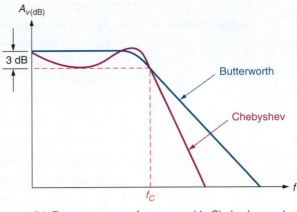

(a)  A Chebyshev response curve

(b)  Response curves for comparable Chebyshev and Butterworth filters

FIGURE 17.8    Chebyshev frequency response.

but at the expense of its high initial roll-off rate. In other words, if a Chebyshev filter is designed for a lower ripple width, its roll-off characteristics come closer to those of a Butterworth filter.

Figure 17.8b shows the response curves of comparable Chebyshev and Butterworth filters. Note that the Chebyshev has a higher initial roll-off rate than the Butterworth filter. However, as the circuit operating frequency moves futher outside the pass band, the two circuits eventually reach the same roll-off rate.

Even though the Chebyshev filter has a higher initial roll-off rate, the Butterworth filter provides consistent gain across its pass band. As a result, the Butterworth is by far the more commonly used of the two. Even so, the Butterworth filter has its own drawback: The time delay (from input and output) produced by the Butterworth filter is not constant across its pass band. This means that two (or more) frequencies applied to a Butterworth filter do not experience the same phase shift from input to output. The gain provided by the filter is constant, but the phase shift is not. This can produce severe signal distortion from input to output.

The **Bessel filter** is designed to provide a constant phase shift across its pass band. The constant phase shift of the Bessel filter results in greater **fidelity** (ability to reproduce a waveform accurately) than either the Butterworth or Chebyshev filter. However, the circuit has the disadvantage of a lower initial roll-off rate (as shown in Figure 17.9).

Of the three filter types introduced in this section, the Butterworth is the most commonly used. For this reason, we will concentrate on this type of circuit in the upcoming sections.

> **Bessel filter**
> A filter designed to provide a constant phase shift across its pass band.
>
> **Fidelity**
> The ability of a circuit to accurately reproduce a waveform.

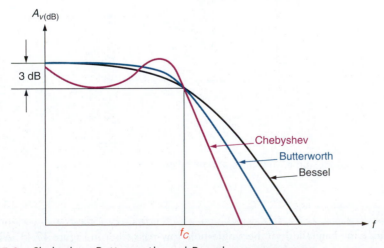

FIGURE 17.9    Chebyshev, Butterworth, and Bessel response curves.

1. What is an *active filter*?

2. Describe the frequency-response characteristics of low-pass, high-pass, band-pass, and notch filters.

3. What is a *pole*?

4. What is the relationship between the *order* of a filter and its poles?

5. Why is the number of poles in an active filter important?

6. What are the frequency-response characteristics of a *Butterworth filter*?

7. Compare the frequency response of a *Chebyshev filter* to that of a Butterworth filter.

8. What is *ripple width*? How does ripple width relate to gain roll-off?

9. What is the inherent drawback of a Butterworth filter? What can result from this drawback?

10. What is a *Bessel filter*? What are its primary strengths and weaknesses?

11. Write a brief paragraph comparing the frequency-response characteristics of Butterworth, Chebyshev, and Bessel filters.

## 17.3   Low-Pass and High-Pass Filters

### 17.3.1   The Single-Pole Low-Pass Filter

**OBJECTIVE 5 ▶** The single-pole Butterworth filter is designed as either a *variable-gain* circuit or a *voltage follower*. The variable-gain circuit is shown in Figure 17.10. As you can see, this circuit is simply a noninverting amplifier with an *RC* circuit added to the active input. Since the reactance of the capacitor decreases as frequency increases, the *RC* circuit limits the high-frequency response of the circuit. The cutoff frequency for this type of circuit is found as

$$f_C = \frac{1}{2\pi RC} \tag{17.5}$$

This is the same basic relationship that we used to find $f_C$ for the circuits we discussed in Chapter 14. Example 17.4 demonstrates the process for determining the bandwidth of a single-pole low-pass filter.

> Many of the circuits in the next several sections contain op-amps that are drawn with the noninverting input at the *top* of the symbol (like the one in Figure 17.10). They are drawn like this to simplify the illustrations.

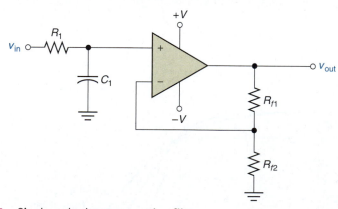

FIGURE 17.10   Single-pole, low-pass active filter.

### EXAMPLE 17.4

Determine the bandwidth of the single-pole low-pass filter in Figure 17.11. Also (as a review) determine the maximum value of $A_{CL}$ for the circuit.

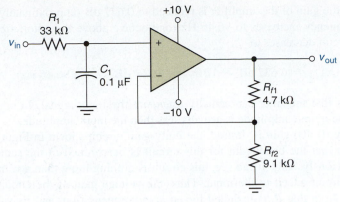

FIGURE 17.11

***Solution:*** Using the circuit values of $R_1$ and $C_1$, the upper cutoff frequency is found as

$$f_C = \frac{1}{2\pi R_1 C_1} = \frac{1}{2\pi (33 \text{ k}\Omega)(0.1 \text{ }\mu\text{F})} = \textbf{48.23 Hz}$$

Since the amplifier is capable of working as a dc amplifier (0 Hz), there is no lower cutoff frequency for the circuit. Thus, the bandwidth is equal to the value of $f_C$.

Because the circuit is a noninverting amplifier, its value of $A_{CL}$ is found using a form of equation (15.16) as follows:

$$A_{CL} = \frac{R_{f1}}{R_{f2}} + 1 = \frac{4.7 \text{ k}\Omega}{9.1 \text{ k}\Omega} + 1 = \textbf{1.52}$$

**PRACTICE PROBLEM 17.4**

A single-pole low-pass filter like the one in Figure 17.11 has values of $R_1 = 47 \text{ k}\Omega$, $C_1 = 0.033 \text{ }\mu\text{F}$, $R_{f1} = 10 \text{ k}\Omega$, and $R_{f2} = 10 \text{ k}\Omega$. Determine the values of $f_C$ and $A_{CL}$ for the circuit.

The frequency-response curve for the circuit in Example 17.4 can be drawn as shown in Figure 17.12. The dB value of midband voltage gain shown in the figure was determined as follows:

$$A_{CL(dB)} = 20 \log A_{CL} = 20 \log (1.52) = \textbf{3.637 dB}$$

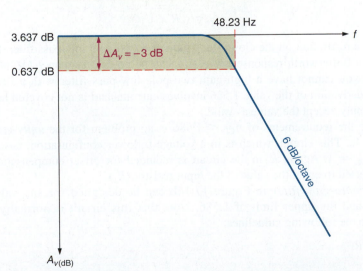

FIGURE 17.12   The frequency response of the circuit shown in Example 17.4.

At 48.23 Hz, the gain of the amplifier is reduced to 0.637 dB (approximately 1.08). If the operating frequency increases to 96.46 Hz (one octave above the cutoff frequency), the gain of the circuit decreases to

$$A_{CL} = 0.637 \text{ dB} - 6 \text{ dB} = -5.363 \text{ dB} \qquad \text{(at 96.46 Hz)}$$

This indicates that the filter can actually *attenuate* the input signal. In other words, at some point, the output amplitude is actually less than the input amplitude.

When unity (0 dB) gain is desired, the unity-gain circuit shown in Figure 17.13 may be used. Note that the bandwidth for this circuit is determined in the same manner as described previously. As you can see, this circuit is nothing more than a *voltage follower* with an *RC* circuit added to the input. Thus, the voltage gain of the circuit is approximately 0 dB. Note that $R_f$ is included for op-amp compensation, and its value does not affect any circuit calculations. The value of $R_f$ is normally selected to equal the value of the input resistor ($R_1$).

**Lab Reference**: The operation of a unity-gain filter is demonstrated in Exercise 26.

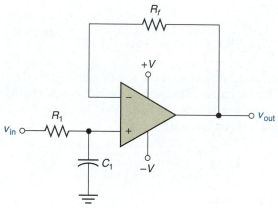

**FIGURE 17.13** Unity-gain circuit.

## 17.3.2 The Two-Pole Low-Pass Filter

The two-pole circuits shown in Figure 17.14 are also referred to as *Sallen-Key* filters (after their developers) and *VCVS* (voltage-controlled voltage-source) filters.

As stated earlier, a two-pole low-pass filter has a roll-off rate of 40 dB/decade. Two commonly used two-pole low-pass filter configurations are shown in Figure 17.14. As you can see, each circuit has two *RC* circuits, $R_1$–$C_1$ and $R_2$–$C_2$. As the operating frequency increases beyond $f_C$, each *RC* circuit reduces $A_{CL}$ by 20 dB/decade, giving a total roll-off rate of 40 dB/decade. The cutoff frequency for each circuit in Figure 17.14 is found as

$$f_C = \frac{1}{2\pi\sqrt{R_1 R_2 C_1 C_2}} \tag{17.6}$$

There is a restriction on the closed-loop gain of a two-pole low-pass filter. For the filter to have a Butterworth response curve, $A_{CL}$ can be no greater than *1.586 (4 dB)*. This means that you cannot have a high-gain two-pole low-pass filter with a flat response curve. The derivation of the value 1.586 involves calculus and is not covered here. In this case, we simply accept the value as valid.

**Lab Reference**: The operation of a two-pole unity-gain filter is demonstrated in Exercise 26.

Fulfilling the requirements of $A_{CL} < 1.586$ is no problem for the *unity-gain filter* in Figure 17.14a. This circuit, which is in a voltage-follower configuration, has a standard value of $A_{CL} = 1$. Again, $R_f$ in the circuit is included for offset compensation, and its value is selected to equal the value of the input resistor ($R_1$).

The *variable-gain filter* in Figure 17.14b can be designed for any value of $A_{CL}$ between 1 and the upper limit of 1.586. Note that this circuit is normally designed according to the following guidelines:

1. $R_2 = R_1$
2. $C_2 = 2C_1$
3. $R_{f1} \leq 0.586 R_{f2}$

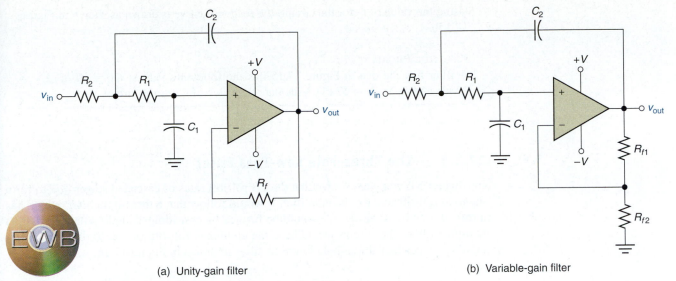

(a) Unity-gain filter             (b) Variable-gain filter

**FIGURE 17.14** Common two-pole low-pass filter configurations.

Example 17.5 illustrates these relationships along with the process for determining the bandwidth of a two-pole low-pass filter.

## EXAMPLE 17.5

Verify that the circuit in Figure 17.15a conforms to the component value relationships listed for a two-pole Butterworth filter. Also, determine the bandwidth of the filter, and draw its response curve.

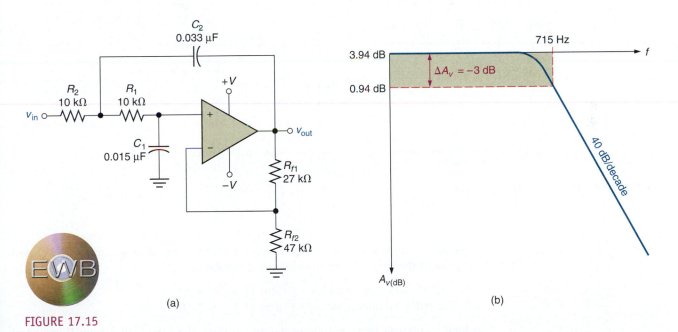

**FIGURE 17.15**

**Solution:** First, let's verify the component relationships. $R_1$ and $R_2$ are equal in value, while $C_2$ is approximately twice the value of $C_1$. Also note that the value of $R_{f1}$ is approximately equal to $0.586R_{f2}$.

The cutoff frequency (and, thus, the bandwidth) of the filter is found as

$$f_C = \frac{1}{2\pi\sqrt{R_1 R_2 C_1 C_2}} = \frac{1}{2\pi\sqrt{(10\text{ k}\Omega)(10\text{ k}\Omega)(0.015\text{ μF})(0.033\text{ μF})}} = \textbf{715 Hz}$$

The value of 3.94 dB in Figure 17.15b was found by solving

$$A_{CL} = \frac{R_{f1}}{R_{f2}} + 1$$

and converting the result to dB form.

Using the value of $f_C$ for the circuit, the response curve is drawn as shown in Figure 17.15b.

**PRACTICE PROBLEM 17.5**

A filter like the one in Figure 17.15 has the following values: $C_1 = 10$ nF, $C_2 = 22$ nF, and $R_1 = R_2 = 12$ k$\Omega$. Calculate the value of $f_C$ for the circuit.

### 17.3.3    The Three-Pole Low-Pass Filter

In Chapter 14, you were shown that the dB roll-off rates of cascaded stages *add* to form the total roll-off rate for the amplifier. It would follow that a three-pole filter, which has a roll-off rate of 60 dB/decade, could be formed by cascading a single-pole filter with a two-pole filter. This circuit would have the combined roll-off rate of 20 dB + 40 dB = 60 dB per decade. A three-pole low-pass filter is shown in Figure 17.16.

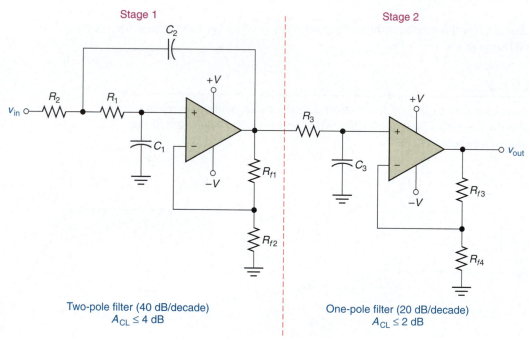

**FIGURE 17.16**   Three-pole low-pass filter.

For the three-pole low-pass filter to have a Butterworth response curve, the closed-loop gains of the filter stages must be limited as follows:

**1.** The two-pole filter must have a value of $A_{CL} \leq 4$ dB.
**2.** The one-pole filter must have a value of $A_{CL} \leq 2$ dB.

As long as both of these requirements are fulfilled, the filter will have a Butterworth response curve. Also, note that the two filters would be tuned to the same value of $f_C$. All circuit calculations for this amplifier are the same as those performed earlier.

Can active filters have more than three poles? Yes. If you see an active filter with more than three poles, just remember the following points:

**1.** The filter will have a 20 dB/decade roll-off *for each pole*. For example, a five-pole filter would have a roll-off rate of 5 × 20 dB/decade = 100 dB/decade.

| How is a multipole active filter analyzed? |

2. All stages will be tuned to the same cutoff frequency. Thus, the overall cutoff frequency is approximately equal to the value of $f_C$ for any stage.

## 17.3.4 High-Pass Filters

Figure 17.17 shows several typical high-pass filters. The high-pass filter differs from the low-pass filter in two respects: First, and most obvious, is that the resistors and capacitors have swapped positions. Second, the multipole circuits are designed to fulfill the following conditions:

What are the differences between low-pass and high-pass active filters?

1. $C_1 = C_2$.
2. $R_1 = 2R_2$.

Since the capacitors are in series with the amplifier input, they limit the *low-frequency* operation of the circuit. Note that the value of $f_C$ for each circuit is found using the same equation that we used to find $f_C$ for the equivalent low-pass filter.

As frequency decreases, the reactance of a given series capacitor increases. This causes a larger portion of the input signal to be dropped across the capacitor. When the

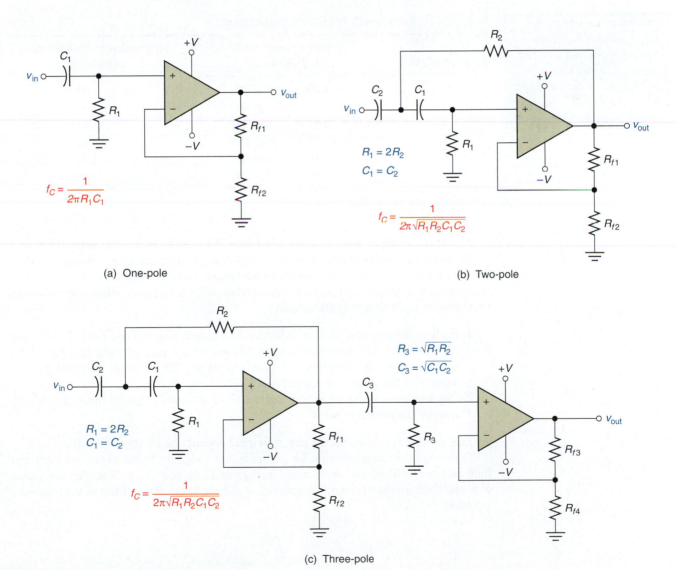

(a) One-pole

(b) Two-pole

(c) Three-pole

**FIGURE 17.17** Typical high-pass active filters.

**Lab Reference:** The operating characteristics of one-pole and two-pole high-pass filters are demonstrated in Exercise 28.

operating frequency reaches the value of $f_C$, the series capacitors reduce the gain of the amplifier by 3 dB.

We will not spend any time on these circuits since their operating principles should be familiar to you by now. Just remember that these circuits abide by all the rules established up to this point. The only difference is that they pass frequencies *above* their cutoff frequencies.

### 17.3.5    Filter Gain Requirements

In this section, you have been provided with the gain requirements for several low-pass and high-pass active filters. Fulfilling these requirements provides a Butterworth response curve; that is, each filter has relatively constant gain until its cutoff frequency is reached. Then, the gain rolls off at an approximate rate of 20 dB/decade per pole.

The gain requirements for a variety of Butterworth active low-pass and high-pass filters are summarized in Table 17.1. The derivations of these gain requirements are way beyond the scope of this text. However, you should be aware that almost every type of low-pass or high-pass active filter has gain requirements that must be fulfilled if the circuit is to have a Butterworth response curve.

Note that the maximum gain for low-pass and high-pass active filters increases by 2 dB for each pole added to the circuit. Also note that the approximated roll-off rate is equal to 20 dB/decade per pole.

**TABLE 17.1**    Butterworth Filter Gain Requirements

| Number of Poles | Maximum Overall Voltage Gain[a] | | Approximate Roll-Off Rate (dB/Decade) |
|---|---|---|---|
| 2 | 1.586 | (4 dB) | 40 |
| 3 | 2 | (6 dB) | 60 |
| 4 | 2.58 | (8 dB) | 80 |
| 5 | 3.29 | (10 dB) | 100 |
| 6 | 4.21 | (12 dB) | 120 |
| 7 | 5.37 | (14 dB) | 140 |

[a]Decibel values are rounded off to the nearest whole number.

Any of the multiple filters described in Table 17.1 can be constructed using the appropriate number of two-pole and one-pole cascaded stages. For example, a five-pole circuit can be constructed by cascading two two-pole circuits and one one-pole circuit. To obtain the appropriate frequency-response characteristics for a five-pole filter, the following requirements must be met by the circuit:

1. Each of the two-pole circuits is limited to a closed-loop voltage gain of 4 dB, and the one-pole circuit is limited to a closed-loop voltage gain of 2 dB. This provides a maximum overall voltage gain of 4 dB + 4 dB + 2 dB = 10 dB, the value given in Table 17.1 for a five-pole active filter.
2. The resistor and capacitor values must to be chosen so that all three stages have identical cutoff frequencies.

As long as these requirements are met, the circuit operates as a Butterworth filter.

The component requirements for one-pole and two-pole Butterworth low-pass and high-pass active filters are summarized in Figures 17.18 and 17.19. Note that the values of $R$ and $C$ determine the cutoff frequencies according to the proper form of the following equation:

$$f_C = \frac{1}{2\pi RC}$$

Also, note that any of the circuits shown can be converted to unity-gain (0 dB) circuits by removing the lower (grounded) resistor in the feedback path, $R_{f2}$.

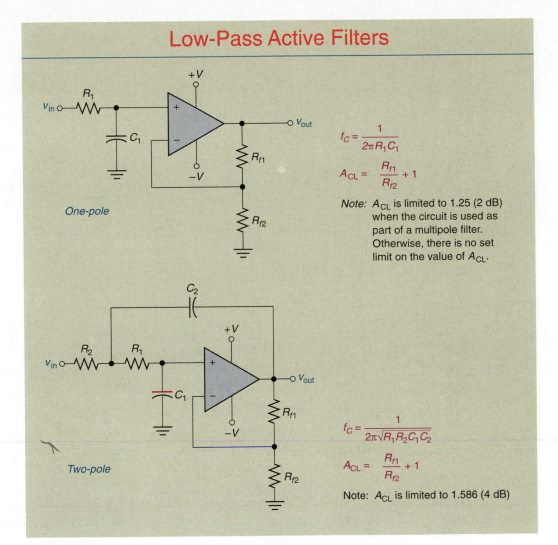

**Low-Pass Active Filters**

$$f_C = \frac{1}{2\pi R_1 C_1}$$

$$A_{CL} = \frac{R_{f1}}{R_{f2}} + 1$$

*Note:* $A_{CL}$ is limited to 1.25 (2 dB) when the circuit is used as part of a multipole filter. Otherwise, there is no set limit on the value of $A_{CL}$.

One-pole

$$f_C = \frac{1}{2\pi \sqrt{R_1 R_2 C_1 C_2}}$$

$$A_{CL} = \frac{R_{f1}}{R_{f2}} + 1$$

*Note:* $A_{CL}$ is limited to 1.586 (4 dB)

Two-pole

**FIGURE 17.18**

---

1. What steps are involved in analyzing active one-pole, two-pole, and three-pole low-pass filters?

2. What is the difference between low-pass and high-pass active filters?

3. Describe the dB gain and roll-off characteristics of low-pass and high-pass filters.

◀ **Section Review**

---

## 17.4 Band-Pass and Notch Circuits

You may recall that *band-pass filters* are designed to *pass* all frequencies within their bandwidths, while *notch filters (band-stop filters)* are designed to *block* all frequencies within their bandwidths. In this section, we will take a look at several band-pass and notch filters.

◀ *OBJECTIVE 6*

### 17.4.1    The Two-Stage Band-Pass Filter

A band-pass filter can be constructed by cascading a high-pass filter and a low-pass filter. Such a circuit is shown in Figure 17.20a. The first stage of the amplifier passes all frequencies that are below its cutoff frequency. The frequencies passed by the first stage are coupled to the second stage, which passes all frequencies above its cutoff frequency. The result of this circuit action is shown in Figure 17.20b. Note that the only frequencies that pass through the amplifier are those that fall within the pass band of *both* amplifiers.

*A Practical Consideration:*
The order of the low-pass and high-pass filters is unimportant. We could have a high-pass first stage and a low-pass second stage. The results would be the same.

# High-Pass Active Filters

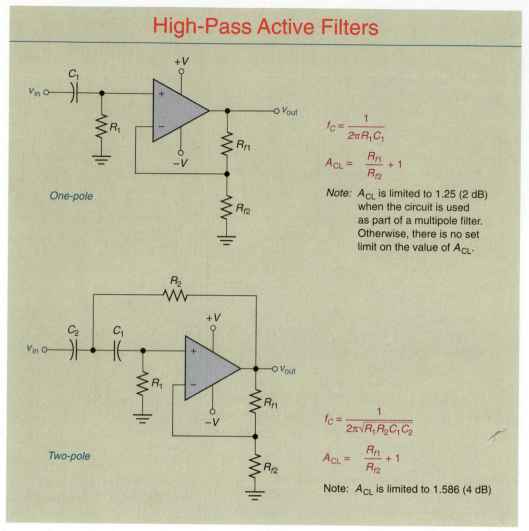

One-pole

$$f_C = \frac{1}{2\pi R_1 C_1}$$

$$A_{CL} = \frac{R_{f1}}{R_{f2}} + 1$$

*Note:* $A_{CL}$ is limited to 1.25 (2 dB) when the circuit is used as part of a multipole filter. Otherwise, there is no set limit on the value of $A_{CL}$.

Two-pole

$$f_C = \frac{1}{2\pi\sqrt{R_1 R_2 C_1 C_2}}$$

$$A_{CL} = \frac{R_{f1}}{R_{f2}} + 1$$

*Note:* $A_{CL}$ is limited to 1.586 (4 dB)

**FIGURE 17.19**

The frequency analysis of a circuit like the one in Figure 17.20a is relatively simple. The cutoff frequencies are found as shown in Section 17.3. Once the values of $f_{C1}$ and $f_{C2}$ are known, the circuit values of bandwidth, geometric center frequency, and $Q$ are found as follows:

$$BW = f_{C2} - f_{C1}$$
$$f_0 = \sqrt{f_{C1} f_{C2}}$$
$$Q = \frac{f_0}{BW}$$

The following example demonstrates a complete frequency analysis of the two-stage band-pass filter.

### EXAMPLE 17.6

Perform the complete frequency analysis of the amplifier in Figure 17.21.

*Solution:*   The value of $f_{C2}$ is determined by the first stage of the circuit (a *low-pass* filter) as follows:

$$f_{C2} = \frac{1}{2\pi\sqrt{R_1 R_2 C_1 C_2}} = \frac{1}{2\pi\sqrt{(10\text{ k}\Omega)(10\text{ k}\Omega)(0.01\text{ }\mu\text{F})(0.02\text{ }\mu\text{F})}} = \textbf{1.13 kHz}$$

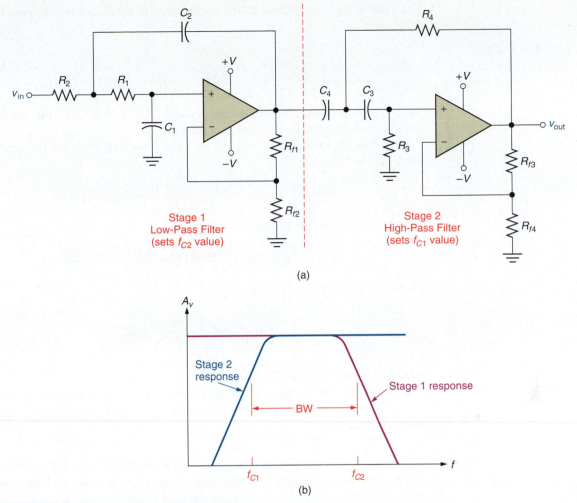

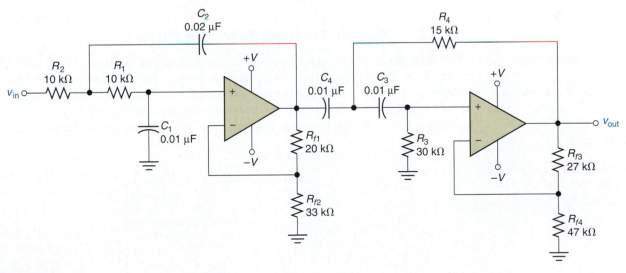

**FIGURE 17.20** A two-stage band-pass filter and response curve.

**FIGURE 17.21**

The value of $f_{C1}$ is determined by the second stage of the circuit (a *high-pass* filter) as follows:

$$f_{C1} = \frac{1}{2\pi\sqrt{R_3 R_4 C_3 C_4}} = \frac{1}{2\pi\sqrt{(30 \text{ k}\Omega)(15 \text{ k}\Omega)(0.01 \text{ }\mu\text{F})(0.01 \text{ }\mu\text{F})}} = \mathbf{750 \text{ Hz}}$$

Thus, the circuit passes all frequencies between 750 Hz and 1.13 kHz. All others are effectively blocked by the band-pass filter.

Now that the values of $f_{C1}$ and $f_{C2}$ are known, we can solve for the amplifier bandwidth as follows:

$$\mathbf{BW} = f_{C2} - f_{C1} = 1.13 \text{ kHz} - 750 \text{ Hz} = \mathbf{380 \text{ Hz}}$$

The geometric center frequency is found as

$$f_0 = \sqrt{f_{C1} f_{C2}} = \sqrt{(750 \text{ Hz})(1.13 \text{ kHz})} = \mathbf{921 \text{ Hz}}$$

Finally, the value of $Q$ is found as

$$Q = \frac{f_0}{\text{BW}} = \frac{921 \text{ Hz}}{380 \text{ Hz}} = \mathbf{2.42}$$

**PRACTICE PROBLEM 17.6**

A filter like the one in Figure 17.21 has values of $R_1 = R_2 = 12 \text{ k}\Omega$, $C_1 = C_3 = C_4 = 0.01 \text{ }\mu\text{F}$, $C_2 = 0.02 \text{ }\mu\text{F}$, $R_3 = 39 \text{ k}\Omega$, and $R_4 = 20 \text{ k}\Omega$. Perform the frequency analysis of the filter.

---

The two-stage band-pass filter is the easiest of the band-pass filters to analyze. However, it has the disadvantage of requiring two op-amps and a relatively large number of resistors and capacitors. As you will see, the physical construction of the *multiple-feedback band-pass filter* is much simpler than that of the two-stage filter. At the same time, the frequency analysis of the multiple-feedback band-pass filter is a bit more difficult than that of the two-stage filter.

### 17.4.2    Multiple-Feedback Band-Pass Filters

**Multiple-feedback band-pass filter**

A band-pass filter that has a single op-amp and two feedback paths, one resistive and one capacitive.

**Lab Reference**: The operation of a circuit like the one in Figure 17.22 is demonstrated in Exercise 27.

The **multiple-feedback band-pass filter**, which is shown in Figure 17.22, derives its name from the fact that it has two feedback networks, one capacitive and one resistive. Note the presence of the input series capacitor ($C_1$) and the input shunt capacitor ($C_2$). The series capacitor affects the low-frequency response of the filter, while the shunt capacitor affects its high-frequency response. This point is illustrated in Figure 17.23.

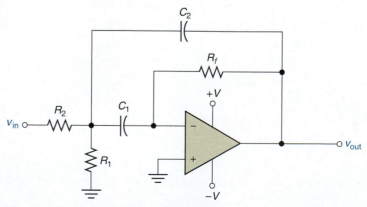

**FIGURE 17.22**    A multiple-feedback band-pass filter.

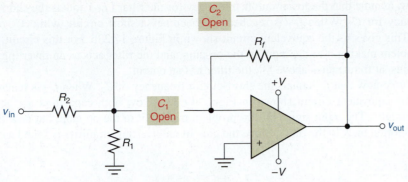

(a) An idealized equivalent for the bandpass filter when $f_{in} < f_{C1}$

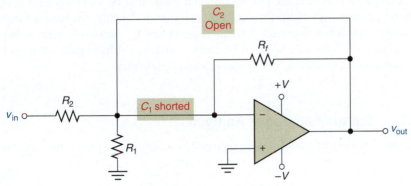

(b) An idealized equivalent for the bandpass filter when $f_{C1} < f_{in} < f_{C2}$

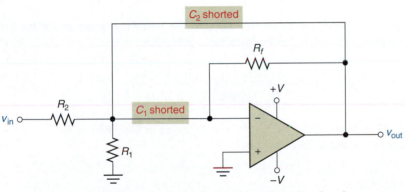

(c) An idealized equivalent for the bandpass filter when $f_{in} > f_{C2}$

**FIGURE 17.23** Ideal band-pass active filter circuit operation.

Let's begin our discussion of the circuit operation by establishing some ground rules. To simplify our discussion, we will assume that:

1. $C_2 < C_1$.
2. A given capacitor acts as an open circuit until a *short-circuit frequency* is reached. At that point, the capacitor is represented as a short circuit.

(Granted, the operation of the capacitors is more complicated than this, but we want to start by getting the overall picture of how the circuit works.)

When the input frequency is below the short-circuit frequency for $C_1$, *both capacitors act as open circuits*. Since $C_1 > C_2$, we know that the short-circuit frequency for $C_1$ is *lower* than that of $C_2$. Therefore, as long as $C_1$ is an open, $C_2$ is also an open. Having both capacitors acting as opens gives us the equivalent circuit shown in Figure 17.23a. As you can see, the input signal is completely isolated from the op-amp. Therefore, the circuit has no output.

Now, assume that the lower cutoff frequency for the filter ($f_{C1}$) equals the short-circuit frequency for $C_1$. When $f_{C1}$ is reached, $C_1$ becomes a short circuit, while $C_2$ remains open. This gives us the equivalent circuit shown in Figure 17.23b. For this circuit, $v_{in}$ has no problem making it to the input of the op-amp, and the filter acts as an inverting amplifier. Thus, at frequencies above $f_{C1}$, the filter has an output.

Assume now that $f_{C2}$ equals the short-circuit frequency of $C_2$. When $f_{C2}$ is reached, we have the equivalent circuit shown in Figure 17.23c. Now both capacitors are acting as short circuits. The input signal has no problem making it to the op-amp, but now the output is shorted back to the input. Since the gain of an inverting amplifier is found as

The actual equation used to find $A_{CL}$ for the multiple-feedback filter is a variation on the equation shown here. However, the equation shown is valid for the point being made.

$$A_{CL} = \frac{R_f}{R_{in}}$$

and $R_f$ is effectively shorted, the closed-loop voltage gain of the circuit is zero at frequencies above $f_{C2}$. If we put these three equivalent circuits together, we have a circuit with zero voltage gain when operated at frequencies below $f_{C1}$, relatively high voltage gain when operated at frequencies between $f_{C1}$ and $f_{C2}$, and zero voltage gain when operated at frequencies above $f_{C2}$. This, by definition, is a band-pass filter.

### 17.4.3 Circuit Frequency Analysis

As you know, the cutoff frequency for a given low-pass or high-pass active filter is found using

$$f_C = \frac{1}{2\pi \sqrt{R_1 R_2 C_1 C_2}}$$

This equation is modified to provide us with the following equation for the geometric center frequency of a multiple-feedback band-pass filter:

$$f_0 = \frac{1}{2\pi \sqrt{(R_1 \parallel R_2)R_f C_1 C_2}} \tag{17.7}$$

Example 17.7 demonstrates the use of this equation in determining the geometric center frequency of a multiple-feedback filter.

### EXAMPLE 17.7

Determine the value of $f_0$ for the filter in Figure 17.24.

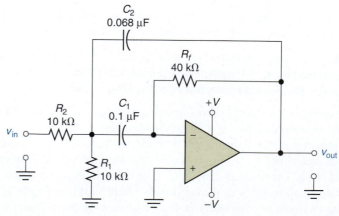

FIGURE 17.24

*Solution:* The center frequency of the circuit is found as

$$f_0 = \frac{1}{2\pi\sqrt{(R_1 \parallel R_2)R_fC_1C_2}} = \frac{1}{2\pi\sqrt{(10 \text{ k}\Omega \parallel 10 \text{ k}\Omega)(40 \text{ k}\Omega)(0.1 \text{ }\mu\text{F})(0.068 \text{ }\mu\text{F})}}$$

$$= \frac{1}{2\pi(0.001166)} = \textbf{136.5 Hz}$$

### PRACTICE PROBLEM 17.7
A multiple-feedback filter has the following values: $R_1 = R_2 = 30$ k$\Omega$, $R_f = 68$ k$\Omega$, $C_1 = 0.22$ $\mu$F, and $C_2 = 0.1$ $\mu$F. Determine the value of $f_0$ of the circuit.

Once the center frequency of a multiple-feedback filter is known, we can calculate the value of $Q$ for the circuit using the following equations:

$$C = \sqrt{C_1C_2} \qquad (17.8)$$

and

$$Q = \pi f_0 R_f C \qquad (17.9)$$

Example 17.8 shows how equations (17.8) and (17.9) are used to determine the $Q$ of a multiple-feedback filter.

### EXAMPLE 17.8

Determine the values of $Q$ and bandwidth for the filter in Figure 17.24.

*Solution:* In Example 17.7, the center frequency of the filter was found to be 136.5 Hz. Using this value, the $Q$ of the circuit is found as

$$C = \sqrt{C_1C_2} = \sqrt{(0.1 \text{ }\mu\text{F})(0.068 \text{ }\mu\text{F})} = \textbf{0.082 }\mu\textbf{F}$$

and

$$Q = \pi f_0 R_f C = \pi(136.5 \text{ Hz})(40 \text{ k}\Omega)(0.082 \text{ }\mu\text{F}) = \textbf{1.41}$$

The bandwidth of the circuit is found as

$$\text{BW} = \frac{f_0}{Q} = \frac{136.5 \text{ Hz}}{1.41} = \textbf{97 Hz}$$

### PRACTICE PROBLEM 17.8
Determine the values of $Q$ and bandwidth for the filter described in Practice Problem 17.7.

Once the values of center frequency, $Q$, and bandwidth are known, the values of $f_{C1}$ and $f_{C2}$ for the multiple-feedback filter can be determined. The equations used to determine the cutoff frequencies of the filter will vary with the $Q$ of the circuit; that is, we will use one set of equations to find $f_{C1}$ and $f_{C2}$ when $Q \geqslant 2$ and another set to find $f_{C1}$ and $f_{C2}$ when $Q < 2$.

You may recall that the value of $f_0$ approaches the *algebraic* average ($f_{ave}$) of $f_{C1}$ and $f_{C2}$ when $Q \geqslant 2$, as was discussed in Section 17.1. Since $f_{ave}$ is halfway between the cutoff frequencies, we can use the following equations to approximate the values of $f_{C1}$ and $f_{C2}$ when $Q \geqslant 2$:

$$f_{C1} \cong f_0 - \frac{\text{BW}}{2} \qquad \text{when } Q \geqslant 2 \qquad (17.10)$$

### 17.5.5 Active Filter Troubleshooting

Once you know the common fault symptoms for active filters, the troubleshooting procedure for these circuits is easy:

1. Determine the type of filter you are dealing with.
2. Verify that the filter has an input signal.
3. Verify that the load is not the cause of the problem by isolating the filter output from the load.
4. Verify that both supply inputs to the filter are working properly.
5. Using your observations and Tables 17.2 through 17.4, determine the source of the fault within the filter.
6. If none of the passive components or IC sockets are faulty, replace the op-amp.

---

**Section Review ▶**

1. What is a *crossover network*?
2. Describe the operation of the crossover network in Figure 17.29.
3. What is a *graphic equalizer*?
4. Describe the operation of the graphic equalizer in Figure 17.30.
5. Describe how high-pass filters can be used to eliminate low-frequency noise in audio systems.
6. List the fault symptoms that can occur in active filters (those described in the text) and the possible cause(s) of each.
7. List the steps involved in troubleshooting an active filter.

## 17.6 Discrete Tuned Amplifiers

While many tuned circuit applications can be filled by active filters, some applications exceed the power-handling and/or high-frequency limits of active filters. In these applications, *discrete tuned amplifiers* are still used.

Discrete component circuits are tuned using parallel $LC$ (inductive-capacitive) circuits in place of a collector (or drain) resistor. A typical BJT tuned amplifier is shown in Figure 17.32. The collector of the transistor is coupled to $V_{CC}$ via the parallel $LC$ circuit ($C_T$ and $L_T$). The parallel $LC$ circuit determines the frequency-response characteristics of the

**Lab Reference:** The operation of a tuned BJT amplifier is demonstrated in Exercise 28.

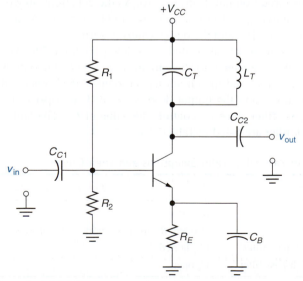

**FIGURE 17.32** A typical BJT discrete tuned amplifier.

amplifier. This being the case, we will start our discussion of the circuit by briefly review-ing the operating characteristics of parallel *LC* circuits.

## 17.6.1  Parallel *LC* Circuits

If we were to graph the *reactance versus frequency* characteristics of a capacitor and an inductor on the same chart, the plot would look like the one in Figure 17.33. The *origin* of the graph represents 0 Ω on the *y*-axis and 0 Hz on the *x*-axis. Note that at 0 Hz, the reactance of the inductor is 0 Ω, while the value of $X_C$ approaches ∞ Ω. As frequency increases, the value of $X_L$ *increases*, while the value of $X_C$ *decreases*. At some fre-quency, the values of $X_C$ and $X_L$ are *equal* for a given capacitor-inductor pair. This fre-quency, called the *resonant frequency*, is found as

$$f_r = \frac{1}{2\pi\sqrt{LC}} \tag{17.18}$$

You may recall from your study of basic electronics that:

**1.** The current in a capacitive branch *leads* the voltage across that branch by 90°.
**2.** The current in an inductive branch *lags* the voltage across that branch by 90°.

In a parallel *LC* circuit, the voltage across the capacitor is in phase with (and equal to) the voltage across the inductor. Combining this fact with the statements above, it is easy to see that $I_C$ and $I_L$ are 180° out of phase in a parallel *LC* tank. Thus,

$$I_{net} = |I_C - I_L| \tag{17.19}$$

where $I_{net}$ is the net current entering (and leaving) the tank circuit.

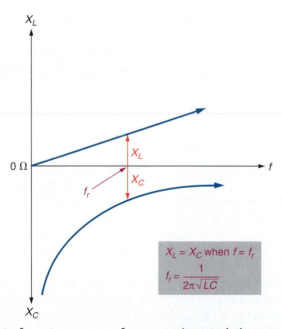

$$X_L = X_C \text{ when } f = f_r$$
$$f_r = \frac{1}{2\pi\sqrt{LC}}$$

**FIGURE 17.33**  Plot of reactance versus frequency characteristics.

At resonance, the values of $X_C$ and $X_L$ are equal. Since the voltages across the parallel components are equal, $I_C$ and $I_L$ are ideally equal. Thus, at resonance,

$$\mathbf{I_{net} = |I_C - I_L| = 0\ A}$$

Since the net current through the *LC* tank circuit is 0 A at resonance, the overall impedance of the tank circuit is (ideally) infinite. At frequencies below resonance, $X_L < X_C$. Thus, $I_L > I_C$, and the circuit is inductive. At frequencies above resonance, $X_C < X_L$.

> Below resonance, a tank circuit is inductive. Above resonance, it is capacitive.

Thus, $I_C > I_L$, and the circuit is capacitive. These points are illustrated in Figure 17.34. Now, let's apply these principles to the tuned amplifier in Figure 17.32.

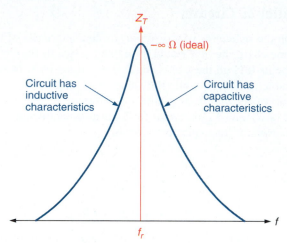

FIGURE 17.34 Frequency response of a parallel *LC* tank circuit.

## 17.6.2 AC Circuit Conditions

**OBJECTIVE 8** ▶ The ac equivalent of Figure 17.32 is derived in the usual fashion. The only difference is that the tank circuit components are not shorted. This ac equivalent circuit is shown in Figure 17.35. The load resistance is included to aid our discussion.

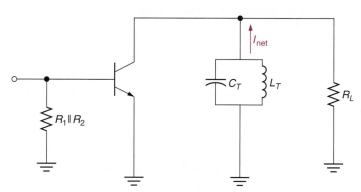

FIGURE 17.35 An equivalent circuit for a discrete tuned amplifier.

To understand the circuitry completely, we need to look at three frequency conditions:

**1.** $f_{in} = f_r$.
**2.** $f_{in} < f_r$.
**3.** $f_{in} > f_r$.

When $f_{in} = f_r$, the tank circuit acts as an open, as was discussed earlier in this section. Since $R_L$ represents the only path to ground in the collector circuit, it provides the only path for the ac output current. Thus, at $f_{in} = f_r$, the ac load current and amplifier efficiency reach their maximum possible values.

As $f_{in}$ begins to drop below the value of $f_r$, the inductor current ($I_L$) *increases* and the capacitor current ($I_C$) *decreases*. As a result, the combination of the two ($I_{net}$) *increases*. As the input frequency continues to decrease, $I_{net}$ also continues to increase. At some point, the transistor loading caused by the increase in $I_{net}$ causes the load voltage to drop by 3 dB. The frequency at which this occurs is the value of $f_{C1}$ for the circuit.

As $f_{in}$ begins to climb above the value of $f_r$, $I_C$ *increases* and $I_L$ *decreases*. Again, this results in an increase in $I_{net}$. As the input frequency continues to increase, another point is reached where the transistor loading caused by the increase in $I_{net}$ causes the load voltage to drop by 3 dB. The frequency at which this occurs is the value of $f_{C2}$ for the circuit.

When $f_{in} = f_r$.

When $f_{in} < f_r$.

When $f_{in} > f_r$.

As you may have figured out by now, the resonant frequency of the tank circuit is the geometric center frequency of the tuned amplifier. By formula,

◀ OBJECTIVE 9

$$f_0 = f_r \qquad (17.20)$$

or

$$f_0 = \frac{1}{2\pi\sqrt{LC}} \qquad (17.21)$$

## EXAMPLE 17.12

Determine the value of $f_0$ for the tuned amplifier shown in Figure 17.36.

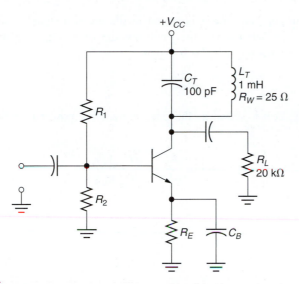

**FIGURE 17.36**

*Solution:* The center frequency equals the resonant frequency of the *LC* tank circuit. Using equation (17.21) and the circuit values shown, $f_0$ is found as

$$f_0 = \frac{1}{2\pi\sqrt{LC}} = \frac{1}{2\pi\sqrt{(1\text{ mH})(100\text{ pF})}} = \frac{1}{1.987\times10^{-6}}\text{ Hz} = \mathbf{503.29\text{ kHz}}$$

### PRACTICE PROBLEM 17.12

A tuned BJT amplifier has values of $L = 33$ mH and $C = 0.1$ μF. Calculate the value of $f_0$ for the circuit.

Once you have determined the value of $f_0$ for a tuned amplifier, the next step is to determine the value of $Q$ for the circuit.

## 17.6.3 The Figure of Merit, Q

Earlier in this chapter, we stated that the $Q$ of a tuned amplifier is equal to the ratio of center frequency to bandwidth. Although this statement helps to explain the use of $Q$ in bandwidth calculations, it does not really tell you what $Q$ is or where it comes from.

The $Q$ of a discrete tuned amplifier is a property of the *LC* circuit; more specifically, it is a property of the inductor. Strictly speaking, the **Q of an LC circuit** is *the ratio of energy stored in the circuit to energy lost per cycle by the circuit.* By formula,

$$Q = \frac{\text{energy stored}}{\text{energy lost}}$$

**Q (LC circuit)**
The ratio of energy stored in the circuit to energy lost per cycle by the circuit.

Another way of writing this is

$$Q = \frac{\text{reactive power}}{\text{resistive power}}$$

**Apparent power**

The combination of reactive (imaginary) power and resistive (true) power.

If you take the voltage across an inductor and multiply it by the current through the component, you obtain an **apparent power** value. We say *apparent* because only a small portion of the power value is actually dissipated by the component. Most of it is stored in the electromagnetic field that surrounds the component.

Apparent power is made up of *reactive (or imaginary) power* and *resistive (or "true")* *power*. Reactive power, which is actually *energy stored* by the component, can be found as

$$P_X = (I_L^2)(X_L)$$

While most of the apparent power is stored in the electromagnetic field, some of the power is actually dissipated across the small amount of winding resistance in the coil. This *resistive power*, or *energy lost*, is found as

$$P_{RW} = (I_L^2)(R_W)$$

Now, the $Q$ of the coil is found as the ratio of reactive power to resistive power. By formula,

$$Q = \frac{P_X}{P_{RW}} = \frac{(I_L^2)(X_L)}{(I_L^2)(R_W)}$$

or, simply,

$$Q = \frac{X_L}{R_W} \tag{17.22}$$

where $Q$ = the quality of the coil, measured at the geometric center frequency
$X_L$ = the reactance of the coil, measured at the geometric center frequency
$R_W$ = the resistance of the winding

The following example demonstrates the process for determining the $Q$ of a tank circuit.

### EXAMPLE 17.13

Determine the $Q$ of the tank circuit shown in Figure 17.36.

*Solution:* In Example 17.12, we determined the geometric center frequency of the amplifier to be 503.29 kHz. Using this value, $X_L$ for the circuit is determined as

$$X_L = 2\pi f L = 2\pi(503.29 \text{ kHz})(1 \text{ mH}) = \textbf{3.16 k}\boldsymbol{\Omega}$$

Using this value and the value of $R_W = 25 \ \Omega$ shown in the figure, the $Q$ of the tank circuit is found as

$$Q = \frac{X_L}{R_W} = \frac{3.16 \text{ k}\Omega}{25 \ \Omega} = \textbf{126.4}$$

### PRACTICE PROBLEM 17.13
The inductor in Practice Problem 17.12 has a value of $R_W = 18 \ \Omega$. Determine the $Q$ of the tank circuit.

The value of $Q$ obtained in Example 17.13 is accurate for the tank circuit. However, before we can use this value for bandwidth calculations, we have to take into account the effects of *circuit loading*, that is, the effect that $R_L$ has on the $Q$ of the tank.

## 17.6.4    Loaded-Q

To see the effects of loading on circuit $Q$, we have to take another look at the ac equivalent of the collector circuit. The ac equivalent is shown in Figure 17.37a. In this figure, we have included the tank circuit capacitor, $C_T$. However, since it does not really weigh into this discussion, we will not refer to it.

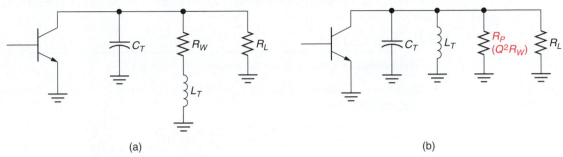

(a)                                                                   (b)

FIGURE 17.37    Effects of loading on circuit $Q$.

If you look closely at Figure 17.37a, you'll notice that the load resistance is in parallel with the inductor and its winding resistance ($R_W$). Therefore, we must include the resistance of the load in our calculations if we want an accurate value of $Q$.

The first step in determining the loaded-$Q$ of the circuit is to replace $R_W$ with an equivalent parallel resistance, $R_P$. This resistance is shown in Figure 17.37b. The value of $R_P$ is derived by determining the values of $X_L$ and $R_W$ and then solving for the equivalent parallel resistance value. This parallel equivalent resistance is found as

$$R_P = Q^2 R_W \qquad \text{(17.23)}$$

The derivation of equation (17.23) is shown in Appendix D. The total ac resistance of the collector circuit can now be found as

$$r_C = R_P \parallel R_L \qquad \text{(17.24)}$$

and the loaded-$Q$ can be found as

$$Q_L = \frac{r_C}{X_L} \qquad \text{(17.25)}$$

Equation (17.25) is also derived in Appendix D. Example 17.14 demonstrates the procedure for determining the loaded-$Q$ of a tuned amplifier.

### EXAMPLE 17.14

Determine the loaded-$Q$ for the circuit shown in Figure 17.36.

*Solution:*    The $Q$ of the tank circuit was found to be 126.4 in Example 17.13. Using this value of $Q$ and the value of $R_W = 25\ \Omega$, the value of $R_P$ is found as

$$R_P = Q^2 R_W = (126.4)^2 (25\ \Omega) \cong 400\ \text{k}\Omega$$

Since $R_P \gg R_L$, we can approximate the parallel combination of the two resistors ($r_C$) to be equal to $R_L$, 20 k$\Omega$. Using this value in equation (17.24), along with the value of $X_L = 3.16$ k$\Omega$ (obtained in Example 17.13), we can find $Q_L$ as follows:

$$Q_L = \frac{r_C}{X_L} \cong \frac{20\ \text{k}\Omega}{3.16\ \text{k}\Omega} = 6.33$$

The circuit described in Practice Problems 17.12 and 17.13 has a 10 kΩ load. Determine the value of $Q_L$ for the circuit.

As you can see, the loaded-$Q$ value for the circuit is considerably lower than the original value of $Q$ found for the tank circuit. Once we have determined the loaded-$Q$ value for the circuit, the bandwidth of the amplifier can be found as shown in Example 17.15.

### EXAMPLE 17.15

Determine the bandwidth for the amplifier in Figure 17.36.

*Solution:*   Using the values of $f_0 = 503.29$ kHz and $Q_L = 6.33$, the bandwidth of the amplifier is found as

$$\textbf{BW} = \frac{f_0}{Q_L} = \frac{503.29 \text{ kHz}}{6.33} = \textbf{79.5 kHz}$$

**PRACTICE PROBLEM 17.15**
Determine the bandwidth of the amplifier in Practice Problem 17.14.

Once the values of $f_0$ and bandwidth are known, we can calculate the cutoff frequencies of the amplifier using the guidelines established in our discussion on active filters. For circuits with $Q_L \geq 2$, the cutoff frequencies can be approximated as

$$f_{C1} \cong f_0 - \frac{\text{BW}}{2}$$

and

$$f_{C2} \cong f_0 + \frac{\text{BW}}{2}$$

When $Q_L < 2$, we need to use the more exact equations:

$$f_{C1} = f_0 \sqrt{1 + \left(\frac{1}{2Q_L}\right)^2} - \frac{\text{BW}}{2}$$

and

$$f_{C2} = f_0 \sqrt{1 + \left(\frac{1}{2Q_L}\right)^2} + \frac{\text{BW}}{2}$$

The only difference between these equations and those presented earlier in the text is the use of the *loaded-Q* ($Q_L$).

**Section Review ▶**

1. What is the circuit recognition feature of the discrete tuned amplifier?
2. Why is the net current in a parallel resonant tank circuit approximately equal to zero?
3. Describe the tuned discrete amplifier response to frequencies that are less than, equal to, and greater than the value of $f_r$.
4. What is the $Q$ of a tank circuit?
5. How does the presence of a load affect the $Q$ of the tank circuit?

## 17.7 Discrete Tuned Amplifiers: Practical Considerations and Troubleshooting

It is common for a technician to build a tuned amplifier and find that the center frequency is not what was expected. There are several reasons for this:

◀ **OBJECTIVE 10**

1. Capacitors and inductors usually have large tolerances, so there can be a significant difference between the *rated* value of an inductor (or capacitor) and its actual value. This can throw your calculations off by a significant margin.
2. Amplifiers tend to have many "natural" capacitances that are not accounted for in the frequency relationships. Values of stray capacitance and junction capacitance, for example, are always present in a discrete amplifier. If the capacitor used in the tank circuit is in the picofarad range, the other circuit capacitances can have a drastic effect on the actual geometric center frequency of the circuit.

> Why are variable components used?

To overcome these problems, discrete tuned amplifiers can be constructed using either a variable capacitor or a variable inductor in the tank circuit. By adjusting the variable component, the value of $f_0$ can be adjusted (up or down) to its design value. Note that variable inductors are used more commonly than variable capacitors because they are less expensive.

> Tuning an amplifier.

Another method of adjusting the tuning of a circuit is shown in Figure 17.38. This type of tuning is referred to as **electronic tuning** because the tuning of the amplifier is *voltage controlled*. The reverse bias on each of the *varactor diodes* ($D_1$ and $D_2$) is adjusted to set the capacitance of the component to the desired value. You may recall from Chapter 5 that the varactor is used as an electronically variable capacitance when reverse biased. The circuit in Figure 17.38 shows exactly this type of application for the component.

> **Electronic tuning**
> Using voltage-controlled or programmable devices to control the tuning of a circuit.

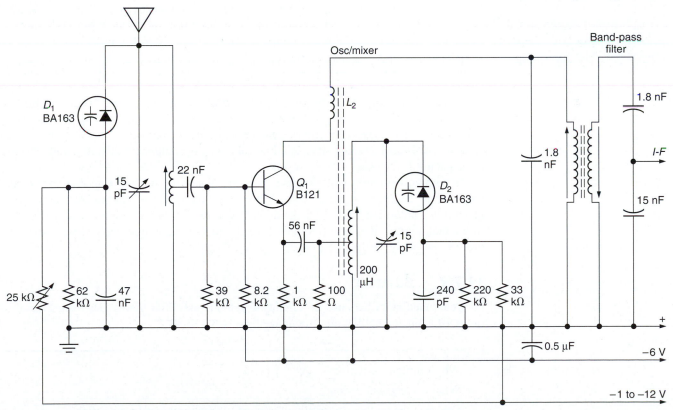

Broadcast band AM receiver front end with electronic tuning.

FIGURE 17.38 Electronic (varactor) tuning.

When the biasing of $D_1$ is adjusted, its junction capacitance changes. This causes the total capacitance in the tuned circuit to change. Changing the total capacitance in the tuned circuit causes the resonant frequency of the circuit to change. Thus, the tuning of the amplifier is changed by varying the reverse bias applied to the varactor.

Don't get caught up in trying to figure out all of Figure 17.38. This circuit, or one like it, will be covered in detail when you study communications electronics. It is introduced here solely to show you the concept of electronic tuning.

### 17.7.1    Circuit Troubleshooting

**Drift**

A change in tuning caused by component aging.

The most common problem that develops in a tuned amplifier is frequency **drift**. Component aging can cause the tuning of an amplifier to change. A simple solution to the problem is to retune the amplifier. If this does not solve the problem, one (or both) of the components in the tank circuit may need to be replaced. Frequency drift can also result from value changes in the input and output coupling components. As such, these components must be checked as well.

If either the capacitor or the inductor fails (opens or shorts), the effects are much more drastic than those of a drift problem. Possible "fatal" faults that can occur in the tank circuit and their respective symptoms are listed in Table 17.5.

TABLE 17.5    Tuned Amplifier Troubleshooting

| Fault | Symptoms |
|---|---|
| $L_T$ open | $V_{CC}$ is effectively removed from the transistor collector circuit (capacitors block dc voltages). Therefore, $I_C$ and $V_C$ both drop to zero. |
| $L_T$ shorted | The ac output from the amplifier is developed across the tank circuit. Shorting $L_T$ effectively removes the tank circuit from the picture. The biasing is now like that of an emitter follower, and $V_C$ equals $V_{CC}$. |
| $C_T$ open | Capacitors are "open" circuits by nature. However, the capacitor *connections* can open, effectively removing the capacitor from the circuit. When this happens, the amplifier gain decreases and varies directly with frequency. |
| $C_T$ shorted | The symptoms for this condition are the same as those that occur when $L_T$ shorts. |

*Don't Forget:*
A leaky capacitor acts (to some degree) like a shorted component.

Since either component shorting will cause the same symptoms, further testing is required when $V_C$ stays at $V_{CC}$. The simplest thing to do is to measure the resistance of the capacitor. If the capacitor shows signs of charging, the inductor is the problem. If the capacitor resistance reading stays at a low value or the component fails to hold a charge, the capacitor is the cause of the problem.

As with any circuit troubleshooting, you should start by making sure that your tuned amplifier is the cause of the problem. Once you have narrowed the problem down to a given tuned amplifier, Table 17.5 should help you to establish whether or not the fault is in the tank circuit. If it is not, troubleshoot the rest of the circuit components using the procedures taught earlier in the book.

### 17.7.2    Summary

Discrete tuned amplifiers use *LC* circuits to provide amplifier tuning. The characteristics of the BJT tuned amplifier are summarized in Figure 17.39.

Section Review ▶

1. Why is amplifier tuning needed?
2. What is *electronic tuning*?
3. Describe the methods normally used to adjust the tuning of an amplifier.
4. What is *drift*? How is drift corrected?
5. What are the symptoms of a major fault in the tank circuit of a tuned amplifier?

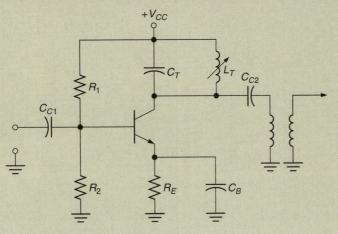

## Discrete Tuned Amplifiers

| Range of $Q$ values: | Normally greater than 2 |
|---|---|
| Geometric center frequency: | $f_0 = \dfrac{1}{2\pi\sqrt{LC}}$ |
| Practical considerations: | • Circuits are commonly transformer-coupled to reduce $Q$-loading. |
| | • Variable components are commonly included to allow adjustments of the circuit center frequency. |
| | • Low $C_T$ values can cause significant errors between calculated and measured center frequency values due to effects of circuit stray and internal capacitances. |
| | • Subject to center frequency drift. When this occurs, readjustment of the tuning is required. |
| | • Many other circuit configurations are possible. The common factor is that all of them contain at least one parallel $LC$ circuit. |

FIGURE 17.39

## 17.8  Class C Amplifiers

Earlier in the book, you were exposed to class A, class B, and class AB amplifiers. In this section, we take a look at **class C amplifiers**. Class C amplifiers were described briefly in Chapter 11 as *tuned amplifiers*.

Class C amplifiers are circuits containing transistors that conduct for *less than 180° of the input cycle*. A basic class C amplifier is shown in Figure 17.40.

> **Class C amplifier**
> An amplifier containing one transistor that conducts during less than 180° of the input cycle.

### 17.8.1  DC Operation

The transistor in a class C amplifier is biased deeply into *cutoff*. This is often done by connecting the base resistor to a negative supply voltage (for *npn* circuits) as shown in Figure 17.40.

◄ **OBJECTIVE 11**

Since a class C amplifier is biased in cutoff, the value of $V_{CEQ}$ is approximately equal to $V_{CC}$. This results in a dc load line like the one shown in Figure 17.41. Note that this dc load line can be somewhat misleading. You see, with the amplifier biased in cutoff, $I_C \cong 0$ A. The dc load line for the circuit would lead you to believe that the value of $I_C$ can be varied, as long as $V_{CEQ} = V_{CC}$. This is not the case. As long as there is no input signal to the transistor, $V_{CEQ} = V_{CC}$ and $I_{CQ} = 0$ A. (Perhaps it would be more appropriate to have a dc load *point* for this circuit rather than a dc load *line*. However, the dc operation of the circuit is traditionally represented by a load line.)

> Why is the dc load line for the class C amplifier misleading?

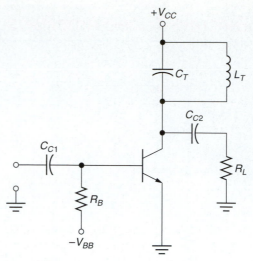

FIGURE 17.40   Class C amplifier.

**Lab Reference:** The dc characteristics of a class C amplifier are observed in Exercise 29.

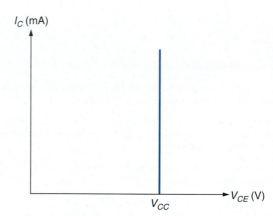

FIGURE 17.41   Class C amplifier dc load line.

When a negative supply voltage is used for $V_{BB}$, it is normally set at a value that fulfills the following relation:

$$V_{\text{in(pk)}} + (-V_{BB}) = 1 \text{ V}$$

or

$$-V_{BB} = 1 \text{ V} - V_{\text{in(pk)}} \qquad \text{(17.26)}$$

Thus, if the amplifier in Figure 17.40 had an input peak value of +4 V, the base biasing voltage would be set to approximately −3 V. This would ensure that the transistor would turn on only at the positive peaks of the input cycle. The purpose served by this type of biasing will be discussed later in this section.

## 17.8.2   AC Operation

The ac operation of a class C amplifier is based on the characteristics of the parallel $LC$ circuit. These characteristics are illustrated in Figure 17.42, which shows a parallel $LC$ circuit in series with a switch ($SW_1$). When $SW_1$ is closed for an instant, a current is generated in the tank circuit. This current initiates a charge/discharge cycle that goes back and forth between the capacitor and the inductor. This charge/discharge action produces the waveform shown in Figure 17.42b. Note that the tank circuit loses some amount of power during each cycle, so the waveform eventually dies out.

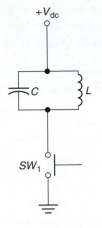

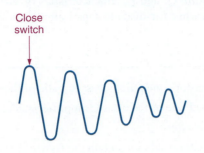

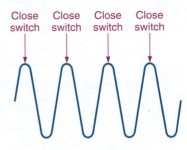

(a) A parallel *LC* circuit

(b) Current waveform produced by the *LC* circuit after a single switch closing

(c) Current waveform produced by the *LC* circuit when the switch is closed at each current peak

FIGURE 17.42   *LC* tank operation.

The charge/discharge cycle described here is known as the **flywheel effect.** Note that the waveform produced by the flywheel effect has a frequency equal to the resonant frequency of the tank circuit.

If we want to keep the flywheel waveform from dying out, we can do so by closing $SW_1$ for an instant at each positive peak, as illustrated in Figure 17.42c. Each time the switch is closed, power is returned to the circuit from the dc power supply. By closing the switch at the intervals shown, additional power is supplied to the tank circuit, and the amplitude of the waveform is maintained. With this in mind, let's take another look at the original class C amplifier circuit. The operation of this circuit is illustrated in Figure 17.43.

> **Flywheel effect**
> A term used to describe the ability of a parallel *LC* circuit to self-oscillate for a brief period of time.

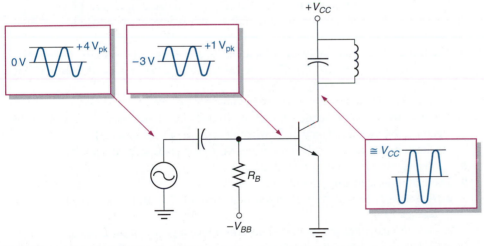

> **Lab Reference:** Class C circuit action is demonstrated in Exercise 29.

FIGURE 17.43   Class C circuit action.

Now, consider the transistor in this circuit to be the switch we talked about earlier. The positive peaks of the input signal cause the transistor to conduct, while the component is in cutoff the rest of the time. Each time the transistor conducts, it is like closing the switch in Figure 17.42. In other words, on each positive peak of the input, the transistor switch is closed, and the tank circuit gets the current pulse it requires to continue producing the output waveform.

There is an important relationship we have yet to establish: For the class C amplifier to work properly, the tank circuit must be tuned to either the same frequency as $v_{in}$ or to some *harmonic* of that frequency. A **harmonic** is a whole number multiple of a given frequency. For example, a 2 kHz signal would have harmonics of 4 kHz, 6 kHz, 8 kHz, and so on.

> **Harmonic**
> A whole number multiple of a given frequency.

When a class C amplifier is tuned to a harmonic of $v_{in}$, the output frequency of the circuit is equal to that harmonic. For example, a class C amplifier that has an input frequency of 3 kHz and a tank circuit tuned to 6 kHz produces a 6 kHz output. Thus, the class C amplifier can be used as a type of *frequency multiplier*.

Since the bandwidth, $Q$, and $Q_L$ characteristics of the class C amplifier are the same as those described earlier for discrete tuned amplifiers, we will not elaborate on them further.

**Section Review ▶**

1. What is the transistor conduction characteristic of the class C amplifier?
2. Why is the dc load line of the class C amplifier misleading?
3. Describe *flywheel effect*.
4. How is the transistor used to maintain the flywheel effect in a class C amplifier?
5. What is the input frequency requirement for proper class C operation?

## CHAPTER SUMMARY

Here is a summary of the major points made in this chapter:

1. The *tuned amplifier* is the heart of any communications system.
2. A tuned amplifier is designed for a specific bandwidth.
   a. For tuned amplifiers, bandwidth is a desired characteristic that is achieved through circuit design.
   b. The bandwidth of a tuned amplifier is usually narrower than that of a standard amplifier having the same geometric center frequency ($f_0$).
3. The *ideal* tuned amplifier provides:
   a. Zero ($-\infty$ dB) gain for all frequencies outside its pass band.
   b. Midband voltage gain for all frequencies within its pass band.
   Ideal and practical pass-band characteristics are compared in Figure 17.3.
4. The lower the gain roll-off of an amplifier, the greater its bandwidth (see Figure 17.5).
5. The roll-off rates and bandwidth of a tuned amplifier are related to its *quality* ($Q$).
   a. The quality ($Q$) of an amplifier is a figure of merit that equals the ratio of center frequency ($f_0$) to bandwidth (BW).
   b. The $Q$ of a tuned circuit is determined by circuit component values.
6. The geometric center frequency ($f_0$) of a tuned circuit approaches the value of the algebraic average frequency ($f_{ave}$) when $Q \geq 2$ (see Example 17.3).
7. Op-amp circuits are normally tuned using *RC* (resistive-capacitive) circuits.
8. An *active filter* is a tuned op-amp circuit.
9. There are four general types of active filters:
   a. A *low-pass* filter passes all frequencies from 0 Hz (dc) to an upper cutoff frequency ($f_C$).
   b. A *high-pass* filter passes all frequencies above a lower cutoff frequency ($f_C$).
   c. A *band-pass* filter passes all frequencies that fall between its values of $f_{C1}$ and $f_{C2}$.
   d. A *band-stop* (or *notch*) filter blocks all frequencies that fall between its values of $f_{C1}$ and $f_{C2}$. (In other words, it passes all frequencies *below* $f_{C1}$ and *above* $f_{C2}$.) See Figure 17.6.
10. A *pole* is a single *RC* circuit.
11. The *order* of an active filter indicates the number of poles it contains. A first-order filter contains one pole, a second-order filter contains two poles, and so on.
12. The more poles in an active filter, the higher the gain roll-off rate when the circuit is operated outside its pass band.
13. The *Butterworth filter* has relatively constant gain across its pass band.
    a. The term *flat response* is used to describe a constant-gain characteristic.
    b. Butterworth filters are commonly referred to as *maximally flat* or *flat-flat* filters.

**14.** The *Chebyshev filter* has higher initial roll-off (per pole) than a Butterworth filter. However:

**a.** The gain of a Chebyshev filter is not constant across its pass band.

**b.** The Chebyshev filter has a greater-roll off rate only for frequencies just outside the pass band. As frequency continues outside the pass band, the roll-off rates of Chebyshev and Butterworth filters become equal.

See Figure 17.8.

**15.** The variation in gain that occurs across the pass band of a Chebyshev filter is called *ripple width* (see Figure 17.8).

**a.** Ripple width is measured in dB.

**b.** Ripple width can be reduced (by design). However, reduced ripple width results in a reduction of the high initial roll-off rate.

**16.** The primary drawback of the Butterworth filter is that the *phase shift* produced by the circuit is not constant across the pass band.

**a.** Two (or more) frequencies applied to a Butterworth filter do not experience the same phase shift.

**b.** This phase shift variation can produce severe signal distortion from input to output.

**17.** The *Bessel filter* is designed to provide a constant phase shift across its pass band.

**a.** The Bessel filter provides better *fidelity*. (Fidelity is the ability to reproduce a waveform accurately.)

**b.** The Bessel filter has the disadvantage of lower initial roll-off rates than the Butterworth filter.

The frequency-response curves for the Butterworth, Chebyshev, and Bessel filters are compared in Figure 17.9.

**18.** A single-pole low-pass filter is designed as either a *variable-gain* circuit or a *voltage follower*.

**19.** The single-pole low-pass filter is simply a noninverting amplifier with an *RC* circuit connected between the signal source and the op-amp (see Figure 17.10).

**20.** A *unity-gain* one-pole low-pass filter is shown in Figure 17.13.

**21.** The two-pole low-pass filter is designed as either a *unity-gain* filter or a *variable-gain* filter (see Figure 17.14). There is a restriction on the closed-loop gain of the variable gain filter. To produce a Butterworth response curve, the closed-loop voltage gain is limited to 4 dB.

**22.** A three-pole low-pass filter can be made by cascading a two-pole circuit with a one-pole circuit (see Figure 17.16).

**23.** You can determine the operating characteristics of a multipole filter if you remember the following:

**a.** A filter has a 20 dB/decade roll-off *for each pole*.

**b.** All stages are tuned to the same cutoff frequency.

**24.** The first-order, second-order, and third-order *high-pass* filters are shown (along with their cutoff frequency equations) in Figure 17.17.

**25.** A *band-pass* filter can be made by cascading a low-pass filter with a high-pass filter (see Figure 17.20).

**26.** The *multiple-feedback band-pass* filter has a single op-amp and two feedback paths, one resistive and one capacitive (see Figure 17.22).

**27.** A *multistage notch filter* contains a low-pass filter, a high-pass filter, and a summing amplifier (see Figure 17.25).

**a.** The cutoff frequency of the high-pass filter is above the cutoff frequency of the low-pass filter.

**b.** The actual circuitry is shown in Figure 17.26.

**28.** A *multiple-feedback notch filter* has one op-amp and two feedback paths (see Figure 17.27).

**29.** Some typical active filter applications are as follows:

**a.** Low-pass and high-pass filters can be combined to form an audio *crossover network* (see Figures 17.28 and 17.29).

**b.** Band-pass filters can be combined to form a simple *graphic equalizer* (see Figure 17.30).

30. When troubleshooting an active filter:
    a. Determine the type of filter you are dealing with.
    b. Verify that the filter has an input signal.
    c. Verify that the load is not the cause of the problem.
    d. Verify that both supply inputs to the filter are working properly.
    e. Using your observations and the faults listed in Tables 17.2 through 17.4, determine the source of the fault.
    f. If none of the passive components or the IC socket are faulty, replace the op-amp.
31. The operation of a discrete tuned amplifier is based on the characteristics of a parallel *LC* circuit.
    a. The geometric center frequency ($f_0$) of the circuit is approximately equal to the resonant frequency of the *LC* circuit.
    b. The relationship between the component reactances is illustrated in Figure 17.33.
32. The bandwidth of the tuned BJT amplifier is determined by the *quality* (*Q*) of its *LC* circuit.
    a. The *Q* of a tank circuit equals the ratio of energy stored in the circuit to the energy lost per cycle by the circuit.
    b. The *Q* of a tank circuit equals the ratio of inductive reactance to inductor winding resistance.
33. The load on a tuned amplifier reduces the overall *Q* of the tank circuit.
34. Variable components are often used in the *LC* circuit of a discrete tuned amplifier. Variable components are needed because:
    a. Capacitors and inductors can have large tolerances, allowing for significant differences between the calculated and actual operating frequencies.
    b. Amplifiers have "natural" capacitances that can lower their operating frequencies.
    c. In many circuits, *electronic* (varactor) *tuning* is used (see Figure 17.38).
35. A *class C amplifier* contains a transistor that conducts for less than 180° of the input cycle.
36. The transistor in a class C amplifier is biased deeply into cutoff, normally using a negative base supply voltage ($V_{BB}$) (see Figure 17.40).
37. The ac operation of a class C amplifier is based on the *flywheel effect*.
    a. *Flywheel effect* is a term used to describe the ability of a parallel *LC* circuit to self-oscillate for a brief period of time.
    b. The waveform produced by the flywheel effect loses energy over each cycle and eventually dies out.
    c. The key to keeping the waveform produced by the flywheel effect constant is to provide a brief current spike during each cycle (see Figure 17.42c).

## EQUATION SUMMARY

| Equation Number | Equation | Section Number |
|---|---|---|
| (17.1) | $Q = \dfrac{f_0}{BW}$ | 17.1 |
| (17.2) | $BW = \dfrac{f_0}{Q}$ | 17.1 |
| (17.3) | $f_0 = \sqrt{f_{C1}f_{C2}}$ | 17.1 |
| (17.4) | $f_{ave} = \dfrac{f_{C1} + f_{C2}}{2}$ | 17.1 |
| (17.5) | $f_C = \dfrac{1}{2\pi RC}$ | 17.3 |

| Equation Number | Equation | Section Number |
|---|---|---|
| **(17.6)** | $f_C = \dfrac{1}{2\pi\sqrt{R_1 R_2 C_1 C_2}}$ | 17.3 |
| **(17.7)** | $f_0 = \dfrac{1}{2\pi\sqrt{(R_1 \parallel R_2)R_f C_1 C_2}}$ | 17.4 |
| **(17.8)** | $C = \sqrt{C_1 C_2}$ | 17.4 |
| **(17.9)** | $Q = \pi f_0 R_f C$ | 17.4 |
| **(17.10)** | $f_{C1} \cong f_0 - \dfrac{\text{BW}}{2}$   (when $Q \geq 2$) | 17.4 |
| **(17.11)** | $f_{C2} \cong f_0 + \dfrac{\text{BW}}{2}$   (when $Q \geq 2$) | 17.4 |
| **(17.12)** | $f_{\text{ave}} = f_0\sqrt{1 + \left(\dfrac{1}{2Q}\right)^2}$ | 17.4 |
| **(17.13)** | $f_{C1} = f_0\sqrt{1 + \left(\dfrac{1}{2Q}\right)^2} - \dfrac{\text{BW}}{2}$   (when $Q < 2$) | 17.4 |
| **(17.14)** | $f_{C2} = f_0\sqrt{1 + \left(\dfrac{1}{2Q}\right)^2} + \dfrac{\text{BW}}{2}$   (when $Q < 2$) | 17.4 |
| **(17.15)** | $A_{\text{CL}} = \dfrac{R_f}{2R_{\text{in}}}$ | 17.4 |
| **(17.16)** | $f_0 = \dfrac{1}{2\pi\sqrt{R_1 R_f C_1 C_2}}$ | 17.4 |
| **(17.17)** | $A_{\text{CL}} \cong 1$   (0 dB) | 17.4 |
| **(17.18)** | $f_r = \dfrac{1}{2\pi\sqrt{LC}}$ | 17.6 |
| **(17.19)** | $I_{\text{net}} = |I_C - I_L|$ | 17.6 |
| **(17.20)** | $f_0 = f_r$ | 17.6 |
| **(17.21)** | $f_0 = \dfrac{1}{2\pi\sqrt{LC}}$ | 17.6 |
| **(17.22)** | $Q = \dfrac{X_L}{R_W}$ | 17.6 |
| **(17.23)** | $R_P = Q^2 R_W$ | 17.6 |
| **(17.24)** | $r_C = R_P \parallel R_L$ | 17.6 |
| **(17.25)** | $Q_L = \dfrac{r_C}{X_L}$ | 17.6 |
| **(17.26)** | $-V_{BB} = 1\text{ V} - V_{\text{in(pk)}}$ | 17.8 |

| | | |
|---|---|---|
| active filter  698 | class C amplifier  735 | multiple-feedback band- |
| apparent power  730 | crossover network  722 | pass filter  712 |
| band-pass filter  699 | drift  734 | pass band  694 |
| band-stop (notch) | electronic tuning  733 | pole  699 |
| filter  699 | fidelity  701 | $Q$ ($LC$ circuit)  729 |
| Bessel filter  701 | flywheel effect  737 | quality ($Q$)  696 |
| biomedical | graphic equalizer  722 | ripple width  700 |
| electronics  722 | harmonic  738 | stop band  694 |
| Butterworth filter  700 | high-pass filter  699 | tuned amplifier  694 |
| Chebyshev filter  700 | low-pass filter  699 | |

## PRACTICE PROBLEMS

### Section 17.1

**1.** An amplifier has values of $f_0 = 14$ kHz and BW = 2 kHz. Calculate the value of $Q$ for the circuit.

**2.** An amplifier has values of $f_0 = 1200$ kHz and BW = 300 kHz. Calculate the value of $Q$ for the circuit.

**3.** An amplifier with a geometric center frequency of 800 kHz has a $Q$ of 6.2. Calculate the bandwidth of the circuit.

**4.** An amplifier with a geometric center frequency of 1100 kHz has a $Q$ of 25. Calculate the bandwidth of the circuit.

**5.** Complete the following table.

| $f_0$ (kHz) | BW | $Q$ |
|---|---|---|
| 740 | ———— | 2.4 |
| 388 | 40 kHz | ———— |
| 1050 | ———— | 5.6 |
| 920 | 600 kHz | ———— |

**6.** An amplifier has cutoff frequencies of 1180 and 1300 kHz. Show that $f_0 = f_{ave}$ for the circuit. Also, determine the $Q$ of the circuit.

### Section 17.3

**7.** Calculate the bandwidth and closed-loop voltage gain for the filter in Figure 17.44.

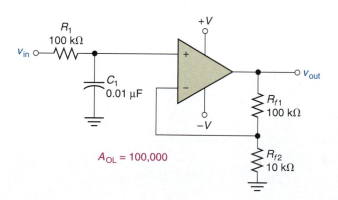

**FIGURE 17.44**

**8.** A filter like the one in Figure 17.44 has values of $R_1 = 82$ kΩ, $C_1 = 0.015$ μF, $R_{f1} = 150$ kΩ, and $R_{f2} = 20$ kΩ. Calculate the bandwidth and closed-loop voltage gain of the circuit.

**9.** Calculate the bandwidth and closed-loop voltage gain of the filter in Figure 17.45.

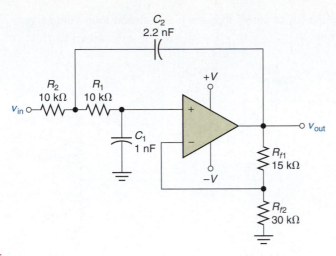

FIGURE 17.45

**10.** A filter like the one in Figure 17.14a has values of $R_1 = R_2 = 33 \text{ k}\Omega$, $C_1 = 100 \text{ pF}$, $C_2 = 200 \text{ pF}$, and $R_f = 220 \text{ k}\Omega$. Calculate the bandwidth and closed-loop voltage gain of the circuit.

**11.** Calculate the bandwidth and closed-loop voltage gain of the filter in Figure 17.46.

**12.** Calculate the bandwidth and closed-loop voltage gain of the filter in Figure 17.47.

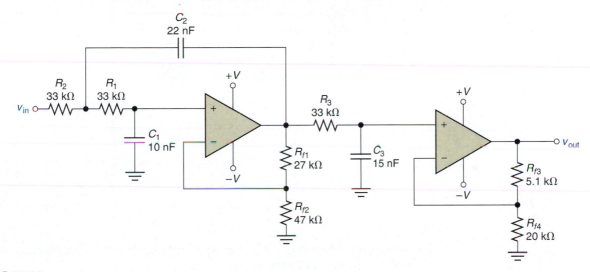

FIGURE 17.46

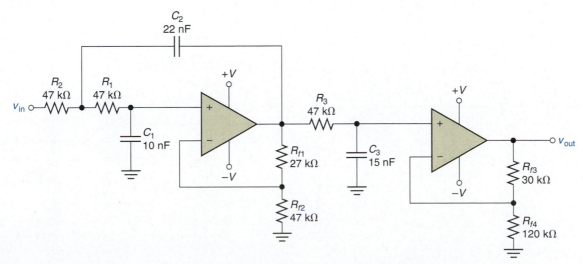

FIGURE 17.47

**13.** Calculate the lower cutoff frequency and closed-loop voltage gain of the filter in Figure 17.48.

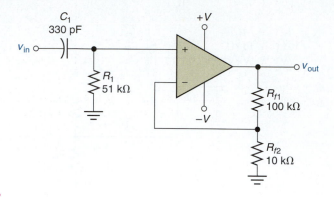

**FIGURE 17.48**

**14.** Calculate the lower cutoff frequency and closed-loop voltage gain of the filter in Figure 17.49.

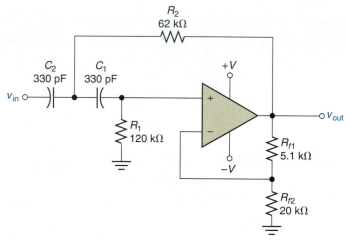

**FIGURE 17.49**

**Section 17.4**

**15.** Calculate the values of $f_{C1}$, $f_{C2}$, bandwidth, geometric center frequency, and $Q$ for the band-pass filter in Figure 17.50.

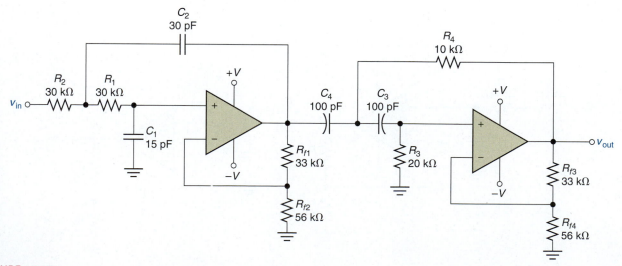

**FIGURE 17.50**

**16.** Calculate the values of $f_{C1}$, $f_{C2}$, bandwidth, geometric center frequency, and $Q$ for the band-pass filter in Figure 17.51.

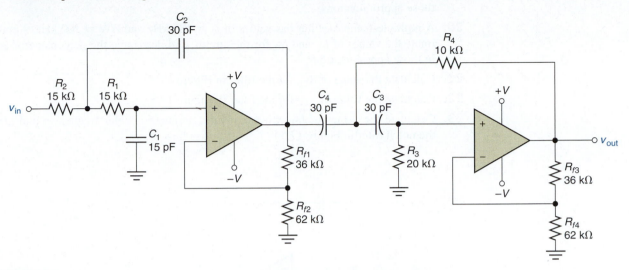

FIGURE 17.51

**17.** Calculate the values of $f_{C1}$, $f_{C2}$, bandwidth, geometric center frequency, and $Q$ for the band-pass filter in Figure 17.52.

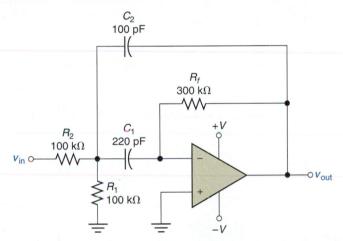

FIGURE 17.52

**18.** Calculate the values of $f_{C1}$, $f_{C2}$, bandwidth, geometric center frequency, and $Q$ for the band-pass filter in Figure 17.53.

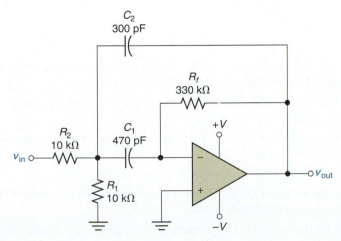

FIGURE 17.53

**19.** A multiple-feedback filter has values of $f_0$ = 62 kHz and BW = 40 kHz. Approximate the values of $f_{C1}$ and $f_{C2}$ for the circuit and determine the percentage of error in those approximations.

**20.** A multiple-feedback filter has values of $f_0$ = 482 kHz and BW = 200 kHz. Approximate the values of $f_{C1}$ and $f_{C2}$ for the circuit, and determine the percentage of error in those approximations.

**21.** Calculate the value of $A_{CL}$ for the filter in Figure 17.52.

**22.** Calculate the value of $A_{CL}$ for the filter in Figure 17.53.

**23.** Calculate the values of $f_{C1}$, $f_{C2}$, bandwidth, geometric center frequency, and $Q$ for the notch filter in Figure 17.54.

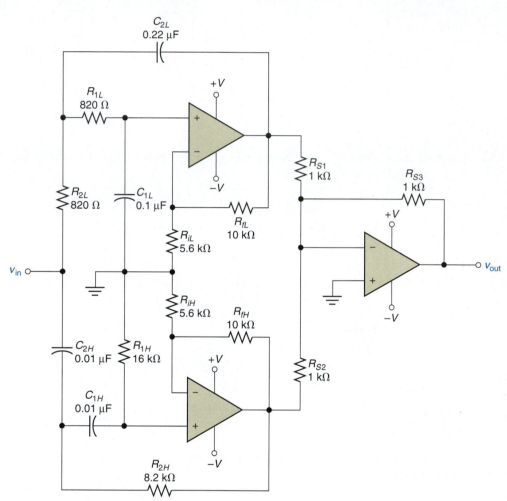

FIGURE 17.54

**24.** Calculate the values of $f_{C1}$, $f_{C2}$, bandwidth, geometric center frequency, and $Q$ for the notch filter in Figure 17.55.

**25.** Calculate the values of $f_{C1}$, $f_{C2}$, bandwidth, geometric center frequency, and $Q$ for the notch filter in Figure 17.56.

**26.** Calculate the values of $f_{C1}$, $f_{C2}$, bandwidth, geometric center frequency, and $Q$ for the notch filter in Figure 17.57.

**Section 17.7**

**27.** Calculate the values of $Q$, $Q_L$, geometric center frequency, bandwidth, $f_{C1}$, and $f_{C2}$ for the circuit shown in Figure 17.58a.

**28.** Calculate the values of $Q$, $Q_L$, geometric center frequency, bandwidth, $f_{C1}$, and $f_{C2}$ for the circuit shown in Figure 17.58b.

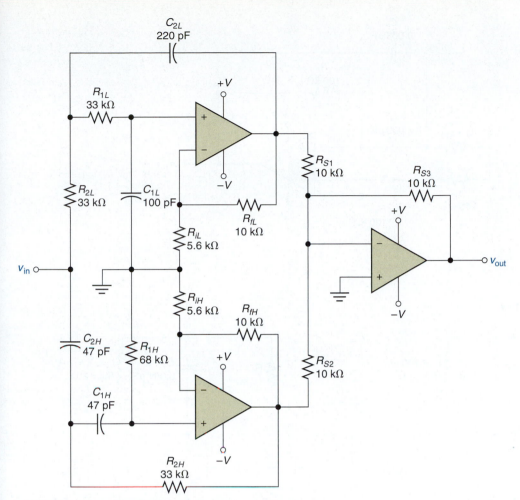

FIGURE 17.55

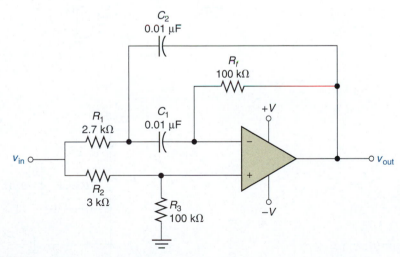

FIGURE 17.56

## Section 17.9

**29.** Calculate the values of $V_{CEQ}$ and $I_{CQ}$ for the class C amplifier in Figure 17.59.

**30.** Calculate the values of $V_{CEQ}$ and $I_{CQ}$ for the amplifier in Figure 17.60.

**31.** Calculate the minimum acceptable peak input voltage for the amplifier in Figure 17.59.

**32.** Calculate the minimum acceptable peak input voltage for the amplifier in Figure 17.60.

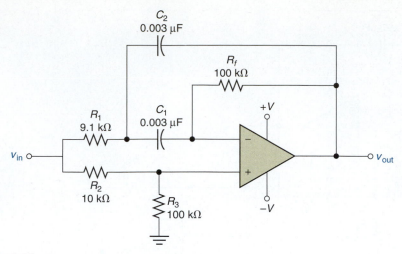

FIGURE 17.57

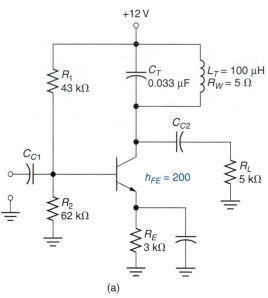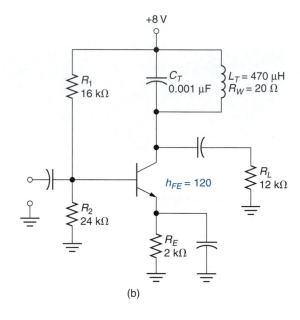

(a)                                          (b)

FIGURE 17.58

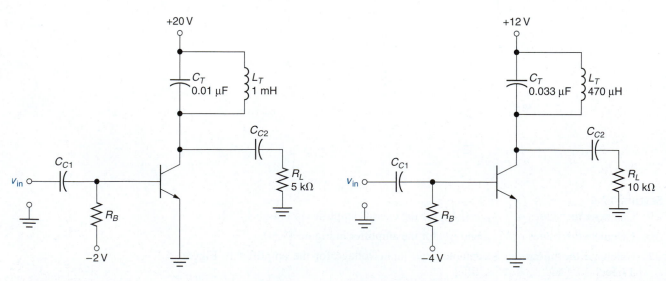

FIGURE 17.59                                          FIGURE 17.60

**33.** What fault is indicated by the readings in Figure 17.61? Explain your answer.

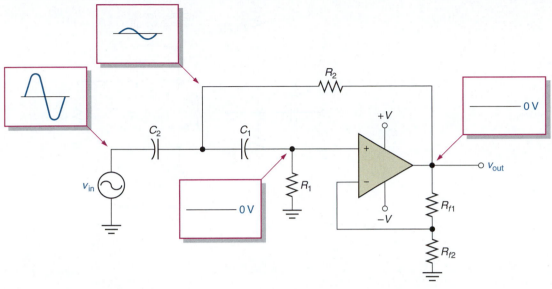

FIGURE 17.61

**34.** What fault is indicated by the readings in Figure 17.62? Explain your answer.

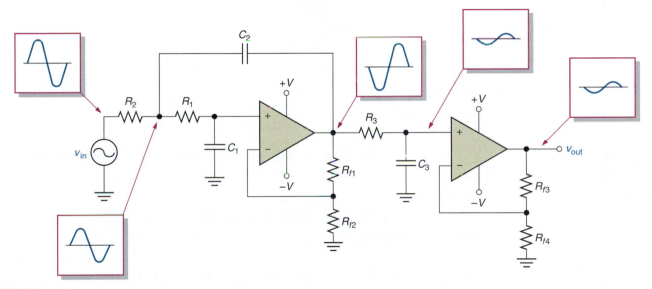

FIGURE 17.62

**PUSHING THE ENVELOPE**

**35.** The inductor in Figure 17.63 is replaced by one with a winding resistance of 17 Ω. Calculate the shift that occurs in each of the cutoff frequencies as a result of this change.

**36.** Figure 17.64 shows a *notch* filter that consists of a multiple-feedback band-pass filter and a summing amplifier. Using your knowledge of these circuits, describe the response of the filter to frequencies below, within, and above its bandwidth.

**SUGGESTED COMPUTER APPLICATIONS PROBLEMS**

**37.** Write a program that will determine the values of $f_0$, $Q$, and bandwidth for a multiple-feedback band-pass active filter.

**38.** Write a program that will determine the values of $f_0$, $Q$, $Q_L$, and BW for a discrete tuned amplifier.

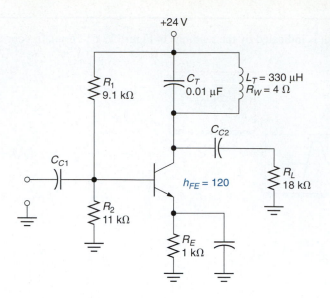

FIGURE 17.63

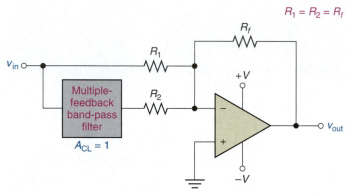

FIGURE 17.64

**ANSWERS TO THE EXAMPLE PRACTICE PROBLEMS**

**17.1**  60

**17.2**  56 kHz

**17.3**  $f_0 = 1029$ kHz, $f_{ave} = 1030$ kHz, $Q = 10.29$

**17.4**  $f_C = 102.6$ Hz, $A_{CL} = 2$

**17.5**  894.2 Hz

**17.6**  $f_{C1} = 570$ Hz, $f_{C2} = 938$ Hz, BW = 368 Hz, $f_0 = 731$ Hz, $Q = 1.99$

**17.7**  33.6 Hz

**17.8**  $Q = 1.08$, BW = 31.1 Hz

**17.9**  17.5 kHz, 22.5 kHz, % of error $(f_{C1}) = 0.91\%$, % of error $(f_{C2}) = 0.71\%$

**17.11**  1.13

**17.12**  2.77 kHz

**17.13**  31.9

**17.14**  11.27

**17.15**  245.8 Hz

## chapter 18

# Oscillators

## Objectives

*After studying the material in this chapter, you should be able to:*

1. State the function of the oscillator.

2. Describe *positive feedback*, how it is produced, and how it maintains oscillations after an oscillator is triggered.

3. Explain the *Barkhausen criterion* and its effect on oscillator operation.

4. List the three requirements for proper oscillator operation.

5. Describe the operating characteristics of the *phase-shift* and *Wien-bridge RC oscillators*.

6. Describe the operation and perform the complete frequency analysis of a *Colpitts oscillator*.

7. Describe the operation of the *Hartley*, *Clapp*, and *Armstrong LC* oscillators.

8. Describe the operating principles of *crystals* and *crystal-controlled oscillators*.

9. Describe the overall approach to (and the difficulties involved in) oscillator troubleshooting.

## Outline

# DC to AC?

In Chapter 3, we went through the analysis of a dc power supply. As you recall, these circuits are used to convert ac to dc.

In this chapter, we will discuss the operation of *oscillators*, circuits used to convert dc to ac. It would be reasonable to wonder why we want to convert dc to ac after going through the trouble of having a power supply convert ac to dc. Why don't we just use the ac supplied by the wall outlet to begin with?

The line frequency in the United States is 60 Hz standard. However, electronic systems depend on internally generated frequencies that typically extend well into the megahertz (MHz) range. In fact, microwave systems depend on internally generated frequencies in the gigahertz (GHz) range. These frequencies must somehow be developed using the dc voltages present in the system. This function is performed by circuits called *oscillators*.

How common are oscillators? You will see throughout your career that there are very few electronic systems that do not contain one or more oscillators.

**OBJECTIVE 1** ▶

**Oscillator**

An ac signal generator. A circuit that converts dc to a sinusoidal (or other) waveform.

An **oscillator** is a circuit that produces an output waveform without an external signal source. The only input to an oscillator is the dc power supply. As such, an oscillator can be viewed as a *signal generator*.

There are several types of oscillators, each classified according to the type of output waveform it produces. In this chapter, we will cover the operation of *sine-wave oscillators*, which have sinusoidal outputs. Then, in Chapter 19, we will discuss (along with a variety of circuits) the operation of a *square-wave oscillator*.

## 18.1 Introduction

**OBJECTIVE 2** ▶

**Positive feedback**

A type of feedback signal that is in phase with the circuit input signal.

In Chapter 15, you were told that **positive-feedback** is the key to oscillator operation. As shown in Figure 18.1, a positive-feedback network produces a feedback voltage ($v_f$) that is *in phase* with the amplifier input signal. Here's how this circuit works: An input signal ($v_{in}$) is applied to the amplifier, which introduces a 180° voltage phase shift. The output signal is applied to the input of the feedback network, which introduces another 180° voltage phase shift. As a result, the signal voltage is shifted 360° as it travels around the loop. As you know, a 360° shift is the same as a 0° shift. Therefore, the feedback and input signals are in phase. (Of course, the same result can be achieved using an amplifier and feedback network that each produce a 0° phase shift.)

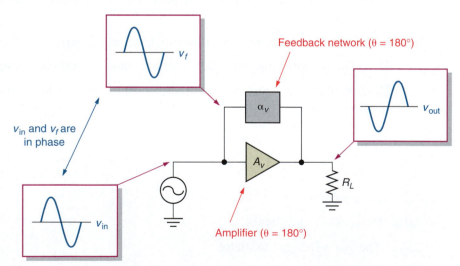

**FIGURE 18.1** Positive feedback.

### 18.1.1 Oscillators: The Basic Idea

Even though the circuit in Figure 18.1 is useful in explaining positive feedback, having an input signal is inconsistent with our definition of an oscillator. However, by modifying Figure 18.1, we can develop a circuit that is very useful for showing you the basic operating principle of the oscillator. This modified circuit is shown in Figure 18.2.

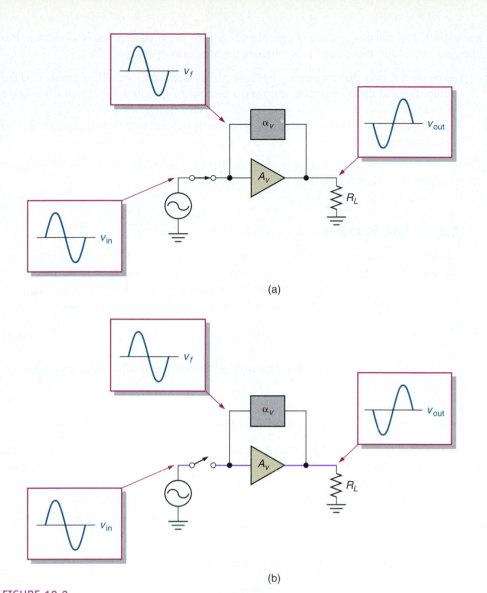

In Chapter 15, you were told that $\alpha_v$ is the *attenuation factor* of a feedback network, equal to the ratio of feedback voltage ($v_f$) to output voltage ($v_{out}$). The value of $\alpha_v$ is always less than 1.

(a)

(b)

**FIGURE 18.2**

In Figure 18.2, we have added a switch in series with the amplifier input. When the switch is closed, the circuit waveforms are as shown in the figure. Now, assume that the switch is opened while the circuit is in operation. If this happens, $v_{in}$ is isolated from the circuit. However, $v_f$ (which is in phase with the original input) is still applied to the amplifier input. The amplifier responds to this signal in the same way that it did to $v_{in}$. In other words, $v_f$ is amplified and sent to the output. Since the feedback network sends a portion of the output back to the input, the amplifier receives another input cycle, and another output cycle is produced. This process continues as long as the amplifier is turned on, and the amplifier produces a sinusoidal output with no external signal source.

The feedback network in an oscillator *generates* an input to the amplifier, which in turn *generates* an input to the feedback network. Since positive feedback produces this circuit action, it is often referred to as **regenerative feedback**. Regenerative feedback is the basis of operation for all oscillators. This point will be demonstrated throughout this chapter.

It should be noted that an oscillator needs only a quick *trigger* signal to start the oscillating circuit action. In other words, anything that causes a slight signal variation at any point in the circuit will start the oscillator. It is not necessary for us to provide a complete input cycle from an external source. In fact, most oscillators will provide their own trig-

**Regenerative feedback**
Another name for *positive feedback*.

ger signals. The sources of these trigger signals will be made clear in the next section. For now, just remember these basic requirements for oscillator operation:

1. The circuit must have a *positive* (regenerative) feedback loop. This means that the amplifier and its feedback circuitry must combine to produce a 360° (or 0°) voltage phase shift.
2. The circuit must receive some trigger, either internally or externally generated, to start the oscillations.

There is one more requirement that must be fulfilled for an oscillator to work: The circuit must fulfill a condition called the *Barkhausen criterion*.

## 18.1.2    The Barkhausen Criterion

In Chapter 16, you were shown that the active component in a feedback amplifier provides voltage gain ($A_v$) while the feedback network introduces a voltage loss, or *attenuation* ($\alpha_v$). In other words, a feedback amplifier has values of $A_v > 1$ and $\alpha_v < 1$.

**OBJECTIVE 3 ▶**

For an oscillator to operate properly, the following relationship must be fulfilled:

$$\alpha_v A_v = 1 \tag{18.1}$$

**Barkhausen criterion**

The relationship between the circuit feedback factor ($\alpha_v$) required and voltage gain ($A_v$) required for proper oscillator operation.

This relationship is known as the **Barkhausen criterion**. The results of *not* fulfilling this criterion are as follows:

1. If $\alpha_v A_v < 1$, the oscillations will fade out within a few cycles.
2. If $\alpha_v A_v > 1$, the oscillator may drive itself into saturation and cutoff clipping.

These points are easy to understand if we apply them to a couple of circuits using several different combinations of $A_v$ and $\alpha_v$. As a reference, we will use the circuits shown in Figure 18.3. In Figure 18.3a, the output voltage is shown to be the product of the amplifier gain ($A_v$) and the input voltage ($v_f$). By formula,

$$v_{out} = A_v v_f$$

The value of $v_f$ depends on the values of $\alpha_v$ and $v_{out}$ as follows:

$$v_f = \alpha_v v_{out}$$

Now, we will use these relationships to demonstrate what happens in Figure 18.3a as the circuit progresses through several cycles of operation. We will assume that the initial input to the amplifier is a 100 mV$_{pk}$ signal. Starting with this value and using the values of $A_v$ and $\alpha_v$ shown in the figure, the circuit cycles as follows:

| Cycle | $v_{in}$ | $v_{out}$ | $v_f$ |
|---|---|---|---|
| 1 | 100 mV$_{pk}$ | 10 V$_{pk}$ | 10 mV$_{pk}$ |
| 2 | 10 mV$_{pk}$ | 1 V$_{pk}$ | 1 mV$_{pk}$ |
| 3 | 1 mV$_{pk}$ | 100 mV$_{pk}$ | 0.1 mV$_{pk}$ |

Note how the value of $v_f$ produced by each cycle is used as $v_{in}$ for the next. This is consistent with the basic operating principle of the oscillator. Now, take a look at the progression in the $v_{out}$ column. As you can see, $v_{out}$ decreases from each cycle to the next. If this progression continues, $v_{out}$ eventually reaches 0 V for all practical purposes. This output deterioration is illustrated in Figure 18.3a. Note that this circuit has an $\alpha_v A_v$ product that is less than 1. As you can see, when $\alpha_v A_v < 1$, the oscillations lose amplitude on each

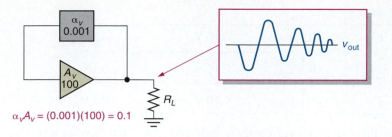

$\alpha_v A_v = (0.001)(100) = 0.1$

(a) The output fades out when $\alpha_v A_v < 1$.

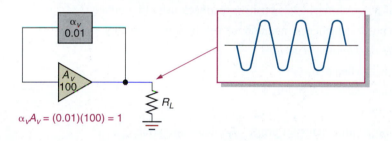

$\alpha_v A_v = (0.1)(100) = 10$

(b) The output is driven into clipping when $\alpha_v A_v > 1$.

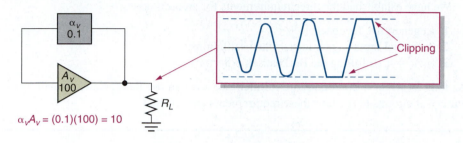

$\alpha_v A_v = (0.01)(100) = 1$

(c) A constant-amplitude output is produced when $\alpha_v A_v = 1$.

FIGURE 18.3   Effects of $\alpha_v A_v$ on oscillator operation.

progressive cycle and will eventually fade out. This loss of signal amplitude is called **damping**.

When $\alpha_v A_v$ is greater than 1, the output amplitude *increases* with each cycle. This problem is illustrated in Figure 18.3b. Using the same initial 0.1 $V_{pk}$ input as before and the circuit values shown, the circuit voltages in Figure 18.3b progress as follows:

**Damping**

The fading and loss of oscillations that occur when $\alpha_v A_v < 1$.

| Cycle | $v_{in}$ | $v_{out}$ | $v_f$ |
|-------|----------|-----------|-------|
| 1 | 0.1 $V_{pk}$ | 10 $V_{pk}$ | 1 $V_{pk}$ |
| 2 | 1 $V_{pk}$ | 100 $V_{pk}$ | 10 $V_{pk}$ |
| 3 | 10 $V_{pk}$ | 1000 $V_{pk}$ | 100 $V_{pk}$ |

For this circuit, it took only two cycles to start hitting some ridiculous values for $v_{out}$. The point is that $v_{out}$ is increasing from each cycle to the next. After several cycles, the output starts to experience saturation and cutoff clipping. Thus, $\alpha_v A_v$ cannot be greater than 1.

The only condition that provides a constant sinusoidal output from an oscillator is to have a product of $\alpha_v A_v = 1$. This results in the circuit operation illustrated in Figure 18.3c. Using the same 0.1 $V_{pk}$ initial input, let's construct the cycle chart for Figure 18.3c.

| Cycle | $v_{in}$ | $v_{out}$ | $v_f$ |
|-------|----------|-----------|-------|
| 1 | $0.1\ V_{pk}$ | $10\ V_{pk}$ | $0.1\ V_{pk}$ |
| 2 | $0.1\ V_{pk}$ | $10\ V_{pk}$ | $0.1\ V_{pk}$ |
| 3 | $0.1\ V_{pk}$ | $10\ V_{pk}$ | $0.1\ V_{pk}$ |

It is obvious that the cycle will continue over and over. Thus, having a product of $\alpha_v A_v = 1$ causes the oscillator to have a consistent sinusoidal output, which is the only acceptable outcome.

OBJECTIVE 4 ▶ We have established the three requirements for an oscillator:

What is required for a circuit to oscillate?

1. Regenerative feedback.
2. An initial input trigger to start the oscillations.
3. $\alpha_v A_v = 1$ (fulfilling the Barkhausen criterion).

As long as all these conditions are fulfilled, we have an oscillator. Now, we will look at a variety of circuits to see how these requirements are fulfilled.

**Section Review ▶**

1. What is an *oscillator*?
2. What is *positive feedback*?
3. In terms of phase shifts, how is positive feedback usually produced?
4. How does positive feedback maintain the oscillations started by a trigger?
5. What is the *Barkhausen criterion*?
6. What happens when $\alpha_v A_v > 1$?
7. What happens when $\alpha_v A_v < 1$?
8. What is *damping*?
9. List the three requirements for proper oscillator operation.

## 18.2  Phase-Shift Oscillators

OBJECTIVE 5 ▶ Probably the easiest oscillator to understand is the *phase-shift oscillator*. This circuit contains *three RC* circuits in its feedback network, as shown in Figure 18.4.

*A Circuit Variation:*
Figure 18.4 shows a phase-shift oscillator that contains series resistors and shunt capacitors. You can also construct a phase-shift oscillator using shunt resistors and series capacitors, that is, by reversing the capacitor and resistor positions.

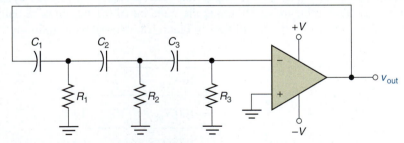

FIGURE 18.4  Basic phase-shift oscillator.

In your study of basic electronics, you were introduced to the *phase shift* that is produced by an *RC* circuit at a given frequency. As you may recall, the phase shift produced by a given *RC* circuit is found as

$$\theta = \tan^{-1}\frac{-X_C}{R} \qquad (18.2)$$

where $\theta$ = the phase angle of the circuit
$\tan^{-1}$ = the *inverse* tangent of the fraction

At this point, we are not interested in the exact value of $\theta$. We *are* interested in the fact that $\theta$ changes with $X_C$ and, thus, with frequency. In other words, a given $RC$ circuit can be designed to produce a specific phase shift at a given frequency.

Now, assume that the **phase-shift oscillator** is designed so that the three $RC$ circuits produce a combined phase shift of 180° at a resonant frequency, $f_r$. This allows the circuit to oscillate at that frequency, provided that the Barkhausen criterion has been met. For example, let's assume that the feedback circuit in Figure 18.4 is designed to produce a 180° voltage phase shift when $f = 10$ kHz, and that $\alpha_v A_v = 1$ at the same frequency. As a result, the circuit oscillates, generating a 10 kHz sine-wave output.

You would think that each $RC$ circuit in the phase-shift oscillator would be designed to produce a 60° phase shift, with the three 60° shifts combining to produce the 180° shift needed for regenerative feedback. However, this is not the case. Each $RC$ circuit in the phase-shift oscillator acts as a *load* on the previous $RC$ circuit. Just as the loaded-$Q$ of an amplifier differs from the unloaded-$Q$, the phase shift of a loaded $RC$ circuit differs from that of an unloaded $RC$ circuit. Thus, the exact phase shift of each $RC$ circuit in the phase-shift oscillator varies from the next. However, the overall phase shift of the three still adds up to 180°.

**Phase-shift oscillator**
An oscillator that uses three $RC$ circuits in its feedback network to produce a 180° phase shift.

## 18.2.1 Practical Considerations

You should be aware that phase-shift oscillators are rarely used because they are extremely unstable. **Oscillator stability** is a measure of its ability to maintain an output that is constant in *frequency* and *amplitude*. Phase-shift oscillators are extremely difficult to stabilize in terms of frequency. Thus, they cannot be used in any application where timing is critical.

Even though the phase-shift oscillator is rarely used, it is introduced here for two reasons. First, as stated earlier, the phase-shift oscillator is easy to understand. Thus, it serves as a valuable learning tool. Second, it is very easy to build a phase-shift oscillator *accidentally*. For example, consider the circuit shown in Figure 18.5. In this circuit, we have three $RC$ networks. Each $RC$ circuit is made up of a coupling capacitor and the input resistance of the following stage. The first $RC$ circuit consists of $C_{C1}$ and the input resistance of the first stage, the second consists of $C_{C2}$ and the input resistance of the second stage, and the third consists of $C_{C3}$ and the input resistance of the third stage. These three $RC$ circuits combine to produce a phase shift of 180°. This phase shift is added to a combined transistor phase shift of 180° (the first two 180° shifts cancel out). The feedback path is provided through the dc power supply ($V_{CC}$). Remember that $V_{CC}$ is simply a dc source connected between the "high" side of the circuit and all the ground connections. If the internal resistance of the dc power supply is high enough, any alternating current coupled to the power supply can cause a significant alternating voltage to be developed across the resistance. This voltage, fed back to the first stage through its biasing circuit, can convert the circuit to an unintentional oscillator.

**Oscillator stability**
A measure of an oscillator's ability to maintain constant output amplitude and frequency.

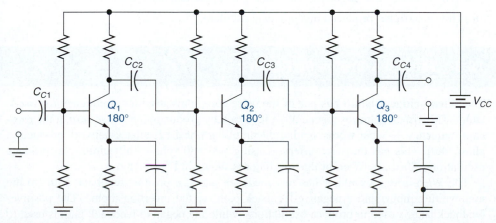

FIGURE 18.5

One solution to this potential problem is to use a *regulated* dc power supply. As you were told in Chapter 10, the internal resistance of a regulated power supply is extremely low. This low internal resistance keeps the voltage produced by changes in current to a minimum. Thus, any changing voltage generated appears as a small ripple in the dc supply voltage. Another solution is to connect a low-value bypass capacitor across the dc power supply to short any alternating current around the internal resistance of the supply. We discussed the use of a decoupling capacitor for this purpose in Chapter 10.

There are several other possible sources of unwanted oscillations. *Stray capacitance* between the stages of an amplifier can cause high-frequency oscillations at the output. Oscillations caused by stray capacitance can be eliminated by increasing the distance between stages or by placing metal shielding over each stage. High-frequency oscillations, called *loop oscillations*, can also be caused by improper ground connections between amplifier stages. Loop oscillations can be prevented by connecting all the stages to the same ground point (if possible) and by keeping the lengths of the ground connections as short as possible.

In practice, $\alpha_v A_v$ must be *slightly* greater than 1.

Another practical consideration involves the Barkhausen criterion. The relationship $\alpha_v A_v = 1$ holds true only for *ideal* circuits. In any practical circuit, the product of $\alpha_v A_v$ must be *slightly greater than 1*. Since each resistive component in an oscillator dissipates *some* amount of power, there is a power loss in the overall circuit. By making $\alpha_v A_v$ slightly greater than 1, the power lost during each cycle is returned to the circuit. How much greater than 1? Just enough to sustain oscillations.

How are practical oscillators triggered?

Finally, we must address the question of how the oscillations start. Consider what happens when power is first applied to the circuit shown in Figure 18.4. Since the output is fed back to the *inverting* input, there must exist a 180° phase shift between the two ends of the feedback network. When the power is applied, the circuit establishes this phase shift in one of two ways:

1. The input remains stable, and the output changes to the opposite-polarity extreme.
2. The output remains stable, and the input voltage changes to the opposite-polarity voltage.

Which of these two possibilities happens is of no consequence. Either way, a *transition* occurs between the output and input. This transition is enough to trigger the oscillating process.

---

**Section Review ▶**

1. What is a *phase-shift oscillator*?
2. How is regenerative feedback produced by the phase-shift oscillator?
3. What is *oscillator stability*?
4. Why are phase-shift oscillators rarely used?
5. Explain how a three-stage BJT amplifier can become a phase-shift oscillator.
6. Explain how the problem in Question 5 is prevented.
7. Why must $\alpha_v A_v$ be slightly greater than 1 in a practical oscillator?
8. How is a trigger produced in a practical oscillator?

## 18.3  The Wien-Bridge Oscillator

**Wien-bridge oscillator**
An oscillator that achieves regenerative feedback by producing no phase shift at $f_r$.

The **Wien-bridge oscillator** is one of the more commonly used low-frequency *RC* oscillators. This circuit achieves regenerative feedback by producing *no phase shift* at the resonant frequency. In other words, neither the amplifier nor the feedback network produces a phase shift. This has the same effect as using two 180° phase-shift circuits to produce oscillations. The basic Wien-bridge oscillator is shown in Figure 18.6.

Why are there *two* feedback paths?

The Wien-bridge oscillator has *two* feedback paths: a positive-feedback path (to the noninverting input) and a negative-feedback path (to the inverting input). The positive-feedback path is used to produce oscillations, while the negative-feedback path is used to control the $A_{CL}$ of the circuit.

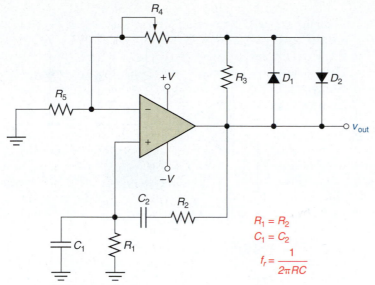

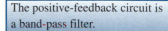

**Lab Reference**: The operation of the oscillator in Figure 18.6 is demonstrated in Exercise 30.

$R_1 = R_2$
$C_1 = C_2$
$$f_r = \frac{1}{2\pi RC}$$

EWB

FIGURE 18.6 Wien-bridge oscillator.

### 18.3.1 The Positive-Feedback Path

The positive-feedback network contains two $RC$ circuits. $R_1C_1$ forms a *low-pass* filter, while $R_2C_2$ forms a *high-pass* filter. As you learned in Chapter 17, the series combination of a low-pass filter and a high-pass filter forms a *band-pass* filter. The resonant frequency of this band-pass filter determines the oscillating frequency of the circuit.

A typical Wien-bridge oscillator is designed so that $R_1C_1 = R_2C_2$. As a result, the two circuits have the same cutoff frequency, as illustrated in Figure 18.7. Note that the frequency-response curves cross at the cutoff frequency ($f_C$) of each circuit. As a result, the circuit oscillates at this frequency.

> The positive-feedback circuit is a band-pass filter.

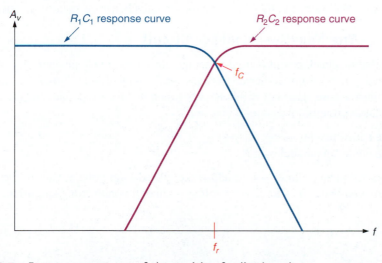

FIGURE 18.7 Frequency response of the positive-feedback path.

As you know, a band-pass filter does not introduce a phase shift when it is operated at its resonant frequency. Also, there is no phase shift between the noninverting input of the op-amp and the component output terminal. Therefore, the output and input signals are in phase, and the feedback is regenerative.

One final note on this circuit: You will often see *trimmer potentiometers* added in series with $R_1$ and $R_2$. These trimmers allow the feedback circuit (and, thus, the resonant frequency) to be "fine-tuned." The circuit shown in Figure 18.8 shows the added trimmer pots. The pots are adjusted to set the frequency of oscillations to the precise value desired.

> Why are trimmer pots used?

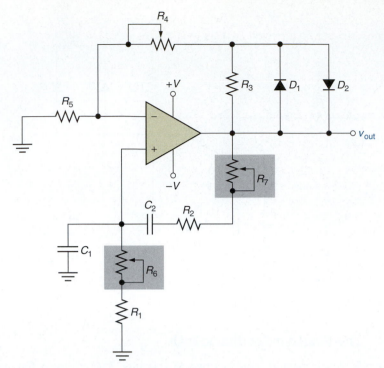

**FIGURE 18.8** Trimmer pots used for oscillator tuning.

Why add two potentiometers? Why not make $R_1$ and $R_2$ potentiometers and save the extra components? The answer to these questions is based on the fact that $R_1$ and $R_2$ are commonly selected to be in the neighborhood of 10 to 200 k$\Omega$. For the circuit to be *fine tuned*, you want the potentiometers to have a range that is equal to approximately 10% of the total series resistance or less. For example, if $R_1 = R_2 = 100$ k$\Omega$, you want the potentiometers to have a maximum range of 10 k$\Omega$ or less. (In all probability, you would want them to be as low as 1 k$\Omega$.) This gives you the ability to vary $f_c$ easily across a relatively narrow range of frequencies.

### 18.3.2  The Negative-Feedback Circuit

The negative-feedback circuit is used to set the closed-loop voltage gain ($A_{CL}$) of the circuit. You may recall from earlier discussions that voltage-divider circuits are commonly used for this purpose. The only differences between this feedback path and the other negative-feedback circuits covered until now are:

1. The added potentiometer, $R_4$.
2. The diodes in parallel with $R_3$.

**Lab Reference**: The effects of the negative-feedback path on circuit operation are observed in Exercise 30.

The potentiometer allows you to adjust the closed-loop voltage gain of the circuit as follows. As you know, the closed-loop voltage gain of a noninverting amplifier is found as

$$A_{CL} = \frac{R_f}{R_{in}} + 1$$

For the circuit shown in Figure 18.8, $R_{in} = R_5$ and $R_f = (R_3 + R_4)$ when the diodes are *off*. Substituting these values into the above equation, we get

$$A_{CL} = \frac{R_3 + R_4}{R_5} + 1$$

As this equation indicates, adjusting the value of $R_3$ has the effect of increasing or decreasing the value of $A_{CL}$ for the circuit.

The diodes are contained in the feedback path to prevent the circuit output from exceeding a predetermined value. If the oscillator output tries to exceed the magnitude of

$(V_{R4} + V_{R5})$ by more than 0.7 V, one of the diodes turns on, shorting out $R_3$. This effectively reduces the closed-loop voltage gain to approximately

$$A_{CL} = \frac{R_4}{R_5} + 1$$

Note that the diodes are essentially acting as *clippers* in this application.

### 18.3.3 Frequency Limits

As the operating frequency of an op-amp increases, a point may be reached where a significant *voltage phase shift* is introduced between the noninverting input and the output terminal. This phase shift is caused by the **propagation delay** of the op-amp. The propagation delay of any component is the time required for a signal to pass from its input to its output.

When the op-amp starts to introduce a phase shift, the oscillating action of the circuit starts to lose its stability. Remember, the operation of the circuit depends heavily on the phase relationship between the input and output. When you change this relationship, you affect the operation of the oscillator.

**Propagation delay**
The time required for a signal to pass through a component or circuit.

## Wien-Bridge Oscillator Characteristics

*A Practical Consideration:*
The frequency stability of the Wien-bridge oscillator is much higher than that of the phase-shift oscillator. However, it still experiences some frequency drift due to component tolerances and heating. Thus, it is normally used in low-frequency applications where the *exact* operating frequency isn't critical.

Schematic diagram:

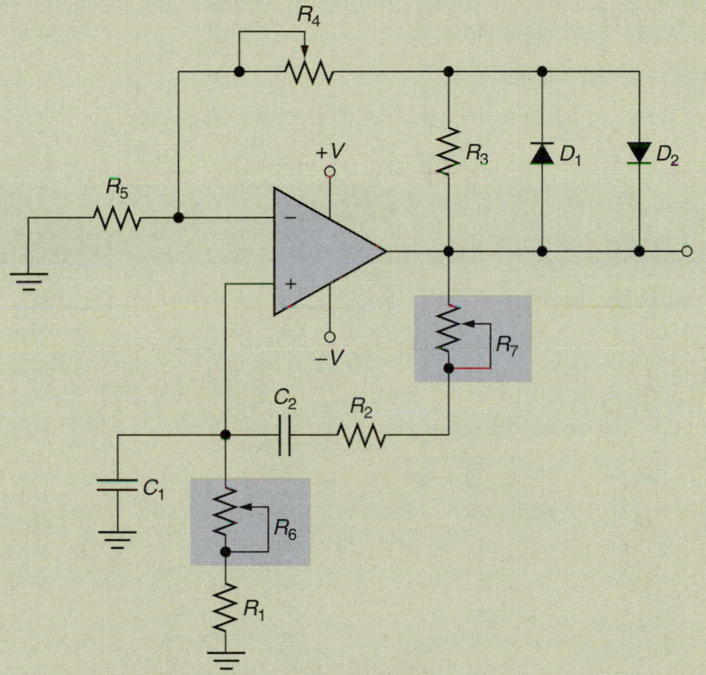

| Circuit recognition: | Dual-feedback networks with a band-pass filter in the positive-feedback network. |
|---|---|
| Gain: | Controlled by the negative feedback network. A potentiometer ($R_4$) is included to allow adjustment. |
| Operating frequency: | • Controlled by the band-pass circuit in the positive-feedback network. Added potentiometers ($R_6$ and $R_7$) allow fine tuning of the output frequency.<br><br>• Operating frequency is limited by the propagation delay of the op-amp; normally restricted to frequencies below 1 MHz. |
| Applications: | Relatively low-frequency systems where the oscillator frequency stability is not critical, that is, where some amount of frequency drift can be tolerated. |

FIGURE 18.9

For most Wien-bridge oscillators, the upper frequency limit is below 1 MHz. Above this frequency, the stability of the oscillator starts to drop. For any oscillator application, this is an unacceptable situation. For higher-frequency applications, discrete $LC$ oscillators are used. We will discuss these circuits through the next two sections.

### 18.3.4 Summary

The Wien-bridge oscillator contains an op-amp and two feedback networks. The positive-feedback network is used to control the operating frequency of the circuit, while the negative-feedback network is used to control its gain. By including variable components in the positive- and negative-feedback networks, both the gain and the frequency of operation for the circuit can be adjusted.

Because of the propagation delay of the op-amp, Wien-bridge oscillators are restricted to operating frequencies below 1 MHz. The Wien-bridge oscillator characteristics are summarized in Figure 18.9.

---

**Section Review ▶**

1. Describe the construction of the Wien-bridge oscillator.

2. How does the positive-feedback network of the Wien-bridge oscillator control its operating frequency?

3. Explain why potentiometers are normally included in the positive-feedback network of a Wien-bridge oscillator.

4. Explain why a potentiometer is normally included in the negative-feedback network of a Wien-bridge oscillator.

5. What is *propagation delay*?

6. How does propagation delay limit the operating frequency of a Wien-bridge oscillator?

---

## 18.4 The Colpitts Oscillator

**OBJECTIVE 6 ▶**

The **Colpitts oscillator** is a discrete $LC$ amplifier that uses a pair of *tapped capacitors* and an inductor to produce the regenerative feedback necessary for oscillations. The Colpitts oscillator is shown in Figure 18.10. As you can see, the transistor is in a common-emitter configuration. Since this configuration produces a 180° voltage phase shift from base to collector, the feedback network must also produce a 180° voltage phase shift if the feedback is to be regenerative.

**Lab Reference**: The operation of an oscillator like the one in Figure 18.10 is demonstrated in Exercise 30.

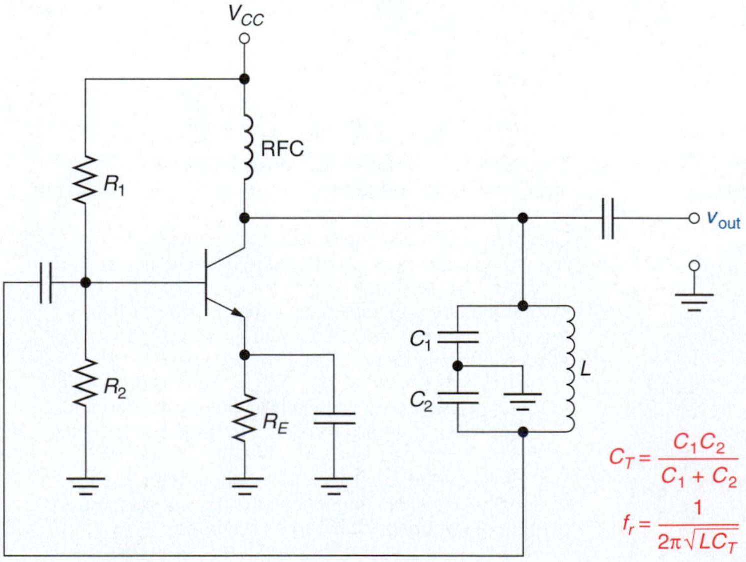

$$C_T = \frac{C_1 C_2}{C_1 + C_2}$$

$$f_r = \frac{1}{2\pi\sqrt{LC_T}}$$

**FIGURE 18.10** Colpitts oscillator (common-emitter configuration).

## 18.4.1    The Feedback Network

The key to understanding the *Colpitts oscillator* is knowing how the feedback network produces a 180° voltage phase shift. The feedback network in Figure 18.10 is made up of $C_1$, $C_2$, and $L$.

The operation of the feedback network is based on the following key points:

1. The amplifier output voltage is developed across $C_1$.
2. The feedback voltage is developed across $C_2$.
3. The voltage across $C_2$ is 180° out of phase with the voltage across $C_1$. Therefore, the feedback voltage is 180° out of phase with the output voltage.

Points 1 and 2 are easier to see if we simplify the circuit as shown in Figure 18.11. In Figure 18.11a, we have dropped the inductor ($L$) from the circuit. Then, in Figure 18.11b, we split the capacitors at the ground connection. As you can see, the input voltage to the amplifier is developed across $C_2$. Therefore, $C_2$ must be the source of the feedback voltage. $C_1$ is obviously across the output of the amplifier. Therefore, $v_{\text{out}}$ is measured across this component. Now, the trick is to see how the voltages across these two components are always 180° out of phase.

> **Colpitts oscillator**
> An oscillator that uses a pair of tapped capacitors and an inductor to produce a 180° voltage phase shift in the feedback network.

> Circuit operation key points.

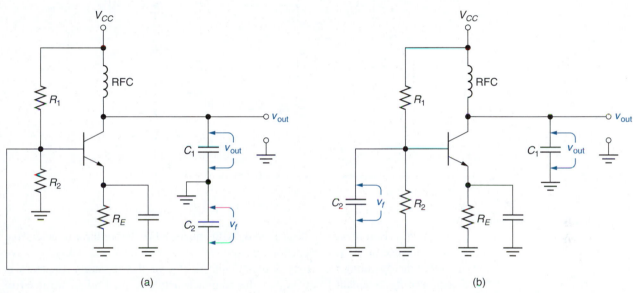

(a)                              (b)

FIGURE 18.11

The phase relationship between $v_{\text{out}}$ and $v_f$ is caused by the *current* action in the feedback network. This action can be seen if we redraw the feedback network as shown in Figure 18.12. By comparing the circuit shown in Figure 18.12 with the feedback network in Figure 18.10, you can see that the circuits are identical. Now, let's assume for a moment that $L$ has a voltage across it with the polarities shown. If we view $L$ as the voltage source, it is easy to see that it produces a current in the circuit. This current develops voltages across $C_1$ and $C_2$ with the polarities shown. As you can see, these voltages are 180° out of phase. If you reverse the voltage polarity across $L$, all polarity signs and current directions reverse, but the voltages across $C_1$ and $C_2$ are still 180° out of phase.

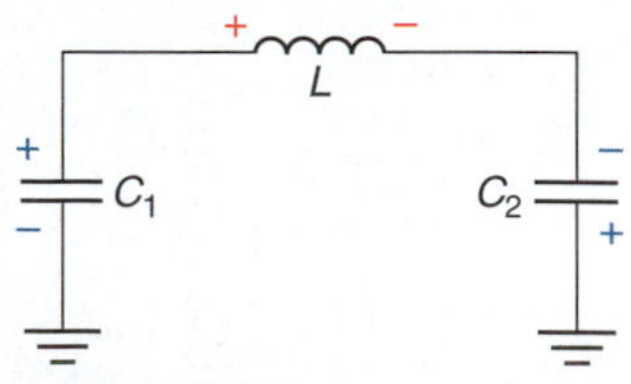

The value of the feedback voltage in the Colpitts oscillator depends on the $\alpha_v$ of the circuit. For this oscillator, $\alpha_v$ is the ratio of $X_{C2}$ to $X_{C1}$. By formula,

FIGURE 18.12

$$\alpha_v = \frac{X_{C2}}{X_{C1}} \qquad\qquad \textbf{(18.3)}$$

Since $X_{C2}$ and $X_{C1}$ are inversely proportional to the values of $C_2$ and $C_1$ at a given frequency, equation (18.3) can be rewritten as

$$\alpha_v = \frac{C_1}{C_2}$$ (18.4)

The validity of equation (18.4) is demonstrated in Example 18.1.

### EXAMPLE 18.1

Determine the value of $\alpha_v$ for the circuit shown in Figure 18.13 using both equations (18.3) and (18.4).

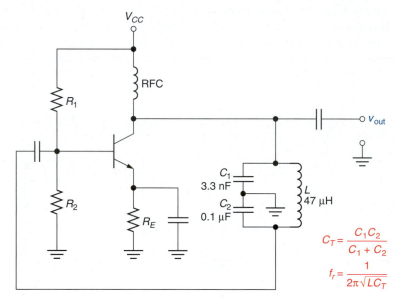

$$C_T = \frac{C_1 C_2}{C_1 + C_2}$$

$$f_r = \frac{1}{2\pi\sqrt{LC_T}}$$

FIGURE 18.13

***Solution:*** Before we can determine $\alpha_v$ with equation (18.3), we have to determine the reactance of the two capacitors. Before we can determine $X_{C1}$ and $X_{C2}$, we have to find the operating frequency of the circuit. As always, the operating frequency is equal to the resonant frequency of the feedback network. To find $f_r$, we need to determine the total capacitance in the $C_1$–$C_2$–$L$ circuit.

Figure 18.11 demonstrated the fact that $C_1$ and $C_2$ are in series. Therefore,

$$C_T = \frac{C_1 C_2}{C_1 + C_2} = \frac{(3.3 \text{ nF})(0.1 \text{ }\mu\text{F})}{3.3 \text{ nF} + 0.1 \text{ }\mu\text{F}} = \textbf{3.19 nF}$$

Now, we can use this value to find $f_r$ as follows:

$$f_r = \frac{1}{2\pi\sqrt{LC_T}} = \frac{1}{2\pi\sqrt{(47 \text{ }\mu\text{H})(3.19 \text{ nF})}} = \textbf{411 kHz}$$

Using the value of $f = 411$ kHz, we can now calculate values of $X_{C1} = 117.34$ $\Omega$ and $X_{C2} = 3.87$ $\Omega$. Using these values in equation (18.3), we get

$$\alpha_v = \frac{X_{C2}}{X_{C1}} = \frac{3.87 \text{ }\Omega}{117.34 \text{ }\Omega} = 0.03298 \cong \textbf{0.033}$$

Now, if we determine the value of $\alpha_v$ using equation (18.4), we obtain

$$\alpha_v = \frac{C_1}{C_2} = \frac{3.3 \text{ nF}}{0.1 \text{ }\mu\text{F}} = 0.033$$

**PRACTICE PROBLEM 18.1**

A Colpitts oscillator like the one in Figure 18.13 has the following values: $C_1 = 10$ nF, $C_2 = 1.5$ $\mu$F, and $L = 10$ $\mu$H. Calculate the value of $\alpha_v$ for the circuit using both equations (18.3) and (18.4).

Because none of us enjoys doing things the hard way, equation (18.4) is the preferred approach to calculating the value of $\alpha_v$.

### 18.4.2 Circuit Gain

As with any other oscillator, the product of $\alpha_v A_v$ for the Colpitts oscillator must be slightly greater than 1. This prevents loss of oscillations due to power losses within the amplifier and, at the same time, prevents the circuit from driving itself into saturation and cutoff clipping.

As described earlier, $v_{\text{out}}$ is developed across $C_2$, and $v_f$ is developed across $C_1$. The voltage gain of any circuit equals the ratio of output voltage to input voltage. Since $v_{\text{in}} = v_f$, the value of $A_v$ for the Colpitts amplifier can be found as

$$A_v = \frac{v_{\text{out}}}{v_f} \cong \frac{C_2}{C_1}$$

### 18.4.3 Amplifier Coupling

As with any parallel resonant tank circuit, the feedback network loses some efficiency when loaded down. To reduce the loading effects of $R_L$, Colpitts oscillators are commonly transformer coupled to the load. A transformer-coupled Colpitts oscillator is shown in Figure 18.14. For this circuit, the primary winding of $T_1$ is the feedback network inductance. Remember that the transformer reduces circuit loading because of the effects of the turns ratio on the reflected load impedance.

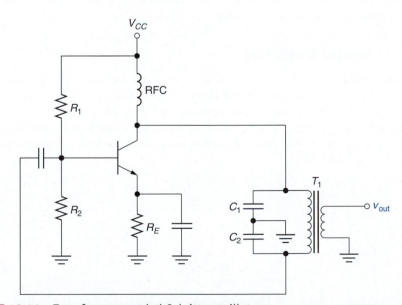

**FIGURE 18.14** Transformer-coupled Colpitts oscillator.

Also, it is acceptable to use capacitive coupling for the Colpitts oscillator, as shown in Figure 18.13, provided that the following relationship is fulfilled:

$$C_C \ll C_T \qquad (18.5)$$

where $C_T$ is the total series capacitance of the feedback network. By fulfilling this relationship, you ensure that the reactance of the coupling capacitor is much greater than that of the series combination of $C_1$ and $C_2$. This prevents circuit loading almost as well as using transformer coupling. At the same time, the relatively high reactance of the coupling capacitor is in series with the load resistance. This causes a significant reduction in load voltage.

### 18.4.4 Summary

The Colpitts oscillator uses a pair of tapped capacitors and an inductor (all in its feedback network) to produce the 180° feedback phase shift required for oscillations. The frequency of operation for the circuit is approximately equal to the resonant frequency of the feedback circuit.

---

**Section Review ▶**

1. What is the *Colpitts oscillator*?

2. Explain how the feedback network in the Colpitts oscillator produces a 180° voltage phase shift.

3. List the key points to remember about the operation of the Colpitts oscillator.

4. How do you calculate the value of $A_v$ for a Colpitts oscillator?

5. Why is transformer coupling often used in Colpitts oscillators?

**Critical Thinking ▶**

6. Identify the following statement as true or false and explain your reasoning: *The higher the Q of the feedback network in Figure 18.13, the more stable the oscillator output frequency.*

---

## 18.5 Other *LC* Oscillators

**OBJECTIVE 7 ▶**

Although the Colpitts is the most commonly used *LC* oscillator, several others are worth mentioning here. In this section, we will take a brief look at three other *LC* oscillator circuits, starting with the Hartley oscillator.

> **Hartley oscillator**
> An oscillator that uses a pair of tapped inductors and a parallel capacitor in its feedback network to produce the 180° voltage phase shift required for oscillation.

### 18.5.1 Hartley Oscillators

The **Hartley oscillator** is almost identical to the Colpitts oscillator. The primary difference is that the feedback network of the Hartley oscillator uses *tapped inductors* and a single capacitor. Two Hartley oscillators are shown in Figure 18.15.

For the oscillator in Figure 18.15a, the output voltage is developed across $L_1$ and the feedback voltage is developed across $L_2$. Therefore, the attenuation caused by the feedback network ($\alpha_v$) is found as

> *A Practical Consideration:*
> Each circuit in Figure 18.15 contains a *blocking capacitor* ($C_3$). In each case, the blocking capacitor is included to keep the RF choke (RFC) and feedback network inductor ($L_1$ or $T_1$) from shorting $V_{CC}$ to ground. By design, the value of $C_3$ is too high to consider in any circuit frequency analysis.

$$\alpha_v = \frac{X_{L2}}{X_{L1}} \qquad (18.6)$$

or

$$\alpha_v = \frac{L_2}{L_1} \qquad (18.7)$$

Since the inductors in Figure 18.15a are in series, the total inductance is found as

$$L_T = L_1 + L_2$$

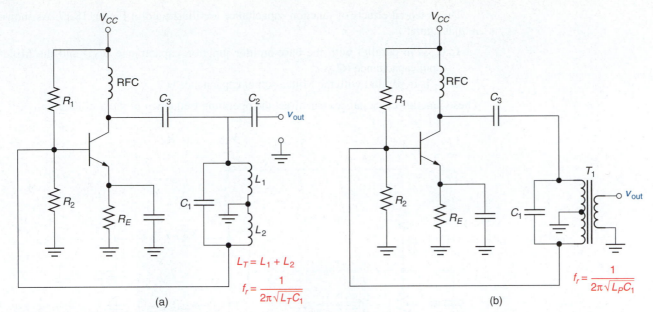

FIGURE 18.15 Hartley oscillators.

This value of $L_T$ is used in the calculation of the circuit's resonant frequency.

The *transformer-coupled* Hartley oscillator (Figure 18.15b) uses a transformer with a tapped primary to accomplish two things:

1. The transformer provides coupling from the circuit to the load.
2. The tapped primary forms the inductor pair for the feedback network.

For this circuit, the total inductance of the primary is used to calculate the circuit's operating frequency.

## 18.5.2 The Clapp Oscillator

The **Clapp oscillator** is simply a Colpitts oscillator with an added capacitor in its feedback network. This capacitor is included to reduce (or eliminate) the effects of *junction capacitance* on operating frequency. As shown in Figure 18.16, the added capacitor ($C_3$) is placed in series with the inductor.

> **Clapp oscillator**
> A Colpitts oscillator with an added capacitor (in series with the feedback inductor) used to reduce the effects of transistor junction capacitance.

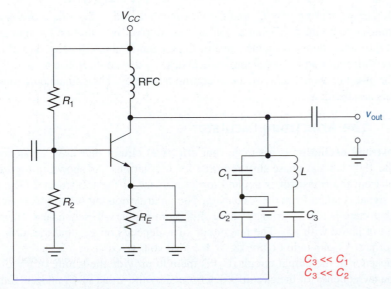

FIGURE 18.16 Clapp oscillator.

The potential effects of junction capacitance are illustrated in Figure 18.17. As shown in that figure:

1. $C_1$ is in parallel with the base-emitter junction capacitance ($C_{be}$) and the Miller input capacitance ($C_{in(M)}$).
2. $C_2$ is in parallel with the Miller output capacitance ($C_{out(M)}$).

These junction capacitances can affect the operating frequency of the oscillator.

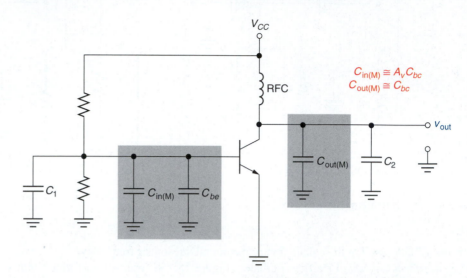

$$C_{in(M)} \cong A_v C_{bc}$$
$$C_{out(M)} \cong C_{bc}$$

**Note:** This is the circuit developed in Figure 18.11 with added BJT capacitances. The coupling capacitors and emitter circuit have been shorted as they typically are in ac equivalent circuits.

FIGURE 18.17   Junction capacitances in a Colpitts oscillator.

The Clapp oscillator eliminates the effects of transistor capacitance by adding the capacitor $C_3$. This capacitor is normally much lower in value than $C_1$ or $C_2$. Because its value is so low, $C_3$ becomes the dominant component in all frequency calculations for the circuit. For the Clapp oscillator,

$$f_r \cong \frac{1}{2\pi\sqrt{LC_3}} \tag{18.8}$$

You may be wondering why $C_1$ and $C_2$ are even included in the Clapp oscillator if $C_3$ is the dominant component. The answer to this is simple. Even though the operating frequency of the circuit depends on the value of $C_3$, $C_1$ and $C_2$ are still needed to provide the 180° phase shift required for regenerative feedback. $C_3$ does not replace $C_1$ and $C_2$; it just reduces the effect of their values on the operating frequency. The gain calculations for the two circuits are identical.

### 18.5.3   The Armstrong Oscillator

**Armstrong oscillator**

An oscillator that uses a transformer in its feedback network to achieve the required 180° voltage phase shift.

The **Armstrong oscillator** is a simple (but effective) circuit that uses a transformer to achieve the 180° voltage phase shift required for oscillations. As shown in Figure 18.18, the output from the transistor is applied to the primary of the transformer ($T_1$), and the feedback signal is taken from the secondary. The polarity dots on the transformer symbol indicate that there is a 180° voltage phase shift from primary to secondary. Therefore, $v_f$ is 180° out of phase with $v_{out}$. The magnitude of $v_f$ depends on $v_{out}$ and the turns ratio of the transformer. As shown in Figure 18.18, the load voltage equals $v_f$.

The capacitor in the output circuit ($C_1$) is there to provide the tuning of the oscillator. The resonant frequency of the circuit is determined by the value of $C_1$ and the inductance of the transformer primary.

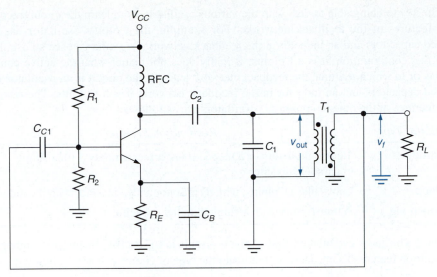

$C_2$ is a blocking capacitor used to prevent the RFC and the primary of $T_1$ from shorting out the dc supply.

FIGURE 18.18   Armstrong oscillator.

### 18.5.4   One Final Note

You have been introduced to several *LC* oscillators, each containing a BJT. Any of the *LC* oscillators we have discussed can be made using either an FET or an op-amp. For example, consider the circuits shown in Figure 18.19. Figure 18.19a shows a Colpitts oscillator that utilizes an op-amp. Figure 18.19b shows an FET Hartley oscillator. These circuits work according to the principles introduced in this section, despite the changes in the active components used. Several other common variations for the circuits covered in this section are as follows:

1. There are several two-stage oscillator circuits where the feedback path is tied between the output of the second stage and the input of the first. The principles of operation for this type of circuit are the same as those covered in this section.
2. Many of the discrete amplifier oscillators covered in this section can also be constructed using a common-base (or common-gate) configuration. In this case, the feedback network is returned from the collector (or drain) to the emitter (or source). The wiring of the feedback network is modified so that it produces a 0° voltage phase shift. This modification, along with the 0° phase shift introduced by the BJT (or FET), results in regenerative feedback.

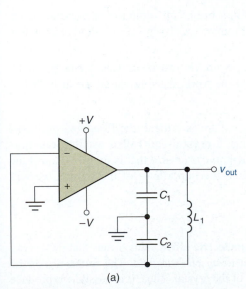

(a)

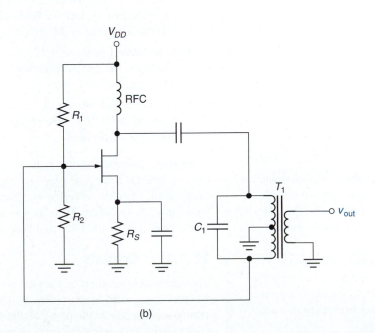

(b)

FIGURE 18.19

What is the key to recognizing oscillator circuits?

The key to being able to deal with the various oscillators is to learn the circuit recognition features of the feedback networks. For example, the Colpitts oscillator has two tapped capacitors and an inductor in the feedback network. Anytime you see an oscillator with this configuration, it is a Colpitts. It really does not matter what the active component is or to which terminal the feedback network leads. If the circuit is an oscillator with tapped capacitors and an inductor in the feedback network, it is a Colpitts. The recognition features of the other common *LC* oscillators are as follows:

| Oscillator Type | Recognition Feature(s) |
|---|---|
| Hartley | Tapped inductors or a tapped transformer primary with a single parallel capacitor |
| Clapp | Looks like a Colpitts, with a capacitor added in series with the inductor |
| Armstrong | A transformer with a single parallel capacitor |

Once you have established that a given circuit is an oscillator, just remember the recognition features listed. They will indicate the type of circuit you are dealing with.

**Section Review ▶**

1. Where are the feedback and output voltages measured in the feedback network of a Hartley oscillator?

2. How does the construction of the Clapp oscillator differ from that of the Colpitts oscillator? What purpose does this difference serve?

3. How is the required 180° voltage phase shift accomplished in the feedback network of an Armstrong oscillator?

4. List the circuit recognition features of the *LC* oscillators covered in this section.

## 18.6 Crystal-Controlled Oscillators

**OBJECTIVE 8 ▶**

Most communications and digital applications require the use of oscillators with *extremely* stable outputs. This requirement can pose a problem for "conventional" oscillators, which can experience output fluctuations for a variety of reasons. For example, the Colpitts oscillator in Figure 18.10 can experience a change in gain or frequency under any of the following conditions:

- *The transistor is replaced.* A replacement transistor may not have the same ac emitter resistance ($r_e'$) as the original. This can result in a change of voltage gain ($A_v$) and the loop gain of the circuit ($\alpha_v A_v$).

- *A reactive component is changed.* When either an inductor or a capacitor is changed, the output frequency may change, requiring one or more frequency adjustments.

- *The circuit warms up.* As a circuit warms up, its resistance values can fluctuate. This can affect the load on the feedback network, causing variations in both frequency and amplitude.

In any system where the stability of the oscillator is critical, the foregoing problems are intolerable. For these types of applications, a **crystal-controlled oscillator** is normally used. Crystal-controlled oscillators have a *quartz crystal* that is used to control the frequency of operation. To help you understand the importance of the crystal, we will take a brief look at what it is and how it works.

### 18.6.1 Crystals

The key to the operation of a crystal is the **piezoelectric effect**. This means that the *crystal vibrates at a constant rate when it is exposed to an electric field*. The frequency of the vibrations depends on the physical dimensions of the crystal. Thus, it is possible to produce crystals with very exact frequency ratings by simply cutting them to the right dimensions.

**Crystal-controlled oscillator**
An oscillator that uses a quartz crystal to produce an extremely stable output frequency.

**Piezoelectric effect**
The tendency of a crystal to vibrate at a fixed frequency when exposed to an electric field.

Three commonly used crystals exhibit a piezoelectric effect. When one of these crystals is placed between two metal plates, it vibrates at its *resonant frequency* as long as a voltage is applied. Crystals can also be made to vibrate by applying a signal to them at a specific frequency. When this is done, the crystal vibrates at that frequency.

The three crystals used in oscillators are *Rochelle salt, quartz,* and *tourmaline*. The best of the three is Rochelle salt because it has the best piezoelectric activity. However, it is also the easiest to break. The toughest of these crystals is tourmaline, but this crystal does not have a very constant vibration rate. The quartz crystal falls between the two extremes: It has a good piezoelectric activity and is strong enough to withstand the vibrations. Quartz is also the least expensive of the three to use.

## 18.6.2   Quartz Crystals

A quartz crystal is made of silicon dioxide, $SiO_2$. This is the same compound used as the insulating layer in the gate of a MOSFET. Quartz crystals are very common. They develop as six-sided crystals, as shown in Figure 18.20. When used in an electronic component, a thin slice of crystal is placed between two conducting plates, like those of a capacitor.

As stated earlier, the physical dimensions of a crystal determine its operating frequency. The only factor that normally alters the physical dimensions of a crystal (and, thus, its operating frequency) is *temperature*. However, using cooling methods to hold the crystal temperature constant eliminates this potential problem.

The electrical operation of the crystal is based on its mechanical properties. However, we can still represent the crystal with an equivalent circuit, as shown in Figure 18.21. Figure 18.21a shows the schematic symbol for the crystal. The equivalent circuit for the device is shown in Figure 18.21b. The components shown represent specific characteristics of the crystal as follows:

$C_C$ = the capacitance of the crystal itself
$C_M$ = the *mounting capacitance* (the capacitance between the crystal and the parallel conducting plates that hold it)
$L$ = the inductance of the crystal
$R$ = the resistance of the crystal

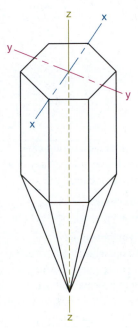

FIGURE 18.20   Quartz crystal.

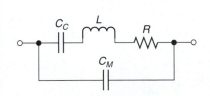

(a) Symbol

(b) Equivalent circuit

(c) Frequency-response curve

FIGURE 18.21   Crystal symbol, equivalent circuit, and frequency response.

*A Practical Consideration:* There is only a slight difference between the values of $f_s$ and $f_p$ for a given crystal. The space left between these frequencies in Figure 18.21c is for illustrative purposes only.

The graph in Figure 18.21c appears strange at first, but it tells us quite a bit about the crystal. At the low-frequency end of the curve, the impedance of the crystal is controlled by the extremely high reactances of $C_M$ and $C_C$. As frequency increases, the combination of $C_C$ and $L$ approaches resonance. At the *series-resonant* frequency of the crystal ($f_s$), the crystal reactance drops to near zero, and the total crystal impedance equals $R$.

When the frequency increases beyond $f_s$, the net reactance rapidly increases. At the same time, the combination of $C_M$ and $L$ approaches resonance. At the *parallel-resonant* frequency $(f_p)$, the impedance of the crystal approaches infinity. At frequencies above $f_p$, the continuing drop in the reactance of $C_M$ causes the crystal impedance to drop off. The bottom line of all this is as follows:

1. At $f_s$, the crystal acts as a *series-resonant* circuit.
2. At $f_p$, the crystal acts as a *parallel-resonant* circuit.

This means that we can use a crystal in place of either a series $LC$ circuit or a parallel $LC$ circuit. If we use it in place of a series $LC$ circuit, the oscillator operates at $f_s$. If we use it in place of a parallel $LC$ circuit, the oscillator operates at $f_p$.

### 18.6.3    Overtone Mode

A given crystal produces outputs at its *resonant frequency* and *harmonics* of that frequency. The *resonant frequency* is sometimes referred to as the **fundamental frequency**, while the *harmonics* are sometimes referred to as **overtones**. (You were first introduced to the concept of harmonics in our discussion on class C amplifiers.)

Crystals are limited by their physical dimensions to fundamental outputs of 10 MHz and less. However, by tuning circuits to the overtones, we can produce usable signals at much higher frequencies. For example, tuning a circuit to the tenth harmonic of a 10 MHz crystal produces a relatively stable 100 MHz signal. This **overtone mode** of operation is commonly used in digital and communications electronics.

### 18.6.4    CCO Circuits

The Colpitts oscillator can be modified into a *crystal-controlled oscillator* (CCO), as shown in Figure 18.22. As you can see, a crystal ($Y_1$) has been added in series between the $LC$ circuit and the amplifier input. The crystal acts as a *series-resonant* circuit (at $f_s$), allowing the feedback voltage to pass to the amplifier input with minimum attenuation. In other words, when operated at $f_s$, the impedance of the crystal drops to nearly zero, and the $LC$ circuit is coupled directly to the amplifier input.

The frequency stability of the CCO comes from the extremely high $Q$ of the crystal, which is typically in the thousands. Because of this high $Q$, the component acts as a series resonant circuit over an extremely limited band of frequencies. Thus, the circuit can oscillate only at the value of $f_s$, and the circuit output frequency is extremely stable. Note that

**Fundamental frequency**
The resonant frequency of a crystal or circuit.

**Overtone**
Another word for *harmonic*.

**Overtone mode**
An operating mode where a circuit is tuned to a harmonic of a fundamental frequency for the purpose of producing a higher output frequency.

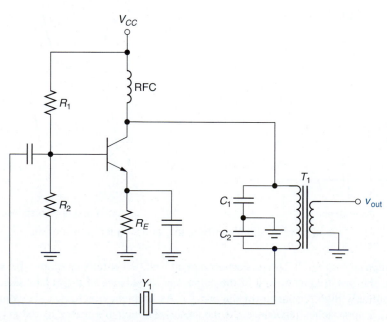

**FIGURE 18.22**   Crystal-controlled Colpitts oscillator.

placing a crystal in the same relative position in the Hartley or Clapp oscillators has the same result.

The simplest of the crystal oscillators is the *Pierce oscillator*, shown in Figure 18.23. It has only four components and a crystal. This oscillator is not highly stable, but its simplicity has kept it around for years.

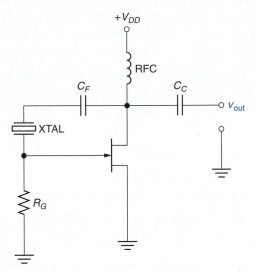

**FIGURE 18.23** Pierce oscillator.

---

1. What is a *crystal-controlled oscillator*?

2. What is the *piezoelectric effect*?

3. Explain how a crystal will act as a series resonant circuit at one frequency $(f_s)$ and as a parallel resonant component at another $(f_p)$.

4. Explain how the addition of a crystal to an *LC* oscillator stabilizes the operating frequency of the circuit.

5. What prevents the circuit shown in Figure 18.22 from operating at the parallel-resonant frequency of the crystal $(f_p)$?

◀ **Section Review**

◀ **Critical Thinking**

## 18.7  Oscillator Troubleshooting

Of all the circuits we have discussed, oscillators can be the most challenging and time-consuming circuits to troubleshoot. The reason is simple: With the exception of the biasing resistors, every component is directly involved in producing the output signal. If any one of these components fails, nothing works. For example, refer to Figure 18.14. Whether a failure of the circuit is caused by the coupling capacitor, $C_2$, $T_1$, or $C_1$, the results are exactly the same: The oscillator does not have an output.

◀ *OBJECTIVE 9*

As with any circuit, you start to troubleshoot an oscillator by making sure that it is the source of the problem. If the circuit does not have an output, check the power supply connections in the circuit. If there is no problem here, the oscillator has a fault. (You do not need to check the oscillator's signal source—it doesn't have one.)

When you have verified that an oscillator is bad, there are a few points to remember that may make the troubleshooting process go a little more smoothly:

1. Remember that for dc analysis, the oscillator is biased like any other type of amplifier. Thus, your approach to dc troubleshooting should be the same as it is for any amplifier with the same biasing circuit.

2. In the case of the oscillator, it is often faster (and, therefore, less costly) to go ahead and replace the reactive components in the feedback network.

*A Practical Consideration:*
Most oscillators contain one or more variable components for output tuning. When you replace any component in the circuit, you must check the output frequency and, if necessary, retune the circuit.

If circumstances require you to test the components in the feedback network, use an analog VOM (set as an ohmmeter) to test them, keeping in mind the following:

1. A good capacitor initially provides a low resistance reading. As the ohmmeter charges the capacitor, its resistance reading increases toward $\infty$ $\Omega$.
2. Inductors always provide low resistance readings.
3. When a transformer is used (as in Figure 18.14), you should obtain a low resistance reading across the primary terminals and the secondary terminals. However, you should obtain a very *high* resistance reading from either primary terminal to either secondary terminal.

If you decide to go the component-replacement (or *swapping*) route, replace the reactive components one at a time. After you replace each component, apply power to the circuit and retest its operation. When the circuit starts to operate at the proper frequency, your task is complete.

**Section Review ▶**

1. Why are oscillators difficult to troubleshoot?
2. What points should be remembered when troubleshooting an oscillator?
3. When testing the reactive components in an oscillator, what points should be kept in mind?

## CHAPTER SUMMARY

Here is a summary of the major points made in this chapter:

1. An *oscillator* is an ac signal generator.
2. A *positive-feedback* network produces a feedback voltage ($v_f$) that is in phase with the amplifier input. Positive feedback can be accomplished using:
   a. An amplifier and feedback network that each produce a 180° voltage phase shift, for a 360° total phase shift.
   b. An amplifier and a feedback network that each produce a 0° phase shift.
   The feedback network generates the input for the amplifier, which in turn generates an input to the feedback network. For this reason, positive feedback is referred to as *regenerative feedback*.
3. For an oscillator to operate properly:
   a. The circuit must have a positive (regenerative) feedback loop.
   b. The circuit must receive an initial trigger to begin the oscillations.
   c. The circuit must fulfill the *Barkhausen criterion*.
4. The Barkhausen criterion states that the amplifier and feedback network must fulfill the relationship $\alpha_v A_v = 1$.
   a. When $\alpha_v A_v < 1$, the output from an oscillator fades out within a few cycles.
   b. When $\alpha_v A_v > 1$, an oscillator drives itself into saturation and cutoff clipping.
   The effects of $\alpha_v A_v$ on oscillator operation are illustrated in Figure 18.3.
5. A *phase-shift* oscillator can be made using an inverting amplifier with a series of *RC* circuits in its feedback path (see Figure 18.4).
   a. The inverting amplifier produces a 180° voltage phase shift.
   b. The *RC* circuits combine to produce a 180° voltage phase shift.
   c. The two 180° voltage phase shifts result in an overall 360° shift, which by definition is regenerative feedback.
6. *Oscillator stability* is a measure of the circuit's ability to maintain constant output amplitude and frequency.
7. Phase-shift oscillators are rarely used because of frequency stability problems.
8. A three-stage amplifier can (inadvertently) act as a phase-shift oscillator (see Figure 18.5). Oscillations in an amplifier (like the one in Figure 18.5) can be prevented using:
   a. A *regulated dc power supply*.
   b. A *decoupling capacitor* connected across the dc power supply.

9. In practice, the value of $\alpha_v A_v$ for an oscillator must be *slightly greater than 1* to make up for the power lost during each cycle of operation.

10. The *Wien-bridge oscillator* is an op-amp circuit that achieves regenerative feedback by producing no phase shift at its resonant frequency ($f_r$) (see Figure 18.6).

    a. The feedback signal is applied to its *noninverting* input, so the op-amp does not produce a voltage phase shift.

    b. The $RC$ circuits in the signal feedback path do not produce a voltage phase shift.

11. The Wien-bridge oscillator has two feedback paths.

    a. The positive-feedback path is a band-pass filter that sets the frequency of operation (see Figure 18.7).

    b. The negative-feedback path controls the gain of the op-amp.

12. Potentiometers are often added to the positive-feedback network of a Wien-bridge oscillator for adjusting the operating frequency (see Figure 18.9).

13. The negative-feedback network of a Wien-bridge oscillator may contain a potentiometer and several diodes (see Figure 18.9).

    a. The potentiometer is used to adjust the gain of the op-amp.

    b. The diodes limit the output amplitude by reducing the circuit voltage gain ($A_{CL}$) if the output exceeds a specified value.

14. Op-amp oscillators are limited to frequencies below approximately 1 MHz because of the *propagation delay* of the component.

15. The *Colpitts* oscillator is a discrete $LC$ amplifier that uses a pair of tapped capacitors and an inductor to produce a 180° voltage phase shift in the feedback network (see Figure 18.10).

    a. $C_1$ is the *output capacitor*. (The output voltage is felt across this component).

    b. $C_2$ is the *feedback capacitor*. (The feedback voltage is felt across this component).

    c. The voltages across $C_1$ and $C_2$ are 180° out of phase.

16. The attenuation factor ($\alpha_v$) of the feedback network equals the ratio of the output capacitor to the feedback capacitor.

17. The voltage gain ($A_v$) equals the ratio of the feedback capacitor to the output capacitor.

18. The *Hartley* oscillator is a variation on the Colpitts. It uses a pair of tapped inductors and a single capacitor in its feedback network to produce the 180° voltage phase shift required for regenerative feedback (see Figure 18.15).

19. The attenuation factor ($\alpha_v$) of the Hartley oscillator equals the ratio of the feedback inductor ($L_2$) to the output inductor ($L_1$).

20. The *Clapp* oscillator is a Colpitts oscillator with an added capacitor that is used to reduce the effects of transistor junction capacitance.

    a. The transistor junction capacitances are in parallel with the output and feedback capacitors and, therefore, can affect the operating frequency (see Figure 18.17).

    b. A low-value capacitor is added in series with the inductor. This capacitor (along with the inductor) determines the circuit operating frequency.

21. An *Armstrong* oscillator uses a transformer in its feedback network to achieve the 180° voltage phase shift required for regenerative feedback (see Figure 18.18).

    a. The output from the transistor is applied to the transformer primary.

    b. The feedback signal is taken from the transformer secondary.

    c. The transformer produces a 180° voltage phase shift, so the feedback voltage is 180° out of phase with the amplifier output voltage.

22. Any of the $LC$ oscillators covered can be built around an FET or an op-amp. The key to recognizing an oscillator lies in the feedback network.

23. $LC$ oscillators are not stable enough for many communications and digital applications.

    a. Replacing the active component can affect the gain of the circuit.

    b. Replacing any reactive component in the feedback network can affect the operating frequency.

    c. Changes in temperature can affect both gain and operating frequency.

24. A *crystal-controlled oscillator* uses a quartz crystal to produce an extremely stable output frequency.

25. The *piezoelectric effect* is the tendency of a crystal to vibrate at a fixed frequency when exposed to an electric field. Three commonly used crystals exhibit this effect:
    a. *Rochelle salt* exhibits the strongest piezoelectric effect, but it is the easiest to break.
    b. *Tourmaline* has a very strong structure, but it has problems with frequency stability.
    c. *Quartz* falls between the other two crystals. It is strong and vibrates at extremely stable frequencies.
26. A quartz crystal is made of *silicon dioxide* ($SiO_2$), the same material used as the insulating layer in MOSFET gates (see Figure 18.20).
27. A quartz crystal has two resonant frequencies: a *series-resonant* frequency ($f_s$) and a *parallel-resonant* frequency ($f_p$) (see Figure 18.21).
    a. There is little difference between the values of $f_s$ and $f_p$.
    b. A quartz crystal can be used as either a series or a parallel resonant circuit.
    c. When used in place of a series *LC* circuit, the crystal operates at $f_s$.
    d. When used in place of a parallel *LC* circuit, the crystal operates at $f_p$.
28. A given crystal produces outputs at its resonant frequency and multiples of that frequency.
    a. The resonant frequency is called the *fundamental* frequency.
    b. The harmonics are often referred to as *overtones*.
29. *Overtone mode* operation involves using tuned circuits to select and amplify a harmonic produced by a crystal, producing stable oscillations at the harmonic frequency.
30. Any *LC* oscillator can be converted to a *crystal-controlled oscillator* (CCO) by adding a crystal to the feedback network (see Figure 18.22).
31. The *Pierce* oscillator is the simplest CCO (see Figure 18.23).
32. Oscillators can be the most challenging and time-consuming circuits to troubleshoot.
    a. Except for the biasing resistors, all components are directly involved in producing the output signal.
    b. All component failures have the same primary symptom: The circuit has no output signal and, therefore, no input signal.
    c. The approach to troubleshooting the dc biasing circuit should be the same as the approach you would take for any amplifier having the same biasing circuit.
    d. When the dc biasing circuit is good, it is often faster (and, therefore, cheaper) simply to replace all the components in the feedback network.

## EQUATION SUMMARY

| Equation Number | Equation | Section Number |
|:---:|:---:|:---:|
| **(18.1)** | $\alpha_v A_v = 1$ | 18.1 |
| **(18.2)** | $\theta = \tan^{-1} \dfrac{-X_C}{R}$ | 18.2 |
| **(18.3)** | $\alpha_v = \dfrac{X_{C2}}{X_{C1}}$ | 18.4 |
| **(18.4)** | $\alpha_v = \dfrac{C_1}{C_2}$ | 18.4 |
| **(18.5)** | $C_C \ll C_T$ | 18.4 |
| **(18.6)** | $\alpha_v = \dfrac{X_{L2}}{X_{L1}}$ | 18.5 |
| **(18.7)** | $\alpha_v = \dfrac{L_2}{L_1}$ | 18.5 |
| **(18.8)** | $f_r \cong \dfrac{1}{2\pi \sqrt{LC_3}}$ | 18.5 |

**PRACTICE PROBLEMS**

### Section 18.3

**1.** Calculate the operating frequency of the circuit shown in Figure 18.24.

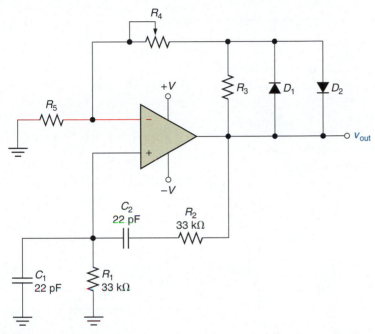

**FIGURE 18.24**

**2.** Calculate the operating frequency of the circuit shown in Figure 18.25.

### Section 18.4

**3.** Calculate the operating frequency for the circuit shown in Figure 18.26.

**4.** Calculate the value of $\alpha_v$ for the circuit shown in Figure 18.27.

**5.** A circuit like the one shown in Figure 18.26 has values of $C_1 = 10$ nF, $C_2 = 1$ μF, and $L_1 = 3.3$ mH. Calculate the operating frequency, the value of $\alpha_v$, and the value of $A_v$ for the circuit.

**6.** A circuit like the one shown in Figure 18.26 has values of $C_1 = 1$ μF, $C_2 = 10$ μF, and $L_1 = 3.3$ mH. Calculate the operating frequency, the value of $\alpha_v$, and the value of $A_v$ for the circuit.

**7.** A circuit like the one shown in Figure 18.26 has values of $C_1 = 0.22$ μF, $C_2 = 3.3$ μF, and $L_1 = 330$ μH. Calculate the operating frequency, the value of $\alpha_v$, and the value of $A_v$ for the circuit.

### Section 18.5

**8.** Calculate the operating frequency, the value of $\alpha_v$, and the value of $A_v$ for the circuit shown in Figure 18.27.

**9.** A circuit like the one shown in Figure 18.27 has values of $L_1 = 0.1$ H, $L_2 = 1$ mH, and $C_1 = 22$ nF. Calculate the operating frequency, the value of $\alpha_v$, and the value of $A_v$ for the circuit.

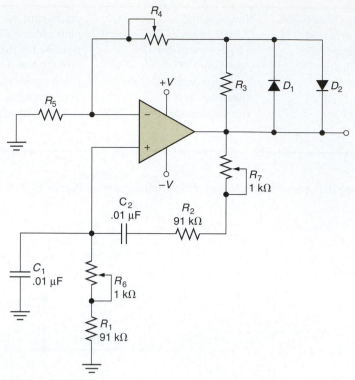

FIGURE 18.25

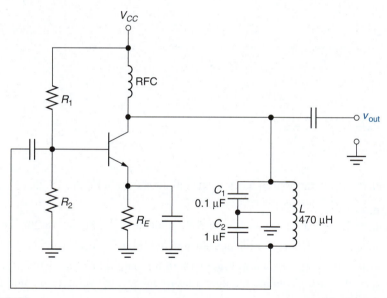

FIGURE 18.26

**10.** For the circuit shown in Figure 18.28, calculate the operating frequency, the value of $\alpha_v$, and the value of $A_v$.

**11.** A circuit like the one shown in Figure 18.28 has values of $C_1 = 0.1\ \mu F$, $C_2 = 1\ \mu F$, $C_3 = 100\ pF$, and $L = 3.3\ mH$. Calculate the operating frequency, the value of $\alpha_v$, and the value of $A_v$ for the circuit.

**12.** The circuit shown in Figure 18.29 does not oscillate. A check of the dc voltages provides the readings indicated. Discuss the possible cause(s) of the problem.

**13.** The circuit shown in Figure 18.30 does not oscillate. A check of the dc voltages provides the readings indicated. Discuss the possible cause(s) of the problem.

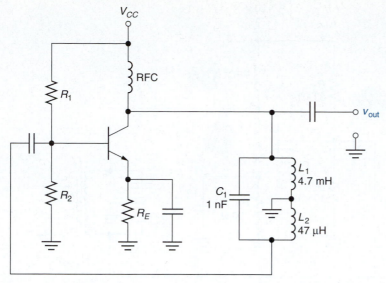

FIGURE 18.27

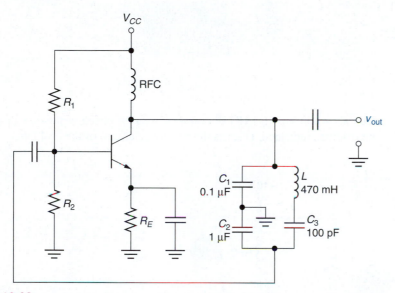

FIGURE 18.28

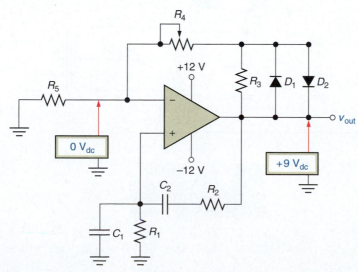

FIGURE 18.29

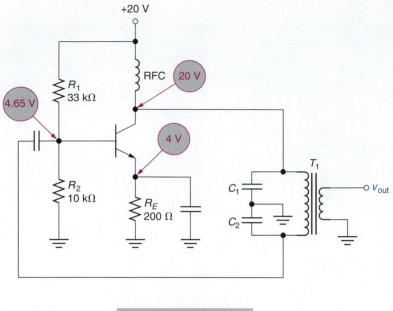

FIGURE 18.30

**14.** The circuit shown in Figure 18.31 does not oscillate. A check of the dc voltages provides the readings indicated. Discuss the possible cause(s) of the problem.

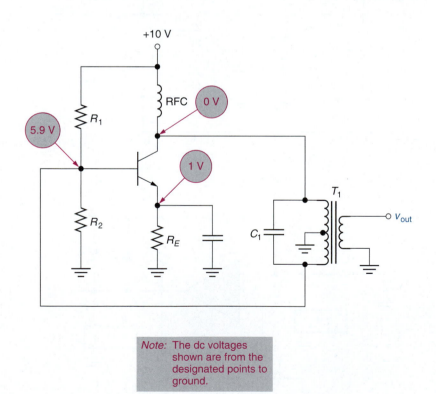

FIGURE 18.31

**15.** Refer to Figure 18.27. Use circuit calculations to show that the circuit fulfills the Barkhausen criterion.

**16.** Despite its appearance, you've already seen the dc biasing circuit shown in Figure 18.32. Determine the type of biasing used, and perform a complete dc analysis of the oscillator.

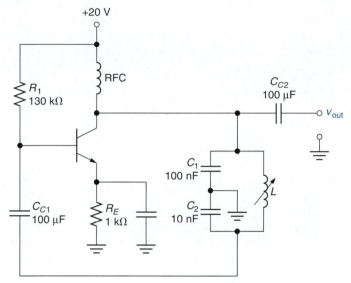

**FIGURE 18.32**

**17.** The feedback inductor in Figure 18.32 is shown to be a variable inductor. Determine the adjusted value of $L$ that would be required for an operating frequency of 22 kHz.

**18.** Write a program to determine the operating frequency of a Wien-bridge oscillator given the positive feedback values of $R$ and $C$.

**19.** Write a program to determine the operating frequency and $\alpha_v$ values for a Colpitts oscillator given the feedback network values of $L$ and $C$.

**18.1** $\alpha_v = 6.7 \times 10^{-3}$ (for both solutions)

## chapter 19

# Solid-State Switching Circuits

## Objectives

*After studying the material in this chapter, you should be able to:*

1. Describe and analyze the operation of the basic BJT, JFET, and MOSFET switches.

2. List the common time and frequency characteristics of rectangular waveforms and describe how each is measured.

3. List and describe the factors that affect BJT and FET switching time.

4. Discuss the methods used to improve the switching times of basic BJT and FET circuits.

5. Compare and contrast the time measurement techniques for *inverters* and *buffers*.

6. Describe and analyze the operation of *inverting Schmitt triggers* and *noninverting Schmitt triggers*.

7. List and describe the output characteristics of the three types of *multivibrators*.

8. Describe the internal construction and operation of the 555 timer.

9. Describe, analyze, and troubleshoot 555 timer *astable* and *monostable* multivibrators.

10. Describe the operation of the 555 timer *voltage-controlled oscillator* (*VCO*).

## Outline

# Computers Anyone?

Solid-state switching circuits are the fundamental components of modern computer systems. It seems appropriate then that we should take a minute to look at the effects that developments in solid-state electronics have had on the digital computer.

The first general-purpose digital computer, called ENIAC, was developed in the 1940s at the University of Pennsylvania. This computer was just about as large as a small building, weighed about 50 tons, and contained more than 18,000 vacuum tubes. As the story is told, several college students spent their days running around the computer shop with shopping carts full of vacuum tubes because ENIAC burned out several tubes each hour. Amazingly, ENIAC was capable of doing less than your calculator.

When the first transistor computer circuits were developed, the *minicomputer* was born. Minicomputers were capable of performing more advanced functions and were much smaller than any vacuum tube computer.

The development of the solid-state *microprocessor*, a complete data-processing unit constructed on a single chip, in the 1970s led to the birth of the *microcomputer*. Since the birth of the microcomputer, computers have been accessible and affordable for almost everyone.

One interesting point: As computer chips became smaller and more powerful, they also became much less expensive. It has been said that if automobiles had gone through the same evolution as computers, you could now buy a Rolls Royce for less than $25.00. Of course, the car would be smaller than a pencil eraser!

---

Throughout the text, we have concentrated almost entirely on the *linear* operation of solid-state devices and circuits. Another important aspect of solid-state electronics is the way in which devices and circuits respond to *nonlinear* waveforms, such as *square waves*. Circuits designed to respond to (or generate) nonlinear waveforms are referred to as **switching circuits**. In this chapter, we will discuss the most basic switching circuits. As you will see, these circuits are viewed quite differently from linear circuits. They are also easier to understand and troubleshoot.

> **Switching circuits**
> Circuits designed to respond to (or generate) nonlinear waveforms, such as square waves.

## 19.1  Introductory Concepts

For you to understand some of the more complex switching circuits, you must be comfortable with the idea of using a BJT, FET, or MOSFET as a switch. In this section, we will discuss some of the switching characteristics of discrete solid-state devices.

### 19.1.1  The BJT as a Switch

OBJECTIVE 1 ▶    A BJT can be used as a switch simply by driving the component back and forth between *saturation* and *cutoff*. A basic transistor switching circuit is shown in Figure 19.1a. As the figure shows, a rectangular input produces a rectangular output. This relationship is easy to understand if you refer to the dc load line shown in Figure 19.1b.

When the input to the transistor is at $-V_{pk}$, the emitter-base junction of the transistor is biased *off*. When the transistor is biased off, the following (*ideal*) conditions exist:

$$V_{CE} = V_{CC} \quad \text{and} \quad I_C = 0$$

As Figure 19.2 shows, these conditions can be represented using an *open* switch.

When the input to the transistor is at $+V_{pk}$, current is generated in the base circuit. Assuming that the value of $I_B$ is high enough, the transistor is driven into *saturation*. When the transistor saturates, the following (*ideal*) conditions exist:

$$V_{CE} = 0\,\text{V} \quad \text{and} \quad I_C = \frac{V_{CC}}{R_C}$$

As Figure 19.2 shows, these conditions can be represented using a *closed* switch.

> Ideal cutoff conditions are similar to those of an *open* switch.

> Ideal saturation conditions are similar to those of a *closed* switch.

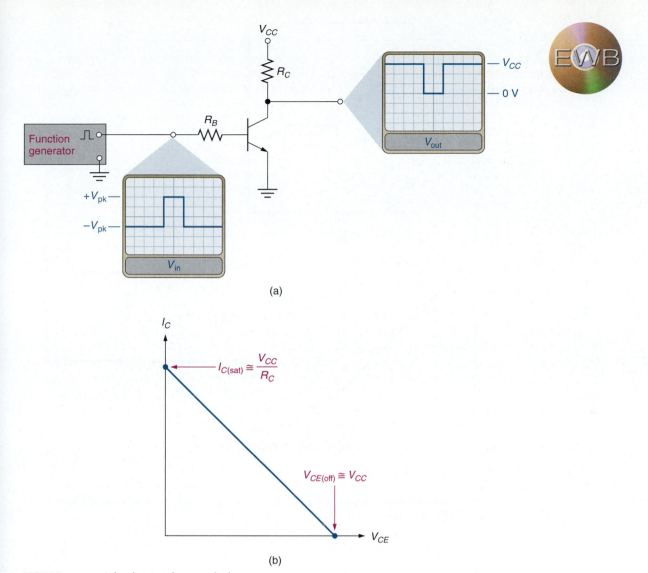

FIGURE 19.1   A basic transistor switch.

Some assumptions were made about the circuit shown in Figure 19.1 to simplify the discussion. These assumptions are:

1. $-V_{pk}$ is low enough to drive $Q_1$ into cutoff.
2. $+V_{pk}$ generates sufficient base current to drive $Q_1$ into saturation.
3. The transistor is an ideal component.

In practice, the value of $-V_{pk}$ must be equal to (or more negative than) the emitter supply voltage ($V_{EE}$) to ensure that $Q_1$ is driven into cutoff. Since the emitter in Figure 19.1 is returned to ground, $V_{EE}$ is 0 V. Therefore, a value of $-V_{pk}$ that is equal to (or more negative than) 0 V ensures that the transistor is driven into cutoff.

Practical input signal characteristics.

The high input voltage ($+V_{pk}$) must drive the transistor into saturation. In practice, this can be accomplished by:

1. Designing the circuit so that $+V_{pk} = V_{CC}$.
2. Using a value of $R_B$ that is low enough to ensure that $I_B > \dfrac{I_{C(sat)}}{h_{FE}}$.

Example 19.1 uses several relationships from Chapter 7 to show how meeting these conditions ensures that a BJT is saturated when $+V_{pk}$ is applied to its input.

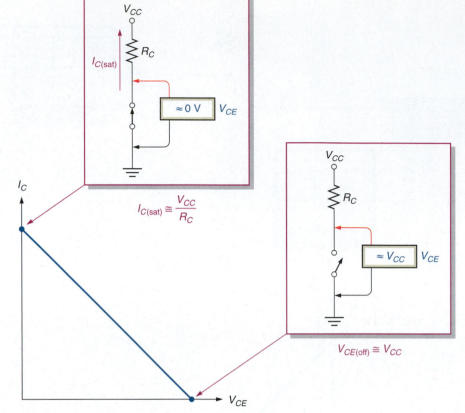

**FIGURE 19.2** The "open" and "closed" transistor switch.

**EXAMPLE 19.1**

Determine the minimum high input voltage $(+V_{pk})$ required to saturate the transistor switch shown in Figure 19.3.

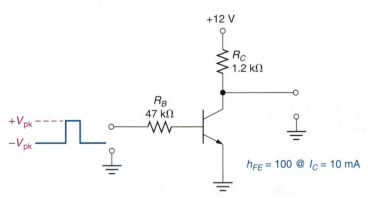

**FIGURE 19.3**

***Solution:*** Assuming that the transistor is an ideal component ($V_{CE} = 0$ V at saturation), the value of $I_{C(sat)}$ is found as

$$I_{C(sat)} = \frac{V_{CC}}{R_C} = \frac{12 \text{ V}}{1.2 \text{ k}\Omega} = 10 \text{ mA}$$

As shown, the value of $h_{FE}$ for the transistor is 100 when $I_C$ equals 10 mA. Using this value, the base current required to set $I_{C(sat)}$ equal to 10 mA is found as

$$I_B = \frac{I_{C(sat)}}{h_{FE}} = \frac{10 \text{ mA}}{100} = 100 \text{ μA}$$

In Chapter 7, you learned that the base current generated in a base bias circuit is found as

$$I_B = \frac{V_{CC} - V_{BE}}{R_B}$$

The base current generated in Figure 19.3 is found by using $+V_{pk}$ in place of $V_{CC}$ in the above equation as follows:

$$I_B = \frac{+V_{pk} - V_{BE}}{R_B}$$

This equation can be rewritten as

$$+V_{pk} = I_B R_B + V_{BE}$$

Using $I_B = 100 \ \mu A$ and $R_B = 47 \ k\Omega$, the *minimum* value of $+V_{pk}$ required to saturate the transistor is found as

$$+V_{pk} = I_B R_B + V_{BE} = (100 \ \mu A)(47 \ k\Omega) + 0.7 \ V$$
$$= 4.7 \ V + 0.7 \ V = \mathbf{5.4 \ V}$$

Since a value of $+V_{pk} = 5.4$ V is sufficient to saturate the transistor, setting the value of $+V_{pk}$ equal to $V_{CC}$ is more than enough to do the trick.

**PRACTICE PROBLEM 19.1**

A BJT switch like the one in Figure 19.3 has the following values: $V_{CC} = +10$ V, $R_C = 1 \ k\Omega$, $R_B = 51 \ k\Omega$, and $h_{FE} = 70$ at $I_C = 10$ mA. Determine the minimum value of $+V_{pk}$ required to saturate the transistor.

In our discussion, we have assumed that the transistor is an *ideal* component; that is, we have assumed that $V_{CE} = V_{CC}$ when the transistor is in cutoff and that $V_{CE} = 0$ V when the transistor is saturated. In practice, the outputs from a transistor switch typically fall within the following ranges:

| Condition | Output Voltage ($V_{CE}$) Range |
|-----------|-------------------------------|
| Cutoff | Within 1 V of $V_{CC}$ |
| Saturation | Between 0.2 and 0.4 V |

When a transistor is in cutoff, there is still some leakage current through the component. This leakage current causes some voltage to be dropped across the collector resistor. Thus, the output voltage is slightly lower than $V_{CC}$. When the transistor is saturated, you still have some voltage developed across the internal resistance of the component. Thus, the output is slightly greater than 0 V. However, the range of output voltages for a saturated transistor is so low (when compared to the value of $V_{CC}$) that we can easily idealize the situation and say that the output equals 0 V.

## 19.1.2    The JFET as a Switch

The basic JFET switch differs from the BJT switch in a couple of ways:

1. A typical JFET switch has much higher input impedance.
2. A *negative* input rectangular wave (or *pulse*) to an *n*-channel JFET is used to produce a *positive* output pulse.

Both of these points are illustrated in Figure 19.4. The high input impedance of the JFET switch is due to the relatively high value of $R_G$ (typically in megohms) and the extremely

*A Practical Consideration:* A circuit like the one shown in Figure 19.3 must be designed so that the collector current never meets or exceeds the $I_{C(max)}$ rating of the BJT. As long as

$$\frac{V_{CC}}{R_C} < I_{C(max)}$$

the BJT will be able to handle an input of $+V_{pk} = V_{CC}$.

What are the practical output values for a BJT switch?

What are the differences between BJT and JFET switches?

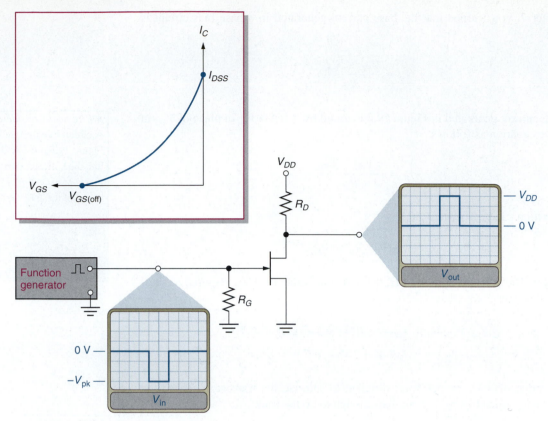

**FIGURE 19.4**   A JFET switch.

JFET switch input/output
relationships.

high input impedance of the JFET gate. For the circuit shown in Figure 19.4, the input impedance of the switch equals the value of $R_G$.

To understand the input/output voltage relationship of the circuit, we need to refer to the transconductance curve shown in Figure 19.4. Recall that $I_D$ reaches its maximum value ($I_{DSS}$) when $V_{GS} = 0$ V. Thus, when the input is at 0 V, the JFET current reaches its maximum value, and the output is found as

$$V_{out} = V_{DD} - I_{DSS}R_D$$

Assuming that the value of $R_D$ has been selected properly, the output from the JFET switch is approximately 0 V when the input is at 0 V.

When the input goes to $-V_{pk}$, the current through the JFET drops nearly to zero when $-V_{pk}$ is more negative than, or equal to, $V_{GS(off)}$. When $I_D \cong 0$ A, little voltage is dropped across $R_D$ and the output voltage is close to $V_{DD}$. By formula,

$$V_{out} \cong V_{DD}$$

when $V_{in}$ is equal to (or more negative than) $V_{GS(off)}$. Using these two relationships for $V_{out}$, it is easy to see that the input pulse shown in Figure 19.4 produces the output pulse shown. Example 19.2 demonstrates the analysis of a basic JFET switch.

## EXAMPLE 19.2

Determine the high and low output voltage values for the circuit shown in Figure 19.5.

*Solution:*   When the input is at 0 V, the JFET has minimum resistance between the source and drain and $I_D = I_{DSS}$. Thus,

$$V_{out} = V_{DD} - I_{DSS}R_D = 5\text{ V} - (5\text{ mA})(1\text{ k}\Omega) = 5\text{ V} - 5\text{ V} = \mathbf{0\ V}$$

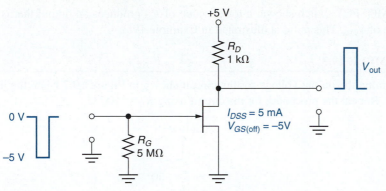

**FIGURE 19.5**

When the input is at $-5$ V, it is equal to $V_{GS(off)}$. Therefore, $I_D = 0$ A. Using this value in place of $I_{DSS}$ in the above equation yields

$$V_{out} = V_{DD} - I_D R_D = +5\,V - 0\,V = \mathbf{+5\,V}$$

**PRACTICE PROBLEM 19.2**

A JFET switch like the one in Figure 19.5 has the following values: $V_{DD} = +10$ V, $R_D = 1.8$ k$\Omega$, $I_{DSS} = 5$ mA, and $V_{GS(off)} = -3$ V. The input signal to the circuit is a rectangular wave that has low and high output voltages of $-4$ V and 0 V, respectively. Determine the high and low output voltage values for the circuit.

*A Practical Consideration:* Example 19.2 assumes that the JFET has only one value of $I_{DSS}$ and $V_{GS(off)}$. In practice, JFETs typically have minimum and maximum values of $I_{DSS}$ and $V_{GS(off)}$. When this is the case, you must use both values to determine the output voltage ranges.

In Example 19.2, we have once again assumed that the active component is *ideal*. In practice, the output values vary slightly from the ideal values of 0 V and $V_{DD}$. These discrepancies are caused by the *drain cutoff current* ($I_{D(off)}$), and the *drain-source on-state voltage* ($V_{DS(on)}$), of the JFET.

What are the practical output values for a JFET switch?

### 19.1.3    The MOSFET as a Switch

A MOSFET switch has the input impedance advantage of a JFET switch and the input/output polarity relationship of a BJT switch. A basic MOSFET switch is shown in Figure 19.6.

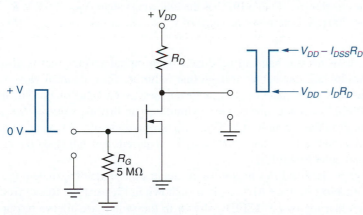 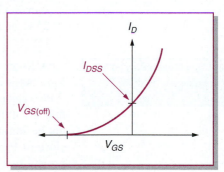

**FIGURE 19.6    The MOSFET switch.**

Since the D-MOSFET can be operated in the enhancement mode, it can produce a positive output signal with a positive input signal. While there is still a 180° voltage phase shift from input to output, both signals vary between approximately 0 V and some positive peak value. (In contrast, the JFET switch uses a *negative* input signal to produce a *positive* output signal, as shown in Figure 19.4.)

When the input to the circuit shown in Figure 19.6 is at 0 V, $I_D = I_{DSS}$ and

$$V_{out} = V_{DD} - I_{DSS} R_D$$

How do MOSFET and JFET switches differ?

For the MOSFET switch shown, a lower value of $R_D$ produces an output that is closer to the value of $V_{DD}$. This point is illustrated in Example 19.3.

### EXAMPLE 19.3

Determine the output voltage for the circuit shown in Figure 19.7 when the input is at 0 V. Repeat the procedure for the circuit using $R_D = 100\ \Omega$.

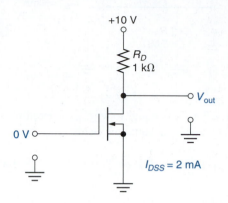

**FIGURE 19.7**

**Solution:** When the input is at 0 V, the current through the MOSFET equals $I_{DSS}$. Therefore, the output voltage is found as

$$V_{out} = V_{DD} - I_{DSS}R_D = 10\ V - 2\ V = 8\ V$$

If the resistor in the circuit is replaced with a 100 $\Omega$ resistor, the output changes to

$$V_{out} = V_{DD} - I_{DSS}R_D = 10\ V - 200\ mV = 9.8\ V$$

Thus, using a smaller value of $R_D$ results in a high output voltage that is much closer to the ideal value, $V_{DD}$.

### PRACTICE PROBLEM 19.3

A circuit like the one shown in Figure 19.7 has the following values: $V_{DD} = +8\ V$, $R_D = 2\ k\Omega$, and $I_{DSS} = 500\ \mu A$. Determine the value of $V_{out}$ for the circuit when $V_{in} = 0\ V$.

> What is the effect of $R_D$ on the output voltage of a MOSFET switch?

A lower-value drain resistor helps to produce a high output voltage that is closer to $V_{DD}$, but it also results in a *low* output voltage that is further from its ideal value (0 V). With a lower value of $R_D$, the value of $I_D R_D$ is lower for a given value of $I_D$. This means that when the MOSFET is conducting to its maximum capability, the value of $I_D R_D$ may not be sufficient to drop the output to 0 V. Thus, we have a trade-off. If having an output close to $V_{DD}$ is important, a low value of $R_D$ *may* be required, depending on the current capability of the particular MOSFET.

The CMOS switch discussed in Chapter 13 eliminates the trade-off problem that can be caused by $R_D$. As a review, the CMOS switch is shown in Figure 19.8. Recall from our previous discussion that the two MOSFETs shown in the switch are always in opposite operating states. Thus, when $Q_1$ is *on*, $Q_2$ is *off*, and vice versa. The following table is based on the relationship between the operating states of $Q_1$ and $Q_2$.

| $Q_1$ State | $Q_2$ State | $Q_1$ Resistance | $Q_2$ Resistance |
|---|---|---|---|
| On | Off | Low | High |
| Off | On | High | Low |

When $Q_1$ is *on*, $Q_2$ is *off*. In this case, the high resistance of $Q_2$ and the low resistance of $Q_1$ combine to produce an output that is very close to 0 V. When $Q_2$ is *on*, $Q_1$ is *off*. In this case, the low resistance of $Q_2$ and the high resistance of $Q_1$ combine to produce an output

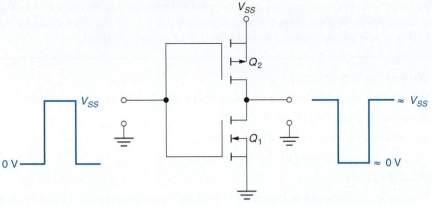

**FIGURE 19.8** The CMOS switch.

that is very close to $V_{SS}$. As a result, the trade-off caused by $R_D$ in the basic MOSFET inverter is eliminated. This is one of the reasons that CMOS switches are almost always used in place of other MOS switches.

### 19.1.4  Basic Switching Applications

A typical application for a BJT switch is illustrated in Figure 19.9. In this figure, the BJT is being used as an LED *driver*. A **driver** is a circuit used to couple a *low-current signal source* to a relatively high-current device. Assuming that the output current from the signal source is insufficient to light the LED, a driver is required to provide the current needed to light the LED.

> **Driver**
> A circuit used to couple a low-current output to a relatively high-current device.

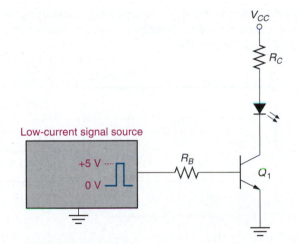

**FIGURE 19.9** A typical application for the simple BJT switch.

How does the circuit work? When the output from the signal source is at 0 V, $Q_1$ is in cutoff (acting as an open switch). As a result, $I_C \cong 0$ mA, and the LED doesn't light. When the output from the signal source is at +5 V, $Q_1$ is in saturation (acting as a closed switch). As a result, $I_C$ increases to a level sufficient to light the LED. Note that the value of $I_C$ depends on the value of $V_{CC}$, the forward voltage across the LED, the value of $V_{CE(sat)}$ for the transistor, and the value of $R_C$ (which serves as a current-limiting resistor).

Even though the driver shown in Figure 19.9 uses a BJT, drivers are commonly made with each of the active devices covered in this section. The device used in any driver application is determined by the system engineer, and no single active device seems to dominate this application.

### 19.1.5  Summary

BJT, JFET, and MOSFET switches are commonly used as *drivers* in switching applications. A driver is a circuit used to couple a low-current circuit output to a relatively high-current device, such as an LED.

The basic switching circuits are used to provide high and low dc outputs when driven by an input signal. A high output is one that is within 1 V of the circuit supply voltage (typically). A low voltage is one that is within 1 V of ground (typically).

---

**Section Review ▶**

1. What is a switching circuit?
2. What is the primary difference between linear circuits and switching circuits?
3. Describe the operation of the BJT switch.
4. Describe the operation of the JFET switch.
5. Describe the operation of the MOSFET switch.
6. What is a *driver*?

**Critical Thinking ▶**

7. When does the transistor in Figure 19.1 dissipate the greatest amount of power? Explain.
8. Using only the components shown, how could the circuit shown in Figure 19.9 be altered to light the LED when the output from the signal source is 0 V?

## 19.2 Basic Switching Circuits: Practical Considerations

**OBJECTIVE 2 ▶** In this section, we will take a look at some of the factors that affect the operation and analysis of basic switching circuits. Most of these factors explain the differences between *ideal* and *practical* time and frequency measurements. We will also look at two switching circuit classifications and several devices that are specially designed for switching applications.

### 19.2.1 Practical Measurements

Before we can discuss any practical time or frequency measurements, we must define some of the terms commonly used to describe rectangular waves. We will define these terms with the help of Figure 19.10.

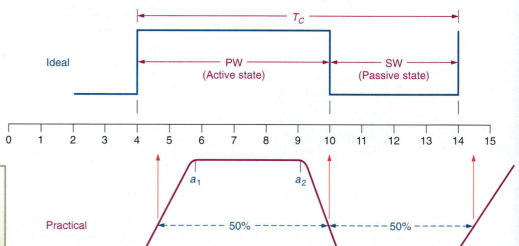

**FIGURE 19.10** Measuring pulse width, space width, and cycle time.

---

**Rectangular waveform**
A waveform made up of alternating (high and low) dc voltages.

**Pulse width (PW)**
The time spent in the active (high) dc voltage state.

**Space width (SW)**
The time spent in the passive (low) dc voltage state.

**Cycle time ($T_C$)**
The sum of pulse width and space width.

**Square wave**
A rectangular waveform that has equal PW and SW values.

---

A waveform made up of alternating (high and low) dc voltages is generally classified as a **rectangular waveform**. The time a rectangular waveform spends in the high dc voltage state is generally called the **pulse width (PW)** of the waveform. The time spent in the low dc voltage state is generally called the **space width (SW)** of the waveform. The sum of pulse width and space width gives you the **cycle time ($T_C$)** of the waveform.

The **square wave** is a rectangular waveform that has equal pulse width and space width values. When PW ≠ SW, the waveform is simply referred to as a rectangular waveform.

When discussing the *ideal* waveform, it is easy to determine the value of pulse width, space width, or cycle time. For example, take a look at the ideal waveform in Figure 19.10. The scale shown between the ideal and practical waveforms is used as a time reference. Assume (for the sake of discussion) that the scale is measuring time in microseconds (μs). For the ideal waveform, it is easy to see that the value of PW is

$$PW = 10\ \mu s - 4\ \mu s = \textbf{6}\ \boldsymbol{\mu s}$$

Similarly, the value of SW can be found as

$$SW = 14\ \mu s - 10\ \mu s = \textbf{4}\ \boldsymbol{\mu s}$$

and the cycle time can be found as

$$T_C = 14\ \mu s - 4\ \mu s = \textbf{10}\ \boldsymbol{\mu s}$$

> *A Practical Consideration:*
> The value of $T_C$ can also be found as $T_C = PW + SW$.

When we wish to make the time measurements of a practical rectangular waveform, we run into a problem. You see, the transitions from low to high and from high to low are not perfectly vertical. (The reasons for this are discussed later in this section.) For example, take a look at the *practical* waveform in Figure 19.10. If we measure the PW of this waveform from $a_1$ to $a_2$, its value is found to be approximately 3.2 μs. However, if we measure the PW from $b_1$ to $b_2$, its value is found to be approximately 6.5 μs. The question here is: Where do we measure the values of PW, SW, and $T_C$ so that everyone gets the same results?

To eliminate the problem of measurement variations, the values of PW, SW, and $T_C$ are *always* measured at the 50% points (where the voltage is halfway between the high and low values). For example, if the practical waveform in Figure 19.10 has peak values of 0 V (low) and +5 V (high), we measure the values of PW, SW, and $T_C$ at the +2.5 V points on the waveform. This point is illustrated further in Example 19.4.

> *A Practical Consideration:*
> Time measurements are always made at the 50% points on the waveform.

## EXAMPLE 19.4

Figure 19.11a represents the output from an inverter as viewed with an oscilloscope. With the oscilloscope set to a horizontal calibration of 50 μs/div, what are the values of PW and $T_C$ for the waveform?

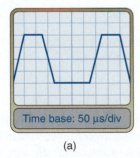

Time base: 50 μs/div

(a)

Time base: 2 μs/div

(b)

**FIGURE 19.11**

*Solution:*   The waveform is positioned so that it varies equally above and below the 0 V reference on the grid. Therefore, we measure PW and $T_C$ at the 0 V points (since these are halfway between the peak values).

The PW is measured during the positive alternation. There are approximately 2.5 major divisions between the two 0 V points. Since each major division represents 50 μs, the PW is found as

$$PW = 2.5 \times 50\ \mu s = \textbf{125}\ \boldsymbol{\mu s}$$

Now, looking at the illustration, we can see that there are approximately 6.5 divisions from the 0 V point on one rising edge to the 0 V point on the next rising edge. Since these points correspond to one complete cycle, the cycle time is found as

$$T_C = 6.5 \times 50 \ \mu s = \mathbf{325 \ \mu s}$$

**PRACTICE PROBLEM 19.4**
Determine the values of PW and $T_C$ for the waveform in Figure 19.11b.

Another time-related measurement commonly made with switching circuits is *duty cycle*. **Duty cycle** is the ratio of pulse width to total cycle time, *measured as a percentage*. By formula,

**Duty cycle**
The ratio of pulse width (PW) to cycle time ($T_C$), measured as a percentage.

$$\text{duty cycle} = \frac{\text{pulse width}}{\text{cycle time}} \times 100 \qquad \textbf{(19.1)}$$

Example 19.5 demonstrates the use of this equation in determining the duty cycle of a given waveform.

**EXAMPLE 19.5**

Determine the duty cycle for the waveform in Figure 19.11a.

*Solution:* In Example 19.4, we determined the waveform to have values of PW = 125 $\mu s$ and $T_C$ = 325 $\mu s$. Using these values and equation (19.1), the duty cycle of the waveform is found as

$$\textbf{duty cycle} = \frac{\text{PW}}{T_C} \times 100 = \frac{125 \ \mu s}{325 \ \mu s} \times 100 = \mathbf{38.5\%}$$

**PRACTICE PROBLEM 19.5**
Determine the duty cycle of the waveform in Figure 19.11b.

The duty cycle of a waveform indicates *the percentage of each cycle that is taken up by the pulse width*. If a given waveform has a duty cycle of 35%, the pulse width takes up 35% of each cycle. A square wave always has a duty cycle of 50% since, by definition, its pulse width is one-half of its cycle time.

Now that you are familiar with basic switching circuit measurements, we will examine the switching characteristics of the BJT and FET individually.

### 19.2.2 BJT Switching Time

**OBJECTIVE 3** ▶ The output from an *ideal* BJT switch would change at the exact instant that the circuit input changes, with the output resembling the ideal waveform you saw in Figure 19.10. In practice, the output from a BJT switch more closely resembles the waveforms shown in Figure 19.12. The $V_B$ waveform represents an ideal input that changes states *instantly* at $T_0$ and $T_3$. Note that this waveform can represent either $V_B$ or $I_B$ since they are always in phase. The $I_C$ and $V_C$ waveforms differ from their ideal characteristics in several respects:

1. There is a delay between each input transition and the start of the output transitions.
2. The output transitions are not vertical, implying that they require some measurable amount of time to occur.

**Propagation delay**
The time delay between input and output transitions, as measured at the 50% point on the two waveforms.

The overall delay between the input and output transitions (as measured at the 50% points on the two waveforms) is referred to as **propagation delay**.

There are actually *four* sources of BJT propagation delay, each measured during one of the time periods shown in Figure 19.12. The first contributor to the delay is called

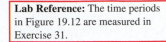

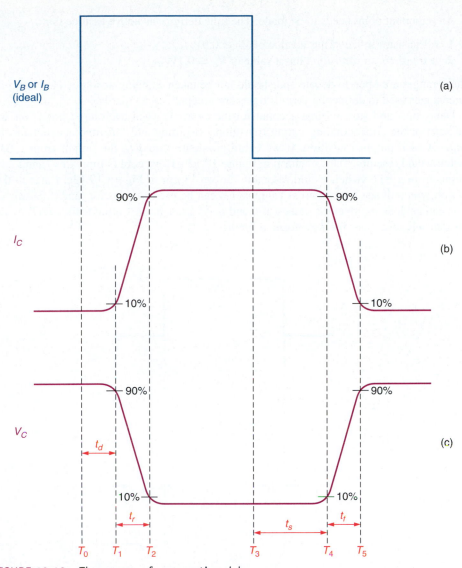

FIGURE 19.12    The causes of propagation delay.

**delay time** ($t_d$). This is the time required for the transistor to come out of cutoff. As shown in Figure 19.12, $t_d$ is measured between $T_0$ and $T_1$ on the output waveforms. In terms of $I_C$, the delay time is *the time required for $I_C$ to reach 10% of its maximum value*.

Oscilloscopes are designed to display *voltage* (as it relates to time). As such, we need to define delay time in terms of $V_C$ so that we can apply the concept to oscilloscope measurements. In terms of $V_C$, delay time is *the time required for $V_C$ to drop to 90% of its maximum value* (in most cases, to 90% of $V_{CC}$).

**Rise time** ($t_r$) is the time required for the BJT to make the transition from cutoff to saturation (in effect, the time required for the transistor to pass through its active region of operation). In Figure 19.12, rise time is measured between $T_1$ and $T_2$. In terms of $I_C$, rise time is *the time required for $I_C$ to rise from 10% to 90% of its maximum value*. In terms of $V_C$, it is *the time required for $V_C$ to drop from 90% to 10% of its maximum value*.

**Storage time** ($t_s$) is the time required for a BJT to come out of saturation. In terms of $I_C$, it is *the time required for $I_C$ to drop from 100% to 90% of its maximum value*, measured between $T_3$ and $T_4$. In terms of $V_C$, *it is the time required for $V_C$ to rise to 10% of its maximum value*. Note that storage time (like delay time) is shown to begin at the point when the input signal changes.

Finally, **fall time** ($t_f$) is the time required for the BJT to make the transition from saturation to cutoff, measured between $T_4$ and $T_5$. In terms of $I_C$, it is the time required for $I_C$ to drop from 90% to 10% of its maximum value. In terms of $V_C$, it is the time required for $V_C$ to increase from 10% to 90% of its maximum value.

**Delay time ($t_d$)**
The time required for a BJT to come out of cutoff.

**Rise time ($t_r$)**
The time required for the BJT to go from cutoff to saturation.

**Storage time ($t_s$)**
The time required for a BJT to come out of saturation.

**Fall time ($t_f$)**
The time required for the BJT to make the transition from saturation to cutoff.

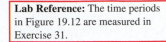

Lab Reference: The time periods in Figure 19.12 are measured in Exercise 31.

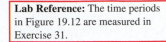

SECTION 19.2    Basic Switching Circuits: Practical Considerations    795

An important point needs to be made: The time definitions above assume that:

**1.** A transistor in *cutoff* has a value of $V_C \geq 0.9V_{CC}$.
**2.** A transistor in *saturation* has a value of $V_C \leq 0.1V_{CC}$.

These ranges are open to debate and should not be taken as being absolute. However, the percentages used to define the various times are accepted as standards.

Delay time and storage time account for the delay between each input transition and the *start* of the corresponding output transition. Rise time and fall time account for the *slope* of each output transition. If we could eliminate $t_r$ and $t_f$, the output from a BJT switch would resemble the waveform in Figure 19.13b. If we could eliminate $t_d$ and $t_s$, the output from a BJT switch would resemble the waveform in Figure 19.13c. Later in this section, you will see how external components can be used to reduce $t_d$ and $t_s$. However, little can be done to affect the values of $t_r$ and $t_f$. As a result, the output from a BJT switch most closely resembles the waveform in Figure 19.13c.

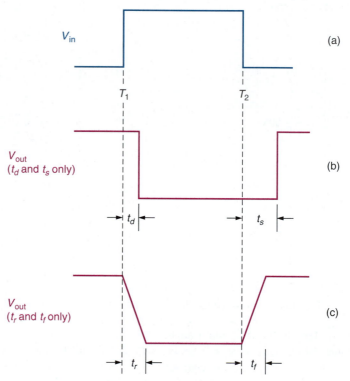

**FIGURE 19.13** The effects of switching times on the output from a BJT switch.

The values of $t_d$, $t_r$, $t_s$, and $t_f$ are all provided on the specification sheet of a given BJT. For example, the 2N3904 spec sheet (Figure 7.17) lists the following maximum values under *switching characteristics*:

$$t_d = 35 \text{ ns} \quad t_r = 35 \text{ ns} \quad t_s = 200 \text{ ns} \quad t_f = 50 \text{ ns}$$

If we add the values of delay time and rise time, we get the maximum time required for the 2N3904 to make the transition from a high output state to a low output state, 70 ns. If we add the values of storage time and fall time, we get the maximum time required for the 2N3904 to make the transition from a low output state to a high output state, 250 ns. As you can see, it takes more time for the device to go from high to low than it takes to go from low to high. The reason for this is explained later in this section.

If we add the four switching times listed, we get a theoretical minimum switching time of 320 ns. It would seem that this minimum switching time could be used to calculate a maximum switching frequency as follows:

$$f_{max} = \frac{1}{t_d + t_r + t_s + t_f} = \frac{1}{320 \text{ ns}} = \mathbf{3.125 \text{ MHz}}$$

In practice, 3.125 MHz is far beyond the frequency capability of a 2N3904 switching circuit. The reason lies in the relationship between the upper cutoff frequency ($f_{C2}$) of the BJT and the rise time of its input waveform. In Appendix D, the following equation is derived for calculating the value of $f_{C2}$ for a discrete switching circuit:

$$f_{C2} = \frac{0.35}{t_r} \qquad\qquad (19.2)$$

A Practical Consideration:
In switching circuit applications, frequency is often referred to as pulse repetition rate (PRR) or pulse repetition frequency (PRF). You may find either of these terms used in some references and system technical manuals.

where $t_r$ = the *rise time* of the active device. Using this equation, the value of $f_{C2}$ for a 2N3904 switching circuit would be found as

$$f_{C2} = \frac{0.35}{35 \text{ ns}} = 10 \text{ MHz} \qquad (t_r = 35 \text{ ns for the 2N3904})$$

Thus, the 2N3904 has an upper cutoff frequency of 10 MHz.

To pass a square wave with minimum distortion, *the upper cutoff frequency of a switching circuit should be at least 100 times the frequency of the input square wave.* Thus, the practical limit on the frequency of a switching circuit input signal is one-hundredth the value of $f_{C2}$ for the circuit. By formula,

$$f_{max} = \frac{0.35}{100 t_r} \quad \text{(practical limit)} \qquad\qquad (19.3)$$

In the case of the 2N3904 switch, this means that the practical limit on $f_{in}$ for a circuit is 100 kHz.

Figure 19.14 shows what happens if the practical limit on $f_{in}$ for a circuit is exceeded. When $f_{in} = f_{C2}/100$, there is virtually no distortion in the output waveform. As the input frequency increases beyond this limit:

1. The leading and trailing edges of the circuit output become more and more rounded.
2. The delays between the input and output transitions increase.
3. The pulse width and space width become *asymmetrical*, meaning that PW ≠ SW.

Example 19.6 demonstrates the complete frequency analysis of a BJT switching circuit.

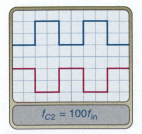
$f_{C2} = 100 f_{in}$

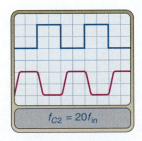
$f_{C2} = 20 f_{in}$

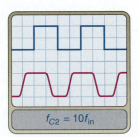
$f_{C2} = 10 f_{in}$

**FIGURE 19.14** Wave shape versus operating frequency.

**EXAMPLE 19.6**

A transistor has the following maximum values listed on its spec sheet: $t_d = 10$ ns, $t_r = 40$ ns, $t_s = 80$ ns, and $t_f = 30$ ns. Determine the value of $f_{C2}$ for the transistor and the practical limit on its input frequency.

**Solution:**  With a rise time of 40 ns, the value of $f_{C2}$ for the component is found as

$$f_{C2} = \frac{0.35}{t_r} = \frac{0.35}{40 \text{ ns}} = \textbf{8.75 MHz}$$

To avoid distortion in the output waveform, the input frequency to the component should be limited to approximately

$$f_{\max} = \frac{0.35}{100 t_r} = \frac{0.35}{(100)(40 \text{ ns})} = \textbf{87.5 kHz}$$

**PRACTICE PROBLEM 19.6**

A transistor has the following maximum values listed on its spec sheet: $t_d = 20$ ns, $t_r = 25$ ns, $t_s = 120$ ns, and $t_f = 20$ ns. Determine the value of $f_{C2}$ for the transistor and the practical limit on its input frequency.

### 19.2.3    Improving BJT Switching Time

**OBJECTIVE 4 ▶**     Before you can understand the methods used to improve BJT switching time, we need to take a moment to discuss the causes of switching time. The various times are explained easily with the help of an illustration that appeared earlier in the text. Figure 6.6 is repeated as Figure 19.15.

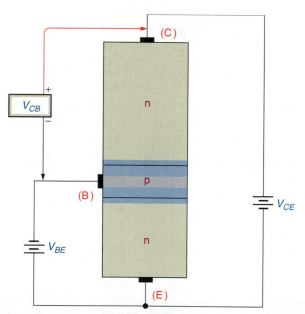

**FIGURE 19.15**    A BJT in cutoff.

| What is the cause of delay time? |
| --- |

When the BJT is in cutoff, the base-emitter depletion layer is at its maximum width, and $I_C$ is essentially at zero. When the input to the BJT goes positive, the depletion layer starts to "dissolve," allowing $I_C$ to begin to increase. Delay time is the time required for the depletion layer to dissolve to the point where 10% of the maximum value of $I_C$ is allowed to pass through the component. How long is this period of time? That depends on three factors:

**1.** The physical characteristics of the particular BJT.
**2.** The amount of reverse bias initially applied to the component.
**3.** The amount of $I_B$ that the input signal generates when it goes positive.

We cannot do anything about the physical characteristics of the BJT, but we can do several things to improve $t_d$. *By keeping the initial reverse bias at a minimum*, we can keep the depletion layer at a minimum width. The narrower the depletion layer, the less time it will take for it to dissolve. Also, you should recall that the BJT is a *current-controlled* device; that is, the width of the depletion layer is determined by the amount of $I_B$. Therefore, *by providing a very high initial value of $I_B$*, delay time is further reduced. Later in this section, we will discuss one method by which a large initial value of $I_B$ is generated.

How is delay time reduced?

After delay time has passed, the depletion layer continues to dissolve, allowing $I_C$ to continue to increase. Rise time is the time required for the depletion layer to dissolve to the point where $I_C$ reaches 90% of its maximum value. Rise time is strictly a function of the physical characteristics of the BJT, and nothing can be done to reduce it.

The biggest overall delay is *storage time*. Referring to the specifications of the 2N3904, you can see that storage time is at least four times as long as any of the other switching times. Therefore, reducing $t_s$ has a major impact on the overall switching time of a given BJT.

When a transistor is in saturation, the base region is flooded with charge carriers. When the input voltage goes low, it takes a great deal of time for these charge carriers to leave the base region, allowing a new depletion layer to begin to form. The time required for this to happen is storage time.

What is the cause of storage time?

The duration of storage time depends on three factors:

1. The physical characteristics of the BJT.
2. The initial value of $I_C$.
3. The initial value of reverse bias applied to the base.

Again, we cannot do anything about the physical characteristics of the BJT, but we *can* do something about the other two variables. Storage time can be greatly reduced *by keeping $I_C$ low enough so that the BJT never completely saturates*. In other words, if we keep the BJT operating point *just below* saturation, the amount of current in the base region of the BJT is greatly reduced, and storage time is also reduced. Also, if we *apply a very large initial value of reverse bias*, the charge carriers are forced rapidly out of the base region and storage time is further reduced.

How is storage time reduced?

Like rise time, fall time is a function of the physical characteristics of the BJT and cannot be reduced by any practical means. Fall time is the time required for the depletion layer to expand to the point where $I_C$ drops to 10% of its maximum value.

Now, let's summarize the steps we can take to reduce the overall switching time of a given BJT:

1. By applying a high *initial* value of $I_B$, delay time is reduced.
2. By using the *minimum* value of reverse bias required to hold the BJT in cutoff, delay time is further reduced.
3. By limiting $I_B$ to a value lower than that required to completely saturate the BJT, storage time is reduced.
4. By applying an *initial* reverse bias that is very large, storage time is further reduced.

*A Practical Consideration:* We can reduce $t_d$ and $t_s$ as described in this section. However, since the practical frequency limit is determined by rise time, improving these times does not increase the operating frequency limit of the switch.

Take another look at statements 1 and 3. These statements indicate that we want $I_B$ to be very high *initially* (to reduce delay time) and then settle down to some level below that required for saturation (to reduce storage time). Statements 2 and 4 can be combined in a similar fashion. We want a very high *initial* value of reverse bias (to reduce storage time) and then a minimum reverse bias (to reduce delay time).

The desired $I_B$ and reverse-bias characteristics described above are both achieved by using a **speed-up capacitor**. A basic BJT switch with an added speed-up capacitor is shown in Figure 19.16. The speed-up capacitor will perform all the required functions when its value is properly selected. The means by which it accomplishes these functions are illustrated in Figure 19.17.

**Speed-up capacitor**
A capacitor used to reduce propagation delay by reducing $t_d$ and $t_s$.

The speed-up capacitor in Figure 19.16 effectively short circuits the resistor ($R_B$) during the time that the input is making its transitions. As indicated in Figure 19.17, a positive spike is coupled to the base of the transistor when $V_{in}$ makes its transition from 0 V to +5 V. This spike generates a high initial value of $I_B$. Then, as the capacitor charges, the

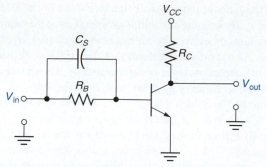

FIGURE 19.16   A speed-up capacitor ($C_S$) improves switching time.

Lab Reference: The effects of a speed-up capacitor on BJT switching characteristics are demonstrated in Exercise 31.

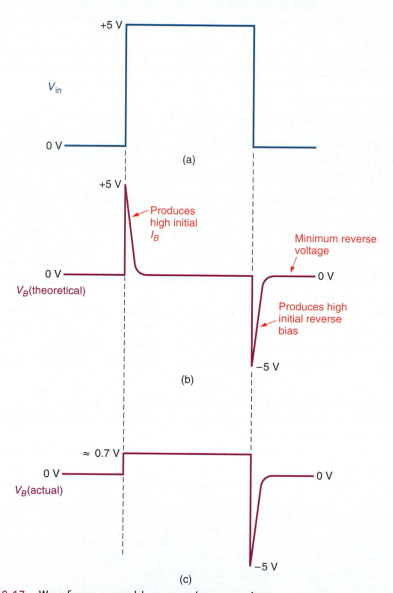

FIGURE 19.17   Waveforms caused by a speed-up capacitor.

spike returns to the 0 V level, and $I_B$ decreases to some value slightly less than that required to fully saturate the transistor. As a result, delay time is reduced while ensuring that $I_B$ returns to a value that is low enough to prevent saturation.

When the input signal returns to 0 V, the charge on $C_S$ drives the output of the $RC$ circuit to $-5$ V. This is the high initial reverse bias needed to reduce storage time. Since the transistor was prevented from saturating by $I_B$, the capacitor has done all that can be done to reduce storage time. Also, since $V_B$ returns to 0 V at the end of the negative spike, the final value of reverse bias is at a minimum. This, coupled with the high initial $I_B$ spike, ensures that delay time is held to a minimum.

The value of $C_S$ is normally chosen so that the time constant of the $RC$ circuit is very short (as compared to the pulse width) at the maximum operating frequency. This requirement is fulfilled by using the following equation to select $C_S$:

$$C_S < \frac{1}{20 R_B f_{max}} \qquad \text{(19.4)}$$

As long as the speed-up capacitor fulfills the requirement of equation (19.4), the values of $t_d$ and $t_s$ for the BJT will be greatly reduced.

In Figure 19.17b, we simplified things by ignoring the effects of the emitter-base junction of the transistor. The actual waveform produced at the base of the transistor is shown in Figure 19.17c. On the positive transition, the emitter-base junction turns on and clips off the positive spike produced by the speed-up capacitor. Since the transistor remains on during the entire pulse width of the input, $V_B$ does not actually return to 0 V. Rather, it remains at the value of $V_{BE}$ (approximately 0.7 V).

The value of $f_{max}$ used in equation (19.4) is the practical $f_{max}$ of the circuit, found as

$$f_{max} = \frac{0.35}{100 t_r}$$

### 19.2.4 JFET Switching Time

Many JFET spec sheets list values of $t_d$, $t_r$, $t_s$, and $t_f$. The causes of these times in the JFET are similar to those of the BJT. When the value of $t_r$ is listed on the JFET spec sheet, the practical limit on $f_{in}$ can be determined in the same fashion as it is for the BJT.

Some JFET spec sheets list only **turn-on time ($t_{on}$)** and **turn-off time ($t_{off}$)**. Turn-on time is the sum of delay time and rise time. Turn-off time is the sum of storage time and fall time. When these are the only values listed in the JFET spec sheet, it is difficult to determine the exact practical limit on $f_{in}$ for the device. This is because you cannot determine the portion of turn-on time that is actually taken up by the rise time of the device. In other words, there is no conversion formula for calculating rise time when turn-on time is known.

When only $t_{on}$ is given, use this value in place of $t_r$ in equation (19.3) to determine the practical operating frequency limit. While the actual limit on $f_{in}$ will be significantly higher than the calculated value, you will be able to approximate its minimum value.

Since JFETs are voltage-controlled devices, speed-up capacitors reduce only the turn-off time of the component. By initially supplying a gate-source voltage that is greater than $V_{GS(off)}$, the speed-up capacitor decreases the time required to turn off the JFET. However, the capacitor does little to improve the turn-on time of the component.

**Turn-on time ($t_{on}$)**
The sum of $t_d$ and $t_r$.

**Turn-off time ($t_{off}$)**
The sum of $t_s$ and $t_f$.

*A Practical Consideration:*
In some cases, the spec sheet of a BJT will list $t_{on}$ and $t_{off}$ in place of the standard parameters. When this is the case, you must follow the guideline discussed here for approximating the practical limit on $f_{in}$.

### 19.2.5 Switching Devices

Groups of BJTs and JFETs have been developed specifically for switching applications. These devices, called **switching transistors**, are designed to have extremely short values of $t_d$, $t_r$, $t_s$, and $t_f$. For example, the 2N2369 has the following maximum values listed on its data sheet:

**Switching transistors**
Devices with extremely low switching times.

$$t_d = 5 \text{ ns} \quad t_r = 18 \text{ ns} \quad t_s = 13 \text{ ns} \quad t_f = 15 \text{ ns}$$

With the extremely short delay and storage time values, a 2N2369 switching circuit probably wouldn't need a speed-up capacitor. Also, with the given rise time of 18 ns, the practical limit on $f_{in}$ for the device would be found as

$$f_{max} = \frac{0.35}{100 t_r} = \frac{0.35}{(100)(18 \text{ ns})} = \textbf{194 kHz}$$

This is nearly twice the acceptable $f_{max}$ limit of the 2N3904. Thus, the 2N2369 can be used at much higher switching frequencies than the 2N3904.

## 19.2.6  Switching Circuit Classifications

*OBJECTIVE 5* ▶

**Inverter**

A basic switching circuit that produces a 180° voltage phase shift.

**Buffer**

A switching circuit that does not produce a voltage phase shift.

Buffer time measurements.

Our discussion on basic switching circuits has been limited to a specific type of switching circuit called an *inverter*. An **inverter** has a 180° voltage phase shift from input to output. Looking back at the various illustrations of BJT and FET circuits, you will see that every circuit had a high output voltage when the input voltage was low, and vice versa.

Another basic switching circuit is the *buffer*. A **buffer** is a switching circuit that does not introduce a 180° voltage phase shift. The BJT and FET buffers are shown in Figure 19.18. As you can see, these circuits are simply the emitter follower and source follower. When a square wave is applied to any of these circuits, the output is a square wave that is in phase with the input signal. Because of this, we must redefine the $V_{out}$ descriptions of $t_d$, $t_r$, $t_s$, and $t_f$. For the buffer:

1. $t_d$ is the time required for $V_{out}$ to reach 10% of its maximum value.
2. $t_r$ is the time required for $V_{out}$ to rise from 10% to 90% of its maximum value.
3. $t_s$ is the time required for $V_{out}$ to drop to 90% of its maximum value.
4. $t_f$ is the time required for $V_{out}$ to drop from 90% to 10% of its maximum value.

The changes in the $V_{out}$ definitions for the buffer are necessary because it does not produce a 180° voltage phase shift. Except for this difference, the operation of the buffer and inverter circuits are nearly identical.

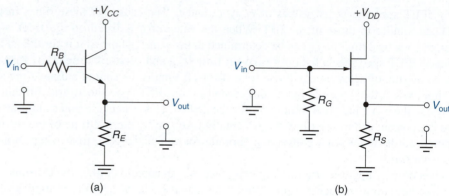

(a)                                                      (b)

**FIGURE 19.18**  Some basic buffer circuits.

## 19.2.7  Summary

A waveform made up of alternating (high and low) dc voltages is classified as a *rectangular* waveform. The square wave is a special-case rectangular waveform that has equal *pulse-width* and *space-width* values.

The ideal rectangular waveform would have instantaneous transitions from high to low and from low to high. However, these transitions are not perfectly vertical in practice. To ensure that pulse-width, space-width, and cycle-time measurements are consistent, these values are always measured at the 50% points on rectangular waveforms, that is, at the points where the amplitude of the waveform is halfway between the high and low output voltage levels.

The duty cycle of a rectangular waveform indicates the percentage of the waveform taken up by the pulse width. Since pulse width and space width are always equal for a square wave, the duty cycle of a square wave is always 50%.

The time between the input and output changes of a switching circuit is referred to as *propagation delay*. For discrete devices, four switching times contribute to propagation delay:

1. Delay time ($t_d$): The time required for the device to come out of cutoff.
2. Rise time ($t_r$): The time required for the device to make the transition from cutoff to saturation.
3. Storage time ($t_s$): The time required for the device to come out of saturation.
4. Fall time ($t_f$): The time required for the device to make the transition from saturation to cutoff.

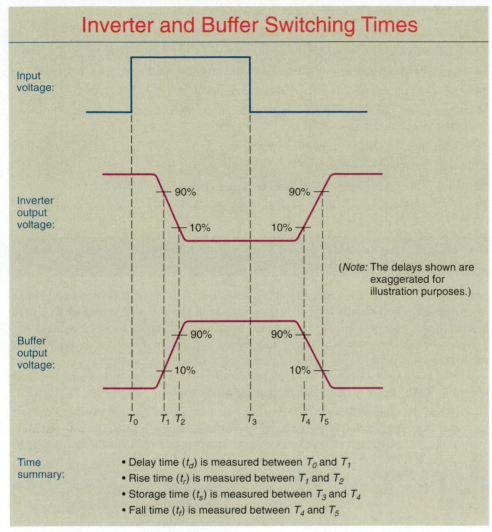

## Inverter and Buffer Switching Times

Input voltage:

Inverter output voltage:

90%        90%

10%        10%

(*Note:* The delays shown are exaggerated for illustration purposes.)

Buffer output voltage:

90%        90%

10%        10%

$T_0$    $T_1$ $T_2$        $T_3$        $T_4$ $T_5$

Time summary:
- Delay time ($t_d$) is measured between $T_0$ and $T_1$
- Rise time ($t_r$) is measured between $T_1$ and $T_2$
- Storage time ($t_s$) is measured between $T_3$ and $T_4$
- Fall time ($t_f$) is measured between $T_4$ and $T_5$

**FIGURE 19.19**

The times listed are measured using an oscilloscope. Figure 19.19 lists these times and shows where they are measured on the output voltage waveforms for the inverter and the buffer. As you can see, the inverter has a 180° voltage phase shift from input to output, while the buffer does not.

The propagation delay of a discrete switch can be reduced by the use of a *speed-up capacitor*. The speed-up capacitor reduces delay and storage time in the BJT and storage time (only) in an FET switch. The rise time of a device determines the maximum practical operating frequency of a switching circuit. Since a speed-up capacitor does not affect rise time, the practical limit on $f_{in}$ is not increased when a speed-up capacitor is added.

---

1. What is a *rectangular waveform*?                     ◀ **Section Review**
2. Define each of the following terms: *pulse width*, *space width*, and *cycle time*.
3. What is a *square wave*?
4. Where is each of the values in Question 2 measured on a practical rectangular waveform?
5. Why do we need a standard for rectangular waveform time measurements?
6. What is *duty cycle*?
7. Why do all square waves have the same duty cycle value? What is this value?
8. What is *propagation delay*?
9. What is delay time? Describe delay time in terms of $I_C$ and $V_C$.

10. What is *rise time*? Describe rise time in terms of $I_C$ and $V_C$.

11. What is *storage time*? Describe storage time in terms of $I_C$ and $V_C$.

12. What is *fall time*? Describe fall time in terms of $I_C$ and $V_C$.

13. Which of the switching times determines the practical limit on $f_{in}$ for the device?

14. How does the use of a *speed-up capacitor* reduce $t_d$ and $t_s$?

15. What effect does a speed-up capacitor have on the practical frequency limit of a BJT? Explain your answer.

16. Define *turn-on time* ($t_{on}$) and *turn-off time* ($t_{off}$).

17. What is a *switching transistor*?

18. What is the primary difference between the *inverter* and the *buffer*?

19. List the points where $t_d$, $t_r$, $t_s$, and $t_f$ are measured on the output voltage of a buffer.

## 19.3 Schmitt Triggers

OBJECTIVE 6 ▶

**Schmitt trigger**
A voltage-level detector.

In Chapter 17, you were introduced to the *comparator*: a circuit that is used as a voltage-level detector. The **Schmitt trigger** is a voltage-level detector that is similar to a comparator in many aspects. However, there are some key differences between Schmitt trigger and comparator operation. These differences (which are discussed throughout this section) result in the two circuits being used under different circumstances.

Figure 19.20 shows a typical input/output combination for a Schmitt trigger. Here's an overview of the circuit's response to the input waveform:

1. When the input makes a *positive-going* transition past a specified voltage, the output from the Schmitt trigger changes from one voltage level ($-V_{out}$) to the other ($+V_{out}$). The voltage at which this change occurs is referred to as the **upper trigger point (UTP)**.

**Upper trigger point (UTP)**
A reference voltage. When a positive-going input passes the UTP, the Schmitt trigger output changes state.

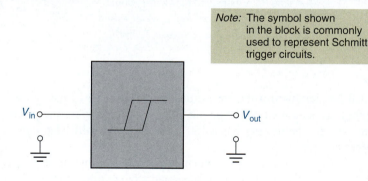

Note: The symbol shown in the block is commonly used to represent Schmitt trigger circuits.

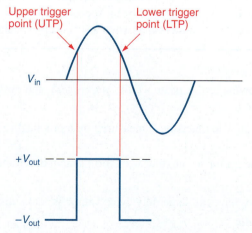

FIGURE 19.20  Schmitt trigger input and output signals.

**2.** When the input makes a *negative-going* transition past a specified voltage, the output from the Schmitt trigger changes from $+V_{out}$ to $-V_{out}$. The voltage at which this change occurs is referred to as the **lower trigger point (LTP)**.

**Lower trigger point (LTP)**
A reference voltage. When a negative-going input passes the LTP, the Schmitt trigger output changes state.

The UTP and LTP values are determined by component values in the circuit. Depending on the circuit design, the UTP and LTP values may or may not be equal. However, the LTP (by definition) can never be greater than the UTP.

The concept of unequal UTP and LTP values is illustrated in Figure 19.21. When the positive-going transition of the input passes the UTP (at $v_1$), the output changes from $-V_{out}$ to $+V_{out}$. After the sine wave peaks and begins its negative-going transition, it passes the UTP again (at $v_2$). However, the output remains $+V_{out}$ until the LTP is reached. When the negative-going transition of the input passes the LTP, the Schmitt trigger output returns to $-V_{out}$.

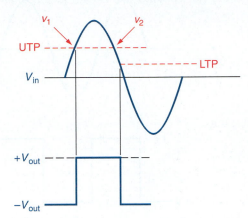

**FIGURE 19.21** Unequal UTP and LTP values.

The waveforms in Figure 19.21 illustrate that the circuit is unaffected by input voltages that lie between the UTP and the LTP. This concept is illustrated further in Figure 19.22. Although the input is going through some significant changes, the output changes only when:

**1.** The UTP is reached by a *positive-going* transition.
**2.** The LTP is reached by a *negative-going* transition.

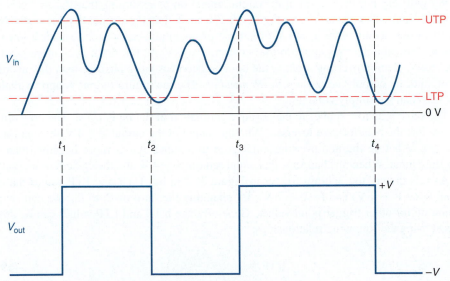

**FIGURE 19.22** Schmitt trigger hysteresis.

No input voltage between the UTP and the LTP affects the output of the circuit. Note that the range of voltages between the UTP and LTP is often referred to as **hysteresis**. For example, if Figure 19.22 has values of UTP = +10 V and LTP = +2 V, the hysteresis of the circuit is the voltage range of $+2\text{ V} < V_{in} < +10\text{ V}$. No voltage in this range will cause the output of the circuit to change *states* (dc voltage levels).

**Hysteresis**
A term that is sometimes used to describe the range of voltages between the UTP and the LTP.

The fact that the Schmitt trigger can have different UTP and LTP values is one of the primary differences between this circuit and a common comparator. The comparator has a single reference voltage used as a basis of comparison for its input waveform. However, the Schmitt trigger may compare its input waveform to *two distinct* reference voltages.

### 19.3.1 Noninverting Schmitt Triggers

The noninverting Schmitt trigger uses a simple feedback resistor connected as shown in Figure 19.23a. Note the similarity between this circuit and the linear inverting amplifiers described earlier. The difference is that the input signal is applied to the *noninverting* op-amp input.

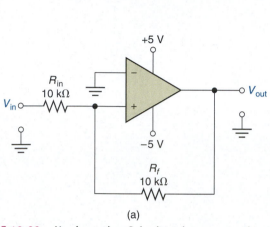

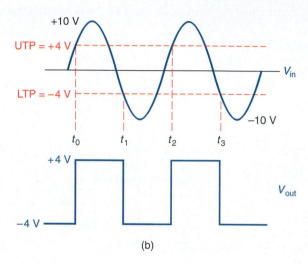

(a)                                         (b)

**FIGURE 19.23** Noninverting Schmitt trigger operation.

> Noninverting Schmitt trigger operation.

Using the component values shown, we will analyze the operation of the noninverting Schmitt trigger in Figure 19.23a. Note that $R_{in} = R_f = 10$ kΩ. As you know, the high input impedance of the op-amp causes the current through the two resistors to be equal. Therefore, the voltages across the components are also equal. This is a key point in the operation of the circuit.

We will start our discussion of the circuit operation by assuming that $V_{out} = -4$ V (its negative limit). The waveforms in Figure 19.23b show that $V_{in}$ reaches a value of $+4$ V at $t_0$. When this occurs, there is a 4 V difference of potential across each resistor in the circuit, and the noninverting input to the op-amp is at 0 V. As the input goes more positive than $+4$ V, the noninverting input to the op-amp becomes *more positive* than the inverting input, and the output goes to $+4$ V. (*Remember*: The maximum output is approximately 1 V less than the supply voltage.)

The same principle applies to the negative alternation of the input cycle. The only difference is that the polarities are reversed. With the output of the circuit at $+4$ V, the input must pass $-4$ V before the noninverting input goes to a value that is more negative than the inverting input. When this happens, the output returns to $-4$ V, and the cycle repeats itself.

As it is drawn, the Schmitt trigger in Figure 19.23a has UTP and LTP values that are equal to $(+V - 1$ V) and $(-V + 1$ V). By changing the ratio of $R_f$ to $R_{in}$, we can set the circuit up for other trigger point values. Generally, the UTP and LTP values can be determined using the following relationships:

> *Remember:*
> The value of $V_{out}$ depends on the load resistance and the supply voltages. As long as the load resistance is greater than 10 kΩ, we can assume that $+V_{out} = +V - 1$ V and $-V_{out} = -V + 1$ V.

$$\text{UTP} = -\frac{R_{in}}{R_f}(-V_{out}) \qquad (19.5)$$

and

$$\text{LTP} = -\frac{R_{in}}{R_f}(+V_{out}) \qquad (19.6)$$

The following example demonstrates the use of these two equations.

## EXAMPLE 19.7

Determine the UTP and LTP values for the circuit shown in Figure 19.24. Also, verify that the output changes states when the UTP and LTP values are exceeded.

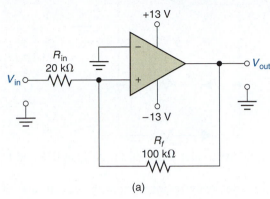

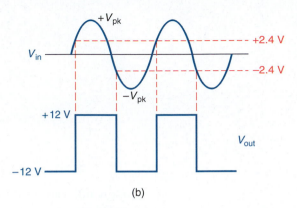

(a)            (b)

**FIGURE 19.24**

*Solution:* Using equation (19.5), the UTP is found as

$$\mathbf{UTP} = -\frac{R_{in}}{R_f}(-V_{out}) = (-0.2)(-12\ V) = \mathbf{2.4\ V}$$

The LTP is found using equation (19.6) as follows:

$$\mathbf{LTP} = -\frac{R_{in}}{R_f}(+V_{out}) = (-0.2)(+12\ V) = \mathbf{-2.4\ V}$$

The UTP value can be verified as follows: When the input is at $+2.4$ V, the output is still at $-12$ V. Using the standard voltage-divider formula, we can calculate the voltage drop across $R_i$. This voltage drop is found as

$$V_{R_{in}} = (V_{in} - V_{out})\frac{R_{in}}{R_{in} + R_f} = (14.4\ V)(0.167) = \mathbf{2.4\ V}$$

With 2.4 V applied to the circuit and a 2.4 V drop across $R_{in}$, it is clear that the voltage at the op-amp input is 0 V. Therefore, if the input goes any higher than $+2.4$ V, the noninverting input to the op-amp becomes more positive than the inverting input, and the output goes to $+12$ V. The same method can be used to validate the value of LTP $= -2.4$ V. The output produced by the circuit for a sine wave input is shown in Figure 19.24b. Note that the output changes states when the input passes the UTP and LTP values calculated in this example.

### PRACTICE PROBLEM 19.7

A noninverting Schmitt trigger like the one in Figure 19.24 has values of $R_f = 33$ kΩ and $R_{in} = 11$ kΩ. Determine the UTP and LTP values for the circuit.

---

The noninverting Schmitt trigger in Figure 19.24 is restricted because the magnitudes of the UTP and the LTP must be equal. For example, the UTP is $+2.4$ V, so the LTP must be $-2.4$ V. If the UTP were 4 V, the LTP would have to be $-4$ V. However, the noninverting Schmitt trigger can be modified so that the magnitudes of the UTP and the LTP are not equal. A circuit that contains this modification is shown in Figure 19.25.

The key to the operation of this circuit is that $D_1$ and $D_2$ conduct on opposite transitions of the output. When the output from the Schmitt trigger is *negative*, $D_1$ conducts and

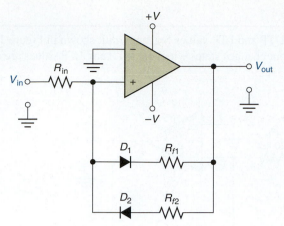

**FIGURE 19.25**  A noninverting Schmitt trigger designed for unequal UTP and LTP values.

$D_2$ is off. This means that $R_{f2}$ is effectively removed from the circuit, and the UTP is found as

$$\text{UTP} = -\frac{R_{in}}{R_{f1}}(-V_{out} + 0.7\text{ V}) \tag{19.7}$$

*Remember:*
The value of $V_{out}$ depends on the load resistance and the supply voltages. As long as the load resistance is greater than 10 kΩ, we can assume that $+V_{out} = +V - 1$ V and $-V_{out} = -V + 1$ V.

When the output is *positive*, $D_1$ is off and $D_2$ conducts. This means that $R_{f1}$ is effectively removed from the circuit, and the LTP is found as

$$\text{LTP} = -\frac{R_{in}}{R_{f2}}(+V_{out} - 0.7\text{ V}) \tag{19.8}$$

Since the UTP and LTP are determined by separate feedback resistors, these values are completely independent of each other. This point is illustrated in Example 19.8.

### EXAMPLE 19.8

Determine the UTP and LTP values for the noninverting Schmitt trigger shown in Figure 19.26a.

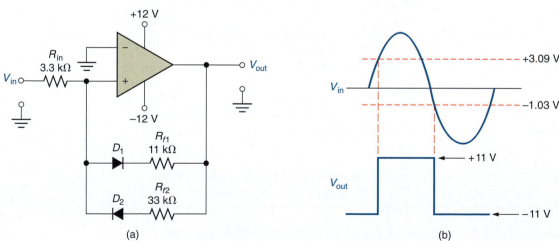

**FIGURE 19.26**

*Solution:*  The circuit has a value of $V_{out} = 11$ V. The UTP is found as

$$\textbf{UTP} = -\frac{R_{in}}{R_{f1}}(-V_{out} + 0.7\text{ V}) = -\frac{3.3\text{ k}\Omega}{11\text{ k}\Omega}(-10.3\text{ V}) = \textbf{3.09 V}$$

The LTP is found as

$$\text{LTP} = -\frac{R_{\text{in}}}{R_{f2}}(+V_{\text{out}} + 0.7\text{ V}) = -\frac{3.3\text{ k}\Omega}{33\text{ k}\Omega}(10.3\text{ V}) = -1.03\text{ V}$$

With these UTP and LTP values, the circuit in Figure 19.26a would have the input/output relationship shown in Figure 19.26b.

**PRACTICE PROBLEM 19.8**

A noninverting Schmitt trigger like the one in Figure 19.26 has $\pm 15$ V supply voltages and the following resistor values: $R_{\text{in}} = 2$ k$\Omega$, $R_{f1} = 20$ k$\Omega$, and $R_{f2} = 14$ k$\Omega$. Determine the UTP and LTP values for the circuit.

---

The circuits in this section are noninverting circuits; that is, their outputs go *high* when the UTP is passed by a *positive-going* input signal and negative when the LTP is passed by a *negative-going* input signal. It is possible to wire an op-amp as an *inverting Schmitt trigger* that has the opposite input/output relationship from the one described here; that is, the output goes *negative* when the UTP is passed and *positive* when the LTP is passed. We will take a look at this type of circuit now.

## 19.3.2 Inverting Schmitt Triggers

A basic inverting Schmitt trigger is shown in Figure 19.27. In this circuit, the input ($V_{\text{in}}$) is applied to the inverting terminal. A feedback signal is applied to the noninverting terminal from the junction of $R_{f1}$ and $R_{f2}$.

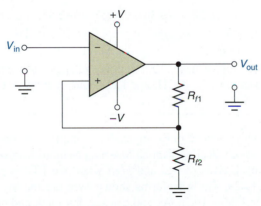

**FIGURE 19.27** An inverting op-amp Schmitt trigger.

To understand the operation of the circuit, we must first establish the equations for the UTP and LTP. The UTP for the circuit in Figure 19.27 is found as

$$\text{UTP} = \frac{R_{f2}}{R_{f1} + R_{f2}}(+V_{\text{out}}) \qquad (19.9)$$

and the LTP is found as

$$\text{LTP} = \frac{R_{f2}}{R_{f1} + R_{f2}}(-V_{\text{out}}) \qquad (19.10)$$

Example 19.9 demonstrates the procedure for determining the UTP and LTP values for an inverting Schmitt trigger circuit.

**Lab Reference:** The operation of an inverting Schmitt trigger is demonstrated in Exercise 32.

*Remember:*
The value of $V_{\text{out}}$ depends on the load resistance and the supply voltages. As long as the load resistance is greater than 10 k$\Omega$, we can assume that $+V_{\text{out}} = +V - 1$ V and $-V_{\text{out}} = -V + 1$ V.

## EXAMPLE 19.9

Determine the UTP and LTP values for the circuit shown in Figure 19.28.

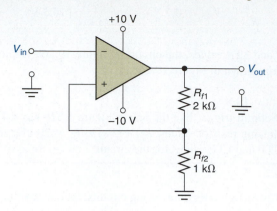

**FIGURE 19.28**

*Solution:*   The quickest approach is to solve the resistance ratio in the circuit equations first. This ratio is found to be

$$\frac{R_{f2}}{R_{f1} + R_{f2}} \cong 0.33$$

Now the UTP is found as

$$\text{UTP} = (0.33)(9\text{ V}) = \textbf{3 V}$$

and the LTP is now found as

$$\text{LTP} = (0.33)(-9\text{ V}) = \textbf{-3 V}$$

### PRACTICE PROBLEM 19.9

An inverting Schmitt trigger has $\pm 10$ V supply voltages and values of $R_{f1} = 3$ k$\Omega$ and $R_{f2} = 1$ k$\Omega$. Determine the UTP and LTP values for the circuit. (Assume that $R_L > 10$ k$\Omega$.)

We will use the values obtained in Example 19.9 to explain the operation of the circuit. First, remember that we are dealing with an *inverting* Schmitt trigger. Therefore, the output goes *low* when the UTP is passed and *high* when the LTP is passed. This point is illustrated in Figure 19.29. The waveforms shown here are the input and output waveforms for the circuit in Figure 19.28. As you can see, the input and output waveforms are out of phase by 180°.

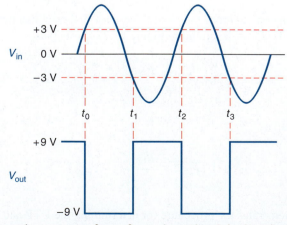

**FIGURE 19.29**   Input/output waveforms for an inverting Schmitt trigger.

When the output of the circuit is at +9 V, the voltage across $R_{f2}$ is +3 V. As the input passes +3 V (at $t_0$), the output switches to −9 V. Now, the voltage across $R_{f2}$ is −3 V. With this voltage applied to the noninverting input of the op-amp, the output does not switch again until the input passes −3 V (at $t_1$).

The basic inverting Schmitt trigger has one limitation. If you look at Figure 19.28, it is not very difficult to figure out that the UTP and LTP values must always be equal in magnitude for this circuit. This is due to the constant resistance ratio and the equal magnitudes of the two output voltages.

The inverting Schmitt trigger can be modified as shown in Figure 19.30 to allow different UTP and LTP values. In this circuit, the UTP is determined by $D_1$, $R_{f1}$, and $R_{f3}$. The LTP is determined by $D_2$, $R_{f2}$, and $R_{f3}$. For example, when the op-amp output is *positive*, $D_1$ is forward biased, and the UTP voltage is found as

$$\text{UTP} = \frac{R_{f3}}{R_{f1} + R_{f3}}(+V_{\text{out}} - 0.7 \text{ V}) \qquad \textbf{(19.11)}$$

When the op-amp output is *negative*, $D_2$ is foward biased, and the LTP voltage is found as

$$\text{LTP} = \frac{R_{f3}}{R_{f2} + R_{f3}}(-V_{\text{out}} + 0.7 \text{ V}) \qquad \textbf{(19.12)}$$

By using different feedback resistor values, the two trigger points can be set to unequal (asymmetrical) values.

> UTP and LTP values that are equal in magnitude are called *symmetrical* trigger points. Trigger-point values that are *not* equal in magnitude are called *asymmetrical* trigger points.

> *Remember:*
> The value of $V_{\text{out}}$ depends on the load resistance and the supply voltages. As long as the load resistance is greater than 10 kΩ, we can assume that $+V_{\text{out}} = +V - 1$ V and $-V_{\text{out}} = -V + 1$ V.

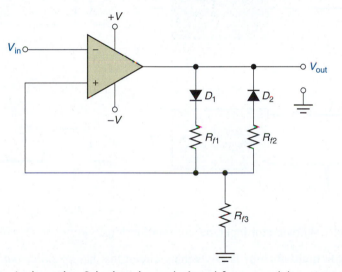

FIGURE 19.30   An inverting Schmitt trigger designed for unequal (asymmetrical) UTP and LTP values.

> **Lab Reference:** The operation of an asymmetrical Schmitt trigger is demonstrated in Exercise 32.

### 19.3.3   One Final Note

You have seen that the Schmitt trigger is a voltage-level detector that can be constructed in noninverting or inverting form. When the input signal to a Schmitt trigger makes a positive-going transition past the UTP, the output of the circuit goes to one dc output level. The output remains at that level until the input signal makes a negative-going transition that passes the LTP. At that time, the output goes to the other dc output level. The dc output level for a specific input condition depends on whether the Schmitt trigger is an inverting or a noninverting circuit.

1. What is a *Schmitt trigger*?

2. What is the *upper trigger point* (UTP)?

3. What is the *lower trigger point* (LTP)?

◀ **Section Review**

4. What is *hysteresis*?

5. Explain the operation of the noninverting Schmitt trigger in Figure 19.23.

6. Explain the operation of the noninverting Schmitt trigger in Figure 19.25.

7. Explain the operation of the inverting Schmitt trigger in Figure 19.27.

8. Explain the operation of the inverting Schmitt trigger in Figure 19.30.

9. Compare and contrast the Schmitt trigger and the comparator.

## 19.4 Multivibrators: The 555 Timer

**OBJECTIVE 7 ▶**

**Multivibrators**
Circuits designed to have zero, one, or two stable output states.

**Multivibrators** are circuits designed to have zero, one, or two *stable* output states. The concept of *stable* output states is illustrated in Figure 19.31.

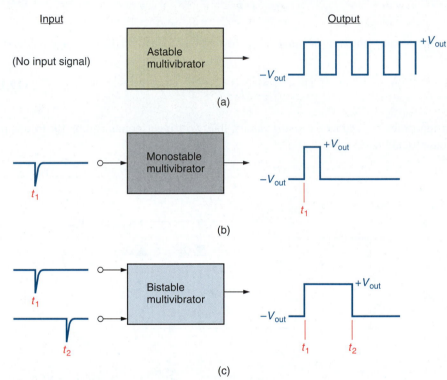

FIGURE 19.31 Multivibrator input/output relationships.

**Astable multivibrator**
A switching circuit that has no stable output state.
A rectangular-wave oscillator.
Also called a **free-running multivibrator**.

**Monostable multivibrator**
A switching circuit with one stable output state. Also called a **one-shot**.

**Bistable multivibrator**
A switching circuit with two stable output states. Also called a **flip-flop**.

The **astable multivibrator** is a switching circuit that has *no stable output state*. This means that output switches back and forth constantly between two dc levels as long as it has a dc supply voltage input. In effect, the circuit acts as a *rectangular-wave oscillator*. Because it produces an output continually, the astable multivibrator is often referred to as a **free-running multivibrator**.

The **monostable multivibrator** has *one* stable output state. For the circuit represented in Figure 19.31b, $-V_{out}$ is the stable output state. When the circuit receives an input pulse, or *trigger*, it switches to the $+V$ output state for a predetermined period of time, then automatically reverts back to its stable output state ($-V_{out}$). The duration of the $+V$ output is determined by component values in the circuit. Since the monostable multivibrator produces a single output pulse for each input trigger, it is generally referred to as a **one-shot**.

The **bistable multivibrator** has two output states. When the proper input trigger is received, it switches from one output state to the other. It then stays at the new output state until another trigger is received, returning the output to its original output state. The bistable multivibrator is generally referred to as a **flip-flop**.

As you have probably figured out by now, the term *astable* means *not stable*, the term *monostable* means *one stable* state, and the term *bistable* means *two stable* states. If you take another look at Figure 19.31, you might notice something else that is appropriate about these names. The astable multivibrator has no signal input. The one-shot generally has one signal input. The flip-flop generally has two signal inputs. Although there are some exceptions to this, the input relationships shown often hold true.

Flip-flops are covered extensively in any digital electronics course, so they are not covered in depth here. Although the astable and monostable multivibrators are also covered in digital electronics, they are discussed here to give you an introductory idea of how these circuits operate.

At one time, multivibrators were commonly constructed using discrete components. However, this is no longer the case. More often than any type of circuit we have discussed thus far, multivibrators are constructed using ICs. One of these ICs is the *555 timer*, an 8-pin IC that can be used for a variety of switching applications. Among its many applications are the astable and monostable multivibrator.

A single 555 timer contains all the active components required to produce a multivibrator. By connecting the proper passive components to the IC, you determine whether it operates as a free-running circuit or as a one-shot. In this section, we will look at the operation of the 555 timer and how it can be made to operate as either of these two circuits.

## 19.4.1   The 555 Timer

The **555 timer** is a switching circuit contained in an 8-pin IC. Its internal circuitry can be represented using two comparators, a flip-flop, an inverter, and several resistors and transistors. The block diagram of the 555 timer is shown in Figure 19.32.   ◀ *OBJECTIVE 8*

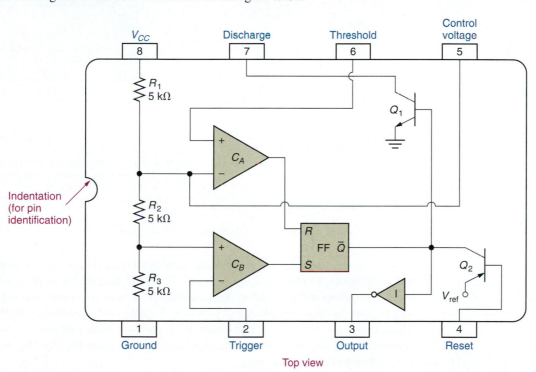

FIGURE 19.32   The 555 timer.

Since you may never have worked with an IC of this type before, there are a few points that should be made. In Figure 19.32, the block diagram is shown inside the IC casing. The small numbered blocks that border the case are the IC pins. Looking at the left side of the case, you will see that there is a small indentation. This indentation is used to identify pin 1. When you hold a 555 timer with the indentation on the left, pin 1 is the lower-left pin. The rest of the pins are set in counterclockwise order around the case. This is identical to the pin-numbering scheme for the 741 op-amp.

Inside the chip are two comparators ($C_A$ and $C_B$), a flip-flop (FF), an inverter (I), two transistors, and a voltage divider ($R_1$ through $R_3$). The voltage divider is used to set the comparator reference voltages. Since $R_1 = R_2 = R_3$, the reference voltages for $C_A$ and $C_B$ are normally equal to $\frac{2}{3}V_{CC}$ and $\frac{1}{3}V_{CC}$, respectively. The only exception to this condition

> **555 timer**
> An 8-pin IC designed for use in a variety of switching applications.

occurs when the *control voltage* input to the timer (pin 5) is used. This point is demonstrated later in this section.

If the input to pin 6 is greater than the reference voltage for $C_A$, the output from that comparator goes *high*. Otherwise, it remains at its low output state. If the input at pin 2 is lower than the reference voltage for $C_B$, the output from that comparator goes *high*. Otherwise, it remains at its low output state.

Under normal circumstances, the comparators operate in one of three input/output combinations, as shown in Figure 19.33. Each of the output combinations depends on the inputs at pins 2 and 6 of the timer. Note that it *is* possible to drive the outputs from both comparators *high* at the same time, but this is normally avoided because of the effect it will have on the flip-flop.

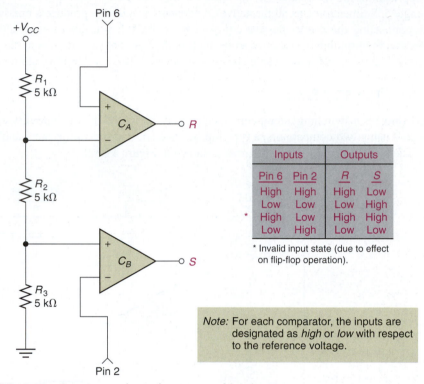

| Inputs | | Outputs | |
|---|---|---|---|
| Pin 6 | Pin 2 | $R$ | $S$ |
| High | High | High | Low |
| Low | Low | Low | High |
| High | Low | High | High |
| Low | High | Low | Low |

\* Invalid input state (due to effect on flip-flop operation).

*Note:* For each comparator, the inputs are designated as *high* or *low* with respect to the reference voltage.

FIGURE 19.33   Comparator input/output combinations.

The flip-flop responds to its inputs as shown in Figure 19.34. As you can see, having two *high* inputs is considered to be "invalid." This is because there is no way to predict how the flip-flop will respond to that condition. If it were to receive two *high* inputs, the output from the flip-flop (labeled $\overline{Q}$) could be high, low, or anywhere in between the two. Thus, the comparator input combination of *pin 6 = high and pin 2 = low* is never allowed. By disallowing this comparator input combination, the flip-flop never receives two high inputs at the same time.

We need to clear up only a few more points and we will be ready to look at some circuits. First, the flip-flop output is tied to the *discharge* output (pin 7) via an *npn* transistor. This transistor is merely acting as a current-controlled switch. When the output from the flip-flop is *high*, pin 7 is grounded through the transistor. When the output from the flip-flop is *low*, pin 7 appears as an open to any external circuitry. The purpose served by this pin will be made clear when we discuss some basic 555 timer circuits. The inverter (pin 3) and the transistor (pin 7) both invert the flip-flop output. The reason that different symbols are used for the two circuits is that the inverter is used to provide some current gain for the flip-flop output while the transistor is used only as a switch.

What is the purpose of the *reset* input?

The *reset* input (pin 4) is used to disable the 555 timer. Even though the timer functions (internally) as usual, the output (pin 3) is held *low* whenever pin 4 is grounded. This input is used only in special applications and is usually held at an inactive level by being tied to $V_{CC}$. When pin 4 is tied to $V_{CC}$, the *pnp* transistor is biased *off* and the 555 timer output

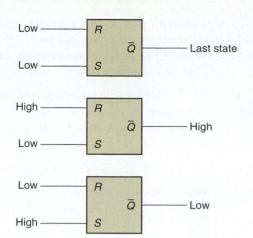

When the flip-flop inputs are not equal, the output follows the input at *R*.

| S | R | $\bar{Q}$ |
|---|---|---|
| Low | Low | L.S.* |
| Low | High | High |
| High | Low | Low |
| High | High | Invalid |

\* Last state. The $\bar{Q}$ output remains in the state it was in before the input combination occurred.

**FIGURE 19.34** Flip-flop input/output combinations.

operates normally. When the *reset* input is *low*, the *pnp* transistor is biased *on*, and the output from the flip-flop is shorted out before it reaches the inverter. In all our circuits, pin 4 will be tied to $V_{CC}$, thus disabling the reset circuit.

When the reset input isn't used, it should be connected to $V_{CC}$.

### 19.4.2   The Monostable Multivibrator

The 555 timer can be converted into a one-shot with the addition of a single resistor and a single capacitor, as shown in Figure 19.35. Before we get into the operation of this circuit, a few practical observations should be made:

◀ *OBJECTIVE 9*

1. The *reset* input to the 555 timer (pin 4) is connected to $V_{CC}$, which disables the *reset* circuit.
2. Pins 8 and 1 are tied to $V_{CC}$ and ground, respectively, as they must be for the IC to operate.

What are the initial circuit conditions?

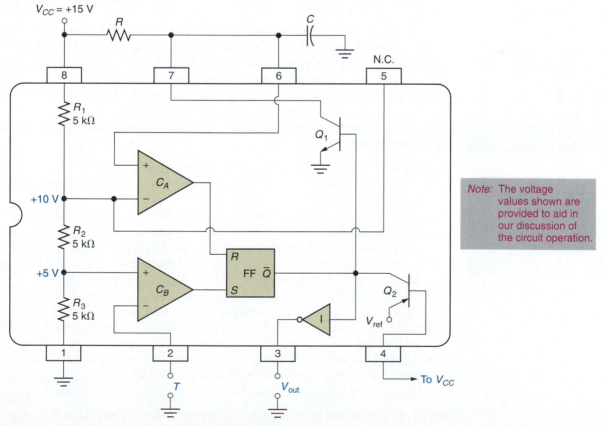

*Note:* The voltage values shown are provided to aid in our discussion of the circuit operation.

**FIGURE 19.35** The 555 timer one-shot (monostable multivibrator).

3. The *control voltage* input (pin 5) is *not connected* (N.C.). This means that this pin is left open, which is acceptable when the control voltage input is not used.
4. Pins 6 and 7 are tied together between $R$ and $C$. Thus, the capacitor voltage ($V_C$) is applied to *both* of these pins.

**Lab Reference:** The operation of a 555 timer one-shot is demonstrated in Exercise 33.

Now, let's look at the operation of the circuit. The appropriate waveforms are shown in Figure 19.36. To start with, we will assume that something caused the output of the flip-flop to be *high* before $t_0$. You will see what caused this condition in a minute. However, we need to assume this high flip-flop output to understand the waveforms at $t_0$.

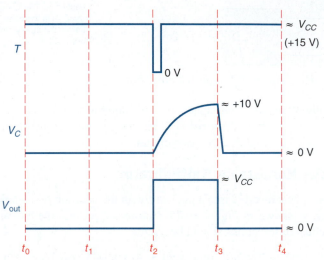

**FIGURE 19.36** The 555 timer one-shot waveforms.

Between $t_0$ and $t_2$, the inactive circuit voltages are shown. Normally, the trigger input ($T$) is approximately equal to $V_{CC}$ when in its inactive state. This ensures that the output from $C_B$ is low.

Assuming that the output from the flip-flop is high, we know that pins 3 and 7 are both tied *low*. The low output at pin 7 ensures that the voltage across the capacitor ($V_C$) is approximately 0 V, as shown in Figure 19.36. This low voltage is applied to $C_A$ via pin 6. With a low input to $C_A$, its output is low. Since both inputs to the flip-flop are low, its output does not change from the *high* voltage that was already present. The circuit voltages will remain as shown until something happens to change them.

At $t_2$, the trigger ($T$) input is driven *low*. When this occurs, the following sequence of events takes place:

1. The low input to $C_B$ drives the output of the comparator *high*.
2. The inputs to the flip-flop are now unequal, and $\bar{Q}$ goes to the same level as the $R$ input, which is *low*.
3. The low flip-flop output is inverted and pin 3 goes *high*, as shown in Figure 19.36.
4. The low flip-flop output biases the *npn* transistor *off*. Thus, the capacitor starts to charge toward $V_{CC}$. The charge time of the capacitor depends on the values of $R$ and $C$.
5. Eventually, the capacitor charges to the point where $V_C > +10$ V. At this point, the input to pin 6 is *high* (as compared to the voltage at the other $C_A$ input).
6. The high input to $C_A$ causes the comparator's output to go *high*. Since the trigger signal is no longer present, the output from $C_B$ has returned to a low.
7. Since the inputs to the flip-flop are unequal again, $\bar{Q}$ switches to the level of $R$, which is *high*.
8. The *high* output from the flip-flop causes pin 3 to go low again (shown at $t_3$).
9. The high flip-flop output turns the *npn* transistor back on, which in turn discharges the capacitor. This returns all circuit potentials to their original values (shown at $t_4$).

As you can see, our initial assumption of $\bar{Q}$ = high is based on the fact that the flip-flop is always set to this value at the end of a trigger cycle.

What happens when ($T$) goes low?

*Remember:*
Whenever the inputs to the flip-flop are not equal, $\bar{Q}$ goes to the same state as the $R$ input. (This is verified in Figure 19.34.)

At this point, you may be thinking that this circuit is very complex. In a way it is, but in a far more practical sense, it is not. In this discussion, we have considered all the details of the internal 555 timer response to a trigger input. When working with this circuit in a practical situation, you are concerned only with the input/output function of the timer. In fact, look at the diagram shown in Figure 19.37. This is what you will see when working with the circuit we just discussed. All you have is one IC, one resistor, and one capacitor. If the circuit does not work, all you need to do is make a few basic measurements. If the passive components check out, you replace the 555 timer. How is *that* for simple? (We will discuss the entire troubleshooting process later in this section.)

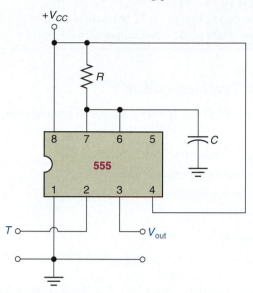

FIGURE 19.37   A 555 timer one-shot.

The pulse width of a 555 timer one-shot is determined by the values of $R$ and $C$ used in the circuit. The charge time of the capacitor in an $RC$ circuit can be found as

$$t = RC\left[\ln\frac{V_S - V_I}{V_S - V_C}\right] \tag{19.13}$$

where $t$ = the time for the capacitor to charge to $V_C$
  $V_S$ = the source voltage used to charge the capacitor
  $V_I$ = the initial charge on the capacitor, usually 0 V
  $V_C$ = the capacitor voltage of interest

Note that "ln" indicates that we are interested in the *natural log* of the fraction. For the 555 timer one-shot, the charging voltage ($V_S$) for the capacitor is $V_{CC}$ and the initial charge ($V_I$) on the capacitor is 0 V. Since the pulse width is determined by the time required for the capacitor to charge to $\frac{2}{3}V_{CC}$, we use this value as $V_C$ in the charge time equation. With these values, we can derive a useful equation that allows us to determine the pulse width of the one-shot for given values of $R$ and $C$ as follows:

$$t = RC\left[\ln\frac{V_S - V_I}{V_S - V_C}\right] = RC\left[\ln\frac{V_{CC} - 0\,V}{V_{CC} - \frac{2}{3}V_{CC}}\right]$$

$$= RC\left[\ln\frac{V_{CC}}{\frac{1}{3}V_{CC}}\right] = RC\left[\ln(3)\right]$$

or

$$PW = 1.1RC \tag{19.14}$$

Example 19.10 demonstrates the use of this equation.

## EXAMPLE 19.10

A one-shot like the one in Figure 19.37 has values of $R = 1.2 \text{ k}\Omega$ and $C = 0.1 \text{ μF}$. Determine the pulse width of the output.

*Solution:* The pulse width is found as

$$\text{PW} = 1.1RC = (1.1)(1.2 \text{ k}\Omega)(0.1 \text{ μF}) = \textbf{132 μs}$$

**Lab Reference:** The effect of changing $C$ on the PW of a one-shot is demonstrated in Exercise 33.

### PRACTICE PROBLEM 19.10

A one-shot like the one in Figure 19.37 has values of $R = 18 \text{ k}\Omega$ and $C = 1.5 \text{ μF}$. Determine the pulse width of the timer output signal.

### 19.4.3 Circuit Troubleshooting

The one-shot in Figure 19.37 is extremely simple to troubleshoot. The process involves the following checks:

Circuit checks.

- Are the $V_{CC}$ and ground connections good?
- Is the *reset* input held at an inactive level?
- Is the circuit receiving a *valid* trigger signal?
- Is the resistor good?
- Is the capacitor good?

If the answer to all these questions is *yes*, replace the 555 timer. If any question has a *no*, you have probably found the problem.

What is a *valid* trigger signal?

The troubleshooting procedure shown involves checking for a *valid* trigger signal. Just what is a *valid* trigger signal? A valid trigger signal is an input to pin 2 that is *low enough* to cause $C_B$ to have a high output. The value of input voltage required to do this depends on whether the *control voltage* input is used. When the control voltage input is *not* used, a valid input trigger is one that fulfills the following relationship:

$$V_T < \tfrac{1}{3}V_{CC} \tag{19.15}$$

This equation is based on the fact that the (+) input of $C_B$ is at $\tfrac{1}{3}V_{CC}$ when the control voltage input is not used. Therefore, the ($T$) input must go *below* this voltage level to cause the output of $C_B$ to go high.

When a control voltage is applied to pin 5, it is dropped equally across $R_2$ and $R_3$ in the resistive ladder. Therefore, the voltage at the (+) input of $C_B$ equals one-half the control voltage. In this case, a valid input trigger is one that fulfills the following relationship:

$$V_T < \tfrac{1}{2}V_{\text{con}} \tag{19.16}$$

where $V_{\text{con}}$ is the voltage at the *control voltage* input. Example 19.11 demonstrates the process for determining the input trigger voltages that are valid for a given one-shot.

### EXAMPLE 19.11

Determine the values of $V_T$ that are valid for each of the one-shots in Figure 19.38.

*Solution:* The circuit shown in Figure 19.38a uses a voltage divider to provide a control voltage input. Using the standard voltage-divider formula, the control voltage is found to be $V_{\text{con}} = +6 \text{ V}$. Therefore, any valid trigger has a value of

$$V_T < \tfrac{1}{2}V_{\text{con}} = \textbf{+3 V}$$

For the circuit shown in Figure 19.38a, the *low* trigger voltage must be lower than $+3 \text{ V}$ for the circuit to trigger properly.

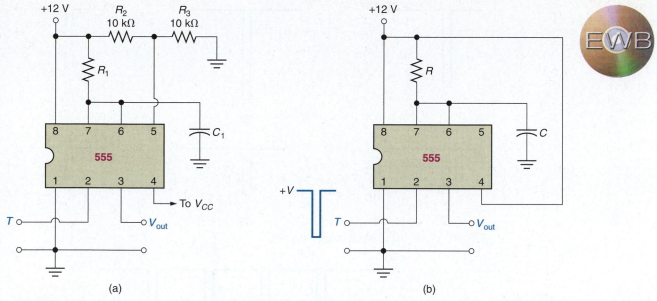

FIGURE 19.38

Since the circuit shown in Figure 19.38b has an open control voltage input, equation (19.13) is used to determine the valid trigger voltages for the circuit as follows:

$$V_T < \tfrac{1}{3}V_{CC} = +4 \text{ V}$$

### PRACTICE PROBLEM 19.11

A circuit like the one shown in Figure 19.38a has values of $R_1 = 2 \text{ k}\Omega$, $R_2 = 10 \text{ k}\Omega$, and $R_3 = 30 \text{ k}\Omega$. If $V_{CC}$ for the circuit is +15 V, what values of $V_T$ are valid for the circuit?

The circuits used in Example 19.11 demonstrate the advantage of using the control voltage input to the 555 timer. In both of these circuits, the output is a pulse that has peak voltages of 0 and +12 V (approximately). However, the circuit shown in Figure 19.38a can be triggered by a lower voltage than the one in Figure 19.38b. If we had wanted a +3 V trigger without using the control voltage input, we would have had to use a +9 V supply. In other words, by using the control voltage input, we can select any trigger signal level for the circuit without having to change $V_{CC}$ and, thus, the maximum possible peak output value from the timer.

Using the values obtained in Example 19.11, we can determine which trigger signals are valid and which are not. For example, consider the trigger signals shown in Figure 19.39. All three of the trigger signals in Figure 19.39a are valid for either circuit shown in Figure 19.38. Since they are all less than the upper limit for $V_T$, they will all cause the two circuits to trigger properly.

The first trigger signal in Figure 19.39b ($V_T = +3.2$ V) will work for the circuit shown in Figure 19.38b, as will the second trigger signal ($V_T = +3.5$ V). However, neither of these trigger signals will work for the circuit shown in Figure 19.38a because their low levels are greater than the maximum value of $V_T$ for that circuit. The third waveform shown will not work for either circuit.

An important point needs to be made at this time. The first waveform in Figure 19.39b may cause the circuit shown in Figure 19.38a to operate *intermittently*. This means that it is close enough to the value of $V_T$ to work one time but not work another. An example of intermittent operation is illustrated in Figure 19.40. Note that the output triggers on two of the input signals but not on the other two. Any time that you have an intermittent problem in a switching circuit, the trigger signal should be suspect. In most cases, the intermittent operation is caused by a trigger signal that is not going low (or high) enough.

> What is the advantage of using the *control input* (pin 5)?

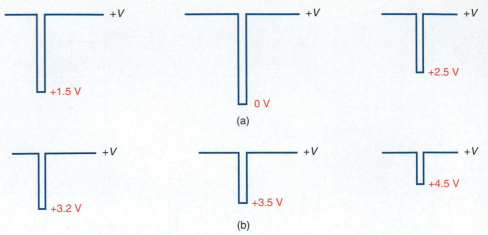

FIGURE 19.39   One-shot trigger signals.

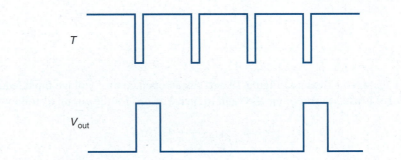

FIGURE 19.40   One-shot intermittent operation.

Sometimes an intermittent problem can develop when a one-shot is operated at a relatively high frequency. In this case, the problem is the switching rate of the 555 and not the trigger signal. You see, when the 555 timer switches from one output state to another, there is an instant when an effective short circuit *may* develop inside the chip between the $V_{CC}$ and ground pins. The cause of this condition is the active components switching through their active regions. The cure for this problem is to use a *decoupling capacitor* between the $V_{CC}$ and ground pins of the chip, as shown in Figure 19.41.

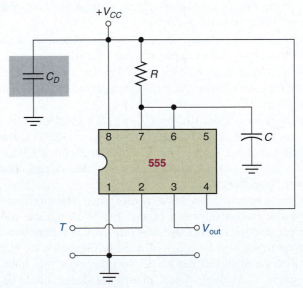

FIGURE 19.41   Use of a decoupling capacitor ($C_D$) can eliminate intermittent output problems.

If a one-shot circuit has a frequency-related intermittent problem, the decoupling capacitor helps to eliminate the problem by holding the supply current ($I_{CC}$) relatively constant during the switching time. Note that $C_D$ is typically around 0.1 μF.

How do you know when an intermittent problem is caused by the circuit operating frequency? Check the trigger signal. If the trigger signal is valid, odds are that the problem is related to the input frequency of the circuit.

### 19.4.4 The Astable Multivibrator

The 555 timer can be wired for free-running operation, as shown in Figure 19.42. Note that the circuit contains *two resistors and one capacitor* and *does not have an input trigger from any other circuit*. The lack of a trigger signal from an external source is the circuit recognition feature of the free-running multivibrator.

Circuit recognition.

**Lab Reference:** The operation of a 555 timer free-running multivibrator is demonstrated in Exercise 33.

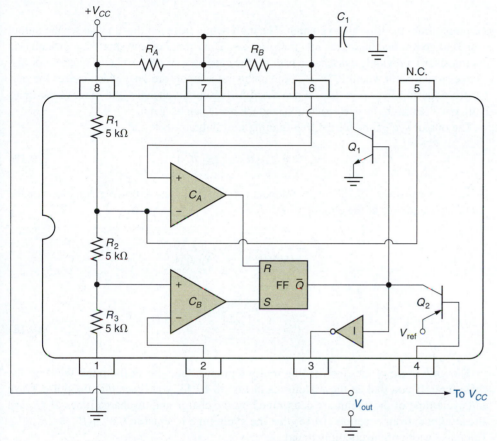

FIGURE 19.42   The 555 timer astable (free-running) multivibrator.

The key to the operation of this circuit is that the capacitor is connected to *both* the trigger input (pin 2) and the threshold input (pin 6). Thus, if the capacitor is caused to charge and discharge between the threshold voltage ($\frac{2}{3}V_{CC}$) and the trigger voltage ($\frac{1}{3}V_{CC}$), the timer produces a steady train of pulses. This point is illustrated in Figure 19.43. Since pins 6 and 2 are tied together, their voltage values are always equal. When $V_C$ charges to $\frac{2}{3}V_{CC}$, both inputs are high and the output from the timer goes *low*. When $V_C$ discharges to $\frac{1}{3}V_{CC}$, both inputs are low and the output from the timer goes *high*.

The charge–discharge action of $C_1$ is caused by the 555 timer internal circuitry. To see how this is accomplished, we will start by making a few assumptions:

Circuit operation.

1. The capacitor is just starting to charge toward the threshold voltage, $V_{th}$.
2. The output from the flip-flop is *low*, causing pin 3 to be high and pin 7 to be *open*.

With pin 7 open, the capacitor is charged by $V_{CC}$ via $R_A$ and $R_B$. As the capacitor charges, $V_C$ eventually reaches $V_{th}$. At this point, both inputs to the 555 timer are *high*, which causes

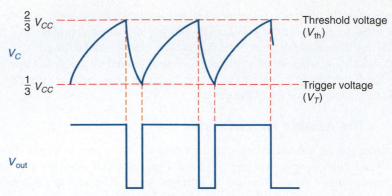

**FIGURE 19.43** Free-running multivibrator waveforms.

the output from the flip-flop to go high. The high output from the flip-flop causes the output (pin 3) to go low and biases the *npn* transistor *on*. Thus, pin 7 is now shorted to ground (via the transistor), providing a discharge path for the capacitor (via $R_B$). When the capacitor discharges to the point where $V_C = V_T$, both comparator inputs are *low*, which causes the output from the flip-flop to go *low*. When the flip-flop output changes, the timer output goes high, pin 7 returns to its open condition, and the cycle begins again.

The output cycle time for the free-running multivibrator is found as

$$T = 0.693[(R_A + 2R_B)C_1] \tag{19.17}$$

Converting equation (19.15) into a frequency form yields the following equation for the operating frequency of the circuit:

$$f_0 = \frac{1}{0.693[(R_A + 2R_B)C_1]}$$

or

$$f_0 = \frac{1.44}{(R_A + 2R_B)C_1} \tag{19.18}$$

The duty cycle of the free-running multivibrator is the ratio of the pulse width to the cycle time. If you look at the waveforms in Figure 19.43, you'll see that the pulse width and cycle time of the circuit are determined by the charge and discharge times of $C_1$. We already know from equation (19.15) that the cycle time is equal to $0.693[(R_A + 2R_B)C_1]$. The pulse width of the circuit is found as

$$PW = 0.693[(R_A + R_B)C_1] \tag{19.19}$$

If we use the charge/discharge equations in place of PW and cycle time in the duty cycle equation, we obtain the following useful equation:

$$\text{duty cycle} = \frac{0.693[(R_A + R_B)C_1]}{0.693[(R_A + 2R_B)C_1]} \times 100$$

or

$$\text{duty cycle} = \frac{R_A + R_B}{R_A + 2R_B} \times 100 \tag{19.20}$$

Example 19.12 demonstrates the basic analysis of a free-running multivibrator.

## EXAMPLE 19.12

Determine the values of $f_0$ and duty cycle of the circuit shown in Figure 19.44. Also, determine the pulse width of the circuit.

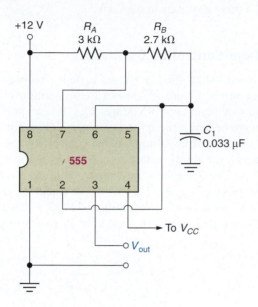

**FIGURE 19.44**

**Solution:** The circuit operating frequency is found as

$$f_0 = \frac{1.44}{(R_A + 2R_B)C_1} = \frac{1.44}{(3\ k\Omega + 5.4\ k\Omega)(0.033\ \mu F)} = \textbf{5.19 kHz}$$

The duty cycle of the circuit is found as

$$\textbf{duty cycle} = \frac{R_A + R_B}{R_A + 2R_B} \times 100 = \frac{5.7\ k\Omega}{8.4\ k\Omega} \times 100 = \textbf{67.9\%}$$

Finally, the pulse width of the circuit is found as

$$\textbf{PW} = 0.693[(R_A + R_B)C_1] = 0.693(5.7\ k\Omega)(0.033\ \mu F) = \textbf{130 }\boldsymbol{\mu}\textbf{s}$$

### PRACTICE PROBLEM 19.12

A circuit like the one shown in Figure 19.44 has the following values: $V_{CC} = +10$ V, $R_A = 5.1$ k$\Omega$, $R_B = 2.2$ k$\Omega$, and $C_1 = 0.022$ $\mu$F. Determine the values of $f_0$, duty cycle, and output pulse width for the circuit.

---

At this point, you might be wondering why the free-running multivibrator requires two resistors in its *RC* network. After all, we required only a single resistor for the one-shot. To answer this question, let's take a look at what would happen if either $R_A$ or $R_B$ were shorted.

> Why are two resistors used in the *RC* circuit?

If we short $R_A$, we have a single resistor, $R_B$, between pins 6 and 7. This situation is unacceptable because the 555 timer will short circuit the power supply when pin 7 goes low. Recall that pin 7 goes low during every discharge cycle. If $R_A$ is shorted, this low output is connected directly to $V_{CC}$. The result is a "well-done" 555 timer. Thus, we must have a resistor between pin 7 and $V_{CC}$.

If $R_B$ were shorted, there would be no *RC* circuit during the discharge cycle of $C_1$. As you have been shown, $C_1$ charges through $(R_A + R_B)$ and discharges through $R_B$. If $R_B$ is replaced with a short circuit, the discharge cycle time will be near zero, and the duty cycle will approach 100% (an unacceptable situation). Thus, $R_B$ is also required for proper operation.

### 19.4.5    Circuit Troubleshooting

The process for troubleshooting the free-running circuit is nearly the same as the procedure used on the one-shot. The only difference is that you have two resistors that must be verified as good. All other tests remain the same.

### 19.4.6    Voltage-Controlled Oscillators

**OBJECTIVE 10** ▶

**Voltage-controlled oscillator (VCO)**

A free-running oscillator whose output frequency is controlled by a dc input voltage.

*Don't Forget:*

$V_{th}$ is the voltage that triggers comparator A. $V_T$ is the input that triggers comparator B.

The 555 can be modified to form a circuit called a **voltage-controlled oscillator (VCO)**. A VCO (shown in Figure 19.45) is a free-running oscillator whose output frequency depends on a dc input control voltage, $V_{con}$. When applied to pin 5 of the timer, the dc control voltage becomes the reference voltage ($V_{ref}$) for $C_A$. The reference voltage for $C_B$ is half the dc control voltage. As the control voltage is increased (by adjusting the potentiometer), the circuit output frequency *decreases* (and vice versa) because the dc control voltage directly affects the values of $V_{th}$ and $V_T$. For example, when $V_{con}$ is set to $+10$ V, the value of $V_{th}$ also becomes $+10$ V, and $V_T$ becomes one-half of $V_{con}$, or $+5$ V. Capacitor $C_1$ must charge to the value of $V_{con}$ ($+10$ V) to change the output of the comparator, $C_A$, and discharge to the value of $V_T$ ($+5$ V) to cause the output of $C_B$ to change states.

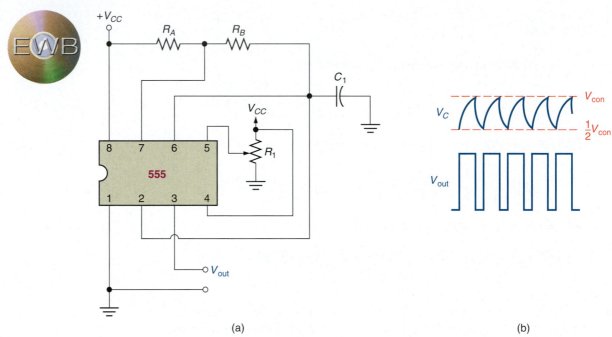

(a)                                      (b)

FIGURE 19.45    A voltage-controlled oscillator (VCO).

When the dc control voltage is changed, the values of $V_{th}$ and $V_T$ also change. If $V_{con}$ is changed to $+7$ V, $C_1$ only needs to change to $+7$ V and discharge to $+3.5$ V to produce an output cycle. This reduces the time required to charge and discharge the capacitor, which causes the output state of both comparators to switch at a faster rate. The faster the comparators' switch rates, the higher the output frequency for the circuit.

### 19.4.7    One Final Note

In this chapter, we have barely scratched the surface of the entire subject of switching circuits. The study of digital electronics provides complete coverage of the subject. However, this chapter has provided you with the fundamental principles of switching circuits in such a way as to aid in your study of digital electronics.

1. What is a *multivibrator*?

2. Describe the output characteristics of each of the following circuits:
   a. *Astable multivibrator (free-running multivibrator).*
   b. *Monostable multivibrator (one-shot).*
   c. *Bistable multivibrator (flip-flop).*

3. What is the *555 timer*?

4. List the input/output relationships for the 555 timer.

5. Briefly describe the operation of the monostable multivibrator (one-shot).

6. What checks are made when troubleshooting a one-shot?

7. Briefly describe the operation of the astable (free-running) multivibrator.

8. Describe the operation of the voltage-controlled oscillator (VCO).

## CHAPTER SUMMARY

Here is a summary of the major points made in this chapter:

1. Circuits designed to respond to (or generate) rectangular waveforms are referred to as *switching circuits*.

2. A BJT can be used as a switch by driving the component back and forth between saturation and cutoff.
   a. Ideally, $V_{CE} = V_{CC}$ and $I_C = 0$ A when a BJT switch is in cutoff. These conditions match those of an open switch. (See Figure 19.2.)
   b. Ideally, $V_{CE} = 0$ V and $I_C = V_{CC}/R_C$ when a BJT switch is in saturation. These conditions match those of a closed switch. (See Figure 19.2.)

3. In practice, $V_{CE}$ is usually within 1 V of $V_{CC}$ when a BJT is in cutoff and somewhere between 0.2 and 0.4 V when the device is saturated.
   a. When in cutoff, $V_{CE} \neq V_{CC}$ because transistor leakage current causes some voltage to be dropped across the collector resistor.
   b. When saturated, $V_{CE} \neq 0$ V because the device current develops some voltage across the internal resistance of the transistor.

4. JFET switches differ from BJT switches because:
   a. They have higher input impedance.
   b. An input voltage of one polarity is used to produce an output of the opposite polarity (see Figure 19.4).

5. A MOSFET switch has the input impedance advantage of a JFET switch and the input/output polarity relationship of a BJT switch.

6. Since the D-MOSFET can be operated in the enhancement mode, it can produce a positive output signal with a positive input signal. The input and output signals are 180° out of phase.

7. A trade-off exists in a MOSFET switch that involves the required high and low output values and the value of the drain resistor (see Example 19.3). The trade-off is eliminated by using CMOS switching circuits.

8. A CMOS switch uses *n*-channel and *p*-channel MOSFETs connected in series between the dc power supply and ground (see Figure 19.8).
   a. A low input turns the upper MOSFET on and the lower MOSFET off, coupling the output to $V_{SS}$.
   b. A high input turns the upper MOSFET off and the lower MOSFET on, coupling the output to ground.

9. A *driver* is a circuit used to couple a low-current output to a relatively high-current device (see Figure 19.9).

10. Any waveform made up of alternating (high and low) dc voltages is generally referred to as a *rectangular* waveform (see Figure 19.10).
    a. The time spent in the high output state is generally referred to as the *pulse width* (PW).

    **b.** The time spent in the low output state is generally referred to as the *space width* (SW).

    **c.** The sum of pulse width (PW) and space width (SW) equals the *cycle time* of the waveform.

**11.** A *square wave* is a rectangular waveform with equal PW and SW values.

**12.** The *ideal* rectangular waveform switches from low to high and high to low instantly.

    **a.** In practice, it takes some finite amount of time for the transitions to occur.

    **b.** As a standard, all time-related measurements are taken at the 50% point on the waveform, that is, halfway between the low and high voltage levels.

    See Example 19.4.

**13.** The *duty cycle* of a waveform is the ratio of PW to cycle time, measured as a percentage.

**14.** By definition, the duty cycle of a square wave is 50%.

**15.** *Propagation delay* is a time delay between input and output transitions, measured at the 50% points on the two waveforms.

**16.** There are four sources of propagation delay in a BJT switch: delay time, rise time, storage time, and fall time (see Figure 19.12).

**17.** *Delay time* ($t_d$) is the time required for a BJT to come out of cutoff. It is the time required for:

    **a.** $I_C$ to increase to 10% of its maximum value.

    **b.** $V_C$ to decrease to 90% of its maximum value.

**18.** *Rise time* ($t_r$) is the time required for a BJT to make the transition from cutoff to saturation (through the active region). It is the time required for:

    **a.** $I_C$ to increase from 10% to 90% of its maximum value.

    **b.** $V_C$ to decrease from 90% to 10% of its maximum value.

**19.** *Storage time* ($t_s$) is the time required for a BJT to come out of saturation. It is the time required for:

    **a.** $I_C$ to decrease to 90% of its maximum value.

    **b.** $V_C$ to increase to 10% of its maximum value.

**20.** *Fall time* ($t_f$) is the time required for a BJT to make the transition from saturation to cutoff (through the active region). It is the time required for:

    **a.** $I_C$ to decrease from 90% to 10% of its maximum value.

    **b.** $V_C$ to increase from 10% to 90% of its maximum value.

**21.** Delay time and storage time account for the delay between an input transition and the *start* of the output transition (see Figure 19.13).

**22.** Rise time and fall time account for the *slope* in the output transitions of a BJT switch (see Figure 19.13).

**23.** The values of $t_d$, $t_r$, $t_s$, and $t_f$ are usually listed in the spec sheet for a given BJT.

**24.** Rise time and fall time are functions of the physical construction of a BJT and cannot be decreased.

**25.** Delay time and storage time can be reduced by using a *speed-up capacitor* (see Figure 19.16).

    **a.** The capacitor effectively bypasses the base resistor during both transitions of the input signal.

    **b.** On the positive-going transition, this generates a high initial base current that helps reduce delay time (see Figure 19.17).

    **c.** On the negative-going transition, this generates a high initial reverse bias that helps reduce storage time (see Figure 19.17).

**26.** Since JFETs are voltage-controlled devices, a speed-up capacitor improves only the *turn-off* time of the component.

**27.** *Switching transistors* are designed for extremely low switching times.

**28.** An *inverter* is a basic switching circuit that produces a 180° voltage phase shift from input to output.

**29.** A *buffer* is a switching circuit that does not produce a voltage phase shift.

**30.** The switching times for a BJT buffer are measured as shown in Figure 19.19.

**31.** A *Schmitt trigger* is a voltage-level detector that is similar to a comparator in many respects.

 **a.** When the input makes a *positive-going* transition past a specified voltage, called the *upper trigger point* (UTP), the output switches from one dc voltage level to another.

 **b.** When the input makes a *negative-going* transition past a specified voltage, called the *lower trigger point* (LTP), the output switches back to the original output voltage level.

 **c.** The UTP and LTP values are determined by circuit component values.

**32.** The range of voltages between the UTP and LTP values is sometimes referred to as *hysteresis*.

**33.** The fact that the Schmitt trigger has unequal UTP and LTP values distinguishes the circuit from a common comparator, which is restricted (by design) to a single reference voltage.

**34.** A *noninverting Schmitt trigger* can be constructed as shown in Figure 19.23.

 **a.** The circuit shown has UTP and LTP values that are equal in magnitude and opposite in polarity.

 **b.** The UTP and LTP values for the circuit are calculated as shown in Example 19.7.

**35.** A noninverting Schmitt trigger can be modified to allow for UTP and LTP values that differ in magnitude (see Example 19.8).

**36.** An inverting Schmitt trigger can be modified to allow for unequal UTP and LTP values, as shown in Figure 19.30.

**37.** A *multivibrator* is a circuit designed to have zero, one, or two *stable* output states (see Figure 19.31).

 **a.** The *astable multivibrator* (or *free-running multivibrator*) has no stable output state. It is essentially a rectangular-wave oscillator.

 **b.** The *monostable multivibrator* (or *one-shot*) has one stable output state. When the circuit receives a trigger, it switches to its high output state for a set period of time, then automatically reverts to its stable (low) output state.

 **c.** The *bistable multivibrator* (or *flip-flop*) has two stable output states. When the proper trigger is received, it switches from one output state to the other. It remains at this output state until it receives another trigger, which returns it to its original output state.

**38.** The 555 timer is an 8-pin IC that can be used in a variety of switching applications.

 **a.** The internal circuitry is typically represented as two comparators, a flip-flop, an inverter, and several resistors and transistors (see Figure 19.32).

 **b.** When you hold a 555 timer with the indentation on the left, pin 1 is located in the lower-left corner. The remaining pins are set in counterclockwise order around the IC.

**39.** Comparator operation is illustrated in Figure 19.33.

 **a.** The reference voltages for the comparators are set by the input voltage divider, ground, and $V_{CC}$ (except when the *control voltage* input is used).

 **b.** The input combination of *pin 6 = high and pin 2 = low* is avoided because it provides an unacceptable combination of inputs to the flip-flop.

**40.** Flip-flop operation is illustrated in Figure 19.34.

 **a.** An input combination of *S = high* and *R = high* is invalid because there is no way of predicting the output response of the component.

 **b.** Whenever the inputs (*S* and *R*) are *not* equal, the output voltage equals the voltage at the *R* input.

**41.** The internal transistor connected to pin 7 of the timer is used to provide a low-resistance path to ground when it receives a *high* input from the flip-flop.

**42.** The *reset* input, when low, blocks the output from the 555 timer. The internal circuit works (up to the flip-flop), but the output pin is held *high*.

**43.** The 555 timer can be wired as a monostable multivibrator, or *one-shot* (see Figure 19.35).

**44.** Example 19.11 demonstrates the effect of using the *control voltage* input on the operation of the 555 timer one-shot.

**45.** *Intermittent* operation of a 555 timer one-shot can be caused by:

**a.** A trigger input that approaches, but does not drop below, the upper limit for a valid input trigger.

**b.** Operating the circuit at a relatively high frequency. This problem is resolved by placing a decoupling capacitor between $V_{CC}$ and ground (see Figure 19.41).

**46.** The 555 timer can be wired as an astable (*free-running*) multivibrator (see Figure 19.42). The output is developed by the charge/discharge action of an external capacitor.

**47.** The 555 can be modified to form a *voltage-controlled oscillator* (VCO). This circuit is essentially a free-running multivibrator whose output frequency is controlled by a dc input control voltage (see Figure 19.45).

**a.** The potentiometer in the circuit is used to adjust the dc control voltage.

**b.** When the dc control voltage *increases*, the output frequency *decreases*, and vice versa.

## EQUATION SUMMARY

| Equation Number | Equation | Section Number |
|---|---|---|
| (19.1) | $\text{duty cycle} = \dfrac{\text{pulse cycle}}{\text{cycle time}} \times 100$ | 19.2 |
| (19.2) | $f_{C2} = \dfrac{0.35}{t_r}$ | 19.2 |
| (19.3) | $f_{\max} = \dfrac{0.35}{100 t_r}$ (practical limit) | 19.2 |
| (19.4) | $C_S < \dfrac{1}{20 R_B f_{\max}}$ | 19.2 |
| (19.5) | $\text{UTP} = -\dfrac{R_{\text{in}}}{R_f}(-V_{\text{out}})$ | 19.3 |
| (19.6) | $\text{LTP} = -\dfrac{R_{\text{in}}}{R_f}(+V_{\text{out}})$ | 19.3 |
| (19.7) | $\text{UTP} = -\dfrac{R_{\text{in}}}{R_{f1}}(-V_{\text{out}} + 0.7 \text{ V})$ | 19.3 |
| (19.8) | $\text{LTP} = -\dfrac{R_{\text{in}}}{R_{f2}}(+V_{\text{out}} - 0.7 \text{ V})$ | 19.3 |
| (19.9) | $\text{UTP} = \dfrac{R_{f2}}{R_{f1} + R_{f2}}(+V_{\text{out}})$ | 19.3 |
| (19.10) | $\text{LTP} = \dfrac{R_{f2}}{R_{f1} + R_{f2}}(-V_{\text{out}})$ | 19.3 |
| (19.11) | $\text{UTP} = \dfrac{R_{f3}}{R_{f1} + R_{f3}}(+V_{\text{out}} - 0.7 \text{ V})$ | 19.3 |
| (19.12) | $\text{LTP} = \dfrac{R_{f3}}{R_{f2} + R_{f3}}(-V_{\text{out}} + 0.7 \text{ V})$ | 19.3 |

| Equation Number | Equation | Section Number |
|---|---|---|
| (19.13) | $t = RC\left[\ln\dfrac{V_S - V_I}{V_S - V_C}\right]$ | 19.4 |
| (19.14) | $PW = 1.1RC$ | 19.4 |
| (19.15) | $V_T < \frac{1}{3}V_{CC}$ | 19.4 |
| (19.16) | $V_T < \frac{1}{2}V_{\text{con}}$ | 19.4 |
| (19.17) | $T = 0.693[(R_A + 2R_B)C_1]$ | 19.4 |
| (19.18) | $f_0 = \dfrac{1.44}{(R_A + 2R_B)C_1}$ | 19.4 |
| (19.19) | $PW = 0.693[(R_A + R_B)C_1]$ | 19.4 |
| (19.20) | duty cycle $= \dfrac{R_A + R_B}{R_A + 2R_B} \times 100$ | 19.4 |

**KEY TERMS**

astable multivibrator 812
bistable multivibrator 812
buffer 802
cycle time ($T_C$) 792
delay time ($t_d$) 795
driver 791
duty cycle 794
fall time ($t_f$) 795
555 timer 813
flip-flop 812
free-running
   multivibrator 812
hysteresis 805

inverter 802
lower trigger point
   (LTP) 805
monostable
   multivibrator 812
multivibrator 812
one-shot 812
propagation delay 794
pulse width (PW) 792
rectangular waveform 792
rise time ($t_r$) 795
Schmitt trigger 804
space width (SW) 792

speed-up capacitor 799
square wave 792
storage time ($t_s$) 795
switching circuits 784
switching transistors 801
turn-off time ($t_{\text{off}}$) 801
turn-on time ($t_{\text{on}}$) 801
upper trigger point
   (UTP) 804
voltage-controlled
   oscillator (VCO) 824

**PRACTICE PROBLEMS**

### Section 19.1

**1.** Determine the minimum high voltage required to saturate the transistor in Figure 19.46.

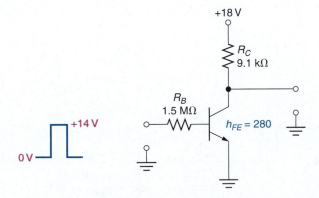

**FIGURE 19.46**

**2.** A switch like the one in Figure 19.46 has the following values: $V_{CC} = +12$ V, $h_{FE} = 150$, $R_C = 1.1$ kΩ, and $R_B = 47$ kΩ. Determine the minimum high input voltage required to saturate the transistor.

3. Determine the minimum high input voltage required to saturate the transistor in Figure 19.47.

4. Determine the minimum high input voltage required to saturate the transistor in Figure 19.48.

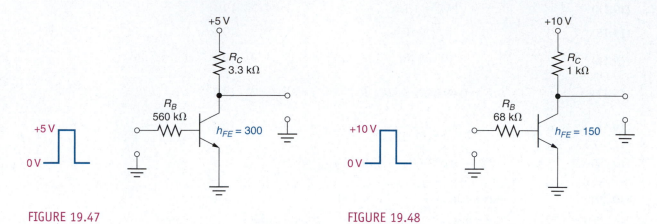

FIGURE 19.47                                   FIGURE 19.48

5. Determine the high and low output voltages for the circuit shown in Figure 19.49.
6. Determine the high and low output voltages for the circuit shown in Figure 19.50.

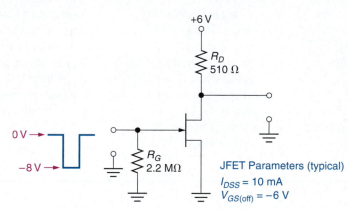

FIGURE 19.49

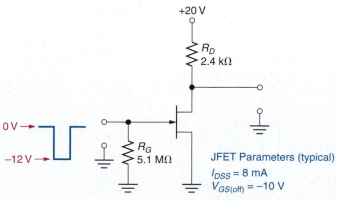

FIGURE 19.50

**Section 19.2**

7. Determine the values of PW and $T_C$ for the waveform in Figure 19.51.

8. Determine the values of PW and $T_C$ for the waveform in Figure 19.52.

9. Calculate the duty cycle of the waveform in Figure 19.51.

10. Calculate the duty cycle of the waveform in Figure 19.52.

11. The 2N4264 BJT has the following parameters: $t_d = 8$ ns, $t_r = 15$ ns, $t_s = 20$ ns, and $t_f = 15$ ns. Determine the value of $f_{C2}$ and the practical limit on $f_{in}$ for the device.

12. The 2N4400 BJT has the following parameters: $t_d = 15$ ns, $t_r = 20$ ns, $t_s = 225$ ns, and $t_f = 30$ ns. Determine the value of $f_{C2}$ and the practical limit on $f_{in}$ for the device.

13. The BCX587 BJT has the following parameters: $t_d = 16$ ns, $t_r = 29$ ns, $t_s = 475$ ns, and $t_f = 40$ ns. Determine the value of $f_{C2}$ and the practical limit on $f_{in}$ for the device.

14. The MPS2222 BJT has the following parameters: $t_d = 10$ ns, $t_r = 25$ ns, $t_s = 225$ ns, and $t_f = 60$ ns. Determine the value of $f_{C2}$ and the practical limit on $f_{in}$ for the device.

15. The 2N3971 JFET has the following parameters: $t_d = 15$ ns, $t_r = 15$ ns, and $t_{off} = 60$ ns. Determine the value of $f_{C2}$ and the practical limit on $f_{in}$ for the device.

16. The 2N4093 JFET has the following parameters: $t_d = 20$ ns, $t_r = 40$ ns, and $t_{off} = 80$ ns. Determine the value of $f_{C2}$ and the practical limit on $f_{in}$ for the device.

17. The 2N4351 MOSFET has the following parameters: $t_d = 45$ ns, $t_r = 65$ ns, $t_s = 60$ ns, and $t_f = 100$ ns. Determine the value of $f_{C2}$ and the practical limit on $f_{in}$ for the device.

18. The 2N4393 JFET has the following parameters: $t_{on} = 15$ ns and $t_{off} = 50$ ns. Determine the value of $f_{C2}$ and the practical limit on $f_{in}$ for the device.

19. Determine the values of $t_d$, $t_r$, $t_s$, and $t_f$ for the waveform in Figure 19.53.

20. Determine the values of $t_d$, $t_r$, $t_s$, and $t_f$ for the waveform in Figure 19.54.

Time base: 10 µs/div

FIGURE 19.51

Time base: 1 µs/div

FIGURE 19.52

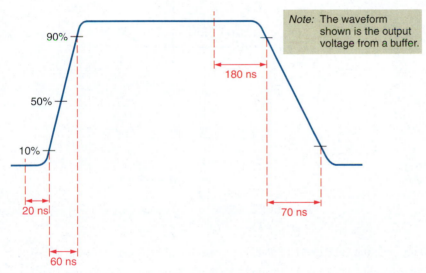

Note: The waveform shown is the output voltage from a buffer.

180 ns

90%

50%

10%

20 ns

60 ns

70 ns

FIGURE 19.53

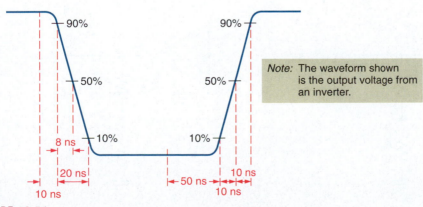

90%

50%

10%

8 ns

20 ns

10 ns

Note: The waveform shown is the output voltage from an inverter.

90%

50%

10%

50 ns

10 ns

10 ns

FIGURE 19.54

**21.** Determine the UTP and LTP values for the Schmitt trigger in Figure 19.55.

**22.** Determine the UTP and LTP values for the Schmitt trigger in Figure 19.56.

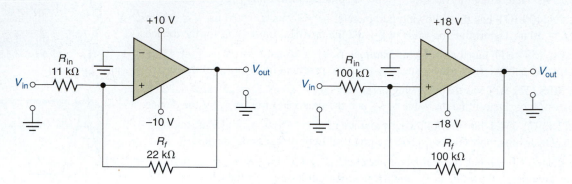

FIGURE 19.55                     FIGURE 19.56

**23.** Determine the UTP and LTP values for the Schmitt trigger in Figure 19.57.

**24.** Determine the UTP and LTP values for the Schmitt trigger in Figure 19.58.

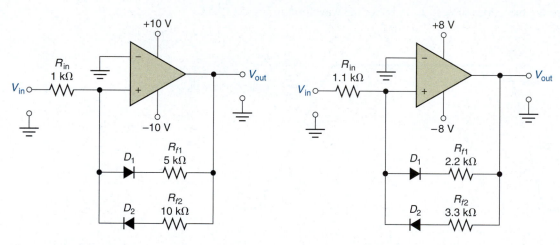

FIGURE 19.57                     FIGURE 19.58

**25.** Determine the UTP and LTP values for the Schmitt trigger in Figure 19.59.

**26.** Determine the UTP and LTP values for the Schmitt trigger in Figure 19.60.

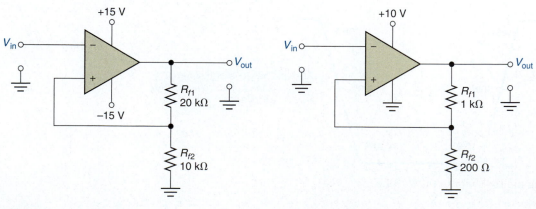

FIGURE 19.59                     FIGURE 19.60

**27.** Determine the UTP and LTP values for the Schmitt trigger in Figure 19.61.

**28.** Determine the UTP and LTP values for the Schmitt trigger in Figure 19.62.

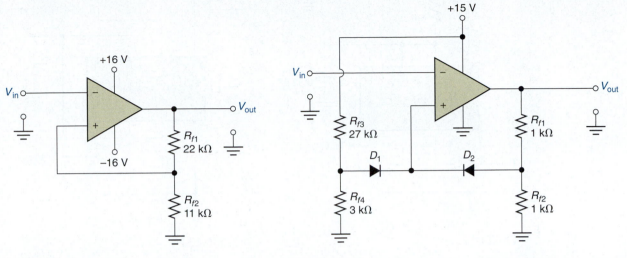

FIGURE 19.61

FIGURE 19.62

## Section 19.4

**29.** Calculate the output pulse width of the one-shot in Figure 19.63.

**30.** Calculate the output pulse width of the one-shot in Figure 19.64.

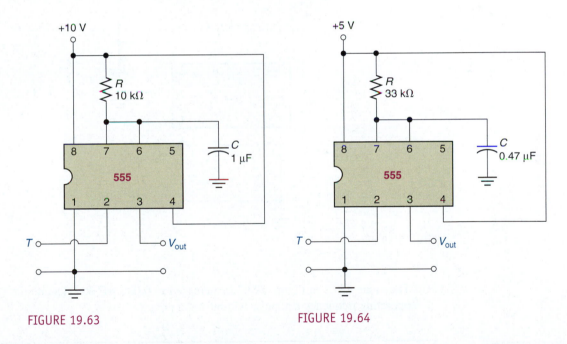

FIGURE 19.63

FIGURE 19.64

**31.** Calculate the value of $V_{T(max)}$ for the switch in Figure 19.65.

**32.** The resistor values in Figure 19.65 are changed to $R_1 = 33$ k$\Omega$ and $R_2 = 11$ k$\Omega$. Recalculate the value of $V_{T(max)}$ for the circuit.

**33.** Calculate the operating frequency, duty cycle, and pulse-width values for the free-running multivibrator in Figure 19.66.

**34.** Calculate the operating frequency, duty cycle, and pulse-width values for the free-running multivibrator in Figure 19.67.

**35.** The value of $R_B$ in Figure 19.66 is increased to 10 k$\Omega$. Recalculate the operating frequency, duty cycle, and pulse-width values for the circuit.

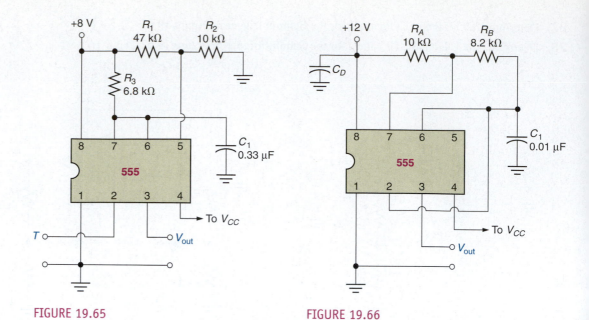

FIGURE 19.65

FIGURE 19.66

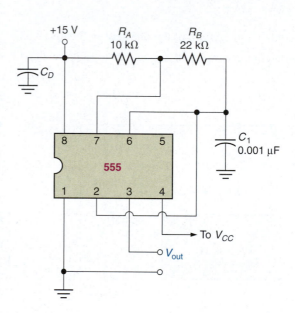

FIGURE 19.67

**36.** The value of $C_1$ in Figure 19.67 is increased to 0.033 μF. Recalculate the values of operating frequency and pulse width for the circuit.

## TROUBLESHOOTING PRACTICE PROBLEMS

**37.** The free-running multivibrator in Figure 19.68 has the dc readings indicated when power is applied. Discuss the possible cause(s) of the problem.

**38.** The free-running multivibrator in Figure 19.69 has the dc readings indicated when power is applied. Discuss the possible cause(s) of the problem.

## PUSHING THE ENVELOPE

**39.** The BJT in Figure 19.70a is faulty. Determine which (if any) of the transistors listed in the selector guide can be used as a substitute component. (Be careful; this one is more difficult than it appears!)

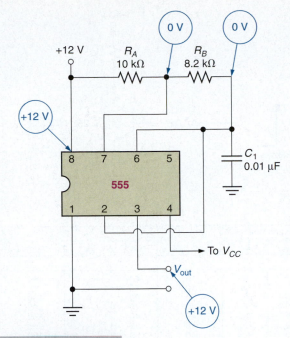

FIGURE 19.68

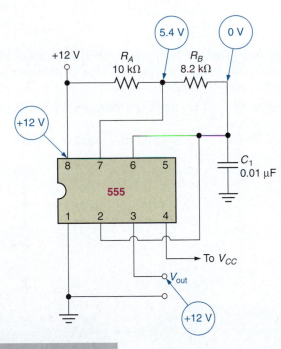

FIGURE 19.69

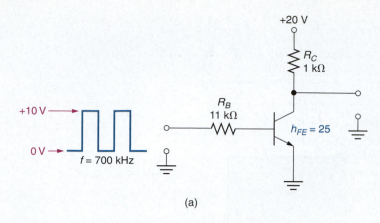

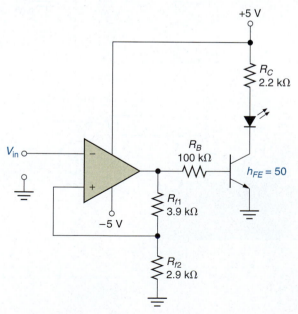

| Device | Marking | Switching Time (ns) | | $V_{(BR)CEO}$ | $h_{FE}$ | | | $f_T$ |
| | | $t_{on}$ | $t_{off}$ | | Min | Max | @ $I_{C (mA)}$ | Min (MHz) |
|---|---|---|---|---|---|---|---|---|
| **NPN** | | | | | | | | |
| MMBT2369 | 1J | 12 | 18 | 15 | 20 | — | 100 | — |
| BSV52 | B2 | 12 | 18 | 12 | 40 | 120 | 10 | 400 |
| MMBT2222 | 1B | 35 | 385 | 30 | 30 | — | 500 | 250 |
| MMBT2222A | 1P | 35 | 385 | 40 | 40 | — | 500 | 200 |
| MMBT4401 | 2X | 35 | 255 | 40 | 40 | — | 500 | 250 |
| MMBT3903 | 1Y | 70 | 225 | 40 | 15 | — | 100 | 250 |
| MMBT3904 | 1A | 70 | 250 | 40 | 30 | — | 100 | 200 |
| **PNP** | | | | | | | | |
| MMBT3638A | BN | 75 | 170 | 25 | 20 | — | 300 | — |
| MMBT3638 | AM | 75 | 170 | 25 | 20 | — | 300 | — |
| MMBT3640 | 2J | 25 | 35 | 12 | 20 | — | 50 | 500 |
| MMBT4403 | 2T | 35 | 225 | 40 | 90 | 180 | 1 | 150 |
| MMBT2907 | 2B | 45 | 100 | 40 | 30 | — | 500 | 200 |
| MMBT2907A | 2F | 45 | 100 | 60 | 50 | — | 500 | 200 |
| MMBT3906 | 2A | 70 | 300 | 40 | 100 | 300 | 10 | 250 |

(b)

**FIGURE 19.70**

**40.** Using a 555 timer, design a free-running multivibrator that produces a 50 kHz output with a 60% duty cycle. The supply voltage for the timer is $+10$ V, and it should have an inactive reset input pin.

**41.** Refer to Figure 19.71. The circuit was intended (by design) to fulfill the following requirements:

  **a.** The LED should light when a positive-going input signal passes a UTP value of $+1.7$ V.

**FIGURE 19.71**

**b.** The LED should turn off when a negative-going input signal passes an LTP of −1.7 V.

There is a flaw in the circuit design. Find the flaw, and suggest a correction that will make the circuit work properly.

**42.** Write a program that will determine the UTP and LTP values for both of the following:
  **a.** A noninverting Schmitt trigger like the one in Figure 19.25.
  **b.** An inverting Schmitt trigger like the one in Figure 19.27.

**43.** Write a program that will determine the PRR, duty cycle, and pulse-width values for a free-running multivibrator.

**19.1**  7.99 V

**19.2**  When $V_{GS} = 0$ V, $V_{out} = 1$ V. When $V_{GS} = -4$ V, $V_{out} = 10$ V.

**19.3**  7 V

**19.4**  PW = 6 μs, $T_C = 18$ μs

**19.5**  33.3%

**19.6**  $f_{C2} = 14$ MHz, $f_{max} = 140$ kHz

**19.7**  UTP = +4 V, LTP = −4 V

**19.8**  UTP = +1.33 V, LTP = −1.9 V

**19.9**  UTP = +2.25 V, LTP = −2.25 V

**19.10**  29.7 ms

**19.11**  $V_T < 5.625$ V

**19.12**  $f_0 = 6.89$ kHz, duty cycle = 76.8%, PW = 111.3 μs

## chapter 20

# Thyristors and Optoelectronic Devices

## Objectives

*After studying the material in this chapter, you should be able to:*

1. Describe the operation of the *silicon unilateral switch* (*SUS*).

2. List and describe the methods commonly used to drive an SUS into cutoff.

3. Describe the construction of the *silicon-controlled rectifier* (*SCR*) and the methods used to trigger the device into conduction.

4. Discuss the problem of SCR *false triggering* and the methods commonly used to prevent it.

5. Discuss the use of the SCR as a *crowbar circuit* and a *phase controller*.

6. Discuss the similarities and differences between *diacs* and *triacs*.

7. List and describe the circuits commonly used to control triac triggering.

8. Describe the construction and operation of *unijunction transistors* (*UJTs*).

9. Compare and contrast *light emitters* and *light detectors*.

10. Discuss *wavelength* and *light intensity*.

11. Describe the operation and critical parameters of the basic discrete photodetectors.

12. Discuss the operation and applications of *optoisolators* and *optointerrupters*.

## Outline

In this chapter, we will take a look at two independent groups of electronic devices. First, we will discuss a group of devices called *thyristors*. **Thyristors** are devices designed specifically for high-power switching applications. Unlike BJTs, FETs, and op-amps (which can also be used as switches), thyristors are not designed to be used as linear amplifying devices.

Second, we will discuss a group known as **optoelectronic devices**. These devices are controlled by or emit (generate) light. In Chapter 2, we discussed the most basic of the optoelectronic devices, the *light-emitting diode*, or *LED*. In this chapter, we will discuss some other devices that belong to this group.

No central theme ties thyristors and optoelectronic devices together. These two groups of components are covered in the same chapter simply because it is convenient to do so.

**Thyristors**
Devices designed specifically for high-power switching applications.

**Optoelectronic devices**
Devices that are controlled by or emit (generate) light.

## 20.1 Introduction to Thyristors: The Silicon Unilateral Switch (SUS)

**OBJECTIVE 1** ▶

The simplest of the thyristors is the **silicon unilateral switch (SUS)**. The SUS is a two-terminal, four-layer device that can be triggered into conduction by applying a specified forward voltage across its terminals. The construction of the SUS (along with its schematic symbol) is shown in Figure 20.1. As you can see, the device contains two *n*-type regions and two *p*-type regions. Because of its construction, the SUS is often referred to as a *pnpn diode* or a *four-layer diode*. Note that the term *diode* is considered to be appropriate because the device has two terminals: an *n*-type cathode (K) and a *p*-type anode (A).

The SUS is actually known by a variety of names. In addition to those listed above, the device is sometimes referred to as a *Schockley diode*, a *current latch*, or a *reverse blocking diode thyristor* (the JEDEC designated name).

**Silicon unilateral switch (SUS)**
A two-terminal, four-layer device that can be triggered into conduction by applying a specified forward voltage across its terminals.

### 20.1.1 SUS Operation

The overall operation of the SUS is easy to understand if we discuss it in terms of a two-transistor equivalent circuit. The derivation of this equivalent circuit is illustrated in Figure 20.2. If we split the SUS as shown in Figure 20.2a, we can see that the device is effectively made up of two transistors: one *pnp* ($Q_1$) and one *npn* ($Q_2$). Note that the collector of each transistor is connected to the base of the other.

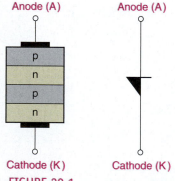

Anode (A)          Anode (A)

Cathode (K)        Cathode (K)

**FIGURE 20.1**
SUS construction and schematic symbol.

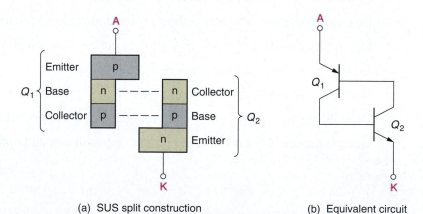

(a) SUS split construction          (b) Equivalent circuit

**FIGURE 20.2** SUS split construction and equivalent circuit.

**Forward breakover voltage ($V_{BR(F)}$)**
The value of forward voltage that forces an SUS into conduction.

Figure 20.3a illustrates the response of an *off* (nonconducting) SUS to an increase in supply voltage. Assuming that both transistors are in cutoff, the total forward current through the device ($I_F$) is approximately equal to zero, as is the total circuit current. Since there is no current through the series resistor ($R_S$), $V_{RS} = 0$ V (as shown in the figure). Note that these circuit conditions exist as long as $V_{AK}$ is less than the **forward breakover**

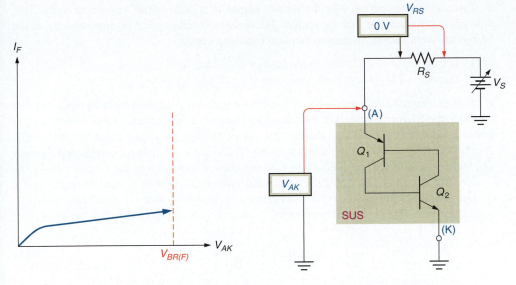

(a) As $V_S$ increases, $V_{AK}$ increases (from 0 V) until it reaches $V_{BR(F)}$.

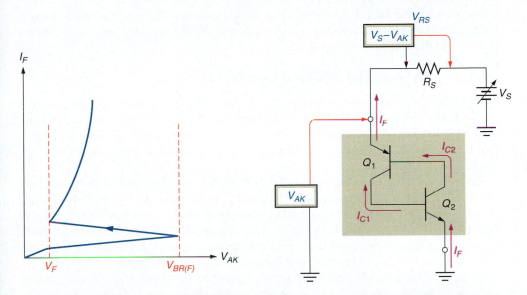

(b) When $V_{AK}$ reaches $V_{BR(F)}$, the SUS breaks over into full conduction.

**FIGURE 20.3** The response of an off (nonconducting) SUS to an increase in supply voltage.

**voltage ($V_{BR(F)}$)** rating of the SUS. This rating indicates the value of forward voltage that will force the device into conduction.

Figure 20.3b shows what happens when the value of $V_{AK}$ reaches the forward breakover voltage of the device. When this occurs, the two transistors break down and begin to conduct. As soon as conduction starts, both transistors in the SUS equivalent circuit are driven into *saturation*. This causes the voltage across the SUS to drop, as shown in the operating curve. Thus, when the SUS is forced into conduction, two things happen:

1. The device current ($I_F$) rapidly increases as the device is driven into saturation.
2. The value of $V_{AK}$ rapidly decreases because of the low resistance of the saturated device materials. (The relatively low value of $V_{AK}$ is identified as $V_F$ in the operating curve.)

As you can see, the SUS acts as an open circuit, dropping the applied voltage until $V_{BR(F)}$ is reached. At that time, the device breaks down and becomes a low-resistance conductor.

You may be wondering why the start of conduction causes both transistors in the SUS equivalent circuit to go into saturation. The answer is simple. If you look closely at the SUS equivalent circuit, you'll see that two conditions exist:

1. The collector of $Q_2$ is tied to the base of $Q_1$. Thus, $I_{B1} = I_{C2}$.
2. The base of $Q_2$ is tied to the collector of $Q_1$. Thus, $I_{B2} = I_{C1}$.

Now, consider what happens when $V_{AK}$ increases. As $V_{AK}$ increases, the leakage current through the SUS increases (as indicated by the curve in Figure 20.3a). When $V_{AK}$ is approximately equal to $V_{BR(F)}$, the current drawn through $Q_2$ is sufficient to turn on $Q_1$. As $Q_1$ begins to conduct, conduction through $Q_2$ increases, increasing conduction through $Q_1$, and so on. This cycle rapidly drives the component into conduction.

The fact that the SUS forces itself into saturation is important. You may recall that the current through a saturated device is controlled by external values of resistance and voltage. Thus, the forward current through the saturated SUS in Figure 20.3b is found as

$$I_F = \frac{V_S - V_{AK}}{R_S} \qquad (20.1)$$

where $V_{AK}$ is the forward voltage drop across the component. Equation (20.1) is important for several reasons. It provides us with the means of determining the forward current through a saturated SUS. It also shows that the SUS requires the use of a *series current-limiting resistor* ($R_S$). Without such a resistor, $I_F$ may exceed the rated limit on the device current, and the SUS may be destroyed.

### 20.1.2  Driving the SUS into Cutoff

**OBJECTIVE 2 ▶**

**Holding current ($I_H$)**
The minimum value of $I_F$ required to maintain conduction.

**Anode current interruption**
A method of driving an SUS into cutoff by breaking the diode current path or shorting the circuit current around the diode.

Once an SUS is driven into conduction, the device continues to conduct as long as $I_F$ is greater than a specified value, called the **holding current ($I_H$)**. The holding current rating of an SUS indicates the minimum forward current required to maintain conduction. Once $I_F$ drops below $I_H$, the device is driven back into cutoff and remains there until $V_{AK}$ reaches $V_{BR(F)}$ again.

Several methods are used to drive $I_F$ below the value of $I_H$. The first of these methods, called **anode current interruption**, is illustrated in Figure 20.4. In Figure 20.4a, a series switch is opened to interrupt the path for $I_F$, causing the device current to drop to zero. In Figure 20.4b, a shunt switch is closed to divert the circuit current ($I_T$) around the SUS, again causing the device current to drop to zero. In both cases, $I_F$ is forced to a value that is less than the $I_H$ rating of the SUS, and the device is driven into cutoff.

In practice, *series interruption* is generally caused by the *opening of a fuse*. As you'll see later in this chapter, thyristors are often used to protect voltage-sensitive loads from overvoltage conditions. When such a circuit is activated, the thyristor triggers and protects the load just long enough for a fuse to blow. The blown fuse then acts as the open switch in Figure 20.4a. *Shunt interruption* is also used for load protection. However, the switch in the shunt interruption circuit is normally a *reset* switch that is pressed when the overvoltage situation has passed. By closing the reset switch, the SUS is driven into cutoff, and normal circuit operation is restored.

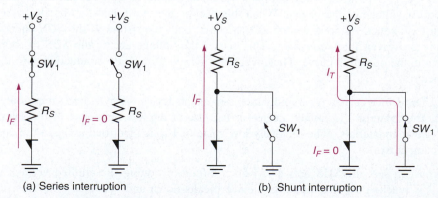

(a) Series interruption                    (b) Shunt interruption

FIGURE 20.4   Anode current interruption.

Another method for driving an SUS into cutoff is called **forced commutation**. This method involves applying a reverse voltage to the component, as shown in Figure 20.5. In this case, a positive pulse is applied to the cathode of the device. The reverse voltage causes the device current to decrease. When $I_F < I_H$, the SUS turns off.

**Forced commutation**
Driving an SUS into cutoff by applying a reverse voltage to the device. The SUS turns off when $I_F < I_H$.

### 20.1.3 SUS Specifications

Figure 20.6 shows the operating curve of the SUS. As you can see, the reverse region of the curve is nearly identical to that of the *pn*-junction diode and the zener diode. The part of the curve that falls between 0 V and the *reverse breakdown voltage* ($V_{BR(R)}$) is called the *reverse blocking region*. When the SUS is operating in this region, it has the same characteristics as a reverse-biased *pn* junction; that is, reverse current ($I_R$) is in the low μA range and the reverse voltage across the device ($V_R$) equals the applied voltage. If the reverse voltage exceeds the value of $V_{BR(R)}$, the device breaks down and conducts in the reverse direction. Like the *pn*-junction diode, the SUS is normally damaged or destroyed by this reverse conduction.

The forward operating curve of the SUS is enlarged in Figure 20.6b. Forward operation is divided into two regions. The **forward blocking region** (shaded) is the area of the curve that illustrates the *off-state* (nonconducting) operation of the device. The **forward operating region** is the area of the curve that illustrates the *on-state* (conducting) operation of the device.

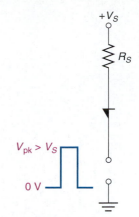

**FIGURE 20.5** Forced commutation.

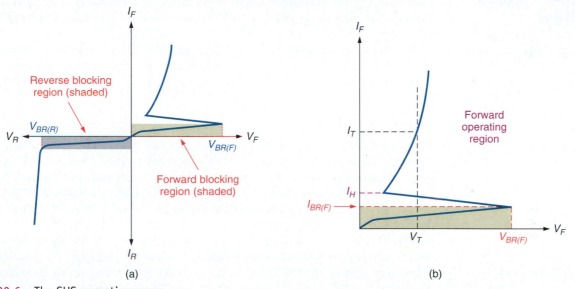

(a)            (b)

**FIGURE 20.6** The SUS operating curve.

The forward blocking region is defined by two parameters: $V_{BR(F)}$ and $I_{BR(F)}$. As you know, $V_{BR(F)}$ is the value of forward voltage that causes the SUS to break down and conduct in the forward direction. The **forward breakover current ($I_{BR(F)}$)** rating of the SUS is the value of $I_F$ at the point where breakover occurs. For example, the ECG6404 has ratings of $V_{BR(F)} = 10$ V (maximum) and $I_{BR(F)} = 500$ μA. This means that the forward current through the device is approximately 500 μA when $V_F$ reaches the breakover rating of 10 V.

The importance of the $I_{BR(F)}$ rating is illustrated in Figure 20.7. The curve shown is for an SUS with an anode-to-cathode voltage ($V_{AK}$) that is lower than the forward breakover rating of the device. As temperature increases, the leakage current through the device also increases. When the temperature reaches 150°C, the leakage current reaches the value of $I_{BR(F)}$ and the device is triggered into its on-state despite the fact that $V_{AK} < V_{BR(F)}$. Thus, *an SUS can be triggered into conduction by a significant increase in operating temperature.*

Once an SUS turns on, the device enters the forward operating region of the curve. The holding current ($I_H$) is the minimum forward current required to maintain the on-state condition. When $I_F$ falls below the value of $I_H$, the device returns to its off-state, as was stated earlier in this section.

**Forward blocking region**
The forward *off-state* (nonconducting) region of operation.

**Forward operating region**
The forward *on-state* (conducting) region of operation.

**Forward breakover current ($I_{BR(F)}$)**
The value of $I_F$ at the point where breakover occurs.

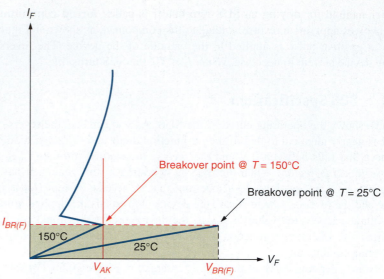

**FIGURE 20.7** The effects of temperature on SUS forward operation.

**Average on-state current ($I_T$)**
The maximum average (dc) forward current.

**Average on-state voltage ($V_T$)**
The value of $V_F$ when $I_F = I_T$.

The **average on-state current ($I_T$)** is the maximum average forward current for the SUS. The series current-limiting resistor for an SUS is selected to ensure that the dc forward current through the device is held below this value. The **average on-state voltage ($V_T$)** rating is the value of $V_F$ when $I_F = I_T$.

In addition to the ratings discussed here, the SUS has the standard maximum power dissipation and voltage ratings. Since we have covered these parameters for other components, we will not discuss them here.

### 20.1.4 Summary

The *silicon unilateral switch* (*SUS*) is a thyristor that is forced into conduction when the forward voltage across the device reaches a specified *forward breakover voltage* ($V_{BR(F)}$). Once triggered, the device becomes a low-resistance conductor. It remains in this *on-state* until its forward current ($I_F$) drops below the holding current ($I_H$) rating for the device. At that time, the component returns to the *off-state* (nonconducting) region of operation.

The SUS can be driven from the on-state to the off-state by one of two methods. *Anode current interruption* involves blocking or diverting the diode current so that $I_F$ falls below the value of $I_H$. *Forced commutation* involves applying a reverse voltage to the device to drop the value of $I_F$ below the value of $I_H$. In either case, the device is driven into its off-state and remains there until triggered again.

The SUS is rarely used in modern circuit design. However, you will see in the next two sections that many of the commonly used thyristors are nothing more than variations on the basic SUS. For this reason, it is important that you understand the SUS operating principles and ratings.

**Section Review ▶**

1. What are *thyristors*?
2. What distinguishes thyristors from other switching devices, such as BJTs, FETs, and op-amps?
3. What are *optoelectronic* devices?
4. What is the *silicon unilateral switch* (*SUS*)?
5. Describe the construction of the SUS.
6. What is forward breakover voltage ($V_{BR(F)}$)?
7. Describe what happens when the voltage applied to an SUS reaches the $V_{BR(F)}$ of the device.

8. Describe the two methods of *anode current interruption*.

9. Explain how *forced commutation* drives an SUS into cutoff.

10. What is the *forward blocking region* of an SUS?

11. What is the *forward operating region* of an SUS?

12. Explain how temperature can drive an SUS into its forward operating region.

## 20.2 Silicon-Controlled Rectifiers (SCRs)

The **silicon-controlled rectifier (SCR)** is a three-terminal device that is very similar in construction and operation to the SUS. The construction of the SCR is shown in Figure 20.8, along with its schematic symbol. As you can see, the construction of the SCR is identical to that of the SUS, except for the addition of a third terminal, called the *gate*. As you will see, the gate provides an additional means of triggering the device into the on-state. Except for that, the operation of the SCR is identical to that of the SUS.

◀ OBJECTIVE 3

> **Silicon-controlled rectifier (SCR)**
> A three-terminal device very similar in construction and operation to the SUS. The third terminal, called the *gate*, provides an additional method for triggering the device.

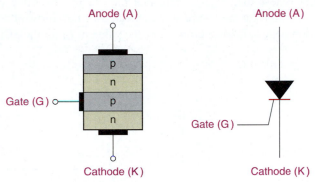

FIGURE 20.8  SCR construction and schematic symbol.

### 20.2.1 SCR Triggering

Figure 20.9a shows the two-transistor equivalent circuit for the SCR. Again, note the similarities between the SCR and the SUS. Like the SUS, the SCR can be triggered into conduction by applying an anode-to-cathode voltage ($V_{AK}$) that meets or exceeds the $V_{BR(F)}$ rating of the device.

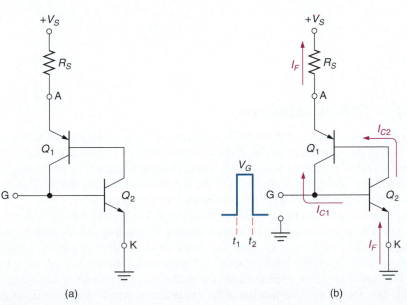

FIGURE 20.9

The addition of the gate terminal provides an additional means of triggering the device, as shown in Figure 20.9b. Here, we have a positive pulse applied to the gate terminal. When the gate pulse goes positive ($t_1$), $Q_2$ is forced into conduction. At that point, the device takes itself into saturation, just like the SUS.

When the gate pulse is removed ($t_2$), the device continues its forward conduction. The forward conduction continues until the *forward current* ($I_F$) drops below the *holding current* ($I_H$) rating of the device. Note that the SCR is driven into the off-state using the same two methods we use for the SUS: *anode current interruption* or *forced commutation*.

## 20.2.2    SCR Operating Curve

Figure 20.10a shows the operating curve for the SCR. This curve is identical to the SUS operating curve in Figure 20.6a. The enlarged forward operating curve in Figure 20.10b shows the effect of gate current on forward conduction. When $I_G = 0$ mA, the device breaks over at $V_{AK} = V_{BR(F)}$. When $I_G > 0$ mA, the device breaks over at some value of $V_{AK}$ that is less than $V_{BR(F)}$. Note that the greater the value of $I_G$, the lower the value of $V_{AK}$ required for the component to break over into forward conduction. Once forward conduction begins, $I_F$ must drop below the value of $I_H$ for the SCR to return to its off-state.

**Lab Reference:** The forward operating characteristics of the SCR are demonstrated in Exercise 34.

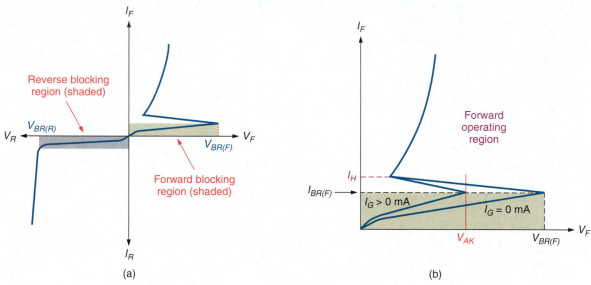

FIGURE 20.10    The SCR operating curve.

## 20.2.3    SCR Specifications

Many of the specifications for the SCR are identical to those for the SUS. For example, take a look at the spec sheet for the 2N6400 series SCRs in Figure 20.11. Table 20.1 lists the specifications contained in this spec sheet that have already been introduced. One important point: As shown in Table 20.1, the forward and reverse blocking voltages and currents are equal in magnitude. The same holds true for the SUS.

In addition to the specifications listed in Table 20.1, several other parameters warrant discussion. The first of these is the **circuit fusing ($I^2t$)** rating. This rating indicates the *maximum forward surge current capability* of the device. For example, the circuit fusing ($I^2t$) rating of the 2N6400 series SCRs is 145 ampere-square-seconds ($A^2s$). If the product of the square of surge current times the duration (in seconds) of the surge exceeds 145 $A^2s$, the device may be destroyed by excessive power dissipation. Example 20.1 demonstrates the determination of $I^2t$ for a given circuit.

**Circuit fusing ($I^2t$)**
A rating that indicates the maximum forward surge current capability of an SCR.

# 2N6400 Series

**Preferred Device**

# Silicon Controlled Rectifiers

## Reverse Blocking Thyristors

Designed primarily for half-wave ac control applications, such as motor controls, heating controls, and power supplies; or wherever half-wave silicon gate-controlled, solid-state devices are needed.

- Glass Passivated Junctions with Center Gate Geometry for Greater Parameter Uniformity and Stability
- Small, Rugged Thermowatt Construction for Low Thermal Resistance, High Heat Dissipation, and Duribility
- Blocking Voltage to 800 Volts
- Device Marketing: Logo, Device type, e.g., 2N6400, Date Code

## ON Semiconductor™

**http://onsemi.com**

**SCRs**
**16 AMPERES RMs**
**50 thru 800 VOLTS**

**MARKING DIAGRAM**

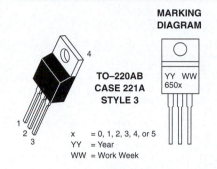

**TO–220AB CASE 221A STYLE 3**

YY WW
650x

x = 0, 1, 2, 3, 4, or 5
YY = Year
WW = Work Week

### MAXIMUM RATINGS ($T_J = 25°C$ unless otherwise noted)

| Rating | Symbol | Value | Unit |
|---|---|---|---|
| *Peak Repetitive Off-State Voltage (Note 1.) ($T_J = -40$ to $125°C$, Sine Wave 50 to 60 Hz; Gate Open) | $V_{DRM}$, $V_{RRM}$ | | Volts |
|     2N6400 | | 50 | |
|     2N6401 | | 100 | |
|     2N6402 | | 200 | |
|     2N6403 | | 400 | |
|     2N6404 | | 600 | |
|     2N6405 | | 800 | |
| On-State RMS Current (180° Conduction Angles; $T_C = 100°C$) | $I_{T(RMS)}$ | 16 | A |
| Average On-State Current (180° Conduction Angles; $T_C = 100°C$) | $I_{T(AV)}$ | 10 | A |
| Peak Non-Repetitive Surge Current (1/2 Cycle, Sine Wave 60 Hz, $T_J = 90°C$) | ITSM | 160 | A |
| Circuit Fusing (t = 8.3 ms) | $I^2t$ | 145 | $A^2s$ |
| Forward Peak Gate Power (Pulse Width ≤ 1.0 µs, $T_C = 100°C$) | $P_{GM}$ | 20 | Watts |
| Forward Average Gate Power (t = 8.3 ms, $T_C = 100°C$) | $P_{G(AV)}$ | 0.5 | Watts |
| Forward Peak Gate Current (Pulse Width ≤ 1.0 µs, $T_C = 100°C$) | $I_{GM}$ | 2.0 | A |
| Operating Junction Temperature Range | $T_J$ | −40 to +125 | °C |
| Storage Temperature Range | $T_{stg}$ | −40 to +150 | °C |

### PIN ASSIGNMENT

| 1 | Cathode |
|---|---|
| 2 | Anode |
| 3 | Gate |
| 4 | Anode |

*Indicates JEDEC Registered Data

1. $V_{DRM}$ and $V_{RRM}$ for all types can be applied on a continuous basis. Ratings apply for zero or negative gate voltage; however, positive gate voltage shall not be applied concurrent with negative potential on the anode. Blocking voltages shall not be tested with a constant current source such that the voltage ratings of the devices are exceeded.

(a)

**FIGURE 20.11** The 2N6400 specification sheet. (Copyright of Semiconductor Component Industries, LLC. Used by permission.)

## TABLE 20.1

| Specification | Symbol | Definition |
|---|---|---|
| Peak repetitive off-state voltage | $V_{DRM}$ or $V_{RRM}$ | The maximum forward or reverse blocking voltage. The same thing as $V_{BR(F)}$ and $V_{BR(R)}$. |
| Peak repetitive forward or reverse blocking current | $I_{DRM}$ or $I_{RRM}$ | The maximum forward or reverse blocking current. The same thing as $I_{BR(F)}$ and $I_{BR(R)}$. |
| Holding current | $I_H$ | The minimum forward current required to maintain conduction. |

$V_{RRM}$ is listed under the *maximum ratings* heading on the spec sheet. $I_{RRM}$ and $I_H$ are listed under the *electrical characteristics* heading.

**THERMAL CHARACTERISTICS**

| Characteristic | Symbol | Max | Unit |
|---|---|---|---|
| Thermal Resistance, Junction to Case | $R_{\theta JC}$ | 1.5 | °C/W |
| Maximum Lead Temperature for Soldering Purposes 1/8″ from Case for 10 Seconds | $T_L$ | 260 | °C |

**ELECTRICAL CHARACTERISTICS** ($T_J$ = 25°C unless otherwise noted.)

| Characteristic | | Symbol | Min | Typ | Max | Unit |
|---|---|---|---|---|---|---|
| **OFF CHARACTERISTICS** | | | | | | |
| *Peak Repetitive Forward or Reverse Blocking Current ($V_{AK}$ = Rated $V_{DRM}$ or $V_{RRM}$, Gate Open) | $T_J$ = 25°C | $I_{DRM}$, $I_{RRM}$ | — | — | 10 | µA |
| | $T_J$ = 125°C | | — | — | 2.0 | mA |
| **ON CHARACTERISTICS** | | | | | | |
| *Peak Forward On-State Voltage ($I_{TM}$ = 32 A Peak, Pulse Width ≤ 1 ms, Duty Cycle ≤ 2%) | | $V_{TM}$ | — | — | 1.7 | Volts |
| *Gate Trigger Current (Continuous dc) ($V_{AK}$ = 12 Vdc, $R_L$ = 100 Ohms) | $T_C$ = 25°C | $I_{GT}$ | — | 9.0 | 30 | mA |
| | $T_C$ = −40°C | | — | — | 60 | |
| *Gate Trigger Voltage (Continuous dc) ($V_D$ = 12 Vdc, $R_L$ = 100 Ohms, $T_C$ = −40°C) | | $V_{GT}$ | | | | Volts |
| | $T_C$ = 25°C | | — | 0.7 | 1.5 | |
| | $T_C$ = −40°C | | — | — | 2.5 | |
| Gate Non-Trigger Voltage ($V_D$ = 12 Vdc, $R_L$ = 100 Ohms) | $T_C$ = +125°C | $V_{GD}$ | 0.2 | — | — | Volts |
| *Holding Current ($V_D$ = 12 Vdc, Initiating Current = 200 mA, Gate Open) | $T_C$ = 25°C | $I_H$ | — | 18 | 40 | mA |
| | *$T_C$ = −40°C | | — | — | 60 | |
| Turn-On Time ($I_{TM}$ = 16 A, $I_{GT}$ = 40 mAdc, $V_D$ = Rated $V_{DRM}$) | | $t_{gt}$ | — | 1.0 | — | µs |
| Turn-Off Time ($I_{TM}$ = 16 A, $I_R$ = 16 A, $V_D$ = Rated $V_{DRM}$) | | $t_q$ | | | | µs |
| | $T_C$ = 25°C | | — | 15 | — | |
| | $T_J$ = +125°C | | — | 35 | — | |
| **DYNAMIC CHARACTERISTICS** | | | | | | |
| Critical Rate–of–Rise of Off-State Voltage ($V_D$ = Rated $V_{DRM}$, Exponential Waveform) | | dv/dt | — | 50 | — | V/µs |

*Indicates JEDEC Registered Data

(b)

FIGURE 20.11  *(continued)*

## EXAMPLE 20.1

In a given circuit, the 2N6400 is subjected to a 50 A current surge that lasts for 18 ms. Determine whether this surge will destroy the device.

*Solution:*  The $I^2t$ value for the circuit is found as

$$I^2t = (50\ \text{A})^2(18\ \text{ms}) = (2500\ \text{A})(18\ \text{ms}) = \mathbf{45\ A^2s}$$

Since this value is well below the maximum rating of 145 $A^2$s, the device can easily handle the surge.

### PRACTICE PROBLEM 20.1

An SCR has a circuit fusing rating of 40 $A^2$s. Determine whether the device will survive a 60 A surge that lasts for 15 ms.

When the circuit fusing rating for an SCR is known, we can determine the maximum allowable duration of a surge with a known current value as follows:

$$t_{max} = \frac{I^2t\ (\text{rated})}{I_S^2} \tag{20.2}$$

where $I_S$ = the known value of surge current. Example 20.2 demonstrates the use of equation (20.2) in determining the maximum allowable duration (time) of a surge.

## EXAMPLE 20.2

The 2N1843A has a circuit fusing rating of 60 A²s. The device is used in a circuit where it could be subjected to a 100 A surge. Determine the limit on the duration of such a surge.

*Solution:* The maximum allowable duration of the surge is found as

$$t_{max} = \frac{I^2t \text{ (rated)}}{I_S^2} = \frac{60 \text{ A}^2\text{s}}{(100 \text{ A})^2} = \textbf{6 ms}$$

Thus, if the 100 A surge is present for more than 6 ms, the 2N1843A may not survive the current.

### PRACTICE PROBLEM 20.2

The C35 series SCRs each have a circuit fusing rating of 75 A²s. Determine the maximum allowable duration of a 150 A surge that passes through one of these devices.

We can also use a variation on equation (20.2) to determine the maximum allowable surge current value for a given period of time as follows:

$$I_{S(max)} = \sqrt{\frac{I^2t \text{ (rated)}}{t_S}} \qquad \textbf{(20.3)}$$

*A Practical Consideration:* If the value obtained using equation (20.3) exceeds the *peak non-repetitive surge current* ($I_{TSM}$) rating of the SCR, the surge will damage the component, regardless of its duration.

where $t_S$ = the time duration of the current surge. Example 20.3 demonstrates the use of equation (20.3) in determining the maximum allowable value of a current surge.

## EXAMPLE 20.3

Determine the highest surge current value that the 2N1843A can withstand for a period of 20 ms.

*Solution:* The peak value for the surge is found as

$$I_{S(max)} = \sqrt{\frac{I^2t \text{ (rated)}}{t_S}} = \sqrt{\frac{60 \text{ A}^2\text{s}}{20 \text{ ms}}} = \textbf{54.77 A}$$

The value of 60 A²s for the 2N1843A was given in Example 20.2.

Thus, a 20 ms current surge would have to exceed 54.77 A to damage the 2N1843A.

### PRACTICE PROBLEM 20.3

Refer to Practice Problem 20.2. Determine the highest surge current value that a C35 series SCR can withstand for a duration of 50 ms.

Later in this section, we will look at several SCR applications. In one of these applications, the SCR is used as a surge-protection device. When the SCR is used in a surge-protection application, the circuit fusing rating becomes critical.

The SCR has another maximum current rating called the **nonrepetitive surge current** ($I_{TSM}$) rating. This is the absolute limit on the surge current through the device. If this rated value is exceeded for *any length of time*, the device may be destroyed.

If you look under the *electrical characteristics* heading in Figure 20.11b, you'll see the **gate nontrigger voltage** ($V_{GD}$) rating. This rating indicates the maximum gate voltage that can be applied without triggering the SCR into conduction. If $V_G$ exceeds this rating, the SCR may be triggered into the on-state.

**Nonrepetitive surge current ($I_{TSM}$)**
The absolute limit on the forward surge current through an SCR.

**Gate nontrigger voltage ($V_{GD}$)**
The maximum gate voltage that can be applied without triggering the SCR into conduction.

The $V_{GD}$ rating is important because it points out one of the potential causes of **false triggering**. False triggering is a situation where the SCR is accidentally triggered into conduction, usually by some type of noise. For example, the 2N6400 series of SCRs has a $V_{GD}$ rating of 200 mV. If a noise signal with a peak value greater than 200 mV appears at the gate, the device may be triggered into conduction.

A more common cause of false triggering is a rise in anode voltage that exceeds the **critical rise rating (*dv/dt*)** of the SCR. This rating indicates the *maximum rate of increase in anode-to-cathode voltage that the SCR can handle without false triggering occurring*. For example, the *dv/dt* rating of the 2N6400 series SCRs is 50 V/µs. This means that a noise signal in $V_{AK}$ with a rate of rise equal to 50 V/µs may cause false triggering.

Obviously, 50 V of noise would never occur under normal circumstances. However, even a relatively small amount of noise can have a sufficient *rate of increase* to cause false triggering. To determine the rate of increase of a given noise signal in V/µs, the following conversion formula can be used:

$$\Delta V = \frac{dv}{dt} \Delta t \qquad (20.4)$$

where $\dfrac{dv}{dt}$ = the critical rise rating of the component

$\Delta t$ = the rise time of the increase in $V_{AK}$

$\Delta V$ = the amount of change in $V_{AK}$

As Example 20.4 demonstrates, even a small amount of noise in $V_{AK}$ can cause false triggering if the rise time of the noise signal is short enough.

### EXAMPLE 20.4

A 2N6400 SCR is used in a circuit where 2 ns (rise time) noise signals occur randomly. Determine the noise amplitude required to cause false triggering.

***Solution:*** With a critical rise rating of 50 V/µs, the amplitude that may cause false triggering is found as

$$\Delta V = \frac{dv}{dt} \Delta t = (50 \text{ V/µs})(2 \text{ ns}) = \textbf{100 mV}$$

As you can see, with a 2 ns rise time, only 100 mV of noise is required to cause false triggering.

### PRACTICE PROBLEM 20.4
The C35D SCR has a critical rise rating of 25 V/µs. Determine the amplitude of noise needed to cause false triggering when the noise signal rise time is 100 ps.

As you can see, many of the critical SCR parameters deal with the current limits and false triggering limits of the device. At this point, we'll move on to look at some methods commonly used to prevent false triggering.

## 20.2.4 Preventing False Triggering

You have been shown that false triggering is usually caused by one of two conditions:

**1.** A noise signal at the gate terminal.
**2.** A short rise-time noise signal in $V_{AK}$.

Noise from the gate signal source is generally reduced using one of the two methods shown in Figure 20.12. In Figure 20.12a, the gate is connected (via a gate resistor) to a negative dc power supply. For the sake of discussion, let's assume that the gate supply voltage is −2 V. If this is the case, the noise at the gate would have to be greater than (2 V + $V_{GD}$) to cause false triggering. This is not likely to occur under any normal circumstances.

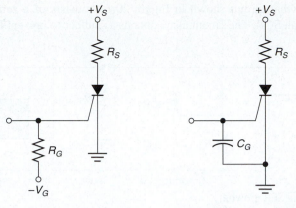

(a) Gate biasing voltage  (b) Gate bypass capacitor

FIGURE 20.12   Preventing false triggering.

The circuit in Figure 20.12b uses a bypass capacitor connected between the gate terminal and ground. This capacitor, which shorts any gate noise to ground, is normally between 0.1 and 0.01 µF. Of the two methods shown in Figure 20.12, the bypass capacitor in Figure 20.12b is by far the more commonly used because the capacitor provides a very inexpensive solution to the gate noise problem.

Noise in $V_{AK}$ is normally eliminated by the use of a **snubber network**. A snubber is an $RC$ circuit connected between the anode and cathode terminals of the SCR. A snubber network is shown in Figure 20.13. The capacitor ($C$) effectively shorts any noise in $V_{AK}$ to ground. The resistor ($R$) is included to limit the capacitor discharge current when the SCR turns on.

**Snubber network**
An $RC$ circuit that is connected between the $SCR$ anode and cathode to eliminate false triggering.

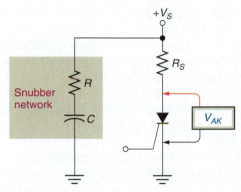

FIGURE 20.13   An $RC$ snubber network.

It should be noted that noise in $V_{AK}$ (Figure 20.13) would originate in the dc voltage source ($V_S$). Assuming that $V_S$ comes from a standard dc power supply, any noise generated in the supply is coupled to the SCR circuit. This noise is reduced by the snubber to a level that prevents it from causing false triggering.

False triggering is most commonly caused by exceeding the critical rise ($dv/dt$) rating of a given SCR. For this reason, snubber networks are included in most SCR circuits.

## 20.2.5   SCR Applications

The SCR is the most commonly used of the thyristors. Most SCR applications are found in the area of **industrial electronics**, which deals with the devices, circuits, and systems used in manufacturing. Although we cannot cover all the possible applications for the SCR, we will touch briefly on a couple of them in this section.

◄ *OBJECTIVE 5*

**Industrial electronics**
The area of electronics that deals with the devices, circuits, and systems used in manufacturing.

## 20.2.6   The SCR Crowbar

One common application for the SCR is in a type of circuit called a **crowbar**. A crowbar is a circuit used to protect a voltage-sensitive load from excessive dc power supply output

**Crowbar**

A circuit used to protect a voltage-sensitive load from excessive dc power supply output voltages.

voltages. The crowbar circuit shown in Figure 20.14 consists of a zener diode, a gate resistor ($R_G$), and an SCR. The circuit also contains a snubber to prevent false triggering.

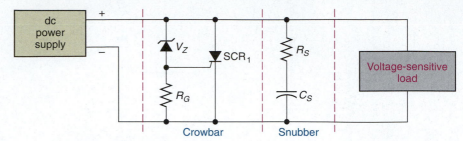

**FIGURE 20.14** The SCR crowbar.

In the crowbar shown, the zener diode and the SCR are normally off. With the zener diode in cutoff, there is no path for current through $R_G$, and no voltage is developed across the resistor. With no voltage developed across $R_G$, the gate of the SCR is at 0 V, holding the device in the off-state. As long as the zener diode is off, the SCR is acting as an open and does not affect either the dc power supply or the load.

The SCR in a crowbar is turned off by a blown primary fuse in the dc power supply. This is an example of *series anode current interruption*.

If the power supply output voltage *increases* beyond a certain point, the zener diode conducts. The current through the zener diode causes a voltage to be developed across $R_G$. This voltage forces the SCR into conduction. When the SCR conducts, the output from the dc power supply is shorted through the device and the load is protected.

Several points should be made regarding this circuit:

*A Practical Consideration:*
The $I^2t$ rating of the SCR is especially critical when the dc power supply contains a slow-blow fuse.

1. The SCR would be activated only by some major fault within the dc power supply that causes its output voltage to increase drastically. Once activated, the SCR continues to conduct until the power supply fuse opens.
2. The *circuit fusing* ($I^2t$) rating of the SCR must be high enough to ensure that the SCR isn't destroyed before the power supply fuse has time to blow. If the SCR gives out before the power supply fuse, the power supply will continue to operate and the load may be damaged after all.
3. The SCR in the crowbar is considered to be expendable, which is why there is no series current-limiting resistor in the circuit. The object of the circuit is to protect the load, not the SCR.
4. The SCR in a crowbar should be replaced after activation of the circuit as a precaution against future component failure from possible internal damage.

Circuit protection is not the only application for the SCR. A completely different type of application can be found in the *SCR phase controller*.

**Phase controller**

A circuit used to control the conduction angle through a load and, thus, the average load voltage.

**Conduction angle**

The portion of the input waveform (in degrees) that is coupled to the load.

**Firing angle**

The point on the input waveform (in degrees) at which the SCR triggers, allowing load conduction to begin.

### 20.2.7 The SCR Phase Controller

A **phase controller** is a circuit that is used to control the *conduction angle* through a load and, thus, the average load voltage. For example, by varying the setting of the potentiometer ($R_2$) in the phase controller in Figure 20.15a, we can vary the conduction angle (θ) through the load as shown in Figure 20.15b. The **conduction angle** of a phase controller indicates the portion of the input waveform (in degrees) that is coupled to the load. For example, the load ($v_L$) waveforms in Figure 20.15b show three conduction angles as shaded areas. When the conduction angle (θ) is 180°, the entire alternation is coupled to the load. When the conduction angle is 90°, half the alternation is coupled to the load, and so on. Note that the point on the waveform at which the SCR triggers (allowing load conduction to begin) is often referred to as the **firing angle**. For example, assume the middle $v_L$ waveform represents a phase controller that allows conduction to begin at 45°. In this case, the firing angle is 45°, and the conduction angle (θ) is found as θ = 180° − 45° = 135°.

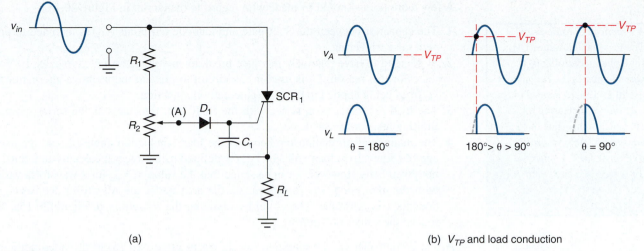

(a)

(b) $V_{TP}$ and load conduction

FIGURE 20.15    The SCR phase controller.

Conduction through the load in Figure 20.15a is controlled by the SCR. When $v_{in}$ causes the voltage at point A to reach a specified level, called the **trigger-point voltage** ($V_{TP}$), the SCR is triggered into conduction. Once triggered, the SCR allows $v_{in}$ to pass through to the load and continues to conduct for the remainder of the positive alternation of $v_{in}$. The relationship among the voltage at point A, the value of $V_{TP}$, and load conduction can be seen in the three waveforms in Figure 20.15b.

For the SCR to be triggered into conduction, the voltage at point A must reach the level specified by the following equation:

$$V_{TP} = V_F + V_{GT}$$

where $V_F$ = the forward voltage drop across the gate diode ($D_1$)
 $V_{GT}$ = the **gate trigger voltage** rating of the SCR

The gate trigger voltage of the SCR is the value of $V_{GK}$ that will cause $I_G$ to reach a level sufficient to trigger the SCR into conduction.

When the voltage at point A reaches the value of $V_{TP}$, it is sufficient to overcome the forward voltage drop across $D_1$ and the value of $V_{GT}$ for the SCR. This triggers the SCR into conduction for the remainder of the positive alternation of $v_{in}$. When $v_{in}$ goes negative, the anode voltage of the SCR also goes negative. This causes the SCR current to drop below the holding current ($I_H$) value of the device, and it returns to the off-state.

An important point needs to be made regarding the circuit and waveforms shown in Figure 20.15: The waveforms shown in the figure are somewhat misleading. You see, the value of $V_{TP}$ for a given phase controller is a fixed value, while the amplitude of the sine wave at point A is variable. For example, let's say that the value of $V_{GT}$ for the SCR in Figure 20.15a is 1.5 V. This gives the circuit a trigger-point voltage found as

$$V_{TP} = V_F + V_{GT} = 0.7\text{ V} + 1.5\text{ V} = \textbf{2.2 V}$$

This is the voltage required at point A to trigger the SCR into conduction. The voltage divider ($R_1$ and $R_2$) applies a sine wave to point A that is *proportional to* $v_{in}$. As the setting of $R_2$ is increased, the amplitude of the sine wave at point A is also increased. Thus, by increasing the setting of $R_2$, we decrease the value of $v_{in}$ required to cause the voltage at point A to reach the value of $V_{TP}$. In the same fashion, decreasing the setting of $R_2$ increases the value of $v_{in}$ required to cause the voltage at point A to reach the value of $V_{TP}$. In other words, adjusting $R_2$ changes the amplitude of $V_A$, not the value of $V_{TP}$ for the circuit.

**Trigger-point voltage ($V_{TP}$)**
The voltage required at the anode of the gate diode to trigger the SCR.

**Lab Reference:** The operation of an SCR phase controller is demonstrated in Exercise 34.

**Gate trigger voltage ($V_{GT}$)**
The value of $V_{GK}$ that causes $I_G$ to reach a level sufficient to trigger the SCR into conduction.

Turning an SCR off by reversing the anode voltage polarity is an example of *forced commutation.*

A few more points should be made with regard to the circuit in Figure 20.15a:

A Practical Consideration:
The fact that the value of $V_{ave}$ for a phase controller is variable makes the circuit very useful as a *speed controller* for a dc motor. Increasing (or decreasing) the value of $V_{ave}$ across a dc motor causes its rotational speed to increase (or decrease). A phase controller is well suited for this type of application.

1. The capacitor between the SCR gate and cathode terminals ($C_1$) is included to prevent false triggering.
2. SCRs tend to have relatively low gate breakdown voltage ratings, typically 10 V or less. For this reason, $D_1$ is used in the circuit to prevent the negative alternation of $v_{in}$ from being applied to the SCR gate-cathode junction.
3. The load resistance is in series with the SCR and, thus, acts as the series current-limiting resistor for the device.
4. The mathematical relationship between the load conduction angle ($\theta$) and the average (dc equivalent) load voltage ($V_{ave}$) is defined using integral calculus and, thus, is not given here. However, it can be stated that the value of $V_{ave}$ for a waveform varies with *the area under the waveform*. As the area under the waveform decreases, so does its average value. Therefore, by reducing the $R_2$ setting in Figure 20.15a, we reduce the values of $\theta$ and $V_{ave}$.

A phase controller can be modified (as shown in Figure 20.16) to provide additional control over the firing angle. In this circuit, the added capacitor ($C_1$) forms a *delay circuit* with $R_1$ and $R_2$. By adjusting the setting of $R_2$, the firing angle can be set beyond the 90° point on the input waveform (which would not be possible otherwise).

A Practical Consideration:
SCR phase controllers are often used to control the rotating speed of a *dc motor*. In this application, a shorted SCR causes the motor to *vibrate rapidly* (rather than rotate).

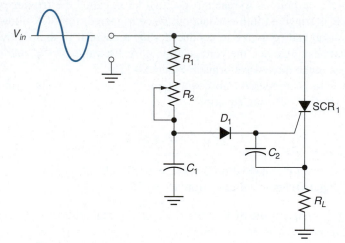

**FIGURE 20.16** Modified phase-angle controller.

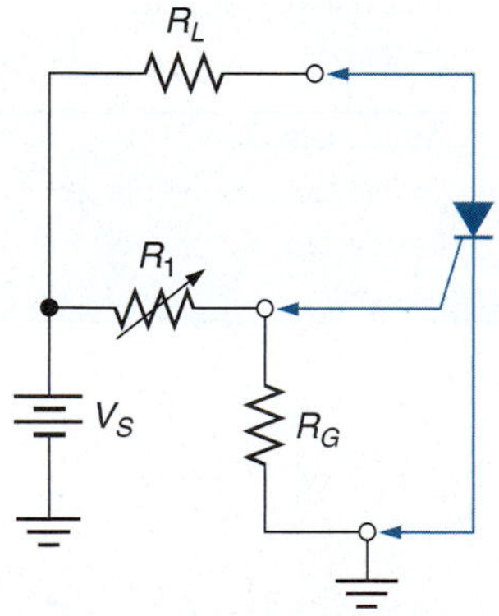

## 20.2.8  SCR Testing

Whenever possible, in-circuit testing on an SCR is performed. An open SCR remains in the nonconducting state when triggered. A shorted SCR usually conducts regardless of the polarity of the anode-cathode voltage ($V_{AK}$). In either case, the SCR fault can be diagnosed by checking the gate, anode, and cathode voltages of the device with a voltmeter or an oscilloscope.

In some circuits (like crowbars), it is impractical to perform any type of in-circuit tests on the SCR. In this case, the device must be tested using a circuit like the one shown in Figure 20.17. The circuit is initially set up so that $V_{GK}$ is approximately 0 V. The value of $R_L$ is selected to equal the value used in the $V_{GT}$ listing on the spec sheet of the device. For example, the 2N6400 series SCRs list a value of $R_L = 100\ \Omega$ for the testing of $V_{GT}$ (see the spec sheet in Figure 20.11). Thus, when testing one of these SCRs, $R_L$ would be a 100 $\Omega$ resistor. The value of $V_S$ is set to equal the sum of $V_{BR(F)}$ and $I_H R_L$ (as shown in the figure).

To test the SCR, the setting of $R_1$ is decreased until $V_G$ equals the value of $V_{GT}$. At this point, the SCR should trigger, and $V_{AK}$ should drop to a relatively low value. If the device fails to trigger, $R_1$ is varied across its entire range. If the device continues in the nonconducting state, it is open and must be replaced.

$V_S = V_{BR(F)} + I_H R_L$

**FIGURE 20.17** An SCR test circuit.

## 20.2.9 Summary

The SCR is a three-terminal device that acts as a gated SUS. The device is normally triggered into conduction by applying a gate-cathode voltage ($V_{GK}$) that causes a specific level of gate current ($I_G$). This gate current triggers the device into conduction.

Like the SUS, the SCR is returned to its nonconducting state by either *anode current interruption* or *forced commutation*. When turned off, the device remains in its nonconducting state until another trigger is received.

Two critical SCR ratings are the *circuit fusing* and *critical rise* ratings. If the *circuit fusing* ($I^2t$) *rating* is exceeded, the device is destroyed by excessive power dissipation. If the *critical rise* ($dv/dt$) rating is exceeded by noise in the anode voltage, the device may experience *false triggering*. False triggering, a condition where the device is unintentionally triggered into conduction, can also be caused by excessive noise in the gate circuit biasing voltage. False triggering is normally prevented by the use of *snubber networks* and *gate bypass capacitors*.

SCR applications generally fall into one of two categories: load overvoltage protection and limiting the conduction through a load. The characteristics of the SCR are summarized in Figure 20.18.

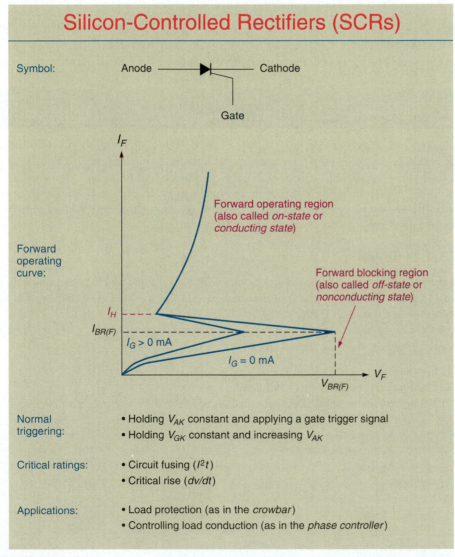

FIGURE 20.18

**Section Review** ▶

1. Describe the construction of the SCR.
2. What is the *forward blocking region* of the SCR?
3. What is the *forward operating region* of the SCR?
4. List the methods used to trigger an SCR.
5. List the methods used to force an SCR into cutoff.
6. Using the operating curve in Figure 20.10, describe the forward operation of the SCR.
7. What is the *circuit fusing* ($I^2t$) *rating* of an SCR? Why is it important?
8. What is *false triggering*?
9. Which SCR ratings are tied directly to false triggering problems?
10. How does a gate bypass capacitor prevent false triggering?
11. What is a *snubber network*? What is it used for?
12. What is the function of the SCR *crowbar*?
13. Describe the operation of the crowbar.
14. What is the function of the SCR *phase controller*?
15. Describe the operation of the SCR phase controller.
16. What are the typical in-circuit symptoms of *open* and *shorted* SCRs?

**Critical Thinking** ▶

17. In Figure 20.14, why not use a fuse in series with the load rather than the more expensive crowbar circuit?
18. Why would it be impractical to perform in-circuit tests on the SCR in a crowbar circuit?

## 20.3   Diacs and Triacs

**OBJECTIVE 6** ▶

**Bidirectional thyristor**
A thyristor capable of conducting in two directions.

*Diacs* and *triacs* are components that are classified as **bidirectional thyristors**. The term *bidirectional* means that they are capable of conducting in two directions. For all practical purposes, the diac can be viewed as a bidirectional SUS, and the *triac* can be viewed as a bidirectional SCR. Since these components are extremely similar in operation to the thyristors we have discussed, most of the material in this section will seem more like a review.

### 20.3.1   Diacs

**Diac**
A two-terminal, three-layer device with forward and reverse characteristics that are identical to the forward characteristics of the SUS.

The **diac** is a two-terminal, three-layer device with forward and reverse operating characteristics that are identical to the forward characteristics of the SUS. The diac symbol and construction are shown in Figure 20.19.

As you can see, the construction of the diac is extremely similar to that of the *npn* (or *pnp*) transistor. However, there are several crucial differences:

1. There is no base connection.
2. The three regions are nearly identical in size.
3. The three regions have nearly identical doping levels.

Diacs are often referred to as *bidirectional diodes*.

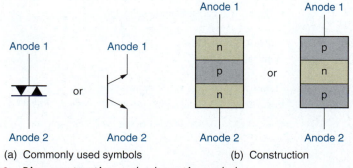

(a) Commonly used symbols          (b) Construction

**FIGURE 20.19**   Diac construction and schematic symbols.

In contrast, the BJT has an extremely narrow and lightly doped base (center) region, as you were shown in Chapter 6. Note that the diac can be constructed in either *npn* or *pnp* form. The operating principles are identical for the two.

The operating curve of the diac is shown in Figure 20.20. As you can see, the forward and reverse operating characteristics are identical to the forward characteristics of the SUS. In either direction, the device acts as an open until the breakover voltage ($V_{BR}$) is reached. At that time, the device is triggered and conducts in the appropriate direction. Conduction then continues until the device current drops below its holding current ($I_H$) value. Note that the breakover voltage and holding current values are usually identical for the forward and reverse regions of operation.

Both *anode current interruption* and *forced commutation* are used to drive a diac into cutoff.

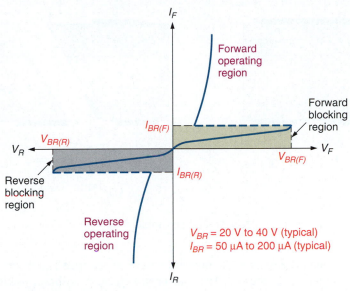

**FIGURE 20.20**   The diac operating curve.

A *Practical Consideration:* Most diacs have symmetrical forward and reverse curves (as shown here). However, some diacs have asymmetrical (unequal) forward and reverse curves.

The diac can be triggered into conduction by an increase in temperature. The most common application for the diac is as a triac triggering device in ac control circuits. This application will be discussed later in this section.

### 20.3.2   Triacs

The **triac** is a three-terminal, five-layer device with forward and reverse characteristics that are identical to the forward characteristics of the SCR. The triac symbol and construction are shown in Figure 20.21.

The primary conducting terminals are identified as *main terminal 1 (MT1)* and *main terminal 2 (MT2)*. Note that MT1 is the terminal on the same side of the device as the gate.

**Triac**
A bidirectional thyristor whose forward and reverse characteristics are identical to the forward characteristics of the SCR.

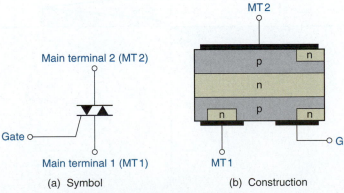

(a) Symbol          (b) Construction

**FIGURE 20.21**   Triac construction and schematic symbol.

Triacs are also referred to as *triodes* and *bidirectional triode thyristors* (the JEDEC designated name).

If we split the triac as shown in Figure 20.22, it is easy to see that we have complementary SCRs that are connected in parallel. When MT2 is *positive* and MT1 is *negative*, a triggered triac has the conduction path shown on the left in Figure 20.22b. When MT2 is *negative* and MT1 is *positive*, a triggered triac has the conduction path shown on the right. Note that the triggered triac conducts in the appropriate direction as long as the current through the device is greater than its *holding current* ($I_H$) rating.

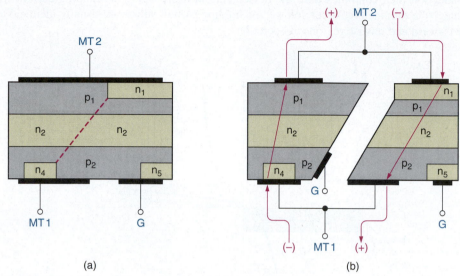

(a)                                             (b)

**FIGURE 20.22**   Triac split construction and currents.

**OBJECTIVE 7 ▶**

Because of its unique structure, the operating curve of the triac is a bit more complicated than any we have seen thus far. The triac operating curve is shown in Figure 20.23. The graph in Figure 20.23 is divided into four *quadrants*, labeled *I*, *II*, *III*, and *IV*. Each quadrant has the MT2 and G (gate) polarities shown. For example, quadrant I indicates the operation of the device when MT2 is positive (with respect to MT1) and a positive potential is applied to the gate. Quadrant III indicates the operation of the device when MT2 is negative (with respect to MT1) and a negative potential is applied to the gate, and so on.

The triac operating curve is normally plotted as shown in Figure 20.23 (to keep it as simple as possible). The curve for quadrants II and IV is simply a mirror image of the one shown.

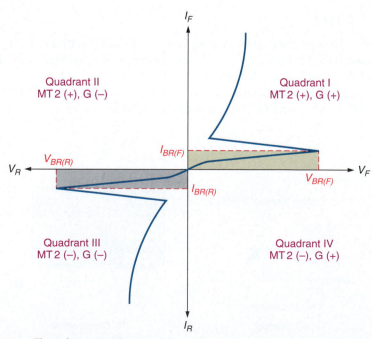

**FIGURE 20.23**   The triac operating curve.

Operation in quadrant I is identical to the operation of the SCR. All the SCR characteristics apply to the triac operated in the first quadrant. For example:

1. The device remains in the off-state until $V_{BR(F)}$ is reached or a positive gate trigger voltage ($V_{GT}$) generates a gate current ($I_G$) sufficient to trigger the device.
2. The device may be false triggered by increases in temperature, gate noise, or anode (MT2) noise.
3. Once triggered, the device continues to conduct until the forward current drops below its holding current ($I_H$) rating.

Note that when operated in quadrant I, the current through the triac is in the direction shown on the left side of Figure 20.22b.

## 2N6344, 2N6349

**Preferred Device**

## Triacs

### Silicon Bidirectional Thyristors

Designed primarily for full-wave ac control applications, such as light dimmers, motor controls, heating controls, and power supplies; or wherever full-wave silicon gate controlled solid-state devices are needed. Triac type thyristors switch from a blocking to a conducting state for either polarity of applied main terminal voltage with positive or negative gate triggering.

- Blocking Voltage to 800 Volts
- All Diffused and Glass Passivated Junctions for Greater Parameter Uniformity and Stability
- Small, Rugged, Thermowatt Construction for Low Thermal Resistance, High Heat Dissipation and Durability
- Gate Triggering Guaranteed in all Four Quadrants
- For 400 Hz Operation, Consult Factory
- Device Marking: Logo, Device Type, e.g., 2N6344, Date Code

**ON Semiconductor™**

**http://onsemi.com**

**TRIACS**
**8 AMPERES RMs**
**600 thru 800 VOLTS**

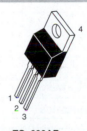

**TO–220AB**
**CASE 221A**
**STYLE 4**

| PIN ASSIGNMENT | |
|---|---|
| 1 | Main Terminal 1 |
| 2 | Main Terminal 2 |
| 3 | Gate |
| 4 | Main Terminal 2 |

**MAXIMUM RATINGS** ($T_J$ = 25°C unless otherwise noted)

| Rating | Symbol | Value | Unit |
|---|---|---|---|
| *Peak Repetitive Off-State Voltage (1) ($T_J$ = −40 to +110 °C, Sine Wave 50 to 60 Hz, Gate Open)       2N6344       2N6349 | $V_{DRM}$, $V_{RRM}$ | 600 800 | Volts |
| *On-State RMS Current ($T_C$ = +80°C) Full Cycle Sine Wave 50 to 60 Hz ($T_C$ = +90°C) | $I_{T(RMS)}$ | 8.0 4.0 | Amps |
| *Peak Non-Repetive Surge Current (One Full Cycle, Sine Wave 60 Hz, $T_C$ = +25°C) Preceded and followed by rated current | $I_{TSM}$ | 100 | Amps |
| Circuit Fusing Consideration (t = 8.3 ms) | $I^2 t$ | 40 | $A^2 s$ |
| *Peak Gate Power ($T_C$ = +80°C, Pulse Width = 2.0 μs) | $P_{GM}$ | 20 | Watts |
| *Average Gate Power ($T_C$ = +80°C, t = 8.3 ms) | $P_{G(AV)}$ | 0.5 | Watts |
| *Peak Gate Current ($T_C$ = +80°C, Pulse Width = 2.0 μs) | $I_{GM}$ | 2.0 | Amps |
| *Peak Gate Voltage ($T_C$ = +80°C, Pulse Width = 2.0 μs) | $V_{GM}$ | 10 | Volts |
| *Operating Junction Temperature Range | $T_J$ | −40 to +125 | °C |
| *Storage Temperature Range | $T_{stg}$ | −40 to +150 | °C |

1. $V_{DRM}$ and $V_{RRM}$ for all types can be applied on a continuous basis. Blocking voltages shall not be tested with a constant current source such that the voltage ratings of the devices are exceeded.

(a)

**FIGURE 20.24** The 2N6344 and 2N6349 triac specification sheet. (Copyright of Semiconductor Component Industries, LLC. Used by permission.)

## THERMAL CHARACTERISTICS

| Characteristic | Symbol | Max | Unit |
|---|---|---|---|
| *Thermal Resistance, Junction to Case | $R_{\theta JC}$ | 2.2 | °C/W |
| Maximum Lead Temperature for Soldering Purposes 1/8″ from Case for 10 Seconds | $T_L$ | 260 | °C |

## ELECTRICAL CHARACTERISTICS ($T_J$ = 25°C unless otherwise noted; Electricals apply in both directions)

| Characteristic | Symbol | Min | Typ | Max | Unit |
|---|---|---|---|---|---|
| **OFF CHARACTERISTICS** | | | | | |
| *Peak Repetitive Blocking Current ($V_D$ = Rated $V_{DRM}$, $V_{RRM}$, Gate Open)     $T_J$ = 25°C | $I_{DRM}$ $I_{RRM}$ | — | — | 10 | μA |
| $T_J$ = 100°C | | — | — | 2.0 | mA |
| **ON CHARACTERISTICS** | | | | | |
| *Peak On–State Voltage ($I_{TM}$ = ± 11 A Peak; Pulse Width = 1 to 2 ms, Duty Cycle ≤ 2%) | $V_{TM}$ | — | 1.3 | 1.55 | Volts |
| Gate Trigger Current (Continuous dc) ($V_D$ = 12 Vdc, $R_L$ = 100 Ohms) | $I_{GT}$ | | | | mA |
| MT2(+), G(+) | | — | 12 | 50 | |
| MT2(+), G(−) | | — | 12 | 75 | |
| MT2(−), G(−) | | — | 20 | 50 | |
| MT2(−), G(+) | | — | 35 | 75 | |
| *MT2(+), G(+); MT2(−), G(−) $T_C$ = −40°C | | — | — | 100 | |
| *MT2(+), G(−); MT2(−), G(+) $T_C$ = −40°C | | — | — | 125 | |
| Gate Trigger Voltage (Continuous dc) ($V_D$ = 12 Vdc, $R_L$ = 100 Ohms) | $V_{GT}$ | | | | Volts |
| MT2(+), G(+) | | — | 0.9 | 2.0 | |
| MT2(+), G(−) | | — | 0.9 | 2.5 | |
| MT2(−), G(−) | | — | 1.1 | 2.0 | |
| MT2(−), G(+) | | — | 1.4 | 2.5 | |
| *MT2(+), G(+); MT2(−), G(−) $T_C$ = −40°C | | — | — | 2.5 | |
| *MT2(+), G(−); MT2(−), G(+) $T_C$ = −40°C | | — | — | 3.0 | |
| Gate Non-Trigger Voltage (Continous dc) ($V_D$ = Rated $V_{DRM}$, $R_L$ = 100 k Ohms, $T_J$ = 100°C) *MT2(+), G(+); MT2(−), G(−); MT2(+), G(−); MT2(−), G(−) | $V_{GT}$ | 0.2 | — | — | Volts |
| *Holding Current ($V_D$ = 12 Vdc, Gate Open)     $T_C$ = 25°C (Initiating Current = ± 200 mA)     *$T_C$ = −40°C | $I_H$ | — — | 6.0 — | 40 75 | mA |
| *Turn-On Time ($V_D$ = Rated $V_{DRM}$, $I_{TM}$ = 11 A, $I_{GT}$ = 120 mA. Rise Time = 0.1 μs, Pulse Width = 2 μs) | $t_{gt}$ | — | 1.5 | 2.0 | μs |
| **DYNAMIC CHARACTERISTICS** | | | | | |
| Critical Rate of Rise of Commutation Voltage $V_D$ = Rated $V_{DRM}$, $I_{TM}$ = 11 A, Commutating di/dt = 4.0 A/ms (Gate Unenergized, $T_C$ = 80°C) | dv/dt(c) | — | 5.0 | — | V/μs |

*Indicates JEDEC Registered Data

(b)

FIGURE 20.24  *(continued)*

Quadrant III operation is simply a reflection of quadrant I operation. In other words, when MT2 is negative (with respect to MT1) and a negative gate trigger voltage is applied to the triac, it is triggered into reverse conduction, as shown on the right side of Figure 20.22b. Again, all the SCR characteristics apply to the triac; the direction of conduction has simply been reversed.

Quadrants II and IV indicate the unique triggering combinations of the triac. The triac can be triggered into conduction by either of the following conditions:

1. Applying a negative gate trigger while MT2 is positive (quadrant II).
2. Applying a positive gate trigger while MT2 is negative (quadrant IV).

Quadrants I and II have the same MT2 polarities. The same is true for quadrants III and IV.

These triggering combinations are possible because the gate is physically connected to both the $p_2$ and $n_5$ regions, as shown in Figure 20.22a. When quadrant II triggering occurs, the current through the device is in the direction indicated on the left side of Figure 20.22b. When quadrant IV triggering occurs, the current through the device is in the direction shown on the right side of Figure 20.22b. Since the polarity of MT2 (with respect to MT1) determines the direction of current through the triac, quadrant II triggering allows current in the same direction as quadrant I triggering. The same can be said for quadrant III triggering and quadrant IV triggering.

Confusing? Let's take a look at the $V_{GT}$ parameters for a typical triac. The spec sheet for the 2N6344 triac shown in Figure 20.24 lists the following values:

| $V_{GT}$ (Maximum Value) | Listed Conditions | Indicated Quadrant |
|---|---|---|
| 2 V | MT2 (+), G (+) | I |
| 2.5 V | MT2 (+), G (−) | II |
| 2 V | MT2 (−), G (−) | III |
| 2.5 V | MT2 (−), G (+) | IV |

Note that these values of $V_{GT}$ were measured with 12 V applied across the component's main terminals. Now, let's apply the values listed to the triacs shown in Figure 20.25. The triacs in Figure 20.25 illustrate the polarities required for triggering in each quadrant. Using the spec sheet values:

- The triac drawn in quadrant I triggers when MT2 = +12 V and $V_{GT}$ = +2 V.
- The triac drawn in quadrant II triggers when MT2 = +12 V and $V_{GT}$ = −2.5 V.
- The triac drawn in quadrant III triggers when MT2 = −12 V and $V_{GT}$ = −2 V.
- The triac drawn in quadrant IV triggers when MT2 = −12 V and $V_{GT}$ = +2.5 V.

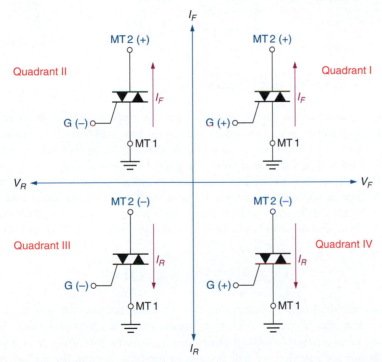

**FIGURE 20.25** Triac triggering.

Note that the values of $V_{GT}$ listed above are *maximum* values and that the typical values of $V_{GT}$ are less than 1 V. Also, note the direction of triac current for each operating quadrant. Two points should be made at this time:

1. Regardless of the triggering method used, triacs are forced into cutoff in the same fashion as the SCR. Either *device current interruption* or *forced commutation* must be used to decrease the device current below its holding current ($I_H$) rating. At that point, the device enters its nonconducting state and remains in cutoff until another trigger is received.

2. Except for the quadrant triggering considerations, the spec sheet for a triac is nearly identical to that of an SCR. This can be seen by comparing the triac spec sheet in Figure 20.24 with the SCR spec sheet in Figure 20.11.

### 20.3.3 Controlling Triac Triggering

The low values of $V_{GT}$ required to trigger a triac can cause problems in many circuit applications. For example, let's assume that the triac shown in Figure 20.26a is a 2N6344. According to its spec sheet, the 2N6344 typically triggers into quadrant I or II operation at $V_{GT} = \pm0.9$ V. Three questions arise when you consider these gate triggering voltages:

**1.** What if we want the 2N6344 to be triggered *only* by a *positive* gate voltage?
**2.** What if we want the 2N6344 to be triggered *only* by a *negative* gate voltage?
**3.** What if we want positive *and* negative triggering but at higher gate voltage values?

The circuit shown in Figure 20.26b illustrates the method commonly used to provide *single-polarity* triggering. In the circuit shown, $D_1$ is forward biased when $V_{GT}$ is *positive* and reverse biased when $V_{GT}$ is *negative*. Since $D_1$ conducts only when $V_{GT}$ is positive, the triac can be triggered only by a positive gate signal. The potential required to trigger the device equals the sum of $V_F$ for $D_1$ and $V_{GT}$. By reversing the direction of the diode, we limit the triggering of the triac to *negative* gate voltages.

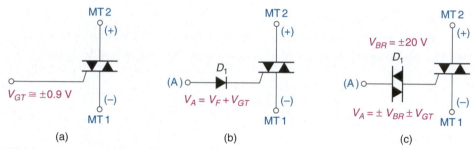

**FIGURE 20.26** Controlling triac triggering.

In many applications, the dual-polarity triggering characteristic of the triac is desirable but the rated values of $V_{GT}$ for the device are too low. When this is the case, the triggering levels can be raised by using a *diac* in the gate circuit, as shown in Figure 20.26c. With the diac present, the triac can still be triggered by both positive and negative gate voltages. However, to get conduction in the gate terminal (which is required for triggering), the voltage at (A) must overcome the values of $V_{BR}$ (for the diac) and $V_{GT}$. For example, in Figure 20.26c, we have placed a diac with $V_{BR}$ values of $\pm20$ V in the gate circuit of the triac. For the 2N6344 to be triggered into conduction, $V_A$ will have to reach a value of

$$V_A = \pm V_{BR} \pm V_{GT} \cong \pm21 \text{ V}$$

We still have the dual-polarity triggering that we want, but now the gate trigger signal ($V_A$) must reach approximately $\pm21$ V to trigger the triac into conduction. Note that *thyristor triggering control is the primary application for the diac.* It is also one of the applications for a device we will discuss in the next section, the *unijunction transistor (UJT).*

### 20.3.4 Triac Applications

A typical triac application can be seen in the *phase controller* in Figure 20.27. The primary difference between the triac phase controller and the SCR phase controller is the triac's ability to conduct during both alternations of the input signal.

With the input polarity indicated in Figure 20.27, $D_1$ is forward biased and $D_2$ is reverse biased. This puts a positive voltage at point A in the circuit. When $V_A$ is high enough to overcome the $V_{BR}$ rating of the diac and the $V_{GT}$ rating of the triac, the triac is triggered into conduction, providing the positive output cycle shown. When the input polarity reverses, $D_1$ becomes reverse biased and $D_2$ becomes forward biased. In this case, point A is still positive, but MT2 is now negative. When the triac is trig-

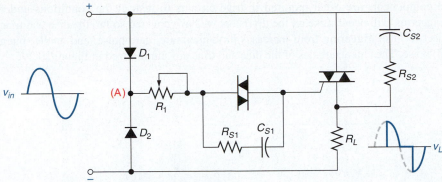

FIGURE 20.27 A triac phase controller.

gered, the current direction through the load is reversed and the negative output cycle is produced. Note that the triac in Figure 20.27 is quadrant I triggered during the positive alternation of the input and quadrant IV triggered during the negative alternation of the input. Also, note the presence of the diac and triac *snubber networks*. These are included because the diac and triac are subject to the same false triggering problems as the SCR.

Figure 20.28 shows an improved circuit that can extend the firing angle beyond 90°. It uses a variable $RC$ network to replace the diodes and diac. The capacitor also reduces the risk of false triggering. $R_3$ is included to prevent any surge current from destroying the triac gate. This control circuit is similar in operation and construction to the SCR circuit shown in Figure 20.16, except that this circuit can be used for phase control over a range of approximately 30° to 320°.

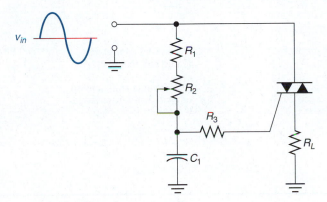

FIGURE 20.28 Large-angle triac phase controller.

### 20.3.5 Component Testing

Diacs and triacs are almost always in-circuit tested. An oscilloscope can be used to check the operation of each of these components. When shorted, both components act as low-value resistors. When open, the components do not conduct, regardless of the presence of a triggering input.

### 20.3.6 Summary

The *diac* acts as a bidirectional SUS. The device conducts in either direction (*anode 1 to anode 2* or *anode 2 to anode 1*) when the terminal voltage reaches the breakover potential of the device. The device then continues to conduct until its current is driven below the holding current ($I_H$) rating by either anode current interruption or forced commutation.

The *triac* acts as a bidirectional SCR. The triac can be triggered into conduction by any one of four biasing and triggering combinations. Because of the ease of triac triggering, external components are often required in triac circuits to restrict the conditions under which triggering will occur. Like all the thyristors we have covered so far, the diac and triac are subject to false triggering from increases in temperature, gate noise, and anode (main terminal) noise. The characteristics of the diac and triac are summarized in Figure 20.29.

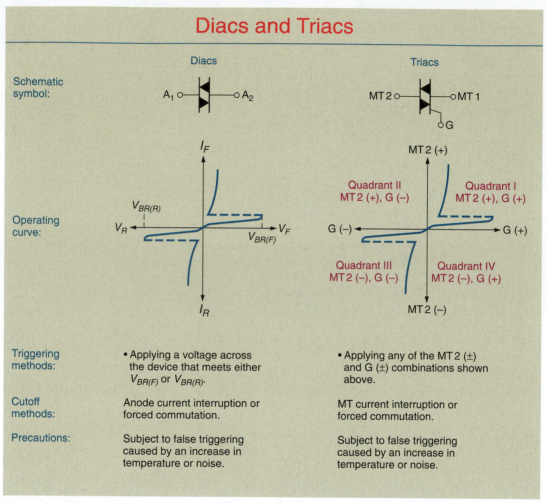

FIGURE 20.29

**Section Review ▶**

1. What is a *bidirectional thyristor*?

2. What is a *diac*?

3. What are the two terminals of the diac called?

4. How is a diac triggered into conduction?

5. How is a diac driven into cutoff?

6. What is the most common diac application?

7. What is a *triac*?

8. Draw the schematic symbol for a triac, and identify the component terminals.

9. Describe the four-quadrant triggering characteristics of the triac.

10. What methods are used to drive a triac into cutoff?

11. Describe the operation of each of the trigger control circuits shown in Figure 20.26.

12. Describe the operation of the triac phase controller in Figure 20.27.

13. What are the symptoms of a *shorted* diac or triac?

14. What are the symptoms of an *open* diac or triac?

The **unijunction transistor (UJT)** is a three-terminal switching device whose trigger voltage is proportional to its applied biasing voltage. The schematic symbol for the UJT is shown in Figure 20.30a. As you can see, the terminals of the UJT are called the *emitter* (E), *base 1* (B1), and *base 2* (B2). The switching action of the UJT is explained with the help of the circuit shown in Figure 20.30b. A biasing voltage ($V_{BB}$) is applied across the two base terminals. The emitter–base 1 junction acts as an open until the voltage across the terminals ($V_{EB1}$) reaches a specified value, called the **peak voltage** ($V_P$). When $V_{EB1} = V_P$, the emitter–base 1 junction triggers into conduction and will continue to conduct until the emitter current drops below a specified value, called the **peak current** ($I_P$). At that time, the device returns to its nonconducting state. Note that the peak voltage (emitter–base 1 triggering voltage) is proportional to the value of $V_{BB}$. Thus, when we change $V_{BB}$, we change the value of $V_{EB1}$ needed to trigger the device.

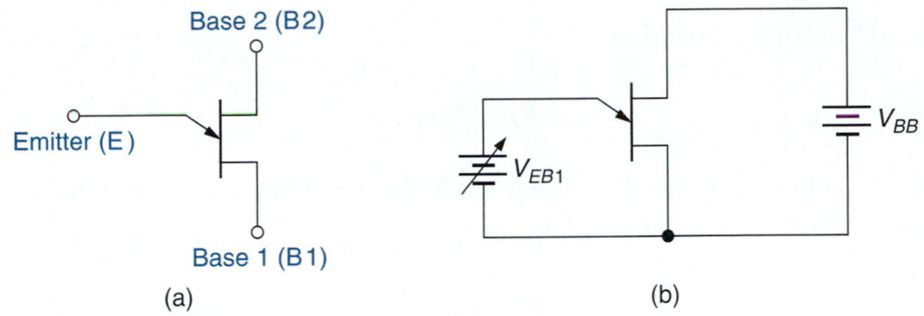

◄ **OBJECTIVE 8**

**Unijunction transistor (UJT)**
A three-terminal device whose trigger voltage is proportional to its applied biasing voltage.

**Peak voltage ($V_P$)**
The value of $V_{EB1}$ that triggers the UJT into conduction.

**Peak current ($I_P$)**
The minimum value of $I_E$ required to maintain emitter conduction.

Base 2 (B2)

Emitter (E)

Base 1 (B1)

(a)

$V_{EB1}$

$V_{BB}$

(b)

**FIGURE 20.30**    UJT schematic symbol and biasing.

## 20.4.1    UJT Construction and Operation

Figure 20.31 shows the construction and equivalent circuit of the UJT. As you can see, the UJT structure is very similar to that of the *n*-channel JFET. In fact, the only difference between the two components is that the *p*-type (gate) material of the JFET *surrounds* the *n*-type (channel) material, as was shown in Chapter 12. The similarity in the construction of the UJT and the *n*-channel JFET is the reason that their schematic symbols are so similar.

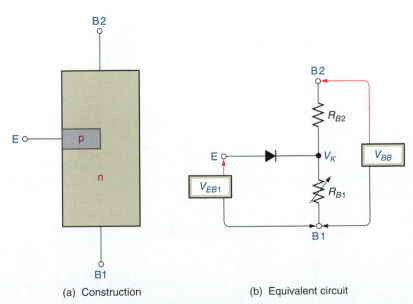

(a) Construction                    (b) Equivalent circuit

**FIGURE 20.31**    UJT construction and equivalent circuit.

The UJT equivalent circuit (Figure 20.31b) represents the component as a diode connected to a voltage divider. The diode represents the junction formed by the emitter ($p$-type) material and the base ($n$-type) material. The voltage divider ($R_{B1}$ and $R_{B2}$) represents the resistance of the $n$-type material between the emitter junction and each base terminal.

The overall principle of operation is simple. For the emitter-base diode to conduct, its anode voltage must be approximately 0.7 V more positive than its cathode. The cathode potential of the diode ($V_K$) is determined by the combination of $V_{BB}$, $R_{B1}$, and $R_{B2}$. Using the standard voltage-divider equation, the value of $V_K$ is found as

$$V_K = V_{BB} \frac{R_{B1}}{R_{B1} + R_{B2}} \tag{20.5}$$

**Intrinsic standoff ratio ($\eta$)**
The ratio of emitter–base 1 resistance to the total *interbase resistance* in a UJT. The **interbase resistance** is the total resistance between base 1 and base 2 when the device is not conducting.

The resistance ratio in equation (20.5) is called the **intrinsic standoff ratio**, or $\eta$ (Greek letter *eta*) of the UJT. Since

$$\eta = \frac{R_{B1}}{R_{B1} + R_{B2}}$$

equation (20.5) is often written as

$$V_K = \eta V_{BB} \tag{20.6}$$

Adding 0.7 V to the value found in equation (20.6), we obtain an equation for the triggering voltage ($V_P$) for the UJT, as follows:

$$V_P = \eta V_{BB} + 0.7 \text{ V} \tag{20.7}$$

The intrinsic standoff ratio for a given UJT is listed on its spec sheet. For example, the spec sheet for a 2N5431 UJT lists an intrinsic standoff ratio of $\eta = 0.8$ (maximum). Example 20.5 shows how this value is used to calculate the peak voltage for the 2N5431 at a given value of $V_{BB}$.

### EXAMPLE 20.5

Determine the maximum peak voltage ($V_P$) value for the 2N5431 UJT in Figure 20.32.

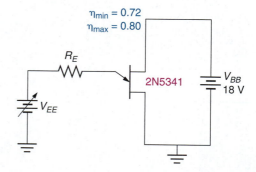

**FIGURE 20.32**

*Solution:*   With $\eta_{max} = 0.8$ and $V_{BB} = +18$ V, the maximum peak voltage of the circuit is found as

$$V_P = \eta V_{BB} + 0.7 \text{ V} = (0.8)(+18 \text{ V}) + 0.7 \text{ V} = \textbf{15.1 V}$$

### PRACTICE PROBLEM 20.5
The 2N4870 UJT has a rating of $\eta_{max} = 0.75$. Determine the maximum value of $V_P$ for the device when it is used in a circuit with $V_{BB} = +12$ V.

The peak voltage is applied across the emitter–base 1 terminals. Because of this, almost all of $I_E$ is drawn through $R_{B1}$ when the device is triggered, as shown in Figure 20.33. Since $I_E$ passes through $R_{B1}$, this resistance value drops off as $I_E$ increases. The value of $R_{B2}$ is not affected very much by the presence of $I_E$. The results of this component operation are shown in the UJT characteristic curve in Figure 20.34. As you can see, $V_{EB1}$ can be increased until the *peak point* is reached. When the peak point is reached, $V_{EB1} = V_P$, and the emitter diode begins to conduct. The initial value of $I_E$ is referred to as the *peak emitter current* ($I_P$). This term can be confusing since we tend to think of a peak current as a maximum value. However, for the UJT, the peak current ($I_P$) is the *minimum* rating because it is associated with the first point of conduction. Note that when $I_E < I_P$ the UJT is said to be in its *cutoff* region of operation.

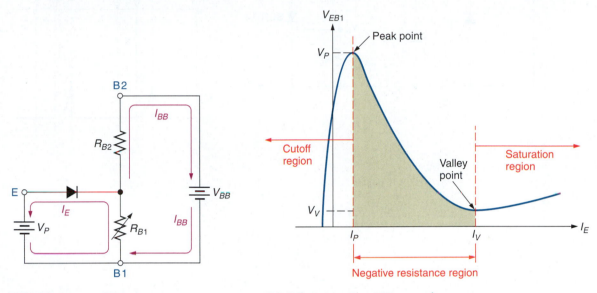

FIGURE 20.33   UJT currents.          FIGURE 20.34   The UJT operating curve.

When $I_E$ increases above the value of $I_P$, the value of $R_{B1}$ drops drastically. As a result, the value of $V_{EB1}$ actually decreases as $I_E$ increases. The decrease in $V_{EB1}$ continues until $I_E$ reaches the *valley current* ($I_V$) rating of the UJT. When $I_E$ is increased beyond $I_V$, the device goes into *saturation*.

As the characteristic curve in Figure 20.34 shows, $V_{EB1}$ and $I_E$ are inversely related between the peak point and valley point of the curve. The region of operation between these two points is called the **negative resistance region**. Note that the term **negative resistance** is used to describe any device with current and voltage values that are *inversely* related.

Once the UJT is triggered, the device continues to conduct as long as $I_E$ is greater than $I_P$. When $I_E$ drops below $I_P$, the device returns to the cutoff region of operation and remains there until another trigger is received.

**Negative resistance region**
The region of operation that lies between the peak and valley points of the UJT curve.

**Negative resistance**
A term used to describe any device with current and voltage values that are *inversely* related.

## 20.4.2   UJT Applications

UJTs are used almost exclusively as *thyristor-triggering devices*. An example of this application can be seen in Figure 20.35. The circuit shown in Figure 20.35a is called a *relaxation oscillator*. A **relaxation oscillator** is a circuit that uses the charge/discharge characteristics of a capacitor or inductor to produce a pulse output. This pulse output is normally used as the triggering signal for a thyristor, such as an SCR or a triac. In a practical relaxation oscillator, the dc supply voltage ($V_S$) is derived from the ac line voltage to control the timing of the oscillator.

The ac input to the circuit is applied to a bridge rectifier. The full-wave rectified signal is then applied to the rest of the circuit. The zener diode ($D_1$) clips the rectified signal, producing the $V_S$ waveform shown in Figure 20.35b. Note that $V_S$ is the biasing voltage for the UJT and its trigger circuit ($R_T$ and $C_T$).

**Relaxation oscillator**
A circuit that uses the charge/discharge characteristics of a capacitor or inductor to produce a pulse output.

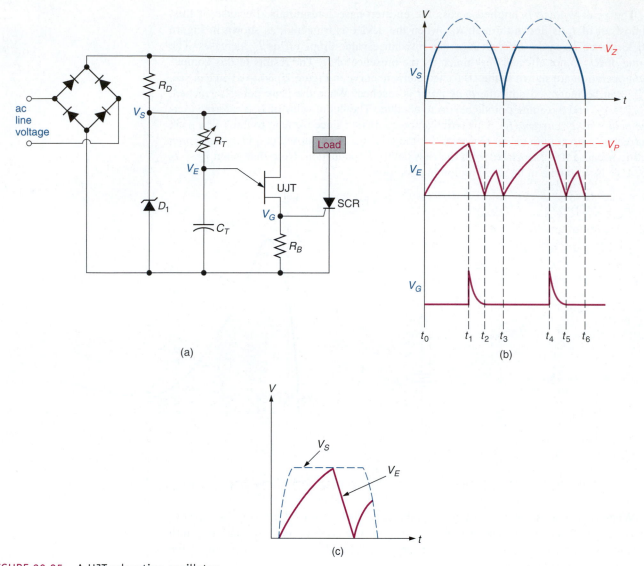

FIGURE 20.35   A UJT relaxation oscillator.

*Don't Forget:*

The time required for a capacitor to charge or discharge is found as $T = 5RC$. Since $(R_{B1} + R_B)$ (during conduction) is much less than $R_T$, the capacitor takes less time to discharge than it does to charge.

From $t_0$ to $t_1$, $C_T$ charges through $R_T$. When the emitter voltage of the UJT ($V_E$) reaches $V_P$, the UJT fires, causing conduction through $R_B$. The current through $R_B$ then produces the gate trigger pulse ($V_G$) required to trigger the SCR into conduction.

During the time period from $t_1$ to $t_2$, $C_T$ is discharging. When the UJT returns to its cutoff state (at $t_2$), the capacitor begins to charge again through $R_T$. However, this time the capacitor charge doesn't make it to $V_P$. The reason for this is shown in Figure 20.35c. As $V_E$ is starting to increase for the second time in the alternation, $V_S$ is decreasing in value. Before $V_E$ can reach the original value of $V_P$, the supply voltage is effectively removed from the UJT. Thus, the second charge cycle of $C_T$ is cut short, producing the $V_E$ waveform shown.

Even if $C_T$ could recharge completely during a given alternation, it wouldn't have any effect on the load. The reason for this is simple: The SCR is triggered into conduction during the discharge cycle of $C_T$ and continues to conduct for the remainder of the alternation. Since the SCR is already conducting when the second $C_T$ charge cycle occurs, the production of another trigger signal by the UJT would have no effect on the SCR and, thus, none on the load.

Since the initial charge cycle of $C_T$ is controlled by the cycles of $V_S$, the relaxation oscillator is synchronized to the ac load cycle. This method of synchronizing the oscillator to the ac load cycle is used extensively in practice.

### 20.4.3  One Final Note

The UJT is technically classified as a *thyristor-triggering device* rather than as a thyristor. Since triggering thyristors (such as SCRs and triacs) is the only common application for the UJT, the device is normally covered along with thyristors. Also, you will normally find the spec sheets for UJTs in the same section of a parts manual as the thyristors.

Another thyristor trigger that is commonly used is the *programmable unijunction transistor (PUT)*. We will conclude this section with a brief introduction to this device.

### 20.4.4  The Programmable UJT (PUT)

The PUT is a four-layer device that is very similar to the SCR. In this case, however, the gate is used as a *reference* terminal rather than as a triggering terminal. The schematic symbol for the PUT is shown, along with an application circuit, in Figure 20.36. Note the similarity between the PUT schematic symbol and the SCR schematic symbol. The similarity in schematic symbols between the two devices indicates the similarity in their operation.

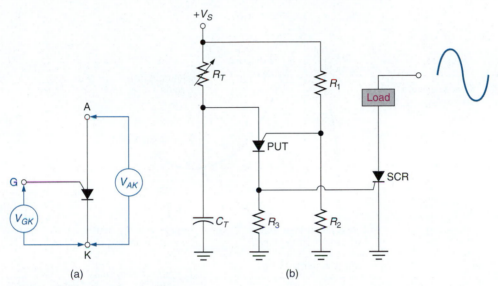

**FIGURE 20.36**  A simple PUT relaxation oscillator.

The PUT blocks anode current until the value of $V_{AK}$ reaches the value of $V_{GK}$. For example, if $V_{GK}$ is $+5$ V, the device blocks anode conduction until $V_{AK}$ reaches $+5$ V. At that time, the device triggers and anode current increases rapidly. Once triggered, the PUT continues to conduct until the anode current drops below the $I_P$ rating of the device. At that point, the device returns to cutoff and blocks anode current until triggered again.

The circuit shown in Figure 20.36b demonstrates the use of the PUT in a relaxation oscillator. $R_1$ and $R_2$ are used to establish the value of $V_{GK}$ for the device. $C_T$ is charged through $R_T$ until $V_{AK}$ reaches the value of $V_{GK}$. At that point, the PUT is triggered, allowing conduction through $R_3$. The conduction through $R_3$ produces a voltage pulse at the gate of the SCR that is similar to the $V_G$ waveform in Figure 20.35.

When the PUT is triggered, $C_T$ discharges through the device. When the current supplied by the capacitor drops below the $I_P$ rating of the PUT, the device returns to cutoff, and the charge cycle begins again.

It should be noted that the PUT relaxation oscillator must be synchronized to the SCR anode signal, as was the case with the UJT relaxation oscillator. Thus, the practical PUT oscillator is actually a bit more complex than the one shown in Figure 20.36b.

### 20.4.5  Summary

The UJT and PUT are devices that are not classified as thyristors but, rather, as *thyristor triggers*. These devices are usually used to control the triggering of SCRs and triacs.

The most common UJT and PUT triggering circuits are *relaxation oscillators*. These oscillators are usually synchronized to the ac signal that is applied to the circuit thyristor (SCR or triac) so that proper phase control is provided. The characteristics of the UJT and PUT are summarized in Figure 20.37.

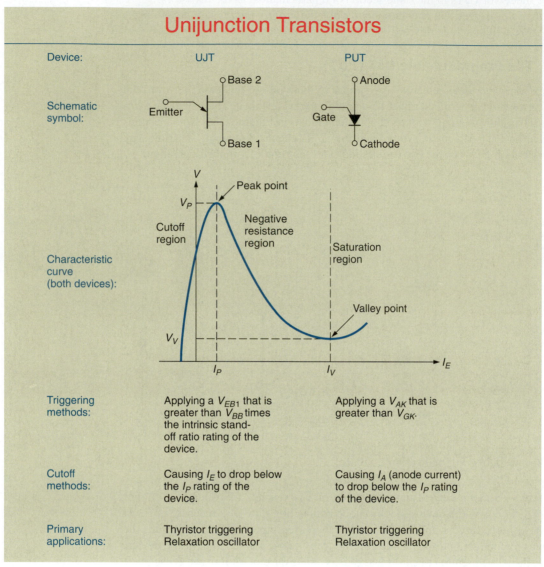

FIGURE 20.37

**Section Review** ▶

1. What is the *unijunction transistor (UJT)*?

2. What is the *peak voltage* ($V_P$) value of a UJT?

3. What is the *peak current* ($I_P$) value of a UJT?

4. What is the *intrinsic standoff ratio* ($\eta$) rating of a UJT?

5. How does the value of $\eta$ for a UJT relate to the value of $V_P$?

6. What is meant by the term *negative resistance*?

7. What current and voltage values define the limits of the negative resistance region of UJT operation?

8. What is the relationship between $I_E$ and $V_{EB1}$ in the negative resistance region of UJT operation?

9. Describe the operation of the *UJT relaxation oscillator* in Figure 20.35.
10. Explain the operation of the *PUT relaxation oscillator* in Figure 20.36.
11. What is the primary difference between the PUT and the SCR?

## 20.5 Discrete Photodetectors

In Chapter 2, we discussed the operation of the *light-emitting diode* (*LED*). The LED is classified as a **light emitter** because it gives off light when biased properly. In this section, we will take a look at the operation of several **light detectors**. Light detectors are components with electrical output characteristics that are controlled by the amount of light they receive. In other words, while *emitters* produce light, *detectors* respond to light.

When we discuss photodetectors, we need to consider a variety of parameters related to light. For this reason, we're starting this section with a brief discussion on light and its properties.

◀ *OBJECTIVE 9*

**Light emitters**
Optoelectronic devices that produce light.

**Light detectors**
Optoelectronic devices that respond to light.

### 20.5.1 Characteristics of Light

**Light** is *electromagnetic energy* that falls within a specific range of frequencies, as shown in Figure 20.38. The entire light spectrum falls within the range of 30 THz to 3 PHz and is further broken down as shown in Table 20.2.

◀ *OBJECTIVE 10*

**Light**
Electromagnetic energy that falls within a specific range of frequencies.

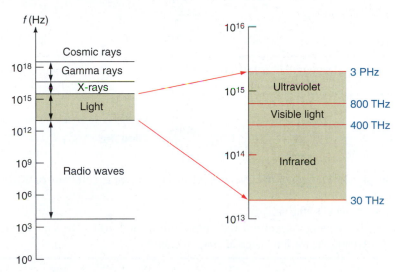

FIGURE 20.38   The light spectrum.

TABLE 20.2   Light Frequency Ranges

| Type of Light | Frequency Range (Approximate)[a] |
|---|---|
| Infrared | 30 THz to 400 THz |
| Visible | 400 THz to 800 THz |
| Ultraviolet | 800 THz to 3 PHz |

[a]P stands for *peta* ($10^{15}$) and T for *tera* ($10^{12}$).

Two characteristics are commonly used to describe light. The first of these is **wavelength** (symbolized by **λ**, the Greek letter *lambda*). Wavelength is the physical length of one cycle of a transmitted electromagnetic wave.

The concept of wavelength is illustrated in Figure 20.39. The wavelength for a signal at a given frequency is found as

$$\lambda = \frac{c}{f}$$

(20.8)

**Wavelength (λ)**
The physical length of one cycle of a transmitted electromagnetic wave.

*A Practical Consideration:*
The wavelength ratings of most optoelectronic devices are given in *nanometers* (*nm*).

where $\lambda$ = the wavelength of the signal (in *nanometers*)
$c$ = the *speed of light*, given as $3 \times 10^{17}$ nm/s
$f$ = the frequency of the transmitted signal

The following example demonstrates the calculation of wavelength for a given signal.

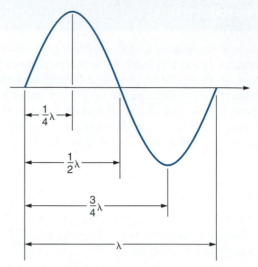

**FIGURE 20.39**   Wavelength.

### EXAMPLE 20.6

Determine the wavelength of a 150 THz light signal.

*Solution:*   The wavelength of the signal is found as

$$\lambda = \frac{c}{f} = \frac{3 \times 10^{17} \text{ nm/s}}{150 \times 10^{12} \text{ Hz}} = \textbf{2000 nm}$$

Thus, the physical length of one cycle of a 150 THz signal is 2000 nm, or 2 $\mu$m.

**PRACTICE PROBLEM 20.6**
The infrared frequency spectrum falls between 30 and 400 THz. Determine the range of wavelengths for this range of frequencies.

Wavelength is an important characteristic of light because photoemitters and photodetectors are rated for specific wavelengths. You see, emitters are commonly used to drive detectors, as shown in Figure 20.40. For a specific emitter to be used with a specific detector, the two must be rated for the same approximate wavelength. Otherwise, the detector will not respond to the photoemitter output. This point is demonstrated later in this chapter.

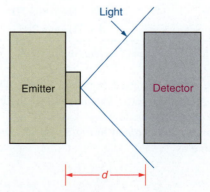

**FIGURE 20.40**   Emitter-detector combination.

The second important light characteristic is *intensity*. **Light intensity** is the amount of light per unit area received by a given photodetector. For example, consider the emitter-detector combination shown in Figure 20.40. The light intensity is a measure of the amount of light received by the detector. Note that *light intensity decreases as the distance (d) between the emitter and detector increases*. As you will see, the output characteristics of most photodetectors are controlled by the light intensity at their inputs as well as the input wavelength.

**Light intensity**
The amount of light per unit area received by a given photodetector; also called *irradiance*.

### 20.5.2 The Photodiode

The **photodiode** is a diode whose *reverse* conduction is light-intensity controlled. When the light intensity at the optical input to a photodiode increases, the reverse current through the device also increases. This point is illustrated in Figure 20.41.

◄ *OBJECTIVE 11*

**Photodiode**
A diode whose *reverse* conduction is light-intensity controlled.

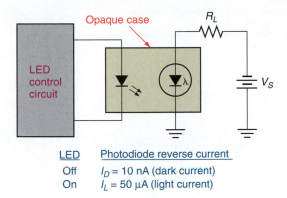

| LED | Photodiode reverse current |
|-----|----------------------------|
| Off | $I_D = 10$ nA (dark current) |
| On  | $I_L = 50$ μA (light current) |

**FIGURE 20.41**

In Figure 20.41, a photodiode is enclosed with an LED in an *opaque case*. The term **opaque** is used to describe anything that blocks light. When in the opaque case, the photodiode receives a light input only when the LED is turned on by its control circuit. When the LED is off, the photodiode reverse current is shown to be 10 nA. When the LED lights, the photodiode reverse current increases to 50 μA. Note that the **light current ($I_L$)** is shown to be approximately 5000 times as great as the **dark current ($I_D$)**. This ratio of light current to dark current is not uncommon.

The spec sheet for a photodiode lists its values of $I_D$ and $I_L$, as can be seen in Figure 20.42. Note that the value of $I_D$ for the MRD500 photodiode is 2 nA (maximum), while the value of $I_L$ for the device is 6 μA (minimum). These two values show the device to have a minimum ratio of light to dark current of 3000.

Several other photodiode ratings listed under the *optical characteristics* heading are important. The first of these is **wavelength of peak spectral response ($\lambda_S$)**. This rating indicates the wavelength that causes the strongest response in the photodiode. For the MRD500, the optimum input wavelength is 0.8 μm (800 nm). As Example 20.7 shows, this value can be used to determine the optimum input frequency for the device.

**Opaque**
A term used to describe anything that blocks light.

**Light current ($I_L$)**
The reverse current through a photodiode with an active light input.

**Dark current ($I_D$)**
The reverse current through a photodiode with no active light input.

**Wavelength of peak spectral response ($\lambda_S$)**
A rating that indicates the wavelength that will cause the strongest response in a photodetector.

#### EXAMPLE 20.7

Determine the optimum input frequency for the MRD500 photodiode.

*Solution:* Rearranging equation (20.8) gives us the following equation for frequency:

$$f = \frac{c}{\lambda}$$

Using the given optimum wavelength of 800 nm, the optimum input frequency for the MRD500 is found as

$$f = \frac{3 \times 10^{17} \text{ nm/s}}{800 \text{ nm}} = \textbf{375 THz}$$

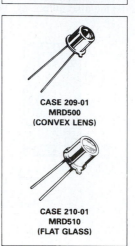

## SEMICONDUCTOR
### TECHNICAL DATA

## Photo Detectors
### Diode Output

... designed for application in laser detection, light demodulation, detection of visible and near infrared light-emitting diodes, shaft or position encoders, switching and logic circuits, or any design requiring radiation sensitivity, ultra high-speed, and stable characteristics.

- Ultra Fast Response — (<1 ns Typ)
- High Sensitivity — MRD500 (1.2 $\mu$A/mW/cm$^2$ Min)
  MRD510 (0.3 $\mu$A/mW/cm$^2$ Min)
- Available With Convex Lens (MRD500) or Flat Glass (MRD510) for Design Flexibility
- Popular TO-18 Type Package for Easy Handling and Mounting
- Sensitive Throughout Visible and Near Infrared Spectral Range for Wide Application
- Annular Passivated Structure for Stability and Reliability

**MRD500**
**MRD510**

**PHOTO DETECTORS
DIODE OUTPUT
PIN SILICON
250 MILLIWATTS
100 VOLTS**

**CASE 209-01
MRD500
(CONVEX LENS)**

**CASE 210-01
MRD510
(FLAT GLASS)**

**MAXIMUM RATINGS**  ($T_A$ = 25°C unless otherwise noted)

| Rating | Symbol | Value | Unit |
|---|---|---|---|
| Reverse Voltage | $V_R$ | 100 | Volts |
| Total Power Dissipation @ $T_A$ = 25°C<br>Derate above 25°C | $P_D$ | 250<br>2.27 | mW<br>mW/°C |
| Operating Temperature Range | $T_A$ | −55 to +125 | °C |
| Storage Temperature Range | $T_{stg}$ | −65 to +150 | °C |

**STATIC ELECTRICAL CHARACTERISTICS** ($T_A$ = 25°C unless otherwise noted)

| Characteristic | Fig. No. | Symbol | Min | Typ | Max | Unit |
|---|---|---|---|---|---|---|
| Dark Current ($V_R$ = 20 V, $R_L$ = 1 megohm) Note 2<br>$T_A$ = 25°C<br>$T_A$ = 100°C | 2 and 3 | $I_D$ | —<br>— | —<br>14 | 2<br>— | nA |
| Reverse Breakdown Voltage ($I_R$ = 10 $\mu$A) | — | $V_{(BR)R}$ | 100 | 200 | — | Volts |
| Forward Voltage ($I_F$ = 50 mA) | — | $V_F$ | — | — | 1.1 | Volts |
| Series Resistance ($I_F$ = 50 mA) | — | $R_S$ | — | — | 10 | Ohms |
| Total Capacitance ($V_R$ = 20 V, f = 1 MHz) | 5 | $C_T$ | — | — | 4 | pF |

**OPTICAL CHARACTERISTICS** ($T_A$ = 25°C unless otherwise noted)

| Characteristic | | Fig. No. | Symbol | Min | Typ | Max | Unit |
|---|---|---|---|---|---|---|---|
| Light Current<br>($V_R$ = 20 V) Note 1 | MRD500<br>MRD510 | 1 | $I_L$ | 6<br>1.5 | 9<br>2.1 | —<br>— | $\mu$A |
| Sensitivity at 0.8 $\mu$m<br>($V_R$ = 20 V) Note 3 | MRD500<br>MRD510 | — | $S_{(\lambda = 0.8\ \mu m)}$ | —<br>— | 6.6<br>1.5 | —<br>— | $\mu$A/mW/<br>cm$^2$ |
| Response Time ($V_R$ = 20 V, $R_L$ = 50 Ohms) | | — | $t_{(resp)}$ | — | 1 | — | ns |
| Wavelength of Peak Spectral Response | | 5 | $\lambda_S$ | — | 0.8 | — | $\mu$m |

NOTES: 1. Radiation Flux Density (H) equal to 5 mW/cm$^2$ emitted from a tungsten source at a color temperature of 2870 K.
2. Measured under dark conditions. (H ≈ 0).
3. Radiation Flux Density (H) equal to 0.5 mW/cm$^2$ at 0.8 $\mu$m.

(a)

**FIGURE 20.42**  The MRD500–510 photodetector specification sheet. (Copyright of Semiconductor Component Industries, LLC. Used by permission.)

Since 375 THz is in the *infrared* frequency spectrum, the MRD500 is classified as an *infrared detector*.

### PRACTICE PROBLEM 20.7
The Motorola MRD821 photodiode has an optimum wavelength of 940 nm. Determine the optimum input frequency for the device.

**Sensitivity**
A rating that indicates the response of a photodetector to a specified light intensity.

The **sensitivity** rating of the photodiode indicates the response of the device to a specified light intensity. For the MRD500, the sensitivity rating (listed under *optical characteristics*) is given as 6.6 $\mu$A/mW/cm$^2$. This means that the reverse current of the device

**MDR500**                                                    **MDR510**

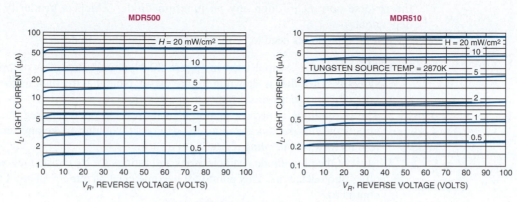

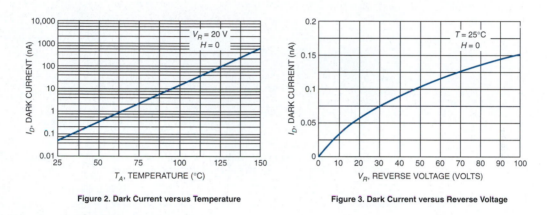

**Figure 1. Irradiated Voltage — Current Characteristic**

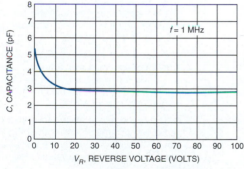

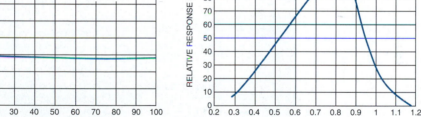

**Figure 2. Dark Current versus Temperature**          **Figure 3. Dark Current versus Reverse Voltage**

**Figure 4. Capacitance versus Voltage**               **Figure 5. Relative Spectral Response**

(b)

**FIGURE 20.42** *(continued)*

increases by 6.6 μA for every 1 mW/cm² of light applied to the device. For example, refer to Figure 20.41. If the light intensity at the photodiode is 1 mW/cm², the value of $I_R$ for the device is approximately 6.6 μA. If the light intensity at the photodiode increases to 2 mW/cm², $I_R$ increases to approximately 13.2 μA, and so on.

The **spectral response** of a photodiode is a measure of the device's response to a change in input wavelength. The spectral response of the MRD500 is shown in the *relative spectral response* curve in Figure 20.42b. As you can see, the response of the photodiode peaks at 800 nm, the value given on the spec sheet. If the input wavelength goes as low as 520 nm or as high as 950 nm, the *sensitivity* of the device drops to 50% of its rated value. Thus, at these two frequencies, the sensitivity of the device drops to approximately 3.3 μA/mW/cm².

The *irradiated voltage–current characteristic* curves in Figure 20.42b show that the light current ($I_L$) of the device is relatively independent of the amount of reverse bias across the diode. The dark current ($I_D$) curves show that the amount of dark cur-

> **Spectral response**
> A measure of a photodetector's response to a change in input wavelength. Note that response is measured in terms of the device *sensitivity*.

rent is far more dependent on temperature than on the value of $V_R$. This is similar to the reverse current through any *pn*-junction diode, which is primarily temperature dependent.

Several final points need to be made regarding the photodiode:

1. The schematic symbol used in Figure 20.41 is only one of two commonly used symbols. The other is shown in Figure 20.43. Note that the light arrows in the symbol point *toward* the diode rather than away from it.
2. The photodiode is always operated in its reverse operating region. (Since light affects the *reverse conduction* of the device, it wouldn't make any sense to operate it in its forward operating region.)
3. Most of the photodiode characteristics and ratings we have discussed apply to many other photodetectors. This point will become clear as our coverage of these devices continues.

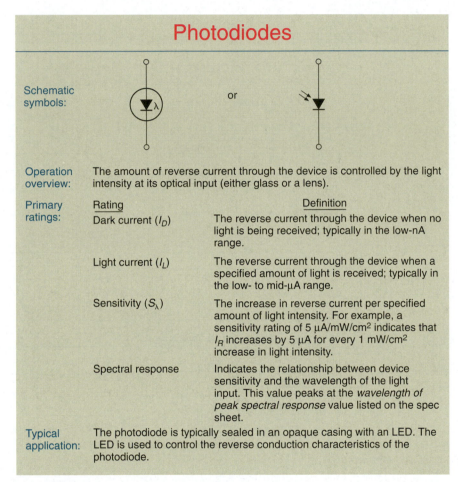

FIGURE 20.43

### 20.5.3   The Phototransistor

The **phototransistor** is a three-terminal photodetector whose collector current is controlled by the intensity of the light at its optical input (base). Figure 20.44 shows one possible configuration for a phototransistor amplifier.

In Figure 20.44, the phototransistor is shown to be enclosed with an LED in an opaque casing. When the light from the LED is varied by its control circuit, the change in light causes a proportional change in base current in the phototransistor. This change in base current causes a change in emitter (and collector) current that causes a proportional voltage to be developed across $R_E$. This voltage is applied to the inverting amplifier, which responds accordingly.

**Phototransistor**

A three-terminal photodetector whose collector current is controlled by the intensity of the light at its optical input (base).

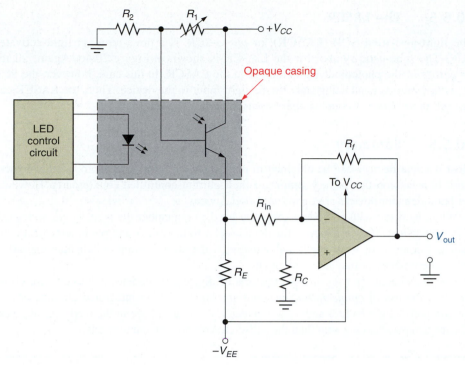

**FIGURE 20.44**   Optocoupling.

The phototransistor is identical to the standard BJT in every respect, except that its collector current is (ultimately) light-intensity controlled. Thus, the phototransistor in Figure 20.44 could be a linear amplifier or a switch, depending on the output of the LED control circuit. The biasing circuit ($R_1$ and $R_2$) is used to adjust the dark (quiescent) operation of the phototransistor. When $R_1$ is varied, the dark emitter current of the phototransistor varies. Thus, by adjusting $R_1$, the input voltage to the op-amp is varied, allowing the output of the op-amp to be set to 0 V. For this reason, $R_1$ in this circuit is called a *zero adjust*.

The arrangement in Figure 20.44 is referred to as **optocoupling** because the output from the LED control circuit is coupled via light to the phototransistor/op-amp circuit. The advantage of optocoupling is that it provides complete electrical isolation between the source (the LED control circuit) and the load (the phototransistor/op-amp circuit).

Several points should be made regarding the phototransistor: First, the spec sheet for the phototransistor contains the same basic parameters and ratings as those found on the spec sheet for a photodiode. The phototransistor is subject to the same wavelength and sensitivity characteristics as the photodiode. The difference is that the optical characteristics of the phototransistor relate the light input to *collector current* (rather than diode reverse current). Second, the base terminal of the phototransistor is normally returned to ground through some type of biasing circuit, as shown in Figure 20.44. The biasing circuit not only allows for adjusting the output from the device; it also adds *biasing stability* to the circuit. You may recal that *voltage-divider bias* is stable against changes in beta that may result from a change in temperature. By using a voltage divider ($R_1$ and $R_2$) to bias the transistor, the emitter and collector circuits are made relatively stable against the effects of any increase in operating temperature.

### 20.5.4   The Photo-Darlington

The **photo-Darlington** is a phototransistor constructed in a Darlington configuration, as shown in Figure 20.45. Except for the increased output current capability (as is always the case with Darlingtons), the operation and characteristics of the device are identical to those of the standard phototransistor.

> **Optocoupling**
> A method of coupling the output of one circuit to the input of another using light.

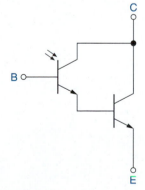

**FIGURE 20.45**   The photo-Darlington schematic symbol.

> **Photo-Darlington**
> A phototransistor with a Darlington pair output.

**Light-activated SCR (LASCR)**
A three-terminal light-activated SCR.

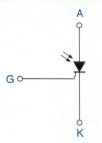

FIGURE 20.46 The LASCR schematic symbol.

### 20.5.5 The LASCR

The **light-activated SCR (LASCR)**, or *photo-SCR*, is a three-terminal light-activated SCR. The schematic symbol for the LASCR is shown in Figure 20.46. Again, all the properties of the photodiode can be found in the LASCR. In this case, however, the SCR is *latched* into its conducting state by the light input to the device. Thus, the LASCR can be used in an optically coupled phase-control circuit.

### 20.5.6 Summary

Most discrete devices can be obtained in *photodetector* form. The photodetector devices work in a fashion that is very similar to their current-controlled counterparts. However, the photodetector devices are *light-controlled* devices.

When working with photodetectors, you need to remember the *wavelength* and *sensitivity* (intensity) characteristics of the devices. If a light emitter and photodetector are not matched in terms of wavelength and sensitivity, the optocoupling circuit may operate at reduced efficiency or may not work at all.

The wavelength and sensitivity concerns of discrete photodetector circuits are eliminated by the use of *optoisolators* and *optointerrupters*. These integrated circuits, or ICs, contain both a light emitter and a light detector. As you will see in the next section, these ICs are far easier to deal with than their discrete component counterparts.

---

**Section Review ▶**

1. What is the difference between a light *emitter* and a light *detector*?
2. What is *wavelength*?
3. Why is the concept of wavelength important when working with optoelectronic devices?
4. What is *light intensity*?
5. What is a *photodiode*?
6. In terms of the photodiode, what is *dark current* ($I_D$)? What is *light current* ($I_L$)?
7. What is the relationship between photodiode light current and dark current?
8. What is the *wavelength of peak spectral response* for a photodiode?
9. What is *sensitivity*?
10. Explain the $\mu A/mW/cm^2$ unit of measure for sensitivity.
11. What is *spectral response*? Which photodiode unit of measure is affected by spectral response?
12. What is a *phototransistor*?
13. What is the relationship between input light and collector current for a phototransistor?
14. What is the difference between the optical ratings of the phototransistor and those of the photodiode?
15. What is the difference between the *photo-Darlington* and the standard phototransistor?
16. What is the *LASCR*? How does it differ from the standard SCR?
17. List the characteristics that must be considered whenever you are working with any discrete photodetector.

## 20.6 Optoisolators and Optointerrupters

**OBJECTIVE 12 ▶**

*Optoisolators* and *optointerrupters* are ICs that contain both a photoemitter and a photodetector. While the photoemitter is always an LED, the photodetector can be any one of the discrete photodetectors we discussed in the last section.

## 20.6.1 Optoisolators

The **optoisolator** is an optocoupler, that is, a device that uses light to couple a signal from its input device (a photoemitter) to its output device (a photodetector). For example, an optoisolator with a transistor output contains the optical circuitry shown in Figure 20.44.

The typical optoisolator comes in a 6-pin *dual in-line package* (DIP), as shown on the spec sheet in Figure 20.47. As the schematic diagram shows, this transistor-output optoisolator contains an LED and a phototransistor. Thus, the 4N35–7 series optoisolators could be used in the circuit shown in Figure 20.44.

Most optoisolator parameters and ratings are the standard LED and phototransistor specifications. However, several new ratings warrant discussion.

> **Optoisolator**
> An optocoupler. A device that uses light to couple a signal from its input device (a photoemitter) to its output device (a photodetector).

---

■ **SEMICONDUCTOR** ■
**TECHNICAL DATA**

## 6-Pin DIP Optoisolators
### Transistor Output

**4N35**
**4N36**
**4N37**

6-PIN DIP
OPTOISOLATORS
TRANSISTOR
OUTPUT

These devices consist of a gallium arsenide infrared emitting diode optically coupled to a monolithic silicon phototransistor detector.

- Convenient Plastic Dual-In-Line Package
- High Current Transfer Ratio — 100% Minimum at Spec Conditions
- Guaranteed Switching Speeds
- High Input-Output Isolation Guaranteed — 7500 Volts Peak
- UL Recognized. File Number E54915
- VDE approved per standard 0883/6.80 (Certificate number 41853), with additional approval to DIN IEC380/VDE0806, IEC435/VDE0805, IEC65/VDE0860, VDE0110b, covering all other standards with equal or less stringent requirements, including IEC204/VDE0113, VDE0160, VDE0832, VDE0833, etc.
- Meets or Exceeds All JEDEC Registered Specifications
- Special lead form available (add suffix "T" to part number) which satisfies VDE0883/6.80 requirement for 8 mm minimum creepage distance between input and output solder pads.
- Various lead form options available. Consult "Optoisolator Lead Form Options" data sheet for details.

CASE 730A-02
PLASTIC

**MAXIMUM RATINGS**  ($T_A$ = 25°C unless otherwise noted)

| Rating | Symbol | Value | Unit |
|---|---|---|---|
| **INPUT LED** | | | |
| Reverse Voltage | $V_R$ | 6 | Volts |
| Forward Current — Continuous | $I_F$ | 60 | mA |
| LED Power Dissipation @ $T_A$ = 25°C with Negligible Power in Output Detector | $P_D$ | 120 | mW |
| Derate above 25°C | | 1.41 | mW/°C |
| **OUTPUT TRANSISTOR** | | | |
| Collector-Emitter Voltage | $V_{CEO}$ | 30 | Volts |
| Emitter-Base Voltage | $V_{EBO}$ | 7 | Volts |
| Collector-Base Voltage | $V_{CBO}$ | 70 | Volts |
| Collector Current — Continuous | $I_C$ | 150 | mA |
| Detector Power Dissipation @ $T_A$ = 25°C with Negligible Power in Input LED | $P_D$ | 150 | mW |
| Derate above 25°C | | 1.76 | mW/°C |
| **TOTAL DEVICE** | | | |
| Isolation Source Voltage (1) (Peak ac Voltage, 60 Hz, 1 sec Duration) | $V_{ISO}$ | 7500 | Vac |
| Total Device Power Dissipation @ $T_A$ = 25°C | $P_D$ | 250 | mW |
| Derate above 25°C | | 2.94 | mW/°C |
| Ambient Operating Temperature Range | $T_A$ | −55 to +100 | °C |
| Storage Temperature Range | $T_{stg}$ | −55 to +150 | °C |
| Soldering Temperature (10 seconds, 1/16" from case) | — | 260 | °C |

(1) Isolation surge voltage is an internal device dielectric breakdown rating.
For this test, Pins 1 and 2 are common, and Pins 4, 5 and 6 are common.

**SCHEMATIC**

1. LED ANODE
2. LED CATHODE
3. N.C.
4. EMITTER
5. COLLECTOR
6. BASE

(a)

**FIGURE 20.47**  The 4N35–7 optoisolator specification sheet. (Copyright of Semiconductor Component Industries, LLC. Used by permission.)

| Characteristic | | Symbol | Min | Typ | Max | Unit |
|---|---|---|---|---|---|---|
| **COUPLED** | | | | | | |
| Output Collector Current     $T_A$ = 25°C | | $I_C$ | 10 | 30 | — | mA |
| ($I_F$ = 10 mA, $V_{CE}$ = 10 V)    $T_A$ = −55°C | | | 4 | — | — | |
|                            $T_A$ = 100°C | | | 4 | — | — | |
| Collector-Emitter Saturation Voltage ($I_C$ = 0.5 mA, $I_F$ = 10 mA) | | $V_{CE(sat)}$ | — | 0.14 | 0.3 | V |
| Turn-On Time | | $t_{on}$ | — | 7.5 | 10 | µs |
| Turn-Off Time | ($I_C$ = 2 mA, $V_{CC}$ = 10 V, | $t_{off}$ | — | 5.7 | 10 | |
| Rise Time | $R_L$ = 100 Ω, Figure 11) | $t_r$ | — | 3.2 | — | |
| Fall Time | | $t_f$ | — | 4.7 | — | |
| Isolation Voltage (f = 60 Hz, t = 1 sec) | | $V_{ISO}$ | 7500 | — | — | Vac(pk) |
| Isolation Current ($V_{I-O}$ = 3550 Vpk)      4N35 | | $I_{ISO}$ | — | — | 100 | µA |
|                     ($V_{I-O}$ = 2500 Vpk)      4N36 | | | — | — | 100 | |
|                     ($V_{I-O}$ = 1500 Vpk)      4N37 | | | — | 8 | 100 | |
| Isolation Resistance (V = 500 V) | | $R_{ISO}$ | $10^{11}$ | — | — | Ω |
| Isolation Capacitance (V = 0 V, f = 1 MHz) | | $C_{ISO}$ | — | 0.2 | 2 | pF |

(b)

**FIGURE 20.47** *(continued)*

> **Lab Reference:** Optoisolator operation is demonstrated in Exercise 35.

> **Isolation source voltage ($V_{ISO}$)**
> The voltage that, if applied across the input and output pins, will destroy an optoisolator.

> **Soldering temperature**
> The maximum amount of heat that can be applied to any pin of the IC without causing internal damage to the chip.

> **Isolation current ($I_{ISO}$)**
> The amount of current that can be forced between the input and output at the rated voltage.

> **Isolation resistance ($R_{ISO}$)**
> The total resistance between the device input and output pins.

> **Isolation capacitance ($C_{ISO}$)**
> The total capacitance between the device input and output pins.

> **Solid-state relay (SSR)**
> A circuit that uses a dc input voltage to pass or block an ac signal.

The **isolation source voltage ($V_{ISO}$)** rating indicates the input-to-output voltage that will cause the optoisolator to break down and conduct. In other words, it is the voltage that will destroy the device if applied across the input and output pins. For the 4N35–7 series chips, the value of $V_{ISO}$ is 7500 $V_{pk}$.

The **soldering temperature** is the maximum amount of heat that can be applied to any pin of the IC without causing internal damage to the chip. If the total heat applied to any pin of the 4N35 reaches 260°C for a duration of 10 s, internal damage may result.

The **isolation current ($I_{ISO}$)** rating indicates the amount of current that can be forced between the input and output at the rated voltage. For example, if 3550 $V_{pk}$ is applied across the 4N35, it will generate 100 µA (maximum) between the input and output pins of the device. The 4N36 requires 2500 $V_{pk}$ (from input to output) to generate 100 µA (maximum), and so on.

The **isolation resistance ($R_{ISO}$)** rating indicates the total resistance between the input pins and output pins of the device. For the 4N35–7 series, this rating is $10^{11}$ Ω (100 GΩ).

Finally, the **isolation capacitance ($C_{ISO}$)** rating indicates the total capacitance between the device input and output pins. For the 4N35–7 series, this rating is 2 pF.

Except for the ratings listed here, all the parameters and ratings for the optoisolator are fairly standard. This can be seen by quickly looking through the spec sheet. As you glance through the spec sheet, you'll also notice two standard light ratings are missing: *wavelength* and *sensitivity*. Since the photoemitter and photodetector are contained in the same chip, the wavelength and sensitivity values of the devices are unimportant.

As stated earlier, optoisolators come with a variety of output devices. Among these are the transistor output, Darlington output, triac output, and SCR output. For each of these devices, the input/output parameters and ratings vary only to reflect the characteristics of the photodetector used. Other than that, they operate exactly as described here and in Section 20.5.

### 20.6.2 An Optoisolator Application: The Solid-State Relay

A **solid-state relay (SSR)** is a circuit that uses a dc input voltage to pass or block an ac signal. The diagram in Figure 20.48a illustrates the input/output relationship for the SSR. When the dc input has the polarity shown, the waveform at the input (ac 1) is coupled to the ac output (ac 2) and, thus, to the load. As long as the dc potential remains, the ac signal will be coupled to the load. If the dc potential is removed, the output goes open circuit, and the ac signal is blocked from the load.

The schematic for the SSR is shown in Figure 20.48b. The dc circuit is coupled to the ac circuit via the diac-output optoisolator. As long as the dc input voltage has the polarity shown, the LED in the optoisolator is forward biased and emits light. This light keeps the diac photodetector conducting. As long as the diac photodetector is conducting, the triac in the ac circuit continues to conduct, passing the ac signal from the input terminal (ac 1) to the output terminal (ac 2).

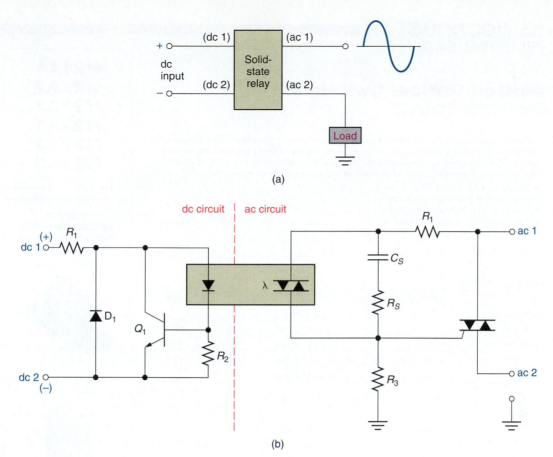

**FIGURE 20.48** The solid-state relay.

If the dc input voltage is removed, the LED turns off and the diac photodetector shuts off. This prevents a gate trigger voltage from being developed across $R_3$, and the output triac shuts off. When the output triac shuts off, the ac input is blocked from the ac output terminal.

The SSR is typically used to allow a digital circuit to control the operation (on or off) of a motor. This application is easy to understand if you picture the input to Figure 20.48a as coming from a digital system and the load as being a motor. Note that SSRs are also available in integrated form. While the exact circuitry of the integrated SSR varies from that shown in Figure 20.48b, the overall principle of operation is the same.

### 20.6.3 Optointerrupters

The **optointerrupter** or **optical switch** is an IC optocoupler designed to allow an external object to block the light path between the photoemitter and the photodetector. Two of the many common optointerrupter configurations are shown on the spec sheet in Figure 20.49. Note that the photoemitter is on one side of the slot, while the photodetector is on the other side.

The open gap between the photoemitter and the photodetector is the primary difference between the optointerrupter and the optoisolator. This gap allows the optointerrupter to be activated by some external object, such as a piece of paper. For example, let's say that the optointerrupter in Figure 20.49 is used in a photocopying machine and that the device is placed so that the edge of a given copy passes through the slot as it (the paper) passes through the machine. When no paper is present in the optointerrupter gap, the light from the emitter reaches the phototransistor, causing it to saturate. Thus, the output from the phototransistor is *low*. When a piece of paper passes through the gap, the emitter light is blocked (by the paper) from the phototransistor. This causes the phototransistor to go into cutoff, causing its collector voltage to go *high*. The low and high output voltages from the phototransistor are used to tell the machine whether a piece of paper has passed the point where the optointerrupter is located.

> **Optointerrupter or optical switch**
> An IC optocoupler designed to allow an external object to block the light path between the photoemitter and the photodetector.

■■ **SEMICONDUCTOR**
TECHNICAL DATA

## Slotted Optical Switches
### Transistor Output

Each device consists of a gallium arsenide infrared emitting diode facing a silicon NPN phototransistor in a molded plastic housing. A slot in the housing between the emitter and the detector provides the means for mechanically interrupting the infrared beam. These devices are widely used as position sensors in a variety of applications.

- Single Unit for Easy PCB Mounting
- Non-Contact Electrical Switching
- Long-Life Liquid Phase Epi Emitter
- 1 mm Detector Aperture Width

**H21A1**
**H21A2**
**H21A3**
**H22A1**
**H22A2**
**H22A3**

**SLOTTED OPTICAL SWITCHES TRANSISTOR OUTPUT**

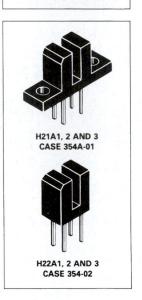

**H21A1, 2 AND 3 CASE 354A-01**

**H22A1, 2 AND 3 CASE 354-02**

**MAXIMUM RATINGS**

| Rating | Symbol | Value | Unit |
|---|---|---|---|
| **INPUT LED** | | | |
| Reverse Voltage | $V_R$ | 6 | Volts |
| Forward Current — Continuous | $I_F$ | 60 | mA |
| Input LED Power Dissipation @ $T_A$ = 25°C  Derate above 25°C | $P_D$ | 150  2 | mW  mW/°C |
| **OUTPUT TRANSISTOR** | | | |
| Collector-Emitter Voltage | $V_{CEO}$ | 30 | Volts |
| Output Current — Continuous | $I_C$ | 100 | mA |
| Output Transistor Power Dissipation @ $T_A$ = 25°C  Derate above 25°C | $P_D$ | 150  2 | mW  mW/°C |
| **TOTAL DEVICE** | | | |
| Ambient Operating Temperature Range | $T_A$ | − 55 to + 100 | °C |
| Storage Temperature | $T_{stg}$ | − 55 to + 100 | °C |
| Lead Soldering Temperature (5 seconds max) | — | 260 | °C |
| Total Device Power Dissipation @ $T_A$ = 25°C  Derate above 25°C | $P_D$ | 300  4 | mW  mW/°C |

**FIGURE 20.49**  The H21–2 series optointerrupters specification sheet. (Copyright of Semiconductor Component Industries, LLC. Used by permission.)

The type of application discussed here is the primary application for the optointerrupter. Any other type of optical-coupling application is usually performed by a discrete optocoupler or an optoisolator.

**Section Review ▶**

1. What is an *optoisolator*?
2. List and define the common *isolation* parameters and ratings of the optoisolator.
3. Describe the operation of the *solid-state relay* (*SSR*).
4. Contrast the *optointerrupter* with the *optoisolator*.

## CHAPTER SUMMARY

Here is a summary of the major points made in this chapter:

1. *Thyristors* are devices designed specifically for high-power switching applications.
2. Unlike other switching devices (such as BJTs and FETs), thyristors are not designed for use in linear amplifiers.
3. *Optoelectronic devices* are controlled by or emit (generate) light.
4. A *silicon unilateral switch* (*SUS*) is a two-terminal, four-layer device that can be triggered into conduction by applying a specified potential across its terminals (see Figure 20.1).

5. The SUS is commonly called by any of the following names: *pnpn diode, four-layer diode, Schockley diode, current latch*, or *reverse blocking diode thyristor* (the JEDEC designated name).

6. The SUS can be represented as cross-connected complementary transistors (see Figure 20.2).

   a. The circuit does not conduct as long as the anode-to-cathode voltage ($V_{AK}$) is lower than the rated *forward breakover voltage* ($V_{BR(F)}$).

   b. When $V_{AK}$ reaches the value of $V_{BR(F)}$, the device is driven into saturation. Device current ($I_F$) rapidly increases, and $V_{AK}$ rapidly decreases.

   c. The saturation value of $V_{AK}$ is identified as $V_F$ on the SUS operating curve (see Figure 20.3).

7. Since the SUS drives itself into saturation when $V_{AK} = V_{BR(F)}$, a series current-limiting resistor must be used.

8. Once it saturates, an SUS conducts as long as $I_F$ is greater than the *holding current* ($I_H$) rating of the component.

9. Once $I_F$ decreases below the value of $I_H$, the device remains in cutoff until $V_{AK}$ again reaches the value of $V_{BR(F)}$.

10. Two methods are commonly used to drive $I_F$ below the value of $I_H$.

    a. *Anode current interruption* drives an SUS into cutoff by breaking the diode current path or shorting current around the component (see Figure 20.4).

    b. In practice, anode current interruption is caused by the opening of a fuse (series) or by pushing a reset switch (parallel).

    c. *Forced commutation* drives an SUS into cutoff by applying a reverse voltage across the device. When $I_F < I_H$, the device cuts off (see Figure 20.5).

11. The forward operating curve of the SUS is divided into two parts:

    a. The *forward block region* falls between $0\text{ V} \leq V_{AK} < V_{BR(F)}$. This region represents the *off-state* (nonconducting) operation of the component.

    b. The *forward operating region* is the area of the curve that represents the *on-state* (conducting) characteristics of the device.

12. The limits of the SUS forward operating region are defined by $V_{BR(F)}$ and $I_{BR(F)}$.

    a. The *forward breakover current* ($I_{BR(F)}$) is the value of $I_F$ at the point where breakover occurs.

    b. The relationship among operating temperature, $V_{BR(F)}$, and $I_{BR(F)}$ is shown in Figure 20.7. Note that an increase in temperature can cause the device to trigger at values of $V_{AK}$ that are lower than $V_{BR(F)}$.

13. The *average on-state current* ($I_T$) rating is the maximum allowable forward current for the device. This value is used when selecting the value of a series current-limiting resistor.

14. The SUS is rarely used anymore. However, the commonly used thyristors are all variations on the SUS and can be explained in terms of SUS characteristics.

15. The *silicon-controlled rectifier* (*SCR*) is a three-terminal device that is very similar to the SUS. The primary difference is the addition of a *gate* terminal.

    a. The added gate terminal provides another method of component triggering (see Figure 20.8).

    b. Except for gate triggering, the operating characteristics of the SCR are identical to those of the SUS.

16. Gate triggering is accomplished by applying a *positive* pulse to the gate terminal (see Figure 20.9).

    a. The positive pulse forces $Q_2$ (in the equivalent circuit) into conduction.

    b. Once driven into conduction, the device takes itself into saturation (just like the SUS).

    c. Once saturated, the gate becomes ineffective until the device is driven back into cutoff by *anode current interruption* or *forced commutation*.

17. The operating curve of the SCR is shown in Figure 20.10.

    a. At a given value of $V_{AK}$, the current generated through the gate can cause the component to break into its forward operating region.

**b.** As long as $I_F > I_H$, the device continues to conduct, regardless of the potential applied to the gate.

18. The *circuit fusing* ($I^2t$) rating of an SCR indicates its maximum forward surge current-handling capability. If the $I^2t$ rating of an SCR is exceeded, the device may be damaged or destroyed by excessive heat.

19. *False triggering* is a situation where an SCR is accidentally triggered into conduction. False triggering can be caused by:
    **a.** A noise signal at the SCR gate terminal.
    **b.** A short rise-time noise signal in $V_{AK}$.
    **c.** A sufficient increase in temperature (as is the case with the SUS).

20. The *critical rise* ($dv/dt$) rating of an SCR indicates the maximum rate of increase in $V_{AK}$ that can occur without causing false triggering.

21. A relatively small amount of noise can have a sufficient *rate of increase* to cause false triggering (see Example 20.4).

22. False triggering from gate noise can be prevented by:
    **a.** Connecting the gate terminal (via a resistor) to a negative supply voltage.
    **b.** Connecting a bypass capacitor across from the SCR gate to ground.
    Both of these methods are illustrated in Figure 20.12.

23. False triggering from noise in $V_{AK}$ is normally prevented by the use of a *snubber network*.
    **a.** A snubber network is a simple *RC* circuit connected across the SCR (see Figure 20.13).
    **b.** The capacitor effectively shorts any noise to ground. The resistor limits the capacitor discharge current when the SCR is triggered.

24. One common application for the SCR is found in a *crowbar* (see Figure 20.14).
    **a.** A crowbar is a circuit designed to protect a voltage-sensitive load from excessive power supply voltages.
    **b.** If the output from a voltage source exceeds a specified limit, the SCR (which is in parallel with the load) triggers, shorting out the excessive voltage.
    **c.** Once activated, the SCR in a crowbar conducts until the power supply fuse gives out (which is an example of anode current interruption).

25. A *phase controller* is a circuit used to control the *conduction angle* through a load and, thus, the average load voltage. Phase controller operation is illustrated in Figure 20.15.

26. Whenever possible, an SCR is tested in-circuit.
    **a.** When open, the device won't trigger.
    **b.** When shorted, the device conducts continually.
    **c.** When an SCR cannot be tested in-circuit, it can be checked using a test circuit like the one shown in Figure 20.17.

27. *Diacs* and *triacs* are *bidirectional* thyristors; that is, they are capable of conducting in two directions.

28. The *diac* is a two-terminal, three-layer device with forward and reverse characteristics that are identical to the forward characteristics of an SUS. (The diac is essentially a bidirectional SUS.)

29. The construction of the diac is similar to that of the BJT (see Figure 20.19). However:
    **a.** There is no base terminal.
    **b.** The three regions are nearly identical in size.
    **c.** The three regions have nearly identical doping levels.
    The BJT base is very small and very lightly doped compared to the other materials.

30. The forward and reverse curves for the diac are identical to the forward operating curve of the SUS (see Figure 20.20).

31. The *triac* is a three-terminal, five-layer device with forward and reverse characteristics that are identical to the forward characteristics of the SCR (see Figure 20.21).

32. The triac is essentially a bidirectional SCR.

33. There are four combinations of component biasing and triggering that can be used to force a triac to conduct (see Figure 20.25).

34. Most triac specifications are identical to those for the SCR.

**35.** The triggering requirements for each operating quadrant are listed on the component spec sheet.

**36.** Other devices are often connected to the gate of a triac to provide:

   **a.** *Positive* gate voltage triggering only.

   **b.** *Negative* gate voltage triggering only.

   **c.** Bidirectional triggering at voltages higher than the component's $V_{GT}$ rating (see Figure 20.26).

**37.** Triacs are typically used in bidirectional phase control circuits (see Figure 20.27).

**38.** Diac and triac operating characteristics are summarized in Figure 20.29.

**39.** The *unijunction transistor (UJT)* is a three-terminal device whose trigger voltage is proportional to its applied biasing voltage.

   **a.** The terminals of a UJT are called the *emitter*, *base 1*, and *base 2*.

   **b.** The schematic symbol for the UJT is similar to that of the JFET because of similarities in their construction (see Figures 20.30 and 20.31).

**40.** The *intrinsic standoff ratio* ($\eta$) is the ratio of emitter–base 1 resistance to the total *interbase resistance*. The intrinsic standoff ratio is a component rating that is provided on the UJT spec sheet.

**41.** As UJT current increases from the *peak current* ($I_P$) to the *valley current* ($I_V$), the voltage across the component *decreasees* (see Figure 20.34). As a result, the UJT is referred to as a *negative resistance* component (as is the tunnel diode).

**42.** UJTs are often used in *relaxation oscillators*, that is, circuits that use the charge/discharge characteristics of a capacitor or inductor to produce a pulse output (see Figure 20.35).

**43.** The *programmable unijunction transistor (PUT)* is very similar to an SCR.

   **a.** A *reference voltage* is applied to the component's gate.

   **b.** When $V_K$ exceeds the reference voltage, the device is triggered into conduction.

   **c.** A relaxation oscillator can be constructed using a PUT, as shown in Figure 20.36.

**44.** UJT and PUT characteristics are summarized in Figure 20.37.

**45.** *Light* is electromagnetic energy that falls within a specific range of frequencies (see Figure 20.38).

**46.** *Optoelectronic devices* are classified as *emitters*, *detectors*, or a combination of the two.

   **a.** A *photoemitter* generates light. (The LED is a photoemitter.)

   **b.** A *photodetector* responds to light at its input.

**47.** Two characteristics of light play important roles in the operation of photoemitters and photodetectors:

   **a.** *Wavelength* ($\lambda$) is the physical length of one cycle of a transmitted electromagnetic wave.

   **b.** *Light intensity* is the amount of light per unit area received by a given photodetector. Light intensity decreases as the distance between a source and detector increases.

**48.** A *photodiode* is a diode whose *reverse* conduction is light-intensity controlled.

   **a.** An increase in light causes an increase in the reverse diode current.

   **b.** The photodiode is a detector that is typically housed in an opaque case with an LED (see Figure 20.41).

   **c.** The *light current* ($I_L$) rating indicates the amount of reverse current through a photodiode with an active light input.

   **d.** The *dark current* ($I_D$) rating indicates the amount of reverse current with no active light input.

   **e.** The value of $I_L$ is typically thousands of times as great as the value of $I_D$.

**49.** The *wavelength of peak spectral response* ($\lambda_S$) is a rating that indicates the wavelength that produces the strongest response from a photodetector, usually measured in nanometers (nm).

**50.** The *sensitivity* of a photodetector is a measure of its response to a specified light *intensity* (as opposed to frequency).

**51.** *Spectral response* is a measure of a photodetector's response to a change in wavelength. It describes how device *sensitivity* changes in response to a change in wavelength.

52. The primary photodiode ratings are summarized in Figure 20.43.
53. A *phototransistor* is a three-terminal photodetector whose collector current is controlled by the intensity of light at its optical input (base).
54. A *photo-Darlington* is a phototransistor that has a higher current capability than the standard phototransistor.
55. A *light-activated SCR* (*LASCR*), or *photo-SCR*, is triggered into conduction by an optical input. As such, it can be used as an optically coupled phase-control circuit.
56. An *optoisolator* is an optocoupler, a device that uses light to couple a signal from its input (a photo emitter) to its output (a photodetector).
   a. Optoisolators house the emitter and detector in a single IC.
   b. An input generates an internal optical response, which in turn causes a change in the circuit output.
   c. The *isolation source voltage* ($V_{ISO}$) rating of the component indicates the maximum difference of potential that can exist across its input and output circuits without damaging the IC.
57. An *optointerrupter* is an optocoupler designed to allow an external object to block the light path between the photoemitter and photodetector (see Figure 20.49).

## EQUATION SUMMARY

| Equation Number | Equation | Section Number |
|---|---|---|
| (20.1) | $I_F = \dfrac{V_S - V_{AK}}{R_S}$ | 20.1 |
| (20.2) | $t_{max} = \dfrac{I^2 t \text{ (rated)}}{I_S^2}$ | 20.2 |
| (20.3) | $I_{S(max)} = \sqrt{\dfrac{I^2 t \text{ (rated)}}{t_S}}$ | 20.2 |
| (20.4) | $\Delta V = \dfrac{dv}{dt} \Delta t$ | 20.2 |
| (20.5) | $V_K = V_{BB} \dfrac{R_{B1}}{R_{B1} + R_{B2}}$ | 20.4 |
| (20.6) | $V_K = \eta V_{BB}$ | 20.4 |
| (20.7) | $V_P = \eta V_{BB} + 0.7 \text{ V}$ | 20.4 |
| (20.8) | $\lambda = \dfrac{c}{f}$ | 20.5 |

## KEY TERMS

anode current interruption  842
average on-state current ($I_T$)  844
average on-state voltage ($V_T$)  844
bidirectional thyristor  856
circuit fusing ($I^2t$)  846
conduction angle  852
critical rise (*dv/dt*)  850
crowbar  852
dark current ($I_D$)  873
diac  856
false triggering  850

firing angle  852
forced commutation  843
forward blocking region  843
forward breakover current ($I_{BR(F)}$)  843
forward breakover voltage ($V_{BR(F)}$)  840
forward operating region  843
gate nontrigger voltage ($V_{GD}$)  849
gate trigger voltage ($V_{GT}$)  853

holding current ($I_H$)  842
industrial electronics  851
interbase resistance  866
intrinsic standoff ratio ($\eta$)  866
irradiance  873
isolation capacitance ($C_{ISO}$)  880
isolation current ($I_{ISO}$)  880
isolation resistance ($R_{ISO}$)  880
isolation source voltage ($V_{ISO}$)  880
light  871

## PRACTICE PROBLEMS

### Section 20.2

1. The 2N6237 SCR has a circuit fusing rating of 2.6 A²s. Determine whether the component can withstand a 25 A surge that lasts for 120 ms.

2. The 2N6342 SCR has a circuit fusing rating of 40 A²s. Determine whether the device can withstand an 80 A surge that lasts for 8 ms.

3. The 2N6237 SCR has a circuit fusing rating of 2.6 A²s. Determine the maximum allowable duration of a 50 A surge through the device.

4. The 2N6342 has a circuit fusing rating of 40 A²s. Determine the maximum allowable duration of a 95 A surge through the device.

5. The 2N6237 has a circuit fusing rating of 2.6 A²s. Determine the maximum surge current that the device can withstand for 50 ms.

6. The 2N6342 has a circuit fusing rating of 40 A²s. Determine the maximum surge current that the device can withstand for 100 ms.

7. The 2N6237 has a critical rise rating of $dv/dt = 10$ V/μs. Determine the anode noise amplitude at $t_r = 10$ ns required to cause false triggering.

8. The 2N6342 has a critical rise rating of $dv/dt = 5$ V/μs. Determine the anode noise amplitude at $t_r = 25$ ns required to cause false triggering.

### Section 20.4

9. The 2N2646 has a range of $\eta = 0.56$ to $0.75$. Determine the range of $V_P$ values for the device when $V_{BB} = +14$ V.

10. The 2N4871 UJT has a range of $\eta = 0.7$ to $0.85$. Determine the range of $V_P$ values for the device when $V_{BB} = +12$ V.

11. The 2N4948 UJT has a range of $\eta = 0.55$ to $0.82$. Determine the range of $V_P$ values for the device when $V_{BB} = +16$ V.

12. The 2N4949 UJT has a range of $\eta = 0.74$ to $0.86$. Determine the range of $V_P$ values for the device when $V_{BB} = +26$ V.

### Section 20.5

13. Determine the wavelength (in nm) for a 650 THz signal.

14. Determine the wavelength (in nm) for a 220 THz signal.

15. Determine the wavelength (in nm) for a 180 THz signal.

16. The visible light spectrum includes all frequencies between approximately 400 and 800 THz. Determine the range of wavelength values for this band of frequencies.

17. A photodetector has a value of $\lambda_S = 0.94$ nm. Determine the optimum operating frequency for the device.

**18.** A photodetector has a rating of $\lambda_S = 0.84$ nm. Determine the optimum operating frequency for the device.

**19.** The spec sheet for the SCR in Figure 20.50 is given in Figure 20.11. The output short-circuit current for the dc power supply is 4 A and will remain at that level until its primary fuse blows. For the circuit, determine the following:

 **a.** The maximum value of $V_{dc}$ required to trigger the SCR.
 **b.** The maximum value of zener current at the instant that the SCR triggers.
 **c.** Whether the fuse or the SCR will open first.

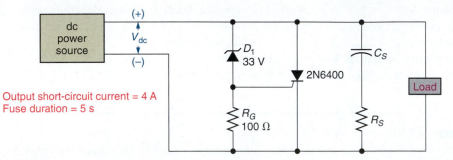

**FIGURE 20.50**

**20.** Refer to Figure 20.51. $R_2$ in the circuit is set to the value shown. The spec sheet for the 2N6400 is given in Figure 20.11. Determine the conduction angle for the circuit load. (*Hint*: Start by finding the maximum value of $v_{in}$ required to trigger the SCR.)

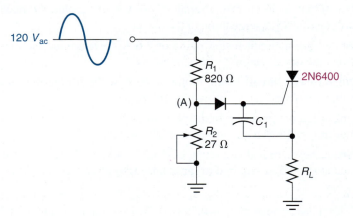

**FIGURE 20.51**

**21.** Refer to Figure 20.35. The relaxation oscillator shown contains a 51 V zener diode. The UJT has a range of $\eta = 0.7$ to $0.82$. Determine the minimum value of $V_E$ required to trigger the UJT.

**20.1** For the application, $I^2t = 54$ A$^2$s. The device cannot handle the surge.

**20.2** 3.3 ms

**20.3** 38.73 A

**20.4** 2.5 mV

**20.5** 9.7 V

**20.6** 750 to 10,000 nm (10 μm)

**20.7** 319.2 THz

## chapter 21

# Discrete and Integrated Voltage Regulators

## Objectives

*After studying the material in this chapter, you should be able to:*

1. List the purposes served by a voltage regulator.

2. Define *line regulation* and its commonly used units of measure.

3. Define *load regulation* and its commonly used units of measure.

4. Describe and analyze the operation of the *pass-transistor regulator*.

5. Describe the means by which *short-circuit protection* is provided for a pass-transistor regulator.

6. Describe and analyze the operation of the *shunt-feedback regulator*.

7. Discuss the need for *overvoltage protection* in a shunt-feedback regulator and the means by which it is provided.

8. List the reasons why the series regulator is preferred over the shunt regulator.

9. List and describe the various types of linear IC voltage regulators.

10. List and describe the commonly used linear IC regulator parameters and ratings.

11. Describe the fundamental operating difference between the linear regulator and the switching regulator.

12. List the four parts of the basic switching regulator, and describe the function of each.

13. Describe the response of a switching regulator to a change in load demand.

14. Describe the two most commonly used methods for controlling power switch conduction.

15. List the circuit recognition features for each of the common switching regulator configurations.

16. Describe the functions performed by a typical switching regulator IC.

17. In terms of their advantages, disadvantages, and applications, compare and contrast linear and switching regulators.

# Outline

In Chapter 3, you were introduced to the operation of the basic power supply. At that time, you were shown that a *rectifier* is used to convert the ac output of a transformer to *pulsating dc*. This pulsating dc is then applied to a *filter*, which reduces the variations in the rectifier output. Finally, the filtered dc is applied to a *voltage regulator*. This regulator serves two purposes:

**OBJECTIVE 1 ▶**

**1.** It reduces the variations (ripple) in the filtered dc.
**2.** It maintains a relatively constant output voltage regardless of minor changes in load current demand and/or input voltage.

The block diagram of the basic power supply is shown (along with its waveforms) in Figure 21.1.

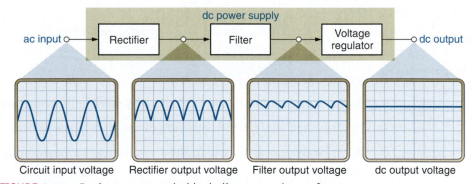

**FIGURE 21.1** Basic power supply block diagram and waveforms.

The voltage regulator we used in Chapter 3 was a *zener diode* connected in parallel with the load. In practice, this type of regulator is rarely used. The primary problem with the simple zener regulator is that it wastes a tremendous amount of power.

Practical voltage regulators contain a number of discrete and/or integrated active devices. In this chapter, we will look at the operation of many of these more practical voltage regulators.

## 21.1 Voltage Regulation: An Overview

The *ideal* voltage regulator maintains a constant dc output voltage regardless of changes in its input voltage or its load current demand. For example, consider the $+10$ $V_{dc}$ regulator shown in Figure 21.2. A change in the regulator input voltage (shown as $\Delta V_{in}$ in Figure 21.2a) would not be coupled to the regulator output. Note that $\Delta V_{in}$ could be either a change in the steady-state (dc) value of $V_{in}$ or could be some ripple voltage. In either case, the change in $V_{out}$ for the ideal regulator would be 0 V. This assumes, of course, that the value of $V_{in}$ does not decrease below the value required to maintain the operation of the regulator.

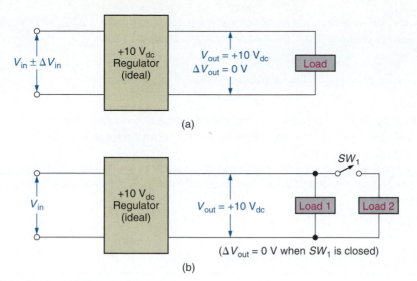

(a)

(b)

FIGURE 21.2   A +10 $V_{dc}$ regulator.

Figure 21.2b shows how the ideal voltage regulator would respond to a change in load current demand. Assume that *load 1* and *load 2* have the same resistance. When $SW_1$ is closed, the added load causes the current demand on the regulator to *double*. The ideal voltage regulator maintains a constant output voltage ($\Delta V_{out} = 0$ V) despite the change in load current demand.

### 21.1.1   Line Regulation

In practice, a change in the input voltage to a regulator *does* cause a change in its output voltage. The **line regulation** rating of a voltage regulator indicates *the change in output voltage that occurs per unit change in the input voltage*. The line regulation of a voltage regulator is found as

$$\text{line regulation} = \frac{\Delta V_{out}}{\Delta V_{in}} \qquad \textbf{(21.1)}$$

◄ *OBJECTIVE 2*

**Line regulation**
A rating that indicates the change in regulator output voltage that occurs per unit change in input voltage.

where $\Delta V_{out}$ = the change in output voltage (usually in microvolts or millivolts)
$\Delta V_{in}$ = the change in input voltage (usually in volts)

Example 21.1 illustrates the concept of line regulation.

#### EXAMPLE 21.1

A voltage regulator experiences a 10 μV change in output voltage when its input voltage changes by 5 V. Determine the value of line regulation for the circuit.

*Solution:*   The line regulation of the circuit is found as

$$\textbf{line regulation} = \frac{\Delta V_{out}}{\Delta V_{in}} = \frac{10 \ \mu V}{5 \ V} = \textbf{2} \ \boldsymbol{\mu} \textbf{V/V}$$

The 2 μV/V rating means that the output voltage will change by 2 μV for every 1 V change in the regulator's input voltage.

#### PRACTICE PROBLEM 21.1

The change in output voltage for a voltage regulator is measured at 100 μV when the input voltage changes by 4 V. Calculate the line regulation rating for the regulator.

As was stated earlier, the *ideal* voltage regulator would have a $\Delta V_{out}$ of 0 V when the input voltage changes. Thus, for the ideal voltage regulator:

$$\text{line regulation} = \frac{\Delta V_{out}}{\Delta V_{in}} = \frac{0 \text{ V}}{\Delta V_{in}} = 0$$

Based on the fact that the ideal line regulation rating is *zero*, we can make the following statement: *The lower the line regulation rating of a voltage regulator, the higher the quality of the circuit.*

Line regulation is commonly expressed in a variety of units. The $\mu$V/V rating used in Example 21.1 is only one of them. The commonly used line regulation units are identifiied in Table 21.1.

*A Practical Consideration:*
Technically, line regulation should have no unit of measure because it is the ratio of one voltage value to another. However, the units listed in Table 21.1 are commonly used on spec sheets, so you need to know what they are.

TABLE 21.1   Commonly Used Line Regulation Units

| Unit | Meaning |
|------|---------|
| $\mu$V/V | The change in output voltage (in microvolts) per 1 V change in input voltage. |
| ppm/V | *Parts-per-million per volt.* Another way of saying *microvolts per volt.* |
| %/V | The percentage of change in the output voltage that may occur per 1 V change in the input voltage. |
| % | The total percentage of change in output voltage that can occur over the rated range of input voltages. |
| mV (or $\mu$V) | The actual change in output voltage that may occur over the rated range of input voltages. |

### 21.1.2   Load Regulation

**OBJECTIVE 3** ▶

The practical voltage regulator also experiences a slight change in output voltage when there is a change in load current demand. The **load regulation** rating indicates the change in regulator output voltage that occurs per unit change in load *current*. The load regulation of a voltage regulator is found as

**Load regulation**
A rating that indicates the change in regulator output voltage per unit change in load *current.*

$$\text{load regulation} = \frac{V_{NL} - V_{FL}}{\Delta I_L} \qquad \textbf{(21.2)}$$

where $V_{NL}$ = the no-load output voltage (the output voltage when the load is open)
$V_{FL}$ = the full-load output voltage (the output voltage when the load current demand is at its maximum value)
$\Delta I_L$ = the *change in* load current demand

Another way of expressing this is

$$\text{load regulation} = \frac{\Delta V_{out}}{\Delta I_L}$$

Example 21.2 illustrates the concept of load regulation.

### EXAMPLE 21.2

A voltage regulator is rated for an output current of $I_L$ = 0 to 20 mA. Under no-load conditions, the output voltage from the circuit is 5 V. Under full-load conditions, the output voltage from the circuit is 4.9998 V. Determine the value of load regulation for the circuit.

*Solution:* The load regulation of the circuit is found as

$$\text{load regulation} = \frac{V_{NL} - V_{FL}}{\Delta I_L} = \frac{5 \text{ V} - 4.9998 \text{ V}}{20 \text{ mA}} = 10 \ \mu\text{V/mA}$$

The 10 $\mu$V/mA rating indicates that the output voltage changes by 10 $\mu$V for each 1 mA change in load current.

### PRACTICE PROBLEM 21.2

A voltage regulator is rated for an output current of $I_L = 0$ to 40 mA. Under no-load conditions, the output voltage from the circuit is 8 V. Under full-load conditions, the output voltage from the circuit is 7.996 V. Determine the value of load regulation for the circuit.

Like the line regulation rating, load regulation is commonly expressed in a variety of units. The commonly used load regulation units are identified in Table 21.2.

TABLE 21.2   Commonly Used Load Regulation Units

| Unit | Meaning |
|------|---------|
| $\mu$V/mA | The change in output voltage (in microvolts) per 1 mA change in load current. |
| %/mA | The percentage of change in the output voltage per 1 mA change in load current. |
| % | The total percentage of change in output voltage over the rated range of load current values. |
| mV (or $\mu$V) | The actual change in output voltage that may occur over the rated range of load current values. |
| $\Omega$ | V/mA, expressed as a resistance value; the rating times the value of $\Delta I_L$ gives you the corresponding value of $\Delta V_L$ for the regulator. |

The ideal voltage regulator would not experience a change in output voltage when the load current demand increases. Thus, for the ideal regulator, $V_{NL} = V_{FL}$ and load regulation equals *zero*. Based on this fact, we can state that *the lower the load regulation rating of a voltage regulator, the higher the quality of the circuit.*

## 21.1.3   Line and Load Regulation: Some Practical Considerations

Some manufacturers combine the *line regulation* and *load regulation* ratings into a single **regulation** rating. This rating indicates the *maximum* change in output voltage that can occur when input voltage *and* load current are varied over their entire rated ranges. For example, one voltage regulator has the following ratings:

$$V_{in} = 12 \text{ to } 24 \text{ V}_{dc}$$

$$I_L = 40 \text{ mA (maximum)}$$

$$\text{regulation} = 0.33\%$$

> **Regulation**
> A rating that indicates the maximum change in regulator output voltage that can occur when input voltage *and* load current are varied over their entire rated ranges.

These ratings indicate that the output voltage of the regulator will vary by no more than 0.33% as long as $V_{in}$ remains between 12 and 24 $V_{dc}$ and load current does not exceed 40 mA. A single regulation rating is always given either as a percentage or in millivolts or microvolts.

There is one other important consideration we need to discuss. The unit of measure used for line regulation and/or load regulation is often a good indicator of the quality of that regulator. For example, let's say that we need to choose between two voltage regulators with the following ratings:

| Regulator | dc Output Voltage | Line Regulation |
|-----------|-------------------|-----------------|
| A | +10 V | 0.12 %/V |
| B | +10 V | 40 μV/V |

If the input voltage to regulator A increases by 5 V, the output voltage of the circuit is

$$V_{out} = V_{dc} + (0.0012 V_{dc})(\Delta V_{in}) = 10 \text{ V} + 0.06 \text{ V} = \textbf{10.06 V}$$

If the input voltage to regulator B increases by 5 V, the output voltage of the circuit is

$$V_{out} = 10 \text{ V} + (40 \text{ μV})(\Delta V_{in}) = 10 \text{ V} + 200 \text{ μV} = \textbf{10.0002 V}$$

As you can see, the regulator with the μV/V rating experiences a smaller change in output voltage than the regulator with the %/V rating. Since the ideal voltage regulator would have no change in output voltage when the input voltage changed, regulator B comes much closer to the ideal regulator than regulator A.

### 21.1.4 Types of Regulators

There are two basic types of discrete voltage regulators: the *series regulator* and the *shunt regulator*. These two types of regulators are represented by the blocks in Figure 21.3. The **series regulator** is placed in *series* with the load, as shown in Figure 21.3a. The **shunt regulator** is placed in *parallel* with the load, as shown in Figure 21.3b. As you will see, series and shunt regulators each have their own advantages and disadvantages.

**Series regulator**
A voltage regulator in series with the load.

**Shunt regulator**
A voltage regulator in parallel with the load.

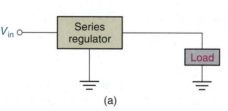

(a)      (b)

**FIGURE 21.3** Series and shunt regulators.

**Section Review ▶**

1. What two purposes are served by a voltage regulator?
2. How would the *ideal* voltage regulator respond to a change in input voltage?
3. How would the *ideal* voltage regulator respond to a change in load current demand?
4. What is *line regulation*?
5. What is the line regulation value of an ideal voltage regulator?
6. What is the relationship between the quality of a voltage regulator and its line regulation rating?
7. List and define the commonly used line regulation ratings.
8. What is *load regulation*?
9. List and define the commonly used load regulation ratings.
10. What is the load regulation value of an ideal voltage regulator?
11. What is the relationship between the quality of a voltage regulator and its load regulation rating?
12. Explain the single *regulation* rating.
13. What is a *series regulator*?
14. What is a *shunt regulator*?

Series voltage regulators can take a variety of forms. However, they all have one or more active devices placed in series with the load. In this section, we will take a look at several common series voltage regulators.

### 21.2.1 Pass-Transistor Regulator

The **pass-transistor regulator** uses a series transistor, called a *pass-transistor*, to regulate load voltage. The term *pass-transistor* is used because the load current passes through the series transistor, $Q_1$, as shown in Figure 21.4.

◀ *OBJECTIVE 4*

**Pass-transistor regulator**
A regulator that uses a series transistor to maintain a constant load voltage.

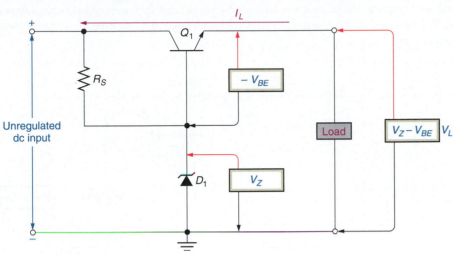

FIGURE 21.4   Pass-transistor regulator.

The key to the operation of the pass-transistor regulator is the fact that the base voltage is held to the relatively constant voltage across the zener diode. For example, if a 9.1 V zener diode is used, the base voltage of $Q_1$ is held at approximately 9.1 V. Since $Q_1$ is an *npn* transistor, $V_L$ is found as

$$V_L = V_Z - V_{BE} \qquad (21.3)$$

If we rearrange this equation, we get

$$V_{BE} = V_Z - V_L \qquad (21.4)$$

Equation (21.4) can be used to explain the response of the pass-transistor to a change in load resistance. If load resistance *increases*, load voltage also increases. Since the zener voltage is constant, the increase in $V_L$ causes $V_{BE}$ to decrease [as given in equation (21.4)]. The decrease in $V_{BE}$ reduces conduction through the pass-transistor, so load current *decreases*. This offsets the increase in load resistance, and a relatively constant load voltage is maintained.

In a similar fashion, a *decrease* in load resistance causes $V_L$ to decrease. The reduction in $V_L$ causes $V_{BE}$ to increase [as given in equation (21.4)]. The increase in $V_{BE}$ increases conduction through the pass-transistor, so load current *increases*. This offsets the decrease in load resistance, and again, a relatively constant load voltage is maintained.

The pass-transistor regulator in Figure 21.4 has relatively good line and load regulation characteristics, but there is a problem with the circuit. If the input voltage or load current values increase, the zener diode may have to dissipate a relatively high amount of power. You see, increases in $V_{in}$ or $I_L$ result in an increase in zener conduction. The increased zener conduction results in higher power dissipation by the device. This problem is reduced by the use of the *Darlington pass-transistor regulator.*

*What Purpose Is Served by $R_S$?*
In many cases, the zener requires more current (to maintain regulation) than the transistor base can allow. $R_S$ provides an alternate path for this additional zener current.

## 21.2.2 Darlington Pass-Transistor Regulator

**Darlington pass-transistor regulator**

A series regulator that uses a Darlington pair in place of a single pass-transistor.

The **Darlington pass-transistor regulator** uses a Darlington pair ($Q_1$ and $Q_2$), in place of a single pass-transistor, as shown in Figure 21.5. The load voltage for the Darlington circuit is found as

$$V_L = V_Z - 2V_{BE} \qquad (21.5)$$

Since the total current gain of a Darlington pair is equal to $h_{FE1}h_{FE2}$, an increase in load current causes very little, if any, increase in zener current. Therefore, the regulator in Figure 21.5 is not subject to the same power concerns as the regulator in Figure 21.4. However, zener conduction *is* affected by temperature and the zener current is applied to the high-current-gain Darlington pair. As a result, the load current in the Darlington pass-transistor regulator can be severely affected by significant increases in operating temperature. For this reason, the Darlington pass-transistor regulator must be kept at a relatively constant temperature.

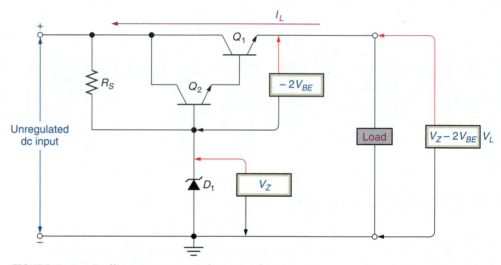

**FIGURE 21.5**   Darlington pass-transistor regulator.

## 21.2.3 Series Feedback Regulator

**Series feedback regulator**

A series regulator that uses an error-detection circuit to improve the line and load regulation characteristics of other pass-transistor regulators.

The **series feedback regulator** uses an *error detector* to improve the line and load regulation characteristics of the other pass-transistor regulators. The block diagram for the series feedback regulator is shown in Figure 21.6.

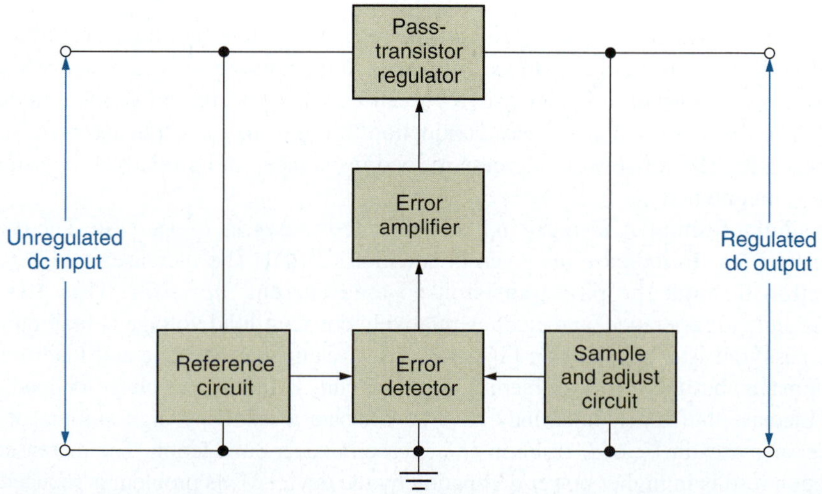

**FIGURE 21.6**   Block diagram for the series feedback regulator.

The error detector receives two inputs: a *reference voltage* derived from the unregulated dc input voltage and a *sample voltage* from the regulated output voltage. The error detector compares the reference and sample voltages and provides an output voltage that is proportional to the difference between the two. This output voltage is amplified and used to drive the pass-transistor regulator.

The series feedback regulator is capable of responding very quickly to differences between its sample and reference input voltages. This gives the circuit much better line and load regulation characteristics than the other circuits we have discussed in this section.

The schematic diagram for a series feedback regulator is shown in Figure 21.7. The *sample and adjust circuit* is the voltage divider that consists of $R_3$, $R_4$, and $R_5$. The reference voltage is set by the zener diode. $Q_2$ detects and amplifies the difference between the reference and sample voltages and adjusts the conduction of the pass-transistor accordingly.

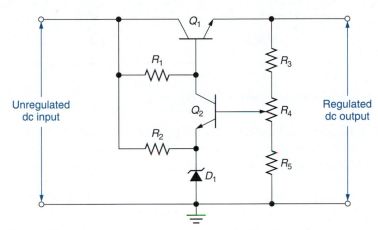

**FIGURE 21.7**  Schematic diagram for a basic series feedback regulator.

The operation of the circuit shown in Figure 21.7 is best illustrated by describing its response to a change in load resistance. If load resistance *increases*, load voltage starts to increase. The voltage divider is in parallel with the load, so the voltage at the base of $Q_2$ also increases. The increase in $V_{B(Q2)}$ increases the conduction through the transistor and its collector resistor ($R_1$). The increased conduction causes $V_{C(Q2)}$ to decrease, which reduces the value of $V_{B(Q1)}$. The reduction in $V_{B(Q1)}$ reduces the conduction through the pass-transistor, causing load current to decrease. The decrease in load current offsets the initial increase in load resistance.

If the load resistance decreases, $V_L$ starts to decrease. The decrease in $V_L$ causes $V_{B(Q2)}$ to decrease, which reduces the conduction through the transistor. The reduced conduction through $Q_2$ causes $V_{C(Q2)}$ to increase, increasing $V_{B(Q1)}$. The increase in $V_{B(Q1)}$ causes the pass-transistor conduction to increase. This increases the value of load current, offsetting the initial decrease in load resistance. Again, load regulation is maintained.

## 21.2.4  Short-Circuit Protection

One weakness of a standard series regulator is the possibility of the pass-transistor being destroyed by excessive load current if the load is shorted. To prevent a shorted load from destroying the pass-transistor, a *current-limiting circuit* can be added to the regulator, as shown in Figure 21.8.

◀ *OBJECTIVE 5*

The current-limiting circuit consists of a transistor ($Q_3$) and a series resistor ($R_S$) that is connected between its base and emitter terminals. For $Q_3$ to conduct, the voltage across $R_S$ must reach approximately 0.7 V. This happens when

$$I_L = \frac{0.7\text{ V}}{1\ \Omega} = 700\text{ mA}$$

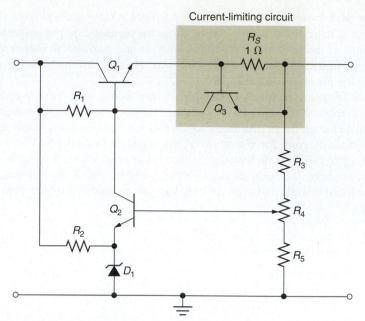

**FIGURE 21.8** Series regulator current-limiting circuit.

Thus, when load current is less than 700 mA, $Q_3$ is in cutoff and the circuit acts exactly as described earlier. If the load current increases above 700 mA, $Q_3$ conducts, decreasing the voltage at the base of $Q_1$. The decreased $V_{B(Q1)}$ reduces the conduction of the pass-transistor, preventing any further increases in load current. Thus, the load current for the circuit is limited to approximately 700 mA. In fact, for the type of current-limited regulator shown, the maximum load current can be found as

$$I_{L(\text{max})} \cong \frac{V_{BE(Q3)}}{R_S} \qquad \textbf{(21.6)}$$

Thus, the maximum allowable load current for the regulator can be set to any value by using the appropriate value of $R_S$.

You may be wondering why we don't simply use a single current-limiting resistor to protect the pass-transistor. If a current-limiting resistor is used, it must have a fairly significant value to protect the pass-transistor. The voltage drop across such a resistor would be much higher than the voltage across a 1 Ω resistor. The current-limiting circuit is used to provide short-circuit protection without causing a significant decrease in the regulator output voltage.

### 21.2.5   One Final Note

The possibility of a shorted load damaging a pass-transistor is not the only drawback to the series regulator. For one thing, the fact that $I_L$ passes through the pass-transistor means that this transistor dissipates a fairly significant amount of power. Also, there is a voltage drop across the pass-transistor. This voltage drop reduces the maximum possible output voltage.

Even with its power dissipation and short-circuit protection problems, the series regulator is used more commonly than the shunt regulator. As you will see in the next section, the shunt regulator has several problems that make the series regulator the more desirable of the two.

---

**Section Review ▶**

1. Describe the basic *pass-transistor regulator*.

2. Describe how the pass-transistor regulator in Figure 21.4 responds to a change in load resistance.

3. Explain how the *Darlington pass-transistor regulator* reduces the problem of excessive zener diode power dissipation.

**4.** Describe the response of the *series feedback regulator* to a change in load resistance.

**5.** Describe the operation of the current-limiting circuit shown in Figure 21.8.

**6.** List the disadvantages of using series voltage regulators.

**7.** How would the circuit shown in Figure 21.4 respond to slight variations in line voltage?   ◄ **Critical Thinking**

**8.** How would the circuit shown in Figure 21.7 respond to slight variations in line voltage?

## 21.3   Shunt Voltage Regulators

The most basic shunt regulator is the simple zener regulator we discussed in Chapter 3. As you know, this regulator wastes far too much power for most applications and, thus, is rarely used. In this section, we will discuss the *shunt feedback regulator*. As you will see, this regulator is similar to the series feedback regulator we covered in the last section.

### 21.3.1   Shunt Feedback Regulator

The **shunt feedback regulator** uses an error detector to control the conduction of a *shunt transistor*. This shunt transistor is shown as $Q_1$ in the shunt feedback regulator in Figure 21.9.

◄ **OBJECTIVE 6**

**Shunt feedback regulator**
A circuit that uses an error detector to control the conduction of a *shunt* transistor.

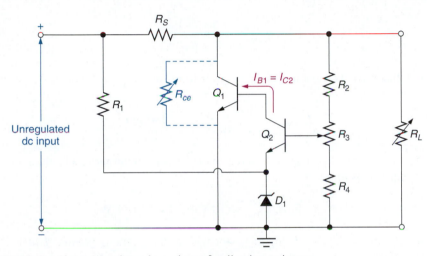

**FIGURE 21.9**   Shunt transistor in a shunt feedback regulator.

The *sample circuit* is (once again) a simple voltage divider. The *reference circuit* is made up of $D_1$ and $R_1$. The outputs from the sample and reference circuits are applied to the *error detector*, $Q_2$. The output from $Q_2$ is then used to control the conduction of the shunt transistor, $Q_1$.

The operation of the shunt feedback regulator is easiest to understand if we view the shunt transistor ($Q_1$) as a variable resistor, $R_{ce}$. When $Q_1$ is not conducting, $R_{ce}$ is at its maximum value. When $Q_1$ is saturated, $R_{ce}$ is at its minimum value. Thus, we can say that *the value of $R_{ce}$ decreases with increases in the conduction of $Q_1$, and vice versa.*

Under normal circumstances, $R_3$ is set so that $Q_2$ is conducting. Since $I_{B1} = I_{C2}$, $Q_1$ is biased somewhere in its active region of operation. The conduction of the two-transistor circuit is set so that $R_{ce}$ is approximately midway between its two extremes in value.

If the load resistance *decreases*, $V_L$ starts to decrease. Since the voltage divider is in parallel with the load, $V_{B(Q2)}$ also decreases. The decrease in $V_{B(Q2)}$ reduces the conduction through $Q_2$, thereby reducing the value of $I_{B(Q1)}$. This decrease in its base current causes $R_{ce}$ to increase, which increases the voltage at the collector of $Q_1$. Since $V_L = V_{C(Q1)}$, the increase in $V_{C(Q1)}$ offsets the initial decrease in $V_L$.

If $R_L$ *increases*, $V_L$ starts to increase. The increase in $V_L$ increases $V_{B(Q2)}$. This increases the conduction through $Q_2$, causing an increase in $I_{B(Q1)}$. The increase in $I_{B(Q1)}$

causes the shunt transistor's conduction to increase, reducing the value of $R_{ce}$. When $R_{ce}$ decreases, $V_{C(Q1)}$ decreases, offsetting the initial increase in $V_L$.

The series resistor ($R_S$) combines with $R_{ce}$, the voltage divider, and the load to form a variable voltage divider that provides regulation control. In addition, $R_S$ provides shorted-load protection for the power supply circuitry. Normally, this resistor must be a high-wattage resistor, since the regulator current and $I_L$ must pass through it.

### 21.3.2  Overvoltage Protection

**OBJECTIVE 7 ▶** Just as the series regulator must be protected against shorted-load conditions, the shunt regulator must be protected from input overvoltage conditions. If the unregulated dc input voltage to the regulator increases, the conduction of the shunt transistor increases to maintain the constant output voltage. Assuming that $V_{CE}$ of $Q_1$ remains relatively constant, the increased conduction through the transistor causes an increase in its power dissipation.

Several things can be done to ensure that any increase in unregulated dc input voltage will not destroy the shunt transistor. First, you can use a transistor whose $P_{D(max)}$ rating is far greater than the maximum power dissipation that would ever be required in the circuit. For example, assume that the regulator in Figure 21.9 has an input voltage that can go as high as 20 V and that the regulator is designed to provide a regulated $10 \text{ V}_{dc}$ output. The worst-case value of transistor current can be approximated using

$$I_{C(max)} = \frac{V_{in} - V_{dc}}{R_S}$$

Assuming that a 100 $\Omega$ resistor is used, $I_{C(max)}$ is found as

$$I_{C(max)} = \frac{20 \text{ V} - 10 \text{ V}}{100} = \textbf{100 mA}$$

Since the transistor is dropping 10 V across its collector-emitter terminals, the maximum power dissipation is found as

$$\boldsymbol{P_D} = V_{CE}I_{C(max)} = (10 \text{ V})(100 \text{ mA}) = \textbf{1 W}$$

Thus, for the circuit described, any transistor with a $P_{D(max)}$ rating *greater than* 1 W can be used.

Of course, there are always circumstances where the unregulated dc input voltage can exceed its maximum *rated* value. To protect the circuit from such a circumstance, a *crowbar* circuit can be added to the input of the regulator. You may recall (from Chapter 20) that a *crowbar* is a circuit that uses an SCR to protect its load from an overvoltage condition. If a crowbar is added to the regulator input, the circuit is protected from any extreme input overvoltage condition that may arise.

Another practical consideration involves the potentiometer, $R_3$. This potentiometer is included in the circuit to provide an adjustment for the dc output voltage. Adjusting $R_3$ varies the condition of $Q_1$ and, thus, the value of $R_{ce}$. Since $R_{ce}$ affects the value of $V_L$, varying $R_3$ sets the value of the regulated dc output voltage.

### 21.3.3  One Final Note

**OBJECTIVE 8 ▶** The series voltage regulator is preferred over the shunt regulator for several reasons. The primary problem with the shunt regulator is that a fairly significant portion of the total current through $R_S$ passes through the shunt transistor rather than to the load. Another problem involves the voltage drop across $R_S$ and the resulting power dissipation. Finally, an input overvoltage condition is far more likely to occur than a shorted-load condition. In other words, the fault that can damage the shunt regulator is more

*Don't Forget:*
High-voltage problems are more common than shorted-component problems.

likely to occur than the fault that can damage the series regulator. When you consider all these drawbacks, it is easy to understand why the series regulator is preferred over the shunt regulator.

◀ Section Review

1. Describe the response of the shunt feedback regulator to a decrease in load resistance.
2. Describe the response of the shunt feedback regulator to an increase in load resistance.
3. Explain how an input overvoltage condition can destroy the shunt transistor in a shunt feedback regulator.
4. Discuss the means by which a shunt regulator can be protected from an input overvoltage problem.
5. Why are series regulators preferred over shunt regulators?
6. Which of the transistors in Figure 21.9 requires a higher power dissipation rating than the other? Why?

◀ Critical Thinking

## 21.4 Linear IC Voltage Regulators

A **linear IC voltage regulator** is a device used to hold the output voltage from a dc power supply relatively constant over a specified range of line and load variations. Most of the commonly used IC voltage regulators are three-terminal devices, though some types require more than three terminals. The schematic symbol for a three-terminal regulator is shown in Figure 21.10.

◀ OBJECTIVE 9

> **Linear IC voltage regulator**
> A device used to hold the output voltage from a dc power supply relatively constant over a specified range of line and load variations.

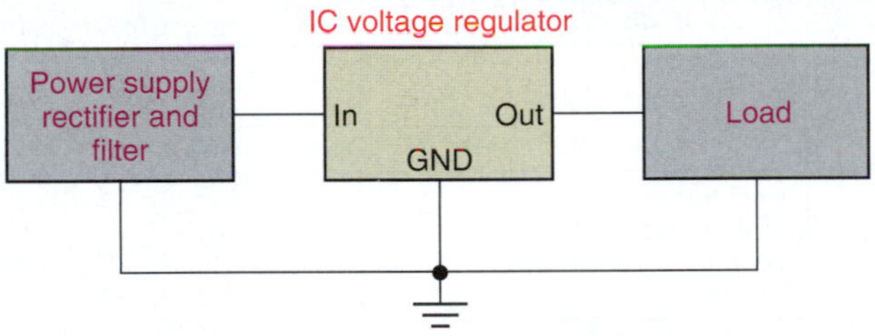

**FIGURE 21.10** Schematic symbol for a three-terminal regulator.

There are basically four types of IC voltage regulators: *fixed positive*, *fixed negative*, *adjustable*, and *dual tracking*. The **fixed-positive** and **fixed-negative** IC voltage regulators are designed to provide specific output voltages. For example, the LM309 (fixed positive) provides a +5 V output (as long as the regulator input voltages and load demand stay within their specified ranges). The **adjustable regulator** can be adjusted to provide any dc output voltage within its two specified limits. For example, the LM317L output can be adjusted to any value between its limits of +1.2 and +37 V. Both positive and negative variable regulators are available. The **dual-tracking regulator** provides equal positive and negative output voltages. For example, the RC4195 provides outputs of +15 and −15 V. Adjustable dual-tracking regulators are also available. These regulators have outputs that can be varied between their two rated limits. A single control varies both outputs so that they are always equal in magnitude. For example, if an adjustable dual-tracking regulator is adjusted for a positive output of +2 V, the negative output is automatically adjusted to −2 V.

Regardless of the type of linear regulator used, the regulator input polarity must match the device's rated output polarity. In other words, positive regulators must have positive input voltages, and negative regulators must have negative input voltages. Dual-tracking regulators require *both* positive and negative input voltages.

IC voltage regulators are *series regulators*. They contain internal pass-transistors and transistor control components. Generally, the internal circuitry of an IC voltage regulator resembles that of the series feedback regulator.

> **Fixed-positive regulator**
> A regulator with a predetermined +V output.
>
> **Fixed-negative regulator**
> A regulator with a predetermined −V output.
>
> **Adjustable regulator**
> A regulator whose output voltage can be set to any value within specified limits.
>
> **Dual-tracking regulator**
> A regulator that provides equal +V and −V outputs.

### 21.4.1  IC Regulator Specifications

**Input/output voltage differential**

The maximum allowable difference between $V_{in}$ and $V_{out}$ for an IC voltage regulator.

For our discussion on common IC regulator specifications, we'll use the spec sheet for the LM317L voltage regulator. As was stated earlier, this device is an adjustable regulator whose output can be varied between +1.2 and +37 V. The spec sheet for the LM317L is shown in Figure 21.11.

The **input/output voltage differential** rating (which is listed in the *absolute maximum ratings* section of the spec sheet) indicates the maximum difference between $V_{in}$ and $V_{out}$ that can occur without damaging the device. For the LM317L, this rating is 40 V. The

# LM317L

## Absolute Maximum Ratings

| Parameter | Symbol | Value | Unit |
|---|---|---|---|
| Input-Output Voltage Differential | $V_I - V_O$ | 40 | V |
| Power Dissipation | $P_D$ | Internally limited | W |
| Operating Junction Temperature Range | $T_j$ | 0 ~ +125 | °C |
| Storage Temperature Range | $T_{STG}$ | -65 ~+125 | °C |

## Electrical Characteristics

($V_I - V_O$ = 5V, $I_O$ = 40mA, 0°C ≤ $T_J$ ≤ +125°C, $P_{DMAX}$ = 625mW, unless otherwise specified)

| Parameter | Symbol | Conditions | Min. | Typ. | Max. | Unit |
|---|---|---|---|---|---|---|
| *Line Regulation | $R_{line}$ | $T_A$ = +25°C<br>3V ≤ $V_I - V_O$ ≤ 40V | - | 0.01 | 0.04 | %/V |
| | | 3V ≤ $V_I - V_O$ ≤ 40V | - | 0.02 | 0.07 | |
| *Load Regulation | $R_{load}$ | $T_A$ = +25°C<br>10mA ≤ $I_O$ ≤100mA<br>$V_O$ ≤ 5V<br>$V_O$ ≥5V | - | 5<br>0.1 | 25<br>0.5 | mV<br>%/ $V_O$ |
| | | 10mA ≤ $I_O$ ≤ 100mA<br>$V_O$ ≤ 5V<br>$V_O$ ≥ 5V | - | 20<br>0.3 | 70<br>1.5 | mV<br>%/ $V_O$ |
| Adjustment Pin Current | $I_{ADJ}$ | - | - | 50 | 100 | μA |
| Adjustment Pin Current Change | $\Delta I_{ADJ}$ | 3V ≤ $V_I - V_O$ ≤ 40V<br>10mA ≤ $I_O$ ≤ 100mA<br>$P_D < P_{DMAX}$ | - | 0.2 | 5 | μA |
| Reference Voltage | $V_{REF}$ | 3V < $V_I - V_O$ <40V<br>10mA ≤ $I_O$ ≤100mA<br>$P_D ≤ P_{DMAX}$ | 1.20 | 1.25 | 1.30 | V |
| Temperature Stability | $ST_T$ | - | - | 0.7 | - | % |
| Minimum Load Current to Maintain Regulation | $I_{L(MIN)}$ | $V_I - V_O$ = 40V | - | 3.5 | 10 | mA |
| Maximum output Current | $I_{O(MAX)}$ | $V_I - V_O$ ≤ 15V, $P_D < P_{DMAX}$ | 100 | 200 | - | mA |
| | | $V_I - V_O$ ≤ 40V<br>$P_D < P_{DMAX}$, $T_A$ = +25°C | 25 | 50 | - | |
| RMS Noise, % of $V_{OUT}$ | $e_N$ | $T_A$ =+ 25°C, 10Hz < f <10KHz | - | 0.003 | - | %/ $V_O$ |
| Ripple Rejection | RR | $V_O$ = 10V, f = 120Hz<br>without $C_{ADJ}$<br>$C_{ADJ}$ = 10uF | 66 | 65<br>80 | - | dB |
| Long-Term Stability | ST | $T_J$ = +125 °C, 1000 Hours | - | 0.3 | - | % |

• Load and Line regulation are specified at constant junction temperature. Change in $V_O$ due to heating effects must be taken into account separately. Pulse testing with low duty cycle is used.

**FIGURE 21.11**  LM317L specification sheet. (Courtesy of Fairchild Semiconductor).

differential voltage rating can be used to determine the maximum allowable value of $V_{in}$ as follows:

$$V_{in(max)} = V_{out(adj)} + V_d \qquad (21.7)$$

where $V_{in(max)}$ = the maximum allowable unrectified dc input voltage
$V_{out(adj)}$ = the *adjusted* output voltage of the regulator
$V_d$ = the input/output voltage differential rating of the regulator

### EXAMPLE 21.3

The LM317L is adjusted to provide a +8 V regulated output. Determine its maximum allowable input voltage.

*Solution:*    With a $V_d$ rating of 40 V, the maximum allowable value of $V_{in}$ is found as

$$V_{in(max)} = V_{out(adj)} + V_d = 8\text{ V} + 40\text{ V} = \textbf{48 V}$$

### PRACTICE PROBLEM 21.3

An adjustable IC voltage regulator has a $V_d$ rating of 32 V. Determine its maximum allowable input voltage when it is adjusted for a +6 V output.

The *line regulation* rating of the LM317L is 0.04%/V (maximum). This rating is measured under the following conditions: $T_A = 25°C$ and $3\text{ V} \le V_{in} - V_{out} \le 40\text{ V}$. This means that the difference between the input and output voltages can be no less than 3 V and no greater than 40 V. For example, if we set the LM317L for a +10 V output, the input voltage to the device must be between +13 and +50 V. As long as these conditions are met, the output will vary by no more than $10\text{ V} \times 0.04\% = 4\text{ mV}$.

The *load regulation* rating for the LM317L depends on whether the device is operated for a $V_{out}$ less than or greater than +5 V. If $V_{out}$ is less than +5 V, the output voltage will not vary by more than 25 mV when the load current is varied within the given range of values. If the output voltage is set to a value greater than (or equal to) +5 V, the same change in load current will cause a maximum output voltage change of 0.5%. Note that the rated range of output currents for the device is 10 mA to approximately 100 mA.

The **minimum load current** rating indicates the minimum allowable load current demand for the regulator. For the LM317L, this rating is 10 mA. If the load current drops below 10 mA (such as when the load is open), regulation of the output voltage is lost.

The **ripple rejection ratio** is the ability of the regulator to block any ripple voltage at its input. For the LM317L, any input ripple is reduced by 65 dB at the output. Using the decibel conversion process discussed in Chapter 8, we find that any input ripple is reduced by a factor of approximately 1780 by the regulator.

> **Minimum load current**
> The value of $I_L$ below which regulation is lost.
>
> **Ripple rejection ratio**
> The ratio of regulator input ripple to maximum output ripple.

## 21.4.2    Output Voltage Adjustment

The output voltage of the LM317L is adjusted using a voltage divider, as shown in Figure 21.12. The *complete* LM317 spec sheet (not shown) contains the following equation for determining the dc output voltage of the circuit:

$$V_{dc} = 1.25\left(\frac{R_2}{R_1} + 1\right) \qquad (21.8)$$

where $V_{dc}$ is the regulated dc output voltage of the regulator. Example 21.4 shows how equation (21.8) is used to determine the regulated dc output voltage for an LM317L regulator circuit.

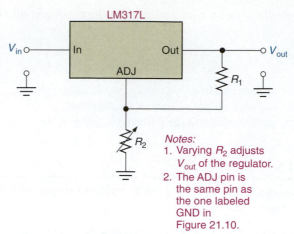

**Lab Reference:** The operation of an LM317 regulator is demonstrated in Exercise 36.

**FIGURE 21.12** An LM317L voltage-adjust circuit.

### EXAMPLE 21.4

$R_2$ in Figure 21.13 is adjusted to 2.4 k$\Omega$. Determine the regulated dc output voltage for the circuit.

**Lab Reference:** The operation of an LM317-regulated power supply (like the one represented in Figure 21.13) is demonstrated in Exercise 37.

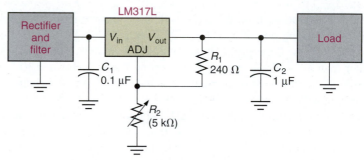

**FIGURE 21.13** LM317L circuit connections.

*Solution:* The regulated dc output voltage is found as

$$V_{dc} = 1.25\left(\frac{R_2}{R_1} + 1\right) = 1.25\left(\frac{2.4 \text{ k}\Omega}{240 \text{ }\Omega} + 1\right) = (1.25)(11) = \mathbf{13.75 \text{ V}}$$

**PRACTICE PROBLEM 21.4**

$R_2$ in Figure 21.13 is adjusted to 1.68 k$\Omega$. Determine the regulated dc output voltage for the LM317L.

Note that the output voltage equation for a given adjustable regulator is always provided on its spec sheet.

Figure 21.13 shows two shunt capacitors connected to the regulator input and output pins. The input capacitor is used to prevent the input ripple from driving the regulator into self-oscillations. The output capacitor is used to improve the ripple reduction of the regulator.

### 21.4.3 Linear IC Regulator Applications: A Complete Dual-Polarity Power Supply

Figure 21.14 shows a complete dual-polarity power supply. The circuit uses *matched* fixed-positive and fixed-negative regulators to provide equal $+V_{dc}$ and $-V_{dc}$ outputs.

As shown in Figure 21.14, the bridge rectifier is connected to a *center-tapped* transformer. When connected in this fashion, the bridge acts as two separate full-wave recti-

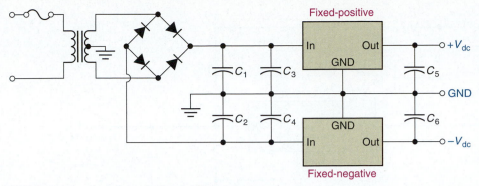

FIGURE 21.14  Complete dual-polarity power supply.

fiers. The diodes on the left form a *negative* full-wave rectifier, while the diodes on the right form a *positive* full-wave rectifier. The rectifier is wired in this fashion to prevent an *imbalance* from being caused when the load demands differ from each other. For example, if the load demand on the positive regulator increases, the center-tapped transformer provides a path for this current. (Otherwise, the current would be drawn through the ground connection of the negative regulator, causing an imbalance in the power supply output.)

Capacitors $C_1$ and $C_2$ are *filter capacitors*. These capacitors have values in the mid- to high-microfarad range. Capacitors $C_3$ and $C_4$ are the regulator input shunt capacitors, typically less than 1 μF in value. Capacitors $C_5$ and $C_6$ are output ripple reduction capacitors and have values in the neighborhood of 1 μF.

### 21.4.4  Circuit Variations

Figure 21.15 shows how a *pnp pass-transistor* can be added in parallel with an IC voltage regulator to increase the maximum possible output current. The value of $R_S$ in the circuit is selected to bias $Q_1$ on as follows:

$$R_S = \frac{V_{BE(Q1)}}{I_{in}} \tag{21.9}$$

where $I_{in}$ is the regulator input current. With $Q_1$ on, an increase in load current demand (above the capability of the regulator output) results in increased $Q_1$ conduction. This provides the needed load current.

As is the case with any series regulator, the pass-transistor in Figure 21.15 may be damaged or destroyed if the load demand exceeds the current capability of the component. To protect the pass-transistor from excessive load current demand, a *current-limiting* circuit ($Q_2$ and $R_{S2}$) can be added as shown in Figure 21.16. Note that this circuit limits the load current in the same fashion as the current-limiting circuit shown in Figure 21.8.

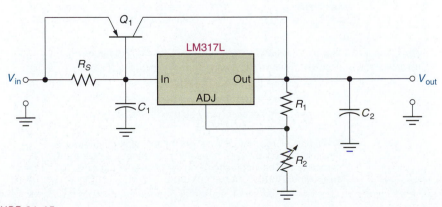

FIGURE 21.15

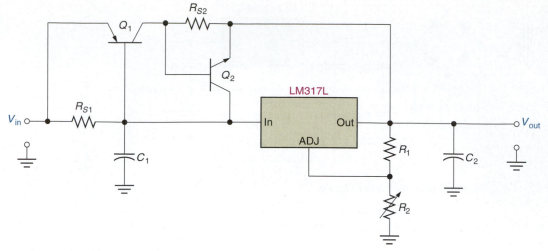

FIGURE 21.16   A complete linear IC voltage regulator.

### 21.4.5   One Final Note

Linear IC voltage regulators are extremely common. While we cannot possibly hope to cover the operation of every type of IC voltage regulator, you should be able to deal with common IC voltage regulators. The characteristics of IC voltage regulators are summarized for you in Figure 21.17.

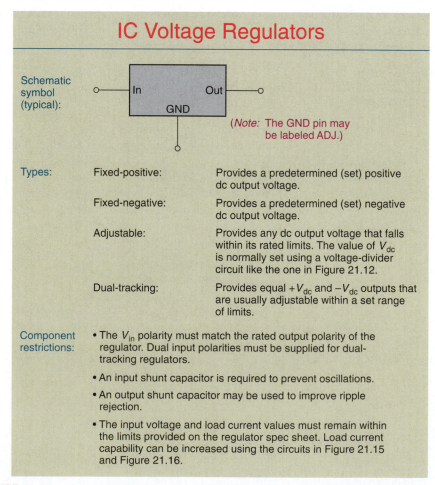

FIGURE 21.17

1. What is an *IC voltage regulator*?
2. List and describe the four types of IC voltage regulators.
3. What input polarity (or polarities) is required for each type of voltage regulator?
4. What is the *input/output differential* rating?
5. What is the *minimum load current* rating?
6. What is the *ripple rejection ratio* rating?
7. What type of circuit is normally used to provide the adjustment of the dc output voltage from an adjustable regulator?
8. Where can you find the $V_{dc}$ equation for a given adjustable regulator?
9. What purpose is served by the circuit shown in Figure 21.15?
10. Why does the circuit shown in Figure 21.15 need to be modified as shown in Figure 21.16?

## 21.5  Switching Regulators

There are two fundamental types of voltage regulators: *linear regulators* and *switching regulators*. The basic block diagram for each type of regulator is shown in Figure 21.18.

The *linear regulator* is designed to provide a continuous path for current between the regulator input and the load. For example, the linear regulator in Figure 21.18a contains a *pass-transistor* ($Q_1$) *that is always conducting*. All the regulators we have covered up to this point have been linear regulators.

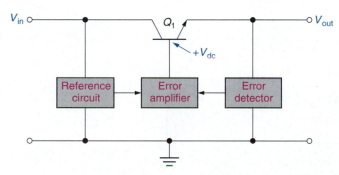

(a)  A series-linear regulator

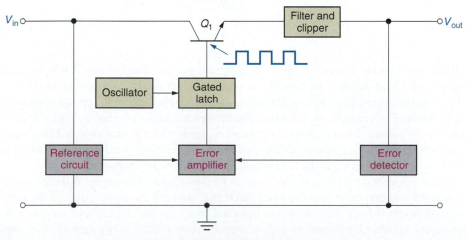

(b)  A series-switching regulator

**FIGURE 21.18**  Typical linear and switching regulators.

The *switching regulator* is designed so that the current path between the regulator input and the load is *not* continuous. For example, the switching regulator in Figure 21.18b contains a pass-transistor ($Q_1$) that is rapidly switched back and forth between saturation and cutoff. Note that the pass-transistor in a series switching regulator is often referred to as a **power switch**.

When saturated, the power switch provides a current path between the regulator's input and output. When in cutoff, the power switch breaks the conduction path between the input and the load. As you will see later in this section, this circuit action results in:

**1.** Higher regulator efficiency.
**2.** Higher regulator power-handling capability.

> **Power switch**
> A term used to describe the pass-transistor in a switching regulator.

### 21.5.1    Switching Regulator Operation

**OBJECTIVE 12 ▶**      The basic switching regulator in Figure 21.19 is divided into four circuit groups.

- The *switch driver* consists of an oscillator and a gated latch.
- The *power switch* is the series (pass) transistor.
- The *clipper and filter* consists of $D_1$, $L_1$, and $C_1$.
- The *control circuit* (not labeled) contains the remaining components: $R_1$, $R_2$, the comparator, and the reference circuit.

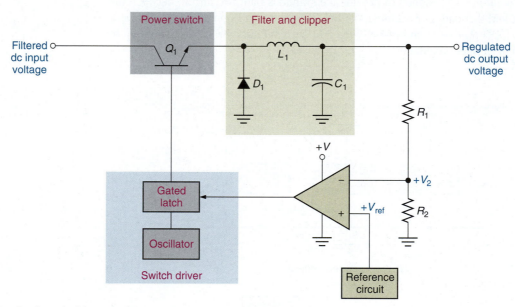

**FIGURE 21.19**    Basic switching regulator.

The control circuit is used to control the output from the switch driver. When it senses a change in the load voltage, the control circuit sends a signal to the switch driver. This signal causes the output from the switch driver to vary according to the type of change that has occurred. For example, the control circuit responds to a *decrease* in load voltage by directing the switch driver to *increase* conduction through the power switch. Likewise, an *increase* in load voltage results in the switch driver decreasing conduction through the power switch. In either case, the load voltage is returned to its proper value.

The *switch driver* contains an oscillator and a *gated latch*. The output from the oscillator is fixed and constant. It is the gated latch that controls the conduction of the power switch. The gated latch accepts inputs from the oscillator and the control circuit. These two inputs are then combined in such a way as to produce a driving signal that either increases or decreases conduction through the power switch. (We will discuss the methods used to produce the driving signal more thoroughly later in this section.)

Since the power switch is constantly changing output states, the voltage at its emitter is (more or less) a *rectangular waveform*. The *filter and clipper* circuit is designed to respond to this waveform as follows:

1. The *capacitor* opposes any change in voltage and, therefore, keeps the load voltage relatively constant.
2. The *inductor* opposes any change in current and, therefore, keeps the load current relatively constant.
3. The *diode* clips the counter emf produced by the *LC* circuit response to a rectangular input. In other words, it provides *transient protection*. (You were first introduced to this application in Chapter 4.)

Now, let's take a look at the overall response of the regulator to a change in load demand. Assume for a moment that the circuit experiences an *increased load*. When a regulator load is increased, $V_{out}$ starts to *decrease*. The decrease in $V_{out}$ causes $+V_2$ in the control circuit to *decrease*. This, in turn, causes the voltage at the comparator output to *decrease*. The decreased output from the control circuit signals the switch driver to *increase* the conduction of the power switch. The increased power switch conduction returns $V_{out}$ to its normal value.

◄ **OBJECTIVE 13**

*Remember:*
The term *increased load* means that the load *current demand* has increased; that is, the load *resistance* has decreased.

When the circuit experiences a *decreased load*, the results are the opposite of those just given. A decreased load demand causes the output voltage to *increase*. The *increase* in output voltage causes $+V_2$ in the control circuit to *increase*. This increase leads to an *increase* in the output from the control circuit. The increased output from the control circuit signals the switch driver to *decrease* the conduction of the power switch. The decrease in power switch conduction returns $V_{out}$ to its normal value.

At this point, you should have a pretty good idea of how the switching regulator in Figure 21.19 operates. To complete the picture, we need to take a closer look at the operation of the *switch driver* and the means by which it controls the conduction of the power switch.

## 21.5.2 Controlling Power Switch Conduction

As you know, the power switch is constantly driven back and forth between saturation and cutoff. When saturated, the transistor couples $V_{in}$ to the load. When in cutoff, the transistor isolates the load from the input.

◄ **OBJECTIVE 14**

The average (dc) value of the waveform produced at the emitter of the power switch can be found as

$$V_{ave} = V_{in}\left(\frac{T_{on}}{T_{on} + T_{off}}\right)$$

(21.10)

The fraction in equation (21.10) is the *duty cycle* of the waveform (written as a ratio rather than a percentage). Therefore, the average value can be described as the product of input voltage and duty cycle.

where $T_{on}$ = the time that the transistor spends in saturation (per cycle)
$T_{off}$ = the time that the transistor spends in cutoff (per cycle)

Example 21.5 demonstrates the use of this equation.

### EXAMPLE 21.5

The regulator shown in Figure 21.19 has the following values: $V_{in} = 24$ V, $T_{on} = 5$ μs, and $T_{off} = 10$ μs. Calculate the dc average of the load voltage.

*Solution:* Using the values given, the average (dc) load voltage can be found as

$$V_{ave} = V_{in}\left(\frac{T_{on}}{T_{on} + T_{off}}\right) = (24\text{ V})\left(\frac{5\text{ μs}}{15\text{ μs}}\right) = 8\text{ V}$$

### PRACTICE PROBLEM 21.5

A regulator like the one shown in Figure 21.19 has the following values: $V_{in} = 36$ V, $T_{on} = 6$ μs, and $T_{off} = 9$ μs. Calculate the value of $V_{ave}$ for the circuit.

Equation (21.10) is important because it demonstrates that *the average output voltage from a switching regulator can be controlled by varying the conduction of the power switch*. At this point, we will take a look at two methods commonly used to control the conduction of the power switch and, thus, the output from the switching regulator.

### 21.5.3    Pulse-Width Modulation (PWM)

**Pulse-width modulation (PWM)**

Using a signal to vary the pulse width of a rectangular wave without affecting its cycle time.

One method commonly used to control the conduction of the power switch (and, therefore, the average emitter voltage) is referred to as **pulse-width modulation**, or **PWM**. In a PWM system, *a signal is used to vary the pulse width of a rectangular waveform while not affecting its total cycle time*.

The control circuit shown in Figure 21.20a is designed to provide PWM for the power switch. The oscillator in this circuit generates a *triangular* waveform ($V_o$). The *error voltage* ($V_{error}$) is a dc voltage that varies directly with the input to the control circuit. The *gated latch* (which actually performs the modulation) *produces a high output whenever the following relationship is fulfilled*:

$$V_o \geq V_{error}$$

For example, take a look at the first set of waveforms in Figure 21.20b. As you can see, the control voltage ($V_C$) goes high as $V_o$ increases beyond the value of $V_{error}$. Then, as $V_o$ drops below the value of $V_{error}$, the control voltage drops back to 0 V. Note that the control voltage shown is nearly a square wave.

Now, take a look at the second set of waveforms in the figure. If you compare the voltages shown to those in the first set of waveforms, you'll see that:

1. The error voltage has increased in value.
2. The *pulse width* ($T_{on}$) of the control voltage has decreased significantly.
3. The total cycle time ($T_{on} + T_{off}$) has not changed.

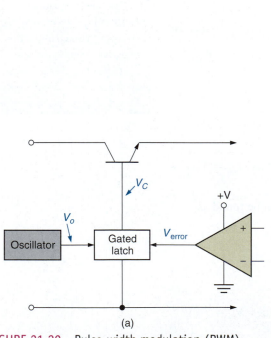

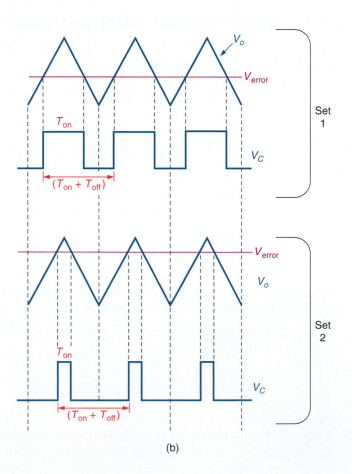

**FIGURE 21.20**    Pulse-width modulation (PWM).

Since the value of $T_{on}$ has decreased, the *average value* of $V_C$ has also decreased. Likewise, a decrease in the error voltage causes an *increase* in the value of $V_C$. This is how PWM (as it applies to switching regulators) works.

### 21.5.4    Variable Off-Time Modulation

Another method commonly used to control the conduction of a power switch is illustrated in Figure 21.21. In this circuit, *the pulse width of the control voltage is fixed and the total cycle time is variable*. The gated latch in this circuit accepts inputs from the error detector and an *astable multivibrator*. As long as the error voltage is high, the *square wave output* from the oscillator is gated to the power switch. When the error voltage goes low, the output from the oscillator is blocked.

**Variable off-time modulation**
Using a signal to vary the cycle time of a rectangular wave without affecting its pulse width.

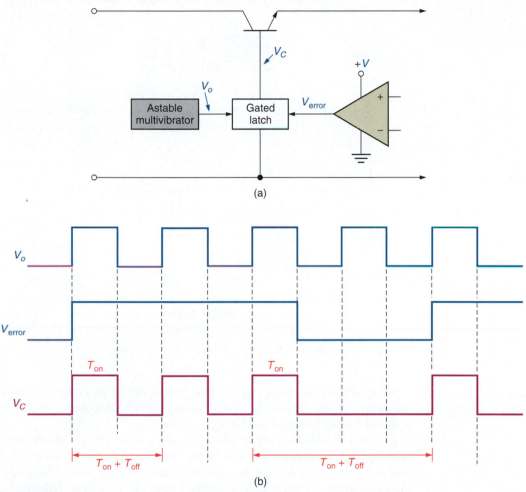

**FIGURE 21.21**    Variable off-time modulation.

The effect that the error voltage has on the cycle time of $V_C$ is shown in Figure 21.21b. The pulse width ($T_{on}$) has not changed from the first highlighted waveform to the second. However, the total cycle time ($T_{on} + T_{off}$) of the second waveform has been altered by the $V_{error}$ input. As the total cycle time varies, so does the average output from the power switch. Again, the average output voltage from the regulator is controlled.

### 21.5.5    Switching Regulator Configurations

One of the advantages of using a switching regulator is that it can be designed in a variety of configurations. So far, we have discussed only the **step-down regulator**. This configuration, which is represented in Figure 21.22a, produces a dc load voltage that is less than (or equal to) the rectified input voltage.

◄ *OBJECTIVE 15*

**Step-down regulator**
A regulator where $V_{out} \leq V_{in}$.

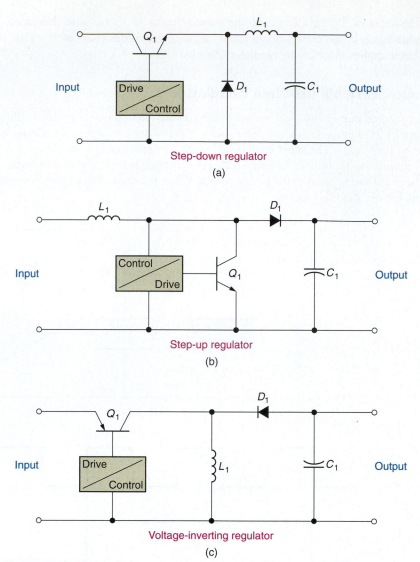

Step-down regulator

(a)

Step-up regulator

(b)

Voltage-inverting regulator

(c)

**FIGURE 21.22**   Basic switching regulator configurations.

**Step-up regulator**

A regulator where $V_{out} \geq V_{in}$.

By modifying the basic configuration shown in Figure 21.22b, we obtain a **step-up regulator**. As you can see, the power switch is now a shunt component, and the inductor is placed directly in series with the input source ($V_{in}$).

The step-up switching regulator can provide output voltages greater than the rectified value of $V_{in}$. This is made possible by the positioning of the inductor ($L_1$). In a nutshell, here is how it works: During the *on-time* of the transistor, the current drawn through $L_1$ causes an induced voltage to be developed across its terminals. This voltage adds to the value of $V_{in}$, making it possible for the output voltage to be greater than the input voltage.

**Voltage-inverting regulator**

A regulator that reverses the polarity of its regulated dc input.

The **voltage-inverting regulator** shown in Figure 21.22c reverses the polarity of the rectified input voltage. For example, with a rectified input voltage of $+7$ V, the regulator in Figure 21.22c could be designed to be any *negative value* less than, equal to, or greater than $-7$ V (within limits). Note the series power switch and shunt inductor connections characteristic of this configuration.

Several points should be made regarding the configurations represented in Figure 21.22. First, the *control* and *switch driver* circuits have been lumped into a single block for simplicity. While the control and switch driver circuits may vary from one switching regulator to another, you should be able to determine the type of regulator you're dealing with by noting the position of the power switch as follows:

1. If the power switch is in series with the input and the inductor, it is a *step-down* regulator.

**2.** If the power switch is a *shunt component* placed *after* the inductor, it is a *step-up* regulator.

**3.** If the power switch is in series with the input and the inductor is a *shunt* component, it is a *voltage-inverting* regulator.

Another important point is that there are no "cookbook" equations to help with the analysis of these circuits. Switching regulators are complex circuits that operate on complex principles. Just as the component placement varies from one switching regulator to another, so do most of the equations used in circuit analysis.

Because switching regulators can be designed for several different input/output voltage relationships, they are sometimes referred to as **dc-to-dc converters**. As the name implies, they can effectively convert one rectified dc voltage to another value.

**dc-to-dc converter**
Another name for a switching voltage regulator.

### 21.5.6 IC Switching Regulators

◄ *OBJECTIVE 16*

In many cases, the control functions and power switching of a switching regulator are handled by a single IC. For example, the MC34063A is an IC that contains the entire control, switch drive, and power switch circuitry. The spec sheet for the MC34063A is shown in Figure 21.23. Note that the transistor labeled ($Q_1$) in the IC is the power switch.

## MC34063A, MC33063A, NCV33063A

## 1.5 A, Step-Up/Down/ Inverting Switching Regulators

The MC34063A Series is a monolithic control circuit containing the primary functions required for DC–to–DC convertors. These devices consist of an internal temperature conpensated reference, comparator, controlled duty cycle oscillator with an active current limit circuit, driver and high current output switch. This series was specifically designed to be incorporated in Step–Down and Step–Up and Voltage–Inverting applications with a minimum number of external components. Refer to Application Notes An920A/D and AN954/D for additional design information.

- Operation from 3.0 V to 40 V Input
- Low Standby Current
- Current Limiting
- Output Switch Current to 1.5 A
- Output Voltage Adjustable
- Frequency Operation to 100 kHz
- Precision 2% Reference

**ON Semiconductor™**

http://onsemi.com

PDIP–8
P, P1 SUFFIX
CASE 626

SO–8
D SUFFIX
CASE 751

**PIN CONNECTIONS**

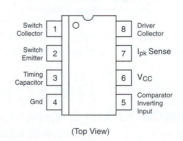

| Switch Collector | 1 | | 8 | Driver Collector |
| Switch Emitter | 2 | | 7 | $I_{pk}$ Sense |
| Timing Capacitor | 3 | | 6 | $V_{CC}$ |
| Gnd | 4 | | 5 | Comparator Inverting Input |

(Top View)

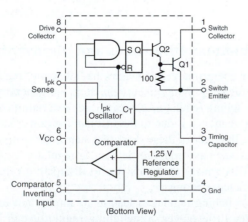

**FIGURE 21.23** The MC34063A switching regulator. (Copyright of Semiconductor Component Industries, LLC. Used by permission.)

If you compare the internal circuitry of the MC34063A to the block diagram in Figure 21.19, it is easy to see that the IC contains the control, driver, and power switch circuitry. The *control circuit* consists of the 1.25 V reference regulator and the comparator. The control circuit for this IC is used to provide *variable off-time* modulation of the oscillator output.

The MC34063A can be used in any standard switching regulator configuration. For example, the circuit shown in Figure 21.24 is an IC-based *step-down* regulator. This particular circuit was designed to produce a 5 V/500 mA output with a 25 V rectified input. Note the positioning of the power switch and the inductor. This is characteristic of a step-down regulator. Also, note the excellent efficiency rating of this circuit.

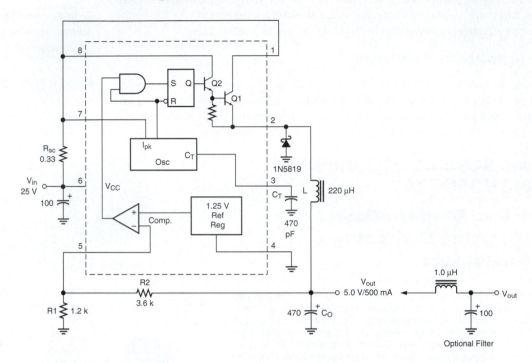

| Test | Conditions | Results |
|---|---|---|
| Line Regulation | $V_{in}$ = 15 V to 25 V, $I_O$ = 500mA | 12 mV = ±0.12% |
| Load Regulation | $V_{in}$ = 25 V, $I_O$ = 50mA to 500 mA | 12 mV = ±0.12% |
| Output Ripple | $V_{in}$ = 25 V, $I_O$ = 500mA | 120 mVpp |
| Short Circuit Current | $V_{in}$ = 25 V, $R_L$ = 0.1 Ω | 1.1 A |
| Efficiency | $V_{in}$ = 25 V, $I_O$ = 500mA | 83.7% |
| Out Ripple With Optional Filter | $V_{in}$ = 25 V, $I_O$ = 500mA | 40 mVpp |

FIGURE 21.24   An MC34063A-based step-down regulator. (Copyright of Semiconductor Component Industries, LLC. Used by permission.)

### 21.5.7   Switching Regulators: Advantages and Disadvantages

OBJECTIVE 17 ▶

As Figure 21.24 indicates, one of the primary advantages that switching regulators have over linear regulators is *higher efficiency*. While linear regulators are generally limited to efficiency ratings below 60%, switching regulators can easily achieve ratings of 90%. The higher efficiency of the switching regulator is due to the fact that the power switch is usually in either saturation or cutoff. In either of these operating states, the component dissipates very little power. In contrast, the pass-transistor in a linear regulator is usually operating within its *active region*. This causes the transistor to dissipate a relatively high amount of power, dropping its efficiency rating.

The power characteristics of switching regulators lead us to another advantage. Since the power switch dissipates very little power, a switching regulator can actually be designed for output power values that are far greater than the maximum power rating of the power switch. For example, a 2 W transistor could easily work in a 20 W switching regulator because the transistor would not be required to dissipate any significant amount of power. The same cannot be said for the linear regulator.

Finally, switching regulators can be built in a variety of configurations (as shown earlier in this section). In contrast, linear regulators can be designed only as *step-down* regulators.

In spite of the above advantages, switching regulators *do* have some distinct disadvantages. First, the operation of the power switch and filter can generate a very significant amount of *noise* that can be transmitted into the surrounding environment. Therefore, switching regulators cannot be used in any low-noise application unless shielding is provided.

Another disadvantage is that switching regulators have a longer *transient response time* than linear regulators; that is, they are slower to respond to a change in the load demand. This is due to the time required for the feedback loop (the control and drive circuits) to respond to a change in the error detection circuitry.

Finally, the design of a switching regulator is far more complex and time-consuming than that of a similar linear regulator. This adds to the production cost of the switching regulator.

## 21.5.8    Regulator Applications

Linear regulators are generally used in low-power applications. For example, most digital systems will have at least one linear regulator for each board-level supply voltage required. These supply voltages are generally not required to handle a significant amount of power.

Switching regulators are generally used in *high-power* (above 10 W) applications. They are also used in some lightweight, battery-operated units because of their minimal size requirements and dc-to-dc conversion capabilities.

---

1. In terms of conduction characteristics, what is the primary difference between *linear* regulators and *switching* regulators?      ◄ **Section Review**

2. List the circuits that make up the basic switching regulator, and describe the function performed by each.

3. Describe how the switching regulator in Figure 21.19 responds to a change in load demand.

4. Describe the relationship between the duty cycle of a power switch and its average output voltage.

5. Compare and contrast *pulse-width modulation* with *variable off-time modulation*.

6. List the circuit recognition features for each of the switching regulators shown in Figure 21.22.

7. Why are switching regulators often referred to as *dc-to-dc converters*?

8. Discuss the advantages, disadvantages, and applications for linear and switching regulators.

---

**CHAPTER SUMMARY**

Here is a summary of the major points made in this chapter:
1. The voltage regulator of a dc power supply serves two purposes:
   a. It reduces the variations (ripple) in the filtered dc.
   b. It maintains a relatively constant output voltage despite minor changes in load current demand and/or input voltage.
2. A zener regulator is rarely used because it wastes a tremendous amount of power.
3. *Line regulation* is a rating that indicates the change in regulator output voltage that occurs per unit change in input voltage.

a. Line regulation equals the ratio of $\Delta V_{\text{out}}$ to $\Delta V_{\text{in}}$.

b. The *ideal* voltage regulator has a value of $\Delta V_{\text{out}} = 0$ V for any value of $\Delta V_{\text{in}}$. Therefore, the *ideal* value of line regulation is zero.

c. The commonly used line regulation units are identified in Table 21.1.

4. *Load regulation* is a rating that indicates the change in regulator output voltage per unit change in load *current*.

a. Load regulation equals the ratio of $\Delta V_{\text{out}}$ to $\Delta I_L$ (where $\Delta V_{\text{out}}$ is the difference between the *no-load* output voltage and the *full-load* output voltage).

b. The *ideal* voltage regulator has a value of $\Delta V_{\text{out}} = 0$ V over the full range of load current values. Therefore, the *ideal* value of load regulation is zero.

c. The commonly used load regulation units are identified in Table 21.2.

5. Some manufacturers combine the line regulation and load regulation ratings into a single *regulation* rating.

a. This rating indicates the maximum change in output voltage that can occur when input voltage *and* load current are varied over their entire rated ranges.

b. The unit of measure used for line and/or load regulation is often a good indicator of the quality of that regulator. Generally, the lower the ratio in the unit of measure, the higher the quality of the regulator.

6. A *series regulator* is connected in series with its load.

7. A *shunt regulator* is connected in parallel with its load.

8. A *pass-transistor* regulator uses a series transistor to regulate load voltage. The term *pass-transistor* is used because the load current passes through the series transistor.

9. The schematic of a pass-transistor regulator is shown in Figure 21.4.

a. The zener diode establishes the base voltage of the pass-transistor.

b. $R_S$ provides a path for zener current.

c. The pass-transistor offsets a change in load resistance by adjusting load current (so that $I_L R_L$ remains relatively constant).

10. The *Darlington pass-transistor regulator* uses a Darlington pair to reduce the amount of power dissipated by the zener diode (see Figure 21.5).

11. A *series feedback regulator* uses an *error detection* circuit to improve the load and line regulation characteristics of the pass-transistor regulator (see Figure 21.6).

a. The error detector receives inputs from a *sample and adjust* circuit and a *reference* circuit.

b. The error detector responds to any difference between its inputs and provides an output that is proportional to that difference. This output is applied to the pass-transistor via an *error amplifier*.

12. The schematic of a basic series feedback regulator is shown in Figure 21.7.

a. The voltage divider ($R_3$, $R_4$, and $R_5$) acts as the *sample and adjust* circuit.

b. $D_1$ and $R_2$ form the *reference* circuit.

c. $Q_2$ and $R_1$ act as both the *error detector* and the *error amplifier*.

13. One weakness of a standard pass-transistor regulator is the possibility of the pass-transistor being destroyed by excessive load current if the load is shorted.

a. A *current-limiting* circuit can be added to protect the circuit (see Figure 21.8).

b. The current through the pass-transistor is limited to $V_{BE}/R_S$.

14. A *shunt feedback regulator* uses an error detector to control the conduction of a *shunt* transistor.

a. In Figure 21.9, the shunt transistor is $Q_1$.

b. The voltage divider ($R_2$, $R_3$, and $R_4$) is the *sample and adjust* circuit.

c. $D_1$ and $R_2$ form the *reference* circuit.

d. $Q_2$ acts as the error detector/amplifier.

15. Overvoltage protection in the shunt feedback regulator is normally accomplished by:

a. Using a shunt transistor with a higher $P_{D(\text{max})}$ rating than would be required in an overvoltage situation.

b. Adding a *crowbar* to the regulator input circuit. (Crowbar circuits were introduced in Chapter 20.)

**16.** A *linear IC voltage regulator* is a device used to hold the output from a dc power supply relatively constant over a specified range of line and load variations (see Figure 21.10).

   **a.** Most linear IC voltage regulators are three-terminal *series* regulators.

   **b.** *Fixed-positive* and *fixed-negative* regulators provide specific output voltages.

   **c.** An *adjustable* regulator can be adjusted to provide any output between two specified output voltage limits. These regulators may provide either positive or negative output voltages.

   **d.** A *dual-tracking* regulator provides equal-magnitude positive and negative output voltages. These regulators may be variable, with a single control adjusting both output voltages.

   **e.** Regardless of the type of regulator used, the input voltage polarity must match the output voltage polarity.

**17.** Here are some of the common IC voltage regulator ratings:

   **a.** The *input/output differential* rating indicates the maximum difference between $V_{in}$ and $V_{out}$ that can occur without damaging the component.

   **b.** The *minimum load current* rating indicates the minimum allowable current demand on the component. (If $I_L$ drops below this rating, regulation is lost.)

   **c.** The *ripple rejection ratio* is the ability of the regulator to block ripple. It is equal to the ratio (in dB) of input ripple to output ripple.

**18.** The current capability of an IC voltage regulator can be increased by adding a *pass-transistor* (see Figure 21.15).

**19.** When a pass-transistor is added to an IC voltage regulator, short-circuit current protection should also be added to the circuit (see Figure 21.16).

**20.** IC voltage regulator types and restrictions are summarized in Figure 21.17.

**21.** *Linear regulators* are designed to provide a constant current path between the source and load. *Switching regulators* are designed so that this current path is *not* continuous.

**22.** The pass-transistor in a switching regulator is constantly switched back and forth between saturation and cutoff (see Figure 21.18). This results in:

   **a.** Higher regulator efficiency.

   **b.** Higher regulator power-handling capability.

**23.** The pass-transistor is commonly referred to as a *power switch*.

**24.** The basic switching regulator can be divided into four circuit groups (see Figure 21.19):

   **a.** The *control circuit* is used to control the output from the switch driver.

   **b.** The *switch driver* controls the conduction of the power switch (pass-transistor).

   **c.** The *power switch* makes (and breaks) the connection between the source and the load.

   **d.** The *filter and clipper* reduces the variations in the power switch output and protects the power switch from transients.

**25.** The *control circuit* in the switching regulator:

   **a.** Samples the output voltage.

   **b.** Provides an output indicating whether the output voltage is above or below a set reference voltage.

   **c.** Couples output from its *comparator* to the switch driver.

**26.** The *switch driver* in the switching regulator:

   **a.** Contains an oscillator and a gated latch.

   **b.** Couples the output from the oscillator to the power switch when enabled by the signal from the control circuit.

**27.** The *filter and clipper* in the switching regulator perform the following functions:

   **a.** The filter capacitor reduces the changes in power switch output *voltage*.

   **b.** The filter inductor reduces the changes in power switch output *current*.

   **c.** The clipper diode protects the power switch from transients produced by the *LC* circuit when the input changes polarity.

**28.** The dc average of the power switch output is determined by the input voltage and the duty cycle of the power switch (see Example 21.5).

**29.** *Pulse-width modulation* (*PWM*) is one method commonly used to control the conduction of the power switch (see Figure 21.20).

a. PWM uses the *error voltage* to control the pulse width of the switch driver output while not affecting the overall cycle time.

b. PW increases when the error voltage decreases, and vice versa.

30. *Variable off-time modulation* is another method of controlling power switch conduction (see Figure 21.21):

a. This method uses a signal to vary the cycle time of the rectangular waveform without affecting its pulse width.

b. When an error voltage is present, the output from the astable multivibrator (oscillator) is coupled to the power switch. This decreases the cycle time of the multivibrator (and power switch) output.

31. Switching regulators can be designed as *step-up*, *step-down*, and *voltage-inverting* circuits (see Figure 21.22).

32. The type of switching regulator can be determined from component placement as follows:

a. If the power switch is in series with the input and the inductor, it is a *step-down* regulator.

b. If the power switch is a *shunt component* placed *after* the inductor, it is a *step-up regulator*.

c. If the power switch is in series with the input and the inductor is a shunt component, it is a *voltage-inverting* regulator.

33. Switching regulators are available in IC form (see Figure 21.23).

34. Switching regulators have the advantages listed in item 22 of this chapter summary. They also have the following disadvantages:

a. The operation of a switching regulator can generate a significant amount of *noise*.

b. Switching regulators have longer *transient response* times. In other words, it takes a switching regulator longer to respond to a change in load than it takes for a linear regulator to respond to the same change in load.

35. Switching regulators are commonly used in *high-power* circuits and in some low-power battery-operated units (because of the small size of the circuit).

## EQUATION SUMMARY

| Equation Number | Equation | Section Number |
|---|---|---|
| (21.1) | $\text{line regulation} = \dfrac{\Delta V_{\text{out}}}{\Delta V_{\text{in}}}$ | 21.1 |
| (21.2) | $\text{load regulation} = \dfrac{V_{NL} - V_{FL}}{\Delta I_L}$ | 21.1 |
| (21.3) | $V_L = V_Z - V_{BE}$ | 21.2 |
| (21.4) | $V_{BE} = V_Z - V_L$ | 21.2 |
| (21.5) | $V_L = V_Z - 2V_{BE}$ | 21.2 |
| (21.6) | $I_{L(\text{max})} \cong \dfrac{V_{BE(Q3)}}{R_S}$ | 21.2 |
| (21.7) | $V_{\text{in(max)}} = V_{\text{out(adj)}} + V_d$ | 21.4 |
| (21.8) | $V_{\text{dc}} = 1.25 \left( \dfrac{R_2}{R_1} + 1 \right)$ | 21.4 |
| (21.9) | $R_S = \dfrac{V_{BE(Q1)}}{I_{\text{in}}}$ | 21.4 |
| (21.10) | $V_{\text{ave}} = V_{\text{in}} \left( \dfrac{T_{\text{on}}}{T_{\text{on}} + T_{\text{off}}} \right)$ | 21.5 |

| | | |
|---|---|---|
| adjustable regulator 901 | linear IC voltage | series feedback |
| Darlington pass-transistor | regulator 901 | regulator 896 |
| regulator 896 | line regulation 891 | series regulator 894 |
| dc-to-dc converter 913 | load regulation 892 | shunt feedback |
| dual-tracking | minimum load current 903 | regulator 899 |
| regulator 901 | pass-transistor | shunt regulator 894 |
| fixed-negative | regulator 895 | step-down regulator 911 |
| regulator 901 | power switch 908 | step-up regulator 912 |
| fixed-positive | pulse-width modulation | variable off-time |
| regulator 901 | (PWM) 910 | modulation 911 |
| input/output voltage | regulation 893 | voltage-inverting |
| differential 902 | ripple rejection ratio 903 | regulator 912 |

### Section 21.1

**1.** A voltage regulator experiences a 20 $\mu$V change in its output voltage when its input voltage changes by 4 V. Determine the line regulation rating of the circuit.

**2.** A voltage regulator experiences a 14 $\mu$V change in output voltage when its input voltage changes by 10 V. Determine the line regulation rating of the circuit.

**3.** A voltage regulator experiences a 15 $\mu$V change in its output voltage when its input voltage changes by 5 V. Determine the line regulation rating of the circuit.

**4.** A voltage regulator experiences a 12 mV change in output voltage when its input voltage changes by 12 V. Determine the line regulation rating of the circuit.

**5.** A voltage regulator is rated for an output current of 0 to 150 mA. Under no-load conditions, the output voltage of the circuit is 6 V. Under full-load conditions, the output from the circuit is 5.98 V. Determine the load regulation rating of the circuit.

**6.** A voltage regulator experiences a 20 mV change in output voltage when the load current increases from 0 to 50 mA. Determine the load regulation rating of the circuit.

**7.** A voltage regulator experiences a 1.5 mV change in output voltage when the load current increases from 0 to 20 mA. Determine the load regulation rating of the circuit.

**8.** A voltage regulator experiences a 14 mV change in output voltage when the load current increases from 0 to 100 mA. Determine the load regulation rating of the circuit.

### Section 21.2

**9.** Determine the approximate value of $V_L$ for the circuit shown in Figure 21.25.

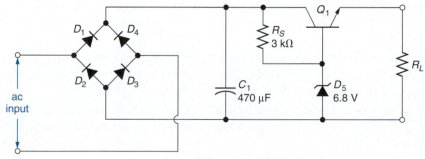

FIGURE 21.25

**10.** Determine the approximate value of $V_L$ for the circuit shown in Figure 21.26.

**11.** Determine the maximum possible value of load current for the circuit shown in Figure 21.27.

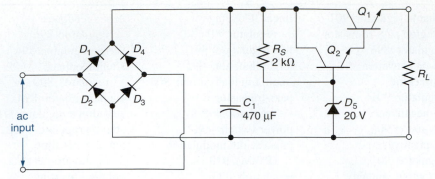

FIGURE 21.26

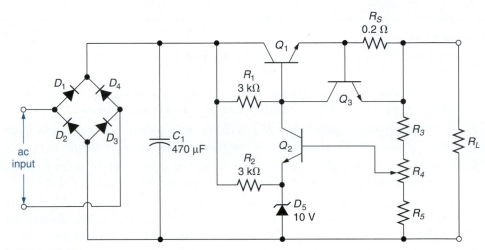

FIGURE 21.27

**12.** $R_S$ in Figure 21.27 is changed to 1.2 Ω. Determine the maximum possible load current for the new circuit.

**13.** An adjustable IC voltage regulator is set for a $+3 \ V_{dc}$ output. The $V_d$ rating for the device is 32 V. Determine the maximum allowable input voltage for the device.

**14.** An adjustable IC voltage regulator is set for a $+6 \ V_{dc}$ output. The $V_d$ rating for the device is 24 V. Determine the maximum allowable input voltage for the device.

**15.** The LM317L is used in a circuit with adjustment resistance values of $R_1 = 330$ Ω and $R_2 = 2.848$ kΩ (adjusted potentiometer value). Determine the output voltage for the circuit.

**16.** The LM317L is used in a circuit with adjustment resistance values of $R_1 = 510$ Ω and $R_2 = 6.834$ kΩ (adjusted potentiometer value). Determine the output voltage for the circuit.

**17.** A power switch like the one in Figure 21.19 has the following values: $V_{in} = 36$ V, $T_{on} = 12$ μs, and $T_{off} = 48$ μs. Determine the average output voltage from the power switch.

**18.** A power switch like the one in Figure 21.19 has the following values: $V_{in} = 24$ V, $T_{on} = 10$ μs, and $T_{off} = 40$ μs. Determine the average output voltage from the power switch.

## PUSHING THE ENVELOPE

**19.** A $+12 \ V_{dc}$ regulator has a line regulation rating of 420 ppm/V. Determine the output voltage for the circuit when $V_{in}$ increases by 10 V.

**20.** A voltage regulator has the following measured values: $V_{out} = 12.002 \ V_{dc}$ when $V_{in} = +20$ V (rated maximum allowable input), and $V_{out} = 12 \ V_{dc}$ when $V_{in} =$

+10 V (rated minimum allowable input). Determine the line regulation rating of the device in mV, %, and %/V.

21. A +5 V regulator has a load regulation rating of 2 $\Omega$ over a range of $I_L = 0$ to 100 mA (maximum). Express the load regulation of the circuit in V/mA, %, and %/mA.

22. A +15 $V_{dc}$ regulator has a 0.02%/mA load regulation rating for a range of $I_L = 0$ to 50 mA (maximum). Assuming that the load current stays within its rated limits, determine the maximum load power that can be delivered by the regulator. (Assume that $V_{out}$ increases as $I_L$ increases.)

21.1  25 µV/V

21.2  100 µV/mA

21.3  38 V

21.4  10 V

21.5  14.4 $V_{dc}$

# Additional Specification Sheets and Resistor Tables

## 2N3906

**Preferred Device**

## General Purpose Transistors

**PNP Silicon**

**ON Semiconductor™**

http://onsemi.com

TO–92
CASE 29
STYLE 1

**MARKING DIAGRAMS**

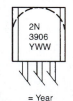

2N
3906
YWW

Y = Year
WW = Work Week

**MAXIMUM RATINGS**

| Rating | Symbol | Value | Unit |
|---|---|---|---|
| Collector–Emitter Voltage | $V_{CEO}$ | 40 | Vdc |
| Collector–Base Voltage | $V_{CBO}$ | 40 | Vdc |
| Emitter–Base Voltage | $V_{EBO}$ | 5.0 | Vdc |
| Collector Current – Continuous | $I_C$ | 200 | mAdc |
| Total Device Dissipation @ $T_A = 25°C$ <br> Derate above 25°C | $P_D$ | 625 <br> 5.0 | mW <br> mW/°C |
| Total Power Dissipation @ $T_A = 60°C$ | $P_D$ | 250 | mW |
| Total Device Dissipation @ $T_C = 25°C$ <br> Derate above 25°C | $P_D$ | 1.5 <br> 12 | Watts <br> mW/°C |
| Operating and Storage Junction Temperature Range | $T_J, T_{stg}$ | −55 to +150 | °C |

**THERMAL CHARACTERISTICS (Note 1.)**

| Characteristic | Symbol | Max | Unit |
|---|---|---|---|
| Thermal Resistance, Junction to Ambient | $R_{\theta JA}$ | 200 | °C/W |
| Thermal Resistance, Junction to Case | $R_{\theta JC}$ | 83.3 | °C/W |

1. Indicates Data in addition to JEDEC Requirements.

**ELECTRICAL CHARACTERISTICS** ($T_A = 25°C$ unless otherwise noted)

| Characteristic | Symbol | Min | Max | Unit |
|---|---|---|---|---|
| **OFF CHARACTERISTICS** | | | | |
| Collector–Emitter Breakdown Voltage (Note 2.) ($I_C = 1.0$ mAdc, $I_B = 0$) | $V_{(BR)CEO}$ | 40 | – | Vdc |
| Collector–Base Breakdown Voltage ($I_C = 10$ μAdc, $I_E = 0$) | $V_{(BR)CBO}$ | 40 | – | Vdc |
| Emitter–Base Breakdown Voltage ($I_E = 10$ μAdc, $I_C = 0$) | $V_{(BR)EBO}$ | 5.0 | – | Vdc |
| Base Cutoff Current ($V_{CE} = 30$ Vdc, $V_{EB} = 3.0$ Vdc) | $I_{BL}$ | – | 50 | nAdc |
| Collector Cutoff Current ($V_{CE} = 30$ Vdc, $V_{EB} = 3.0$ Vdc) | $I_{CEX}$ | – | 50 | nAdc |

(Copyright of Semiconductor Component Industries, LLC. Used by permission.)

# 2N3906

## ON CHARACTERISTICS (Note 2.)

| | | | | |
|---|---|---|---|---|
| DC Current Gain<br>($I_C$ = 0.1 mAdc, $V_{CE}$ = 1.0 Vdc)<br>($I_C$ = 1.0 mAdc, $V_{CE}$ = 1.0 Vdc)<br>($I_C$ = 10 mAdc, $V_{CE}$ = 1.0 Vdc)<br>($I_C$ = 50 mAdc, $V_{CE}$ = 1.0 Vdc)<br>($I_C$ = 100 mAdc, $V_{CE}$ = 1.0 Vdc) | $h_{FE}$ | 60<br>80<br>100<br>60<br>30 | –<br>–<br>300<br>–<br>– | – |
| Collector–Emitter Saturation Voltage<br>($I_C$ = 10 mAdc, $I_B$ = 1.0 mAdc)<br>($I_C$ = 50 mAdc, $I_B$ = 5.0 mAdc) | $V_{CE(sat)}$ | –<br>– | 0.25<br>0.4 | Vdc |
| Base–Emitter Saturation Voltage<br>($I_C$ = 10 mAdc, $I_B$ = 1.0 mAdc)<br>($I_C$ = 50 mAdc, $I_B$ = 5.0 mAdc) | $V_{BE(sat)}$ | 0.65<br>– | 0.85<br>0.95 | Vdc |

## SMALL-SIGNAL CHARACTERISTICS

| | | | | |
|---|---|---|---|---|
| Current–Gain – Bandwidth Product<br>($I_C$ = 10 mAdc, $V_{CE}$ = 20 Vdc, f = 100 MHz) | $f_T$ | 250 | – | MHz |
| Output Capacitance ($V_{CB}$ = 5.0 Vdc, $I_E$ = 0, f = 1.0 MHz) | $C_{obo}$ | – | 4.5 | pF |
| Input Capacitance ($V_{EB}$ = 0.5 Vdc, $I_C$ = 0, f = 1.0 MHz) | $C_{ibo}$ | – | 10 | pF |
| Input Impedance ($I_C$ = 1.0 mAdc, $V_{CE}$ = 10 Vdc, f = 1.0 kHz) | $h_{ie}$ | 2.0 | 12 | kΩ |
| Voltage Feedback Ratio<br>($I_C$ = 1.0 mAdc, $V_{CE}$ = 10 Vdc, f = 1.0 kHz) | $h_{re}$ | 0.1 | 10 | X $10^{-4}$ |
| Small–Signal Current Gain<br>($I_C$ = 1.0 mAdc, $V_{CE}$ = 10 Vdc, f = 1.0 kHz) | $h_{fe}$ | 100 | 400 | – |
| Output Admittance ($I_C$ = 1.0 mAdc, $V_{CE}$ = 10 Vdc, f = 1.0 kHz) | $h_{oe}$ | 3.0 | 60 | µmhos |
| Noise Figure<br>($I_C$ = 100 µAdc, $V_{CE}$ = 5.0 Vdc, $R_S$ = 1.0 kΩ, f = 1.0 kHz) | NF | – | 4.0 | dB |

## SWITCHING CHARACTERISTICS

| | | | | | |
|---|---|---|---|---|---|
| Delay Time | ($V_{CC}$ = 3.0 Vdc, $V_{BE}$ = 0.5 Vdc,<br>$I_C$ = 10 mAdc, $I_{B1}$ = 1.0 mAdc) | $t_d$ | – | 35 | ns |
| Rise Time | | $t_r$ | – | 35 | ns |
| Storage Time | ($V_{CC}$ = 3.0 Vdc, $I_C$ = 10 mAdc,<br>$I_{B1}$ = $I_{B2}$ = 1.0 mAdc) | $t_s$ | – | 225 | ns |
| Fall Time | ($V_{CC}$ = 3.0 Vdc, $I_C$ = 10 mAdc,<br>$I_{B1}$ = $I_{B2}$ = 1.0 mAdc) | $t_f$ | – | 75 | ns |

2. Pulse Test: Pulse Width ≤ 300 µs; Duty Cycle ≤ 2%.

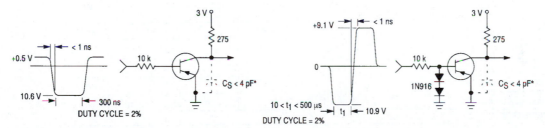

* Total shunt capacitance of test jig and connectors

**Figure 1. Delay and Rise Time Equivalent Test Circuit**

**Figure 2. Storage and Fall Time Equivalent Test Circuit**

# 2N4013

**Preferred Device**

## Switching Transistors

### PNP Silicon

**ON Semiconductor™**

http://onsemi.com

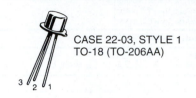

CASE 22-03, STYLE 1
TO-18 (TO-206AA)

**MAXIMUM RATINGS**

| Rating | Symbol | 2N4013 | 2N4014 | Unit |
|---|---|---|---|---|
| Collector-Emitter Voltage | $V_{CEO}$ | 30 | 50 | Vdc |
| Collector-Base Voltage | $V_{CBO}$ | 50 | 80 | Vdc |
| Emitter-Base Voltage | $V_{EBO}$ | 6.0 | | Vdc |
| Collector Current — Continuous — Peak | $I_C$ | 1.0 2.0 | | Adc |
| Total Device Dissipation @ $T_A$ = 25°C Derate above 25°C | $P_D$ | 0.5 28.6 | | Watt mW/°C |
| Total Device Dissipation @ $T_C$ = 25°C Derate above 25°C | $P_D$ | 1.4 6.8 | | Watts mW/°C |
| Operating and Storage Junction Temperature Range | $T_J$, $T_{stg}$ | −65 to +200 | | °C |

**ELECTRICAL CHARACTERISTICS** ($T_A$ = 25°C unless otherwise noted)

| Characteristic | | Symbol | Min | Typ | Max | Unit |
|---|---|---|---|---|---|---|
| **OFF CHARACTERISTICS** | | | | | | |
| Collector-Emitter Breakdown Voltage(1) ($I_C$ = 10 mAdc, $I_B$ = 0) | 2N4014 | $V_{(BR)CEO}$ | 50 | — | — | Vdc |
| | 2N4013 | | 30 | — | — | |
| Collector-Emitter Breakdown Voltage ($I_C$ = 10 μAdc, $V_{BE}$ = 0) | 2N4014 | $V_{(BR)CES}$ | 80 | — | — | Vdc |
| | 2N4013 | | 50 | — | — | |
| Collector-Base Breakdown Voltage ($I_C$ = 10 μAdc, $I_E$ = 0) | 2N4014 | $V_{(BR)CBO}$ | 80 | — | — | Vdc |
| | 2N4013 | | 50 | — | — | |
| Emitter-Base Breakdown Voltage ($I_E$ = 10 μAdc, $I_C$ = 0) | | $V_{(BR)EBO}$ | 6.0 | — | — | Vdc |
| Collector Cutoff Current ($V_{CB}$ = 60 Vdc, $I_E$ = 0) ($V_{CB}$ = 40 Vdc, $I_E$ = 0) ($V_{CB}$ = 60 Vdc, $I_E$ = 0, $T_A$ = 100°C) ($V_{CB}$ = 40 Vdc, $I_E$ = 0, $T_A$ = 100°C) | 2N4014 2N4013 2N4014 2N4013 | $I_{CBO}$ | — — — — | 0.12 0.12 — — | 1.7 1.7 120 120 | μAdc |
| Collector Cutoff Current ($V_{CE}$ = 80 Vdc, $V_{EB}$ = 0) ($V_{CE}$ = 50 Vdc, $V_{EB}$ = 0) | 2N4014 2N4013 | $I_{CES}$ | — — | 0.15 0.15 | 10 10 | μAdc |
| **ON CHARACTERISTICS** | | | | | | |
| DC Current Gain ($I_C$ = 10 mAdc, $V_{CE}$ = 1.0 Vdc) ($I_C$ = 100 mAdc, $V_{CE}$ = 1.0 Vdc) ($I_C$ = 100 mAdc, $V_{CE}$ = 1.0 Vdc, $T_A$ = −55°C) ($I_C$ = 300 mAdc, $V_{CE}$ = 1.0 Vdc) ($I_C$ = 500 mAdc, $V_{CE}$ = 1.0 Vdc) ($I_C$ = 500 mAdc, $V_{CE}$ = 1.0 Vdc, $T_A$ = −55°C) ($I_C$ = 800 mAdc, $V_{CE}$ = 2.0 Vdc) | 2N4014 2N4013 | $h_{FE}$ | 30 60 30 40 35 20 20 25 | — — — — — — | — 150 — — — — | — |
| ($I_C$ = 1.0 Adc, $V_{CE}$ = 5.0 Vdc) | 2N4014 2N4013 | | 25 30 | — | — | |

(Copyright of Semiconductor Component Industries, LLC. Used by permission.)

# 2N5457, 2N5458

**Preferred Device**

## JFETs – General Purpose
### N–Channel – Depletion

N–Channel Junction Field Effect Transistors, depletion mode (Type A) designed for audio and switching applications.

- N–Channel for Higher Gain
- Drain and Source Interchangeable
- High AC Input Impedance
- High DC Input Resistance
- Low Transfer and Input Capacitance
- Low Cross–Modulation and Intermodulation Distortion
- Unibloc Plastic Encapsulated Package

**ON Semiconductor™**

http://onsemi.com

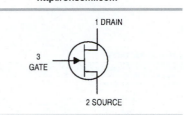

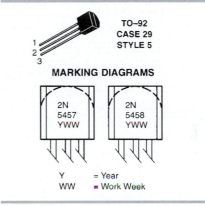

TO–92
CASE 29
STYLE 5

**MARKING DIAGRAMS**

2N
5457
YWW

2N
5458
YWW

Y = Year
WW = Work Week

**MAXIMUM RATINGS**

| Rating | Symbol | Value | Unit |
|---|---|---|---|
| Drain–Source Voltage | $V_{DS}$ | 25 | Vdc |
| Drain–Gate Voltage | $V_{DG}$ | 25 | Vdc |
| Reverse Gate–Source Voltage | $V_{GSR}$ | −25 | Vdc |
| Gate Current | $I_G$ | 10 | mAdc |
| Total Device Dissipation<br>@ $T_A$ = 25°C<br>Derate above 25°C | $P_D$ | 310<br>2.82 | mW<br>mW/°C |
| Operating Junction Temperature | $T_J$ | 135 | °C |
| Storage Temperature Range | $T_{stg}$ | −65 to +150 | °C |

**ELECTRICAL CHARACTERISTICS** ($T_A$ = 25°C unless otherwise noted)

| Characteristic | | Symbol | Min | Min | Max | Unit |
|---|---|---|---|---|---|---|
| **OFF CHARACTERISTICS** | | | | | | |
| Gate–Source Breakdown Voltage | ($I_G$ = −10 μAdc, $V_{DS}$ = 0) | $V_{(BR)GSS}$ | −25 | −25 | – | Vdc |
| Gate Reverse Current | ($V_{GS}$ = −15 Vdc, $V_{DS}$ = 0)<br>($V_{GS}$ = −15 Vdc, $V_{DS}$ = 0, $T_A$ = 100°C) | $I_{GSS}$ | –<br>– | –<br>– | 1.0<br>−200 | nAdc |
| Gate–Source Cutoff Voltage<br>($V_{DS}$ = 15 Vdc, $i_D$ = 1 nAdc)  2N5457<br>2N5458 | | $V_{GS(off)}$ | −1.0<br>−2.0 | –<br>– | −6.0<br>−7.0 | Vdc |
| Gate–Source Voltage<br>($V_{DS}$ = 15 Vdc, $i_D$ = 100 μAdc)  2N5457<br>($V_{DS}$ = 15 Vdc, $i_D$ = 200 μAdc)  2N5458 | | $V_{GS}$ | –<br>– | −2.5<br>−3.5 | −6.0<br>−7.0 | Vdc |
| **ON CHARACTERISTICS** | | | | | | |
| Zero–Gate–Voltage Drain Current (Note 1.)  2N5638<br>($V_{DS}$ = 20 Vdc, $V_{GS}$ = 0)  2N5639 | | $I_{DSS}$ | 1.0<br>2.0 | 3.0<br>6.0 | 5.0<br>9.0 | mAdc |
| **DYNAMIC CHARACTERISTICS** | | | | | | |
| Forward Transfer Admittance (Note 1.)  2N5638<br>($V_{DS}$ = 15 Vdc, $V_{GS}$ = 0, f = 1 kHz)  2N5639 | | $|Y_{fs}|$ | 1000<br>1500 | 3000<br>4000 | 5000<br>5500 | μmhos |
| Forward Transfer Admittance (Note 1.) | ($V_{DS}$ = 15 Vdc, $V_{GS}$ = 0, f = 1 kHz) | $|Y_{os}|$ | – | 10 | 50 | μmhos |
| Input Capacitance | ($V_{DS}$ = 15 Vdc, $V_{GS}$ = 0, f = 1 kHz) | $C_{iss}$ | – | 4.5 | 7.0 | pF |
| Reverse Transfer Capacitance | ($V_{DS}$ = 15 Vdc, $V_{GS}$ = 0, f = 1 kHz) | $C_{rss}$ | – | 1.5 | 3.0 | pF |

1. Pulse Width ≤ 630 ms, Duty Cycle ≤ 10%.

(Copyright of Semiconductor Component Industries, LLC. Used by permission.)

# 2N6426, 2N6427

Preferred Device

# Darlington Transistors

## PNP Silicon

## ON Semiconductor™

http://onsemi.com

## 2N6426*
## 2N6427

### CASE 29-04, STYLE 1
### TO-92 (TO-226AA)

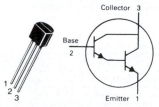

## DARLINGTON TRANSISTORS

NPN SILICON

### MAXIMUM RATINGS

| Rating | Symbol | Value | Unit |
|---|---|---|---|
| Collector-Emitter Voltage | $V_{CEO}$ | 40 | Vdc |
| Collector-Base Voltage | $V_{CBO}$ | 40 | Vdc |
| Emitter-Base Voltage | $V_{EBO}$ | 12 | Vdc |
| Collector Current — Continuous | $I_C$ | 500 | mAdc |
| Total Device Dissipation @ $T_A$ = 25°C<br>Derate above 25°C | $P_D$ | 625<br>5.0 | mW<br>mW/°C |
| Total Device Dissipation @ $T_C$ = 25°C<br>Derate above 25°C | $P_D$ | 1.5<br>12 | Watts<br>mW/°C |
| Operating and Storage Junction<br>Temperature Range | $T_J$, $T_{stg}$ | −55 to +150 | °C |

### THERMAL CHARACTERISTICS

| Characteristic | Symbol | Max | Unit |
|---|---|---|---|
| Thermal Resistance, Junction to Case | $R_{\theta JC}$ | 83.3 | °C/W |
| Thermal Resistance, Junction to Ambient | $R_{\theta JA}(1)$ | 200 | °C/W |

(1) $R_{\theta JA}$ is measured with the device soldered into a typical printed circuit board.

### ELECTRICAL CHARACTERISTICS ($T_A$ = 25°C unless otherwise noted.)

| Characteristic | | Symbol | Min | Typ | Max | Unit |
|---|---|---|---|---|---|---|
| **OFF CHARACTERISTICS** | | | | | | |
| Collector-Emitter Breakdown Voltage(2)<br>($I_C$ = 10 mAdc, $V_{BE}$ = 0) | | $V_{(BR)CES}$ | 40 | — | — | Vdc |
| Collector-Base Breakdown Voltage<br>($I_C$ = 100 μAdc, $I_E$ = 0) | | $V_{(BR)CBO}$ | 40 | — | — | Vdc |
| Emitter-Base Breakdown Voltage<br>($I_E$ = 10 μAdc, $I_C$ = 0) | | $V_{(BR)EBO}$ | 12 | — | — | Vdc |
| Collector Cutoff Current<br>($V_{CE}$ = 25 Vdc, $I_B$ = 0) | | $I_{CEO}$ | — | — | 1.0 | μAdc |
| Collector Cutoff Current<br>($V_{CB}$ = 30 Vdc, $I_E$ = 0) | | $I_{CBO}$ | — | — | 50 | nAdc |
| Emitter Cutoff Current<br>($V_{BE}$ = 10 Vdc, $I_C$ = 0) | | $I_{EBO}$ | — | — | 50 | nAdc |
| **ON CHARACTERISTICS** | | | | | | |
| DC Current Gain(2)<br>($I_C$ = 10 mAdc, $V_{CE}$ = 5.0 Vdc) | 2N6426<br>2N6427 | $h_{FE}$ | 20,000<br>10,000 | —<br>— | 200,000<br>100,000 | — |
| ($I_C$ = 100 mAdc, $V_{CE}$ = 5.0 Vdc) | 2N6426<br>2N6427 | | 30,000<br>20,000 | —<br>— | 300,000<br>200,000 | |
| ($I_C$ = 500 mAdc, $V_{CE}$ = 5.0 Vdc) | 2N6426<br>2N6427 | | 20,000<br>14,000 | —<br>— | 200,000<br>140,000 | |
| Collector-Emitter Saturation Voltage<br>($I_C$ = 50 mAdc, $I_B$ = 0.5 mAdc)<br>($I_C$ = 500 mAdc, $I_B$ = 0.5 mAdc) | | $V_{CE(sat)}$ | —<br>— | 0.71<br>0.9 | 1.2<br>1.5 | Vdc |
| Base-Emitter Saturation Voltage<br>($I_C$ = 500 mAdc, $I_B$ = 0.5 mAdc) | | $V_{BE(sat)}$ | — | 1.52 | 2.0 | Vdc |
| Base-Emitter On Voltage<br>($I_C$ = 50 mAdc, $V_{CE}$ = 5.0 Vdc) | | $V_{BE(on)}$ | — | 1.24 | 1.75 | Vdc |
| **SMALL-SIGNAL CHARACTERISTICS** | | | | | | |
| Output Capacitance<br>($V_{CB}$ = 10 Vdc, $I_E$ = 0, f = 1.0 MHz) | | $C_{obo}$ | — | 5.4 | 7.0 | pF |
| Input Capacitance<br>($V_{BE}$ = 1.0 Vdc, $I_C$ = 0, f = 1.0 MHz) | | $C_{ibo}$ | — | 10 | 15 | pF |

(Copyright of Semiconductor Component Industries, LLC. Used by permission.)

## 2N6426, 2N6427

ELECTRICAL CHARACTERISTICS (continued) ($T_A = 25°C$ unless otherwise noted.)

| Characteristic | | Symbol | Min | Typ | Max | Unit |
|---|---|---|---|---|---|---|
| Input Impedance<br>($I_C = 10$ mAdc, $V_{CE} = 5.0$ Vdc, f = 1.0 kHz) | | $h_{ie}$ | | — | | k Ω |
| | 2N6426 | | 100 | — | 2000 | |
| | 2N6427 | | 50 | — | 1000 | |
| Small-Signal Current Gain<br>($I_C = 10$ mAdc, $V_{CE} = 5.0$ Vdc, f = 1.0 kHz) | | $h_{fe}$ | | | | — |
| | 2N6426 | | 20,000 | — | — | |
| | 2N6427 | | 10,000 | — | — | |
| Current Gain — High Frequency<br>($I_C = 10$ mAdc, $V_{CE} = 5.0$ Vdc, f = 100 MHz) | | $|h_{fe}|$ | | | | — |
| | 2N6426 | | 1.5 | 2.4 | — | |
| | 2N6427 | | 1.3 | 2.4 | — | |
| Output Admittance<br>($I_C = 10$ mAdc, $V_{CE} = 5.0$ Vdc, f = 1.0 kHz) | | $h_{oe}$ | — | — | 1000 | $\mu$mhos |
| Noise Figure<br>($I_C = 1.0$ mAdc, $V_{CE} = 5.0$ Vdc, $R_S = 100$ kΩ,<br>f = 1.0 kHz) | | NF | — | 3.0 | 10 | dB |

(2) Pulse Test: Pulse Width $\leq$ 300 $\mu$s, Duty Cycle $\leq$ 2.0%.

## Standard Resistor Values

### 10% Tolerance

| Ω | | | | kΩ | | | MΩ | |
|------|-----|----|-----|-----|----|-----|-----|------|
| 0.10 | 1.0 | 10 | 100 | 1.0 | 10 | 100 | 1.0 | 10.0 |
| 0.12 | 1.2 | 12 | 120 | 1.2 | 12 | 120 | 1.2 | 12.0 |
| 0.15 | 1.5 | 15 | 150 | 1.5 | 15 | 150 | 1.2 | 15.0 |
| 0.18 | 1.8 | 18 | 180 | 1.8 | 18 | 180 | 1.8 | 18.0 |
| 0.22 | 2.2 | 22 | 220 | 2.2 | 22 | 220 | 2.2 | 22.0 |
| 0.27 | 2.7 | 27 | 270 | 2.7 | 27 | 270 | 2.7 | |
| 0.33 | 3.3 | 33 | 330 | 3.3 | 33 | 330 | 3.3 | |
| 0.39 | 3.9 | 39 | 390 | 3.9 | 39 | 390 | 3.9 | |
| 0.47 | 4.7 | 47 | 470 | 4.7 | 47 | 470 | 4.7 | |
| 0.56 | 5.6 | 56 | 560 | 5.6 | 56 | 560 | 5.6 | |
| 0.68 | 6.8 | 68 | 680 | 6.8 | 68 | 680 | 6.8 | |
| 0.82 | 8.2 | 82 | 820 | 8.2 | 82 | 820 | 8.2 | |

In addition to the values listed in the table, multiples of the following values are available with 2% and 5% tolerance only:

| | | |
|-----|-----|-----|
| 1.1 | 2.4 | 5.1 |
| 1.3 | 3.0 | 6.2 |
| 1.6 | 3.6 | 7.5 |
| 2.0 | 4.3 | 9.1 |

## Precision Resistor Values

Precision resistors are commonly available in 0.1%, 0.25%, 0.5%, and 1% tolerances. The values listed below are the standard digit combinations for the resistors in this tolerance range. Resistors with 1% tolerance are available only in the magnitudes shown in **bold**.

| | | | | | | | | | | | |
|------|-----|---------|-----|---------|-----|---------|-----|---------|-----|---------|-----|
| **100** | 101 | **102** | 104 | **105** | 106 | **107** | 109 | **110** | 111 | **113** | 114 |
| **115** | 117 | **118** | 120 | **121** | 126 | **127** | 129 | **130** | 132 | **133** | 135 |
| **137** | 138 | **140** | 142 | **143** | 145 | **147** | 149 | **150** | 152 | **154** | 156 |
| **158** | 160 | **162** | 164 | **165** | 167 | **169** | 172 | **174** | 176 | **178** | 180 |
| **182** | 184 | **187** | 189 | **191** | 193 | **196** | 198 | | | | |
| **200** | 203 | **205** | 208 | **210** | 213 | **215** | 218 | **221** | 223 | **226** | 229 |
| **232** | 234 | **237** | 240 | **243** | 246 | **249** | 252 | **255** | 258 | **261** | 264 |
| **267** | 271 | **274** | 277 | **280** | 284 | **287** | 291 | **294** | 298 | | |
| **301** | 305 | **309** | 312 | **316** | 320 | **324** | 328 | **332** | 336 | **340** | 344 |
| **348** | 352 | **357** | 361 | **365** | 370 | **374** | 379 | **383** | 388 | **392** | 397 |
| **402** | 407 | **412** | 417 | **422** | 427 | **432** | 437 | **442** | 448 | **453** | 459 |
| **464** | 470 | **475** | 481 | **487** | 493 | **499** | | | | | |
| 505 | **511** | 517 | **523** | 530 | **536** | 542 | **549** | 562 | **566** | 569 | **576** |
| 583 | **590** | 599 | | | | | | | | | |
| **604** | 612 | **619** | 626 | **634** | 642 | **649** | 657 | **665** | 673 | **681** | 690 |
| **698** | | | | | | | | | | | |
| 706 | **715** | 723 | **732** | 741 | **750** | 759 | **768** | 777 | **787** | 796 | |
| **806** | 816 | **825** | 835 | **845** | 856 | **866** | 876 | **887** | 898 | | |
| **909** | 920 | **931** | 942 | **953** | 965 | **976** | 988 | | | | |

# Approximating Circuit Values

The analysis of a given circuit can often be simplified by approximating many of its resistance and current values. In this appendix, we will look at the methods by which you can approximate circuit values and the circumstances that make circuit approximations valid.

## What Is Meant by Approximating Circuit Values?

When you approximate circuit values, you ignore any resistance and/or current values that have little impact on the analysis of a given circuit. For example, consider the simple series circuit shown in Figure B.1a. What is the total resistance in this circuit? Obviously, it is 10,000,010 $\Omega$. However, *for all practical purposes*, couldn't we just say that it is *approximately* 10 M$\Omega$? In this circuit, the value of $R_1$ has very little noticeable effect on the values of $R_T$, $I_T$, or $V_{R2}$. Therefore, the value of $R_1$ can be dropped from the circuit with very little loss in the accuracy of our calculations.

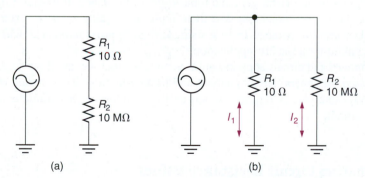

FIGURE B.1

Now, take a look at the circuit shown in Figure B.1b. In terms of circuit *current*, which component can be ignored? In this case, it may be valid to ignore the value of $R_2$ since there is very little current through this branch. In fact, if we assume the voltage source to be 10 V, we can use Ohm's law to calculate the current values of $I_1 = 1$ A and $I_2 = 1$ $\mu$A. In this case, the value of $I_2$ has virtually no effect on the value of total circuit current and, thus, can be ignored in circuit calculations.

Later, we'll establish the guidelines for approximating circuit values. At this point, we'll take a look at *when* it is valid to use circuit approximations.

## When Can You Use Circuit Approximations?

Any time that you are troubleshooting a given circuit, you can assume that it is all right to use approximated values. Generally, when you are troubleshooting, you are interested only in whether a voltage, current, or resistance is *close to* its rated value. For example, consider the circuit shown in Figure B.2. Assume that you are troubleshooting this circuit. As the figure shows, $V_B$ should be a 1 $V_{PP}$ sine wave, $V_E$ should be a 1 $V_{PP}$ sine wave, and $V_C$ should be an 8 $V_{PP}$ sine wave that is 180° out of phase with the other two voltages. When checking this circuit with an oscilloscope, we are not concerned with whether a given value of $V_{PP}$ is off by 0.001 V. We are concerned only with whether the value of $V_{PP}$ is *close* to the given value and whether the phase relationships are correct. Therefore, any approximations of circuit $V_{PP}$ values are completely acceptable.

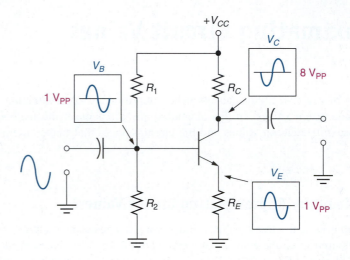

FIGURE B.2

There are two circumstances when you should not use the circuit approximations covered here. The first is when you are performing an exact circuit analysis involving *h*-parameters. *h*-Parameter equations, such as those introduced in Appendix C, lead to very exact values for transistor gain and input/output impedance. It wouldn't make much sense to use these exact equations and then approximate all the values external to the transistor. The second instance for not using circuit approximations is when you are designing a circuit for a specific application.

As a summary, the circuit approximation techniques that we are about to discuss can usually be used for general circuit analysis and troubleshooting. They should not be used when an exact analysis is required (such as those involving *h*-parameters) or when designing a circuit.

## Approximating Circuit Resistance Values

*A Practical Consideration:* In this discussion, we are assuming that the tolerance of the resistors used is 10%. For this reason, a percentage of error of 10% in our calculations is considered acceptable. *In practice, the acceptable margin of error should be no greater than the tolerance of the components used.*

Generally, you can ignore the value of a resistor in a series circuit *when the value of that component is less than 10% of the value of the next smallest resistance value.* For example, take a look at the voltage dividers shown in Figure B.3.

In Figure B.3a, the value of $R_2$ is equal to $0.1R_1$. Therefore, we can approximate the total resistance in the circuit as being equal to the value of $R_1$, 4.7 k$\Omega$. In Figure B.3b, the value of $R_3$ can be ignored because it is less than $0.1R_1$, and $R_1$ is the next lowest resistance value in the circuit. Note that the value of $R_2$ is not considered because it is greater than the value of $R_1$. In Figure B.3c, none of the resistors can be dropped from the circuit. Why? Even though $R_1$ is less than $0.1R_2$, it is not less than $0.1R_3$. Since it is not less than (or equal to) 10% of the *next lowest individual resistance value*, it must be kept in the circuit.

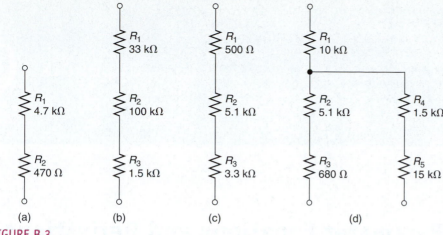

FIGURE B.3

Which resistor (if any) can be dropped from Figure B.3d? If you picked $R_4$, you are correct. The value of $R_4$ is 10% of the value of $R_5$, so it can be dropped from the circuit. Why doesn't the value of $R_1$ (10 kΩ) prevent us from dropping $R_4$? The circuit shown in Figure B.3d is a *series-parallel* circuit. While $R_1$ is in series with the *combination* of $R_2$, $R_3$, $R_4$, and $R_5$, it is not considered to be in series with $R_4$ directly. This is because the current through $R_1$ is not necessarily equal to the current through $R_4$. By definition, two components are in series only when the same current passes through the two components.

For parallel circuits, the 10% rule still holds. In this case, however, *the larger resistance value is the one that can be dropped*. The reason for this can be seen by referring to Figure B.1b. For the circuit shown, the larger resistor provided less than 10% of the total circuit current. Thus, its value can easily be ignored.

# *h*-Parameter Equations and Derivations

In Chapter 9, you were introduced to the four transistor *h*-parameters and their use in basic circuit analysis. In this appendix, we'll take a more in-depth look at *h*-parameter equations and the transistor hybrid equivalent circuit.

## The Transistor Hybrid Equivalent Circuit

Using the *h*-parameter values described in Chapter 9, the *hybrid equivalent circuit* of a transistor is developed. To understand this ac equivalent circuit, the transistor must be viewed as a four-terminal device. This representation of the transistor is shown in Figure C.1. Since the emitter terminal of the transistor is common to both the input and output circuits in a common-emitter amplifier, it is represented as two separate terminals. Realize, however, that both these terminals represent the single transistor emitter. The input and output voltages are represented as $v_1$ and $v_2$, respectively.

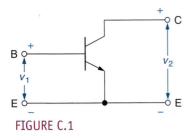

FIGURE C.1

## Input Circuit

The ac equivalent of the input circuit is derived using Thevenin's theorem. Thevenin's theorem states that any circuit can be represented as a single voltage source in series with a single resistance. The hybrid equivalent circuit for the transistor input is derived as illustrated in Figure C.2.

The Thevenin voltage for the input is equal to the value of $v_1$ with the base circuit open. As Figure C.2 shows, this voltage is found as

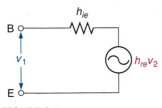

FIGURE C.2

$$v_1 \cong h_{ie}v_2 \tag{C.1}$$

This equation is simply another form of equation (9.30). Since $v_{be} = v_1$ and $v_{ce} = v_2$, equation (9.30) can be rewritten as above.

The Thevenin resistance of the transistor input is equal to $h_{ie}$. This value is found as shown in Chapter 9.

So why is the input represented as a Thevenin equivalent circuit? Since the voltage across the base-emitter junction remains fairly constant, it is convenient to define the operation of the input circuit in terms of this voltage.

## Output Circuit

The hybrid equivalent of the output circuit is derived using Norton's theorem. Norton's theorem represents a circuit as a current source in parallel with a single resistance. The ac equivalent of the output is represented in Figure C.3.

The value of the Norton current source is written as the product of the transistor current gain ($h_{fe}$) and the base input current ($i_1$). This current source value is consistent with the value of $h_{fe}i_b$ used in Chapter 9. The Norton resistance equals the reciprocal of the *transistor output admittance* ($h_{oe}$).

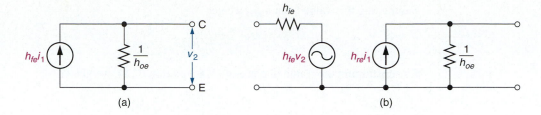

**FIGURE C.3**

The output circuit is represented as a Norton equivalent circuit because $i_c$ is dependent on factors outside the collector-emitter circuit under normal circumstances. For example, changing the value of $R_C$ will not cause $i_c$ to change significantly. However, changing $i_b$ *will*. Since the transistor output current is relatively independent of the external component values, it is the reference value for output calculations.

The complete hybrid equivalent circuit for the transistor is shown in Figure C.3b. This circuit will be used throughout our discussions on *h*-parameter calculations.

## Calculations Involving *h*-Parameters

Very exact values of transistor current gain ($A_i$), voltage gain ($A_v$), input impedance ($Z_{in(base)}$), and output impedance ($Z_{out}$) can be obtained using *h*-parameters. *Note that these calculations provide values of $A_i$, $A_v$, $Z_{in}$, and $Z_{out}$ for specific values of $I_C$. If $I_C$ changes, the values of $A_v$, $A_i$, $Z_{in(base)}$, and $Z_{out}$ obtained with the h-parameters will change as well.*

## Current Gain ($A_i$)

The current gain of a transistor equals the ratio of output current to input current. By formula,

$$A_i = \frac{i_2}{i_1} \tag{C.2}$$

Now, let's apply this equation to the circuit shown in Figure C.4. This circuit is the small-signal equivalent of a voltage-divider biased amplifier. The only difference between this circuit and those discussed earlier is that the transistor has been replaced by its hybrid equivalent circuit. The boundaries of the transistor are represented by the shaded area.

> *A Practical Consideration:*
> The discussion here applies *only* to the current gain of the *transistor* in a common-emitter amplifier. The current gain of a common-emitter amplifier is found using equation (9.21). The derivation of this equation appears in Appendix D.

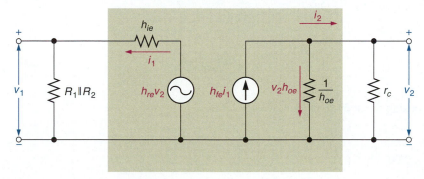

**FIGURE C.4**

The output current is equal to the difference between $h_{fe}i_1$ and the current through the Norton parallel resistance. Using Ohm's law, the current through this resistance can be found as

$$i = \frac{v_2}{1/h_{oe}} = v_2 h_{oe}$$

Subtracting this value from $h_{fe}i_1$, we get

$$i_2 = h_{fe}i_1 - v_2h_{oe} \qquad \text{(C.3)}$$

By substituting this value in place of $i_2$ in equation (C.2), we get

$$A_i = \frac{h_{fe}i_1 - v_2h_{oe}}{i_1}$$

which can be rewritten as

$$A_i = \frac{h_{fe}i_1}{i_1} - \frac{v_2h_{oe}}{i_1}$$

or

$$A_i = h_{fe} - \frac{v_2h_{oe}}{i_1}$$

Since $v_2 = i_2r_C$,

$$A_i = h_{fe} - \frac{i_2r_Ch_{oe}}{i_1}$$

which can be rewritten as

$$A_i = h_{fe} - \left(\frac{i_2}{i_1}\right)r_Ch_{oe}$$

This equation is now rewritten as

$$A_i = h_{fe} - A_ir_Ch_{oe}$$

Dividing both sides of the equation by $A_i$, we get

$$\frac{A_i}{A_i} = \frac{h_{fe}}{A_i} - \frac{A_ir_Ch_{oe}}{A_i}$$

or

$$1 = \frac{h_{fe}}{A_i} - r_Ch_{oe}$$

Transposing the above equation, we get

$$1 + r_Ch_{oe} = \frac{h_{fe}}{A_i}$$

or

$$A_i = \frac{h_{fe}}{1 + r_Ch_{oe}} \qquad \text{(C.4)}$$

Equation (C.4) is our final goal: a formula that defines $A_i$ strictly in terms of the transistor $h$-parameters and the ac resistance of the collector circuit. The following example shows how equation (C.4) is used.

**EXAMPLE C.1**

Determine the value of $A_i$ for the transistor in Figure C.5. Assume that the $h$-parameter values for the transistor are as follows:

$$h_{ie} = 1 \text{ k}\Omega$$
$$h_{re} = 2.5 \times 10^{-4}$$
$$h_{fe} = 50$$
$$h_{oe} = 25 \text{ μS}$$

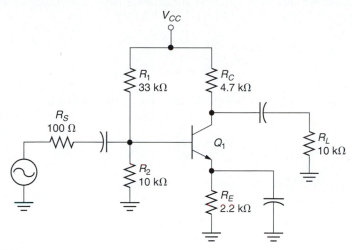

**FIGURE C.5**

**Solution:** To determine the value of $A_i$, we need to determine $r_C$. For the circuit shown,

$$r_C = R_C \parallel R_L = 4.7 \text{ k}\Omega \parallel 10 \text{ k}\Omega = \textbf{3.2 k}\Omega$$

Now, $A_i$ is determined as

$$A_i = \frac{h_{fe}}{1 + r_C h_{oe}} = \frac{50}{1 + (3.2 \text{ k}\Omega)(25 \text{ μS})} = \textbf{46.3}$$

If you take a look at the final result in Example C.1, you'll notice that it is very close to the original value of $h_{fe}$. The reason for this is the fact that $r_C h_{oe} \ll 1$. Since this is normally the case, equation (C.4) can be approximated as

$$A_i \cong h_{fe} \qquad \text{(C.5)}$$

## Voltage Gain ($A_v$)

The voltage gain of an amplifier equals the ratio of output voltage to input voltage. By formula,

$$A_v = \frac{v_2}{v_1} \qquad \text{(C.6)}$$

Now, take a look at the circuit shown in Figure C.6. As shown, $v_1$ is the sum of $h_{re}v_2$ and the voltage developed across $h_{ie}$. The voltage across $h_{ie}$ is found as $h_{ie}i_1$; therefore,

$$v_1 = h_{re}v_2 + h_{ie}i_1 \qquad \text{(C.7)}$$

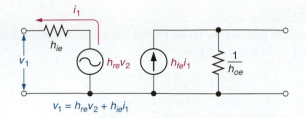

$$v_1 = h_{re}v_2 + h_{ie}i_1$$

**FIGURE C.6**

It was already shown that $v_2 = i_2 r_C$, and substituting these two equations into equation (C.6), we get

$$A_v = \frac{v_2}{v_1} = \frac{i_2 r_C}{h_{re}v_2 + h_{ie}i_1}$$

If we divide both the numerator and the denominator of the fraction by $i_2$, we get

$$A_v = \frac{r_C}{h_{re}r_C + h_{ie}/A_i} \qquad \textbf{(C.8)}$$

Confused? When you divide the numerator by $i_2$, you get

$$\frac{i_2 r_C}{i_2} = r_C$$

When you divide $h_{re}v_2$ by $i_2$, you get

$$\frac{h_{re}v_2}{i_2} = h_{re}\left(\frac{v_2}{i_2}\right) = h_{re}r_C$$

Finally, when you divide $h_{ie}i_1$ by $i_2$, you get

$$\frac{h_{ie}i_1}{i_2} = h_{ie}\left(\frac{i_1}{i_2}\right) = h_{ie}\left(\frac{1}{A_i}\right) = \frac{h_{ie}}{A_i}$$

These values were substituted into equation (C.8).

The final equation for $A_v$ is derived by substituting equation (C.4) for $A_i$ in equation (C.8). This gives us

$$A_v = \frac{h_{fe}r_C}{h_{ie} + (h_{ie}h_{oe} - h_{ie}h_{fe})r_C} \qquad \textbf{(C.9)}$$

Again, we have defined a gain value strictly in terms of the transistor $h$-parameters and the ac collector resistance. The following example demonstrates the use of this equation.

### EXAMPLE C.2

Determine the voltage gain for the circuit shown in Figure C.5 (Example C.1).

*Solution:* Using the $h$-parameter values listed in Example C.1, we get

$$A_v = \frac{h_{fe}r_C}{h_{ie} + (h_{ie}h_{oe} - h_{re}h_{fe})r_C}$$

$$= \frac{(50)(3.2\ k\Omega)}{1\ k\Omega + [(1\ k\Omega)(25\ \mu S) - (2.5 \times 10^{-4})(50)](3.2\ k\Omega)}$$

$$= \frac{1.6 \times 10^5\ \Omega}{1.04\ k\Omega}$$

$$= 153.85$$

Anyone care for an aspirin? Don't give up yet . . . it gets better. You see, the input impedance of a transistor can be approximated as being equal to $h_{ie}$. In Chapter 9, we defined $Z_{in(base)}$ as being equal to $h_{fe}r'_e$. Based on these two relationships, we defined $r'_e$ as follows:

$$r'_e = \frac{h_{ie}}{h_{fe}} \qquad\qquad \textbf{(C.10)}$$

Now, let's attack the problem in Example C.2 from another angle.

### EXAMPLE C.3

Determine the voltage gain for the circuit shown in Figure C.5 (Example C.1).

*Solution:*  First, using the *h*-parameters listed in Example C.1, we determine the value of $r'_e$ as

$$r'_e = \frac{1\ k\Omega}{50} = 20\ \Omega$$

The voltage gain can now be calculated as

$$A_v = \frac{r_C}{r'_e} = \frac{3.2\ k\Omega}{20\ \Omega} = 160$$

You will find that the method of determining $A_v$ used in Example C.3 will always give you a result that is well within 10% of the exact value. For almost every analysis problem, this value of $A_v$ will work just fine.

## Input Impedance ($Z_{in(base)}$)

The input impedance to a transistor is defined as the ratio of input voltage ($v_1$) to input current ($i_1$). By formula,

$$Z_{in(base)} = \frac{v_1}{i_1} \qquad\qquad \textbf{(C.11)}$$

If you refer to Figure C.6, you'll recall that $v_1 = h_{re}v_2 + h_{ie}i_1$. Substituting this value of $v_1$ into equation (C.11) gives us

$$Z_{in(base)} = \frac{h_{re}v_2 + h_{ie}i_1}{i_1}$$

which simplifies to

$$Z_{in(base)} = h_{ie} + \frac{h_{re}v_2}{i_1}$$

This equation can be rewritten as

$$Z_{\text{in(base)}} = h_{ie} + h_{re}\left(\frac{v_2}{i_1}\right)$$

and

$$Z_{\text{in(base)}} = h_{ie} + h_{re}\left(\frac{i_2 r_c}{i_1}\right)$$

Since $\dfrac{i_2}{i_1} = A_i$, the above equation can be rewritten as

$$Z_{\text{in(base)}} = h_{ie} + h_{re}A_i r_C$$

Finally, substituting equation (C.4) into the above equation gives us

$$Z_{\text{in(base)}} = h_{ie} + \frac{h_{re}h_{fe}r_C}{1 + r_C h_{oe}} \qquad \textbf{(C.12)}$$

Remember earlier when we assumed that $Z_{\text{in(base)}} \cong h_{ie}$? The following example will show this assumption to be valid.

### EXAMPLE C.4

Determine the exact value of $Z_{\text{in}}$ for the amplifier shown in Example C.1.

*Solution:*  Using the $h$-parameter values from Example C.1, the value of $Z_{\text{in(base)}}$ is found as

$$Z_{\textbf{in(base)}} = h_{ie} + \frac{h_{re}h_{fe}r_C}{1 + r_C h_{oe}} = 1\,\text{k}\Omega + \frac{(2.5 \times 10^{-4})(50)(3.2\,\text{k}\Omega)}{1 + (3.2\,\text{k}\Omega)(25\,\mu\text{S})}$$

$$= 1\,\text{k}\Omega + \frac{40}{1.08}\,\Omega = \textbf{1037}\,\boldsymbol{\Omega}$$

As you can see, there was only a 37 $\Omega$ difference between the values of $Z_{\text{in(base)}}$ and $h_{ie}$ in Example C.4.

## Output Impedance ($Z_{\text{out}}$)

The output impedance of a transistor is defined as the ratio of output voltage ($v_2$) to output current ($i_2$). By formula,

$$Z_{\text{out}} = \frac{v_2}{i_2} \qquad \textbf{(C.13)}$$

Substituting equation (C.5) in place of $i_2$ gives us

$$Z_{\text{out}} = \frac{v_2}{h_{fe}i_1 - v_2 h_{oe}} \qquad \textbf{(C.14)}$$

Now, take a look at the circuit shown in Figure C.7. The *Kirchhoff*'s voltage equation for the input is

$$v_{\text{in}} = i_1 r_s + i_1 h_{ie} + h_{re}v_2$$

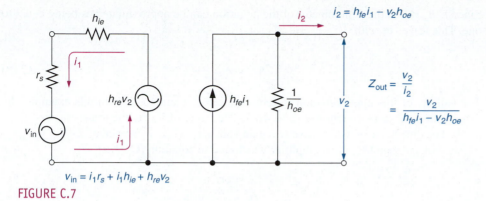

$$v_{in} = i_1 r_s + i_1 h_{ie} + h_{re} v_2$$

**FIGURE C.7**

If we short out the voltage source (a normal step in thevenizing a multisource circuit), we get

$$h_{re} v_2 = i_1 (r_s + h_{ie})$$

And finally,

$$i_1 = \frac{h_{re} v_2}{r_s + h_{ie}} \qquad \textbf{(C.15)}$$

Now, if you substitute equation (C.15) for the value of $i_1$ in equation (C.14), you get

$$Z_{out} = \frac{r_s - h_{ie}}{(r_s + h_{ie})h_{oe} - h_{re} h_{fe}} \qquad \textbf{(C.16)}$$

or

$$Z_{out} = \frac{1}{h_{oe} - \left( \dfrac{h_{fe} h_{re}}{h_{ie} + r_s} \right)} \qquad \textbf{(C.17)}$$

## EXAMPLE C.5

Determine the value of the output impedance for the transistor used in Example C.1.

*Solution:* The source resistance is shown to be 100 Ω. Using this value and the parameters given in Example C.1, $Z_{out}$ is found as

$$Z_{out} = \frac{1}{h_{oe} - \left( \dfrac{h_{fe} h_{re}}{h_{ie} + r_s} \right)} = \frac{1}{(25\mu S) - \dfrac{(50)(2.5 \times 10^{-4})}{(1\,k\Omega) + (100\,\Omega)}} = 73.3\,k\Omega$$

## *h*-Parameter Approximations

Most of the *h*-parameter formulas covered in this section can be approximated to a form that is easier to handle. While these formula approximations will not produce results that are as accurate as the original equations, they may be used for most applications.

Equation (C.4) shows *transistor* current gain ($A_i$) to be

$$A_i = \frac{h_{fe}}{1 + r_C h_{oe}}$$

Since $r_C h_{oe} \ll 1$, the denominator of the equation can be approximated as being equal to one. This leaves us with

$$A_i \cong h_{fe} \qquad \text{(C.18)}$$

The validity of this approximation is demonstrated in Example C.1. In this example, the exact value of $A_i$ was determined to be 46.3. The value of $h_{fe}$ for the transistor was given as 50. The difference between these two gain values is 3.7, a difference of 7.4%.

The voltage gain ($A_v$) of an amplifier was found in equation (C.9) as

$$A_v = \frac{h_{fe} r_C}{h_{ie} + (h_{ie} h_{oe} - h_{re} h_{fe}) r_C}$$

The approximation of this equation is based on several factors. First, we originally defined $A_v$ as

$$A_v = \frac{r_C}{r_e'}$$

This equation can also be written as

$$A_v = r_C \left( \frac{1}{r_e'} \right)$$

Now, equation (C.10) defined $r_e'$ as

$$r_e' = \frac{h_{ie}}{h_{fe}}$$

Therefore,

$$\frac{1}{r_e'} = \frac{h_{fe}}{h_{ie}}$$

Thus, $A_v$ can be approximated as

$$A_v \cong \frac{h_{fe} r_e}{h_{ie}} \qquad \text{(C.19)}$$

The validity of this approximation was demonstrated in Examples C.2 and C.3. Example C.2 provided a value of $A_v$ equal to 153.85. Example C.3 approximated the value of $A_v$ for the same circuit as 160. The percentage of error for the approximated value was about 4%.

Example C.4 calculated a $Z_{in}$ for an amplifier transistor of 1037 $\Omega$. The value of $h_{ie}$ for the transistor was 1 k$\Omega$. Thus, we can approximate the value of $Z_{in}$ as

$$Z_{in} \cong h_{ie} \qquad \text{(C.20)}$$

Unfortunately, there is no quick approximation for the output impedance of a transistor. Whenever you need to calculate this value, you need to use equation (C.17). However, there is some good news here. In most cases, there is no reason to calculate the output impedance of a transistor. This is because it rarely weighs into any common-emitter circuit calculations. When you do need a general idea of how large $Z_{out}$ is for a given transistor, you can use

$$Z_{out} > \frac{1}{h_{oe}} \qquad \text{(C.21)}$$

While equation (C.21) will not tell you the exact value of $Z_{out}$, it will give you an idea of how large the value is.

## Other Transistor Configurations

$h$-Parameters apply not only to common-emitter circuits but to common-base and common-collector circuits as well. However, you must use several conversion factors to obtain the values needed to calculate the $A_i$, $A_v$, $Z_{in}$, and $Z_{out}$ values for these circuit configurations. In this section, you will be introduced to the $h$-parameter conversions as well as the common-base and common-collector formulas used for circuit calculations. We will not analyze these formulas and conversions. They are listed simply for future reference.

The common-base $h$-parameters are identified by the use of the subscript letter $b$ in place of $e$. The following chart illustrates this point:

| Parameter | Common Emitter | Common Base |
|---|---|---|
| Transistor current gain | $h_{fe}$ | $h_{fb}$ |
| Voltage feedback ratio | $h_{re}$ | $h_{rb}$ |
| Input impedance | $h_{ie}$ | $h_{ib}$ |
| Output admittance | $h_{oe}$ | $h_{ob}$ |

To convert the common-emitter parameters to common-base parameters, use the following conversion formulas:

$$h_{fb} = \frac{h_{fe}(1 - h_{re}) - h_{ie}h_{oe}}{(1 + h_{fe})(1 - h_{re}) + h_{ie}h_{oe}} \cong \frac{h_{fe}}{h_{fe} + 1} \qquad \text{(C.22)}$$

$$h_{rb} = \frac{h_{ie}h_{oe} - h_{re}(1 + h_{oe})}{(1 + h_{fe})(1 - h_{re}) + h_{ie}h_{oe}} \cong \frac{h_{ie}h_{oe}}{h_{fe} + 1} - h_{re} \qquad \text{(C.23)}$$

$$h_{ib} = \frac{h_{ie}}{(1 + h_{fe})(1 - h_{re}) + h_{ie}h_{oe}} \cong \frac{h_{ie}}{h_{fe} + 1} \qquad \text{(C.24)}$$

$$h_{ob} = \frac{h_{oe}}{(1 + h_{fe})(1 - h_{re}) + h_{ie}h_{oe}} \cong \frac{h_{oe}}{h_{fe} + 1} \qquad \text{(C.25)}$$

Once the parameter conversions are complete, the circuit gain and impedance values can be found as

$$A_i = \frac{h_{fb}}{1 + h_{ob}r_C} \qquad \text{(C.26)}$$

$$A_v = \frac{h_{fb}r_C}{h_{ib} + (h_{ib}h_{ob} - h_{fb}h_{rb})r_C} \qquad \text{(C.27)}$$

$$Z_{in} = h_{ib} - \frac{h_{rb}h_{fb}r_C}{1 + h_{ob}r_C} \qquad \text{(C.28)}$$

$$Z_{out} = \frac{1}{h_{ob} - \dfrac{h_{fb}h_{rb}}{h_{ib} + r_s}} \qquad \text{(C.29)}$$

The *common-collector h*-parameters are identified by the letter $c$ in the subscript in place of $e$. The conversions from common-emitter parameters to common-collector parameters are as follows:

$$h_{fc} = h_{fe} + 1 \tag{C.30}$$

$$h_{rc} = 1 - h_{re} \tag{C.31}$$

$$h_{ic} = h_{ie} \tag{C.32}$$

$$h_{oc} = h_{oe} \tag{C.33}$$

Using the converted parameters, the gains and impedances of the common-collector amplifier can be found as

$$*A_i = \frac{h_{fc}}{1 + h_{oc}r_E} \tag{C.34}$$

$$A_v = \frac{h_{fc}r_E}{h_{ic} + (h_{ic}h_{oc} - h_{fc}h_{rc})r_E} \tag{C.35}$$

$$Z_{in} = h_{ic} - \frac{h_{rc}h_{fc}r_E}{1 + h_{oc}r_E} \tag{C.36}$$

$$Z_{out} = \frac{1}{h_{oe} + \dfrac{h_{fc}h_{rc}}{h_{ic} + r_s}} \tag{C.37}$$

The actual gain and impedance equations shown in this section are the same as those used for the common-emitter circuit. However, before they can be used, the common-emitter *h*-parameters must be converted to the necessary form.

---

* For the common-collector circuit, the output is taken from the emitter. Therefore, the output ac resistance is the ac resistance of the emitter circuit, $r_E$.

# Selected Equation Derivations

## Equation (3.11)

The average value of any curve equals the area under the curve divided by its length. For example, the average value of the sine-wave alternation shown in Figure D.1 can be found as

$$V_{ave} = \frac{A}{\ell}$$

where $A$ is the value of a shaded area under the curve and $\ell$ is the difference between the zero points on the curve.

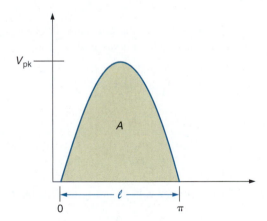

FIGURE D.1

To find the area under the curve, we start with the curve equation. The equation of the curve in Figure D.1 is given as

$$v = V_{pk} \sin \omega t \qquad \text{(D.1)}$$

where $\omega t$ = the phase angle, given as a product of angular velocity ($\omega$), in radians, and time

Using calculus, the area under the curve is found as the integral of equation (D.1). By formula,

$$A = \int_0^\pi V_{pk} \sin \omega t \, d(\omega t) \qquad \text{(D.2)}$$

Solving equation (D.2), we get

$$A = \int_0^\pi V_{\text{pk}} \sin \omega t \; d(\omega t) = -V_{\text{pk}}[-\cos \omega t]_0^\pi$$
$$= -V_{\text{pk}}[-\cos \pi - \cos 0] = -V_{\text{pk}}[-1 - 1] = 2V_{\text{pk}}$$

Now we know that the area under the curve equals $2V_{\text{pk}}$. As shown in Figure D.1, the difference between the zero crossings of the waveform equals $\pi$ (radians). Using these two values in equation (D.1),

$$V_{\text{ave}} = \frac{A}{\ell} = \frac{2V_{\text{pk}}}{\pi} \tag{D.3}$$

Equation (D.3) gives us the average value of the positive alternation produced by a half-wave rectifier. The positive alternation makes up only *half* the cycle time, with the other half having a value of 0 V. Therefore, the average *over one complete output cycle* is half the above value, or

$$V_{\text{ave}} = \frac{V_{\text{pk}}}{\pi}$$

which is equation (3.11).

## Equation (7.15)

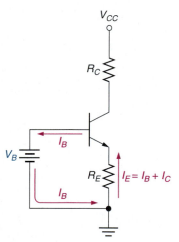

**FIGURE D.2**

Refer to Figure D.2. According to Ohm's law, the base current ($I_B$) generated by the voltage source ($V_B$) can be found as

$$I_B = \frac{V_B}{R_{IN(\text{base})} + R_E}$$

where $R_{IN(\text{base})}$ is the base input resistance of the transistor. When $R_{IN(\text{base})} \gg R_E$ (which is normally the case), $I_B$ can be assumed to have a value of

$$I_B = \frac{V_B}{R_{IN(\text{base})}}$$

or

$$V_B = I_B R_{IN(\text{base})} \tag{D.4}$$

As implied by equation (7.9), the emitter voltage in Figure D.2 can be found as

$$V_E = I_E R_E \tag{D.5}$$

If the forward voltage across the base-emitter junction of the transistor is ignored, then the circuit in Figure D.2 can be assumed to have a value of

$$V_B = V_E$$

Substituting equations (D.4) and (D.5) into the above equation, we have

$$I_B R_{IN(\text{base})} = I_E R_E \tag{D.6}$$

As given in equation (6.5),

$$I_E = (h_{FE} + 1)I_B$$

Substituting this relationship into equation (D.6), we get

$$I_B R_{IN(\text{base})} = (h_{FE} + 1) I_B R_E$$

Assuming that $h_{FE} \gg 1$, the above relationship can be rewritten as

$$I_B R_{IN(\text{base})} = h_{FE} I_B R_E$$

Finally, factoring out $I_B$, we have

$$R_{IN(\text{base})} = h_{FE} R_E$$

which is equation (7.15).

# Equation (9.2)

Schockley's equation for the total current through a *pn* junction is

$$I_T = I_S(e^{Vq/kT} - 1) \tag{D.7}$$

where $I_S$ = the reverse saturation current through the diode
$\quad e$ = the *exponential constant*; approximately 2.71828
$\quad V$ = the voltage across the depletion layer
$\quad q$ = the charge on an electron; approximately $1.6 \times 10^{-19}$ V
$\quad k$ = Boltzmann's constant; approximately $1.38 \times 10^{-23}$ J/°K
$\quad T$ = the temperature of the device, in degrees Kelvin (°K = °C + 273)

We can solve for the value of $q/kT$ at room temperature (approximately 21°C) as

$$q/kT = \frac{1.6 \times 10^{-19}\,\text{J}}{(1.38 \times 10^{-23})\text{J/°K}(294°\text{K})} = 40$$

Using this value, the equation for $I_T$ is rewritten as

$$I_T = I_S(e^{40\,V} - 1) \tag{D.8}$$

Now, differentiating the above equation gives us

$$\frac{dI}{dV} = 40 I_S e^{40\,V} \tag{D.9}$$

If we rearrange equation (D.8) as

$$I_S e^{40\,V} = (I_T + I_S)$$

we can use this equation to rewrite equation (D.9) as

$$\frac{dI}{dV} = 40(I_T + I_S) \tag{D.10}$$

Now, if we take the reciprocal of equation (D.10), we will have an equation for the ac resistance of the junction, $r_e'$, as follows:

$$\frac{dV}{dI} = \frac{1\,\text{V}}{40(I_T + I_S)} = \frac{1\,\text{V}}{40}\frac{1}{I_T + I_S} = 25\,\text{mV}\frac{1}{I_T + I_S} = \frac{25\,\text{mV}}{I_T + I_S}$$

In this case, $(I_T + I_S)$ is the current through the emitter-base junction of the transistor, $I_E$. Therefore, we can rewrite the above equation as

$$r'_e = \frac{25 \text{ mV}}{I_E}$$

## Equation (9.21)

According to Ohm's law, the input current to a common-emitter amplifier can be found as

$$i_{in} = \frac{v_s}{Z_{in}}$$

As Figure D.3 shows, the input current to a common-emitter amplifier is divided between the resistors in the biasing network and the base of the transistor. Using the current-divider relationship, the value of $i_b$ can be found as

$$i_b = i_{in}\left(\frac{Z_{in}}{Z_{in(base)}}\right) \tag{D.11}$$

where $Z_{in}$ = the parallel combination of the biasing resistors and the base input impedance, $Z_{in(base)}$.

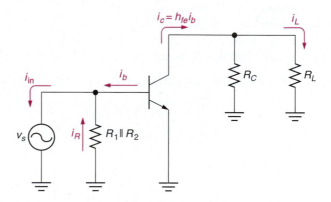

**FIGURE D.3**

Figure D.3 shows the transistor output current $(i_c)$ to be equal to the product of the transistor ac current gain $(h_{fe})$ and the ac base current. By formula,

$$i_c = h_{fe}i_b$$

Substituting equation (D.11) for the value of $i_b$ in the above equation, we get

$$i_c = h_{fe}i_{in}\left(\frac{Z_{in}}{Z_{in(base)}}\right) \tag{D.12}$$

As is the case with the input circuitry, the output circuitry forms a current divider. The output current divider is made up of the collector resistor $(R_C)$ and the load. Once we know the value of $i_c$ for a given common-emitter amplifier, the value of the ac load current can be found as

$$i_L = i_c\left(\frac{r_C}{R_L}\right) \tag{D.13}$$

where $r_C$ is the parallel combination of the collector resistor and the load resistance.

The *effective* current gain of any amplifier equals the ratio of *ac load current* to *ac input current*. By equation,

$$A_i = \frac{i_L}{i_{in}}$$

Substituting the relationships we have established for the values of $i_L$ and $i_{in}$, we get

$$A_i = \frac{i_c \dfrac{r_C}{R_L}}{\dfrac{v_s}{Z_{in}}}$$

or

$$A_i = h_{fe} i_b \left(\frac{r_C}{R_L}\right)\left(\frac{Z_{in}}{v_s}\right)$$

Since $\dfrac{Z_{in}}{v_s} = \dfrac{1}{i_{in}}$, the above equation can be rewritten as

$$A_i = h_{fe}\left(\frac{i_b}{i_{in}}\right)\left(\frac{r_C}{R_L}\right) \qquad \textbf{(D.14)}$$

According to equation (D.11),

$$\frac{i_b}{i_{in}} = \frac{Z_{in}}{Z_{in(base)}}$$

Substituting this equation for the current ratio in equation (D.14), we get

$$A_i = h_{fe}\left(\frac{Z_{in} r_C}{Z_{in(base)} R_L}\right)$$

This is equation (9.21).

## Equation (10.15)

The derivation of equation (10.15) is best understood by looking at the circuits shown in Figure D.4. Figure D.4a shows the ac equivalent of the emitter follower. Note that the input circuit consists of the source resistance ($R_S$) in parallel with the combination of $R_1$ and $R_2$. Thus, the total resistance in the base circuit (as seen from the base of the transistor) can be found as

$$R'_{in} = R_1 \parallel R_2 \parallel R_S$$

This resistance is shown as a single resistor in Figure D.4b.

To determine the value of $Z_{out}$, we start by writing the Kirchhoff's voltage equation for the circuit and solving that equation for $i_e$. The voltage equation is

$$v_{in} = i_b R'_{in} + i_e(r'_e + R_E)$$

Since $i_b = \dfrac{i_e}{h_{fc}}$, the voltage equation can be rewritten as

$$v_{in} = \frac{i_e R'_{in}}{h_{fc}} + i_e(r'_e + R_E)$$

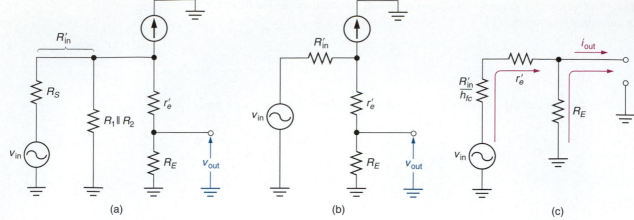

FIGURE D.4

Solving for $i_e$ yields

$$i_e = \frac{v_{in}}{r'_e + R_E + \dfrac{R'_{in}}{h_{fc}}}$$

This equation indicates that the emitter "sees" a three-resistor circuit, as shown in Figure D.4c. If we thevenize the circuit, we find that the load sees $R_E$ in parallel with the series combination of the other two resistors. Thus, $Z_{out}$ is found as

$$Z_{out} = R_E \parallel \left( r'_e + \frac{R'_{in}}{h_{fc}} \right)$$

This is equation (10.15).

## Equation (10.21)

Equation (10.13) gives the value of the base input impedance as

$$Z_{in(base)} = h_{fc}(r'_e + r_E)$$

where $r_E$ is the parallel combination of $R_E$ and $R_L$. Expanding the right side of the equation, we obtain

$$Z_{in(base)} = h_{fc}r'_e + h_{fc}r_E$$

Now, since $h_{ic} = h_{fc}r'_e$, we can rewrite the above equation as

$$Z_{in(base)} = h_{ic} + h_{fc}r_E$$

Thus, for the Darlington amplifier shown in Figure D.5, the value of $Z_{in(base)}$ for $Q_2$ is

$$Z_{in(base)2} = h_{ic2} + h_{fc}r_{E2}$$

and the value of $Z_{in(base)}$ for $Q_1$ is

$$Z_{in(base)1} = h_{ic1} + h_{fc1}r_{E1}$$

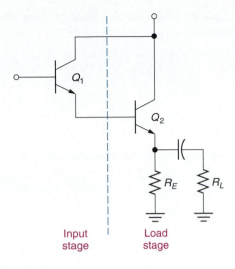

**FIGURE D.5**

As Figure D.5 shows, $Q_2$ is the ac load on $Q_1$. Thus, $r_{E1} = Z_{in(base)2}$; thus,

$$Z_{in(base)1} = h_{ic1} + h_{fc1}r_{E1} = h_{ic1} + h_{fc1}(h_{ic2} + h_{fc}r_{E2})$$

or

$$Z_{in(base)} = h_{ic1} + h_{fc1}(h_{ic2} + h_{fc}r_E)$$

where $Z_{in(base)}$ = the input impedance of $Q_1$
$r_E$ = the parallel combination of $R_E$ and $R_L$

## *RC*-Coupled Class A Efficiency

The *ideal* class A amplifier would have the following characteristics:

$$\textbf{1.} \quad V_{CEQ} = \frac{V_{CC}}{2}$$

$$\textbf{2.} \quad V_{PP} = V_{CC}$$

$$\textbf{3.} \quad I_{CQ} = \frac{I_{C(sat)}}{2}$$

$$\textbf{4.} \quad I_{PP} = I_{C(sat)}$$

These ideal characteristics will be used in our derivation of the maximum ideal efficiency rating of 25%.

Equation (11.10) gives us the following equation for load power:

$$P_L = \frac{V_{PP}^2}{8R_L}$$

Since $R_L = V_{PP}/I_{PP}$, we can rewrite the above equation as

$$P_L = \left(\frac{V_{PP}^2}{8}\right)\left(\frac{1}{R_L}\right) = \left(\frac{V_{PP}^2}{8}\right)\left(\frac{I_{PP}}{V_{PP}}\right) = \frac{V_{PP}I_{PP}}{8} \qquad \textbf{(D.15)}$$

Now, we will use the characteristics listed on page 949 and equation (D.15) to determine the ideal maximum load power as follows:

$$P_L = \frac{V_{PP}I_{PP}}{8} = \frac{V_{CC}I_{C(sat)}}{8} = \frac{(2V_{CEQ})(2I_{CQ})}{8}$$

$$= \frac{4(V_{CEQ}I_{CQ})}{8} = \frac{V_{CEQ}I_{CQ}}{2} \quad \text{(maximum, ideal)}$$

Now, recall that the power drawn from the supply of an amplifier is found as

$$P_S = V_{CC}I_{CC}$$

where

$I_1$ is being used to designate the current through the base biasing circuit.

$$I_{CC} = I_{CQ} + I_1$$

Since $I_{CQ}$ is normally *much greater than* $I_1$, we can approximate the equation above to

$$P_S \cong V_{CC}I_{CQ}$$

Note that this equation is valid for most amplifier power analyses. We can rewrite the above for the ideal amplifier as follows:

$$P_S = 2V_{CEQ}I_{CQ}$$

We can now use the derived values of $P_L$ and $P_S$ to determine the *maximum ideal value* of $\eta$ as follows:

$$\eta = \frac{P_L}{P_S} \times 100 = P_L \frac{1}{P_S} \times 100 = \frac{V_{CEQ}I_{CQ}}{2} \frac{1}{2V_{CEQ}I_{CQ}} \times 100$$

$$= \frac{V_{CEQ}I_{CQ}}{4V_{CEQ}I_{CQ}} \times 100 = \frac{1}{4} \times 100 = 25\%$$

## Transformer-Coupled Class A Efficiency

Assume for a minute that we are dealing with an *ideal* transformer-coupled amplifier. This amplifier would have characteristics of $V_{CEQ} = V_{CC}$ and $V_{CE(off)} = 2V_{CC}$. These characteristics ignore the fact that there would be some voltage dropped across the emitter resistor of the amplifier. Now, recall that the maximum power that can be delivered to a load from a class A amplifier is found as

$$P_{L(max)} = \frac{V_{CEQ}I_{CQ}}{2}$$

This equation was derived in our initial discussion on class A amplifiers. Since the ideal transformer-coupled amplifier would have a value of $V_{CEQ} = V_{CC}$, the equation above can be rewritten as

$$P_{L(max)} = \frac{V_{CC}I_{CQ}}{2} \quad \text{(maximum, ideal)}$$

You may also recall that the power drawn from the supply can be approximated as

$$P_S = V_{CC}I_{CQ}$$

We can use these two equations to calculate the maximum *ideal* efficiency of the transformer-coupled amplifier as follows:

$$\eta = \frac{P_L}{P_S} \times 100 = P_L \frac{1}{P_S} \times 100 = \frac{V_{CC}I_{CQ}}{2} \frac{1}{V_{CC}I_{CQ}} \times 100 = \frac{1}{2} \times 100 = 50\%$$

## Class B Amplifier Efficiency

Equation (11.31) defines $I_{C1(ave)}$ as

$$I_{C1(ave)} = \frac{V_{CC}}{2\pi R_L}$$

This equation is based on the relationship

$$I_{C1(ave)} = \frac{I_{pk}}{\pi}$$

In the practical class B amplifier, $I_{C1(ave)} \gg I_1$. Therefore, we can assume that $I_{CC} = I_{C1(ave)}$ in the *ideal* class B amplifier. Based on this point, the determination of the ideal maximum efficiency for the class B amplifier proceeds as follows:

$$\eta = \frac{P_L}{P_S} \times 100 = P_L\left(\frac{1}{P_S}\right) \times 100 = \left(\frac{V_{CC}^2}{8R_L}\right)\left(\frac{1}{V_{CC}I_{C1(ave)}}\right) \times 100$$

$$= \left(\frac{V_{CC}^2}{8R_L}\right)\left(\frac{2\pi R_L}{V_{CC}^2}\right) \times 100 = \frac{2\pi}{8} \times 100 = 78.54\%$$

The final result can, of course, be assumed to be equal to 78.5%. Remember that this maximum efficiency rating is ideal. Any practical value of efficiency for a class B amplifier must be less than this value.

## Equation (11.37)

The instantaneous power dissipation in a class B amplifier is found as

$$p = V_{CEQ}(1 - k \sin \theta)I_{C(sat)}(k \sin \theta) \qquad \text{(D.16)}$$

where $V_{CEQ}$ = the quiescent value of $V_{CE}$ for the amplifier
$I_{C(sat)}$ = the saturation current for the amplifier, as determined by the ac load line for the circuit
$\theta$ = the phase angle of the output at the instant that $p$ is measured
$k$ = a constant factor, representing the percentage of the load line that is actually being used by the circuit; $k$ is represented as a decimal value between 0 and 1

The average power dissipation can be found by integrating equation (D.16) for one half-cycle. Note that a half-cycle is represented as being between 0 and $\pi$ *radians* in the integration

$$P_{ave} = \frac{1}{2\pi} \int_0^\pi p \, d\theta \qquad \text{(D.17)}$$

Performing the integration yields

$$P_{\text{ave}} = \left(\frac{V_{CEQ}I_{C(\text{sat})}}{2\pi}\right)\left(2k - \frac{\pi k^2}{2}\right) \tag{D.18}$$

The next step is to find the maximum value of $k$ in equation (D.18). The first step in finding this maximum value is to differentiate $P_{\text{ave}}$ with respect to $k$ as follows:

$$\frac{dP_{\text{ave}}}{dk} = \frac{V_{CEQ}I_{C(\text{sat})}}{2}(2 - \pi k) \tag{D.19}$$

Next, we set the right-hand side of equation (D.19) equal to zero. This provides us with the following equations:

$$\frac{V_{CEQ}I_{C(\text{sat})}}{2} = 0 \quad \text{and} \quad 2 - \pi k = 0$$

Now, we can solve the equation on the right to find the maximum value:

$$k = \frac{2}{\pi} \tag{D.20}$$

At this point, we want to take the value of $k$ obtained in equation (D.20) and plug it into equation (D.17). Reducing this new equation gives us

$$P_{\text{ave}} = 0.1V_{CEQ}I_{C(\text{sat})} \tag{D.21}$$

Finally, we replace the values of $V_{CEQ}$ and $I_{C(\text{sat})}$ in equation (D.21) using the following relationships:

$$V_{CEQ} = \frac{\text{PP}}{2} \quad \text{and} \quad I_{C(\text{sat})} = \frac{V_{CEQ}}{R_L}$$

This gives us

$$P_D = 0.1\left(\frac{V_{CEQ}\text{PP}}{2R_L}\right)$$

or

$$P_D = \left(\frac{V_{CEQ}\text{PP}}{20R_L}\right)$$

Now, since $V_{CEQ}$ is equal to $\dfrac{\text{PP}}{2}$ we can rewrite the above equation as

$$P_D = (V_{CEQ})\left(\frac{\text{PP}}{20R_L}\right) = \left(\frac{\text{PP}}{2}\right)\left(\frac{\text{PP}}{20R_L}\right)$$

or

$$P_D = \frac{\text{PP}^2}{40R_L}$$

This is a form of equation (11.37).

## Equation (14.17)

The proof of this equation starts by taking a look at the $RC$ circuit shown in Figure D.6. Since the current through a series circuit is constant, the ratio of $v_{out}$ to $v_{in}$ is equal to the ratio of $X_C$ (the output shunt component) to the total impedance of the circuit. By formula,

$$\frac{v_{out}}{v_{in}} = \frac{X_C}{Z_T} \tag{D.22}$$

or

$$\frac{v_{out}}{v_{in}} = \frac{\dfrac{1}{j\omega C}}{\dfrac{1}{j\omega C} + R}$$

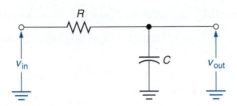

FIGURE D.6

Multiplying both the numerator and the denominator in equation (D.22) by $j\omega C$, we get

$$\frac{v_{out}}{v_{in}} = \frac{1}{1 + j\omega RC} \tag{D.23}$$

Now, since $f_C = \dfrac{1}{2\pi RC}$ and $\omega = 2\pi f$, equation (D.23) can be rewritten as

$$\frac{v_{out}}{v_{in}} = \frac{1}{1 + j\dfrac{1}{f_C}} \tag{D.24}$$

Now, the ratio of $X_C$ to $Z_T$ varies with the ratio of operating frequency to $f_C$. Thus, equation (D.23) can be rewritten as

$$\frac{v_{out}}{v_{in}} = \frac{1}{1 + j(f_{C1}/f)} \tag{D.25}$$

Finally, we use the relationships

$$A_v = \frac{v_{out}}{v_{in}}$$

and

$$1 + jX = \sqrt{1 + X^2}$$

(where $X$ = any quantity) to rewrite equation (D.25) as

$$\Delta A_v = \frac{1}{\sqrt{1 + (f_{C1}/f)^2}}$$

(D.26)

Rewriting equation (D.26) in $dB$ form, we obtain

$$\Delta A_v = 20 \log \left( \frac{1}{\sqrt{1 + (f_{C1}/f)^2}} \right)$$

(D.26)

## Equations (14.42) and (14.43)

In this derivation, we will concentrate on the *low-frequency* response of a multistage amplifier. We will then expand the derivation to include the *high-frequency* response of a given multistage amplifier.

To minimize any confusion, we now define two variables that will be used extensively in this derivation:

$$A_{v(mid)} = \text{the } \textit{midband} \text{ gain of an amplifier}$$

$$A_v = \text{the gain of an amplifier at any specified frequency}$$

Note that $A_v = A_{v(mid)}$ when the amplifier is operated above $f_{C1}$ and that $A_v < A_{v(mid)}$ when the amplifier is operated below $f_{C1}$.

Earlier in the text, you were shown that the ratio of amplifier gain to its midband gain $(A_v')$ is found as

$$A_v' = \frac{A_v}{A_{v(mid)}} = \frac{1}{\sqrt{1 + (f_{C1}/f)^2}}$$

(D.27)

It follows that two amplifier stages with identical values of $f_{C1}$ and operated at the same value of $f$ will have values of $A_v'$ that are equal. By formula,

$$A_{v1}' = A_{v2}'$$

when the two circuits have the same cutoff frequency and are at the same operating frequency.

Now, recall that the total gain of $n$ cascaded amplifier stages is equal to the product of the individual stage gain values. By formula,

$$A_{vT} = (A_{v1})(A_{v2}) \cdot \cdot \cdot (A_{vn})$$

The same relationship holds true for the gain ratios of the individual stages. By formula,

$$A_{vT}' = (A_{v1}')(A_{v2}') \cdot \cdot \cdot (A_{vn}')$$

(D.28)

Now, assuming that we have $n$ stages with equal values of $A_v'$, the value of $A_{vT}'$ is found as

$$A_{vT}' = (A_v')^n$$

(D.29)

where $A_v'$ = the gain ratio of any individual stage. Substituting equation (D.27) into equation (D.28) gives us

$$A_{vT}' = (A_v')^n = \left( \frac{1}{\sqrt{1 + (f_{C1}/f)^2}} \right)^2$$

(D.30)

or

$$A'_{vT} = \left( \frac{1}{\sqrt{1 + (f_{C1}/f)^2}} \right)^n \qquad \text{(D.31)}$$

Now, when $f = f_{C1}$ (the lower 3 dB point), equation (D.31) simplifies as follows:

$$\left( \frac{1}{\sqrt{1 + (f_{C1}/f_{C1})^2}} \right) = \left( \frac{1}{\sqrt{1 + 1}} \right) = \left( \frac{1}{\sqrt{2}} \right) = 2^{1/2}$$

If we set $A'_{vT}$ to the value of $2^{1/2}$, equation (D.31) can be rewritten as

$$2^{1/2} = \left\{ \left[ 1 + (f_{C1}/f)^2 \right]^{1/2} \right\}^n$$

Since the exponent value of $(1/2)$ appears on both sides of the equation, it can be dropped completely, leaving

$$2 = \left[ 1 + (f_{C1}/f)^2 \right]^n$$

or

$$2^{1/n} = 1 + (f_{C1}/f)^2 \qquad \text{(D.32)}$$

Finally, solving equation (D.32) for $f$ yields

$$f = \frac{f_{C1}}{\sqrt{2^{1/n} - 1}}$$

or, in another form,

$$f_{C1(T)} = \frac{f_{C1}}{\sqrt{2^{1/n} - 1}} \qquad \text{(D.33)}$$

This is equation (14.43). The same process is used to derive equation (14.42).

## Equation (15.7)

The equation for the instantaneous voltage at any point on a sine wave is given as

$$v = V_{pk} \sin \omega t$$

where $V_{pk}$ = the peak input voltage
    $\omega = 2\pi f$
    $t$ = the time from the start of the cycle to the instant that $v$ occurs

If we differentiate $v$ with respect to $t$, we get

$$\frac{dv}{dt} = \omega V_{pk}(\cos \omega t)$$

Since the slew rate is a maximum rating of a given op-amp and the rate of change in a sine wave varies, we need to determine the point at which the rate of change in the sine wave is at its maximum value. This point occurs when the sine wave passes the reference voltage, or at $t = 0$. The value of $dv/dt$ at $t = 0$ determines the maximum operating frequency of the op-amp as follows:

$$\frac{dv}{dt}\text{max} = \omega_{max} V_{pk}$$

or

$$\text{slew rate} = 2\pi f_{max} V_{pk}$$

Finally, this equation is rearranged into the form of equation (15.7):

$$f_{max} = \frac{\text{slew rate}}{2\pi V_{pk}}$$

## Equation (15.24)

The value of $v_{in}$ for an amplifier with feedback can be given as

$$v_{in} = v_s - v_f$$

According to equation (15.22),

$$v_f = \alpha_v v_{out}$$

Substituting this equation for $v_f$ in the $v_{in}$ equation, we get

$$v_{in} = v_s - \alpha_v v_{out} \tag{D.34}$$

Now, equation (D.34) can be rewritten to solve for $v_s$ as

$$v_s = v_{in} + \alpha_v v_{out} \tag{D.35}$$

Since $v_{out} = A_v V_{in}$, we can rewrite equation (D.35) as

$$v_s = v_{in} + \alpha_v A_v v_{in}$$

or

$$v_s = v_{in}(1 + \alpha_v A_v) \tag{D.36}$$

Now, $A_v$ is the total gain of the feedback amplifier from source to output. By formula,

$$A_{vf} = \frac{v_{out}}{v_s} \tag{D.37}$$

Substituting equation (D.36) for $v_s$ in equation (D.37) gives us

$$A_{vf} = \frac{v_{out}}{v_{in}(1 + \alpha_v A_v)}$$

or

$$A_{vf} = \frac{A_v}{1 + \alpha_v A_v}$$

## Equations (17.23) and (17.25)

We start these proofs with an assumption that can be seen intuitively; that is, for any given series reactive-resistive circuit, there exists an equivalent parallel reactive-resistive circuit. These two circuits are represented in Figure D.7. All this assumption does is let us assume that it *is* possible to derive a parallel equivalent circuit for a series reactive-resistive circuit.

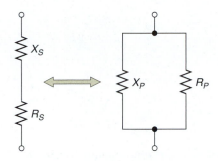

FIGURE D.7

For the series circuit shown in Figure D.7,

$$Q = \frac{X_S}{R_S} \tag{D.38}$$

and

$$Z_S = R_S \pm jX_S \tag{D.39}$$

For the parallel circuit shown in Figure D.7, the total impedance ($Z_P$) is found as

$$Z_P = \frac{(R_P)(\pm jX_P)}{R_P \pm jX_P} \tag{D.40}$$

If we multiply the right-hand side of equation (D.40) by 1 in the form of

$$\frac{R_P \pm jX_P}{R_P \pm jX_P}$$

we get

$$Z_P = \frac{R_P X_P^2}{R_P^2 + X_P^2} \pm j\frac{X_P R_P^2}{R_P^2 + X_P^2} \tag{D.41}$$

Now, refer to our original assumption regarding the two circuits shown in Figure D.7. If these two circuits are equivalent circuits, then the total impedance of the two circuits must be equal. By formula,

$$Z_S = Z_P$$

Substituting this relationship into equation (D.41) gives us

$$Z_S = \frac{R_P X_P^2}{R_P^2 + X_P^2} \pm j\frac{X_P R_P^2}{R_P^2 + X_P^2} \tag{D.42}$$

Now, if we substitute equation (D.39) for $Z_S$ in equation (D.42), we get

$$R_S \pm jX_S = \frac{R_P X_P^2}{R_P^2 + X_P^2} \pm j \frac{X_P R_P^2}{R_P^2 + X_P^2} \qquad \textbf{(D.43)}$$

For the two sides of equation (D.43) to be equal, the real components on each side must be equal, and the imaginary components on each side must also be equal. Therefore,

$$R_S = \frac{R_P X_P^2}{R_P^2 + X_P^2} \qquad \textbf{(D.44)}$$

and

$$X_S = \frac{X_P R_P^2}{R_P^2 + X_P^2} \qquad \textbf{(D.45)}$$

Using these two equations, we can rewrite equation (D.38) as

$$Q = \frac{\dfrac{X_P R_P^2}{R_P^2 + X_P^2}}{\dfrac{R_P X_P^2}{R_P^2 + X_P^2}} = \frac{X_P R_P^2}{R_P X_P^2}$$

and

$$Q = \frac{R_P}{X_P} \qquad \textbf{(D.46)}$$

Equation (D.46) will be used to prove equation (17.25) at the end of this proof. For now, we'll concentrate on equation (D.45). Using the relationship

$$Q = \frac{R_P}{X_P}$$

we can rewrite equation (D.45) as

$$X_S = \frac{X_P}{1 + (1/Q)^2} \qquad \textbf{(D.47)}$$

Since tuned amplifiers are *high-Q* circuits, it is safe to assume that $(1/Q)^2 \ll 1$. Based on this assumption, equation (D.47) simplifies to

$$X_S = X_P \qquad \textbf{(D.48)}$$

Now, we can rewrite equation (D.46) as

$$R_P = QX_P$$

or, using equation (D.48),

$$R_P = QX_S$$

Since $X_S = QR_S$ [from equation (D.38)], we can solve the above equation as

$$R_P = Q(QR_S)$$

or

$$R_P = Q^2 R_S \qquad \text{(D.49)}$$

Now, let's relate equations (D.46) and (D.49) to the *LC* tuned circuit. In an *LC* tuned circuit, the series reactance is $X_L$ and the series resistance is $R_W$. Replacing $R_S$ in equation (D.49) with $R_W$ gives us

$$R_P = Q^2 R_W$$

Finally, replacing $X_P$ in equation (D.46) with $X_L$ gives us

$$Q = \frac{R_P}{X_L}$$

or

$$Q_L = \frac{R_P \parallel R_L}{X_L}$$

when a load is connected to the equivalent parallel circuit. Of course, this equation is the equivalent of equation (17.25).

## Equation (19.2)

The time required for a capacitor to charge to a given value is found as

$$t = RC \left[ \ln \frac{V - V_0}{V - V_C} \right] \qquad \text{(D.50)}$$

where $V$ = the charging voltage
$V_0$ = the initial charge on the capacitor
$V_C$ = the capacitor charge of interest
$\ln$ = an operator, indicating that we are to take the *natural log* of the fraction

We can use equation (D.50) to find the time required for $V_C$ to reach 0.1 V (10% of $V$) as follows:

$$t = RC \left[ \ln \frac{V - 0\,V}{V - 0.1V} \right] = RC \left[ \ln \frac{V}{0.9V} \right] = RC\,[\ln(1.11)] = 0.1RC$$

Now, we can use equation (D.50) to find the time required for $V_C$ to reach 0.9$V$ (90% of $V$) as follows:

$$t = RC \left[ \ln \frac{V - 0\,V}{V - 0.9V} \right] = RC \left[ \ln \frac{V}{0.1V} \right] = RC\,[\ln(10)] = 2.3RC$$

Since rise time is defined as the time for $V_C$ to fall from 90% of its maximum value to 10% of its maximum value, it can be found as

$$t_r = t_{(90\%)} - t_{(10\%)} = 2.3RC - 0.1RC = 2.2RC \qquad \text{(D.51)}$$

Now, the upper cutoff frequency of a given *RC* circuit is found as

$$f_{C2} = \frac{1}{2\pi RC}$$

Rewriting the $f_{C2}$ equation, we get

$$RC = \frac{1}{2\pi f_{C2}}$$

Substituting this equation in equation (D.51), we get

$$t_r = 2.2RC = 2.2\left(\frac{1}{2\pi f_{C2}}\right) = \frac{2.2}{2\pi f_{C2}} = \frac{0.35}{f_{C2}}$$

or

$$f_{C2} = \frac{0.35}{t_r}$$

# Glossary

**ac beta ($\beta_{ac}$ or $h_{fe}$).** The ratio of ac collector current to ac base current; listed on specification sheets as $h_{fe}$.

**Acceptor atoms.** Another name for *trivalent atoms*.

**ac emitter resistance ($r'_e$).** The dynamic resistance of the transistor base-emitter junction, used in BJT amplifier voltage gain and input impedance calculations.

**ac equivalent circuit.** A representation of a circuit that shows how the circuit appears to an ac source.

**ac load line.** A graph that represents all possible combinations of $i_c$ and $v_{ce}$.

**Active filter.** A tuned op-amp circuit.

**Active region.** The BJT operating region between saturation (maximum $I_C$) and cutoff (minimum $I_C$).

**Adjustable regulator.** A regulator whose output dc voltage can be set to any value within specified limits.

**AM detector.** A diode clipper that converts a varying amplitude ac input to a varying dc level.

**Amplification.** The process of increasing the power of an ac signal.

**Amplifier.** A circuit used to increase the strength of an ac signal.

**Amplifier input impedance.** The impedance of an amplifier as seen by its source.

**Amplifier output impedance.** The impedance of an amplifier as seen by its load.

**Amplitude.** The maximum value of a changing voltage or current.

**Anode.** The *p*-type terminal of a diode.

**Anode current interruption.** A method of driving an SUS into cutoff by breaking the diode current path or shorting the current around the diode.

**Apparent power.** The combination of reactive (imaginary) power and resistive (true) power. The geometric sum of resistive and reactive power; measured in *volt-amperes* (VA).

**Armstrong oscillator.** An oscillator that uses a transformer in its feedback network to achieve the required 180° voltage phase shift.

**Astable multivibrator.** A switching circuit that has no stable output state. A square-wave oscillator. Also called a *free-running* multivibrator.

**Attenuation.** Any reduction in the amplitude of a signal.

**Attenuation factor ($\alpha_v$).** The ratio of feedback voltage to output voltage. The value of $\alpha_v$ is always less than 1.

**Audio amplifier.** The final audio stage in communications receivers; used to drive the speakers.

**Avalanche current.** The current that occurs when $V_{RRM}$ is reached. Avalanche current can generate sufficient heat to destroy a *pn*-junction diode.

**Average current ($I_{ave}$).** The dc equivalent of an alternating current. $I_{ave}$ is measured with a dc ammeter.

**Average forward current.** The maximum allowable value of dc forward current for a diode.

**Average on-state current ($I_T$).** The maximum average (dc) forward current for a silicon-unilateral switch (SUS).

**Average on-state voltage ($V_T$).** The voltage across a silicon-unilateral switch (SUS) when $I_F = I_T$.

**Average voltage ($V_{ave}$).** The dc equivalent of an alternating voltage. $V_{ave}$ is measured with a dc voltmeter.

**Averaging amplifier.** A summing amplifier that provides an output proportional to the *average* of the input voltage.

**Back-to-back suppressor.** A single package that contains two transient suppressors in either a common-cathode or common-anode configuration.

**Band.** Another name for an orbital shell (in atomic theory) or a range of frequencies (in frequency analysis).

**Band-pass filter.** A filter designed to pass all frequencies that fall between its lower and upper cutoff frequencies ($f_{C1}$ and $f_{C2}$).

**Band-stop (notch) filter.** A filter designed to block (attenuate) all frequencies that fall between its lower and upper cutoff frequencies ($f_{C1}$ and $f_{C2}$).

**Bandwidth (BW).** The range of frequencies over which gain is relatively constant.

**Barkhausen criterion.** The relationship between the circuit feedback factor ($\alpha_v$) and voltage gain ($A_v$) for proper oscillator operation.

**Barrier potential.** The natural potential across a *pn* junction.

**Base.** One of three bipolar junction transistor (BJT) terminals. The other two are the *collector* and the *emitter*.

**Base bias.** A BJT biasing circuit that consists of a single base resistor between the base terminal and $V_{CC}$ and no emitter resistor. Also known as *fixed bias*.

**Base current ($I_B$).** Current that can be varied to control the emitter current ($I_E$) and collector current ($I_C$).

**Base curve.** A curve illustrating the relationship between $I_B$ and $V_{BE}$.

**Base cutoff current ($I_{BL}$).** The maximum amount of current through a reverse-biased emitter-base junction.

**Base-emitter junction.** One of two *pn* junctions that make up the bipolar junction transistor. The other is the *collector-base* junction.

**Base-emitter saturation voltage ($V_{BE(sat)}$).** The rated value of $V_{BE}$ for a transistor in saturation.

**Bessel filter.** A filter designed to provide a constant phase shift across its pass band.

**Beta ($\beta$).** A Greek letter often used to represent the base-to-collector current gain of a bipolar junction transistor.

**Beta curve.** A curve that shows the relationship between beta and temperature and/or collector current.

**Beta-dependent circuit.** A circuit with $Q$-point values that are affected by changes in $h_{FE}$.

**Beta-independent circuit.** A circuit with $Q$-point values that are not affected significantly by changes in $h_{FE}$.

**Bias.** A potential applied to a *pn* junction to obtain a desired mode of operation.

**Biased clamper.** A clamper that allows a waveform to be shifted above or below a dc reference other than 0 V.

**Biased clipper.** A shunt clipper that uses a dc biasing source to set the limit(s) on the circuit output voltage.

**Bidirectional thyristor.** A thyristor capable of conducting in two directions.

**Biomedical electronics.** The area of electronics that deals with medical test and treatment equipment.

**Bipolar junction transistor (BJT).** A three-terminal solid-state device whose output current, voltage, and/or power are normally controlled by its input current.

**Bistable multivibrator.** A switching circuit with two stable output states. Also called a *flip-flop*.

**Bode plot.** An idealized frequency-response curve that assumes $\Delta A_{p(mid)}$ is zero until the cutoff frequency is reached.

**Breakdown voltage ($V_{BR}$).** The peak or dc reverse voltage that will drive a transient suppressor into its reverse breakdown (zener) operating region. The value of reverse voltage that forces a *pn* junction to conduct in the reverse direction.

**Bridge rectifier.** A four-diode rectifier used to control the direction of conduction through a load. It is the most commonly used of the various power-supply rectifier circuits.

**Buffer.** A circuit used to compensate for an impedance mismatch between a source and its load. A switching circuit that does not produce a voltage phase shift.

**Bulk resistance.** The natural resistance of a forward-biased diode.

**Butterworth filter.** An active filter characterized by a flat pass-band response and 20 dB/decade/pole roll-off rate.

**Bypass capacitor.** A capacitor used to establish an ac ground at a specific point in a circuit.

**Calibration resistor.** A variable resistor used to adjust for minor variations in gain.

**Capacitance ratio ($C_R$).** The factor by which the capacitance of a varactor varies from one value of $V_R$ to another.

**Capacitive filter.** A power-supply filter that uses one or more capacitors to oppose any variations in load voltage. *See also* Inductive filter.

**Cascaded.** A term meaning *connected in series*.

**Cascaded amplifiers.** Amplifiers connected in series. Each amplifier in the circuit is referred to as a *stage*.

**Cascode amplifier.** A low-$C_{in}$ amplifier used in high-frequency applications.

**Cathode.** The *n*-type terminal of a diode.

**Center-tapped transformer.** A transformer with an output lead connected to the center of the secondary winding.

**Channel.** The material that connects the source and drain terminals of an FET.

**Chebyshev filter.** An active filter characterized by an irregular pass-band response and high initial roll-off rates.

**Circuit fusing ($I^2t$).** A rating that indicates the maximum forward surge current capability of an SCR.

**Clamper.** A diode circuits that is used to set (or restore) the dc reference of a waveform. Also referred to as a *dc restorer*.

**Clamping voltage ($V_C$).** The rated reverse voltage across a transient suppressor when it is conducting.

**Clapp oscillator.** A Colpitts oscillator with an added capacitor (in series with the feedback inductor) used to reduce the effects of stray capacitance.

**Class A amplifier.** An amplifier with a single transistor that conducts throughout the entire input cycle.

**Class AB amplifier.** An amplifier that contains two (or more) transistors, each conducting for slightly more than 180° of the input cycle.

**Class B amplifier.** An amplifier with two transistors that each conduct for 180° of the input cycle.

**Class C amplifier.** An amplifier with one transistor that conducts during less than 180° of the input cycle.

**Clipper.** A diode circuit that is used to eliminate some portion(s) of a waveform. Also referred to as a *limiter*.

**Closed-loop voltage gain ($A_{CL}$).** The voltage gain of an op-amp with a feedback path; always lower than the value of $A_{OL}$.

**Coarse tuning.** A control that allows you to tune a circuit or system over a wide range of frequencies. *See also* Fine tuning.

**Collector.** One of the three terminals of a bipolar junction transistor (BJT). The other two are the *base* and the *emitter*.

**Collector-base junction.** One of two *pn* junctions that make up the bipolar junction transistor (BJT). The other is the *base-emitter* junction.

**Collector biasing voltage ($V_{CC}$).** A dc supply voltage connected to the collector of a bipolar junction transistor (BJT) as part of its biasing circuitry.

**Collector bypass capacitor.** *See* Decoupling capacitor.

**Collector characteristic curve.** A curve that relates the values of $I_C$, $I_B$, and $V_{CE}$ for a given transistor.

**Collector current ($I_C$).** One of the three terminal currents of a bipolar junction transistor (BJT).

**Collector cutoff current ($I_{CO}$ or $I_{CEX}$).** The maximum value of $I_C$ through a cutoff transistor.

**Collector-emitter saturation voltage ($V_{CE(sat)}$).** The rated value of $V_{CE}$ when the transistor is in saturation.

**Collector-feedback bias.** A bias circuit that is constructed so that the collector voltage ($V_C$) has a direct effect on the base voltage ($V_B$).

**Colpitts oscillator.** An oscillator that uses a pair of tapped capacitors and an inductor to produce the 180° phase shift in the feedback network.

**Common-anode display.** A display in which all LED anodes are tied to a single pin.

**Common-base amplifier.** A BJT amplifier that provides voltage and power gain.

**Common-cathode display.** A display in which all LED cathodes are tied to a single pin.

**Common-collector amplifier.** A BJT amplifier that provides current and power gain. Also called an *emitter-follower*.

**Common-drain amplifier.** The JFET counterpart of a common-collector amplifier. It is also referred to as a *source follower*.

**Common-emitter amplifier.** The only BJT amplifier with a 180° voltage phase shift from input to output.

**Common-gate amplifier.** The JFET counterpart of a common-base amplifier.

**Common-mode gain ($A_{CM}$).** The gain a differential amplifier provides to common-mode signals; typically less than 1.

**Common-mode rejection ratio (CMRR).** The ratio of differential gain to common-mode gain.

**Common-mode signals.** Identical signals that appear simultaneously at the two inputs of an op-amp.

**Common-source amplifier.** The JFET counterpart of a common-emitter amplifier.

**Comparator.** A circuit used to compare two voltages.

**Compensating diodes.** Diodes used in the bias circuit of a class AB amplifier with forward voltage ratings that match the transistor $V_{BE}$ ratings.

**Compensating resistor.** A resistor connected to the noninverting input to an op-amp to compensate for differences in the input currents.

**Complementary MOS (CMOS).** A logic family made up of MOSFETs.

**Complementary-symmetry amplifier.** A class B circuit configuration using complementary transistors. Also called a *push-pull emitter follower*.

**Complementary transistors.** Two bipolar junction transistors, one *npn* and one *pnp*, that have nearly identical electrical characteristics and ratings.

**Compliance.** The limit that the output circuit of an amplifier places on its peak-to-peak output voltage.

**Conduction angle.** For a phase controller, the portion of the input waveform (in degrees) that is coupled to the load.

**Conduction band.** The energy band outside the valence shell.

**Constant-current diode.** A diode that maintains a relatively constant device current over a wide range of forward operating voltages.

**Constant-current region.** The region of the JFET operating curve (between $V_P$ and $V_{BR}$) where drain current remains constant for fixed values of $V_{GS}$.

**Counter emf.** A voltage developed across a coil as its magnetic field collapses. The polarity of the voltage is the opposite of the voltage that originally generated the magnetic field.

**Coupling capacitor.** A capacitor connected between amplifier stages to provide dc isolation between the stages while allowing the ac signal to pass without distortion.

**Covalent bonding.** A means by which atoms are held together. In a covalent bond, shared electrons hold the atoms together.

**Critical rise ($dv/dt$).** The maximum rate of increase in $V_{AK}$ for an SCR that will not cause false triggering.

**Crossover distortion.** Distortion caused by class B transistor biasing. It occurs during the zero-crossing of the input, when neither of the transistors is fully conducting.

**Crossover network.** A circuit designed to separate high-frequency audio from low-frequency audio.

**Crowbar.** An SCR circuit used to protect a voltage-sensitive load from excessive dc power supply output voltages.

**Crystal-controlled oscillator.** An oscillator that uses a quartz crystal to produce an extremely stable output frequency.

**Current-controlled resistance.** Any resistance whose value is determined by the value of a controlling current.

**Current feedback.** A type of feedback that uses a portion of the output current from a circuit to control the circuit's operating characteristics.

**Current gain.** The factor by which current increases from the base of a BJT to its collector. The ratio of output current to input current for a component or circuit.

**Current gain–bandwidth product.** The frequency at which BJT current gain drops to unity. Also referred to as *beta cutoff frequency*.

**Current-limiting resistor ($R_S$).** A resistor place in series with a component to limit the current through that component.

**Current regulator diode.** Another name for a constant-current diode.

**Current-source bias.** A JFET biasing circuit that uses a BJT to maintain a constant value of drain current ($I_D$).

**Cutoff.** A BJT operating state where $I_C$ is nearly zero.

**Cutoff clipping.** A type of distortion caused by driving a transistor into cutoff.

**Cutoff frequencies.** The frequencies at which the power gain ($A_p$) of an amplifier is reduced to 50% of its midband value.

**Cycle time ($T_C$).** For a sine wave, the time required for one 360° rotation. For a rectangular wave, the sum of pulse width and space width.

**Damping.** The fading and loss of oscillations that occurs when $\alpha_v A_v < 1$. *See* Barkhausen criterion.

**Dark current ($I_D$).** The reverse current through a photodiode with no active light input.

**Darlington emitter-follower.** An emitter follower that uses a specific two-transistor configuration to provide higher values of current gain and input impedance than can be obtained using a single transistor.

**Darlington pair.** A two-transistor configuration used to provide extremely high current gain and input impedance. The emitter of the input transistor is connected to the base of the output transistor, and the collectors of the two are tied together.

**Darlington pass-transistor regulator.** A series regulator that uses a Darlington pair in place of a single pass-transistor.

**dBm reference.** The ratio of a power value to one milliwatt (1 mW), given as 10 times the common log of the ratio.

**dB power gain.** The ratio of circuit output power to input power, equal to 10 times the common log of that ratio.

**dB voltage gain.** The ratio of output voltage to input voltage, given as 20 times the common log of the ratio.

**dc alpha ($\alpha$).** The ratio of dc collector current to dc emitter current.

**dc amplifier.** Any amplifier that exhibits midband gain at 0 Hz.

**dc beta ($\beta_{dc}$ or $h_{FE}$).** The ratio of dc collector current to dc base current.

**dc blocking voltage.** The maximum allowable value of reverse voltage for a given *pn*-junction diode. Also referred to as *peak reverse voltage*.

**dc load line.** A graph of all possible combinations of $I_C$ and $V_{CE}$ for a given amplifier.

**dc power dissipation rating.** *See* Maximum steady-state power dissipation rating.

**dc reference voltage.** A dc voltage used to determine the input signal level that will trigger a circuit such as a comparator or Schmitt trigger.

**dc restorer.** *See* Clamper.

**dc-to-dc converter.** A switching voltage regulator used to change one dc voltage level to another.

**Decade.** A frequency multiplier equal to 10.

**Decibel.** A logarithmic unit used to express the ratio of one value to another.

**Decoder-driver.** An IC used to drive a multisegment display.

**Decoupling capacitor.** A capacitor connected between $V_{CC}$ and ground (in parallel with the dc power supply).

**Delay time ($t_d$).** The time required for a BJT to come out of cut-off when provided with a pulse input. In terms of $I_C$, it is the time required for $I_C$ to reach 10% of its maximum value.

**Depletion layer.** The area around a *pn* junction that is depleted of free carriers. Also called the *depletion region*.

**Depletion-mode operation.** Using an input voltage to effectively decrease the channel size of an FET.

**D-MOSFET.** A MOSFET that can be operated in both the depletion mode and the enhancement mode.

**Designator code.** An IC code that indicates circuit type and operating temperature range.

**Diac.** A two-terminal, three-layer device with forward and reverse characteristics that are identical to the forward characteristics of the silicon unilateral switch (SUS).

**Differential amplifier.** A circuit that amplifies the difference between two input voltages. The input circuit of the op-amp, driven by the inverting and noninverting inputs.

**Differentiator.** A circuit whose output is proportional to the rate of change of its input signal.

**Diffusion current.** The current that "wanders" across a *pn* junction in the process of increasing or decreasing in size. The value of junction forward current when diode voltage is less than $V_F$.

**Digital circuit.** A circuit designed to respond to specific alternating dc voltage levels.

**Digital communications.** A method of transmitting and receiving information in digital form.

**Digital-to-analog (D/A) converter.** A circuit that converts digital circuit outputs to equivalent analog voltages.

**Diode.** A two-terminal device that acts as a one-way conductor.

**Diode bias.** A complementary-symmetry amplifier biasing circuit that uses two diodes in place of the resistor(s) between the two transistors. *See* Class AB amplifier.

**Diode capacitance ($C_t$).** The rated value (or range) of capacitance for a varactor at a specified value of $V_R$.

**Diode capacitance temperature coefficient ($TC_C$).** A multiplier used to determine the amount by which varactor capacitance changes when temperature changes.

**Discrete components.** A term used to describe devices that are packaged in individual casings.

**Distortion.** Any unwanted change in the shape of an ac signal.

**Donor atoms.** Another name for *pentavalent atoms*.

**Doping.** Adding impurity elements to intrinsic semiconductors to improve their conductivity.

**Drain.** The JFET counterpart of the BJT collector.

**Drain-feedback bias.** The MOSFET counterpart of collector-feedback bias.

**Drift.** A change in tuning caused by component aging.

**Driver.** A circuit used to couple a low-current output to a relatively high-current device.

**Dual-gate MOSFET.** A MOSFET constructed with two gates to reduce gate input capacitance.

**Dual-polarity power supply.** A dc supply that provides both positive and negative dc output voltages.

**Dual-tracking regulator.** A regulator that provides equal +dc and −dc output voltages.

**Duty cycle.** The ratio of pulse width (PW) to cycle time ($T_C$), measured as a percentage.

**Efficiency ($\eta$).** The percentage of the power drawn from the dc power supply that an amplifier actually transfers to its load. The ratio of ac load power to dc input power for an amplifier (or other circuit).

**Electrical characteristics.** The guaranteed operating characteristics of a device, listed on its specification sheet.

**Electron-hole pair.** A free electron and its matching valence band hole.

**Electronic tuning.** Using voltage-controlled or programmable devices to control the tuning of a circuit.

**Electron-volt (eV).** The energy absorbed by an electron when it is subjected to a 1 V difference of potential.

**Emitter.** One of the three terminals of a bipolar junction transistor (BJT). The other two are the *base* and the *collector*.

**Emitter bias.** A BJT bias circuit that contains a dual-polarity power supply and a grounded base resistor.

**Emitter current ($I_E$).** One of the three terminal currents of a bipolar junction transistor (BJT).

**Emitter-feedback bias.** A biasing circuit that is designed so that the emitter voltage ($V_E$) has a direct effect on the base voltage ($V_B$).

**Emitter follower.** *See* Common-collector amplifier.

**E-MOSFET.** A MOSFET that is restricted to enhancement-mode operation.

**Energy gap.** The difference between the energy levels of any two orbital shells.

**Enhancement-mode operation.** Using an input voltage to effectively increase the channel size of an FET.

**Epicap.** Another name for a *varactor*.

**Extrinsic.** Another word for *impure*.

**Fall time ($t_f$).** The time required for a BJT to make the transition from saturation to cutoff. The time required for $I_C$ to drop from 90% to 10% of its maximum value.

**False triggering.** When a noise signal triggers an SCR into conduction.

**Fanout.** For a given logic family, the number of inputs that can be driven by a single output.

**Feedback.** A circuit connection that "feeds" a portion of the output voltage or current back to the input to control the circuit's operating characteristics.

**Feedback bias.** A term used to describe a circuit that "feeds" a portion of the output voltage or current back to the input to control the circuit operation.

**Feedback factor ($1 + \alpha_v A_{OL}$).** A value used in the gain and impedance calculations for a given negative-feedback amplifier.

**Feedback path.** A signal path from the output of an amplifier back to the input.

**Fidelity.** The ability of a circuit to accurately reproduce a waveform.

**Field-effect transistor (FET).** A three-terminal, voltage-controlled device used in amplification and switching applications.

**Filter.** A circuit that reduces the variations in the output of a rectifier.

**Fine tuning.** A control that allows you to tune a circuit or system within a narrow band of frequencies. It is used to select one out of a limited range of frequencies.

**Firing angle.** For a phase controller, the point on the input waveform (in degrees) at which the SCR (or triac) fires, allowing load conduction to begin.

**555 timer.** An 8-pin IC designed for use in a variety of switching applications.

**Fixed bias.** *See* Base bias.

**Fixed-negative regulator.** A regulator with a set negative dc output voltage.

**Fixed-positive regulator.** A regulator with a set positive dc output voltage.

**Flip-flop.** A bistable multivibrator; a circuit with *two* stable output states.

**Flywheel effect.** A term used to describe the ability of a parallel *LC* circuit to self-oscillate for a brief period of time.

**Forced commutation.** Driving an SUS into cutoff by applying a reverse voltage to the device.

**Forward bias.** A potential used to reduce the resistance of a *pn* junction.

**Forward blocking region.** The *off-state* (nonconducting) region of operation for a thyristor.

**Forward breakover current ($I_{BR(F)}$).** The value of $I_F$ at the point where breakover occurs for a given thyristor.

**Forward breakover voltage ($V_{BR(F)}$).** The value of forward voltage that forces an SUS into conduction. The value of $V_F$ at the point where breakover occurs.

**Forward gate current ($I_{G(F)}$).** The maximum amount of current that can be drawn through the JFET gate without damaging the device.

**Forward operating region.** The *on-state* (conducting) region of device operation.

**Forward power dissipation ($P_{D(max)}$).** A rating that indicates the maximum allowable power dissipation of a forward-biased device.

**Forward voltage ($V_F$).** The voltage across a forward-biased *pn* junction.

**Free-running multivibrator.** Another name for the *astable multivibrator*.

**Frequency-response curve.** A curve showing the relationship between gain and operating frequency.

**Full load.** A term used to describe a *minimum* load resistance that draws maximum current.

**Full-wave rectifier.** A power supply circuit that converts ac to pulsating dc. The full-wave rectifier provides two half-cycle outputs for each full cycle input.

**Full-wave voltage doubler.** A circuit that produces a dc output that is twice the peak value of a sinusoidal input. The term *full-wave* indicates that the entire input cycle is used to produce the dc output.

**Fundamental frequency.** The lowest frequency in any group of harmonically related frequencies. The resonant frequency of a crystal or circuit.

**Gain.** A multiplier that exists between the input and output of a circuit. The ratio of an output value to its corresponding input value.

**Gain-bandwidth product.** A constant equal to the unity-gain frequency of an op-amp.

**Gain-stabilized amplifier.** An amplifier that uses a partially bypassed emitter (or source) resistance to stabilize its voltage gain against changes in device parameters. Also referred to as a *swamped amplifier*.

**Gate.** The JFET counterpart of the BJT base.

**Gate bias.** The JFET counterpart of BJT base bias.

**Gate nontrigger voltage ($V_{GD}$).** The maximum gate voltage that can be applied to an SCR without triggering the device into conduction.

**Gate reverse current ($I_{GSS}$).** The maximum amount of current that can be generated through a reverse-biased gate-source junction at the rated value of $V_{GS}$.

**Gate-source breakdown voltage ($V_{(BR)GSS}$).** The value of $V_{GS}$ that can cause the gate-source junction to break down.

**Gate-source cutoff voltage ($V_{GS(off)}$).** The value of $V_{GS}$ that reduces $I_D$ to approximately zero.

**Gate trigger voltage ($V_{GT}$).** The value of $V_{GK}$ that generates sufficient gate current ($I_G$) to trigger an SCR into conduction.

**General amplifier model.** An amplifier model that represents the circuit as one having gain, input impedance, and output impedance. It is used to show how the circuit is affected by its source and load.

**General-class equation.** An equation derived for a summing amplifier that is used to predict the output voltage for any combination of input voltages.

**Geometric center frequency ($f_0$).** The geometric average of $f_{C1}$ and $f_{C2}$.

**Germanium.** One of the semiconductor materials commonly used to produce solid-state devices. Another is *silicon*.

**Graphic equalizer.** A circuit or system designed to allow you to control the amplitude of different audio-frequency ranges.

**Half-wave rectifier.** A power supply circuit that converts ac to pulsating dc. The half-wave rectifier produces one half-cycle output for each full-cycle input.

**Half-wave voltage doubler.** A circuit that produces a dc output approximately equal to twice its peak input voltage. The name is derived from the fact that the output capacitor is charged during one-half of the input cycle.

**Harmonic.** A whole number multiple of a given frequency.

**Hartley oscillator.** An oscillator that uses a tapped inductor (pair) and a parallel capacitor in its feedback network to produce the 180° voltage phase shift required for oscillation.

**Heat sink.** A large metallic object that helps to cool components by increasing their effective surface area.

**Heat-sink compound.** A compound used to aid in the transfer of heat from a component to a heat sink.

**$h_{fe}$.** A label used to represent the base-to-collector ac current gain (ac beta) of a BJT.

**$h_{FE}$.** A label used to represent the dc current gain of a BJT. The ratio of BJT collector current to base current.

**$h_{ie}$.** A label used to represent the input (base-to-emitter) impedance of a BJT.

**High-current transistor.** A BJT with a high maximum $I_C$ rating.

**High-pass filter.** A filter designed to pass all frequencies *above* a given cutoff frequency.

**High-power transistor.** A BJT with a high power dissipation rating.

**High-voltage transistor.** A BJT with a high reverse breakdown voltage rating.

**$h_{oe}$.** A label used to represent the output (collector-to-emitter) impedance of a BJT.

**Holding current ($I_H$).** The minimum value of $I_F$ required to maintain conduction through a thyristor.

**Hole.** A gap in a covalent bond.

**Hot-carrier diode.** Another name for the *Schottky diode*.

**$h$-Parameters.** *See* Hybrid parameters.

**Hybrid parameters ($h$-parameters).** Transistor specifications that describe the operation of small-signal circuits under full-load or no-load conditions.

**Hysteresis.** A term that is sometimes used to describe the range of voltages between the UTP and LTP values of a Schmitt trigger.

**$I_{CQ}$.** The $Q$-point value of $I_C$.

**Ideal voltage amplifier.** A theoretical voltage amplifier having infinite gain, input impedance, and bandwidth, and zero output impedance.

**Inductive filter.** A power supply filter that uses one or more inductors to oppose any variations in load current. *See also* Capacitive filter.

**Industrial electronics.** The area of electronics that deals with the devices, circuits, and systems used in manufacturing.

**Input bias current.** The average value of quiescent dc biasing current drawn by the signal inputs of an op-amp.

**Input impedance.** The load that an amplifier places on its source. The opposition to current provided by the input to a circuit.

**Input offset current.** A slight difference in op-amp input currents, caused by differences in the transistor beta ratings.

**Input offset voltage ($V_{io}$).** The difference between the base-emitter voltages in a differential amplifier that produces an output offset voltage when the signal inputs are grounded.

**Input/output voltage differential.** The maximum allowable difference between $V_{in}$ and $V_{out}$ for a voltage regulator.

**Input voltage range.** The maximum differential input that an op-amp can accept without risking damage to its input differential amplifier.

**Instrumentation amplifier.** A circuit used to amplify low-level signals in process control or measurement applications.

**Integrated circuit (IC).** An entire circuit (or group of components) constructed on a single piece of semiconductor material. It contains any number of active and/or passive components.

**Integrated rectifier.** A power supply rectifier that is constructed on a single piece of semiconductor material and housed in a single package.

**Integrated transistor.** A group of transistors having identical characteristics that are constructed on a single piece of semiconductor material and housed in a single package.

**Integrator.** A circuit whose output is proportional to the area of the input waveform.

**Interbase resistance ($\eta$).** The resistance between the base terminals of a *unijunction transistor* (UJT), measured with the emitter-base junction reverse biased.

**Intrinsic.** Another word for *pure*.

**Intrinsic standoff ratio ($\eta$).** The ratio of emitter–base 1 resistance to the total *interbase resistance* of a UJT.

**Inverter.** A basic switching circuit that produces a 180° voltage phase shift. A logic-level converter.

**Inverting amplifier.** A basic op-amp circuit that produces a 180° signal phase shift. The op-amp counterpart of the common-emitter and common-source circuits.

**Inverting input.** The op-amp input that produces a 180° voltage phase shift (from input to output) when used as a signal input.

**Irradiance.** *See* Light intensity.

**Isolation capacitance ($C_{ISO}$).** The total capacitance between the input and output pins of an optoisolator.

**Isolation current ($I_{ISO}$).** The amount of current that can be forced between the input and output pins of an optoisolator at the rated voltage.

**Isolation resistance ($R_{ISO}$).** The total resistance between the input and output pins of an optoisolator.

**Isolation source voltage ($V_{ISO}$).** The voltage that, if applied across the input and output pins, will destroy an optoisolator.

**Isolation transformer.** A transformer with a 1:1 turns ratio that is used to provide electrical isolation between two electronic circuits or systems.

**Junction capacitance.** The capacitance of a reverse-biased *pn* junction. This capacitance decreases when the amount of reverse bias increases, and vice versa.

**Kick emf.** *See* Counter emf.

**Knee voltage ($V_k$).** The point on a current-versus-voltage graph where current suddenly increases or decreases.

**Large-signal voltage gain.** Another term for the *open-loop voltage gain* of an op-amp.

**Lateral double-diffused MOSFET (LDMOS).** A high-power MOSFET that uses a short channel and a heavily doped *n*-type region to obtain high $I_D$ and low $r_{d(on)}$.

**Leaky.** A term used to describe a component that is partially shorted.

**Level detector.** Another name for a comparator used to compare an input voltage to a fixed dc reference voltage.

**Lifetime.** The time between electron-hole pair generation and recombination.

**Light.** Electromagnetic energy that falls within a specific range of frequencies.

**Light-activated SCR (LASCR).** A three-terminal light-activated SCR.

**Light current ($I_L$).** The reverse current through a photodiode with an active light input.

**Light detectors.** Optoelectronic devices that respond to light.

**Light emitters.** Optoelectronic devices that produce light.

**Light-emitting diode (LED).** A diode designed to produce one or more colors of light under specific biasing conditions.

**Light intensity.** The amount of light per unit area received by a given photodetector; also called *irradiance*.

**Limiter.** *See* Clipper.

**Linear.** A change that occurs at a constant rate.

**Linear IC voltage regulator.** A device used to hold the output voltage from a dc power supply relatively constant over a specified range of line voltages and load current demands.

**Linear power supply.** A power supply whose output is a linear function of its input. Any power supply whose output is controlled by a linear voltage regulator.

**Line regulation.** The ability of a voltage regulator to maintain a constant load voltage despite anticipated variations in rectifier output voltage. A rating that indicates the change in regulator output voltage that occurs per unit change in input voltage.

**Liquid-crystal display (LCD).** A display made of segments that can be made to reflect (or not reflect) ambient light.

**Loaded-Q.** The *Q* (*quality*) of an *LC* tank circuit when a resistive load is connected to that circuit. The loaded-*Q* of a tank circuit is always *lower* than its unloaded *Q*.

**Load regulation.** The ability of a regulator to maintain a constant load voltage despite anticipated variations in load current demand. A rating that indicates the change in regulator output voltage that occurs per unit change in load current.

**Logic families.** Groups of digital circuits with nearly identical characteristics.

**Logic levels.** A term used to describe the two dc levels used to represent information in a digital circuit or system.

**Lower trigger point (LTP).** A Schmitt trigger reference voltage. When a negative-going input reaches the LTP, the Schmitt trigger output changes state.

**Low-pass filter.** A filter designed to pass all frequencies *below* a given cutoff frequency.

**Majority carrier.** In a given semiconductor material, the charge carrier that is introduced by doping. The electrons in a pentavalent material. The holes in a trivalent material.

**Matched transistors.** Transistors that have the same operating characteristics.

**Maximum limiting voltage ($V_L$).** The voltage at which a constant-current diode starts to limit current.

**Maximum ratings.** A spec sheet table which lists device parameters that must not be exceeded under any circumstances.

**Maximum reverse stand-off voltage ($V_{RWM}$).** The maximum allowable peak or dc reverse voltage that will not turn on a transient suppressor.

**Maximum reverse surge current.** In a bidirectional thyristor, the maximum surge current that the device can handle when biased in its reverse operating region.

**Maximum steady-state power dissipation rating.** The maximum allowable average power dissipation for a zener diode that is operating in reverse breakdown.

**Maximum temperature coefficient.** The percentage of change in $V_{BR}$ per 1°C rise in operating temperature.

**Maximum zener current ($I_{ZM}$).** A zener diode rating indicating the maximum allowable value of reverse current through the device.

**Metal-oxide-semiconductor FET (MOSFET).** A field-effect transistor designed to operate in the depletion and enhancement modes, or the enhancement mode only. The gate construction allows for enhancement-mode operation.

**Midpoint bias.** Having a $Q$-point that is centered on the load line.

**Miller's theorem.** Allows a feedback capacitor to be represented as separate input and output capacitances.

**Minimum dynamic impedance ($Z_T$).** The minimum forward impedance of a constant-current diode when operated in its current-limiting region of operation.

**Minimum knee impedance ($Z_K$).** The minimum forward impedance of a constant-current diode when operated at the knee voltage of its operating curve.

**Minimum load current.** For a voltage regulator, the value of $I_L$ below which regulation is lost.

**Minority carrier.** In a given semiconductor material, the charge carrier that is not introduced by doping. The holes in a pentavalent material. The electrons in a trivalent material.

**Model.** A representation of a component or circuit that demonstrates one (or more) of its characteristics.

**Modulator.** A circuit that combines two signals of different frequencies into a single signal.

**Monostable multivibrator.** A switching circuit with one stable output state. Also called a *one-shot*.

**MOSFET.** An FET that can be operated in the enhancement mode.

**Multicolor LED.** An LED that emits different colors when the polarity of its biasing voltage changes.

**Multiple-feedback band-pass filter.** A band-pass filter that has a single op-amp and two feedback paths, one resistive and one capacitive.

**Multiple-feedback notch filter.** A notch filter that has a single op-amp and two feedback paths, one resistive and one capacitive.

**Multisegment display.** An LED circuit used to display alphanumeric symbols (numbers, letters, and punctuation marks).

**Multistage amplifier.** A term used to describe two or more amplifiers connected in series. Each amplifier is called a *stage*.

**Multivibrators.** Switching circuits designed to have zero, one, or two stable output states.

**Negative clamper.** A circuit that shifts an entire input signal *below* a dc reference voltage.

**Negative feedback.** A type of feedback in which the feedback signal is 180° out of phase with the input signal.

**Negative resistance.** Any device whose current and voltage are inversely related.

**Negative resistance oscillator.** An oscillator whose operation is based on a negative resistance device, such as a tunnel diode or a unijunction transistor.

**Negative resistance region.** A term used to describe any region of device operation where an *increase* in current results in a *decrease* in voltage, or vice versa. The region of UJT operation that lies between the *peak* and *valley* points on the operating curve.

**Noise.** Any undesired voltage or current generated within, or externally to, an electronic system. Noise sources can be natural or man-made.

**Nominal zener voltage.** The rated value of $V_Z$ for a given zener diode.

**Noninverting amplifier.** An op-amp circuit with no signal phase shift.

**Noninverting input.** The op-amp input that does not produce a voltage phase shift (from input to output) when used as a signal input.

**Nonlinear distortion.** A type of distortion caused by driving the base-emitter junction of a transistor into its nonlinear operating region.

**Nonrepetitive surge current ($I_{SM}$).** The absolute limit on the forward surge current through a device, such as a *pn*-junction diode or an SCR.

**Notch filter.** A filter designed to reject (or block) a band of frequencies while allowing the frequencies outside its bandwidth to pass. Also referred to as a *band-stop filter*.

**npn transistor.** A BJT with *n*-type emitter and collector materials and a *p*-type base.

**n-type material.** A semiconductor that has added pentavalent impurities.

**Octave.** A frequency multiplier equal to 2.

**Off characteristics.** The characteristics of a transistor that is in cutoff.

**Offset null.** Input pins on an op-amp that are used to eliminate an output offset voltage.

**Ohmic region.** The portion of the JFET operating curve that lies below $V_P$.

**On characteristics.** The characteristics of a transistor in the active and saturation regions of operation.

**One-shot.** Another name for the *monostable multivibrator*.

**Opaque.** A term used to describe anything that blocks light.

**Open-loop voltage gain ($A_{OL}$).** The maximum possible gain of a given op-amp. The voltage gain of an op-amp with no feedback path.

**Operational amplifier (op-amp).** A high-gain dc amplifier that has high input impedance and low output impedance.

**Optical switch.** An IC optocoupler designed to allow an external object to block the light path between the photoemitter and the photodetector.

**Optocoupler.** A device that uses light to couple a signal from its input (a photoemitter) to its output (a photodetector).

**Optocoupling.** A method of coupling the output of one circuit to the input of another using light.

**Optoelectronic devices.** Devices that are controlled by and/or emit (generate) light.

**Optointerrupter.** An IC optocoupler designed to allow an external object to block the light path between the photoemitter and the photodetector.

**Optoisolator.** An optocoupler. A device that uses light to couple a signal from its input (a photoemitter) to its output (a photodetector).

**Oscillator.** An ac signal generator. A circuit that converts dc to a sinusoidal (or other) waveform.

**Oscillator stability.** A measure of an oscillator's ability to maintain constant output amplitude and frequency.

**Output admittance ($h_{oe}$).** The admittance of the collector-emitter circuit, measured under no-load conditions. This $h$-parameter is used mainly in amplifier design.

**Output impedance.** The "source impedance" that a circuit presents to its load. The opposition to current provided by the output of an amplifier.

**Output offset voltage ($V_{out(offset)}$).** A voltage that may appear at the output of an op-amp; caused by an imbalance in the differential amplifier.

**Output short-circuit current.** The maximum output current for an op-amp, measured under shorted-load conditions.

**Overtone.** Another word for *harmonic*. A whole-number multiple of a given frequency.

**Overtone mode.** An operating mode where a circuit is tuned to a harmonic of a fundamental frequency for the purpose of producing a higher output frequency.

**Parameter.** A limit.

**Pass band.** The range of frequencies passed (amplified) by a tuned amplifier.

**Pass-transistor regulator.** A regulator that uses a series transistor to maintain a constant load voltage.

**Peak current ($I_P$).** One of two current values that define the limits of a negative resistance region of operation for a device such as a tunnel diode or unijunction transistor (UJT). The other is the *valley current* ($I_V$).

**Peak inverse voltage (PIV).** The maximum reverse bias that a diode will be exposed to in a rectifier.

**Peak operating voltage (POV).** The maximum allowable value of $V_F$ for a constant-current diode.

**Peak power dissipation rating ($P_{pk}$).** Indicates the amount of surge power that a transient suppressor can dissipate.

**Peak repetitive reverse voltage ($V_{RRM}$).** The diode rating that indicates the maximum allowable value of diode reverse voltage.

**Peak reverse voltage ($V_{RRM}$).** The maximum reverse voltage that won't force a *pn* junction to conduct.

**Peak voltage ($V_P$).** One of two voltage values that define the limits of a negative resistance region of operation for a device such as a tunnel diode or unijunction transistor (UJT). The other is the *valley voltage* ($V_V$).

**Pentavalent.** Elements with five valence shell electrons.

**Phase controller.** A circuit used to control the conduction phase angle through a load, and thus the average load voltage.

**Phase-shift oscillator.** An oscillator that uses three *RC* circuits in its feedback network to produce a 180° phase shift.

**Photo-Darlington.** A phototransistor with a Darlington pair output.

**Photodiode.** A diode whose *reverse* conduction is light-intensity controlled.

**Phototransistor.** A three-terminal photodetector whose collector current is controlled by the intensity of the light at its optical input (base).

**Piezoelectric effect.** The tendency of a crystal to vibrate at a fixed frequency when subjected to an electric field.

**Pinch-off voltage ($V_P$).** The value of drain-source voltage ($V_{DS}$) that allows maximum JFET drain current ($I_D$), measured at $V_{GS} = 0$ V.

**PIN diode.** A diode made up of *p*-type and *n*-type materials that are separated by an insulating material.

***pnp* transistor.** A BJT with *p*-type emitter and collector materials and an *n*-type base.

**Pole.** A single *RC* circuit.

**Positive clamper.** A circuit that shifts an entire input signal *above* a dc reference voltage.

**Positive feedback.** A type of feedback where the feedback signal is *in phase* with the circuit input signal. Also referred to as *regenerative feedback*.

**Power consumption.** A rating that indicates the amount of power dissipated by an op-amp when operated in its quiescent state.

**Power derating factor.** The rate at which the maximum power rating decreases per 1°C rise above a specified temperature.

**Power gain ($A_p$).** The ratio of circuit output power to circuit input power.

**Power rectifier.** A diode with extremely high forward current and/or power dissipation ratings.

**Power supply.** A group of circuits that combine to convert ac to dc.

**Power supply rejection ratio.** The ratio of a change in op-amp output voltage to a change in supply voltage.

**Power switch.** The pass-transistor in a switching regulator that controls conduction through the load.

**Precision rectifier.** A clipper that consists of a diode and an op-amp; characterized by the ability to clip extremely low-level input signals.

**Programmable unijunction transistor (PUT).** A four-layer thyristor trigger whose construction and operation are similar to that of the SCR. However, the PUT gate is used to provide a reference for the triggering voltage rather than as a trigger input.

**Propagation delay.** The time required for a signal to pass through a component or circuit; measured at the 50% point on the two waveforms.

***p*-type material.** Silicon or germanium that has trivalent impurities.

**Pulse width (PW).** The time spent in the active (high) dc output state.

**Pulse-width modulation (PWM).** Modulating a rectangular waveform so that its pulse width varies while its overall cycle time remains unchanged. Essentially, it varies the duty cycle of the modulated waveform.

**Push-pull emitter follower.** *See* Complementary-symmetry amplifier.

**Q (*LC* circuit).** A measure of the *quality* of an *LC* circuit. The ratio of energy stored in the circuit to energy lost per cycle. The higher the *Q* of an *LC* circuit, the closer it comes to being an *ideal* circuit.

**Q-point.** A point on the dc load line of a BJT amplifier that indicates the values of $V_{CE}$ and $I_C$ for the circuit when it is at rest (has no active input). A point on the dc load line for a circuit that indicates its inactive voltage and current values.

**Q-point shift.** A condition where a change in operating temperature (or device substitution) indirectly causes a change in the Q-point values of $I_C$ and $V_{CE}$.

**Quality (Q).** A figure of merit for a tuned amplifier that is equal to the ratio of center frequency to bandwidth.

**Quiescent.** At rest.

**Rail.** A term use to describe either supply voltage connection to an op-amp.

**Ramp.** Another name for a voltage that changes at a constant (linear) rate.

**RC-coupled class A amplifier.** A class A amplifier that uses a capacitor to couple its output signal to a load.

**Recombination.** A term used to describe the process of a free electron giving up its energy and returning to a valence band orbit.

**Rectangular waveform.** A waveform made up of alternating (high and low) dc voltages.

**Rectifier.** A circuit that converts ac to pulsating dc.

**Regenerative feedback.** Another name for *positive feedback*.

**Regulated dc power supply.** A dc power supply with extremely low internal resistance.

**Regulation.** A rating that indicates the maximum change in regulator output voltage that can occur when input voltage and load current are varied over their entire rated ranges.

**Regulator current ($I_P$).** For a constant-current diode, the regulated value of forward current for forward voltages between its $V_L$ and POV ratings.

**Relaxation oscillator.** A circuit that uses the charge/discharge characteristics of a capacitor or inductor to produce a pulse output.

**Reverse bias.** A potential that causes a *pn* junction to have a high resistance.

**Reverse blocking region.** The thyristor off-state (nonconducting) region of operation between 0 V and $V_{BR(R)}$.

**Reverse breakdown voltage ($V_{BR}$).** The minimum reverse voltage that causes a device to break down and conduct in the reverse direction.

**Reverse current ($I_R$).** The current through a reverse-biased *pn* junction.

**Reverse saturation current ($I_S$).** A current caused by thermal activity in a reverse-biased diode. $I_S$ is temperature dependent.

**Reverse voltage ($V_R$).** The voltage across a reverse-biased *pn* junction.

**Reverse voltage feedback ratio ($h_{re}$).** An *h*-parameter that equals the ratio of $v_{be}$ to $v_{ce}$ for a given transistor, measured under no-load conditions.

**RF amplifier.** Radio frequency amplifier; the input circuit of a communications receiver.

**Ripple rejection ratio.** The ratio of regulator input ripple to maximum output ripple.

**Ripple voltage.** The variation in the output voltage from a filter.

**Ripple width.** The maximum variation in gain over the pass band of an active filter, measured in dB.

**Rise time ($t_r$).** The time required for a BJT to go from cutoff to saturation. In terms of $I_C$, the time required for the 10% to 90% transition.

**Roll-off rate.** The rate of gain reduction for a circuit when operated beyond its cutoff frequencies.

**Saturation.** The BJT operating region where $I_C$ reaches its maximum value.

**Saturation clipping.** A type of distortion caused by driving a transistor into saturation.

**Schmitt trigger.** A voltage-level detector.

**Schottky diode.** A high-speed diode with very little junction capacitance.

**Selector guide.** A publication that groups components according to their critical ratings and/or characteristics.

**Self-bias.** A JFET biasing circuit that uses a source resistor to establish a negative $V_{GS}$.

**Semiconductor.** An element that is neither an insulator nor a conductor.

**Sensitivity.** A rating that indicates the response of a photodetector to a specified light intensity.

**Series clipper.** A clipper that is in series with its load.

**Series current regulator.** A circuit that uses a constant-current diode to maintain a constant circuit input current over a wide range of input voltages.

**Series feedback regulator.** A series regulator that uses an error-detection circuit to improve the line and load regulation characteristics of other pass-transistor regulators.

**Series regulator.** A voltage regulator that is in series with its load.

**Seven-segment display.** A display made up of seven LEDs shaped in a figure-8 that is used to display numbers.

**Shorted-gate drain current ($I_{DSS}$).** The maximum possible value of $I_D$ for a JFET. Also called *zero-gate-voltage drain current*.

**Shunt clipper.** A clipper that is in parallel with its load. The circuit provides a waveform output when the diode is reverse biased (not conducting).

**Shunt feedback regulator.** A circuit that uses an error detector to control the conduction of a shunt transistor.

**Shunt regulator.** A voltage regulator that is in parallel with its load.

**Silicon.** One of the semiconductor materials that are commonly used to produce solid-state devices. Another is *germanium*.

**Silicon-controlled rectifier (SCR).** A three-terminal device very similar in construction and operation to the SUS. A third terminal, called the *gate*, provides another method for triggering the device.

**Silicon unilateral switch (SUS).** A two-terminal, four-layer device that can be triggered into conduction by applying a specified forward voltage across its terminals.

**Slew rate.** The maximum rate at which op-amp output voltage can change.

**Snubber network.** An *RC* circuit that is connected between an SCR anode and cathode to eliminate false triggering.

**Soldering temperature.** The maximum amount of heat that can be applied to any pin of the IC without causing damage to the chip.

**Solid-state relay (SSR).** A circuit that uses a dc input voltage to pass or block an ac signal.

**Source.** The JFET counterpart of the BJT emitter.

**Source follower (common-drain amplifier).** The JFET counterpart of the emitter follower.

**Space width (SW).** The time spent in the passive (low) dc output stage. The time between pulses in switching circuits.

**Specification sheet.** A listing of all the important parameters and operating characteristics of a device or circuit.

**Spectral response.** A measure of a photodetector's response to a change in input wavelength. Spectral response is measured in terms of the device *sensitivity*.

**Speed-up capacitor.** A capacitor used in the base circuit of a BJT to allow the device to switch rapidly between saturation and cutoff. It reduces propagation delay by reducing $t_d$ and $t_s$.

**Square wave.** A rectangular waveform that has equal pulse width (PW) and space width (SW) values.

**Stage.** Each amplifier in a cascaded group of amplifiers.

**Standard push-pull amplifier.** A class B circuit that uses two identical transistors and a center-tapped transformer.

**Static reverse current ($I_R$).** The reverse current through a diode when $V_R$ is less than the reverse breakdown value.

**Step-down regulator.** A switching regulator whose dc output voltage is lower than its dc input voltage.

**Step-down transformer.** One with a secondary voltage that is less than the primary voltage.

**Step-recovery diode.** A heavily doped diode with an ultrafast switching time.

**Step-up regulator.** A switching regulator whose dc output voltage is greater than its dc input voltage.

**Step-up transformer.** One with a secondary voltage greater than the primary voltage.

**Stop band.** The range of frequencies outside an amplifier's pass band.

**Storage time ($t_s$).** The time required for a BJT to come out of saturation. In terms of $I_C$, the time required for $I_C$ to drop to 90% of its maximum value.

**Substrate.** The foundation material of a MOSFET.

**Subtractor.** A summing amplifier that provides an output proportional to the difference between two input voltages.

**Summing amplifier.** An op-amp circuit that produces an output voltage proportional to the sum of its input voltages.

**Supply current.** A rating that indicates the value of quiescent (inactive) current that an op-amp draws from its power supply.

**Surface leakage current ($I_{SL}$).** A current along the surface of a reverse-biased diode. $I_{SL}$ is $V_R$ dependent.

**Surface-mount package (SMP).** An IC package that is much smaller and lighter than its dual in-line package (DIP) counterpart.

**Surge current.** The high initial current in a power supply. Any current caused by a nonrepetitive high-voltage condition.

**Swamped amplifier.** An amplifier that uses a partially bypassed emitter resistance to increase the ac emitter resistance.

**Switching circuits.** Circuits designed to respond to (or generate) nonlinear waveforms, such as square waves.

**Switching power supply.** A power supply that contains a switching regulator.

**Switching time constants.** A term used to describe a condition where a capacitor has different charge and discharge times.

**Switching transistors.** Devices with extremely low switching times.

**Switching voltage regulator.** Voltage regulator that alternates between being fully *on* and fully *off*, resulting in very high efficiency.

**Thermal contact.** Placing two or more components in physical contact with each other (or a common surface) so that their operating temperatures are equal.

**Thermal resistance ($R_\theta$).** Any opposition to the flow of heat (power dissipation).

**Threshold voltage ($V_{GS(th)}$).** The value of $V_{GS}$ that turns an E-MOSFET on.

**Thyristors.** Devices designed specifically for high-power switching applications. Also called *breakover devices*.

**Transconductance ($g_m$).** A ratio of a change in drain current ($I_D$) to a change in $V_{GS}$, measured in microsiemens ($\mu$S) or micromhos ($\mu$mhos).

**Transconductance curve.** A plot of all possible combinations of $V_{GS}$ and $I_D$ for a JFET or MOSFET.

**Transducer.** A device that converts energy from one form to another.

**Transformer-coupled class A amplifier.** A class A amplifier that uses a transformer to couple its output signal to a load.

**Transient.** An abrupt current or voltage spike. A surge.

**Transient suppressor.** A zener diode with extremely high surge-handling capabilities.

**Transistor.** A three-terminal device whose output current, voltage, and/or power are controlled (under normal circumstances) by its input current or voltage.

**Triac.** A bidirectional thyristor whose forward and reverse characteristics are identical to the forward characteristics of the SCR.

**Trigger-point voltage ($V_{TP}$).** The voltage required at the anode of the gate diode to trigger the SCR.

**Trivalent.** Elements with three valence electrons.

**Troubleshooting.** The process of locating faults in electronic equipment.

**Tuned amplifier.** A circuit designed to have a specific value of gain over a specified range of frequencies. An amplifier designed to have a specific bandwidth.

**Tunnel diode.** A heavily doped diode used in high-frequency communications circuits.

**Tunnel diode oscillator.** A signal generator whose operation is based on the negative resistance characteristics of a tunnel diode.

**Turn-off time ($t_{off}$).** The sum of storage time ($t_s$) and fall time ($t_f$).

**Turn-on time ($t_{on}$).** The sum of delay time ($t_d$) and rise time ($t_r$).

**Ultrahigh frequency (UHF).** The band of frequencies between 300 MHz and 3 GHz.

**Unijunction transistor (UJT).** A three-terminal device whose trigger voltage is proportional to its applied biasing voltage.

**Unity-gain frequency.** The maximum possible operating frequency for an op-amp, measured at $A_{OL} = 0$ dB.

**Universal bias.** Another name for *voltage-divider bias*.

**Upper trigger point (UTP).** A Schmitt trigger reference voltage. When a positive-going input reaches the UTP, the Schmitt trigger output changes state.

**Valence shell.** The outermost shell of an atom. The number of electrons in the valence shell determines the conductivity of the atom.

**Valley current ($I_V$).** One of two current values that define the limits of a negative resistance region of operation for a device such as a tunnel diode or unijunction transistor (UJT). The other is the *peak current ($I_P$)*.

**Valley voltage ($V_V$).** One of two voltage values that define the limits of a negative resistance region of operation for a device such as a tunnel diode or unijunction transistor (UJT). The other is the *peak voltage ($V_P$)*.

**Varactor.** A diode designed for relatively high junction capacitance when reverse biased.

**Variable comparator.** A comparator with an adjustable reference voltage.

**Variable off-time modulation.** A method of modulating a rectangular waveform so that cycle time varies while pulse width remains constant.

**Varicap.** Another name for a *varactor*.

**$V_{CEQ}$.** The $Q$-point value of $V_{CE}$.

**Very high frequency (VHF).** The band of frequencies between 30 and 300 MHz.

**V-MOSFET (VMOS).** An E-MOSFET designed to handle high values of drain current.

**Voltage-controlled current source.** A circuit with a constant-current output controlled by the circuit input voltage.

**Voltage-controlled oscillator (VCO).** A free-running oscillator whose output frequency is controlled by a dc input voltage.

**Voltage-divider bias.** A biasing circuit that contains a voltage divider. This type of bias is sometimes referred to as *universal bias*.

**Voltage doubler.** A circuit whose dc output is twice the peak value of its sinusoidal input.

**Voltage follower.** The op-amp counterpart of the emitter follower and the source follower.

**Voltage gain ($A_v$).** The factor by which ac signal voltage increases from the input of an amplifier to its output.

**Voltage-inverting regulator.** A switching regulator that produces a negative output when its input voltage is positive, and vice versa.

**Voltage multiplier.** A circuit used to produce a dc output voltage that is some multiple of an ac peak input voltage.

**Voltage quadrupler.** Produces a dc output voltage that is four times its peak input voltage.

**Voltage regulator.** A circuit designed to maintain a constant power supply output voltage despite anticipated variations in line voltage and load current demand.

**Voltage tripler.** Produces a dc output voltage that is three times its peak input voltage.

**Wavelength ($\lambda$).** The physical length of one cycle of a transmitted electromagnetic wave.

**Wavelength of peak spectral response ($\lambda_s$).** A rating that indicates the wavelength that will cause the strongest response in a photodetector.

**Wien-bridge oscillator.** An oscillator that achieves regenerative feedback by producing no phase shift at $f_r$.

**Working peak reverse voltage ($V_{RWM}$).** The maximum peak or dc reverse voltage that will not drive a transient suppressor into its reverse breakdown (zener) region of operation.

$y_{os}$. A label used to represent the *output admittance* of a JFET.

**Zener breakdown.** A type of reverse breakdown that occurs at relatively low reverse voltages.

**Zener clamper.** A clamper that uses a zener diode to establish the dc reference voltage.

**Zener diode.** Diodes that are designed to work in the reverse operating region.

**Zener impedance ($Z_Z$).** A zener diode's opposition to a change in current.

**Zener knee current ($I_{ZK}$).** The minimum value of zener reverse current required to maintain voltage regulation.

**Zener test current.** The value of zener current at which the nominal values of the component are measured.

**Zener voltage ($V_Z$).** The approximate voltage across a zener diode when it is operated in its reverse breakdown region of operation.

**Zero bias.** A term used to describe the biasing of a *pn* junction at room temperature with no potential applied. A D-MOSFET biasing circuit that has quiescent values of $V_{GS} = 0$ V and $I_D = I_{DSS}$.

# Transistor Amplifier Design

The complete common-emitter design process starts with knowing the requirements of the amplifier and the values you have to work with. Normally, you will be provided with the following information:

- The available supply voltage and ac input voltage.
- The load resistance or impedance.
- The required output voltage or the required value of $A_v$.
- The specifications of the transistor being used.

The step-by-step process for designing a common-emitter amplifier is as follows:

Equation (11.6):

$$PP = 2V_{CEQ}$$

1. The required output voltage swing is the minimum compliance (PP) of the amplifier. Using this value and equation (11.6), determine the minimum allowable value of $V_{CEQ}$. Whenever possible, set $V_{CEQ}$ to half the value of $V_{CC}$.
2. Determine the desired value of $I_{CQ}$. The exact value is not usually critical in terms of input and output voltages. However, remember that a lower value of $I_{CQ}$ means that the amplifier will dissipate less power. Typical values of $I_{CQ}$ fall between 1 and 10 mA for small-signal amplifiers.
3. For the value of $I_{CQ}$ used, check the specification sheet to determine the value of $h_{FE}$.
4. Once you have determined the proper value of $h_{FE}$, calculate the value of $I_E$ using

$$I_E = I_{CQ}\left(1 + \frac{1}{h_{FE}}\right) \tag{F.1}$$

5. Set $V_E$ to approximately $0.1V_{CC}$.
6. Using the values from steps 4 and 5, find $R_E$ as

$$R_E = \frac{V_E}{I_E}$$

7. Determine $V_{RC}$ as

$$V_{RC} = V_{CC} - (V_{CEQ} + V_E)$$

8. Find $R_C$ as

$$R_C = \frac{V_{RC}}{I_C}$$

**9.** Find $V_B$ as

$$V_B = V_E + V_{BE}$$

**10.** Find $I_B$ as

$$I_B = \frac{I_{CQ}}{h_{FE(min)}}$$

**11.** Set $I_2$ as

$$I_2 \geqslant 10(I_B)$$

**12.** Select the value of $R_2$ as

$$R_2 = \frac{V_B}{I_2}$$

**13.** Now, determine the value of current through $R_1$ as

$$I_1 = I_2 + I_B$$

**14.** Determine the voltage across $R_1$ as

$$V_1 = V_{CC} - V_B$$

**15.** Find the value of $R_1$ as

$$R_1 = \frac{V_1}{I_1}$$

At this point, the dc design process is complete. Now, you have to analyze the ac operation of the circuit. Using the value of $I_{CQ}$, you need to determine the $h$-parameter values for the transistor. Then:

**16.** Determine the value of $r_C$.
**17.** Determine the value of $r'_e$ using equation (9.32).
**18.** Determine the value of $A_v$ for the circuit. If the value of $A_v$ is too low, you must redesign the circuit using a higher value of $I_{CQ}$. If it is too high, you may want to consider swamping the emitter circuit to reduce the gain.

Equation (9.32):
$$r'_e = \frac{h_{ie}}{h_{fe}}$$

The entire process is illustrated in the following example.

## EXAMPLE F.1

Design an amplifier to the following specifications: The voltage gain is to be 50 with an input voltage of 40 mV. The available supply voltage is $+10$ V. The transistor used is the 2N3904. Use an $I_{CQ}$ of 1 mA. $R_L = 12$ k$\Omega$.

*Solution:* A model circuit for this example is shown in Figure F.1. Along with the circuit, this figure contains the $h$-parameters for the 2N3904 at $I_C = 1$ mA. First, the required output compliance from the amplifier is found by converting the value of $v_{out}$ to a peak-to-peak value. The value of $v_{out}$ is found as

$$v_{out} = A_v v_{in} = 2 \text{ V}$$

The peak-to-peak output voltage is found as

$$V_{PP} = 2.828(v_{out}) = 5.66 \text{ V}$$

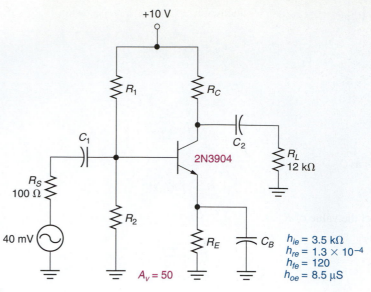

FIGURE F.1

Therefore, the minimum value of $V_{CEQ}$ is

$$V_{CEQ} = \frac{V_{PP}}{2} = 2.83 \text{ V}$$

Since 2.83 V < $\frac{1}{2}V_{CC}$, we will use $V_{CEQ} = 5$ V.

At $I_{CQ} = 1$ mA, the 2N3904 has a value of $h_{FE(min)} = 70$. Using this value, $I_E$ is found as

$$I_E = I_{CQ}\left(1 + \frac{1}{h_{FE}}\right) = 1.01 \text{ mA}$$

The value of $V_E$ is found as

$$V_E = 0.1V_{CC} = 1 \text{ V}$$

and the value of $R_E$ is found as

$$R_E = \frac{V_E}{I_E} = 990 \text{ } \Omega \qquad \text{(use 910 } \Omega \text{ standard)}$$

Using $R_E = 910$ $\Omega$, the value of $V_E$ is now recalculated as

$$V_E = I_E R_E = 920 \text{ mV}$$

Now, $V_{RC}$ is found as

$$V_{RC} = V_{CC} - (V_{CEQ} + V_E) = 4 \text{ V}$$

and $R_C$ is found as

$$R_C = \frac{V_{RC}}{I_C} = 4 \text{ k}\Omega \qquad \text{(use 3.9 k}\Omega \text{ standard)}$$

This completes the emitter-collector circuit. Now, the first step in designing the base circuit is to find $V_B$ as

$$V_B = V_E + V_{BE} = 1.62 \text{ V}$$

Next, $I_B$ is found as

$$I_B = \frac{I_{CQ}}{h_{FE(min)}} = 14.3 \text{ } \mu\text{A}$$

and $I_2$ is set as

$$I_2 = 10 I_B = 143 \text{ } \mu\text{A}$$

Now, $R_2$ is determined as

$$R_2 = \frac{V_B}{I_2} = 11.3 \text{ k}\Omega \qquad \text{(use 11 k}\Omega \text{ standard)}$$

Using $R_2 = 11$ k$\Omega$, the value of $I_2$ is now recalculated as

$$I_2 = \frac{V_B}{R_2} = 147.3 \text{ } \mu\text{A}$$

The value of $I_1$ is found as

$$I_1 = I_2 + I_B = 161.6 \text{ } \mu\text{A}$$

and the value of $V_1$ is found as

$$V_1 = V_{CC} - V_B = 8.4 \text{ V}$$

Finally, the value of $R_1$ is found as

$$R_1 = \frac{V_1}{I_1} = 51.9 \text{ k}\Omega \qquad \text{(use 51 k}\Omega \text{ standard)}$$

At this point, we need to take a look at the ac operation of our biasing circuit. The first step is to determine the total ac resistance in the collector circuit. This resistance is found as

$$r_C = R_C \| R_L = 3 \text{ k}\Omega$$

From the specification sheet, $r'_e$ is found as

$$r'_e = \frac{h_{ie}}{h_{fe}} = 29 \text{ } \Omega$$

and $A_v$ is found as

$$A_v = \frac{h_{fe} r_C}{h_{ie}} = 103$$

**33.** 24.1 Hz  **35.** 2.04 MHz, 7.96 MHz  **37.** 9 pF, 5 pF, 1 pF
**39.** 795.8 Hz, 1.24 MHz  **41.** 77.23 kHz  **43.** 537.5 kHz

**Chapter 15:**
**1.** $(+), (-), (+)$  **3.** 16 $V_{PP}$  **5.** 133.3 m$V_{PP}$  **7.** 150 m$V_{PP}$
**9.** 12 mV  **11.** 143.2 kHz  **13.** 35.4 kHz  **15.** $A_{CL} = 120$, $Z_{in} \cong$
1 kΩ, $Z_{out} < 50$ Ω, CMRR = 6000, $f_{max} = 79.6$ kHz  **17.** $A_{CL} = 1$,
$Z_{in} \cong 100$ kΩ, $Z_{out} < 75$ Ω, CMRR = 200, $f_{max} = 159$ kHz  **19.** $A_{CL} =$
101, $Z_{in} \geq 1$ MΩ, $Z_{out} < 40$ Ω, CMRR = 50,500, $f_{max} = 78.78$ kHz
**21.** $A_{CL} = 41$, $Z_{in} > 2$ MΩ, $Z_{out} < 100$ Ω, CMRR = 1242, $f_{max} =$
291.1 kHz  **23.** $A_{CL} = 1$, $Z_{in} = 5$ MΩ, $Z_{out} = 40$ Ω, CMRR = 1000,
$f_{max} = 318.3$ kHz  **25.** 30 kHz  **27.** 198.6 kHz  **29.** The
component can be used.  **31.** 12 MHz  **33.** 25 MHz  **35.** 4.53
**37.** 124.9  **39.** $Z_{in(f)} = 48$ kΩ, $Z_{out(f)} = 995$ mΩ  **41.** $A_{CL} = 100$,
$Z_{in(f)} = 1$ kΩ, $Z_{out(f)} = 49.98$ mΩ  **43.** $R_f$ is open.  **45.** The input
op-amp is faulty.  **47.** The input op-amp is open.  **49.** 500,000

**Chapter 16:**
**1.** 0 V  **3.** +4 V  **5.** 1.59 kHz  **7.** 72.34 Hz  **9.** 32.15 Hz
**11.** 159 Hz, 1.59 kHz  **13.** $-V_{out} = V_1 + V_2 + V_3$, −5 V
**15.** $-V_{out} = 1.5V_1 + 3V_2 + 1.5V_3$, −4.35 V  **17.** 3 kΩ  **19.** −2 V
**21.** $R_2$ is likely open  **23.** $R_1$ is open  **25.** 4.1 V, 1.9 V

**Chapter 17:**
**1.** 7  **3.** 129 kHz  **5.** 308.3 kHz, 9.7, 187.5 kHz, 1.53
**7.** 159.2 Hz, 11  **9.** 10.7 kHz, 1.5  **11.** 321.5 Hz, 1.98
**13.** 9.46 kHz, 11  **15.** 112.5 kHz, 250.1 kHz, 137.6 kHz, 167.7 kHz,
1.22  **17.** 5.88 kHz, 13.04 kHz, 7.16 kHz, 8.76 kHz, 1.224
**19.** $f_{C1}$: 6.98%, $f_{C2}$: 3.7%.  **21.** 1.5  **23.** 1.31 kHz, 1.39 kHz, 80 Hz,
1.35 kHz, 16.88  **25.** 809.3 Hz, 1128 Hz, 318.6 Hz, 968.6 Hz, 3.04

**27.** 11, 9.82, 87.6 kHz, 8.92 kHz, 83.14 kHz, 92.06 kHz  **29.** 20 V, 0 A
**31.** +3 V  **33.** $C_1$ is open.  **35.** $\Delta f_{C1} = \Delta f_{C2} = 3.1$ kHz

**Chapter 18:**
**1.** 219.2 kHz  **3.** 24.35 kHz  **5.** 27.84 kHz, 0.01, 100
**7.** 19.29 kHz, 0.067, 15  **9.** 3.38 kHz, 0.01, 100  **11.** 277.05 kHz,
0.1, 10  **17.** 5.76 mH

**Chapter 19:**
**1.** 11.3 V  **3.** 3.54 V  **5.** 0.9 V, 6 V  **7.** 20 μs, 50 μs
**9.** 40%  **11.** 23.3 MHz, 233 kHz  **13.** 12.07 MHz, 120.7 kHz
**15.** 23.3 MHz, 233 kHz  **17.** 5.38 MHz, 53.8 kHz  **19.** 20 ns, 60 ns,
180 ns, 70 ns  **21.** 4.5 V, −4.5 V  **23.** 1.66 V, −830 mV
**25.** 4.67 V, −4.67 V  **27.** 5 V, −5 V  **29.** 11 ms  **31.** 0.7 V
**33.** 5.45 kHz, 68.9%, 126 μs  **35.** 4.8 kHz, 66.7%, 138.6 μs
**37.** $R_A$ is open or pin 7 is shorted.  **39.** MMBT2222, MMBT2222A,
MMBT4401

**Chapter 20:**
**1.** It cannot withstand the surge.  **3.** 1.04 ms  **5.** 7.21 A
**7.** 100 mV  **9.** 8.54 V to 11.20 V  **11.** 9.5 V to 13.82 V
**13.** 461.5 nm  **15.** 1666.7 nm  **17.** 319 PHz (P = $10^{15}$)
**19.** 34.5 V, 15 mA, the fuse.  **21.** 36.4 V

**Chapter 21:**
**1.** 5 μV/V  **3.** 3 μV/V  **5.** 133.3 μV/mA  **7.** 75 μV/mA
**9.** 6.1 V  **11.** 3.5 A  **13.** 35 V  **15.** 12.0 V  **17.** 7.2 V
**19.** 12.0042 V  **21.** 0.0417%/mA

Filters, *continued*
  and diode PIV, 109
  faults in, 120, 121
  inductive, 109
  output voltages of, 105–106
  in rectifier analysis, 107–108
  surge current in, 103–105
  in switching regulators, 908
Fine tuning circuits, 178–180
Fixed bias. *See* Base bias
Fixed-negative IC voltage regulators, 904
Fixed-positive IC voltage regulators, 904
Flat-flat filters, 700–701
Flat response, 700–701
Flip-flops, 814
Flywheel effect, 737
Forced commutation
  for silicon-controlled rectifiers, 846
  for silicon unilateral switches, 843
  for triacs, 859
Forward bias, 12–13
  for BJTs, 206, 207, 208, 225, 227
  for diodes, 20–21, 24
Forward blocking region
  for silicon-controlled rectifiers, 846
  for silicon unilateral switches, 842
Forward breakover current, 843
Forward breakover voltage, 840
Forward characteristics of diodes, 23, 25, 35
Forward current gain, 204, 212–215
Forward current ratings, 32, 39, 42, 86
Forward operating region, 23, 35
  for diacs, 857
  for silicon-controlled rectifiers, 846
  for silicon unilateral switches, 841–842
Forward power dissipation ratings, 32–34
Forward surge current rating
  for silicon-controlled rectifiers, 846–847
  for transient suppressors, 182
Forward transconductance ratings, 505–506
Forward transfer admittance, 505–506
Forward voltage, 13
  of diodes, 25–29, 43
  of light-emitting diodes, 54
Four-layer diodes
  cutting off, 841–843
  operation of, 840–841
  specifications for, 843–844
Free-running multivibrators, 812
  555 timers as, 821–824
  troubleshooting, 824
Frequency
  of astable multivibrators, 822–823
  of crystals, 770–772
  of differentiators, 670–672
  in integrators, 669–670
  of light, 871
  of operational amplifiers, 615–616, 617, 632–635
  in oscillator stability, 757
  and reactance, 727–728
  in Wien-bridge oscillators, 758–761
Frequency multipliers, 738
Frequency response, 550–586. *See also* Active filters

and bandwidth, 550–554
of BJT amplifiers, 558–575
Bode plots for, 557–558
capacitance in, 568–570, 583–584
Chebyshev, 700–701
and decibel power gain, 555–557
gain roll-off rate in, 557–558, 562–565, 574–575, 695–697
high-frequency, 568–575, 579–582
of JFETs, 576–582
low-frequency, 558–562, 576–579
measurement units for, 557–558
of multistage amplifiers, 584–586
of operational amplifiers, 632–633
practical considerations in, 567–568, 575, 584
of single-pole low-pass filters, 702–704
of tuned amplifiers, 726–732
of two-pole low-pass filters, 704–706
Frequency-response curves, 550, 555, 558, 694
Full-cycle average forward voltage drop ratings, 43
Full-cycle average reverse current ratings, 43
Full-load current gain, 345–346
Full-load input impedance, 345
Full loads with zener voltage regulators, 114–115
Full-wave bridge rectifiers, 92–96
  vs. full-wave rectifiers, 94–95
  load voltage and current calculations for, 93–96
  operation of, 93
  peak inverse voltage in, 95
Full-wave rectifiers, 87–91
  faults in, 119
  vs. full-wave bridge rectifiers, 94–95
  load voltage and current calculations for, 88–90
  negative, 91
  operation of, 88
  ripple voltage from, 106–107
Full-wave voltage doublers, 152
Fundamental frequency of crystals, 772
Fuses
  in power supplies, 118–123
  for silicon unilateral switches, 841

Gain, 284–286, 290, 1035
  of active filters, 708–709, 710
  of audio amplifiers, 683
  of BJTs, 212–215
  of class B amplifiers, 427–428
  in Colpitts oscillators, 765
  in common-base amplifiers, 295
  in common-collector amplifiers, 294, 370–372, 374
  in common-drain amplifiers, 493–494
  in common-emitter amplifiers, 294, 318, 329–334, 335, 337–338, 339–340, 341–344
  in common-gate amplifiers, 497
  in common-source amplifiers, 487–489
  in Darlington amplifiers, 383–386
  in decibels, 305

equation for, 946–947
with feedback, 638–641, 956
and frequency response. *See* Frequency response
$h$-parameters for, 345–351, 933–937
in instrumentation amplifiers, 681–682
in inverting amplifiers, 623–624
of multiple-feedback band-pass filters, 718
in multistage amplifiers, 338–340, 556–557
of noninverting amplifiers, 626
of operational amplifiers, 602, 623–624, 626, 628–629, 637, 639–643
as ratio, 285–287
of single-pole low-pass filters, 702–704
of voltage followers, 628–629
Gain-bandwidth product
  of BJTs, 569
  of operational amplifiers, 633–635
Gain roll-off rate, 557
  vs. bandwidth, 695
  equation for, 953–954
  high-frequency, 574–575
  low-frequency, 562–565
Gain-stabilized amplifier. *See* Swamping amplifiers
Gallium, 7
Gate bias, 470–472, 491–493
  in E-MOSFETs, 531
Gate current, 465, 504–505
Gated latches in switch drivers, 908
Gate-drain current, 463
Gate impedance, 464
Gate nontrigger voltage ratings, 849
Gate reverse current ratings, 504–505
Gates
  in JFETs, 460–461
  in silicon-controlled rectifiers, 845
  in triacs, 857–858
Gate-source voltage
  in common-source amplifiers, 483–484
  in JFETs, 462–464, 504–505
Gate trigger voltages, 853
Gate voltage, 483–484
General amplifier model, 284–285
General-class output equations, 675–676
General voltage amplifier model, 287
Geometric averages
  for center frequencies, 552, 697–698
Geometric progression in frequency response, 557
Germanium, 3
  forward voltage for, 13
  limitations of, 74
  vs. silicon, 14
Glossary, 961–971
Graphic equalizers, 722–723
Ground connections
  in class AB amplifiers, 440, 441
  in common-base amplifiers, 295
  in common-collector amplifiers, 294
  in common-emitter amplifiers, 293, 325
  for integrators, 668
  for inverting amplifiers, 621